FLORA EUROPAEA

FLORA EUROPAEA

VOLUME 3
DIAPENSIACEAE TO MYOPORACEAE

EDITED BY

T. G. TUTIN   V. H. HEYWOOD

N. A. BURGES   D. M. MOORE   D. H. VALENTINE

S. M. WALTERS   D. A. WEBB

WITH THE ASSISTANCE OF

P. W. BALL   A. O. CHATER   R. A. DeFILIPPS

I. K. FERGUSON   I. B. K. RICHARDSON

CAMBRIDGE

AT THE UNIVERSITY PRESS

1972

Published by the Syndics of the Cambridge University Press
Bentley House, 200 Euston Road, London NW1 2DB
American Branch: 32 East 57th Street, New York, N.Y.10022

© Cambridge University Press 1972

Library of Congress Catalogue Card Number: 64–24315

ISBN: 0 521 08489 X

Printed in Great Britain
at the University Printing House, Cambridge
(Brooke Crutchley, University Printer)

CONTENTS

THE FLORA EUROPAEA ORGANIZATION

SPONSOR

The Linnean Society of London

EDITORIAL COMMITTEE

T. G. TUTIN, University of Leicester; *Chairman*
V. H. HEYWOOD, University of Reading; *Secretary*
N. A. BURGES, New University of Ulster
D. M. MOORE, University of Reading
D. H. VALENTINE, University of Manchester
S. M. WALTERS, University of Cambridge
D. A. WEBB, University of Dublin (Trinity College)

RESEARCH ASSISTANTS

P. W. BALL, *Flora Europaea*, at University of Liverpool (until September 1969)
A. O. CHATER, *Flora Europaea*, at University of Leicester
R. A. DeFILIPPS, *Flora Europaea*, at University of Reading
I. K. FERGUSON, *Flora Europaea*, at British Museum (Natural History), London (until September 1968)
I. B. K. RICHARDSON, *Flora Europaea*, at University of Reading

REPRESENTATIVES OF THE LINNEAN SOCIETY

R. D. MEIKLE, Royal Botanic Gardens, Kew
A. MELDERIS, British Museum (Natural History), London

ADVISORY EDITORS

T. W. BÖCHER, Institut for Planteanatomi og Cytologi, København
A. R. CLAPHAM, University of Sheffield
J. E. DANDY, British Museum (Natural History), London
J. DOSTÁL, Universita Palackého, Olomouc
ANDREY A. FEDOROV, Akademija Nauk S.S.S.R., Leningrad
H. GAUSSEN, Faculté des Sciences, Toulouse
†N. HYLANDER, Uppsala Universitet
K. H. RECHINGER, Naturhistorisches Museum, Wien (resigned May 1970)
SIR GEORGE TAYLOR, Royal Botanic Gardens, Kew

REGIONAL ADVISERS

Albania	F. MARKGRAF, Universität Zürich
	I. MITRUSHI, Universiteti Shtetëror i. Tiranës, Tirana
Austria	F. EHRENDORFER, Universität Wien
Belgium	A. LAWALRÉE, Jardin Botanique Nationale de Belgique, Bruxelles

Bulgaria	B. KUZMANOV, Bălgarska Akademija na Naukite, Sofija
	†N. STOJANOV, Bălgarska Akademija na Naukite, Sofija
Czechoslovakia	J. DOSTÁL, Universita Palackého, Olomouc
	J. HOLUB, Československa Akademie Ved, Průhonice u Prahy
Denmark	A. HANSEN, Universitetets Botaniske Museum, København
Finland	J. JALAS, Yliopiston kasvitieteen laitos, Helsinki
France	P. DUPONT, Université de Nantes
	P. JOVET, Muséum National d'Histoire Naturelle, Paris
	R. DE VILMORIN, Université de Paris-Sud
Germany	H. MERXMÜLLER, Botanische Staatssammlung, München
	K. WERNER, Martin-Luther-Universität, Halle-Wittenberg
Greece	W. GREUTER, Conservatoire et Jardin Botaniques, Genève
	D. PHITOS, Panepistimion Patron, Patrai
	K. H. RECHINGER, Naturhistorisches Museum, Wien (resigned May 1970)
	H. RUNEMARK, Lunds Universitet, Lund
Hungary	†Z. E. KÁRPÁTI, Kertészeti és Szölészeti Föiskola, Budapest
	R. DE SOÓ, Egyetemi Botanikus Kert, Budapest
Iceland	E. EINARSSON, Náttúrufrædistofnun Íslands, Reykjavík
Italy	G. MOGGI, Università di Firenze
	R. E. G. PICHI-SERMOLLI, Università di Genova
Jugoslavia	V. BLEČIĆ, Beogradski Univerzitet, Beograd
	E. MAYER, Universa Ljubljana
Netherlands	S. J. VAN OOSTSTROOM, Rijksherbarium, Leiden
Norway	R. NORDHAGEN, Universitet i Oslo Botanisk Museum, Oslo
Poland	†B. PAWŁOWSKI, Polska Akademia Nauk, Kraków
Portugal	J. DO AMARAL FRANCO, Instituto Superior de Agronomia, Lisboa (assisted by M. DA LUZ DA ROCHA AFONSO)
	A. R. PINTO DA SILVA, Estação Agronómica Nacional, Oeiras
Romania	†A. BORZA, Academia Republicii Socialiste România, Bucureşti
	C. ZAHARIADI, Academia Republicii Socialiste România, Bucureşti (resigned September 1969)

Russia	T. V. Egorova, Akademija Nauk S.S.S.R., Leningrad
	A. I. Pojarkova, Akademija Nauk S.S.S.R., Leningrad
	V. N. Tikhomirov, Moskovskij Ordena Lenina i Ordena Trudovogo Krasnogo Znameni Gosudarstvennij Universitet im. M. V. Lomonosova, Moskva
Spain	O. de Bolós, Instituto Botánico de Barcelona
	E. Fernández-Galiano, Universidad de Sevilla
	E. Guinea López, Jardín Botánico, Madrid
Sweden	B. Jonsell, Uppsala Universitet
Switzerland	E. Landolt, Eidgenössische Technische Hochschule, Zürich
Turkey	P. H. Davis, Royal Botanic Garden, Edinburgh

TECHNICAL CONSULTANT
Á. Löve, University of Colorado, Bouldei

GEOGRAPHICAL CONSULTANT
H. Meusel, Martin-Luther-Universität, Halle-Wittenberg

The above lists refer to the Organization as it was constituted during the preparation of volume 3, which was completed in April 1971

LIST OF CONTRIBUTORS TO VOLUME 3

The following is a list of authors who have contributed accounts of genera or parts of them.

P. W. BALL, University of Toronto

J. BORJA CARBONELL, Jardín Botánico, Madrid

D. BRAMWELL, University of Reading

R. K. BRUMMITT, Royal Botanic Gardens, Kew

J. M. CAMARASA, Instituto Botánico, Barcelona

S. J. CASPER, Friedrich-Schiller Universität, Jena

A. O. CHATER, University of Leicester

J. CULLEN, University of Liverpool

R. A. DeFILIPPS, University of Reading

R. DOMAC, Sveučilište u Zagrebu

J. M. EDMONDS, University of Cambridge

N. FEINBRUN, Hebrew University of Jerusalem

I. K. FERGUSON, Royal Botanic Gardens, Kew

L. F. FERGUSON, British Museum (Natural History), London

R. B. FERNANDES, Universidade de Coimbra

J. DO AMARAL FRANCO, Instituto Superior de Agronomia, Lisboa

J. W. FRANKS, Manchester Museum, University of Manchester

F. M. GETLIFFE, University of Natal, Durban

P. E. GIBBS, University of St Andrews

J. GRAU, Universität München

P. S. GREEN, Royal Botanic Gardens, Kew

E. GUINEA LÓPEZ, Jardín Botánico, Madrid

A. HANSEN, Universitetets Botaniske Museum, København

R. M. HARLEY, Royal Botanic Gardens, Kew

J. G. HAWKES, University of Birmingham

I. C. HEDGE, Royal Botanic Garden, Edinburgh

V. H. HEYWOOD, University of Reading

J. JALAS, Yliopiston kasvitieteen laitos, Helsinki

M. KOVANDA, Československa Akademie Ved, Průhonice u Prahy

S. KOŽUHAROV, Bǎlgarska Akademija na Naukite, Sofija

A. KRESS, Botanischer Garten, München-Nymphenburg

B. KŘÍSA, Universita Karlova, Praha

A. LAWALRÉE, Jardin Botanique Nationale de Belgique, Bruxelles

F. MARKGRAF, Universität Zürich

E. MAYER, Universa Ljubljana

R. D. MEIKLE, Royal Botanic Gardens, Kew

A. MELDERIS, British Museum (Natural History), London

H. MERXMÜLLER, Botanische Staatssammlung, München

D. M. MOORE, University of Reading

A. PATZAK, Naturhistorisches Museum, Wien

S. PAWŁOWSKA, Polska Akademia Nauk, Kraków

†B. PAWŁOWSKI, Polska Akademia Nauk, Kraków

D. PHILCOX, Royal Botanic Gardens, Kew

S. PIGNATTI, Università degli Studi di Trieste

A. R. PINTO DA SILVA, Estação Agronómica Nacional, Oeiras

T. N. POPOVA, Akademija Nauk S.S.S.R., Leningrad

N. M. PRITCHARD, University of Aberdeen

I. B. K. RICHARDSON, University of Reading

H. RIEDL, Naturhistorisches Museum, Wien

E. M. RIX, University of Cambridge

M. DA LUZ DA ROCHA AFONSO, Instituto Superior de Agronomia, Lisboa

W. SAUER, Universität München

H. D. SCHOTSMAN, Muséum National d'Histoire Naturelle, Paris

N. H. SINNOTT, Royal Botanic Gardens, Kew

A. R. SMITH, Royal Botanic Gardens, Kew

R. DE SOÓ, Egyetemi Botanikus Kert, Budapest

C. A. STACE, University of Manchester

W. T. STEARN, British Museum (Natural History), London

P. TAYLOR, Royal Botanic Gardens, Kew

C. C. TOWNSEND, Royal Botanic Gardens, Kew

C. TURNER, University of Cambridge

T. G. TUTIN, University of Leicester

B. VALDÉS CASTRILLÓN, Universidad de Sevilla

D. H. VALENTINE, University of Manchester

S. M. WALTERS, University of Cambridge

D. A. WEBB, Trinity College, Dublin

D. WOOD, Royal Botanic Garden, Edinburgh

P. F. YEO, University of Cambridge

PREFACE

The publication of Volume 3 of *Flora Europaea* marks another stage in the progress of this international cooperative venture. A work of this nature is the outcome of a vast fund of goodwill and collaboration which we have continued to enjoy from our friends and advisers throughout Europe.

It is a matter of great satisfaction to the Editorial Committee to see the many ways in which the previous two volumes of *Flora Europaea* have stimulated taxonomic and floristic research in many parts of Europe. Furthermore, the project has been a source of inspiration to botanists in other continents when contemplating the production of Floras of their own regions.

The United Kingdom Science Research Council has continued its generous financial support during the period of preparation of this volume and has enabled us to maintain our secretariat and staff, as well as permitting us to invite a further number of visiting bursars, who have assisted us in preparing accounts. These visiting bursars have been Professor J. Dostál (Olomouc), Dr K. Werner (Halle), Dr Viera Feráková (Bratislava), Dr S. Snogerup (Lund), Professor L. Boulos (Cairo), Dr M. Kovanda (Praha), and Dr Delphine Cartier (Orsay). The total amount of the Science Research Council grant for the period 1968–72 is £36,560, and we wish to record our deep gratitude for this support.

The sixth Flora Europaea symposium was held in Switzerland and France in 1970, and we are grateful to Professors J. Miège (Genève), M. Guinochet (Orsay), J.-P. Barry (Nice) and their staff for the local organization of this highly successful meeting. Financial support was received from The Royal Society, the International Union of Biological Sciences, Hoffmann–La Roche & Cie S.A., Bâle, Zyma S.A., Nyon, Ciba S.A., Bâle, and Geigy S.A., Bâle. Publication in 1971 of the *Actes du VIe Symposium de Flora Europaea* (as *Boissiera* Volume 19), under the editorship of J. Miège and W. Greuter, was made possible by a grant from the Fonds National Suisse pour la Récherche Scientifique.

Although it was our original intention to publish the whole of the Sympetalae as a single volume, it became evident at an early stage of preparation that such a volume would run to some 1,000 pages and would be too large to handle conveniently either editorially or physically. Accordingly it was decided to spread the Sympetalae over two volumes, of which this is the first. Preparation of the other part (Volume 4) covering the Plantaginales–Campanulales is already well advanced.

The Editorial Committee has remained the same for Volume 3 as for Volume 2 but there have been changes in our team of Research Associates. Dr P. W. Ball left us in 1969 after having worked for *Flora Europaea* for more than ten years, during which he made many contributions to the Flora and assisted with editorial and secretarial matters. Dr I. K. Ferguson also resigned, in 1968, and was replaced by Mr I. B. K. Richardson the same year. Dr Ball's successor was Dr R. A. DeFilipps who joined us in 1969.

The Linnean Society of London have continued to act as our sponsor and have generously agreed to the publication of our series *Notulae Systematicae ad Floram Europaeam spectantes* in their Botanical Journal, following a decision to discontinue publication in *Feddes Repertorium*.

Our thanks are due to the Keeper of Botany and Staff of the Department of Botany,

British Museum (Natural History) for continuing to provide us and our collaborators with many facilities. We are also grateful to the Director and Staff of the Herbarium and Library, Royal Botanic Gardens, Kew, for similar assistance and advice. Other European institutions have assisted with the loan of material and given us hospitality during our visits, especially the Naturhistorisches Museum, Wien, the Conservatoire et Jardin botaniques, Genève, and the Istituto Botanico dell'Università di Firenze. To these and many other institutions we would like to express our thanks.

Once again Mr J. E. Dandy, one of our Advisory Editors, has performed a most valuable service in checking the nomenclature of the great majority of the manuscripts and given freely of his expert advice. As in the previous volumes Mr P. D. Sell (Cambridge) has prepared the index for the press and we are indebted to him for undertaking this complex and painstaking work.

The names of the authors primarily responsible for the accounts of families and genera are given in footnotes. The Editorial Committee accepts full responsibility for the text as published, which may in some cases have been substantially altered in form and presentation.

We wish to make acknowledgement to the Universities of Cambridge, Dublin, Leicester, Manchester, Reading and Ulster for kindly providing facilities for members of the Editorial Committee and their assistants. In 1968, following the acceptance of an appointment at the University of Reading by our Secretary, Professor V. H. Heywood, the Secretariat moved to Reading and is housed in the University Department of Botany.

Finally we wish to record our gratitude to the secretarial staff who play a key role in the production of the Flora. In Liverpool the team comprised Mrs Joan Beck, Mrs Margaret Donnelly, and Mrs Margaret Pollard. The staff at Reading are Mrs Rosa Husain and Mrs Linda Bennett. We greatly appreciate their devoted service.

INTRODUCTION

The aim of the Flora is in general diagnostic, and the descriptions, while brief, are as far as possible comparable for related species. The Floras listed on pp. xix–xxi, and the monographs or revisions given when appropriate after the descriptions of families and genera, may assist the reader in obtaining more detailed information. Other references to published work are occasionally given in cases of special taxonomic difficulty.

All available evidence, morphological, geographical, ecological and cytogenetical, has been taken into consideration in delimiting species and subspecies, but they are in all cases definable in morphological terms. (Taxa below the rank of subspecies are not normally included.) The delimitation of genera is often controversial and the solution adopted in the Flora may be a somewhat arbitrary choice between conflicting opinions. We have endeavoured to weigh as fairly as possible the various opinions available, but there has been no consistent policy of 'lumping' or 'splitting' genera (or, for that matter, species). The order and circumscription of the families is that of Melchior in Engler, *Syllabus der Pflanzenfamilien* ed. 12 (1964). Since, however, this edition of the *Syllabus* did not appear until Volume 1 of the Flora had gone to press, there are some small discrepancies between the two with regard to the sequence of families. In particular, the Cactaceae and Guttiferae, which should have been in Volume 1, were inserted in Volume 2. The main text of the Rubiaceae will appear in Volume 4, as noted on p. 73.

All descriptions of taxa refer only to their representatives in Europe. In practice, we have relaxed this rule slightly for families and genera to avoid giving taxonomically misleading information, particularly in those cases where a large family or genus has only one or few, somewhat atypical, members in Europe. In such cases we have occasionally added 'in European members' or a similar phrase to emphasize the atypical representation. It should, however, never be assumed that the description is valid for all non-European taxa.

For the purpose of this Flora, we have tried as far as possible to interpret Europe in its traditional sense. The area covered is shown on the maps at the end of the volume.

Place-names used in the summaries of geographical distribution have been given in their English form when they refer to independent states (including the constituent republics of the U.S.S.R.) or to such geographical features of Europe as transcend national boundaries. All other place-names are given in the language of the country concerned. Thus we write *Sweden, Ukraine, Danube, Alps, Mediterranean* but *Corse, Kriti, Slovenija, Rodopi Planina, Ahvenanmaa*.

In *transliteration* from Cyrillic characters we have followed the ISO system recommended in the UNESCO *Bulletin for Libraries* **10**: 136–137 (1956) for place-names and titles of journals. With personal names, however, we have followed the list of transliterations given in the index-volume (1962) to *Not. Syst. (Leningrad)*, and have transliterated personal names which do not occur in this list according to the conventions used there.

In transliterating place-names from Greek characters, we have, except for omitting the accents, followed *The Times Atlas of the World*, Mid-Century Edition, vol. 4 (London, 1956).

On pp. xix–xxi, we give a list of *Basic and Standard Floras*. The reason behind the choice of these Floras was not made clear in Volume 1. Basic Floras have been chosen as widely known Floras covering large or important parts of Europe. Standard Floras are considered to represent those Floras in current use and likely to be familiar to a large number of people in the particular country concerned; the list has been revised since the publication of Volume 2.

Synonyms, whether full or partial, are given in parentheses in the text only when they are used in one of the Basic Floras or when they are necessary to prevent confusion. (For primarily Iberian and Mediterranean species, synonyms used in the *Prodromus* of Willkomm & Lange, and the *Supplementum* by Willkomm (p. xxi) are also included.) Synonyms (or the basionym) are also usually given in the text when the combination has not previously been used in a Flora or monograph, or when the nomenclature is otherwise unfamiliar or in need of explanation. Otherwise, synonyms are given in the Index only; but it is important to note that no attempt has been made to give a complete synonymy. Even at the binomial level, the number of names for European plants is four or five times the number of accepted species, and to include all these would be impracticable. Thus, in addition to the binomials in the text, the Index contains all synonyms at specific rank which are used in the Basic and Standard Floras, or in cited monographs, with an indication of the species in the text under which they have been relegated to synonymy. Some subspecific names also appear in the Index. In this way, we hope that users of any Basic or Standard Flora will be able to relate the names used in their own Floras to those in *Flora Europaea*. In cases where the name of a familiar species has been changed, an explanation of this is usually published as a Notula (see p. xviii).

Citations have been abbreviated, and the abbreviations used for authors and places of publication have been standardized; lists of these abbreviations are given in Appendices I, II and III. These lists apply only to the abbreviations used in Volume 3.

Species descriptions attempt to give, within the limits of length set by the Flora, both the diagnostic characters of the plant and a general idea of its appearance. Where dimensions are given, a measurement without qualification refers to length. Two measurements connected by × indicate length followed by width. Further measurements in parentheses indicate exceptional sizes outside the normal ranges. In order to save space and facilitate identification, descriptions may sometimes take the form of a comparison with another description. The conventional way of setting this out is, to give an example (p. 233):

44. Linaria lilacina Lange...Like **43** but...

This implies that the description with which it is being compared (in this example **43. L. anticaria** Boiss. & Reuter) applies to this taxon but for the differences noted. It does not necessarily mean that the two taxa are similar in general appearance. Additional descriptive information is sometimes also given, but in separate sentences.

The *diploid chromosome number* ($2n =$) is given where it has been possible to verify that the count was made on material of known wild European origin. For naturalized and cultivated species, the count is from material which is naturalized or is cultivated in the way which justifies its inclusion in the Flora. It is hoped to publish separately a list of references to the data on which the published numbers are based.

Ecological information is given sparingly, and only where the ecological characteristics of a species are clearly and concisely definable for its total European range. Sometimes a general statement, applicable to a whole genus or to a group of species, is made. There is an inevitable irregularity of treatment, as in a great many cases reliable ecological information is not available.

The description of each species is followed by an indication of its *distribution within Europe*. This falls into two parts: (1) a summary in a short phrase; (2) a list of abbreviations of 'territories' in which the species occurs. The summary phrase makes use of everyday geographical phrases and concepts such as 'W. Europe', 'the Mediterranean region', 'the Balkan peninsula', etc. Maps IV and V and the legends accompanying them indicate the interpretation which is to be put on these phrases. We would emphasize that they are to be interpreted in a simple geographical sense, and do not attempt in any way to divide Europe phytogeographically.

Species believed to be endemic to Europe are distinguished by a symbol (●) before the summary of geographical distribution.

A more precise indication of distribution is given by the enumeration of the 'territories' (indicated by a two-letter abbreviation) in which the plant is believed to occur. The limits of these territories follow, with very few exceptions, existing political boundaries (see Map I). The territories, of course, vary greatly in size, and Ga, Hs or Ju gives very much less information than does Fa, Rs(K) or Tu. In all cases, however, the lists provide a guide to which national Floras should be searched for further detailed information, whether on taxonomy or on distribution. Occasionally, the list of territories is followed by a brief indication, in parentheses, of extra-European distribution. This is done only for plants of which the European range is but a small fraction of the total and for species not native in Europe.

In general the only infraspecific taxa described and keyed in the Flora are subspecies. Any formal treatment of variation below the level of subspecies would have been impossible in a Flora of this kind; the known variation of taxa is, however, covered in the descriptions. No 'experimental' categories, such as ecotypes, are used in the Flora in a formal systematic sense, though they are sometimes mentioned in notes.

Where it is difficult to distinguish between a number of closely similar species in a genus, an *ad hoc* 'group' has been made, and these groups, not the individual species, are keyed out in the main species key. They will serve for at least a partial identification. Following the description of a group in the text, a key to the component species is given, and they are then numbered and described, so that a more detailed study, or the availability of more adequate material, may enable the user to take the identification further. For example, in *Linaria* there is the *L. arvensis* group, which comprises the species *L. arvensis* (L.) Desf., *L. simplex* (Willd.) DC. and *L. micrantha* (Cav.) Hoffmanns. & Link. Such groups have no nomenclatural status.

For some of the difficult hemiparasitic genera in the Scrophulariaceae, other *ad hoc* treatments have been devised; these are described in the notes following the descriptions of the genera concerned.

Only those few *hybrids* which reproduce vegetatively and are frequent over a reasonably large area (e.g. *Mentha × verticillata*) are described and keyed as for species. Other common hybrids may be mentioned individually in notes (e.g. in *Euphrasia*), or collectively for the whole genus (e.g. in *Armeria*).

We have attempted to include the following categories of *alien species*:

(i) Aliens which are effectively naturalized. These include garden plants which have escaped to situations not immediately adjacent to those in which they are cultivated, as well as weeds and other plants which have been accidentally introduced; provided, in both cases, that the plant has been established in a single station for at least 25 years, or is reported as naturalized in a number of widely separated localities.

(ii) Trees or crop-plants which are planted or cultivated in continuous stands on a fairly extensive scale.

Casual aliens, i.e. those which do not persist without repeated re-introduction, are not included unless they have often been mistaken for a native or established species, or are for any other reason of special interest. In assessing the status of a species in any part of Europe we have, however, been dependent very largely on the information contained in the national Floras, and it is clear that the criteria used by different authors vary widely. All data on native, naturalized or casual status relating to synanthropic plants must, therefore, be regarded only as approximate.

It is the policy of the Committee not to publish new names in the Flora itself. To deal with the publication of much of this material, an arrangement has been made with our sponsor The Linnean Society of London, by which taxonomic and nomenclatural notes are being published as part of a series entitled *Notulae Systematicae ad Floram Europaeam spectantes* in the *Botanical Journal of the Linnean Society*. The *Notulae* corresponding to Volumes 1 and 2 were published in *Feddes Repertorium*.

LISTS OF BASIC
AND STANDARD FLORAS

BASIC FLORAS

COSTE, H. *Flore descriptive et illustrée de la France, de la Corse et des Contrées limitrophes.* Vols. 1–3. Paris, 1900–1906.

HAYEK, A. VON. *Prodromus Florae Peninsulae balcanicae* (in *Feddes Repert. (Beih.)* 30). Vols. 1–3. Berlin-Dahlem, 1924–1933.

HEGI, G. *Illustrierte Flora von Mittel-Europa*, ed. 1. Vols. 1–7. München, 1906–1931. Ed. 2. Vols. 1– . München, 1936– . Ed. 3. Vols. 2– . München, 1966– .

HYLANDER, N. *Nordisk Kärlväxtflora.* Vols. 1– . Stockholm, 1953– .

KOMAROV, V. L. *et al.* (ed.). *Flora URSS.* Vols. 1–30. Leningrad & Moskva, 1934–1964.

STANDARD FLORAS

ARCANGELI, G. *Compendio della Flora italiana*, ed. 1. Torino, 1882.

BARCELÓ Y COMBIS, F. *Flora de las Islas Baleares.* Palma de Mallorca, 1879–1881.

BECK VON MANNAGETTA, G. *Flora Bosne, Hercegovine i Novipazarskog Sandžaka.* Vols. 1–4(1). Beograd & Sarajevo, 1903–1950.

BINZ, A. *Schul- und Exkursionsflora für die Schweiz*, ed. 13 by A. Becherer. Basel, 1968.

BINZ, A. & THOMMEN, E. *Flore de la Suisse*, ed. 2. Lausanne, 1953.

BOISSIER, E. *Flora orientalis.* Vols. 1–5. Genève, Bâle & Lyon, 1867–1884. *Supplementum.* 1888.

BORNMÜLLER, J. Beiträge zur Flora Mazedoniens (in *Bot. Jahrb.* 59 (Beibl. 136), 60 (Beibl. 140), 61). Leipzig, 1925–1928.

BORZA, A. *Conspectus Florae Romaniae.* Cluj, 1947–1949.

BRIQUET, J. *Prodrome de la Flore corse.* Vols. 1–3. Genève, Bâle, Lyon & Paris, 1910–1955.

CADEVALL I DIARS, J. *Flora de Catalunya.* Vols. 1–6. Barcelona, 1913–1937.

CLAPHAM, A. R., TUTIN, T. G. & WARBURG, E. F. *Flora of the British Isles*, ed. 2. Cambridge, 1962.

COUTINHO, A. X. PEREIRA. *Flora de Portugal*, ed. 2 by R. T. Palhinha. Lisboa, 1939.

DEGEN, A. VON. *Flora velebitica.* Vols. 1–4. Budapest, 1936–1938.

DIAPOULIS, K. A. *Ellenike Khloris.* Vols. 1–3. Athenai, 1939–1949.

DOMAC, R. *Flora za odredivanje i upoznavanje Bilja.* Zagreb, 1950.

DOSTÁL, J. *Květena ČSR.* Praha, 1948–1950.

EHRENDORFER, F. *Liste der Gefässpflanzen Mitteleuropas.* Graz, 1967.

FIORI, A. *Nuova Flora analitica d'Italia.* Vols. 1–2. Firenze, 1923–1929.

FIORI, A. & PAOLETTI, G. *Iconographia Florae italicae*, ed. 3. San Casciano, Val di Pesa, 1933.

FOMIN, A. V. *et al.* (ed.). *Flora RSS Ucr.*, ed. 1. Vols. 1–12. Kijiv, 1936–1965. Ed. 2. Vol. 1. Kijiv, 1938.

FOURNIER, P. *Les quatre Flores de la France, Corse comprise.* Poinsin-les-Grancey, 1934–1940. (Reprints with additions and corrections, Paris, 1946 and 1961.)

FRITSCH, K. *Exkursionsflora für Österreich und die ehemals österreichischen Nachbargebiete*, ed. 3. Wien & Leipzig, 1922.

GEIDEMAN, T. S. *Opredelitel' Rastenij Moldavskoj SSR.* Moskva & Leningrad, 1954.

GOFFART, J. *Nouveau Manuel de la Flore de Belgique et des Régions limitrophes,* ed. 3. Liège, 1945.

GORODKOV, B. N. & POJARKOVA, A. I. (ed.). *Flora Murmanskoj Oblasti.* Vols. 1–5. Moskva & Leningrad, 1953–1966.

HALÁCSY, E. VON. *Conspectus Florae graecae.* Vols. 1–3. Leipzig, 1900–1904. *Supplementum* 1. Leipzig, 1908. *Supplementum* 2 (in *Magyar Bot. Lapok* 11). Budapest, 1912.

HEUKELS, H. *Flora van Nederland,* ed. 16 by S. J. van Ooststroom. Groningen, 1970.

HIITONEN, H. I. A. *Suomen Kasvio.* Helsinki, 1933.

HULTÉN, E. *Atlas of the Distribution of vascular Plants in N.W. Europe.* Stockholm, 1950.

HYLANDER, N. *Förteckning over Nordens Växter.* 1. *Kärlvaxter.* Lund, 1955. *Tillägg och Rättelser* (in *Bot. Not.* 112). Lund, 1959.

JANCHEN, E. *Catalogus Florae Austriae.* Vol. 1. Wien, 1956–1960. *Ergänzungsheft.* Wien, 1963. *Zweites Ergänzungsheft.* Wien, 1964. *Drittes Ergänzungsheft.* Wien, 1966. *Viertes Ergänzungsheft.* Wien & New York, 1968.

JORDANOV, D. (ed.). *Flora na Narodna Republika Bălgarija.* Vol. 1– . Sofija, 1963– .

KNOCHE, H. *Flora balearica.* Vols. 1–4. Montpellier, 1921–1923.

LID, J. *Norsk og svensk Flora.* Oslo, 1963. —— *The Flora of Jan Mayen.* Oslo, 1964.

LINDMAN, C. A. M. *Svensk Fanerogamflora,* ed. 2. Stockholm, 1926.

LÖVE, Á. *Íslenzkar Jurtir.* København, 1945.

MAEVSKIJ, P. F. *Flora srednej Polosy evropejskoj Časti SSSR,* ed. 9 by B. K. Schischkin. Leningrad, 1964.

MAYER, E. *Seznam praprotnic in Cvetnic Slovenskega Ozemlja* (in *Razpr. Mat.-*

Prir. Akad. Ljubljani. Dela 5. *Inšt. Biol.* 3). Ljubljana, 1952.

MERINO Y ROMÁN, P. B. *Flora descriptiva é illustrada de Galicia.* Vols. 1–3. Santiago, 1905–1909.

MULLENDERS, W. (ed.). *Flora de la Belgique, du Nord de la France et des Régions voisines.* Liège, 1967.

NORDHAGEN, R. *Norsk Flora.* Oslo, 1940.

NYMAN, C. F. *Conspectus Florae europaeae.* Örebro, 1878–1882. *Supplementum 1,* 1883–1884. *Additamenta,* 1886. *Supplementum 2,* 1889–1890.

OSTENFELD, C. E. H. & GRÖNTVED, J. *The Flora of Iceland and the Faeroes.* København, 1934.

PALHINHA, R. T. *Catálogo das Plantas vasculares dos Açores.* Lisboa, 1966.

RACIBORSKI, M., SZAFER, W. & PAWŁOWSKI, B. (ed.). *Flora polska.* Vols. 1– . Kraków & Warszawa, 1919– .

RASMUSSEN, R. *Föroya Flora,* ed. 2. Tórshavn, 1952.

RAUNKIÆR, C. *Dansk Ekskursions-Flora,* ed. 7 by K. Wiinstedt. København, 1950.

RECHINGER, K. H. *Flora aegaea* (in *Denkschr. Akad. Wiss. Math.-Nat. Kl.* (*Wien*) 105(1)). Wien, 1943. *Supplementum* (in *Phyton* (*Austria*) 1). Horn, 1949.

ROBYNS, W. (ed.). *Flore Générale de Belgique. Spermatophytes.* Vols. 1– . Bruxelles, 1952– .

ROHLENA, J. *Conspectus Florae montenegrinae* (in *Preslia* 20 and 21). Praha, 1942.

RØNNING, O. I. *Svalbards Flora.* Oslo, 1964.

ROSTRUP, F. G. E. *Den danske Flora,* ed. 19 by C. A. Jørgensen. København, 1961.

ROTHMALER, W. *Exkursionsflora von Deutschland.* 2: *Gefässpflanzen.* Berlin, 1962. 4: *Kritischer Ergänzungsband*: *Gefässpflanzen.* Berlin, 1963. *Atlas der Gefässpflanzen.* Berlin, 1966.

ROUY, G. C. C. *Conspectus de la Flore de France.* Paris, 1927.

ROUY, G. C. C. *et al. Flore de France.* Vols. 1–14. Asnières, Paris & Rochefort, 1893–1913.

SAMPAIO, G. A. DE SILVA FERREIRA. *Flora portuguesa,* ed. 2 by A. Pires de Lima. Porto, 1947.

SĂVULESCU, T. (ed.). *Flora Republicii Populare Române.* Vols. 1– . Bucureşti, 1952– .

SCHMEIL, O. & FITSCHEN, J. *Flora von Deutschland,* ed. 67/68 by H. Voerkel & G. Müller. Jena, 1957.

Soó, R. *A Magyar Flóra es Vegetáció Rendszertani-Növényföldrajzi Kézikönyve.* Vols. 1– . Budapest, 1964– .

Soó, R. & KÁRPÁTI, Z. *Növényhatározo 2. Magyar Flóra.* Budapest, 1968.

STANKOV, S. S. & TALIEV, V. I. *Opredelitel' vysšikh Rastenij Evropejskoj Časti SSSR,* ed. 2. Moskva, 1957.

STEFÁNSSON, S. *Flóra Íslands,* ed. 3 by S. Steindórsson. Akureyri, 1948.

STOJANOV, N., STEFANOV, B. & KITANOV, B. *Flora na Bǎlgarija,* ed. 4. Vols. 1–2. Sofija, 1966–67.

SZAFER, W., KULCZYŃSKI, S. & PAWŁOWSKI, B. *Rósliny polskie.* Warszawa, 1953.

TRELEASE, W. *Botanical Observations on the Azores* (in *Ann. Rep. Missouri Bot. Gard.* **8**). St Louis, 1897.

WEBB, D. A. *An Irish Flora,* ed. 5. Dundalk, 1967.

—— *The Flora of European Turkey* (in *Proc. Roy. Irish Acad.* **65B**: 1–100). Dublin, 1966.

WEEVERS, T. *et al.* (ed.). *Flora neerlandica.* Vols. 1– . Amsterdam, 1948– .

WILLKOMM, H. M. *Supplementum Prodromi Florae hispanicae.* Stuttgart, 1893.

WILLKOMM, H. M. & LANGE, J. *Prodromus Florae hispanicae.* Vols. 1–3. Stuttgart, 1861–1880.

WULF, E. V. *Flora Kryma.* Vols. 1–3. Yalta, Leningrad & Moskva, 1927–1969.

ZEROV, D. K. *et al.* (ed.). *Viznačnyk Roslyn Ukrajiny,* ed. 2. Kijiv, 1965.

SYNOPSIS OF FAMILIES

KEY TO FAMILIES OF ANGIOSPERMAE

This key covers all the families of Angiospermae in volumes 1–3 and the great majority of those to be included in volumes 4–5, though some introduced families and, doubtless, some anomalous genera, have been omitted. A comprehensive key will be included in volume 5

1 Plant free-floating on or below surface of water, not rooted in mud
2 Plant with small bladders on leaves or on apparently leafless stems; leaves divided into filiform segments   **CLXI. Lentibulariaceae**
2 Not as above
3 Plant without obvious differentiation into stems and leaves   **Lemnaceae**
3 Plant with obvious stems and leaves
4 Leaves with a cuneate basal part, 4–6 setaceous segments and a terminal orbicular lobe   **LXXI. Droseraceae**
4 Leaves not as above
5 Floating leaves sessile   **Hydrocharitaceae**
5 Floating leaves long-petiolate
6 Floating leaves cordate-orbicular, entire   **Hydrocharitaceae**
6 Floating leaves rhombic, dentate in upper $\frac{2}{3}$ **CXX. Trapaceae**
1 Land-plants or aquatics rooted in mud
7 2- to 4-fid coloured staminodes present inside the sepals; leaves often fasciculate   **LIII. Molluginaceae**
7 Not as above
8 Perianth of 2 (rarely more) whorls differing markedly from each other in shape, size or colour
9 Petals not all united into a tube at base, very rarely cohering at apex, or else flowers papilionate
10 Ovary superior
11 Carpels 2 or more, free, or united at the base only
12 Sepals and petals 3
13 Carpels more than 3
14 Leaves lobed   **LXI. Ranunculaceae**
14 Leaves entire   **Alismataceae**
13 Carpels 3
15 Leaves palmately divided; petioles spiny   **Palmae**
15 Leaves simple, sessile   **LXXII. Crassulaceae**
12 Sepals or petals more than 3
16 Flowers zygomorphic; petals deeply divided   **LXIX. Resedaceae**
16 Flowers actinomorphic; petals entire
17 Stamens more than twice as many as petals
18 Shrubs or herbs with stipulate leaves; flowers perigynous   **LXXX. Rosaceae**
18 Herbs; stipules 0, though leaf-bases sometimes sheathing; flowers hypogynous
19 Fruit a head of achenes; sepals deciduous   **LXI. Ranunculaceae**
19 Fruit of 2–5 follicles; sepals persistent   **LXII. Paeoniaceae**
17 Stamens not more than twice as many as petals
20 Leaves 3-foliolate   **LXXX. Rosaceae**
20 Leaves simple
21 Carpels spirally arranged on an elongated receptacle   **LXI. Ranunculaceae**
21 Carpels in 1 whorl
22 Trees with palmately lobed leaves; flowers in globose capitula   **LXXIX. Platanaceae**
22 Herbs or shrubs; leaves not palmately lobed; flowers not in globose capitula
23 Herbs or dwarf shrubs with terete stems; leaves ± succulent   **LXXII. Crassulaceae**
23 Shrubs with angular stems; leaves not succulent   **XCIII. Coriariaceae**
11 Carpels obviously united for c. $\frac{1}{2}$ their length or more, or carpel solitary
24 Flowers actinomorphic
25 Corona of long filaments present inside the petals   **CXI. Passifloraceae**

25 Flowers without a corona
26 Petals more than 10
27 Aquatic herbs with petiolate leaves
28 Leaves floating, usually with a deep basal sinus   **LVIII. Nymphaeaceae**
28 Leaves not floating, peltate   **LIX. Nelumbonaceae**
27 Terrestrial herbs or shrubs with sessile or subsessile leaves
29 Stamens 4–6   **LXIII. Berberidaceae**
29 Stamens numerous   **LII. Aizoaceae**
26 Petals fewer than 10
30 Stamens more than twice as many as petals
31 Stamens with their filaments united into a tube   **CVI. Malvaceae**
31 Stamens free or united into bundles
32 Perianth-segments persistent in fruit, 2 large and 2 small   **XLVII. Polygonaceae**
32 Perianth-segments not as above
33 Ovary on a long gynophore   **LXVII. Capparaceae**
33 Ovary sessile or nearly so
34 Ovary surrounded by a cup-shaped perigynous zone; ovule 1   **LXXX. Rosaceae**
34 No cup-shaped perigynous zone; ovules 2 or more
35 Leaves 2-pinnate or simple phyllodes present   **LXXXI. Leguminosae**
35 Leaves not as above
36 Carpel 1; leaves 2-ternate, lower leaflets stalked   **LXI. Ranunculaceae**
36 Carpels 2 or more; leaves not as above
37 Large trees; inflorescence with a conspicuous bract partly adnate to peduncle   **CV. Tiliaceae**
37 Not as above
38 Styles more than 1, free
39 All or most leaves alternate; outer perianth-segments petaloid   **LXI. Ranunculaceae**
39 All leaves opposite or verticillate; outer perianth-segments sepaloid   **CIX. Guttiferae**
38 Style 1 or 0
40 Petals 4   **LXVI. Papaveraceae**
40 Petals 5
41 Ovary 1-locular or septate at base only; stamens numerous   **CXII. Cistaceae**
41 Ovary 3-locular; stamens 15   **LXXXV. Zygophyllaceae**
30 Stamens not more than twice as many as petals
42 Trees, shrubs or woody climbers
43 Flowers on tough leaf-like cladodes; leaves scale-like, brownish   **Liliaceae**
43 Not as above
44 Leaves small, scale-like or ericoid
45 Perianth-segments in 2 whorls of 3; stamens 3   **CXXXIII. Empetraceae**
45 Perianth-segments and stamens more than 3 in a whorl
46 Leaves opposite   **CXIV. Frankeniaceae**
46 Leaves alternate   **CXIII. Tamaricaceae**
44 Leaves neither scale-like nor ericoid
47 Peduncles adnate to petioles; ovary on a short gynophore   **LXXXIX. Cneoraceae**

47 Not as above
48 All leaves opposite
49 Leaves pinnate
50 Shrubs; fruit a capsule   **CI. Staphyleaceae**
50 Trees; fruit of 2 single-seeded samaras
   XCV. Aceraceae
49 Leaves entire or palmately lobed
51 Fruit of 2 single-seeded samaras; leaves usually palmately lobed.   **XCV. Aceraceae**
51 Fruit a fleshy capsule; leaves not palmately lobed   **C. Celastraceae**
48 At least some leaves alternate
52 Stamens 6   **LXVIII. Cruciferae**
52 Stamens 4, 5, 10 or 12
53 Stamens 4 or 5
54 Stamens opposite petals
55 Shrubs or small trees; petals shorter than sepals.   **CIII. Rhamnaceae**
55 Woody climbers; petals longer than sepals   **CIV. Vitaceae**
54 Stamens alternating with petals
56 Bark resinous; ovule 1   **XCIV. Anacardiaceae**
56 Bark not resinous; ovules several   **LXXVIII. Pittosporaceae**
53 Stamens 10 or 12
57 Leaves entire   **CXXXII. Ericaceae**
57 Leaves pinnate
58 Spiny trees   **LXXXI. Leguminosae**
58 Unarmed shrubs or small trees
59 Stamens free   **XCIV. Anacardiaceae**
59 Stamens with connate filaments   **XCI. Meliaceae**
42 Herbs, sometimes ± woody at base
60 Sepals 2, petals 5
61 Stems erect or procumbent, not twining   **LV. Portulacaceae**
61 Stems twining   **LVI. Basellaceae**
60 Sepals as many as the petals
62 Flowers 3-merous
62 Flowers 4- or more-merous   **Commelinaceae**
63 Leaves forming long pitchers; stigma very large, peltate   **LXX. Sarraceniaceae**
63 Not as above
64 Flowers strongly perigynous with a long tubular or campanulate receptacle   **CXIX. Lythraceae**
64 Flowers hypogynous or perigynous with a flat or weakly concave receptacle
65 Cauline leaves opposite or whorled
66 Leaves deeply divided, rarely only serrate
67 Petals 4   **LXVIII. Cruciferae**
67 Petals 5
68 Stamens without scales on the inner side of the filaments   **LXXXIII. Geraniaceae**
68 Stamens with scales on the inner side of the filaments   **LXXXV. Zygophyllaceae**
66 Leaves simple and entire
69 Leaves in 1 whorl; flower solitary, terminal   **Liliaceae**
69 Leaves opposite or in more than 1 whorl
70 Stipules present
71 Stipules scarious; land-plants   **LVII. Caryophyllaceae**
71 Stipules not scarious; usually submerged aquatics   **CXV. Elatinaceae**
70 Stipules absent
72 Sepals united to more than half-way
73 Styles connate; placentation parietal   **CXIV. Frankeniaceae**
73 Styles free; placentation free-central   **LVII. Caryophyllaceae**
72 Sepals free or united at base only

74 Ovary 1-celled; placentation free-central   **LVII. Caryophyllaceae**
74 Ovary 4- to 5-celled; placentation axile   **LXXXVI. Linaceae**
65 Leaves alternate or all basal
75 Leaves ternate   **LXXXII. Oxalidaceae**
75 Leaves not ternate
76 Sepals and petals 2–3   **XLVII. Polygonaceae**
76 Sepals and petals 4–5
77 Both whorls of perianth-segments green   **LXXX. Rosaceae**
77 Inner whorl of perianth-segments not green
78 Sepals and petals 4; stamens 4 or 6
79 Stipules absent; stamens usually 6   **LXVIII. Cruciferae**
79 Stipules present; stamens 4   **LVII. Caryophyllaceae**
78 Sepals and petals 5; stamens 5 or 10
80 Leaves with conspicuous, red, viscid, glandular hairs   **LXXI. Droseraceae**
80 Not as above
81 Leaves with numerous pellucid glands, strongly scented when crushed   **LXXXVIII. Rutaceae**
81 Leaves without pellucid glands
82 Style 1; stigma entire or shallowly lobed; anthers opening by pores   **CXXXI. Pyrolaceae**
82 Style or stigmas more than 1; anthers opening by longitudinal slits
83 Stigmas 5
84 Leaves lobed or pinnate   **LXXXIII. Geraniaceae**
84 Leaves entire
85 Sepals united; leaves basal   **CXXXVI. Plumbaginaceae**
85 Sepals free; leaves cauline   **LXXXVI. Linaceae**
83 Stigmas 2–4
86 Flowers with conspicuous glandular-fimbriate staminodes   **LXXIV. Parnassiaceae**
86 Glandular-fimbriate staminodes absent
87 Stamens 5   **LVII. Caryophyllaceae**
87 Stamens 10   **LXXIII. Saxifragaceae**
24 Flowers zygomorphic
88 Flowers saccate or spurred at base
89 Sepals 2, small   **LXVI. Papaveraceae**
89 Sepals 3 or 5
90 Sepals 3, very unequal, 1 spurred; petals 3, not spurred   **XCVIII. Balsaminaceae**
90 Sepals 5; petals 5
91 Leaves peltate   **LXXXIV. Tropaeolaceae**
91 Leaves not peltate
92 Leaves alternate   **CX. Violaceae**
92 Leaves opposite   **LXXXIII. Geraniaceae**
88 Flowers not saccate or spurred at base
93 All, or all but one, of the stamens united into a tube   **LXXXI. Leguminosae**
93 All stamens free
94 Trees or shrubs
95 Leaves simple
96 Ovary on a long gynophore   **LXVII. Capparaceae**
96 Ovary sessile
97 Petals 4   **LXVIII. Cruciferae**
97 Petals 5   **LXXXI. Leguminosae**
95 Leaves compound
98 Leaves trifoliolate or pinnate   **LXXXI. Leguminosae**
98 Leaves palmate with more than 3 leaflets   **XCVII. Hippocastanaceae**
94 Herbs

99 Ovary and fruit deeply 5-lobed
  100 Flowers in umbellate cymes; fruit with a long beak   **LXXXIII. Geraniaceae**
  100 Flowers in racemes; fruit not beaked   **LXXXVIII. Rutaceae**
99 Ovary and fruit not deeply 5-lobed
  101 Petals fimbriate or lobed.   **LXIX. Resedaceae**
  101 Petals entire or emarginate
    102 Stamens 10   **LXXXI. Leguminosae**
    102 Stamens not more than 6
      103 Sepals inserted on a cup-like perigynous zone   **LVII. Caryophyllaceae**
      103 Sepals free
        104 Ovary 2-locular; gynophore short or 0   **LXVIII. Cruciferae**
        104 Ovary 1-locular; gynophore long   **LXVII. Capparaceae**
10 Ovary inferior or partly so
105 Petals numerous
  106 Aquatic plants; leaves not succulent   **LVIII. Nymphaeaceae**
  106 Land-plants; leaves succulent   **LII. Aizoaceae**
105 Petals 5 or fewer
  107 Petals and sepals 3
    108 Flowers zygomorphic
      109 Style and filaments obvious   **Iridaceae**
      109 Stigma and stamens sessile   **Orchidaceae**
    108 Flowers actinomorphic
      110 Outer perianth-whorl sepaloid   **Hydrocharitaceae**
      110 Both perianth-whorls petaloid
        111 Stamens 6   **Amaryllidaceae**
        111 Stamens 3   **Iridaceae**
  107 Petals and sepals 2, 4 or 5
    112 Stamens numerous
      113 Leaves opposite, with pellucid glands   **CXXI. Myrtaceae**
      113 Leaves alternate, without pellucid glands
      114 Leaves entire; seeds covered with pulp   **CXXII. Punicaceae**
      114 Leaves serrulate; seeds dry
        115 Styles free; fruit fleshy   **LXXX. Rosaceae**
        115 Styles united, except at the top; fruit a capsule   **LXXV. Hydrangeaceae**
    112 Stamens 10 or fewer
      116 Aquatic; leaves pinnate, segments filiform; flowers in spikes   **CXXIV. Haloragaceae**
      116 Not as above
      117 Trees, shrubs or woody climbers
        118 Flowers in umbels
          119 Climbers   **CXXVIII. Araliaceae**
          119 Erect shrubs
            120 Evergreen; umbels flat   **CXXIX. Umbelliferae**
            120 Deciduous; umbels globose   **CXXVII. Cornaceae**
        118 Flowers not in umbels
          121 Leaves palmately lobed   **LXXVII. Grossulariaceae**
          121 Leaves not lobed
            122 Both perianth-whorls petaloid   **CXXIII. Onagraceae**
            122 Outer perianth-whorl sepaloid
              123 Calyx-teeth very small; ovules 1 in each carpel; fruit a drupe   **CXXVII. Cornaceae**
              123 Calyx-teeth large; ovules numerous; fruit a capsule
                124 Stamens 10   **LXXV. Hydrangeaceae**
                124 Stamens 5   **LXXVI. Escalloniaceae**
      117 Herbs
        125 Both perianth-whorls sepaloid   **LXXX. Rosaceae**
        125 Inner perianth-whorl petaloid
          126 Petals 5
            127 Stamens 5   **CXXIX. Umbelliferae**
            127 Stamens 10   **LXXIII. Saxifragaceae**
          126 Petals 4 or 2

128 Flowers in umbels surrounded by 4 conspicuous white bracts   **CXXVII. Cornaceae**
128 Flowers not in umbels; no conspicuous white bracts   **CXXIII. Onagraceae**
9 Petals all united at base into a longer or shorter tube
129 Ovary superior
  130 Flowers papilionate
    131 Sepals free; stamens 8   **XCII. Polygalaceae**
    131 Sepals connate; stamens 10   **LXXXI. Leguminosae**
  130 Flowers not papilionate
    132 Stamens at least twice as many as corolla-lobes
      133 Herbs with succulent leaves   **LXXII. Crassulaceae**
      133 Shrubs or trees
        134 Flowers unisexual   **CXXXVII. Ebenaceae**
        134 Flowers hermaphrodite
          135 Anthers opening by pores; hairs simple or scale-like   **CXXXII. Ericaceae**
          135 Anthers opening by longitudinal slits; hairs stellate   **CXXXVIII. Styracaceae**
    132 Stamens as many as or fewer than corolla-lobes
      136 Plant without chlorophyll; leaves scale-like
        137 Flowers zygomorphic; stem stout, erect   **CLX. Orobanchaceae**
        137 Flowers actinomorphic; stem slender, twining   **CXLVI. Convolvulaceae**
      136 Green plants
        138 Sepals 2; flowers actinomorphic
          139 Petals 2; leaves in a rosette   **Eriocaulaceae**
          139 Petals 5; leaves not in a rosette   **LV. Portulacaceae**
        138 Sepals more than 2, or flowers zygomorphic
          140 Ovary deeply 4-lobed with 1 ovule in each lobe
            141 Leaves alternate   **CXLVIII. Boraginaceae**
            141 Leaves opposite   **CLI. Labiatae**
          140 Ovary not 4-lobed
            142 Flowers actinomorphic or nearly so
              143 Carpels free
                144 Leaves peltate; carpels 5   **LXXII. Crassulaceae**
                144 Leaves not peltate; carpels 2
                  145 Corolla with a corona; styles 2, free but united by the stigma   **CXLIII. Asclepiadaceae**
                  145 Corolla without a corona; styles 2, united except at the very base   **CXLII. Apocynaceae**
              143 Carpels united
                146 Stamens fewer than corolla-lobes
                  147 Herbs   **CLIV. Scrophulariaceae**
                  147 Shrubs or trees
                    148 Leaves opposite   **CXXXIX. Oleaceae**
                    148 Leaves alternate
                      149 Leaves with numerous pellucid glands   **CLXII. Myoporaceae**
                      149 Leaves without pellucid glands
                        150 Flowers yellow   **CXXXIX. Oleaceae**
                        150 Flowers not yellow   **CLIV. Scrophulariaceae**
              146 Stamens as many as corolla-lobes
                151 Stamens opposite the corolla-lobes
                  152 Styles or stigmas more than 1; ovule 1   **CXXXVI. Plumbaginaceae**
                  152 Style 1; stigma 1; ovules numerous
                    153 Herbs   **CXXXV. Primulaceae**
                    153 Shrubs   **CXXXIV. Myrsinaceae**
              151 Stamens alternating with the corolla-lobes
                154 Leaves opposite
                  155 Shrubs
                    156 Large, erect; leaves deciduous   **CLIII. Buddlejaceae**
                    156 Small, procumbent; leaves evergreen
                      157 Leaves elliptical or oblong; flowers pink   **CXXXII. Ericaceae**
                      157 Leaves spathulate; flowers white   **CXXX. Diapensiaceae**
                  155 Herbs
                    158 Land-plants; leaves sessile   **CXL. Gentianaceae**

158 Aquatic plants; leaves petiolate
    CXLI. Menyanthaceae
154 Leaves alternate or all basal
159 Sepals, petals and stamens 4
160 Shrubs   **XCIX. Aquifoliaceae**
160 Herbs
  161 Corolla not violet-blue   **Plantaginaceae**
  161 Corolla violet-blue   **CLIX. Gesneriaceae**
159 Sepals, petals and stamens 5 (rarely sepals fewer)
162 Ovary 3-celled; stigmas 3 or 3-lobed
  163 Leaves pinnate   **CXLV. Polemoniaceae**
  163 Leaves simple   **CXXX. Diapensiaceae**
162 Ovary 2-celled; stigmas 2 or 1
  164 Ovules 4 or fewer
    165 Flowers numerous, in scorpioid cymes; corolla-lobes distinct
      CXLVIII. Boraginaceae
    165 Flowers solitary or few, not in scorpioid cymes; corolla not or scarcely lobed   **CXLVI. Convolvulaceae**
  164 Ovules numerous
    166 Aquatic or bog-plants; corolla fimbriate
      CXLI. Menyanthaceae
    166 Land-plants; corolla not fimbriate
      167 Leaves all basal   **CLIX. Gesneriaceae**
      167 Some leaves cauline
        168 Style deeply divided
          CXLVII. Hydrophyllaceae
        168 Style undivided
          169 Corolla-tube much shorter than lobes; stamens patent **CLIV. Scrophulariaceae**
          169 Corolla-tube long, or anthers connivent   **CLII. Solanaceae**
142 Flowers strongly zygomorphic
170 Anthers opening by pores   **CXXXII. Ericaceae**
170 Anthers opening by slits
  171 Calyx with patent spines and erect, membranous, usually dark-spotted lobes
    CXXXV. Primulaceae
  171 Calyx not as above
    172 Flowers small, crowded in capitula
      CLV. Globulariaceae
    172 Flowers not in capitula
      173 Ovary 1-celled; carnivorous plants
        CLXI. Lentibulariaceae
      173 Ovary 2-celled; not carnivorous plants
        174 Ovules numerous
          175 Capsule not more than twice as long as wide   **CLIV. Scrophulariaceae**
          175 Capsule many times longer than wide
            176 Capsule with a short beak
              CLVII. Pedaliaceae
            176 Capsule with a horn 8–20 cm
              CLVIII. Martyniaceae
        174 Ovules 4
          177 Bracts shorter than calyx
            CXLIX. Verbenaceae
          177 Bracts or bracteoles much longer than calyx   **CLVI. Acanthaceae**
129 Ovary inferior
178 Stamens 8–10, or 4–5 with filaments divided to base
179 Herb; anthers opening by slits; leaves ternate
    Adoxaceae
179 Woody; anthers opening by pores; leaves simple
    CXXXII. Ericaceae
178 Stamens 5 or fewer; filaments not divided
180 Leaves in whorls of 4 or more   **CXLIV. Rubiaceae**
180 Leaves not in whorls
  181 Stamens opposite corolla-lobes   **CXXXV. Primulaceae**
  181 Stamens alternating with corolla-lobes
    182 Leaves opposite; stipules interpetiolar
      CXLIV. Rubiaceae

    182 Leaves alternate, or stipules not interpetiolar
      183 Flowers in capitula surrounded by an involucre of more than 2 bracts
        184 Anthers coherent in a ring round the style
          185 Ovule 1; calyx, if present, represented by hairs or scales   **Compositae**
          185 Ovules numerous; calyx-lobes conspicuous, green   **Campanulaceae**
        184 Anthers free
          186 Ovules numerous; corolla-lobes longer than tube   **Campanulaceae**
          186 Ovule 1; corolla-lobes much shorter than tube
            Dipsacaceae
      183 Flowers not in capitula, or bracts 2
        187 Anthers cohering in a tube round the style
          Lobeliaceae
        187 Anthers not cohering to one another
          188 Anthers sessile; pollen-grains cohering in pollinia   **Orchidaceae**
          188 Stamens with filaments; pollen-grains free
            189 Stamens 1–3
              190 Perianth 4- to 5-merous   **Valerianaceae**
              190 Perianth 3-merous
                191 Stamens 5–6; flowers usually unisexual; fruit indehiscent   **Musaceae**
                191 Stamen 1; flowers usually hermaphrodite; fruit dehiscent
                  192 Sepals connate into a tube   **Zingiberaceae**
                  192 Sepals free   **Cannaceae**
             189 Stamens 4–5
              193 Shrubs (sometimes small and creeping), or woody climbers   **Caprifoliaceae**
              193 Herbs
                194 Tendrils present   **CXVII. Cucurbitaceae**
                194 Tendrils absent
                  195 Leaves pinnate   **Caprifoliaceae**
                  195 Leaves not pinnate
                    196 Flowers hermaphrodite; fruit a capsule
                      Campanulaceae
                  196 Flowers unisexual; fruit fleshy
                    CXVII. Cucurbitaceae
8 Perianth not of 2 or more markedly different whorls
197 Perianth entirely petaloid
198 Parasites or saprophytes without chlorophyll
  199 Flowers mostly unisexual; stamen 1
    XLVI. Balanophoraceae
  199 Flowers hermaphrodite; stamens 6–16
    200 Filaments free   **CXXXI. Pyrolaceae**
    200 Filaments united into a column   **XLV. Rafflesiaceae**
198 Green plants
  201 Perianth-segment 1, bract-like   **Aponogetonaceae**
  201 Perianth-segments more than 1, or perianth tubular
    202 Stems succulent, leafless but with groups of spines
      CXVIII. Cactaceae
    202 Not as above
      203 Stamens more than 12
        204 Herbs, or, rarely, woody climbers with pinnate leaves   **LXI. Ranunculaceae**
        204 Trees with simple leaves   **LXIV. Magnoliaceae**
      203 Stamens 12 or fewer
        205 Flowers in ovoid capitula without an involucre
          LXXX. Rosaceae
        205 Flowers not in capitula, or capitula with an involucre
          206 Ovary superior
            207 Perianth-segments 4
              208 Flowers zygomorphic   **XLI. Proteaceae**
              208 Flowers actinomorphic
                209 Perianth tubular below   **CVII. Thymelaeaceae**
                209 Perianth-segments free
                  210 Herbs   **Liliaceae**
                210 Shrubs   **XLVII. Polygonaceae**
            207 Perianth-segments more than 4

211 Carpels more than 1, free or nearly so
  212 Leaves triquetrous, all basal   **Butomaceae**
  212 Leaves flat, cauline   **LI. Phytolaccaceae**
211 Carpel 1, or carpels obviously united
  213 Perianth-segments 6
    214 Terrestrial or marsh plants; inflorescence not subtended by a spathe-like leaf-sheath   **Liliaceae**
    214 Aquatic herbs; inflorescence subtended by a spathe-like leaf-sheath   **Pontederiaceae**
  213 Perianth-segments 5
    215 Stigmas 2–3; stipules sheathing, scarious   **XLVIII. Polygonaceae**
    215 Stigma 1; stipules absent
      216 Ovules numerous; perianth divided almost to base   **CXXXV. Primulaceae**
      216 Ovule 1; perianth with a long tube   **L. Nyctaginaceae**
206 Ovary inferior, or flowers male
217 Leaves in whorls of 4 or more   **CXLIV. Rubiaceae**
217 Leaves not in whorls
  218 Flowers in capitula surrounded by an involucre
    219 Anthers cohering in a tube round the style, or flowers unisexual   **Compositae**
    219 Anthers free; flowers hermaphrodite   **Dipsacaceae**
  218 Flowers not in capitula, though sometimes shortly pedicellate in compact umbels
    220 Ovules numerous
      221 Perianth-segments 3, or perianth tubular with a unilateral entire limb   **XLIV. Aristolochiaceae**
      221 Perianth-segments 6
        222 Stamens 6   **Amaryllidaceae**
        222 Stamens 3   **Iridaceae**
    220 Ovules 1 or 2
      223 Leaves opposite   **Valerianaceae**
      223 Leaves alternate
        224 Flowers in simple cymes or solitary   **XLII. Santalaceae**
        224 Flowers in umbels or superposed whorls   **CXXIX. Umbelliferae**
197 Perianth not petaloid, often absent, if brightly coloured then dry and scarious
225 Trees or shrubs, sometimes small
226 Parasitic on branches of trees or shrubs   **XLIII. Loranthaceae**
226 Not parasitic
227 Stems creeping or climbing with adventitious roots; evergreen   **CXXVIII. Araliaceae**
227 Not as above
  228 Flowers borne on flattened evergreen cladodes; leaves small, brownish, scale-like   **Liliaceae**
  228 Not as above
    229 Most leaves opposite or subopposite
      230 Stems green and fleshy or leaves fleshy   **XLVIII. Chenopodiaceae**
      230 Neither leaves nor stems fleshy
        231 Styles 3   **CII. Buxaceae**
        231 Styles 4, or 1
          232 Flowers in catkins   **XXXI. Salicaceae**
          232 Flowers not in catkins
            233 Leaves pinnate; stamens 2   **CXXXIX. Oleaceae**
            233 Leaves simple; stamens 4 or more
              234 Stamens 5, alternating with sepals   **CIII. Rhamnaceae**
              234 Stamens 8; sepals 5   **XCV. Aceraceae**
    229 Most leaves alternate
      235 Leaves pinnate
        236 Ovary inferior; styles 2; pith septate   **XXXIII. Juglandaceae**
        236 Ovary superior; styles 3 or 1; pith not septate

237 Style 1; fruit a lomentum   **LXXXI. Leguminosae**
237 Styles 3; fruit a dry, 1-seeded drupe   **XCIV. Anacardiaceae**
235 Leaves simple
  238 Leaves not more than 2 mm wide, oblong or linear
    239 Stigma 1   **CVII. Thymelaeaceae**
    239 Stigmas 2–9
      240 Stamens 3   **CXXXIII. Empetraceae**
      240 Stamens 5   **XLVIII. Chenopodiaceae**
  238 Leaves more than 2 mm wide
    241 Petiole with dilated base, enclosing the bud   **LXXIX. Platanaceae**
    241 Petiole-base not enclosing the bud
      242 Anthers opening by transverse valves   **LXV. Lauraceae**
      242 Anthers opening by longitudinal slits
        243 Flowers not in catkins or dense heads
          244 Inflorescence of several male flowers, each of 1 stamen, and a female flower, appearing as a stalked ovary, all surrounded by 4 or 5 conspicuous glands; latex present   **LXXXVII. Euphorbiaceae**
          244 Inflorescence not as above; no latex
            245 Flowers unisexual
              246 Peltate scale-like silvery or ferrugineous hairs present beneath the leaves and often elsewhere; ovary 1-locular; fruit fleshy   **CVIII. Elaeagnaceae**
              246 No scale-like hairs; ovary 3-locular; fruit dry   **LXXXVII. Euphorbiaceae**
            245 Flowers hermaphrodite
              247 Trees; perianth-tube short, with stamens inserted near its base.   **XXXVII. Ulmaceae**
              247 Shrubs; perianth-tube long, with stamens inserted near its apex   **CVII. Thymelaeaceae**
        243 Flowers in catkins or dense heads
          248 Latex present; fruit or false fruit fleshy   **XXXVIII. Moraceae**
          248 Latex absent; fruit dry
            249 Dioecious; perianth absent
              250 Bracts (catkin-scales) fimbriate or lobed at apex; flowers with a cup-like disc   **XXXI. Salicaceae**
              250 Bracts (catkin-scales) entire; disc absent
                251 Leaves without pellucid glands; stamens with long filaments; ovules numerous   **XXXI. Salicaceae**
                251 Leaves with pellucid glands; stamens with short filaments; ovule 1   **XXXII. Myricaceae**
            249 Monoecious; perianth present in male or female flowers or both
              252 Styles 3 or more; flowers of both sexes with perianth   **XXXVI. Fagaceae**
              252 Styles 2; perianth present in flowers of 1 sex only
                253 Male flowers 3 to each bract; perianth present   **XXXIV. Betulaceae**
                253 Male flowers 1 to each bract; perianth absent   **XXXV. Corylaceae**
225 Herbs
254 Perianth absent or represented by scales or bristles, minute in flower; flowers in the axils of bracts, a number of which are usually closely imbricate on a rhachis, forming a spikelet; leaves usually linear, grass-like, sheathing below
  255 Flowers usually with a bract above and below; sheaths usually open; stems usually with hollow internodes   **Gramineae**

255 Flowers with a bract below only; sheaths usually closed; stems usually with solid internodes **Cyperaceae**

254 Perianth present, or flowers not arranged in spikelets

256 Aquatic plants; leaves submerged or floating; inflorescence sometimes emergent

257 Leaves divided into numerous filiform segments

258 Leaves pinnately divided; flowers in a terminal spike **CXXIV. Haloragaceae**

258 Leaves dichotomously divided; flowers solitary, axillary **LX. Ceratophyllaceae**

257 Leaves entire or dentate

259 Flowers in spikes

260 Rhizome densely covered with stiff fibres; spikes subtended by a group of leaf-like bracts (marine) **Posidoniaceae**

260 Not as above

261 Flowers unisexual, arranged on one side of a flat rhachis (marine) **Zosteraceae**

261 Flowers hermaphrodite, arranged all round or on 2 sides of a terete rhachis (fresh or brackish water)

262 Spikes 2-flowered; carpels in fruit with stalks several times their own length **Ruppiaceae**

262 Spikes more than 2-flowered; carpels sessile in fruit **Potamogetonaceae**

259 Flowers not in spikes

263 Flowers (at least the fertile) solitary or few, sessile or shortly pedicellate

264 Leaves in whorls of 8 or more **CXXVI. Hippuridaceae**

264 Leaves not in whorls of 8 or more

265 Carpels 2 or more, free **Zannichelliaceae**

265 Carpels united, or solitary

266 Perianth-segments 4–6; stamens 4 or more; leaves ovate to obovate

267 Perianth-segments 4; ovary inferior **CXXIII. Onagraceae**

267 Perianth-segments 6; ovary superior **CXIX. Lythraceae**

266 Perianth-segments fewer than 4, or perianth absent, stamens 1–3; leaves linear to lanceolate

268 Perianth present, of 3 segments; ovary inferior; stamens 2–3 **Hydrocharitaceae**

268 Perianth absent or 2-lipped; ovary superior; stamen 1

269 Leaves entire, without sheathing base; ovary compressed, deeply 4-lobed **CL. Callitrichaceae**

269 Leaves spinulose-dentate, with sheathing base; ovary terete, not lobed **Najadaceae**

263 Flowers in heads on long peduncles or in compound inflorescences

270 Flowers hermaphrodite; heads few-flowered **Juncaceae**

270 Flowers unisexual; heads many-flowered

271 Leaves all basal; heads solitary on long scapes **Eriocaulaceae**

271 Some leaves cauline; inflorescence with female heads below and male heads above **Sparganiaceae**

256 Terrestrial plants or, if aquatic, with inflorescence and either stems or leaves emergent

272 Climbing plants with unisexual flowers

273 Leaves opposite; perianth-segments 5 **XXXIX. Cannabaceae**

273 Leaves alternate; perianth-segments 6 **Dioscoreaceae**

272 Not climbing, or rarely climbers with hermaphrodite flowers

274 Leaves linear

275 Flowers unisexual

276 Female flowers solitary; male flowers solitary or in short cymes **XLVIII. Chenopodiaceae**

276 Male and female flowers numerous, in dense heads or spikes

277 Male and female (and some hermaphrodite) flowers mixed together in the same spike; stamen 1 **Lilaeaceae**

277 Male and female flowers separate in the inflorescence; stamens 2 or more

278 Male and female flowers in separate globose heads **Sparganiaceae**

278 Flowers in a dense cylindrical spike, male above, female below **Typhaceae**

275 Flowers hermaphrodite

279 Plant densely pubescent **XLVIII. Chenopodiaceae**

279 Plant glabrous or sparsely hairy

280 Flowers in dense spikes; spikes apparently lateral on a flattened leaf-like stem **Araceae**

280 Not as above

281 Carpel 1

282 Leaves not subverticillate, exstipulate **XLVIII. Chenopodiaceae**

282 Leaves subverticillate, with minute stipules **LVII. Caryophyllaceae**

281 Carpels more than 1

283 Carpels free (except at base); leaves with a conspicuous pore at apex **Scheuchzeriaceae**

283 Carpels ± completely united; leaves without a conspicuous pore at apex

284 Flowers in unbranched racemes; styles short or 0 **Juncaginaceae**

284 Flowers in cymes in a branched inflorescence; styles 3, distinct **Juncaceae**

274 Leaves lanceolate or wider, or sometimes small and scale-like, but never linear

285 Leaves compound

286 Flowers in compound umbels **CXXIX. Umbelliferae**

286 Flowers not in compound umbels

287 Flowers in capitula

288 Leaves simply pinnate; styles 1 or 2 **LXXX. Rosaceae**

288 Leaves ternate; styles 3–5 **Adoxaceae**

287 Flowers not in capitula

289 Stamens numerous **LXI. Ranunculaceae**

289 Stamens 4 or 5(–10)

290 Epicalyx present **LXXX. Rosaceae**

290 Epicalyx absent **LXXXIII. Geraniaceae**

285 Leaves simple or apparently absent

291 Flowers numerous, small, crowded on an axis (spadix) subtended and often ± enclosed by a conspicuous bract (spathe) **Araceae**

291 Not as above

292 Inflorescence of several male flowers, each of 1 stamen, and a female flower, appearing as a stalked ovary, all surrounded by 4 or 5 conspicuous glands; latex present **LXXXVII. Euphorbiaceae**

292 Not as above

293 Leaves apparently absent; stem green and succulent **XLVIII. Chenopodiaceae**

293 Leaves obvious; stem not succulent

294 Lower leaves opposite, upper alternate; monoecious; male flowers with 2-partite perianth, female with tubular perianth **CXXV. Theligonaceae**

294 Not as above

295 Plant densely clothed with stellate hairs; ovary 3-locular with 1 ovule in each loculus **LXXXVII. Euphorbiaceae**

295 Not as above

296 Densely papillose annuals

297 Leaves oblong-lanceolate, never hastate; fruit opening by 5 valves **LII. Aizoaceae**

297 Leaves ovate-rhombic, often hastate; fruit indehiscent **LIV. Tetragoniaceae**

296 Not densely papillose annuals
298 Leaves whorled
  299 Stigma 1; stems hollow
                  CXXVI. Hippuridaceae
  299 Stigmas 3; stems solid   **LIII. Molluginaceae**
298 Leaves not in whorls
  300 Leaves alternate or all basal (rarely the
        lower opposite)
    301 Stamens numerous; carpels free except
        sometimes at base   **LXI. Ranunculaceae**
    301 Stamens 12 or fewer; carpels not free,
        or one only
      302 Carpels attached to a central axis,
          otherwise free   **LI. Phytolaccaceae**
      302 Carpels united, or one only
        303 Stamens 12   **XLIV. Aristolochiaceae**
        303 Stamens 10 or fewer
          304 Stipules united into a sheath
                      XLVII. Polygonaceae
          304 Stipules free or absent
            305 Leaves very large, palmately lobed,
               all basal; inflorescence of dense
               many-flowered spikes much
               shorter than the leaves
                     CXXIV. Haloragaceae
            305 Not as above
              306 Epicalyx present; stipules leaf-
                like   **LXXX. Rosaceae**
              306 Epicalyx 0; stipules small or 0
                307 Ovary superior
                  308 Perianth tubular below
                    309 Ovule basal
                        XLVIII. Chenopodiaceae
                    309 Ovule pendent
                        CVII. Thymelaeaceae
                308 Perianth-segments free or near-
                  ly so, rarely absent in female
                  flowers
                  310 Perianth-segments 4
                    311 Flowers in ebracteate racemes
                        LXVIII. Cruciferae
                    311 Flowers in axillary clusters
                        XL. Urticaceae
                  310 Perianth-segments 5

312 Perianth herbaceous, rarely
      absent in female flowers
             XLVIII. Chenopodiaceae
312 Perianth scarious
             XLIX. Amaranthaceae
307 Ovary inferior
313 Leaves reniform, cordate
             LXXIII. Saxifragaceae
313 Leaves subulate to linear-
    lanceolate   **XLII. Santalaceae**
300 Leaves opposite (rarely a few upper
    apparently alternate)
  314 Leaves toothed or lobed
    315 Flowers hermaphrodite
      316 Ovary inferior; stigmas 2
            LXXIII. Saxifragaceae
      316 Ovary superior; stigmas 5
            LXXXIII. Geraniaceae
    315 Flowers unisexual
      317 Perianth-segments 4 or 2; style 1
               XL. Urticaceae
      317 Perianth-segments 3; styles 2
            LXXXVII. Euphorbiaceae
  314 Leaves entire
    318 Perianth 0; ovary compressed, 4-lobed
               CL. Callitrichaceae
    318 Perianth present; ovary not com-
      pressed and 4-lobed
      319 Perianth-segments 3
               XLVII. Polygonaceae
      319 Perianth-segments 4 or more
      320 Ovary inferior   **CXXIII. Onagraceae**
      320 Ovary superior
        321 Perianth-segments 6 or 12; style
        and stigma 1   **CXIX. Lythraceae**
        321 Perianth-segments 4 or 5; styles or
        stigmas 2 or more
          322 Leaves without a long spinose
          apex; fruit unwinged
             LVII. Caryophyllaceae
        322 Leaves with a long spinose apex;
        fruit transversely winged
             XLVIII. Chenopodiaceae

EXPLANATORY NOTES ON THE TEXT

General notes

The sequence of families is that of Melchior in Engler, *Syllabus der Pflanzenfamilien* ed. 12 (1964).

Descriptions of taxa refer only to the European populations of the taxon in question. If extra-European representatives differ substantially, an explanatory note is sometimes added.

Groups of species have been used in some genera where the species are very difficult to separate. These groups have no formal nomenclatural status and are simply a device to enable a partial identification to be made.

Taxa below the rank of subspecies are neither keyed nor described, and varieties are mentioned only when there are special reasons.

Aliens are included only when they appear to be effectively naturalized or when planted in continuous stands on a fairly large scale.

Hybrids are mentioned only when they occur frequently.

A measurement given without qualification refers to length. Two measurements connected by × indicate length followed by width. Further measurements in parentheses indicate exceptional cases outside the normal range.

Synonyms given in the text are principally those names under which the species or subspecies is described in the Basic Floras listed on p. xix. The index contains (in addition to these) names which occur in any of the Standard Floras (p. xix) or in well-known monographs.

Chromosome numbers are given only when the editors are satisfied that the count has been made on correctly identified material known to be of wild European origin. For naturalized and cultivated species the count is from material which is naturalized or is cultivated in the way which justifies its inclusion in the Flora.

Ecological information is provided only when the habitat-preference of a species is sufficiently uniform over its European range to permit it to be summed up in a short phrase.

Geographical terms such as 'W. Europe', 'Mediterranean region', etc., are to be interpreted as shown on maps IV and V. The statement that a plant occurs in one or more of these regions does not necessarily imply that it occurs throughout the region.

Extra-European distribution is indicated only for those plants whose European range is small and whose range outside Europe is considerably greater, or for species which are not native in Europe.

SPERMATOPHYTA

ANGIOSPERMAE

DICOTYLEDONES

(continued)

DIAPENSIALES

CXXX. DIAPENSIACEAE[1]

Perennial herbs or dwarf shrubs. Leaves simple, exstipulate. Flowers actinomorphic, 5-merous. Petals connate towards the base. Stamens alternating with the petals; anthers opening by slits; pollen not in tetrads. Ovary superior, 3-locular; placentation axile; style 1. Fruit a loculicidal capsule.

1. Diapensia L.[2]

Flowers solitary, pedunculate, erect. Calyx deeply 5-lobed, coriaceous. Anthers broadly ovate, with divergent loculi. Ovary deeply 3-lobed; style slender; stigma 3-lobed. Seeds numerous, small.

1. **D. lapponica** L., *Sp. Pl.* 141 (1753). Evergreen, densely caespitose herb up to 6 cm. Leaves 5–10 mm, obovate-spathulate, rounded at apex, coriaceous, crowded. Peduncles 5–40 mm, each with usually 1 bract and 2 bracteoles, the latter close below the flower. Calyx-lobes *c.* 5 mm, obtuse. Petals up to 10 mm, obovate, white. Stamens with wide, flattened filaments. Capsule *c.* 4 mm, ovoid. $2n=12$. *Rock-crevices and stony ground. Arctic Europe, extending southwards in the mountains to c. 56° 40′ N. in Scotland and c. 60° N. in Norway.* Br Fe Is No Rs (N) Su.

ERICALES

CXXXI. PYROLACEAE[3]

Small perennial herbs (rarely dwarf shrubs). Leaves simple, alternate or opposite. Flowers hermaphrodite, actinomorphic, (4-)5-merous. Sepals free or united at base. Petals free. Stamens twice as many as petals, obdiplostemonous. Stigma (4-)5-lobed. Ovary superior, 4- to 5-locular, with parietal placentae; fruit a loculicidal capsule; seeds numerous, small, with the testa prolonged into two appendages.

Literature: H. Andres, *Verh. Bot. Ver. Brandenb.* 56: 1–76 (1914).

1 All leaves scale-like, without chlorophyll   **5. Monotropa**
1 Green leaves present
  2 Leaves mostly cauline; flowers in corymbs or umbels   **4. Chimaphila**
  2 Leaves mostly basal; flowers solitary or in racemes
   3 Flowers solitary   **3. Moneses**
   3 Flowers in racemes
    4 Racemes secund; pollen-grains free   **2. Orthilia**
    4 Racemes not secund; pollen-grains in tetrads   **1. Pyrola**

1. Pyrola L.[4]

Glabrous, rhizomatous herbs. Leaves green, somewhat coriaceous, alternate, in lax basal rosettes; cauline leaves reduced to small scales. Flowers in terminal racemes, not secund. Anthers opening by pores at the end of short tubes. Pollen-grains in tetrads.

Literature: B. Křísa, *Nov. Bot. Inst. Bot. Univ. Carol. Prag.* 1965: 31–35 (1965). G. Knaben & T. Engelskjön, *Årbok Univ. Bergen* 1967(4): 1–71 (1968).

1 Style not expanded into a disc below the stigma
  2 Style 1–2 mm, included   **1. minor**
  2 Style at least 5 mm, exserted
   3 Bracts lanceolate, equalling or shorter than pedicels   **6. norvegica**
   3 Bracts ovate-lanceolate, longer than pedicels   **7. carpatica**
1 Style expanded into a disc below the stigma
  4 Style straight; petiole not longer than lamina   **2. media**
  4 Style curved; petiole longer than lamina
   5 Flowers 15–20 mm in diameter; leaves subentire   **5. grandiflora**
   5 Flowers 8–12 mm in diameter; leaves crenulate
    6 Calyx-lobes 1·5–2 mm, deltate, acute; petals yellowish-green   **3. chlorantha**
    6 Calyx-lobes 2–4·5 mm, lanceolate, acuminate; petals white   **4. rotundifolia**

1. **P. minor** L., *Sp. Pl.* 396 (1753). Stem 5–20(–30) cm. Leaves 2·5–6 × 1·5–5 cm, broadly elliptical, obtuse to subacute, crenulate-serrulate; petiole shorter than lamina. Bracts linear-lanceolate, at least as long as pedicels. Flowers 5–7 mm in diameter, globose. Calyx-lobes 1·5–2 mm, deltate, acute to subobtuse. Petals whitish to lilac-pink. Anthers 0·8–1·2 mm. Style 1–2 mm, not expanded below the stigma, straight, included. $2n=46$. *Woods, moors and other humus-rich soils; occasionally on sand-dunes. Most of Europe, but rare in the south.* All except Az Bl Cr Lu Sa Sb Si Tu.

2. **P. media** Swartz, *Kungl. Svenska Vet.-Akad. Handl.* nov. ser., 25: 257 (1804). Stem (5–)15–30 cm. Leaves 3–5 × 2–5 cm, ovate to orbicular, obtuse, crenulate; petiole shorter than or equalling lamina. Bracts ovate to lanceolate, acuminate, about equalling or longer than pedicels. Flowers 7–10 mm in diameter, globose. Calyx-lobes *c.* 2·5 mm, triangular-ovate, obtuse or subacute. Petals white or pink. Anthers *c.* 2·5 mm. Style 4–6 mm, expanded into a disc below the stigma, exserted, straight. $2n=92$. *Woods, moors and heaths. N. & C. Europe, extending southwards locally, and mainly in the mountains, to N. Italy, Macedonia, Krym and S. Ural.* Au Br Bu Cz Da Fe Ga Ge Hb He Hu It Ju No Po Rm Rs (N, B, C, W, K, E) Su.

3. **P. chlorantha** Swartz, *op. cit.* 31: 190 (1810) (*P. virens* Koerte, *P. virescens* auct.). Stem 10–30 cm. Leaves 0·5–2(–3) × 0·5–2·5 cm, pale green above, darker beneath, oblong-obovate

[1] Edit. T. G. Tutin.
[2] By T. G. Tutin.
[3] Edit. D. A. Webb.
[4] By B. Křísa.

to orbicular, crenulate; petiole longer than lamina. Bracts linear-lanceolate, equalling pedicels. Flowers 8–12 mm in diameter, broadly campanulate. Calyx-lobes 1·5–2 mm, deltate, acute. Petals yellowish-green. Anthers 2·3–2·7 mm. Style 6–7 mm, expanded into a disc below the stigma, exserted, curved. $2n=46$. *Mainly in coniferous woods. Most of Europe, but rare in the west and absent from most of the islands.* Au Be Bu Co Cz Da Fe Ga Ge Gr He Hs Hu It Ju No Po Rm Rs (N, B, C, W, K, E) Su.

4. **P. rotundifolia** L., *Sp. Pl.* 396 (1753). Stem 20–30(–35) cm. Leaves 2–4·5(–6) × 1·5–5 cm, orbicular to ovate, crenulate; petiole longer than lamina. Cauline scales oblong-lanceolate. Bracts lanceolate to ovate-lanceolate, amplexicaul, equalling or up to twice as long as pedicels. Flowers 8–12 mm in diameter, campanulate. Calyx-lobes 2–4·5 mm, lanceolate, acuminate. Petals pure white. Style 4–10 mm, expanded into a disc below the stigma, exserted, curved. *Europe southwards to N. Spain, N. Italy, Bulgaria and Krym.* Au Be Br Bu Cz Da Fe Ga Ge Hb He Ho Hs Hu It Ju No Po Rm Rs (N, B, C, W, K, E) Su.

(a) Subsp. **rotundifolia**: Leaves ovate, cuneate to rounded at base. Cauline scales 1–2. Pedicels 4–8 mm. Calyx-lobes 3·5–4·5 mm, linear-lanceolate, acuminate. Anthers 2·3–2·8 mm. Style 6–10 mm. $2n=46$. *Woods, bogs and fens. Almost throughout the range of the species.*

(b) Subsp. **maritima** (Kenyon) E. F. Warburg in Clapham, Tutin & E. F. Warburg, *Fl. Brit. Is.* 789 (1952): Leaves orbicular, rounded to truncate at base. Cauline scales 2–5. Pedicels not more than 5 mm. Calyx-lobes 2–3 mm, ovate to ovate-lanceolate, obtuse. Anthers 1·9–2·4 mm. Style 4–6 mm. $2n=46$. *Damp hollows in sand-dunes.* ● *N.W. Europe.*

5. **P. grandiflora** Radius, *Diss.* 27 (1821). Stem 8–20 cm. Leaves 1·5–3 × 1–2 cm, shiny, suborbicular, almost entire, pale beneath; petiole longer than lamina. Cauline scales elliptical to ovate. Bracts ovate-lanceolate, longer than pedicels. Flowers 15–20 mm in diameter. Calyx-lobes 2·5–3·5 mm, elliptical to ovate-lanceolate, obtuse. Petals white, greenish-white or pink. Anthers 1·7–2·2 mm. Style 3·5–5·5 mm, expanded into a disc below the stigma, exserted, curved. $2n=46$. *Tundra. Iceland; Arctic Russia.* Is Rs (N).

6. **P. norvegica** Knaben, *Bergens Mus. Årbok* **1943**(6): 5 (1943). Stem 10–20(–23) cm. Leaves 1–3(–5) × 1–3 cm, orbicular-ovate to elliptical; petiole equalling or shorter than lamina. Cauline scales broadly ovate. Bracts lanceolate, equalling or shorter than pedicels, reddish to purple. Flowers 10–25 mm in diameter, broadly campanulate. Calyx-lobes 2–3 mm, ovate-lanceolate, acuminate. Petals white or lilac. Anthers 1–2·5 mm. Style 6·5–8·5 mm, not expanded below the stigma, exserted, curved. $2n=46$. *Dry mountain slopes; calcicole.* ● *W. & N. Fennoscandia.* Fe No Rs (N) Su.

7. **P. carpatica** J. Holub & Křísa, *Folia Geobot. Phytotax.* (*Praha*) **6**: 82 (1971). Stem (8–)12–20 cm, slightly winged in upper part. Leaves 2–3 × 1·5–3 cm, elliptical to suborbicular, obtuse, cuneate at base; petiole shorter than lamina. Cauline scales ovate-lanceolate. Bracts broadly ovate-lanceolate, longer than pedicels. Flowers 10–15 mm in diameter, campanulate. Calyx-lobes 3–3·5 mm, oblong to linear-lanceolate, obtuse. Petals white (rarely pinkish). Anthers 2–2·5 mm. Style 5·5–7·5 mm, not expanded below the stigma, exserted, curved. $2n=46$. *Usually calcicole.* ● *Carpathians.* Cz Rm Rs (W).

[1] By B. Křísa.

2. Orthilia Rafin.[1]

(*Ramischia* Opiz ex Garcke)

Like *Pyrola* but racemes secund; anthers without tubes; pollen-grains free.

Literature: H. Andres, *Feddes Repert.* **19**: 209–224 (1923).

1. **O. secunda** (L.) House, *Amer. Midl. Nat.* **7**: 134 (1921) (*Pyrola secunda* L., *Ramischia secunda* (L.) Garcke). Stem 5–25 cm. Leaves 2–4 cm, ovate or ovate-elliptical to orbicular, crenulate to serrulate, light green. Petiole shorter than lamina. Cauline scales numerous, lanceolate. Bracts linear-lanceolate, about equalling pedicels. Flowers 5–6 mm in diameter. Calyx-lobes triangular-ovate, obtuse. Petals greenish-white, erect. Style somewhat thickened below the stigma, straight, exserted. *Woods and moors. Most of Europe, but rare in the Mediterranean region.* Al Au Br Bu Cz Da Fe Ga Ge Gr Hb He †Ho Hs Hu Is It Ju No Po Rm Rs (N, B, C, W, K, E) Si Su [Be].

(a) Subsp. **secunda**: Leaves 2–5 cm, ovate to ovate-elliptical, acute. Anthers *c.* 1·5 mm. Style 4·5–6 mm. $2n=38$. *Almost throughout the range of the species.*

(b) Subsp. **obtusata** (Turcz.) Böcher, *Bot. Tidsskr.* **57**: 31 (1961) (*Ramischia obtusata* (Turcz.) Freyn): Leaves 1–2 cm, orbicular to orbicular-ovate, obtuse. Anthers *c.* 1 mm. Style 3–4·5 mm. *N. Russia, from c. 62° to c. 70° N.*

3. Moneses Salisb.[1]

Like *Pyrola* but leaves opposite; flowers solitary; anthers with longer tubes.

1. **M. uniflora** (L.) A. Gray, *Man. Bot.* 273 (1848) (*Pyrola uniflora* L.). Stem 5–15 cm. Leaves 1–3 cm, orbicular to ovate-elliptical, crenulate to serrulate, light green; petiole shorter than lamina. Cauline scale 1. Flowers 13–20 mm in diameter. Calyx-lobes 1·5–3 mm, ovate, obtuse, erose-denticulate. Petals ovate-lanceolate, white. Anthers 2·5–3·2 mm. Style 5–7 mm, not expanded below the stigma, straight. $2n=26$. *Woods, usually coniferous. Most of Europe, but absent from many of the islands and the extreme south.* Al Au Br Bu Co Cz Da Fe Ga Ge He Ho Hs Hu It Ju No Po Rm Rs (N, B, C, W, K, E) Su [Be].

4. Chimaphila Pursh[1]

Glabrous, rhizomatous dwarf shrubs. Leaves green, coriaceous, alternate or pseudoverticillate, mostly cauline. Flowers in terminal corymbs or umbels. Anthers opening by pores at the end of short tubes. Pollen-grains in tetrads.

Leaves oblanceolate, without white veins; flowers 7–12 mm in diameter   **1. umbellata**
Leaves ovate-lanceolate, with white veins; flowers 20 mm in diameter   **2. maculata**

1. **C. umbellata** (L.) W. Barton, *Veg. Mat. U.S.* **1**: 17 (1817) (*Pyrola umbellata* L.). Stem 10–30 cm. Leaves 4–7 × 0·5–2·5 cm, oblanceolate, serrate; petiole 3–5 mm. Pedicels glandular. Flowers 7–12 mm in diameter. Calyx-lobes 1·5–2·5 mm, ovate, obtuse, erose-denticulate. Petals ovate-orbicular, pinkish. Anthers *c.* 2 mm. Style very short. $2n=26$. *Coniferous woods. N., C. & E. Europe, westwards to Norway, E. France and Switzerland and southwards to C. Romania and Krym.* Au Cz Da Fe Ga Ge He Hu Ju No Po Rm Rs (N, B, C, W, K, E) Su.

2. C. maculata (L.) Pursh, *Fl. Amer. Sept.* 300 (1814). Like **1** but leaves ovate-lanceolate, with white veins; flowers 20 mm in diameter; petals white. *Naturalized in woods near Paris (Fontainebleau).* [Ga.] (*E. North America.*)

5. Monotropa L.[1]

Wholly saprophytic herbs, without chlorophyll. Leaves scale-like, all cauline, alternate, sessile. Flowers in terminal racemes. Petals saccate at base. Anthers opening by longitudinal slits. Pollen-grains free.

Literature: K. Domin, *Sitz.-Ber. Böhm. Ges. Wiss.* (*Math.-Nat. Kl.*) **1915**(1): 3–46 (1915).

1. M. hypopitys L., *Sp. Pl.* 387 (1753) (*Hypopitys monotropa* Crantz). Whole plant creamy-white to yellow, sometimes slightly tinged with brown or pink. Stem 5–25 cm. Cauline scales 5–13 mm, ovate-oblong, entire to fimbriate. Sepals obovate-spathulate. Petals 6–13 mm, oblong to oblong-spathulate, erect, usually recurved at apex. Stamens shorter than petals. Capsule subglobose. $2n=16, 48$. *Damp woods, especially of conifers or Fagus.* *Most of Europe, but ran in the extreme south.* All except Az Cr Fa Is Sa Sb Tu.

Some plants are wholly glabrous; others are pubescent, especially in the inflorescence. The glabrous plants have been distinguished as var. *glabra* Roth, or as a separate species, **M. hypophegea** Wallr., *Sched. Crit.* 191 (1822), which is said to be further distinguished by its short petals not recurved at the apex, its shorter capsule, and its occurrence mainly in *Fagus*-woods. Recent work, however, suggests that there is little correlation between pubescence on the one hand and other morphological features, habitat or chromosome-number on the other.

CXXXII. ERICACEAE[2]

Dwarf shrubs to small trees; leaves simple, exstipulate, usually evergreen. Flowers actinomorphic (rarely slightly zygomorphic), in racemes or umbels, or solitary or in clusters in the leaf-axils. Petals usually connate. Stamens twice as many as petals (rarely fewer), inserted on the receptacle or disc, but sometimes adnate to the extreme base of the corolla and falling with it; anthers usually opening by pores; pollen grains usually in tetrads. Ovary superior (rarely inferior); placentation axile; style single. Fruit a capsule, berry or drupe.

All European members of the family which have been investigated are mycorrhizal; the majority are calcifuge, and are especially characteristic of peaty soils.

1 Corolla persistent in fruit
2 Calyx petaloid, larger than corolla; leaves opposite, closely imbricate on non-flowering shoots **3. Calluna**
2 Calyx smaller than corolla; leaves whorled, usually ± patent
3 Sepals free; bracteoles present **1. Erica**
3 Sepals connate, with calyx-tube about as long as lobes; bracteoles absent **2. Bruckenthalia**
1 Corolla deciduous
4 Ovary inferior **18. Vaccinium**
4 Ovary superior
5 Fruit a berry or drupe, or a capsule enclosed in a succulent, accrescent calyx
6 Leaves spine-tipped **15. Pernettya**
6 Leaves not spine-tipped
7 Inflorescence glandular-hirsute; anther-appendages 2-fid **14. Gaultheria**
7 Inflorescence glabrous or slightly puberulent; anther-appendages simple
8 Tree or erect shrub; fruit verrucose or sulcate **12. Arbutus**
8 Procumbent dwarf shrub; fruit smooth **13. Arctostaphylos**
5 Fruit a capsule; calyx not succulent
9 Anthers with appendages
10 Leaves not more than 5 mm, crowded and ± imbricate **4. Cassiope**
10 Larger leaves at least 10 mm, not very crowded
11 Leaves whitish-glaucous beneath; flowers pink, in condensed, umbel-like clusters **16. Andromeda**
11 Leaves brownish beneath; flowers white, in elongate, secund, leafy racemes **17. Chamaedaphne**
9 Anthers without appendages
12 Corolla urceolate; tube at least 3 times as long as lobes
13 Leaves oblong-lanceolate to elliptical, white-tomentose beneath; flowers 4-merous **11. Daboecia**
13 Leaves linear-oblong, not tomentose beneath; flowers 5-merous **10. Phyllodoce**
12 Corolla not urceolate; tube not more than $1\frac{1}{2}$ times as long as lobes
14 Bracts scarious, deciduous at anthesis
15 Petals free **6. Ledum**
15 Petals connate at base **5. Rhododendron**
14 Bracts leaf-like or coriaceous, persistent
16 Stamens 5; corolla 4–5 mm in diameter **9. Loiseleuria**
16 Stamens 10; corolla at least 8 mm in diameter
17 Corolla lobed almost to base; leaves ciliate **7. Rhodothamnus**
17 Corolla lobed less than half-way to base; leaves not ciliate **8. Kalmia**

1. Erica L.[3]

Dwarf to medium-sized evergreen shrubs; leaves whorled, often linear, or apparently linear on account of revolute margins; petiole short, appressed. Flowers 4–(5)-merous, in terminal umbels or racemes, or in axillary clusters or umbels more or less aggregated into terminal or intercalary panicles; pedicel with 2 or more bracteoles. Sepals free, green or pinkish, shorter than the corolla. Corolla cylindrical, campanulate or urceolate, with lobes shorter than or equalling the tube, persistent in fruit. Stamens 8(–10), inserted between the lobes of a nectariferous disc; anthers with or without appendages. Fruit a loculicidal capsule.

A large genus, centred in South Africa, but with a secondary centre in S. & W. Europe.

1 Anthers with basal appendages
2 Leaves with contiguous revolute margins, which entirely conceal the lower surface
3 Bracteoles confined to basal half of pedicel, not overlapping the sepals
4 Some of the hairs on young twigs echinate; corolla 2·5–4 mm, pure white; stigma broadly capitate, white **9. arborea**
4 All hairs smooth; corolla 4–5 mm, usually tinged with pink; stigma obconical, red **10. lusitanica**
3 Some or all of the bracteoles on apical half of pedicel overlapping the sepals

[1] By B. Křísa.
[2] Edit. D. A. Webb.
[3] By D. A. Webb and E. M. Rix.

5 Leaves in whorls of 3; sepals and ovary glabrous   **6. cinerea**
5 Leaves in whorls of 4; sepals and ovary pubescent
                                                 7. australis
2 Lower surface of leaf visible, at least towards the base
6 Sepals villous beneath   **4. tetralix**
6 Sepals glabrous to puberulent beneath (though sometimes
    with stout marginal cilia)
7 Stem 15–60 cm; leaves white beneath; ovary glabrous
                                        3. mackaiana
7 Stem 100–250 cm; leaves green beneath; ovary pubescent
                                        5. terminalis
1 Anthers without basal appendages
8 Anthers included
9 Sepals *c.* 6 mm; flowers usually 5-merous   **1. sicula**
9 Sepals not more than 2·5 mm; flowers 4-merous
10 Leaves ovate to oblong-lanceolate   **2. ciliaris**
10 Leaves linear
11 Stems at least 100 cm; corolla not more than 3 mm,
    green, often tinged with red   **16. scoparia**
11 Stems not more than 80 cm; corolla more than 3 mm,
    bright pinkish-purple   **8. umbellata**
8 Anthers exserted, at least for part of their length
12 Pedicel about as long as sepals
13 Stems procumbent, ± glabrous; ridges decurrent from leaf-
    bases continuing throughout the internode without much
    change in width or height   **14. herbacea**
13 Stems erect, puberulent when young; ridges decurrent from
    leaf-bases much lower and narrower in lower part of
    internode than in upper   **15. erigena**
12 Pedicel at least twice as long as sepals
14 Flowers in terminal umbels; pedicels pubescent   **8. umbellata**
14 Flowers in axillary clusters or lateral umbels, usually
    aggregated into panicles; pedicels glabrous
                                     (11–13). vagans group

1. E. sicula Guss., *Cat. Pl. Boccad.* 74 (1821) (*Pentapera sicula* (Guss.) Klotzsch). Compact, bushy dwarf shrub 20–50 cm; branches erect or ascending; young twigs densely pubescent. Leaves 6–9 mm, in whorls of 4(–5), linear, patent, densely pubescent when young; margins revolute, contiguous, concealing the lower surface. Flowers (4–)5-merous, in terminal umbels of 4–10; pedicels with 3 bracteoles near the middle. Sepals 6 mm, ovate-lanceolate, pubescent, pink; corolla 7–8 mm, white or very pale pink, urceolate, pubescent; anthers included, without appendages. Ovary pubescent; stigma capitate. *Calcareous cliffs by the sea. W. Sicilia, Malta.* Si. (*Libya and E. Mediterranean region.*)

2. E. ciliaris L., *Sp. Pl.* 354 (1753). Dwarf shrub 30–80 cm; stems erect, usually from a decumbent base, with patent branches; young twigs pubescent. Leaves 2–4 mm, in whorls of 3(–4), ovate to oblong-lanceolate, patent, pubescent above at least when young, and with stout, usually gland-tipped cilia on the apparent margin; margins revolute, but not concealing more than $\frac{1}{3}$ of the whitish lower surface. Flowers in terminal racemes; pedicels with 2–3 bracteoles above or near the middle. Sepals *c.* 2 mm, more or less pubescent, ciliate like the leaves; corolla 8–12 mm, bright reddish-pink, urceolate, somewhat curved or gibbous, with sub-erect lobes; anthers included, without appendages. Ovary glabrous; stigma obconical. $2n=24$. *Bogs, heaths, scrub and open woods; calcifuge. W. Europe, northwards to W. Ireland and eastwards to c. 3° E. in C. France.* Br Ga Hb Hs Lu.

3. E. mackaiana Bab. in Mackay, *Fl. Hibern.* 181 (1836) (*E. mackaii* Hooker). Compact, bushy dwarf shrub 15–60 cm; stems decumbent to erect, with suberect branches; young twigs glandular-puberulent and often also hirsute. Leaves 2–4·5 mm, in whorls of 4, patent in all parts of the plant, oblong, glabrous except for long, stout, gland-tipped cilia on true and apparent

margins; margins slightly revolute, leaving most of the white lower surface exposed. Internodes on flowering stems all about equal in length. Flowers in terminal umbels; pedicels with 1–2 bracteoles in apical half. Sepals 2–3 mm, glabrous except for long cilia; corolla 5–7 mm, bright purplish-pink, urceolate, with patent or revolute lobes; anthers included, with long, narrow basal appendages. Ovary glabrous; stigma capitate. $2n=24$. *Calcifuge.* ● *N.W. Spain; three stations in W. Ireland.* Hb Hs.

Sometimes confused with *E.* × *watsonii* Bentham (2×4), which differs in having leaves, sepals and ovary pubescent, flowers in a condensed raceme, and anther-appendages much shorter.

4. E. tetralix L., *Sp. Pl.* 353 (1753). Straggling dwarf shrub 20–70 cm, with weak, ascending stems and rather few suberect branches; young twigs pubescent to villous. Leaves 3–6 mm, in whorls of 4, appressed below the inflorescence, patent elsewhere, linear-oblong to lanceolate, pubescent at least when young, usually with long, stout, often glandular cilia on true and apparent margins; margins variably revolute, but always leaving part of lower surface visible. Internodes immediately below inflorescence longer than those elsewhere on flowering stems. Flowers in terminal umbels; pedicels with 2–3 bracteoles near the top. Sepals 2–3 mm, villous, usually ciliate like the leaves; corolla 5–9 mm, pale pink, urceolate, with patent or revolute lobes; anthers included, with long, narrow basal appendages. Ovary pubescent; stigma capitate. $2n=24$. *Bogs, wet heaths and pinewoods; calcifuge.* ● *W. & N. Europe, eastwards to Latvia and C. Finland.* Be Br Da Fe Ga Ge Hb Ho Hs Lu No Po Rs (B) Su [Cz He].

Variable, especially in indumentum and leaf-shape. The hybrids *E.* × *watsonii* Bentham in DC., *Prodr.* 7: 665 (1839) (2×4), and *E.* × **praegeri** Ostenf., *New Phytol.* 11: 120 (1912) (3×4), are rather frequent where the parent species occur together. Both are intermediate between the parents in most characters and are sterile.

5. E. terminalis Salisb., *Prodr.* 296 (1796) (*E. stricta* Donn ex Willd.). Bushy shrub 1–2·5 m, with suberect branches; young twigs puberulent. Leaves 3–5·5 mm, in whorls of 4(–6), broadly linear, patent, puberulent at least when young; margins revolute, sometimes contiguous near apex, but leaving $\frac{1}{3}$ of lower surface exposed near base. Flowers in terminal umbels of 3–8; pedicels with 3 narrow bracteoles, usually above the middle. Sepals 2–3 mm, ovate, puberulent; corolla 5–7 mm, bright pink, urceolate, with recurved lobes; anthers included, with triangular basal appendages. Ovary pubescent; stigma capitate. *River-banks, wooded ravines and other damp, shady places. W. Mediterranean region, from Corse to S.W. Spain and S. Italy.* Co Hs It Sa [Hb]

6. E. cinerea L., *Sp. Pl.* 352 (1753). Dwarf shrub 15–75 cm, of rather lax habit; branches ascending; young twigs pubescent. Leaves 4–5(–7) mm, in whorls of 3, linear, erecto-patent, glabrous; margins strongly revolute, contiguous, completely concealing the lower surface; apparent margin denticulate. Flowers in terminal racemes or umbels; flowering stems with numerous short, leafy lateral shoots below the inflorescence; pedicels pubescent, with 3 small bracteoles appressed to sepals. Sepals 2–3 mm, lanceolate, glabrous, with scarious margin; corolla 4–7 mm, bright reddish-purple, urceolate, with erect to patent lobes; anthers included, with short, broad, dentate basal appendages. Ovary glabrous; stigma capitate. $2n=24$. *Heaths, rocky ground, woods and dry moorland; calcifuge. W. Europe, northwards to 63° N. in Norway and eastwards to N. Italy.* Be Br Fa Ga Ge Hb Ho Hs It Lu No.

7. **E. australis** L., *Diss. Erica* 9 (1770) (incl. *E. aragonensis* Willk.). Slender shrub up to 2 m, with suberect branches; young twigs puberulent. Leaves 3·5–6 mm, in whorls of 4, linear, erecto-patent, glabrous at maturity; margins revolute, contiguous, completely concealing the lower surface. Flowers in clusters of 4–8, terminating short lateral branches, often grouped into lax, terminal panicles; pedicels 2–3 mm, pubescent, largely concealed by bract, with 3 bracteoles resembling the sepals and overlapping them. Sepals 2 mm, ovate, keeled, pubescent at least on keel and margins; corolla 6–9 mm, reddish-pink, tubular to narrowly campanulate, with revolute lobes; anthers included or partly exserted, with triangular, often fimbriate basal appendages. Ovary pubescent; stigma capitate. *Heaths, scrub and open woods; calcifuge. Portugal, N., C. & W. Spain.* Hs Lu.

8. **E. umbellata** L., *Sp. Pl.* 352 (1753). Dwarf shrub 20–80 cm; young twigs pubescent. Leaves 3–4 mm, in whorls of 3, linear, suberect, more or less pubescent or glandular-ciliate when young, later glabrous; margins revolute, contiguous, completely concealing the lower surface. Flowers in terminal umbels of 3–6; pedicels 2–4 mm, pubescent, with 2–3 bracteoles immediately below the sepals. Sepals 2 mm, ovate, shortly ciliate; corolla 3·5–5·5 mm, bright pinkish-purple, subglobose to broadly campanulate, with erect lobes; anthers usually exserted, without appendages. Ovary glabrous; stigma obconical. *Heaths, scrub and open woods; calcifuge. W. half of Iberian peninsula.* Hs Lu.

Plants generally similar to **8**, but with pedicels up to 7 mm, corolla 6–6·5 mm, and anthers abortive and included, occur here and there throughout its range. They may be hybrids, but the identity of the other parent is obscure.

9. **E. arborea** L., *Sp. Pl.* 353 (1753). Shrub or small tree 1–4(–7) m; young twigs densely pubescent with short, smooth hairs and longer, strongly echinate hairs. Leaves 3–5 mm, in whorls of (3–)4, linear, erecto-patent, glabrous, minutely denticulate, rather dark green; margins revolute, contiguous, completely concealing lower surface. Flowers in lateral racemes, usually aggregated to form a thyrsoid, terminal or intercalary panicle; pedicels glabrous, with 2–3 minute bracteoles below the middle. Sepals 1·5 mm, ovate, glabrous, saccate at base; corolla 2·5–4 mm, white, broadly campanulate, with erect lobes; anthers included, with flat, ciliate basal appendages. Ovary glabrous; style stout; stigma broadly capitate, white. *Woods, evergreen scrub and by streams. Mediterranean region and S.W. Europe.* Al Bl Bu Co Cr Ga Gr Hs It Ju Lu Sa Si Tu.

10. **E. lusitanica** Rudolphi in Schrader, *Jour. für die Bot.* 1799(2): 286 (1800). Like **9** but a shrub 1–3·5 m, with erect branches; all hairs smooth; leaves 5–7 mm, light green; sepals 1 mm, not saccate; corolla 4–5 mm, tinged with pink, especially in bud; style slender; stigma obconical, red. *Damp heaths and wood-margins.* ● *S.W. Europe, from S. Portugal to 44° 30′ N. in S.W. France.* Ga Hs Lu [Br].

(11–13). **E. vagans** group. Shrubs 30–80(–250) cm, with decumbent to erect stems; young twigs glabrous or puberulent. Leaves in whorls of 3–5, linear to oblong, glabrous or slightly puberulent; margins revolute, contiguous, concealing the lower surface. Flowers in small axillary clusters, aggregated into terminal or axillary, raceme-like panicles; pedicels much longer than sepals, with 3 small bracteoles near the base. Sepals 1·5–2 mm, ovate to ovate-lanceolate, glabrous or finely ciliate, pinkish; corolla 3–7 mm, cylindrical to campanulate; anthers exserted, without appendages. Ovary glabrous; stigma scarcely wider than style.

The distinct geographical ranges of the three species constituting this group have helped to disguise their great morphological similarity. The distinctive characters of **13**, however, though inconspicuous, appear to be constant; and although on purely morphological criteria **12** might be reduced to a subspecies of **11**, its different habitat, wide geographical separation, and usually rather distinct habit probably justify its retention as an independent species.

1 Anthers *c.* 1·5 mm; lobes 3–5 times as long as wide, usually parallel and contiguous   **13. multiflora**
1 Anthers 0·5–1 mm; lobes 1½–2½ times as long as wide, widely separated and usually divergent apically
 2 Leaves 6–11 mm, patent or somewhat deflexed on older shoots; anthers 0·5–0·8 mm   **11. vagans**
 2 Leaves 4–8 mm, mostly erect or erecto-patent; anthers 0·7–1 mm   **12. manipuliflora**

11. **E. vagans** L., *Diss. Erica* 10 (1770). Stems up to 60(–80) cm, decumbent to ascending, flexuous. Leaves 6–11 mm, in whorls of 4–5, mostly patent or somewhat deflexed. Inflorescences up to 10 cm, dense, terminal or intercalary. Sepals 1 mm; corolla 2·5–3·5 mm, lilac-pink or white, broadly campanulate, with erect lobes; anthers 0·5–0·8 mm, with widely separated, divergent lobes about twice as long as wide. $2n=24$. *Heaths and woods; calcifuge.* ● *W. Europe, from C. Spain to S.W. England (? and N.W. Ireland), and eastwards to c. 6° E. in France.* Br Ga *Hb Hs [He].

12. **E. manipuliflora** Salisb., *Trans. Linn. Soc. London* 6: 344 (1802) (*E. verticillata* Forskål, non Bergius). Stems up to 50(–75) cm, decumbent to ascending, rarely erect, straight or somewhat flexuous. Leaves 4–8 mm, in whorls of 3–4, mostly erect or erecto-patent. Inflorescences not more than 6 cm, usually lax and intercalary or terminal and subcapitate. Sepals 1 mm; corolla 3–3·5 mm, pink, broadly campanulate, with erect lobes; anthers 0·7–1 mm, with widely separated, divergent lobes 2–2½ times as long as wide. *Evergreen scrub and dry, rocky places. E. & C. Mediterranean region.* Al ?Bu Cr Gr It Ju Si Tu.

13. **E. multiflora** L., *Sp. Pl.* 355 (1753). Stems up to 80(–250) cm, usually more or less erect and rigid. Leaves 6–11(–14) mm, in whorls of 4–5. Inflorescences up to 5 cm but often much shorter, terminal or intercalary. Sepals 1·5–2 mm; corolla (3–)4–5(–7) mm, pink, cylindrical to narrowly campanulate, with erect to patent lobes; anthers *c.* 1·5 mm, with contiguous, usually parallel lobes 3–5 times as long as wide. *Rocky hillsides, dry woods and thickets; usually calcicole. Mediterranean region, eastwards to Jugoslavia.* ?Al Bl Co Ga Hs It Ju Sa Si.

14. **E. herbacea** L., *Sp. Pl.* 352 (1753) (*E. carnea* L.). Dwarf shrub with procumbent stems and ascending flowering branches not more than 25 cm high; young twigs almost glabrous, bearing conspicuous ridges running from the base of each leaf downwards throughout an internode with nearly constant height and width. Leaves 5–8 mm, in whorls of 4, linear, more or less patent, acute to apiculate, with apparent margin minutely denticulate; margins revolute, closely contiguous, completely concealing the lower surface. Flowers in short, leafy, terminal racemes; pedicels 2–4 mm, glabrous, with 3 small bracteoles near the middle; corolla 5–6 mm, cylindrical, with erect lobes; anthers without appendages, usually completely exserted. Ovary glabrous; stigma scarcely wider than style. $2n=24$. *Coniferous woods and stony slopes, mainly in the mountains; usually calcicole.* ● *Alps and S.C. Europe, northwards to E.C. Germany and eastwards to E. Austria, and extending locally southwards to C. Italy and Macedonia.* Al Au Cz Ga Ge He It Ju ?Rm.

15. E. erigena R. Ross, *Watsonia* 7 : 164 (1969) (*E. mediterranea* auct., non L., *E. hibernica* (Hooker & Arnott) Syme, non Utinet). Like **14** but an erect shrub 60–120(–200) cm; young twigs distinctly puberulent, and with the ridges decurrent from the leaf-bases rapidly becoming narrower and less prominent when traced downwards; leaves obtuse to subacute, with apparent margin entire and with true margins not quite contiguous, leaving a wider white line beneath; racemes often grouped into a panicle; anthers usually only partly exserted. 2n=24. *Damp places.* ● *Iberian peninsula and S.W. France; W. Ireland.* Ga Hb Hs Lu.

16. E. scoparia L., *Sp. Pl.* 353 (1753). Slender, erect shrub 1–6 m; young twigs glabrous or puberulent. Leaves 4–7 mm, in whorls of 3–4, linear, erecto-patent, glabrous; margins revolute, concealing about ⅔ of lower surface. Flowers in narrow, often interrupted, terminal racemes, sometimes grouped into lax panicles; pedicels 2 mm, glabrous, with 3 bracteoles near the middle. Sepals 0·5–1·5 mm, ovate, glabrous; corolla 1·5–3 mm, green, variably tinged with red, broadly campanulate, with erect, deltate lobes about as long as the tube; anthers included, without appendages. Ovary glabrous; stigma conspicuously capitate, reddish-purple. *Woods and heaths; calcifuge. S.W. Europe, extending eastwards to W.C. Italy and northwards to N.C. France.* Az Bl Co Ga Hs It Lu Sa [Ho].

(a) Subsp. **scoparia**: Not more than 2·5 m; corolla (2–)2·5–3 mm, about twice as long as sepals; stigma included or slightly exserted. *Throughout the range of the species except Açores.*

(b) Subsp. **azorica** (Hochst.) D. A. Webb, *Bot. Jour. Linn. Soc.* **65** : 259 (1972) (*E. azorica* Hochst.): Up to 6 m; corolla 1·5–1·75 mm, *c.* 3 times as long as sepals; stigma conspicuously exserted. ● *Açores.*

2. Bruckenthalia Reichenb.[1]

Like *Erica* but with sepals united to form a campanulate calyx; pedicel without bracteoles; disc absent.

1. B. spiculifolia (Salisb.) Reichenb., *Fl. Germ. Excurs.* 414 (1831). Dwarf shrub with ascending stems 10–15 cm; young twigs puberulent. Leaves 4–5 mm, crowded, in rather irregular whorls of 4–5, linear-oblong, minutely ciliate and often with a stout, terminal, gland-tipped hair; margins revolute, leaving ¼–⅓ of the lower surface exposed. Flowers 4-merous, in dense, shortly cylindrical or globose, terminal racemes; pedicels 2–3 mm. Calyx 1·5 mm, with broadly deltate lobes scarcely as long as the tube, pink; corolla 3 mm, campanulate, with ovate lobes about as long as the tube, bright reddish-pink. Anthers included, without appendages. Ovary and capsule glabrous. *Woods and subalpine pastures; calcifuge. Mountains of Romania and of the Balkan peninsula, southwards to N. Greece.* ?Al Bu Gr Ju Rm.

3. Calluna Salisb.[1]

Small evergreen shrubs; leaves opposite, sessile. Flowers 4-merous, in terminal or intercalary racemes or panicles; pedicels with several bracteoles. Sepals free, petaloid, exceeding the corolla; corolla deeply lobed, persistent in fruit; stamens 8; anthers acuminate, with basal appendages, opening by slits along their whole length, not by pores. Fruit a septifragal capsule.

1. C. vulgaris (L.) Hull, *Brit. Fl.* ed. 2, **1** : 114 (1808). Subglabrous to densely grey-pubescent. Stems 15–80(–150) cm,

erect, freely branched. Leaves 2·5–3·5 mm (including auricles), closely appressed, widely spaced on leading shoots but densely imbricate on lateral, non-flowering branches, oblong-lanceolate, concave on adaxial, keeled on abaxial surface, amplexicaul, with proximally directed auricles. Flowers shortly pedicellate, in narrow racemes which are sometimes grouped into panicles; bracteoles usually 6–8, crowded together beneath the flower, the upper 4 simulating sepals. Sepals 3–4 mm, oblong, pinkish-lilac; corolla lobed nearly to the base, its lobes like the sepals but smaller; anthers included. 2n=16. *Moors, heaths, open woods, sand-dunes, etc., often dominant over large areas; calcifuge. Most of Europe, but rare in much of the Mediterranean region and in the south-east.* All except Al Bl ?Co Cr Gr Rs (K) Sa Sb Si.

4. Cassiope D. Don[1]

Evergreen dwarf shrubs with procumbent to ascending stems; leaves alternate or opposite, small, sessile, imbricate. Flowers 5-merous, axillary or terminal, nodding, on slender pedicels; bracteoles absent. Sepals free, shorter than corolla. Corolla campanulate to hemispherical, lobed about half-way to the base, deciduous. Stamens 10; anthers included, with filiform, deflexed apical appendages. Fruit an erect, loculicidal capsule.

Leaves decussate, obtuse, deeply sulcate on abaxial surface, puberulent, closely appressed to stem **1. tetragona**
Leaves alternate, acute, not sulcate, glabrous except for minute cilia, not closely appressed to stem **2. hypnoides**

1. C. tetragona (L.) D. Don, *Edinb. New Philos. Jour.* **17** : 158 (1834). Stems 10–30 cm, decumbent to ascending. Leaves 3–5 mm, decussate, oblong-lanceolate, obtuse, concave adaxially, rounded but deeply sulcate abaxially, closely appressed to stem, puberulent. Flowers axillary, forming short, lax, intercalary racemes; pedicel *c.* 10 mm in flower, up to 25 mm in fruit, subglabrous. Sepals 2·5 mm, yellowish, scarious. Corolla 6–8 mm, campanulate, creamy-white. Capsule *c.* 4 mm. *Dryish, stony or sandy heaths or tundra; somewhat calcicole. Arctic Europe, mainly in the mountains.* Fe No Rs (N) Sb Su.

2. C. hypnoides (L.) D. Don, *loc. cit.* (1834) (*Harrimanella hypnoides* (L.) Coville). Stems 5–15 cm, procumbent, branched, forming a mat. Leaves 2–4 mm, alternate, linear-oblong, acute, not sulcate, erecto-patent or loosely appressed, sometimes somewhat secund, minutely ciliate but otherwise glabrous. Flowers terminal; pedicel 7–10 mm in flower, slightly longer in fruit, puberulent. Sepals 2 mm, crimson, with scarious margin. Corolla 4–5 mm, broadly campanulate to hemispherical, white. Capsule *c.* 3 mm. 2n=32. *By streams, or in damp, mossy tundra, usually in late snow-patches. Arctic Europe, Iceland, mountains of Fennoscandia.* Fe Is No Rs (N) Sb Su.

5. Rhododendron L.[2]

(incl. *Azalea* L. pro parte)

Shrubs or dwarf shrubs, usually evergreen; leaves alternate, shortly petiolate. Flowers 5-merous, actinomorphic or slightly zygomorphic, in short terminal racemes; bracts scarious, deciduous. Sepals connate at the base, or almost throughout their length, often very small. Corolla campanulate to infundibuliform. Stamens 10, rarely 5; anthers without appendages. Ovary 5-locular; fruit a septicidal capsule; seeds numerous, minute.

Literature: J. B. Stevenson (ed.), *The Species of Rhododendron,* ed. 2. London. 1947.

1 Deciduous; stamens 5; ovary and immature capsule strigose
1. luteum
1 Evergreen; stamens usually 10; ovary and immature capsule glabrous or lepidote
2 Leaves at least 6 cm, glabrous; ovary and immature capsule glabrous
2. ponticum
2 Leaves not more than 4 cm, hairy or lepidote; ovary and immature capsule lepidote
3 Procumbent; corolla glabrous; stamens 5–8   **6. lapponicum**
3 Erect; corolla lepidote outside; stamens 10
4 Leaves ciliate   **5. hirsutum**
4 Leaves not ciliate
5 Leaves acute or mucronate; pedicels glabrous; style about twice as long as ovary   **3. ferrugineum**
5 Leaves obtuse; pedicels pubescent; style about as long as ovary   **4. myrtifolium**

1. R. luteum Sweet, *Hort. Brit.* ed. 2, 343 (1830) (*Azalea pontica* L., *R. flavum* G. Don fil.). Erect, deciduous shrub 2–4 m, spreading by suckers; young twigs glandular-pubescent. Leaves 6–12 × 2–4 cm, oblong-lanceolate, cuspidate, ciliate-serrulate, shortly strigose above and on the veins beneath, somewhat glaucous. Flowers in crowded, umbel-like racemes terminating leafless shoots; pedicels glandular-pubescent. Calyx 5–7 mm, deeply lobed; lobes narrowly oblong, ciliate. Corolla yellow, pubescent outside; tube 10–15 mm, narrow; limb 50 mm in diameter. Stamens 5. Ovary glandular-strigose. *Woods, usually coniferous, or wet, peaty ground. E. & E.C. Europe, in a few scattered localities from N.W. Jugoslavia to S. White Russia.* Au Ju Po Rs (C, W) [Br]. (*Anatolia and Caucasus.*)

2. R. ponticum L., *Sp. Pl.* ed. 2, 562 (1762). Erect, evergreen shrub 2–5(–8) m, with spreading branches. Leaves 8–25 cm, entire, coriaceous, dark shining green above, paler beneath, glabrous. Racemes with 8–15 flowers; pedicels 2–6 cm. Calyx 1–2 mm; lobes rounded. Corolla 40–60 mm, campanulate, violet-purple; tube somewhat shorter than lobes. Stamens 10. Ovary glabrous. *In woods or by streams in the mountains; calcifuge. E. part of Balkan peninsula; S. Spain, C. & S. Portugal; cultivated elsewhere for ornament and extensively naturalized in parts of N.W. Europe.* Bu Hs Lu Tu [Be Br Ga Hb].

(a) Subsp. **ponticum**: Leaves 12–18(–25) cm, obovate, 2½–3½ times as long as wide; inflorescence-axis more or less glabrous. *Turkey-in-Europe and S.E. Bulgaria (Istrancadağlari).* (*W. Caucasus, Anatolia, Lebanon.*)
(b) Subsp. **baeticum** (Boiss. & Reuter) Hand.-Mazz., *Ann. Naturh. Mus.* (*Wien*) **23**: 53 (1909) (*R. baeticum* Boiss. & Reuter): Leaves 6–12(–16) cm, elliptic-oblong, 3–5 times as long as wide; inflorescence-axis tomentose. ● *S. Spain, C. & S. Portugal.*

3. R. ferrugineum L., *Sp. Pl.* 392 (1753). Evergreen shrub 50–120 cm; branches numerous, erect; young twigs ferruginous-lepidote. Leaves 2–4 cm, elliptic-oblong, acute or mucronate, entire, dark shining green above, densely lepidote beneath with ferruginous scales (yellowish when young); margin revolute. Racemes with 6–10 flowers; pedicels *c.* 8 mm, glabrous. Calyx *c.* 1·5 mm, ciliate. Corolla 15 mm, campanulate-infundibuliform, with lobes about as long as tube, deep pinkish-red, lepidote outside. Stamens 10. Style twice as long as ovary; ovary lepidote. *Mountain slopes and open woods or scrub; often dominant over large areas in dwarf-shrub zone; somewhat calcifuge.* ● *Alps, Pyrenees, Jura, mountains of W. Jugoslavia; rarely in adjacent lowlands.* Au Ga Ge He Hs It Ju.

4. R. myrtifolium Schott & Kotschy, *Bot. Zeit.* **9**: 17 (1851) (*R. kotschyi* Simonkai, *R. ferrugineum* subsp. *kotschyi* (Simonkai)

Hayek). Like **3** but usually not more than 50 cm; leaves 1–2 cm, obtuse, crenulate, rather less densely lepidote and with lower surface often greenish; pedicels pubescent; corolla clear pink, very sparsely lepidote; style about as long as ovary. *Shady rocks and stony slopes.* ● *E. & S. Carpathians; mountains of Bulgaria and Makedonija.* Bu Ju Rm Rs (W).

5. R. hirsutum L., *Sp. Pl.* 392 (1753). Like **3** but less densely lepidote throughout and with paler scales; young twigs hairy; leaves 1–3 cm, ciliate, obscurely crenulate, with scarcely revolute margin; pedicels up to 15 mm; calyx 4 mm, with acute, lanceolate lobes; corolla bright pink; style scarcely longer than ovary. *Open woods and scrub, screes and stony slopes; calcicole.* ● *C. & E. Alps, mountains of N.W. Jugoslavia; planted in Jura and W. Carpathians but scarcely naturalized.* Au Ga Ge He It Ju.

6. R. lapponicum (L.) Wahlenb., *Fl. Lapp.* 104 (1812). Procumbent, evergreen dwarf shrub; stems up to 50 cm, freely branched; young twigs densely lepidote. Leaves 0·6–2 cm, obtuse, sparsely lepidote above and densely so beneath with brownish scales, otherwise glabrous; margins revolute. Racemes with 3–6 flowers; pedicels 6–12 mm. Calyx 1·5 mm; lobes triangular, ciliate. Corolla *c.* 8 × 15 mm, broadly campanulate, purple, glabrous. Stamens 5–8. Style much longer than ovary; ovary lepidote. *Dry heaths and stony slopes; calcicole. Mountains of Fennoscandia.* Fe No Su. (*North America.*)

6. Ledum L.[1]

Evergreen shrubs; leaves alternate, shortly petiolate. Flowers 5-merous, in terminal, umbel-like, more or less corymbose racemes; bracts scarious, deciduous; bracteoles absent. Sepals small, connate for most of their length. Petals free. Stamens (5–)8–10(–14); anthers without appendages. Fruit a septicidal capsule; seeds with very loose testa.

1. L. palustre L., *Sp. Pl.* 391 (1753). Stems up to 120 cm, decumbent to erect; young twigs ferruginous-tomentose. Leaves 12–50 × 1·5–12 mm, linear to elliptic-oblong, ferruginous-tomentose beneath, deflexed in winter; margins revolute. Flowers numerous; pedicels 5–25 mm, persistently glandular-verrucose and often also ferruginous-tomentose at first, erect in flower, deflexed in fruit. Petals 4–8 mm, obovate, patent, white. Ovary and capsule verrucose-glandular. *Bogs, heaths and coniferous woods. N. & C. Europe, southwards to S. Germany, N.E. Austria and N. Ukraine; extinct in several stations further south.* Au Cz Fe Ge No Po †Rm Rs (N, B, C, W) Su [*Br].

A polymorphic, circumpolar species. The two taxa recognized below as subspecies are often treated as separate species, but intermediates are frequent in N.E. Asia. Some populations in Arctic Europe have very narrow leaves with strongly revolute margins, and have been assigned to subsp. **decumbens** (Aiton) Hultén, *Kungl. Svenska Vet.-Akad. Handl.* ser. 3, **8**(2): 8 (1930), which is widespread in arctic and subarctic Asia and America, but it is probably better to regard them as minor variants of subsp. *palustre.*

(a) Subsp. **palustre** (incl. *L. decumbens* sensu N. Busch, vix (Aiton) Loddiges ex Steudel): Leaves narrowly oblong to linear, 4–12 times as long as wide; midrib usually visible from beneath; stamens (7–)10(–11). $2n = c.$ 52. *Throughout the range of the species.*
(b) Subsp. **groenlandicum** (Oeder) Hultén, *Fl. Alaska Yukon* 1219 (1948) (*L. groenlandicum* Oeder): Leaves elliptic-oblong,

---
[1] By D. A. Webb.

2·5–5 times as long as wide; midrib usually concealed by tomentum on lower surface; stamens (5–)8(–14). *Naturalized locally in Britain and Germany.* (*North America, W. Greenland.*)

7. Rhodothamnus Reichenb.[1]

Evergreen dwarf shrubs; leaves alternate, subsessile. Flowers 5-merous, on long pedicels in the axils of foliage-leaves. Sepals connate at the base. Corolla rotate, with short tube and deeply lobed limb. Stamens 10; anthers without appendages. Fruit a septicidal capsule; seeds with close-fitting testa.

1. **R. chamaecistus** (L.) Reichenb. in Moessler, *Handb.* ed. 2, **1**: 688 (1827). Dwarf shrub up to 40 cm, with numerous ascending branches. Leaves 5–10 × 2–4 mm, elliptic-oblanceolate, entire or obscurely crenulate, usually glabrous except for stiff, white, marginal cilia. Flowers in subterminal groups of 1–3; pedicels glandular-pubescent. Calyx-lobes 5–6 mm, oblong-lanceolate; corolla 20–30 mm in diameter, pale pink; stamens *c.* 12 mm; style 15–20 mm. $2n=24$. *Dry, stony slopes and screes, mostly 1000–2000 m; calcicole.* ● *E. Alps.* Au Ge It Ju.

8. Kalmia L.[1]

Evergreen shrubs; leaves alternate to whorled. Flowers 5-merous, in small, umbel-like corymbs; bracts small, herbaceous, persistent. Sepals connate at the base. Corolla with a short tube and cup-shaped limb, only slightly lobed, plicate towards the base so as to form 10 pouches in which the anthers lie before dehiscence. Stamens 10, shorter than corolla; anthers without appendages. Fruit a septicidal capsule; seeds with loose-fitting testa.

Corymbs axillary; leaves ± flat, green or ferruginous beneath
         1. angustifolia
Corymbs terminal; leaves with revolute margins, glaucous or whitish beneath   **2. polifolia**

1. **K. angustifolia** L., *Sp. Pl.* 391 (1753). Stems up to 150 cm but often much less, erect or ascending; young twigs terete, glabrous. Leaves 20–60 × 8–18 mm, opposite or in whorls of 3, elliptical to oblong, petiolate, entire, sparsely puberulent, ferruginous beneath when young; margin flat or slightly revolute. Corymbs axillary, on upper part of previous year's growth; pedicels glandular-puberulent. Sepals herbaceous, acute; corolla 7–12 mm in diameter, bright reddish-pink. *Cultivated for ornament and naturalized in Germany and N.W. England.* [Br Ge.] (*E. North America.*)

2. **K. polifolia** Wangenh., *Schr. Ges. Naturf. Freunde Berlin* **8**(2): 130 (1788). Stems up to 70 cm, ascending or erect, slender; young twigs with 2 angular ridges, pubescent. Leaves 10–30 × 2–8 mm, mostly opposite, linear- to elliptic-oblong, subsessile, entire, glabrous, glaucous or whitish beneath; margin usually strongly revolute. Corymbs terminal; pedicels glabrous. Sepals scarious, obtuse; corolla 10–20 mm in diameter, purplish-pink. *Cultivated for ornament, and naturalized in a bog in S.E. England.* [Br.] (*North America.*)

9. Loiseleuria Desv.[1]

Evergreen dwarf shrubs; leaves opposite, very shortly petiolate. Flowers 5-merous, solitary or in small, terminal umbels. Bracts coriaceous, persistent in fruit. Sepals almost free. Corolla

broadly campanulate to cup-shaped, moderately deeply lobed, deciduous. Stamens 5, alternating with corolla-lobes; anthers without appendages, opening by slits along their whole length, not by pores. Fruit a septicidal (sometimes also partly loculicidal) capsule.

1. **L. procumbens** (L.) Desv., *Jour. Bot. Appl.* **1**: 35 (1813) (*Azalea procumbens* L.). Glabrous; stems procumbent, rooting, freely branched, forming extensive mats. Leaves 4–7 × 2 mm, oblong, obtuse; margins revolute; midrib prominent beneath. Bracts like the leaves but smaller and subacute; pedicels 2–8 mm. Sepals lanceolate, red. Corolla *c.* 4 × 6 mm, pink. Stamens shorter than corolla. Capsule 3–4 mm, red. $2n=24$. *Dry, stony or peaty places, mainly on mountains, often in very exposed situations; calcifuge. N. Europe, southwards to c. 56° N. in Scotland; mountains of France, Pyrenees, Alps, Carpathians.* Au Br Fa Fe Ga Ge He Hs Is It Ju No Rm Rs (N, W) Su.

10. Phyllodoce Salisb.[1]

Evergreen dwarf shrubs; leaves alternate, subsessile. Flowers 5-merous, in short, subterminal racemes; bracts similar to foliage-leaves; bracteoles absent. Sepals almost free. Corolla urceolate, with short lobes, deciduous. Stamens 10; anthers without appendages. Fruit a septicidal capsule.

1. **P. caerulea** (L.) Bab., *Man. Brit. Bot.* 194 (1843). Stems 10–35 cm, erect or ascending from a rooting base, freely branched. Leaves 6–12 mm, linear- to linear-oblong, obtuse, coriaceous, crowded; margins strongly revolute; apparent margins denticulate. Flowers pendent, on long, slender, glandular-pubescent pedicels. Sepals 4 mm, lanceolate, glandular-pubescent. Corolla 7–12 mm, lilac to pinkish-purple; lobes short, patent to erect. Stamens included. Capsule 4 mm, glandular-pubescent. $2n=24$. *Heathy ground, on mountains except in the extreme north. N. Europe, southwards to S. Norway; isolated stations in Scotland and the Pyrenees.* Br Fe Ga Is No Rs (N) Su.

11. Daboecia D. Don[1]

Evergreen dwarf shrubs; leaves alternate, petiolate. Flowers 4-merous, in bracteate racemes terminal on the current year's growth; bracteoles absent. Sepals free almost to base. Corolla urceolate, with short lobes, deciduous. Stamens 8; anthers without appendages. Ovary hairy; fruit a septicidal capsule.

Larger leaves at least 9 mm; corolla 9–14 mm, more or less glandular-hairy   **1. cantabrica**
Leaves not more than 8 mm; corolla not more than 8 mm, glabrous   **2. azorica**

1. **D. cantabrica** (Hudson) C. Koch, *Dendrologie* 2 (1): 132 (1872) (*D. polifolia* D. Don). Stems decumbent to ascending, rather weak and straggling, up to 70 cm if supported by other shrubs, but not more than 35 cm if isolated. Leaves variable in size, the largest (9–)11–14 × 3–5(–7) mm, narrowly lanceolate to ovate-elliptical, dark green and glandular-hirsute above; margins revolute, but leaving most of the white-tomentose lower surface visible. Racemes lax, with 3–9(–12) pendent flowers. Corolla 9–14 × 5–8 mm, more or less glandular-hairy, reddish-purple; lobes patent. $2n=24$. *Heaths, open woods and rocky ground; calcifuge.* ● *W. Europe, from N. Portugal to W. Ireland.* Ga Hb Hs Lu.

2. **D. azorica** Tutin & E. F. Warburg, *Jour. Bot.* (*London*) **70**: 12 (1932). Like **1** but not more than 15(–20) cm, with procum-

---

[1] By D. A. Webb.

bent leafy stems and erect flowering shoots; leaves 5–8 × (1–)2·5–3·5 mm, broadly elliptical; racemes with 3–7 more or less patent flowers; corolla 7–8 × 5–7 mm, glabrous, pink to crimson, with revolute lobes. *Moorland and stony slopes above 500 m; calcifuge.* ● *Açores.* Az.

12. Arbutus L.[1]

Evergreen shrubs or trees; leaves alternate, petiolate. Flowers 5-merous, in terminal, bracteate panicles; bracteoles absent. Sepals connate at base. Corolla urceolate, with short, revolute lobes, deciduous. Stamens 10, with villous filaments; anthers with subulate, apical appendages. Fruit a globose berry.

Young twigs glandular-setose, at least in part; flowers produced in autumn; berry covered with conical papillae   **1. unedo**
Young twigs glabrous; flowers produced in spring; berry reticulate-sulcate, but without papillae   **2. andrachne**

1. A. unedo L., *Sp. Pl.* 395 (1753). Usually a bushy shrub 1·5–3 m, occasionally a small tree up to 12 m. Bark fissured, peeling off in small flakes, mostly dull brown; young twigs glandular-setose, at least in part. Leaves 4–11 × 1·5–4 cm, oblong-lanceolate, usually 2–3 times as long as wide, serrate to subentire, glabrous except at extreme base; petiole 10 mm or less. Panicle 4–5 cm, drooping, appearing in autumn. Calyx 1·5 mm, with suborbicular lobes; corolla *c.* 9 × 7 mm, white, often tinged with green or pink. Berry *c.* 20 mm, covered with conical papillae, ripening through yellow and scarlet to deep crimson. $2n=26$. *Evergreen scrub, wood-margins and rocky slopes. Mediterranean region and S.W. Europe, extending northwards locally to N.W. Ireland.* Al Bl Co Cr Ga Gr Hb Hs It Ju Lu Sa Si Tu.

2. A. andrachne L., *Syst. Nat.* ed. 10, **2**: 1024 (1759). Like **1** but bark smooth, peeling off in papery sheets, orange-red; young twigs glabrous; leaves 3–6 cm wide, less than twice as long as wide, often entire; petiole 15–30 mm; panicle erect, appearing in spring; calyx 2·5 mm, with ovate-rhombic, acute lobes; berry 8–12 mm, reticulate-sulcate, orange. *Evergreen scrub and rocky slopes. S. Albania, Greece, Aegean region; S. Krym.* Al Cr Gr Rs (K) Tu.

A. × andrachnoides Link, *Enum. Hort. Berol. Alt.* **1**: 395 (1821), the hybrid between **1** and **2**, is frequent wherever the parent species grow together; it is fertile, and introgression appears to take place. It usually has the brightly-coloured bark of **2**, but at least some glandular-hairs on the young twigs.

13. Arctostaphylos Adanson[1]

(incl. *Arctous* (A. Gray) Niedenzu)

Dwarf shrubs; leaves alternate, shortly petiolate. Flowers 5-merous, in small, terminal racemes or clusters; bracteoles 2, at base of pedicel. Sepals small, connate for much of their length. Corolla urceolate, with short lobes. Stamens 10; anthers with apical appendages, opening by pores or short slits. Fruit a small, berry-like drupe with 5–10 pyrenes.

Leaves entire, evergreen; ripe fruit red   **1. uva-ursi**
Leaves serrate, withering in autumn; ripe fruit black   **2. alpinus**

1. A. uva-ursi (L.) Sprengel, *Syst. Veg.* **2**: 287 (1825) (*Arbutus uva-ursi* L.). Stems up to 150 cm long, procumbent, freely branched, forming large, dense mats. Leaves 12–30 × 4–12 mm, obovate to oblanceolate-cuneate, obtuse to emarginate, entire,

coriaceous, evergreen, usually pubescent on midrib and margins, dark and shining above, paler beneath, conspicuously reticulate-veined. Bracts similar but smaller. Calyx-lobes suborbicular. Corolla 5–6 mm, greenish-white to pink; lobes revolute. Stamens included; appendages about as long as anthers. Fruit 10 mm, globose, bright red; flesh rather dry; pyrenes 5–7(–10). $2n=52$. *Heaths, rocky ground and open woods. Most of Europe except the extreme south.* Al Au Br Bu Cz Da Fe Ga Ge Gr Hb He Ho Hs Is It Ju No Po Rm Rs (N, B, C, W) Su.

2. A. alpinus (L.) Sprengel, *loc. cit.* (1825) (*Arbutus alpina* L., *Arctous alpinus* (L.) Niedenzu). Like **1** but stems not more than 60 cm; leaves up to 50 mm, subacute, serrate, not coriaceous, withering in autumn but not falling till spring, ciliate in basal half; bracts whitish, truncate, densely ciliate-fimbriate at apex; appendages much shorter than anthers; ripe fruit black, succulent; pyrenes 5. $2n=26$. *Heaths and stony slopes, on mountains except in the extreme north. N. Russia, N. & C. Fennoscandia, Scotland; mountains of C. & S. Europe southwards to the Pyrenees, C. Appennini and N. Albania.* Al Au Br Cz †Da Fe Ga Ge He Hs It Ju No Rs (N, W) Su.

14. Gaultheria L.[1]

Evergreen shrubs; leaves alternate, shortly petiolate. Flowers 5-merous, in lax, axillary, bracteate racemes; bracteoles 2, at base of pedicel. Sepals connate at base. Corolla urceolate, with revolute lobes, deciduous. Stamens 10; anthers with 2-fid apical appendages. Fruit a septicidal capsule, enclosed by the accrescent calyx.

1. G. shallon Pursh, *Fl. Amer. Sept.* 283 (1814). Stems 50–175 cm, erect, multiplying freely by underground rhizomes to form a dense thicket; twigs and leaves glandular-hirsute when young, glabrescent. Leaves 6–9 × 3–7 cm, shortly cuspidate, serrulate, coriaceous. Racemes subterminal on the previous year's growth. Calyx and corolla pinkish-white, glandular-hairy; corolla 8–10 mm. Capsule closely invested by the succulent, accrescent calyx, which forms a subglobose, dark purple, berry-like pseudocarp. *Cultivated for ornament and as cover for game, and naturalized in N.W. Europe.* [Br Ga.] (*W. North America.*)

15. Pernettya Gaud.-Beaupré[1]

Small evergreen shrubs; leaves alternate, subsessile. Flowers 5-merous, axillary or in small terminal corymbs; bracteoles 2–4. Sepals connate at base. Corolla urceolate, with acute, revolute lobes, deciduous. Stamens 10; anthers with short apical appendages. Fruit a globose berry.

There is some evidence to suggest that this genus, defined primarily by the succulent fruit, is polyphyletic, succulence of fruit having apparently arisen independently in different sections of *Gaultheria*. Until a full monographic study of *Gaultheria* has been made it seems best to maintain *Pernettya* as a separate genus.

1. P. mucronata (L. fil.) Gaud.-Beaupré ex Sprengel, *Syst. Veg.* **4**(2): 158 (1827). Stems 30–100 cm, erect or spreading, much-branched; young twigs glabrescent. Leaves 8–20 mm, elliptic-lanceolate, serrate to subentire, spine-tipped, glabrous, coriaceous. Flowers drooping; pedicel shorter than the subtending leaf. Corolla 5–6 mm, white. Berry *c.* 12 mm in diameter, purple, pink or white. *Cultivated for ornament and locally naturalized in Britain and Ireland.* [Br Hb.] (*S. Chile.*)

---

[1] By D. A. Webb.

16. Andromeda L.[1]

Evergreen dwarf shrubs; leaves alternate, subsessile. Flowers 5-merous, in very short, umbel-like, terminal racemes; bracteoles 2, at extreme base of pedicel, concealed by bract. Sepals connate at base. Corolla urceolate; lobes short. Stamens 10; anthers with apical appendages. Fruit an erect, loculicidal capsule.

1. **A. polifolia** L., *Sp. Pl.* 393 (1753). Flowering stems 15–35 cm, decumbent to erect, sparingly branched, arising from a creeping, woody rhizome. Leaves 10–40 × 2–8 mm, linear to oblong, acute to apiculate, entire, dark green above, silvery-glaucous beneath; margins usually revolute. Flowers inclined or pendent, in racemes of 2–7(–12). Pedicels 7–15 mm, slender. Calyx-lobes 1·5 mm, triangular, reddish. Corolla 5–8 mm, urceolate, bright pink at first, fading almost to white; lobes revolute. Appendages as long as anthers, slightly curved. Capsule 4–5 mm, depressed-globose. $2n = c. 48$. Sphagnum-*bogs and other wet, acid habitats. N. & N.C. Europe, extending locally southwards to the S. Alps, E. Carpathians and S.C. Russia.* Au Be Br Cz Da Fe Ga Ge Hb He Ho It Ju No Po Rm Rs (N, B, C, W, E) Su.

17. Chamaedaphne Moench[1]

(*Cassandra* D. Don)

Evergreen shrubs; leaves alternate, very shortly petiolate. Flowers 5-merous, solitary in the axils of leaf-like bracts, forming terminal, secund racemes; bracteoles 2, inserted immediately below the calyx and overlapping it. Sepals connate at base. Corolla cylindrical, lobed for $\frac{1}{4}$–$\frac{1}{3}$ of its length. Stamens 10; anthers without appendages but with the lobes prolonged apically into a tube with terminal pore. Fruit a pendent, loculicidal capsule.

1. **C. calyculata** (L.) Moench, *Meth.* 457 (1794) (*Cassandra calyculata* (L.) D. Don). Stems 20–50 cm, erect; young twigs puberulent. Leaves 10–40(–50) × 4–10(–15) mm, elliptic-oblong, entire to obscurely serrate, covered with brownish scales beneath. Bracts similar, but progressively smaller upwards. Racemes inclined or horizontal, with 5–20 pendent flowers. Bracteoles and calyx *c.* 2 mm. Corolla 5–6 mm, white. Capsule 2–2·5 mm, depressed-globose. $2n = 22$. Bogs, marshes and wet woods. N.E. Europe, extending southwards to N. Poland and to c. 52° N. in S.C. Russia.* Fe Po Rs (N, B, C, W, E) Su.

18. Vaccinium L.[2]

Deciduous or evergreen dwarf shrubs, shrubs or small trees; leaves alternate, subsessile or very shortly petiolate. Flowers 4- or 5-merous, in axillary or terminal, usually bracteate racemes, sometimes reduced to a single flower. Sepals largely, often almost completely connate. Corolla rotate to globose. Stamens 8 or 10; anthers with or without appendages, each lobe prolonged apically into a tube with a pore at the apex. Ovary inferior; fruit a berry.

All European species are more or less calcifuge.

Literature: W. Gleisberg, *Bot. Arch.* (*Berlin*) **2**: 1–34 (1922). H. Sleumer, *Bot. Jahrb.* **71**: 375–510 (1941). W. H. Camp, *Bull. Torrey Bot. Club* **71**: 426–437 (1944).

V. corymbosum L., *Sp. Pl.* 350 (1753), the blueberry, from E. North America, is cultivated locally in W. & C. Europe for its edible fruits. It is a deciduous shrub with elliptical leaves *c.* 40 × 15 mm, white, cylindrical corolla *c.* 8 mm, and blue berries 7–10 mm.

1 Corolla divided almost to the base; leafy stems procumbent; pedicels filiform, ± erect (Subgen. *Oxycoccus*)
  2 Bracteoles 1–1·5 mm wide, ovate-elliptical; inflorescence usually intercalary   **3. macrocarpon**
  2 Bracteoles not more than 0·5 mm wide, linear-oblong; inflorescence terminal
    3 Pedicels puberulent; larger leaves 3–6 mm wide   **1. oxycoccos**
    3 Pedicels glabrous; leaves not more than 2·5 mm wide   **2. microcarpum**
1 Corolla not divided more than half-way to the base; leafy stems usually erect; pedicels relatively short and stout, patent or deflexed (Subgen. *Vaccinium*)
  4 Leaves entire; margin ± revolute
    5 Leaves evergreen, not glaucous; filaments hairy; berry red   **4. vitis-idaea**
    5 Leaves deciduous, glaucous; filaments glabrous; berry bluish-black   **5. uliginosum**
  4 Leaves serrulate, flat
    6 Small shrub with angular twigs; calyx not or scarcely lobed; filaments glabrous   **6. myrtillus**
    6 Large shrub or small tree with terete twigs; calyx distinctly 5-lobed; filaments hairy
      7 Corolla 5–7 mm, broadly campanulate; racemes bracteate   **7. arctostaphylos**
      7 Corolla 8–12 mm, cylindrical to narrowly campanulate; racemes ebracteate   **8. cylindraceum**

Subgen. **Oxycoccus** (Hill) A. Gray. Evergreen; aerial stems slender, procumbent or decumbent, with short, erect or ascending flowering branches. Flowers 4-merous, in terminal or intercalary racemes of (1–)2–5; bracts scale-like. Pedicels filiform, more or less erect, with 2 bracteoles. Corolla rotate, divided almost to the base; lobes revolute at full anthesis. Anthers without appendages.

1. **V. oxycoccos** L., *Sp. Pl.* 351 (1753) (*Oxycoccus quadripetalus* Br.-Bl., *O. palustris* Pers.). Stems up to 80 cm, procumbent. Leaves 6–10(–15) × 2·5–5(–6) mm, ovate to oblong, subglabrous, dark shining green above, whitish beneath; margin revolute. Racemes terminal; pedicels puberulent; bracteoles *c.* 1·5 × 0·4 mm, linear-oblong. Calyx-lobes deltate to semicircular, ciliate. Corolla pinkish-red; lobes 5–6 mm, ovate-oblong. Filaments puberulent, much shorter than anthers. Berry (6–)8–10(–15) mm, globose to pyriform, red, sometimes speckled with white or brown. $2n = 48$. Peat-bogs, usually in the wetter parts. N. & C. Europe, extending locally southwards to S.C. France, N. Italy and S.E. Russia.* Au Be Br Cz Da Fe Ga Ge Hb He Ho Hu It Ju No Po Rm Rs (N, B, C, W, E) Su.

Plants like **1** but more robust, with larger leaves and fruit, and with $2n = 72$, are known from Scandinavia. They have been interpreted as hybrids between **1** and **2**, which have not achieved fertility in spite of doubling of the chromosome number. Fertile plants intermediate between **1** and **2** have been reported from Sweden and Finland; their chromosome number is not known and their status is obscure.

2. **V. microcarpum** (Turcz. ex Rupr.) Schmalh., *Trudy Imp. S. Peterb. Obšč. Estestv.* **2**: 149 (1871) (*Oxycoccus microcarpus* Turcz. ex Rupr.). Like **1** but stems not more than 30 cm; leaves 3–8 × 1–2·5 mm, triangular-ovate; pedicels, and sometimes calyx, glabrous; corolla smaller; berry 5–8 mm wide, pyriform to ellipsoid. $2n = 24$. Peat-bogs, sometimes in rather drier parts than **1**. N. & N.C. Europe, extending southwards to N.W. Ukraine; Alps; Carpathians.* Au Br Cz Fe Ge He Is It Ju No Po Rm Rs (N, B, C, W, ?E) Su.

[1] By D. A. Webb.   [2] By T. N. Popova.

3. V. macrocarpon Aiton, *Hort. Kew.* **2**: 13 (1789). Like **1** but more robust and taller; leaves 8–18 × 3·5–5 mm, oblong, with margins only slightly revolute; axis continued beyond raceme as leafy shoot of current year, so that raceme is intercalary; bracteoles *c.* 5 × 1–1·5 mm, ovate-elliptical; corolla-lobes 6–10 mm; berry 10–20 mm wide. *Cultivated in parts of N. & C. Europe for its edible fruit (cranberries) and locally naturalized.* [Br Ge He Ho.] (*E. North America.*)

Subgen. **Vaccinium**. Evergreen or deciduous; aerial stems relatively stout, usually more or less erect. Flowers 4- or 5-merous, in axillary (rarely terminal) racemes; bracts leaf-like. Pedicels patent or deflexed, not filiform, with or without bracteoles. Corolla campanulate to globose, divided not more than halfway to the base. Anthers often with appendages.

4. V. vitis-idaea L., *Sp. Pl.* 351 (1753). Evergreen. Stems up to 30 cm but often much less, erect, arching or decumbent, arising from a creeping rhizome; twigs terete, puberulent when young. Leaves elliptical to oblong or obovate, entire, obtuse, glabrous, coriaceous, dark green above, paler and dotted with dark glands beneath; margin revolute. Flowers 4- or 5-merous, in short, crowded, terminal racemes. Bracteoles reddish. Calyx-lobes deltate. Corolla 5–8 mm, campanulate, white or pink; lobes about half as long as tube. Filaments hairy; anthers without appendages. Berry 5–10 mm, globose, red, acid. 2*n*=24. *Moors, heaths, coniferous woods, subalpine pastures and tundra. N. & C. Europe, extending southwards in the mountains to the N. Appennini, Albania and Bulgaria.* Al Au Be Br Bu Cz Da Fa Fe Ga Ge Hb He Ho Hu Is It Ju No Po Rm Rs (N, B, C, W, E) Su.

(a) Subsp. **vitis-idaea**: Stems 8–30 cm; leaves 10–25 × 6–15 mm, with conspicuous lateral veins; racemes with 3–8(–15) flowers; corolla white or pale pink; style long-exserted; berry 8–10 mm. *Throughout the range of the species.*
(b) Subsp. **minus** (Loddiges) Hultén, *Fl. Aleut. Isl.* 268 (1937): Stems 3–8 cm; leaves 4–8 × 2·5–5 mm, with inconspicuous lateral veins; racemes with 2–5 flowers; corolla bright pink; style scarcely exserted; berry 5–8 mm. *Arctic Russia.* (*N. Asia, North America.*)

5. V. uliginosum L., *Sp. Pl.* 350 (1753). Deciduous. Stems up to 75(–100) cm, erect, freely branched, arising from a creeping rhizome; twigs terete, usually glabrous, brownish. Leaves 6–25(–35) × 4–12(–20) mm, obovate, entire, obtuse to subacute, glabrous, glaucous; margin slightly revolute. Flowers 4- or 5-merous, in racemes of 1–3 which terminate short branches bearing only scale-leaves. Bracteoles absent. Calyx-lobes rounded, scarious, reddish. Corolla 4–6 mm, urceolate, white, usually tinged with pink; lobes short, revolute. Filaments glabrous; anthers with small, subulate appendages. Style included. Berry 7–10 mm, globose to ellipsoid, bluish-black, sweetish. *Moors, heaths, coniferous woods, subalpine pastures and tundra. N. & C. Europe, extending southwards in the mountains to the Sierra Nevada, N. Appennini, Albania and Bulgaria.* Al Au Be Br Bu Cz Da Fa Fe Ga Ge He Ho Hs Is It Ju No Po Rm Rs (N, B, C, W) Sb Su.

(a) Subsp. **uliginosum**: Stems up to 75(–100) cm, erect; leaves 10–25(–35) mm; pedicel as long as corolla; lobes of corolla revolute. 2*n*=48. *Almost throughout the range of the species.*
(b) Subsp. **microphyllum** Lange, *Meddel. Grønl.* **3**: 91 (1880): Stems not more than 15 cm, procumbent to decumbent; leaves 6–15 mm; pedicel 1–3 mm, much shorter than corolla; lobes of corolla scarcely revolute. 2*n*=24. *Arctic and subarctic Europe.*

6. V. myrtillus L., *Sp. Pl.* 349 (1753). Deciduous. Stems 15–35(–60) cm, erect, freely branched, arising from a creeping rhizome; twigs acutely 3-angled, glabrous, green. Leaves 10–30 × 6–18 mm, ovate, acute, serrulate, glabrous, bright green, flat. Flowers 4- or 5-merous, in axillary racemes, which are reduced to 1(–2) flowers; pedicel arising from between 2 scale-like bracts; bracteoles absent. Calyx scarcely lobed, with sinuate or subentire margin. Corolla 4–6 mm, depressed-globose, pale green tinged with pink; lobes short, revolute. Filaments glabrous, very short; anthers with subulate appendages. Style usually included. Berry 6–10 mm, globose, bluish-black, sweet. 2*n*=24. *Heaths, moors and woods. Most of Europe, but only on mountains in the south.* All except Az Bl Cr Gr Rs (K) Sa Sb Si Tu.

Hybrids with **4** (*V.* × *intermedium* Ruthe) are found locally with the parents.

7. V. arctostaphylos L., *Sp. Pl.* 351 (1753). Deciduous shrub or small tree 2–3(–5) m. Leaves 40–60(–100) mm, oblong, acuminate, serrulate. Flowers 5-merous, in pendent axillary racemes of 5–8, each in the axil of a somewhat leaf-like bract; pedicel with 1–2 small bracteoles at the base. Sepals deltate. Corolla 5–7 mm, broadly campanulate, greenish-white tinged with pink; lobes short, slightly revolute. Anthers with small, subulate appendages. Berry 6–8 mm, globose, purplish-black. *Wooded ravines. Turkey-in-Europe and S.E. Bulgaria (Istrancadağları).* Bu Tu. (*N. Anatolia, Caucasus.*)

8. V. cylindraceum Sm. in Rees, *Cyclop.* **36**: no. 23 (1817). Like **7** but leaves 20–50 mm; racemes ebracteate, with 10–20 flowers; bracteoles deciduous; corolla 8–12 mm, cylindrical to narrowly campanulate, with erect lobes; anthers without appendages; berry 8–10 mm, ovoid. *Woods.* ● *Açores.* Az.

CXXXIII. EMPETRACEAE[1]

Small, evergreen shrubs of ericoid habit. Leaves mostly in somewhat irregular whorls of 3–4 but often alternate on some of the stronger shoots; margins strongly revolute; petiole adnate to stem, at least in part. Flowers small, actinomorphic, often unisexual, (2-) 3-merous, in small terminal heads, or in short axillary racemes which are often reduced to a single flower. Sepals, petals and stamens (2–)3, free. Ovary superior, 6- to 9-locular, with one ovule in each loculus; fruit a small, berry-like drupe with 2–9 pyrenes.

Stems erect; young twigs conspicuously pubescent; flowers in terminal heads; fruit white or pink **1. Corema**
Stems procumbent to ascending; young twigs glabrous or minutely glandular-puberulent; flowers in axillary racemes of 1–3; fruit black **2. Empetrum**

1. Corema D. Don[2]

Dioecious. Flowers in capitate, bracteate racemes, terminal at anthesis, but in the female intercalary in fruit from continued growth of the main axis.

---

[1] Edit. D. A. Webb.   [2] By D. A. Webb.

1. C. album (L.) D. Don in Sweet, *Hort. Brit.* ed. 2, 460 (1830). Erect shrub; stems 30–75 cm, with numerous branches; young twigs pubescent; older twigs rough with persistent petioles. Leaves 6–10 × 1 mm, linear, obtuse, covered when young with minute, subsessile glands, glabrous at maturity, patent, with a deep, narrow groove on the apparent abaxial surface. Heads with 5–9 subsessile flowers. Sepals 2 mm, suborbicular, pubescent. Petals in male flowers 3–4 mm, obovate, pinkish; in female flowers reduced or absent. Stamens 5–6 mm; anthers red. Style present; stigma red, with 3 linear-oblong lobes. Drupe (5–)6–8 mm in diameter, white (rarely pink). $2n = 26$. *Maritime (or rarely inland) sands.* ● *W. half of Iberian peninsula; Açores.* Az Hs Lu [Ga].

Male plants have straight, erect branches and a strict habit; in female plants the branches are shorter, more patent and somewhat tortuous, so that the habit is more diffuse.

2. Empetrum L.[1]

Dioecious or hermaphrodite. Flowers in short lateral racemes which bear 3 bracts, of which 1 (rarely 2–3) bears in its axil a sessile flower.

1. E. nigrum L., *Sp. Pl.* 1022 (1753). Stems procumbent to ascending; young twigs minutely glandular-puberulent, soon glabrous. Leaves 3–7 × 1–2 mm, linear-oblong to narrowly elliptical, glandular when young, glabrous later; margins closely approximated but not or only slightly involute, whitish, so that the apparent abaxial surface of the leaf bears a pale stripe or shallow groove. Sepals and petals *c.* 1·5 mm, oblong, greenish-pink to reddish-purple. Stamens up to 5 mm, long-exserted. Stigma sessile, with 6–9 rays. Drupe 5 mm in diameter, black. *Heaths, moors, coniferous forests and stony mountain-slopes. N. Europe, and on mountains southwards to the Pyrenees, C. Italy, Bulgaria and S. Ural.* Al Au Be Br Bu Cz Da Fa Fe Ga Ge Hb He Ho Hs Is It Ju No Po Rm Rs (N, B, C, W) Sb Su.

The genus consists of a number of closely related taxa (for the most part American) which are variously treated as species, subspecies or varieties. The two European taxa, here treated as subspecies, are often regarded as separate species, largely because of their difference in chromosome number; but the correlation between chromosome number and reproductive pattern, though good, is not perfect (the latter being possibly susceptible to environmental influences), and the differences in vegetative characters show some overlap.

(a) Subsp. **nigrum**: Normally dioecious. Stems up to 120 cm, procumbent, sparingly branched, usually rooting; leaves mostly linear-oblong, 2½–5 times as long as wide, usually not grooved beneath. $2n = 26$. *Throughout most of the range of the species, but rare in the extreme north and absent from most of the Alps.*

(b) Subsp. **hermaphroditum** (Hagerup) Böcher, *Meddel. Grønl.* 147(9): 81 (1952) (*E. hermaphroditum* Hagerup, *E. eamesii* subsp. *hermaphroditum* (Hagerup) D. Löve): Flowers normally hermaphrodite, with stamens usually persistent in fruit. Stems seldom more than 50 cm, freely branched, decumbent to ascending, not rooting; leaves mostly elliptical to elliptic-oblong, 2–4 times as long as wide, often with a shallow groove beneath. $2n = 52$. *N. Europe, southwards to c. 60° N., and on some of the higher mountains further south.*

PRIMULALES

CXXXIV. MYRSINACEAE[2]

Shrubs or trees. Leaves alternate, simple, exstipulate. Flowers actinomorphic. Calyx 4- to 6-fid. Corolla usually sympetalous; lobes 4–6. Stamens equal in number to the corolla-lobes and opposite to them, epipetalous. Ovary superior, 1-locular. Fruit an indehiscent, dry or fleshy drupe.

1. Myrsine L.[3]

Usually dioecious. Flowers in axillary fascicles. Calyx small, persistent. Corolla-lobes imbricate; tube usually very short. Fruit 1-seeded.

1. M. africana L., *Sp. Pl.* 196 (1753). Evergreen shrub usually 1–2 m; young twigs and petioles puberulent. Leaves 0·5–1·5 cm, obovate to suborbicular, serrulate, obtuse, emarginate or subacute, cuneate to rounded at base, coriaceous, gland-dotted. Inflorescence of 3–8 sub-sessile flowers. Corolla *c.* 1·5 mm in diameter, pale brown, rotate. Fruit 4–5 mm, globose, bluish-purple. *Evergreen scrub, 300–900 m. Açores.* Az. (*C. & S. Africa; Afghanistan to Nepal.*)

CXXXV. PRIMULACEAE[4]

Herbs, rarely dwarf shrubs. Leaves exstipulate, usually simple. Flowers 5(to 7)-merous, actinomorphic (rarely zygomorphic), sympetalous. Stamens epipetalous, opposite the corolla-lobes; staminodes sometimes present. Ovary 1-locular, superior (rarely semi-inferior), with 1-numerous ovules and basal, free-central placenta; style simple. Fruit a capsule.

1 Aquatic; leaves pinnate   **6. Hottonia**
1 Terrestrial; leaves simple

  2 Corolla-lobes sharply deflexed; stock tuberous   **7. Cyclamen**
  2 Corolla-lobes not sharply deflexed; stock not tuberous
    3 Flowers zygomorphic; calyx spiny   **14. Coris**
    3 Flowers actinomorphic; calyx not spiny
      4 Leaves basal, often in a rosette; flowers solitary or on a leafless scape
        5 Corolla-lobes laciniate   **5. Soldanella**
        5 Corolla-lobes not laciniate
          6 Filaments with dilated base, connate; leaves lobed   **4. Cortusa**
          6 Filaments without dilated base, not connate; leaves not lobed

---

[1] By D. A. Webb.   [2] Edit. T. G. Tutin.
[3] By T. G. Tutin.   [4] Edit. D. H. Valentine.

7　Throat of corolla constricted; plants not heterostylous　**3. Androsace**
7　Throat of corolla not or scarcely constricted; plants often heterostylous
　8　Seeds numerous; flowers usually in umbels, rarely solitary　**1. Primula**
　8　Seeds 2–5; flowers solitary　**2. Vitaliana**
4　Stems leafy, bearing flowers
　9　Flowers with petaloid calyx and no corolla　**11. Glaux**
　9　Flowers with calyx and corolla
　10　Flowers perigynous to epigynous　**13. Samolus**
　10　Flowers hypogynous
　11　Corolla shorter than calyx
　12　Leaves opposite; capsule opening by valves　**10. Asterolinon**
　12　Leaves alternate; capsule circumscissile　**12. Anagallis**
　11　Corolla equalling or longer than calyx
　13　Flowers 7-merous
　14　Flowers white, solitary　**9. Trientalis**
　14　Flowers yellow, in racemes　**8. Lysimachia**
　13　Flowers 5-merous
　15　Capsule circumscissile　**12. Anagallis**
　15　Capsule opening by valves
　16　Heterostylous; corolla-tube longer than calyx　**2. Vitaliana**
　16　Homostylous; corolla-tube equalling or shorter than calyx
　17　Throat of corolla constricted　**3. Androsace**
　17　Throat of corolla open, not constricted **8. Lysimachia**

1. Primula L.[1]

Perennial scapigerous herbs with a basal rosette of leaves. Flowers 5-merous, bracteate, in umbels or rarely solitary, usually heterostylous. Calyx campanulate or cylindrical, 5-toothed. Corolla-tube cylindrical, not constricted at throat; limb 5-lobed, infundibuliform or rotate. Stamens included. Stigma capitate. Capsule valvate; seeds numerous.

Literature: W. Wright Smith & H. R. Fletcher, *Trans. Roy. Soc. Edinb.* **61**: 1–69 (1943), 631–686 (1948); *Trans. Proc. Bot. Soc. Edinb.* **34**: 402–468 (1946).

1　Leaves with revolute vernation, not fleshy, sometimes rugose; calyx angled
2　Leaves distinctly farinose
　3　Corolla-tube 2–3 times as long as calyx　**7. halleri**
　3　Corolla-tube 1–1½ times as long as calyx
　4　Homostylous
　5　Corolla 5–8(–10) mm in diameter; stigma ±5-lobed　**8. scotica**
　5　Corolla 9–12(–15) mm in diameter; stigma globose　**9. scandinavica**
　4　Heterostylous
　6　Calyx campanulate; teeth acute　**5. frondosa**
　6　Calyx cylindrical or urceolate; teeth obtuse, rarely acute　**4. farinosa**
2　Leaves efarinose
7　Corolla yellow
　8　Flowers arising individually from the centre of the leaf-rosette　**1. vulgaris**
　8　Flowers in an umbel at the apex of a scape
　9　Calyx-lobes acuminate; mature capsule equalling or exceeding calyx　**2. elatior**
　9　Calyx-lobes acute; mature capsule shorter than calyx **3. veris**
7　Corolla lilac, pink, purple or violet
　10　Flowers arising individually from the centre of the leaf-rosette　**1. vulgaris**
　10　Flowers in an umbel at the apex of a scape
　11　Homostylous

[1] By D. H. Valentine and A. Kress.

12　Corolla 10–15 mm in diameter　**5. frondosa**
12　Corolla 4–9 mm in diameter
　13　Leaves spathulate; lamina tapering gradually into petiole　**10. stricta**
　13　Leaves oblong-ovate to elliptical; lamina ±abruptly contracted into petiole　**12. egaliksensis**
11　Heterostylous
　14　Bracts distinctly auriculate at base　**11. nutans**
　14　Bracts not or only slightly auriculate at base
　15　Scapes up to 30 cm, stout; corolla not more than 10 mm in diameter　**6. longiscapa**
　15　Scapes up to 20 cm, slender; corolla usually more than 10 mm in diameter
　16　Calyx campanulate; teeth acute　**5. frondosa**
　16　Calyx cylindrical or urceolate; teeth obtuse, rarely acute　**4. farinosa**
1　Leaves with involute vernation, ±fleshy, not rugose; calyx rounded, not angled
17　Corolla yellow
　18　Smaller bracts broadly ovate　**28. auricula**
　18　Smaller bracts lanceolate　**24. palinuri**
17　Corolla lilac, violet, pink or red
　19　Leaves obtriangular; sides entire, apex deeply and narrowly dentate　**19. minima**
　19　Leaves not obtriangular; apex not sharply distinct from the sides
　20　Leaf-margin and interior of calyx farinose　**26. marginata**
　20　Leaf-margin and interior of calyx efarinose
　21　Leaf-margin glabrous, or glands, if present, minute, subsessile
　22　Bracts enveloping the calyx　**14. glutinosa**
　22　Bracts not enveloping the calyx
　23　Leaves with minute dots (sunken glands) above
　24　Longer leaves lanceolate, not pressed to the ground　**13. deorum**
　24　Longer leaves rhombic-ovate, pressed to the ground　**15. spectabilis**
　23　Leaves without minute dots above
　25　Corolla farinose at the throat　**27. carniolica**
　25　Corolla efarinose
　26　Umbels mostly 3- to 5-flowered　**16. glaucescens**
　26　Umbels mostly 1- to 2-flowered　**17. wulfeniana**
　21　Leaf-margin puberulent (strong lens); some glands at least with a distinct, short stalk
　27　Tips of hairs mostly blackish or brown
　28　At most single hair-tips blackish; shrunken apical cells ±apically depressed　**29. hirsuta**
　28　Many hair-tips blackish; shrunken apical cells ±laterally depressed
　29　Leaves mostly 6–12 mm wide; longest pedicels mostly 1–6 mm　**32. daonensis**
　29　Leaves mostly 9–16 mm wide; longest pedicels mostly 4–15 mm
　30　Longest hairs at least 0·3 mm　**33. villosa**
　30　Longest hairs not more than 0·25 mm
　31　Leaves usually subglabrous above　**30. pedemontana**
　31　Leaves puberulent above　**31. apennina**
　27　Tips of hairs colourless or pale
　32　Scape usually not more than 5 mm　**23. allionii**
　32　Scape usually more than 5 mm
　33　At least half the bracts ±ovate
　34　Pedicels not more than 3 mm, much shorter than bracts　**22. tyrolensis**
　34　Pedicels up to 20 mm, usually longer than bracts
　35　Corolla violet; umbels at anthesis ±unilateral　**25. latifolia**
　35　Corolla lilac, pink or red; umbels erect　**29. hirsuta**
　33　More than half the bracts linear-lanceolate
　36　Leaf-margins ciliate; cilia in some places subappressed or semi-patent　**21. integrifolia**
　36　Leaf-margins puberulent; hairs uniformly patent
　37　Leaves suborbicular　**22. tyrolensis**
　37　Leaves obovate to lanceolate

38 Leaves puberulent at margins and above
20. kitaibeliana
38 Leaves densely puberulent at margins, usually glabrous above **18. clusiana**

Subgen. **Primula.** Heterostylous. Efarinose. Leaves hairy, with revolute vernation, rugose, neither fleshy nor coriaceous. Bracts not gibbous or saccate. Calyx angled.

The 3 species in this subgenus all hybridize; the 2 most frequent and widely distributed hybrids are *P. × digenea* A. Kerner (*P. elatior × vulgaris*) and *P. × tommasinii* Gren. & Godron (*P. × variabilis* Goupil, non Bast., *P. veris × vulgaris*).

1. **P. vulgaris** Hudson, *Fl. Angl.* 70 (1762) (*P. acaulis* (L.) Hill). Leaves 5–25 × 2–6 cm, in a lax rosette, oblanceolate to obovate, irregularly dentate to crenate, glabrous above. Scape absent. Bracts 1–1·5 cm, linear to linear-lanceolate. Pedicels 6–20 cm, ascending in flower, decumbent in fruit, with long hairs. Calyx 1–2 cm, tubular, lobes lanceolate, acuminate. Corolla 2–4 cm in diameter. Capsule ovoid, enclosed in calyx. Seeds viscid, with oily caruncle. *Moist, shady places. W., S. & S.C. Europe, extending to E. Denmark and N. Ukraine.* All except Az Fe Is Po Rs (N, B, E) Sa Sb †Su.

1 Flowers white **(b) subsp. balearica**
1 Flowers yellow, red or purple
  2 Leaves tapering gradually into petiole **(a) subsp. vulgaris**
  2 Leaves cuneate, contracted ± abruptly into petiole
(c) subsp. sibthorpii

(a) Subsp. **vulgaris**: Leaves tapering gradually into the short, winged petiole, hairy on the veins beneath. Flowers pale yellow, with 5 yellow-green to orange areas at throat, fragrant. $2n = 22$. *Almost throughout the range of the species.*
(b) Subsp. **balearica** (Willk.) W. W. Sm. & Forrest, *Notes Roy. Bot. Gard. Edinb.* **16**: 42 (1928): Petiole longer than lamina, narrowly winged; lamina glabrescent beneath. Flowers white, very fragrant. ● *Mountains of Mallorca.*
(c) Subsp. **sibthorpii** (Hoffmanns.) W. W. Sm. & Forrest, *loc. cit.* (1928) (*P. sibthorpii* Hoffmanns.): Lamina cuneate, contracted more or less abruptly into petiole, slightly pubescent beneath. Flowers usually red or purple. *E. part of Balkan peninsula; Krym; locally naturalized in C. Europe.*

White-flowered populations (**P. komarovii** Losinsk., *Bull. Acad. Sci. URSS* ser. 7, **1933**(2): 301 (1933)), which may be related to (b) and (c), are known from parts of S.E. Europe.

2. **P. elatior** (L.) Hill, *Veg. Syst.* **8**: 25 (1765). Leaves 5–20 × 2–7 cm, in a lax rosette, suborbicular to oblong or elliptical, usually widest at about the middle, abruptly contracted or gradually narrowed into the long, winged, hairy petiole but not cordate, crenate to denticulate to entire, pubescent or glabrescent above, grey-green and sparsely hairy or tomentose or glabrous beneath. Scape 10–30 cm, usually hairy. Umbel secund; bracts 3–11 mm, linear to ovate; pedicels 5–20 mm. Calyx 6–15 mm, lobes lanceolate or triangular, acuminate. Corolla 15–25 mm in diameter, yellow, sometimes orange at throat. Capsule cylindrical or tapering, usually exceeding calyx. Seeds dry, without caruncle. *Meadows and woods. S., W. & C. Europe, extending northwards to N. Denmark and eastwards to C. Ukraine.* Al Au Be Br Bu Cz Da Ga Ge He Ho Hs Hu It Ju Po Rm Rs (C, W), Su [Fe No].

5 subspecies, largely allopatric, may be distinguished.

1 Leaves glabrous **(e) subsp. pallasii**
1 Leaves ± hairy
  2 Leaves grey-tomentose beneath; calyx 6–8 mm
(b) subsp. leucophylla

2 Leaves ± pubescent beneath; calyx usually more than 8 mm
3 Leaves abruptly contracted into petiole; capsule tapering slightly **(a) subsp. elatior**
3 Leaves gradually narrowing into petiole; capsule tapering markedly
  4 Corolla pale yellow; leaves slightly rugose **(c) subsp. intricata**
  4 Corolla yellow; leaves rugose **(d) subsp. lofthousei**

(a) Subsp. **elatior** (incl. subsp. *carpatica* (Griseb. & Schenk) W. W. Sm. & Forrest, *P. carpatica* (Griseb. & Schenk) Fuss, *P. poloninensis* (Domin) Fedorov): Leaves suborbicular or ovate, abruptly contracted into petiole, crenate to denticulate, sparsely hairy beneath, rugose. Inflorescence many-flowered. Calyx c. 10 mm, pubescent; lobes triangular. Corolla 20–25 mm in diameter; limb more or less flat, pale yellow. Capsule up to 15 mm, tapering only slightly, usually longer than calyx. $2n = 22$. ● *Throughout most of the range of the species, but rarer in the south.*
(b) Subsp. **leucophylla** (Pax) H.-Harrison ex W. W. Sm. & Fletcher, *Trans. Proc. Bot. Soc. Edinb.* **34**: 422 (1946) (*P. leucophylla* Pax): Leaves oblong or elliptical, gradually narrowed into petiole, crenulate or entire, grey-tomentose beneath, rugose. Calyx 6–8(–11) mm, hairy; lobes short, lanceolate. *Calcicole.* ● *E. Carpathians.*
(c) Subsp. **intricata** (Gren. & Godron) Lüdi in Hegi, *Ill. Fl. Mitteleur.* **5**(3): 1748 (1927): Leaves oblong, gradually narrowed into petiole, entire when young, crenulate when mature, pubescent above and beneath, only slightly rugose. Inflorescence few-flowered. Calyx 10–12(–15) mm, puberulent; lobes triangular. Corolla up to 20 mm in diameter; limb more or less flat, pale yellow. Capsule up to 15 mm, tapering distinctly, equalling or slightly exceeding calyx. *Moist grassland.* ● *Mountains of S. and S.C. Europe.*
(d) Subsp. **lofthousei** (H.-Harrison) W. W. Sm. & Fletcher, *Trans. Proc. Bot. Soc. Edinb.* **34**: 423 (1946): Leaves oblong, gradually narrowed into petiole, crenulate, rugose, pubescent above and beneath. Inflorescence many-flowered. Calyx 12–17 mm in fruit, puberulent; lobes triangular. Corolla 14–18 mm in diameter; limb saucer-shaped, yellow. Capsule up to 15 mm, tapering, often exceeding calyx. $2n = 22$. ● *S. Spain (Sierra Nevada).*
(e) Subsp. **pallasii** (Lehm.) W. W. Sm. & Forrest, *Notes Roy. Bot. Gard. Edinb.* **16**: 42 (1928): Leaves obovate-oblong, gradually narrowed into petiole, dentate, glabrous. Inflorescence 3- to 6-flowered. Calyx 12 mm, glabrous; lobes triangular. Corolla 20–25 mm in diameter; limb more or less flat, pale yellow. Capsule up to 17 mm, tapering slightly, usually exceeding calyx. *Alpine meadows. C. Ural.* (*C. & S.W. Asia.*)

3. **P. veris** L., *Sp. Pl.* 142 (1753) (*P. officinalis* (L.) Hill). Leaves 5–20 × 2–6 cm in a lax rosette, enlarging in fruit; lamina ovate to ovate-oblong, sometimes cordate, obtuse, usually widest at or near the base, abruptly contracted or narrowed into the sometimes winged petiole, crenate, pubescent or glabrescent above, grey- or white-tomentose to glabrescent beneath. Scape 10–30 cm, hairy. Umbel secund; bracts 2–7 mm, linear-lanceolate; pedicels 3–20 mm. Calyx 8–20 mm, campanulate, lobes triangular, acute, sometimes mucronate. Corolla 8–28 mm in diameter, concave or almost flat, bright yellow, with orange spots at the base of the lobes. Capsule ovate, shorter than calyx. Seeds dry, without caruncle. *Meadows and pastures. Most of Europe except the extreme north and much of the Mediterranean region.* Al Au Be Br Bu Cz Da Fe Ga Ge Gr Hb He Ho Hs Hu It Ju ?Lu No Po Rm Rs (N, B, C, W, K, E) Su.

Variable; 4 subspecies may be distinguished but there are many transitional plants.

1 Lamina weakly hairy or glabrescent beneath; hairs usually
  straight, unbranched
2 Calyx 8–15 mm; corolla 9–12 mm in diameter   **(a) subsp. veris**
2 Calyx 15–20 mm; corolla 18–28 mm in diameter
                           (d) subsp. macrocalyx
1 Lamina hairy beneath; hairs twisted, often branched
3 Lamina gradually decurrent, grey-tomentose beneath; petiole
  winged   **(b) subsp. canescens**
3 Lamina usually cordate, thickly white-hairy beneath; petiole
  not or slightly winged   **(c) subsp. columnae**

(a) Subsp. **veris**: Lamina usually abruptly contracted at base, weakly hairy to glabrescent; hairs 0·15–0·3 mm, stiff, unbranched; petiole winged. Calyx 8–15 mm, not widely divergent at apex. Corolla 9–12 mm in diameter; limb concave; tube usually longer than calyx. 2*n*=22. ● *Throughout most of the range of the species, southwards to the Alps.*

(b) Subsp. **canescens** (Opiz) Hayek ex Lüdi in Hegi, *Ill. Fl. Mitteleur.* **5**(3): 1752 (1927): Lamina grey-tomentose beneath, gradually decurrent into winged petiole; hairs 0·5–0·75 mm, twisted, often branched or interwoven. Calyx 16–20 mm. Corolla 8–20 mm in diameter; limb slightly concave; tube usually equalling calyx. ● *Lowlands of C. Europe, extending to S. France and N. Spain.*

(c) Subsp. **columnae** (Ten.) Lüdi, *loc. cit.* (1927) (*P. suaveolens* Bertol.): Lamina ovate, usually cordate, thickly white-hairy beneath; hairs 1 mm, twisted, often branched and interwoven; petiole not or slightly winged. Calyx 16–20 mm, widened at apex, usually rather hairy. Corolla 10–22 mm in diameter; limb almost flat; tube longer than calyx. *Mountains of S. Europe.*

(d) Subsp. **macrocalyx** (Bunge) Lüdi, *op. cit.* 1753 (1927) (*P. macrocalyx* Bunge): Lamina narrowed to a long, winged petiole, somewhat hairy to glabrescent beneath, the hairs 0·3–0·5 mm, usually straight, unbranched. Calyx 15–20 mm, widened at apex, often hairy. Corolla 18–28 mm in diameter; tube longer than calyx. *S.E. Russia, Krym.*

Subgen. **Aleuritia** (Duby) Wendelbo. Hetero- or homostylous. Usually farinose or with farina-producing hairs, otherwise glabrous. Leaves with revolute vernation, not fleshy, coriaceous or rugose. Bracts usually saccate or gibbous. Calyx angled.

4. **P. farinosa** L., *Sp. Pl.* 143 (1753). Farinose or efarinose; heterostylous. Leaves 1–10 × 0·3–2 cm, oblanceolate to elliptical, obtuse, entire or finely denticulate, green above, usually farinose beneath; petiole variable in length, winged. Scape 3–20 cm, 2- to many-flowered. Bracts 2–8 mm, linear-lanceolate to subulate, gibbous at the base. Pedicels up to twice as long as bracts, lengthening in fruit. Calyx 3–6 mm, often tinged with black or purple, cylindrical or urceolate; teeth obtuse, rarely acute. Corolla 8–16 mm in diameter, lilac-pink, rarely purple or white, yellow in the throat. Capsule 5–9 mm, equalling or up to twice as long as calyx. *Marshes and damp meadows, usually on base-rich soils. From Scotland and C. Sweden southwards to C. Spain and Bulgaria, mainly in the mountains.* Au Br Bu Cz Da Fe Ga Ge He Hs Hu It Ju Po Rm Rs (N, B, C, W) Su.

(a) Subsp. **farinosa**: Usually a stout plant, with up to 20 flowers on short pedicels. Scape, bracts and calyx farinose; leaves usually densely farinose beneath, rarely efarinose (var. *denudata* Koch). 2*n*=18, 36. *Throughout the range of the species.*

(b) Subsp. **exigua** (Velen.) Hayek, *Prodr. Fl. Penins. Balcan.* **2**: 25 (1928): Usually a smaller and more slender plant than subsp. (a), with 2–6(–10) flowers on longer pedicels. Scape, bracts and calyx efarinose; leaves not or sparingly farinose. ● *Mountains of Bulgaria.*

5. **P. frondosa** Janka, *Österr. Bot. Zeitschr.* **23**: 204 (1873). Farinose or efarinose; heterostylous or homostylous. Leaves

3–9 × 1–2 cm, spathulate, oblong or obovate, denticulate or crenate, thickly farinose or sometimes efarinose. Scape 4–12 cm, slightly farinose or glandular, 10- to 30-flowered. Bracts 4–5 mm, linear or linear-lanceolate, wider at the base, with or without farina. Pedicels 5–35 mm. Calyx 4–6 mm, sometimes farinose, campanulate; teeth acute. Corolla 10–15 mm in diameter, pinkish-lilac to reddish-purple. In homostylous flowers, stamens and stigma both at apex of corolla-tube. Capsule cylindrical, equalling or just exceeding calyx. *Shady cliffs, near melting snow.* ● *C. Bulgaria* (*C. Stara Planina*). Bu.

6. **P. longiscapa** Ledeb., *Mém. Acad. Sci. Pétersb.* **5**: 520 (1815). Heterostylous. Leaves 3–13 × 0·7–2 cm, oblong-obovate to lanceolate, usually rounded at apex, entire or crenulate, glaucescent, efarinose. Scape 10–30 cm, stout, often farinose, 10- to 100-flowered. Bracts 3–10 mm, linear-lanceolate, somewhat gibbous or slightly auriculate at base. Pedicels 3–10 mm. Calyx 3–5 mm; teeth sometimes slightly farinose. Corolla 6–10 mm in diameter, purple. Capsule 5–8 mm, oblong, nearly twice as long as calyx. *Damp meadows. S.E. Russia.* Rs (C, E). (*C. Asia.*)

7. **P. halleri** J. F. Gmelin, *Onomat. Bot. Compl.* **7**: 407 (1775) (*P. longiflora* All.). Farinose; homostylous. Leaves 2–8 × 0·5–3 cm, oblong-obovate to oblanceolate, subentire or minutely denticulate, densely covered beneath with yellow farina. Scape 8–18 cm, stout, farinose, 2- to 12-flowered. Bracts 5–10 mm, narrowly lanceolate, subsaccate or slightly auriculate at base, sparingly farinose. Pedicels 5–10 mm. Calyx 8–12 mm, farinose. Corolla 15–20 mm in diameter, lilac; tube 20–30 mm, 2–3 times as long as calyx. Stamens inserted at apex of tube; style exserted. Capsule cylindrical, equalling or exceeding calyx. *Alpine meadows and rock-crevices; somewhat calcicole.* ● *Alps, Carpathians, mountains of Balkan peninsula.* Al Au Bu Cz Ga He It Ju Po Rm Rs (W).

8. **P. scotica** Hooker in Curtis, *Fl. Lond.* **4**: t.133 (1821). Farinose; homostylous. Leaves 1–5 × 0·4–1·5 cm, elliptical, oblong or spathulate, entire or remotely crenulate-denticulate, usually abundantly farinose beneath. Scapes 0·5–6(–9) cm, 1–2(–4) in number, farinose, 1- to 6-flowered. Bracts 2–5 mm, lanceolate, subsaccate at base, farinose. Pedicels 1–5 mm. Calyx 4–6 mm; teeth obtuse, farinose. Corolla 5–8(–10) mm in diameter, dark purple with a yellow throat, rarely white; tube 1½ times as long as calyx. Stigma obscurely 5-lobed. Capsule equalling or exceeding calyx. 2*n*=54. *Moist pastures near the sea.* ● *N. Scotland.* Br.

9. **P. scandinavica** Bruun, *Svensk Bot. Tidskr.* **32**: 249 (1938). Like **8** but scapes 4–10(–15) cm, 2- to 8(–10)-flowered; bracts 5–8 mm; calyx 5–6 mm; corolla 9–12(–15) mm in diameter; stigma globose, not 5-lobed. 2*n*=72. *Mountain slopes; calcicole.* ● *Norway, N.W. Sweden.* No Su.

10. **P. stricta** Hornem., *Fl. Dan.* **8**(24): 3 (1810). Homostylous. Leaves 0·5–5 × 0·2–1·5 cm, narrowly obovate to oblanceolate, entire or denticulate, efarinose or sometimes sparingly farinose beneath. Scape 2–30 cm, efarinose except sometimes at apex, 1- to 15-flowered. Bracts 3–8 mm, lanceolate to subulate, saccate at the base. Pedicels 1–3 times as long as bracts. Calyx 4–6 mm, farinose between the ribs, and almost always farinose within; teeth often glandular-ciliolate. Corolla 4–9 mm in diameter, violet to lilac, lobes more or less oblong, emarginate; tube 1–1½ times as long as calyx. Capsule 5·5–8 mm, ovoid, a little longer than calyx. 2*n*=126. *Meadows and cliffs, up to 750 m. Iceland and Fennoscandia, southwards to 62° 30′ N.* Fe Is No Rs (N) Su.

11. P. nutans Georgi, *Bemerk. Reise* **1**: 200 (1775) (*P. sibirica* Jacq., *P. finmarchica* Jacq.). Efarinose; heterostylous. Leaves 1–12 × 0·5–1·5 cm, ovate, oblong or elliptic-orbicular, entire or obscurely denticulate, glabrous, somewhat fleshy; lamina not or scarcely decurrent, ½ as long to as long as petiole. Scape 2–30 cm, 1- to 10-flowered. Bracts 5–12 mm, oblong, saccate at base, with auricles up to 1·5 mm. Pedicels 5–45 mm. Calyx 4–8 mm, with 5 distinct ribs. Corolla 10–20 mm in diameter, lilac to pink, with a yellow centre; tube 1¼–2 times as long as calyx. Capsule 10–16 mm, usually 1½ times as long as calyx. 2*n*=22. *Damp meadows, usually near the sea. N. Russia, N.E. Fennoscandia.* Fe No Rs (N, C) Su.

12. P. egaliksensis Wormsk. in Hornem., *Fl. Dan.* **9**(26): 4 (1816). Efarinose; homostylous. Leaves 0·5–5 × 0·2–1·5 cm, elliptical to oblong-ovate, entire or obscurely denticulate, glabrous; lamina more or less abruptly contracted into petiole. Scape 1–15 cm, 1- to 9-flowered. Bracts 3–7 mm, subulate, gibbous at the base. Pedicels 1–3 times as long as bracts. Calyx 4–6 mm; teeth often glandular-ciliolate. Corolla 5(–9) mm in diameter, white or lilac; tube slightly longer than calyx. Capsule 5–10 mm, narrowly cylindrical, 1½–2 times as long as calyx. *Moist pastures near the sea. N. Iceland.* Is. (*Greenland, Canada, Alaska.*)

Subgen. **Auriculastrum** Schott. Heterostylous. Farinose or efarinose. Hairs glandular, sometimes minute. Leaves with involute vernation, fleshy or coriaceous, not rugose. Bracts not gibbous or saccate. Calyx rounded, not angled.

Numerous hybrids are known; many are fertile and some are frequent, especially **14** × **19, 25** × **29, 28** × **29**. The last, *P.* × *pubescens* Jacq., is occasionally found outside the area of **29**, *P. hirsuta*.

13. P. deorum Velen., *Sitz.-Ber. Böhm. Ges. Wiss.* (*Math.-Nat. Kl.*) **1890**(1): 55 (1890). Longest leaves 2–15 × 0·4–2 cm, narrowly oblanceolate, erect or erecto-patent; lamina coriaceous, shining, with very short, pale, glandular hairs embedded in the upper surface; margin cartilaginous, entire. Scape 5–20 cm, fragrant and viscid, more or less dark violet towards the apex, 3- to 18-flowered. Bracts 3–10 mm, narrowly triangular-lanceolate. Pedicels 3–10 mm. Calyx 3–6 mm. Corolla deep reddish- to violet-purple. 2*n*=66. *Wet alpine grassland.* ● *S.W. Bulgaria* (*Rila Planina*). Bu.

14. P. glutinosa Wulfen in Jacq., *Fl. Austr.* **5**: 41 (1778). Plant viscid, the secretion rarely drying to a white crust. Longest leaves 1·5–5 × 0·3–0·8 cm, more or less erect, narrowly oblanceolate, more or less rounded at the apex, coriaceous, with minute, pale glandular hairs embedded in the upper surface; margin somewhat cartilaginous, usually finely crenate or dentate towards the apex. Scape 1·5–8 cm, 2- to 8-flowered. Bracts 4–12 mm, broadly ovate to oblong, viscid. Pedicels up to 3 mm. Calyx 5–9 mm. Corolla deep violet (reddish-violet when dry). 2*n*=66, 67. *Wet, heavy soils; calcifuge.* ● *E. Alps; C. Jugoslavia* (*Bosna*). Au He It Ju.

15. P. spectabilis Tratt., *Ausgem. Taf. Arch. Gewächsk.* **4**: 34 (1814). Longest leaves 2–9 × 1–3 cm, pressed to the ground, rhombic-ovate, coriaceous, somewhat viscid, shining, with scattered, very short, pale glandular hairs embedded in the upper surface (darkish dots, if illuminated from the side); margin cartilaginous, minutely erose. Scape 2–15 cm, 2- to 5-flowered. Bracts 2–15 mm, triangular-lanceolate to linear. Pedicels 3–30 mm. Calyx 6–12 mm. Corolla pinkish-red to lilac. 2*n*=66. *Rocks and stony slopes; calcicole.* ● *S. Alps between 10° 15′ and 11° 50′ E.* It.

16. P. glaucescens Moretti, *Gior. Fis.* (*Brugnat.*) ser. 2, **5**: 249 (1822). Longest leaves 2–7 × 0·7–2 cm, broadly lanceolate to obovate, patent when mature, acute, coriaceous, stiff, shining, glabrous; margin cartilaginous, minutely doubly crenulate. Scape 3–12 cm, 2- to 6-flowered. Bracts 4–30 mm, triangular-lanceolate to linear. Pedicels 3–20 mm. Calyx 6–18 mm. Corolla pinkish-red to lilac. 2*n*=66. *Rocks and stony slopes; calcicole.* ● *S. Alps between 9° 15′ and 10° 45′ E.* It.

17. P. wulfeniana Schott, *Wilde Blendl. Österr. Primeln* 17 (1852). Longest leaves 1·5–4 × 0·5–1·2 cm, oblanceolate to obovate, patent when mature, shining, coriaceous; margin entire, cartilaginous, densely covered with pale glandular hairs up to 0·1 mm, which are more or less concealed by non-farinose secretions; leaves otherwise glabrous. Scape 0·5–7 cm, 1- or 2-flowered. Bracts 3–15 mm, triangular-lanceolate to linear. Pedicels 2–8 mm. Calyx 6–12 mm. Corolla pinkish-red to lilac. 2*n*=66. *Rocks and stony slopes; calcicole.* ● *S.E. Alps, westwards to 12° 15′ E.; S. Carpathians.* Au It Ju Rm.

Plants from the S. Carpathians (subsp. **baumgarteniana** (Degen & Moesz) Lüdi in Hegi, *Ill. Fl. Mitteleur.* **5**(3): 1776 (1927)) are said to differ from those of the Alps in their wider, more obtuse leaves, but are doubtfully distinct.

18. P. clusiana Tausch, *Flora* (*Regensb.*) **4**: 364 (1821). Longest leaves 1·5–6 × 0·5–2·5 cm, lanceolate to obovate, patent when mature, shining, coriaceous; margins entire or nearly so, narrowly cartilaginous, densely covered with pale glandular hairs up to 0·25 mm, leaves otherwise glabrous. Scape 1–8 cm, 1- to 4-flowered. Bracts 3–15 mm, triangular-lanceolate to linear. Pedicels 2–12 mm. Calyx 8–15 mm. Corolla pinkish-red to lilac. 2*n*=c. 198. *Rocks and stony slopes; calcicole.* ● *N.E. Alps, westwards to c. 13° E.* Au Ge.

19. P. minima L., *Sp. Pl.* 143 (1753). Longest leaves 0·5–3 × 0·3–1 cm, cuneate, shining, coriaceous; the margins entire, the apex truncate or more or less rounded, deeply and densely dentate; teeth cartilaginous at apex; glandular hairs c. 0·05 mm, pale, scattered or very few. Scape 0·2–4 cm, 1- to 2-flowered. Bracts 3–8 mm, linear-lanceolate. Pedicels up to 3 mm. Calyx 5–9 mm. Corolla bright pink, with deeply divided, V-shaped lobes; throat with many-celled glandular hairs, the longer 0·4–0·8 mm. 2*n*=66, 67, 68, 69, 70, 73. *Alpine pastures, snow-patches; calcifuge.* ● *E.C. & S.E. Europe, extending westwards to c. 10° 30′ E. in the Alps.* Al Au Bu Cz Ge It Ju Po Rm Rs (W).

20. P. kitaibeliana Schott, *Österr. Bot. Wochenbl.* **2**: 268 (1852). Plant viscid. Longest leaves 2–9 × 1–4 cm, not reaching full length until after anthesis, broadly lanceolate to obovate at maturity, mostly with shallow, distant teeth, sometimes entire, fleshy, more or less densely covered with pale 3- to 5-celled glandular hairs, the longest 0·1–0·5 mm. Scape 1·5–8 cm, 1- to 3-flowered. Bracts 2–11 mm, linear. Pedicels 2–12 mm. Calyx 7–13 mm. Corolla pink to lilac. 2*n*=66. *Rocks and stony soil; calcicole.* ● *W.C. Jugoslavia.* Ju.

21. P. integrifolia L., *Sp. Pl.* 144 (1753). Slightly viscid, the secretion sometimes drying to a white crust. Longest leaves

1–4 × 0·6–1·2 cm, often not reaching full length until after anthesis, lanceolate to obovate, entire or nearly so, fleshy to coriaceous, slightly viscid; glandular hairs many-celled, pale-tipped; leaves with scattered, inconspicuous hairs above; leaf-margin ciliate, the cilia variously deflexed, patent or more or less erecto-patent, the longest 0·2–1 mm. Scape 0·5–5 cm, 1- to 3-flowered. Bracts 2–10 mm, linear to lanceolate. Pedicels up to 3·5 mm at anthesis, up to 6 mm in fruit. Calyx 6–10 mm. Corolla reddish-purple or pinkish-lilac. $2n=66$, c. 68, c. 70. *Snow-patches and stony alpine soils; calcifuge.* ● *C. Alps; E. & C. Pyrenees.* Au Ga He Hs It.

22. P. tyrolensis Schott, *Sippen Österr. Primeln* 13 (1851). Plant viscid. Longest leaves 0·8–2·5 × 0·5–1·2 cm, suborbicular to broadly obovate, with fine cartilaginous teeth at margin, fleshy, densely puberulent; withered leaves persistent for some time; glandular hairs pale, 3-celled, those at apex up to 0·1 mm, those at base up to 0·4 mm. Scape 0·5–2 cm, 1- or 2-flowered. Bracts 2–8 mm, linear-lanceolate to obovate. Pedicels up to 2·5 mm. Calyx 5–8 mm. Corolla bright purplish-pink. $2n=66$. *Shady rock-crevices and screes.* ● *Alpi dolomitiche.* It.

23. P. allionii Loisel. in Desv., *Jour. Bot. Rédigé* 2: 262 (1809). Plant viscid. Longest leaves 1·5–4·5 × 0·7–1·2 cm, oblanceolate to suborbicular, entire or crenulate or finely serrate, fleshy, very viscid, densely puberulent; withered leaves often persistent for some years; glandular hairs pale-tipped, 3- or 4-celled, the longest 0·15–0·4 mm. Scape very short, up to 8 mm in fruit, 1- to 5-flowered. Bracts 1–3 mm, broadly ovate. Pedicels 1–5 mm. Calyx 3–6 mm. Corolla pale pink to reddish-purple. $2n=66$. *Shady cliffs.* ● *Maritime Alps.* Ga It.

24. P. palinuri Petagna, *Inst. Bot.* 2: 332 (1787). Leaves efarinose, the longest 4–16 × 2–7 cm, obovate, more or less dentate, fleshy, viscid, fragrant, densely puberulent or almost glabrous above and beneath, slightly cartilaginous at margin; glandular hairs colourless, the longest 0·1–0·25 mm. Scape 8–20 cm, 5- to 25-flowered. Bracts 5–25 mm, farinose, the outer broadly ovate, acute, the inner lanceolate. Pedicels 5–20 mm, farinose. Calyx 6–10 mm, densely farinose. Corolla deep yellow with a white farinose ring at the throat. $2n=66$. *Rock-crevices by the sea.* ● *S.W. Italy, from 39° 45′ to 40° 5′ N.* It.

25. P. latifolia Lapeyr., *Hist. Abr. Pyr.* 97 (1813) (*P. viscosa* All., non Vill.). Leaves efarinose, the longest 3–15 × 1–4·5 cm, oblanceolate to obovate, rarely dentate, usually with wide teeth towards apex or sometimes entire, not cartilaginous at margin, fleshy, fragrant, viscid, covered with many pale glandular hairs, the longer 0·1–0·4 mm, with shrunken apical cells apically depressed. Scape 3–18 cm, 2- to 20-flowered; umbel secund. Bracts 2–9 mm, more or less scarious, ovate to transversely ovate. Pedicels 2–20 mm. Calyx 2–6 mm. Corolla purplish- or dark violet, more or less farinose at the throat. $2n=64, 65, 66, 67$. *Rocks; calcifuge.* ● *S.W. & C. Alps; E. Pyrenees.* Ga He Hs It.

26. P. marginata Curtis, *Bot. Mag.* 6: t. 191 (1792). Longest leaves 2–9 × 1–3·5 cm, obovate to lanceolate, fleshy, serrate, the teeth more or less slender and obtuse; surface with short, more or less farinose hairs or glabrous; margin not cartilaginous, but usually densely glandular and strongly farinose; glandular hairs up to 0·15 mm, colourless, with shrunken apical cells laterally depressed. Scape 2–9 cm, 3- to 15-flowered. Bracts 2–9 mm, ovate to transversely ovate, more or less scarious. Pedicels 3–20 mm. Calyx 2–6 mm, farinose. Corolla bluish- (rarely

pinkish-) lilac with a white farinose ring at the throat. $2n=62$, 63, ?124, 126, 127, 128. *Rock-crevices.* ● *S.W. Alps.* Ga It.

The tetraploids are restricted to the E. Maritime Alps and the Alpi Ligure, the diploids to the northern and western parts of the area of the species.

27. P. carniolica Jacq., *Fl. Austr.* 5: 28 (1778). Leaves efarinose, the longest 3–12 × 1–4 cm, obovate to lanceolate, almost entire or shallowly toothed towards the apex, fleshy, with a narrow cartilaginous margin, glabrous. Scape 5–20 cm, 2- to 15-flowered. Bracts 1–7 mm, ovate to transversely ovate, more or less scarious. Pedicels 2–20 mm. Calyx 3·5–7 mm. Corolla purplish-pink, white-farinose at the throat. $2n=62$, c. 68. *Pastures; shady rocks in wet ravines.* ● *Slovenija.* Ju.

28. P. auricula L., *Sp. Pl.* 143 (1753). Longest leaves 1·5–12 × 1–6 cm, suborbicular to lanceolate, usually obovate, with or without farina on margins and surface, entire to dentate, fleshy, the surface glabrous or with scattered short glandular hairs; margin narrowly cartilaginous, puberulent, the hairs 0·05–0·5 mm, glandular, pale. Scape 1–16 cm, 2- to 30-flowered, often farinose. Bracts 1–8 mm, ovate to transversely ovate, more or less scarious. Pedicels 2–20 mm. Calyx 3–7·5 mm, usually farinose. Corolla deep yellow, white-farinose at the throat. $2n=62, 63, 64, 65, 66$. *Rock-crevices or wet alpine grassland.* ● *Alps, Carpathians, Appennini.* Au Cz Ga Ge He Hu It Ju Po Rm.

Size, leaf-shape, indumentum, dentation, farina and odour vary, occasionally even within a single population; it is therefore difficult to delimit subspecies.

29. P. hirsuta All., *Auct. Syn. Stirp. Horti Taur.* 10 (1773) (*P. viscosa* Vill.). Efarinose. Longest leaves 1–9 × 0·8–3·5 cm, ovate, obovate or suborbicular, rather abruptly narrowed into the winged petiole, finely or coarsely dentate, fleshy, viscid, covered with glandular hairs which produce a pale or reddish-brown, rarely blackish secretion; longest hairs 0·1–0·5 mm, with shrunken apical cells depressed apically. Scape 1–7 cm, usually shorter than the leaves, even in fruit, 1- to 15-flowered. Bracts 1–8 mm, ovate to transversely ovate, more or less scarious. Pedicels 2–16 mm. Calyx 3·5–9 mm. Corolla pale lilac to deep purplish-red, usually with a white centre. Capsule ⅓ as long to as long as calyx. $2n=62, 63, 64, 67$. *Rocks and stony alpine pastures; calcifuge.* ● *Alps; C. Pyrenees.* Au Ga He Hs It.

30. P. pedemontana Thomas ex Gaudin, *Fl. Helv.* 2: 91 (1828). Efarinose. Longest leaves 2–8 × 0·8–2·5 cm, broadly obovate to broadly lanceolate, entire or with shallow teeth towards the apex, fleshy, usually subglabrous above and beneath, with scattered glands towards the margin; margin densely puberulent; hairs producing a reddish-black or reddish-brown secretion, the longest 0·1–0·25 mm; shrunken apical cells depressed laterally. Scape 2·5–12 cm, usually longer than the leaves in fruit, 1- to 15-flowered. Bracts 1–6 mm, ovate to transversely ovate, more or less scarious. Pedicels 2–15 mm. Calyx 3–7·5 mm. Corolla purple or deep pink, usually with a white centre. Capsule somewhat shorter to slightly longer than calyx. $2n=62$. *Rocks and stony alpine pastures; calcifuge.* ● *S.W. Alps; Cordillera Cantábrica.* Ga Hs It.

Subsp. **iberica** Losa & P. Monts., *Anal. Inst. Bot. Cavanilles* 10(2): 482 (1952), from Spain, is doubtfully distinct.

31. P. apennina Widmer, *Eur. Arten Primula* 48, 140 (1891). Like **30** but somewhat smaller; leaves 2·5–6·5 × 1–2 cm; scapes

3–8 cm; leaf-surfaces not subglabrous, the glandular hairs 0·25 mm, less dark. 2n=62. *Sandstone cliff-ledges.* ● *N. Appennini (Monte Orsaro).* It.

32. **P. daonensis** (Leybold) Leybold, *Flora (Regensb.)* **38**: 345 (1855) (*P. oenensis* Thomas ex Gremli). Efarinose. Longest leaves 1·5–7 × 0·6–2 cm, narrowly cuneate-obovate, more rarely ovate, dentate towards the apex, fleshy, viscid, covered with glandular hairs; hairs producing a reddish-black or brown secretion, the longest usually 0·15–0·5 mm; shrunken apical cells depressed laterally. Scape 1·5–9 cm, usually longer than the leaves in fruit, 2- to 8-flowered. Bracts 1–6 mm, ovate to transversely ovate, more or less scarious. Pedicels 1–9 mm. Calyx 2·5–6 mm. Corolla pink to lilac, usually white at the throat. Capsule about as long as calyx. 2n=62, 63, 64. *Rocks and stony alpine pastures; calcifuge.* ● *E. Alps from 9° 45′ to 11° E.* Au He It.

33. **P. villosa** Wulfen in Jacq., *Fl. Austr.* **5**: 41 (1778). Efarinose. Longest leaves 2–15 × 1–3·5 cm, broadly to narrowly obovate or ovate, dentate towards the apex or sometimes entire, fleshy, viscid, covered with glandular hairs; hairs producing a reddish-black or brown secretion, the longest 0·3–1 mm; shrunken apical cells depressed laterally. Scape 2–15 cm, usually longer than the leaves in fruit, 2- to 5-flowered. Bracts 1–5 mm, ovate to transversely ovate, more or less scarious. Pedicels 2–15 mm. Calyx 3–6·5 mm. Corolla pink to lilac with a white centre. Capsule as long as or a little longer than calyx. 2n=62, 63, 64. *Rocks and stony alpine pastures; calcifuge.* ● *E. Alps westwards to 13° 30′ E.; very locally in S.W. Alps.* Au Ga It Ju.

2. Vitaliana Sesler[1]

Trailing or caespitose perennial herbs. Leaves basal, entire. Flowers 5-merous, solitary; heterostylous. Calyx obconical with linear-lanceolate lobes. Corolla yellow; tube at least 1½ times as long as calyx, cylindrical, scarcely constricted at the throat. Filaments short, without dilated base, not connate. Capsule dehiscing by valves to the middle; seeds 2–4(–5).

Literature: A. Chiarugi, *Nuovo Gior. Bot. Ital.* nov. ser., **37**: 319–368 (1930). O. Schwarz, *Feddes Repert.* **67**: 16–41 (1963). A. Kress, *Androsace* in Hegi, *Ill. Fl. Mitteleur.* 5(3), *Nachtr.*: 2248c–2249b (1968).

1. **V. primuliflora** Bertol., *Fl. Ital.* **2**: 368 (1835) (*Androsace vitaliana* (L.) Lapeyr., *Gregoria vitaliana* (L.) Duby). Stems more or less creeping and forming mats or cushions. Leaves 4–12 × 1–2 mm, linear to oblong-lanceolate, acute, acuminate or rounded. Calyx 4–10 mm, glabrous or pubescent. Corolla 6–22 mm. ● *Screes, rocks and stony pastures, usually above 2000 m. Mountains of S. & W.C. Europe, eastwards to S.E. Alps and C. Appennini.* Au Ga He Hs It.

1 Leaves oblong to oblong-lanceolate, rounded and distinctly tomentose at apex (e) **subsp. praetutiana**
1 Leaves linear to linear-lanceolate, acute to obtuse, not tomentose at apex
  2 Corolla 6–9(–12) mm; leaves usually distinctly keeled towards apex (d) **subsp. assoana**
  2 Corolla 10–22 mm; leaves usually flat
    3 Leaves densely stellate-puberulent above (c) **subsp. cinerea**
    3 Leaves glabrous or sparsely pubescent above
      4 Calyx and leaves glabrous or ciliate (a) **subsp. primuliflora**
      4 Calyx and lower surface of leaves pubescent (b) **subsp. canescens**

---

[1] By I. K. Ferguson.

(a) Subsp. **primuliflora**: Stems trailing or laxly caespitose; leaves flat, acuminate, glabrous or sometimes ciliate; calyx glabrous or ciliate; corolla 10–16 mm. *S.E. Alps.*

(b) Subsp. **canescens** O. Schwarz, *Feddes Repert.* **67**: 24 (1963): Stems trailing or laxly caespitose; leaves flat, acute to obtuse, pubescent or more or less tomentose beneath, usually glabrous above; calyx pubescent; corolla 12–18 mm. *S.W. Alps; Pyrenees.*

(c) Subsp. **cinerea** (Sünd.) I. K. Ferguson, *Taxon* **18**: 303 (1969) (*Gregoria vitaliana* var. *cinerea* Sünd.): Stems trailing to densely caespitose; leaves flat, acuminate, densely pubescent to grey-tomentose; calyx pubescent; corolla 12–22 mm. 2n=40. *C. & S.W. Alps; E. Pyrenees.*

(d) Subsp. **assoana** Laínz, *Bol. Inst. Estud. Astur. (Supl.Ci.)* **10**: 199 (1964) (*V. congesta* O. Schwarz, *V. intermedia* O. Schwarz): Densely caespitose; leaves acuminate, usually keeled, pubescent to grey-tomentose; calyx 4–6 mm, tomentose; corolla 6–12 mm. *Mountains of S. & E. Spain.*

(e) Subsp. **praetutiana** (Buser ex Sünd.) I. K. Ferguson, *Taxon* **18**: 303 (1969) (*Gregoria vitaliana* var. *praetutiana* Buser ex Sünd.): Stems trailing or laxly caespitose; leaves flat, oblong to oblong-oblanceolate, obtuse or rounded at apex, densely tomentose on the lower surface and at the apex above; calyx pubescent; corolla 8–12 mm. *C. Appennini.*

3. Androsace L.[1]

Annual or perennial herbs; when perennial, often caespitose. Leaves usually basal. Flowers bracteate, borne singly or in umbels; homostylous. Calyx campanulate or subglobose; lobes triangular. Corolla-tube somewhat inflated, usually shorter than calyx; throat constricted, closed by a ring of scales. Filaments very short, without dilated base, not connate. Capsule spherical, dehiscing almost to the base; seeds 2–many.

Literature: F. Pax & R. Knuth in Engler, *Pflanzenreich* 22(IV. 237): 172–220 (1905). A. Kress in Hegi, *Ill. Fl. Mitteleur.* 5(3), *Nachtr.*: 2248c–2249b (1968).

1 Annual, without non-flowering rosettes
  2 Involucral bracts large, leaf-like, equalling or longer than pedicels at anthesis **1. maxima**
  2 Involucral bracts small, much shorter than pedicels
    3 Corolla and capsule shorter than calyx **5. elongata**
    3 Corolla and capsule equalling or longer than calyx
      4 Leaves ovate or elliptical, distinctly petiolate **4. filiformis**
      4 Leaves oblong-lanceolate, sessile or very shortly petiolate
        5 Inflorescence 5- to 8(–10)-flowered; corolla 5–7 mm; seeds 3–5 **3. chaixii**
        5 Inflorescence 5- to 30-flowered; corolla 4–6 mm; seeds numerous **2. septentrionalis**
1 Perennial, usually caespitose, with non-flowering rosettes
  6 Flowers bracteate, usually in umbels or pairs
    7 Leaves ± densely villous or ciliate with long, simple hairs
      8 Young leaves densely villous, at least near the apex **12. villosa**
      8 Young leaves ± glabrous but distinctly ciliate **11. chamaejasme**
    7 Leaves glabrous or shortly pubescent with short, stellate or hooked hairs
      9 Leaves persistent for many years, forming columnar shoots after withering **10. pyrenaica**
      9 Leaves not persistent after withering
        10 Leaves widest above the middle
          11 Calyx glabrous **8. hedraeantha**
          11 Calyx pubescent **7. obtusifolia**
        10 Leaves linear or linear-lanceolate, parallel-sided or widest at the base
          12 Scapes glabrous **6. lactea**
          12 Scapes pubescent or puberulent **9. carnea**
  6 Flowers usually ebracteate, solitary
    13 Pedicels usually 20 mm or more, glabrous **6. lactea**
    13 Pedicels usually less than 20 mm, pubescent

14 Leaves persistent for many years after withering, forming columnar shoots; plants forming deep, dense cushions
  15 Flowers with 2–3 bracts below the calyx   **10. pyrenaica**
  15 Flowers ebracteate
    16 Leaves densely white-tomentose   **14. vandellii**
    16 Leaves pubescent, not white-tomentose
      17 Pedicels 6–20 mm   **15. cylindrica**
      17 Pedicels less than 6 mm   **13. helvetica**
14 Leaves not persistent for long after withering, forming short rosettes at the end of the branches; plants forming shallow cushions
  18 Leaves pubescent; hairs *c.* 1 mm, simple or forked   **17. pubescens**
  18 Leaves glabrous or hairy; hairs *c.* 0·25 mm, usually stellate
    19 Leaves ovate-spathulate or oblong-ovate, shortly ciliate; leaf-surfaces glabrous or sparsely puberulent   **18. ciliata**
    19 Leaves linear, linear-lanceolate, oblong-lanceolate or spathulate, rarely ciliate
      20 Leaves glabrous, or shortly ciliate when young   **16. mathildae**
      20 Leaves puberulent or pubescent
        21 Branches shortly capitate, not forming a cushion; corolla 4–5 mm in diameter   **19. hausmannii**
        21 Branches many-headed, forming a cushion; corolla 7–12 mm in diameter
          22 Leaves ± keeled, acute; calyx not divided to middle; corolla 10–12 mm in diameter   **21. wulfeniana**
          22 Leaves flat, obtuse or subacute; calyx divided at least to middle; corolla 7–9 mm in diameter
            23 Leaves spathulate-lanceolate, with forked hairs; pedicels 2–3(–4) times as long as leaves   **22. brevis**
            23 Leaves oblong-lanceolate, with stellate hairs; pedicels 1–2 times as long as leaves   **20. alpina**

1. A. maxima L., *Sp. Pl.* 141 (1753) (*A. turczaninowii* Freyn). Annual. Leaves 6–30 × 3–11 mm, ovate, obovate or oblong-lanceolate, sessile or very shortly petiolate, dentate, glabrous or sparsely hairy. Scapes 1–several, 3–15 cm, with short, glandular and eglandular patent hairs. Inflorescence many-flowered; involucral bracts 6–12 × 2–7 mm, ovate, obovate or oblong-lanceolate, connate at the base. Pedicels equal to or shorter than bracts, elongating in fruit. Calyx 8–10 mm, more or less densely hairy, accrescent. Corolla *c.* 6 mm, white or pink, shorter than the calyx. *Dry open habitats; somewhat calcicole. Europe, northwards to c. 50° N. in W. Germany and to c. 55° N. in E. Russia.* Au Bu Cz Ga Ge Gr He Hs Hu It Ju ?Po Rm Rs (C, W, K, E) Tu.

2. A. septentrionalis L., *Sp. Pl.* 142 (1753). Annual, somewhat pubescent below with simple, forked and stellate hairs. Leaves 9–35 × 3–10 mm, oblong-lanceolate or elliptical, dentate, sessile or very shortly petiolate. Scapes 8–30 cm, 1–several. Inflorescence 5- to 30-flowered; involucral bracts 2–4 mm, oblong or lanceolate. Pedicels at least 3 times as long as bracts. Calyx 2·5–3·5 mm, divided to about the middle. Corolla 4–5(–6) mm, white or pink. Seeds 0·5–1 mm, numerous. *Dry places. N. & C. Europe, extending southwards to the S.W. Alps and S. Russia.* Au Cz Da Fe Ga Ge He It No Po Rm Rs (N, B, C, W, E) Su.

3. A. chaixii Gren. & Godron, *Fl. Fr.* 2: 458 (1853) (*A. lactiflora* sensu Coste, non Pallas). Like **2** but inflorescence lax; flowers 5–8(–10); bracts ovate or ovate-lanceolate; corolla 5–7 mm; seeds *c.* 3 mm, 3–5. 2n = 34. *Woods, rocks and pastures.* ● *Mountains of S.E. France.* Ga.

4. A. filiformis Retz., *Obs. Bot.* 2: 10 (1781). Annual, glabrous below. Leaves 25–40 × 3–7 mm; lamina ovate to elliptical, dentate, distinctly petiolate. Scapes 10–15 cm, usually several, glabrous or sparsely glandular-pubescent above. Inflorescence many-flowered, lax; involucral bracts 1·5–3 mm, lanceolate. Pedicels 15–50 mm, at least 10 times as long as bracts, glandular-pubescent. Calyx 1·5–2·5 mm, glabrous. Corolla 2–3 mm, white. Seeds *c.* 0·5 mm, numerous. *Damp places. U.S.S.R., northwards from c. 53° N.* Rs (N, B, C, W, E).

5. A. elongata L., *Sp. Pl.* ed. 2, 1668 (1763). Annual. Leaves 6–14 × 1·5–5 mm, lanceolate or oblanceolate, entire or dentate, finely ciliate. Scapes 4–10 cm, usually numerous, densely pubescent with forked hairs. Inflorescence many-flowered, lax; involucral bracts 3–8 mm, lanceolate to ovate-lanceolate. Pedicels 2–8 times as long as bracts, glabrous. Calyx 3·5–5 mm, glandular-pubescent. Corolla *c.* 3 mm, white, shorter than the calyx-lobes. Seeds *c.* 1 mm, *c.* 5. *Dry open habitats. From C. Germany eastwards to E.C. Russia and Bulgaria; Sicilia; Pyrenees.* Au Bu Cz Ga Ge Hs Hu Ju Po Rm Rs (C, W, K, E) Si [?No].

6. A. lactea L., *Sp. Pl.* 142 (1753). Perennial, laxly caespitose or rosettes solitary, shortly stoloniferous; stolons up to 3 cm. Leaves 8–20(–30) × 0·5–2 mm, linear, sessile, glabrous or very sparsely hairy usually on the margins. Scapes 1–4, (2–)3–15 cm, glabrous. Inflorescence 1- to 4(–6)-flowered, often arising from axils of basal leaves only; involucral bracts 2·5–4 mm, linear-lanceolate, glabrous or sparsely hairy. Pedicels 2–10 times as long as bracts, glabrous. Calyx 3–5 mm, glabrous. Corolla up to 12 mm in diameter, white. Seeds 2–3 mm, 4–5. 2n = 76. *Rocks and screes; calcicole.* ● *Mountains of C. Europe, extending to the French Jura and C. Jugoslavia.* Au Cz Ga Ge He It Ju Po Rm.

7. A. obtusifolia All., *Fl. Pedem.* 1: 90 (1785). Perennial, laxly caespitose or rosettes solitary. Leaves 6–20 × 1·5–4 mm, lanceolate or oblong-lanceolate, tapered to the base, scarcely petiolate, more or less sparsely hairy and ciliate. Scapes 1–several, 2–12 cm, more or less densely tomentose with stellate or forked hairs. Inflorescence 1- to 7-flowered; involucral bracts 2–3·5 mm, linear-lanceolate, puberulent. Pedicels 1–4(–7) times as long as bracts, densely hairy with stellate hairs. Calyx 3–4 mm, densely pubescent. Corolla 6–9 mm in diameter, white or pink. 2n = 36, 38. *Rocks and screes; calcifuge.* ● *Mountains of C. Europe, extending southwards to the S.W. Alps and N. Appennini; mountains of Balkan peninsula southwards to Macedonia.* Au Bu Cz Ga Ge He It Ju Po Rm.

8. A. hedraeantha Griseb., *Spicil. Fl. Rumel.* 2: 3 (1844). Like **7** but inflorescence subcapitate, with (2–)5–10 flowers; scapes often short, 1–2(–12) cm; pedicels shorter than to 1–2 times as long as bracts; calyx and bracts glabrous. *Schistose rocks.* ● *Mountains of C. part of Balkan peninsula.* Al Bu Ju.

9. A. carnea L., *Sp. Pl.* 142 (1753). Perennial, laxly caespitose. Leaves 4–20(–35) × 1–2 mm, linear, sometimes fleshy and keeled, pubescent or glabrous on the surfaces, ciliate. Scapes 1–4, 1–6 cm, more or less densely pubescent with stellate or forked hairs. Inflorescence 2- to many-flowered; involucral bracts 2–5 mm, linear-lanceolate, ciliate or glabrous. Pedicels 1–2 times as long as bracts, densely hairy with stellate hairs. Calyx 2·5–6 mm, sparsely hairy. Corolla 5–8 mm, white or pink. *Rocks and screes; calcifuge.* ● *Mountains of W. Europe from the C. Pyrenees to the Vosges and eastwards to 8° 10′ E. in the W.C. Alps.* Ga He Hs It.

A very variable species in which the following subspecies can be recognized.

1 Leaves more or less glabrous on the surfaces, usually ciliate
2   Leaves 10–25 mm, flat or recurved   (c) subsp. rosea
2   Leaves 4–6(–10) mm, caniculate   (b) subsp. laggeri
1 Leaves densely pubescent or puberulent, at least beneath
3   Leaves entire   (a) subsp. carnea
3   Leaves ± dentate   (d) subsp. brigantiaca

(a) Subsp. **carnea**: More or less caespitose. Leaves 6–12(–20) × 1–2 mm, entire, more or less flat or recurved, more or less densely pubescent. Corolla 5–6 mm in diameter, pink or white. $2n=38$. *C. & W. Alps; E. Pyrenees.*

(b) Subsp. **laggeri** (Huet) Nyman, *Consp.* 607 (1881): Rather small, usually densely caespitose, with rather numerous small rosettes. Leaves 4–6(–10) × *c*.1 mm, entire, caniculate and somewhat fleshy, glabrous on the surfaces, shortly ciliate. Corolla 6–8 mm in diameter, usually pink. $2n=38$. *C. Pyrenees.*

(c) Subsp. **rosea** (Jordan & Fourr.) Rouy, *Fl. Fr.* **10**: 211 (1908): Rather robust, often with few rather large rosettes. Leaves 10–25 × 1–2 mm, entire, flat, glabrous on the surfaces or sparsely puberulent towards the apex above, shortly ciliate, somewhat deflexed. Corolla 6–8(–9) mm in diameter, pink. $2n=38$. *Mountains of E. & S.C. France, E. Pyrenees.*

(d) Subsp. **brigantiaca** (Jordan & Fourr.) I. K. Ferguson, *Bot. Jour. Linn. Soc.* **64**: 376 (1971): Rather robust with few rather large rosettes. Leaves 10–30 × 1–2·5 mm, flat, puberulent, at least some dentate. Corolla 5–8 mm in diameter, white. $2n=78$. *S.W. Alps.*

10. A. pyrenaica Lam., *Tabl. Encycl. Méth. Bot.* **1**: 432 (1792). Densely caespitose perennial; stems regularly branched forming deep, dense, more or less hemispherical cushions. Leaves 3–7(–9) mm, linear-oblong, pubescent and ciliate, densely imbricate, persistent after withering and forming columnar shoots. Peduncles 2–22 mm, usually longer than leaves. Flowers solitary but subtended by 2–3 ovate-lanceolate, glabrous involucral bracts. Calyx 2–3·5 mm, subglabrous. Corolla 4–5 mm in diameter. Seeds 1–1·5 mm. $2n=38$. *Rocks and screes.* ● *E. & C. Pyrenees.* Ga Hs.

11. A. chamaejasme Wulfen in Jacq., *Collect. Bot.* **1**: 194 (1787) (incl. *A. bungeana* Schischkin & Bobrov). Laxly caespitose perennial, shortly stoloniferous; stolons 1–3 cm. Leaves 5–16 × 2–3 mm, oblong-lanceolate, in flattened rosettes, entire, ciliate with long silky hairs on the margins and sometimes on the midveins. Scapes 1–7 cm, densely villous and also interspersed with short glandular hairs. Inflorescence 3- to 7-flowered; involucral bracts *c*. 4 mm, lanceolate or ovate-lanceolate, densely villous. Pedicels as long as or a little longer than bracts. Calyx 2–3 mm, villous. Corolla 5–7 mm, white or pink. $2n=20$. *Alps, Carpathians; one station in C. Pyrenees; Arctic Russia, N. & C. Ural.* Au Cz Ga Ge He It Ju Po Rm Rs (N, C).

12. A. villosa L., *Sp. Pl.* 142 (1753) (incl. *A. taurica* Ovcz., *A. koso-poljanskii* Ovcz.). Like **11** but usually more densely caespitose, with numerous cushion-forming rosettes; leaves in hemispherical rosettes, more or less densely villous, at least near the apex. $2n=20$. *Rocks and screes; calcicole. Mountains of S. & C. Europe, from the Swiss Jura and E. Carpathians to N.W. Spain, C. Italy, C. Greece, Krym; also at low altitude in S.E. Russia and E. Ukraine.* Al Au Bu Ga Gr He Hs It Ju Rm Rs (C, K, E).

Plants from S.E. Europe and Italy are usually more densely villous and tend to have shorter scapes; they have been described as **A. arachnoidea** Schott, Nyman & Kotschy, *Analect. Bot.* 17 (1854) but probably only deserve varietal status.

13. A. helvetica (L.) All., *Fl. Pedem.* **1**: 91 (1785) (*A. imbricata* Lam.). Densely caespitose perennial; stems regularly branched, forming deep, dense, more or less hemispherical cushions. Leaves 2–6(–7·5) × 0·5–1·5 mm, lanceolate, oblong or spathulate, more or less densely pubescent with simple hairs, densely imbricate, persistent after withering and forming columnar shoots. Flowers solitary. Pedicels very short or absent; involucral bracts absent. Calyx 2–3·5 mm, pubescent. Corolla 4–6 mm in diameter, white with a yellow throat. Seeds 1–1·5 mm. $2n=40$. *Mountain rocks; calcicole.* ● *Alps; ?Pyrenees.* Au Ga Ge He It.

14. A. vandellii (Turra) Chiov., *Nuovo Gior. Bot. Ital.* nov. ser., **26**: 27 (1919). (*A. imbricata* auct., non Lam., *A. multiflora* (Vandelli) Moretti, non Lam.). Like **13** but leaves linear-lanceolate, densely whitish-tomentose with stellate hairs; pedicels 1–8(–12) mm. $2n=40$. *Mountain rocks; calcifuge.* ● *Alps, Pyrenees, Sierra Nevada.* Au Ga He Hs It.

15. A. cylindrica DC. in Lam. & DC., *Fl. Fr.* ed. 3, **3**: 439 (1805). Densely caespitose perennial; stems somewhat irregularly branched, forming cushions. Leaves (4–)5–8(–10) × (1–)1·5–2·5 mm, oblong-lanceolate, pubescent with simple or forked hairs, densely imbricate, persistent after withering and forming columnar shoots. Flowers solitary. Pedicels 6–20 mm, densely pubescent; involucral bracts absent. Calyx 3–5 mm, pubescent. Corolla 7–9 mm in diameter, pink with a yellow throat. Seeds 2–3 mm. $2n=40$. ● *C. Pyrenees.* Ga Hs.

Plants from the same area forming more even hemispherical cushions, with leaves 1–1·5 mm wide and shorter pedicels, and with $2n=40$, have been called **A. hirtella** Dufour, *Actes Soc. Linn. Bordeaux* **8**: 100 (1856) and need further investigation.

16. A. mathildae Levier, *Nuovo Gior. Bot. Ital.* **9**: 43 (1877). Caespitose perennial. Leaves 10–15 × 1–2·5 mm, linear, glabrous, or sometimes shortly ciliate when young, imbricate. Flowers solitary. Pedicels 1–1½ times as long as the leaves, pubescent with stellate hairs, sometimes glabrous. Calyx 3–4 mm, more or less pubescent. Corolla *c*. 5 mm in diameter, white or pink. $2n=40$. *Mountain rocks, c. 2500 m.* ● *C. Appennini.* It.

17. A. pubescens DC. in Lam. & DC., *Fl. Fr.* ed. 3, **3**: 438 (1805). Caespitose perennial forming more or less lax cushions. Leaves 4–10 × 1–2·5 mm, spathulate or oblong-lanceolate with short, simple or forked hairs *c*. 1 mm, somewhat imbricate in rosettes at the ends of the shoots, not usually persistent after withering. Flowers solitary. Pedicels *c*. 5 mm, pubescent. Calyx 2–5 mm, pubescent. Corolla 4–6 mm in diameter, white or pink with a yellow throat. Seeds 1–2 mm. $2n=40$. *Rocks and screes above* 2000 *m; somewhat calcicole.* ● *C. & S.W. Alps, Pyrenees.* Ga He Hs It.

18. A. ciliata DC. in Lam. & DC., *op. cit.*: 441 (1805). Like **17** but leaves 6–15 × 2–4 mm, ovate-spathulate or oblong-ovate, ciliate, more or less pubescent; pedicels 5–15 mm; corolla 5–8 mm in diameter, pink or violet with an orange or yellow throat. $2n=c$. 80. *Screes at high altitude.* ● *C. Pyrenees.* Ga Hs.

19. A. hausmannii Leybold, *Flora* (Regensb.) **35**: 401 (1852). Caespitose perennial; branches shortly capitate, rosettes few, not forming a cushion. Leaves 5–10 × 1–1·5 mm, linear-lanceolate, densely pubescent with stellate or forked hairs, not persistent after withering. Flowers solitary. Pedicels 1–12 mm, pubescent. Calyx 3–4 mm, pubescent. Corolla 4–5 mm in diameter, pink with a yellow throat. Seeds 1–1·5 mm. $2n=40$. *Rocks and screes above* 1900 *m; calcicole.* ● *E. Alps, westwards to* 10° 50′ *E.* Au Ge It Ju.

20. A. alpina (L.) Lam., *Fl. Fr.* **3**: 642 (1778) (incl. *A. tiroliensis* F. Wettst.). Laxly caespitose perennial; branches many-headed forming flat, lax cushions. Leaves 5–10 × 1–2 mm, in rosettes at the ends of the branches, laxly imbricate, oblong or oblanceolate, pubescent with short, stellate hairs, not persistent after withering. Flowers solitary. Pedicels 5–10 mm, more or less pubescent. Calyx 3–4 mm, pubescent, divided to middle or beyond. Corolla 7–9 mm in diameter, white or pink with a yellow throat. $2n = 40$. *Rocks and screes, usually above* 2000 *m; calcifuge.* ● *Alps.* Au Ga He It.

21. A. wulfeniana Sieber ex Koch, *Syn. Fl. Germ.* ed. 2, 670 (1844). Like **20** but cushions usually more dense; leaves 3–5 mm, lanceolate, acute, somewhat fleshy and keeled, sparsely pubescent; pedicels 4–12 mm; calyx 4–5 mm, not divided to middle; corolla 10–12 mm in diameter, deep pink with a yellow throat. $2n = 40$. *Rocks and screes above* 2000 *m; calcifuge.* ● *E. Alps.* Au It.

22. A. brevis (Hegetschw.) Cesati, *Sagg. Geogr. Bot. Fl. Lomb.* 58 (1844). Like **20** but more densely caespitose and cushions smaller; leaves 3–5(–6) mm, spathulate, more imbricate and forming denser rosettes; pedicels 4–23 mm. $2n = 40$. *Mountain rocks,* 1700–2600 *m; calcifuge.* ● *S. Alps (mountains around Lago di Como).* He It.

4. Cortusa L.[1]

Perennial herbs. Leaves basal, long-petiolate. Inflorescence scapose, umbellate, bracteate. Corolla much longer than calyx, divided to $\frac{1}{3}$; limb infundibuliform, divided to $\frac{1}{2}$ or more. Filaments with dilated base, connate to form a ring. Capsule ovoid to subcylindrical; valves 5. Seeds numerous.

1. C. matthioli L., *Sp. Pl.* 144 (1753). Plant 20–40 cm, pubescent and with ferruginous hairs throughout. Leaves 10–25 cm; petiole up to 17 cm. Lamina reniform to suborbicular, cordate at base, up to 12 cm wide, lobed to $\frac{1}{4}$ of radius; lobes coarsely and irregularly dentate. Scape 15–35 cm. Bracts 5–15 mm, laciniate. Flowers 5–20. Calyx divided to $\frac{1}{2}$-way, lobes acute to acuminate. Corolla *c.* 1 cm, $1\frac{1}{2}$–2 times as long as calyx, divided to $\frac{1}{3}$, purple-violet. Stamens included, apiculate. Style exserted. Capsule (4–)5–7(–9) mm. Seeds *c.* 40, polyhedral, with narrow wings on the angles. $2n = 24$. *Mountain woods and other damp places; calcicole. C. & E. Russia; mountains of S.E. & S.C. Europe from Bulgaria and E. Carpathians to S.W. Alps.* Au Bu Cz Ga Ge He It Ju Po Rm Rs (N, C, W, E).

Small plants from Ural, subglabrous and lacking ferruginous hairs, with dentate bracts, deeply lobed leaves and few-flowered umbels have been called **C. altaica** Losinsk., *Acta Inst. Bot. Acad. Sci. URSS* (Ser. 1) **3**: 243 (1936), but they merge into and are not clearly separated from **1**.

5. Soldanella L.[2]

Perennial herbs. Leaves in a basal rosette, entire, coriaceous, evergreen, long-petiolate. Inflorescence scapose. Flowers nodding. Corolla much longer than calyx, infundibuliform to campanulate, with 5 wide, more or less incised lobes alternating with 5 linear segments (laciniae); throat naked, or with scales which alternate with the stamens. Capsule with a deciduous operculum and 5–10 teeth. Seeds small.

Literature: F. Vierhapper, *Pflanzenareale* **1**(1): Karte 7–8 (1926). G. Cristofolini & S. Pignatti, *Webbia* **16**: 443–475 (1962). S. Pawłowska, *Fragm. Fl. Geobot.* **9**: 3–30 (1963).

---

[1] By L. F. Ferguson.   [2] By S. Pawłowska.

1 Corolla divided to not more than $\frac{1}{3}$ of its length; throat usually naked
  2 Petioles, scape and pedicels with sessile glands   **1. pusilla**
  2 Petioles (when young), scape and pedicels with glandular hairs
    3 Glandular hairs 0·15–0·2 mm, the stalk longer than the head   **2. minima**
    3 Glandular hairs 0·05–0·1 mm, the stalk equalling or shorter than the head   **3. austriaca**
1 Corolla divided usually to $\frac{1}{2}$ or more of its length; throat with scales which are usually 2-dentate
  4 Petioles with sessile glands, most of which do not persist to maturity
    5 Leaves bluish-green beneath, pruinose   **6. pindicola**
    5 Leaves green or violet beneath, not pruinose
      6 Scape and pedicels with sessile glands   **4. alpina**
      6 Scape and pedicels with shortly stalked glands   **5. carpatica**
  4 Petioles with glandular hairs, most of which persist to maturity
    7 Glandular hairs of the petiole 0·1–0·4 mm, the stalk up to 5 times as long as the head; cells of the stalk short and wide
      8 Leaves green or violet beneath, not pruinose   **7. hungarica**
      8 Leaves bluish-grey beneath, pruinose   **8. dimoniei**
    7 Glandular hairs of the petiole 0·4–1·6 mm, the stalk more than 5 times as long as the head; cells of the stalk elongated
      9 Hairs of the petiole 0·4–0·8 mm; sepals 1-veined   **9. montana**
      9 Hairs of the petiole 0·8–1·6 mm; sepals 3-veined   **10. villosa**

Sect. TUBIFLORES Borbás. Corolla campanulate or narrowly infundibuliform, usually naked at the throat. Staminal appendages small or absent. Style shorter than corolla.

1. S. pusilla Baumg., *Enum. Stirp. Transs.* **1**: 138 (1816). Petioles, scape and pedicels with sessile glands, most of which do not persist to maturity. Lamina (4–)10(–20) mm wide, somewhat coriaceous, orbicular to orbicular-reniform, with a wide basal sinus, with prominent veins above and small glandular pits beneath. Scape 2–10 cm, 1- (to 2-) flowered. Corolla 8–15 mm, narrowly campanulate, divided to $\frac{1}{4}$–$\frac{1}{3}$ of its length, reddish-violet. Anthers hastate at the base. $2n = 40$. *Wet, alpine soils,* 1800–3100 *m; calcifuge.* ● *C. & E. Alps; S. Carpathians; isolated stations in N. Appennini and mountains of S.W. Bulgaria.* Au Bu Ge He It Ju Rm.

2. S. minima Hoppe in Sturm, *Deutschl. Fl.* Abt. 1, Band 5, Heft 20 (1806). Petioles and scape densely covered with glandular hairs 0·15–0·2 mm, most of which do not persist to maturity, their stalks 2- to 4- or more-celled, longer than the head. Lamina usually less than 10 mm wide, often longer than wide, suborbicular, usually without a basal sinus, veins not prominent above, with rather large glandular pits beneath; stomata only on the lower surface of the leaf. Scape 2–10 cm, usually 1-flowered. Corolla 8–15 mm, campanulate, divided to $\frac{1}{4}$ or $\frac{1}{5}$ of its length, pale violet to almost white; lobes not deeply cut. Anthers rounded at the base. *Wet, calcareous soils.* ● *E. Alps; C. Appennini.* Au Ge It Ju.

(a) Subsp. **minima**: Corolla divided to *c.* $\frac{1}{4}$ of its length, broadly campanulate. $2n = 40$. *E. Alps.*

(b) Subsp. **samnitica** Cristofolini & Pignatti, *Webbia* **16**: 464 (1962): Corolla divided to *c.* $\frac{1}{5}$ of its length, narrowly campanulate. *C. Appennini (Monte Majella).*

3. S. austriaca Vierh. in Urban & Graebner, *Festschr. Ascherson* 503 (1904). Like **2** but petioles, scape and pedicels with glandular hairs 0·05–0·1 mm, their stalk 1- to 2-celled, the head larger and wider than the stalk; lamina often with a slight basal sinus; stomata on both surfaces of the leaf; corolla less deeply divided. *Wet, calcareous soils.* ● *Austrian Alps, from* 13° 30′ *E. eastwards.* Au.

Sect. SOLDANELLA (Sect. *Crateriflores* Borbás). Corolla broadly infundibuliform, with 2-dentate scales at the throat. Staminal appendages long, at least ⅓ as long as the anther. Style longer than corolla.

4. S. alpina L., *Sp. Pl.* 144 (1753). Petioles when young with sessile glands, becoming almost glabrous at maturity. Lamina up to 40 mm wide, orbicular-reniform, dark green on both sides, not concentrically rugose when dry; basal sinus wide and shallow, or lacking. Scape 5–15 cm, up to 30 cm in fruit, (1–)2- to 4-flowered; pedicels with sessile glands. Corolla 8–13(–15) mm, divided to ½ or more of its length, violet or blue-violet; lobes divided into 4–5(–6) linear lobules which are almost as long as the laciniae, so that the whole corolla appears to be uniformly incised. Filaments glabrous; staminal appendages ½ or less the length of the anther. Capsule-teeth rounded at apex. 2*n*=40. *Wet pastures and rocks, 450–3000 m.* ● *Mountains of Europe, from the C. Pyrenees to Crna Gora, northwards to S.W. Germany (Schwarzwald) and southwards to Calabria.* Al Au Ga Ge He Hs It Ju.

5. S. carpatica Vierh. in Urban & Graebner, *Festschr. Ascherson* 504 (1904). Petioles when young with rather sparse, sessile glands, most of which do not persist to maturity. Lamina 8–50 mm wide, suborbicular, dark green, usually violet beneath, concentrically rugose above when dry, veins not visible; basal sinus narrow, shallow. Scape (3–)5–15(–20) cm, (1–)2- to 5-flowered; pedicels with glandular hairs, the stalk equalling or up to 3 times as long as the head. Corolla 8–15 mm, divided to ⅔–¾ of its length, violet; lobes 3- to 4(5)-partite or 3-dentate, more or less distinctly longer than the laciniae. Filaments glabrous, shorter than or almost equalling the anthers; staminal appendages ⅔–¾ the length of the anther. Capsule-teeth rounded at apex. Seeds acutely angled, often with a narrow wing along one side. 2*n*=40. ● *W. Carpathians.* Cz Po.

6. S. pindicola Hausskn., *Mitt. Thür. Bot. Ver.* **5**: 61 (1886). Like **5** but glands on the scape and pedicels sessile or almost sessile; lamina with a wide basal sinus and blue-grey, pruinose beneath. *By mountain springs.* ● *N.W. Greece (Zygos, near Metsovon).* Gr.

Details of the floral characters are not available and further study is needed. This species was thought by Markgraf, *Notizbl. Bot. Gart. Berlin* **11**: 219–223 (1931), to be related to **8**.

7. S. hungarica Simonkai, *Enum. Fl. Transs.* 461 (1887). Petioles more or less densely covered with rather stout, persistent, glandular hairs 0·2–0·4(0·5) mm; stalks of the hairs 2–5 times as long as the head; cells of the stalks 1–3 times as long as wide. Lamina orbicular-reniform with a narrow basal sinus, often violet beneath. Glandular hairs on the scape and pedicels 0·15–0·25 mm, the stalk 2–3 times as long as the head. Corolla violet. Filaments glandular. Capsule-teeth truncate. ● *Mountains of E.C. Europe and of the Balkan peninsula; two outlying stations in Calabria.* Al Au Bu Cz It Ju Po Rm Rs (W).

(a) Subsp. **hungarica**: Scape 3–10 cm, (1–)2- to 4-flowered. Lamina 10–20 mm wide, thick, entire, irregularly rugose. Corolla 8–10(–12) mm, divided to ⅓–½ of its length; lobes digitately divided into 3–4 short lobules. Staminal appendages ⅓–⅔ the length of the anthers. Capsule less than 10 mm. *Alpine meadows. E. & S. Carpathians; Balkan peninsula.*

(b) Subsp. **major** (Neilr.) S. Pawł., *Fragm. Fl. Geobot.* **9**: 11 (1963): Scape up to 20–25 cm, 2- to 8(–10)-flowered. Lamina

20–50(–60) mm wide, rather thin, occasionally remotely crenulate, not rugose. Corolla (8–)10–15 mm, divided to ½–⅔ of its length; lobes divided to up to ½ of their length into 3 or 4 lobules. Staminal appendages about ½ the length of the anthers. Capsule 10–15 mm. 2*n*=40. *Woodland and scrub. Throughout the range of the species except Macedonia.*

8. S. dimonici Vierh., *Österr. Bot. Zeitschr.* **59**: 148 (1909). Like **7** but lamina blue-grey and pruinose beneath, densely covered with glandular pits; scape 3–9 cm, 2- to 3-flowered; corolla divided to more than ½ its length, violet; lobes of corolla crenately divided into 3 lobules, the apical up to 0·5 mm. ● *Mountains of Macedonia, E. Albania and Bulgaria.* Al Bu Ju.

9. S. montana Willd., *Enum. Pl. Hort. Berol.* 192 (1809). Petioles more or less densely covered with slender, persistent glandular hairs 0·4–0·8(–1) mm; stalks of the hairs 8–10 times longer than the head; cells of the stalk 3–6 times as long as wide. Lamina 2–6(–7) cm wide, rather thin, orbicular or orbicular-reniform, sometimes shallowly dentate or crenate, not rugose when dry, bright green, often violet beneath; basal sinus deep, narrow or closed. Scape 5–25(–30) cm, 6-(8-) flowered. Sepals 1-veined. Corolla 10–15(–18) mm, divided to ⅜–¾ of its length; lobes usually divided to more than ½ their length into 3–4 narrow lobules. Filaments more or less densely glandular-hairy; staminal appendages up to ⅘ of the length of the anthers. Capsule-teeth truncate. 2*n*=40. *Woods and meadows, 700–1600 m; calcifuge.* ● *C. Europe, westwards to c. 9° 45′ E. in the Italian Alps; mountains of Bulgaria.* Au Bu Cz Ge It Po Rm Rs (W) [Fe].

10. S. villosa Darracq, *Actes Soc. Linn. Bordeaux* **16**, *Mélanges*: 2 (1850). Rhizome long and stolon-like. Petioles densely covered with slender, persistent glandular hairs 0·8–1·6 mm; stalks of the hairs 10–15 times longer than the head; cells of the stalk 3–6 times as long as wide. Lamina up to 7 cm wide, rather thin, orbicular or orbicular-reniform, more or less entire, pale green, not violet beneath; basal sinus narrow. Scape 10–30 cm, usually 3- to 4-flowered, densely hairy; hairs 0·4–1 mm. Sepals ovate-lanceolate, up to 2·5 mm wide, 3-veined. Corolla 10–16 (–18) mm, violet, divided to ⅔–⅘ of its length; lobes divided to more than ½ their length into 3–5 lobules. Filaments more or less densely glandular-hairy; staminal appendages up to ⅔ the length of the anthers. *Damp, shady places in the foothills on siliceous, base-rich soils.* ● *W. Pyrenees.* Ga Hs.

6. Hottonia L.[1]

Aquatic perennial herbs. Leaves pinnate. Flowers aerial, heterostylous. Corolla-tube equalling calyx; limb rotate. Stamens 5, included. Capsule globose; valves 5, remaining coherent at apex. Seeds numerous.

1. H. palustris L., *Sp. Pl.* 145 (1753). Subglabrous, with creeping stolons and erect leafy stems 30–90 cm. Leaves 2–13 cm, alternate or verticillate, submerged and floating, 1- to 2-pinnate; lobes linear. Inflorescence (15–)30–40(–60) cm, bracteate, with 3–9 whorls of flowers; pedicels 12–30 mm, glandular-pubescent, deflexed in fruit. Calyx 4–6 mm, equalling corolla-tube, lobed nearly to base; lobes linear. Corolla 20–25 mm in diameter, violet; throat yellow. Capsule 3–6 mm; seeds brown, polyhedral, coarsely reticulate. 2*n*=20. *Shallow fresh water. From C. Sweden southwards to C. Italy and Romania.* Au Be Br †Bu Cz Da Ga Ge He Ho Hu It Ju Po Rm Rs (B, C, W, E) Su [Hb].

---

[1] By L. F. Ferguson.

7. Cyclamen L.[1]

Perennial tuberous herbs. Leaves simple, cordate, long-petiolate, glabrous, rather fleshy, usually blotched or marbled above and purple beneath. Flowers axillary, solitary, nodding; pedicel long, usually coiling spirally in fruit. Calyx campanulate, deeply 5-lobed. Corolla with a short, hemispherical basal tube and 5 lobes, the lobes at first contorted, later strongly deflexed. Anthers large, connivent and forming a cone. Capsule globose or broadly ovoid, splitting either irregularly or into 5 teeth. Seeds large, sticky.

All the species occur in woods and thickets or on shaded, rocky hillsides, usually on limestone.

Literature: R. Knuth in Engler, *Pflanzenreich* **22 (IV. 237)**: 246–256 (1905). O. Schwarz, *Feddes Repert.* **58**: 234–283 (1955); **69**: 73–103 (1964). D. E. Saunders, *Bull. Alp. Gard. Soc.* **27**: 18–76 (1959).

1 Corolla-lobes auriculate at base; flowers appearing in late summer or autumn, usually before the leaves
  2 Tuber usually rooting from upper surface, without thick retractile roots; fruiting pedicel coiling from apex; leaves often distinctly angled or lobed   **1. hederifolium**
  2 Tuber rooting only from base, with thick retractile roots; fruiting pedicel coiling from middle or base; leaves seldom angled or lobed   **2. graecum**
1 Corolla-lobes not auriculate at base; flowers appearing in winter, spring or summer when leaves are fully developed
  3 Flowers appearing in summer and early autumn, very fragrant; leaves persistent, broadly cordate or reniform, subentire or obscurely denticulate; tuber irregular, often with corky protuberances   **3. purpurascens**
  3 Flowers appearing in winter or spring; leaves not persistent; tubers subglobose or depressed-globose
    4 Corolla white or veined pale pink, without a darker zone at the base of the lobes; leaves usually speckled with white above
      5 Corolla veined pale pink; corolla-lobes 10–18 mm; leaf-margins obtusely toothed or undulate, sometimes almost entire   **6. balearicum**
      5 Corolla white; corolla-lobes 16–25 mm; leaf-margins rather acutely dentate or sometimes lobed, rarely undulate or subentire   **5. creticum**
    4 Corolla pink or purple, or occasionally white with a darker zone or blotch at base of lobes; leaves sometimes marbled above but rarely speckled with white
      6 Corolla-lobes 7–15 mm, with a prominent, dark violet, pale-centred blotch at base; leaves reniform or suborbicular with entire or obscurely undulate-denticulate margins   **7. coum**
      6 Corolla-lobes 15–45 mm, with a darker zone at base but without distinct blotches; leaves cordate, sometimes angular
        7 Leaves coarsely dentate or shallowly lobed, often angled; corolla-lobes 15–30 mm; fruiting pedicel coiled   **4. repandum**
        7 Leaves closely and obtusely denticulate, rarely angled or lobed; corolla-lobes 25–45 mm; fruiting pedicel arcuate but not coiled   **8. persicum**

1. C. hederifolium Aiton, *Hort. Kew.* **1**: 196 (1789) (*C. neapolitanum* Ten.). Tuber 3–15 cm or more in diameter, subglobose or depressed-globose, corky, rooting mainly from the upper surface. Leaves 3–14 × 2–10 cm, appearing in autumn after anthesis, cordate, acute or obtuse; margins obscurely and obtusely denticulate, often distinctly angled or lobed. Corolla

pale pink or white with dark purple bifurcate blotches at the throat; lobes *c.* 20 mm, auriculate at base. Style scarcely exserted. Fruiting pedicel coiling from the apex. Flowering from August to November. $2n = 24$. *S. Europe, from France to Bulgaria.* Al Bu Co Ga Gr He It Ju Sa Si ?Tu [Br].

Records for Kriti appear to be erroneous.

2. C. graecum Link, *Linnaea* **9**: 573 (1835) (incl. *C. mindleri* Heldr.). Tuber 3–10 cm in diameter, globose or depressed-globose, corky, rooting only from the base, with retractile roots. Leaves 3–7(–10) × 2·5–6(–8) cm, appearing in autumn or winter after anthesis, cordate, acute or obtuse; margins minutely and obtusely denticulate, rarely angled. Corolla white or pale pink with dark purple bifurcate or trifurcate blotches at the throat; lobes *c.* 20 mm, auriculate at base. Style included or slightly exserted. Fruiting pedicel coiling from the middle or base. Flowering from September to December. $2n = 78–80$. *Greece and Aegean region.* Cr Gr.

3. C. purpurascens Miller, *Gard. Dict.* ed. 8, no. 2 (1768) (*C. europaeum* auct.). Tuber 2–3 cm in diameter, irregularly globose or depressed-globose, often with corky protuberances, rooting all over. Leaves 2·5–8 × 2–6 cm, persistent, broadly cordate or reniform, rounded or subacute; margins subentire or obscurely and obtusely denticulate, not angled. Flowers very fragrant. Corolla without auricles, reddish-pink or purplish with a darker blotch at the throat; lobes 15–20 mm. Style exserted for *c.* 1 mm. Fruiting pedicel coiling from the apex. Flowering from June to October. $2n = 34$. *From S.E. France to the W. Carpathians and C. Jugoslavia.* Au Cz Ga Ge He Hu It Ju Po [Rm Rs(W)].

4. C. repandum Sibth. & Sm., *Fl. Graec. Prodr.* **1**: 128 (1806) (*C. vernale* sensu O. Schwarz, non Miller). Tuber 1·5–3·5(–6) cm in diameter, subglobose or depressed-globose, pubescent, rooting from the base. Leaves 4–13 × 3–13 cm, with or without marbling, broadly cordate, acute or subacute; margins deeply dentate or shallowly lobed. Corolla without auricles, uniformly purplish-pink, occasionally white or pale pink with a darker zone around the throat; lobes 15–30 mm. Style exserted for 1·5–3 mm. Fruiting pedicel coiling from the apex. Flowering from March to May. *C. & E. Mediterranean region.* Co Ga Gr It Ju Sa Si.

5. C. creticum Hildebr., *Beih. Bot. Centr.* **19**(2): 367 (1906). Tuber 1·5–3 cm, depressed-globose, pubescent, rooting from the base. Leaves 3·5–9(–12) × 3–7(–11) cm, white-marbled above, broadly ovate, usually acute; margins rather sharply dentate, rarely undulate or subentire. Corolla without auricles, uniformly white; lobes 16–25 mm. Style included in the corolla-tube or very shortly exserted. Fruiting pedicel coiling from the apex. Flowering from March to May. $2n = 24$. ● *Kriti.* Cr.

6. C. balearicum Willk., *Österr. Bot. Zeitschr.* **25**: 111 (1875). Tuber 1·5–2·5 cm in diameter, depressed-globose, pubescent, rooting from base. Leaves 2–9 × (1·3–)2–5·5(–9) cm, white-marbled above, broadly ovate, obtuse or subacute; margins obtusely toothed or undulate, sometimes almost entire. Corolla without auricles, white, veined pale pink; lobes 10–18 mm. Style included in corolla-tube. Fruiting pedicel coiling from the apex. Flowering from February to April. $2n = 18, 20$ ● *S. France; Islas Baleares.* Bl Ga.

7. C. coum Miller, *Gard. Dict.* ed. 8, no. 6 (1768) (incl. *C. durostoricum* Panţu & Solac.). Tuber 2–3·5 cm in diameter, subglobose or depressed-globose, pubescent, rooting from base. Leaves (1·2–)3–7 × (1·3–)3–7 cm, with or without marbling,

---

[1] By R. D. Meikle and N. H. Sinnott.

reniform or suborbicular, rarely broadly ovate, rounded or obtuse, rarely subacute; margins entire or obscurely undulate-denticulate. Corolla without auricles, magenta with a prominent, dark violet, pale-centred blotch at the base of each corolla-lobe; lobes 7–15 mm. Style very shortly exserted. Fruiting pedicel coiling from the apex. Flowering from January to April. *S.E. Europe.* Bu Rs (K) Tu.

8. C. persicum Miller, *op. cit.* no. 3 (1768). Tuber 4–15 cm or more in diameter, subglobose or depressed-globose, corky, rooting from base. Leaves 3–14 × 3–14 cm, cordate, acute or obtuse; margins somewhat thickened, closely and obtusely denticulate, but rarely angled or lobed. Corolla without auricles, white or pink with a darker purple zone around the throat; lobes 25–45 mm. Styles exserted for 1–2 mm. Fruiting pedicel arcuate but not coiling. Flowering from January to May. *Aegean region (Athos, Karpathos).* Gr. (*S.W. Asia.*)

Records for Kriti appear to be erroneous.

8. Lysimachia L.[1]

Erect or procumbent herbs, rarely dwarf shrubs. Leaves opposite, whorled or rarely alternate, entire or crenate. Flowers 5(7)-merous, axillary, solitary or clustered, or in terminal bracteate panicles or racemes. Corolla rotate; lobes entire or dentate, often glandular within. Capsule more or less globose, with 5(–7) valves. Seeds rugose, numerous.

Literature: F. Pax & R. Knuth in Engler, *Pflanzenreich* 22(IV. 237): 256–313 (1905). H. Handel-Mazzetti, *Notes Roy. Bot. Gard. Edinb.* **16**: 51–122 (1928). J. D. Ray, *Illinois Biol. Monogr.* **24**(3–4): 1–160 (1956).

1 Stems procumbent
  2 Leaves gland-dotted; calyx-lobes ovate, overlapping; corolla 8–18 mm   **6. nummularia**
  2 Leaves not gland-dotted; calyx-lobes subulate, not overlapping; corolla 6–8·5 mm
    3 Leaves 5–30 mm wide, petiolate; pedicels ±equalling the subtending leaf   **1. nemorum**
    3 Leaves 3–8 mm wide, subsessile; pedicels at least twice as long as the subtending leaf   **2. serpyllifolia**
1 Stems erect
  4 Flowers 7-merous, in dense axillary racemes   **13. thyrsiflora**
  4 Flowers 5-merous, not in dense axillary racemes
    5 At least the lower flowers subtended by ordinary leaves; corolla completely or partly yellow
      6 Corolla 4–5·5 mm, pink at base, yellowish at apices of lobes   **8. minoricensis**
      6 Corolla more than 5·5 mm, yellow, sometimes streaked or dotted with red or black
        7 Flowers terminal
          8 Flowers in a panicle; corolla without streaks or dots; axillary bulbils never present   **3. vulgaris**
          8 Flowers in a raceme; corolla streaked and dotted with red and black; axillary bulbils usually present   **4. terrestris**
        7 Flowers axillary
          9 Flowers solitary, rarely paired; corolla with red basal blotches; leaf-surfaces glabrous   **5. ciliata**
          9 Flowers usually in clusters of 2 or more; corolla without blotches; leaf-surfaces puberulent   **7. punctata**
    5 Flowers subtended by minute bracts; corolla never yellow
      10 Corolla white
        11 Leaves linear-lanceolate, opposite   **9. ephemerum**
        11 Leaves broadly lanceolate to ovate, alternate   **10. clethroides**

---

      10 Corolla pink or purple
        12 Corolla purple; flowers sessile; style indurate in fruit
              11. atropurpurea
        12 Corolla pink; flowers pedicellate; style filiform in fruit
              12. dubia

1. L. nemorum L., *Sp. Pl.* 148 (1753). Evergreen, glabrous. Stems 10–45 cm, procumbent. Leaves (10–)15–30(–50) × (5–)10–20(–30) mm, opposite, ovate to ovate-lanceolate, acute, shortly petiolate. Flowers solitary in the axils of the middle leaves; pedicels filiform, slightly shorter than to 1½ (rarely 3, in young plants) times as long as the subtending leaf. Calyx 3·5–6 mm; lobes linear-lanceolate or subulate. Corolla 6–8·5 mm, yellow. Capsule *c.* 3 mm, globose; style deciduous. 2*n*=16, 18, 28. *Damp or shady places. W. & C. Europe, extending locally to S. Sweden, C. Jugoslavia and Sicilia.* Au Az Be Br Cz Da Fa Ga Ge Hb He Ho Hs It Ju Lu No Po Rm Rs (W) Si Su.

Specimens from the Açores have been recognized as subsp. **azorica** (Hooker) Palhinha, *Bol. Soc. Brot.* ser. 2, **30**: 77 (1956); the characters used are those of first-year flowering material in cultivation, and mature specimens, apart from having revolute leaf-margins and narrower corolla-lobes, are not clearly separable from **1**. The taxon does not warrant more than varietal status.

2. L. serpyllifolia Schreber, *Nova Acta Acad. Leop.-Carol.* **4**: 144 (1770). Like **1** but dwarf shrub with a short stock; leaves 6–11 × 3–6(–8) mm, subsessile, ovate, obtuse; pedicels at least twice as long as subtending leaves. *Stony ground.* ● *Mountains of C. & S. Greece and Kriti.* Cr Gr.

3. L. vulgaris L. *Sp. Pl.* 146 (1753). Pubescent, stoloniferous perennial. Stems 50–160 cm, erect. Leaves 5·5–9(–12) × 2–3·5(–5) cm, opposite or in whorls of 3–4, ovate to lanceolate, acuminate, dotted with black or orange glands; petiole 2–10 mm. Inflorescence a terminal panicle, leafy below; upper bracts linear, subulate. Calyx ½–¾ as long as corolla; lobes triangular-lanceolate; margins red, minutely ciliate. Corolla 8–11(–13) mm, yellow. Capsule 3·5–5·5(–6) mm, subglobose; style deciduous. 2*n*=56, 84. *Fens, wet woods, lake-shores and river-banks. Almost throughout Europe.* All except Az Bl Cr Fa Is Sa Sb.

4. L. terrestris (L.) Britton, E. E. Sterns & Poggenb., *Prelim. Cat.* 34 (1888) (*L. stricta* Aiton). Glabrous perennial. Stems 30–80 cm, erect. Leaves up to 9 × 1·5 cm, lanceolate, opposite, acute, dotted with black glands, subsessile; usually with moniliform bulbils in the axils. Inflorescence a terminal raceme, rarely produced; pedicels up to 15 mm. Calyx 2–4 mm. Corolla 5·5–7 mm; lobes lanceolate, yellow, streaked and dotted with red or black. Capsule globose; style persistent. *Naturalized on lake-margins. N.W. England (Windermere).* [Br.] (*North America.*)

5. L. ciliata L., *Sp. Pl.* 147 (1753). Rhizomatous. Stems 50–130 cm, erect, glabrous. Leaves 7–14 × 2·5–6·5 cm, opposite or in whorls of 4, ovate to lanceolate, glabrous except for ciliate margins; petiole *c.* 1 cm. Flowers solitary or in pairs, in the axils of the upper leaves; pedicels up to 9 times as long as the calyx. Calyx 3·5–8 mm; lobes lanceolate. Corolla 9–13 mm; lobes obovate, yellow with red basal blotches. Capsule 4–5·5 mm, ovoid or globose; style *c.* 4 mm, persistent, filiform. *Cultivated for ornament and locally naturalized in N.W. Europe.* [Be Br.] (*North America.*)

6. L. nummularia L., *Sp. Pl.* 148 (1753). Evergreen, glabrous. Stems 10–50(–60) cm, creeping. Leaves 12–20(–30) × 11–17(–28) mm, opposite, shortly petiolate, broadly ovate to suborbicular, obtuse, rounded to almost cordate at base, dotted

---

[1] By L. F. Ferguson.

with glands. Flowers solitary (rarely in pairs) in the axils of the middle leaves; pedicels stout, somewhat shorter than to twice as long as the subtending leaf. Calyx 7–11 mm; lobes ovate, overlapping, acuminate. Corolla 8–18 mm, yellow; lobes ob-ovate, dotted with black glands and minutely glandular-puberulent. Capsule *c*. 3 mm, globose, rarely produced in N. and C. Europe. 2*n* = 32, 36, 43, 45. *Ditches, lake-shores and wet grass-land. Most of Europe, northwards to c. 62° N. in Fennoscandia and Russia; cultivated for ornament and perhaps only naturalized in the northernmost part of its range.* Al Au Be Br Bu Cz Da Ga Ge Gr Hb He Ho Hs Hu It Ju No Po Rm Rs (N, B, C, W, K, E) Su Tu [?Co Fe].

7. **L. punctata** L., *Sp. Pl.* 147 (1753) (incl. *L. verticillaris* Sprengel). Puberulent, with a few thin rhizomes. Stems 35–90(–130) cm, erect. Leaves 4·5–7·5(–11) × 2–3·5(–4) cm, opposite or in whorls of 3–4, petiolate, lanceolate to elliptical, acute, rounded at the base, ciliate, puberulent, dotted with glands beneath. Flowers in axillary clusters of (1–)2(–4); pedicels *c*. 3 times as long as calyx. Calyx 5–8 mm; lobes narrowly lanceolate, ciliate. Corolla 12–16(–18) mm, yellow; lobes ovate to lanceolate, glandular-puberulent. Capsule *c*. 4 mm, subglobose; style *c*. 3 mm, persistent, filiform. *Marshes, river-banks and other wet places. S.E. & E.C. Europe, extending to W. Austria and N. Italy; cultivated for ornament and widely naturalized elsewhere.* Al Au Bu Cz Gr Hu It Ju *Po Rm Rs (W, K, E) Tu [Be Br Da Fe Ga Ge He Ho No Rs (B) Su].

8. **L. minoricensis** Rodr., *Bull. Soc. Bot. Fr.* 25: 240 (1878). Glabrous, rhizomatous. Stems 30–60 cm, erect, rarely branched from the base. Leaves 4–8·5 × 1·5–2·5 cm, alternate to suboppo-site, sessile, lanceolate to ovate, entire. Flowers solitary in the axils of the upper leaves, subsessile or the lower with pedicels up to 6 mm. Calyx 3·5–4·5 mm; lobes linear-lanceolate. Corolla 4–5·5 mm, subcampanulate, pinkish at the base and greenish-yellow at the apices of the lobes. Capsule 4·5–5·5 mm, globose; style less than 2 mm, persistent. *Wooded gorge on limestone.* ● *Menorca (Sa Vall).* †Bl.

9. **L. ephemerum** L., *Sp. Pl.* 146 (1753). Glabrous, rhizoma-tous. Stems 40–110 cm, erect, simple. Leaves 9·5–14·5(–17) × 1–2·5(–3) cm, opposite, linear-lanceolate to linear-spathulate, amplexicaul, entire, sometimes glaucous. Flowers in a terminal raceme, rarely with smaller axillary racemes at the base; pedicels as long as the flowers. Calyx 3–5 mm; lobes ovate, obtuse, with white margins. Corolla 8–11 mm, white; lobes oblanceolate to spathulate, obtuse. Stamens more or less equalling the corolla. Capsule 4–5 mm, globose; style 3 mm, eventually 7 mm, per-sistent, filiform. *Damp grassland, especially around springs.* ● *Iberian peninsula and S.W. France.* Ga Hs Lu.

10. **L. clethroides** Duby in DC., *Prodr.* 8: 61 (1844). Peren-nial. Stems up to 100 cm, erect, simple. Leaves 7–13 × 2–4·5 cm, alternate, ovate-lanceolate, subsessile, sparsely pubescent, dotted with black glands. Flowers in a terminal raceme; pedicels 1½–3 times as long as the flowers. Calyx 2–3 mm; lobes ovate, obtuse, with white margins. Corolla 5–7·5 mm, white; lobes lanceolate to spathulate, obtuse. Stamens half as long as the corolla. Capsule *c*. 2 mm, subglobose; style stout, persistent. *Naturalized in shady places in the Netherlands.* [Ho.] (*E. Asia.*)

11. **L. atropurpurea** L., *Sp. Pl.* 147 (1753). Puberulent, rhizomatous. Stems 20–65 cm, erect, with a few branches at the base. Leaves 3·5–8(–10) × 0·5–1(–1·5) cm, alternate, the upper

sessile, the lower with short petioles, linear-lanceolate to spathu-late, irregularly crenate and undulate, glaucous. Flowers in a terminal spike. Calyx *c*. 4 mm. Corolla 6–7 mm, dark purple; lobes oblong, spathulate. Stamens longer than the corolla. Capsule 3–5 mm; style 4–5 mm, persistent, strongly indurated. *Waste ground and damp, sandy places. Balkan peninsula.* Al Bu Gr Ju Tu.

12. **L. dubia** Aiton, *Hort. Kew.* 1: 199 (1789). Annual or biennial, glabrous. Stems 30–65 cm, erect. Leaves 4–8·5 × 0·5–1·5 cm, alternate or more or less opposite, subsessile, linear-lanceolate, entire, margins subundulate. Flowers pedicellate in a loosely branched, terminal, spike-like raceme. Calyx 3·5–5 mm; lobes oblong with white, ciliate margins. Corolla 4·5–6 mm, pink; lobes cuneate to spathulate. Stamens equalling or shorter than corolla. Capsule 3–4 mm; style persistent, filiform. *Ditches and other damp places. S. & E. parts of the Balkan peninsula.* Al Bu Gr Ju ?Rs (K) Tu.

13. **L. thyrsiflora** L., *Sp. Pl.* 147 (1753) (*Naumburgia thyrsi-flora* (L.) Reichenb.). Rhizomatous, usually glabrous, but pubescent under dry conditions. Stems 30–70 cm, erect. Leaves 7–9·5(–12) × 0·5–2(–3) cm, sessile, lanceolate, with numerous black glands. Flowers 7-merous, in dense axillary racemes; bracts *c*. 6 mm, linear-lanceolate; pedicels shorter than bracts. Calyx 3–4 mm; lobes linear-oblong. Corolla 4–6 mm; lobes linear-lanceolate, yellow; stamens longer than the corolla. Capsule 1·5–2·5 mm; style *c*. 8 mm, persistent, filiform. 2*n* = 54. *Bogs and marshes. Most of Europe northwards from C. France, N. Jugoslavia, C. Romania and S.E. Russia; one isolated station in N.C. Bulgaria.* Au Be Br Bu Cz Da Fe Ga Ge He Ho Hu Ju No Po Rm Rs (N, B, C, W, E) Su.

9. Trientalis L.[1]

Perennial herbs with slender rhizomes. Leaves in a whorl at the top of the stem; occasionally a few alternate leaves below. Flowers (5–)7(–9)-merous, axillary, solitary on capillary pedicels, ebracteate. Corolla rotate. Capsule globose; valves 5(–7), deciduous. Seeds few, reticulate.

1. **T. europaea** L., *Sp. Pl.* 344 (1753). Glabrous. Stem 5–30 cm, usually unbranched. Leaves 10–90 × 5–15 mm, obovate to lanceolate, subsessile, entire to minutely dentate at apex. Pedicel 1–7 cm. Calyx 4–7 mm, teeth linear. Corolla 11–19 mm in diameter, white or tinged with pink; segments ovate, acuminate or acute. Stamens included; filaments adnate to top of tube. Seeds 7–11, black, dull, minutely reticulate. 2*n* = c. 160. *Damp grassy places and coniferous woods. N. Europe, extending south-wards, mainly in the mountains, to the S. Alps and N. Romania.* Au Be Br Cz Da Fe Ga Ge He Ho Is It No Po Rm Rs (N, B, C, W, ?E) Su.

10. Asterolinon Hoffmanns. & Link[1]

Annual. Leaves opposite. Flowers 5-merous, solitary in the axils of the upper leaves. Corolla shorter than calyx. Capsule with 5 valves.

1. **A. linum-stellatum** (L.) Duby in DC., *Prodr.* 8: 68 (1844) (*A. stellatum* Hoffmanns. & Link). Glabrous. Stems 3–12(–18) cm, erect, freely branched. Leaves 3–7 × 0·5–2 mm, lanceolate, sessile, acute, entire, equalling or longer than the internodes. Pedicels 2–4 mm. Calyx 3–6 mm; lobes lanceolate to linear-lanceolate, acute to acuminate. Corolla 0·2–2 mm, white, rotate; segments ovate, imbricate; filaments adnate to corolla, anthers

---

[1] By L. F. Ferguson.

exserted. Seeds 2–5, blackish, coarsely rugose. *Dry, open habitats. S. & W. Europe, northwards to N.W. France.* Al Bl Bu Co Cr Ga Gr Hs It Ju Lu Rs (K) Sa Si Tu.

11. Glaux L.[1]

Perennial herbs. Lower leaves opposite, upper alternate. Flowers 5-merous, subsessile, axillary. Calyx petaloid. Corolla absent. Capsule globose; valves 5. Seeds few.

1. G. maritima L., *Sp. Pl.* 207 (1753). Fleshy herb. Stems 5–30 cm, erect or procumbent, rooting at nodes. Leaves on rhizomes broadly triangular, more or less membranous; leaves on erect shoots 4–15 × 1·5–6 mm, decussate, sessile, entire, elliptic-oblong to obovate, obtuse or acute. Flowers 3–6 mm in diameter. Calyx white to purple or pink; lobes obtuse, broadly hyaline. Style 2–5 mm, persistent. Capsule 2–5 mm. Seeds 3–11, trigonous, minutely reticulate. 2n=30. *Maritime habitats and on saline soils inland. Atlantic, Baltic and Arctic coasts of Europe, and very locally in the W. Mediterranean region; inland in C. & E. Europe and Spain.* Au Be ?Bl Br Cz Da Fe Ga Ge Hb Ho Hs Hu Is Lu No Po Rm Rs (N, B, C, W, K, E) Su [?It].

12. Anagallis L.[1]

(incl. *Centunculus* L.)

Glabrous herbs. Leaves opposite or alternate. Flowers 5-merous, solitary, axillary, pedicellate or rarely subsessile. Corolla rotate or infundibuliform. Capsule globose, circumscissile. Seeds 6–45, papillose.

Literature: F. Kollmann & N. Feinbrun, *Notes Roy. Bot. Gard. Edinb.* 28: 173–186 (1968). P. Taylor, *Kew Bull.* [10]: 321–350 (1955).

1 Stems 1–4(–10) cm; calyx greatly exceeding corolla   **1. minima**
1 Stems 5–50(–70) cm; calyx equalling or shorter than corolla
  2 Stems procumbent, rooting at the nodes; corolla infundibuliform, pink, white or cream
    3 Pedicels shorter than subtending leaves; corolla white or cream   **2. crassifolia**
    3 Pedicels exceeding subtending leaves; corolla pink, rarely white   **3. tenella**
  2 Stems decumbent, ascending or erect; corolla rotate, usually blue or red
    4 Perennial; corolla 5–12 mm; stamens (when dry) 1–2 mm   **6. monelli**
    4 Usually annual; corolla (2–)4–7(–10) mm; stamens (when dry) 0·2–1 mm
      5 Marginal hairs of petals numerous, always 3-celled   **4. arvensis**
      5 Marginal hairs of petals absent or few, usually 4-celled   **5. foemina**

1. A. minima (L.) E. H. L. Krause in Sturm, *Deutschl. Fl.* ed. 2, 9: 251 (1901) (*Centunculus minimus* L.). Erect annual. Stems 1–4(–10) cm. Leaves 3–5 × 2–4 mm, alternate, ovate, entire, subsessile. Flowers subsessile in axils of upper leaves. Calyx 1·5–2·5 mm, divided to ¾; lobes lanceolate. Corolla much shorter than calyx, white or pink. Capsule 1·5 mm. Seeds 5–13. 2n=22. *Damp, open habitats, especially on sandy soils. Most of Europe except the north-east and extreme north.* All except Bl Cr Fa Is Rs (N, K) Sb Si Tu.

2. A. crassifolia Thore, *Essai Chlor. Land.* 62 (1803). Procumbent perennial, rooting at the nodes. Stems 5–20 cm. Leaves 8–10 × 5–9 mm, alternate or subopposite, orbicular to shortly

spathulate, entire, subsessile. Flowers in axils of middle leaves; pedicels 2–5 mm, always shorter than subtending leaf. Calyx 3–4 mm, divided to base; lobes linear-lanceolate, acute. Corolla 3–6 mm, white or cream, infundibuliform; tube short; lobes ovate. Capsule 1·5–2·5 mm; style 2–4 mm, filiform, persistent. Seeds 8–12. *Wet places. S.W. Europe.* Ga Hs Lu Sa.

3. A. tenella (L.) L., *Syst. Veg.* ed. 13, 165 (1774). Procumbent perennial, rooting freely at the nodes. Stems (3–)5–15 cm. Leaves 4–9 × 3–6 mm, opposite or rarely alternate, suborbicular to broadly elliptical, entire; petiole 0–2 mm. Flowers in axils of middle leaves; pedicels (5–)15–35 mm, capillary, always much exceeding subtending leaf. Calyx 3–4 mm, divided to base; lobes lanceolate, acuminate. Corolla 6–10 mm, pink, rarely white, infundibuliform; tube short; lobes lanceolate. Capsule 2·5–4 mm. Seeds 6–12. 2n=22. *Damp turf, bogs, ditches and lake-shores. W. Europe, northwards to Færöer; a few isolated stations in W.C. Europe, N. Italy, Greece and Kriti.* Au Az Be Bl Br Cr Fa Ga Ge Gr Hb He Ho Hs It Lu Sa.

4. A. arvensis L., *Sp. Pl.* 148 (1753) (*A. phoenicea* Scop., *A. platyphylla* Baudo, *A. parviflora* Hoffmanns. & Link). Annual or biennial, rarely short-lived perennial. Stems 6–50(–90) cm, quadrangular, diffuse, ascending to erect. Leaves 8–18(–25) × 4–10(–18) mm, opposite or rarely whorled, sessile, both surfaces dotted with glands; margins scarious; upper leaves narrower than lower, ovate to lanceolate. Flowers in axils of upper leaves; pedicels 3–35 mm, at anthesis usually exceeding subtending leaves, recurved in fruit. Calyx 3·5–5 mm, divided nearly to base; lobes narrowly lanceolate, apiculate, not concealing corolla in bud. Corolla (2–)4–7(–10) mm, blue, red or various paler colours; lobes ovate to elliptical, entire or dentate or crenate; marginal hairs usually numerous, always 3-celled, with a globose terminal cell. Stamens 0·5–1·5 mm, when dry 0·2–1 mm. Capsule 4–6 mm. Seeds 12–45. 2n=40. *Cultivated ground, waste places and maritime sands. Almost throughout Europe.* All except Fa Is Sb.

This is a species in which many variants have been described, for example *A. parviflora* Hoffmanns. & Link, which is sporadic and seems to be little more than a small-flowered variety. Blue-flowered plants predominate in the Mediterranean region; they become rarer northwards, being replaced by red-flowered plants. Plants with flowers of other colours occur throughout the range of the species.

5. A. foemina Miller, *Gard. Dict.* ed. 8, no. 2 (1768) (*A. caerulea* Schreber, non L., *A. arvensis* subsp. *caerulea* Hartman, *A. arvensis* subsp. *foemina* (Miller) Schinz & Thell.). Like 4 but upper leaves usually lanceolate; pedicels at anthesis usually shorter than subtending leaves; calyx concealing corolla in bud; corolla blue; lobes cuneate, crenate-dentate to denticulate; marginal hairs absent or few, 4- or rarely 3-celled, terminal cell elongate. 2n=40. *Cultivated ground, waste places and maritime sands. S., W. & C. Europe, but only casual or locally naturalized in most of the north.* All except Fa Hb Is Rs (N, B, C, E) ?Sa Sb ?Si.

6. A. monelli L., *Sp. Pl.* 148 (1753) (*A. collina* Schousboe, *A. linifolia* L.). Perennial, erect or ascending. Stems 10–50(–70) cm, terete, freely branched. Leaves (8–)10–15(–20) × 2–6 mm, linear-lanceolate to lanceolate, or 6–10 × 3–6 mm, elliptical, opposite or verticillate, sessile; often with small axillary, sterile shoots. Flowers in axils of upper leaves; pedicels 20–50 mm, always exceeding subtending leaves. Calyx 4–6 mm; lobes narrowly lanceolate, subulate; margins scarious. Corolla 5–12 mm, blue or red, rarely of various paler colours; lobes

---

[1] By L. F. Ferguson.

obovate to elliptical, always exceeding calyx; stamens 1·75–3·5 mm, when dry 1–2 mm, with numerous purple or yellow basal hairs. Capsule 5–7 mm. Seeds 20–45. $2n=22$. *Dry, open habitats. S.W. Europe.* Hs Lu Sa Si.

Variable in several characters, including size of leaf and corolla; though subspecific taxa have been described, they intergrade, and it is difficult to recognize them over the whole range of the species. Further investigation is required.

13. Samolus L.[1]

Perennial herbs with short stock and fibrous roots. Flowers bracteolate, 5-merous. Calyx campanulate. Corolla twice as long as calyx. Stamens alternating with clusters of 1–2(–3) staminodes. Ovary semi-inferior. Capsule globose; valves 5. Seeds numerous.

1. **S. valerandi** L., *Sp. Pl.* 171 (1753). Glabrous, with a basal rosette; stems 5–60 cm, leafy, erect. Leaves 10–90 × 5–25 mm, obovate to spathulate, obtuse, entire. Inflorescence racemose, simple or branched; pedicels becoming geniculate after anthesis. Calyx-teeth obtuse. Corolla 2–3 mm in diameter. Capsule 2–3 mm. Seeds polyhedral, reddish-brown, minutely tuberculate. $2n=$?24, 26. *Wet, usually saline or calcareous habitats. S., W. & C. Europe, extending northwards to S. Finland and eastwards to Krym.* All except Fa Is No Rs (N, ?C, E) Sb.

14. Coris L.[1]

Perennial or biennial. Stems woody at the base. Leaves alternate, subsessile. Flowers zygomorphic, in a terminal spike-like raceme. Calyx campanulate, with 2 rows of teeth. Corolla-tube short; limb unequally 5-partite to c. ⅔; lobes unequally bifid.

Stamens exserted. Capsule globose; valves 5. Seeds 4–6, papillose.

Literature: F. Pax & R. Knuth in Engler, *Pflanzenreich* **22**(IV. 237): 344–346 (1905). F. Masclans, *Collect. Bot. (Barcelona)* **7** (2, n.42): 749–758 (1968).

Up to 8 taxa have been recognized in Europe, mainly on the basis of indumentum and habit. Only the 2 following show consistent ecological and geographical separation.

Flowers pink, purple or blue; outer calyx-teeth 6–21, unequal **1. monspeliensis**
Flowers white or pale pink; outer calyx-teeth 0–4, equal **2. hispanica**

1. **C. monspeliensis** L., *Sp. Pl.* 177 (1753). Stems 10–30 cm, ascending to erect. Leaves 2–20 × 1–3 mm, glabrous to densely puberulent, numerous, entire or sinuate; margin subrevolute or, particularly in upper leaves, spinose-dentate; often with two rows of black spots on either side of the midrib. Racemes dense. Calyx 5–7 mm, membranous, tinged purple; the outer teeth 6–21, spinose, unequal, prominent in bud; the inner 5 broadly triangular with central red or black patch, connivent in fruit. Corolla 9–12 mm, pink, purple or blue, drying darker. Capsule 0·5–2 mm. $2n=18, 56$. *Dry places, particularly near the coast. W. & C. Mediterranean region, extending to N.W. Spain.* Al Bl Ga Hs It Ju Sa Si.

2. **C. hispanica** Lange, *Vid. Meddel. Dansk Naturh. Foren. Kjøbenhavn* **1863**: 53 (1863). Like **1** but stem 12–35 cm, erect, more robust; leaves 4–12 × 1–3 mm, sinuate; margin revolute, rarely spinose-dentate; racemes lax; calyx green; outer teeth 0–4, developed dorsally only, subequal, not prominent in bud; inner row of teeth with 3(–5) dark patches; corolla white or pale pink; limb subequally 5-partite. *Clay soils and maritime rocks.* ● *S.E. Spain (S. of Vera, prov. Almería).* Hs.

PLUMBAGINALES

CXXXVI. PLUMBAGINACEAE[2]

Herbs or shrubs. Leaves alternate or in basal rosettes, exstipulate. Inflorescence usually cymose, often contracted into a capitulum, rarely spike-like. Flowers actinomorphic, 5-merous, usually in bracteate spikelets. Calyx tubular below, dentate or lobed and at least slightly scarious and often plicate distally, persistent. Petals connate only at the base, or the corolla with a usually short tube. Stamens epipetalous. Styles 5, or 1 with 5 stigma-lobes. Ovary superior, 1-locular. Fruit dry, membranous, 1-seeded, surrounded by calyx, indehiscent or with circumscissile or irregular dehiscence.

Two tribes are traditionally recognized in the family, but most of the distinguishing characters are unsatisfactory. Heteromorphic incompatibility systems are known in both tribes. In the *Plumbagineae* (*Plumbago, Ceratostigma*) heterostyly is associated with monomorphism of pollen and stigma. In the *Staticeae* heterostyly is known only in some species of *Limonium*; *Acantholimon* and *Goniolimon* have dimorphic pollen and monomorphic stigmas, while dimorphism of both pollen and stigma occurs in *Armeria, Limonium* and *Limoniastrum* (cf. H. G. Baker, *Evolution* **20**: 349–368 (1966).)

1 Flowers in capitula; top of scape enclosed in a tubular scarious sheath **4. Armeria**
1 Flowers in spikes or panicles; top of scape not enclosed in a tubular sheath
2 Leaves in a basal rosette or absent
3 Spikes not secund, the terminal much the largest; corolla-tube much longer than lobes **8. Psylliostachys**
3 Spikes secund, the terminal not distinctly larger; corolla-tube much shorter than lobes
4 Stigmas filiform **5. Limonium**
4 Stigmas capitate **6. Goniolimon**
2 Stems leafy
5 Leaves pungent; stigmas capitate; fruit indehiscent **3. Acantholimon**
5 Leaves not pungent; stigmas filiform or lobed; fruit dehiscent
6 Calyx tubular; stamens free; style 1; fruit circumscissile near base, dehiscing upwards by 5 valves
7 Leaves auriculate-amplexicaul; corolla with narrow tube and rotate limb **1. Plumbago**
7 Leaves cuneate at base; corolla infundibuliform **2. Ceratostigma**
6 Calyx infundibuliform; stamens inserted in base of corolla; styles 5, free or connate in basal half; fruit circumscissile towards apex or with irregular dehiscence
8 Corolla-tube much shorter than lobes **5. Limonium**
8 Corolla-tube about as long as lobes **7. Limoniastrum**

---

[1] By L. F. Ferguson.
[2] Edit. D. M. Moore.

29

1. Plumbago L.[1]

Shrubs or perennial herbs. Leaves alternate, simple. Inflorescence of (2–)3-bracteate, 1-flowered spikelets grouped into terminal spikes. Calyx tubular, 5-ribbed, scarious; limb deeply 5-dentate. Corolla with narrow tube and rotate limb. Stamens free. Style 1, with 5 stigma-lobes. Fruit circumscissile near base, dehiscing upwards into 5 valves.

1. **P. europaea** L., *Sp. Pl.* 151 (1753). Perennial herb; stems 30–100 cm, erect, much-branched. Leaves up to 5–8 × 3–4·5 cm, the lower long-petiolate, ovate to oblong, cordate, the upper sessile, oblanceolate to lanceolate or linear, acute, auriculate-amplexicaul, glandular-dentate, glabrous, farinose beneath, rarely densely stipitate-glandular. Calyx 5–7 mm, densely stipitate-glandular on ribs, the teeth triangular. Corolla-tube 1½ times as long as calyx, the lobes obovate, violet to lilac-pink. Fruit 5–8 mm, oblong to oblong-ovoid. $2n=12$. *Roadsides, maritime sands and other dry, open habitats. S. Europe.* Al Bu Co Cr Ga Gr Hs It Ju Lu Rm Sa Si Tu [Cz].

P. auriculata Lam., *Encycl. Méth. Bot.* **2**: 270 (1786) (*P. capensis* Thunb.), a shrub native to South Africa, distinguished from **1** by its larger, pale blue corolla *c.* 25 mm in diameter, is frequently cultivated for ornament in the warmer parts of Europe and is occasionally found as an escape from gardens.

2. Ceratostigma Bunge[1]

Perennial herbs. Leaves alternate, simple. Inflorescence of 3-bracteate, 1-flowered spikelets grouped into compact terminal and axillary clusters. Calyx tubular, 5-ribbed, slightly scarious and plicate distally. Corolla infundibuliform, with patent limb. Stamens free. Style 1, with 5 stigma-lobes. Fruit circumscissile near base, dehiscing upwards into 5 valves.

1. **C. plumbaginoides** Bunge, *Enum. Pl. Chin. Bor.* 55 (1833). Stems 15–30 cm, ascending, much-branched, slightly hispid, reddish. Leaves 2–5 cm, obovate, acute to obtuse, rarely emarginate, cuneate at base, entire, appressed-ciliate, sessile. Calyx 10–12 mm, the teeth 1·5–2 mm, linear-triangular. Corolla violet-blue, the tube 15 mm, the limb 16–20 mm in diameter, with obcordate, emarginate, mucronate lobes. *Locally naturalized in N.W. France and N.W. Italy.* [Ga It.] (*China.*)

3. Acantholimon Boiss.[1]

Spiny, cushion-forming shrubs. Leaves alternate, simple. Inflorescence of 3-bracteate, 1-flowered spikelets grouped into a simple, spike-like panicle. Calyx infundibuliform; tube slender and 10-ribbed at base; limb scarious, plicate, shortly 5-dentate, with slightly excurrent ribs. Petals connate at base. Stamens free. Styles 5; stigmas capitate. Fruit indehiscent.

1. **A. androsaceum** (Jaub. & Spach) Boiss., *Diagn. Pl. Or. Nov.* 1(7): 73 (1846) (*A. echinus* auct., non (L.) Bunge). Leaves 10–20 × *c.* 1 mm, linear, subulate-triquetrous distally, pungent, glabrous or puberulent. Peduncle up to 4 cm, shorter than or somewhat exceeding the leaves, puberulent; inflorescence (1–)3- to 7-flowered. Calyx 14–16 mm, the tube about twice as long as the limb, puberulent towards base, white with purplish veins. Petals shorter than or exceeding the calyx by *c.* 3 mm, long-clawed, purple. Fruit narrowly cylindrical. *Mountains of S. part of Balkan peninsula and Kriti.* Al Cr Gr Ju.

---

¹ By D. M. Moore.   ² By A. R. Pinto da Silva.

4. Armeria Willd.[2]

Caespitose perennial herbs or cushion-forming dwarf shrubs, with a branched woody stock. Leaves simple, in basal rosettes or on densely leafy branches. Inflorescence a capitulum, with an involucre of imbricate bracts and a tubular scarious sheath at the base enclosing the top of the scape. Flowers usually bracteolate, in usually bracteate, cymose spikelets. Calyx infundibuliform, inserted obliquely, with a basal spur; tube 5- or 10-ribbed; limb scarious; lobes usually awned. Corolla infundibuliform; petals connate at the extreme base, persistent. Stamens inserted at base of corolla. Styles 5, connate at the base; stigmas plumose. Fruit with circumscissile or irregular dehiscence.

Most species in temperate Europe show dimorphism of pollen and stigma and are self-incompatible. In Europe only **1(b)** is known to be monomorphic and self-compatible.

The extent of hybridization in the genus is not clear, but the following hybrids seem worthy of mention: **1** × **32**, **18** × **31(a)**, **35** × **37** and **40** × **38**.

This account follows Bernis (1953) in considering the calyx-spur as the unveined basal extremity of the calyx, adnate to the pedicel at least before maturity, and including the insertion-scar; the tube is measured from the apex of the pedicel to the base of the limb. Measurements of tube and spur refer to the flower with the longest pedicel in the spikelet, i.e. the oldest one, which is opposite the bract.

Literature: F. A. Novák, *Mém. Soc. Sci. Bohême* **1938**: 1–25 (1938); **1939**: 1–24 (1939). G. H. M. Lawrence, *Gentes Herb.* **4**: 389–418 (1940). F. Bernis, *Bol. Soc. Brot.* ser. 2, **23**: 225–263 (1950); *Anal. Inst. Bot. Cavanilles* **11**(2): 5–287 (1953); **12**(2): 77–252 (1954); **14**: 259–432 (1956).

1 Calyx-tube not more than 2(–2½) times as long as spur; insertion-scar of calyx narrowly elliptical to linear
  2 Dwarf shrub; stock with many long branches
    3 Outer involucral bracts at least as long as the inner, acuminate, mucronate to cuspidate; spikelet-bract at least as long as calyx   **34. welwitschii**
    3 Outer involucral bracts shorter than the inner, obtuse, muticous; spikelet-bract absent or rudimentary   **35. pungens**
  2 Herb; stock short
    4 Leaves obovate- to linear-lanceolate, usually at least 2 mm wide
      5 Leaves velutinous   **39. velutina**
      5 Leaves glabrous or puberulent
        6 Leaves (10–)15–35 mm wide; outer involucral bracts at least as long as the inner, acuminate to cuspidate   **40. gaditana**
        6 Leaves not more than 10 mm wide; outer involucral bracts shorter than the inner, obtuse, rarely mucronate   **41. hirta**
    4 Leaves (at least the inner) filiform to linear, sometimes linear-lanceolate, usually not more than 1 mm wide
      7 Leaves not more than 30 mm   **42. hispalensis**
      7 Leaves at least 40 mm
        8 Spikelet-bract present
          9 Spikelet-bract longer than calyx, coriaceous; leaves all similar; awns absent or not more than 0·4 mm   **37. rouyana**
          9 Spikelet-bract shorter than calyx, scarious; leaves dimorphic; awns 0·7–2 mm   **43. cariensis**
        8 Spikelet-bract absent (but bracteoles sometimes rudimentary, scarious)
          10 Involucral bracts puberulent, the outer ±cuspidate; calyx-tube usually hirsute   **36. pinifolia**

10 Involucral bracts glabrous, the outer mucronulate; calyx-
    tube hirsute only on ribs   **38. macrophylla**
1 Calyx-tube 2–8 times as long as spur; insertion-scar of calyx
  orbicular to ovate-oblong
11 Leaves at least 5 mm wide, lanceolate to obovate-lanceolate
12 Leaves mostly mucronate; scapes 10–20 cm **33. berlengensis**
12 Leaves muticous; scapes 20–80 cm
13 Dwarf shrub; stock large, stout; capitula 30–40 mm wide
    10. pseudarmeria
13 Herb; stock rather short, slender; capitula 10–30 mm wide
14 Involucral bracts spinescent, the inner with narrow
    scarious margins; calyx-awns 1·5–3 mm **13. villosa**
14 Involucral bracts not spinescent, the inner with wide
    scarious margins; calyx-awns 0·3–1·5 mm
15 Sheath 8–20 mm; spikelet-bract at least as long as calyx
16 Calyx 7–8 mm, the awns (0·8–)1–1·5 mm; spikelets
    stipitate   **15. morisii**
16 Calyx 6–6·5 mm, the awns 0·3–0·5 mm; spikelets
    sessile or subsessile   **16. colorata**
15 Sheath 20–60 mm; spikelet-bract shorter than calyx
17 Capitula 25–30 mm wide; calyx 8–10 mm; leaves
    (60–)100–200 mm; involucral bracts greenish
    17. macropoda
17 Capitula 8–20(–30) mm wide; calyx 5–9 mm; leaves
    13–130 mm; involucral bracts brownish or reddish
18 Outer involucral bracts usually longer than the inner,
    with narrow scarious margins; spikelet-bract almost
    as long as calyx   **8. alliacea**
18 Outer involucral bracts shorter than the inner, with
    wide scarious margins; spikelet-bract shorter than
    calyx   **12. langei**
11 At least the inner leaves not more than 4(–7) mm wide,
  filiform or linear to linear-spathulate
19 At least the inner leaves less than 0·7 mm wide, filiform,
    usually canaliculate, conduplicate or convolute
20 Inner leaves muticous
21 Capitula soft, not becoming hard in fruit; outer involucral
    bracts long-acuminate; corolla purple **6. vandasii**
21 Capitula compact, becoming hard in fruit; outer involu-
    cral bracts obtuse, mucronate to shortly cuspidate;
    corolla pinkish or white
22 Calyx-awns 0·3–0·6 mm; leaves subobtuse; corolla white
    or pale pinkish   **20. littoralis**
22 Calyx-awns at least 0·7 mm; leaves usually acute or sub-
    acute; corolla pinkish, rarely white   **21. duriaei**
20 Inner leaves usually mucronate or mucronulate
23 Sheath (10–)15–35 mm
24 Calyx-awns not more than 1 mm; scapes erect or some-
    what divergent; spikelet-bract puberulent, brown
    12. langei
24 Calyx-awns 1·2–1·5 mm; scapes patent-ascending; spike-
    let-bract glabrous, white-scarious **24. eriophylla**
23 Sheath 4–15 mm
25 Calyx-awns (0·5–)1–1·5 mm
26 Outer leaves mucronulate; bracteoles reduced
    25. humilis
26 Outer leaves muticous; bracteoles well-developed
    19. trachyphylla
25 Calyx-awns 0·3–0·9 mm
27 Scapes 18–32 cm, straight, erect; outer involucral bracts
    much shorter than the inner   **18. filicaulis**
27 Scapes 4–15 cm, arcuate-ascending; outer involucral
    bracts almost as long as the inner
28 Spikelet-bract shorter than calyx, white-scarious;
    scapes 10–15 cm   **23. girardii**
28 Spikelet-bract as long as calyx, dark brown, with
    scarious margin and fimbriate apex; scapes 4–5 cm
    22. arcuata
19 Leaves more than 0·7 mm wide, at least the inner linear to
  linear-spathulate, mostly flat
29 At least the inner leaves mucronate or mucronulate
30 Leaves pungently mucronate; scapes not more than 5 cm;
    bracteoles absent or reduced   **30. juniperifolia**

30 Leaves not pungent; scapes more than 5 cm; bracteoles
    ± well-developed
31 Sheath not more than 15 mm
32 Capitula (12–)20–25 mm wide; leaves uniform; spike-
    let-stipe up to 2 mm; corolla usually white
    27. leucocephala
32 Capitula not more than 20 mm wide; leaves dimorphic;
    spikelets sessile to shortly stipitate; corolla usually
    pinkish
33 Calyx-awns 0·3–0·6 mm; scapes 18–32 cm; outer
    leaves 10–30 mm, subacute, mucronulate **18. filicaulis**
33 Calyx-awns (0·5–)1–1·5 mm; scapes 4–20 cm; outer
    leaves 30–50 mm, subobtuse, muticous **19. trachyphylla**
31 Sheath 15–65 mm
34 Calyx-awns not more than 1 mm
35 Capitula 8–10(–13) mm wide, becoming hard in
    fruit; outer involucral bracts with wide scarious
    margins   **12. langei**
35 Capitula 10–20(–30) mm wide, not hard in fruit;
    outer involucral bracts with narrow scarious
    margins   **8. alliacea**
34 Calyx-awns 1–4 mm
36 Calyx-awns 2–4 mm; outer involucral bracts ± con-
    tracted, cuspidate-spinescent **14. transmontana**
36 Calyx-awns not more than 2 mm; outer involucral
    bracts ± acuminate
37 Spikelets usually subsessile; leaves usually more than
    3 mm wide, lanceolate to linear-spathulate; capi-
    tula 10–20(–30) mm wide; bract almost as long as
    calyx   **8. alliacea**
37 Spikelets stipitate; inner leaves 1·5–3 mm wide,
    linear; capitula 20–35 mm wide; bract at least as
    long as calyx   **4. rumelica**
29 At least the inner leaves muticous
38 Sheath not more than 15 mm
39 Outer involucral bracts longer than the inner, acuminate-
    cuspidate
40 Calyx-awns 1–1·5 mm   **9. ruscinonensis**
40 Calyx-awns not more than 1 mm
41 Capitula 15–25 mm; leaves all similar, entire
    1. maritima
41 Capitula 10–12 mm; leaves dimorphic, the outer
    ± undulate-denticulate   **7. denticulata**
39 Outer involucral bracts shorter than or as long as the
    inner, usually mucronate to muticous
42 Herb; stock short, with short branches
43 Leaves all similar, or somewhat unequal but not di-
    morphic, with at most very narrow scarious
    margins; outer involucral bracts with narrow
    scarious margins
44 Calyx-awns absent or not more than 1 mm; spikelet-
    bract nearly as long as or longer than calyx; leaves
    not becoming coriaceous   **1. maritima**
44 Calyx-awns 1–1·5 mm; spikelet-bract shorter than
    calyx; leaves becoming coriaceous **9. ruscinonensis**
43 Leaves dimorphic, at least the outer with wide
    scarious margins; outer involucral bracts with
    wide scarious margins
45 Outer involucral bracts acuminate-cuspidate; scapes
    up to 50 cm; corolla usually white **3. undulata**
45 Outer involucral bracts ± acute, muticous or
    mucronate; scapes not more than 20(–25) cm;
    corolla usually pink to purple **2. canescens**
42 Dwarf shrub; stock woody, stout, dense, much-
    branched
46 Scapes not more than 15 cm
47 Calyx-awns 1–1·5 mm; outer involucral bracts as
    long as the inner   **9. ruscinonensis**
47 Calyx-awns not more than 1 mm; outer involucral
    bracts shorter than the inner
48 Plant usually velutinous; outer involucral bracts
    with narrow, scarious margins, mucronate to
    ± cuspidate (coastal)   **32. pubigera**

48 Plant glabrous or ±puberulent; outer involucral bracts with wide scarious margins, muticous or mucronate (mountains)
  49 Spikelets usually sessile or subsessile; outer involucral bracts suborbicular to ovate-acuminate **31. splendens**
  49 Spikelets stipitate; outer involucral bracts ovate to lanceolate
    50 Calyx-awns not more than 0·5 mm, usually scarcely distinct from the lobe; leaves 0·7–1·5(–2) mm wide; capitula 10–15(–18) mm wide **28. multiceps**
    50 Calyx-awns 0·5–1 mm; leaves 1·5–5 mm wide; capitula (15–)18–30 mm wide **1. maritima**
46 Scapes more than 15 cm
  51 Outer involucral bracts muticous or mucronulate; calyx 5–6 mm; spikelets sessile or subsessile
    52 Leaves not papillose, the inner 1–1·2 mm wide **26. sardoa**
    52 Leaves papillose, the inner *c.* 2 mm wide **29. soleirolii**
  51 Outer involucral bracts acuminate-cuspidate or mucronate; calyx 6–8 mm; spikelets ±stipitate
    53 Calyx-awns 1–1·5 mm **9. ruscinonensis**
    53 Calyx-awns not more than 1 mm
      54 Plant usually velutinous **32. pubigera**
      54 Plant glabrous or puberulent **1. maritima**
38 Sheath at least 15 mm
55 Outer involucral bracts longer than the inner
  56 Involucral bracts and leaves velutinous **11. vestita**
  56 Involucral bracts glabrous; leaves glabrous, ciliate or puberulent
    57 Outer involucral bracts obtuse, with wide scarious margins; capitula becoming hard in fruit; calyx-tube with secondary ribs glabrous or less hirsute than primary ribs **20. littoralis**
    57 Outer involucral bracts acuminate, with narrow scarious margins; capitula not becoming hard in fruit; calyx-tube pubescent, or hirsute on both primary and secondary ribs
      58 Leaves acute or subacute, usually few **8. alliacea**
      58 Leaves obtuse or subobtuse, numerous
        59 Calyx-awns absent or not more than 1 mm; scapes straight or arcuate only at base **1. maritima**
        59 Calyx-awns 1–1·5 mm; scapes arcuate or flexuous **9. ruscinonensis**
55 Outer involucral bracts not longer than the inner
  60 Calyx-awns at least 1 mm
    61 Capitula not more than 14 mm wide, becoming hard in fruit; calyx 4·5–6·5 mm; outer involucral bracts with wide scarious margins **21. duriaei**
    61 Capitula usually more than 14 mm wide, not becoming hard in fruit; calyx 6–12 mm; outer involucral bracts with narrow scarious margins
      62 Scapes not more than 20 cm; outer involucral bracts at least as long as the inner; usually a dwarf shrub with the stock often long-branched; leaves becoming coriaceous **9. ruscinonensis**
      62 Scapes at least 25 cm; outer involucral bracts not longer than the inner; herb with shortly branched stock; leaves not becoming coriaceous
        63 Inner leaves 1–2 mm wide, linear, subobtuse; outer involucral bracts usually acuminate-cuspidate, the inner mucronulate or muticous, with wide scarious margins **2. canescens**
        63 Leaves at least 3 mm wide, lanceolate to linear-spathulate, subacute; outer involucral bracts ±contracted at apex, cuspidate-spinescent, the inner subacute to emarginate, mucronate-spinescent, with narrow scarious margins **13. villosa**
60 Calyx-awns not more than 1 mm
  64 Capitula at least 15 mm wide
    65 Dwarf shrub, usually velutinous **32. pubigera**
    65 Herb, glabrous, ciliate or puberulent

      66 Calyx-tube hirsute only on primary ribs; leaves dimorphic, the outer 3–10 mm wide; outer involucral bracts much shorter than the inner **5. sancta**
      66 Calyx-tube pubescent throughout, or hirsute only on ribs; leaves all similar or slightly dimorphic; outer involucral bracts slightly shorter than to as long as the inner
        67 Leaves usually not more than 4 mm wide, usually obtuse, linear to linear-spathulate; outer involucral bracts with narrow scarious margins **1. maritima**
        67 Leaves (3–)4–14 mm wide, acute, the outer lanceolate-spathulate; outer involucral bracts with wide scarious margins **8. alliacea**
  64 Capitula not more than 15 mm wide
    68 Capitula not becoming hard in fruit; outer involucral bracts with narrow scarious margins **1. maritima**
    68 Capitula becoming hard in fruit; outer involucral bracts with wide scarious margins
      69 Sheath 40–60 mm, lax (on serpentine and dolomite) **12. langei**
      69 Sheath not more than 30 mm, appressed (on schist and granite)
        70 Calyx-awns 0·3–0·6 mm; leaves subobtuse; corolla white or pale pinkish **20. littoralis**
        70 Calyx-awns at least 0·7 mm; leaves usually acute or subacute; corolla pink, rarely white **21. duriaei**

Sect. ARMERIA. Calyx-tube 2–8 times as long as spur; insertion-scar of calyx orbicular to ovate-oblong.

1. **A. maritima** (Miller) Willd., *Enum. Pl. Hort. Berol.* 333 (1809). Herb, or sometimes a dwarf shrub; stock shortly branched. Leaves linear to linear-spathulate, obtuse, muticous, with very narrow scarious margins, flat. Scapes straight or arcuate only at base. Sheath 10–20(–30) mm. Capitula 10–30 mm wide; outer involucral bracts more or less acuminate-subulate, mucronate or shortly cuspidate, usually with narrow scarious margins; inner scarious or almost so. Spikelet-bract almost as long as or longer than calyx. Calyx 5–10 mm; tube pubescent, or hirsute only on ribs; awns absent or up to 1 mm. *Most of Europe, but absent from much of the south-east and many of the islands.* Al Au Be Br Bu Cz Da Fa Fe Ga Ge Hb He Hs Hu Is It Ju Lu No Po Rm Rs (N, B, C) Su.

A. maritima, even in the present restricted sense, is the most polymorphic species of the genus. The following eight subspecies can usually be readily distinguished.

1 Capitula not more than 15(–17) mm wide
  2 Leaves 0·7–1·8 mm wide; scapes not more than 30 cm **(e) subsp. halleri**
  2 Leaves 2–3(–4) mm wide; scapes 40–60 cm **(h) subsp. barcensis**
1 Capitula more than 15 mm wide
  3 Outer involucral bracts usually at least as long as the inner, long-acuminate-subulate or shortly cuspidate to acute
    4 Leaves linear, usually ciliate **(d) subsp. elongata**
    4 Leaves linear-spathulate, not or sparsely ciliate **(c) subsp. miscella**
  3 Outer involucral bracts shorter than the inner, mucronate
    5 Leaves mostly at least 2 mm wide **(g) subsp. alpina**
    5 Leaves not more than 2 mm wide
      6 Corolla purplish **(f) subsp. purpurea**
      6 Corolla pinkish or white
        7 Leaves 40–120(–180) × 0·8–1·2 mm, ±linear **(a) subsp. maritima**
        7 Leaves 30–50(–80) × 1·5–2 mm, linear-spathulate **(b) subsp. sibirica**

(a) Subsp. **maritima** (*A. vulgaris* Willd., *Statice armeria* L.): Leaves 40–120(–180) × 0·8–1·2 mm, more or less linear. Scapes up to 20 cm, usually pubescent. Capitula (15–)17–25 mm wide; outer involucral bracts shorter than the inner, mucronate. Corolla pink, red or white. $2n=18$. *Coastal salt-marshes, cliffs and pastures.* ● *N. Europe, eastwards to N.W. Russia.*

(b) Subsp. **sibirica** (Turcz. ex Boiss.) Nyman, *Consp.* 616 (1881) (incl. *A. arctica* sensu Sternb., non (Cham.) Wallr.): Leaves 30–50(–80) × 1·5–2 mm, linear-spathulate. Scapes up to 20 cm. Capitula 16–20 mm wide; outer involucral bracts shorter than the inner, mucronate. Spikelets stipitate. Corolla pale pink. *Dry stony places, heaths and alpine meadows. N. Fennoscandia.* (*Siberia, Arctic America.*)

(c) Subsp. **miscella** (Merino) Malagarriga, *Acta Phytotax. Barcinon.* 1: 23 (1969): Leaves 1·5–3 mm wide, linear-spathulate, glabrous or sparsely ciliate. Scapes 25 cm or more. Capitula (15–)17–25 mm wide; outer involucral bracts as long as the inner, acute to shortly cuspidate. Corolla pink to pinkish-purple. *Estuarine meadows.* ● *S.W. France, N. Spain, N.W. Portugal.*

(d) Subsp. **elongata** (Hoffm.) Bonnier, *Fl. Compl. Fr.* 9: 54 (1927): Leaves 50–150 × 1·1–2·5 mm, linear, usually ciliate. Scapes mostly more than 25 cm, glabrous. Capitula (15–)18–25 mm wide; outer involucral bracts up to 25 mm, as long as or longer than the inner, long-acuminate-subulate. Corolla pale pink. $2n=18$. *Sandy heaths and grassy open places.* ● *N. & C. Europe.*

Intermediates between this subspecies and **1(a)**, which occur where their ranges overlap, mainly in the Baltic region, have been described as **A. vulgaris** subsp. **intermedia** (Marsson) Nordh., *Norsk Fl.* 498 (1940), and **A. ambifaria** Focke, *Abh. Naturw. Ver. Bremen* 17: 445 (1903).

A. pocutica Pawł., *Fragm. Fl. Geobot.* 8: 399 (1962), from Topicz (N.E. Carpathians), is said to differ from **1(d)** mainly by the outer leaves being wider than the inner, but this is also sometimes observed in plants of **1(d)** from N. & C. Europe.

(e) Subsp. **halleri** (Wallr.) Rothm., *Feddes Repert.* 67: 9 (1963) (*A. halleri* Wallr.): Leaves 30–100 × 0·7–1·8 mm, linear, ciliate. Scapes up to 30 cm. Capitula 10–15(–17) mm wide; outer involucral bracts not longer than the inner. Corolla bright pink to red. $2n=18$. *Dry pastures and gravelly soils rich in heavy metals or over serpentine.* ● *Scattered through W. & C. Europe, from the Pyrenees and the Netherlands to Poland.*

(f) Subsp. **purpurea** (Koch) Á. & D. Löve, *Bot. Not.* 114: 54 (1961): Leaves up to 80 × c. 2 mm, linear, ciliate. Scapes 20–35 cm, glabrous. Capitula 20–25 mm wide; outer involucral bracts shorter than the inner, mucronate. Corolla purplish. $2n=18$. *Marshy meadows by lakes.* ● *S. Germany and formerly in Switzerland; ? N. Italy.*

(g) Subsp. **alpina** (Willd.) P. Silva, *Bot. Jour. Linn. Soc.* 64: 376 (1971) (*A. alpina* Willd., *A. cantabrica* Boiss. & Reuter ex Willk., *A. pubinervis* Boiss., *Statice montana* Miller): Leaves 30–80 × (1·5–)2–4(–5) mm, linear to linear-lanceolate, usually glabrous. Scapes up to 25(–30) cm, usually glabrous. Capitula (15–)18–30 mm wide; outer involucral bracts not longer than the inner, mucronate. Corolla deep pink, purplish or red, rarely white. $2n=18, 36$. ● *High mountains, from N. Spain to W. Carpathians and Balkan peninsula.*

(h) Subsp. **barcensis** (Simonkai) P. Silva, *loc. cit.* (1971) (*A. barcensis* Simonkai): Leaves up to 150 × 2–3(–4) mm, linear, ciliate. Scapes 40–60 cm. Capitula 10–15 mm wide; outer involucral bracts shorter than the inner, brown at the apex. Corolla pink to purplish. ● *C. Romania (near Braşov).*

2. **A. canescens** (Host) Boiss. in DC., *Prodr.* 12: 686 (1848) (incl. *A. dalmatica* G. Beck). Herb; stock shortly branched. Leaves muticous, dimorphic; outer up to 2–3(–6) mm wide, linear-subspathulate, flat; inner 1–2 mm wide, linear or linear-

lanceolate, flat or revolute. Scapes erect or arcuate-ascending. Sheath 7–30 mm. Capitula (10–)15–25 mm wide; outer involucral bracts shorter than or as long as the inner; inner obtuse. mucronulate or muticous, with wide scarious margins. Spikelets sessile or subsessile; bract as long as calyx. Calyx 6–10 mm; tube usually hirsute only on ribs; awns (0·3–)1–2 mm. Corolla usually pinkish to reddish-purple. *C. & E. Mediterranean region.* Al Gr It Ju Si.

A very polymorphic species (cf. Novák (1938)).

(a) Subsp. **canescens**: Leaves with narrow scarious margins, glabrous, usually subobtuse. Scapes 25–70 cm. Sheath 18–30 mm. Outer involucral bracts with very narrow scarious margins, usually acuminate-cuspidate. $2n=18$. *Mountain grassland; calcicole.* ● *Almost throughout the range of the species.*

(b) Subsp. **nebrodensis** (Guss.) P. Silva, *Bot. Jour. Linn. Soc.* 64: 376 (1971) (*A. majellensis* Boiss., *Statice nebrodensis* Guss.): At least the outer leaves usually with wide scarious margins, sometimes ciliate and puberulent on veins, obtuse. Scapes 7–20(–25) cm. Sheath 7–13 mm. Outer involucral bracts with wide scarious margins, mostly subacute, muticous or mucronate. $2n=18$. *Sunny mountain rocks.* ● *Throughout the range of the species.*

3. **A. undulata** (Bory) Boiss. in DC., *Prodr.* 12: 685 (1848). Like **2** but leaves all subobtuse, the outer with wide scarious margins; scapes 10–50 cm; sheath 8–15 mm; outer involucral bracts more or less acuminate-cuspidate, with wide scarious margins; calyx-awns 0·8–1·2 mm; corolla usually white. *Mountain pastures. Greece.* Gr. (*Lebanon.*)

4. **A. rumelica** Boiss., *op. cit.* 677 (1848). Like **2** but leaves all linear, subacute, the outer acuminate, the inner 1·5–3 mm wide, mucronulate; sheath 15–45 mm; capitula 20–35 mm wide; outer involucral bracts with narrow scarious margins; spikelets stipitate; bract as long as or longer than calyx. $2n=18$. *Mountain grassland.* ● *C. & S.E. parts of Balkan peninsula.* Bu Gr Ju Tu.

3 and **4** are close to **2**, but there is insufficient evidence to justify their inclusion in it.

5. **A. sancta** Janka, *Term. Füz.* 6: 165 (1882). Like **2(a)** but outer leaves 70–150 × 3–10 mm, acuminate; outer involucral bracts much shorter than the inner, usually shortly acuminate; calyx hirsute only on primary ribs; awns 0·4–0·7 mm. *Pastures and calcareous rocks up to 1200 m.* ● *N. Greece (Makedhonia).* Gr.

6. **A. vandasii** Hayek, *Feddes Repert.* 21: 257 (1925). Herb. Leaves 30–60 × 0·4–0·5 mm, filiform, acute, muticous, flat, glabrous. Scapes 10–15 cm, glabrous. Sheath 10–20 mm. Capitula 10–16 mm wide; outer involucral bracts shorter than the inner, long-acuminate; inner obtuse, mucronulate, with scarious margins. Spikelets sessile. Calyx 5–6 mm, pubescent at least on ribs; awns 0·5–0·8 mm. Corolla purple. *Mountain rocks.* ● *S. Jugoslavia (near Prilep).* Ju.

A little-known species needing further investigation. Perhaps related to **2**, despite its uniform leaves.

7. **A. denticulata** (Bertol.) DC., *Cat. Pl. Horti Monsp.* 7 (1813). Herb; stock shortly branched. Leaves linear, acuminate,

subobtuse, muticous, with narrow scarious margins, dimorphic; outer 15–40 × 1·2–1·5(–2) mm, undulate-denticulate, otherwise flat; inner 30–70(–130) × 0·7–1 mm, usually flat. Scapes 25–35 cm, straight or slightly arcuate, glabrous. Sheath 10–15 mm. Capitula 10–12 mm wide; involucral bracts with wide scarious margins; outer longer than the inner, long-acuminate-cuspidate; inner obtuse, mucronate or muticous. Spikelets subsessile; bract shorter than calyx. Calyx 4·5–6·5 mm; tube hirsute on ribs; awns 0·5–0·8 mm, dark purplish. Corolla white. *Serpentine rocks.* ● *C. Italy (Toscana).* It.

8. **A. alliacea** (Cav.) Hoffmanns. & Link, *Fl. Port.* 1: 441 (1813–1820) (*A. allioides* Boiss., *A. bupleuroides* Gren. & Godron, *A. castellana* Boiss. & Reuter ex Leresche, *A. plantaginea* Willd., ?*Statice plantaginea* All.). Herb; stock with few branches. Leaves 50–130(–200) × (1–)3–14 mm, usually few, sometimes unequal, lanceolate to linear-spathulate, acute or subacute, usually muticous, usually with narrow scarious margins. Scapes 20–50(–70) cm. Sheath 20–60 mm. Capitula 10–20(–30) mm wide; involucral bracts brownish or reddish; outer usually longer than the inner, acuminate, cuspidate or rarely mucronate, with narrow scarious margins; inner emarginate, with wide scarious margins. Spikelets usually subsessile; bract slightly shorter than calyx. Calyx 5–9 mm; tube usually hirsute only on ribs; awns 0·7–1·5(–2) mm. Corolla purplish to white. $2n=18$. *Dry grassland, mainly in the mountains.* ● *W. Europe, extending to S.W. Germany and N. Italy.* Ga Ge He Hs It Lu.

A polymorphic species within which many populations have been given specific, subspecific or varietal status. They do not seem consistently separable and further investigation is required.

Plants from C. Spain described as **A. alliacea** subsp. **matritensis** (Pau) Borja, Rivas Goday & Rivas Martínez, *Anal. Inst. Bot. Cavanilles* 25: 154 (1969), sometimes have shorter leaves (20 mm or more) and sheath (17 mm or more) and the outer involucral bracts are shorter than the inner, are usually mucronate, and have wide scarious margins.

9. **A. ruscinonensis** Girard, *Ann. Sci. Nat.* ser. 3, 2: 323 (1844) (*A. majellensis* auct. gall. et hisp., non Boiss.). Dwarf shrub, with a usually long-branched, rather stout stock, sometimes a herb with a shortly branched stock. Leaves 40–90(–170) × 2–4(–6) mm, numerous, linear or linear-spathulate, flat or some canaliculate or more or less convolute, subobtuse, muticous, with narrow scarious margins, becoming coriaceous. Scapes (7–)10–20(–40) cm, arcuate or flexuous. Sheath 10–24(–30) mm. Capitula 15–20 mm wide; outer involucral bracts as long as or longer than the inner, usually acuminate-cuspidate, with narrow scarious margins. Spikelets more or less stipitate; bract shorter than calyx. Calyx 6–8 mm; tube hirsute on ribs; awns 1–1·5 mm. Corolla pinkish or white. $2n=18$. *Maritime cliffs and mountains near the sea, up to 1300 m; calcifuge.* ● *S. France, N.E. Spain.* Ga Hs.

Although intermediates are found between 8 and 9, the species are usually distinct.

10. **A. pseudarmeria** (Murray) Mansfeld, *Feddes Repert.* 47: 140 (1939) (*A. latifolia* Willd.). Compact dwarf shrub; stock much-branched, stout. Leaves 100–230 × 14–22 mm, obovate-lanceolate to lanceolate-spathulate, subacute, muticous, with distinct scarious margins, flat, glabrous. Scapes 25–50(–70) cm, glabrous, straight or arcuate, erect. Sheath 50–60(–100) mm, lax. Capitula 30–40 mm wide; involucral bracts with wide scarious margins; outer not longer than the inner, shortly cuspidate or mucronate; inner mucronate. Spikelets sessile to stipitate; bract not longer than calyx. Calyx 10–11 mm; tube hirsute on ribs; awns 1·2–1·7 mm. Corolla usually white. $2n=18$. *Pastures and scrub on granite.* ● *W.C. Portugal (Cabo da Roca).* Lu [Br].

11. **A. vestita** Willk. in Willk. & Lange, *Prodr. Fl. Hisp.* 2: 366 (1868). Herb; stock somewhat lax with few branches. Leaves acuminate, subacute, muticous, with narrow scarious margins, velutinous, erect, dimorphic; outer 50–70 × 4–7 mm, lanceolate; inner 80–100 × 1·5–3 mm, linear-spathulate. Scapes 37–45 cm, stout, puberulent mainly at base. Sheath 35–45 mm. Capitula *c.* 17 mm wide; involucral bracts velutinous; outer longer than the inner, acuminate, long-cuspidate-spinescent, with very narrow scarious margins; inner mucronate to muticous, with wide scarious margins. Spikelets subsessile; bract nearly as long as calyx, velutinous. Calyx 6·5–7 mm; tube hirsute on ribs; awns 1–1·3 mm. Corolla pink. *Mountain pastures.* ● *W.C. Spain.* Hs.

12. **A. langei** Boiss. ex Lange, *Vid. Meddel. Dansk Naturh. Foren. Kjøbenhavn* **1861**: 59 (1861). Herb; stock short, with few, short branches. Leaves 13–130 × 0·3–8(–13) mm, more or less unequal, lanceolate or linear-spathulate to filiform, acute, glaucous-violet to glaucescent. Scapes 15–50(–80) cm, often tinged with violet near the base. Capitula becoming hard in fruit; involucral bracts light yellowish-brown to purplish, with wide scarious margins; outer shorter than the inner, mucronate to cuspidate; inner mucronate to muticous. Spikelets sessile or subsessile; bract coriaceous, broadly scarious at the apex, puberulent. Calyx 5–6 mm; tube pubescent or hirsute on 5 or 10 ribs; awns 0·5–1 mm. Corolla white to pink. *Stony places and rock-crevices, usually on serpentine or dolomite.* ● *N.W. Spain, N.E. Portugal.* Hs Lu.

(a) Subsp. **langei**: Leaves 13–80 × 0·3–3 mm, numerous, mucronate, canaliculate or more or less convolute, 1(–3)-veined. Scapes 15–40 cm, 4–6 times as long as leaves, numerous, slender, erect or somewhat divergent. Sheath 15–35 mm, appressed. Capitula 8–10(–13) mm wide. Bract equalling calyx. *Throughout the range of the species.*

(b) Subsp. **daveaui** (Coutinho) P. Silva, *Agron. Lusit.* 30: 219 (1970) (*A. villosa* sensu Coutinho, non Girard): Leaves 30–130 × (2–)4–8(–13) mm, few, muticous, flat, 3- or 5-veined. Scapes 30–50(–80) cm, 5–11 times as long as leaves, few, stout, erect. Sheath usually 40–60 mm, rather lax. Capitula 10–15 mm wide. Bract shorter than calyx. *N. León, N.E. Trás-os-Montes.*

Both subspecies occur in the serpentine areas of N.E. Portugal, but they are vicarious and never sympatric there.

13. **A. villosa** Girard, *Ann. Sci. Nat.* ser. 3, 2: 323 (1844). Herb; stock short, with few, short branches. Leaves 50–120 × 3–12 mm, lanceolate to linear-spathulate, subacute, muticous, with narrow scarious margins, flat. Scapes 25–45(–60) cm. Sheath 15–27 mm. Capitula 20–28 mm wide; outer involucral bracts not longer than the inner, more or less contracted, cuspidate-spinescent, with narrow scarious margins; inner usually mucronate-spinescent, with narrow scarious margins. Spikelets subsessile or stipitate; bract not longer than calyx. Calyx 7–12 mm; tube hirsute on ribs; awns 1·5–3 mm. Corolla white or pink. $2n=18$. *Gravelly soils in mountains.* ● *S. Spain.* Hs.

Populations intermediate between this species and 8 are found in the northern and western parts of its range.

14. **A. transmontana** (Samp.) Lawrence, *Agron. Lusit.* 12: 303, 380 (1950). Like 13 but leaves 1–6(–9) mm wide, acute,

mucronate, flat, more or less caniculate or convolute, 1- or 3-veined; sheath (25-)40–65 mm; capitula 15–25 mm wide; inner involucral bracts obtuse, shortly mucronate or muticous, with wide scarious margins; spikelets usually sessile; calyx-awns 2–4 mm. *Mountain pastures and rock-crevices.* ● *W.C. & N.W. Spain, N. & C. Portugal.* Hs Lu.

Plants intermediate between this species and **8** occur near its southern limit in Portugal, where the species are sympatric.

15. **A. morisii** Boiss. in DC., *Prodr.* 12: 687 (1848) (incl. *A. gussonei* Boiss.). Herb; stock short, with long branches. Leaves 20–50(–100) × 5–12 mm, lanceolate-spathulate, subacute, muticous, with narrow scarious margins, flat, 3-veined, puberulent or glabrous, subcoriaceous. Scapes 20–40 cm, erect or somewhat arcuate, more or less verrucose, glabrous. Sheath 8–20 mm. Capitula 20–25 mm wide; involucral bracts with wide scarious margins; outer shorter than the inner, obtuse or acute, mucronulate; inner muticous. Spikelets stipitate; bract at least as long as calyx. Calyx 7–8 mm; tube hirsute on ribs; awns (0·8–)1–1·5 mm. Corolla pale pink. *Mountain rocks; calcicole.* ● *Sardegna, Sicilia.* Sa Si.

16. **A. colorata** Pau, *Mem. Mus. Ci. Nat. Barcelona (Bot.)* 1(1): 66 (1922). Like **15** but at least some leaves obovate-lanceolate, somewhat rigid, often undulate on margins; scapes straight; spikelets sessile or subsessile; calyx 6–6·5 mm; awns 0·3–0·5 mm. *Mountains, on base-rich rocks.* ● *S.W. Spain (Sierra Bermeja).* Hs.

17. **A. macropoda** Boiss. in DC., *Prodr.* 12: 688 (1848). Like **15** but leaves (60–)100–200 × 8–12 mm, lanceolate, acute, 3- or 5-veined; scapes (30–)40–60 cm; sheath 20–50 mm; capitula 25–30 mm wide; involucral bracts bright greenish; spikelet-bract somewhat shorter than calyx; calyx 8–10 mm. *Mountains; somewhat calcicole.* ● *C. & S. Italy.* It.

18. **A. filicaulis** (Boiss.) Boiss., *Voy. Bot. Midi Esp.* 2: 527 (1841). Herb; stock short, often with long branches. Leaves subacute, mucronulate, dimorphic; outer 10–30 × 0·7–2 mm, linear or linear-lanceolate, with narrow scarious margins, flat; inner (7–)15–60(–90) × 0·4–0·8 mm, linear to filiform, caniculate-triquetrous. Scapes 18–32 cm, straight, erect. Sheath 4–15 mm. Capitula 8–12(–16) mm; involucral bracts muticous or mucronulate, with wide scarious margins; outer much shorter than the inner, obtuse or subacute. Spikelets subsessile or shortly stipitate; bract nearly equalling calyx, exserted. Calyx 4–7 mm; tube pubescent or sometimes hirsute on ribs; awns 0·3–0·6 mm. Corolla pink or white. 2n=18. *Mountain pastures and scrub; usually calcicole.* ● *S.E. Spain, S.E. France.* Ga Hs.

A. quichiotis (González-Albo) Lawrence, *Gentes Herb.* 4: 412 (1940), described from S.E. Spain (Sierra de Alcaraz), where **18** is also found, can be separated by its wider, glaucous-violet leaves, the outer up to 3·5 mm wide, the inner up to 1·1 mm wide, its arcuate-ascending scapes, capitula 14–18(–20) mm wide, bract exceeding calyx, and calyx-lobes truncate or slightly acuminate.

A. maritima var. *profilicaulis* Bernis (*A. belgenciensis* Donadille), from S.E. France (Var), differs from **18** in having glabrous leaves, the inner being 80–140 × 0·8–1 mm and linear, and occurs on dolomite.

19. **A. trachyphylla** Lange, *Overs. Kong. Danske Vid. Selsk. Forh.* 1893: 199 (1893) (incl. *A. duriaei* auct. hisp., non Boiss.). Like **18** but outer leaves 30–50 × 0·8–2·5(–4) mm, subobtuse, muticous, sometimes withering early, the inner mucronate;

scapes 4–20 cm, usually arcuate-ascending, often verrucose; capitula 10–15(–20) mm wide; outer involucral bracts not longer than the inner, acute, shortly cuspidate or mucronate, sometimes nearly muticous; inner shortly cuspidate or muticous; spikelets sessile or subsessile; calyx-awns (0·5–)1–1·5 mm. *Dry, stony pastures; calcicole.* ● *Mountains of E. & S.E. Spain.* Hs.

Polymorphic, even within a single population; further investigation is required.

20. **A. littoralis** Willd., *Enum. Pl. Hort. Berol.* 333 (1809) (*A. duriaei* auct., non Boiss.). Herb; stock stout, with short branches having swollen nodes. Leaves 20–120 × 0·3–1·5 mm, uniform or obscurely dimorphic, linear-spathulate to filiform, subobtuse, usually muticous, with narrow scarious margins. Scapes 9–40 cm, arcuate or flexuous, often verrucose. Sheath usually 15–25 mm. Capitula 8–15 mm wide, becoming hard in fruit; involucral bracts with wide scarious margins, glabrous; outer shorter than to slightly exceeding the inner, obtuse, mucronate to shortly cuspidate; inner mucronulate. Spikelets sessile or subsessile; bract not longer than calyx. Calyx 4·5–6·5 mm; tube hirsute at least on primary ribs; awns 0·3–0·6 mm. Corolla white or pinkish. 2n=18. *Dry, stony slopes on schist; lowland.* ● *S.W. Spain, S. Portugal.* Hs Lu.

Closely related to **18**.

21. **A. duriaei** Boiss. in DC., *Prodr.* 12: 684 (1848). Like **20** but stock with few branches; leaves usually acute or subacute, usually dimorphic; outer 20–30 × 2–4 mm, linear-lanceolate, flat; inner 60–80(–160) × 0·3–1·5(–2) mm, filiform to linear-lanceolate, acuminate, more or less caniculate-convolute; scapes straight; inner involucral bracts mucronulate or muticous; spikelet-bract as long as calyx, exserted, scarious distally; calyx-tube on ribs or more or less pubescent in the upper part; awns 0·7–1·2(–1·7) mm; corolla pinkish, rarely white. *Calcifuge.* ● *Mountains of N.W. Spain and N. & C. Portugal.* Hs Lu.

22. **A. arcuata** Welw. ex Boiss. & Reuter, *Pugillus* 101 (1852). Like **20** but stock very reduced and unbranched; leaves 24–30 × 0·2–0·7 mm, filiform, subacute, mostly mucronulate, caniculate; scapes 4–5 cm, decumbent-arcuate; sheath *c.* 8 mm; outer involucral bracts mucronate, the inner obtuse, muticous; spikelet-bract as long as calyx, dark brown, with wide scarious margins, fimbriate-scarious at apex; calyx-awns 0·8–0·9 mm. *Moist pastures.* ● *Formerly in S.W. Portugal (Vila Nova de Milfontes).* †Lu.

23. **A. girardii** (Bernis) Litard. in Briq., *Prodr. Fl. Corse* 3(2): 4 (1955) (*A. juncea* Girard, non Wallr.). Herb; stock short, with short branches. Leaves usually glabrous, dimorphic; outer 12–17 × 1–1·2 mm, linear or linear-triangular, subacute, muticous, with narrow scarious margins, withering early; inner 15–40 × 0·3–0·7 mm, filiform, acute, white-pungent-mucronate, arcuate or flexuous. Scapes 10–15 cm, arcuate-ascending, glabrous. Sheath 6–10 mm. Capitula 10–15 mm; involucral bracts with wide scarious margins; outer shorter than the inner, subobtuse, shortly cuspidate; inner muticous. Spikelets subsessile; bract shorter than calyx, glabrous, white-scarious. Calyx 5–6 mm; tube densely hirsute throughout or only on ribs; awns 0·4–0·7 mm. Corolla pink. 2n=18. *Rock-crevices and stony slopes on dolomite.* ● *S.C. France.* Ga.

24. **A. eriophylla** Willk., *Bol. Soc. Brot.* 2: 145 (1884). Like **23** but outer leaves 10–35 × 1·5–2 mm, linear-spathulate, subobtuse, flat; inner often ciliate, puberulent or hispidulous at

least beneath; scapes decumbent to ascending; sheath (10–)15–28 mm; outer involucral bracts usually at least as long as the inner, acute, more or less long-cuspidate; inner mucronate; calyx-tube hirsute on ribs; awns 1·2–1·5 mm; corolla pink or white. *Dry pastures and rock-crevices on serpentine.* ● *N.E. Portugal, ?N.W. Spain.* ?Hs Lu.

25. A. humilis (Link) Schultes in Roemer & Schultes, *Syst. Veg.* **6**: 772 (1820). Dwarf shrub; stock usually much-branched. Leaves mucronulate, with narrow scarious margins, mostly ciliate, somewhat rigid, dimorphic; outer 20–30 × 0·8–2 mm, linear or linear-lanceolate, subacute or obtuse, flat; inner 10–30(–50) × 0·2–0·7 mm, filiform, subacute, becoming more or less canaliculate. Scapes 3–12 cm, ascending or flexuous. Capitula 13–16(–20) mm wide. Spikelets subsessile; bract as long as calyx, scarious; bracteoles very small. Calyx 5–7·5 mm; tube hirsute on ribs; awns (0·8–)1–1·5 mm. Corolla pale pink or white. *Pastures and rock-crevices on granite.* ● *Mountains of N.W. Portugal.* Lu.

(a) Subsp. **humilis**: Outer leaves (1–)2 mm wide, the inner 0·3–0·7 mm wide. Sheath 5–7(–10) mm. Outer involucral bracts shorter than the inner, mucronate to cuspidate, with narrow scarious margins; inner mucronulate or muticous. Calyx 6·5–7·5 mm; awns (0·8–)1–1·2 mm. $2n=18$. *Serra Amarela, Serra do Gerês.*
(b) Subsp. **odorata** (Samp.) P. Silva, *Bot. Jour. Linn. Soc.* **64**: 377 (1971) (*Statice humilis* var. *odorata* Samp.): Outer leaves 0·8–1·2 mm wide, the inner 0·2–0·4 mm wide. Sheath 8–12 mm. Involucral bracts acuminate-cuspidate, the outer equalling or longer than the inner, with wide scarious margins. Calyx 5(–7·5) mm; awns 1·3–1·5 mm. *Serra de Arga, Serra Amarela.*

26. A. sardoa Sprengel, *Syst. Veg.* **4**(2): 127 (1827). Dwarf shrub. Leaves acute or subacute, glabrous or puberulent, dimorphic; outer 25–30 × 2–4 mm, wider than scapes, linear to linear-lanceolate, flat; inner up to 40 × 1–1·2 mm, linear, somewhat convolute or canaliculate. Scapes 15–30 cm. Sheath shorter than capitulum. Capitula up to 15 mm wide, becoming hard in fruit; involucral bracts mucronulate or muticous; outer half as long as the inner, acuminate, subobtuse or acute, with narrow scarious margins; inner subobtuse, with wide scarious margins. Spikelets sessile or subsessile; bract as long as calyx. Calyx 5–6 mm; tube hirsute on ribs; awns 0·3–0·8 mm. Corolla pale pinkish to purple. $2n=18$. *Mountain pastures and rocks.* ● *Sardegna.* Sa.

27. A. leucocephala Salzm. ex Koch, *Flora* (*Regensb.*) **6**: 712 (1823). Dwarf shrub; stock stout, with short branches. Leaves 30–60(–120) × 1–2·5(–4) mm, linear or linear-lanceolate, subacute, mucronate, with narrow scarious margins, flat or somewhat canaliculate. Scapes 15–25(–40) cm, straight, sometimes verrucose. Sheath 10–15 mm. Capitula (12–)20–25 mm wide; involucral bracts few, with wide scarious margins; outer shorter than the inner, subacute, mucronate or shortly cuspidate; inner obtuse, muticous. Spikelet-stipe up to 2 mm; bract as long as calyx. Calyx (5–)6–8 mm; tube hirsute on ribs; awns (0·5–)0·8–2 mm. Corolla usually white. $2n=18$. *Mountain pastures and rocks; calcifuge.* ● *Corse.* Co.

28. A. multiceps Wallr., *Beitr. Bot.* **1**: 196 (1844). Like **27** but leaves 10–25(–60) × 0·7–1·5(–2) mm, linear, muticous; scapes 2–9(–15) cm, arcuate or flexuous; sheath 5–8 mm; capitula 10–15(–18) mm wide; outer involucral bracts ovate, obtuse, muticous or sometimes mucronulate; inner obtuse or emarginate; spikelets more shortly stipitate; calyx 5–6 mm; awns

0·3–0·5 mm; corolla usually pink. $2n=18$. *Mountain pastures and rock-crevices.* ● *Corse.* Co.

27 and **28** are recorded from many of the same localities and intermediates occur.

29. A. soleirolii (Duby) Godron in Gren. & Godron, *Fl. Fr.* **2**: 737 (1853). Dwarf shrub; stock with long branches. Leaves acute, muticous, the margins not scarious, dimorphic, papillose; outer 40–50 × 2–4 mm, linear-spathulate, flat or canaliculate, glabrous; inner 70–80 × c. 2 mm, linear or linear-lanceolate, canaliculate, glabrous or puberulent. Scapes 20–40 cm, straight, glabrous or pubescent, somewhat verrucose. Sheath 9–12 mm. Capitula (10–)15–20 mm wide; involucral bracts with narrow scarious margins; outer shorter than the inner, obtuse, muticous or mucronulate; inner muticous. Spikelets usually subsessile; bract almost as long as calyx. Calyx c. 6 mm; tube hirsute on ribs; awns 0·5–1(–1·2) mm. Corolla pinkish or white. $2n=18$. *Maritime rocks or sands.* ● *Corse* (*near Calvi*). Co.

30. A. juniperifolia (Vahl) Hoffmanns. & Link, *Fl. Port.* **1**: 442 (1813–1820) (*A. cespitosa* (Cav.) Boiss.). Dwarf shrub; stock much-branched. Leaves 4–15(–20) × 0·7–1 mm, linear, acute, white-pungent-mucronate, ciliate, puberulent on midrib, flat above but triquetrous. Scapes 0·6–2·5(–5) cm. Sheath 3–9 mm. Capitula 10–14(–17) mm wide; outer involucral bracts shorter than the inner, acuminate, mucronate or cuspidate, with narrow scarious margins; inner mucronulate or muticous, with wide scarious margins. Spikelets sessile; bract as long as calyx; bracteoles absent or very small. Calyx 5–6·5 mm; tube hirsute on ribs; awns 1 mm. Corolla pale purplish or pink. $2n=18$. *Mountain pastures and rock-crevices; calcifuge.* ● *C. Spain.* Hs.

31. A. splendens (Lag. & Rodr.) Webb, *Iter Hisp.* 18 (1838). Dwarf shrub; stock with many short branches. Leaves linear or linear-lanceolate, obtuse to acute, muticous, usually glabrous, with narrow scarious margins, flat. Sheath 3–11 mm. Involucral bracts with wide scarious margins; outer much shorter than the inner, ovate-acuminate or suborbicular, muticous or mucronulate; inner muticous or mucronulate. Spikelets usually sessile or subsessile; bract as long as calyx. Calyx-tube pubescent or hirsute on ribs only; awns 0·5–0·8 mm. Corolla bright purplish. *Mountain pastures and rock-crevices; calcifuge.* ● *N., W. & S. Spain.* Hs.

(a) Subsp. **splendens**: Leaves 5–15(–25) × 0·7–1·3 mm. Scape 3–8 cm, sinuate. Capitula 8–12 mm wide. Calyx 4·5–5·5 mm. $2n=18$. *S. Spain* (*Sierra Nevada*).
(b) Subsp. **bigerrensis** (C. Vicioso & Beltrán) P. Silva, *Bot. Jour. Linn. Soc.* **64**: 377 (1971) (*A. cespitosa* var. *bigerrensis* C. Vicioso & Beltrán): Leaves 10–30 × 1–2 mm. Scape 4–15 cm, arcuate. Capitula 11–19 mm wide. Calyx 5–7·5 mm. *N. & W. Spain.*

32. A. pubigera (Desf.) Boiss. in DC., *Prodr.* **12**: 678 (1848). Dwarf shrub, usually velutinous; stock woody, stout, much-branched. Leaves (5–)10–40(–80) × (0·7–)1·5–3(–7) mm, linear-spathulate, obtuse, muticous, flat or canaliculate. Scapes 4–15(–30) cm, erect or ascending. Sheath 6–20 mm. Capitula 15–20 mm wide; outer involucral bracts shorter than the inner, obtuse or acuminate, mucronate to cuspidate, with narrow scarious margins; inner muticous or mucronate, with wide scarious margins. Spikelets usually stipitate; bract at least as long as calyx. Calyx 6–8 mm; tube villous, sometimes only on ribs; awns up to 0·8 mm. Corolla pink to white. *Maritime rocks.* ● *N.W. Spain, N.W. Portugal.* Hs Lu.

A species showing great variation, even within small populations, in the pubescence of the leaves, scapes, sheath and bracts, and in the size of the leaves, scapes and capitula. It has been reported from S.W. France, Ireland and Scotland, but the records require confirmation.

Intermediates between **32** and **1** occur in intermediate habitats when the species are sympatric.

33. A. berlengensis Daveau, *Bol. Soc. Geogr. Lisboa* 4(9): 426 (1884). Compact dwarf shrub, glabrous or velutinous. Leaves 50–60 × 5–10(–13) mm, lanceolate to obovate-lanceolate, acute, mostly mucronate, with narrow scarious margins, flat. Scapes 10–20 cm. Sheath 20–30 mm. Capitula 20–30 mm wide; outer involucral bracts at least as long as the inner, acuminate, mucronate, with narrow scarious margins; inner mucronate, with wide scarious margins. Spikelets stipitate; bract at least as long as calyx. Calyx 8–9 mm; tube hirsute throughout or only on ribs; awns up to 0·6 mm. Corolla deep pink to white. *Maritime slopes; calcifuge.* ● *W.C. Portugal (Berlengas Is.).* Lu.

Sect. MACROCENTRON Boiss. Calyx-tube up to 2(–2½) times as long as spur; insertion-scar of calyx narrowly elliptical to linear.

34. A. welwitschii Boiss. in DC., *Prodr.* **12**: 676 (1848). Compact dwarf shrub; stock with many long branches. Leaves 50–100(–150) × (1–)2–7 mm, linear-lanceolate to linear-spathulate, acute or subacute, usually muticous, flat or somewhat convolute. Scapes 13–25(–35) cm, straight, sometimes verrucose. Sheath 25–35(–55) mm. Capitula 20–30 mm wide; outer involucral bracts as long as to much longer than the inner, acuminate, mucronate to cuspidate; inner mucronulate. Spikelets subsessile; bract at least as long as calyx, glabrous; bracteoles more or less developed. Calyx 8–10 mm; tube hirsute on ribs; awns up to 0·5(–0·8) mm. Corolla pink to white. 2*n*=18. *Maritime calcareous rocks and sands.* ● *W.C. Portugal (from Cabo Mondego to Cascais).* Lu.

Polymorphic mainly in width of leaves, size of outer involucral bracts and pubescence. Velutinous forms described as **A. cinerea** Boiss. & Welw. in Boiss. & Reuter, *Pugillus* 101 (1852), found only in the northern part of the range of the species (from Nazaré to Cabo Carvoeiro), alone or with the glabrous form, may perhaps be worthy of subspecific rank.

Plants of *A. welwitschii* var. *platyphylla* Daveau, from the northern part of the range of the species, look like small plants of **33**.

35. A. pungens (Link) Hoffmanns. & Link, *Fl. Port.* **1**: 439 (1813–1820) (*A. fasciculata* (Vent.) Willd.). Like **34** but involucral bracts usually muticous, the outer shorter than the inner, obtuse; spikelet-bract absent or rudimentary; calyx-tube usually hirsute throughout. 2*n*=18. *Maritime sands.* ● *S. Portugal and S.W. Spain; Corse and Sardegna.* Co Hs Lu Sa.

Var. *velutina* Coutinho, with longer, narrower, more or less puberulent leaves and longer scapes, known from S.W. Portugal, could perhaps have originated by introgression with **36**.

36. A. pinifolia (Brot.) Hoffmanns. & Link, *op. cit.* 437 (1813–1820). Puberulent herb; stock short, rather stout. Leaves 90–160 × 0·5–1 mm, filiform, canaliculate, acute, mucronate. Scapes 30–50 cm, straight. Sheath 30–40 mm. Capitula 20–25 mm wide; involucral bracts thin, with narrow scarious margins, puberulent; outer much shorter than the inner, more or less cuspidate; inner mucronate. Spikelets sessile; bract and bracteoles absent or rudimentary. Calyx 9–11 mm; tube usually hirsute throughout; awns 0·8–1 mm. Corolla white or pink. 2*n*=18. *Scrub on sandy or gravelly soil; calcifuge.* ● *C. & S.W. Portugal.* Lu.

37. A. rouyana Daveau, *Bol. Soc. Brot.* **6**: 166 (1889). Like **36** but leaves sometimes up to 2·5 mm wide, linear-lanceolate; involucral bracts coriaceous, with wide scarious margins, the outer mucronate; spikelet-bract longer than calyx, coriaceous, broadly scarious at the apex; calyx 6·5–7·5(–8·5) mm; awns absent or up to 0·4 mm. 2*n*=18. *Scrub on sandy soil; calcifuge.* ● *S.W. Portugal.* Lu.

Dwarf shrubs with a long-branched stock, linear-lanceolate leaves 40–70 × 1·3–2·5 mm, scapes 20–35 cm, sheath 15–20 mm and calyx 8–8·5 mm, which are known only from maritime sands near Sines, were described as f. *littorea* (Bernis) Bernis and perhaps originated by introgression with **35**.

38. A. macrophylla Boiss. & Reuter, *Pugillus* 100 (1852). Like **36** but leaves (80–)120–250 mm, filiform to sublinear; scapes glabrous, or puberulent at base; involucral bracts with wide scarious margins, glabrous, the outer obtuse or emarginate, mucronulate, the inner muticous; calyx-tube hirsute on ribs. 2*n*=18. *Scrub on sandy or gravelly soil.* ● *S. Portugal, S.W. Spain.* Hs Lu.

39. A. velutina Welw. ex Boiss. & Reuter, *loc. cit.* (1852). Velutinous herb; stock short, stout, with few branches. Leaves 80–120 × 2·5–8 mm, lanceolate-spathulate, acute, mucronate, flat. Scapes 30–50 cm, straight, velutinous or glabrous above, numerous. Sheath 18–30 mm. Capitula 20–30 mm wide; outer involucral bracts shorter than the inner, rather shortly cuspidate, with narrow scarious margins; inner mucronate or muticous, with wider scarious margins. Spikelets more or less stipitate; bract as long as calyx, with scarious margins; bracteoles 4–8 mm. Calyx 8–9·5 mm; tube pubescent; awns up to 0·2 mm. Corolla pink or white. *Scrub on dry acid sand.* ● *S.E. Portugal, S.W. Spain.* Hs Lu.

40. A. gaditana Boiss. in DC., *Prodr.* **12**: 675 (1848). Glabrous herb; stock short, stout. Leaves 100–250 × (10–)15–35 mm, lanceolate to obovate-spathulate, acute, with narrow scarious margins. Scapes 50–100 cm, straight. Sheath 40–80 mm. Capitula 30–40 mm wide; involucral bracts with wide scarious margins; outer at least as long as the inner, acuminate to long-cuspidate; inner mucronate. Spikelets sessile or subsessile; bract and bracteoles absent or rudimentary. Calyx 10–12 mm; tube hirsute on ribs; awns *c.* 1 mm. Corolla usually deep pink. *Estuarine meadows.* ● *S. Portugal, S.W. Spain.* Hs Lu.

41. A. hirta Willd., *Enum. Pl. Hort. Berol.* 333 (1809) (*A. boetica* Boiss., *A. boissierana* Cosson). Herb; stock short. Leaves 50–130 × (1–)2–10 mm, lanceolate-spathulate to linear-lanceolate, obtuse, glabrous or puberulent. Scapes up to 65 cm, straight, glabrous, smooth or verrucose. Sheath 25–50 mm, glabrous. Capitula 20–30 mm wide; outer involucral bracts shorter than the inner, obtuse, rarely mucronate; inner with wide scarious margins, mucronulate. Spikelets sessile or subsessile; bract absent or rudimentary; bracteoles more or less developed. Calyx 10–11 mm; tube hirsutulous on the ribs; awns 0·8 mm. Corolla pink or white. *Sandy lowlands.* ● *S. Spain.* Hs.

42. A. hispalensis Pau, *Not. Bot. Fl. Esp.* **6**: 89 (1895). Like **41** but leaves *c.* 30 × 0·7(–1·5) mm, filiform; scapes up to 25 cm, somewhat arcuate; sheath 13–18 mm; capitula *c.* 15 mm wide;

involucral bracts not mucronate, paler, thinner; calyx-tube as long as spur, scarcely pubescent on ribs; awns 0·4 mm. *Calcareous hills.* ● *S.W. Spain (between Sevilla and Jerez de la Frontera).* Hs.

43. A. cariensis Boiss. in DC., *Prodr.* **12**: 677 (1848). Herb; stock short. Leaves dimorphic, linear, acute, with very narrow scarious margins, glabrous; outer 50 × 3 mm, flat; inner 70 × 1 mm. Scapes 20–35 cm, straight, glabrous. Sheath *c.* 22 mm. Capitula *c.* 22 mm wide; outer involucral bracts shorter than the inner, acuminate, with narrow scarious margins; inner with wide scarious margins, mucronate or muticous. Spikelets long-stipitate; bract shorter than calyx, scarious; bracteoles well-developed. Calyx 10–12 mm; tube hirsute on primary ribs, rarely also on secondary ones; awns 0·7–2 mm. Corolla white or pinkish. *N.E. Greece (Thraki, between Arvas and Aisymi).* Gr. (*W. Anatolia.*)

5. Limonium Miller[1]

Perennial, rarely annual, herbs or dwarf shrubs. Leaves simple, usually in a basal rosette, but densely leafy branches sometimes present; leaves often absent at anthesis. Inflorescence a corymbose panicle, with terminal, secund spikes, often with non-flowering branches, usually with a reddish scale at the base of each branch. Spikes of 3-bracteate, 1- to 5-flowered spikelets; inner and outer bracts external to the spikelet, the middle one internal and often inconspicuous. Calyx infundibuliform; limb scarious, usually coloured, sometimes shortly dentate between the lobes. Corolla with a short tube, or the petals connate only at the base. Stamens inserted at the base of the corolla. Styles 5, glabrous, free or connate at the base; stigmas filiform. Fruit with circumscissile or irregular dehiscence.

Measurements of scales refer to the largest, which subtends the lowest branch of the inflorescence. Descriptions give the average number of spikelets per cm, but generally spikes are described as dense if the spikelets are closely set and contiguous, and lax if the spikelets are more or less separated. The corolla colour is violet unless otherwise indicated.

The taxonomically important non-flowering branches, which are present in many species, completely lack flowers and have only a single, very reduced scale (usually 0·5–1 mm) at the apex. They should not be confused with the sterile branches, which occur in a few species at the end of the summer or under unfavourable conditions, and which have abortive spikelets and bear numerous, imbricate, sterile floral bracts (usually 1–3 mm) at the apex.

Most species of *Limonium* exhibit dimorphism in pollen and stigmas associated with self-incompatibility. From this system there has apparently been derived monomorphic self-compatibility, known in Europe for **18** and **87**, and monomorphic agamospermy, known for **36**, **51**, **60**, **61**, **64–70** and **77** (H. G. Baker, *Evolution* **20**: 349–368 (1966)). Agamospermic plants have high pollen-sterility and some populations of **66**, **70** and **72** are completely male-sterile. The occurrence of agamospermy in parts of subgen. *Limonium* undoubtedly accounts for some of its taxonomic difficulty, particularly in view of the occurrence in disjunct coastal habitats of populations showing relatively small morphological differences.

1 Leaves sinuate or lobed; spikes usually subtended by leaf-like wings
  2 Calyx-limb yellow   **2. bonduellei**
  2 Calyx-limb pale blue, violet or white

¹ By S. Pignatti.

3 Annual; stem angular or subcylindrical, not winged; calyx-limb with acute lobes   **3. thouinii**
3 Perennial; stem with wings up to 20 mm wide; calyx-limb entire or crenate
  4 Inner bracts with 2–3 almost spiny apical mucros; wings of the stem not more than 3 mm wide   **1. sinuatum**
  4 Inner bracts acute, not mucronate; wings of the stem 5–20 mm wide   **4. brassicifolium**
1 Leaves entire or absent; flowering branches not winged
  5 Corolla with short tube; leaves usually absent at anthesis
    6 Calyx completely enclosed by inner bract; stems green, not more than 1·5 mm in diameter, flexible
      7 Basal scales of the flowering spikes 4–5 mm, with reddish limb and apical awn 1–2 mm   **5. ferulaceum**
      7 Basal scales of the flowering spikes 1–2·5 mm, with scarious or silvery limb, muticous or mucronulate   **6. diffusum**
    6 Calyx-limb exceeding inner bract by 2–3 mm; stems farinose or white, 1·7–3 mm in diameter, rigid
      8 Corolla 7–8 mm, the limb 5–7 mm in diameter   **7. caesium**
      8 Corolla 10–14 mm, the limb 8–10 mm in diameter   **8. insigne**
  5 Petals connate at base only; at least dead leaves present at anthesis
    9 Annuals, entirely herbaceous
      10 Plant flowering in April–June; inner bract tuberculate dorsally   **87. echioides**
      10 Plant flowering later; inner bract smooth   **59. duriusculum**
    9 Perennials with woody stock
      11 Leaves pinnately veined
        12 Stems, at least the upper part, and leaf-midribs hairy
          13 Lower branches of the inflorescence sterile; spikes 1·8–2·5 cm, lax   **11. sareptanum**
          13 Inflorescence with few or no sterile branches; spikes not more than 1·2(–2) cm, dense
            14 Leaves 250–600 × 80–150 mm; outer bract entirely hyaline   **9. latifolium**
            14 Leaves 50–130(–250) × 10–40(–90) mm; outer bract with narrow hyaline margin
              15 Leaves not more than 13 mm wide, the lamina 6–8 times as long as wide   **10. asterotrichum**
              15 Leaves 20–40(–90) mm wide, the lamina 2–5 times as long as wide   **16. tomentellum**
        12 Stems and leaves glabrous
          16 Lower branches of the inflorescence sterile   **12. bungei**
          16 Inflorescence with few or no sterile branches
            17 Leaves 250–600 × 80–150 mm; outer bract hyaline   **9. latifolium**
            17 Leaves 70–170(–300) × 13–60 mm; outer bract herbaceous, at least on the midrib
              18 Calyx 3·6–6·5 mm; corolla 6–8 mm; inner bract 3–5 mm; outer bract overlapping 1/6–1/3 of the inner
                19 Spikes 1–2 cm, patent, with 6–8 spikelets per cm   **17. vulgare**
                19 Spikes 3–5 cm, erect, with 2–3 spikelets per cm   **18. humile**
              18 Calyx 3–4·5 mm; corolla 5–5·5 mm; inner bract (1·5–)2–3 mm; outer bract overlapping 1/4–2/3 of the inner
                20 Spikes 0·6–1 cm   **13. gmelinii**
                20 Spikes 1·5–3 cm
                  21 Calyx densely hairy on the veins; scales 10–16 mm   **15. hirsuticalyx**
                  21 Calyx sparsely hairy on the veins or subglabrous; scales 6–9 mm   **14. meyeri**
      11 Leaves with single or parallel veins
        22 Inflorescence with non-flowering branches
          23 Outer bract hyaline, herbaceous only on the keel; inner bract hyaline for at least (1/4–)1/3 of its length   **19. bellidifolium**
          23 Bracts herbaceous, with or without narrow (0·3–0·5 mm) hyaline margin
            24 Segments of the inflorescence-branches strongly constricted at the nodes

25 Inflorescence 5–15(–20) cm, the branches coarsely tuberculate; leaves mostly withered at anthesis
    29. articulatum

25 Inflorescence 20–40 cm, the branches smooth; leaves green at anthesis   **41. melium**

24 Segments of the inflorescence-branches not constricted at the nodes

26 Stems, branches and bracts hairy

27 Calyx *c.* 5·5 mm   **27. calcarae**

27 Calyx 3·5–5 mm

  28 Leaves withered at anthesis

    29 Inner bract 2–2·5 mm; calyx 3·5–4(–4·5) mm   **26. dichotomum**

    29 Inner bract 4–4·5 mm; calyx 4·5–5 mm

      30 Inflorescence hemispherical, with patent to deflexed branches; spikelets erecto-patent   **24. furfuraceum**

      30 Inflorescence corymbose, with patent branches; spikelets patent   **25. lucentinum**

  28 Leaves green at anthesis

    31 Inflorescence with reticulate branching, the branches mostly diverging at an angle of 90°–180°

      32 Leaves 5–20, 3–5(–8) mm wide, forming a lax rosette   **20. cancellatum**

      32 Leaves at least 20, 5–10 mm wide, forming a dense, cushion-like rosette   **21. vestitum**

    31 Inflorescence with acutely dichotomous branching

      33 Inflorescence 3–14 cm; leaves not more than 15 mm

        34 Inner bract acute; calyx 5–5·5 mm, exceeding the inner bract by *c.* 2 mm   **31. minutum**

        34 Inner bract truncate; calyx 3·5–5 mm, exceeding the inner bract by *c.* 1 mm   **32. caprariense**

      33 Inflorescence 15–50 cm; leaves 15–60 mm

        35 Outer bract overlapping ⅓–½ of the inner; spikelets 3- to 4-flowered   **66. dufourei**

        35 Outer bract overlapping ¼–⅓ of the inner; spikelets 1- to 2-flowered

          36 Inflorescence pyramidal; spikes rudimentary or 1–2 cm, with 2–4 spikelets per cm   **22. johannis**

          36 Inflorescence corymbose; spikes not more than 1 cm, with 7 spikelets per cm   **23. cordatum**

26 Stems, branches and bracts glabrous, sometimes tuberculate

37 Calyx 3–3·5 mm

  38 Leaves acute, green at anthesis

    39 Leaves linear, 0·5–1 mm wide   **35. aragonense**

    39 Leaves spathulate, 5–6 mm wide   **(75–79). delicatulum** group

  38 Leaves obtuse, green or withered at anthesis

    40 Leaves 1–7 mm wide, 1-veined

      41 Leaves 1–1·5(–4) mm wide; spikes not more than 4 cm   **45. hermaeum**

      41 Leaves 4–7 mm wide; spikes not more than 1 cm   **44. tenuiculum**

    40 Leaves 7–14 mm wide, 1- to 3-veined

      42 Inner bract *c.* 2 mm; inflorescence hemispherical; leaves usually withered at anthesis   **85. supinum**

      42 Inner bract *c.* 3 mm; inflorescence pyramidal; leaves green at anthesis   **(75–79). delicatulum** group

37 Calyx (3·8–)4–7 mm

43 Inflorescence 2–12 cm; spikes 0·5–1·8(–3) cm

  44 Leaves on densely leafy, woody basal branches 4–12 cm long

    45 Leaves 2–3 mm wide, narrowly spathulate; inner bract cucullate   **33. parvifolium**

    45 Leaves *c.* 1 mm wide, linear; inner bract not cucullate   **34. acutifolium**

  44 Leaves in cushion-like rosettes on somewhat woody branches 1–2 cm long

46 Branches tuberculate, with segments diverging at 90°; bracts tuberculate; calyx *c.* 4 mm   **30. japygicum**

46 Branches smooth, with segments diverging at an acute angle; bracts smooth; calyx 5–5·5 mm   **31. minutum**

43 Inflorescence 10–60 cm; spikes, at least the terminal, (1–)2–9(–12) cm

47 Inner bract with the midrib not or slightly excurrent, the margin reddish-brown, not or only narrowly hyaline

  48 Leaves 3-veined   **60. ramosissimum**

  48 Leaves 1-veined

    49 Spikes 1–2 cm or ±rudimentary, with 6–10 spikelets per cm; inner bract truncate   **(67–71). binervosum** group

    49 Spikes 2–5(–8) cm, with 1–4 spikelets per cm; inner bract obtuse

      50 Spikes 3–5(–8) cm, with 1–3 spikelets per cm; spikelets straight, cylindrical; calyx narrowly conical after pollination, with straight tube   **55. graecum**

      50 Spikes 2–4 cm, with 4 spikelets per cm; spikelets arcuate, compressed; calyx infundibuliform after pollination, with curved tube   **56. oleifolium**

47 Inner bract with the midrib excurrent into a conspicuous acute point, the margin hyaline

  51 Inner bract obtuse; leaves usually 3- to 5-veined

    52 Spikes 1–2 cm, with 4–5 spikelets per cm; calyx *c.* 4·5 mm   **63. parvibracteatum**

    52 Spikes 2–6(–12) cm, with 1–3 spikelets per cm; calyx 5–6 mm   **28. catalaunicum**

  51 Inner bract acute; leaves 1-veined

    53 Spikes 2–6(–10) cm, with 1–2 spikelets per cm; inner bract 3–3·5 mm   **46. remotispiculum**

    53 Spikes 1–5 cm, with 3–7 spikelets per cm; inner bract 3·5–4·5 mm

      54 Calyx 5·5 mm, exceeding the inner bract by 2–2·5 mm; plant tuberculate   **43. calaminare**

      54 Calyx 4–5 mm, exceeding the inner bract by 1·5 mm; plant not tuberculate

        55 Calyx-limb ⅓–⅔ as long as the tube; non-flowering branches numerous   **(36–40). cosyrense** group

        55 Calyx-limb as long as the tube; non-flowering branches few

          56 Spikelets 1-flowered; outer bract overlapping ⅓ of the inner; calyx-teeth acute, conspicuously veined   **42. tremolsii**

          56 Spikelets 2-flowered; outer bract overlapping ⅓ of the inner; calyx-teeth obtuse, indistinctly veined   **48. inarimense**

22 Inflorescence without non-flowering branches, sometimes with a few basal sterile branches

57 Leaves with 3–7(–9) parallel veins near base

  58 Inner bract 2–3·5 mm; calyx 2·5–4 mm

    59 Petiole 4–6 mm wide   **74. ovalifolium**

    59 Petiole 1–4 mm wide

      60 Plant 80–120 cm; leaves 80–125 mm, coriaceous   **81. eugeniae**

      60 Plant 8–80 cm; leaves usually not more than 80 mm, usually herbaceous

        61 Spikes not more than 1 cm   **84. cymuliferum**

        61 Spikes at least 1 cm

          62 Leaves whitish   **83. album**

          62 Leaves green or glaucous

            63 Spikes 2–3·2 cm, lax, with 3(–4) spikelets per cm; inner bract 2–2·2 mm   **82. coincyi**

            63 Spikes 1–2·5(–3) cm, dense, with 4–10 spikelets per cm; inner bract 2·5–3·5 mm   **(75–79). delicatulum** group

58 Inner bract usually 4–7 mm; calyx (4–)4·5–8 mm

64 Spikes (2–)3–10 cm, lax, with 1–5 spikelets per cm
65 Leaves linear-spathulate, 5–10 times as long as wide; scale 6–7(–20) mm **58. emarginatum**
65 Leaves obovate-spathulate to cuneate, $1\frac{1}{2}$–3 times as long as wide; scale 2–3 mm **(52–54). albidum** group
64 Spikes 0·7–3(–4) cm, dense, with 4–15 spikelets per cm
66 Leaves 50–100 mm, 4–5 mm wide at base, often with resinous secretions **72. auriculae-ursifolium**
66 Leaves 20–50(–100) mm, 1–2(–3) mm wide at base, without resin
67 Outer bract not more than 2 mm, overlapping not more than $\frac{1}{4}$ of the inner **60. ramosissimum**
67 Outer bract at least 2 mm, overlapping at least $\frac{1}{3}$ of the inner
68 Leaves with revolute margin **73. majoricum**
68 Leaves not revolute
69 Calyx conical or slightly infundibuliform, the limb $\frac{2}{3}$ as long to as long as the tube **(67–71). binervosum** group
69 Calyx infundibuliform, the limb $\frac{1}{3}$ as long as the tube
70 Plant 8–20 cm; spikes 0·7–1·3 cm; inner bract as long as wide, the midrib excurrent for 0·2–0·4 mm **64. girardianum**
70 Plant 15–40 cm; spikes up to 2·5 cm; inner bract $1\frac{1}{2}$ times as long as wide, the midrib excurrent for 0·8–1·6 mm **65. densiflorum**
57 Leaves 1-veined near base
71 Leaves enlarged at base into an auriculate sheath; spikes sessile **86. suffruticosum**
71 Leaf-sheath without auricles; spikes pedunculate
72 Leaves acute
73 Spikes 3–5 cm, with 2–3 spikelets per cm **55. graecum**
73 Spikes 0·6–2·5(–4) cm, with 4–15 spikelets per cm
74 Inner bract not more than 3 mm; calyx not more than 3·5 mm
75 Spikelets 2- to 3-flowered; leaves 5–6 mm wide, green **(75–79). delicatulum** group
75 Spikelets 1-flowered; leaves *c.* 10 mm wide, glaucous **80. sibthorpianum**
74 Inner bract at least 3·5 mm; calyx at least 4 mm
76 Leaves not more than 5 mm wide
77 Spikes 0·6–0·8 cm, with 2–3 spikelets **(67–71). binervosum** group
77 Spikes 1–1·5 cm, usually with 8–12 spikelets **50. laetum**
76 Leaves at least 8 mm wide
78 Spikes with 4–8 spikelets per cm; outer bract overlapping not more than $\frac{1}{4}$ of the inner **60. ramosissimum**
78 Spikes with 10–15 spikelets per cm; outer bract overlapping $\frac{1}{3}$–$\frac{1}{2}$ of the inner **64. girardianum**
72 Leaves obtuse
79 Leaves 1–2 mm wide, linear **57. carpathum**
79 Leaves 3–20 mm wide, ±spathulate
80 Leaves on woody basal branches 2–8 cm
81 Spikes 0·9–1·3 cm, with 10–13 spikelets per cm **49. tenoreanum**
81 Spikes 1–4 cm, with 4–6 spikelets per cm
82 Calyx-limb $1\frac{1}{2}$ times as long as the tube; inflorescence-branches straight **47. minutiflorum**
82 Calyx-limb as long as the tube; inflorescence-branches arcuate **48. inarimense**
80 Leaves in basal rosettes
83 Spikes 2–8 cm, with 2–5 spikelets per cm
84 Leaves obovate-spathulate, $1\frac{1}{2}$–3 times as long as wide; calyx-teeth not uncinate in fruit **(52–54). albidum** group
84 Leaves oblanceolate-spathulate, 4–6 times as long as wide; calyx-teeth becoming uncinate in fruit **59. duriusculum**
83 Spikes 1–4 cm, with 4–9 spikelets per cm
85 Calyx 3·2–3·5 mm
86 Leaves obovate-spathulate, $1\frac{1}{2}$–2 times as long as wide; calyx-limb $\frac{2}{3}$ as long as the tube **(75–79). delicatulum** group
86 Leaves oblanceolate-spathulate, 3–5 times as long as wide; calyx-limb $1\frac{1}{2}$ times as long as the tube
87 Leaves green; spikes with 4 spikelets per cm **61. ocymifolium**
87 Leaves whitish-glaucous; spikes with 6–7 spikelets per cm **62. densissimum**
85 Calyx at least 4 mm
88 Inner bract *c.* 3 mm **63. parvibracteatum**
88 Inner bract at least 4 mm
89 Calyx conical, gradually widening from the base; outer bract overlapping $\frac{1}{3}$–$\frac{1}{2}$ of the inner **(67–71). binervosum** group
89 Calyx infundibuliform, abruptly widening in apical $\frac{1}{4}$–$\frac{1}{3}$; outer bract overlapping $\frac{1}{6}$–$\frac{1}{4}$ of the inner
90 Plant 10–20 cm; leaves 3–9 mm wide **51. gougetianum**
90 Plant 20–50 cm; leaves 9–20 mm wide **60. ramosissimum**

Subgen. **Pteroclados** (Boiss.) Pignatti. Annual or perennial herbs. Leaves sinuate or lobed. Ultimate branches of the inflorescence usually expanded into a leaf-like wing subtending the spike. Fruit with distal circumscissile dehiscence.

1. L. sinuatum (L.) Miller, *Gard. Dict.* ed. 8, no. 6 (1768) (*Statice sinuata* L.). Perennial 20–40 cm. Leaves 30–100 × *c.* 15 mm, with 4–7 rounded lobes on each side, hispid. Flowering stems with 3–4 wings; wings 1–3 mm wide, sometimes narrower. Inflorescence compact. Spikes dense; wing 15 × 4 mm, 3-dentate; spikelets 2-flowered. Inner bract 7 mm, with 2–3 pungent, apical spines. Calyx 10–12 mm; tube *c.* 7 mm, glabrous; limb *c.* 4 mm, entire, white or pale violet. Corolla yellowish or pink, purple when dry. *Dry, sandy places. Mediterranean region, S. Portugal.* Al Bl Co Cr Ga Gr Hs It Ju Lu Sa Si.

2. L. bonduellei (Lestib.) O. Kuntze, *Rev. Gen. Pl.* **2**: 395 (1891). Like **1** but annual 10–30 cm; basal leaves sparsely hairy; flowering stems subcylindrical, not winged or the wings less than 1 mm wide; calyx *c.* 5 mm, the tube hispid, the limb 5–5·5 mm, yellow. *Dry places. N. Spain (prov. Burgos).* Hs [It]. (*N. Africa.*)

3. L. thouinii (Viv.) O. Kuntze, *op. cit.* 396 (1891) (*Statice thouinii* Viv.). Annual 10–20 cm. Leaves 30–60 × 7–20 mm, with 5–7 lobes on each side. Flowering stems angular or subcylindrical. Inflorescence compact. Spikes with a 3-dentate wing *c.* 20 × 5 mm; spikelets 2-flowered. Inner bract *c.* 6 mm. Calyx *c.* 10 mm; tube *c.* 5 mm, glabrous; limb *c.* 5 mm, *c.* 9 mm in diameter, with acute lobes, pale blue or whitish. Corolla yellow. *Dry places near the sea. S. Spain.* Hs ?Lu [Ga]. (*N. Africa, S.W. Asia.*)

4. L. brassicifolium (Webb & Berth.) O. Kuntze, *op. cit.* 395 (1891). Perennial 20–40 cm. Stems with 2 wings, at least above; wings 5–20 mm wide. Leaves 10–30 cm, with 3–4 lobes on each side; lateral lobes *c.* 1 cm, the terminal 5–15 × 4–11 cm, ovate. Spikelets 1-flowered. Inner bracts *c.* 4 mm, cylindrical, truncate. Calyx *c.* 10 mm; tube *c.* 4 mm, glabrous; limb *c.* 6 mm, crenate, pale bluish-violet. Corolla white. *Cultivated for ornament and occasionally naturalized in S.W. Europe.* [?Ga Hs.] (*Islas Canarias.*)

Subgen. **Myriolepis** (Boiss.) Pignatti. Densely branched dwarf shrubs with many non-flowering branches. Leaves entire, usually absent at anthesis. Corolla with short tube. Fruit with irregular dehiscence.

5. L. ferulaceum (L.) O. Kuntze, *Rev. Gen. Pl.* **2**: 395 (1891) (*Statice ferulacea* L.). Stems 10–40 cm, up to 1·5 mm in diameter, green, flexible, with reddish-brown scales at the base. Leaves absent at anthesis. Flowering branches regularly alternate, with flowers only towards apex; scales 4–5 mm, reddish, with apical awn 1–2 mm. Spikes clavate; spikelets 1-flowered. Inner bract 4–5 mm, truncate, surrounded by numerous shorter, aristate bracts. Calyx 3–3·5 mm, cylindrical, enclosed by inner bract. Corolla 5–6 mm, pink. *W. & C. Mediterranean region, C. & S. Portugal.* Bl Ga Hs Ju Lu Si.

6. L. diffusum (Pourret) O. Kuntze, *loc. cit.* (1891) (*Statice diffusa* Pourret). Like **5** but with some leaves present at anthesis; flowering branches with flowers along their length; lower branches often entirely covered by regularly alternate scales, the scales scarious or silvery, muticous or mucronulate; spikes with more or less distant spikelets; inner bract *c.* 3·5 mm; outer bract *c.* 2 mm, acute, muticous; corolla pale violet. *Maritime sands and salt-marshes.* ● *S. Portugal, S.W. Spain; S. France.* Ga Hs Lu.

7. L. caesium (Girard) O. Kuntze, *loc. cit.* (1891) (*Statice caesia* Girard). Leaves spathulate, 1(–3)-veined. Inflorescence 20–60 cm; non-flowering branches 5–15 cm, 2 mm in diameter, white, often farinose, rigid; scales 2–3 mm, the upper white. Spikes *c.* 1 cm; spikelets 1-flowered. Inner bract *c.* 2·5 mm; outer bract *c.* 0·5 mm. Calyx 4·5–5 mm, exceeding inner bract by 2–3 mm. Corolla 7–8 mm, pinkish; tube shorter than the calyx; limb 5–7 mm in diameter. $2n=16$. *Dry places.* ● *S.E. Spain.* Hs.

8. L. insigne (Cosson) O. Kuntze, *loc. cit.* (1891) (*Statice insignis* Cosson). Like **7** but more robust, rigid and fragile; leaves 40–50 × 15 mm, 1(3–5)-veined, emarginate; scales brown; spikes up to 3 cm, linear; corolla 10–14 mm, purple, the tube exceeding the calyx by 1–3 mm, the limb 8–10 mm in diameter. $2n=18$. *Dry places.* ● *S.E. Spain.* Hs.

(a) Subsp. **insigne**: Plant 20–40(–60) cm, greenish. Spikelets (1–)2-flowered. Calyx 6·5(–8) mm. Inner bract 4–5 mm. *Provinces of Almería and Granada.*

(b) Subsp. **carthaginiense** Pignatti, *Collect. Bot.* (*Barcelona*) **6**: 295 (1962): Plant 20–60 cm, whitish. Spikelets 1(–2)-flowered. Calyx 5–6 mm. Inner bract *c.* 3 mm. *Near Cartagena.*

Subgen. **Limonium.** Perennial herbs, rarely annuals or dwarf shrubs. Leaves entire, present at anthesis. Petals connate at base only. Fruit with irregular dehiscence.

9. L. latifolium (Sm.) O. Kuntze, *Rev. Gen. Pl.* **2**: 395 (1891) (*Statice latifolia* Sm.). Plant 50–80(–100) cm, usually hairy. Leaves 250–600 × 80–150 mm, spathulate to elliptical, sparsely hairy, pinnately veined. Sterile branches few or absent; scales 5–6 mm. Spikes up to 1·5(–2) cm, dense; spikelets 1(–2)-flowered. Inner bract 2 mm, with hyaline margin 0·7–1 mm wide; outer bract hyaline, overlapping $\frac{1}{6}$ of the inner. Calyx 3–3·5 mm, with 5 short, rounded lobes. Corolla *c.* 6 mm, pale violet. *Steppes and dry grassland. S.E. Europe, from E. Bulgaria to S.E. Russia.* Bu Rm Rs (C, W, K, E).

10. L. asterotrichum (Salmon) Salmon, *Jour. Bot.* (*London*) **62**: 336 (1924) (*Statice asterotricha* Salmon). Plant 40–65 cm, densely pubescent. Leaves 100–180 × 10–13 mm, linear-spathulate, acute, mucronate, pinnately veined; petiole $\frac{2}{3}$–$\frac{3}{4}$ as long as lamina, with stellate hairs. Sterile branches absent; scales 6–12 mm. Spikes up to 1·5 cm, dense; spikelets 1(–2)-flowered. Bracts with narrow hyaline margin; inner bract 2·2–2·8 mm, 2-fid. Calyx *c.* 4 mm, with 5 acute lobes and shorter intermediate ones. Corolla pale violet. *Dry, saline grassland.* ● *C. & S.E. Bulgaria.* Bu.

11. L. sareptanum (A. Becker) Gams in Hegi, *Ill. Fl. Mitteleur.* **5**(3): 1880 (1927). Plant 20–60 cm, densely puberulent. Leaves 100–150 × 20–50 mm, obovate-spathulate, pinnately veined. Sterile branches usually present; scales 6–8 mm. Spikes 1·8–2·5 cm, lax; spikelets 1(–4)-flowered. Inner bract *c.* 2·5 mm, with hyaline margin *c.* 0·7 mm wide. Calyx *c.* 4·5 mm. Corolla reddish. *Steppes. S. & E. Ukraine, S.E. & S.C. Russia, W. Kazakhstan.* Rs (C, W, K, E).

12. L. bungei (Claus) Gamajun., *Izv. Kazakhsk. Fil. Akad. Nauk SSSR* (*Ser. Bot.*) **1**: 10 (1944). Like **11** but glabrous; leaves 50–150 × 30–70 mm; spikes 0·8–1·8 cm, spikelets (1–)2- to 3-flowered. *Steppes. E. Ukraine, S.E. Russia, W. Kazakhstan.* Rs (C, W, E).

13. L. gmelinii (Willd.) O. Kuntze, *Rev. Gen. Pl.* **2**: 395 (1891) (*Statice gmelinii* Willd.). Plant 20–60 cm, glabrous. Leaves 70–110(–300) × 20–30(–60) mm, spathulate, indistinctly tuberculate, pinnately veined. Sterile branches few or absent; scales up to 25 mm. Spikes 0·6–1 cm, dense; spikelets 1(–2)-flowered. Bracts densely tuberculate; inner bracts *c.* 3 mm, emarginate; outer bract hyaline except on the midrib, overlapping $\frac{1}{3}$–$\frac{2}{5}$ of the inner. Calyx *c.* 4 mm, sparsely pubescent. Corolla 5–5·5 mm, reddish. $2n=18, 27, 36$. *Alkaline soils. E.C. & S.E. Europe.* Bu Cz ?Gr Hu Ju Rm Rs (C, W, K, E) Tu.

L. hypanicum Klokov in Kotov & Barbarich, *Fl. RSS Ucr.* **8**: 524 (1957), from S.W. Ukraine and Moldavia, with subsessile leaves and calyx-hairs up to 0·3 mm, and **L. hungaricum** Klokov, *op. cit.* 525 (1957), from Hungary and Romania, with subsessile leaves and calyx-hairs 0·3–0·6 mm, are here included in **13.** They have been separated from the remaining populations of **13**, which have a subglabrous calyx-tube and a petiole $\frac{1}{6}$–$\frac{1}{3}$ as long as the lamina, but their status requires further examination.

14. L. meyeri (Boiss.) O. Kuntze, *Rev. Gen. Pl.* **2**: 395 (1891). Like **13** but scales 6–9 mm; spikes 2–3 cm; spikelets somewhat contiguous; inner bract *c.* 2 mm, rounded; outer bract with narrow hyaline margin; calyx *c.* 3·5 mm, with sparsely hairy veins or subglabrous. *Alkaline soils. S. & E. Ukraine, S.E. Russia, W. Kazakhstan.* Rs (W, K, E).

15. L. hirsuticalyx Pignatti, *Bot. Jour. Linn. Soc.* **64**: 361 (1971) (*Statice gmelinii* var. *limonioides* Wangerin, non *Statice limonioides* Biv.). Like **13** but leaves densely tuberculate; scales 10–16 mm; spikes 1·5–3 cm; inner bract *c.* 2 mm, rounded, with hyaline margin *c.* 0·2 mm wide; outer bract almost entirely herbaceous, overlapping $\frac{1}{4}$–$\frac{1}{3}$ of the inner; calyx *c.* 3 mm, with densely hairy veins. *Maritime salt-marshes. S.E. Greece, Kikladhes.* Gr.

16. L. tomentellum (Boiss.) O. Kuntze, *Rev. Gen. Pl.* **2**: 396 (1891). Plant 20–30(–80) cm; stems densely tomentose to sparsely hairy in the upper half. Leaves 50–130(–250) × 20–40(–90) mm, hairy on midrib, tuberculate, pinnately veined. Sterile branches few or absent; scales 9–14 mm or more. Spikes up to 1·2 cm, dense; spikelets 1- to 3-flowered. Inner bract 2·5–3·3(–4·5) mm; outer bract 1–2 mm, with narrow hyaline margin, overlapping $\frac{1}{6}$–$\frac{1}{5}$ of the inner. Calyx 3·5–5·5 mm, subglabrous. Corolla 5–6 mm, pinkish. *Alkaline soils and salt-marshes. S. & E. Ukraine, S.E. Russia, W. Kazakhstan.* ?Rm Rs (W, K, E).

L. czurjukiense (Klokov) Lavrenko, in Lavrenko & Soczava, *Descr. Veg. URSS* **2**: 669 (1956), from the coast of Ukraine, having the petiole $\frac{1}{3}-\frac{1}{2}$ as long as the lamina, **L. alutaceum** (Steven) O. Kuntze, *Rev. Gen. Pl.* **2**: 395 (1891), from W. Ukraine, having subsessile leaves 2–3 times as long as wide, and **L. donetzicum** Klokov in Kotov & Barbarich, *Fl. RSS Ucr.* **8**: 523 (1957), from E. Ukraine and adjacent parts of Russia, having subsessile leaves $2\frac{1}{2}$–5 times as long as wide, probably belong here but their status requires further examination.

17. **L. vulgare** Miller, *Gard. Dict.* ed. 8, no. 1 (1768) (*Statice limonium* L.). Plant 15–70 cm, glabrous. Leaves 100–150 × 15–40 mm, oblanceolate-spathulate, usually mucronate, pinnately veined, usually erect, in a sparse basal rosette; petiole 2 mm wide, $\frac{1}{2}$ as long to as long as lamina. Flowering stems cylindrical; non-flowering branches absent; sterile branches few or absent; scales 6–18 mm. Spikes 1–2 cm, patent, with 6–8 spikelets per cm; spikelets 2-flowered. Inner bract 2·7–5 mm; outer bract 0·9–2·7 mm, herbaceous, overlapping $\frac{1}{3}$–$\frac{1}{2}$ of the inner. Calyx 3·6–6 mm. Corolla 6–8 mm, reddish. *Maritime salt-marshes. S. & W. Europe, extending north-eastwards to S.W. Sweden.* Al Az Be Br Bu Co Cr Da Ga Ge Gr Ho Hs It Ju Lu Rm Sa Si Su Tu.

(a) Subsp. **vulgare**: Plant 15–30(–40) cm. Flowering stems usually branched only in upper half; branches short, crowded. Inner bract 4–5 mm; outer bract 2–2·7 mm, subequalling middle bract. Calyx 4·7–6 mm, 5-lobed. $2n=36$. ● *W. & N. Europe.*
(b) Subsp. **serotinum** (Reichenb.) Gams in Hegi, *Ill. Fl. Mitteleur.* **5**(3): 1882 (1927): Plant 30–70 cm. Flowering stems branched in the upper $\frac{1}{2}$–$\frac{2}{3}$; branches long, patent, often recurved. Inner bract 2·7–4·1 mm; outer bract 0·9–2 mm, $\frac{1}{2}$ as long as middle bract. Calyx 3·6–4·7 mm, 5-lobed, with smaller intermediate lobes. *S. Europe.*

Populations in Portugal frequently appear to be intermediate between **17**(a) and **17**(b).

18. **L. humile** Miller, *Gard. Dict.* ed. 8, no. 4 (1768) (*Statice bahusiensis* Fries). Like **17** but plant 20–40 cm; leaves 80–150 × 10–20 mm; flowering stems branched in the upper half, compressed, subangular; branches long, erect or erecto-patent; spikes 3–5 cm, erect, with 2–3 spikelets per cm; inner bract *c.* 5 mm; outer bract *c.* 3 mm, about equalling middle bract; calyx *c.* 6·5 mm. $2n=36$. *Maritime salt-marshes.* ● *N.W. Europe and W. part of Baltic region.* Br Da Ga Ge Hb No Su.

17 and **18** show variation in size and in the extent of branching. (Cf. W. Wangerin, *Zeitschr. Naturwiss.* (*Halle*) **82**: 401–443 (1911).)

19. **L. bellidifolium** (Gouan) Dumort., *Fl. Belg.* 27 (1827) (*Statice bellidifolia* (Gouan) DC.). Plant 9–30(–40) cm. Leaves 14–40 × 3–6(–15) mm, spathulate, acute, (1–)3(–5)-veined, usually withered and often completely absent at anthesis. Inflorescence with tuberculate branches; non-flowering branches numerous. Spikes dense, patent; spikelets 1- to 3-flowered. Inner bract 1·6–3·9 mm, hyaline for at least $(\frac{1}{4}-)\frac{1}{3}$ of its length; outer bract hyaline except on the keel, overlapping $\frac{1}{4}$–$\frac{1}{2}$ of the inner. Calyx 2·6–4·3 mm; limb $1\frac{1}{2}$ times as long as the tube. Corolla 4–5·5 mm, pale violet. $2n=18$. *Maritime salt-marshes; inland on saline soils in U.S.S.R. Coasts of Mediterranean and Black Seas; S. part of U.S.S.R.; E. England.* Br ?Co Ga Gr Hs It Ju Rm Rs (W, K, E) Sa.

Procumbent plants with the calyx 3·6–4·1 mm and with bracts having a narrow hyaline margin are recorded from the coasts of W. France and England as **L. dubyi** (Gren. & Godron) O. Kuntze, *Rev. Gen. Pl.* **2**: 395 (1891); they may represent a distinct

species. **L. danubiale** Klokov in Kotov & Barbarich, *Fl. RSS Ucr.* **8**: 526 (1957), from S.W. Ukraine, is a larger variant with leaves up to 120 mm and calyx 4–4·5 mm.

20. **L. cancellatum** (Bernh. ex Bertol.) O. Kuntze, *Rev. Gen. Pl.* **2**: 395 (1891) (*Statice cancellata* Bernh. ex Bertol.; incl. *S. cordata* sensu Hayek, non L.). Plant 10–18(–28) cm, densely pubescent. Leaves 5–20, 15–27 × 3–5(–8) mm, spathulate, obtuse or emarginate, forming a lax rosette. Inflorescence reticulate, the branches mostly diverging at an angle of 90°–180°; non-flowering branches numerous. Spikes up to 2·5 cm, with 4–6 spikelets per cm; spikelets 1- to 2-flowered. Bracts with narrow hyaline margin; inner bract (3–)4 mm; outer bract overlapping $\frac{1}{3}$ of the inner. Calyx *c.* 4·5 mm; limb $\frac{1}{2}$ as long as the tube. Corolla 5·5 mm. *Rocks near the sea.* ● *Coasts of the Adriatic sea and W. Greece.* Gr It Ju.

21. **L. vestitum** (Salmon) Salmon, *Jour. Bot.* (*London*) **62**: 336 (1924). Like **20** but plant 4–8(–13) cm, puberulent; leaves at least 20, 5–10 mm wide, forming a dense, cushion-like rosette; spikelets 1-flowered; inner bract *c.* 3·5 mm; calyx 4·7–5 mm. *Maritime cliffs.* ● *Adriatic islands of Jabuka* (*Pomo*) *and Kamik* (*W.S.W. of Split*). Ju.

22. **L. johannis** Pignatti, *Bot. Jour. Linn. Soc.* **64**: 362 (1971). Plant 15–25 cm. Leaves 20–50 × 6–9 mm. Inflorescence pyramidal, with numerous non-flowering and few flowering branches; branches pubescent, erect or erecto-patent, diverging at an acute angle. Spikes rudimentary or 1–2 cm, scorpioid, lax, with 2–4 spikelets per cm; spikelets (1–)2-flowered. Bracts with narrow hyaline margin; inner bract 3–3·5 mm; outer bract overlapping $\frac{1}{8}$ of the inner. Calyx 4·5–5 mm; tube not longer than the limb. Corolla 5–6 mm. *Volcanic maritime cliffs.* ● *S.W. Italy, Sicilia.* It Si.

23. **L. cordatum** (L.) Miller, *Gard. Dict.* ed. 8, no. 10 (1768) (*Statice cordata* L.). Plant 15–30 cm, greyish-pubescent. Leaves 15–30 × 2–6 mm, linear-spathulate. Inflorescence corymbose, with erect to erecto-patent branches diverging at an acute angle; non-flowering branches numerous. Spikes up to 1 cm, dense, with 7 spikelets per cm; spikelets 1-flowered, in short clusters. Bracts with narrow hyaline margin; inner bract *c.* 3·5 mm, puberulent; outer bract overlapping $\frac{1}{8}$ of the inner. Calyx *c.* 4·5 mm, exceeding bracts by *c.* 1·5 mm, with acute lobes; limb $\frac{2}{3}$ as long as the tube. *Calcareous maritime cliffs.* ● *W. Italy and S.E. France.* Ga It.

24. **L. furfuraceum** (Lag.) O. Kuntze, *Rev. Gen. Pl.* **2**: 395 (1891) (*Statice furfuracea* Lag.). Plant 18–25 cm, whitish-pubescent. Leaves 20–26 × 4–6 mm, usually withered at anthesis. Inflorescence hemispherical, with patent to deflexed branches; non-flowering branches present. Spikes 2–3 cm, the terminal up to 4–5 cm, often scorpioid or divaricate, with 2–4 spikelets per cm; spikelets 2-flowered, erecto-patent. Bracts with narrow hyaline margin; inner bract *c.* 4·5 mm; outer bract overlapping $\frac{1}{8}$ of the inner. Calyx *c.* 4·5 mm; limb $\frac{1}{2}$ as long as the tube. Corolla 6–7 mm, pale violet. $2n=18$. *Calcareous maritime cliffs.* ● *S.E. Spain* (*Alicante prov.*). Hs.

25. **L. lucentinum** Pignatti & Freitag, *Bot. Jour. Linn. Soc.* **64**: 363 (1971). Plant 15–33 cm, greyish-pubescent. Leaves 30–40 × 11–15 mm, withered at anthesis. Inflorescence corymbose, with patent branches; non-flowering branches present. Spikes 1–1·5 cm, sometimes forming irregular clusters; spikelets 1- to 2-flowered, patent. Bracts with narrow hyaline margin; inner bract 4–4·5 mm, forming a narrowly cylindrical envelope around

the flowers. Calyx *c.* 5 mm; limb $\frac{2}{3}$ as long as the tube. Corolla 5–6 mm, blue-violet. *Maritime salt-marshes.* ● *S.E. Spain (Alicante prov.).* Hs.

26. L. dichotomum (Cav.) O. Kuntze, *Rev. Gen. Pl.* **2**: 395 (1891) (*Statice dichotoma* Cav.). Plant (15–)30–50 cm; stems rugose, pubescent, glaucous. Leaves 30–40 × 8–9 mm, spathulate, withered at anthesis. Inflorescence usually hemispherical, with dense branches diverging at an obtuse angle; non-flowering branches present. Lateral spikes rudimentary, the terminal up to 3 cm; spikelets 1- to 2-flowered. Bracts with narrow hyaline margin; inner bract 2–2·5 mm; outer bract overlapping $\frac{1}{4}$ of the inner. Calyx 3·5–4(–4·5) mm; limb as long as the tube. Corolla *c.* 5 mm. *Saline and gypsaceous soils.* ● *C. Spain.* Hs.

27. L. calcarae (Tod. ex Janka) Pignatti, *Bot. Jour. Linn. Soc.* **64**: 364 (1971) (*Statice calcarae* Tod. ex Janka). Plant 20–30 cm; stems rough, greyish-pubescent. Leaves *c.* 25 × *c.* 4 mm, linear-spathulate, acute, usually withered in summer; sheaths reddish-brown. Inflorescence with branches diverging at an angle of 90°–120°; non-flowering branches numerous. Lateral spikes rudimentary, the terminal 5–6 cm, with 3 spikelets per cm; spikelets 2- to 3-flowered. Bracts with narrow hyaline margin; inner bract 3·5–4 mm; outer bract overlapping $\frac{1}{8}$ of the inner. Calyx *c.* 5·5 mm; limb as long as the tube. *Gypsaceous soils.* ● *C. Sicilia (near Caltanissetta).* Si.

28. L. catalaunicum (Willk. & Costa) Pignatti, *Collect. Bot. (Barcelona)* **6**: 300 (1962). Stems glabrous, smooth. Leaves 30–65 × 5–17 mm, oblanceolate-spathulate, usually withered or absent at anthesis. Inflorescence 20–60 cm, usually hemispherical, the branches with long, divaricate segments; non-flowering branches numerous. Spikes 2–6(–12) cm, very lax, with 1–3 spikelets per cm; spikelets 1- to 3-flowered. Bracts with narrow hyaline margin; inner bract 2·5–3(–4) mm, obtuse, the midrib excurrent into a conspicuous point; outer bract *c.* 0·3 mm. Calyx 5–6 mm, exceeding the inner bract by 2–3 mm; tube as long as the limb. Corolla *c.* 8·5 mm. *Saline and gypsaceous soils.* ● *N.E. Spain (Ebro basin).* Hs.

1 Spikelets 1(–2)-flowered   **(b) subsp. procerum**
1 Spikelets 2- to 3-flowered
 2 Inflorescence 20–40 cm; leaves not more than 40 × 5–8 mm
                       (a) subsp. catalaunicum
 2 Inflorescence 40–60 cm; leaves 50–65 × 14–17 mm
                       (c) subsp. viciosoi

(a) Subsp. **catalaunicum**: Leaves up to *c.* 40 × 5–8 mm, 1(–3)-veined. Inflorescence 20–40 cm; spikelets 2- to 3-flowered. *Near Logroño and Lérida.*
(b) Subsp. **procerum** (Willk.) Pignatti, *loc. cit.* (1962): Leaves 30–40 × 8–11 mm, 1(–3)-veined. Inflorescence 20–40 cm; spikelets 1(–2)-flowered. *From Lérida to Huesca.*
(c) Subsp. **viciosoi** (Pau) Pignatti, *loc. cit.* (1962): Leaves 50–65 × 14–17 mm, 5-veined. Inflorescence 40–60 cm; spikelets 2- to 3-flowered. *Near Calatayud and Logroño.*

29. L. articulatum (Loisel.) O. Kuntze, *Rev. Gen. Pl.* **2**: 395 (1891) (*Statice articulata* Loisel.). Plant 5–15(–20) cm. Leaves 30–35 × 4–5 mm, linear-spathulate, 1-veined, usually withered at anthesis. Flowering stems branched from near the base; branches coarsely tuberculate, often divaricate; segments 2–5 mm, strongly constricted at the nodes; non-flowering branches present. Spikes rudimentary, the terminal 1–2·5 cm, lax; spikelets 1- to 2-flowered. Bracts with narrow hyaline margin; inner bract 3·2–4 mm; outer bract 0·5–1 mm. Calyx

c. 4·5 mm; limb $\frac{1}{3}$ as long as the tube. *Rocks near the sea.* ● *Corse, N. Sardegna.* Co Sa.

30. L. japygicum (Groves) Pignatti, *Bot. Jour. Linn. Soc.* **64**: 364 (1971) (*Statice cancellata* var. *japygica* Groves). Plant 5–10(–20) cm, glabrous. Leaves 10–15 × *c.* 3 mm, linear-spathulate, bullate, in cushion-like rosettes on somewhat woody branches 1–2 cm. Inflorescence 4–12 cm; branches tuberculate, with segments diverging at 90°; non-flowering branches numerous. Spikes 0·5–1·8(–3) cm, with 5–6 spikelets per cm; spikelets 2-flowered. Bracts with narrow hyaline margin; inner bract 3–3·5 mm, tuberculate; outer bract up to 2 mm, overlapping $\frac{1}{4}$ of the inner. Calyx *c.* 4 mm; limb $\frac{1}{2}$ as long as the tube, the veins divaricate after anthesis. *Calcareous maritime cliffs.* ● *S.E. Italy (Taranto to Otranto).* It.

31. L. minutum (L.) Fourr., *Ann. Soc. Linn. Lyon* nov. ser., **17**: 141 (1869) (*Statice minuta* L.). Plant 3–14 cm, usually glabrous. Leaves 8–10 × 2–3 mm, spathulate, revolute, bullate, 1-veined, in cushion-like rosettes at apex of somewhat woody branches 1–2 cm. Inflorescence 2–12 cm; branches smooth, with straight, rigid segments diverging at an acute angle; non-flowering branches few to numerous; scales 3–5 mm. Lateral spikes 0·5–1 cm, irregularly subglobose, the apical 2–3 cm, with 6–7 spikelets per cm; spikelets 1- to 4-flowered. Bracts with narrow hyaline margin; inner bract 3·5–4 mm, acute; outer bract up to 2 mm, slightly overlapping the base of the inner. Calyx 5–5·5 mm; limb $\frac{2}{3}$ as long as the tube. *Calcareous maritime cliffs.* ● *S.E. France.* Ga.

Densely pubescent plants recorded from Martigues possibly result from introgression with **23**.

32. L. caprariense (Font Quer & Marcos) Pignatti, *Arch. Bot. (Forlì)* **31**: 77 (1955) (*Statice minuta* auct. balear., non L.). Like **31** but plant densely puberulent; leaves up to 7 mm wide, spathulate, obtuse, in a hemispherical, cushion-like basal rosette; non-flowering branches usually numerous; spikelets contiguous; outer bract overlapping $\frac{1}{4}$–$\frac{1}{3}$ of the inner, the inner truncate; calyx 3·5–5 mm, the limb as long as the tube. *Calcareous maritime cliffs.* ● *Islas Baleares.* Bl.

(a) Subsp. **caprariense**: Spikelets usually 1(–2)-flowered. Inner bract 4–4·5 mm. Calyx 4–5 mm. *Menorca, Mallorca and Cabrera.*
(b) Subsp. **multiflorum** Pignatti, *op. cit.* 81 (1955): Spikelets usually 3- to 4-flowered. Inner bract 3–4 mm. Calyx 3·5–4·5 mm. *Throughout the range of the species.*

33. L. parvifolium (Tineo) Pignatti, *Bot. Jour. Linn. Soc.* **64**: 364 (1971) (*Statice parvifolia* Tineo). Plant 10–22 cm, glabrous, with densely leafy, somewhat woody, basal branches 4–12 cm. Leaves 10–16 × 2–3 mm, narrowly spathulate. Inflorescence 6–10 cm; non-flowering branches numerous, incurved. Spikes 0·5–2 cm, with 1–3 spikelets per cm; spikelets 1(–2)-flowered. Bracts with narrow hyaline margin; inner bract *c.* 3·5 mm, cucullate; outer bract *c.* 0·8 mm, overlapping $\frac{1}{4}$ of the inner. Calyx 3·8–4·2 mm; limb as long as the tube. *Rocky slopes up to 600 m.* ● *Pantelleria.* Si.

34. L. acutifolium (Reichenb.) Salmon, *Jour. Bot. (London)* **62**: 336 (1924) (*Statice rupicola* Badaro ex Reichenb.). Like **33** but leaves 10–18 × *c.* 1 mm, linear, acute; inflorescence 3–6 cm, with few non-flowering branches; spikelets 2-flowered; inner bract 4–4·5 mm, flat; calyx 4·5–5 mm. *Rocky slopes near the sea.* ● *S. Corse, N. Sardegna.* Co Sa.

Plants intermediate between **29** and **34** occur in N. Sardegna.

35. L. aragonense (Debeaux) Pignatti, *Collect. Bot. (Barcelona)* **6**: 301 (1962) (*Statice aragonensis* Debeaux). Like **33** but leafy basal branches 1–3 cm; leaves 10×0·5–1 mm, linear, acute, 1-veined; non-flowering branches few or almost absent; spikes with 5–6 spikelets per cm; spikelets 2- to 3-flowered; inner bract *c.* 2·8 mm; calyx *c.* 3 mm, the limb $\frac{1}{3}$–$\frac{1}{2}$ as long as the tube. *Gypsaceous soils.* ● E. Spain (*prov. Teruel*). Hs.

(36–40). L. cosyrense group. Plant (10–)15–50 cm, glabrous, usually smooth, often with woody basal branches. Leaves linear-spathulate, flat to convolute, 1-veined, in dense basal rosettes. Inflorescence with numerous non-flowering branches. Spikes 1–5 cm, with 3–5 spikelets per cm. Bracts with narrow hyaline margin; inner bract 3·5–4·3 mm, shorter than the calyx, acute, with midrib excurrent in conspicuous acute point. Corolla *c.* 8 mm.

37–38 are represented by many local variants scattered on small islands and rocky slopes along the W. coast of Italy.

1 Calyx-limb $\frac{2}{3}$ as long to as long as the tube
  2 Calyx-lobes 0·6–0·8 mm, obtuse   **38. multiforme**
  2 Calyx-lobes 1–2 mm, acute   **40. anfractum**
1 Calyx-limb $\frac{1}{3}$–$\frac{1}{2}$ as long as the tube
  3 Inflorescence-segments very fragile, dark when dry **39. bocconei**
  3 Inflorescence-segments not fragile, greenish
    4 Leaves flat or with slightly revolute margin; calyx *c.* 4 mm, the lobes *c.* 0·8 mm **36. cosyrense**
    4 Leaves with revolute margin or entirely convolute; calyx 4–5 mm, the lobes 1–1·5 mm **37. pontium**

36. L. cosyrense (Guss.) O. Kuntze, *Rev. Gen. Pl.* **2**: 395 (1891) (*Statice cosyrensis* Guss.). Plant 15–50 cm. Leaves 20–25 × 2·5–3 mm, flat or slightly revolute, obtuse, in rosettes on somewhat woody basal branches 1–3 cm. Inflorescence up to 25 cm; segments up to 2 cm, flexuous or rigid, diverging at an acute angle, greenish. Spikes 2–5 cm, lax, with 3–4 spikelets per cm; spikelets 1-flowered. Inner bract 3·5(–4) mm; outer bract *c.* 1 mm, overlapping $\frac{1}{6}$–$\frac{1}{4}$ of the inner. Calyx *c.* 4 mm, exceeding the bract by *c.* 1·5 mm; limb campanulate, $\frac{1}{3}$–$\frac{1}{2}$ as long as the tube; lobes *c.* 0·8 mm. *Rocky slopes.* ● Malta, Pantelleria. Si.

37. L. pontium Pignatti, *Bot. Jour. Linn. Soc.* **64**: 364 (1971). Like **36** but inflorescence with the segments diverging at an angle of 60°–90°; leaves revolute or convolute; spikelets sometimes 3- to 4-flowered; calyx 4–5 mm, the lobes 1–1·5 mm. *Calcareous maritime cliffs.* ● Isole Ponziane (*W. of Napoli*). It.

A variant with long, slender segments resembles **44**.

38. L. multiforme (U. Martelli) Pignatti, *loc. cit.* (1971) (*Statice minuta* var. *multiformis* U. Martelli). Like **36** but plant 20–40 cm, without woody basal branches; leaves usually revolute or convolute; spikelets 1- to 4-flowered; calyx-limb $\frac{2}{3}$ as long to as long as the tube, the lobes 0·6–0·8 mm, obtuse. *Calcareous maritime cliffs.* ● W.C. Italy (*Gaeta to Livorno*), Arcipelago Toscano. It.

A polymorphic species with more than twenty local variants.

39. L. bocconei (Lojac.) Litard. in Briq., *Prodr. Fl. Corse* **3**(2): 16 (1955) (*Statice cordata* sensu Guss., non L.). Plant (10–)20–40 cm. Leaves 20–50 × 5–10 mm, obtuse or emarginate. Inflorescence-segments up to 2 cm, rigid, usually arcuate and very fragile, smooth, dark when dry. Spikes 1·5–4 cm, lax, with 4 spikelets per cm; spikelets 1-flowered. Inner bract 3·5–4 mm; outer bract *c.* 1·5 mm. Calyx 4–5 mm; limb $\frac{1}{3}$ as long as the tube. *Maritime cliffs.* ● N.W. Sicilia, Isole Egadi. Si.

Plants with larger spikelets, having the inner bract 4·5–5 mm and the calyx *c.* 6 mm, occur near Trapani and were described as **Statice ambigua** Lojac., *Fl. Sic.* **2**(2): 26 (1907) (*S. ponzoi* Fiori); they may be intermediate between **39** and **55**.

40. L. anfractum (Salmon) Salmon, *Jour. Bot. (London)* **62**: 336 (1924) (*Statice anfracta* Salmon, *S. cosyrensis* sensu Hayek, non Guss.). Plant 20–30 cm, slightly rough. Leaves 20–25 × 3–4 mm. Inflorescence-segments usually diverging at 90°–120°. Spikes 1–2 cm, with 4–5 spikelets per cm; spikelets 1- to 2(–3)-flowered. Inner bract 4–4·3 mm; outer bract overlapping $\frac{1}{4}$ of the inner. Calyx *c.* 5 mm; limb $\frac{2}{3}$ as long as the tube; lobes 1–2 mm, acute. *Calcareous rocks.* ● Coasts of Albania and S. Jugoslavia. Al Ju.

41. L. melium (Nyman) Pignatti, *Bot. Jour. Linn. Soc.* **64**: 365 (1971) (*Statice cosyrensis* subsp. *melia* Nyman). Plant 20–40 cm, glabrous, smooth. Leaves 25–50 × 4–7 mm, linear-spathulate, revolute. Inflorescence branched from the base; segments constricted at the nodes; non-flowering branches few. Spikes up to 1·5 cm, with 3–5 spikelets per cm; spikelets 1-flowered. Bracts with narrow hyaline margin; inner bract *c.* 3 mm; outer bract overlapping $\frac{1}{4}$ of the inner one. Calyx 5–5·5 mm, exceeding bracts by *c.* 2 mm; limb as long as to $1\frac{1}{2}$ times as long as the tube. Corolla *c.* 6 mm. *Calcareous cliffs near the sea.* ● S. Greece, Kikhlades. Gr.

42. L. tremolsii (Rouy) P. Fourn., *Quatre Fl. Fr.* 721 (1937) (*Statice tremolsii* Rouy). Plant 10–25 cm, glabrous, slightly rough. Leaves 15–35 × 3–8 mm, linear-lanceolate, bullate, 1-veined, in rosettes on somewhat woody basal branches 2–3 cm. Flowering stems branched from the base; branches erect; segments arcuate, flexible; non-flowering branches few. Spikes up to 2 cm, with 4–5 spikelets per cm; spikelets 1-flowered. Bracts with narrow hyaline margin; inner bract *c.* 4 mm, acute, with midrib excurrent in conspicuous acute point; outer bract overlapping $\frac{1}{8}$ of the inner. Calyx 4–4·5 mm, exceeding bracts by *c.* 1·5 mm; limb as long as the tube, the teeth acute and conspicuously veined. Corolla *c.* 6 mm, exceeding calyx by at least 1 mm. $2n=18$. *Calcareous cliffs near the sea.* ● N.E. Spain, Islas Baleares, S.E. France. Bl Ga Hs.

43. L. calaminare Pignatti ex Pignatti, *Bot. Jour. Linn. Soc.* **65**: 353 (1972). Like **42** but plant 20–40 cm, tuberculate; spikes with 3–5 spikelets per cm; spikelets 2(–3)-flowered; inner bract *c.* 3·5 mm, forming an open envelope around the flowers, usually completely open after anthesis; calyx *c.* 5·5 mm, exceeding inner bract by 2–2·5 mm, the limb $\frac{2}{3}$ as long as the tube; corolla *c.* 5·5 mm, pale pink, slightly exceeding the calyx. $2n=18$. *Zinc-rich soils.* ● S.E. Spain (*near Cartagena*). Hs.

44. L. tenuiculum (Tineo ex Guss.) Pignatti, *op. cit.* **64**: 365 (1971) (*Statice tenuicula* Tineo ex Guss.). Plant 10–15 cm, slender, glabrous, smooth. Leaves 20–30 × 4–7 mm, linear-spathulate, obtuse, revolute, smooth, 1-veined, on somewhat woody basal branches 2–3 cm. Inflorescence-segments 2–2·5 cm, flexible, sometimes very slender, straight or arcuate, diverging at an angle of 15°–45°; non-flowering branches numerous. Spikes up to 1 cm, with 3 spikelets per cm; spikelets 1- to 2-flowered. Bracts with narrow hyaline margin; inner bract *c.* 2·5 mm; outer bract up to 2 mm, overlapping $\frac{1}{10}$ of the inner. Calyx *c.* 3·5 mm; limb as long as the tube. Corolla *c.* 5 mm. *Hillsides near the sea.* ● Isole Egadi (*Marettimo*). Si.

45. L. hermaeum (Pignatti) Pignatti, *loc. cit.* (1971) (*L. tenuiculum* subsp. *hermaeum* Pignatti). Like **44** but plant rough;

leaves 10–20 × 1–1·5(–4) mm, linear-lanceolate, usually withered at anthesis, on woody basal branches 2–5 cm; spikes few, up to 4 cm, with 1(–2) spikelets per cm; spikelets 2-flowered; inner bract 3–3·5 mm, cucullate. *Calcareous hillsides.* ● *N.E. Sardegna (island of Tavolara).* Sa.

46. L. remotispiculum (Lacaita) Pignatti, *loc. cit.* (1971) (*Statice remotispicula* Lacaita). Plant 20–30 cm, glabrous, smooth. Leaves 30–40 × 4–5 mm, linear-spathulate, revolute, 1-veined, in rosettes on somewhat woody basal branches 1–2 cm. Inflorescence-segments up to 3 cm, flexible, straight or arcuate, diverging at an angle of 15°–45°; non-flowering branches numerous. Spikes 2–6(–10) cm, usually arcuate, with 1–2 spikelets per cm; spikelets 2- to 3(–4)-flowered. Bracts with narrow hyaline margin; inner bract 3–3·5 mm, acute, with mid-rib excurrent in conspicuous acute point; outer bract *c.* 1·2 mm, overlapping $\frac{1}{4}$ of the inner. Calyx 4–4·5 mm; limb as long as the tube. Corolla 5–6 mm. *Maritime cliffs.* ● *S.W. Italy, from 39° 40′ to 40° 40′ N.* It.

47. L. minutiflorum (Guss.) O. Kuntze, *Rev. Gen. Pl.* **2**: 395 (1891). Plant 15–25(–35) cm, glabrous, often slightly rough. Leaves 20–25 × 5–9 mm, obovate-spathulate, obtuse, 1-veined, on somewhat woody basal branches 2–3 cm. Inflorescence pyramidal; non-flowering branches absent. Spikes usually lax, with 4–6 spikelets per cm; spikelets 1- to 2(–3)-flowered. Inner bract 4–4·5 mm; outer bract overlapping $\frac{1}{4}$ of the inner. Calyx 4·5–5 mm, exceeding bracts by *c.* 1 mm; limb 1½ times as long as the tube. *Rocky slopes near the sea.* ● *Islands of W. Mediterranean region.* Bl Si.

(a) Subsp. **minutiflorum**: Inflorescence-branches more or less rigid. Leaves flat or with somewhat revolute margin. Spikes 1·5–4 cm; spikelets (1–)2-flowered. *Sicilia.*

(b) Subsp. **balearicum** Pignatti, *Arch. Bot.* (*Forlì*) **31**: 85 (1955): Inflorescence-branches slender and flexible. Leaves revolute. Spikes 1·2–1·5(–2) cm; spikelets (1–)2(–3)-flowered. *Islas Baleares.*

48. L. inarimense (Guss.) Pignatti, *op. cit.* 84 (1955). Plant 10–20 cm, glabrous, smooth. Leaves 30–40 × 4–10 mm, obtuse, 1-veined, on somewhat woody basal branches 2–3 cm. Inflorescence pyramidal; segments flexible, arcuate; non-flowering branches few or absent; scales 3–5 mm. Spikes 2–2·5 cm, with 5, rarely more, often contiguous spikelets per cm; spikelets 2-flowered. Bracts with narrow hyaline margin; inner bract *c.* 4·5 mm, acute, with midrib excurrent in conspicuous point; outer bract overlapping $\frac{1}{3}$ of the inner. Calyx 4·5–5 mm; limb as long as the tube, the teeth obtuse and indistinctly veined. *Maritime rocks and cliffs.* ● *W. Mediterranean region.* Bl It.

(a) Subsp. **inarimense**: Leaves 4–5 mm wide, linear-lanceolate to linear. *S.W. Italy (Golfo di Napoli).*

(b) Subsp. **ebusitanum** (Font Quer) Pignatti, *loc. cit.* (1955): Leaves 7–10 mm wide, lanceolate-spathulate. *Islas Baleares.*

In the Islas Baleares a hybrid swarm is known involving this species, **32** and **56**.

49. L. tenoreanum (Guss.) Pignatti, *Bot. Jour. Linn. Soc.* **64**: 365 (1971) (*Statice tenoreana* Guss.). Plant glabrous, smooth. Leaves 25–40 × 4–7 mm, lanceolate-spathulate, obtuse, revolute, 1-veined, on woody basal branches 4–8 cm. Inflorescence 10–15 cm, dense, sparingly branched only in the upper half; non-flowering branches absent; scales 1·5–2·5 mm. Spikes 0·9–1·3 cm, with 10–13 contiguous spikelets per cm; spikelets 2-flowered. Inner bract 3·5–4 mm; outer bract overlapping $\frac{1}{3}$–$\frac{1}{2}$ of

the inner. Calyx 5–5·5 mm; limb as long as the tube. Corolla *c.* 6·5 mm. *Calcareous cliffs.* ● *S.W. Italy (near Castellammare).* It.

50. L. laetum (Nyman) Pignatti, *loc. cit.* (1971) (*Statice duriuscula* subsp. *laeta* Nyman, *S. laeta* Moris, non Salisb.). Plant 10–20 cm, glabrous, slightly rough. Leaves 20–25 × 2–3 mm, linear-spathulate, acute, 1-veined, on woody basal branches 3(–8) cm. Inflorescence dense, branched only in the upper half; non-flowering branches absent. Spikes 1–1·5 cm, with 8 contiguous spikelets per cm; spikelets 2-flowered. Inner bract *c.* 4·5 mm; outer bract overlapping $\frac{1}{4}$–$\frac{1}{3}$ of the inner. Calyx *c.* 4 mm; limb $\frac{1}{2}$ as long as the tube. Corolla *c.* 6 mm. ● *Coast of W. Sardegna.* Sa.

51. L. gougetianum (Girard) O. Kuntze, *Rev. Gen. Pl.* **2**: 395 (1891). Plant 10–20 cm, glabrous, slightly rough. Leaves 15–30 × 3–9 mm, obovate-spathulate, obtuse, 1-veined, in dense basal rosettes. Inflorescence with straight, rigid segments; non-flowering branches usually absent, rarely 1–4; basal branches sterile; scales 3–5 mm. Spikes 1·5–2 cm, with 7–9 contiguous spikelets per cm; spikelets 2-flowered. Inner bract *c.* 4 mm; outer bract overlapping $\frac{1}{6}$–$\frac{1}{4}$ of the inner. Calyx 4·5–5 mm; limb 1½ times as long as the tube. Corolla 5–6 mm, pale violet. *Salt-marshes.* Islas Baleares. Bl. (*Algeria.*)

(52–54). **L. albidum** group. Plant 7–25 cm, glabrous. Leaves obovate-spathulate to cuneate, 1- to 7-veined, in hemispherical basal rosettes. Flowering stems branched only in the upper part; non-flowering branches absent, sometimes a few basal branches sterile. Spikes 2–8 cm, lax. Calyx 4·5–7 mm, about equalling or exceeding the bracts.

1 Inner bract *c.* 7 mm; calyx about equalling bracts **53. frederici**
1 Inner bract 3–4·5 mm; calyx exceeding bracts
  2 Inner bract *c.* 4·5 mm; calyx exceeding bracts by *c.* 1 mm; spikelets 2- to 4(–6)-flowered **52. albidum**
  2 Inner bract 3–3·5 mm; calyx exceeding bracts by 2–2·5 mm; spikelets (1–)2-flowered **54. panormitanum**

52. L. albidum (Guss.) Pignatti, *Bot. Jour. Linn. Soc.* **64**: 365 (1971) (*Statice albida* Guss.). Plant 7–25 cm, glabrous. Leaves 10–22 × 5–9(–13) mm, obovate-spathulate, obtuse or emarginate, revolute. Inflorescence-branches smooth; scales 2–3 mm. Spikes with 2–5 spikelets per cm; spikelets 2- to 4(–6)-flowered. Inner bract *c.* 4·5 mm; outer bract overlapping $\frac{1}{8}$ of the inner. Calyx *c.* 5 mm, exceeding the bracts by *c.* 1 mm; limb as long as the tube. *Rocky slopes near the sea.* Isole Pelagie. Si.

Intermediates between this species and **47(a)** occur in S. Sicilia.

53. L. frederici (W. Barbey) Rech. fil., *Denkschr. Akad. Wiss. Math.-Nat. Kl.* (*Wien*) **105(1)**: 427 (1943) (*Statice frederici* W. Barbey). Like **52** but leaves up to 18 mm wide, cuneate, canaliculate, on woody basal branches 7–12 cm; inflorescence-branches sometimes rough; spikes with 1–2 spikelets per cm; spikelets 2-flowered; inner bract *c.* 7 mm; outer bract *c.* 3 mm, overlapping at least $\frac{1}{3}$ of the inner; calyx *c.* 7 mm, about equalling the bracts, the limb $\frac{1}{4}$ as long as the tube. *Calcareous cliffs.* S. Aegean region. Cr Gr.

54. L. panormitanum (Tod.) Pignatti, *Bot. Jour. Linn. Soc.* **64**: 365 (1971) (*Statice panormitana* Tod.). Like **52** but leaves 25–45 × 9–17 mm, obtuse; spikelets (1–)2-flowered; inner bract 3–3·5 mm; outer bract overlapping $\frac{1}{4}$ of the inner; calyx 4·5–5 mm, exceeding the bracts by 2–2·5 mm. *Calcareous cliffs.* ● *N.W. Sicilia (near Palermo).* Si.

55. L. graecum (Poiret) Rech. fil., *Denkschr. Akad. Wiss. Math.-Nat. Kl. (Wien)* **105(1)**: 427 (1943). Plant 10–35 cm, glabrous, usually rough. Leaves 20–40 × 4–12 mm, linear-lanceolate to obovate-spathulate, 1-veined, usually on long woody basal branches. Inflorescence usually with numerous non-flowering branches; segments 1–2 cm, more than 0·7 mm in diameter; scales 3–4 mm. Spikes 3–5(–8) cm, with 1–3 spikelets per cm; spikelets 1- to 5-flowered, straight, cylindrical. Inner bract 5·5–6(–8) mm, obtuse, rounded dorsally, usually tuberculate, with reddish-brown margin, sometimes hyaline near apex, dark when dry, forming a wide, subpatent envelope around the flowers; outer bract 1·5–2·5(–3) mm, overlapping $\frac{1}{6}$–$\frac{1}{3}$ of the inner. Calyx 6–7 mm, narrowly conical after pollination; tube straight; limb $\frac{2}{3}$ as long as the tube. Corolla 7–8 mm. *Maritime rocks and sands. Mediterranean region, mainly in the islands*. Bl Co Cr Gr Sa Si. (*W. Asia.*)

(a) Subsp. **graecum** (*Statice graeca* Poiret, *S. sieberi* Boiss., *S. hyssopifolia* Girard): Inflorescence-branches mostly diverging at not more than 45°. Leaves green in summer; sometimes with clusters of reduced leaves on the lower branches. *S. Greece, S. Aegean region.*

A very variable subspecies, containing two widespread variants having linear-spathulate and obovate-spathulate leaves respectively, and differing in the arrangement of the leaves and the number of flowers per spikelet. Further study is required to clarify their status. Plants intermediate between **55** and **29**, **53** and **56** occur frequently.

(b) Subsp. **divaricatum** (Rouy) Pignatti, *Bot. Jour. Linn. Soc.* **64**: 366 (1971) (*Statice virgata* var. *divaricata* Rouy): Inflorescence-branches diverging at an angle of (45–)60–90(–120)°. Leaves usually withered in summer; cauline leaves absent. *Islands of W. & C. Mediterranean region.*

56. L. oleifolium Miller, *Gard. Dict.* ed. 8, no. 3 (1768) (*Statice oleifolia* (Miller) Scop.). Plant 5–50 cm, glabrous, smooth. Leaves 30–55 × 4–9 mm, linear-spathulate, 1-veined, on long basal branches. Inflorescence 10–40 cm; non-flowering branches numerous. Spikes up to 4 cm, with 4 spikelets per cm; spikelets 1- to 2(–5)-flowered, arcuate, compressed. Inner bract *c.* 4–5 mm, obtuse, smooth, with reddish-brown margin, sometimes hyaline near apex, keeled dorsally, incurved, dark when dry, forming a narrow envelope around the flowers; outer bract 0·8–2 mm, overlapping $\frac{1}{4}$ of the inner. Calyx infundibuliform after pollination; tube curved. Corolla *c.* 8 mm. *Coastal habitats. Mediterranean region, C. & S. Portugal.* Al Bl Co Cr Ga Gr Hs It Ju Lu Sa Si Tu.

1 Inflorescence-branches regularly reticulate, diverging at an angle of (45–)90(–120)°; segments usually 2–4 mm, black when dry **(c) subsp. sardoum**
1 Inflorescence-branches not reticulate, diverging at an angle of not more than 45°; segments 15–35 mm, the lower longer than the upper, not black when dry
2 Spikelets 1-flowered; inner bract *c.* 5 mm; calyx 6–7 mm, the limb as long as the tube **(b) subsp. dictyocladum**
2 Spikelets 1- to 5-flowered; inner bract 4–4·5 mm; calyx 4·5–5 mm, the limb $\frac{1}{2}$ as long as the tube
3 Stems and branches smooth; segments cylindrical **(a) subsp. oleifolium**
3 Stems and branches rough; segments slightly constricted at nodes **(d) subsp. pseudodictyocladum**

(a) Subsp. **oleifolium** (*Statice virgata* Willd.): Plant 15–45 cm. Leaves in basal rosettes. Inflorescence-branches 0·5–0·7 mm in diameter when dry, smooth, diverging at not more than 45°; segments 15–35 mm, cylindrical, the lower longer than the upper. Spikelets 1- to 5-flowered. Inner bract 4–4·5 mm, about equalling the calyx. Calyx 4·5–5 mm; limb $\frac{1}{2}$ as long as the tube. *Throughout the range of the species.*

A highly polymorphic subspecies. Plants with 3- to 5-flowered spikelets (var. *majus* (Guss.) Pignatti) are frequent on calcareous cliffs in the southern part of the range and have been described as a distinct species (**Statice dubia** Andrews ex Guss., *Fl. Sic. Prodr., Suppl.* 89 (1832), *S. smithii* Ten.). Intermediates between this subspecies and **30**, **31**, **42**, **47**, **48**, **51** and **85** occur sporadically in areas of contact.

(b) Subsp. **dictyocladum** Arcangeli, *Comp. Fl. Ital.* ed. 2, 459 (1894): Plant 15–45 cm. Leaves in basal rosettes. Inflorescence-branches 0·5–0·7 mm in diameter when dry, diverging at not more than 45°; lower segments longer than the upper. Spikelets 1-flowered. Inner bract *c.* 5 mm. Calyx 6–7 mm; limb as long as the tube. ● *Corse, ?Sardegna.*

(c) Subsp. **sardoum** (Pignatti) Pignatti, *Bot. Jour. Linn. Soc.* **64**: 366 (1971) (*L. virgatum* subsp. *sardoum* Pignatti): Plant 5–15(–35) cm. Leaves on somewhat woody basal branches. Inflorescence-branches regularly reticulate, diverging at an angle of (45°–)90°(–120°); segments usually 2–4 mm, subequal, black when dry. Spikelets 2- to 5-flowered. Inner bract up to 4 mm. Calyx *c.* 4·5 mm; limb as long to $1\frac{1}{2}$ times as long as the tube. ● *S. Sardegna* (*near Cagliari*).

(d) Subsp. **pseudodictyocladum** (Pignatti) Pignatti, *loc. cit.* (1971) (*L. virgatum* subsp. *pseudodictyocladum* Pignatti): Plant 15–30 cm. Leaves in basal rosettes. Inflorescence-branches 0·5–0·7 mm in diameter when dry, rough, diverging at not more than 45°; segments slightly constricted at the nodes, the lower longer than the upper. Spikelets 2- to 3-flowered. Inner bract 4–4·5 mm. Calyx 4·5–5 mm; limb $\frac{1}{2}$ as long as the tube. ● *Islas Baleares* (*S. Mallorca*).

57. L. carpathum (Rech. fil.) Rech. fil., *Denkschr. Akad. Wiss. Math.-Nat. Kl. (Wien)* **105(1)**: 427 (1943). Plant 20–40 cm, glabrous, rough. Leaves 40–90 × 1–2 mm, linear, fleshy, somewhat subcylindrical, obtuse, 1-veined. Inflorescence without non-flowering branches. Spikes 4–6 cm, with 2–4 spikelets per cm; spikelets (1–)2-flowered. Inner bract *c.* 5·5 mm; outer bract *c.* 1·5 mm, overlapping $\frac{1}{5}$ of the inner. Calyx *c.* 5·5 mm. Corolla pale violet. *Calcareous cliffs.* ● *Karpathos* (*near Vrondi*). Cr.

58. L. emarginatum (Willd.) O. Kuntze, *Rev. Gen. Pl.* **2**: 395 (1891) (*Statice spathulata* var. *emarginata* (Willd.) Boiss.). Plant 30–50 cm, glabrous. Leaves 40–60 × 7–15 mm, linear-spathulate, 1- to 5-veined, truncate to emarginate. Inflorescence cylindrical; non-flowering branches absent; scales 6–7 mm, sometimes the lowest up to 20 mm and leaf-like. Spikes 2–3 cm, lax, with 4 spikelets per cm; spikelets 1(–2)-flowered. Inner bract 5·5–7 mm; outer bract 2–2·5 mm, overlapping $\frac{1}{5}$–$\frac{1}{4}$ of the inner. Calyx 7–8 mm; limb $\frac{1}{2}$ as long as the tube. Corolla 11–12 mm. *Rocks near the sea.* ● *Gibraltar.* Hs.

L. spathulatum (Desf.) O. Kuntze, *op. cit.* 396 (1891) (*Statice spathulata* Desf.), from N. Africa, is like **58** but has acute leaves *c.* 90 mm, 2- to 4-flowered spikelets, and calyx-limb $\frac{1}{5}$ as long as the tube. It has frequently been recorded from S. Spain but probably in error for **58**.

59. L. duriusculum (Girard) Fourr., *Ann. Soc. Linn. Lyon* nov. ser., **17**: 141 (1869) (*Statice duriuscula* Girard). Sometimes annual, 20–30 cm, glabrous, smooth. Leaves 30–40 × 5–9 mm, oblanceolate-spathulate, obtuse to truncate, 1(–3)-veined. Inflorescence with the lower 1–5 branches usually sterile; non-flowering branches absent; scales 2–4 mm. Spikes 2–3(–6) cm, with 2–3(–4) spikelets per cm; spikelets 1- to 3-flowered, incurved. Inner bract 4·5–5 mm, smooth; outer bract overlapping $\frac{1}{4}$ of the inner. Calyx (4·5–)5·5–6 mm; limb $1\frac{1}{2}$ times as long as the tube; teeth becoming uncinate in fruit. *Saline soils and maritime rocks.* ● *W. Mediterranean region.* Bl Ga Hs ?It Sa ?Si.

A very critical species, resembling **56** in its incurved inner bract enclosing the calyx, but of similar habit to **87**, which is smaller and always annual, and to **28**, which is taller and always perennial.

60. L. ramosissimum (Poiret) Maire, *Bull. Soc. Hist. Nat. Afr. Nord* **27**: 244 (1936) (*Statice globulariifolia* Desf.). Plant 20–50 cm. Leaves 30–100 × 7–20 mm, obovate- to oblanceolate-spathulate or subspathulate, 1- to 5-veined. Inflorescence 20–50 cm; scales 3–20 mm, often leaf-like. Spikes 1–4 cm, rather dense, with 4–8 spikelets per cm; spikelets 2- to 5-flowered. Inner bract 4–5·5 mm, with reddish-brown margin, sometimes hyaline near apex; outer bract overlapping up to $\frac{1}{4}$ of the inner. Calyx 4–6 mm; tube subcylindrical, often incurved; limb cupuliform. Corolla 5–7 mm, pale pink. *Salt-marshes. Mediterranean region.* Bl Co Ga Gr Hs It Sa Si.

Subsp. *ramosissimum* occurs in Algeria.

1 Leaves rounded at apex; scales 11–20 mm, often leaf-like (a) subsp. **confusum**
1 Leaves acute or obtuse; scales 3–12 mm, never leaf-like
 2 Leaves obovate-spathulate to subspathulate, abruptly contracted into the petiole (e) subsp. **doerfleri**
 2 Leaves oblanceolate-spathulate, gradually narrowed into the petiole
  3 Non-flowering branches absent; spikes up to 4 cm (b) subsp. **provinciale**
  3 Non-flowering branches present; spikes up to 2 cm
   4 Non-flowering branches few; inner bract and calyx 4–4·5 mm (c) subsp. **tommasinii**
   4 Non-flowering branches numerous; inner bract and calyx 5–6 mm (d) subsp. **siculum**

(a) Subsp. **confusum** (Gren. & Godron) Pignatti, *Bot. Jour. Linn. Soc.* **64**: 366 (1971) (*Statice confusa* Gren. & Godron): Leaves up to 50 × 10 mm, oblanceolate-spathulate, rounded at apex. Inflorescence 20–30 cm; non-flowering branches absent; scales 11–20 mm, often leaf-like. Spikes up to 2 cm, with 6–8 spikelets per cm; spikelets 3- to 5-flowered, subpatent. Inner bract *c.* 4·5 mm. Calyx 5–5·5 mm; limb $\frac{1}{2}$ as long as the tube. *W. Mediterranean region.*

(b) Subsp. **provinciale** (Pignatti) Pignatti, *loc. cit.* (1971) (*L. globulariifolium* subsp. *provinciale* Pignatti, *Statice psiloclada* Boiss.): Leaves 80–100 × 15–20 mm, oblanceolate-spathulate, gradually narrowed into the petiole. Inflorescence 30–50 cm; non-flowering branches absent; scales 3–12 mm. Spikes up to 4 cm, with 4–6 spikelets per cm; spikelets 2- to 3-flowered, erecto-patent. Inner bract *c.* 5 mm. Calyx *c.* 5 mm; limb $\frac{2}{3}$ as long as the tube. ● *S. France and islands of W. Mediterranean.*

(c) Subsp. **tommasinii** (Pignatti) Pignatti, *loc. cit.* (1971) (*L. globulariifolium* subsp. *tommasinii* Pignatti, *Statice confusa* sensu Reichenb. fil., non Gren. & Godron): Leaves 30–40 × 7–12 mm, oblanceolate-spathulate, gradually narrowed into the petiole. Inflorescence 30–40 cm, with few non-flowering branches; scales 3–12 mm. Spikes with 5–6 spikelets per cm; spikelets 2(–3)-flowered, erecto-patent. Inner bract 4–4·5 mm. Calyx 4–4·5 mm; limb $\frac{1}{2}$ as long as the tube. ● *Coast of N.E. Italy.*

(d) Subsp. **siculum** Pignatti, *loc. cit.* (1971): Leaves up to 60 × 15 mm, oblanceolate-spathulate, gradually narrowed into the petiole. Inflorescence 30–50 cm, the lowest branches non-flowering; scales 3–12 mm. Spikes 1–1·5 cm; spikelets 2(–3)-flowered, incurved. Inner bract *c.* 5·5 mm. Calyx 5–6 mm; limb as long as the tube. ● *S. coast of Sicilia.*

This subspecies is similar to **56**, but the leaves are wider, distinctly 3-veined, glaucous and persistent in summer, while the spikes are short and dense (3–7 spikelets per cm), with an incurved secund axis.

(e) Subsp. **doerfleri** (Halácsy) Pignatti, *loc. cit.* (1971) (*Statice doerfleri* Halácsy): Leaves 30–50 × 9–15 mm, obovate-spathulate to subspathulate, abruptly contracted into the petiole; petiole *c.* 2 mm wide. Inflorescence 20–40 cm; scales 3–12 mm. Spikes up to 1·5 cm, with 5–6 spikelets per cm; spikelets 2- to 3-flowered, incurved. Inner bract 4–4·5 mm. Calyx (4·5–)5 mm; limb $\frac{2}{3}$ as long as the tube. ● *Kikladhes.*

61. L. ocymifolium (Poiret) O. Kuntze, *Rev. Gen. Pl.* **2**: 396 (1891) (*L. corinthiacum* (Boiss. & Heldr.) O. Kuntze, *Statice ocymifolia* Poiret). Plant 20–30 cm, glabrous, smooth. Leaves 30–40 × 10 mm, oblanceolate-spathulate, obtuse, 1-veined or indistinctly 3-veined. Inflorescence without non-flowering branches. Spikes up to 2·5 cm, rather dense, with 4 spikelets per cm; spikelets 2-flowered. Inner bract *c.* 3·8 mm, truncate; outer bract overlapping $\frac{1}{4}$ of the inner. Calyx *c.* 3·5 mm, almost completely enclosed by the bract; limb $1\frac{1}{2}$ times as long as the tube. ● *S. Greece.* Gr.

62. L. densissimum (Pignatti) Pignatti, *Bot. Jour. Linn. Soc.* **64**: 367 (1971) (*L. confusum* subsp. *densissimum* Pignatti). Like **61** but leaves whitish-glaucous; spikes up to 1·5 cm, with 6–7 spikelets per cm; outer bract overlapping $\frac{1}{3}$–$\frac{1}{4}$ of the inner. *Salt-marshes.* ● *N.E. Spain* (*mouth of R. Ebro*). Hs.

63. L. parvibracteatum Pignatti, *loc. cit.* (1971). Plant 40–90 cm, puberulent, rough; stems whitish. Leaves 40–60 × *c.* 20 mm, spathulate, mucronulate, (1–)3- to 5-veined. Inflorescence usually with non-flowering branches. Spikes 1–2 cm, with 4–5 spikelets per cm; spikelets 2-flowered. Bracts with narrow hyaline margin; inner bract *c.* 3 mm, obtuse, with midrib excurrent into conspicuous acute point; outer bract overlapping $\frac{1}{4}$ of the inner. Calyx *c.* 4·5 mm, exceeding bracts by almost $\frac{1}{2}$ its length; limb $\frac{2}{3}$ as long as the tube. *Calcareous slopes near the sea.* ● *E. Spain* (*near Calpe*). Hs.

64. L. girardianum (Guss.) Fourr., *Ann. Soc. Linn. Lyon* nov. ser., **17**: 141 (1869) (*Statice girardiana* Guss.). Plant 8–20 cm, glabrous, smooth. Leaves 20–50 × 8–13 mm, obovate-spathulate, acute, 1- to 5-veined, glaucous, in short basal rosettes. Flowering stems branched in upper $\frac{1}{3}$–$\frac{1}{2}$; non-flowering branches absent; a few sterile branches sometimes present. Spikes 0·7–1·3 cm, very dense, with 10–15 spikelets per cm; spikelets 3- to 4-flowered. Inner bract 3·5–4 mm, as long as wide, the midrib excurrent for 0·2–0·4 mm; outer bract overlapping $\frac{1}{3}$–$\frac{1}{2}$ of the inner. Calyx *c.* 4 mm, infundibuliform; limb $\frac{1}{3}$ as long as tube. Corolla *c.* 5 mm. $2n=35$. *Salt-marshes.* ● *Coasts of E. Spain and S. France.* ?Bl Ga Hs.

65. L. densiflorum (Guss.) O. Kuntze, *Rev. Gen. Pl.* **2**: 395 (1891) (*Statice densiflora* Guss.). Plant 15–40 cm, glabrous, smooth. Leaves 50–70 × 13–15 mm, oblanceolate, spathulate, acute, 3(–5)-veined, glaucous, in basal rosettes. Flowering stems branched only in upper $\frac{1}{3}$–$\frac{1}{2}$; non-flowering branches absent; sometimes a few sterile branches at the base. Spikes up to 2·5 cm, very dense, with 10 or more spikelets per cm; spikelets

2- to 3-flowered. Inner bract 5–5·5 mm, 1½ times as long as wide, the midrib excurrent for 0·8–1·6 mm; outer bract *c.* 3 mm, overlapping at least ⅓ of the inner. Calyx *c.* 5·5 mm; teeth 1½ times as long as wide. 2*n*=26. *Salt-marshes. Islands of W. Mediterranean.* ?Co Sa Si.

Records from Spain and France are referable to **64**.

66. L. dufourei (Girard) O. Kuntze, *loc. cit.* (1891) (*Statice dufourei* Girard). Plant 30–50 cm, sparsely hairy on stems and bracts. Leaves 30–60 × *c.* 10 mm, obovate-spathulate, 1(–3)-veined. Inflorescence usually cylindrical, with erect branches diverging at an acute angle; non-flowering branches numerous. Spikes 1·2–1·6 cm, dense, with 10 or more spikelets per cm; spikelets 3- to 4-flowered. Bracts with narrow hyaline margin; inner bract *c.* 5 mm; outer bract overlapping ⅓–½ of the inner. Calyx *c.* 6 mm, subcylindrical. *Saline soils near the sea.* ● *E. Spain (near Valencia).* Hs.

(67–71). L. binervosum group. Leaves 3–12 mm wide, 1- to 3-veined. Spikes 1–2 cm, or more or less rudimentary, dense, with 6–10 spikelets per cm. Inner bract truncate, with reddish-brown margin; outer bract 2–2·5 mm, overlapping ⅓–½ of the inner. Calyx conical or slightly infundibuliform; limb ⅔ as long as the tube, up to 2·7 mm in diameter.

1 Non-flowering branches numerous   **71. salmonis**
1 Non-flowering branches absent; rarely a few sterile branches at base
  2 Spikes reduced to clusters of 2–3 spikelets   **70. paradoxum**
  2 Spikes (at least the terminal of major branches) 1–3 cm, with 6–10 spikelets per cm
    3 Spikelets 6–8 per cm; outer bracts of adjacent spikelets not overlapping   **67. binervosum**
    3 Spikelets 8–10 per cm, closely set so that their outer bracts overlap those of the adjacent spikelets in the same row
      4 Leaves 4–8 mm wide, 3-veined at base   **68. recurvum**
      4 Leaves 3–6 mm wide, 1-veined   **69. transwallianum**

67. L. binervosum (G. E. Sm.) Salmon, *Jour. Bot.* (*London*) **45**: 24 (1907) (*Statice occidentalis* Lloyd, *S. dodartii* Girard). Plant 20–30 cm, glabrous, smooth. Leaves 40–55 × 8–12 mm, acute or obtuse, 1- to 3-veined. Inflorescence pyramidal; non-flowering branches absent, rarely the lower branches sterile; scales 7–10 mm. Spikes 2–2·5 cm, straight, erect, dense, with 6–8 spikelets per cm; spikelets 1- to 3-flowered. Inner bract 4·5–5 mm; outer bract *c.* 2·5 mm, overlapping ⅓–½ of the inner. Calyx 5–5·5 mm. Corolla *c.* 8 mm, violet-blue, with wide, imbricate petals. 2*n*=34–36. *Maritime cliffs and rocks.* ● *W. Europe, northwards to S.W. Scotland.* Br Ga Hb Hs Lu.

A smaller plant, with acute, 1-veined leaves (*Statice legrandii* Gaut. & Timb.-Lagr.), from the Mediterranean coast of France, which has been referred here, seems closer to **60**.

68. L. recurvum Salmon, *op. cit.* **41**: 67 (1903). Like **67** but leaves 4–8 mm wide, obtuse, 3-veined at base; spikes up to 3 cm, patent, often recurved; spikelets 8–10 per cm, closely set so that their outer bracts overlap those of adjacent spikelets in the same row. 2*n*=27. *Maritime cliffs.* ● *S. England.* Br.

69. L. transwallianum (Pugsley) Pugsley, *op. cit.* **62**: 277 (1924). Like **67** but leaves 3–6 mm wide, obtuse, mucronulate, 1-veined; inflorescence pyramidal; spikes patent; spikelets 8–10 per cm, closely set so that their outer bracts overlap those of adjacent spikelets in the same row; petals narrow, not imbricate. 2*n*=27, 35. *Maritime rocks.* ● *S.W. Wales, W. Ireland.* Br Hb.

70. L. paradoxum Pugsley, *op. cit.* **69**: 46 (1931). Like **67** but leaves *c.* 15 × 5 mm, 1-veined; inflorescence 5–15 cm; spikes 0·6–0·8 cm, reduced to clusters of 2–3 spikelets. 2*n*=33. *Maritime cliffs.* ● *S.W. Wales, N.W. Ireland.* Br Hb.

71. L. salmonis (Sennen & Elias) Pignatti, *Collect. Bot.* (*Barcelona*) **6**: 321 (1962). Plant 10–15(–40) cm, glabrous, smooth. Leaves 20–30 × 5–7 mm, acute, mucronate, 1-veined. Inflorescence with numerous non-flowering branches; scales 3–4(–8) mm. Spikes 1–2 cm, with 6–7 spikelets per cm; spikelets 1(–2)-flowered. Inner bract 4–4·5 mm; outer bract *c.* 2 mm, overlapping ⅓ of the inner. Calyx *c.* 5·5 mm. Corolla *c.* 7 mm, violet-blue. *Maritime cliffs.* ● *S.W. France, N.E. Spain.* Ga Hs.

72. L. auriculae-ursifolium (Pourret) Druce, *Brit. Pl. List* ed. 2, 77 (1928) (*Statice lychnidifolia* Girard). Plant 20–40 cm, glabrous, smooth. Leaves 25–100 × 9–20 mm, oblanceolate-spathulate, usually acute, mucronate, 3- to 7-veined, glaucous, often with resinous secretions at the base. Inflorescence pyramidal; non-flowering branches absent; scales up to 7 mm. Spikes 1–2·5 cm, with 5–12 spikelets per cm; spikelets 2- to 6-flowered. Inner bract *c.* 4·5 mm; outer bract 1·5–2·5 mm, overlapping ¼–½ of the inner. Calyx 5–5·5 mm, the teeth as long as wide. Corolla 7–8 mm, violet-blue. 2*n*=25, 26. *Maritime cliffs and salt-marshes. France, Iberian peninsula and Islas Baleares.* Bl Ga Hs Lu.

1 Outer bract *c.* 1·5 mm; spikes with at least 10 spikelets per cm
    (a) subsp. auriculae-ursifolium
1 Outer bract 2–2·5 mm; spikes with 5–8 spikelets per cm
  2 Spikes with 5–6 spikelets per cm; spikelets 2-flowered
    (b) subsp. lusitanicum
  2 Spikes with 8 spikelets per cm; spikelets 5- to 6-flowered
    (c) subsp. multiflorum

(a) Subsp. **auriculae-ursifolium**: Spikes dense, with at least 10 spikelets per cm; spikelets (2–)3-flowered, distichous. Outer bract *c.* 1·5 mm, overlapping ¼ of the inner. *Throughout the range of the species.*

(b) Subsp. **lusitanicum** (Pignatti) Pignatti, *Bot. Jour. Linn. Soc.* **64**: 367 (1971) (*L. globulariifolium* subsp. *lusitanicum* Pignatti): Leaves 35–40 × 10–15 mm, with wide, scarious wings. Spikes with 5–6 spikelets per cm; spikelets 2-flowered, borne at a very acute angle to the axis on main branches. Outer bract 2–2·5 mm, overlapping ⅓–½ of the inner. *C. Portugal* (*Trafaria to S. Martinho do Porto*).

Apparently intermediate between **60** and **70**.

(c) Subsp. **multiflorum** (Pignatti) Pignatti, *loc. cit.* (1971) (*L. binervosum* subsp. *multiflorum* Pignatti): Leaves 25–33 × 9–13 mm, acute. Spikes with 8 spikelets per cm; spikelets 5- to 6-flowered, more or less regularly distichous. Outer bract 2–2·5 mm, overlapping ⅓–½ of the inner. *S.C. Portugal* (*Torres Vedras to Colares*).

Apparently intermediate between **67** and **70**.

73. L. majoricum Pignatti, *Arch. Bot.* (*Forlì*) **31**: 89 (1955). Like **72** but leaves 30–40 × 20 mm, obtuse or emarginate, revolute, 3-veined, dark green; inflorescence 20–30 cm, corymbose; spikes 1–1·5 cm; spikelets 2(–3)-flowered; inner bract *c.* 5 mm; outer bract 2–2·5 mm, overlapping ⅓ of the inner. *Calcareous cliffs.* ● *Islas Baleares.* Bl.

74. L. ovalifolium (Poiret) O. Kuntze, *Rev. Gen. Pl.* **2**: 396 (1891) (*Statice ovalifolia* Poiret). Plant glabrous, smooth.

Leaves spathulate, acute, mucronate, 3- to 7-veined, glaucous; petiole 4–6 mm wide. Inflorescence branched only in upper half; non-flowering branches absent. Spikes *c.* 1 cm, very dense, with 10 or more spikelets per cm. Inner bract 2·8–3·1 mm; outer bract *c.* 1 mm, overlapping ⅕–¼ of the inner. Calyx 3·5–4 mm, conical. Corolla pale pink. 2*n*=16. *Coasts of Portugal and W. France.* Ga ?Hs Lu.

Subsp. *ovalifolium* occurs in Morocco.

(a) Subsp. **gallicum** Pignatti, *Collect. Bot.* (*Barcelona*) **6**: 316 (1962): Leaves *c.* 60 × 20 mm. Inflorescence 10–25 cm; scales 4–6 mm. Spikelets 1- to 3(–4)-flowered. *W. France.*

(b) Subsp. **lusitanicum** Pignatti, *op. cit.* 318 (1962): Leaves 100–150 × 20–30 mm. Inflorescence 30–50 cm; scales 10–25 mm, sometimes leaf-like. Spikelets 2- to 3-flowered. *Portugal.*

(75–79). L. delicatulum group. Plant glabrous, usually smooth. Leaves sometimes on somewhat woody basal branches. Inflorescence with few or no non-flowering branches. Spikes with 4–10 spikelets per cm; spikelets 1- to 3-flowered. Bracts with narrow hyaline margin; outer bract overlapping ¼ of the inner.

This group consists of a very polymorphic species (**77**) with wide ecological amplitude, and four narrowly endemic species. **75**, **76** and **78** occur only in coastal habitats, **79** in the interior. **77** hybridizes frequently with such closely related species as **85** and perhaps **83**.

1 Leaves not more than 14 mm wide, with distinct central and 2, usually indistinct, lateral veins
  2 Leaves 7–14 mm wide, obtuse or emarginate; spikes with 6 spikelets per cm   **78. gibertii**
  2 Leaves 5–6 mm wide, acute; spikes with 4–5 spikelets per cm   **79. costae**
1 Leaves 15–32 mm wide, distinctly 3- to 9-veined
  3 Petiole 3–4 mm wide; spikelets 1(–2)-flowered; inner bract 2·5–2·9 mm   **75. biflorum**
  3 Petiole not more than 2·5 mm wide; spikelets 2- to 5-flowered; inner bract 3–3·5 mm
    4 Leaves on somewhat woody basal branches 10–30 cm; petiole shorter than the lamina   **76. lausianum**
    4 Leaves in basal rosettes; petiole as long as or longer than the lamina   **77. delicatulum**

75. L. biflorum (Pignatti) Pignatti, *Bot. Jour. Linn. Soc.* **64**: 368 (1971) (*L. ovalifolium* f. *biflorum* Pignatti). Plant 40–60 cm. Leaves 60–80 × 25–27 mm, spathulate, 5- to 7-veined; petiole 3–4 mm wide, longer than the lamina. Inflorescence without non-flowering branches, sometimes with a few sterile branches at the base; scales 5–6 mm. Spikes 1–1·3 cm, with *c.* 8 spikelets per cm; spikelets 1(–2)-flowered. Inner bract 2·5–2·9 mm. Calyx 3·2–3·5 mm, subconical. Corolla *c.* 5·5 mm, pale violet. *Maritime salt-marshes.* ● *Islas Baleares (Mallorca).* Bl.

76. L. lausianum Pignatti, *op. cit.*: 369 (1971). Leaves 50–70 × 15–24 mm, oblanceolate-spathulate, 5- to 9-veined, sub-glaucous, on somewhat woody, basal branches 10–30 cm; petiole 2·5 mm wide, shorter than the lamina. Inflorescence 10–20 cm; sterile branches few or absent; scales 3–4 mm. Spikes 1 cm, dense, with 9 spikelets per cm; spikelets 3-flowered. Inner bract *c.* 3·5 mm. Calyx *c.* 3·5 mm, conical. Corolla *c.* 5·5 mm, bluish-violet. *Slopes near the sea.* ● *W. Sardegna (near Oristano).* Sa.

77. L. delicatulum (Girard) O. Kuntze, *Rev. Gen. Pl.* **2**: 395 (1891) (*Statice delicatula* Girard). Plant 40–70 cm, slightly rough. Leaves 35–50 × 18–32 mm, oblanceolate to spathulate, (3–)5- to 7(–9)-veined; petiole as long as or longer than the

lamina. Inflorescence corymbose; non-flowering branches absent, sometimes a few sterile branches at the base; scales 6–15 mm. Spikes 1–3 cm, with 5–10 spikelets per cm; spikelets 2- to 3-flowered. Inner bract *c.* 3 mm; outer bract *c.* 0·8 mm. Calyx *c.* 3·5 mm; limb ⅔ as long as the tube. Corolla *c.* 6 mm, pale pink. 2*n*=27. *Gypsaceous and saline soils. E. & S.E. Spain, Islas Baleares.* Bl Hs.

1 Scales 10–15 mm   (c) subsp. **valentinum**
1 Scales not more than 10 mm
  2 Scales 6–9 mm   (a) subsp. **delicatulum**
  2 Scales 2–3 mm   (b) subsp. **tournefortii**

(a) Subsp. **delicatulum**: Leaves 35–50 × 18–20 mm, oblanceolate, usually acute, gradually narrowed at base, 5- to 7-veined; petiole 1–2 mm wide. Scales 6–9 mm. Spikelets 2-flowered. *Throughout the range of the species.*

(b) Subsp. **tournefortii** Pignatti, *Collect. Bot.* (*Barcelona*) **6**: 305 (1962): Leaves 20–45 × 18–32 mm, spathulate, rounded, truncate or emarginate, abruptly narrowed at base, 3- to 7-veined; petiole *c.* 2 mm wide. Scales 2–3 mm. Spikelets 2- to 3-flowered. ● *E. Spain, Islas Baleares.*

(c) Subsp. **valentinum** Pignatti, *op. cit.* 306 (1962): Leaves 25–32 mm wide, oblanceolate-spathulate, (5–)7(–9)-veined. Scales 10–15 mm. Spikelets 2-flowered. ● *E. Spain (near Valencia).*

78. L. gibertii (Sennen) Sennen, *Diagn. Pl. Esp. Maroc* 271 (1936). Plant 8–30 cm, glabrous, smooth. Leaves 15–22 × 7–14 mm, obtuse or emarginate, 1-veined or obscurely 3-veined. Inflorescence pyramidal; non-flowering branches few or absent; scales 3 mm. Spikes 1·5–2·5 cm, with 6 spikelets per cm; spikelets 1- to 3-flowered. Inner bract *c.* 3 mm; outer bract overlapping ¼ of the inner. Calyx *c.* 3·2 mm. Corolla pale pink. 2*n*=25–27. *Calcareous cliffs.* ● *N.E. Spain (near Tarragona).* Hs.

79. L. costae (Willk.) Pignatti, *Collect. Bot.* (*Barcelona*) **6**: 302 (1962). Like **78** but leaves 20–30 × 5–6 mm, acute, sometimes mucronate; spikes with 4–5 spikelets per cm; spikelets 2- to 3-flowered. *Gypsaceous soils.* ● *N.E. Spain (Teruel to Urgel).* Hs.

80. L. sibthorpianum (Guss.) O. Kuntze, *Rev. Gen. Pl.* **2**: 396 (1891). Like **78** but leaves 20–40 × 10 mm, acute, long-mucronate, glaucous; non-flowering branches absent; scales 6–7 mm; spikes 1–1·3 cm, with 5–6 spikelets per cm; spikelets 1-flowered. *Slopes near the sea.* N.E. Sicilia. Si. (*Libya.*)

81. L. eugeniae Sennen, *Diagn. Pl. Esp. Maroc* 98 (1936). Plant 80–120 cm, glabrous, smooth, with woody stock bearing coriaceous, reddish-brown, triangular leaf-sheaths *c.* 10 × 12 mm. Leaves spathulate, coriaceous, glaucous; lamina *c.* 45 × 30 mm, acute, 7-veined; petiole 60–80 × 2 mm, with resin glands. Inflorescence narrow, little-branched; non-flowering branches absent; a few sterile branches sometimes present; scales *c.* 5 mm. Spikes *c.* 0·6 cm, at apices of short branches; spikelets 2- to 4-flowered, scarcely contiguous. Inner bract *c.* 2 mm; outer bract very small. Calyx *c.* 2·5 mm. Corolla white. 2*n*=16. *Dry grassland.* ● *S.E. Spain (Cuevas de Almanzora, Almería prov.).* Hs.

82. L. coincyi Sennen, *op. cit.* 73 (1936). Like **81** but plant 60–80 cm; leaves very glaucous; lamina 35–45 × 25–35 mm, 7- to 9-veined; petiole 30–40 mm; spikes up to 3·2 cm, with 3(–4) spikelets per cm; spikelets 1- to 2-flowered, not contiguous; inner bract 2–2·2 mm. 2*n*=16. *Dry grassland.* ● *S.E. Spain (near Murcia).* Hs.

Sometimes regarded as a hybrid between **83** and **77(b)** but the chromosome numbers suggest that this is unlikely.

83. L. album (Coincy) Sennen, *op. cit.* 72 (1936). Like **81** but plant 30–50 cm; leaves whitish; lamina *c.* 15 × 20 mm, 5-veined; petiole *c.* 30 mm, canaliculate; spikes up to 1·8 cm, with 4 spikelets per cm; spikelets 1- to 2-flowered; inner bract *c.* 3 mm, truncate; calyx 3·2–3·8 mm; corolla *c.* 4·5 mm. $2n=18$. *Gypsaceous soils.* ● *S.E. Spain (near Lorca).* Hs.

84. L. cymuliferum (Boiss.) Sauvage & Vindt, *Fl. Maroc* **1**: 68 (1952) (*Statice gummifera* Durieu ex Boiss. & Reuter). Plant 20–40 cm, glabrous, smooth. Leaves 30–50 × 14–22 mm, spathulate, whitish; lamina suborbicular, emarginate, often mucronate; petiole 20–40 × 2–4 mm, widening below into a reddish sheath and often with conspicuous resin glands. Inflorescence without non-flowering branches; scales 6–12 mm. Spikes up to 1 cm; spikelets 1- to 6-flowered. Inner bract *c.* 2·5 mm. Calyx 2·8–3·2 mm. Corolla *c.* 4 mm, white. $2n=16$. *Maritime cliffs and salt-marshes, and saline soils inland. S. Spain.* Hs. (*N. Africa.*)

Variants with 1-flowered (var. *uniflorum* Pignatti), 2- to 3-flowered (var. *cymuliferum*) and 4- to 6-flowered (var. *corymbulosum* Cosson, sensu stricto) spikelets occur throughout the range of the species and seem quite distinct, but their status needs further investigation.

85. L. supinum (Girard) Pignatti, *Collect. Bot.* (*Barcelona*) **6**: 309 (1962) (*Statice salsuginosa* Boiss.). Plant glabrous, slightly rough. Leaves 35–50 × 11–14 mm, oblanceolate-spathulate, 3-veined, usually withered at anthesis. Inflorescence 20–30 cm, hemispherical, much-branched; lower branches mostly non-flowering; scales 4–6 mm. Spikes 1–1·5 cm, with 6 spikelets per cm; spikelets 1(–2)-flowered. Bracts with narrow hyaline margin; inner bract *c.* 2 mm; outer bract *c.* 0·5 mm. Calyx *c.* 3 mm. Corolla *c.* 4 mm. $2n=26–27$. *Gypsaceous soils, usually inland.* ● *S.E. Spain (Almería to Alicante).* Hs.

Coastal populations are sometimes intermediate between **56** and **85**, and have well-developed leaves at anthesis, large spikes and more incurved spikelets. Plants intermediate between **77** and **85** are very common. A rather distinct variant (var. *diegoi* (Sennen) Pignatti, *L. diegoi* Sennen) has 7- to 9-veined leaves and a larger inflorescence (up to 60 cm) and is perhaps a hybrid between **85** and **77(b)**.

86. L. suffruticosum (L.) O. Kuntze, *Rev. Gen. Pl.* **2**: 396 (1891). Dwarf shrub with leafy basal branches 2–5 cm. Leaves 15–25(–70) × 2–5 mm, linear-subspathulate, 1-veined, fleshy, glaucous, enlarged at base into an auriculate sheath. Inflorescence 10–20 cm, distichous, spicate, usually simple or little-branched; non-flowering branches absent. Spikes 0·7–1 cm, subglobose, inserted on the main axis, sessile, very dense; spikelets 2- to 3-flowered. Inner bract *c.* 3 mm. Calyx 3–4·5 mm, conical. Corolla whitish. *Saline soils. S.E. part of U.S.S.R.* Rs (W, K, E).

87. L. echioides (L.) Miller, *Gard. Dict.* ed. 8, no. 11 (1768) (*Statice echioides* L.). Annual 5–30(–40) cm. Leaves 20–40 × 10–14 mm, obovate to spathulate, obscurely pinnately veined, with tuberculate glands. Scales (3–)5–6 mm. Spikes 2–10 (–20) cm, with up to 2 spikelets per cm; spikelets 2-flowered, curved. Inner bract 4–5 mm, tuberculate dorsally; outer bract overlapping ⅛ of the inner. Calyx *c.* 5 mm; teeth becoming

uncinate in fruit. Corolla pinkish, evanescent. $2n=18$. *Dry, saline soils. Mediterranean region, S. Portugal.* Bl Co Cr Ga Gr Hs It Lu Sa Si.

6. Goniolimon Boiss.[1]

Like *Limonium* but always perennial herbs; stems usually compressed; leaves coriaceous, few, in a basal rosette; spikelets with mucronate bracts; stigma capitate.

All species, except those for which another habitat is indicated, are found on steppes, dry grassland or stony hillsides.

1 Calyx-lobes wider than long, or obscure   **2. speciosum**
1 Calyx-lobes longer than wide
   2 Inner bract of spikelet entire   **1. elatum**
   2 Inner bract of spikelet 2- to 3-fid
     3 Calyx-tube glabrous
       4 Leaves spathulate, 3–4 times as long as wide, abruptly contracted into the petiole   **11. sartorii**
       4 Leaves linear-spathulate, 6–12 times as long as wide, gradually narrowed into the petiole
         5 Spikes dense, globose; corolla whitish   **9. collinum**
         5 Spikes lax, with 4–5 spikelets per cm; corolla reddish-purple   **10. besseranum**
     3 Calyx-tube pubescent or hirsute
       6 Bracts puberulent   **5. tataricum**
       6 Bracts glabrous
         7 Inner bract 3·5–4 mm   **3. heldreichii**
         7 Inner bract more than 4 mm
           8 Spikes lax, with 2–3 spikelets per cm; inner bract 7–10 mm
             9 Leaves 0·5–1 cm wide; inner and outer bracts subequal   **7. rubellum**
             9 Leaves 0·1–0·5(–0·7) cm wide; outer bract ⅓–½ as long as the inner   **8. graminifolium**
           8 Spikes dense, with 5–8 spikelets per cm; inner bract 5–6 mm
             10 Inner bract with middle mucro much longer than the lateral; spikes with 8 spikelets per cm   **4. dalmaticum**
             10 Inner bract with all mucros subequal; spikes with 5 spikelets per cm   **6. tauricum**

1. G. elatum (Fischer ex Sprengel) Boiss. in DC., *Prodr.* **12**: 634 (1848). Plant 30–80 cm, sparsely pubescent. Leaves 6–15 × 2·5–4 cm, obovate to oblanceolate, obtuse or emarginate, mucronate. Spikes 1·5–2·5 cm; spikelets 2- to 3(4)-flowered, scarcely contiguous. Bracts *c.* 4 mm, entire, with wide, hyaline margin. Calyx 5–6·5 mm; tube hairy, twice as long as the limb. ● *S.E. Russia, W. Kazakhstan.* Rs (C, E).

2. G. speciosum (L.) Boiss., *loc. cit.* (1848). Stems 10–50 cm, stout, subangular. Leaves 3–8 × 1–3 cm, obovate to obcuneate, obtuse, mucronate. Scales 1–2 cm. Spikes 1–2 cm, globose, dense; spikelets 3- to 4-flowered. Inner bract 2-fid, rarely entire; outer bract 7–8 mm, usually puberulent, with wide, hyaline margin. Calyx 7–8 mm; lobes wider than long, or obscure. Corolla reddish-purple. *E. Russia, from c.* 51° *to* 57° *N.* Rs (C, E). (*N. & C. Asia.*)

3. G. heldreichii Halácsy, *Verh. Zool.-Bot. Ges. Wien* **36**: 241 (1886). Plant 10–20 cm, glabrous; stems slender, cylindrical, flexuous, rough. Leaves 5–6 × *c.* 1·5 cm, spathulate, whitish-punctate. Scales *c.* 4 mm; mucro 2–3 mm. Spikes 2–3 cm, lax, with 3 spikelets per cm; spikelets usually 1-flowered. Bracts glabrous; inner bract 3·5–4 mm, 3-fid; outer bract *c.* 2 mm, acuminate. Calyx 6·5 mm; tube pubescent, 4 times as long as the limb; lobes longer than wide. Corolla whitish. ● *E. Greece* (*Thessalia*). Gr.

[1] By S. Pignatti.

4. G. dalmaticum (C. Presl) Reichenb. fil., *Icon. Fl. Germ.* **17**: 61 (1855). Plant 7–30 cm; stems stout, angular to compressed. Leaves 5–8 × 1·2–1·6 cm, lanceolate-spathulate, acute, mucronate, sparsely whitish-punctate. Scales 5–6 mm; mucro 1–2 mm. Spikes 1·5–3 cm, dense, with 8 spikelets per cm. Bracts glabrous, sometimes with stipitate glands; inner bract 5–6 mm, 3-fid, with narrow, hyaline margin, the middle mucro much longer than the lateral; outer bract 5–7 mm. Calyx 7–8 mm; tube sparsely pubescent, 2–2½ times as long as the limb. Corolla reddish-purple. ● *Balkan peninsula.* Al Bu ?Gr Ju ?Tu.

5. G. tataricum (L.) Boiss. in DC., *Prodr.* **12**: 632 (1848). Stems 20–30 cm, 2–3 mm in diameter, subangular. Leaves 2·5–15(–30) × 0·5–3·5 cm, obovate to lanceolate-spathulate, mucronate, 3- to 5-veined, densely whitish-punctate above. Scales 9–12 mm. Spikes 10–13 mm, dense, with 8 spikelets per cm; spikelets 1- to 2-flowered. Bracts puberulent; inner bract 5·5 mm, 3-fid, with narrow, hyaline margin, the mucros subequal; outer bract 5 mm, acute. Calyx 7 mm, hirsute; tube 1½ times as long as the limb; lobes 1·5 mm, longer than wide. Corolla reddish-purple. *S.E. Europe, from S. Jugoslavia to W. Kazakhstan, and extending northwards to 51° N. in S.C. Russia.* Bu Gr Ju Rm Rs (C, W, E) [Hu].

6. G. tauricum Klokov in Kotov & Barbarich, *Fl. RSS Ucr.* **8**: 521 (1957). Like **5** but stems narrowly winged above; spikes with 5 spikelets per cm; bracts glabrous; calyx 7·5–8 mm. *S. Ukraine.* ● Rs (W, K).

7. G. rubellum (S. G. Gmelin) Klokov & Grossh. in Grossh., *Opred. Rast. Kavk.* 593 (1949). Like **5** but stems less than 2 mm in diameter; leaves 3–6 × 0·5–1 cm, linear-spathulate, acute, 1-veined; spikes lax, with 2–3 spikelets per cm; spikelets 1- to 2-flowered; bracts 7–8 mm, glabrous, the inner with wide, hyaline margin, the mucros shortly emergent; calyx 7–8 mm. *S. Ukraine, S.E. Russia, W. Kazakhstan.* Rs (W, K, E).

8. G. graminifolium (Aiton) Boiss. in DC., *Prodr.* **12**: 633 (1848). Stems 20–50 cm, less than 2 mm in diameter. Leaves 6–12 × 0·1–0·5(–0·7) cm, linear-spathulate, acute, 1-veined. Spikes 1·5–3 cm, lax, with 2–3 spikelets per cm; spikelets 1- to 2-flowered. Bracts glabrous; inner bract 8–10 mm, 3-fid, with wide, hyaline margin; outer bract ⅓–½ as long as the inner. Calyx 8–9 mm, pubescent; tube about as long as the limb. *Sandy ground by rivers.* ● *S. Ukraine.* Rs (W).

9. G. collinum (Griseb.) Boiss., *loc. cit.* (1848). Stems 10–20 cm, subangular and narrowly winged towards base. Leaves 3–12 × 0·3–0·8 cm, linear-spathulate, gradually narrowed into the petiole. Inflorescence scorpioid-corymbose. Scales *c.* 8 mm, with hyaline margin. Spikes 1 cm, dense, globose; spikelets 1-flowered. Inner bract 3·5–4 mm, 3-fid, with narrow, hyaline margin, auriculate; outer bract 4·5 mm, rounded, acute. Calyx 7 mm, glabrous; lobes longer than wide. Corolla whitish. *E. half of Balkan peninsula, E. Romania.* Bu Gr Rm Tu.

10. G. besseranum (Schultes ex Reichenb.) Kusn. in Kusn., N. Busch & Fomin, *Fl. Cauc. Crit.* **4**(1): 202 (1902). Like **9** but leaves 3–8 × 0·3–1·5 cm, mucronate, sparsely whitish-glandular; inflorescence pyramidal; spikes 1–2 cm, lax, with 4–5 spikelets per cm; spikelets 1- to 2-flowered; bracts 3–4 mm; calyx 7–8 mm; corolla reddish-purple. 2n = 36. *From N.E. Bulgaria to S.C. Ukraine.* Bu Rm Rs (W).

11. G. sartorii Boiss., *Diagn. Pl. Or. Nov.* **3**(4): 67 (1859). Plant 10–20 cm, glabrous. Stems stout, arcuate, with whitish glands. Leaves 3–4 × 1 cm, spathulate, acute, densely whitish-glandular, 1-veined, abruptly contracted into the petiole. Scales 5–6 mm. Spikes 1–2 cm, dense; spikelets 1-flowered. Inner bract 5–5·5 mm, 3-fid, with wide, hyaline margin; outer bract 4·5 mm, auriculate, mucronate. Calyx 7–7·5 mm, glabrous; lobes longer than wide. Corolla pale pink. *Maritime rocks.* ● *E. Greece, Kikladhes.* Gr.

7. Limoniastrum Heister ex Fabr.[1]

Like *Limonium* but dwarf shrubs; corolla-tube about as long as the lobes; styles connate in basal half.

1. L. monopetalum (L.) Boiss. in DC., *Prodr.* **12**: 689 (1848). Stems 50–120 cm, branched, leafy. Leaves 2–3(–8) × 0·5(–1·5) cm, oblanceolate to linear-spathulate, fleshy, glaucous, expanded below into a wide, amplexicaul sheath. Spikes 5–10 cm, with fragile rhachis. Spikelets 1(–2)-flowered, 5–10 mm apart, nearly parallel to the rhachis; outer bract 4 mm, truncate; inner bract 8 mm, forming an ellipsoid envelope around the flowers. Calyx 9 mm, enclosed by the bracts. Corolla 1–2 cm in diameter, pink, violet when dry. *Maritime sands and salt-marshes. Mediterranean region, S. Portugal.* Cr Ga Hs It Lu Sa Si [Bl].

8. Psylliostachys (Jaub. & Spach) Nevski[2]

Pubescent annual herbs. Leaves in a basal rosette, simple. Inflorescence of simple spikes; spikelets 2- to 4-flowered, bracteate. Calyx infundibuliform; tube 10-ribbed at base; limb scarious, the lobes with excurrent ribs. Corolla infundibuliform, with long tube and short, deflexed lobes. Stamens slightly connate. Styles 5; stigmas filiform. Fruit with irregular dehiscence.

1. P. spicata (Willd.) Nevski, *Acta Inst. Bot. Acad. Sci. URSS* (Ser. 1) **4**: 314 (1937). Plant 10–40(–60) cm. Leaves 5–15 × 1–3·5 cm, oblanceolate in outline, deeply lobed, shortly petiolate; lobes irregularly triangular, obtuse, usually deflexed. Inflorescence up to 30(–40) cm. Terminal spike 2–9 × *c.* 1 cm, the lateral shorter, sessile; bracts 2–3 mm, pubescent. Calyx 3–3·5 mm; limb about as long as the tube; lobes broadly triangular, acuminate. Corolla *c.* 4–4·5 mm, with ovate lobes, pale pink. Fruit oblong-obovoid. *Saline clays or sands. S.E. Russia (near W. coast of Caspian sea).* Rs (?K, E). (*Caucasus, C. Asia.*)

---

[1] By S. Pignatti.   [2] By D. M. Moore.

EBENALES

CXXXVII. EBENACEAE[1]

Trees or shrubs; usually dioecious. Leaves simple, exstipulate, usually alternate. Flowers actinomorphic, usually 4- to 5-merous. Calyx often accrescent in fruit. Corolla sympetalous, urceolate-campanulate; lobes contorted. Male flowers with 14–24 stamens; female flowers with 8 staminodes. Ovary superior, 4- to 12-locular, with axile placentae. Fruit a large berry.

1. Diospyros L.[2]

Dioecious. Flowers in small, axillary cymes or sometimes solitary. Calyx accrescent and with recurved lobes in fruit. Styles 2–4. Seeds large, dorsally convex and laterally compressed.

1. **D. lotus** L., *Sp. Pl.* 1057 (1753). Tree up to 14 m, with furrowed bark. Young twigs pubescent. Leaves 6–12 × 2·5–5 cm, elliptical to oblong, rounded to broadly cuneate at base, acuminate, entire, pubescent when young but usually glabrescent above, shortly petiolate. Male flowers 2–3 together, *c.* 5 mm; female flowers solitary, 8–10 mm. Calyx with 4 short, acute, ciliate lobes, villous within. Corolla reddish- or greenish-white, with recurved, rounded, ciliate lobes *c.* ½ as long as the tube. Fruit *c.* 15 mm in diameter, globose, yellow or blue-black. *Cultivated locally for its edible fruits; widely naturalized in the Balkan peninsula and occasionally elsewhere in S. Europe.* [Al Bl Bu ?Ga Gr He Hs It Ju.] (*Asia.*)

D. kaki L. fil., *Suppl.* 439 (1781), from Japan and Korea, is sometimes cultivated in S. Europe for its edible fruits. It has male flowers *c.* 10 mm in diameter and the orange-yellow fruit is 35–70 mm in diameter.

CXXXVIII. STYRACACEAE[1]

Shrubs or trees. Leaves alternate, simple, exstipulate. Flowers actinomorphic, hermaphrodite. Calyx more or less adnate to the ovary. Corolla usually sympetalous. Stamens usually twice as many as the corolla-lobes. Ovary superior to inferior, 3- to 5-locular, with axile placentae. Fruit a drupe or capsule.

1. Styrax L.[2]

Flowers in short, lax, terminal or axillary racemes. Calyx campanulate. Corolla-lobes imbricate; tube usually very short. Stamens 10(–12), adnate to the base of the corolla. Ovary 3-locular; ovules few in each loculus. Fruit a dry, usually 1-seeded drupe.

1. **S. officinalis** L., *Sp. Pl.* 444 (1753). Stellate-pubescent shrub or small tree 2–7 m. Leaves 3–7 cm, broadly ovate to ovate-oblong, entire, obtuse, rounded or shortly cuneate at base. Inflorescence usually 3- to 6-flowered. Pedicels 1–2 cm. Calyx nearly entire; corolla *c.* 2 cm, white, campanulate. *E. Mediterranean region, extending westwards to W.C. Italy.* Al Cr Gr It Ju Tu [Ga].

OLEALES

CXXXIX. OLEACEAE[1]

Usually glabrous trees or shrubs. Leaves opposite, rarely alternate, exstipulate. Flowers hermaphrodite, rarely unisexual, 4-(5-) to 6-merous, actinomorphic. Calyx campanulate, often small. Corolla sympetalous or polypetalous, rarely absent. Stamens 2, often inserted on the corolla-tube. Ovary superior, 2-locular; style usually short or absent; ovules usually 2 in each loculus. Fruit a loculicidal capsule, a samara, a drupe or a berry; seeds 1–4.

1 Petals free or almost so, or absent
  2 Style absent; fruit a drupe   **9. Picconia**
  2 Style distinct, though sometimes short; fruit a samara
    3 Leaves usually pinnate; fruit with a terminal wing   **4. Fraxinus**
    3 Leaves simple; fruit with lateral wings   **2. Fontanesia**
1 Corolla with a distinct tube
  4 Leaves lepidote beneath   **7. Olea**
  4 Leaves not lepidote beneath
    5 Twigs green, 4-angled; leaves pinnate or trifoliolate   **1. Jasminum**
    5 Twigs brown or grey, terete; leaves simple
      6 Corolla yellow   **3. Forsythia**
      6 Corolla white or lilac
        7 Corolla *c.* 2 mm; tube very short   **8. Phillyrea**
        7 Corolla at least 5 mm; tube about as long to twice as long as lobes
          8 Flowers lilac, rarely white; corolla-tube twice as long as lobes; fruit a capsule   **5. Syringa**
          8 Flowers white; corolla-tube about as long as lobes; fruit a berry   **6. Ligustrum**

1. Jasminum L.[3]

Deciduous or evergreen shrubs, sometimes climbing. Twigs usually angled. Leaves opposite or alternate, imparipinnate; leaflets entire. Flowers hermaphrodite, usually in terminal or axillary cymes. Calyx campanulate, with 4–9 lobes. Corolla hypocrateriform, with 4–9 lobes, convolute in bud. Stamens

---

[1] Edit. T. G. Tutin.   [2] By T. G. Tutin.
[3] By J. do Amaral Franco and M. L. da Rocha Afonso.

epipetalous; anthers large; filaments short. Fruit a black berry, usually didymous.

J. odoratissimum L., *Sp. Pl.* 7 (1753), from Madeira, is cultivated in S.E. France for the perfume extracted from the flowers. It is like **3** but has almost terete branches and somewhat larger flowers and fruit.

1 Leaves opposite
  2 Flowers yellow, solitary   **1. nudiflorum**
  2 Flowers white, in terminal cymes   **2. officinale**
1 Leaves alternate
  3 Leaflets 0·7–2(–3) cm; calyx-lobes at least as long as tube
                                        3. fruticans
  3 Leaflets 2–6 cm; calyx-teeth minute   **4. humile**

1. J. nudiflorum Lindley, *Jour. Hort. Soc.* (*London*) **1**: 153 (1846). Deciduous shrub up to 5 m, with recurved branches. Leaves opposite; leaflets 3, 1–3 cm, ovate to oblong-ovate. Flowers appearing in winter, yellow, solitary, on short peduncles covered with small imbricate bracts. Calyx-lobes linear, about as long as or longer than tube. Corolla-limb 15–30 mm in diameter; lobes usually 6, obovate and undulate, about half as long as tube. *Widely cultivated for ornament, and locally naturalized.* [Ga.] (*W. China.*)

2. J. officinale L., *Sp. Pl.* 7 (1753). Deciduous or half-evergreen climbing shrub up to 10 m. Leaves opposite; leaflets 5–7(–9), 1–6 cm, more or less ovate, the terminal larger and acuminate. Flowers white, or purplish outside (var. *affine* (Royle ex Lindley) Dippel), fragrant, in terminal cymes of 2–10. Calyx-lobes 6–10 mm, subulate, longer than tube. Corolla-limb 20–25 mm in diameter; lobes 4–5, oblong, about as long as tube. *Cultivated for ornament, and widely naturalized in S. Europe.* [Ga ?He Hs It Ju Lu Rm Si.] (*S.W. Asia.*)

J. grandiflorum L., *Sp. Pl.* ed. 2, 9 (1762), from the mountains of S.E. Asia, with calyx-lobes twice as long as tube, is cultivated in S.E. France for the perfume extracted from the flowers.

3. J. fruticans L., *Sp. Pl.* 7 (1753). Evergreen or half-evergreen shrub up to 3 m, with slender, erect or patent, 4-angled branches. Leaves alternate; leaflets 3, rarely 1, 0·7–2(–3) cm, oblong to obovate-oblong. Flowers yellow, in cymes of 1–5 on short lateral branches. Corolla-limb 12–15 mm in diameter; lobes 5, oblong, slightly shorter than tube. Berry 7–9 mm, globose. *Scrub and margins of woods. S. Europe.* Al Bu Ga Gr Hs It Ju Lu Rm Rs (K) Tu [Be ?He Hu.]

4. J. humile L., *Sp. Pl.* 7 (1753). Evergreen shrub up to 1·5 m. Leaves alternate; leaflets 3, rarely 5 or 7, ovate to oblong. Flowers yellow, in axillary cymes of 2–6. Corolla-limb *c.* 10 mm in diameter; lobes half as long as tube. *Cultivated for ornament, and locally naturalized in S. Europe.* [Gr Ju Si.] (*W. China.*)

2. Fontanesia Labill.[1]

Deciduous shrubs. Leaves opposite, entire or minutely serrulate. Inflorescences shortly paniculate, axillary and terminal. Calyx and corolla 4-lobed. Fruit a samara, 1- or 2-seeded.

Literature: A. Lingelsheim in Engler, *Pflanzenreich* **72** (IV. 243): 7–9 (1920).

1. F. philliraeoides Labill., *Icon. Pl. Syr.* **1**: 9 (1791). Densely branched shrub up to 3 m. Leaves 1·5–4 cm, elliptic-lanceolate.

Calyx 0·5 mm. Corolla-lobes 1·5–2 mm, whitish. Filaments 2–2·5 mm; anthers 2–3 mm. Fruit 6–9 mm, ellipsoid, winged. *S.E. Sicilia (S. of Siracusa).* Si [It]. (*Anatolia and Syria.*)

3. Forsythia Vahl[1]

Deciduous shrubs. Leaves opposite, entire or serrate, simple (in most species). Inflorescences axillary. Flowers yellow, heterostylous, appearing before or with the leaves. Calyx and corolla 4-lobed. Ovary with several ovules. Fruit a capsule.

Literature: A. Lingelsheim in Engler, *Pflanzenreich* **72** (IV. 243): 109–113 (1920). F. Markgraf, *Mitt. Deutsch. Dendrol. Ges.* **42**: 1–12 (1930).

1. F. europaea Degen & Bald., *Österr. Bot. Zeitschr.* **47**: 406 (1897). 2 m, with upright branches. Leaves (2·5–)4–5(–7) cm, ovate to ovate-lanceolate, more or less acuminate, entire or serrate. Calyx 3 mm. Corolla-tube 4 mm; lobes 12 × 4 mm, narrowly oblong. Capsule 1·2–1·5 cm, smooth, ovoid, beaked. ● *N. Albania and adjacent parts of Jugoslavia.* Al Ju.

4. Fraxinus L.[2]

Deciduous trees. Twigs flattened at the nodes. Leaves opposite or sometimes in whorls of 3, imparipinnate, rarely reduced to one leaflet. Flowers hermaphrodite or unisexual, small, in terminal or axillary panicles. Calyx campanulate and 4-lobed, or absent. Corolla usually absent; when present with 2–6 more or less free petals; stamens hypogynous; stigmas 2. Fruit a samara, winged distally.

The leaves vary greatly from young to adult plants; in the juvenile stage they commonly bear a larger number of smaller, rounded and more coarsely toothed leaflets; the first leaves of the seedlings are simple.

1 Corolla present; panicles terminal or axillary on leafy twigs
                                        1. ornus
1 Corolla absent; panicles axillary, appearing before the leaves
  2 Leaflets distinctly petiolulate; calyx present; body of samara
    terete   **2. pennsylvanica**
  2 Leaflets sessile or subsessile; calyx absent; body of samara
    flattened
    3 Buds black; leaflets with more serrations than lateral veins
                                    3. excelsior
    3 Buds brown; leaflets usually with as many serrations as
    lateral veins
    4 Twigs and leaf-rhachis glabrous   **4. angustifolia**
    4 Twigs and leaf-rhachis densely pubescent   **5. pallisiae**

1. F. ornus L., *Sp. Pl.* 1057 (1753) (*Ornus europaea* Pers.). Tree up to 20 m, with smooth, grey bark; twigs usually glabrous, yellowish-grey to grey; buds greyish or brownish, more or less pruinose. Leaflets 5–9, 30–80 × 18–45 mm, ovate to lanceolate, cuspidate, distinctly petiolulate, irregularly serrulate. Leaflets of juvenile trees 15–30 × 10–20 mm, subsessile and crenate-serrate. Calyx small, deeply lobed, persistent. Corolla white; petals 4, 5–6 mm, linear. Samara 20–25 × 4–6 mm, obovate-linear to oblong, acute to emarginate. *Mixed woods, thickets and rocky places. Mediterranean region and S.C. Europe, northwards to S. Czechoslovakia and N.E. Romania.* Al Au Bu Co Cz Gr He Hs Hu It Ju Rm Sa Si Tu [Ga].

2. F. pennsylvanica Marshall, *Arbust. Amer.* 51 (1785) (*F. lanceolata* Borkh.). Tree up to 20 m, with brownish-red and slightly furrowed bark; twigs and petioles glabrous to densely pubescent, stout; buds brown. Leaflets 5–7(–9), 80–150 ×

---

[1] By P. S. Green.
[2] By J. do Amaral Franco and M. L. da Rocha Afonso.

25–50 mm, ovate-lanceolate to oblong-lanceolate, acuminate, broadly cuneate, irregularly serrulate to entire, glabrous above, more or less pubescent beneath, at least along the midrib; petiolules 3–6 mm; rhachis glabrous to densely white-pubescent. Calyx persistent; corolla absent. Samara 30–60 × 4–6 mm, the wing lanceolate to oblong-obovate, acuminate to rounded or rarely emarginate, decurrent to below the middle of the slender terete body *c.* 2 mm in diameter. *Cultivated for timber and for shelter in C. & S.E. Europe; sometimes naturalized.* [Au Bu Cz Ge Rm Rs (W, E).] (*E. North America.*)

F. americana L., *Sp. Pl.* 1057 (1753), also from E. North America, is planted on a small scale mainly in E. Europe, though the name is at times wrongly applied to *F. pennsylvanica*. It is like **2** but the petiolules are 5–15 mm and the wing of the samara is not decurrent.

3. **F. excelsior** L., *Sp. Pl.* 1057 (1753). Tree up to 40 m, with grey bark, at first smooth but rough and finally fissured on old trunks; twigs and petioles glabrous to densely pubescent; buds black. Leaflets 7–13(–15) (rarely reduced to a large terminal one, var. *diversifolia* Aiton), (30–)50–110 × 10–30(–40) mm, oblong-ovate to oblong-lanceolate, long-acuminate, tapering to a rounded base, sessile, usually crenate-serrulate (the serrations more numerous than the lateral veins), villous on the midrib and towards the base beneath; rhachis subglabrous to pubescent. Calyx and corolla absent; anthers dark purple. Samara (20–)25–50 × (5–)7–10 mm, the wing oblong-obcordate to lanceolate. 2*n*=46. *Most of Europe, except the northern, southern and eastern margins.* Au Be Br Bu Cz Da Ga Ge Hb He Ho Hs Hu It Ju No Po Rm Rs (B, C, W, K, E) Su ?Tu.

(a) Subsp. **excelsior**: Twigs and petioles glabrous; leaf-rhachis glabrous or puberulent at the leaflet attachments only. 2*n*=46. *Throughout the range of the species.*

(b) Subsp. **coriariifolia** (Scheele) E. Murray in Rech. fil., *Fl. Iran.* **52**: 6 (1968): Twigs, petioles and rhachis densely pubescent. *S.E. Europe.* Rm Rs (K). (*Iran.*)

4. **F. angustifolia** Vahl, *Enum. Pl.* **1**: 52 (1804). Tree up to 25 m, with grey bark, soon finely and deeply reticulate-fissured; twigs and petioles glabrous; buds dark brown. Leaflets 5–13, 30–90(–100) × 8–25(–30) mm, oblong- to linear-lanceolate, acuminate, cuneate and entire at the base, sessile or nearly so, the serrations usually as many as the lateral veins; rhachis glabrous. Leaflets of juvenile trees 7–15, 8–30 × 5–17 mm, obovate to ovate, obtuse to acute at the apex. Calyx and corolla absent. Samara 20–45(–50) × 6–10 mm, glabrous, the wing oblong-obcordate to lanceolate. *River-banks, flood-plains and deciduous woods. S. & E.C. Europe.* Al Au *Bl Bu Co Cz Ga Gr Hs Hu It Ju Lu Rm Rs (W, K) Sa Si Tu.

(a) Subsp. **angustifolia**: Leaflets always glabrous beneath. *W. Mediterranean region and Portugal.*

(b) Subsp. **oxycarpa** (Bieb. ex Willd.) Franco & Rocha Afonso, *Bot. Jour. Linn. Soc.* **64**: 377 (1971) (*F. oxycarpa* Bieb. ex Willd., *F. pojarkoviana* V. Vassil., *F. syriaca* sensu Hayek, non Boiss.): Leaflets pubescent along the proximal part of the midrib beneath. *E.C. Europe, and S. Europe from N.E. Spain eastwards.*

5. **F. pallisiae** Wilmott, *Jour. Linn. Soc. London* (*Bot.*) **43**: 284 (1916) (*F. holotricha* auct., ?an Koehne). Like **4(b)** but up to 30 m, with twigs, petioles and leaf-rhachis densely pubescent, and leaflets densely pubescent on both sides when young, be-

coming glabrescent above; samara pubescent. *River-banks and flood-plains.* ● *S.E. Europe, from Turkey-in-Europe to Moldavia.* Bu Rm Rs (W) Tu.

5. Syringa L.[1]

Deciduous shrubs or small trees. Leaves opposite, entire or serrate, simple. Inflorescences terminal or axillary, paniculate; flowers appearing after the leaves. Calyx small, with 4 shallow lobes. Corolla more or less cylindrical with 4 valvate, more or less cucullate lobes. Fruit a capsule.

Literature: A. Lingelsheim in Engler, *Pflanzenreich* **72** (**IV. 243**): 74–95 (1920). S. D. McKelvey, *The Lilac, a Monograph.* New York. 1928.

Leaves ovate, with subcordate to broadly cuneate base; inflorescences usually paired from apical axillary buds (terminal bud abortive), usually without basal leaves   **1. vulgaris**
Leaves elliptical to elliptic-lanceolate, with acute to broadly cuneate base; inflorescences usually solitary, terminal, leafy at the base   **2. josikaea**

1. **S. vulgaris** L., *Sp. Pl.* 9 (1753). 3(–7) m. Leaves 4–8 (–12) cm, ovate; base subcordate to broadly cuneate. Inflorescences 10–20 cm, usually paired, from apical axillary buds (the terminal bud abortive), usually without basal leaves. Corolla lilac, rarely white; tube 8–12 mm, slender, cylindrical; lobes patent or slightly deflexed. Capsule 8–10(–14) mm, ovoid, acuminate. *In scrub on rocky hill-slopes.* ● *From N.C. Romania to C. Albania and N.E. Greece; widely cultivated for ornament, and naturalized in W. & C. Europe.* Al Bu Gr Ju Rm [Au Be Br Cz Ga Ge Hb He Hu It Rs (K)].

2. **S. josikaea** Jacq. fil. ex Reichenb., *Pl. Crit.* **8**: 32 (1830). 3–4 m. Leaves 6–12 cm, elliptical to elliptic-oblanceolate or elliptic-lanceolate, somewhat glaucous beneath; base acute to broadly cuneate. Inflorescences 10–16 cm, terminal, usually solitary, with basal leaves. Corolla purplish; tube 7–11 mm, slender, infundibuliform; lobes slightly patent. Capsule 9–14 mm, ellipsoid, obtuse, mucronulate. ● *Mountains of Transylvania, and Ukrainian Carpathians.* Rm Rs (W) [Ge].

6. Ligustrum L.[2]

Deciduous or evergreen shrubs or small trees. Twigs terete. Leaves opposite, simple and entire, shortly petiolate. Flowers hermaphrodite, small, mainly in erect terminal panicles. Calyx campanulate, sinuately 4-dentate. Corolla hypocrateriform, white, with 4 lobes. Stamens epipetalous; anthers small, included or exserted; filaments short or long. Fruit a berry, usually blackish.

1. **L. vulgare** L., *Sp. Pl.* 7 (1753). Shrub 1–3(–5) m, with slender, patent branches and smooth grey bark. Young twigs puberulent, brownish. Leaves 3–6 × 1–2 cm, lanceolate, thin, deciduous or semi-evergreen. Panicles 3–6 cm, puberulent. Flowers fragrant; corolla-tube as long as limb; limb 4–6 mm in diameter; stamens with slightly exserted filaments and oblong anthers shorter than corolla-lobes. Berry 6–8 mm, subglobose or ovoid, lustrous black. 2*n*=46. *Wood-margins and scrub; somewhat calcicole. S., W. & C. Europe, northwards to c. 59° N. and extending eastwards to C. Ukraine; also in a small region of S.E. Norway and S.W. Sweden.* Al Au Be Br Bu Cz Ga Ge Gr Hb He Ho Hs Hu It Ju Lu No Po Rm Rs (W, K) Si Su Tu [Az].

L. ovalifolium Hassk., *Cat. Horto Bogor.* 119 (1844), from Japan, which differs from **1** in its glabrous young twigs, broader,

---

usually more evergreen leaves, larger, glabrous panicles, and corolla-tube 2–3 times as long as the limb, is widely planted for hedges (often as a golden-leaved cultivar) and may be locally naturalized.

L. lucidum Aiton fil. in Aiton, *Hort. Kew.* ed. 2, **1**: 19 (1810) (*L. japonicum* auct. eur., non Thunb.), from China and Japan, is planted as a street-tree in S. Europe, and sometimes in hedges. It forms a small tree up to 10 m, with glabrous twigs, evergreen leaves 8–12 cm, panicles 12–20 cm, and a pruinose berry.

7. Olea L.[1]

Evergreen trees or shrubs. Twigs terete or quadrangular, not flattened at the nodes. Leaves opposite, simple, lanceolate to obovate, entire, coriaceous. Flowers hermaphrodite or polygamous, arranged in many-flowered axillary or terminal racemes or panicles. Calyx small, with 4 teeth or shallow lobes. Corolla subrotate, 4-lobed, with the lobes slightly longer than wide. Stamens epipetalous; anthers large; filaments short. Fruit an oleaginous drupe.

1. O. europaea L., *Sp. Pl.* 8 (1753). Tree up to 15 m, with a broad crown and a thick trunk, sometimes a shrub; bark grey, finely fissured; twigs lepidote, grey; buds very small, lepidote-sericeous, greyish. Leaves (10–)20–80 × (3–)5–15(–20) mm, subsessile, mucronate, dark greyish-green and glabrous above, light grey and densely lepidote beneath. Panicles axillary. Corolla white. Drupe 10–35 × 6–20 mm, ellipsoid to subglobose, green when unripe, becoming black, brownish-green or rarely ivory-white. $2n=46$. *Woods and scrub in dry rocky places; commonly cultivated for its fruit. Mediterranean region, Portugal, Krym.* Al Bl Co Cr Ga Gr Hs It Ju Lu Rs (K) Sa Si Tu [He].

Var. *europaea* is the cultivated olive. Many cultivars have been selected for their edible fruit, which is an important source of oil. Wild plants (var. *sylvestris* Brot.) differ from the cultivars mainly in the spiny lower branches and small leaves and drupes.

8. Phillyrea L.[1]

Evergreen trees or shrubs. Twigs terete. Leaves opposite, simple. Flowers hermaphrodite, in short axillary racemes. Calyx small, more or less 4-lobed. Corolla subrotate, greenish-white, 4-lobed, with the lobes longer than wide and imbricate in bud. Stamens epipetalous; anthers large, exserted; filaments short; stigma 2-lobed. Fruit a dry bluish-black drupe with a crustaceous endocarp.

Literature: C. Sébastian, *Trav. Inst. Sci. Chérif. Sér. Bot.*, **6**: 1–100 (1956).

Leaves all similar, with 4–6 pairs of distant, nearly obsolete veins; calyx with short, rounded lobes; drupe apiculate   **1. angustifolia**

Leaves dimorphic, with 7–11 pairs of close, usually distinct veins; calyx with triangular lobes; drupe muticous   **2. latifolia**

1. P. angustifolia L., *Sp. Pl.* 7 (1753). Shrub up to 2·5 m with subfastigiate branches and grey, smooth bark; twigs, buds and petioles glabrous or puberulent. Leaves (20–)30–80(–100) × 3–15 mm, all similar, linear to lanceolate, entire or more rarely remotely serrulate; lateral veins 4–6 pairs, distant, nearly obsolete, making a small angle with the midrib, long, nearly straight, not or very slightly forked distally. Calyx thick, brownish, lobed to ¼ of its length; lobes rounded. Stigma rounded, with obtuse lobes. Drupe 6–8 × 5–7 mm, ovoid while young, later subglobose; style more or less persistent. *Evergreen scrub. W. & C. Mediterranean region, Portugal.* Al Bl Co Ga Hs It Ju Lu Sa Si [Rs (K)].

2. P. latifolia L., *Sp. Pl.* 8 (1753) (incl. *P. media* L.). Shrub or small tree up to 15 m, with subfastigiate branches while young, later spreading; bark grey, smooth, becoming finely reticulate; twigs, buds and petioles puberulent-tomentose. Leaves dimorphic; juvenile leaves 20–70 × 10–40 mm, ovate-cordate to ovate-lanceolate, rarely lanceolate, more or less dentate or serrate; adult leaves 10–60 × 4–20 mm, lanceolate to elliptical, entire or finely serrulate; lateral veins 7–11 pairs, close, usually distinct, making a wide angle with the midrib, sinuous, clearly forked distally. Calyx thin, yellowish, lobed to ¾ of its length; lobes triangular. Stigma elongate, with acute lobes. Drupe 7–10 mm, globose; style caducous. $2n=46$. *Evergreen woods. Mediterranean region, Portugal.* Al Bl Bu Co Cr Ga Gr Hs It Ju Lu Sa Si Tu.

Specimens from dry calcareous places have small and thick leaves, but their taxonomic status is uncertain.

P. media L., *Syst. Nat.* ed. 10, **2**: 847 (1759) represents the adult leaf-stage of **2**, and so cannot be accepted as a distinct taxon.

9. Picconia DC.[2]

Evergreen trees or shrubs, with smooth, whitish bark. Twigs terete. Leaves opposite, simple, entire or rarely serrulate, coriaceous, glabrous. Flowers hermaphrodite, small, in short axillary racemes. Calyx rotate-stellate, 4-lobed. Corolla rotate, white, deeply divided into 4 patent, oblong lobes. Stamens epipetalous; anthers subsessile, exserted. Fruit a dry drupe with a crustaceous endocarp.

1. P. azorica (Tutin) Knobl., *Notizbl. Bot. Gart. Berlin* **11**: 1028 (1934). Shrub or small tree. Twigs fulvous when young, becoming grey. Leaves 3–6(–10) × 1–3(–6) cm, lanceolate to obovate, acute to obtuse, with a prominent midrib and inconspicuous lateral veins beneath. Calyx-lobes *c.* 1 mm, ovate-mucronate, ciliate; anthers mucronate; stigma deeply emarginate. Drupe 10–13 mm, oblong-ovoid, black. *Scrub, up to 700 m.* ● *Açores.* Az.

---

[1] By J. do Amaral Franco and ·M. L. da Rocha Afonso.
[2] By J. do Amaral Franco.

GENTIANALES

CXL. GENTIANACEAE[1]

Glabrous bitter-tasting herbs. Leaves opposite (very rarely some alternate), entire, exstipulate, usually sessile and often connate at the base. Flowers 4- to 5(to 12)-merous, actinomorphic. Calyx more or less deeply lobed. Corolla-lobes contorted in bud. Stamens inserted on the corolla-tube. Ovary superior, 1-locular, with 2 parietal placentae, rarely almost 2-locular owing to the intrusive placentae; ovules numerous. Fruit a septicidal capsule.

1 Flowers 6- to 12-merous   **3. Blackstonia**
1 Flowers 4- to 5-merous
 2 Each corolla-lobe with a slender nectary-spur   **9. Halenia**
 2 Corolla-lobes without spurs
  3 Style slender, distinct, often caducous
   4 Flowers usually 5-merous; calyx-lobes keeled; anthers twisting spirally after dehiscence   **4. Centaurium**
   4 Flowers 4-merous; calyx-lobes not keeled; anthers not twisting spirally after dehiscence
    5 Calyx with short, deltate lobes; corolla-tube ovoid   **1. Cicendia**
    5 Calyx with long, linear lobes; corolla-tube cylindrical   **2. Exaculum**
  3 Ovary narrowed into the stout, persistent style, or stigmas sessile
  6 Each corolla-lobe with 2 fimbriate nectaries near the base
   7 Perennial; flowers in terminal racemes   **8. Swertia**
   7 Annual; flowers solitary   **7. Lomatogonium**
  6 Corolla-lobes without nectaries
   8 Corolla with appendages between the lobes; throat and lobes not ciliate   **5. Gentiana**
   8 Corolla without appendages between the lobes; throat or lobes usually ciliate   **6. Gentianella**

1. Cicendia Adanson[2]

Annual herbs. Calyx campanulate, with 4 short, deltate lobes. Corolla with ovoid tube and 4 short lobes. Anthers basifixed, cordate, not twisting spirally after dehiscence. Stigma peltate.

  1. C. filiformis (L.) Delarbre, *Fl. Auvergne* ed. 2, 29 (1800) (*Microcala filiformis* (L.) Hoffmanns. & Link). Slender annual 3–14(–20) cm. Stems simple or with few branches. Leaves 2–6(–10) mm, linear, few, soon withering. Flowers 3–6 mm, yellow. Capsule 4–5 mm, ovoid. $2n=26$. *Damp, sandy or peaty places. S. & W. Europe.* Az Be Bl Br Co †Da Ga Ge Gr Hb Ho Hs It Ju Lu Sa Si Tu.

2. Exaculum Caruel[2]

Like *Cicendia* but calyx deeply divided into linear lobes; corolla-tube cylindrical; anthers ovate; stigma bifid.

  1. E. pusillum (Lam.) Caruel in Parl., *Fl. Ital.* 6: 743 (1886) (*Cicendia pusilla* (Lam.) Griseb.). Slender, divaricately branched annual 1–12 cm. Leaves 2–12 mm, linear-lanceolate. Flowers 4–5 mm, pink or cream. Capsule 5–6 mm, fusiform. $2n=20$. *Damp, sandy or grassy places. S.W. Europe.* Co Ga Hs It Lu Sa.

3. Blackstonia Hudson[2]

Annual herbs. Calyx deeply divided into 6–12 lobes. Corolla rotate, yellow; tube short; lobes 6–12. Anthers basifixed, oblong to linear, not or scarcely twisting spirally after dehiscence. Style filiform, caducous; stigma 2-lobed.

  1. B. perfoliata (L.) Hudson, *Fl. Angl.* 146 (1762) (*Chlora perfoliata* (L.) L.). Erect, glaucous annual 10–60 cm. Leaves 5–30 mm, ovate to triangular, the basal obtuse, not connate at base, the cauline acute, connate or rarely free at base. Flowers (6–)8–35 mm, yellow. Calyx-tube very short; lobes 6–12, linear to lanceolate. Lobes of the stigma erect at anthesis, becoming patent later. *W., S. & C. Europe.* Al Au Be Bl Br Bu Co Cr Cz Ga Ge Gr Hb He Ho Hs Hu It Ju Lu Rs (K) Sa Si Tu.

1 Flowers 20–35 mm, 8- to 12-merous   **(d) subsp. grandiflora**
1 Flowers (6–)8–15 mm, 6- to 8-merous
 2 Upper cauline leaves not or scarcely narrowed towards the base; calyx-lobes linear   **(a) subsp. perfoliata**
 2 Upper cauline leaves narrowed towards the base; calyx-lobes linear-lanceolate to lanceolate
  3 Calyx-lobes linear-lanceolate, 3–4 times as long as tube; upper cauline leaves narrowly connate at base   **(b) subsp. serotina**
  3 Calyx-lobes lanceolate, twice as long as tube; upper cauline leaves not or scarcely connate at base   **(c) subsp. imperfoliata**

  (a) Subsp. **perfoliata**: Usually robust. Upper cauline leaves triangular, not or scarcely narrowed towards the base. Flowers 8–15 mm, 6- to 8-merous. Calyx-lobes linear, 1-veined. $2n=40$. *Throughout most of the range of the species, but rare in the south.*
  (b) Subsp. **serotina** (Koch ex Reichenb.) Vollmann, *Fl. Bayern* 594 (1914) (*B. serotina* (Koch ex Reichenb.) G. Beck, *B. perfoliata* subsp. *acuminata* (Koch & Ziz) Dostál, *Chlora serotina* Koch ex Reichenb.): Usually slender. Upper cauline leaves distinctly narrowed towards the base. Flowers (6–)8–10 mm, 6- to 8-merous. Calyx-lobes linear-lanceolate, often 3-veined in lower part, 3–4 times as long as tube. $2n=40$. *Usually inland. S. & C. Europe; Netherlands.*
  (c) Subsp. **imperfoliata** (L. fil.) Franco & Rocha Afonso, *Bot. Jour. Linn. Soc.* 64: 378 (1971) (*Chlora imperfoliata* L. fil.). Like (b) but upper cauline leaves not or scarcely connate at base; calyx-lobes lanceolate, distinctly 3-veined, twice as long as tube. $2n=20$. *Damp places near the sea, rarely inland. S.W. Europe extending to S. Italy and Sicilia.*
  (d) Subsp. **grandiflora** (Viv.) Maire in Jahandiez & Maire, *Cat. Pl. Maroc* 3: 578 (1934) (*Chlora grandiflora* Viv.): Robust. Upper cauline leaves slightly narrowed towards the base. Flowers 20–35 mm, 8- to 12-merous. Calyx-lobes linear-lanceolate, 3-veined. *Sicilia and Sardegna; perhaps also in Corse, Islas Baleares and S.E. Spain.*

Plants with small, pale yellow flowers growing with normal *B. perfoliata* near Granada (S. Spain) were described as **Chlora citrina** Boiss. & Reuter, *Pugillus* 77 (1852). This is probably a variant of subsp. *serotina* and does not appear to have been collected again.

4. Centaurium Hill[3]

Flowers in dichasial, corymbiform, capitate or spiciform cymes. Calyx deeply divided into (4)5, keeled, linear, acute lobes. Corolla pinkish-purple, rarely white or yellow, infundibuliform. Stamens usually inserted at apex of corolla-tube; anthers linear or oblong-linear, twisting spirally after dehiscence. Ovary with

---

[1] Edit. T. G. Tutin.   [2] By T. G. Tutin.   [3] By A. Melderis.

a short, filiform, persistent style with 2 caducous stigmas. Nectaries at the base of the ovary absent.

A taxonomically difficult genus. Numerous taxa have been described, but there is no uniformity in their treatment by various authors. Most of the infraspecific taxa hybridize freely in natural habitats, the hybrids being more or less inter-fertile and producing swarms which obscure the limits between the taxa involved.

Parallel variations in type of inflorescence, size of flowers, the scabridity of vegetative organs and calyx, length of the calyx in relation to the corolla-tube and insertion of stamens in the corolla-tube are frequent in the groups of related species. Insertion of stamens lower down inside the corolla-tube is usually associated with dwarf plants, smaller flowers, a more or less capitate inflorescence and more crowded leaves.

In this account the relative lengths of the calyx and corolla-tube refer to the flowers at anthesis.

Literature: K. Ronniger, *Mitt. Naturw. Ver. Steierm.* **52**: 312–321 (1916). A. Melderis, *Acta Horti Bot. Univ. Latv.* **6**: 123–156 (1932). L. Zeltner, *Bull. Soc. Neuchâtel. Sci. Nat.* **92**: 1–164 (1970).

1 Perennial with decumbent non-flowering shoots; leaves of non-flowering shoots petiolate **1. scilloides**
1 Annual or biennial without non-flowering shoots; leaves sessile
2 Corolla yellow, rarely slightly tinged with purple **14. maritimum**
2 Corolla not yellow
3 Flowers in a spiciform cyme **13. spicatum**
3 Flowers in a corymbiform, capitate or dichasial cyme, rarely solitary
4 Cauline leaves usually lanceolate or wider, rarely linear-spathulate, never parallel-sided, usually 3-veined; rosette-leaves more than 5 mm wide, 3- to 7-veined
5 Biennial with a distinct basal leaf-rosette; flowers usually in a corymbiform, rarely a capitate cyme; calyx up to ¾ as long as the corolla-tube; corolla-lobes usually more than 5 mm
6 Calyx usually ½–¾ as long as the corolla-tube; rosette-leaves oblong-obovate or obovate, distinctly attenuate towards the base **2. erythraea**
6 Calyx not more than ⅓ as long as the corolla-tube; rosette-leaves oblong-spathulate, wide at base **3. suffruticosum**
5 Annual usually without, rarely with a weak basal leaf-rosette; flowers in a dichasial cyme; calyx equalling or nearly equalling the corolla-tube; corolla-lobes usually 2–4 mm
7 Stem usually dichotomously branched in the lower part, with patent branches; cauline internodes usually 2–4; flowers long-pedicellate, usually in a lax dichasial cyme **11. pulchellum**
7 Stem branched in the upper part, with strict branches; cauline internodes usually 5–9; flowers shortly pedicellate, usually in a dense dichasial cyme **12. tenuiflorum**
4 Cauline leaves narrow, often parallel-sided, 1-veined; rosette-leaves not more than 5 mm wide, 1- to 3-veined
8 Cauline leaves 5–7 times as long as wide, narrowly elliptical, narrowly obovate or narrowly oblong; flowers usually sessile or subsessile
9 Plant 3–10 cm, densely caespitose, with numerous stems thickened towards the apex; rosette-leaves caducous; cauline-leaves crowded; flowers solitary or in a few-flowered cyme **4. chloodes**
9 Plant 5–25 cm, not caespitose, with several stems, branched in the upper part, not thickened towards the apex; rosette-leaves persistent; cauline leaves more or less distant; flowers solitary or in a more or less dense corymbiform cyme **5. littorale**
8 Cauline leaves 10–20 times as long as wide, linear-oblong, linear-spathulate or linear; flowers usually pedicellate
10 Calyx not more than ⅓ as long as the corolla-tube **10. microcalyx**
10 Calyx at least ½ as long as the corolla-tube
11 Plant tomentose-scabrid, greyish-green
12 Rosette-leaves not attenuate towards the base; flowers pedicellate; corolla-lobes usually 2–4 mm; capsule as long as the calyx **6. favargeri**
12 Rosette-leaves slightly attenuate towards the base; flowers subsessile; corolla-lobes usually 5–8 mm; capsule exceeding the calyx **7. triphyllum**
11 Plant smooth and glabrous, glaucous
13 Annual without a basal leaf-rosette; calyx ½–¾ as long as the corolla-tube; corolla-lobes 5–6 mm **8. rigualii**
13 Biennial with a dense leaf-rosette; calyx ¾ as long as or nearly equalling the corolla-tube; corolla-lobes 10–12 mm **9. linariifolium**

Sect. CENTAURIUM. Flowers pink, rarely white. Cyme corymbiform, capitate or dichasial. Stigmas broadly ovate, wider than long.

1. C. scilloides (L. fil.) Samp., *Lista Esp. Herb. Port.* 106 (1913) (*Erythraea diffusa* J. Woods). Perennial, with numerous decumbent, non-flowering shoots and ascending flowering stems up to 30 cm. Leaves of non-flowering shoots and the lower ones of flowering stems up to 1 cm, suborbicular to rhombic, with a short petiole; upper leaves lanceolate, sessile. Flowers 15–20 mm, pedicellate, solitary or in few-flowered cymes. Calyx *c.* ¾ as long as the corolla-tube; corolla-lobes 8–9 mm. $2n=20$. ● *Atlantic coasts of Europe from N. Portugal to Britain; Açores.* Az Br Ga Hs Lu.

2. C. erythraea Rafn, *Danm. Holst. Fl.* **2**: 75 (1800). Biennial (3–)10–50 cm. Stem usually solitary, sometimes several, usually branched in the upper part. Rosette-leaves 1–5 × 0·8–2 cm, obovate to elliptical, usually obtuse, 3- to 7-veined; cauline leaves usually much smaller and narrower, often acute, 3-veined. Flowers of variable size, pink to pinkish-purple, rarely white, in a dense or lax corymbiform cyme. Calyx ½–¾ as long as, rarely nearly equalling the corolla-tube. *Dry grassland, scrub and mountain slopes. Europe, northwards to c.* 61° *N. in N.W. Russia.* All except Fa Fe Is No Sb.

Variable in branching of the stem, shape and size of the leaves, presence or absence of scabridity on the stem, leaves, bracts and on the calyx, density of the inflorescence, size of the flowers, length of the calyx in relation to the corolla-tube, length of the anthers, etc. Populations showing different combinations of these characters and occupying more or less restricted areas have been treated by different authors as species, subspecies or varieties. In many cases, however, these taxa are linked by transitional forms. The following subspecies appear to be reasonably distinct.

1 Stem branched usually from the base or middle; rosette-leaves oblanceolate or narrowly elliptical, ± acute; flowers usually pedicellate, in a lax corymbiform cyme; corolla-lobes usually acute; calyx usually ⅔–¾ as long as the corolla-tube; capsule distinctly exceeding the calyx
2 Upper part of stem, bracts and calyces scabrid; calyx sometimes ½ as long as the corolla-tube; corolla-lobes usually 5–7 mm, pink, more than ½ as long as the tube **(e) subsp. rhodense**
2 Stem, bracts and calyces smooth; calyx ⅔–¾ as long as the corolla-tube; corolla-lobes usually 8–10 mm, usually equalling or longer than the tube, deep pink **(f) subsp. majus**

1 Stem branched usually in the upper part; rosette-leaves obovate or elliptical, usually obtuse; flowers sessile or subsessile, fasciculate, in a ± dense corymbiform cyme; corolla-lobes obtuse; calyx usually *c.* $\frac{1}{2}$ as long as the corolla-tube; capsule slightly exceeding the calyx

3 Corolla-lobes usually 6–8 mm, usually equalling the tube **(d) subsp. grandiflorum**

3 Corolla-lobes usually less than 6 mm; usually *c.* $\frac{1}{2}$ as long as the tube

4 Corolla-lobes usually 2·5–5 mm, middle cauline leaves ± linear-oblong; inflorescence fairly lax **(c) subsp. rumelicum**

4 Corolla-lobes usually 5–6 mm; middle cauline leaves lanceolate, narrowly lanceolate or narrowly elliptical; inflorescence dense

5 Stem, margin of leaves, bracts and calyx smooth **(a) subsp. erythraea**

5 Stem, margin of leaves, bracts and calyx scabrid **(b) subsp. turcicum**

(a) Subsp. **erythraea** (*C. umbellatum* auct., *Erythraea centaurium* auct., non (L.) Pers., *E. capitata* Willd.): Middle cauline leaves usually narrowly lanceolate or narrowly elliptical; calyx usually *c.* $\frac{1}{2}$ as long as the corolla-tube; stamens inserted at the apex or at the base (var. *capitatum* (Willd.) Melderis) of the corolla-tube. $2n=40$. *Almost throughout the range of the species.*

Hybrids with **5** ($2n=40$, 40–48), with **11** and with **12** ($2n=40$) occur.

Variants with the calyx $\frac{2}{3}$ as long to almost as long as the corolla-tube, occur in C. and S.E. Europe.

(b) Subsp. **turcicum** (Velen.) Melderis, *Bot. Jour. Linn. Soc.* **65**: 232 (1972) (*C. turcicum* (Velen.) Ronniger): Like **2(a)** but stem, margins of leaves, bracts and calyx scabrid; leaves narrower. Calyx more than $\frac{1}{2}$ as long as the corolla-tube. *C. & S.E. Europe.*

(c) Subsp. **rumelicum** (Velen.) Melderis, *loc. cit.* (1972) (*Erythraea centaurium* subsp. *rumelica* Velen.): Stem slender, branched in the upper part. Middle cauline leaves linear-oblong, acute. Flowers deep pinkish-purple, in a lax, many-flowered cyme. Corolla-lobes usually 2·5–5 mm. Calyx *c.* $\frac{1}{2}$ as long as the corolla-tube. $2n=20$. *C. & E. parts of Mediterranean region.*

(d) Subsp. **grandiflorum** (Biv.) Melderis, *op. cit.* 234 (1972) (*Erythraea grandiflora* Biv.): Stem sometimes branched from the base. Cauline leaves elliptical, obtuse, not conspicuously decreasing upwards. Calyx *c.* $\frac{1}{2}$ as long as the corolla-tube. Corolla-lobes usually 6–8 mm, usually equalling the tube. *S.W. Europe.*

The distribution and status of this subspecies is in need of further investigation.

(e) Subsp. **rhodense** (Boiss. & Reuter) Melderis, *loc. cit.* (1972) (*Erythraca rhodensis* Boiss. & Reuter, *Centaurium limoniiforme* W. Greuter): Stem in the upper part and bracts scabrid. Cauline leaves narrowly elliptical, linear-oblong or linear-spathulate, acute, conspicuously decreasing upwards. Flowers pink, pedicellate (the central often sessile or subsessile), in a lax cyme. Calyx scabrid, ($\frac{1}{2}$–)$\frac{2}{3}$–$\frac{3}{4}$ as long as the corolla-tube. Corolla-lobes usually 5–7 mm, narrowly elliptical, acute. $2n=40$. *C. & E. parts of the Mediterranean region; ?S.W. Germany.*

(f) Subsp. **majus** (Hoffmanns. & Link) Melderis, *op. cit.* 235 (1972) (*Erythraea major* Hoffmanns. & Link): Stem as in **2(e)**. Cauline leaves more or less lanceolate, acute, conspicuously decreasing upwards. Calyx $\frac{2}{3}$–$\frac{3}{4}$ as long as the corolla-tube. Flowers deep pink, pedicellate, in a very lax inflorescence. Corolla-lobes usually 8–10 mm, usually equalling or longer than the tube, elliptical, more or less acute. $2n=20$. ● *Mountain regions of S.W. Europe.*

C. latifolium (Sm.) Druce, *Ann. Scott. Nat. Hist.* **1905**: 48 (1905) formerly occurred on sand-dunes in N.W. England. It is like **2(a)** but with broadly ovate, 5- to 7-veined leaves, the cauline being crowded. The flowers are 7–9 mm, sessile, in a dense capitate cyme and the corolla-lobes 3–4 mm. It is probably a mutant of **2(a)**.

C. enclusense O. Bolós, Molinier & P. Monts., *Acta Geobot. Barcinon.* **5**: 60 (1970) (*Erythraea divaricata* Porta, non Schaffner ex Schlecht.), from Islas Baleares (Menorca), with a stem 30 cm, divaricately branched from the base, a large corolla, and calyx $\frac{3}{4}$ as long as the corolla-tube, seems to be related to **2(e)** but is in need of further study.

3. C. suffruticosum (Griseb.) Ronniger, *Mitt. Naturw. Ver. Steierm.* **52**: 321 (1916). Biennial 20–60 cm. Stem stout, branched above. Rosette-leaves 5–8 cm, spathulate-oblong, obtuse, 5-veined; cauline leaves gradually decreasing in size upwards, elliptic-oblong, subacute. Flowers *c.* 15 mm, in a dense corymbose cyme. Calyx $\frac{1}{4}$–$\frac{1}{3}$ as long as the corolla-tube. Corolla-lobes 5–6 mm, *c.* $\frac{1}{2}$ as long as the corolla-tube. $2n=20$. *Mountain regions of S.W. Europe.* ?Ga Hs.

4. C. chloodes (Brot.) Samp., *Lista Esp. Herb. Port.* 106 (1913) (*Erythraea chloodes* (Brot.) Gren. & Godron). Annual or biennial 3–10 cm. Stems numerous, forming dense tufts. Rosette-leaves caducous; cauline leaves up to 1 cm, fleshy, crowded, 1-veined. Flowers usually 10 mm, 3–10, in a more or less lax corymbiform cyme. Calyx equalling the corolla-tube. Corolla-lobes 3–5 mm, obtuse. $2n=40$. *Sea-cliffs and maritime sands.* ● *W. Europe from 40° to 46° N.* Ga Hs Lu.

5. C. littorale (D. Turner) Gilmour, *Kew Bull.* **1937**: 498 (1937) (*C. vulgare* Rafn). Biennial, usually 5–25 cm, with a basal leaf-rosette. Stem solitary or frequently several, branched above. Rosette-leaves 1–2 × 0·3–0·5 cm; cauline leaves shorter and narrower, all obtuse. Flowers sessile, in a more or less dense corymbiform cyme. *From Ireland and C. Fennoscandia southwards to N.W. France, Austria and S.E. Russia.* Au Be Br Cz Da Fe Ga Ge Hb Ho Hu ?It No Po Rm Rs (B, C, W, E) Su.

Hybridizes with **11** (*C.* × *aschersonianum* (Seemen) Hegi).

This species shows a considerable amount of morphological variation, especially in size and shape of the leaves, size of the flowers and in presence or absence of scabridity on vegetative organs and calyx. Several taxa have been described from S. & C. Europe and given specific, subspecific or varietal rank, but intermediates are frequently met with. The following subspecies can be recognized.

(a) Subsp. **littorale** (*Erythraea littoralis* (D. Turner) Fries): Stem, margins of leaves, bracts and calyx usually smooth; corolla-lobes 6–7 mm. Stamens inserted at apex or at the middle of corolla-tube (var. *glomeratum* (Wittrock) Melderis). $2n=38$, 40, 42, *c.* 56. *Maritime grassland and sandy places near the sea. N. Europe and W. part of U.S.S.R.*

(b) Subsp. **uliginosum** (Waldst. & Kit.) Melderis, *Bot. Jour. Linn. Soc.* **65**: 241 (1972) (*C. uliginosum* (Waldst. & Kit.) G. Beck ex Ronniger, *Erythraea tenuifolia* Griseb. pro parte): Stem, margins of leaves, bracts and calyx conspicuously scabrid. Corolla-lobes usually 3–6 mm, shorter than the corolla-tube. $2n=40$. *Saline habitats, mainly inland. E.C. & E. Europe, northwards to N. Germany.*

6. C. favargeri Zeltner, *Bull. Soc. Neuchâtel. Sci. Nat.* **93**: 73 (1970) (*Erythraea tenuifolia* Griseb. pro parte). Biennial 10–20 cm, tomentose-scabrid, with a dense rosette of linear-oblong leaves, not attenuate at the base. Stem solitary to several, branched from the middle or below. Cauline leaves linear. Flowers small, pedicellate, in a contracted pyramidal or subfastigiate cyme. Calyx equalling the corolla-tube. Corolla-lobes usually 2–4 mm, ovate-lanceolate or oblong, obtuse. Capsule as long as the calyx. $2n = 20$. ● *S.W. Europe.* Ga Hs Lu.

7. C. triphyllum (W. L. E. Schmidt) Melderis, *Bot. Jour. Linn. Soc.* **65**: 243 (1972) (*Erythraea gypsicola* Boiss. & Reuter). Biennial 15–30 cm, tomentose-scabrid, with a dense rosette of linear-oblong leaves slightly attenuate at the base. Stem often branched from the lower part. Cauline leaves linear. Flowers large, subsessile in a usually more or less dense cyme. Calyx usually equalling the corolla-tube. Corolla-lobes 5–8 mm, elliptical, acute. $2n = 20$. *Gypsaceous soils.* ● *C. Spain.* Hs.

8. C. rigualii Esteve, *Anal. Inst. Bot. Cavanilles* **23**: 182 (1965). Annual 40–50 cm, without a basal leaf-rosette. Stem branched from near the base. Cauline leaves linear, indistinctly 1- to 3-veined, the upper acute. Flowers long-pedicellate, in a very lax dichasial cyme. Calyx $\frac{1}{2}–\frac{3}{5}$ as long as the corolla-tube. Corolla-lobes 5–6 mm, lanceolate, acute. *Dry grassland.* ● *S. Spain.* Hs.

9. C. linariifolium (Lam.) G. Beck, *Fl. Nieder-Österr.* **2**: 935 (1893) (*Erythraea barrelieri* Dufour). Biennial 20–40 cm, with a basal leaf-rosette. Stem solitary to several, branched from or above the middle. Rosette-leaves usually $1–5 \times 0.3$ cm, linear-spathulate, indistinctly 3-veined, obtuse; cauline leaves linear or linear-lanceolate, decreasing upwards. Flowers 20–30 mm, numerous, in a lax corymbiform cyme. Calyx $\frac{3}{4}$ as long as to nearly as long as the corolla-tube. Corolla-lobes 10–12 mm, oblong-elliptical. $2n = 20$. *Dry pastures and dry, open habitats; calcicole.* ● *S.E. Spain, ?Mallorca.* ?Bl Hs.

10. C. microcalyx (Boiss. & Reuter) Ronniger, *Mitt. Naturw. Ver. Steierm.* **52**: 321 (1916). Biennial 10–30 cm, with a basal leaf-rosette. Stems usually several, usually branched in the upper part. Rosette-leaves *c.* 2×0.2–0.3 cm, linear, 3-veined; cauline leaves narrower and shorter, 1- to 3-veined. Flowers 10–15 mm, with short pedicels, in a lax, corymbiform cyme, with short bracts at the base. Calyx $\frac{1}{4}–\frac{1}{3}$ as long as the corolla-tube. Corolla-lobes 4–6 mm, obtuse. *Wet grassland.* ● *S. Portugal, S.W. Spain.* IIs Lu.

11. C. pulchellum (Swartz) Druce, *Fl. Berks.* 342 (1898) (*Erythraea pulchella* (Swartz) Fries; incl. *E. morierei* Corb., *C. meyeri* (Bunge) Druce). Annual 2–20 cm, without a basal leaf-rosette. Stem with 2–4 internodes, usually dichotomously branched from below the middle; branches erecto-patent. Leaves $0.2–1.5 \times 0.1$–1 cm, ovate to ovate-lanceolate, acute, increasing in length upwards. Flowers 10–12 mm, pinkish-purple, rarely white, in a lax dichasial cyme or solitary (sometimes 4-merous in small, delicate, unbranched plants). Calyx nearly equalling the corolla-tube. Corolla-lobes 2–4 mm. Stamens inserted at the apex or in the upper $\frac{1}{3}$ (var. *morierei* (Corb.) Melderis) of the corolla-tube. $2n = 36$. *Open habitats and damp grassy places, especially near the sea. Most of Europe, northwards to c. 61° N. in Finland.* All except Fa Is Rs (N) Sb.

12. C. tenuiflorum (Hoffmanns. & Link) Fritsch, *Mitt. Naturw. Ver. Wien* **5**: 97 (1907) (*Erythraea latifolia* auct., non Sm.). Annual 15–25(–40) cm, without or with a weak basal leaf-rosette. Stem with 5–9 internodes, branched above; branches strict. Cauline leaves $1–2.5 \times 0.8$–1 cm, ovate to elliptical, obtuse to subacute, increasing in length upwards. Flowers 12–14 mm, in a more or less dense dichasial cyme; pedicels at least 2 mm. Calyx nearly equalling the corolla-tube. Corolla-lobes usually 3–4 mm. *Damp grassy places, especially near the sea. S. & W. Europe northwards to S. England.* Bl Br Co Cr Ga Gr Hs It Ju Lu Rs (K) Sa Si Tu.

(a) Subsp. **tenuiflorum**: Plant with a weak basal leaf-rosette. Inflorescence of 2–90 flowers. Corolla-lobes deep pink, entire at the apex. Corolla-tube not constricted below the limb. $2n = 40$. *Almost throughout the range of the species.*

(b) Subsp. **acutiflorum** (Schott) Zeltner, *Bull. Soc. Neuchâtel. Sci. Nat.* **93**: 94 (1970) (*Erythraea acutiflora* Schott): Plant without a basal leaf-rosette. Inflorescence of 20–180 flowers. Corolla-lobes pale pink, notched at the apex. Corolla-tube constricted below the limb. $2n = 20$. *Mediterranean region.*

Sect. SPICARIA (Griseb.) Ronniger. Flowers pinkish-purple, rarely white, sessile or subsessile in a spike-like cyme. Stigmas subcapitate, wider than long.

13. C. spicatum (L.) Fritsch, *Mitt. Naturw. Ver. Wien* **5**: 97 (1907) (*Erythraea spicata* (L.) Pers.). Annual or biennial 10–55 cm, with a more or less caducous basal leaf-rosette. Stem usually branched from the base or middle. Rosette-leaves broadly ovate. Cauline leaves elliptic-oblong to lanceolate, 3- to indistinctly 5-veined. Flowers 12–14 mm. Calyx nearly equalling the corolla-tube. Corolla-lobes *c.* 4·5 mm. Stamens inserted at or below the apex of the corolla-tube. $2n = 22$. *Damp grassy or sandy places near the sea. S. Europe.* Al Bl Bu Co Cr Ga Gr Hs It Ju Lu Rm Rs (W, K, E) Sa Si Tu.

Variants with scabridulous leaf-margins, bracts and calyx, occurring in Bulgaria, Kriti and Turkey, have been separated from **13** as **C. subspicatum** (Velen.) Ronniger, *Mitt. Naturw. Ver. Steierm.* **52**: 321 (1916). Since many transitions occur between them, the status of this taxon is uncertain.

Sect. XANTHAEA (Reichenb. fil.) Ronniger. Flowers yellow, in a few-flowered, corymbiform cyme. Stigmas oblong, longer than wide.

14. C. maritimum (L.) Fritsch, *Mitt. Naturw. Ver. Wien* **5**: 97 (1907) (*Erythraea maritima* (L.) Pers.). Annual or biennial 10–20 cm, with a caducous basal leaf-rosette. Stem solitary, simple or branched above. Leaves $0.6–2 \times 0.3$–0.6 cm, elliptic-oblong, increasing in length upwards. Flowers 20–25 mm, yellow, rarely tinged with purple, pedicellate. Calyx $\frac{3}{4}$ as long to nearly as long as the corolla-tube. Corolla-lobes elliptical, subacute. Stamens inserted at the apex or in the upper $\frac{1}{3}$ (var. *shuttleworthianum* (Rouy) Melderis) of the corolla-tube. $2n = 20$. *Sandy or grassy places, usually near the sea. S. & W. Europe, northwards to N.W. France.* Al Az Bl Bu Co Cr Ga Gr Hs It Ju Lu Sa Si Tu.

5. Gentiana L.[1]

Glabrous herbs. Calyx (4–)5(–9)-toothed, the teeth partly joined by a membrane which forms the upper part of the tube. Corolla-tube campanulate, obconical or cylindrical; limb sometimes rotate, usually plicate between the lobes; throat naked; lobes usually 5, not fringed, with a usually small appendage in the sinus between each pair. Anthers basifixed. Ovary narrowed into the style, or style absent; stigmas 2, persistent. Nectaries at the base of the ovary.

1 Corolla-lobes much longer than the tube   **1. lutea**
1 Corolla-lobes not more than $\frac{1}{2}$ as long as the tube
  2 Flowers crowded in a terminal head and sometimes also in axillary clusters; leaves usually more than 10 cm

---

[1] By T. G. Tutin.

3 Corolla blue, unspotted
  4 Leaves linear-lanceolate; ovary stipitate   **11. decumbens**
  4 Leaves oblong-ovate; ovary sessile   **10. cruciata**
3 Corolla yellow or purple, usually with dark spots
  5 Calyx split down one side, otherwise entire, or with short
    irregular teeth
    6 Corolla reddish-purple   **4. purpurea**
    6 Corolla yellow   **5. burseri**
  5 Calyx not split down one side, with long teeth
    7 Calyx-teeth erect; corolla greenish-yellow with dark purple
      spots   **2. punctata**
    7 Calyx-teeth recurved; corolla bluish-purple with reddish-
      black spots   **3. pannonica**
2 Flowers solitary or few, not crowded in a terminal head;
  leaves less than 10 cm
  8 Appendage in the sinus between the corolla-lobes more than
    $\frac{1}{2}$ as long as the lobes, the corolla thus appearing 10-lobed
    9 Annual, without non-flowering shoots   **12. prostrata**
    9 Perennial, with non-flowering shoots
      10 Leaves linear-lanceolate; corolla 20–30 mm **13. pyrenaica**
      10 Leaves ovate to suborbicular; corolla 8–10 mm **14. boryi**
  8 Appendage in the sinus between the corolla-lobes much
    shorter than the lobes
    11 Annual, without non-flowering shoots
      12 Calyx-tube 2–3 mm wide in flower, angled but not winged
        28. nivalis
      12 Calyx-tube (3–)4–7 mm wide in flower, with wings 2–3
        mm wide   **29. utriculosa**
    11 Perennial, with non-flowering shoots
      13 Lower leaves not forming a well-marked rosette
        14 Most leaves more than 1 cm; flowering stems usually
          more than 10 cm
          15 Leaves distinctly 3- to 5-veined; flowers secund
            6. asclepiadea
          15 Leaves 1-veined; flowers erect   **7. pneumonanthe**
        14 Most leaves less than 1 cm; flowering stems usually less
          than 10 cm
          16 Leaves about 4 times as long as wide, linear to
            linear-lanceolate; calyx-teeth *c.* $\frac{1}{3}$ as long as tube
            26. rostanii
          16 Leaves about twice as long as wide, ovate-lanceolate to
            obovate or spathulate; calyx-teeth ($\frac{1}{3}$–)$\frac{1}{2}$–$\frac{2}{3}$ as long as
            tube
            17 Leaves obovate to spathulate, not cartilaginous at
              apex; margin smooth; calyx-teeth *c.* $\frac{2}{3}$ as long as
              tube   **25. bavarica**
            17 Leaves ovate-lanceolate, with a cartilaginous apex;
              margin strongly papillose; calyx-teeth $\frac{1}{3}$–$\frac{1}{2}$ as long
              as tube   **27. terglouensis**
      13 Lower leaves in a rosette; cauline leaves few
        18 Corolla yellowish-white with blue stripes and with
          bluish spots in the throat   **8. frigida**
        18 Corolla blue, sometimes with white lines and greenish
          spots in the throat
          19 Corolla (30–)50–70 mm; tube obconical; lobes erecto-
            patent
            20 Stigma-lobes oblong or linear, not fimbriate
              9. froelichii
            20 Stigma-lobes rounded, fimbriate **(15–21). acaulis** group
          19 Corolla 15–25(–30) mm; tube almost cylindrical; lobes
            patent
            21 Longest rosette-leaves about twice as long as cauline
              leaves   **22. verna**
            21 Longest rosette-leaves not or little longer than cauline
              leaves
             22 Rosette-leaves linear-lanceolate   **24. pumila**
             22 Rosette-leaves rhombic to suborbicular
              23 Rosette-leaves usually obtuse, with a flat apex;
                uppermost leaves usually separated from calyx by
                a distinct internode   **23. brachyphylla**
             23 Rosette-leaves usually acute, with an incurved apex;
                uppermost leaves usually arising at base of calyx
                27. terglouensis

Sect. GENTIANA. Leaves up to 30 cm, crowded towards the base of the stem. Calyx more or less membranous, with small, irregular teeth, sometimes split to the base down one side. Corolla rotate or campanulate, 5- to 9-lobed.

1. G. lutea L., *Sp. Pl.* 227 (1753). Perennial 50–120 cm. Stem simple, stout, erect. Leaves lanceolate to elliptical or broadly ovate, acute or obtuse, more or less petiolate, glaucous, with 5–7 strong longitudinal veins. Flowers pedicellate, crowded in terminal and axillary cymes. Corolla yellow, very rarely brick-red; lobes patent. ● *Mountains of C. & S. Europe.* Al Au Bu Co Ga Ge Gr He Hs It Ju Lu Rm Rs (W) Sa [Cz.]

(a) Subsp. **lutea**: Anthers free; stigmas spirally coiled after anthesis. $2n=40$. *Throughout the range of the species, except the Balkan peninsula.*
(b) Subsp. **symphyandra** (Murb.) Hayek, *Prodr. Fl. Penins. Balcan.* 2: 417 (1930): Anthers connate in a tube; stigmas erecto-patent after anthesis. *Balkan peninsula and S.E. Alps.*

Hybrids between **1** and the other species of the section occur, though rather infrequently.

2. G. punctata L., *Sp. Pl.* 227 (1753). Perennial 20–60 cm. Stems simple, stout, erect, with a metallic tinge above. Leaves lanceolate to broadly ovate, acute, petiolate, with 5–7 strong longitudinal veins. Flowers sessile, crowded in terminal and axillary clusters. Calyx with 5–8 green, erect teeth. Corolla 14–35 mm, pale greenish-yellow, usually with dark purple spots; lobes short, erect. Anthers connate at first, later free. $2n=40$. ● *Mountains of C. Europe, extending southwards to S. Bulgaria.* Al Au Bu Cz Ga Ge He It Ju Po Rm Rs (W).

3. G. pannonica Scop., *Fl. Carn.* ed. 2, **1**: 182 (1771). Like **2** but stems without metallic tinge; calyx with recurved teeth; corolla purple with reddish-black spots; anthers connate. $2n=40$. *Mountains of C. Europe, from N.E. Switzerland and S.W. Czechoslovakia to N.W. Jugoslavia.* Au Cz Ge He It Ju.

4. G. purpurea L., *Sp. Pl.* 227 (1753). Perennial 20–60 cm. Stems simple, erect. Leaves lanceolate to broadly ovate, acute, petiolate, with 5–7 strong longitudinal veins. Flowers sessile, in small, terminal clusters, sometimes also in few-flowered axillary whorls. Calyx membranous, split to the base down one side. Corolla 15–25 mm, reddish-purple with dark purple spots. Anthers connate. $2n=40$. ● *Mountains of C. Europe; S. Norway.* Au Ga Ge He It No.

5. G. burseri Lapeyr., *Hist. Abr. Pyr.* 132 (1813). Like **4** but flowers usually in large clusters; corolla up to 40 mm, yellow, often with brown spots. ● *Pyrenees, S.W. Alps.* Ga Hs It.

(a) Subsp. **burseri**: Corolla-lobes acute, the sinuses between them with an acute appendage. $2n=40$. *Pyrenees.*
(b) Subsp. **villarsii** (Griseb.) Rouy, *Fl. Fr.* **10**: 256 (1908): Corolla-lobes obtuse or subacute, the sinuses between them with a truncate appendage. $2n=c.$ 40. *S.W. Alps.*

Sect. PNEUMONANTHE (Gled.) Link. Leaves 1–10 cm, those at the base of the stem scale-like; stems with 1–several flowers. Calyx green, 5-lobed, sometimes split halfway to the base down one side. Corolla obconical, plicate, 5-lobed.

6. G. asclepiadea L., *Sp. Pl.* 227 (1753). Perennial 15–60 (–100) cm. Stems simple, slender, erect. Leaves lanceolate to ovate, acuminate, sessile, with 3–5 strong, longitudinal veins. Flowers 1–3 in the leaf-axils, secund, sessile or subsessile. Calyx-teeth subulate, much shorter than the tube. Corolla

35–50 mm, blue (rarely white), usually with reddish-purple spots within; appendage in the sinus small, obtuse. Anthers connate. Seeds winged. 2*n*=36. *Damp or shady places.* ● *C. Europe, mainly in the mountains, extending locally southwards to C. Italy and C. Greece, and eastwards to N.W. Ukraine.* Al Au Bu Co Cz Ga Ge Gr He Hu It Ju Po Rm Rs (W) Tu.

7. **G. pneumonanthe** L., *Sp. Pl.* 228 (1753). Perennial 5–40 cm. Stems usually simple, slender, erect or ascending. Leaves linear to oblong or ovate-lanceolate, obtuse or subacute, sessile, 1-veined. Stems with a terminal and sometimes 1–6 axillary flowers, which are pedicellate or subsessile. Calyx-teeth linear to linear-lanceolate, acute, about as long as the tube. Corolla 25–50 mm, blue with 5 greenish lines; appendage in the sinus small, acute. Anthers connate. Seeds not winged. 2*n*=26. *Wet places; calcifuge. Much of Europe, but absent from most of the islands, Fennoscandia north of c. 60° N. and much of the south.* Al Au Be Br Bu Cz Da Ga Ge He Ho Hs Hu It Ju Lu No Po Rm Rs (N, B, C, W, E) Su.

Sect. FRIGIDA Kusn. Leaves 1–2 cm, crowded towards the base of the stem; stems with a terminal and sometimes 1–2 axillary flowers. Calyx herbaceous or membranous, 5-lobed. Corolla campanulate or obconical, plicate, 5-lobed.

8. **G. frigida** Haenke in Jacq., *Collect. Bot.* 2: 13 (1789). Perennial 5–10(–15) cm. Stems simple, erect. Leaves linear-lanceolate to ovate-lanceolate, obtuse, somewhat fleshy, 1-veined, the lower shortly petiolate, the upper sessile, shortly sheathing at base. Calyx-teeth obtuse, about as long as the tube. Corolla 20–35 mm, yellowish-white with blue stripes and with bluish spots in the throat; appendage in the sinus triangular, acute. Anthers free. Seeds covered with white, membranous lamellae. 2*n*=24. ● *Stony places in the mountains; calcifuge. Carpathians, Alps of C. Austria; S.W. Bulgaria.* Au Bu Cz Po Rm ?Rs (W).

9. **G. froelichii** Jan ex Reichenb., *Fl. Germ. Excurs.* 865 (1832). Perennial 5–10 cm. Stems simple, erect. Basal leaves linear-lanceolate to oblong-lanceolate, acute, somewhat fleshy, 3-veined, shortly sheathing at base; cauline leaves usually 1–2 pairs. Calyx surrounded by the uppermost pair of leaves; teeth lanceolate, acute, somewhat shorter than tube. Corolla 30–40 mm, clear blue, unspotted; appendage in the sinus triangular, acute. Anthers connate. Seeds covered with white, membranous lamellae. 2*n*=42. ● *S.E. Alps.* Au It Ju.

Sect. CRUCIATA Gaudin. Leaves 5–10 cm, the basal forming a rosette, the cauline evenly spaced; stems with terminal and often axillary clusters of flowers. Calyx membranous, with irregular teeth or split down one side. Corolla nearly cylindrical, plicate, 4-lobed.

10. **G. cruciata** L., *Sp. Pl.* 231 (1753). Perennial 15–40 cm. Stems usually several from the basal rosette, stout, ascending, often purplish. Leaves oblong-ovate, obtuse, 3-veined, sheathing at base; cauline numerous. Corolla 20–25 mm, 4-lobed, bluish to greenish outside, clear blue inside; appendages in the sinus 1–several. Anthers free. Ovary sessile. Seeds not winged. 2*n*=52. *S., C. & E. Europe, extending north-westwards to the Netherlands; local.* Al Au Be Bu Cz Ga Ge Gr He Ho Hs Hu It Ju Po Rm Rs (N, B, C, W, K, E).

(a) Subsp. **cruciata**: Calyx-teeth broadly triangular, usually much shorter than the tube; corolla *c.* 3 times as long as calyx. *Throughout the range of the species.*

(b) Subsp. **phlogifolia** (Schott & Kotschy) Tutin, *Bot. Jour.*

Linn. Soc. 64: 378 (1971) (*G. phlogifolia* Schott & Kotschy): Calyx-teeth linear to linear-lanceolate, about as long as the tube; corolla *c.* twice as long as calyx. *E. & S. Carpathians, at high altitudes.*

11. **G. decumbens** L. fil., *Suppl.* 174 (1781). Like 10 but stems 5–30 cm; basal leaves linear-lanceolate; cauline leaves 2(3) pairs, linear; corolla (22–)30–35(–40) mm; ovary stipitate. *Dry grassland. E. Russia.* Rs (C, E). (*C. & N.E. Asia.*)

Sect. CHONDROPHYLLAE Bunge. Leaves up to 1 cm, not crowded towards the base of the stem; stems with solitary, terminal flowers. Calyx herbaceous, with 5 teeth. Corolla obconical, plicate, 5-lobed, the appendage in the sinus almost as long as the corolla-lobes. Anthers free. Seeds unwinged.

12. **G. prostrata** Haenke in Jacq., *Collect. Bot.* 2: 66 (1789). Annual 2–5(–7) cm. Stem procumbent or ascending, simple or with a few basal branches. Leaves narrowly obovate to oblong, obtuse, shortly sheathing at base, rather fleshy; margin cartilaginous. Calyx-teeth broadly triangular, obtuse. Corolla 10–20 mm, narrowly obconical, steel-blue, greenish-white towards the base. 2*n*=*c.* 36. *Damp, stony slopes, 2200–2720 m. E. Alps.* Au He It.

13. **G. pyrenaica** L., *Mantissa* 55 (1767) (*G. laciniata* Kit.). Perennial 3–10 cm. Stems simple, ascending. Leaves linear-lanceolate, acute or mucronate, shortly sheathing at base, rather coriaceous; margin cartilaginous. Calyx-teeth narrowly triangular, acuminate. Corolla 20–30 mm, obconical, purplish-blue, greenish towards the base. 2*n*=26. *E. Pyrenees; Carpathians; mountains of S.W. Bulgaria.* Bu Ga Hs Rs (W).

14. **G. boryi** Boiss., *Biblioth. Univ. Genève* ser. 2, 13: 410 (1838). Caespitose perennial 2–5 cm. Stems simple, ascending. Leaves ovate to suborbicular, obtuse, rather thick and coriaceous; margin cartilaginous. Calyx-teeth ovate, cuspidate. Corolla 8–10 mm, obconical, blue on the upper side of the lobes, dark bluish-green outside, with white plicae and a pale blue appendage in the sinus. *Damp mountain pastures, 900–3000 m.* ● *N., C. & S. Spain (Cordillera Cantábrica, Sierra de Gredos, Sierra Nevada).* Hs.

G. septemfida Pallas, *Fl. Ross.* 1(2): 101 (1788), from the Caucasian region, with a fimbriate appendage in the sinus of the corolla and usually several flowers on the stem, has been once recorded from Krym, probably erroneously. It is commonly cultivated for ornament.

Sect. MEGALANTHE Gaudin (Sect. *Thylacites* Griseb.). Leaves (0·5–)2–8 cm, crowded towards the base of the stem; flowers solitary, terminal. Calyx herbaceous, with 5 teeth. Corolla obconical, plicate, 5-lobed, the appendage in the sinus small. Anthers connate. Stigma 2-lobed. Seeds not winged.

(15–21). **G. acaulis** group. Perennials. Leaves crowded towards the base of the stem; margin cartilaginous. Stem 1–8(–12) cm, elongating in fruit. Flowers solitary, terminal. Calyx green, with 5 teeth. Corolla 40–70 mm, obconical, plicate, 5-lobed, the appendage in the sinus small, dark blue, with or without green spots within. Anthers connate. Stigma 2-lobed, the lobes rounded, fimbriate. Seeds unwinged.

A group of closely related taxa which have been variously regarded as species, varieties or, sometimes, subspecies. They are treated here as species, in accordance with most modern Floras, but the rank of subspecies might well be more appropriate. The diploid chromosome number, as far as is known, is universally

36, but there appears to be no experimental evidence about the genetic relationships of the various taxa. Occasional specimens have morphological characters intermediate between two taxa, though on geographical and ecological grounds it seems unlikely that they are of hybrid origin.

1 Mature rosette-leaves scarcely longer than wide   **19. alpina**
1 Mature rosette-leaves at least 1½ times as long as wide
  2 Mature rosette-leaves linear-oblanceolate to oblanceolate
                                                21. angustifolia
  2 Mature rosette-leaves lanceolate, elliptical or obovate
    3 Calyx-teeth triangular, widest at base, usually more than ½ as long as the tube   **15. clusii**
    3 Calyx-teeth lanceolate to ovate, narrowed at base
      4 Corolla with green spots in the throat; calyx-teeth usually less than ½ as long as the tube
        5 Calyx-teeth about 1½ times as long as wide; corolla-lobes acute or cuspidate   **18. acaulis**
        5 Calyx-teeth little longer than wide; corolla-lobes acuminate   **17. ligustica**
      4 Corolla nearly or quite without green spots in the throat; calyx-teeth usually at least ½ as long as the tube
        6 Calyx-teeth usually more than ½ as long as the tube; corolla-lobes acute   **16. occidentalis**
        6 Calyx-teeth about ½ as long as the tube; corolla-lobes acuminate   **20. dinarica**

15. G. clusii Perr. & Song., *Bull. Soc. Hist. Nat. Savoie* **1853**: 185 (1855). Leaves elliptical to oblong-lanceolate. Calyx-teeth usually at least ½ as long as the tube, triangular, widest at the base. Corolla with few or no green spots in the throat; lobes obtuse, usually apiculate. $2n=36$. *Calcicole.* ● *Mountains of C. & S. Europe, from S.W. Germany and the Carpathians to S. France, N. Italy and N. Jugoslavia.* Au Cz Ga Ge He It Ju Po Rm.

16. G. occidentalis Jakowatz, *Sitz.-Ber. Akad. Wiss. Wien* **108**(1): 342 (1899). Leaves elliptical to oblong-lanceolate. Calyx-teeth usually more than ½ as long as the tube, ovate to lanceolate, narrowed at the base. Corolla with few or no green spots in the throat; lobes acute to acuminate. *Calcicole.* ● *W. Pyrenees, Cordillera Cantábrica.* Ga Hs.

17. G. ligustica R. de Vilmorin & Chopinet, *Rapports Comm. VIII Congr. Internat. Bot., Sect. 21–27*, 166 (1956). Leaves oblong-obovate to broadly ovate, at most 3 times as long as wide. Calyx-teeth less than ½ as long as the tube, broadly ovate, narrowed at the base. Corolla with green spots in the throat; lobes acuminate or cuspidate. *Calcicole.* ● *Maritime Alps; C. Appennini.* Ga It.

18. G. acaulis L., *Sp. Pl.* 228 (1753) (*G. excisa* C. Presl, *G. kochiana* Perr. & Song.). Leaves lanceolate, elliptical or, less frequently, obovate. Calyx-teeth usually less than ½ as long as the tube, ovate, narrowed at the base. Corolla with green spots in the throat; lobes acute or cuspidate. $2n=36$. *Calcifuge.* ● *From the Alps and Carpathians southwards to N.E. Spain, C. Italy and C. Jugoslavia.* Au Bu Cz Ga Ge He Hs It Ju Rm Rs (W).

19. G. alpina Vill., *Prosp. Pl. Dauph.* 22 (1779). Leaves usually *c.* 1 cm, suborbicular. Calyx-teeth usually about ½ as long as the tube, ovate, narrowed at the base. Corolla with green spots in the throat; lobes obtuse, usually rounded. $2n=36$. *Calcifuge;* 2000–2600 m. ● *S.W. & W.C. Alps, C. Pyrenees, S. Spain (Sierra Nevada).* Ga He Hs It.

20. G. dinarica G. Beck, *Ann. Naturh. Mus. (Wien)* **2**: 128 (1887). Leaves broadly elliptical. Calyx-teeth about ½ as long

as the tube, narrowly lanceolate, narrowed at the base. Corolla unspotted in the throat; lobes acuminate. *Calcicole.* ● *Mountains of S.W. Jugoslavia and Albania; C. Italy (Abruzzesi).* Al It Ju.

21. G. angustifolia Vill., *Hist. Pl. Dauph.* **2**: 526 (1787). Leaves linear-lanceolate to oblanceolate. Calyx-teeth usually less than ½ as long as the tube, lanceolate, narrowed at the base. Corolla with green spots in the throat; lobes acute. $2n=36$. *Calcicole.* ● *S.W. Alps, Jura, Pyrenees.* Ga He It.

Sect. CALATHIANAE Froelich. Leaves up to 2(–3) cm, sometimes crowded towards the base of the stem; stems with usually solitary, terminal flowers. Calyx herbaceous, with 5 teeth. Corolla-tube almost cylindrical, plicate, with 5 patent lobes, the appendage in the sinus small. Anthers free. Stigma-lobes contiguous, forming a circular, slightly concave disc. Seeds not winged.

22. G. verna L., *Sp. Pl.* 228 (1753). Perennial 2–20 cm. Stem from a rosette, erect, elongating after flowering. Leaves lanceolate, elliptical or broadly ovate; margin more or less papillose; cauline leaves 1–3 pairs. Calyx-teeth lanceolate, acuminate; calyx-tube (3–)4–7 mm wide, winged on the angles. Corolla 15–25 mm; lobes deep blue; tube greenish-blue outside, with a white line from each sinus. *C. Europe, extending westwards to C. France; mountains of S. Europe; Arctic Russia; N. England; W. Ireland.* Al Au Br Bu Cz Ga Ge Gr Hb He Hs It Ju Po Rm Rs (N, W).

1 Wings of calyx 1–2 mm wide   **(a) subsp. verna**
1 Wings of calyx 2–4 mm wide
  2 Rosette-leaves broadly ovate, obtuse   **(b) subsp. pontica**
  2 Rosette-leaves narrowly lanceolate, acuminate
                                         (c) subsp. tergestina

(a) Subsp. **verna**: Leaves usually lanceolate or elliptical, usually acute. Wings of calyx 1–2 mm wide. Corolla-lobes obtuse. $2n=28$. *Throughout most of the range of the species.*
(b) Subsp. **pontica** (Soltok.) Hayek, *Prodr. Fl. Penins. Balcan.* **2**: 419 (1930): Rosette-leaves broadly ovate, obtuse, about twice as long as wide. Wings of calyx 2–3 mm wide. Corolla-lobes obtuse. *C. & E. parts of Balkan peninsula.*
(c) Subsp. **tergestina** (G. Beck) Hayek, *op. cit.* 420 (1930): Rosette-leaves narrowly lanceolate, acuminate, *c.* 4 times as long as wide. Wings of calyx *c.* 4 mm wide. Corolla-lobes acute. *N. & W. parts of the Balkan peninsula, S.W. Bulgaria, Italy.*

23. G. brachyphylla Vill., *Prosp. Pl. Dauph.* 23 (1779) (incl. *G. sierrae* Briq.). Perennial 3–6(–15) cm. Stem from a rosette, erect. Leaves rhombic to suborbicular, rarely more than twice as long as wide; margin cartilaginous; cauline leaves 0–1(–2) pairs. Calyx-teeth narrowly triangular, acute; calyx-tube 2–5 mm wide, angled but not or very narrowly winged on the angles. Corolla 15–30 mm; lobes deep blue; tube greenish-blue outside. *Alps, Pyrenees, Carpathians, S. Spain (Sierra Nevada).* Au Ga Ge He Hs It Ju Rm.

(a) Subsp. **brachyphylla**: Rosette-leaves rhombic, obtuse to subacute; calyx-tube not or very narrowly winged on the angles; corolla-lobes about twice as long as wide. $2n=28$. *Usually calcifuge;* 1800–4200 m. *Alps, Pyrenees, Sierra Nevada.*
(b) Subsp. **favratii** (Rittener) Tutin, *Bot. Jour. Linn. Soc.* **64**: 378 (1971) (*G. favratii* (Rittener) Favrat, *G. verna* var. *favratii* Rittener): Rosette-leaves obovate to suborbicular, rounded; calyx-tube distinctly winged on the angles; corolla-lobes usually as wide or wider than long. $2n=28, 30+2B, 32.$ *Calcicole;* 2000–2800 m. ● *Alps, Carpathians.*

24. **G. pumila** Jacq., *Enum. Stirp. Vindob.* 41, 215 (1762). Like **23** but rosette-leaves linear-lanceolate, acute; margin strongly papillose; cauline leaves 1–3 pairs; calyx-teeth linear-lanceolate, acute. $2n=20$. *Calcicole. S.W. & E. Alps, E. Pyrenees.* Au Ga Hs It Ju.

(a) Subsp. **pumila**: Calyx usually 11–14 mm; corolla-lobes acute. *E. Alps.*
(b) Subsp. **delphinensis** (Beauverd) P. Fourn., *Quatre Fl. Fr.* 857 (1938): Calyx usually 16–20 mm; corolla-lobes obtuse, sometimes apiculate. *S.W. Alps, E. Pyrenees.*

25. **G. bavarica** L., *Sp. Pl.* 229 (1753). Perennial 4–20 cm. Stem ascending. Leaves not forming a rosette, about twice as long as wide, obovate to spathulate, obtuse; margin cartilaginous, smooth. Calyx-teeth lanceolate, acute; calyx-tube somewhat infundibuliform, angled, very narrowly winged on the angles. Corolla 20–30 mm; lobes deep blue; tube paler blue. $2n=30$. *Damp places, usually between* 1800 *and* 3600 *m.* ● *Alps.* Au Ga Ge He It Ju.

26. **G. rostanii** Reuter ex Verlot, *Cat. Pl. Dauph.* 242 (1872). Like **25** but lower leaves rather crowded, the upper distant, all about 4 times as long as wide, linear or linear-lanceolate; margin not cartilaginous; calyx-teeth *c.* ⅓ as long as the tube. $2n=30$. *S.W. & S.C. Alps.* Ga It.

27. **G. terglouensis** Hacq., *Pl. Carn.* 9 (1782). Like **25** but 3–6 cm; leaves densely crowded, often imbricate, ovate-lanceolate, thick, with a cartilaginous apex; margin narrowly cartilaginous, strongly papillose; calyx-teeth ⅓–½ as long as the tube. $2n=40$. *Stony slopes,* 1900–2700 *m; usually calcicole.* ● *S. & E. Alps.* Au Ga He It Ju.

(a) Subsp. **terglouensis**: Leaves at base of flowering stems more or less erect; calyx-teeth *c.* ½ as long as tube. *S.E. Alps.*
(b) Subsp. **schleicheri** (Vacc.) Tutin, *Bot. Jour. Linn. Soc.* 64: 378 (1971) (*G. verna* var. *imbricata* f. *schleicheri* Vacc.): Leaves at base of flowering stems usually patent or erecto-patent, forming an apparent rosette; calyx-teeth *c.* ⅓ as long as tube. *Maritime Alps to W. Switzerland.*

28. **G. nivalis** L., *Sp. Pl.* 229 (1753). Annual 1–15(–30) cm. Stems simple or branched. Leaves ovate to elliptical, obtuse; margin weakly papillose. Calyx-teeth ½–⅔ as long as the tube, narrowly triangular, acute; calyx-tube 2–3 mm wide, angled but not winged on the angles. Corolla *c.* 15 mm, intense blue. $2n=14$. *N. Europe, southwards to* 56° 30′ *N. in Scotland; mountains of Europe southwards to the Pyrenees, S. Appennini and Bulgaria.* Au Br Bu Cz Fe Ga Ge He Hs Is It Ju No Po Rm Rs (N, W) Su.

29. **G. utriculosa** L., *Sp. Pl.* 229 (1753). Annual 6–25(–35) cm. Stems simple or branched. Leaves obovate to lanceolate, obtuse or acute; margin weakly papillose. Calyx-teeth ⅓–½ as long as the tube, lanceolate, acute; calyx-tube (3–)4–7 mm wide, angled, the angles with wings 2–3 mm wide. Corolla 15–20 mm, intense blue. *C. Europe, mainly in the mountains; mountains of Italy and Balkan peninsula.* Au Bu Ga Ge He It Ju Rm Rs (W).

6. Gentianella Moench[1]

Biennials or annuals. Stems usually more or less branched from above the basal rosette. Calyx divided to at least ½-way; lobes

---

[1] By N. M. Pritchard and T. G. Tutin, partly from data supplied by H. Merxmüller.

4–5, not joined by an inner membrane. Corolla cylindrical or obconical; lobes 4–5, more or less patent, without a small lobe in the sinus; corolla usually fimbriate in the throat. Anthers 4–5, versatile. Ovary gradually tapered to style, or style absent. Stigmas 2, persistent on the capsule. Nectaries at the base of the corolla.

Some species of Sect. *Gentianella* show variation which is to some extent correlated with time of germination and flowering and sometimes also with altitude. The non-commital term 'ecotypic variant' is used for these; they are listed under the appropriate species by names commonly found in Floras, where such are available, at whatever rank this may happen to be. Such citation does not imply approval of the appropriateness of the rank in question.

Literature: R. von Wettstein, *Denkschr. Akad. Wiss. Math.-Nat. Kl.* (*Wien*) 64: 309–382 (1896).

1 Corolla not fimbriate at base of lobes
  2 Corolla less than 12 mm, not fimbriate at throat or margin; flowers numerous, closely subtended by uppermost leaves **22. aurea**
  2 Corolla more than 25 mm, fimbriate or ciliate at margin; flowers solitary, not closely subtended by uppermost leaves
    3 Calyx-lobes equal; corolla with cilia about as long as width of lobes **3. ciliata**
    3 Two calyx-lobes longer and narrower than the others; corolla with cilia much shorter than width of lobes, occasionally absent
      4 Basal rosette withered at anthesis; stems simple or branched only in upper part **5. barbata**
      4 Basal rosette persistent at fruiting; stems branched from the base **4. detonsa**
1 Corolla fimbriate at base of lobes
  5 Calyx-lobes with strongly crispate, blackish margin **6. crispata**
  5 Calyx-lobes with flat or recurved, but not crispate, green margin
    6 Flowers 4-merous; calyx-lobes very unequal, the 2 larger enclosing the 2 smaller **(8–9). campestris** group
    6 Flowers 4- or 5-merous; calyx-lobes equal or somewhat unequal, but the larger not enclosing the smaller
      7 Calyx divided almost to base, the lobes not appressed to the corolla-tube
        8 Flowers usually 4-merous; pedicels several times as long as flowers; corolla-lobes patent **1. tenella**
        8 Flowers usually 5-merous; pedicels not more than twice as long as flowers; corolla-lobes ± erect **2. nana**
      7 Calyx divided to ¾ or less, the lobes usually appressed to the corolla-tube
        9 Corolla more than 25 mm **(13–21). germanica** group
        9 Corolla less than 25 mm
          10 Corolla more than twice as long as calyx, obconical; plant with long branches from base **(13–21). germanica** group
          10 Corolla less than twice as long as calyx, cylindrical; plant usually without long branches from base
            11 Uppermost internode and terminal pedicel together forming at least ½ height of plant; internodes usually 3 or fewer **(10–12). amarella** group
            11 Uppermost internode and terminal pedicel together forming at most ⅛ height of plant; internodes usually more than 3, all ± equal or the terminal one short
              12 Lowland or subalpine plants up to 60 cm, with short branches from base or middle upwards **(10–12). amarella** group
              12 Alpine plants not more than 15 cm, with spreading, leafy flowering branches from base
                13 Flowers usually 5-merous; corolla white or yellow; calyx-sinus rounded **(13–21). germanica** group
                13 Flowers usually 4-merous; corolla blue; calyx-sinus acute **7. columnae**

Sect. COMASTOMA (Wettst.) Pritchard. Calyx divided nearly to base; lobes 4–5, unequal. Corolla up to 12 mm, narrowly obconical, with a few scales forming a thin fringe at the base of the acute or subacute lobes. Style very short or absent; capsule sessile.

1. G. tenella (Rottb.) Börner, *Fl. Deutsche Volk* 542 (1912) (*Gentiana tenella* Rottb.). Annual or biennial 2–10 cm, rarely more. Basal leaves spathulate; cauline 1–4 pairs, elliptical, more or less acute; all leaves 5–12(–18) mm. Pedicels several times as long as flowers. Calyx and corolla usually 4-merous, with 2 nectaries between each pair of stamens at the base. Corolla sky-blue or dirty violet, rarely white or yellowish; lobes 6–12 mm, more or less patent at anthesis. $2n = 10$. *Pastures, screes and river-gravels, in mountains except in the extreme north; usually calcicole. Europe, southwards to the Sierra Nevada, S. Alps and S. Carpathians.* Au Cz Fe Ga Ge He Hs Is It No Po Rm Rs (N, W) Sb Su.

2. G. nana (Wulfen) Pritchard, *Bot. Jour. Linn. Soc.* **65**: 260 (1972) (*Gentiana nana* Wulfen). Like **1** but 2–3 cm; pedicels not more than twice as long as flowers; calyx and corolla usually 5-merous, with 1 nectary between each pair of stamens; corolla dark blue; lobes more or less erect at anthesis. $2n = 30$. *Damp screes and late snow-patches; calcifuge.* ● *E. Alps.* Au It.

Sect. CROSSOPETALAE (Froelich) Pritchard. Calyx-lobes 4, equal or unequal, not overlapping. Corolla at least 25 mm, obconical, 4-lobed; lobes ovate to obovate, obtuse, usually more or less fringed or ciliate at margin, patent. Style distinct; capsule stipitate.

3. G. ciliata (L.) Borkh., *Arch. Bot. (Roemer)* **1**(1): 29 (1796) (*Gentiana ciliata* L.). Biennial 5–30 cm. Lower leaves spathulate, obtuse; cauline 1–3 cm, lanceolate or linear-lanceolate, acute. Internodes usually 4–6, more or less equal and about equalling the terminal pedicel. Calyx-lobes equal, ½ as long as corolla. Corolla 25–50 mm, blue; lobes ovate, long-fimbriate at margin. *Europe, except the north, the extreme west and most of the islands.* Al Au Be Bu ?Co Cz Ga Ge Gr He Ho Hs Hu It Ju Po Rm Rs (N, C, W, E).

(a) Subsp. **ciliata**: 7–30 cm. Corolla blue, usually more than 35 mm long and more than 10 mm wide below the throat; lobes usually more than 17 mm, oblanceolate, apiculate. $2n = 44$. *Meadows and wood-margins. Throughout the range of the species eastward to Poland and C. Ukraine.*

(b) Subsp. **doluchanovii** (Grossh.) Pritchard, *Bot. Jour. Linn. Soc.* **65**: 260 (1972) (*Gentiana doluchanovii* Grossh.): Like subsp. (a) but somewhat smaller in all its parts, rarely more than 20 cm. Corolla dark blue, usually less than 35 mm long and not more than 10 mm wide below the throat; lobes usually less than 17 mm, triangular, acute. *Wet places. E. Russia.* (*Siberia.*)

4. G. detonsa (Rottb.) G. Don fil., *Gen. Syst.* **4**: 179 (1837) (*Gentiana detonsa* Rottb.). Annual or biennial 5–25 cm. Basal leaves spathulate, obtuse, persistent at fruiting; cauline 1–3 pairs, up to 25 mm, linear-lanceolate, acute. Pedicels up to 12 cm, much longer than the internodes. Calyx-lobes unequal, the longer pair about equalling the corolla-tube. Corolla (25–)35–40 mm, dark blue; lobes 10–15 mm, often with a few short cilia on the margins, the apex usually serrulate. $2n = 44$. *Damp places, mainly by the sea. Arctic and subarctic Europe.* Is No Rs (N).

5. G. barbata (Froelich) Berchtold & J. Presl, *Rostlinář* **1**: 22 (1823) (*Gentiana barbata* Froelich). Like **4** but up to 40(–60) cm; basal leaves withered at anthesis; cauline up to 40(–100) mm;

pedicels usually not more than 6 cm; corolla-lobes with entire or slightly toothed apex. *S. Ural.* Rs (C). (*N. & C. Asia.*)

Sect. GENTIANELLA. Calyx-lobes 4–5, equal, or unequal. Corolla usually more than 1·5 cm, obconical or cylindrical, fimbriate in throat, 4- or 5-lobed, lobes erect or patent. Style absent. Capsule stipitate or sessile.

Throughout the section, variation in leaf-shape, number of internodes, density of branching, and ratio of leaf-length to internode-length show parallel variation. These characters are more or less correlated with flowering time, so that plants flowering before mid-August have in general obtuse middle cauline leaves, (0–)2–6 internodes, few branches, and internodes longer than the leaves; plants flowering after mid-August have acute middle cauline leaves, 6–12(–15) internodes, many branches and leaves longer than the internodes. Much of the confused nomenclature results from the intersection of these morphological gradients with other more definite characters; most of the morphological nodes thus formed have been given specific rank. Confusion has arisen also from the variable duration of the plants; most are overwintering annuals, germinating in late summer and autumn to form rosettes of 4–10 acute, lanceolate leaves. This rosette withers during winter in most species, leaving only a small bud. In spring, the secondary rosette formed from this bud has obtuse, oblanceolate or spathulate leaves. Sometimes spring or early summer germination is followed by flowering in the same year from the primary rosette, and these phenological differences may again transgress the usual taxonomic boundaries. Hybridization is apparently common and hybrid swarms occur in zones of overlap between species.

6. G. crispata (Vis.) J. Holub, *Folia Geobot. Phytotax. (Praha)* **2**: 117 (1967) (*Gentiana crispata* Vis.). Glabrous biennial; stem 2–20 cm, covered at base with brown remains of the previous year's leaves. Basal leaves obovate-spathulate, middle cauline elliptic-lanceolate, subobtuse, the upper ovate-lanceolate, acute. Inflorescence corymbose. Flowers 5-merous. Calyx-lobes subequal, about as long as the tube, with strongly crispate, blackish margin. Corolla 12–20 mm, violet or whitish; tube nearly twice as long as calyx. Capsule distinctly stipitate. *Mountains.* ● *C. part of Balkan peninsula; S. Italy.* Al Bu It Ju.

Ecotypic variation occurs, but further investigation of it is needed.

G. caucasea (Loddiges ex Sims) J. Holub, *loc. cit.* (1967) has been recorded, though not recently, from Turkey-in-Europe; it has 5-merous flowers, equal, narrowly linear, falcate calyx-lobes and the calyx-tube split down one side.

7. G. columnae (Ten.) J. Holub, *op. cit.* 119 (1967) (*Gentiana columnae* Ten.). Biennial 1–15 cm, branched from the base to give a caespitose cushion-like habit. Cauline leaves lanceolate. Flowers 4-merous. Calyx-lobes unequal in width, the two wider not enfolding the others; sinus acute. Corolla 10–20 mm, purple or whitish. Capsule distinctly stipitate. ● *C. Appennini.* It.

(8–9). G. campestris group. Annual or biennial. Flowers 4-merous. Calyx-lobes very unequal, the two outer ovate to ovate-lanceolate, enfolding the two lanceolate inner ones. (One or both the inner lobes may be wanting, especially in white-flowered plants). Corolla (12–)15–25(–40) mm.

Calyx-lobes widest below the middle	**8. campestris**
Calyx-lobes widest above the middle	**9. hypericifolia**

8. G. campestris (L.) Börner, *Fl. Deutsche Volk* 542 (1912) (*Gentiana campestris* L.). Stem 2–35 cm, simple or branched;

branches erect, forming a racemose or subcorymbose inflorescence. Leaves 1–3·5 cm, glabrous, the cauline lingulate to ovate-lanceolate, obtuse or subacute. Calyx-lobes widest below the middle, with a flat, papillose to ciliate margin. Corolla (12–)15–30 mm, bluish-lilac or white; tube up to twice as long as calyx. Capsule sessile or stipitate. *Grassland and heaths.* ● *N. & C. Europe, eastwards to N.W. Russia and C. Austria, and extending southwards to E. Spain and C. Italy.* Au Be Br Cz Da Fa Fe Ga Ge Hb He Ho Hs Is It No Po Rs (N, B) Su.

(a) Subsp. **campestris**: Biennial; base of stem covered with brown remains of the previous year's leaves; basal leaves spathulate; corolla-tube up to twice as long as calyx. 2*n*=36. *Throughout most of the range of the species.*
Ecotypic variants:
Autumnal: Type of the species.
Aestival: **Gentiana campestris** var. **suecica** Froelich, *Gent. Diss.* 92 (1796).
Montane: **Gentiana islandica** (Murb.) Dörfler, *Herb. Norm.* 38: 269 (1898) (*G. campestris* subsp. *islandica* (Murb.) Vollmann).

(b) Subsp. **baltica** (Murb.) Tutin, *Bot. Jour. Linn. Soc.* 65: 260 (1972) (*Gentiana baltica* Murb.): Annual; stem with cotyledons or few green leaves near base; basal leaves ovate or lanceolate; corolla-tube scarcely longer than calyx. *From S. Sweden and N.W. France eastwards to Poland.*

9. **G. hypericifolia** (Murb.) Pritchard, *Bot. Jour. Linn. Soc.* 65: 260 (1972). Like 8 but cauline leaves elliptical to broadly ovate; calyx-lobes widest above the middle; corolla *c.* 25 mm, often whitish. 2*n*=36. *W. & C. Pyrenees.* Ga Hs.

Gentiana laevicalyx (Rohlena) Rohlena, *Mem. Soc. Sci. Bohême* 1933(8): 7 (1934), from S.W. Jugoslavia, may belong to this group or be nearer to *G. columnae*. It requires further investigation.

(10–12). **G. amarella** group. Annual or biennial. Flowers 4- or 5-merous. Calyx-lobes more or less unequal, the two wider ones not enfolding the others. Corolla 10–20(–23) mm.

1 Uppermost internode and terminal pedicel together forming at most ⅛ height of plant; internodes usually more than 3
        10. amarella
1 Uppermost internode and terminal pedicel together forming at least ½ height of plant; internodes usually 3 or fewer
  2 Upper leaves ovate to ovate-lanceolate; calyx-lobes unequal, the largest up to 3 mm wide, patent **11. uliginosa**
  2 Upper leaves lanceolate; calyx-lobes usually subequal, up to 1·5 mm wide, appressed to corolla-tube **12. anglica**

10. **G. amarella** (L.) Börner, *Fl. Deutsche Volk* 542 (1912) (*Gentiana amarella* L.). Stems (3–)5–30(–50) cm, erect, simple or branched from above the base. Branches usually ascending, but the lower not reaching or overtopping the upper and not forming a corymbose inflorescence. Internodes (2–)4–9(–12), the terminal one and the terminal pedicel together forming not more than ⅛ the height of the plant. Cauline leaves 1–2(–3) cm, ovate to linear-lanceolate, acute or more or less obtuse. Calyx-lobes erect. Corolla less than twice as long as calyx. Capsule sessile or rarely shortly stipitate. 2*n*=36. *N. & C. Europe, extending south-eastwards to E. Ukraine.* Au Be Br Cz Da Fe Ga Ge Hb He Hu Is It No Po Rm Rs (N, B, C, W, K, E) Su.

(a) Subsp. **amarella**: Branched from the lowest nodes (very rarely simple). Internodes 4–9(–11). Corolla (14–)16–18 (–22) mm, usually 5-merous, dull purple, blue, pink or whitish; lobes erecto-patent at anthesis. *Throughout the range of the species, except for Iceland and N. Scotland.*

(b) Subsp. **septentrionalis** (Druce) Pritchard, *Watsonia* 4: 235 (1960): Simple or branched from the base. Internodes 2–7. Corolla 12–17 mm, 4- or 5-merous, creamy white within, reddish-purple outside. Lobes more or less erect at anthesis. ● *Iceland and N. Scotland.*

The taxa known as *Gentiana axillaris* (F. W. Schmidt) Reichenb., *G. lingulata* Agardh and *G. uliginosa* sensu Murb., non Willd. are respectively autumnal, aestival and non-over-wintering annual variants of 10.

11. **G. uliginosa** (Willd.) Börner, *Fl. Deutsche Volk* 542 (1912) (*Gentiana uliginosa* Willd.). Stems 1–15 cm, when annual often with 1–2 flowers from a basal rosette, when biennial usually with long, ascending branches from the base. Internodes 0–2(–4), the terminal one and the terminal pedicel together forming at least half the height of the plant. Cauline leaves ovate to ovate-lanceolate. Calyx-lobes unequal, the largest up to 3 mm wide, patent. Corolla about as long as the longest calyx-lobe. Capsule sessile. 2*n*=*c.* 54. *Dune-slacks and damp meadows.* ● *N. & N.C. Europe.* Br Cz Da Fe ?Ga Ge Ho No Po Rs (?N, B, C) Su.

12. **G. anglica** (Pugsley) E. F. Warburg in Clapham, Tutin & E. F. Warburg, *Fl. Brit. Is.* 826 (1952). Biennial; stem 4–20 cm, usually with long branches from the base. Internodes 2–3(–5). Cauline leaves lanceolate to linear. Calyx-lobes more or less equal, not more than 1·5 mm wide. Corolla *c.* 1½ times as long as calyx, dull purple. Capsule sessile. *Grassland and dune-slacks.* ● *S. England.* Br.

(a) Subsp. **anglica**: Uppermost internode *c.* 1½ times as long as the others, or rarely short and then the next long. Basal leaves narrow. Terminal pedicel forming about half the height of the plant. Calyx-lobes somewhat unequal. Corolla 13–16 mm. *Devon to Lincolnshire.*
(b) Subsp. **cornubiensis** Pritchard, *Watsonia* 4: 184 (1959): Uppermost internode not longer than the others. Basal leaves wide. Terminal pedicel less than ⅓ the height of the plant. Calyx-lobes subequal. Corolla (15–)17–20 mm. *Cornwall.*

(13–21). **G. germanica** group. Annual or biennial. Stems simple or branched. Flowers nearly always 5-merous. Calyx-lobes more or less equal in length, equal or unequal in width. Corolla (10–)25–45 mm, usually more than twice as long as calyx, more or less obconical, with patent lobes. Capsule nearly always stipitate.

1 Calyx-sinus obtuse
  2 Corolla 10–20 mm; calyx-tube much shorter than the lobes
    3 Pedicels short **14. ramosa**
    3 Pedicels long **13. bulgarica**
  2 Corolla 18–45 mm; calyx-tube at least half as long as the lobes
    4 Calyx-lobes usually distinctly longer than the tube; corolla 24–45 mm **20. austriaca**
    4 Calyx-lobes as long as or shorter than the tube; corolla 18–25 mm **21. lutescens**
1 Calyx-sinus acute
  5 Calyx-lobes not ciliate **19. germanica**
  5 Calyx-lobes distinctly but very shortly ciliate
    6 Two calyx-lobes much wider than the others; margins strongly revolute; midrib not ciliate
      7 Corolla less than 20 mm, dirty violet or white; capsule nearly or quite sessile **16. engadinensis**
      7 Corolla 21–30 mm, blue-violet (rarely white); capsule distinctly stipitate **17. anisodonta**
    6 Two calyx-lobes only a little wider than the others; margins ± flat; midrib ciliate
      8 Leaves narrowly lanceolate **15. pilosa**
      8 Leaves ovate to ovate-lanceolate **18. aspera**

13. G. bulgarica (Velen.) J. Holub, *Folia Geobot. Phytotax.* (*Praha*) **2**: 117 (1967) (*Gentiana germanica* subsp. *bulgarica* (Velen.) Hayek, *G. bulgarica* Velen.). Stems 5–20 cm, with long, slender, procumbent or ascending branches from the base; internodes 4–8. Cauline leaves oblong-lanceolate, subobtuse. Pedicels longer than calyx-tube. Calyx-lobes subequal, linear, flat or slightly revolute, smooth or nearly so; sinus rounded. Corolla 12–20 mm, whitish or pale violet. Capsule stipitate. ● *C. part of Balkan peninsula, S. Carpathians.* Al Bu Ju Rm.

14. G. ramosa (Hegetschw.) J. Holub, *op. cit.* 118 (1967) (*Gentiana ramosa* Hegetschw.). Biennial 1–15 cm, usually branched from the base to give a caespitose cushion-like habit. Cauline leaves ovate to ovate-lanceolate, only the uppermost more or less acute. Pedicels shorter than calyx-tube. Calyx-lobes subequal, linear-lanceolate, flat, smooth; sinus rounded. Corolla 10–20 mm, pale violet or whitish. Capsule stipitate. *Calcicole.* ● *C. & S.W. Alps.* ?Ga He It.

15. G. pilosa (Wettst.) J. Holub, *loc. cit.* (1967) (*Gentiana pilosa* Wettst., *G. germanica* subsp. *pilosa* (Wettst.) Hayek). Biennial 4–20 cm, simple or with short branches. Lower cauline leaves narrowly lanceolate, upper lanceolate, papillose-ciliate on the margin and midrib beneath. Calyx-lobes somewhat unequal, lanceolate, flat or slightly recurved, shortly hirsute on the margin and midrib beneath; sinus acute. Corolla 15–26 mm, little exceeding the calyx, violet. Capsule stipitate or subsessile. ● *S.E. Alps.* It Ju.

16. G. engadinensis (Wettst.) J. Holub, *op. cit.* 117 (1967) (*Gentiana engadinensis* (Wettst.) Br.-Bl. & Samuelsson). Like **15** but cauline leaves ovate-lanceolate; two calyx-lobes much wider than the others, revolute, with glabrous midrib; corolla not more than 20 mm, 1½ times as long as calyx, vinous-violet or whitish; capsule nearly or quite sessile. 2n=36. ● *C. & E. Alps, from 9° 30' to 10° 45' E.* He It.

G. liburnica E. Mayer & H. Kunz, *Österr. Bot. Zeitschr.* **116**: 397 (1969), from N.W. Jugoslavia, is like **16** but has narrower, more acute cauline leaves, a narrower corolla and a stipitate capsule.

17. G. anisodonta (Borbás) Á. & D. Löve, *Bot. Not.* **114**: 42 (1961) (*Gentiana anisodonta* Borbás, *G. germanica* subsp. *calycina* (Koch) Hayek, *G. calycina* (Koch) Wettst.). Like **15** but lower cauline leaves obovate to spathulate, the upper triangular-lanceolate, glabrous or sparsely ciliate; two calyx-lobes much wider than the others, revolute, strongly ciliate, with glabrous midrib; corolla 20–30(–33) mm, exceeding the calyx; capsule stipitate. 2n=36. *Calcicole.* ● *S.C. & S.E. Alps; Appennini; N.W. Jugoslavia.* Au He It Ju.

Ecotypic variants:
  Autumnal: Type of the species.
  Aestival-montane: var. **antecedens** (Wettst.) E. Mayer, *Österr. Bot. Zeitschr.* **116**: 394 (1969) (*Gentiana anisodonta* subsp. *antecedens* (Pacher) Mansfeld).
  Aestival-alpine: var. **calycina** (Koch) E. Mayer, *loc. cit.* (1969) (*Gentiana anisodonta* subsp. *calycina* (Koch) Mansfeld).

G. insubrica (H. Kunz) J. Holub, *Folia Geobot. Phytotax.* (*Praha*) **2**: 117 (1967), from S. Switzerland, is like **17** but has stout, rigid stems and slightly unequal, papillose, not hairy, calyx-lobes. It is also calcicole.

G. albanica (Jáv.) J. Holub, *op. cit.* 116 (1967), from Albania, has a compact habit and revolute calyx-lobes. It requires further investigation.

18. G. aspera (Hegetschw. & Heer) Dostál ex Skalický, Chrtek & Gill, *Preslia* **38**: 92 (1966) (*Gentiana aspera* Hegetschw. & Heer). Biennial 1–40 cm, usually branched from the base. Cauline leaves ovate to ovate-lanceolate or ovate-triangular, subobtuse, papillose-hirsute on the margin and often on the midrib. Calyx-lobes somewhat unequal, papillose-hirsute on the often recurved margin and on the midrib; sinus acute. Corolla 10–40 mm, distinctly exceeding the calyx, violet, pink or whitish. Capsule stipitate. ● *E. & C. Alps, mountains of S. Germany and W. Czechoslovakia.* Au Cz Ge He Ju.

Ecotypic variants:
  Alpine: Type of the species
  Aestival: **Gentiana norica** A. & J. Kerner, *Sched. Fl. Exs. Austro-Hung.* **6**: 56 (1893) (*G. aspera* subsp. *norica* (A. & J. Kerner) Vollmann)
  Autumnal: **Gentiana sturmiana** A. & J. Kerner, *op. cit.* **2**: 122 (1882) (*G. aspera* subsp. *sturmiana* (A. & J. Kerner) Vollmann)

19. G. germanica (Willd.) E. F. Warburg in Clapham, Tutin & E. F. Warburg, *Fl. Brit. Is.* 824 (1952) (*Gentiana germanica* Willd.). Like **18** but leaves and calyx glabrous; calyx-lobes somewhat scabrid on the margin. 2n=36. ● *W. & C. Europe, from S. England to the S. Alps and E. Carpathians.* Au Be Br Cz Ga Ge He Ho It Ju Rm.

Ecotypic variants:
  Autumnal-montane: Type of the species.
  Autumnal-alpine: **Gentiana rhaetica** A. & J. Kerner, *Sched. Fl. Austro-Hung.* **2**: 124 (1882) (*G. germanica* subsp. *semleri* Vollmann).
  Aestival: **Gentiana solstitialis** Wettst., *Denkschr. Akad. Wiss. Math.-Nat. Kl.* (*Wien*) **64**: 337 (1896) (*G. germanica* subsp. *solstitialis* (Wettst.) Vollmann).

Gentiana rhaetica subsp. *kerneri* Wettst. is intermediate between the autumnal and aestival variants.

20. G. austriaca (A. & J. Kerner) J. Holub, *Preslia* **37**: 102 (1965) (*Gentiana austriaca* A. & J. Kerner, *G. praecox* A. & J. Kerner). Biennial (1–)10–20(–40) cm, usually branched from the base, forming a corymbose inflorescence. Cauline leaves ovate-lanceolate to lanceolate, gradually narrowed from the base, acuminate, glabrous. Calyx-lobes nearly equal, longer than the tube, linear, glabrous; margin flat or revolute; sinus obtuse. Corolla 24–45 mm, purplish or whitish. Capsule stipitate. ● *E.C. Europe.* Au Cz Ge Hu Ju Po Rm.

Ecotypic variants:
  Autumnal: Type of the species.
  Aestival: **Gentiana germanica** subsp. **austrica** var. **neilreichii** (Dörfler & Wettst.) Hayek, *Prodr. Fl. Penins. Balcan.* **2**: 423 (1930) (*G. austriaca* subsp. *neilreichii* (Dörfler & Wettst.) Wettst.).
  Aestival-montane: **Gentiana praecox** A. & J. Kerner, *Verh. Zool.-Bot. Ges. Wien* **38**: 669 (1888), non sensu Wettst. et auct. pl.

G. bohemica Skalický, *Preslia* **41**: 144 (1969) is intermediate between **19** and **20**. It occurs in S.E. Germany, Czechoslovakia and N.E. Austria. There are two ecotypic variants, the type of the species, which is montane, and **G. gabretae** Skalický, *op. cit.* 146 (1969), which is alpine.

21. G. lutescens (Velen.) J. Holub, *Folia Geobot. Phytotax.* (*Praha*) **2**: 117 (1967) (*Gentiana praecox* sensu Wettst., non A. & J. Kerner, *G. lutescens* Velen.). Biennial 3–40 cm, simple or branched above, forming a racemose inflorescence. Cauline leaves narrowly ovate, obtuse, except the uppermost which are acuminate, glabrous. Calyx-lobes shorter than to as long as the tube, 2 wider than the others, linear, glabrous; margin flat or revolute; sinus obtuse or subacute. Corolla 18–25 mm, purplish or yellowish. Capsule stipitate. $2n=36$. ● *E.C. Europe and N. half of Balkan peninsula, mainly in the mountains.* Au Bu Cz Ge Ju Po Rm Rs (W).

Ecotypic variants:
Aestival: Type of the species.
Autumnal: **Gentianella lutescens** subsp. **carpatica** (Hayek) J. Holub, *op. cit.* 119 (1967), non *Gentiana carpatica* Kit. (*G. carpaticola* Borbás, *G. praecox* subsp. *carpatica* Hayek).

Sect. ARCTOPHILA (Griseb.) J. Holub. Stems simple or with a few long branches from the base. Flowers 4- to 5-merous, more or less numerous and crowded at the ends of the branches. Corolla cylindrical, not fringed in the throat or at the margins; lobes acuminate or mucronate. Style absent. Capsule sessile.

22. G. aurea (L.) H. Sm., *Uppsala Univ. Årsskr.* **1945**(7): 259 (1945) (*Gentiana aurea* L.). Annual or biennial 5–15 cm, with long erect branches from base. Cauline leaves 2–3 times as long as wide, the lower obtuse, the upper more or less acute. Flowers 4- to 5-merous, small, crowded at the ends of the branches and closely subtended by the uppermost leaves. Calyx-teeth more or less equal; sinus rounded. Corolla 7–10 mm, pale yellow, rarely blue. $2n=36$. *Sea- and lake-shores. Arctic Europe.* ?Fe Is No Rs (N) Su.

7. Lomatogonium A. Braun[1]

Annual herbs. Calyx deeply divided into (4–)5 lobes. Corolla rotate, pale blue or white; tube short; lobes (4–)5, each with 2 fimbriate nectaries on the inner surface near the base. Anthers basifixed, linear, not twisting spirally after dehiscence. Stigma sessile; lobes decurrent.

Calyx-lobes distinctly shorter than corolla, ovate   **1. carinthiacum**
Calyx-lobes equalling or longer than corolla, linear-lanceolate
     2. rotatum

1. L. carinthiacum (Wulfen) Reichenb., *Fl. Germ. Excurs.* 421 (1831). Glabrous. Stems 1–12 cm, 4-angled. Leaves ovate to oblong, papillose on the margin, the basal narrowed to a short petiole, the others sessile. Flowers solitary, on long, naked branches. Calyx-lobes shorter than the corolla, ovate. Corolla 12–16 mm in diameter; lobes papillose on the outside towards the apex. $2n=40$. *Dry grassland and open habitats. C. & E. Alps, S. Carpathians; very local.* Au Ge He It Rm. (*Mountains of Asia and North America.*)

2. L. rotatum (L.) Fries ex Fernald, *Rhodora* **21**: 194 (1919). Like **1** but up to 25 cm; lower leaves spathulate, upper linear-lanceolate; calyx-lobes as long as or longer than corolla, linear-lanceolate. $2n=10$. *Wet places. Iceland; arctic Russia.* Is Rs (N). (*N. & C. Asia, North America.*)

8. Swertia L.[1]

Glabrous perennial herbs. Calyx 4- to 5-lobed almost to base. Corolla rotate, 4- to 5-lobed almost to base, each lobe with 2 fimbriate nectaries near the base. Anthers basifixed, usually sagittate. Ovary unilocular, ovoid, narrowed to the sessile, 2-lobed stigma.

1. S. perennis L., *Sp. Pl.* 226 (1753). Stem (8–)15–60 cm, 4-angled, brownish-red or purplish. Lower leaves crowded, ovate to elliptical, obtuse, narrowed to a winged petiole; upper leaves remote, opposite or, less frequently, alternate, sessile, semi-amplexicaul. Calyx-lobes narrow, acuminate. Corolla-lobes 6–16 mm, blue or dirty violet, rarely yellowish-green or white, with darker dots or lines. Capsule 10–12 mm, 2-valved. $2n=28$. *Wet places, mainly in the mountains. From N.C. France and N.W. Russia southwards to N. Spain, Bulgaria and S. Ural.* Au Bu Cz Ga Ge He Hs It Ju Po Rm Rs (B, C, E).

Very variable. Three other species are sometimes recognized, but the correlation of the supposedly distinguishing characters is poor and plants with one or more of these characters occur sporadically across Europe. The names applied to these variants are: **S. alpestris** Baumg. ex Fuss, *Mantissa* 19 (1846), with alternate cauline leaves, dark violet flowers with acute corolla-lobes and the fimbriae of the nectary short; **S. obtusa** Ledeb., *Mém. Acad. Sci. Pétersb.* **5**: 526 (1815), with alternate cauline leaves, dark violet flowers with obtuse corolla-lobes and the fimbriae of the nectary 3–4 times as long as the nectary; and **S. punctata** Baumg., *Enum. Stirp. Transs.* **1**: 190 (1816), which has opposite cauline leaves, greenish-yellow corolla, with acute lobes and the nectary with short fimbriae. None of these seems to merit even subspecific rank and the latter is probably no more than a colour variant of typical *S. perennis*.

9. Halenia Borkh.[1]

Glabrous annual herbs. Calyx deeply 4-lobed. Corolla campanulate, 4-lobed half-way to base, each lobe with a slender nectary-spur at its base. Anthers basifixed. Ovary unilocular, ovoid, narrowed to the sessile, 2-lobed stigma.

1. H. corniculata (L.) Cornaz, *Bull. Soc. Sci. Nat. Neuchâtel* **25**: 171 (1897). Stem 12–50 cm, 4-angled, erect, usually branched. Leaves opposite, the lower obovate-lanceolate, petiolate, the upper elliptic-oblong, sessile. Flowers in corymbose clusters; pedicels very unequal. Calyx 6–8 mm; lobes linear-subulate, $\frac{1}{2}$–$\frac{2}{3}$ as long as corolla. Corolla 8–11 mm, pale greenish-yellow, persistent in fruit; lobes ovate to ovate-oblong, shortly acuminate; nectary-spur horizontal or ascending. *Woodland-margins. C. Ural (W. of Sverdlovsk).* Rs (C). (*N. & C. Asia.*)

CXLI. MENYANTHACEAE[2]

Aquatic or semi-aquatic perennial herbs. Leaves alternate, except sometimes on flowering stems. Flowers 5-merous, actinomorphic, heterostylous. Calyx deeply divided. Corolla deeply divided; lobes valvate in bud. Stamens inserted on the corolla-tube or in the sinuses between the lobes. Ovary superior to semi-inferior, 1-locular; placentae parietal; ovules numerous. Fruit a capsule.

Leaves 3-foliolate; flowers pink or white   **1. Menyanthes**
Leaves simple; flowers yellow   **2. Nymphoides**

1. Menyanthes L.[1]

Leaves 3-foliolate, all alternate. Capsule dehiscing by 2 valves.

1. M. trifoliata L., *Sp. Pl.* 145 (1753). Glabrous, with a stout, creeping rhizome; flowering stems 12–35 cm. Leaflets up to

[1] By T. G. Tutin.   [2] Edit. T. G. Tutin.

5-2

10 cm, obovate to rhombic, sinuate and sometimes remotely denticulate, rounded to subacute, shortly petiolulate; petioles 7–20 cm, sheathing below. Raceme 10- to 20-flowered; pedicels 5–10 mm, longer than the ovate bracts. Calyx-lobes ovate, somewhat recurved. Corolla *c.* 15 mm in diameter, pink outside, paler or white inside; lobes fimbriate. $2n = 54$. *In shallow water or the wetter parts of bogs and fens. Most of Europe, but rare in the Mediterranean region.* ?Al Au Be Br Bu Co Cz Da Fa Fe Ga Ge Hb He Ho Hs Hu Is It Ju Lu No Po Rm Rs (N, B, C, W, E) Su.

2. Nymphoides Séguier[1]

Leaves simple, those on the flowering stems opposite. Capsule dehiscing irregularly.

1. **N. peltata** (S. G. Gmelin) O. Kuntze, *Rev. Gen. Pl.* **2**: 429 (1891) (*N. flava* Druce, *Limnanthemum peltatum* S. G. Gmelin). Glabrous; stems up to 160 cm, creeping and floating. Leaves 3–10 cm across, suborbicular, sinuate or entire, with a deep, narrow sinus at base; petioles often sheathing below. Flowers in axillary fascicles of 2–5; pedicels 3–10 cm. Calyx-lobes oblong-lanceolate. Corolla 30–40 mm in diameter, yellow; lobes fimbriate-ciliate. $2n = 54$. *In still water and slow-flowing rivers. Most of Europe, northwards to England, the Baltic and N.C. Russia.* Al Au Be Br Bu Cz Ga Ge Gr Ho Hs Hu It Ju Lu Po Rm Rs (N, B, C, W, E) Sa Tu [Da He Su].

CXLII. APOCYNACEAE[2]

Small trees, shrubs, or herbs, with poisonous latex. Leaves simple, exstipulate, with glands in or above the axils. Flowers solitary or in cymes, hermaphrodite, actinomorphic, 5-merous. Corolla with oblique lobes, contorted in bud. Stamens 5; anthers introrse, closely surrounding the stigma. Ovary superior; carpels 2, usually free, united by a single style above. Fruit of two fusiform follicles, usually opening by the ventral suture; rarely only one follicle developing.

Aestivation is described with reference to side-view of the bud.

1 Corolla infundibuliform or campanulate, the throat wide and open; anthers adhering to the stigma; seeds numerous, comose
  2 Leaves coriaceous, evergreen; flowers at least 2 cm   **1. Nerium**
  2 Leaves thin, deciduous; flowers less than 1 cm **2. Trachomitum**
1 Corolla hypocrateriform, the throat narrow and closed by hairs; anthers not adhering to the stigma; seeds few, not comose
  3 Leaves alternate; flowers in terminal cymes   **3. Rhazya**
  3 Leaves opposite; flowers solitary in leaf-axils   **4. Vinca**

1. Nerium L.[3]

Shrubs or small trees. Leaves opposite or in whorls of 3 or 4, evergreen. Inflorescence terminal, corymbose. Corolla cylindrical below, infundibuliform above, the throat wide and open, with 5 large, laciniate scales; lobes overlapping to the right in bud. Anthers adhering to the stigma, with appendage. Disc absent. Follicles united until just before dehiscence. Seeds numerous, tomentose, apically comose.

1. **N. oleander** L., *Sp. Pl.* 209 (1753) (*N. kotschyi* auct. eur., non Boiss.). Stems up to 4 m. Leaves 6–12(–15) × 1·2–2(–3) cm, linear-lanceolate, acute, coriaceous. Calyx densely glandular inside. Corolla (2–)3–4 cm in diameter, usually pink; tube 2 cm, the lobes 2 cm, obtuse, patent. Anther-appendages long, hairy, twisted. Follicles 8–16 × 0·5–1 cm, erect, reddish-brown. *River-banks and river-gravels. Mediterranean region, S. Portugal; frequently planted for ornament, and perhaps only naturalized in some areas.* Al Bl Co Cr Ga Gr Hs It Ju Lu Sa Si Tu [Az].

2. Trachomitum Woodson[3]

Rhizomatous perennials. Leaves opposite, deciduous. Inflorescence terminal, thyrsiform. Corolla campanulate, the throat wide and open; lobes overlapping to the right in bud, suberect at anthesis. Anthers auriculate, adhering to the stigma. Disc of 5 fleshy scales. Seeds numerous, apically comose, otherwise glabrous.

Literature: R. E. Woodson, *Ann. Missouri Bot. Gard.* **17**: 156–164 (1930).

1 Calyx-lobes 1 mm, ovate, obtuse; corolla-lobes shorter than tube   **2. sarmatiense**
1 Calyx-lobes 1·5–2 mm, ± lanceolate, acute; corolla-lobes as long as tube
  2 Leaves obtuse, mucronate, closely denticulate; corolla 4–5 mm   **1. venetum**
  2 Leaves acute, remotely denticulate; corolla 7 mm   **3. tauricum**

1. **T. venetum** (L.) Woodson, *Ann. Missouri Bot. Gard.* **17**: 158 (1930) (*Apocynum venetum* L.). Stems up to 50 cm. Leaves 20–40(–70) × 5–8(–15) mm, narrowly oblong, closely denticulate, rounded or narrowed at base, obtuse, mucronate, thin, glabrous. Calyx-lobes 1·5–2 mm, lanceolate, rarely ovate-lanceolate, acute, the margins scarious. Corolla 4–5 mm, whitish, the throat *c.* 4 mm wide; lobes as long as the tube. Follicles *c.* 15 × 0·5 cm, pendent. *Maritime sands. Coasts of the Adriatic, Aegean and Black Seas; rare and local.* Bu It Ju ?Rm Tu.

2. **T. sarmatiense** Woodson, *op. cit.* 162 (1930). Stems *c.* 100 cm. Leaves 20–40 × 15–20 mm, ovate-oblong, acute, remotely denticulate, truncate at base, glabrous, glaucous beneath. Calyx-lobes 1 mm, ovate, obtuse. Corolla 3–4 mm, pinkish, the throat 3 mm wide; lobes shorter than the tube. Follicles *c.* 15 × 0·4 cm. *Marshy meadows and salt-marshes. S. & S.E. parts of U.S.S.R.* Rs (C, W, K, E).

T. russanovii (Pobed.) Pobed. in Schischkin & Bobrov, *Fl. URSS* **18**: 654 (1952) (*Apocynum russanovii* Pobed.), from the island of Džarylgač, off the S. coast of Ukraine, seems to be a few-flowered variant of **2.**

3. **T. tauricum** (Pobed.) Pobed., *op. cit.* 657 (1952). Like **2** but leaves 40–70 × 8–15 mm, oblong-lanceolate, narrowed at base; calyx-lobes 2 mm, lanceolate, acute, the margins scarious; corolla 7 mm, yellowish-white, the throat 4 mm wide, the lobes as long as the tube. *Hillsides above the sea.* ● *S.E. Krym (Feodosija).* Rs (K).

3. Rhazya Decne[3]

Erect, perennial herbs, woody at base. Leaves alternate, deciduous. Inflorescence terminal, corymbose. Calyx-lobes *c.* $\frac{1}{4}$ as

[1] By T. G. Tutin.   [2] Edit. V. H. Heywood.   [3] By F. Markgraf.

long as the corolla-tube. Corolla hypocrateriform, the tube slender; throat narrow, hairy, without scales; lobes overlapping to the left in bud. Anthers acute, surrounding the stigma but not adhering to it. Carpels surrounded by a disc; ovules many. Seeds few, glabrous.

1. R. orientalis (Decne) A.DC. in DC., *Prodr.* **8**: 386 (1844). Stems up to 50 cm, numerous, simple. Leaves *c.* 4 × 2 cm, lanceolate, rarely ovate, narrowed or rarely rounded at base, thin, ciliate. Calyx-lobes 2–3 mm, lanceolate, acute, ciliate. Corolla *c.* 1 cm in diameter, blue; tube 1–1·5 cm, the lobes *c.* ⅓ as long as the tube, narrowly ovate, subacute. Follicles *c.* 5–8 × 0·5 cm, erect. *Wet places near the sea. Thrace.* Gr †Tu. (*N.W. Anatolia.*)

4. Vinca L.[1]

Low, creeping dwarf shrubs or herbaceous perennials, usually with trailing vegetative shoots and ascending flowering stems. Leaves opposite. Flowers solitary in leaf-axils, long-pedicellate. Calyx-lobes more than ¼ as long as corolla-tube. Corolla hypocrateriform, usually blue (rarely reddish-purple, pink or white), the tube gradually widened, without conspicuous appendages but with a zone of hairs above the insertion of the stamens and a low ridge connecting the lobes at the mouth; lobes oblique, as long as the tube, overlapping to the left in bud. Stamens inserted half-way up the corolla-tube; filaments bent abruptly at base; anthers with the connective expanded above into a flap-like appendage. Carpels with 4–8 ovules, alternating with 2 disc-scales. Follicles patent. Seeds glabrous.

All species grow in woods, scrub, hedges or other shady places.

Literature: M. Pichon, *Bull. Mus. Nat. Hist. Nat.* (*Paris*) ser. 2, **23**: 439–444 (1951).

1 Plant dying down completely in winter; veins of leaves diverging from midrib at 10°–30° or scarcely perceptible
           2. herbacea
1 Plant evergreen, with previous year's leaves present at flowering time; veins of leaves diverging from midrib at 30°–50°
 2 Calyx-lobes 3–4(–5) mm, narrowly ovate to narrowly triangular, glabrous   **1. minor**
 2 Calyx-lobes (4–)6–18 mm, very narrowly triangular or almost linear
  3 Margin of leaves and calyx-lobes glabrous   **3. difformis**
  3 Margin of leaves and calyx-lobes ciliate
   4 Leaves not more than 3·5 × 2 cm; flowering stems 9–12 cm
           4. balcanica
   4 Leaves up to 9 × 6 cm; flowering stems up to 30 cm
    5 Hairs of calyx 0·5–1 mm   **5. major**
    5 Hairs of calyx 0·2 mm or less   **3. difformis**

1. V. minor L., *Sp. Pl.* 209 (1753). Stems procumbent or ascending, over-wintering. Leaves 1·5–4·5 × 0·5–2·5 cm, mostly lanceolate or elliptical (but often ovate on trailing shoots), rounded or cuneate at base, evergreen, glabrous. Flowering stems up to 20 cm; pedicels often longer than the subtending leaves. Calyx-lobes 3–4(–5) mm, narrowly ovate to narrowly triangular, glabrous. Corolla-tube 9–11 mm; limb 25–30 mm in diameter, usually blue; lobes obliquely truncate. $2n = 46$. *S., W. & C. Europe, extending eastwards to Lithuania and Krym.* Au Be Bu Co Cz Ga Ge ?Gr He Ho Hs Hu It Ju Lu Rm Rs (B, C, W, K) Po Si [Br Da Hb No Su ?Tu].

This species has been cultivated for many centuries and occurs as a relict of cultivation or of deliberate naturalization so often that the limits of its natural distribution are rather uncertain.

By W. T. Stearn.

There exist variants with white, pink, reddish-purple and violet flowers.

2. V. herbacea Waldst. & Kit., *Pl. Rar. Hung.* **1**: 8 (1799) (incl. *V. mixta* Velen.). Stems procumbent or ascending, dying back in winter. Leaves 1·5–4 × 0·4–1·5 cm, very narrowly elliptical or narrowly lanceolate to broadly elliptical or the lowermost sometimes obovate, cuneate at base, with margins sometimes minutely ciliate. Flowers well spaced along the stems; pedicels shorter than to longer than the subtending leaves. Calyx-lobes 4–8 mm, very narrowly triangular, the margins smooth, ciliate or scabrid. Corolla-tube 10–15 mm; limb 20–35 mm in diameter, blue; lobes acute or very obliquely truncate. *E.C. & S.E. Europe, extending to c. 51° N. in S.C. Russia.* Au Bu Cz Gr Hu Ju Rm Rs (C, W, K, E) Tu.

3. V. difformis Pourret, *Mém. Acad. Sci. Toulouse* **3**: 337 (1788) (*V. media* Hoffmanns. & Link). Stems up to 200 cm, procumbent or ascending, overwintering. Leaves 2·5–7 × 1·5–4·5 cm, ovate to lanceolate, mostly narrowly lanceolate, rounded at base then attenuate into the petiole, evergreen, glabrous or with minutely ciliate margin. Flowering stems up to 30 cm; pedicels shorter than the subtending leaves. Calyx-lobes 5–14 mm, very narrowly triangular, glabrous or the margin minutely ciliate with hairs up to 0·2 mm. Corolla-tube 12–18 mm; limb 30–70 mm in diameter, pale blue or almost white; lobes acute or obliquely truncate. $2n = c.\ 46$. *S.W. Europe, extending eastwards to S.C. Italy.* Az Bl Co Ga Hs It Lu Sa.

(a) Subsp. **difformis**: Leaf-margins glabrous. Calyx-lobes glabrous or, very rarely, ciliate with a few sparse hairs. Corolla-limb 30–45 mm in diameter. *Almost throughout the range of the species.*

(b) Subsp. **sardoa** Stearn, *Bot. Jour. Linn. Soc.* **65**: 255 (1972): Leaf-margins scabrid, the hairs up to 0·2 mm. Calyx-lobes minutely ciliate. Corolla-limb 60–70 mm in diameter. ● *Sardegna.*

4. V. balcanica Pénzes, *Acta Bot. Acad. Sci. Hung.* **8**: 329 (1962). Stems procumbent or ascending, overwintering. Leaves 1·2–3·5 × 0·5–2 cm, ovate, rounded at base, the margins ciliate with hairs 0·6–1 mm. Flowering stems 9–12 cm, the pedicels shorter or longer than the subtending leaves. Calyx-lobes 4–8 mm, subulate, the margins ciliate with hairs 0·4–1·2 mm. Corolla-tube 10–11 mm; limb 25–35 mm in diameter, blue; lobes obliquely truncate. ● *Balkan peninsula.* Al Bu Ju.

A little-known species very close to **5**, requiring further study.

5. V. major L., *Sp. Pl.* 209 (1753). Stems up to 100 cm, ascending in the lower part, then arching or procumbent, overwintering. Leaves 2·5–9 × 2–6 cm, mostly ovate or broadly ovate (rarely lanceolate), evergreen, the margins ciliate with hairs 0·1–1 mm. Flowering stems up to 30 cm; pedicels shorter than the subtending leaves. Calyx-lobes 7–17 mm, very narrowly triangular, the margins densely ciliate with hairs 0·5–1 mm. Corolla-tube 12–15 mm; limb 30–50 mm in diameter, usually bluish-purple; lobes obliquely truncate. $2n = 92$. *W. & C. Mediterranean region; frequently naturalized elsewhere.* Ga Hs It Ju Si [Au Br Bu Co Cr Gr Hb He Lu Rs (W, K)].

Cultivated for many centuries and widely naturalized.

Records from Europe of subsp. **hirsuta** (Boiss.) Stearn, *Jour. Bot.* (*London*) **70**: 27 (1932), a native of the Caucasus and Anatolia, are referable to variants of **5** with lanceolate leaves and narrower, more pointed, violet corolla-lobes.

CXLIII. ASCLEPIADACEAE[1]

Shrubs or perennial herbs, sometimes with twining stems. Leaves usually opposite, simple, exstipulate, sometimes scale-like. Flowers in axillary or terminal cymes or umbels, actinomorphic, 5-merous. Corolla-lobes contorted or valvate in bud. Single or double corona, of 5 or 10 free or more or less connate segments, inserted at the base of the filaments. Anthers united in a ring and usually adnate to the stigma, forming a gynostegium. Pollen in pollinia or tetrads, becoming attached to translators (pollen-transfer devices) which consist either of a cochleariform body and an elongate portion ending in an adhesive disc, or of 2 arms and a claw. Ovary superior; carpels 2, free below, united at the stigma. Fruit a pair of follicles, though often only one developing. Seeds comose at apex.

1 Corolla-lobes deflexed at anthesis; follicles usually spiny
  2 Corolla white; leaves 2–10 cm   **3. Gomphocarpus**
  2 Corolla pink or purple; leaves (10–)15–23 cm   **4. Asclepias**
1 Corolla-lobes patent or erect at anthesis; follicles not spiny
  3 Corona double, with 10 segments
    4 Stems strongly twining; leaves herbaceous   **5. Cynanchum**
    4 Stems not twining; leaves scale-like   **8. Caralluma**
  3 Corona single, with 5 segments
    5 Corona-segments awned; anthers opening by longitudinal slits   **1. Periploca**
    5 Corona-segments not awned; anthers opening by apical pores
      6 Corolla 15–20 mm in diameter   **2. Araujia**
      6 Corolla 5–12 mm in diameter
        7 Corona-segments ± united by a membrane; inflorescences few-flowered; pollinia pendent   **6. Vincetoxicum**
        7 Corona-segments free; inflorescences many-flowered; pollinia erect   **7. Cionura**

Subfam. Periplocoideae

Anthers adjacent to, but not adnate to the stigma, opening by longitudinal slits; pollen in tetrads. Translators consisting of a cochleariform body and an elongate portion ending in an adhesive disc.

1. Periploca L.[2]

Shrubs, with erect or twining stems. Leaves opposite. Flowers in axillary cymes. Corolla contorted in bud; lobes more or less erect at anthesis. Corona single, with 5 free, abaxially awned segments. Follicles cylindrical, slightly sulcate, not spiny.

Literature: K. Browicz, *Arboretum Kórnickie* **11**: 5–104 (1966). K. H. Rechinger fil., *Denkschr. Akad. Wiss. Math.-Nat. Kl. (Wien)* **105**(2, 1): 131 (1943).

Stems up to 12 m, twining; leaves 4–12 × 2–7 cm, membranous; corolla *c.* 2 cm in diameter   **1. graeca**
Stems 1·5–3 m, erect; leaves 1–3·5 × 0·2–0·6 cm, somewhat coriaceous; corolla *c.* 1 cm in diameter   **2. laevigata**

1. **P. graeca** L., *Sp. Pl.* 211 (1753). Stems up to 12 m, twining. Leaves 4–12 × 2–7 cm, petiolate, ovate, more or less obtuse, rounded to cuneate at the base, glabrous, rarely pubescent beneath, membranous. Petioles *c.* 1 cm. Cymes many-flowered. Corolla *c.* 2 cm in diameter; lobes oblong, obtuse, green outside, dark purple-brown inside; margins deflexed at anthesis. Awns of corona erect, purple-brown. Anthers spiny. Follicles 10–15 cm. Seeds *c.* 15 × 4 mm; coma *c.* 30 mm. 2*n* = 24. *Woods, thickets and river-banks. Balkan peninsula, extending to S. Romania and*

locally westwards to Italy. Al Bu ?Cr Gr It Ju Rm Tu [Ga Hs].

2. **P. laevigata** Aiton, *Hort. Kew.* **1**: 301 (1789). Stems 1·5–3 m, erect, sometimes twining at apex. Leaves 1–3·5 × 0·2–0·6 cm, subsessile, linear, acute or obtuse, cuneate at base, glabrous, coriaceous. Cymes few-flowered. Corolla *c.* 1 cm in diameter; lobes oblong, obtuse, green outside, green, purple-brown and white inside; margins not deflexed at anthesis. Awns of corona deflexed, green. Anthers smooth. Follicles 5–8 (–10) cm. Seeds *c.* 7 × 3 mm; coma *c.* 40 mm. *Dry, rocky places. S.E. Spain; small islands of C. & E. Mediterranean.* Cr Hs Si. (*Islas Canarias, N. Africa, Syria.*)

The European representatives belong to subsp. **angustifolia** (Labill.) Markgraf, *Bot. Jour. Linn. Soc.* **64**: 375 (1971) (*P. angustifolia* Labill.). The type subspecies occurs in Islas Canarias.

Subfam. Asclepiadoideae

Anthers adnate to the stigma, opening apically; pollen in pollinia. Translators consisting of 2 arms and a claw.

2. Araujia Brot.[2]

Shrubs with twining stems. Leaves opposite. Flowers in axillary cymes. Corolla slightly contorted in bud; lobes more or less erect at anthesis. Corona single; segments 5, free, without awns. Follicles ovoid, not spiny.

Literature: G. O. A. Malme, *Ark. Bot.* **8**(1): 1–30 (1909).

1. **A. sericifera** Brot., *Trans. Linn. Soc. London* **12**: 62 (1818). Stems up to 10 m. Leaves 5–8 × 2–2·5 cm, ovate-oblong, acute, truncate at base, subglabrous above, white-tomentose beneath. Cymes few-flowered; peduncles long. Corolla 15–20 mm in diameter; lobes erecto-patent, ovate-acuminate, white with pink stripes. Follicles *c.* 10 × 6 cm, pruinose. *Formerly cultivated for ornament; locally naturalized in S.W. Europe.* [Az Ga Hs Lu.] (*S.E. South America.*)

3. Gomphocarpus R.Br.[2]

Erect shrubs. Leaves usually opposite, sometimes in whorls of 3 or the upper alternate. Flowers in extra-axillary cymes. Corolla valvate in bud; lobes deflexed at anthesis. Corona single, with 5 free segments, each with 2 lateral auricles. Follicles ovoid, acute, spiny.

1. **G. fruticosus** (L.) Aiton fil. in Aiton, *Hort. Kew.* ed. 2, **2**: 80 (1811). Stems 1–2 m. Leaves 2–10 × 0·3–1 cm, linear-lanceolate, glabrous. Cymes many-flowered; peduncles short, pubescent. Corolla white; lobes ovate-oblong; margins ciliate. Corona-segments cucullate, fleshy. Follicles 4–6 × 2–3 cm. 2*n* = 16. *Naturalized in waste places. S. Europe.* [Al Az Bl Co Cr Ga Gr Hs It Ju Lu Sa Si.] (*S. Africa.*)

4. Asclepias L.[2]

Erect herbs with many straight stems. Leaves opposite. Flowers in axillary and terminal umbels. Corolla valvate in bud; lobes deflexed at anthesis. Corona single, with 5 free segments, each with a central adaxial appendage curved over the anthers. Follicles ovoid, usually spiny.

---

[1] Edit. V. H. Heywood.   [2] By F. Markgraf.

Literature: R. E. Woodson, *Ann. Missouri Bot. Gard.* **41**: 1–211 (1954).

1. A. syriaca L., *Sp. Pl.* 214 (1753) (*A. cornuti* Decne). Glaucous perennial; stems 100–150 cm. Leaves 15–23 × 5–9 cm, ovate-oblong, rounded at base, white-pubescent beneath. Umbels many-flowered, subglobose; peduncles 5–10 cm; pedicels 3–6 cm. Corolla pink; lobes 8 × 3–4 mm, ovate-oblong, pubescent beneath. Corona 4 mm, pink. Follicles 8–11 × 2–3 cm, white-pubescent, sulcate, spiny. *Formerly cultivated for fibre and as a food-plant for bees; naturalized in cultivated ground and dry grassland in various parts of Europe.* [Au Bu ?Co Cz Ga He Hu It Ju Po Rm Rs (W, K).] (*E. North America.*)

A. curassavica L., *Sp. Pl.* 215 (1753), a native of tropical America, is perhaps locally naturalized in damp places in S. Spain (prov. Málaga). It differs from **1** in its shorter stems (up to 50 cm), lanceolate leaves *c.* 10 × 2 cm, which are attenuate at base and subglabrous beneath, pedicels 1–2 cm, purple, glabrous corolla-lobes, orange corona and glabrous, smooth follicles.

5. Cynanchum L.[1]

Woody climbers. Leaves opposite. Flowers in axillary and terminal umbels. Corolla contorted in bud; lobes patent at anthesis. Corona double, with 10 free segments. Follicles fusiform, smooth.

1. C. acutum L., *Sp. Pl.* 212 (1753). Glabrous. Stems up to 3 m, slender, twining. Leaves 2–15 × 1·5–10 cm, petiolate, sagittate, subacute; petioles 1–5 cm. Umbels few- to many-flowered; peduncles 1–6 cm. Corolla 8–12 mm in diameter, white or pink, glabrous; lobes *c.* 5 × 1 mm, linear, subacute. Corona-segments triangular. Follicles *c.* 8 × 1 cm. $2n = 18$. *Scrub and hedges, often near the sea.* S. Europe, extending northwards to 51° N. in S.E. Russia. Al Bl Bu Cr Ga Gr Hs It Ju Lu Rm Rs (W, K, E) Si.

6. Vincetoxicum N. M. Wolf[1]

Rhizomatous herbs or small shrubs, often with more or less twining shoot-apices. Leaves opposite, shortly petiolate, the lower suborbicular, the upper lanceolate. Flowers in axillary, few-flowered cymes. Corolla contorted in bud; lobes patent at anthesis. Corona single, the segments not awned, connate at least at base by a more or less folded or toothed membrane; lobes 5, fleshy, flat. Pollinia pendent. Follicles usually fusiform, smooth.

In the descriptions, leaf-characters refer to middle leaves, and peduncle-lengths to upper inflorescences.

Literature: A. Borhidi & S. Priszter, *Acta Bot. Acad. Sci. Hung.* **12**: 241–254 (1966). R. Ross, *Acta Bot. Neerl.* **15**: 161 (1966). H. P. Fuchs, *Verh. Naturf. Ges. Basel* **72**: 343–349 (1961).

1 Leaves grey-tomentose; corona-lobes stipitate   **1. canescens**
1 Leaves glabrous to velutinous; corona-lobes sessile
 2 Corolla-lobes with straight hairs on upper surface
  3 Corolla 10–12 mm in diameter, the lobes lanceolate
         2. speciosum
  3 Corolla 5–8 mm in diameter, the lobes ± ovate
   4 Corolla dark purple
    5 Peduncles 5–10 mm; leaves 5–7 × 2–3 cm   **3. nigrum**
    5 Peduncles 10–50 mm; leaves 9–13 × 4–8 cm   **4. scandens**

    4 Corolla greenish-yellow
     6 Peduncles 5–10 mm   **5. huteri**
     6 Peduncles 12–30 mm   **6. juzepczukii**
 2 Corolla-lobes glabrous or with curved hairs on upper surface
  7 Corolla white or yellow   **11. hirundinaria**
  7 Corolla dark purple or brownish
   8 Stems twining; corolla dark purple   **7. rossicum**
   8 Stems not twining; corolla brownish
    9 Peduncles 10–20 mm; corolla 5–6 mm in diameter
         8. schmalhausenii
    9 Peduncles 0–5 mm; corolla 8–10 mm in diameter
     10 Corolla-lobes brown on both surfaces   **9. fuscatum**
     10 Corolla-lobes greenish-brown above, yellow with brown margins beneath   **10. pannonicum**

1. V. canescens (Willd.) Decne in DC., *Prodr.* **8**: 523 (1844) (*Cynanchum canescens* (Willd.) K. Schum.). Stems 40–70 cm, more or less twining, tomentose. Leaves 6–8 × 3–6 cm, subsessile, broadly ovate, acute, grey-tomentose. Cymes 8- to 15-flowered, sessile. Calyx-lobes ovate. Corolla 7–8 mm in diameter, yellow; lobes triangular, hirsute above. Corona-membrane *c.* ⅛ as long as the segments; lobes stipitate. Follicles 7–10 × 1·5–2 cm, ovoid, shortly acuminate, pubescent. *Rocks and screes.* Kriti. Cr. (*S.W. Asia.*)

2. V. speciosum Boiss. & Spruner in Boiss., *Diagn. Pl. Or. Nov.* 1(4): 39 (1844) (*Cynanchum speciosum* (Boiss. & Spruner) Nyman). Stems 60–70 cm, erect, puberulent. Leaves 7–10 × 4–6 cm, subsessile, broadly ovate, obtuse or acute, velutinous. Cymes 5- to 15-flowered, subsessile. Calyx-lobes oblong. Corolla 10–12 mm in diameter, dark purple; lobes lanceolate, with straight hairs on upper surface. Corona-membrane *c.* ⅔ as long as the segments; lobes sessile. Follicles *c.* 7 × 0·8 cm, fusiform, tomentose. *Mountain woods and meadows.* S. & C. parts of Balkan peninsula. Al Bu Gr Ju Tu.

3. V. nigrum (L.) Moench, *Meth., Suppl.* 313 (1802) (*Cynanchum nigrum* (L.) Pers., non Cav.). Stems 40–80 cm, mostly twining, puberulent. Leaves 5–7 × 2–3 cm, ovate, long-acuminate; veins and margins pubescent. Petioles 10–15 mm. Cymes 5- to 7-flowered; peduncles 5–10 mm. Calyx-lobes ovate. Corolla 6–8 mm in diameter, dark purple; lobes ovate, with straight hairs on upper surface. Corona-membrane *c.* ⅔ as long as the segments, minutely toothed; lobes sessile. Follicles 5–7 × *c.* 0·8 cm, fusiform, glabrous. ● *S.W. Europe.* Bl Ga Hs It Lu [Iᴼo].

4. V. scandens Sommier & Levier, *Acta Horti Petrop.* **12**: 158 (1892) (*Antitoxicum scandens* (Sommier & Levier) Pobed.). Stems up to 200 cm, twining, puberulent. Leaves 9–13 × 4–8 cm, ovate, acute to acuminate; veins and margins ciliate. Petioles 8–15 mm. Cymes 6- to 10-flowered; peduncles 10–50 mm. Calyx-lobes ovate. Corolla 8 mm in diameter, dark purple; lobes broadly ovate, with straight hairs on upper surface. Corona-membrane *c.* ½ as long as the segments; lobes sessile. Follicles 5–8 × *c.* 0·6 cm, fusiform, sparsely pubescent. *Woodland and scrub.* S. & E. Ukraine, S. Russia. Rs (C, W, K, E).

5. V. huteri Vis. & Ascherson, *Österr. Bot. Zeitschr.* **19**: 67 (1869) (*Cynanchum huteri* (Vis. & Ascherson) K. Schum.). Stems (15–)50–150 cm, slightly twining, puberulent. Leaves 5–9 × 2–3 cm, ovate-lanceolate, long-acuminate; veins and margins ciliate. Petioles 5–10 mm. Cymes 6- to 10-flowered; peduncles 5–10 mm. Calyx-lobes oblong. Corolla 5–7 mm in diameter, greenish-yellow; lobes ovate-oblong, with straight hairs on upper surface. Corona-membrane *c.* ½ as long as the segments, minutely toothed; lobes sessile. Follicles *c.* 5·5 × 0·8 cm,

[1] By F. Markgraf.

fusiform, glabrous. *Scrub.* ● *N.W. part of Balkan peninsula.* Al Ju.

6. **V. juzepczukii** (Pobed.) Privalova, *Trudy Gos. Nikitsk. Bot. Sada* **26**: 140 (1956) (*Antitoxicum juzepczukii* Pobed.). Like **5** but stems 40–60 cm; leaves 3–4 cm wide; petioles 3–6 mm; cymes 5- to 8-flowered; peduncles 12–30 mm. Fagus-*woodland.* ● *Krym.* Rs (K).

7. **V. rossicum** (Kleopow) Barbarich, *Vyzn. Rosl. Ukr.* 346 (1950) (*Antitoxicum rossicum* (Kleopow) Pobed.). Stems up to 150 cm, twining, puberulent. Leaves 9–12 × 5–6 cm, ovate, acute; veins and margins ciliate. Petioles 10 mm. Cymes 5- to 6-flowered; peduncles 15–30 mm. Calyx-lobes ovate-lanceolate. Corolla 7 mm in diameter, dark purple; lobes ovate-lanceolate, glabrous on upper surface. Corona-membrane $c.$ $\frac{1}{2}$ as long as the segments; lobes sessile. Follicles 4–6 × $c.$ 0·5 cm, fusiform, glabrous. *Scrub.* ● *Ukraine, S.E. Russia.* Rs (C, W, E).

8. **V. schmalhausennii** (Kusn.) Markgraf, *Bot. Jour. Linn. Soc.* **65**: 358 (1972) (*Antitoxicum schmalhausenii* (Kusn.) Pobed., *Cynanchum schmalhausenii* Kusn.). Stems 35–80 cm; erect, not twining, puberulent. Leaves 7–8 × 3–3·5 cm, ovate, long-acuminate, the veins and margins pubescent. Petioles 50–100 mm. Cymes 10- to 12-flowered; peduncles 10–20 mm. Calyx-lobes broadly ovate. Corolla 5–6 mm in diameter, dark brown; lobes broadly ovate, glabrous on upper surface. Corona-membrane $c.$ $\frac{2}{3}$ as long as the segments, minutely toothed; lobes sessile. Follicles 5–6 × $c.$ 0·5 cm, fusiform, glabrous. *Limestone slopes. Krym.* Rs (K). (*Caucasus.*)

9. **V. fuscatum** (Hornem.) Reichenb. fil., *Icon. Fl. Germ.* **17**: 17 (1854) (*Cynanchum fuscatum* Link; incl. *C. urumoffii* (Davidov) Hayek, *Vincetoxicum intermedium* Taliev, *Antitoxicum maeoticum* (Kleopow) Pobed., *A. minus* (C. Koch) Pobed., *A. intermedium* (Taliev) Pobed.). Stems 20–40 cm, erect, not twining, puberulent. Leaves 3–6·5 × 1·5–4 cm, more or less broadly ovate, acute; veins pubescent and margins ciliate. Petioles 3–8 mm. Cymes 5- to 6-flowered; peduncles up to 5 mm. Calyx-lobes lanceolate. Corolla 8–10 mm in diameter, brown on both surfaces; lobes oblong, glabrous or with curved hairs on upper surface. Corona-membrane $c.$ $\frac{2}{3}$ as long as the segments, more or less toothed; lobes sessile. Follicles 4–5 × $c.$ 0·7 cm, fusiform, glabrous. *Dry stony places, scrub and steppes. S.E. Europe.* Al Bu Gr Ju Rs (W, K, E) Tu.

10. **V. pannonicum** (Borhidi) J. Holub, *Folia Geobot. Phytotax. (Praha)* **5**: 439 (1970) (*Cynanchum pannonicum* Borhidi). Like **9** but stems (10–)15–30(–40) cm; petioles 8–10 mm; calyx-lobes triangular-lanceolate; corolla greenish-brown above, yellow with brown margins beneath, the lobes triangular-ovate, glabrous on the upper surface; corona-membrane $c.$ $\frac{1}{2}$ as long as the segments. $2n=44$. *Dry grassland, and crevices in calcareous rocks.* ● *Hungary.* Hu.

11. **V. hirundinaria** Medicus, *Hist. Comment. Acad. Elect. Theod. Palat. Mannheim Phys.* **6**: 404 (1790) (*V. officinale* Moench, *Cynanchum vincetoxicum* (L.) Pers., *Antitoxicum officinale* Pobed.). Stems up to 120 cm, erect or slightly twining, puberulent or subglabrous. Leaves 6–10 × 2·5–5 cm, broadly ovate to ovate-lanceolate, acute, more or less pubescent, especially on the veins and margins. Petioles 5–10 mm. Cymes 6- to 8-flowered; peduncles 10–40 mm. Calyx-lobes linear. Corolla 3–10 mm in diameter, white or yellow; lobes ovate, glabrous or with curved hairs on upper surface. Corona-membrane up to $\frac{2}{3}$ as long as the segments; lobes sessile. Follicles $c.$ 6 × 0·8 cm,

fusiform, glabrous. $2n=22$. *Wood-margins and open places. Most of Europe, northwards to 60° 30′ N. in S.W. Finland, but absent from many of the islands.* Al Au Be Bl Bu Co Cz Da Fe Ga Ge Gr He Ho Hs Hu It Ju Lu No Po Rm Rs (B, C, W, K, E) Sa Su Tu.

Very variable, even within populations. Many constituent taxa have been described, often as species. Yellow-flowered taxa occur mainly in the Mediterranean region and white-flowered taxa further north.

1 Corolla yellow
  2 Corolla-lobes hairy on upper surface; leaves acuminate
    3 Stem subglabrous; corolla 3–4 mm in diameter; corona-lobes obtuse   **(a)** subsp. **lusitanicum**
    3 Stem pubescent; corolla 6–9 mm in diameter; corona-lobes acute   **(b)** subsp. **intermedium**
  2 Corolla-lobes glabrous on upper surface; leaves acute
    4 Corolla 8–9 mm in diameter; corona folded or toothed   **(c)** subsp. **nivale**
    4 Corolla 7 mm in diameter; corona not folded or toothed   **(d)** subsp. **adriaticum**
1 Corolla whitish
  5 Corona-membrane $c.$ $\frac{2}{3}$ as long as the segments
    6 Corolla 10 mm in diameter; corona not toothed   **(e)** subsp. **stepposum**
    6 Corolla 8 mm in diameter; corona toothed   **(f)** subsp. **cretaceum**
  5 Corona-membrane not more than $\frac{1}{2}$ as long as the segments
    7 Corona-membrane $c.$ $\frac{1}{8}$ as long as the segments   **(h)** subsp. **contiguum**
    7 Corona-membrane $c.$ $\frac{1}{2}$ as long as the segments
      8 Corona-margin sinuate   **(g)** subsp. **jailicola**
      8 Corona-margin abruptly lobed   **(i)** subsp. **hirundinaria**

(a) Subsp. **lusitanicum** Markgraf, *Bot. Jour. Linn. Soc.* **65**: 358 (1972) (*V. luteum* Hoffmanns. & Link pro parte): Stems up to 50 cm, subglabrous. Leaves acuminate. Corolla 3–4 mm in diameter, yellow; lobes hairy on upper surface. Corona-lobes obtuse; membrane $c.$ $\frac{1}{2}$ as long as the segments. ● *N.W. Portugal, N. Spain.*

(b) Subsp. **intermedium** (Loret & Barrandon) Markgraf, *Bot. Jour. Linn. Soc.* **64**: 374 (1971) (*V. officinale* var. *intermedium* Loret & Barrandon): Stems 50–60 cm, pubescent. Leaves 50–70 mm, acuminate. Corolla 6–9 mm in diameter, yellow; lobes hairy on upper surface. Corona-lobes acute; membrane $c.$ $\frac{1}{2}$ as long as the segments, sometimes toothed. ● *S. France, N.E. Spain.*

(c) Subsp. **nivale** (Boiss. & Heldr.) Markgraf, *op. cit.* 375 (1971) (*Vincetoxicum nivale* Boiss. & Heldr.; incl. *Cynanchum sulfureum* (Velen.) Hayek, *C. vincetoxicum* var. *nivale* (Boiss. & Heldr.) Hayek): Stems 20–30 cm, minutely pubescent. Leaves broadly ovate, acute. Corolla 8–9 mm in diameter, yellow; lobes glabrous on upper surface. Corona-membrane $c.$ $\frac{1}{2}$ as long as the segments, more or less folded or toothed. ● *Balkan peninsula.*

(d) Subsp. **adriaticum** (G. Beck) Markgraf, *op. cit.* 374 (1971) (*V. adriaticum* G. Beck, *Cynanchum adriaticum* (G. Beck) Fritsch): Up to 50 cm, minutely pubescent. Leaves acute. Corolla 7 mm in diameter, yellow; lobes glabrous on upper surface (rarely hairy). Corona-membrane $c.$ $\frac{1}{2}$ as long as the segments, not folded or toothed. ● *N.W. Jugoslavia.*

(e) Subsp. **stepposum** (Pobed.) Markgraf, *op. cit.* 375 (1971) (*Antitoxicum stepposum* Pobed.): Stems up to 70 cm, subglabrous. Leaves ovate-lanceolate. Corolla 10 mm in diameter, whitish; lobes glabrous on upper surface. Corona-membrane $c.$ $\frac{2}{3}$ as long as the segments, not toothed. *C. & S. Russia, Ukraine.*

(f) Subsp. **cretaceum** (Pobed.) Markgraf, *op. cit.* 374 (1971) (*Antitoxicum cretaceum* Pobed.): Stems up to 30 cm, subglabrous. Leaves ovate-lanceolate. Corolla 8 mm in diameter, whitish; lobes glabrous on upper surface. Corona-membrane *c.* ⅔ as long as the segments, minutely toothed. *S.E. Russia, S. & E. Ukraine.*

(g) Subsp. **jailicola** (Juz.) Markgraf, *op. cit.* 375 (1971) (*V. jailicola* Juz.; incl. *Antitoxicum tauricum* (Pobed.) Pobed.): Stems 20–70 cm, subglabrous. Leaves shortly acuminate. Corolla 7–8 mm in diameter, whitish; lobes hairy at the base on upper surface. Corona-membrane toothed, *c.* ½ as long as the segments; margin sinuate. ● *Krym.*

(h) Subsp. **contiguum** (Koch) Markgraf, *op. cit.* 374 (1971) (*V. contiguum* (Koch) Gren. & Godron, *Cynanchum contiguum* Koch): Stems *c.* 70 cm, subglabrous. Leaves broadly ovate, subglabrous. Corolla 7 mm in diameter, whitish; lobes glabrous on upper surface. Corona-membrane *c.* ⅛ as long as the segments; lobes contiguous. ● *W. Jugoslavia.*

(i) Subsp. **hirundinaria**: Stems up to 120 cm, subglabrous or pubescent. Leaves ovate, shortly acuminate. Corolla 5–10 mm in diameter, white; lobes glabrous or hairy on upper surface. Corona-membrane more or less toothed, *c.* ½ as long as the segments; margin abruptly lobed. *Throughout the range of the species, except the Iberian peninsula.*

7. Cionura Griseb.[1]

Glabrous herbs, often with slightly twining shoot-apices. Leaves opposite, long-petiolate. Flowers in axillary, many-flowered cymes. Corolla contorted in bud; lobes patent at anthesis. Corona single; segments 5, free, without awns. Pollinia erect. Follicles narrowly ovoid, acuminate, smooth.

Literature: K. Browicz, *Arboretum Kórnickie* 12: 5–32 (1967).

1. **C. erecta** (L.) Griseb., *Spicil. Fl. Rumel.* 2: 69 (1844) (*Cynanchum erectum* L., *Marsdenia erecta* (L.) R. Br.). Stems up to 100 cm, numerous. Leaves *c.* 7 × 5 cm, petiolate, ovate, acuminate, cordate at base; petioles *c.* 5 cm. Corolla *c.* 1 cm in diameter, white; lobes *c.* 7 × 2 mm, oblong, obtuse, glabrous. Follicles *c.* 8 × 1·5 cm. *Rocky places, river-gravels and maritime sands. Aegean region and S. & E. parts of Balkan peninsula.* Al Bu Cr Gr Ju Tu.

8. Caralluma R. Br.[1]

Succulent, caespitose herbs with subterranean stolons. Stems glabrous, glaucous, 4-angled, the angles with opposite, scale-like caducous leaves above and toothed leaf-scars below. Inflorescence umbellate, subterminal. Corolla valvate in bud; lobes patent at anthesis. Corona double, with 10 segments, the outer 5 connate, 2-fid. Follicles fusiform, smooth.

Literature: A. White & B. L. Sloane, *The Stapelieae* 222–232. Pasadena, 1937.

Stem-faces ± flat; corolla 13–17 mm in diameter, yellow with purple bands　　　　　　　　　　**1. europaea**
Stem-faces sulcate; corolla 6–8 mm in diameter, brown　**2. munbyana**

1. **C. europaea** (Guss.) N. E. Br., *Gard. Chron.* ser. 3, 12: 369 (1892) (*Apteranthes gussoneana* Mikan). Stolons long. Stems 10–15 cm, ascending or procumbent, the faces more or less flat; teeth of leaf-scars less than 1 cm apart. Leaves sessile, orbicular, shortly acuminate. Inflorescences with 10 or more pedicellate flowers; pedicels 3–4 mm. Corolla 13–17 mm in diameter, rotate; lobes suborbicular, yellow with purple bands, more or less ciliate. Follicles *c.* 100 × 7 mm, slightly recurved, patent. *Dry, rocky places. S.E. Spain; Linosa, Lampedusa.* Hs Si. (*N. Africa.*)

Plants from Linosa and Lampedusa differ from those from Spain in their slightly smaller corolla and appendaged corona.

2. **C. munbyana** (Decne) N. E. Br., *op. cit.* 370 (1892). Stolons short. Stems 5–15 cm, ascending, the faces sulcate; teeth of leaf-scars more than 1 cm apart. Leaves sessile, ovate, obtuse. Inflorescence with fewer than 10 sessile flowers. Corolla 6–8 mm in diameter, rotate-campanulate; lobes linear, brown, the margins deflexed, glabrous. Follicles 80–100 × 8 mm, straight, erect. *Dry, rocky places. S.E. Spain (Caravaca, prov. Murcia).* Hs. (*N. Africa.*)

CXLIV. RUBIACEAE

Herbs or dwarf shrubs. Leaves opposite or in whorls of 4–12, stipulate, the stipules sometimes leaf-like, simple, entire. Flowers in spikes, heads, cymes, corymbs or panicles, hermaphrodite or male. Calyx 4- or 5(–6)-toothed or absent. Corolla 3- to 5-lobed, actinomorphic, rotate, infundibuliform or hypocrateriform. Stamens 4–5, epipetalous on the corolla-tube and alternating with the lobes. Ovary inferior, usually 2-locular; ovules 1 to numerous per locule; style simple or 2-fid. Fruit fleshy or dry, dividing into 1–2 mericarps. Seeds endospermic.

The text of this family will be published in Vol. 4.

TUBIFLORAE

CXLV. POLEMONIACEAE[2]

Herbs or rarely shrubs. Leaves usually alternate, entire to pinnate, exstipulate. Flowers actinomorphic, 5-merous. Calyx campanulate, more or less deeply lobed. Corolla infundibuliform, campanulate or rotate; lobes contorted. Stamens epipetalous, alternating with the corolla-lobes, often unequal; anthers versatile. Ovary superior, 3-locular, with axile placentae; ovules usually 2–12 in each loculus; style 1, shortly 3-fid at the apex. Fruit a loculicidal or septicidal capsule.

Literature: A. Brand in Engler, *Pflanzenreich* 27(IV. 250) (1907). J. F. Davidson, *Univ. Calif. Publ. Bot.* 23(5): 209–282 (1960).

Leaves pinnate; stamens exserted　　　**1. Polemonium**
Leaves entire; stamens included　　　　**2. Collomia**

[1] By F. Markgraf.　　[2] Edit. T. G. Tutin.

73

1. Polemonium L.[1]

Usually perennial. Leaves pinnate. Flowers in terminal corymbose inflorescences. Corolla blue, campanulate or with a rotate limb; stamens exserted, all inserted at the same height on the corolla, the downward-curved, hairy bases of the filaments almost closing its throat. Capsule subglobose; seeds 4–14 in each loculus.

1 Stem with 1(–2) leaves   **3. boreale**
1 Stem with several leaves
2 Corolla-lobes ciliate   **2. acutiflorum**
2 Corolla-lobes not or sparsely ciliate   **1. caeruleum**

1. P. caeruleum L., *Sp. Pl.* 162 (1753) (*P. caucasicum* auct. eur., non N. Busch). Stems 30–90 cm, simple, leafy, hairy above, glabrous below, sparsely to densely glandular. Leaves up to 40 cm, imparipinnate, the lower petiolate, with usually 10–12 pairs of leaflets, the upper smaller and subsessile. Flowers numerous, shortly pedicellate in terminal and axillary cymes. Calyx-lobes lanceolate, acute, shorter than to longer than the tube. Corolla 8–15 mm, rotate, 2–2½ times as long as the calyx, blue, rarely white; lobes ovate, rounded to subacute, sometimes sparsely ciliate. Stamens exserted. Capsule subglobose; seeds angled and rugose. $2n=18$. *Rocks and damp meadows, mostly in mountain regions. N. & C. Europe, extending southwards to the Pyrenees, C. Jugoslavia and S. Russia; cultivated for ornament and often naturalized.* Au Br Cz Fe Ga Ge He *Is ?It Ju No Po Rm Rs (N, B, C, W, E) Su [Be Da Ho.]

2. P. acutiflorum Willd. in Roemer & Schultes, *Syst. Veg.* 4: 792 (1819). Like **1** but lower leaves with usually not more than 8 pairs of leaflets; inflorescence few-flowered; corolla 18–22 mm, campanulate; lobes narrowed above, acute, ciliate. $2n=18$. *Damp thickets and river-banks. Arctic and subarctic Europe.* Fe No Rs (N) Su.

3. P. boreale Adams, *Mém. Soc. Nat. Moscou* **5**: 92 (1817). Foetid when bruised. Stems 7–15 cm, erect or ascending, often curved, with 1(–2) leaves, glandular-pubescent. Leaves usually with 6–8 pairs of leaflets. Flowers 3–6 in a terminal cluster; pedicels shorter than the calyx. Calyx-lobes oblong-ovate, obtuse, longer than the tube. Corolla *c.* 15 mm, campanulate, blue; lobes rounded, not ciliate. $2n=18$. *Gravelly and sandy places. Arctic Europe, extending locally southwards to 62° N. in N.W. Russia.* No Rs (N) Sb.

P. pulchellum Ledeb., *Icon. Pl. Fl. Ross.* **1**: 6 (1829) occurs in Ural, but probably not in Europe. It is like **3** but has the pedicels longer than the calyx.

2. Collomia Nutt.[1]

Annual. Leaves entire. Flowers in dense heads. Calyx with a projecting fold below each sinus in fruit; corolla yellow or lilac, infundibuliform; stamens included, unequal. Capsule obovoid; seeds 1 in each loculus, mucilaginous when wet.

1. C. grandiflora Douglas ex Lindley, *Bot. Reg.* **14**: sub t. 1166 (1828) (*C. coccinea* auct., non Lehm.). Stems 20–100 cm, glabrous, puberulent or glandular, leafy throughout. Leaves 3–5 cm, linear to lanceolate, sessile. Bracts ovate, leaf-like. Flowers subsessile. Calyx 7–10 mm; lobes lanceolate. Corolla 15–30 mm, yellow, becoming reddish, narrowly tubular, with more or less patent, lanceolate lobes. Capsule *c.* 5 mm. *Cultivated for ornament and locally naturalized in damp places in W. & C. Europe.* [Cz Ga Ge.] (*California.*)

C. linearis Nutt., *Gen. N. Amer. Pl.* **1**: 126 (1818), with corolla with lilac limb and scarcely exceeding the calyx, from North America, is a frequent casual in Fennoscandia and may be confused with **1**.

CXLVI. CONVOLVULACEAE[2]

Herbs or shrubs. Stems often climbing. Leaves alternate, exstipulate. Flowers actinomorphic, in terminal or axillary inflorescences, sometimes solitary. Sepals 5, usually free. Corolla infundibuliform, tubular or campanulate, usually 5-lobed or -angled. Stamens 5, alternating with corolla-lobes. Ovary 1- to 4-locular, superior; ovules 1 or 2 in each loculus; style terminal. Fruit a capsule.

1 Plant a twining parasite with haustoria, not green, more or less glabrous   **1. Cuscuta**
1 Plant free-living, green or covered with hairs
 2 Corolla up to 5 mm, divided to about halfway
  3 Leaves sessile, lanceolate to ovate   **2. Cressa**
  3 Leaves with long petioles, reniform to orbicular   **3. Dichondra**
 2 Corolla more than 5 mm, scarcely lobed
  4 Bracteoles broad, leaf-like, partly obscuring the sepals   **4. Calystegia**
  4 Bracteoles lanceolate to filiform, not or scarcely obscuring the sepals
   5 Stigma with 1–3 globose lobes; pollen grains pantoporate, spinous   **6. Ipomoea**
   5 Stigma with 2 filiform to cylindric-clavate lobes; pollen grains tricolpate, ±smooth   **5. Convolvulus**

1. Cuscuta L.[3]

Herbaceous parasites, usually annual. Stems twining, with haustoria. Leaves reduced to minute scales. Inflorescences cymose, spike-like, umbellate or capitulate. Flowers (3–)4- to 5-merous, small, white, yellowish or reddish. Stamens inserted in throat of corolla-tube. Hypostaminal scales attached at base of corolla-tube opposite the stamens. Ovary 2-celled, each cell containing 2 anatropous ovules. Styles free or united. Stigmas capitate or elongate. Fruit a circumscissile capsule dehiscing by an irregular or regular line near base, or remaining closed. Seeds 4 or fewer. Embryo filiform, surrounded by a cartilaginous endosperm. Cotyledons rudimentary or absent.

Inflorescences with capitulate cymes are referred to as glomerules. Mature flowers are essential for identification.

Literature: T. G. Yuncker, *Mem. Torrey Bot. Club* **18**: 113–331 (1932). N. Feinbrun, *Israel Jour. Bot.* **19**: 16–29 (1970).

1 Style 1; mostly parasites of trees and shrubs
 2 Style in fruit about as long as stigma; calyx ⅔ as long as to equalling corolla-tube   **17. monogyna**
 2 Style in fruit 2–3 times as long as stigma; calyx ⅓–½ as long as corolla-tube   **16. lupuliformis**
1 Styles and stigmas 2 (rarely more); parasites of trees, shrubs and herbs

[1] By T. G. Tutin.
[2] Edit. D. H. Valentine.
[3] By N. Feinbrun.

3 Stigmas capitate
  4 Flowers in a lax, racemose-paniculate inflorescence; some pedicels at least as long as flowers; capsule globose to ovoid
    5 Corolla-lobes deflexed; calyx-lobes ovate, obtuse; corolla detaching from its base at maturity to form a cap to the capsule   **4. gronovii**
    5 Corolla-lobes erect; calyx-lobes triangular-ovate, subacute; corolla persistent at maturity, surrounding the capsule   **3. suaveolens**
  4 Flowers forming a ± compact globose inflorescence; pedicels usually shorter than flowers; capsule depressed-globose
    6 Capsule 2–3 mm in diameter; corolla-lobes triangular, acute; scales abundantly fimbriate, not bifid; flowers 5-merous   **2. campestris**
    6 Capsule 3·5–4 mm in diameter; corolla-lobes ovate, obtuse or subacute; scales deeply bifid; flowers 4- to 5-merous   **1. australis**
3 Stigmas elongate
  7 Capsule dehiscing irregularly; stigmas subsessile; flowers long-pedicellate, in few-flowered umbel-like glomerules   **5. pedicellata**
  7 Capsule regularly circumscissile near base; styles distinct; flowers mostly sessile or shortly pedicellate, in capitate glomerules
    8 Calyx-tube golden-yellow, shiny and reticulate when dry; calyx-lobes wide and short, abruptly ending in a short fleshy or long swollen, nearly cylindrical appendage   **15. approximata**
    8 Calyx-tube neither golden-yellow nor shiny and reticulate; calyx-lobes without a cylindrical appendage
      9 Styles (including stigmas) shorter than ovary; glomerules 10–15 mm in diameter; scales not reaching the stamens, or absent
        10 Parasitic on *Linum usitatissimum* (rarely on other hosts); calyx-lobes acute; stems nearly simple   **8. epilinum**
        10 Parasitic on various species, often on *Urtica dioica*, rarely on *Linum usitatissimum*; calyx-lobes obtuse, rarely subacute; stems much branched
          11 Flowers 3–4 mm; calyx-lobes obtuse, not longer than tube   **6. europaea**
          11 Flowers 2·5 mm; calyx-lobes acute, longer than tube   **7. pellucida**
      9 Styles (including stigmas) longer than ovary, or if as long as ovary or shorter, glomerules 4–7 mm in diameter; scales often reaching the stamens
        12 Styles (including stigmas) much longer than ovary; flowers (2·5–)3–4(–5) mm; glomerules 7–10 mm in diameter; calyx-lobes membranous
          13 Lobes of calyx and corolla obtuse; pedicels as long as or nearly as long as flowers   **11. triumvirati**
          13 Lobes of calyx and corolla acute to acuminate; flowers sessile or shortly pedicellate   **12. epithymum**
        12 Styles (including stigmas) equalling or only slightly longer than ovary; flowers 1·5–2·5 mm; glomerules 4–6(–7) mm in diameter; calyx-lobes often ± fleshy
          14 Flowers (3–)4-merous; corolla-lobes erect   **9. palaestina**
          14 Flowers (4–)5-merous; corolla-lobes patent
            15 Plant blackening when dry; flowers dark purple   **10. atrans**
            15 Plant not blackening when dry; flowers white or purplish
              16 Calyx-lobes oblong (rarely ovate), longer than tube, swollen, nearly semicircular in cross-section   **14. planiflora**
              16 Calyx-lobes triangular-ovate, acute, about as long as tube, membranous or somewhat fleshy
                17 Styles (including stigmas) shorter than ovary; calyx membranous, usually white   **13. brevistyla**
                17 Styles (including stigmas) as long as or slightly longer than ovary; calyx fleshy, reddish   **12. epithymum**

Subgen. **Grammica** (Lour.) Yuncker. Styles 2. Stigmas capitate.

1. C. australis R. Br., *Prodr. Fl. Nov. Holl.* 491 (1810). Stems moderately stout, greenish-yellow to orange. Flowers *c.* 2 mm, 4- to 5-merous, often glandular, shortly pedicellate, in compact glomerules. Calyx campanulate, about as long as corolla-tube; lobes ovate to orbicular, obtuse, not overlapping. Corolla campanulate; lobes obtuse or subacute, erect to patent, slightly shorter than to longer than tube. Stamens exserted. Scales shorter than corolla-tube, bifid, with a few fimbriae or reduced to mere teeth at base of corolla-tube (rarely reaching the stamens and irregularly fimbriate). Styles shorter than the globose ovary. Capsule 3·5–4 mm in diameter, globose to depressed-globose, with the persistent corolla at its base; intra-stylar opening large. Seeds 1·5 mm. *S. Europe; naturalized in C. Europe.* Al Bu Co Ga Ge Gr Hs It Ju Lu Rm Rs (W, E) Si Tu [?Au Cz He Ho Hu Po].

A mainly Asiatic and Australian species, represented in Europe by two subspecies:

(a) Subsp. **tinei** (Insenga) Feinbrun, *Israel Jour. Bot.* **19**: 19 (1970) (*C. breviflora* Vis., *C. tinei* Insenga): Flowers often 4-merous; scales much reduced, each often in the shape of two short wings near the base of corolla. *On* Polygonum *and other herbs. S. & S.E. Europe, with isolated localities in C. Europe.*

(b) Subsp. **cesattiana** (Bertol.) Feinbrun, *loc. cit.* (1970) (*C. cesattiana* Bertol.): Flowers 5-merous; scales well-developed, nearly reaching the stamens. *Mainly on hydrophilous species of* Polygonum. ● *C. France to Romania and N. Greece.*

2. C. campestris Yuncker, *Mem. Torrey Bot. Club* **18**: 138 (1932) (*C. arvensis* auct., non Beyrich ex Engelm.; incl. *C. basarabica* Buia). Stems moderately stout, yellowish. Flowers 2–3 mm, 5-merous; pedicels short; glomerules 10–12 mm in diameter, compact, globose. Calyx campanulate, about as long as corolla-tube; lobes ovate or orbicular, obtuse, slightly overlapping. Corolla-lobes acute, triangular, patent (often with inflexed apex), about as long as the shortly campanulate tube. Stamens exserted. Scales long, densely fimbriate, exserted. Styles slender, about as long as the globose ovary. Capsule 2–3 mm in diameter, depressed-globose, pale, with the persistent corolla at its base. Seeds 1–1·2 mm. *Mainly on cultivated species of* Trifolium *and* Medicago, *but also on other herbs. Widely naturalized in S., C. & W. Europe.* [Al Au Be Br Bu Cz Ga Ge Gr He Ho Hu It Ju Lu Po Rm Rs (C, W) ?Sa.] (*North America.*)

Introduced to Europe about 1900, and spread mainly with agricultural seed; now a weed in some regions.

3. C. suaveolens Ser., *Ann. Sci. Phys. Nat. Agric. Industr.* **3**: 519 (1840) (*C. corymbosa* sensu Choisy, non Ruiz & Pavón). Stems slender or of medium thickness. Flowers 3–4 mm, oblong, 5-merous, pedicellate, in lax clusters *c.* 10 mm in diameter. Calyx not more than ½ and usually only ⅓ as long as corolla-tube; lobes triangular-ovate, subacute, not overlapping, the sinuses often with revolute margins. Corolla campanulate or infundibuliform; lobes *c.* ½ as long as tube, erect, mostly ovate-triangular, with acute, inflexed apices. Stamens slightly exserted. Scales scarcely reaching the filaments, oblong-ovate, with rather short fimbriae. Styles slender, as long as or sometimes longer than ovary, often unequal. Capsule globose, with the persistent corolla at its base, not circumscissile. Seeds 1·5–2 mm. *On cultivated species of* Medicago *and* Trifolium *and on other herbs. Naturalized in S. & C. Europe.* [Co Cz Ga Ge Gr He Hs It Lu Po Rs (C, W).] (*Chile.*)

4. **C. gronovii** Willd. in Roemer & Schultes, *Syst. Veg.* **6**: 205 (1820). Stems stout. Flowers 3–4 mm, 5-merous, more or less glandular, in lax or dense clusters. Calyx campanulate, shorter than corolla-tube; lobes ovate, obtuse, overlapping, often with uneven edges. Corolla-lobes mostly shorter than the tubular-campanulate tube, ovate, obtuse, deflexed. Stamens somewhat exserted. Scales about as long as tube, narrow, long-fimbriate. Ovary ovoid-conical, thickened at apex. Styles somewhat shorter than ovary, thickened below. Capsule globose-conical or pyriform, capped by the detached corolla. *Mainly on* Salix, Populus *and other trees, especially on river-banks. Naturalized mainly in W. & C. Europe.* [Be Bu Cz Ga Ge Ho It Po Rs (W).] (*C. & E. North America.*)

In Europe only as var. *calyptrata* Engelm.

Subgen. **Cuscuta.** Styles 2. Stigmas elongate, cylindrical.

Sect. EPISTIGMA Engelm. Styles very short; capsule irregularly circumscissile.

5. **C. pedicellata** Ledeb., *Fl. Altaica* **1**: 293 (1829). Stems slender. Flowers 2–3 mm, mostly 4-merous; pedicels usually as long as or longer than flowers; clusters umbellate, few-flowered. Calyx rotate or cupulate, loose about the corolla; lobes triangular, acute or subobtuse, not overlapping. Corolla at first globose-campanulate, later closely investing the capsule; lobes connivent to patent, acute, about as long as the tube. Filaments longer than anthers. Scales truncate, dentate, usually about as long as tube. Ovary globose-conical. Styles very short, flattened. Stigmas slender, shorter than the ovary. Seeds 1·25 mm. *S.E. Russia.* Rs (E). (*C. & W. Asia, North Africa.*)

Records of *C. kotschyana* Boiss. for S.E. Russia are probably misidentifications of **5**.

Sect. CUSCUTA. Styles slender. Capsule regularly circumscissile.

Subsect. *Europaeae* Yuncker. Flowers mostly membranous, often 4-merous. Styles short. Scales mostly small, appressed to the corolla-tube. Corolla-lobes mostly erect.

6. **C. europaea** L., *Sp. Pl.* 124 (1753). Stems stout, much branched, often reddish. Flowers 3–4(–5) mm, 4–(5-)merous, on thickish pedicels. Glomerules 10–15 mm in diameter, numerous, each with a bract at its base. Calyx obconical, shorter than or about as long as corolla-tube; lobes ovate, obtuse, as long as or shorter than tube. Corolla campanulate, urceolate in fruit, detached from its base; lobes erect to patent, triangular to ovate, usually obtuse. Scales thin, small, shorter than corolla-tube and appressed to it, usually bifid and sparsely fimbriate, sometimes almost absent. Stamens often included. Styles (including stigmas) shorter than ovary. Capsule globose, conical or pyriform, capped by the detached corolla. Seeds *c.* 1·5 mm, sometimes united in pairs. *Usually on* Urtica dioica *or* Humulus lupulus *but also on many other hosts. Throughout Europe except the extreme north and some of the islands; only on mountains in the south.* All except Az Bl Cr Fa Hb Is Lu Sb Si.

7. **C. pellucida** Butkov, *Not. Syst. Herb. Inst. Bot. Zool. Acad. Sci. Uzbek.* **11**: 53 (1948) (*C. indica* (Engelm.) Petrov). Stems moderately stout, much branched, yellowish or slightly reddish. Flowers 2·5 mm, mostly 5-merous, almost sessile. Glomerules rather large, numerous, each with a membranous bract at its base. Calyx 1–1·5 mm, pellucid, divided nearly to its base into narrowly ovate, acute lobes. Corolla urceolate, nearly twice as long as calyx, pellucid; lobes erect, half as long as tube, ovate, subacute.

Scales minute, thin, bifid and fimbriate. Styles (including stigmas) shorter than ovary. Capsule circumscissile, almost globose, pellucid. Seeds *c.* 1 mm. *On herbs and dwarf shrubs (e.g.* Alhagi pseudalhagi). *S.E. Russia (near Volgograd).* Rs (E). (*C. Asia.*)

8. **C. epilinum** Weihe, *Arch. Apothekerver. Nördl. Teutschl.* **8**: 50 (1824). Stems slender or moderately stout, simple or sparingly branched. Flowers *c.* 3 mm, 5-merous, sessile; glomerules *c.* 10 mm in diameter, sparse, compact. Calyx as long as corolla-tube and somewhat loose about it; lobes broadly triangular-ovate, acute. Corolla urceolate; lobes ovate-triangular, subobtuse or acute, shorter than tube, mostly erect. Scales shorter than tube, truncate, shortly fimbriate, usually bifid, sometimes reduced to short wings near the base of corolla. Stamens shorter than corolla-lobes, slightly exserted. Styles (including stigmas) shorter than the ovary. Stigmas thickish, short, about as long as styles. Capsule depressed. Seeds *c.* 1–2 mm, often united in pairs. *Flax-fields; mainly on* Linum usitatissimum *and other species of* Linum *and on* Camelina sativa. *Much of Europe, but now rare in the west though formerly widespread there.* [Au Bu Co Cz Fe Ga Ge He Ho Hs Hu It Ju Lu No Po Rm Rs (N, B, C, W, E) Sa Si Su.] (*?S.W. Asia.*)

Introduced into Europe probably in prehistoric times.

9. **C. palaestina** Boiss., *Diagn. Pl. Or. Nov.* **2**(11): 86 (1849) (*C. globularis* Bertol.). Stems very slender, usually reddish, much branched. Flowers 1·5–2 mm, usually 4-merous (rarely 3-merous and single flowers in a glomerule 5-merous), sessile; glomerules 4–6 mm in diameter, globose. Calyx about as long as corolla-tube, usually reddish; lobes triangular, as long as or shorter than wide, acute, somewhat fleshy, keeled. Corolla-lobes usually erect, about as long as the cylindrical tube, acute, often with cucullate apices. Anthers scarcely exserted. Scales usually reaching the filaments. Styles (including stigmas) about as long as ovary. Capsule globose, enveloped by the corolla. Seeds *c.* 1 mm. *On dwarf shrubs. E. Mediterranean region and Sicilia.* Cr Gr Si.

10. **C. atrans** Feinbrun, *Israel Jour. Bot.* **19**: 22 (1970). Plant blackening when dry. Stems filiform, dark purple. Flowers 2–2·5 mm, 5-merous, sessile; glomerules *c.* 5 mm in diameter, globose. Calyx *c.* 2 mm, obconical-campanulate, enclosing the corolla-tube, dark purple, slightly fleshy, densely pustular, divided to the middle or beyond; lobes broadly obovate, shortly mucronate. Corolla purplish, somewhat fleshy, campanulate to infundibuliform; lobes ovate, patent, about as long as tube. Stamens shorter than corolla-lobes. Scales bifid, shortly fimbriate, reaching the stamens. Styles (including stigmas) shorter than the subglobose ovary. Capsule globose. Seeds ovoid. *On* Verbascum spinosum. ● *High mountains of Kriti.* Cr.

Subsect. *Planiflorae* Yuncker. Scales well-developed, connivent and arched above the ovary. Flowers usually 5-merous, membranous or fleshy and turgid. Styles (including stigmas) about as long as or longer than ovary.

11. **C. triumvirati** Lange, *Vid. Meddel. Dansk Naturh. Foren. Kjøbenhavn* **1881**: 98 (1882). Stems slender or moderately stout. Flowers 3–4 mm, 5-merous, reddish; pedicels as long as or nearly as long as flowers. Calyx shorter than corolla-tube; lobes ovate, obtuse, not overlapping. Corolla-lobes broadly ovate, obtuse, patent, about as long as the campanulate tube. Stamens exserted. Scales ovate, not reaching the stamens, shallowly fimbriate. Styles (including stigmas) longer than the depressed ovary, slender. *On* Erinacea anthyllis. *S. Spain (Sierra Nevada).* Hs. (*Morocco.*)

12. C. epithymum (L.) L., *Syst. Veg.* ed. 13, 140 (1774). Stems slender, much-branched, often reddish or purplish; flowers (2·5–)3–4(–5) mm, 5-merous, mostly sessile, sometimes shortly pedicellate; glomerules 5–10 mm in diameter, dense, globose, each with a subacute bract at its base. Calyx usually shorter than corolla-tube, sometimes about as long; lobes triangular, acute, sometimes lanceolate and acuminate; corolla-lobes patent, triangular, acute, rarely acuminate, mostly shorter than the campanulate tube. Stamens exserted; anthers often purple-tinged. Scales spathulate, shorter than tube, fimbriate, connivent over the ovary. Styles (including stigmas) about as long as to much longer than the subglobose ovary, exserted. Capsule enveloped by the corolla. Seeds *c.* 1 mm. 2*n*=14. *Almost throughout Europe, except the extreme north.* All except Az Fa Is Sb.

(a) Subsp. **epithymum** (incl. subsp. *trifolii* (Bab.) Hegi, *C. trifolii* Bab., *C. alba* C. Presl): Flowers 3–4(–5) mm (rarely 2·5 mm); glomerules 7–10 mm in diameter. Calyx shorter than corolla-tube, not fleshy; lobes of calyx and corolla triangular to lanceolate, acute or acuminate to subulate. Styles (including stigmas) much longer than ovary. *On shrubs and herbs, most frequently on* Calluna vulgaris, Ulex europaeus *and* Trifolium *spp. Almost throughout the range of the species.*
C. stenoloba Bornm. & O. Schwarz, *Feddes Repert. (Beih.)* **26**: 56 (1924), known only from Weimar in Germany, was probably an abnormal variant of **12(a)**; it is now extinct.

(b) Subsp. **kotschyi** (Desmoulins) Arcangeli, *Comp. Fl. Ital.* 480 (1882): Flowers *c.* 2·5 mm; glomerules 5–6 mm in diameter. Calyx as long as corolla-tube or slightly shorter, fleshy, usually red or purple; lobes ovate, keeled. *Mainly on dwarf shrubs. S. Europe.*

13. C. brevistyla A. Braun ex A. Richard, *Tent. Fl. Abyss.* **2**: 79 (1851). Stems slender or of medium thickness, much branched. Flowers 2–3 mm, (4–)5-merous, sessile; glomerules 4–7 mm in diameter, globose. Calyx membranous, about as long as corolla-tube; lobes ovate-triangular, subacute, not carinate, about as long as or shorter than tube; apex sometimes slightly fleshy; tube easily splitting to the base. Corolla campanulate; lobes patent, somewhat shorter than tube, triangular-ovate, obtuse or subacute. Stamens exserted. Scales nearly reaching the stamens, mostly truncate, fimbriate at apex. Style shorter than stigma; styles (including stigmas) shorter than or about as long as the globose ovary. Capsule depressed-globose, enveloped by the corolla. Seeds *c.* 0·5 mm. *Mainly on herbs. S. & E. parts of Mediterranean region.* Al Cr Gr IIs It.

14. C. planiflora Ten., *Fl. Nap.* **3**: 250 (1824–1829). Stems slender, much branched. Flowers 1·5–2·5 mm, 5-merous; glomerules 5–6 mm in diameter, small, compact, globose. Calyx white, rarely pink, as long as corolla-tube, deeply divided; lobes longer than tube, distinctly turgid (swollen, nearly semicircular in cross-section), oblong or more rarely ovate. Corolla white; lobes patent, acute, about as long as the campanulate tube, often cucullate at apex. Stamens exserted. Scales oblong, usually reaching the stamens, shortly fimbriate. Styles (including stigmas) about as long as ovary. Capsule depressed-globose, enveloped by the corolla. Seeds less than 1 mm. *On dwarf shrubs and annuals. S. Europe.* Al Bl Bu Co Cr Ga Gr Hs It Ju Lu Rm Sa Si.

15. C. approximata Bab., *Ann. Nat. Hist.* **13**: 253 (1844). Stems moderately stout. Flowers (2·5–)3–4 mm, (4–)5-merous, sessile; glomerules (6–)7–12 mm in diameter, compact, globose.

Calyx turbinate, about as long as corolla-tube, golden-yellow, shiny and finely reticulate when dry, like a fly's wing; lobes short, broader than long, overlapping at base, with rounded sides and ending more or less abruptly in a short and obtuse, or long and subacute whitish fleshy apex. Corolla campanulate, becoming globose or urceolate around the capsule; lobes patent, triangular-ovate, about as long as tube. Stamens exserted. Scales shorter than corolla-tube or reaching the stamens, oblong, shallowly fimbriate. Styles (including stigmas) as long or longer than ovary; stigmas often shorter than styles. Capsule depressed-globose. *Mostly on shrubs. S. Europe and S. half of U.S.S.R.; casual further north.* Al Co Ga Gr He Hs It Ju Lu Rm Rs (C, W, K, E) Sa Si ?Su [Br Bu Ge Hu].

1 Fleshy apex of the calyx-lobe short, obtuse   **(a) subsp. approximata**
1 Fleshy apex of calyx-lobe acute, longer than the lobe itself
2 Flowers 3–4 mm, 5-merous   **(b) subsp. macranthera**
2 Flowers 2·5 mm, often 4-merous   **(c) subsp. episonchum**

(a) Subsp. **approximata** (*C. cupulata* Engelm.): Flowers 3–4 mm. Fleshy apex of the calyx-lobe short, obtuse. *Throughout the range of the species.*
(b) Subsp. **macranthera** (Heldr. & Sart. ex Boiss.) Feinbrun & W. Greuter, *Israel Jour. Bot.* **19**: 27 (1970) (*C. macranthera* Heldr. & Sart. ex Boiss.): Flowers 3–4 mm, 5-merous. Fleshy apex of calyx-lobe acute, longer than the lobe itself. *S.E. Europe; Sicilia.*
(c) Subsp. **episonchum** (Webb & Berth.) Feinbrun, *Israel Jour. Bot.* **19**: 28 (1970) (*C. episonchum* Webb & Berth.): Flowers 2·5 mm, often 4-merous. Fleshy apex of calyx-lobe acute, longer than the lobe itself but shorter than in subsp. (b). Capsule enveloped by the corolla, but with the upper part usually more exposed than in subspp. (a) and (b). *Spain, ?Portugal.*

Subgen. **Monogyna** (Engelm.) Yuncker. Stems stout. Styles united. Stigmas capitate to ovoid. Capsule circumscissile.

16. C. lupuliformis Krocker, *Fl. Siles.* **1**: 261 (1787). Stems often reddish. Flowers 3–4 mm, 5-merous, sessile or shortly pedicellate, in spike-like inflorescences. Calyx campanulate, $\frac{1}{3}$–$\frac{1}{2}$ as long as corolla-tube; lobes oblong-ovate, obtuse, overlapping. Corolla-lobes oblong-ovate, more or less obtuse, slightly crenulate or entire, erect, less than half as long as cylindrical tube. Stamens included. Scales often bifid, dentate, not reaching the stamens. Ovary ovoid. Style twice as long as stigma or longer. Capsule capped by the detached corolla. Seeds 2–3 mm. *On* Salix *spp. and other trees. C. & E. Europe, extending westwards to the Netherlands.* Au Bu Cz Ge Ho Hu Ju Po Rm Rs (C, W, E).

17. C. monogyna Vahl, *Symb. Bot.* **2**: 32 (1791). Flowers 3–4 mm, sessile or shortly pedicellate, in spike-like inflorescences of 1- to 4-flowered cymules. Calyx campanulate, $\frac{2}{3}$ as long as corolla-tube or equalling it in length; lobes orbicular-ovate, obtuse, overlapping, somewhat carinate and with crenulate margins. Corolla-lobes ovate, obtuse, crenulate, erect, *c.* $\frac{1}{3}$ as long as tube; corolla-tube cylindrical. Stamens included. Scales dentate, nearly reaching the stamens. Ovary globose-conical. Style about as long as stigma. Capsule capped by the detached corolla. Seeds 3–3·5 mm. *Mainly on shrubs and trees. S.E. Europe, extending locally westwards to Portugal.* Bu Ga Gr It Ju Lu Rm Rs (C, W, K, E) Tu.

2. Cressa L.[1]

Annual or perennial herbs, often somewhat woody at the base. Leaves sessile, simple, entire. Flowers in short, terminal, conges-

---

[1] By C. A. Stace.

ted racemes; pedicels with 2 small bracteoles. Corolla divided to about halfway into 5 patent, pubescent lobes. Stamens inserted at about the middle of the corolla-tube, glabrous, exserted. Ovary 2-locular; styles 2, exserted; stigmas capitate. Capsule usually 1-seeded. Pollen grains tricolpate; exine not spinose.

1. **C. cretica** L., *Sp. Pl.* 223 (1753). Stems up to 30(–50) cm, greyish-pubescent, spreading, much-branched. Leaves 2–10 mm, lanceolate to ovate, cordate or rounded at base, subacute to acuminate, passing gradually into the bracts. Flowers 3–5 mm; pedicels very short. Sepals as long as corolla-tube, obtuse. Corolla 3–5 mm in diameter, whitish-pink or yellow, persistent. *Sandy or saline soils, usually near the sea. Mediterranean region, extending to C. Portugal and S.E. Bulgaria.* Al Bl Bu Co Cr Ga Gr Hs It Lu Sa Si.

3. Dichondra J. R. & G. Forster[1]

Creeping perennial herb rooting at the nodes. Leaves petiolate, simple, entire. Flowers axillary, solitary. Corolla divided to about halfway into 5 erect lobes. Stamens inserted near the top of the corolla-tube, more or less included. Ovary 2-locular; styles 2; stigmas capitate. Capsule 2-lobed; lobes usually 1-seeded.

Literature: A. Lawalrée, *Acta Bot. Neerl.* 19: 717–721 (1970).

1. **D. micrantha** Urban, *Symb. Antill.* 9: 243 (1924) (*D. repens* auct. eur., non J. R. & G. Forster). Stems up to 50 cm, pubescent, slender. Lamina of leaf 5–30 × 5–30 mm, orbicular to reniform, appressed-pubescent; petiole 5–50 mm. Peduncle 3–20 mm; calyx 1·5–2·5 mm, sericeous, divided to more than halfway; corolla 2–2·5 mm, whitish or greenish. *Naturalized in grassland and open sandy ground in W. Europe.* [Az Br It.] (*E. Asia.*)

4. Calystegia R. Br.[2]

Perennial rhizomatous herbs; stem procumbent or climbing. White latex present. Leaves petiolate, variously lobed at the base. Flowers solitary, axillary, with a pair of large and sometimes inflated bracteoles partly or entirely concealing the sepals. Corolla large, infundibuliform or tubiform, glabrous or ciliate. Ovary glabrous, unilocular; ovules 4. Stigma bilobed; lobes swollen and elongate. Pollen grains pantoporate; exine more or less smooth.

The genus differs constantly from *Convolvulus* in its pollen grains and unilocular ovary, but the European species are most easily distinguished by the pair of broad bracteoles immediately below the sepals. The generic delimitation is discussed by W. H. Lewis & R. L. Oliver, *Ann. Missouri Bot. Gard.* 52: 217–222 (1965).

A *flore pleno* cultivar of *C. pubescens* Lindley, from China, easily recognized by its pink flowers and oblong-hastate leaves, is widely cultivated and has sometimes persisted as a garden escape.

Hybrids may become well established by vegetative reproduction. White-flowered plants from S. Sweden which have been referred to 3 are probably 2(a) × 4; and glabrous pink-flowered plants with saccate bracteoles and a rounded leaf-sinus, well established in C. England (Nottinghamshire), may be 3 × 4. 2(a) × 3 (*C. × lucana* (Ten.) G. Don) may be widespread in S. Europe and in the British Isles, but is often difficult to distinguish from one or other parent.

1 Stems not or weakly twisting, procumbent; leaves reniform
                                        1. soldanella
1 Stems strongly twisting and climbing; leaves not reniform

2 Corolla white (occasionally the centre-band only of each lobe ± pinkish)
  3 Bracteoles rarely more than 15 mm wide, not overlapping, flat or keeled at the base; corolla usually not more than 50 mm; stamens not more than 23(–25) mm   **2. sepium**
  3 Bracteoles 15–40 mm wide when flattened, overlapping, ± saccate at the base; corolla 50–90 mm; stamens 24–37(–40) mm   **3. silvatica**
2 Corolla pink
  4 Bracteoles not or scarcely overlapping, acute or somewhat rounded at the apex, flat or keeled at the base; leaf-sinus acute or rounded, with divergent sides   **2. sepium**
  4 Bracteoles overlapping, obtuse to emarginate at the apex, ± saccate at the base; leaf-sinus oblong   **4. pulchra**

1. **C. soldanella** (L.) R. Br., *Prodr. Fl. Nov. Holl.* 484 (1810) (*Convolvulus soldanella* L.). Glabrous. Stem up to 50(–100) cm, procumbent, not or weakly twisting. Leaves 1–2 times as wide as long, reniform, somewhat fleshy; apex very broadly obtuse or emarginate. Bracteoles ovate to suborbicular, flat or sometimes tending to invest the calyx. Corolla 32–52 mm, pink. Stamens 20–30 mm; anthers 4–6 mm. 2n=22. *Maritime sands. Coasts of S. Europe, and of W. Europe northwards to Scotland and Denmark.* Al Az Be Bl Br Bu Co Cr Da Ga Ge Gr Hb Ho Hs It Ju Lu Rs (K) Sa Si Tu.

2. **C. sepium** (L.) R. Br., *op. cit.* 483 (1810) (*Convolvulus sepium* L.). Glabrous or pubescent. Stems strongly twisting and climbing. Leaves more or less sagittate; sinus with divergent sides. Peduncle without a narrow, repand wing. Bracteoles 10–30 × 5–20 mm, not or scarcely overlapping, the sides not closely investing the calyx, acute or rarely subobtuse at the apex, flat or keeled at the base. Corolla 30–70 mm, white or pink. Stamens 15–30 mm; anthers 4–6·5 mm. 2n=22. *Most of Europe except the extreme north.* All except Fa Is Rs (N) Sb.

1 Corolla white   **(a) subsp. sepium**
1 Corolla pink
2 Leaf-sinus rounded   **(d) subsp. spectabilis**
2 Leaf-sinus acute
  3 Glabrous; leaf usually shortly acute at the apex; corolla usually pale pink   **(a) subsp. sepium**
  3 Stem, petiole and peduncle often pubescent; leaf usually tapered to a long-acute apex; corolla bright pink
    4 Stamens 17–25 mm   **(b) subsp. roseata**
    4 Stamens 25–34 mm   **(c) subsp. americana**

(a) Subsp. **sepium**: Glabrous. Leaves variable; sinus acute. Bracteoles usually 10–15 mm wide. Corolla usually 30–50 mm, white or rarely pale pink. Stamens (15–)17–23(–25) mm; anthers 4–6 mm. 2n=22. *River- and lake-margins, marshes, hedges and waste places. Almost throughout the range of the species.*

(b) Subsp. **roseata** Brummitt, *Watsonia* 6: 298 (1967): Stem, petiole and peduncle usually pubescent. Lamina glabrous or sparsely pubescent, usually long-acute; sinus acute. Bracteoles usually 10–15 mm wide. Corolla usually 40–55 mm, pink. Stamens 17–25 mm; anthers 4–6 mm. 2n=22. *Salt-marshes, maritime sands, waste places. Coastal regions of W. Europe.*

(c) Subsp. **americana** (Sims) Brummitt, *Ann. Missouri Bot. Gard.* 52: 216 (1965): Stems, petioles and peduncles usually pubescent. Lamina often pubescent; sinus acute. Bracteoles usually 10–20 mm wide. Corolla usually 45–65 mm, pink. Stamens 25–34 mm; anthers 4–6 mm. *Salt-marshes and maritime sands. Açores.* (*Atlantic coasts of North and South America.*)

(d) Subsp. **spectabilis** Brummitt, *Bot. Jour. Linn. Soc.* 64: 73 (1971): Glabrous or with stems and petioles pubescent. Leaf-sinus rounded. Bracteoles (12–)14–18(–24) mm wide, usually strongly keeled, occasionally slightly overlapping at the margins; apex sometimes more or less obtuse. Corolla usually 50–60 mm,

¹ By A. Lawalrée.   ² By R. K. Brummitt.

pink. Stamens 20–30 mm; anthers 5·5–6·5 mm. *Naturalized in Fennoscandia, and perhaps elsewhere. (Probably a native of Siberia but perhaps a garden hybrid between 2(a) and 4.)*

3. C. silvatica (Kit.) Griseb., *Spicil. Fl. Rumel.* **2**: 74 (1844) (*Convolvulus silvaticus* Kit., *Calystegia sylvestris* (Waldst. & Kit. ex Willd.) Roemer & Schultes). Glabrous. Stems strongly climbing. Leaves more or less sagittate; sinus rounded, with divergent or sometimes parallel sides. Peduncle without a narrow, repand wing. Bracteoles 14–32(–38) mm wide when flattened, overlapping, closely investing the calyx and almost completely concealing it, weakly to strongly saccate at the base, subacute to broadly rounded or emarginate at the apex. Corolla usually 50–90 mm, white; occasionally the outside of the centre-band only of each lobe suffused with pink. Stamens 24–37(–40) mm; anthers 6–8 mm. 2*n* = 22. *Scrub, hedges and open ground. S. Europe; naturalized from gardens elsewhere.* Al Bl Bu ?Co ?Cr Ga Gr Hs It Ju Rm Rs (K) Sa Si Tu [Br Hb He Lu].

Variable; there are marked differences between plants at the western and eastern ends of the European range. In the east the bracteoles are strongly saccate at the base, broadly rounded to emarginate at the apex and often twice as wide as long, the corolla is 55–90 mm and the stamens (25–)28–36(–39) mm. In the west (N.W. Italy, S.E. France and Spain) the bracteoles are much less strongly saccate, obtuse to subacute and about as long as wide, the corolla is 50–65 mm and the stamens 24–28(–30) mm. There seems, however, to be not quite sufficient morphological or geographical discontinuity to recognize subspecies.

4. C. pulchra Brummitt & Heywood, *Proc. Bot. Soc. Brit. Is.* **3**: 385 (1960). Stems strongly climbing. Stems, petioles and peduncles pubescent, at least when young. Leaves more or less sagittate, rather dull green; sinus oblong, with more or less parallel sides. Peduncle usually with a narrow, repand wing. Bracteoles usually 15–25 mm wide when flattened, overlapping, closely investing and almost completely concealing the calyx, more or less saccate at the base, rounded to emarginate at the apex. Corolla 50–75 mm, pink. Stamens 25–35 mm; anthers 6–7·5 mm. 2*n* = 22. *Naturalized from gardens in N. & C. Europe.* [Au Be Br Cz Da Ga Gc Hb Ho Po Su.] *(Origin uncertain; either introduced from N.E. Asia or originated as a garden hybrid in Europe.)*

5. Convolvulus L.[1]

Annual or perennial herbs or small shrubs up to 100 cm, sometimes with latex. Stems erect, trailing or twining. Leaves simple, usually entire. Flowers solitary or in terminal or axillary inflorescences; pedicels with 2 minute to leaf-like bracteoles. Corolla infundibuliform, glabrous except for five usually pubescent stripes often of a different colour. Stamens inserted at the base of the corolla-tube, included. Ovary 2-locular; stigma 2-lobed, the lobes filiform to cylindric-clavate. Capsule usually 4-seeded, 1- to 2-celled. Pollen grains tricolpate; exine not spinose.

Literature: F. Sa'ad, *The Convolvulus Species of the Canary Isles, the Mediterranean Region, and the Near and Middle East.* Rotterdam. 1967.

1 Dwarf cushion-plant with woody stems
  2 Leaves sericeous above, spathulate to rhombic; lateral veins distinct   **2. boissieri**
  2 Leaves glabrous above, linear to oblanceolate; only the midrib distinct   **3. libanoticus**

[1] By C. A. Stace.

1 Not a cushion-plant
  3 Leaves very abruptly narrowed into a distinct petiole
    4 Lamina cuneate to truncate at base
      5 Plant robust, densely lanate; sepals obtuse   **1. persicus**
      5 Plant slender, pubescent but not lanate; sepals acute to acuminate
        6 Entirely herbaceous; corolla 7–12 mm   **16. siculus**
        6 Woody stock present; corolla 15–30 mm
          7 Lamina 1–1½(–3) times as long as wide, orbicular to oblong   **17. sabatius**
          7 Lamina (2½–)3½–10 times as long as wide, linear to narrowly oblong   **18. valentinus**
    4 Lamina hastate or sagittate, or cordate at base
      8 Upper leaves and bracts very deeply divided   **23. althaeoides**
      8 Upper leaves and bracts not divided, except for basal lobes
        9 Corolla 25–45 mm; peduncles usually much longer than subtending bracts
          10 Corolla usually pink; sepals acute to acuminate; plant pubescent   **21. betonicifolius**
          10 Corolla usually yellow; sepals emarginate and apiculate; plant glabrous   **22. scammonia**
        9 Corolla 7–25 mm; peduncles usually shorter or not much longer than subtending bracts
          11 Annual or short-lived perennial with slender root; corolla up to 12 mm, blue   **16. siculus**
          11 Perennial with stout, often woody stock; corolla white to pink
            12 Leaves crenate; sepals acuminate to mucronate   **20. farinosus**
            12 Leaves more or less entire; sepals obtuse to emarginate, often apiculate   **19. arvensis**
  3 Leaves sessile or gradually tapered into a petiole
    13 Mature stems woody and divaricately branched throughout   **4. dorycnium**
    13 Mature stems herbaceous at least above, not divaricately branched
      14 Annual or short-lived perennial, entirely herbaceous
        15 Peduncle and pedicel together not or scarcely longer than calyx   **12. humilis**
        15 Peduncle and pedicel together several times longer than calyx
          16 Sepals pubescent, herbaceous, with distinct distal and proximal regions; capsule pubescent   **13. tricolor**
          16 Sepals glabrous or sparsely pubescent, with scarious margins, not divided into distinct regions; capsule glabrous
            17 Sepals acute to shortly acuminate; corolla 14–22 mm   **14. meonanthus**
            17 Sepals obtuse and mucronate; corolla 7–10 mm   **15. pentapetaloides**
      14 Perennial; shoots woody below, herbaceous above
        18 Inflorescence with all flowers on each main branch or stem crowded into a compact head
          19 Extreme base of leaves at base of main stems widened and scarious; all sepals long-acuminate to caudate
            20 Ovary and capsule glabrous   **7. lanuginosus**
            20 Ovary and capsule pubescent   **8. calvertii**
           19 Extreme base of leaves at base of main stems scarcely widened, herbaceous; at least outer sepals obtuse to shortly acuminate
            21 Inner sepals obtuse to shortly acuminate; corolla usually white   **5. cneorum**
            21 Inner sepals acuminate to cuspidate; corolla usually pink   **6. oleifolius**
        18 Inflorescence lax, or with at least some branches conspicuous
          22 At least lower parts of stems with mostly patent hairs   **11. cantabrica**
          22 Shoots sericeous, sometimes with some patent hairs
            23 Outer sepals cordate at base and conspicuously gibbous on back   **9. holosericeus**
            23 Outer sepals cuneate to rounded at base; usually convex but scarcely gibbous on back

24 Extreme base of leaves at base of main stems widened
    and scarious   **10. lineatus**
24 Extreme base of leaves at base of main stems scarcely
    widened, herbaceous
  25 Inner sepals obtuse to shortly acuminate; corolla
      usually white   **5. cneorum**
  25 Inner sepals acuminate to cuspidate; corolla usually
      pink   **6. oleifolius**

1. C. persicus L., *Sp. Pl.* 158 (1753). Densely lanate perennial 10–50 cm, with creeping woody stock and stout, erect, simple or sparingly branched stems. Leaves petiolate, firm, elliptic-oblong, rounded to broadly cuneate at base, rounded to apiculate at apex. Peduncles axillary, 1-flowered, shorter than bracts. Sepals obtuse. Corolla 30–45 mm, white. *Maritime sands. W. coast of Black Sea.* Bu Rm Tu. (*S.W. Asia.*)

2. C. boissieri Steudel, *Nomencl. Bot.* ed. 2, **1**: 407 (1840). Dwarf, cushion-like, densely appressed-pubescent perennial up to 10 cm; stems much-branched, woody, often rooting along their length. Leaves up to 20(–30) mm, crowded, thick, with a distinctly narrowed proximal region, linear to obovate; midrib and lateral veins distinct. Peduncles very short, terminal and axillary, 1-flowered. Corolla 15–20 mm, white to red. *Mountain rocks and screes; calcicole. S. Spain; Balkan peninsula.* Al Bu Gr Hs Ju.

(a) Subsp. **boissieri** (*C. nitidus* Boiss., non Desr.): Distal widened portion of leaves oblong-obovate, distinctly longer than wide. ● *S. Spain.*
(b) Subsp. **compactus** (Boiss.) Stace, *Bot. Jour. Linn. Soc.* **64**: 58 (1971) (*C. compactus* Boiss.; incl. *C. parnassicus* Boiss. & Orph., *C. cochlearis* Griseb.): Distal widened portion of leaves rhombic-obovate, about as long as wide. *Balkan peninsula, from Albania and S. Bulgaria to C. Greece.*

Usually considered to be two distinct geographically separated species, but intermediates occur, particularly within the range of subsp. (b). These have been called *C. nitidus* var. *acutifolius* Košanin, and sometimes more closely resemble subsp. (a) than subsp. (b). The European representatives of subsp. (b) have been treated as a separate subspecies or species (*C. parnassicus* Boiss. & Orph.), but both extremes and many intermediates occur in Anatolia. Plants from Slavjanka, S.W. Bulgaria, resemble Spanish material of subsp. (a) very closely; they may be best referable to that taxon or described as a third subspecies of *C. boissieri*.

3. C. libanoticus Boiss., *Diagn. Pl. Or. Nov.* **2**(11): 82 (1849) (*C. radicosus* Heldr. & Sart.). Like **2** but less compact and less densely pubescent; leaves thin, glabrous above, linear to oblanceolate, with only the midrib distinct; corolla 10–15 mm, red. *Mountain slopes and pastures. S. Greece, Kriti.* Cr Gr. (*S.W. Asia.*)

4. C. dorycnium L., *Syst. Nat.* ed. 10, **2**: 923 (1759). Upright, divaricately-branched, densely pubescent shrub 50–100 cm; mature stems and branches woody, rigid. Leaves linear-spathulate to oblanceolate. Peduncles much longer than bracts and flowers, 1- to few-flowered. Sepals obtuse to emarginate, apiculate. Corolla 10–20 mm, pink. *Dry places. S. Greece and Aegean region.* Cr Gr.

C. lanatus Vahl, *Symb. Bot.* **1**: 16 (1790), was recorded for Kriti by Sibthorp and others but no record can be confirmed. It occurs in N.E. Africa and S.W. Asia. It is an irregularly branched, densely pubescent shrub up to 50 cm, often with rigid spine-tipped mature branches, and the flowers are crowded in dense axillary heads. Abnormal forms of **6** have been misidentified as this species.

5. C. cneorum L., *Sp. Pl.* 157 (1753). Densely sericeous, erect or spreading, branched perennial 10–50 cm, woody in much of the lower parts. Leaves linear to oblanceolate, attenuate at base, herbaceous and scarcely widened at the extreme base. Flowers in terminal, usually crowded, heads; peduncles absent or very short, usually hidden. Sepals lanceolate to broadly ovate, obtuse to shortly acuminate. Corolla 15–25 mm, usually white. Capsule longer than calyx when ripe, pubescent. ● *Calcareous rocks near the sea. W. & C. parts of Mediterranean region.* Al *Hs It Ju Si [?Ga].

Plants from Sicilia usually have narrower leaves than those from elsewhere, and have been separated as a variety or subspecies. Their status requires more investigation. **5** is replaced in Malta by **6**.

6. C. oleifolius Desr. in Lam., *Encycl. Méth. Bot.* **3**: 552 (1792). Like **5** but usually less robust and stems often less extensively woody; leaves filiform to linear; inflorescence usually somewhat more diffuse with some branches visible; inner sepals acuminate to cuspidate; corolla usually pink; capsule shorter than calyx when ripe. *Rocky places near the sea; calcicole. S. Greece and Aegean region; Malta.* Cr Gr Si.

Geographically separated from **5**, which it sometimes closely resembles, and perhaps only a subspecies of it.

C. argyrothamnos W. Greuter, *Bauhinia* **3**: 251 (1967), is related to **6** but differs in its arcuate-pendent stems with the longest branches near the apex. It is known only from a calcareous cliff in Kriti, and requires further study.

7. C. lanuginosus Desr. in Lam., *Encycl. Méth. Bot.* **3**: 551 (1792). Densely pubescent perennial 5–30 cm, with branched woody stock; stems usually branched and often woody below, erect. Leaves filiform to linear, attenuate at the base, the basal ones on each stem widened and scarious at the extreme base. Flowers crowded into dense terminal heads; peduncles absent or nearly so. Sepals linear-lanceolate to broadly lanceolate, long-acuminate to cuspidate. Corolla 15–25 mm, pink. Ovary and capsule glabrous. *Dry, calcareous rocks. S. & E. Spain, S. France.* Ga Hs.

Very variable in the degree to which the hairs are appressed or patent.

8. C. calvertii Boiss., *Diagn. Pl. Or. Nov.* **3**(3): 124 (1856) (incl. *C. sericocephalus* Juz., *C. tauricus* (Bornm.) Juz.). Like **7** but stems usually simple and herbaceous; leaves linear to linear-oblanceolate; sepals lanceolate to broadly ovate, long-acuminate to caudate; corolla 12–20 mm, white to pink; ovary and capsule pubescent. *Dry places. Krym.* Rs (K). (*S.W. Asia.*)

Geographically widely separated from **7**, but perhaps only a subspecies of it.

9. C. holosericeus Bieb., *Fl. Taur.-Cauc.* **1**: 147 (1808). Densely sericeous perennial 10–30 cm, with woody stock and usually erect herbaceous stems. Leaves linear to oblanceolate, attenuate at base; basal leaves on each stem widened and scarious at the extreme base. Peduncles terminal and axillary, rarely as long as subtending bract, 1- to 2-flowered. Outer sepals broadly ovate, cordate, cuspidate, conspicuously gibbous on the back. Corolla 20–30 mm, white to yellow. Ovary and capsule pubescent. *Dry, stony slopes. Balkan peninsula, from N. Macedonia to E.C. Greece; Krym.* Bu Gr Ju Rs (K).

10. C. lineatus L., *Syst. Nat.* ed. 10, **2**: 923 (1759). Densely sericeous perennial 3–25 cm, with woody stock and usually procumbent or ascending herbaceous stems. Leaves linear to elliptical or oblanceolate, attenuate at base; basal leaves on each shoot widened and scarious at the extreme base. Peduncles terminal and axillary, much shorter than bracts, 1- to several-flowered. Outer sepals linear to lanceolate, convex on the back below, cuneate to rounded at base, acute to acuminate. Corolla 12–25 mm, pink. Ovary and capsule pubescent. *Dry, open ground. S. Europe, extending northwards to 46° 30′ N. in W. France and to c. 51° N. in S.C. Russia, but rare in the E. Mediterranean region.* Al Bl Bu Co ?Cr Ga ?Gr Hs It ?Ju Lu Rm Rs (C, W, K, E) Si.

C. pilosellifolius Desr. in Lam., *Encycl. Méth. Bot.* **3**: 551 (1792), was once recorded from Turkey-in-Europe but was perhaps only a casual. It differs from **11** mainly in the broadly cuneate to subcordate base of the cauline leaves, and in the glabrous ovary and capsule. It occurs in N.E. Africa and S.W. Asia.

11. C. cantabrica L., *Sp. Pl.* 158 (1753). Pubescent perennial (with many patent hairs at least below) with woody stock and erect herbaceous stems 10–50 cm. Leaves linear to oblanceolate, attenuate at base; basal leaves on each stem widened and scarious at the extreme base. Peduncles terminal and axillary, the lower longer than subtending bracts, 1- to several-flowered. Outer sepals lanceolate to oblanceolate, convex on the back below, cuneate to rounded at base, acute to acuminate. Corolla 15–25 mm, pink. Ovary and capsule pubescent. $2n=30$. *Dry places. S. & S.C. Europe, extending northwards to c. 48° N. in E. France and S. Czechoslovakia.* Al Au Bl Bu Co ?Cr Cz Ga Gr Hs Hu It Ju Rm Rs (W, K, E) Sa Si Tu.

12. C. humilis Jacq., *Collect. Bot.* **4**: 209 (1791) (*C. undulatus* Cav.). Herbaceous annual or short-lived perennial up to 40 cm. Leaves sessile, mostly oblanceolate. Peduncle and pedicel together shorter than or about as long as calyx, 1-flowered. Sepals elliptic-oblong, sparsely pubescent, with scarious margins, mucronulate to acuminate. Corolla 7–12 mm, violet. Capsule pubescent. *Dry, open habitats. C. & S. Spain, S.E. Portugal; isolated stations in N.W. Italy and Sicilia.* Hs It Lu Si.

13. C. tricolor L., *Sp. Pl.* 158 (1753). Herbaceous annual or short-lived perennial up to 60 cm. Leaves sessile, mostly obovate to oblanceolate. Peduncle and pedicel together several times as long as calyx, 1-flowered. Sepals with distinct distal and proximal regions, pubescent, herbaceous, obtuse to acuminate. Corolla 15–40 mm, usually with basipetally-zoned blue, white and yellow bands. Capsule pubescent. *Dry, open habitats. Mediterranean region, Portugal.* Az Bl Co Ga Gr Hs *It Ju Lu Si.

(a) Subsp. **tricolor**: Distal part of sepals broadly ovate-oblong, obtuse, subacute or shortly acuminate, shorter than to about as long as proximal part. *Throughout the range of the species, but perhaps not native in Sicilia.*
(b) Subsp. **cupanianus** (Sa'ad) Stace, *Bot. Jour. Linn. Soc.* **64**: 58 (1971) (*C. tricolor* var. *cupanianus* Sa'ad): Distal part of sepals ovate to lanceolate, acute to long-acuminate, conspicuously longer than proximal part. *Sicilia; naturalized elsewhere.*

This species is often cultivated for ornament, and the natural ranges of the two subspecies have evidently become partly obscured by introductions.

14. C. meonanthus Hoffmanns. & Link, *Fl. Port.* **1**: 369 (1813–1820). Herbaceous annual or short-lived perennial up to 50 cm. Leaves sessile, mostly linear to oblanceolate. Peduncle

and pedicel together several times as long as calyx, 1-flowered. Sepals lanceolate to narrowly ovate, glabrous to sparsely pubescent, acute to shortly acuminate; margins scarious. Corolla 14–22 mm, usually with basipetally-zoned blue, white and yellow bands. Capsule glabrous. *Cultivated ground and other open habitats. S. Spain, C. & S. Portugal; W. Italy, Sicilia.* Hs It Lu Si.

15. C. pentapetaloides L., *Syst. Nat.* ed. 12, **3**: 229 (1768). Like **14** but up to 30 cm; sepals elliptic-ovate, glabrous, obtuse and mucronate; corolla 7–10 mm, usually blue distally and yellow proximally. *Dry places. Mediterranean region, S.E. Portugal.* Bl Gr Hs It ?Ju Lu Sa Si Tu.

16. C. siculus L., *Sp. Pl.* 156 (1753). Sparsely pubescent annual or short-lived perennial; stems 10–60 cm, slender, trailing (rarely twining), herbaceous. Leaves petiolate, lanceolate to ovate, cordate or rarely truncate, entire. Peduncles axillary, usually shorter than the bracts, 1-(to 2-) flowered. Sepals acute to acuminate, pubescent. Corolla 7–12 mm, blue. *Dry, open habitats. Mediterranean region, Portugal.* Bl Co Cr Ga Gr Hs It Lu Sa Si.

(a) Subsp. **siculus**: Bracteoles leaf-like, borne just below the sepals and greatly overlapping them. $2n=44$. *Throughout the range of the species.*
(b) Subsp. **agrestis** (Schweinf.) Verdcourt, *Kew Bull.* [12]: 344 (1957): Bracteoles filiform, borne well below the sepals and not or scarcely overlapping them. *Sardegna. (Africa.)*

17. C. sabatius Viv., *Fl. Lib.* 67 (1824). Sparsely to densely pubescent perennial with branched, woody stock; stems 10–50 cm, slender, trailing or ascending, simple or branched, herbaceous. Leaves petiolate; lamina 1–1½(–3) times as long as wide and 3–8 times as long as petiole, orbicular to oblong, cuneate to truncate, entire. Peduncles axillary, usually as long as or shorter than bracts, 1- to 3-flowered. Outer sepals usually acute, inner sepals usually acuminate. Corolla 15–22 mm, pink to blue. *Dry, calcareous rocks. Coasts of N.W. Italy and Sicilia.* It Si. (*N.W. Africa.*)

Variable, particularly with regard to pubescence.

18. C. valentinus Cav., *Icon. Descr.* **2**: 65 (1793) (*C. suffruticosus* Desf.). Like **17** but lamina linear to narrowly oblong, (2·5–)3·5–10 times as long as wide and 6–13 times as long as petiole; outer sepals usually acuminate; corolla 18–30 mm, white, yellow or pink to blue. *Dry places. E. Spain.* Hs. (*N.W. Africa.*)

Very variable in leaf-shape, pubescence, sepal-shape and colour of corolla. Perhaps only a subspecies of **17**, with which it is, however, sympatric in N. Africa.

19. C. arvensis L., *Sp. Pl.* 153 (1753). Glabrous or pubescent perennial; stems up to 200 cm, slender, trailing or twining, herbaceous. Leaves petiolate, triangular or ovate-oblong to linear, hastate to sagittate, more or less entire. Peduncles axillary, often about as long as bracts, 1- to 2(3)-flowered. Sepals obtuse to emarginate, and often apiculate. Corolla 10–25 mm, white to pink. Ovary glabrous. $2n=50$. *Usually in disturbed habitats. Europe, except the extreme north.* All except Fa Is Sb.

Very variable in pubescence and leaf-shape. Plants with linear to oblong-lanceolate leaves (often with very prominent backwardly-directed auricles) are known as var. *linearifolius* Choisy. They are native in S. Europe and widely naturalized in the north, and may merit subspecific status.

C. mairei Halácsy, *Bull. Soc. Sci. Nancy* sér. 3, **8**: 176 (1907), differs from **19** in its smaller parts (stems not more than 25 cm, corolla 7–10 mm) and dense pubescence on the vegetative parts and ovary. It occurs in S. Greece, and may merit specific or subspecific status.

20. **C. farinosus** L., *Mantissa Alt.* 203 (1771). Appressed-pubescent perennial; stems up to 100 cm, slender, twining, herbaceous. Leaves petiolate, triangular-ovate, sagittate, crenate. Peduncles axillary, mostly about as long as bracts, 1- to 3(5)-flowered. Sepals acuminate to mucronate. Corolla *c.* 10 mm, pinkish. *Naturalized in hedges near Lisboa.* [Lu.] (*C. & S. Africa.*)

21. **C. betonicifolius** Miller, *Gard. Dict.* ed. 8, no. 20 (1768) (*C. hirsutus* Bieb.). Perennial with patent hairs; stems up to 100 cm, slender, trailing or twining, herbaceous. Leaves petiolate, triangular-ovate or -lanceolate, sagittate to cordate, entire. Peduncles axillary, mostly much longer than bracts, frequently with more than 3 flowers. Sepals acute to acuminate. Corolla 30–45 mm, usually pink. *Dry, usually acid soils. S.E. Europe, from C. Jugoslavia and Krym southwards.* Al Bu Gr Ju Rs (K) Tu [Ga It Si].

22. **C. scammonia** L., *Sp. Pl.* 153 (1753). Glabrous perennial; stems up to 75 cm, slender, trailing or twining, herbaceous. Leaves petiolate, triangular-ovate or -lanceolate, sagittate, entire. Peduncles axillary, mostly much longer than bracts, frequently with more than 3 flowers. Sepals emarginate, apiculate. Corolla 25–45 mm, pale yellow. *Scrub and forest-margins. Krym.* ?Gr Rs (K). (*Anatolia and E. Aegean region.*)

23. **C. althaeoides** L., *Sp. Pl.* 156 (1753). Pubescent perennial; stems up to 100 cm, slender, trailing or twining, herbaceous. Leaves petiolate, very variable but at least the upper deeply lobed and cordate to sagittate. Peduncles axillary, usually much longer than bracts, 1- to 3(5)-flowered. Sepals variable, acute to rounded. Corolla 25–40 mm, usually pink. *Dry places. S. Europe.* Al Bl Bu Co Cr Ga Gr Hs It Ju Lu Rm Sa Si Tu.

(a) Subsp. **althaeoides**: Relatively robust, with sparse to dense, mostly patent hairs; leaf-lobes relatively wide and shallow; sepals (7)8–9(10) mm. *Throughout most of the range of the species, but rarer in the east.*

(b) Subsp. **tenuissimus** (Sibth. & Sm.) Stace, *Bot. Jour. Linn. Soc.* **64**: 59 (1971) (*C. tenuissimus* Sibth. & Sm., *C. elegantissimus* Miller, *C. althaeoides* var. *pedatus* Choisy): Relatively slender, with dense, almost entirely appressed, hairs; leaf-lobes relatively narrow and deep; sepals (6–)7–8(–9) mm. *From Italy eastwards; naturalized in France.*

Subsp. (b) is more eastern in distribution, but (a) and (b) are sympatric in many areas. Intermediates occur rather rarely.

6. Ipomoea L.[1]

(incl. *Pharbitis* Choisy)

Annual or perennial herbs with long, trailing or climbing stems. Leaves entire or lobed, alternate; petioles long, distinct. Flowers solitary or in axillary cymes, usually with 2 very small bracteoles. Corolla at least 25 mm, infundibuliform or tubiform, scarcely lobed, more or less glabrous. Ovary 2- to 4-locular; ovules usually 2 in each loculus; style filiform; stigma with 1–3 globose lobes. Capsule usually 4-seeded. Pollen grains pantoporate; exine spinose.

---

[1] By C. A. Stace.

The leaves in all species may vary considerably in shape, usually being entire near the base of the plant and variously lobed towards the apex. The descriptions cover most of this variation, but ignore the lowest leaves which are often very aberrant.

3 and **4** are two of several species grown for ornament and frequently found as casuals over much of Europe; they have been very much confused in the past.

1	Sepals acute to rounded, usually abruptly mucronate, glabrous or rarely hairy	
2	Leaves sagittate	**2. sagittata**
2	Leaves not sagittate	
3	Plant without underground tubers; leaves obtuse to emarginate or 2-lobed at apex	**1. stolonifera**
3	Plant with large underground tubers; leaves acute to acuminate at apex or at apex of lobes	**5. batatas**
1	Sepals acute to long-acuminate, pubescent	
4	Sepals finely pubescent, without bristles	**3. acuminata**
4	Sepals finely pubescent and also with long bristles near the base	**4. purpurea**

Sect. LEIOCALYX Hallier fil. Glabrous, without underground tubers; stems creeping to weakly climbing; sepals acute to rounded, mucronate.

1. **I. stolonifera** (Cyr.) J. F. Gmelin, *Syst. Nat.* **2**: 345 (1791) (*I. littoralis* (L.) Boiss., non Blume). Stems creeping and usually rooting at the nodes. Leaves more or less fleshy; lamina 1–5 cm, usually oblong, entire to deeply 3- to 5-lobed, broadly cuneate to cordate at base, obtuse to emarginate or 2-lobed at apex. Flowers usually solitary; sepals 8–15 mm, acute to obtuse, mucronate; corolla 35–50 mm, white or pale yellow, sometimes with a purple centre. *Sandy places near the sea. S. Europe; very local.* Az ?Bl Cr It. (*Tropics and warm-temperate regions.*)

2. **I. sagittata** Poiret, *Voy. Barb.* **2**: 122 (1789). Stems creeping or climbing. Lamina 3–8 cm, usually sagittate with 3 narrow, acute to obtuse, mucronate lobes. Flowers usually solitary; sepals 5–10 mm, rounded to emarginate, mucronate; corolla 40–70 mm, pink to purple. *Usually damp, often saline places near the sea. Mediterranean region.* Bl Co Gr Hs It Si. (*S. & E. parts of Mediterranean region; Tropical America.*)

Despite the large disjunction in its distribution this species is usually considered to be a native of the Mediterranean region.

Sect. IPOMOEA (Sect. *Pharbitis* (Choisy) Griseb.). Pubescent or hispid, without underground tubers; stems usually strongly climbing; sepals acute to long-acuminate.

3. **I. acuminata** (Vahl) Roemer & Schultes, *Syst. Veg.* **4**: 228 (1819) (*Pharbitis cathartica* (Poiret) Choisy). Usually perennial. Lamina 4–16 cm, ovate or broadly ovate, entire to deeply 3-lobed, acuminate, cordate. Inflorescences few- to many-flowered; sepals 13–22 mm, lanceolate, gradually tapering from near the base, uniformly finely pubescent; corolla 50–85 mm, white, blue, pink, or multicoloured, usually fading to pink. *Naturalized in hedges and on roadsides.* [Az Bl Ga Lu Si.] (*Tropics.*)

4. **I. purpurea** Roth, *Bot. Abh.* 27 (1787) (*Pharbitis purpurea* (Roth) Voigt). Like **3** but usually annual; lamina usually entire; inflorescences sometimes 1-flowered; sepals 10–16 mm, uniformly finely pubescent and with conspicuous bristles below; corolla 40–60 mm. *Naturalized in hedges and on roadsides.* [Au Bl Ga Gr Hs It Rm.] (*Tropical America.*)

Sect. BATATAS (Choisy) Griseb. Glabrous to slightly pubescent, with underground fusiform tubers; stems creeping to weakly climbing; sepals obtuse to acute, mucronate.

5. **I. batatas** (L.) Lam., *Tab. Encycl. Méth. Bot.* **1**: 465 (1793). Lamina 4–14 cm, broadly ovate, entire to deeply 3- to 5(7)-lobed, truncate to cordate. Inflorescences 1- to several-flowered; sepals 7–12 mm, oblong to ovate-oblong, glabrous to hairy; corolla 30–50 mm, white, or violet to purple. *Cultivated for its edible tubers (sweet potatoes) in parts of S. Europe.* [Az Gr Hs Lu.] (*Tropical America.*)

CXLVII. HYDROPHYLLACEAE[1]

Herbs. Leaves usually alternate, entire to pinnatisect, exstipulate. Flowers usually in simple or dichotomous scorpioid inflorescences (sometimes solitary), actinomorphic, usually 5-merous. Calyx deeply divided. Corolla infundibuliform, campanulate or rotate; lobes imbricate. Stamens as many as the corolla-lobes and alternating with them; anthers versatile. Ovary superior, 1- to 2-locular, with 2, parietal placentae; ovules 2 to many on each placenta; styles 1–2; stigmas usually capitate. Fruit a usually loculicidal capsule.

1 Flowers solitary; calyx with sepal-like appendages between the
  lobes **1. Nemophila**
1 Flowers in scorpioid inflorescences; calyx without appendages
  between the lobes
2 Leaves pinnatisect; styles connate below **2. Phacelia**
2 Leaves crenate-serrate; styles free to base **3. Wigandia**

1. Nemophila Nutt.[2]

Annual herbs with pinnatisect, opposite leaves. Flowers solitary, pedicellate. Calyx with sepal-like appendages between the lobes. Corolla rotate to campanulate. Filaments with appendages at their base.

1. **N. menziesii** Hooker & Arnott, *Bot. Beech. Voy.* 152 (1833). Stems 10–30 cm, diffuse, pubescent. Leaves 2–5 cm, petiolate, ovate to oblong, pinnatisect, with usually 5–9 serrate lobes. Calyx-lobes 4–6 mm, lanceolate, the appendages 1·5–2·5 mm. Corolla 15–40 mm in diameter, almost rotate, blue or white. Capsule 5–12 mm, subglobose. *Often cultivated for ornament and sometimes escaping.* [No Su.] (*W. North America.*)

2. Phacelia Juss.[2]

Annual herbs. Leaves alternate. Flowers in scorpioid inflorescences, shortly pedicellate. Calyx without appendages between the lobes. Corolla rotate to campanulate. Filaments with appendages at their base.

1. **P. tanacetifolia** Bentham, *Trans. Linn. Soc. London* **17**: 280 (1835). Stems 20–80 cm, glandular-pubescent and hispid, usually erect and simple or sparingly branched. Leaves 2–10 cm, shortly petiolate, ovate to ovate-oblong in outline, 1- to 2-pinnatisect. Calyx-lobes (5–)6–8, usually linear. Corolla 6–9 mm, blue, persistent. Stamens long-exserted. Styles connate for *c.* 1 mm. Capsule 3–4 mm, ovoid, pubescent at apex. *Cultivated for bees and frequently naturalized.* [Au Be Bu Cz Da Ga He Ho Hu Ju Lu No Po Rm Rs (B, C, W, E) Su.] (*W. North America.*)

Several other species of this large American genus are cultivated and some have been recorded as more or less naturalized. The two commonest of these are: **P. ciliata** Bentham, *loc. cit.* (1835), with a strongly accrescent calyx, and **P. minor** (Harvey) Thell., *Ber. Bayer. Bot. Ges.* **14**: 79 (1914), with pedicels up to 15 mm and the styles connate to above the middle.

3. Wigandia Kunth[2]

Robust perennial herbs or shrubs with stinging hairs. Leaves alternate. Flowers in scorpioid inflorescences, shortly pedicellate. Calyx without appendages between the lobes. Corolla with short, campanulate tube and rotate limb. Filaments with deflexed hairs in the lower part. Styles free to the base.

1. **W. caracasana** Kunth in Humb., Bonpl. & Kunth, *Nov. Gen. Sp.* **3**: 128 (1819). Erect, hispid shrub 200–400 cm. Leaves 20–45 × 12–35 cm, ovate, obtuse, cordate at base, petiolate, coarsely crenate-serrate, yellowish-pubescent beneath. Calyx-lobes linear-lanceolate. Corolla *c.* 20 mm, the limb lilac, the tube white. Stamens shortly exserted. Capsule oblong-conical, slightly greyish-pubescent. *Cultivated for ornament, and locally naturalized on rocks and walls in the W. Mediterranean region.* [Ga Hs It.] (*Mexico to Colombia.*)

CXLVIII. BORAGINACEAE[3]

Herbs or dwarf shrubs, often hispid. Leaves alternate, exstipulate, simple. Flowers usually in scorpioid cymes, usually actinomorphic. Calyx 5-toothed or -lobed. Corolla 5-lobed, cylindrical, campanulate, hypocrateriform or rotate, usually with a distinct tube and limb; tube often with 5 scales, invaginations, or tufts or lines of hairs inside, sometimes with an annulus at the base. Stamens 5, inserted on the corolla and alternating with the lobes. Ovary superior, 2- or 4-locular; style usually simple, arising from between the 4 lobes of the ovary (gynobasic), rarely terminal. Fruit of 2 or 4 nutlets (rarely 1 or 3 by abortion).

1 Corolla distinctly zygomorphic
2 Corolla-tube without scales or invaginations **14. Echium**
2 Corolla-tube with scales or small invaginations
3 Corolla-tube with small invaginations; style exserted
  13. Halacsya
3 Corolla-tube with ovate or oblong scales; style included
  21. Anchusa
1 Corolla more or less actinomorphic
4 Style terminal
5 Corolla infundibuliform; fruit fleshy **1. Argusia**
5 Corolla hypocrateriform; fruit dry **2. Heliotropium**
4 Style gynobasic
6 Nutlets connate in 2 separate pairs at maturity **10. Cerinthe**
6 Nutlets separate at maturity

7 Calyx-lobes dentate, accrescent and forming a 2-lipped covering around the nutlets in fruit   **27. Asperugo**
7 Calyx-lobes entire, ± equal in fruit
  8 Nutlets with a distinct, thickened, collar-like ring around the base
    9 Corolla rotate
      10 Corolla-lobes revolute; scales in 2 series   **24. Trachystemon**
      10 Corolla-lobes not revolute; scales in 1 series
        11 Scales glabrous; stamens long-exserted   **23. Borago**
        11 Scales hairy or papillose; stamens included or only slightly exserted
          12 Racemes ebracteate; corolla 3–4 mm   **20. Brunnera**
          12 Racemes mostly bracteate; corolla more than 4 mm
            13 Nutlets stipitate   **22. Pentaglottis**
            13 Nutlets not stipitate   **21. Anchusa**
    9 Corolla tubular to campanulate
      14 Stamens exserted
        15 Corolla-lobes longer than remainder of corolla   **19. Procopiania**
        15 Corolla-lobes shorter than remainder of corolla
          16 Corolla with flange-like ring at mouth of tube   **17. Elizaldia**
          16 Corolla with scales at mouth of tube   **21. Anchusa**
      14 Stamens included
        17 Style exserted; scales at least 3 times as long as wide   **18. Symphytum**
        17 Style included; scales usually less than 3 times as long as wide, or absent
          18 Corolla-lobes longer than remainder of corolla; scales glabrous   **23. Borago**
          18 Corolla-lobes shorter than remainder of corolla; scales hairy or laciniate, or absent
            19 Nutlets smooth
              20 Corolla with 5 hairy scales near mouth of tube   **16. Nonea**
              20 Corolla with 5 tufts of hairs forming a ring in the throat   **15. Pulmonaria**
            19 Nutlets rugose, reticulate or ribbed
              21 Scales conspicuous, at least as long as wide   **21. Anchusa**
              21 Scales absent or shorter than wide, or replaced by tufts of hairs   **16. Nonea**
  8 Nutlets without a collar-like ring around the base
    22 Nutlets glochidiate
      23 Stamens exserted, or equalling corolla; style exserted   **35. Solenanthus**
      23 Stamens and style included
        24 Nutlets broadly winged   **36. Mattiastrum**
        24 Nutlets unwinged
          25 Nutlets 2   **28. Rochelia**
          25 Nutlets 4
            26 Inflorescence usually ebracteate; nutlets attached to the receptacle for the whole of their length   **34. Cynoglossum**
            26 Inflorescence bracteate; nutlets usually attached to the receptacle for only part of their length   **31. Lappula**
    22 Nutlets not glochidiate
      27 Nutlets winged, the wing often forming an umbilicus
        28 Nutlets more than 10 mm   **37. Rindera**
        28 Nutlets less than 5 mm
          29 Nutlets with erect or incurved wing; sparsely hairy or glabrous annual, biennial or perennial, not caespitose   **33. Omphalodes**
          29 Nutlets with flat wing; densely hairy, caespitose perennial   **30. Eritrichium**
      27 Nutlets not winged, or at most with a narrow rim
        30 Corolla with scales or invaginations in the throat
          31 Corolla rotate   **29. Myosotis**
          31 Corolla not rotate, the limb tubular to campanulate or infundibuliform

            32 Style exserted   **13. Halacsya**
            32 Style included
              33 Nutlets with conical spines   **31. Lappula**
              33 Nutlets smooth to tuberculate
                34 Nutlets granulate or tuberculate   **12. Alkanna**
                34 Nutlets smooth
                  35 Plant glabrous; nutlets fleshy   **25. Mertensia**
                  35 Plant hairy; nutlets dry
                    36 Corolla less than 2 mm, blue; nutlets triquetrous   **32. Trigonotis**
                    36 Corolla more than 2 mm, whitish; nutlets ovoid   **4. Lithospermum**
        30 Corolla without scales or invaginations in the throat, though often with tufts of hairs, and sometimes with a narrow flange at the base
          37 Nutlets conspicuously stipitate   **12. Alkanna**
          37 Nutlets not stipitate
            38 Stamens unequal   **14. Echium**
            38 Stamens equal
              39 Corolla yellow
                40 Style exserted   **9. Onosma**
                40 Style included
                  41 Corolla more than 15 mm   **8. Macrotomia**
                  41 Corolla not more than 15 mm
                    42 Inflorescence bracteate to apex   **5. Neatostema**
                    42 Inflorescence ebracteate, or bracteate only at base   **26. Amsinckia**
              39 Corolla blue, purple, brownish or whitish
                43 Style exserted
                  44 Anthers sagittate at base, usually with a projecting connective at the apex   **9. Onosma**
                  44 Anthers not or scarcely sagittate at base, without a projecting connective at the apex but sometimes apiculate
                    45 Annual; nutlets tuberculate   **3. Arnebia**
                    45 Perennial; nutlets smooth   **11. Moltkia**
                43 Style included
                  46 Corolla with 5 longitudinal bands of hairs inside   **6. Buglossoides**
                  46 Corolla glabrous or hairy inside, but without 5 longitudinal bands of hairs
                    47 Annual; nutlets 4   **3. Arnebia**
                    47 Perennial; nutlets usually 1(–2)   **7. Lithodora**

Subfam. **Heliotropioideae**

Style terminal, with a wide, glandular ring below the apex.

1. **Argusia** Boehmer[1]

Perennial herbs or shrubs. Flowers in terminal, ebracteate, often branched cymes. Calyx lobed almost to the base. Corolla infundibuliform, without scales; limb lobed to ½-way or more. Stamens included. Style included; stigma entire, surrounded by a ring at the base. Fruit dry, splitting into two 2-seeded mericarps.

1. **A. sibirica** (L.) Dandy, *Bot. Jour. Linn. Soc.* **65**: 256 (1972) (*Tournefortia arguzia* Roemer & Schultes, *T. sibirica* L.). Erect perennial 10–40 cm; rhizome creeping. Stems simple or branched, with appressed or patent hairs. Leaves 1–4 cm, entire, oblong or lanceolate, sessile, hairy. Flowers fragrant, in short, few-flowered cymes. Corolla tube 6–7 mm, 2–3 times as long as calyx; calyx and corolla sericeous; limb of corolla 8 mm in diameter, broadly 5-fid, white. Stigma shortly hairy. Fruit *c.* 7 × 6 mm, slightly 4-lobed at apex, with coarse appressed hairs. $2n = 26$. *Base-rich or saline soils. U.S.S.R., southwards from c. 56° N., and shores of the Black Sea.* Bu Rm Rs (C, W, K, E) Tu.

2. **Heliotropium** L.[2]

Annual or perennial herbs, sometimes woody at the base. Flowers in terminal or pseudoaxillary, ebracteate and branched

---

cymes. Calyx lobed usually to the base. Corolla hypocrateriform; tube usually without scales; limb lobed more or less to the base, usually with teeth between the lobes. Stamens included. Style included, usually very short; stigma large and disc-like, or conical to subulate, entire or 2- to 4-lobed. Fruit dry, splitting into 4 or 2 nutlets, or 1-seeded and remaining entire.

All species grow in dry, open habitats, especially in cultivated ground or sandy soils.

Literature: H. Riedl, *Ann. Naturh. Mus. (Wien)* 69: 81–93 (1966).

1 Flowers long-pedicellate   **8. micranthos**
1 Flowers ±sessile
  2 Plant glabrous   **10. curassavicum**
  2 Plant hairy
    3 Perennial; racemes not more than 1 cm, with not more than 8 flowers   **9. arguzioides**
    3 Annual or perennial; racemes more than 1 cm, with more than 8 flowers
      4 Perennial; corolla purple   **11. amplexicaule**
      4 Annual; corolla white or cream
        5 Calyx lobed to less than ¼ of its length, almost concealing the single-seeded fruit and falling with it   **12. supinum**
        5 Calyx lobed to the base, not concealing the 4 free nutlets, and persisting after they fall
          6 Stems densely villous; corolla with scales in the throat   **7. hirsutissimum**
          6 Stems ±appressed-pubescent; corolla without scales in the throat
            7 Corolla 2–5 mm; limb 2–4(–5) mm in diameter   **(1–4). europaeum group**
            7 Corolla 4–7 mm; limb 4·5–8 mm in diameter
              8 Stigma filiform, deeply bifid   **6. halacsyi**
              8 Stigma conical, entire or shallowly 2- to 4-lobed   **5. suaveolens**

(1–4). H. europaeum group. Annual; stems 4–40 cm, erect or ascending, usually branched, appressed- to erecto-patent-pubescent. Leaves up to 5·5(–6·5) × 2·8(–3·5) cm, ovate to elliptical, cuneate to almost rounded at the base, rounded to subacute at the apex, appressed-pubescent; petiole up to 3·5 cm. Sepals 1·8–3·5 mm, lanceolate to linear-oblong. Corolla 2–5 mm, at least the limb white; limb 2–4(–5) mm in diameter. Anthers 0·7–1·5 mm, inserted in middle of corolla-tube. Stigma conical to linear-subulate, deeply or shallowly divided into 2(–4) lobes at apex or rarely entire, glabrous to hairy. Fruit splitting into 4 nutlets, glabrous or pubescent.

The species included here are not well understood and some may perhaps not merit specific status. 1 and 4 are usually fairly easily distinguished from each other in the Mediterranean region, but in Russia and S.W. Asia both may be difficult to separate from 2. It might prove more satisfactory to regard 3 as merely a variant of 2 with pubescent fruits, and possibly 2 is only subspecifically distinct from 1. *H. stevenianum* Andrz., *Enum. Pl. Podol.* 116 (1862), described from W. Ukraine, is probably not specifically distinct from 1 but has also been regarded as synonymous with 2. In the literature 2 and 4 have been much confused and details of their distribution in E. Europe are uncertain.

1 Stigma linear-subulate, usually glabrous   **1. europaeum**
1 Stigma conical, usually papillose or hairy
  2 Fruit hairy   **3. lasiocarpum**
  2 Fruit glabrous
    3 Sepals lanceolate; anthers 1–1·5 mm; stigma usually densely hairy   **4. dolosum**
    3 Sepals oblong or narrowly triangular; anthers 0·7–1 mm; stigma usually papillose   **2. ellipticum**

1. H. europaeum L., *Sp. Pl.* 130 (1753). Flowers scentless. Sepals 0·4–0·8 mm wide, linear-oblong or narrowly triangular, irregularly patent-pubescent, usually patent soon after anthesis. Corolla 2–4·2 mm. Anthers 0·7–1 mm. Stigma linear-subulate, usually glabrous. Fruit glabrous or pubescent, usually rugose. *S., C. & W. Europe, northwards to N. France, Czechoslovakia & C. Ukraine; occasionally as a weed or casual further north.* Al Au Az Bl Bu Co Cr Cz Ga Ge Gr He Hs Hu It Ju Lu Rm Rs (W, K, E) Sa Si Tu.

Variable in flower-size and fruit-pubescence.

2. H. ellipticum Ledeb. in Eichw., *Pl. Nov. It. Casp.-Cauc.* 10 (1831). Like **1** but corolla 2·3–3·4 mm; stigma shorter and wider, usually more or less papillose; fruit glabrous. *S. part of U.S.S.R.* Rs (W, K, E).

3. H. lasiocarpum Fischer & C. A. Meyer, *Ind. Sem. Horti Petrop.* **4**: 38 (1838). Like **1** but stigma shorter and wider, papillose or hairy; fruit densely hairy. *S.E. Russia, W. Kazakhstan.* Rs (E). *(S.W. Asia.)*

4. H. dolosum De Not., *Repert. Fl. Ligust.* 284 (1844). Flowers scented. Sepals 0·8–1·1 mm wide, lanceolate, usually appressed-pubescent, curving upwards round the developing fruit, though often patent later. Corolla 3–5 mm. Anthers 1·1–1·5 mm. Stigma conical, hairy. Fruit glabrous, smooth. *Usually coastal. C. & E. Mediterranean region and coast of Black Sea.* Bu Cr Gr It Rm Rs (W) ?Si Tu.

5. H. suaveolens Bieb., *Fl. Taur.-Cauc.* **3**: 116 (1819). Annual; stems up to 40 cm, erect or ascending, more or less appressed-pubescent. Leaves up to 5·5 × 2·8 cm, ovate to elliptical, cuneate to rounded at the base, rounded to subacute at the apex, appressed-pubescent; petiole up to 3·5 cm. Flowers scented. Sepals 2–3·5 mm, linear-oblong, with dense, patent hairs. Corolla 4–6·5 mm; limb (3·5–)4·5–8 mm in diameter, white. Anthers inserted near base of corolla-tube. Stigma 0·1–0·5 mm, shortly conical or hemispherical, entire or shallowly 2- to 4-lobed, about as long as wide, usually pubescent or papillose. Fruit splitting into 4 nutlets. *S.E. Europe, extending to S. Italy and Sicilia.* Bu ?Co ?Cr Gr It Rm Rs (W, K, E) Si Tu.

(a) Subsp. **suaveolens**: Ripe fruit 1·5–2 mm, light brown, usually hairy. *S.E. Europe.*

(b) Subsp. **bocconei** (Guss.) Brummitt, *Bot. Jour. Linn. Soc.* **64**: 67 (1971) (*H. bocconei* Guss.): Ripe fruit 1–1·7 mm, blackish-brown, usually glabrous. *Italy and Sicilia, perhaps Corse.*

6. H. halacsyi Riedl, *Ann. Naturh. Mus. (Wien)* **69**: 89 (1966) (*H. bocconei* sensu Hayek pro parte, non Guss.). Like **5** but anthers inserted at middle of corolla-tube; stigma c. 2·2 mm, filiform, deeply bifid, glabrous; fruit glabrous. ● *W. Greece.* Gr.

7. H. hirsutissimum Grauer, *Pl. Min. Cogn. Dec.* 1 (1784). Like **5** in habit and vegetative characters but stems and petioles with dense, greyish or yellowish, patent hairs c. 1 mm. Flowers scented. Sepals 3–4 mm, linear-oblong, with dense, patent hairs. Corolla 5–8 mm; limb 4–8 mm in diameter, white, with 5 scales in the throat. Anthers inserted near base of corolla-tube. Stigma shallowly hemispherical, wider than long, pubescent. Fruit splitting into 4 nutlets, glabrous. *C. & S. Greece, S. Aegean region.* Cr Gr.

8. H. micranthos (Pallas) Bunge, *Beitr. Kenntn. Fl. Russl.* 223 (1852). Annual; stems up to 30 cm, glabrous below, sparsely setose above. Leaves up to 4 × 1·2 cm, narrowly elliptical to ovate, cuneate to rounded at the base, subacute, hirsute beneath, subglabrous above; petiole as long as or longer than lamina. Inflorescence very lax, few-flowered; pedicels up to 1·8 cm. Sepals 2 mm, linear-oblong, rather sparsely hairy, accrescent and up to 4 mm in fruit. Corolla *c.* 2·5 mm; limb very small, white. Style almost as long as stigma; stigma conical, densely hairy. Fruit splitting into 4 nutlets, sericeous. *S.E. Russia (near Astrakhan').* Rs (E). *(C. Asia.)*

9. **H. arguzioides** Kar. & Kir., *Bull. Soc. Nat. Moscou* **15**: 406 (1842). Perennial; stems up to 40 cm, erect, branched, more or less appressed-pubescent with deflexed hairs, glabrescent below. Leaves up to 3 × 1·5 cm, ovate, cuneate at the base, subacute, margins crenulate-undulate; petiole up to 1 cm. Racemes very short, each branch up to 1 cm and with not more than 8 flowers. Calyx 2 mm, lobed to *c.* ½ way; lobes oblong. Corolla 3·5–4 mm; lobes very short, white. Anthers inserted in upper part of corolla-tube. Style as long as or longer than the stigma; stigma conical, subentire or lobed, papillose-hairy. Fruit splitting into 4 nutlets, which are white-sericeous dorsally. *S.E. Russia, W. Kazakhstan.* Rs (E). *(S.W. & C. Asia.)*

10. **H. curassavicum** L., *Sp. Pl.* 130 (1753). Perennial, glabrous and rather glaucous and fleshy throughout; stems up to 70 cm, procumbent. Leaves up to 4 × 1 cm, spathulate to linear-oblong; lateral veins scarcely visible. Sepals 1–2 mm, ovate. Corolla 1–2·5 mm; limb *c.* 2 mm in diameter, white. Stigma shortly conical. Fruit splitting into 4 nutlets, glabrous. *Sandy saline places near the sea. Naturalized in S. Europe.* [Az Bl Ga Hs Lu Rm Sa Si.] *(North and South America.)*

11. **H. amplexicaule** Vahl, *Symb. Bot.* **3**: 21 (1794) *(Cochranea anchusifolia* (Poiret) Gürke). Perennial; stems up to 60 cm, woody at the base, glandular-hairy above. Leaves *c.* 5 × 1·5 cm, elliptic- to linear-oblong, sessile and cuneate to slightly amplexicaul at the base, acute, the margins undulate. Sepals 2–4 mm, linear-lanceolate, glandular-hairy. Corolla 5–7 mm; limb 4–8 mm in diameter, purple. Stigma shortly conical or hemispherical, entire or shallowly 2- to 4-lobed, about as long as wide, usually pubescent or papillose. Fruit splitting into 2 nutlets. *Widely cultivated for ornament; naturalized in C. Italy and perhaps elsewhere.* [It.] *(South America.)*

H. arborescens L., *Syst. Nat.* ed. 10, **2**: 913 (1759), which differs from **11** in having leaves more than 1·2 cm wide and 4 nutlets, is widely cultivated for its fragrant flowers and occasionally naturalized for short periods in S. Europe.

12. **H. supinum** L., *Sp. Pl.* 130 (1753). Annual, branched at the base, the central branch erect, the lateral procumbent. Leaves up to 3·5 × 2 cm, narrowly elliptical to suborbicular, rounded or cuneate at the base, whitish-pubescent at least beneath, the veins conspicuously impressed above; petiole 0·3–1·5(–2) cm. Calyx 2–2·5 mm, lobed to less than ¼ of its length, tubular, accrescent in fruit and becoming pyriform, closely enfolding the fruit and falling with it. Corolla 2·5–3 mm, the limb white. Style almost as long as stigma, inserted slightly obliquely on the ovary; stigma conical, hairy, often shallowly bifid. Fruit a single one-seeded nutlet. *S. Europe, extending northwards to Hungary.* Al Bl Bu Co Cr Ga Gr Hs Hu It Ju Lu Rm Sa Si.

Subfam. **Boraginoideae**

Style gynobasic, without a glandular ring.

---

3. **Arnebia** Forskål[1]

Annual, biennial or perennial herbs. Flowers usually dimorphic and heterostylous, in terminal, bracteate cymes. Calyx lobed almost to the base. Corolla yellow, blue or violet, infundibuliform, without scales; tube with a ring of hairs at the base. Stamens included, inserted at mouth or in middle of tube. Style included or exserted, divided into 2 or 4; stigma capitate or bifid. Nutlets erect, ovoid or oblong, acute, tuberculate, with flat base.

1. **A. decumbens** (Vent.) Cosson & Kralik, *Bull. Soc. Bot. Fr.* **4**: 398, 402 (1857). Annual. Stems 5–20 cm, erect, with patent and not very dense hairs. Leaves 20–50 × 3–10 mm, lanceolate or linear-lanceolate, hairy, the lower obtuse, the upper acute. Inflorescence terminal, erect, up to 10 cm in fruit. Bracts linear-lanceolate, slightly exceeding flowers. Calyx 6–8 mm, hairy, with long, linear segments. Corolla 10–15 mm, almost tubular, hairy, yellow; limb 2–4 mm in diameter; lobes 1–2 mm. Nutlets 2 mm, grey, shallowly tuberculate. *Sandy or stony places. S.E. Russia (near Astrakhan').* Rs (E). *(N. Africa, S., W. & C. Asia.)*

4. **Lithospermum** L.[2]

Perennial herbs. Flowers in terminal and axillary, leafy cymes, sometimes heterostylous. Calyx 5-lobed more or less to the base, accrescent. Corolla cylindrical to infundibuliform, whitish; tube with 5 more or less glandular invaginations in the throat and with an annulus at the base. Stamens included, inserted at about the middle of the throat. Style included, simple; stigma 2-lobed. Nutlets 4, detaching completely from the receptacle, smooth, white and shining.

1. **L. officinale** L., *Sp. Pl.* 132 (1753). Perennial; stems 20–100 cm, solitary to many from a stout rhizome, erect, scabrid-pubescent, much-branched above. Leaves up to 10 × 2 cm, lanceolate to ovate-lanceolate, sometimes linear-lanceolate, very acute, the middle and upper sessile, discolorous, scabrid-pubescent mainly above. Cymes fairly dense in flower. Calyx-lobes oblong-linear, obtuse. Corolla (3–)4–6 mm, yellowish- or greenish-white; tube cylindrical, about as long as calyx. Nutlets 2·7–4 mm, ovoid, obtuse or subacute, rounded dorsally, obtusely keeled ventrally. $2n = 28$. *Hedges, thickets and forest-margins. Most of Europe, but rare in parts of the north and west; only as an alien in Finland.* All except Az Cr Fa Is Sb.

5. **Neatostema** I. M. Johnston[2]

Like *Lithospermum* but annual; corolla yellow; tube with a ring or 5 clusters of hairs in the throat; stamens inserted just above the annulus; nutlets light brown, tuberculate.

1. **N. apulum** (L.) I. M. Johnston, *Jour. Arnold Arb.* **34**: 6 (1953) *(Lithospermum apulum* (L.) Vahl). Stems 3–30 cm, solitary to many, erect, hispid, corymbosely branched above. Leaves up to 7 × 0·5 cm, setose mainly on the margins; the cauline linear or oblong-linear, erect, sessile, acute; the basal linear- to oblong-spathulate, attenuate into petiole. Cymes dense. Calyx 3·5–4 mm; lobes linear, acute, hispid outside, softly white-hairy inside, triangular in fruit. Corolla 6–6·5 mm; tube slightly exceeding calyx; lobes of limb glandular-puberulent on both surfaces. Style 0·25–0·5 mm. Nutlets 1·5–1·75 mm, subtetrahedral, contracted at the beak, flat dorsally, keeled ventrally. $2n = 28$. *Dry, open habitats. S. Europe.* Bl Bu Co Cr Ga Gr Hs It Ju Lu Rs (K) Sa Si Tu.

6. Buglossoides Moench[1]

Like *Lithospermum* but annual or perennial herbs or dwarf shrubs; cymes all terminal; corolla white, blue or purple, infundibuliform or hypocrateriform, with 5 longitudinal bands of hairs inside, without invaginations, the annulus sometimes discontinuous; stamens sometimes inserted near base of tube; nutlets sometimes tuberculate.

1 Perennial; corolla 12–20 mm; nutlets smooth or punctate-reticulate, white or yellowish, shining
  2 Flowering stems procumbent, arising from previous year's growth; leaves widest at the middle, not acuminate; corolla blue   **3. calabra**
  2 Flowering stems erect, arising from the rhizome; leaves widest below the middle; corolla at first purple, then blue
    3 Leaves sessile, amplexicaul, the lowest scale-like; corolla 12–14 mm; filaments inserted 1–1·5 mm above the base of the corolla-tube   **2. gastonii**
    3 Lower leaves attenuate into petiole, not scale-like; corolla (12–)14–19 mm; filaments inserted 5–8 mm above the base of the corolla-tube   **1. purpurocaerulea**
1 Annual; corolla not more than 10 mm; nutlets ± tuberculate, brownish or greyish, not shining
  4 Plant very glandular   **6. glandulosa**
  4 Plant eglandular or with some glandular hairs on inflorescence
    5 Indumentum of calyx yellowish; nutlets with a very fragile pericarp, distinctly 2-gibbous, with the base 0·5–0·75 mm wide   **7. tenuiflora**
    5 Indumentum of calyx whitish; nutlets with ± hard pericarp, not or only slightly 2-gibbous, with the base more than 0·75 mm wide
      6 Calyx c. ½ as long as corolla-tube, very densely hispid; lobes obtuse   **5. minima**
      6 Calyx distinctly more than ½ as long as corolla-tube, with long, sparse hairs; lobes acute   **4. arvensis**

1. B. purpurocaerulea (L.) I. M. Johnston, *Jour. Arnold Arb.* **35**: 44 (1954) (*Lithospermum purpurocaeruleum* L.). Perennial; flowering stems 15–60(–70) cm, arising from a rhizome, erect, unbranched, with appressed hairs below and more patent hairs above. Leaves 3·5–8 × 0·7–1·5 cm, lanceolate to narrowly elliptical, very acute. Cymes 2–3. Calyx 6–8·5 mm; lobes linear, acute, setose. Corolla (12–)14–19 mm, at first reddish-purple, then bright blue. Filaments inserted 5–8 mm above the base of the corolla-tube. Style 6·5–9 mm. Nutlets 3·5–5 × 3–3·5 mm, ovoid-globose, convex dorsally, obtusely keeled ventrally, white, smooth, shining. 2n=16. *Scrub and wood-margins. S. & C. Europe, extending to Britain and S.C. Russia.* Al Au Be Br Bu ?Co Cz Ga Ge Gr He Hs Hu It Ju Po Rm Rs (C, W, K, E) Sa Si Tu.

2. B. gastonii (Bentham) I. M. Johnston, *op. cit.* 45 (1954) (*Lithospermum gastonii* Bentham). Perennial; flowering stems 10–30 cm, arising from a rhizome, erect, unbranched, appressed-hairy. Leaves up to 6·5 × 2·5 cm, the middle and upper crowded, ovate-lanceolate or lanceolate, semiamplexicaul, acuminate, appressed-hairy, the lowest scale-like. Cymes solitary or paired; flowers crowded; pedicels nearly as long as the calyx. Calyx-lobes linear-lanceolate, acute. Corolla 12–14 mm, at first purple or violet, then blue. Filaments inserted 1–1·5 mm above the base of corolla-tube. Style 0·5–1 mm. Nutlets 4·5 × 4·5 mm, ovoid, acute, keeled near the apex, yellowish, punctate-reticulate. *Mountain woods and rocky slopes.* ● *W. Pyrenees.* Ga.

3. B. calabra (Ten.) I. M. Johnston, *op. cit.* 45 (1954) (*Lithospermum calabrum* Ten.). Perennial; stems 30–80 cm, procum-

bent, slender, patent-hairy, the flowering stems unbranched, arising from previous year's growth from a rosette of persistent, coriaceous leaves. Leaves up to 4 × 1·5 cm, elliptical or ovate-oblong, acute or obtuse. Cymes solitary, few-flowered; pedicels short. Calyx 6–7 mm; lobes linear, hairy. Corolla 17–20 mm, blue, 2–3 times as long as the calyx. Stamens inserted c. 2 mm above the base of the corolla-tube. Style 1–2 mm. Nutlets 3·5 × 2·5 mm, ellipsoid, convex dorsally, obtuse ventrally, white, smooth, shining. 2n=20. *Rocky places on mountains.* ● *S. Italy.* It.

4. B. arvensis (L.) I. M. Johnston, *op. cit.* 42 (1954) (*Lithospermum arvense* L.). Annual; stems (1·5–)5–50(–90) cm, solitary to many, appressed-setulose-hispid. Lowest leaves oblong-spathulate or obovate, obtuse, the others oblong to linear, acute or subacute, usually erect. Cymes solitary or paired. Calyx-lobes linear or linear-lanceolate, unequal, acute. Corolla (4–)6–9 mm, white, purplish or blue. Stamens inserted below the middle of the corolla-tube. Nutlets brownish, hard. 2n=28, 42. *Most of Europe.* All except Az Fa Is Sb.

1 Nutlets 2·5–4 × 1·5–2·5 mm, with crowded, prominent tubercles; corolla 6–9·5 mm, white (rarely purplish), infundibuliform; calyx usually equalling or longer than corolla-tube in flower; pedicels cylindrical in fruit   **(a) subsp. arvensis**
1 Nutlets 2–3 × 1·5–2 mm; corolla 4–8 mm, blue or white, hypocrateriform; calyx usually distinctly shorter than corolla-tube in flower
  2 Pedicels cylindrical in fruit; receptacle not oblique in fruit; adaxial nutlet not embedded in receptacle and pedicel; nutlets with prominent tubercles   **(d) subsp. sibthorpiana**
  2 Pedicels obconical or clavate in fruit; receptacle ± oblique in fruit; adaxial nutlet ± embedded in the receptacle and pedicel; nutlets with small, low tubercles
    3 Stems 1·5–17(–45) cm, usually many, procumbent or ascending; cauline leaves 1·5–4(–6) mm wide   **(c) subsp. gasparrinii**
    3 Stems 10–65 cm, usually solitary, erect; cauline leaves up to 8(–10) mm wide   **(b) subsp. permixta**

(a) Subsp. **arvensis**: Stems 10–50(–90) cm, solitary, usually erect, sometimes branched, sparsely hispid. Cauline leaves up to 8(–10) mm wide, oblong, obovate-oblong or oblong-spathulate to linear. Corolla white or with a blue zone at about the middle of the tube, rarely purplish. Pedicels cylindrical in fruit. Receptacle not oblique in fruit. 2n=28. *Cultivated ground and dry, open habitats. Almost throughout the range of the species.*
Plants with 2n=42 from Portugal (and perhaps from elsewhere in S. Europe) have the pedicels shorter and thicker in fruit and less expanded corollas than in those of N. Europe. They should probably be distinguished as a separate subspecies.
(b) Subsp. **permixta** (Jordan ex F. W. Schultz) R. Fernandes, *Bot. Jour. Linn. Soc.* **64**: 379 (1971) (*Lithospermum permixtum* Jordan ex F. W. Schultz): Like subsp. (a) but pedicels obconical in fruit; receptacle oblique; corolla 5–7 mm, blue; nutlets 2·75–3 × 2 mm. 2n=28. *S.W. & W.C. Europe.*
(c) Subsp. **gasparrinii** (Heldr. ex Guss.) R. Fernandes, *loc. cit.* (1971) (*Lithospermum gasparrinii* Heldr. ex Guss., *L. incrassatum* Guss.): Stems 1·5–17(–45) cm, usually numerous, the central erect, the others procumbent or ascending, densely leafy, densely hispid. Cauline leaves 1·5–4(–6) mm wide, linear or oblong. Pedicels obliquely clavate. Receptacle very oblique. Corolla 4–6·5(–8) mm, usually blue or blue-violet. Nutlets 1·5–2·5 × (1–)1·5–2 mm, contracted into a short beak. *Mountain grassland. Mediterranean region.*
(d) Subsp. **sibthorpiana** (Griseb.) R. Fernandes, *loc. cit.* (1971) (*Lithospermum sibthorpianum* Griseb., *L. leithneri* Heldr. &

[1] By R. Fernandes.

Sart.): Like subsp. (c) but pedicels cylindrical; receptacle not or only slightly oblique in fruit; corolla usually white; nutlets 2·5–2·75 × 1·5–1·75 mm, more attenuate than in subsp. (c). *S.E. Europe.*

5. B. minima (Moris) R. Fernandes, *loc. cit.* (1971) (*Lithospermum minimum* Moris). Annual; stems 2–10(–27) cm, solitary or few, erect or ascending, simple or sparingly branched, appressed-hispid. Basal leaves obovate or oblong, obtuse, the others oblong to oblong-linear, obtuse or acute. Calyx 1·5–3 mm in flower, 4–6 mm in fruit, very densely appressed-hispid with rather long hairs; lobes oblong, usually obtuse. Corolla 4–6 mm, blue or white, densely hispid outside; tube 2·5–4·5 mm, cylindrical, abruptly dilated at the limb. Nutlets 2–2·75 × 1·25–1·75 mm, subtrigonous-conical, not much constricted below the beak, brownish, tuberculate-echinulate. *Dry places.* ● *S.W. Italy, Sicilia, Sardegna.* It Sa Si.

Considered by some authors as identical with **7**, because their corolla-tubes are very alike; but it approaches rather more **4(d)** (mainly the dwarf littoral variants from Greece), from which it differs in the more densely hispid and relatively longer corolla-tube, the oblong-linear, obtuse calyx-lobes and the smaller base of the nutlets.

6. B. glandulosa (Velen.) R. Fernandes, *loc. cit.* (1971) (*Lithospermum glandulosum* Velen.). Hispid and glandular annual; stems numerous, erect. Basal leaves spathulate, narrowed into petiole, the cauline oblong-spathulate to linear, undulate or dentate. Cymes short; pedicels short, slightly thickened in fruit. Calyx-lobes linear-subulate, setose. Corolla 6–8 mm, white. Nutlets ovoid, acute, brownish, more or less tuberculate, dull. ● *Bulgaria, S.E. Romania.* Bu Rm.

7. B. tenuiflora (L. fil.) I. M. Johnston, *Jour. Arnold Arb.* 35: 42 (1954) (*Lithospermum tenuiflorum* L. fil.). Annual; stems (5–)10–30(–50) cm, simple or branched above, usually erect, appressed-hispid. Cauline leaves 15–40 × (2–)3–8 mm, oblong-spathulate to linear, obtuse. Cymes up to 8(–12) cm in fruit, paired or 3 together; flowers crowded; pedicels very short, not or only slightly thickened in fruit. Calyx 3–4 mm, up to 10 mm in fruit, with dense, yellowish, subappressed setae; lobes linear-lanceolate, obtuse. Corolla 5–6 mm, bluish-violet, rarely white; tube narrowly cylindrical, somewhat expanded at the base. Nutlets *c.* 2 mm, erect or slightly incurved, urceolate in outline, distinctly 2-gibbous laterally on the back, abruptly contracted into a beak, densely and minutely tuberculate; pericarp very fragile. *Dry hills and cultivated ground. S.E. Europe.* Gr Ju Rs (E) ?Si. (*S.W. & C. Asia, N. Africa.*)

7. Lithodora Griseb.[1]

Dwarf shrubs. Flowers in terminal, leafy cymes. Calyx 5-lobed more or less to the base, slightly accrescent. Corolla blue, purple or white, infundibuliform or hypocrateriform; tube without scales or invaginations in the throat, without a basal annulus, sometimes glandular or villous. Stamens usually included. Style simple or branched above. Nutlets usually 1(–2), circumscissile above the base.

1 Corolla glabrous outside
2 Leaves up to 15(–20) × 4·5 mm, oblong-ovate to oblanceolate with flat or slightly inflexed margin; calyx setose, *c.* ½ as long as corolla-tube **3. hispidula**

2 Leaves 20–40 × 2–4 mm, linear or oblong-linear with strongly revolute margin; calyx softly hairy, more than ½ as long as corolla-tube **4. zahnii**
1 Corolla sericeous or setulose outside
3 Leaves obovate to obovate-oblong, not more than 3 times as long as wide
4 Leaves obtuse, sparsely hispidulous above, whitish-sericeous beneath; calyx-lobes not more than 1 mm wide, linear, subacute **5. oleifolia**
4 Leaves shortly acuminate, mucronate, very densely whitish-sericeous on both surfaces; calyx-lobes 1·25–1·5 mm wide, lanceolate-oblong, acuminate **6. nitida**
3 Leaves linear to lanceolate, usually more than 3 times as long as wide
5 Stamens inserted at different levels in the corolla-tube or throat; corolla-throat ± densely hairy **7. diffusa**
5 Stamens all inserted at the same level at the apex or near the base of the corolla-tube; corolla-throat glabrous
6 Corolla setulose throughout; leaves 10–60 × 1–10 mm, sparsely appressed-setose above; anthers *c.* 1·75 mm **1. rosmarinifolia**
6 Corolla setulose only on the lobes (sometimes very sparsely so); leaves not more than 23 × 3·5 mm, patent-hispid; anthers 2·5–3 mm **2. fruticosa**

1. L. rosmarinifolia (Ten.) I. M. Johnston, *Contr. Gray Herb.* nov. ser., 73: 56 (1924) (*Lithospermum rosmarinifolium* Ten.). Caespitose dwarf shrub 30–60 cm; branches erect or pendent, greyish-appressed-hairy above. Leaves 10–60 × 1–10 mm, linear to lanceolate, acute or subacute, rigid, dark green, sparsely appressed-setulose or glabrous above, densely greyish-appressed-setose beneath; margin inflexed or strongly revolute. Calyx *c.* 6 mm, whitish-appressed-setose. Corolla blue, lilac or whitish, setulose on the outside, mainly at the middle; tube *c.* 12 mm; limb *c.* 17 mm in diameter, with oblong lobes, rounded at the apex. Anthers *c.* 1·75 mm. Nutlets whitish, smooth. 2*n*=26. *Rock-crevices. S. Italy, Sicilia.* It Si.

2. L. fruticosa (L.) Griseb., *Spicil. Fl. Rumel.* 2: 531 (1846) (*Lithospermum fruticosum* L.). Caespitose dwarf shrub 15–60 cm; branches erect, tortuous, the oldest dark grey, the younger whitish-setose. Leaves up to 23 × 1–2·5(–3·5) mm, oblong-linear or linear, obtuse, those of the current year concolorous, whitish and appressed-setulose on both surfaces, the older coriaceous, patently tuberculate-hispid mainly on the strongly revolute margin and on the veins beneath. Calyx *c.* 6 mm, hispid. Corolla 12–15 mm, purplish-violet or blue, sparsely setulose on the lobes outside. Anthers 2·5–3 mm. Nutlets *c.* 4 × 2 mm, ovoid, strongly constricted just above the base, whitish, striate. 2*n*=28. *Dry ground; calcicole. S., C. & E. Spain, S. France.* Ga Hs.

3. L. hispidula (Sibth. & Sm.) Griseb., *loc. cit.* (1846) (*Lithospermum hispidulum* Sibth. & Sm.). Much-branched dwarf shrub 10–35 cm; branches short, stiff, whitish-appressed-setulose. Leaves up to 15(–20) × 4·5 mm (usually smaller), oblong-obovate to oblanceolate, obtuse or acute, coriaceous, dark green, appressed-setulose and patently tuberculate-hispid above, strongly whitish-appressed-setulose beneath; margin flat or more or less inflexed. Cymes 1- to 4-flowered. Calyx *c.* 7 mm; lobes appressed-setose. Corolla bluish-violet, glabrous; tube *c.* 12 mm; limb *c.* 10 mm in diameter, with short, rounded lobes. Anthers 2 mm. Nutlets ovoid-trigonous, white, minutely tuberculate. *Rock-crevices and other dry places. Kriti and Karpathos.* Cr. (*E. Mediterranean.*)

4. L. zahnii (Heldr. ex Halácsy) I. M. Johnston, *Contr. Gray Herb.* nov. ser., 73: 56 (1924) (*Lithospermum zahnii* Heldr. ex Halácsy). Much-branched dwarf shrub up to 40(–60) cm, forming dense tufts up to 90 cm in diameter; branches erect, the older black and leafless, the younger densely leafy, sericeous. Leaves 20–40 × 2–4 mm, linear or oblong-linear, more or less obtuse, coriaceous, green or greyish and appressed-hispid above, densely greyish-setose beneath; margin strongly revolute. Cymes 1- to 3-flowered. Calyx (6·5–)8–11 mm; lobes linear, densely and softly white-appressed-hairy. Corolla blue or white, hypocrateriform, glabrous; tube *c.* 10 mm; limb 13–15 mm in diameter with patent, ovate, obtuse lobes. Anthers 2–2·5 mm. Nutlets smooth. ● *Cliffs. S. Greece (S.E. of Kalamai).* Gr.

5. L. oleifolia (Lapeyr.) Griseb., *Spicil. Fl. Rumel.* **2**: 531 (1846) (*Lithospermum oleifolium* Lapeyr.). Laxly branched dwarf shrub with slender, diffuse ascending stems 10–45 cm, leafless for some distance at the base. Leaves up to 4 × 1·5 cm, obovate or obovate-oblong, obtuse, shortly petiolate, dull green and sparsely hispidulous above, whitish-sericeous beneath, not very thick, crowded at the apex of short non-flowering branches or surrounding the base of the flowering ones; leaves of flowering branches smaller and sparser. Cymes 3- to 7-flowered. Calyx 5–8(–10) mm, whitish-sericeous; lobes not more than 1 mm wide, linear, subacute. Corolla at first pale pink, then blue, sericeous outside, glabrous inside; tube 8–12 mm; lobes *c.* 3 mm, rounded. Nutlets *c.* 3 × 2·25 mm, broadly ovoid, contracted into a short beak, greyish-white, smooth. *Rocks.* ● *E. Pyrenees (a small region c. 25 km W.N.W. of Figueras); planted in a few places on the French side of the frontier.* Hs.

6. L. nitida (Ern) R. Fernandes, *Bot. Jour. Linn. Soc.* **64**: 73 (1971) (*Lithospermum oleifolium* subsp. *nitidum* Ern). Like **5** in habit but leaves not more than 2 × 0·8 cm, elliptical-obovate, shortly acuminate, mucronate, sessile, very densely whitish-sericeous on both surfaces, hairs shorter, stiffer and more appressed; cymes 2- to 5-flowered; calyx-lobes 1·25–1·5 mm wide, lanceolate-oblong, acuminate; corolla-tube up to 14 mm; lobes *c.* 5·5 mm, ovate-oblong, obtuse. 2*n* = 50. *Stony places, 1800–1900 m.* ● *S. Spain (Sierra Mágina).* Hs.

7. L. diffusa (Lag.) I. M. Johnston, *Contr. Gray Herb.* nov. ser., **73**: 56 (1924) (*Lithospermum diffusum* Lag.). Stems with appressed and patent setae. Leaves obtuse, more or less discolorous, setose on both surfaces, the older ones tuberculate-hispid above. Calyx (5–)6–8(–10) mm. Corolla up to 21 mm, infundibuliform or hypocrateriform, appressed-hairy outside. Stamens inserted at different levels in the corolla-tube or throat. Nutlets pale brown to greyish, smooth. *Pinewoods, scrub, hedges and maritime sands; usually calcifuge.* *S.W. Europe, extending northwards to N.W. France.* Ga Hs Lu.

(a) Subsp. **diffusa** (*Lithospermum prostratum* Loisel.): Diffuse; branches up to 60 cm, procumbent or climbing, flexuous. Leaves 7–38 × 1–8 mm, patent, linear to oblong or elliptical, flat or more or less inflexed at the margin. Corolla blue, rarely purple; throat with a wide ring of very dense, white, soft, long hairs. Anthers 1·5–1·75 mm, oblong. Nutlets 3·25–4 × 2 mm, oblong, somewhat acute. 2*n* = 24. *Almost throughout the range of the species.*

(b) Subsp. **lusitanica** (Samp.) P. Silva & Rozeira, *Agron. Lusit.* **24**: 170 (1964): Erect, much-branched, forming dense tufts. Leaves up to 23 × 3·5(–6) mm, appressed to the stems, linear, strongly revolute, thick. Corolla blue; throat sparsely or somewhat densely hairy. Anthers *c.* 0·5 mm, suborbicular. Nutlets 2·5–3 × 2 mm, ovoid, obtuse. *C. & S. Portugal, S. Spain.*

8. Macrotomia DC.[1]

Perennial herbs. Flowers in short, terminal, bracteate cymes clustered to form a head. Calyx lobed to the base. Corolla violet or yellow, infundibuliform, without scales or annulus. Stamens included, inserted at or above the middle of the tube. Style included; stigma emarginate or bifid. Nutlets ovoid or trigonous, erect, keeled, slightly concave at base.

1. M. densiflora (Ledeb. ex Nordm.) Macbride, *Contr. Gray Herb.* nov. ser., **48**: 56 (1916) (*M. cephalotes* (A. DC.) Boiss.). Perennial. Stock up to 7 × 4 cm, stout, erect or horizontal, covered with lanate bases of old leaves. Current basal leaves up to 10 × 2 cm, forming a rosette, lanceolate, acute, sessile, entire, with appressed hairs. Stem 15–20 cm, simple or branched above, erect, clothed with linear-lanceolate leaves and terminating in a cluster of cymes to form a dense head 5–10 cm in diameter. Calyx-lobes linear, acute, pectinate-setose. Corolla up to 3 cm; tube twice as long as calyx; limb 2 cm in diameter, patent, golden-yellow. Nutlets minutely pitted. *Mountain rocks. S. Greece (Aroania Oros).* Gr. (*Anatolia.*)

9. Onosma L.[2]

Biennial or perennial, hispid herbs. Flowers in terminal, usually branched, bracteate cymes. Calyx lobed almost to the base, often accrescent. Corolla yellow, whitish or purplish, tubular to tubular-campanulate, with 5 short, patent or deflexed lobes, without scales, but with an annulus near the base. Stamens included or exserted, inserted at about the middle of the corolla; anthers sagittate at base, usually with a projecting connective at the apex. Style exserted; stigma capitate to bifid. Nutlets sometimes fewer than 4 by abortion, ovoid or trigonous, acute or beaked, erect, smooth or tuberculate, with triangular, flat base.

Literature: S. Jávorka, *Ann. Hist.-Nat. Mus. Hung.* **4**: 406–449 (1906); *Jour. Bot.* (London) **66**: 1–9, 57–75 (1928). C. C. Lacaita, *Nuovo Gior. Bot. Ital.* nov. ser., **31**: 18–35 (1924). G. Stroh, *Beih. Bot. Centr.* **59B**: 430–454 (1939). M. G. Popov, *Not. Syst.* (*Leningrad*) **14**: 287–304 (1951).

In addition to the setae with an enlarged base, known from many genera of the Boraginaceae, many species of *Onosma* have setae with usually 4–20 rays arising from the base. These are referred to as stellate setae, and are fully described by Jávorka (*loc. cit.*). The central seta is occasionally absent from these hairs, but usually it is distinctly longer and stouter than the rays. The descriptions of the stellate setae in this account are of those found on the lower leaves.

The genus presents considerable taxonomic difficulty, particularly in C. & S.E. Europe, which cannot be satisfactorily resolved without experimental investigation. The presence or absence of stellate setae is widely used as a major character in the genus, but in a number of species (particularly **13–20**) there may be a wide range of variation in the presence, frequency and length of the stellate setae. There has, in the past, been considerable nomenclatural confusion and the following names have been widely misapplied: *O. echioides* L., *O. setosa* Ledeb., *O. stellulata* Waldst. & Kit. and *O. taurica* Willd.

O. fruticosa Sibth. & Sm., *Fl. Graec. Prodr.* **1**: 122 (1806), a species endemic to Cyprus, has been recorded in error from Kriti.

All species occur in dry, sunny habitats.

---

[1] By D. H. Valentine.
[2] By P. W. Ball, with assistance from H. Riedhl.

1 All setae simple
2 Corolla 8–15 mm
3 Nutlets 6–7 mm, tuberculate
4 Corolla *c.* 1½ times as long as calyx; anthers shorter than filaments **2. graeca**
4 Corolla scarcely exceeding calyx; anthers longer than filaments **3. taygetea**
3 Nutlets 2·5–4(–5) mm, smooth
5 Corolla red or purple, or yellow with a red or purple tinge **10. tinctoria**
5 Corolla pale yellow
6 Corolla 8–12(–15) mm; setae white, sometimes with yellow apex **10. tinctoria**
6 Corolla 12–15 mm; setae yellow or dull brown **15. arenaria**
2 Corolla 15–25 mm
7 Nutlets 4–9 mm, with 3 horns at apex **7. tricerosperma**
7 Nutlets 2–6 mm, not horned
8 Nutlets 4–6 mm, ±tuberculate
9 Corolla puberulent
10 Calyx 12–17 mm in flower, 20–25 mm in fruit; cells in tubercles of setae in 1–2 rows **4. visianii**
10 Calyx 15–20 mm in flower, 25–40 mm in fruit; cells in tubercles of setae in at least 3 rows **5. rhodopea**
9 Corolla glabrous, but sometimes shortly papillose
11 Cauline setae 1–3 mm; plant with long, slender, branched woody stock and simple flowering stems **1. frutescens**
11 Longest cauline setae 3–4 mm; plant caespitose with branched flowering stem
12 Calyx 20–30 mm in fruit; corolla 20–25 mm, not papillose **6. setosa**
12 Calyx not more than 20 mm in fruit; corolla 18–20 mm, papillose **7. tricerosperma**
8 Nutlets 2–4 mm, usually smooth
13 Lower leaves white or grey, at least beneath, with very dense, closely appressed setae
14 Calyx 6–8 mm in flower, 10–13 mm in fruit; setae on stem and calyx appressed **8. simplicissima**
14 Calyx 12–15 mm in flower, 20–25 mm in fruit; setae on stem and calyx patent **9. polyphylla**
13 Lower leaves green or grey-green with patent or semi-patent setae
15 Pedicels 5–8 mm in flower, up to 12 mm in fruit **11. propontica**
15 Pedicels 0–4 mm
16 Corolla *c.* 1⅓ times as long as calyx; flowering stems 30–60 cm, with numerous branches **15. arenaria**
16 Corolla 1½–2 times as long as calyx
17 Flowering stems 10–20 cm, simple or with 1–2 branches; calyx with bright yellow setae when dry **12. bubanii**
17 Flowering stems 20–50 cm, usually with at least 2 branches; calyx with pale yellow or brownish setae when dry
18 Upper surface of leaves densely puberulent between setae **13. fastigiata**
18 Upper surface of leaves sparsely puberulent between setae **14. vaudensis**
1 Stellate setae present, at least on the lower leaves
19 Rays of the stellate setae not more than ⅕ as long as the seta on the upper surface of the leaves, not more than ¼ as long on the lower surface
20 Corolla 10–18 mm
21 Lower leaves 1–4 mm wide
22 Caespitose; calyx (8–)9–12 mm in flower **19. helvetica**
22 Plant with lax, slender, much-branched stock; calyx 6–8 mm in flower **21. thracica**
21 Lower leaves 4–13 mm wide
23 Leaves ±glabrous between setae **17. pseudarenaria**
23 Leaves puberulent between setae
24 Corolla 12–15 mm, *c.* 1⅓ times as long as calyx **15. arenaria**
24 Corolla 15–18 mm, usually 1½–2 times as long as calyx
25 Leaves densely puberulent between setae **13. fastigiata**
25 Leaves sparsely puberulent between setae **15. arenaria**
20 Corolla 18–25 mm
26 Flowering stems simple or with not more than 4 branches at apex
27 Leaves with mainly simple setae; stellate setae few, with rays *c.* 0·1 mm
28 Biennial, with several flowering stems and usually without non-flowering rosettes; upper surface of leaves densely setose and densely puberulent between setae **13. fastigiata**
28 Perennial, usually with non-flowering rosettes; upper surface of leaves sparsely setose and sparsely puberulent between setae **14. vaudensis**
27 Leaves with mainly stellate setae and few simple setae; stellate setae with rays at least 0·2 mm
29 Lower leaves linear-oblong, 20–25 times as long as wide; corolla 18–20 mm **18. tridentina**
29 Lower leaves oblong-spathulate, 10–20 times as long as wide; corolla (18–)20–24 mm **19. helvetica**
26 Flowering stems branched, usually with at least 6 branches in the upper half
30 Setae nearly all stellate; anthers 8–10 mm **20. lucana**
30 Simple setae numerous, particularly on stem; anthers 5–8 mm
31 Upper surface of leaves glabrous between setae; usually biennial with several flowering stems **16. austriaca**
31 Upper surface of leaves sparsely to densely puberulent between setae
32 Biennial without non-flowering rosettes and usually with several flowering stems; leaves usually densely puberulent between the setae **13. fastigiata**
32 Perennial with non-flowering rosettes and usually 1–2 flowering stems; leaves sparsely puberulent between the setae
33 Corolla not more than 19 mm; bracts shorter than calyx **15. arenaria**
33 Corolla 20–24 mm; lower bracts equalling or exceeding calyx **14. vaudensis**
19 Rays of stellate setae at least ⅕ as long as the seta on the upper surface of the leaves, at least ¼ as long on the lower surface
34 Pedicels 6–14 mm in flower; corolla glabrous
35 Leaves densely setose; stellate setae with *c.* 20 rays **31. leptantha**
35 Leaves sparsely setose; stellate setae with 3–10 rays **32. stellulata**
34 Pedicels 0–6 mm in flower
36 Corolla glabrous, but sometimes papillose
37 Lower bracts exceeding calyx
38 Calyx with numerous stellate setae **29. erecta**
38 Calyx with mainly simple setae **30. taurica**
37 Lower bracts shorter than or about equalling calyx
39 Corolla *c.* 1⅓ times as long as calyx; pedicels 4–6 mm in flower **31. leptantha**
39 Corolla about twice as long as calyx; pedicels 0–3 mm in flower
40 Lower leaves *c.* 2 mm wide, narrowly linear; stem and leaves densely appressed-stellate-setose **25. elegantissima**
40 Lower leaves 3–7 mm wide, linear-spathulate or linear-oblong; stem and leaves ±patent-setose
41 Corolla not papillose; lower and upper leaves not markedly different in size **23. tornensis**
41 Corolla densely papillose; lower leaves at least twice as long as upper **27. mattirolii**
36 Corolla puberulent
42 Lower bracts exceeding calyx
43 Setae semipatent; lower leaves 3–5 mm wide, oblong- to linear-lanceolate **29. erecta**
43 Setae closely appressed; lower leaves 10–18 mm wide, lanceolate **33. spruneri**
42 Lower bracts shorter than or about equalling calyx
44 Corolla 10–18 mm

45 Lower leaves 1–3 mm wide, linear, with setae 1–2 mm on lower surface **21. thracica**
45 Lower leaves 3–8 mm wide, linear-spathulate to oblong or oblong-spathulate, with setae 0–1 mm on lower surface
46 Lower leaves linear-spathulate, 7–12 times as long as wide **23. tornensis**
46 Lower leaves oblong or oblong-spathulate, 3–6 times as long as wide **26. euboica**
44 Corolla 18–30 mm
47 Stellate setae on stem with all rays deflexed; corolla cylindrical at apex and then tapering to the base **22. echioides**
47 Stellate setae on stem with rays patent or semipatent; corolla tapering from apex to base
48 Calyx with stellate setae, the seta 1–2 times as long as rays **28. montana**
48 Calyx with simple setae, or with stellate setae and the seta at least 5 times as long as rays
49 Lower surface of leaves with the seta at least 3 times as long as rays **19. helvetica**
49 Lower surface of leaves with the seta not more than twice as long as rays, sometimes absent **24. heterophylla**

1. O. frutescens Lam., *Tabl. Encycl. Méth. Bot.* **1**: 407 (1792). Perennial with slender, branched, woody stock; flowering stems 10–25 cm, erect, simple, puberulent and patent-setose, the setae 1–3 mm. Leaves 20–70 × 4–10 mm, linear- or oblong-lanceolate, puberulent and setose. Inflorescence simple or with few short branches; pedicels 5–8 mm; bracts not exceeding calyx. Calyx 10–15 mm in flower, 12–16 mm in fruit. Corolla 16–21 mm, pale yellow tinged with purple, glabrous, *c.* 1½ times as long as calyx. Nutlets *c.* 5 mm, smooth. $2n = 14$. *C. & S. Greece, Aegean region.* ?Cr Gr.

2. O. graeca Boiss., *Diagn. Pl. Or. Nov.* **2**(11): 106 (1849). More or less caespitose biennial; stems 10–30 cm, erect, much-branched, puberulent and patent-setose, the setae 2–4 mm. Lower leaves up to 100 × 10 mm, linear or linear-lanceolate. Inflorescence much-branched; pedicels 2–5 mm; bracts about equalling calyx. Calyx *c.* 10 mm in flower, up to 15 mm in fruit. Corolla 14–15 mm, pale yellow tinged with purple, puberulent, *c.* 1½ times as long as calyx; anthers shorter than filaments. Nutlets 6–7 mm, tuberculate. $2n = 14$. *S. Greece and S. Aegean region.* Cr Gr.

3. O. taygetea Boiss. & Heldr. in Boiss., *op. cit.* 101 (1849) (*O. visianii* subsp. *taygetea* (Boiss. & Heldr.) Hayek). Like **2** but corolla white, scarcely exceeding calyx; anthers subsessile; nutlets finely and sparsely tuberculate. ● *S. Greece (Taïyetos).* Gr.

4. O. visianii G. C. Clementi, *Atti 3 Riun. Sci. Ital. Firenze* 519 (1841) (incl. *O. calycina* Steven ex Avé-Lall., *O. setosa* auct., non Ledeb.). Biennial; stems 15–60 cm, erect, much-branched, puberulent and patent-setose, the setae 2–4 mm; cells in tubercles of setae in 1–2 rows. Lower leaves 100–250 × 4–8(–12) mm, linear-lanceolate, densely setose, glabrous or sparsely puberulent between setae. Inflorescence much-branched; pedicels 1–5 mm; lower bracts about equalling calyx. Calyx 12–17 mm in flower, 20–25 mm in fruit. Corolla 15–20 mm, cream or pale yellow, puberulent, *c.* 1¼ times as long as calyx. Nutlets 4–6 mm, minutely tuberculate. $2n = 18$. *E.C. & S.E. Europe.* Al Au Bu Cz Gr Hu Ju Rm Rs (K) Tu.

5. O. rhodopea Velen., *Sitz.-Ber. Böhm. Ges. Wiss. (Math.-Nat. Kl.)* **1894**(29): 22 (1895) (*O. visianii* subsp. *rhodopea* (Velen.)

Hayek). Like **4** but setae 3–7 mm, coarse; cells in tubercles of setae in at least 3 rows; calyx 15–20 mm in flower, 25–40 mm in fruit; corolla 18–22 mm, scarcely exceeding calyx. ● *Mountains of S. Bulgaria and N. Greece.* Bu Gr.

6. O. setosa Ledeb., *Beitr. Naturk. (Dorpat)* **1**: 70 (1820). Like **4** but lower leaves 50–100 × 5–10 mm; calyx 20–30 mm in fruit; corolla 20–25 mm, glabrous. *S.E. Russia, W. Kazakhstan.* Rs (E).

7. O. tricerosperma Lag., *Gen. Sp. Nov.* 10 (1816). Perennial; flowering stems 20–40 cm, erect, much-branched, puberulent and patent-setose, the setae 2–4 mm. Lower leaves 60–150 × 4–15 mm, oblong-spathulate, patent-setose, puberulent. Inflorescence much-branched; pedicels up to 10 mm; bracts shorter than or about equalling calyx. Calyx 10–22 mm in flower, up to 32 mm in fruit. Corolla 16–23 mm, pale yellow, glabrous but minutely papillose, usually *c.* 1⅓ times as long as calyx. Nutlets 4–9 mm, rugose-tuberculate, usually with 3 horns. ● *C., S. & S.E. Spain.* Hs.

1 Nutlets without horns (c) subsp. **granatensis**
1 Nutlets with 3 horns
2 Stem usually reddish-purple; nutlets 7–9 mm (a) subsp. **tricerosperma**
2 Stem usually greenish-yellow; nutlets 4–7 mm (b) subsp. **hispanica**

(a) Subsp. **tricerosperma**: Stem usually reddish-purple; setae with large conspicuous tubercle at base; leaves green; nutlets 7–9 mm, with horns up to 5 mm. *S.E. Spain (near Alcaraz).*
(b) Subsp. **hispanica** (Degen & Hervier) P. W. Ball, *Bot. Jour. Linn. Soc.* **65**: 354 (1972) (*O. setosa* f. vel subsp. *hispanica* Degen & Hervier, *O. hispanica* (Degen & Hervier) Lacaita): Stem usually greenish-yellow; setae with small tubercle at base; leaves greygreen; nutlets 4–7 mm, with horns 1–3 mm. *Almost throughout the range of the species.*
(c) Subsp. **granatensis** (Debeaux & Degen) Stroh, *Beih. Bot. Centr.* **59B**: 432 (1939): Like subsp. (b) but nutlets without horns. *S. Spain.*

8. O. simplicissima L., *Sp. Pl.* ed. 2, 196 (1762). Perennial with branched woody stock and numerous non-flowering rosettes; flowering stems 15–30 cm, simple, appressed-setose. Lower leaves 30–50 × 2–4 mm, linear or linear-lanceolate, densely white or grey appressed-setose; upper leaves green above, white or grey appressed-setose beneath. Inflorescence simple or with few short branches; pedicels 1–4 mm; bracts shorter than calyx. Calyx 6–8 mm in flower, 10–13 mm in fruit, appressed-setose. Corolla 16–22 mm, pale yellow, glabrous but shortly papillose, 2–3 times as long as calyx. Nutlets 2–3 mm, smooth. *S.E. Russia, E. Ukraine.* Rs (C, E).

9. O. polyphylla Ledeb., *Beitr. Naturk. (Dorpat)* **1**: 72 (1820). Like **8** but stem, middle and upper leaves and calyx patent-setose; lower leaves 20–70 × 2–5 mm; pedicels up to 10 mm; calyx 12–15 mm in flower, 20–25 mm in fruit; corolla 20–24 mm, about twice as long as calyx; nutlets 3–4 mm. *Krym.* Rs (K).

10. O. tinctoria Bieb., *Beschr. Länd. Terek Casp.* 136 (1800) (incl. *O. polychroma* Klokov). More or less caespitose biennial; stems 20–70 cm, several, erect, much-branched, puberulent and patent-setose, the setae 1–3 mm. Lower leaves 30–150 × 3–15 mm, oblong or oblong-spathulate, puberulent and setose above, the setae with conspicuous tubercle at base, puberulent beneath with setae on midrib and margin. Inflorescence much-branched; pedicels 1–2 mm; bracts about equalling calyx.

Calyx 6–11 mm in flower, 12–20 mm in fruit. Corolla 8–12 (–15) mm, pale yellow, often with purple tinge, glabrous or with the lobes puberulent, *c.* $1\frac{1}{3}$ times as long as calyx. Nutlets 3–4 mm, smooth. *S. part of U.S.S.R., northwards to c. 53° N.* Rs (C, W, K, E).

11. O. propontica Aznav., *Bull. Soc. Bot. Fr.* **46**: 145 (1899). More or less caespitose perennial; flowering stems up to 50 cm, several, erect, much-branched, puberulent and patent-setose, the setae 1–3 mm. Lower leaves 40–50 × 5–7 mm, linear-spathulate, setose above, glabrous or sparsely puberulent between setae, puberulent beneath with setae on midrib and margin. Inflorescence much-branched; pedicels 5–8 mm in flower, up to 12 mm in fruit; bracts shorter than calyx. Calyx 10–15 mm in flower. Corolla *c.* 20 mm, pale yellow, glabrous but minutely papillose, $1\frac{1}{2}$–2 times as long as calyx. Nutlets 3–4 mm, smooth. *Turkey-in-Europe (W. of Istanbul)*. Tu.

12. O. bubanii Stroh, *Beih. Bot. Centr.* **59B**: 432 (1939). Perennial; flowering stems 15–20 cm, erect, simple, puberulent and patent-setose, the setae up to 2–3 mm. Lower leaves 40–50 × 3–5 mm, narrowly oblong. Inflorescence simple; pedicels 0–2 mm; bracts about equalling calyx. Calyx 11–14 mm in flower, with dense, long, yellow setae. Corolla 16–20 mm, pale yellow, glabrous, but sometimes with a few hairs on the lobes. ● *Pyrenees*. Hs.

13. O. fastigiata (Br.-Bl.) Br.-Bl. ex Lacaita, *Nuovo Gior. Bot. Ital.* nov. ser., **31**: 30 (1924). Caespitose biennial, usually with several flowering stems; stems 10–30 cm, branched, puberulent and patent-setose, the setae 1·5–2 mm. Lower leaves 50–100 × 4–8 mm, oblong, often with stellate setae 1·5–2 mm, with rays 0·1–0·2 mm. Inflorescence branched; pedicels 0–2 mm; bracts equalling calyx. Calyx 10–15 mm in flower, 16–25 mm in fruit. Corolla 16–22 mm, pale yellow, sparsely puberulent, $1\frac{1}{2}$–2 times as long as calyx. Nutlets *c.* 3 mm, smooth. ● *C. & S. France, N.W. Italy*. Ga It.

O. echioides var. *catalaunica* Sennen, from N.E. Spain, appears to be intermediate in most characters between **7(b)** and **13**. It is like **13** but with the stem up to 45 cm, with setae 2–4 mm, corolla 15–16 mm, glabrous, and nutlets 3–4·5 mm, sometimes with 2 minute lateral horns.

14. O. vaudensis Gremli, *Beitr. Fl. Schweiz* 83 (1870). Perennial with non-flowering rosettes and 1–2 flowering stems; stems 10–40 cm, usually with 2–4 branches at apex, puberulent and patent-setose, the setae 2–4 mm. Lower leaves 60–150 × 4–12 mm, narrowly oblanceolate, sparsely puberulent, setose, the setae 1·5–2 mm, sometimes stellate, with rays *c.* 0·1 mm. Inflorescence with few branches; pedicels 0–2 mm; bracts about equalling calyx. Calyx 12–15 mm in flower, up to 16 mm in fruit. Corolla 19–22 mm, pale yellow, glabrous or with hairs on the lobes or in lines from the lobes, *c.* $1\frac{1}{2}$ times as long as calyx. Anthers 6·5–8 mm, entire or slightly toothed at apex. Nutlets *c.* 3 mm, smooth. 2*n*=20. ● *W. Switzerland (Rhône valley near Aigle)*. He.

O. arenaria subsp. *pennina* Br.-Bl., *Viert. Naturf. Ges. Zürich* **62**: 606 (1917), from S. Switzerland and N. Italy, is intermediate between **14** and **15**, and may merit specific rank. It is like **14** but with flowering stems 30–50 cm, usually much branched, calyx 10–14 mm, corolla 20–24 mm, often puberulent and anthers toothed at apex.

15. O. arenaria Waldst. & Kit., *Pl. Rar. Hung.* **3**: 308 (1812). Perennial or biennial with non-flowering rosettes and usually a single flowering stem; stem 15–70 cm, with numerous branches, puberulent and semipatent-setose, the setae 2–3 mm. Lower leaves 45–180 × 4–13 mm, oblong-spathulate, sparsely to densely puberulent, setose, the setae 1–2 mm, sometimes stellate. Inflorescence branched; pedicels 0–4 mm; bracts shorter than or about equalling calyx. Calyx 6–12 mm in flower, up to 18 mm in fruit. Corolla 12–19 mm, pale yellow, glabrous or puberulent, $1\frac{1}{3}$–$1\frac{1}{2}$ times as long as the calyx. Nutlets 2·5–3 mm, smooth, shining. 2*n*=12. *S.E. & S.C. Europe.* Al Au Bu Cz Ge Hu Ju Rm Rs (C, W, E).

(a) Subsp. **arenaria** (incl. *O. transrhymnensis* Klokov): Upper surface of leaves densely puberulent; setae simple or some stellate with rays *c.* 0·1 mm; corolla 12–17 mm, *c.* $1\frac{1}{3}$ times as long as calyx. 2*n*=20. *Almost throughout the range of the species*.

(b) Subsp. **fallax** (Borbás) Jáv., *Ann. Hist.-Nat. Mus. Hung.* **4**: 430 (1906): Upper surface of leaves sparsely puberulent; some setae always stellate, with rays 0·1–0·2 mm; corolla 15–19 mm, *c.* $1\frac{1}{2}$ times as long as calyx. 2*n*=26. *N.W. part of Balkan peninsula.*

16. O. austriaca (G. Beck) Fritsch, *Excursionsfl. Österr.* 461 (1897). Like **15(a)** but usually perennial with several stems, each with 1–2 branches at apex; leaves always with some stellate setae, upper surface not puberulent between setae; corolla 18–22 mm, $1\frac{1}{2}$–2 times as long as calyx; nutlets 3–3·5 mm. 2*n*=26. ● *E. Austria.* Au ?Hu.

17. O. pseudarenaria Schur, *Verh. Mitt. Siebenb. Ver. Naturw.* **10**: 76 (1859). Biennial; stems up to 80 cm, robust, much-branched, puberulent and patent-setose, the setae 2–3 mm. Lower leaves 80–200 × 8–13 mm, linear-oblong, not puberulent but densely setose, the setae 1·5–2 mm, some always stellate, with rays 0·1–0·2 mm. Inflorescence branched; pedicels 0–2 mm; bracts shorter than calyx. Calyx 8–10 mm in flower, up to 16 mm in fruit. Corolla 12–16 mm, pale yellow, glabrous or sparsely puberulent, $1\frac{1}{3}$–$1\frac{1}{2}$ times as long as calyx. Nutlets 2·5–3 mm, smooth, shining. ● *Romania.* Rm.

18. O. tridentina Wettst. in A. Kerner, *Sched. Fl. Exsicc. Austro-Hung.* **4**: 64 (1886). Perennial with several, erect flowering stems. Stems 20–40 cm, simple or with up to 6 branches at apex, densely puberulent and setose, the setae 2–3 mm, usually stellate, with rays 0·1–0·2 mm. Lower leaves 60–130 × 3–5 mm, linear-oblong, stellate-setose, the setae 1–2 mm, with rays 0·2–0·3 mm, glabrous or sparsely puberulent between setae. Pedicels 2–4 mm; bracts shorter than calyx. Calyx 10–12 mm in flower, up to 16 mm in fruit, with simple setae. Corolla 18–20 mm, pale yellow, sparsely puberulent. Nutlets *c.* 3 mm, smooth, shining. 2*n*=28. ● *N.E. Italy (Trentino), ?N.W. Jugoslavia*. It ?Ju.

19. O. helvetica (A. DC.) Boiss., *Diagn. Pl. Or. Nov.* **2(11)**: 111 (1849). Perennial, usually with non-flowering rosettes and several erect flowering stems; stems 20–50 cm, simple or with up to 6 branches at apex, puberulent and setose, the setae 2–3 mm, usually stellate, with rays *c.* 0·1 mm. Lower leaves 30–70 × (2–)4–6 mm, oblong-spathulate, puberulent, stellate-setose, the setae 1–1·5 mm, with rays 0·1–0·3 mm. Pedicels 0–2 mm; bracts shorter than calyx. Calyx (8–)9–12 mm in flower, up to 17 mm in fruit, with stellate setae. Corolla (16–)20–24 mm, pale yellow, puberulent, $1\frac{1}{2}$–2 times as long as calyx; anthers 6–8 mm. Nutlets 2–4 mm, smooth, shining. 2*n*=26, 28. ● *S.W. Alps.* Ga He It.

20. O. lucana Lacaita, *Nuovo Gior. Bot. Ital.* nov. ser., **31**: 33 (1924). Like **19** but more robust with stems 40–75 cm, much branched; lower leaves 60–120 × 6–8 mm, the setae on the upper surface 1·5–2 mm; bracts about equalling calyx; calyx 10–11 mm in flower, up to 20 mm in fruit; corolla papillose, shortly puberulent on teeth; anthers 8–10 mm. ● *S. Italy (Basilicata).* It.

21. **O. thracica** Velen., *Sitz.-Ber. Böhm. Ges. Wiss. (Math.-Nat. Kl.)* 1895(37): 8 (1896). Perennial with slender branched stock and several erect flowering stems; stems 15–25 cm, usually simple, densely puberulent and setose, the setae 1–2 mm, usually stellate, with rays 0·1–0·2 mm. Lower leaves 10–25 × 1–3 mm, linear, stellate-setose, the setae 1–2 mm, with rays 0·1–0·3 mm. Pedicels 0–2 mm; bracts shorter than calyx. Calyx 6–8 mm in flower, with stellate setae. Corolla 14–18 mm, pale yellow, puberulent, about twice as long as calyx. Nutlets *c.* 2·5 mm, smooth, shining. ● *S. & E. Bulgaria.* Bu.

22. **O. echioides** L., *Sp. Pl.* ed. 2, 196 (1762) (*O. aucherana* subsp. *javorkae* (Simonkai) Hayek). More or less caespitose perennial with several erect flowering stems; stems 10–30 cm, simple or with few branches, puberulent and stellate-setose, the rays all deflexed. Lower leaves 20–60 × 1–7 mm, linear or linear-oblong, sparsely to densely stellate-setose, the setae 1–1·5 mm, white, grey or yellow, with rays 0·3–0·4 mm. Pedicels 0–2 mm; bracts shorter than or about equalling calyx. Calyx *c.* 10 mm in flower, up to 15 mm in fruit, with simple and some stellate setae with short rays. Corolla 18–25 mm, pale yellow, puberulent, about twice as long as calyx, cylindrical at the apex then tapering towards the base. Nutlets *c.* 2·5 mm, smooth, shining. 2*n* = 14. ● *Italy, Sicilia, W. part of Balkan peninsula.* Al It Ju Si.

Variable in leaf-width and in the density and colour of the setae. Lacaita, *Jour. Linn. Soc. London (Bot.)* 46: 387–400 (1924), gives an account of this variation.

23. **O. tornensis** Jáv., *Ann. Hist.-Nat. Mus. Hung.* 4: 431 (1906). Perennial with several erect flowering stems; stems 15–30 cm, usually simple, stellate-setose. Lower leaves 30–60 × 3–5 mm, linear-spathulate, stellate-setose, above with setae 1–2 mm and rays 0·2–0·4 mm, beneath with setae 0–1 mm and rays 0·2–0·4 mm. Pedicels 0–2 mm; bracts shorter than calyx. Calyx 6–10 mm in flower, up to 12 mm in fruit, with simple and stellate setae. Corolla 14–17 mm, pale yellow, sparsely to densely puberulent, about twice as long as calyx. Nutlets 2·5–3 mm, smooth, shining. 2*n* = 14. ● *S.E. Czechoslovakia, just extending to N.E. Hungary.* Cz Hu.

Possibly not specifically distinct from **22**.

24. **O. heterophylla** Griseb., *Spicil. Fl. Rumel.* 2: 80 (1844) (incl. *O. paradoxa* Janka, *O. tubiflora* Velen.). Perennial, caespitose or with long branched stock, with several flowering stems; stems 15–40 cm, simple or branched, puberulent and stellate-setose. Lower leaves 25–150 × 4–12 mm, oblong or linear-oblong, stellate-setose, above with setae 1–2 mm and rays 0·2–0·4 mm, beneath with setae 0–1 mm and rays 0·2–0·4 mm. Pedicels 0–2 mm; bracts usually shorter than calyx. Calyx 8–12 mm in flower, up to 15 mm in fruit, with simple and some stellate setae with short rays. Corolla 20–30 mm, pale yellow, puberulent, about twice as long as calyx, tapering from apex to base. Nutlets 2·5–3 mm, smooth, shining. *Balkan peninsula, S. Romania.* Al Bu Gr Ju Rm Tu.

O. halacsyi Hayek, *Prodr. Fl. Penins. Balcan.* 2: 88 (1928) (*O. helvetica* sensu Halácsy, non (A. DC.) Boiss.), from N. Greece, is a little-known but probably distinct species. It is a perennial with few, erect flowering stems; stems 15–20 cm, with simple setae with a ring of short hairs around the base arranged

in a stellate manner; leaves 35–50 × 4–6 mm, the indumentum on the lower surface similar to that on the stem, on the upper surface with stellate setae, the setae *c.* 3 times as long as the rays; lower bracts equalling or longer than calyx; calyx 15–18 mm, with simple setae; corolla 20–22 mm, puberulent.

25. **O. elegantissima** Rech. fil. & Goulimy, *Anzeig. Akad. Wiss. (Wien)* 94: 22 (1957). Caespitose perennial with numerous erect flowering stems; stems 20–35 cm, simple, densely appressed-stellate-setose. Lower leaves 50–120 × *c.* 2 mm, narrowly linear, densely appressed-stellate-setose, the setae 1–2 mm, with rays 0·3–0·5 mm. Pedicels 1–3 mm; lower bracts about as long as calyx. Calyx *c.* 10 mm in flower, with simple setae. Corolla *c.* 20 mm, pale yellow, glabrous, about twice as long as calyx. Nutlets *c.* 3 mm, smooth. ● *N. Greece (Vourinos).* Gr.

26. **O. euboica** Rech. fil., *Bot. Jahrb.* 80: 378 (1961). Perennial, lax, caespitose or with elongate woody stock, and several flowering stems; stems 20–30 mm, simple, puberulent and setose, the setae simple or with rays 0·1 mm. Lower leaves 12–50 × 4–8 mm, oblong or oblong-spathulate, stellate-setose, above with setae *c.* 1 mm and rays 0·3–0·4 mm, beneath with setae 0–1 mm and rays 0·3–0·4 mm. Pedicels 0–2 mm; bracts shorter than calyx. Calyx 8–9 mm in flower, up to 10 mm in fruit, with usually stellate setae. Corolla 15–16 mm, pale yellow, puberulent, *c.* 1½ times as long as calyx. Nutlets *c.* 3 mm, smooth, shining. ● *E.C. Greece (N. Evvoia).* Gr.

Plants from E.C. Greece (Pilion Oros, near Portaria) are similar to **26** but have the leaves 7–15 mm wide, oblong or oblong-elliptical and the corolla 23–25 mm, about 3 times as long as the calyx. Their taxonomic position is uncertain.

27. **O. mattirolii** Bald., *Bull. Herb. Boiss.* 3: 226 (1895). Caespitose perennial with several flowering stems; stems 10–30 cm, simple, puberulent and setose, the setae usually stellate. Lower leaves 30–110 × 4–7 mm, linear-oblong, sparsely stellate-setose, the setae 1–2 mm with rays 0·5–0·7 mm; middle and upper leaves 15–50 × 2–5 mm, not more than ½ as large as the lower. Pedicels 2–3 mm in flower; bracts about equalling calyx. Calyx 7–10 mm in flower, 12–15 mm in fruit, setae simple or with very short rays. Corolla 15–20 mm, pale yellow, glabrous but densely papillose, about twice as long as calyx. Nutlets *c.* 3 mm, smooth, shining. ● *C. Albania and N.W. Greece.* Al Gr.

28. **O. montana** Sibth. & Sm., *Fl. Graec. Prodr.* 1: 121 (1806) (*O. aucherana* auct., non DC.). Caespitose perennial with several flowering stems; stems 15–30 cm, simple or with few branches at apex, puberulent and stellate-setose. Lower leaves 30–100 × 3–8 mm, oblong-spathulate, densely stellate-setose; setae 0·7–1·5 mm with rays 0·3–0·5 mm, 0–0·5 mm beneath, with rays 0·3–0·5 mm. Pedicels 0–2 mm; bracts shorter than or about equalling calyx. Calyx 8–11 mm in flower, up to 15 mm in fruit, stellate-setose. Corolla 20–30 mm, pale yellow, puberulent, 2–3 times as long as calyx. Nutlets 2·5–3·5 mm, smooth, shining. 2*n* = 14. *S. part of Balkan peninsula and Aegean region.* Bu Gr ?Rs (K).

It is not certain whether this species occurs frequently in Europe, and many records clearly refer to **24**. The European plants referred to this species differ somewhat from those from Anatolia, and are transitional to **24**. Their taxonomic position must be considered uncertain.

O. rigida Ledeb., *Beitr. Naturk. (Dorpat)* 1: 67 (1820), from S. Ukraine, is probably not distinct from **28**. It is said to differ mainly in having the corolla 15–22 mm, 1½ times as long as the calyx.

29. O. erecta Sibth. & Sm., *Fl. Graec. Prodr.* **1**: 121 (1806) (incl. *O. laconica* Boiss. & Orph.). Caespitose perennial with several flowering stems; stems 15–25 cm, simple, densely stellate-setose. Lower leaves 30–60 × 3–5 mm, oblong or linear-lanceolate, densely stellate-setose, the setae 1–1·5 mm with rays 0·3–0·4 mm. Pedicels 1–5 mm; lower bracts exceeding calyx. Calyx 10–14 mm, densely stellate-setose. Corolla 20–24 mm, pale yellow, glabrous or puberulent, about twice as long as calyx. Nutlets 3–3·5 mm, smooth, shining. ● *S. Greece, Kriti.* Cr Gr.

The plants from Kriti have a glabrous corolla, those from S. Greece a puberulent corolla.

30. O. taurica Pallas ex Willd., *Ges. Naturf. Freunde Berlin Neue Schr.* **2**: 122 (1799). Caespitose perennial with several flowering stems; stems 10–40 cm, simple, puberulent and stellate-setose. Lower leaves 40–120 × 3–8 mm, linear-oblong, stellate-setose, above with setae 1–2 mm and rays 0·3–0·5 mm, beneath with setae *c.* 1 mm and rays 0·4–0·5 mm. Pedicels 0–2 mm; lower bracts exceeding calyx. Calyx 10–13 mm in flower, up to 18 mm in fruit, with usually simple setae. Corolla 20–30 mm, pale yellow, glabrous but often papillose, 2–3 times as long as calyx. Nutlets 3·5–4 mm, smooth, shining. 2*n* = 14. *S.E. Europe.* Bu Gr Ju Rm Rs (W, K) Tu.

31. O. leptantha Heldr., *Sched. Herb. Graec. Norm.* no. 1565 (1898). Caespitose perennial with several flowering stems; stems 10–20 cm, simple, puberulent and stellate-setose. Lower leaves 30–50 × 4–6 mm, linear-oblong, densely stellate-setose, the setae 1–1·5 mm with rays *c.* 0·5 mm. Pedicels 4–8 mm in flower, up to 10 mm in fruit; bracts shorter than calyx. Calyx 9–12 mm in flower, up to 16 mm in fruit, with usually simple setae. Corolla 13–17 mm, pale yellow, glabrous, not more than 1⅓ times as long as calyx. Nutlets smooth, shining. ● *S. Greece (Taïyetos).* Gr.

32. O. stellulata Waldst. & Kit., *Pl. Rar. Hung.* **2**: 189 (1804). Perennial with branched woody stock and several flowering stems; stems 10–25 cm, simple, puberulent and stellate-setose. Lower leaves 40–140 × 7–15 mm, oblong-spathulate, sparsely stellate-setose, the setae 1–1·5 mm with rays *c.* 0·4 mm. Pedicels 6–14 mm in flower; bracts shorter than calyx. Calyx 7–9 mm, with simple setae and sometimes with stellate setae with very short rays. Corolla 15–18 mm, pale yellow, glabrous, about twice as long as calyx. Nutlets 2·5–3 mm, smooth, shining. 2*n* = 22. ● *W. Jugoslavia.* ?Al ?Bu Ju.

Erroneously recorded from many parts of Europe.

33. O. spruneri Boiss., *Diagn. Pl. Or. Nov.* 2(11): 109 (1849). Biennial; stems 25–40 cm, simple, with dense, closely appressed stellate setae. Lower leaves 100–180 × 10–18 mm, lanceolate, densely appressed-stellate-setose, the setae 1–1·5 mm with rays *c.* 0·5 mm. Pedicels 1–6 mm; lower bracts about equalling calyx. Calyx 10–12 mm, with usually simple setae. Corolla 20–30 mm, white, puberulent, 2–3 times as long as calyx. Nutlets *c.* 5 mm, granular-tuberculate, dull. ● *C. & S. Greece.* Gr.

10. Cerinthe L.[1]

Annual, biennial or perennial herbs, glabrous or glabrescent but often with white tubercles, usually glaucous. Flowers in terminal, usually branched, bracteate cymes. Calyx lobed to half-way or more. Corolla yellow, usually tinged with violet or red, more or less tubular, with 5 erect or recurved lobes; lobes shorter than or as long as rest of corolla, without scales. Stamens included or slightly exserted, inserted at about middle of corolla; anthers appendiculate at base. Style usually exserted; stigma capitate or emarginate. Nutlets connate in 2 separate pairs, each pair ovoid, slightly beaked, dark brown or black, smooth, with a flat base.

C. tristis Teyber, *Österr. Bot. Zeitschr.* **63**: 491 (1913), was described from W. Jugoslavia (Biokovo mountains) on the basis of a non-flowering rosette of leaves; it differs from all other species in having a bluish-black margin *c.* 2 mm wide to the broadly elliptical, petiolate leaves. Its affinities are unknown.

1 Corolla-lobes lanceolate, acuminate, erect, almost as long as the rest of the corolla **1. minor**
1 Corolla-lobes ovate, sharply recurved at apex, much shorter than the rest of the corolla
2 Corolla curved above, pale yellow, violet distally **4. retorta**
2 Corolla straight, yellow or dark red, usually with red spots or ring at the throat
3 Corolla 5–8 mm wide, more than twice as long as calyx; annual **3. major**
3 Corolla 3–4 mm wide, not more than twice as long as calyx; perennial or biennial **2. glabra**

1. C. minor L., *Sp. Pl.* 137 (1753). Annual, biennial or perennial; stems 15–60 cm. Lower leaves oblong-ovate, petiolate, glabrescent; the upper ovate, sessile. Bracts cordate. Calyx-lobes obtuse, ciliate. Corolla 10–12 × 3·5–5 mm, yellow, sometimes with 5 violet spots in the throat; lobes lanceolate, acuminate, erect, almost as long as the rest of the corolla. *C. & E. Europe, northwards to c. 53° N., and extending westwards to S.E. France; sometimes casual elsewhere.* Al Au Bu ?Cr Cz Ga Ge Gr He Hu It Ju Po Rm Rs (C, W, K) Si Tu [Bl].

(a) Subsp. **minor**: Annual or biennial. Corolla cylindrical. Pedicels not scabrid. Nutlets dark brown. *Throughout most of the range of the species.*

(b) Subsp. **auriculata** (Ten.) Domac, *Bot. Jour. Linn. Soc.* **65**: 260 (1972) (*C. auriculata* Ten., *C. lamprocarpa* Murb.): Perennial. Corolla swollen. Pedicels scabrid. Nutlets black or blackish-brown. *Alps, Appennini, Sicilia and Balkan peninsula.*

Variable, especially in the colour of the leaves and flowers.

2. C. glabra Miller, *Gard. Dict.* ed. 8, no. 2 (1768). Perennial or biennial; stems 15–50 cm. Lower leaves oblong-cuneate, petiolate; the upper ovate, cordate at base, sessile. Cymes very long in fruit; bracts cordate. Calyx-lobes obtuse, glabrous or slightly ciliate. Corolla 8–13 × 3–4 mm, not more than twice as long as calyx, yellow, usually with 5 dark red spots in the throat; lobes ovate, sharply recurved at apex. Nutlets blackish. *Usually in damp or shady places; somewhat calcicole. C. & S. Europe, from the Jura and Carpathians to the Pyrenees, S. Appennini and S.W. Bulgaria.* Al Au Bu Co Cz Ga Ge He ?Hs It Ju Po Rm.

1 Leaves tuberculate; peduncles curved **(b) subsp. tenuiflora**
1 Leaves almost smooth; peduncles straight
2 Rosette-leaves persistent at flowering time; calyx-lobes lanceolate; anthers included **(a) subsp. glabra**
2 Rosette-leaves withered at flowering time; calyx-lobes ovate; anthers exserted **(c) subsp. smithiae**

(a) Subsp. **glabra** (*C. alpina* Kit.): *Throughout most of the range of the species.*

(b) Subsp. **tenuiflora** (Bertol.) Domac, *Bot. Jour. Linn. Soc.* **65**: 260 (1972) (*C. tenuiflora* Bertol.): 2*n* = 16. ● *Corse.*

(c) Subsp. **smithiae** (A. Kerner) Domac, *loc. cit.* (1972) (*C. smithiae* A. Kerner). ● *N.W. Jugoslavia.*

---

[1] By R. Domac.

3. C. major L., *Sp. Pl.* 136 (1753). Annual; stems 15–60 cm. Lower leaves obovate-spathulate, petiolate, ciliate, tuberculate; the upper ovate-lanceolate, cordate at base. Bracts ovate, subcordate at base, often reddish-glaucous, equalling or longer than calyx. Calyx-lobes acute, ciliate. Corolla 15–30 × 5–8 mm, slightly saccate, straight, yellow, with a reddish-brown ring in the throat, sometimes cream distally, or dark red throughout, more than twice as long as calyx; lobes ovate, sharply recurved at apex. Nutlets blackish. 2n = 16. *Cultivated fields, waste places and meadows. Mediterranean region, C. & S. Portugal.* Al Co Cr Ga Gr Hs It Ju Lu Sa Si Tu.

4. C. retorta Sibth. & Sm., *Fl. Graec. Prodr.* 1: 120 (1806). Annual; stems 10–50 cm. Lower leaves ovate to oblong-spathulate, petiolate, tuberculate; the upper cordate at base, sessile. Bracts ovate, cordate at base, violet, equalling calyx. Corolla 10–15 × 3–5 mm, curved above and constricted at throat, pale yellow, violet distally; lobes ovate, sharply recurved at apex. Nutlets blackish. *Rocky places. S. & W. parts of Balkan peninsula.* Al ?Cr Gr Ju.

11. Moltkia Lehm.[1]

Perennial herbs or dwarf shrubs. Flowers in terminal bracteate cymes. Calyx lobed almost to the base. Corolla blue, purple or yellow, infundibuliform, without scales or annulus. Stamens exserted or included, inserted at or above the middle of the corolla; anthers sometimes apiculate but without a projecting connective at the apex and not sagittate or apiculate at the base. Style exserted; stigma small, entire or emarginate. Nutlets often solitary by abortion, ovoid and curved, usually smooth and shining, with a ventral keel, flat at base.

1 Perennial herb; corolla 19–25 mm; anthers longer than filaments, included in corolla-tube   **3. doerfleri**
1 Dwarf shrub; corolla less than 19 mm; anthers subequalling or shorter than filaments, exserted from corolla-tube
2 Leaves 50–150 × 1–4 mm; anthers about as long as filaments; stamens not or scarcely exceeding corolla-lobes **1. suffruticosa**
2 Leaves 10–50 × 1–6 mm; anthers shorter than filaments; stamens much exceeding corolla-lobes   **2. petraea**

1. M. suffruticosa (L.) Brand in Koch, *Syn. Deutsch. Fl.* ed. 3, 3: 1999 (1902). Caespitose dwarf shrub with trailing stock, producing short, non-flowering, densely leafy stems and taller flowering stems. Flowering stems (2–)6–25 cm, erect, slender, appressed-setose, arising at the apex of the non-flowering stems of previous season. Leaves 50–150 × 1–3(–4) mm, linear, acute, sparsely setose and green above, very densely setose and whitish beneath; margin flat or more or less revolute. Cymes short and dense, grouped in corymbs. Calyx 5–6 mm, hispid, with persistent lobes. Corolla 13–16(–17) mm, gradually widened above, blue. Filaments about as long as anthers; anthers c. 3 mm, apiculate, yellow, exserted from the corolla-tube and more or less reaching the apex of the corolla-lobes. Nutlets c. 3 mm, ovoid, beaked. 2n = 16. *Rocky places; calcicole.* ● *Mountains of N. Italy.* It.

2. M. petraea (Tratt.) Griseb., *Spicil. Fl. Rumel.* 2: 515 (1846). Much-branched dwarf shrub 20–40 cm. Branches slender, rigid, erect, densely whitish-appressed-setose. Leaves 10–50 × 1–6 mm, linear to oblong-oblanceolate, obtuse or subacute, sparsely setulose and green above, very densely setulose and whitish beneath; margin more or less revolute. Cymes short, forming dense heads. Calyx c. 4 mm, with persistent

lobes. Corolla 6–10 mm, cylindric-obconical, deep violet-blue, bright pink in bud. Filaments 6·5–8 mm, somewhat exserted; anthers 1·25–2 mm, not apiculate. Nutlets 2·5–3 mm. *Rock-crevices.* ● *Mountains of Balkan peninsula, from C. Jugoslavia to C. Greece.* Al Gr Ju.

3. M. doerfleri Wettst., *Anzeig. Akad. Wiss.* (*Wien*) 55: 284 (1918). Perennial herb. Stems 30–50 cm, erect, simple, appressed-setulose, produced from a thick, horizontal, sympodial rhizome. Leaves lanceolate, acute, sparsely appressed-setulose and also ciliate, the lowest imperfectly developed. Calyx c. 10 mm; lobes falling off individually before the calyx-base. Corolla 19–25 mm, gradually widened above, deep purple. Filaments 1·5–2 mm; anthers 2·5–3 mm, biapiculate, included in the tube. Nutlets c. 4 mm. ● *Mountains of N.E. Albania.* Al.

12. Alkanna Tausch[2]

Annual or perennial herbs. Flowers in terminal, bracteate cymes. Calyx lobed almost to the base. Corolla variously coloured, with more or less cylindrical tube and infundibuliform limb, with a ring of hairs and sometimes with small invaginations in the throat. Stamens included, inserted at about the middle of the tube. Style included; stigma small, entire. Nutlets usually 2 by abortion, subreniform to obliquely ovoid and curved, granular to tuberculate, strongly stipitate.

All the species grow in dry places.

Literature: K. H. Rechinger fil., *Ann. Naturh. Mus.* (*Wien*) 68: 191–220 (1965).

1 Corolla pubescent or puberulent outside
2 Plant eglandular   **15. scardica**
2 Plant glandular
3 Corolla-tube as long as or slightly longer than calyx
4 Corolla-limb c. 6 mm in diameter   **3. calliensis**
4 Corolla-limb 8–11 mm in diameter   **5. stribrnyi**
3 Corolla-tube at least 1⅓ times as long as calyx
5 Calyx densely glandular-pubescent and sparsely setose; bracts as long as or slightly longer than calyx **6. methanaea**
5 Calyx eglandular or ± glandular but with dense setae; bracts usually twice as long as calyx **(11–14). pindicola** group
1 Corolla glabrous outside
6 Corolla blue
7 Bracts not or only slightly longer than calyx   **7. tinctoria**
7 Bracts nearly twice as long as calyx   **15. scardica**
6 Corolla yellow, orange, whitish, greyish or brownish
8 Bracts twice as long as calyx
9 Leaf-margin erose-sinuate and undulate   **1. orientalis**
9 Leaf-margin entire, flat
10 Corolla-limb 8–10 mm in diameter   **2. graeca**
10 Corolla-limb 5–7 mm in diameter
11 Annual; bracts elliptical, much longer than flowers   **9. lutea**
11 Perennial; bracts lanceolate-triangular, not or only slightly longer than flowers **(11–14). pindicola** group
8 Bracts ± equalling calyx
12 Calyx eglandular
13 Corolla-limb c. 10 mm in diameter; tube slightly longer than calyx   **10. pelia**
13 Corolla-limb c. 4 mm in diameter; tube twice as long as calyx   **16. sieberi**
12 Calyx ± glandular
14 Corolla-tube nearly twice as long as calyx **8. corcyrensis**
14 Corolla-tube as long as or slightly longer than calyx
15 Calyx 4–5 mm in flower, c. 7 mm in fruit **4. primuliflora**
15 Calyx c. 7 mm in flower, 9–15 mm in fruit
16 Nutlets 3·5 mm in diameter; corolla yellow **1. orientalis**
16 Nutlets 2 mm in diameter; corolla whitish **17. sartoriana**

1. A. orientalis (L.) Boiss., *Diagn. Pl. Or. Nov.* **1**(4): 46 (1844). Setose and glandular-pubescent perennial. Stems 30–50(–80) cm, erect or ascending. Leaves erose-sinuate and undulate; basal 10–15 × 1·5–2 cm, lanceolate to oblong-lanceolate; cauline oblong to ovate-lanceolate. Racemes dense or lax in fruit; bracts up to twice as long as calyx. Calyx 6–8 mm in flower, 10–15 mm in fruit, glandular. Corolla yellow, glabrous outside; tube up to 1⅓ times as long as calyx; limb 9–12 mm in diameter. Nutlets *c.* 3·5 mm in diameter, minutely granular, tuberculate. $2n = 28$. *Rocky places. S. Greece.* Gr. (S.W. Asia.)

2. A. graeca Boiss. & Spruner in Boiss., *op. cit.* 47 (1844). Setose-hispid perennial. Stems 15–50(–80) cm, ascending, pubescent, sometimes glandular. Leaves entire; basal 6–10 × 1–1·5 cm, linear-lanceolate to lanceolate; cauline linear-oblong to linear. Bracts narrowly triangular-lanceolate, twice as long as calyx. Calyx 5–6 mm in flower, 10–12 mm in fruit, usually eglandular. Corolla yellow, glabrous outside; tube 1–2 times as long as calyx; limb 8–10 mm in diameter. Nutlets 2·5–3·5 mm in diameter, tuberculate-reticulate. $2n = 30$. *Mountain rocks.* ● *S. part of Balkan peninsula.* Al Gr Ju.

(a) Subsp. **graeca**: Stems glandular. Calyx 6 mm in flower, 11 mm in fruit, long-pedicellate. Corolla-tube as long as calyx. Nutlets tuberculate-reticulate. *Mainly below* 1500 *m.*

(b) Subsp. **baeotica** (DC.) Nyman, *Consp.* 517 (1881) (*A. baeotica* DC.): Stems eglandular. Calyx 5–6 mm in flower, 10–12 mm in fruit, shortly pedicellate. Corolla-tube nearly twice as long as calyx. Nutlets tuberculate. *Mainly above* 1500 *m.*

Many intermediates are known, some of which may deserve subspecific rank.

3. A. calliensis Heldr. ex Boiss., *Fl. Or., Suppl.* 353 (1888). Densely glandular-pubescent and sparsely hispid perennial; stems 20–40 cm, ascending or decumbent. Basal leaves 6–10 × 0·7–1·2 cm, oblong; cauline ovate-lanceolate. Bracts lanceolate, the lower twice, the upper half as long as calyx. Calyx *c.* 6 mm in flower, *c.* 8 mm in fruit, densely glandular. Corolla yellow, pubescent outside; tube slightly exceeding calyx; limb *c.* 6 mm in diameter. Nutlets *c.* 2 mm in diameter, irregularly reticulate-tuberculate. ● *Mountains of S.C. Greece.* Gr.

4. A. primuliflora Griseb., *Spicil. Fl. Rumel.* **2**: 89 (1844). Setose-hispid and glandular-pubescent perennial; stems (5–)10–30 cm, ascending. Basal leaves 3–6 × 0·4–0·6 cm, linear-lanceolate; cauline oblong-lanceolate. Bracts as long as calyx, narrowly lanceolate. Calyx 4(–5) mm in flower, *c.* 7 mm in fruit, shortly glandular-hairy. Corolla yellow, glabrous outside; tube slightly longer than calyx; limb 5–7 mm in diameter. Nutlets *c.* 2 mm in diameter, minutely granular, irregularly tuberculate. *Dry, rocky slopes; calcifuge.* ● *Bulgaria and E. Greece.* Bu Gr.

5. A. stribrnyi Velen., *Fl. Bulg.* 647 (1891). Setose-hispid and glandular-pubescent perennial; stems 30–60 cm, ascending or suberect. Basal leaves not developed; lower cauline 3–6 × 0·5–0·9 cm, oblong-lanceolate. Bracts much longer than calyx, lanceolate. Calyx 8–9 mm in flower, 10–11 mm in fruit, glandular. Corolla deep orange, violet or greyish, hairy outside; tube shorter or slightly longer than calyx; limb 8–11 mm in diameter. Nutlets *c.* 2 mm, minutely scrobiculate-tuberculate. *Dry, stony slopes.* ● *W. & C. Bulgaria, S.E. Jugoslavia.* Bu Ju.

6. A. methanaea Hausskn., *Mitt. Thür. Bot. Ver.* **6**: 32 (1888). Setose-hispid and densely glandular-pubescent perennial; stems 10–30 cm, ascending. Basal leaves 5–12 × 0·7–1·4 cm, lanceolate; cauline oblong-lanceolate. Bracts as long as or slightly longer than calyx, linear-lanceolate. Calyx 6–7 mm in flower, 10–11 mm in fruit, densely glandular, and with sparse long setae. Corolla yellowish, hairy outside; tube at least 1½ times as long as calyx; limb 6–7 mm in diameter. Nutlets *c.* 3 mm in diameter, minutely granular, tuberculate. *Mountain rocks.* ● *S. & C. Greece.* Gr.

Many plants are known which are intermediate between **5** and **6** in presence and shape of basal leaves, indumentum of corolla and in calyx and corolla dimensions. Some of them, with large greyish to violet flowers, stout, woody stems and wider leaves are ecologically well isolated.

7. A. tinctoria (L.) Tausch, *Flora (Regensb.)* **7**: 234 (1824). Setose-hispid perennial; stems (5–)10–20(–30) cm, procumbent or ascending, eglandular. Basal leaves 6–15 × 0·7–1·5 cm, linear-lanceolate; lower cauline oblong-linear, cordate at base. Bracts not or only slightly longer than calyx, oblong-lanceolate. Calyx 4–5 mm in flower, 5–6 mm in fruit, eglandular. Corolla blue, glabrous outside; tube as long as or slightly longer than calyx; limb 6–7(–8) mm in diameter. Nutlets *c.* 2 mm in diameter, irregularly reticulate-tuberculate. $2n = 30$. *Sandy and rocky places. S. Europe, extending northwards to S.E. Czechoslovakia.* Al Bu Cr Cz Ga Gr Hs Hu It Ju Rm Sa Si Tu.

A very variable species, which needs further investigation.

8. A. corcyrensis Hayek, *Prodr. Fl. Penins. Balcan.* **2**: 71 (1928). Setose-hispid, shortly glandular-pubescent perennial; stems 25–40 cm, ascending. Basal leaves 6–8 × 1–2 cm, withering at maturity; lower cauline oblong, obtuse. Bracts as long as or slightly longer than calyx, oblong or lanceolate. Calyx 7–8 mm in flower, 8–9 mm in fruit, more or less glandular. Corolla whitish, the lobes bluish when dry, glabrous outside; tube nearly twice as long as calyx. Nutlets *c.* 2·5 mm in diameter, minutely granular-tuberculate. *Calcareous mountain rocks.* ● *S. Albania, W. Greece.* Al Gr.

9. A. lutea DC., *Prodr.* **10**: 102 (1846). Setose-hispid, glandular-pubescent annual; stems 30–80 cm, erect or ascending. Basal leaves not developed; lower cauline 2–7 × 0·7–1·2 cm, elliptical to elliptic-lanceolate. Bracts at least twice as long as calyx, elliptical to elliptic-lanceolate. Calyx 5–6 mm in flower, 6–8 mm in fruit. Corolla yellow, glabrous outside; tube nearly twice as long as calyx; limb 5–7 mm in diameter. Nutlets 2·5 mm in diameter, scrobiculate-tuberculate. $2n = 28$. *Dry, stony places.* ● *W. Mediterranean region.* Bl Co Ga Hs It ?Lu Sa.

10. A. pelia (Halácsy) Rech. fil., *Ann. Naturh. Mus. Wien* **68**: 206 (1965) (*A. stribrnyi* var. *pelia* Halácsy). Rigidly setose and glandular-pubescent perennial. Basal leaves 6–10 × 2–2·5 cm, lanceolate; cauline cordate. Bracts longer than calyx. Calyx 6–7 mm in flower, 9 mm in fruit, eglandular. Corolla white (brownish when dry), glabrous outside; tube as long as calyx; limb *c.* 10 mm in diameter. Nutlets 2 mm, minutely granular, tuberculate. *Mountain woods.* ● *C. Greece.* Gr.

(11–14). A. pindicola group. Setose-hispid, more or less glandular-pubescent perennials, woody at base. Basal leaves lanceolate to oblong-lanceolate; cauline ovate to ovate-lanceolate, cordate at base. Bracts twice as long as calyx, cordate- to lanceolate-triangular. Corolla hairy, rarely subglabrous. Nutlets minutely granular, scrobiculate-tuberculate.

1 Corolla-tube only slightly longer than calyx   **13. noneiformis**
1 Corolla-tube nearly twice as long as calyx
  2 Corolla-limb *c.* 9 mm in diameter   **14. pulmonaria**
  2 Corolla-limb 5–6 mm in diameter
    3 Bracts cordate-triangular; calyx *c.* 8 mm in flower **11. pindicola**
    3 Bracts lanceolate, acute; calyx 4–5 mm in flower **12. sandwithii**

11. A. pindicola Hausskn., *Mitt. Thür. Bot. Ver.* **6**: 32 (1888). Stems 30–50 cm, ascending. Basal leaves 10–20 × 3 cm. Lower bracts twice as long as calyx, cordate-triangular. Calyx *c.* 8 mm in flower, *c.* 11 mm in fruit. Corolla pale yellowish to brownish, sometimes with blue veins, blackish when dry, hairy outside; tube nearly twice as long as calyx; limb 5–6 mm in diameter. Nutlets *c.* 3 mm in diameter, reticulate-tuberculate. ● *Balkan peninsula, from 39° to 41° N.* Al Gr Ju.

12. A. sandwithii Rech. fil., *Ann. Naturh. Mus.* (*Wien*) **68**: 212 (1965). Stems 20–40 cm. Basal leaves 4–7 × 0·7–1·2 cm. Bracts nearly twice as long as calyx, lanceolate, acute. Calyx 4–5 mm in flower, 5–6 mm in fruit. Corolla yellow, hairy or subglabrous outside; tube nearly twice as long as calyx; limb *c.* 5 mm in diameter. Nutlets *c.* 2·5 mm in diameter, reticulate-tuberculate. *Stony limestone slopes.* ● *S.E. Albania.* Al.

13. A. noneiformis Griseb., *Spicil. Fl. Rumel.* **2**: 90 (1844). Stems 25–40 cm. Basal leaves 10–15 × 1·5 cm. Bracts slightly longer than to twice as long as calyx. Calyx 6–7 mm in flower, 10–11 mm in fruit. Corolla purplish-violet, hairy outside; tube slightly longer than calyx; limb 6–7 mm in diameter. Nutlets 2–2·5 mm in diameter, minutely granular and irregularly reticulate-tuberculate. *Rocks.* ● *S. Jugoslavia* (*Makedonija*). Ju.

14. A. pulmonaria Griseb., *loc. cit.* (1844). Stems 25–40 cm; basal leaves 10–15 × 1·5 cm; bracts slightly longer than to twice as long as calyx. Calyx 5–7 mm in flower, *c.* 15 mm in fruit. Corolla pale orange, hairy outside; tube twice as long as calyx; limb 9 mm in diameter. Nutlets 2–2·5 mm in diameter, granular-tuberculate. *Limestone rocks.* ● *S. Jugoslavia* (*S. Makedonija*). Ju.

15. A. scardica Griseb., *op. cit.* 91 (1844). Shortly setose-hispid, eglandular perennial. Stems 10–30 cm, ascending or erect. Basal leaves 4–6 × 0·8–1·2 cm, oblanceolate; cauline oblong-lanceolate. Bracts nearly twice as long as calyx, lanceolate. Calyx *c.* 5 mm in fruit. Corolla blue, glabrous or hairy outside; tube twice as long as calyx; limb 8–11 mm in diameter. Nutlets *c.* 2 mm in diameter, reticulate, minutely granular. ● *Mountains of N. Albania and S.W. Jugoslavia.* Al Ju.

16. A. sieberi DC., *Prodr.* **10**: 99 (1846). Shortly setose-hispid, eglandular, caespitose perennial. Stems 5–15 cm, erect. Basal leaves 3–4 × 0·5–0·6 cm, oblong-linear; cauline oblong-linear. Bracts as long as or slightly longer than calyx, linear-lanceolate. Calyx 4–5 mm in flower, 6–7 mm in fruit, eglandular. Corolla pale yellowish, later turning blue, glabrous outside; tube twice as long as calyx; limb *c.* 4 mm in diameter. Nutlets *c.* 2 mm in diameter, rugose-tuberculate. *Dry, often sandy soil.* ● *Kriti.* Cr.

17. A. sartoriana Boiss. & Heldr. in Boiss., *Diagn. Pl. Or. Nov.* 3(3): 134 (1856). Setose, glandular-pubescent perennial; stems 10–40 cm, ascending. Basal leaves 5–7 × 0·8–1·2 cm, oblong; cauline oblong-linear. Bracts as long as or slightly longer than calyx, lanceolate. Calyx *c.* 7 mm in flower, *c.* 9 mm in fruit, densely glandular-pubescent. Corolla whitish, yellowish when dry, glabrous outside; tube as long as calyx; limb *c.* 10 mm in diameter. Nutlets *c.* 2 mm in diameter, tuberculate. *Dry, sandy places.* ● *S. Greece* (*Navplion*). Gr.

13. Halacsya Dörfler[1]

Perennial herbs. Flowers in terminal, bracteate cymes. Calyx slightly unequally lobed almost to the base. Corolla yellow, infundibuliform, slightly unequally 5-lobed to *c.* ½ way; tube without annulus or scales, but with 5 small invaginations just below the throat. Stamens included, inserted at the throat. Style exserted; stigma small, entire. Nutlets usually solitary by abortion, ovoid, acute, keeled, rugose, with more or less flat base.

1. H. sendtneri (Boiss.) Dörfler, *Herb. Norm.* **44**: 103 (1902). Appressed-pubescent, with stout, creeping, sometimes branched stock; stems 10–25 cm, ascending to erect, simple. Basal leaves up to 10 × 0·5 cm, linear-lanceolate, narrowed into an indistinct petiole; cauline leaves *c.* 2 cm. Cymes solitary, dense in flower, lax in fruit. Calyx *c.* 5 mm in flower, up to 8 mm in fruit. Corolla *c.* 10 mm, yellow. Nutlets *c.* 3 mm. $2n = 22$. *Serpentine rocks.* ● *C. Jugoslavia, N. Albania.* Al Ju.

14. Echium L.[2]

Annual, biennial or perennial, hispid herbs or shrubs, with tubercle-based setae. Flowers in bracteate cymes, often forming panicles. Calyx lobed almost to the base, sometimes accrescent. Corolla blue, purple, yellow or white, broadly to narrowly infundibuliform, usually with an annulus of 10 minute scales or tufts of hairs, or a flange-like membrane, at the base, but without scales or invaginations above; tube straight; limb usually oblique. Stamens unequal, exserted or included, inserted below middle of corolla. Style exserted; stigma capitate or bifid. Nutlets ovoid-trigonous, erect, rugose, with flat base.

All species, except those for which another habitat is specified, grow in dry, more or less open habitats, such as sand-dunes, roadsides, cultivated fields or dry pastures.

Many taxa are very variable in habit, leaf-shape and indumentum; species limits in a number of cases are still critical.

E. **strictum** L. fil., *Suppl.* 131 (1781) (*E. ambiguum* DC.), an erect shrub with petiolate, hispid leaves with prominent lateral veins, native of Islas Canarias, is reported from Sicilia, but is not known with certainty to be naturalized. E. **candicans** L. fil., *Suppl.* 131 (1781), an erect shrub with densely grey-villous leaves with prominent lateral veins, native of Madeira, has been reported from N.W. Spain, probably as a garden escape; E. **marianum** Boiss., *Diagn. Pl. Or. Nov.* 2(11): 90 (1849) and E. **pavonianum** Boiss., *op. cit.* 91 (1849), also reported from Spain, are probably variants of this species.

Literature: G. Klotz, *Wiss. Zeitschr. Univ. Halle* (*Math.-Nat.*) **9**(3): 363–377 (1960); **11**(2): 293–302; **11**(5): 703–711; **11**(9): 1087–1103 (1962); **12**(2): 137–142 (1963). R. Fernandes, *Bol. Soc. Brot.* ser. 2, **43**: 145–158 (1969); **44**: 146–166 (1970).

1 All stamens completely included in corolla-tube
  2 Calyx 6–8 mm at anthesis, up to 15 mm in fruit with lobes 3–6 mm wide at the base   **17. parviflorum**
  2 Calyx 5–7 mm at anthesis, up to 10 mm in fruit with lobes 2–3 mm wide at the base   **18. arenarium**
1 At least 1–2 stamens exserted from corolla-tube
    3 Stigma capitate, somewhat bilobed   **9. russicum**
    3 Stigma distinctly bifid

---

[1] By D. H. Valentine and E. Mayer.   [2] By P. E. Gibbs.

4 Corolla flesh-coloured or yellowish- or bluish-white (usually brown or blue-grey when dry)
5 Inflorescence intricately branched
6 Corolla 13–18 mm; filaments pinkish-red; basal leaves usually oblanceolate, ± rounded at base, usually somewhat strigose **4. asperrimum**
6 Corolla 10–12 mm; filaments pale; basal leaves usually ± lanceolate, attenuate at base, usually not strigose **5. italicum**
5 Inflorescence ± spike-like
7 Plant usually with a single or a dominant flowering stem, with stout, asperous to stinging setae
8 Corolla 10–12 mm, usually yellowish- or bluish-white; filaments pale; flowering stem 40–100 cm **5. italicum**
8 Corolla 16–18 mm, flesh-coloured; filaments pink-carmine; flowering stem 60–250 cm **8. boissieri**
7 Plant usually with several, ± equal ascending or erect flowering stems with soft setae
9 Corolla 13–16 mm, flesh-coloured; basal leaves 40–150 mm **7. flavum**
9 Corolla 7–10 mm, bluish-white; basal leaves usually more than 250 mm **6. lusitanicum**
4 Corolla blue, reddish-purple, or pink-carmine turning blue-purple (reddish to blue or purple when dry)
10 Corolla 7–10 mm
11 Basal leaves usually more than 250 mm; perennial; stems several to many, arising from beneath basal rosette **6. lusitanicum**
11 Basal leaves usually less than 150 mm; biennial; stems 1 to several, arising from centre of basal rosette **10. vulgare**
10 Corolla 11–40 mm
12 Corolla subglabrous, hairy on veins and margins only; basal leaves usually broadly ovate to spathulate with prominent lateral veins **11. plantagineum**
12 Corolla ± uniformly hairy; basal leaves usually lanceolate to oblanceolate, without prominent lateral veins
13 Calyx 10–17 mm at anthesis, densely villous with long, white hairs and very sparse setae **1. albicans**
13 Calyx 5–10 mm at anthesis, without long, white hairs and with ± dense setae
14 Most flowers with 1–2 exserted stamens
15 Stems ascending, with underlayer of forwardly directed or irregularly patent hairs; upper cauline leaves usually oblanceolate to spathulate, attenuate or petiolate at base **12. sabulicola**
15 Stems erect, with underlayer of appressed, uniformly deflexed hairs at least in the lower third; upper cauline leaves usually narrowly elliptical, lanceolate or obovate, ± abruptly sessile
16 Calyx 7–9 mm at anthesis, enlarging to 12–19 mm in fruit; corolla 15–40 mm, broadly infundibuliform **13. creticum**
16 Calyx 5–8 mm at anthesis, not enlarging in fruit; corolla 15–25 mm, narrowly infundibuliform **14. tuberculatum**
14 Most flowers with 3–5 distinctly exserted stamens
17 Plant whitish or greyish with very dense, patent, stinging setae; leaves usually 1·5–5(–8) mm wide, linear-oblong or narrowly lanceolate to oblanceolate
18 Corolla 13–14 mm, usually bluish-purple; upper cauline leaves 1·5–3 mm wide, ± linear-oblong **3. humile**
18 Corolla 16–22 mm, usually reddish-purple; upper cauline leaves 3–5(–8) mm wide, ± narrowly lanceolate to oblanceolate **2. angustifolium**
17 Plant ± greenish with sparse to dense, appressed to patent setae; basal and lower cauline leaves usually 10–30 mm wide, ovate, lanceolate or oblanceolate
19 Stems several to many, robust, ascending, arising from a stout, ± woody stock; upper cauline leaves ovate to broadly lanceolate
20 Perennial; stems and leaves ± greenish; corolla pinkish-violet, the tube dilating gradually **16. rosulatum**
20 Biennial; stems and leaves ± greyish and pustulate; corolla clear blue to bluish-violet, the tube ± abruptly dilating at the middle **15. gaditanum**
19 Stems 1 to several, ± slender, erect, usually without a stout stock; upper cauline leaves ± narrowly lanceolate or narrowly oblong
21 Corolla 11–19 mm, usually bright blue, with 4–5 long-exserted stamens **10. vulgare**
21 Corolla 15–25 mm, usually dark bluish-purple, with 2–4 variously exserted stamens **14. tuberculatum**

1. E. albicans Lag. & Rodr., *Anal. Ci. Nat.* **5**: 269 (1802). Erect, more or less softly hairy perennial 20–75 cm, with 1 to several flowering stems; leaves and stems with a dense, short, whitish, layer of appressed hairs and more or less sparse setae. Leaves 35–70 × 4–9 mm, linear-oblong to narrowly lanceolate. Calyx 10–17 mm at anthesis, scarcely accrescent, with long, white hairs. Corolla 16–26 mm, infundibuliform, pinkish-red to bluish-purple, with 2–3 exserted stamens. *Rocks and screes; calcicole.* ● *Mountains of S. Spain.* Hs.

2. E. angustifolium Miller, *Gard. Dict.* ed. 8, no. 6 (1768) (incl. *E. diffusum* Sibth. & Sm., *E. sericeum* Vahl). Erect, hispid perennial 25–40 cm with several flowering stems; stems, leaves and calyx densely white-setose. Leaves 20–55 × 3–8 mm, narrowly lanceolate to narrowly oblanceolate. Calyx 7–10 mm at anthesis, up to 15 mm in fruit. Corolla 16–22 mm, infundibuliform, reddish or reddish-purple to purplish-violet, with 4 more or less long-exserted stamens. *Aegean region.* Cr Gr Tu.

Following Klotz, a broad view has been taken of this very variable species.

3. E. humile Desf., *Fl. Atl.* **1**: 165 (1798) (*E. angustifolium* Lam., non Miller; incl. *E. pycnanthum* Pomel). Hispid perennial 10–25 cm, with several flowering stems; stems, leaves and especially the calyx densely white-setose. Leaves 20–40 × 1·5–3 mm, linear-oblong to very narrowly oblanceolate. Calyx 6–8 mm at anthesis, not strongly accrescent. Corolla *c.* 13 mm, more or less infundibuliform, bluish-purple, with 4 more or less long-exserted stamens. *C. & S.E. Spain.* Hs.

4. E. asperrimum Lam., *Tabl. Encycl. Méth. Bot.* **1**: 412 (1792) (*E. italicum* subsp. *pyrenaicum* Rouy). Erect biennial up to 100 cm, much-branched from near the base and with an indumentum of dense, whitish-grey, patent, stinging setae. Basal leaves 120–250 × 30–50 mm, oblanceolate; cauline leaves narrowly lanceolate. Inflorescence intricately branched. Calyx 6–7 mm at anthesis, slightly accrescent. Corolla 13–18 mm, very narrowly infundibuliform, flesh-pink, with 4–5 long-exserted stamens; filaments red. 2n = 14. *W. Mediterranean region.* Bl Ga Hs It.

5. E. italicum L., *Sp. Pl.* 139 (1753). Erect, hispid biennial 40–100 cm, usually with a single or a dominant flowering stem. Basal leaves 200–350 × 15–40 mm, lanceolate, with appressed, soft setae; cauline leaves more or less narrowly elliptical. Inflorescence spike-like or much-branched and pyramidal. Calyx 6–7 mm. Corolla 10–12 mm, very narrowly infundibuliform, yellowish-, pinkish- or bluish-white, with 4–5 long-exserted stamens; filaments pale. 2n = 16, 32. *S. & S.C. Europe.* Al Au Bl Bu Co Cr Cz Ga Gr Hs Hu It Ju Rm Rs (W, K) Sa Si Tu.

Plants with a branched inflorescence approach **4** and intermediates possibly occur in S. France and Islas Baleares.

6. E. lusitanicum L., *Sp. Pl.* 140 (1753). Erect or ascending more or less softly hairy perennial with several to many flowering stems. Basal leaves 250–450 × 15–70 mm, broadly lanceolate, with appressed, soft setae; cauline leaves narrowly lanceolate. Inflorescence spike-like. Calyx 5–7 mm. Corolla 7–10 mm, infundibuliform, bluish-white to dark grey-blue, with all the stamens long-exserted; filaments red. ● *C. Portugal, W. & W.C. Spain.* Hs Lu.

(a) Subsp. **lusitanicum**: Stems shortly and more or less sparsely hairy; corolla bluish-white, very narrowly infundibuliform. *Portugal, W. Spain.*
(b) Subsp. **polycaulon** (Boiss.) P. Gibbs, *Bot. Jour. Linn. Soc.* **64**: 379 (1971) (*E. polycaulon* Boiss., *?E. salmanticum* Lag.): Stems and leaves densely hairy; corolla dark bluish-grey, more or less dilated, infundibuliform. *W.C. Spain.*

7. E. flavum Desf., *Fl. Atl.* **1**: 165 (1798) (*E. fontanesii* DC.). Erect or ascending biennial or perennial 20–80 cm, with 1 or usually several flowering stems. Basal leaves 40–150 × 8–30 mm, narrowly obovate to oblanceolate, with soft, appressed setae; cauline leaves very narrowly elliptical. Inflorescence spike-like. Calyx 5–8 mm. Corolla 13–16 mm, very narrowly infundibuliform, flesh-pink, with 5 long-exserted stamens; filaments red. *Screes and mountain pastures; calcicole. S., C. & E. Spain.* Hs.

8. E. boissieri Steudel, *Nomencl. Bot.* ed. 2, **1**: 540 (1840) (*E. pomponium* Boiss.). Large, erect, hispid biennial 60–250 cm, usually with a single or a dominant flowering stem. Basal leaves 100–250 × 40–50 mm, elliptical to lanceolate, with soft, appressed setae; cauline leaves narrowly elliptical to lanceolate. Inflorescence spike-like. Calyx 8–9 mm. Corolla 16–18 mm, very narrowly infundibuliform, flesh-pink, with 5 long-exserted stamens; filaments red. *S. Spain, Portugal.* Hs Lu.

9. E. russicum J. F. Gmelin, *Syst. Nat.* **2**: 323 (1791) (*E. rubrum* Jacq., non Forskål). Erect, hispid biennial 25–50 cm, with 1 to several flowering stems. Basal and cauline leaves 55–100 × 4–10 mm, lanceolate to narrowly elliptical, attenuate at the base, with soft, appressed setae. Inflorescence spike-like. Calyx 5–6 mm. Corolla 9–12 mm, more or less infundibuliform, dark red, with 4–5 long-exserted stamens; filaments red. $2n=24$. *Meadows and uncultivated slopes, scrub. E.C. & S.E. Europe, extending northwards to c. 55° N. in C. Russia.* Al Au Bu Cz Hu Ju Po Rm Rs (C, W, K, E).

10. E. vulgare L., *Sp. Pl.* 139 (1753). Erect, hispid biennial (or sometimes perennial?) 20–90 cm, with 1 to several flowering stems. Leaves with rather soft, appressed setae; the basal and lower cauline 50–150 × 10–20 mm, elliptical to lanceolate, attenuate and petiolate at base; the upper narrowly lanceolate, sessile. Inflorescence more or less spike-like to paniculate. Calyx 5–7 mm. Corolla 10–19 mm, blue to bluish-violet, with 4–5 long-exserted stamens. $2n=16, 32$. *Almost throughout Europe.* All except Az Bl Fa Is Rs (N) Sb; casual in some of these.

Very variable and requiring further study. Plants with more or less sparse, rigid setae sometimes arising from white tubercles have been separated as **E. pustulatum** Sibth. & Sm., *Fl. Graec. Prodr.* **1**: 125 (1806) but many intermediates occur.

11. E. plantagineum L., *Mantissa Alt.* 202 (1771) (*E. lycopsis* L. pro parte, *E. maritimum* Willd.). Erect, softly hairy, annual or biennial 20–60 cm, with 1 to many flowering stems. Basal leaves 50–140 × c. 15 mm, ovate, with prominent lateral veins

and soft appressed setae; cauline leaves oblong to lanceolate, the uppermost more or less cordate at the base. Inflorescence usually branched. Calyx 7–10 mm at anthesis, up to 15 mm in fruit. Corolla 18–30 mm, infundibuliform, blue becoming pink through purple, hairy on veins and margins only, usually with 2 exserted stamens. $2n=16$. *Roadsides, fields and sandy areas near the sea. S. & W. Europe, northwards to S.W. England; an occasional casual elsewhere.* Al Az Bl Br Bu Co Cr Ga Gr Hs It Ju Lu Rs (C, K) Sa Si Tu.

12. E. sabulicola Pomel, *Nouv. Mat. Fl. Atl.* 90 (1874) (*E. maritimum* auct., non Willd., *E. confusum* Coincy). Procumbent, ascending or erect, hispid biennial or perennial 15–50 cm, with several to many flowering stems. Basal leaves ovate to obovate; cauline leaves oblanceolate to spathulate, attenuate and subpetiolate at base, with more or less dense, white, appressed or patent setae. Inflorescence more or less branched, laxly paniculate. Calyx 6–10 mm at anthesis, up to 16 mm in fruit, densely white-setose. Corolla 12–22 mm, infundibuliform, dark blue or pinkish- to bluish-purple, usually with 1–2 stamens exserted; filaments sometimes sparsely hairy. *Roadsides, fields and sandy areas near the sea. W. Mediterranean region.* Bl Co Ga Hs It Sa Si.

13. E. creticum L., *Sp. Pl.* 139 (1753). Erect, hispid biennial 25–90 cm, with 1 to several flowering stems. Basal and lower cauline leaves 60–180 × 10–25 mm, narrowly oblanceolate, with sparse to dense, patent setae; cauline leaves usually narrowly elliptical or oblong. Inflorescence more or less branched, paniculate. Calyx 7–9 mm at anthesis, 12–19 mm in fruit. Corolla 15–40 mm, infundibuliform, persistently reddish-purple, or pink-carmine turning blue or bluish-purple, with 1–2 stamens exserted; filaments sparsely hairy or glabrous. *Roadsides and grassy slopes. W. Mediterranean region, S. Portugal.* Bl Co Ga Hs ?It Lu Sa.

1 Plant usually with a single, little-branched stem; corolla 23–35 mm, reddish-purple   **(a) subsp. creticum**
1 Plant with several, usually branched, flowering stems; corolla 15–30 mm, usually bluish-purple or bluish
2 Bracts much longer than the calyx   **(c) subsp. algarbiense**
2 Bracts shorter than or equalling the calyx   **(b) subsp. coincyanum**

(a) Subsp. **creticum**: Flowering stems usually solitary; lower leaves ovate-lanceolate with yellowish hairs. Filaments with long, straggling hairs. *S. France, N.E. Spain, Islas Baleares, Corse, Sardegna.*
(b) Subsp. **coincyanum** (Lacaita) R. Fernandes, *Bol. Soc. Brot.* ser. 2, **43**: 153 (1969): Flowering stems usually several; lower leaves oblong or oblanceolate with whitish hairs. Filaments very sparsely hairy or glabrous. *S. & E. Spain, S. Portugal.*
(c) Subsp. **algarbiense** R. Fernandes, *op. cit.* 154 (1969): Like **(b)** but more robust; leaves elliptic to ovate-oblong. Filaments glabrous. $2n=16$. *S. Portugal.*

14. E. tuberculatum Hoffmanns. & Link, *Fl. Port.* **1**: 183 (1810). Like **13** but basal leaves 50–130 × 6–15 mm; upper cauline leaves 25–110 × 4–15 mm, narrowly oblong to lanceolate; calyx enlarging up to c. 8 mm in fruit; corolla 15–25 mm, dark blue-purple, with 2–4 irregularly exserted stamens. $2n=16$. *C. & S. Portugal.* Lu.

15. E. gaditanum Boiss., *Voy. Bot. Midi Esp.* **2**: 422 (1841). Erect or ascending, roughly hispid biennial 20–75 cm, with several to many robust flowering stems arising from a more or less woody stock. Leaves 30–110 × 5–30 mm, the lowermost lanceolate to oblanceolate, attenuate at base and long-petiolate;

the uppermost ovate to broadly lanceolate, sessile. Leaves and stems with rigid setae arising from more or less prominent white tubercles. Inflorescence rather lax. Calyx 6–8 mm at anthesis, scarcely accrescent. Corolla 11–20 mm, blue to bluish-violet, with 3–4 exserted stamens. *Coastal regions of S. Spain and S. Portugal.* Hs Lu.

16. **E. rosulatum** Lange, *Ind. Sem. Horto Haun.* **1857**: 22 (1857). Erect, hispid perennial 30–70 cm, with 1 to several flowering stems. Leaves 20–80 × 5–25 mm, ovate to lanceolate, sessile, with more or less short, appressed to patent setae. Inflorescence very laxly paniculate with prominent leaf-like bracts. Calyx 6–9 mm at anthesis, 10–14 mm in fruit. Corolla 11–25 mm, usually with a narrow tube, shortly oblique to almost regular, pinkish-violet, with 3–4 exserted stamens. ● *Portugal, N.W. Spain.* Hs Lu.

17. **E. parviflorum** Moench, *Meth.* 423 (1794) (*E. calycinum* Viv.). Hispid annual or biennial 10–40 cm, with several to many ascending or erect flowering stems. Leaves with sparse to dense appressed setae; the basal 55–120 × 5–25 mm, spathulate to oblanceolate, long-petiolate; the cauline obovate or oblong, the uppermost sessile. Calyx 6–8 mm at anthesis, up to 15 mm with wide lobes in fruit. Corolla 10–13 mm, pale or dark blue, with all the stamens included. $2n = 16$. *Mediterranean region.* Al Bl Co Cr Ga Gr Hs It Ju ?Lu Sa Si.

18. **E. arenarium** Guss., *Ind. Sem. Horto Boccad.* 5 (1826). Hispid biennial 10–25 cm, with several to many ascending flowering stems. Leaves with whitish, short appressed setae; the basal 30–60 × 5–14 mm, spathulate, long-petiolate; the cauline spathulate, elliptical or oblong, the uppermost sessile. Calyx 5–7 mm at anthesis, up to 10 mm in fruit, with the lobes remaining narrow. Corolla 6–11 mm, dark blue, with all the stamens included. *Dunes and sandy fields near the sea. Mediterranean region.* Bl Cr Gr It ?Lu Sa Si [Ga].

15. Pulmonaria L.[1]

Perennial herbs with creeping rhizome and simple stems. Flowers heterostylous, in terminal, bracteate cymes. Calyx lobed to *c.* ⅓, 5-angled. Corolla red, blue or violet, infundibuliform, without scales or invaginations but with 5 tufts of hairs in the throat meeting to form a ring. Stamens included, inserted in the throat or in the middle of the tube. Style included; stigma capitate or somewhat bifid. Nutlets ovoid, erect, smooth, hairy, sometimes glabrescent, contracted at the base above a distinct collar-like ring.

In most modern Floras, the variability in the genus is attributed to hybridization between 3 or 4 "basic species", but this is unlikely for two reasons. First, it does not agree with the chromosome counts which have been made; and secondly, in this genus, true hybrids between species with different chromosome numbers are always sterile.

Important considerations in determining the species are the rosette leaves which develop during flowering and fruiting (summer leaves), the extent to which the leaves are spotted (which may vary even in a single population), the effect of dimorphic heterostyly on flower-size and stamen-insertion, and the interior of the corolla-tube, which may be glabrous or hairy below the tufts of hairs.

All species are found in deep, humus-rich soil, usually in shade, but some (**12, 13, 14**) also occur in sunny places.

[1] By H. Merxmüller and W. Sauer.

Literature: B. J. C. Dumortier, *Bull. Soc. Bot. Belg.* **7**: 6–36 (1868). A. Kerner, *Monographia Pulmonariarum.* Oeniponte. 1878. I. T. Tarnavschi, *Bul. Fac. Şti. Cernăuţi* **9**: 47–122 (1935). B. Pawłowski, *Acta Soc. Bot. Polon.* **31**: 229–238 (1962). H. Merxmüller & J. Grau, *Rev. Roum. Biol. (Sér. Bot.)* **14**: 57–63 (1969).

1 Summer leaves cordate or truncate, long-petiolate, very rough, with dense, small aculeoli
  2 Summer leaves unspotted or with faint green spots; lamina shorter than petiole   **1. obscura**
  2 Summer leaves white-spotted; lamina longer than petiole   **2. officinalis**
1 Summer leaves not cordate, narrowing abruptly or ± gradually into petiole
  3 Flowers red at maturity
    4 Leaves 1½–3 times as long as wide; lamina abruptly contracted into petiole, roughly hairy   **3. rubra**
    4 Leaves 2½–5 times as long as wide; lamina narrowing gradually into petiole, softly hairy   **4. filarszkyana**
  3 Flowers violet or blue at maturity
    5 Summer leaves soft, with dense indumentum of very short, slender hairs; setae scattered; inflorescence glandular-viscid
      6 Summer leaves usually conspicuously spotted; lamina undulate, rather abruptly contracted into petiole; the short hairs *c.* 0·1 mm   **7. vallarsae**
      6 Summer leaves unspotted; lamina not undulate, gradually narrowed into petiole; the short hairs *c.* 0·3 mm   **6. mollis**
    5 Summer leaves rough, setose, without dense indumentum of very short, slender hairs; inflorescence not viscid
      7 Summer leaves with stiff, uniform setae and without long glandular hairs
        8 Summer leaves 6–9 times as long as wide, unspotted; inflorescence with many setae and few glandular hairs   **12. angustifolia**
        8 Summer leaves 3–6 times as long as wide, usually spotted; inflorescence with equal numbers of setae and long glandular hairs   **14. kerneri**
      7 Summer leaves with both long and short setae and with scattered glandular hairs
        9 Summer leaves 6–9 times as long as wide, usually spotted; inflorescence remaining dense   **11. longifolia**
        9 Summer leaves 2–6 times as long as wide; inflorescence lax after flowering
          10 Summer leaves usually distinctly spotted
            11 Flowers bright blue at maturity   **5. stiriaca**
            11 Flowers reddish-violet or bluish-violet at maturity
              12 Lamina of summer leaves abruptly contracted into petiole; the short setae uniform, thickened at the base, erect   **8. affinis**
              12 Lamina of summer leaves ± gradually contracted into petiole; the shorter setae unequal, slender, erecto-patent   **9. saccharata**
          10 Summer leaves not, or rarely, faintly spotted; flowers blue at maturity
            13 Upper cauline leaves ovate, with wide, almost cordate base; corolla-tube hairy inside below the ring of hairs   **10. montana**
            13 Upper cauline leaves lanceolate, usually with a contracted base; corolla-tube glabrous inside below the ring of hairs   **13. visianii**

1. **P. obscura** Dumort., *Bull. Soc. Bot. Belg.* **4**: 341 (1865) (*P. officinalis* subsp. *obscura* (Dumort.) Murb.). Lamina of summer leaves up to 16 × 8 cm, cordate, unspotted or with faint, green spots; upper surface very rough, with scattered, unequal setae, occasional glandular hairs, and dense, small aculeoli; petiole up to 25 cm × 3 mm. Inflorescence with setae, glandular hairs and short hairs. Corolla reddish- to bluish-violet; corolla-tube glabrous inside below the ring of hairs. Nutlets up to

4 × 3 mm. 2n = 14. *From Belgium, E.C. Sweden and N. Russia southwards to S.E. France, Bulgaria and Krym.* Au Be Bu Cz Da Fe Ga Ge He Hu Ju Po Rm Rs (N, B, C, W, K, E) Su [Br].

2. **P. officinalis** L., *Sp. Pl.* 135 (1753) (*P. officinalis* subsp. *maculosa* (Hayne) Gams). Like 1 but lamina of summer leaves up to 16 × 10 cm, with distinct white spots and with uniform setae; petiole up to 15 cm × 4 mm; glandular hairs on the inflorescence longer and denser. 2n = (14), 16. ● *From the Netherlands and S. Sweden southwards to N. Italy and Bulgaria.* Al Au Be Bu Cz Da Ga Ge He Ho Hu It Ju Po Rm Rs (?B, C, W) Su [Br].

1 and 2 have often been treated as subspecies of one species; the sterility of the hybrids makes treatment as separate species more appropriate.

Populations at the S. margin of the Alps differ from 2 in their larger leaves and flowers and in being more glandular; they have the same chromosome number. They have been described as **P. tridentina** Evers, *Verh. Zool.-Bot. Ges. Wien* 46: 74 (1896) and may deserve subspecific status.

3. **P. rubra** Schott, *Bot. Zeit.* 9: 395 (1851). Lamina of summer leaves up to 15 × 7 cm, contracted abruptly into petiole, unspotted or rarely somewhat spotted; upper surface rough with long setae; short setae and glandular hairs also present; petiole up to 13 cm. Cauline leaves decurrent for up to 0·5 cm, the lowermost distinctly smaller than the summer-leaves. Inflorescence with long, glandular hairs and short setae. Corolla remaining red; tube hairy inside below the ring of hairs. Nutlets up to 4·5 × 3 mm. 2n = 14. ● *Carpathians and mountains of Balkan peninsula.* Al Bu Ju Rm Rs (W).

4. **P. filarszkyana** Jáv., *Bot. Közl.* 15: 51 (1916). Like 3 but lamina longer, narrowed gradually into petiole, always unspotted, with shorter and softer hairs; petiole shorter; leaf as a whole up to 30 × 6 cm; cauline leaves decurrent for 1–2 cm, the lowermost not much smaller than the summer leaves. 2n = 14. ● *E. Carpathians.* Rm Rs (W).

5. **P. stiriaca** A. Kerner, *Monogr. Pulm.* 36 (1878). Lamina of summer leaves up to 22 × 8 cm, narrowed gradually into petiole, almost always white-spotted; upper surface rough, with unequal setae and glandular hairs; petiole up to 12 cm. Inflorescence with shorter setae and long glandular hairs. Corolla bright blue; tube obscurely hairy inside below the ring of hairs. Nutlets 3 × 2 mm. 2n = 18. ● *E. Alps.* Au ?Ju.

6. **P. mollis** Wulfen ex Hornem., *Hort. Hafn.* 1: 179 (1813) (incl. *P. montana* subsp. *mollis* Gams). Summer leaves up to 60 × 12 cm, narrowed gradually into petiole, unspotted; the upper surface distinctly soft, with dense, slender, short hairs and usually scattered slender, unequal setae and glandular hairs; petiole shorter than or as long as lamina. Inflorescence densely glandular-viscid, often also with scattered setae. Corolla violet to blue-violet; tube densely hairy inside below the ring of hairs. Nutlets 5 × 4 mm. 2n = 18. *C. & S.E. Europe.* Al Au Bu Cz ?Ga Ge Gr He Hu ?It Ju Po Rm Rs (C, W, E).

The taxonomy, nomenclature and distribution of this widely distributed complex are not yet clear; the characters of the many taxa that have been described are difficult to define. The very softly hairy **P. mollissima** A. Kerner, *Monogr. Pulm.* 47 (1878) (*P. mollis* subsp. *mollissima* (A. Kerner) Nyman, *P. media* Host pro parte), which occurs mainly in E.C. & E. Europe, as well as **P. auriculata** (Boiss.) Halácsy, *Consp. Fl.*

Graec. 2: 331 (1902), from a few isolated localities in Greece, are not clearly distinguishable from the type from S. Germany. In the mountains of E. & C. Europe, more hispid and less softly hairy plants are found (*P. montana* auct., non Lej.); the most characteristic of these, from the N. Alps, also have 2n = 18. The position of the plants from E. Europe with blue flowers or narrow leaves or both is also obscure; these have been described as **P. montana** subsp. **porciusii** Guşuleac in Săvul., *Fl. Rep. Pop. Române* 7: 643 (1960) (*P. dacica* Simonkai pro parte) and var. *pseudoangustifolia* Guşuleac. A chromosome number of 2n = 14 has been reported for similar plants from the U.S.S.R., and of 2n = 28 for a plant from Romania.

7. **P. vallarsae** A. Kerner, *Monogr. Pulm.* 33 (1878). Lamina of summer leaves up to 20 × 10 cm, usually undulate, rather abruptly narrowed into petiole, often with distinct bright green or whitish, often confluent spots, rarely almost unspotted; surface very soft, with very dense, short, fine hairs, with a few long setae and glandular hairs. Petiole up to 18 cm, broadly winged above. Inflorescence densely glandular-hairy, viscid, with setae and also short hairs. Corolla violet; tube hairy inside below the ring of hairs. Nutlets 4·5 × 3·5 mm. 2n = 22. ● *Italy.* It.

8. **P. affinis** Jordan in F. W. Schultz, *Arch. Fl. Fr. Allem.* 321 (1854). Lamina of summer leaves up to 18 × 9 cm, abruptly narrowed into petiole, with white spots, very rough, with very short broad-based setae, also with scattered longer setae and glandular hairs. Petiole up to 18 cm, narrowly winged. Inflorescence with long glandular hairs and unequal setae. Corolla violet to blue-violet; tube glabrous inside below the ring of hairs. Nutlets up to 4 × 2·5 mm. 2n = 22. ● *C. & W. France, N. Spain.* Ga Hs.

P. alpestris Lamotte, *Prodr. Fl. Centr. Fr.* 535 (1881) from S.C. France has 2n = 22 but differs from 8 in its more or less unspotted summer leaves and red-violet corolla; its status awaits investigation.

9. **P. saccharata** Miller, *Gard. Dict.* ed. 8, no. 3 (1768) (*P. tuberosa* prol. *confusa* Rouy, *P. picta* Rouy). Lamina of the summer leaves up to 27 × 10 cm, narrowing rather gradually into the petiole, usually with conspicuous and confluent white spots, with more or less dense, unequal and rather fine short hairs, as well as long, rather slender setae and glandular hairs. Corolla red-violet to dark violet; tube hairy inside below the ring of hairs. Nutlets 4 × 3 mm. 2n = 22. ● *S.E. France, N. & C. Appennini.* Ga It [Be].

This species appears to be connected by transitional forms with 7 and 8 in the zones of overlap. It differs from 7 and 8 especially in the indumentum and in the size of the leaves. As the three species have the same chromosome number, introgression is a possibility, but this has not been demonstrated experimentally.

10. **P. montana** Lej., *Fl. Spa* 1: 98 (1811) (*P. tuberosa* auct., non Schrank, *P. vulgaris* Mérat pro parte, *P. angustifolia* subsp. *tuberosa* Gams pro parte). Summer leaves up to 50 × 12·5 cm, elongate-lanceolate, gradually narrowing to the base, unspotted or rarely with indistinct green spots; the upper surface with unequal setae and scattered glandular hairs. Cauline leaves cordate, ovate, acute. Inflorescence with many setae and long glandular hairs. Corolla blue; tube hairy inside below the ring of hairs. Nutlets 4·5 × 3·5 mm. 2n = 22. ● *W. & W.C. Europe.* Be Ga Ge He †Ho.

A misunderstood species whose area is not yet clear because of confusion with **6, 9, 11** and **13**. Especially uncertain is the status of plants from S. Germany (whence *P. tuberosa* was described) which have been found to have $2n=18$; similar plants seem to occur in E. France and C. Germany.

11. P. longifolia (Bast.) Boreau, *Fl. Centre Fr.* ed. 3, **2**: 460 (1857) (*P. vulgaris* Mérat pro parte, *P. angustifolia* auct., non L.). Summer leaves up to 50×6 cm, narrowly lanceolate, very gradually narrowed to the base, usually with white spots, rarely (especially in the far west) unspotted or with faint green spots; the upper surface with more or less equal setae and few glands, the lower surface with unequal setae and more glandular. Inflorescence remaining dense throughout the flowering period, with long setae and a few long glandular hairs. Corolla violet to blue-violet (often almost blue in the far west); tube glabrous inside below the ring of hairs. Nutlets 4×3 mm. $2n=14$. ● *W. Europe, northwards to England.* Br Ga Hs Lu.

In W. & C. France, unspotted blue-flowered plants with $2n=14$ are found; they have usually been confused with **12**. Special taxonomic treatment may be necessary.

12. P. angustifolia L., *Sp. Pl.* 135 (1753) (*P. azurea* Besser, *P. angustifolia* subsp. *azurea* (Besser) Gams). Summer leaves up to 40×5 cm, narrowly lanceolate, very gradually narrowed to the base, unspotted, with equal setae, more or less eglandular. Cauline leaves narrowly lanceolate, sessile, with narrow base. Inflorescence with setae and scattered glands. Calyx in fruit very slender and short. Corolla bright blue; tube glabrous inside below the ring of hairs. Nutlets 4.5×3.5 mm. $2n=14$. N.E., E. & E.C. *Europe; isolated stations in W.C. Europe and the S.W. Alps.* Au ?Bu Cz Da Ga Ge Hu It ?Ju Po Rs (B, C, W, E) Su.

Many records of this species from S.E. Europe are of narrow-leaved variants of **6**; records from the E. Alps and neighbouring territory to the south-east are referable to **13**.

13. P. visianii Degen & Lengyel in Degen, *Fl. Veleb.* **2**: 569 (1937) (*P. media* Host pro parte, *P. angustifolia* auct., non L.). Like **12** but summer leaves up to 30×6 cm, unspotted or rarely with indistinct green spots, with unequal setae and scattered glands; inflorescence hispid and glandular; fruiting calyx stout; corolla deep blue-violet to blue; nutlets 3.5×3 mm. $2n=20$. ● *C. & E. Alps, mountains of N.W. Jugoslavia.* Au ?Ga He It Ju.

14. P. kerneri Wettst., *Verh. Zool.-Bot. Ges. Wien* **38**: 559 (1888). Summer leaves up to 60×10 cm, lanceolate, very gradually narrowed to the base, thick, coriaceous, usually with distinct white spots, rarely more or less unspotted, with stiff setae and scattered short glands. Cauline leaves lanceolate, sessile, with broad base. Corolla bright blue, tube glabrous inside below the ring of hairs. Nutlets 5×3 mm. $2n=26$. ● *N.E. Alps.* Au.

16. Nonea Medicus[1]

Annual or perennial, hispid herbs. Flowers in bracteate, terminal cymes. Calyx lobed to $\frac{1}{4}-\frac{2}{3}$, accrescent. Corolla yellow, purple, brown or white, with cylindrical tube and infundibuliform limb, with 5 hairy or laciniate scales, or 5 tufts of hairs, or a ring of hairs in the throat. Stamens included in corolla-tube, inserted at about or above the middle of the tube. Style included; stigma entire to bifid. Nutlets ovoid to reniform, erect or oblique, smooth or ribbed, with a thickened collar-like ring at the base.

All are plants of dry, often waste places.

---

[1] By A. O. Chater.

1 Nutlets longer than wide
  2 Corolla yellow; nutlets rugose and reticulately ribbed   **1. lutea**
  2 Corolla blue, brown or purplish; nutlets smooth or weakly
      longitudinally ribbed   **2. obtusifolia**
1 Nutlets wider than long
  3 Calyx-teeth at least $\frac{3}{4}$ as long as tube
    4 Flowers brownish-purple; nutlets constricted just above the
      base   **9. vesicaria**
    4 Flowers pale purplish, yellowish or white; nutlets not constricted above the base
      5 Corolla purplish; leaves mostly less than 1 cm wide **5. caspica**
      5 Corolla yellowish or white; leaves mostly more than 1 cm
        wide   **4. pallens**
  3 Calyx-teeth c. $\frac{1}{3}$ as long as tube
    6 Nutlets ±symmetrical, reniform   **7. ventricosa**
    6 Nutlets asymmetrical, obliquely ovoid
      7 Limb of corolla more than 9 mm in diameter **6. versicolor**
      7 Limb of corolla less than 9 mm in diameter
        8 Corolla (9–)10–14 mm, usually dark reddish- or blackish-
          purple; nutlets $2.5–3 \times 3.5–5$ mm   **3. pulla**
        8 Corolla 7–8 mm, pale blue or yellowish; nutlets c. 1.5×3
          mm   **8. micrantha**

1. N. lutea (Desr.) DC. in Lam. & DC., *Fl. Fr.* ed. 3, **3**: 626 (1805). Setose-hispid and glandular-pubescent annual; stems 10–60 cm, erect or ascending, branched. Leaves $2–7 \times 0.5–2$ cm, obtuse or acute, entire or dentate. Calyx 6–10 mm at anthesis, 10–20 mm in fruit; teeth $\frac{1}{2}$ as long to almost as long as tube. Corolla 7–12 mm; limb 5–15 mm in diameter, pale yellow, divided to c. $\frac{1}{2}$ into ovate-orbicular lobes. Nutlets 3.5–6 mm, oblong-ellipsoid, rugose, ribbed; collar-like basal ring smooth. $2n=14$. *Ukraine and S.E. Russia; a frequent casual in parts of C. & S.E. Europe and locally naturalized.* Rs (W, E) [Hu Ju Rm].

2. N. obtusifolia (Willd.) DC., *loc. cit.* (1805). Shortly setose-hispid, eglandular annual; stems 5–15 cm, ascending, usually simple. Leaves $3–5 \times c. 1$ cm, oblong to ovate-oblong, subentire, subobtuse. Calyx c. 5 mm at anthesis, 8–10 mm in fruit; teeth c. $\frac{1}{2}$ as long as tube. Corolla c. 8 mm; limb 3–6 mm in diameter, blue, divided to c. $\frac{1}{2}$ into ovate-orbicular lobes. Nutlets c. 3 mm, oblong-ovoid, erect, smooth, shiny, blackish; collar-like basal ring smooth. *Aegean region and S.E. Bulgaria; naturalized in C. Italy.* Bu Gr [It].

N. rosea (Bieb.) Link, *Enum. Hort. Berol. Alt.* **1**: 167 (1821), from the Caucasus, with purplish or brownish corolla 15–18 mm long and often slightly ribbed nutlets, is a frequent casual in parts of W. & C. Europe.

3. N. pulla (L.) DC. in Lam. & DC., *Fl. Fr.* ed. 3, **3**: 626 (1805). Grey-setose and more or less glandular-pubescent annual to perennial; stems 25–50 cm, erect, usually branched above. Leaves $3–12 \times 0.7–2$ cm, lanceolate to linear-lanceolate, acute, entire. Calyx 6–8 mm at anthesis, 9–12 mm in fruit; teeth c. $\frac{1}{3}$ as long as tube. Corolla (9–)10–14 mm; limb 5–8 mm in diameter, dark reddish- or blackish-purple (rarely yellowish-white), divided to c. $\frac{1}{4}$ into broadly ovate lobes. Nutlets $2.5–3 \times 3.5–5$ mm, obliquely ovoid, ribbed; collar-like basal ring ribbed. $2n=18, 20$. *E. & E.C. Europe; locally naturalized further west and progressively extending its area westwards.* Au Bu Cz Ge Gr Hu *Ju Po Rm Rs (*N, *B, C, W, K, E) [Fe Ga He].

Annual or biennial plants with long hairs on the pedicels, linear-lanceolate (not triangular-lanceolate) calyx-teeth, eglandular leaves and small nutlets have been called **N. atra** Griseb., *Spicil. Fl. Rumel.* **2**: 94 (1844), but there is poor correlation of these characters; such plants do not seem to merit formal taxonomic status.

4. N. pallens Petrović, *Add. Fl. Agri Nyss.* 129 (1885). Sparsely setose and glandular-pubescent annual; stems 10–40 cm, erect, branched. Leaves 4–9 × 1–2 cm, linear-lanceolate to elliptical, acute, entire. Cymes long. Calyx 5–8 mm at anthesis, 10–15 mm in fruit; teeth *c.* 1½ times as long as tube. Corolla *c.* 8 mm; limb 4–5 mm in diameter, pale yellow or white, divided to ⅓–¼ into obovate lobes. Nutlets *c.* 2·5 × 4 mm, obliquely ovoid, rugose, ribbed; collar-like basal ring strongly ribbed. ● *S.E. Europe, from Albania to S. Ukraine.* Al Bu Ju Rm Rs (W).

5. N. caspica (Willd.) G. Don fil., *Gen. Syst.* 4: 336 (1837). Weakly setose and shortly glandular-pubescent annual; stems 5–30 cm, usually erect, branched. Leaves 2–10 × 0·5–1 cm, lanceolate to linear-lanceolate, acute, weakly dentate below, usually entire above. Calyx 6–8 mm at anthesis, 7–12 mm in fruit; teeth as long as tube. Corolla 8–12 mm; limb 4–5 mm in diameter, purplish, divided to *c.* ⅕ into lobes which are much wider than long. Nutlets *c.* 2 × 3–4 mm, obliquely ovoid, ribbed; collar-like basal ring ribbed. *S.E. Russia, W. Kazakhstan.* Rs (E).

6. N. versicolor (Steven) Sweet, *Hort. Brit.* 292 (1827). Hispid annual or perennial, glandular-pubescent above; stems 20–40 cm, simple or branched above. Leaves 3–8 × 1–2·5 cm, lanceolate to linear-lanceolate, acute, usually entire. Calyx 6–8 mm at anthesis, 7–17 mm in fruit; teeth *c.* ⅓ as long as tube. Corolla 12–17 mm; limb 10–15 mm in diameter, purplish or violet, divided to *c.* ⅓ into lobes which are wider than long. Nutlets *c.* 3 × 4 mm, obliquely ovoid, ribbed; collar-like basal ring slightly ribbed. *Naturalized in S. Fennoscandia.* [Da No Su.] (*Caucasus and E. Anatolia.*)

7. N. ventricosa (Sibth. & Sm.) Griseb., *Spicil. Fl. Rumel.* 2: 93 (1844) (incl. *N. alba* DC.). Hispid, eglandular annual; stems 10–40 cm, ascending, little-branched. Leaves 2–7 × 0·5–1 cm, lanceolate to oblong-lanceolate, acute to subobtuse. Calyx *c.* 5 mm at anthesis, 8–12 mm and often very strongly inflated in fruit; teeth *c.* ⅓ as long as tube, often connivent in fruit. Corolla 7–8 mm; limb 4–5 mm in diameter, pale yellowish or white, divided to *c.* ⅓ into lobes which are wider than long. Nutlets 1·5–2 × 2·5–3·5 mm, reniform, symmetrical, rugose, ribbed, not constricted above the ribbed collar-like basal ring. *S. Europe.* Bu Ga Gr Hs It Ju Rs (K) Tu.

8. N. micrantha Boiss. & Reuter, *Diagn. Pl. Nov. Hisp.* 21 (1842). Like **7** but corolla-limb pale blue or yellowish, more deeply divided; nutlets *c.* 1·5 × 3 mm, obliquely ovoid, asymmetrical, strongly ribbed. *E., C. & S. Spain.* Hs.

9. N. vesicaria (L.) Reichenb., *Fl. Germ. Excurs.* 338 (1831) (*N. nigricans* (Desf.) DC.). Setose-hispid and shortly glandular-pubescent annual or biennial; stems 15–50 cm, ascending, little-branched. Leaves 3–20 × 0·5–2·5 cm, lanceolate to linear-lanceolate, subacute, sometimes dentate below. Calyx 5–7 mm at anthesis, 10–15 mm in fruit; teeth as long as tube. Corolla 8–12 mm; limb 3–5 mm in diameter, brownish-purple, divided to *c.* ⅓ into lobes which are wider than long. Nutlets 2·5–3·5 × 3·5–4·5 mm, obliquely ovoid, strongly ribbed, constricted just above the strongly ribbed collar-like basal ring. *S.W. Europe, Sicilia.* Bl Hs Lu Si.

17. Elizaldia Willk.[1]

Like *Nonea* but corolla with a flange-like ring in the throat, without scales; stamens exserted from tube and reaching apex of corolla, inserted at apex of tube.

1. E. calycina (Roemer & Schultes) Maire, *Bull. Soc. Hist. Nat. Afr. Nord* 20: 192 (1929) (*E. nonneoides* Willk.). Densely setose-hispid annual; stems *c.* 30 cm, ascending, branched at the base. Leaves 3–7 × 0·5–1·5 cm, linear-oblong or -lanceolate, acute, subentire. Flowers in dense, leafy cymes; pedicels 1·5–3 mm. Calyx *c.* 8 mm in flower, 12–16 mm and violet in fruit; teeth linear-lanceolate, 1½–2 times as long as tube. Corolla 10–12 mm; limb *c.* 5 mm in diameter, violet, with semicircular lobes. Nutlets *c.* 3 × 5 mm, obliquely ovoid, longitudinally and transversely ribbed, keeled, deeply excavate at base with a peg-like attachment; collar-like basal ring strongly ribbed. *Under bushes on maritime sands. S.W. Spain (near Cádiz).* Hs. (*N.W. Africa.*)

Not found recently, and perhaps extinct in Europe; the European plant is subsp. **multicolor** (G. Kunze) Chater, *Bot. Jour. Linn. Soc.* 64: 69 (1971) (*Nonea multicolor* G. Kunze) which also occurs, with the typical subspecies, in N.W. Africa.

18. Symphytum L.[2]

Perennial, usually hispid herbs. Flowers in short ebracteate terminal cymes. Calyx lobed to ¼ or almost to the base, accrescent. Corolla variously coloured, with cylindrical tube and tubular-campanulate limb, with triangular to semicircular lobes much shorter than the rest of the corolla; throat with 5 long scales, usually with marginal papillae. Stamens included, inserted at about the middle of the tube; filaments not more than 1½ times as long as anthers. Style exserted; stigma very small, entire. Nutlets ovoid, erect, sometimes curved, usually verruculose and often rugose, concave at the base with a thickened collar-like ring.

The number of the cauline leaves given in the descriptions does not include the pairs of leaves on the inflorescence-branches. The length of the filaments refers only to the part not concealed by the anther.

Literature: C. Bucknall, *Jour. Linn. Soc. London (Bot.)* 41: 491–556 (1913). B. Pawłowski, *Fragm. Fl. Geobot.* 7: 327–356 (1961); *op. cit.* 17: 17–37 (1971). G. E. Wickens, *Notes Roy. Bot. Gard. Edinb.* 29: 157–180 (1969).

1 Nutlets smooth, shining; connective projecting beyond thecae; filament about as wide as anther   **1. officinale**
1 Nutlets minutely verrucose and ± reticulate-rugose; connective usually not projecting beyond thecae; filament narrower than anther
   2 Scales of corolla exserted
    3 Rhizome creeping, with tubers; corolla (excluding scales) (7–)8–11(–12) mm   **13. bulbosum**
    3 Stock fusiform; corolla (excluding scales) 5–7 mm   **14. ottomanum**
   2 Scales of corolla included
    4 Plant aculeolate at least on stems and on mid-vein of the leaves beneath
     5 Leaves all petiolate, or the upper subsessile but not amplexicaul or decurrent   **2. asperum**
     5 Upper leaves sessile and shortly decurrent or at least amplexicaul   **3. × uplandicum**
    4 Plant pubescent to rigidly hispid, but not aculeolate
     6 Scales of corolla lingulate, not or slightly widened in the lower part, rounded or ± emarginate at apex, the marginal papillae dense above, sparse below
      7 Rhizome creeping, slender, producing ascending or procumbent non-flowering and flowering stems   **8. ibiricum**
      7 Rhizome not creeping, or stock fusiform; non-flowering stems absent
       8 Calyx lobed to ¼–⅖   **12. orientale**
       8 Calyx lobed to at least ⅗

¹ By A. O. Chater.   ² By B. Pawłowski.

9 Leaves shallowly cordate, rounded or truncate at base; corolla pale yellow; plant asperous **11. tauricum**
9 Leaves mostly ±narrowed at base; corolla white or white tinged with pink; plant ±softly hispid
10 Corolla (17–)18–22 mm; scales 5–6 mm **9. davisii**
10 Corolla 13–16·5 mm; scales 1·6–4 mm **10. cycladense**
6 Scales of corolla lanceolate, gradually narrowed from the dilated base to the ±acute apex, the marginal papillae dense throughout
11 Corolla violet, or pink turning blue **3. × uplandicum**
11 Corolla pale yellow or white
12 Corolla white; stock fusiform **7. naxicola**
12 Corolla pale yellow; rhizome creeping
13 Lower leaves cordate; rhizome of ±even thickness throughout **6. cordatum**
13 Leaves not cordate; rhizome with alternate thick tuberous and thin portions
14 Connective not projecting beyond thecae **4. tuberosum**
14 Connective projecting slightly beyond thecae **5. gussonei**

1. S. officinale L., *Sp. Pl.* 136 (1753). Stock stout, vertical, branched. Stem (30–)50–120 cm, stout, erect, often branched. Leaves large, ovate-lanceolate to lanceolate, acuminate; the middle and upper sessile, often decurrent. Cymes many-flowered. Calyx lobed to $\frac{3}{4}$–$\frac{4}{5}$, with lanceolate lobes. Corolla 12–18 mm, purple-violet or dirty pink or white, with deflexed lobes. Scales broadly triangular-lanceolate, the lower marginal papillae shortly cylindric-conical, the upper much smaller and shorter, all dense. Stamens with connective projecting beyond thecae; filaments as wide as anther. Nutlets 5–6 mm, black, very smooth, shining. $2n=24$, $24+4B$, 26, *c.* 36, 40, 48, 54. *River-banks and damp grassland. Much of Europe, but rare in the extreme south and only as a naturalized alien in much of the north.* Au Be Br Bu Cz Ga Ge He Ho Hs Hu It Ju ?Lu Po Rm Rs (C, W, K, E) Sa Si Tu [Da Fe Hb No Rs (N, B) Su].

(a) Subsp. **officinale**: Middle and upper leaves deeply and broadly decurrent, and the stem distinctly winged. Stem, leaves, pedicels and calyces more or less densely hairy and setose. *Throughout the range of the species.*

(b) Subsp. **uliginosum** (A. Kerner) Nyman, *Consp.* 509 (1881): Leaves, even the uppermost, not or shortly and narrowly decurrent. Stem and leaves only sparsely covered with very short, stiff setae and densely verrucose-hispid. Sepals with stiff, apical setae, otherwise more or less glabrous. ● *E.C. Europe.*

S. floribundum R. J. Shuttlew. ex Buckn., *Jour. Linn. Soc. London* (*Bot.*) **41**: 531 (1913) (*S. mediterraneum* F. W. Schultz, non Koch), an asperous-hispid plant with the calyx lobed to $\frac{1}{4}$–$\frac{1}{2}$, dirty white corolla with the lobes not recurved, scales with all the marginal papillae similar, about $1\frac{1}{2}$ times as long as wide, the connective not projecting beyond the thecae and abortive pollen, is probably a hybrid of **1** and some other species; it was known from S. France (Var), but has not been found for 75 years.

2. S. asperum Lepechin, *Nova Acta Acad. Sci. Petrop.* **14**: 442 (1805). Stock vertical. Stem up to 180 cm, branched, aculeolate and with scattered hairs. Leaves ovate to oblong, cuneate to rounded or subcordate at base, petiolate or the uppermost sessile, not decurrent and not amplexicaul, densely setose and aculeolate at least on the mid-vein beneath. Calyx 3–5 mm, lobed to *c.* $\frac{2}{3}$–$\frac{3}{4}$, the lobes rounded. Corolla 11–17 mm, pink at first, turning blue. Scales lingulate, slightly widened at the base; marginal papillae dense, narrowly cylindric-conical. Stamens with connective not projecting beyond thecae; filament narrower than and about as long as anther. Nutlets reticulate-rugose and finely verrucose. *Formerly cultivated for fodder, and naturalized in several parts of Europe.* [Au Be Br Da Fe Ga He No Rs (B, C, W) Su.] (*S.W. Asia.*)

3. S. × uplandicum Nyman, *Syll.* 80 (1854) (*S. asperum* × *officinale*). Intermediate between the parents. Stem up to 2 m. Leaves never cordate, the upper sessile, shortly decurrent or more or less amplexicaul. Calyx 5–7 mm; lobes usually acute. Corolla 12–18 mm, pink at first, turning blue, or persistently violet. Scales as in **1** but less wide at the base. $2n=36$. *Formerly cultivated for fodder, and naturalized; also of spontaneous local origin.* [Au Be Br Da Fe Ga Hb Ho No Su.]

4. S. tuberosum L., *Sp. Pl.* 136 (1753). Rhizome creeping, with alternate thick tuberous and thin portions. Stem (10–)15–40(–60) cm, simple or little-branched; stem and leaves more or less densely hairy; hairs up to 1·5 mm. Basal leaves (which disappear at flowering time) long-petiolate; the lower cauline leaves elliptical to lanceolate, the upper sessile, shortly and narrowly decurrent. Calyx 5–8 mm, lobed to $\frac{1}{3}$–$\frac{9}{10}$. Corolla 13–19 mm, pale yellow; lobes with deflexed apex. Scales triangular-lanceolate; marginal papillae dense, the lower not more than $1\frac{1}{2}$ times as long as wide, the upper very small. Anthers 3–4·5(–5) mm; connective not projecting beyond thecae. *Woods and other damp or shady places. W., C. & S. Europe, northwards to England, C. Germany and N. Ukraine.* Al Au Bl Br Bu Co Cz Ga Ge ?Gr He Hs Hu It Ju Po Rm Rs (C, W) Tu.

(a) Subsp. **tuberosum**: Rhizome stout, up to 12 mm in diameter, the tuberous portions of irregular shape and close together. Cauline leaves usually 6–12. Inflorescence with 8–16(–40) flowers. Corolla-scales 5·5–7·5 mm, not more than 2 mm wide. Stamens with filament $\frac{1}{3}$–$\frac{1}{2}$ as long as anther. Nutlets up to 4 mm. $2n=144$. ● *W. Europe.*

(b) Subsp. **nodosum** (Schur) Soó, *Acta Geobot. Hung.* **4**: 192 (1941): Rhizome slender, the tuberous portions irregular-oblong and separated. Cauline leaves usually 3–7. Inflorescence mostly with 1–9(–20) flowers. Corolla-scales 4·5–6(–6·5) mm, up to 2·5 mm wide. Stamens with filament $\frac{1}{4}$–$\frac{1}{3}$(–$\frac{1}{2}$) as long as anther. Nutlets 2·5–3·5 mm. $2n=18$, 72, 96, 100. *C. & S.E. Europe.*

5. S. gussonei F. W. Schultz, *Arch. Fl. Eur.* 58 (1874). Like **4** but rhizome very slender, the tuberous portions more or less globose and very distant; stem slender; cauline leaves 3–8; inflorescence with 4–10(–20) flowers; corolla 16–22 mm; scales 6–8·5 mm, up to 2 mm wide; stamens with filament $\frac{1}{5}$–$\frac{1}{3}$ as long as anther; anthers (3·5–)4–5(–6) mm; connective projecting slightly beyond thecae. ● *Sicilia.* Si.

6. S. cordatum Waldst. & Kit. ex Willd., *Ges. Naturf. Freunde Berlin Neue Schr.* **2**: 121 (1799). Rhizome creeping, uniformly stout. Stem 15–35(–50) cm, simple, sparsely hairy; hairs up to 1·1 mm. Cauline leaves 2–4(–5), the lower large, cordate, long-petiolate, the upper rounded at the base, shortly petiolate or sessile. Inflorescence with up to 20 flowers. Calyx deeply lobed. Corolla 13–18 mm, pale yellow, $1\frac{1}{2}$–2 times as long as calyx. Scales triangular-lanceolate, the lower marginal papillae $2\frac{1}{2}$–$3\frac{1}{2}$ times as long as wide, the upper smaller but distinctly longer than wide. Stamens with filament $\frac{4}{7}$–$\frac{3}{4}$ as long as anther. $2n=18$. *Deciduous woods.* ● *Carpathians, C. Romania, W. Ukraine.* Cz Po Rm Rs (W).

7. S. naxicola Pawł., *Fragm. Fl. Geobot.* **17**: 21 (1971). Stock fusiform. Plant often more or less sericeous-villous, with rather soft hairs up to 4 mm and shorter straight and hooked hairs and setae. Stem 12–45 cm, unwinged. Lower leaves rather long-petiolate, the upper sessile, not or scarcely decurrent. Calyx

7·5–12 mm, lobed to $\frac{3}{4}-\frac{7}{8}$. Corolla (14–)15–17(–18) mm, white. Scales (3)3·5–4·5(5) mm, triangular-lanceolate, gradually narrowed to the apex, the margin densely papillose throughout, the papillae not more than $1\frac{1}{2}$ times as long as wide. Stamens 4–5 mm; anthers 2·5–4 mm; filament mostly shorter than anther. $2n = 30$. ● *S. Aegean region (Naxos).* Gr.

8. S. ibiricum Steven, *Bull. Soc. Nat. Moscou* 24 (1): 579 (1851). Rhizome creeping, rather slender, funiculiform, producing more or less ascending non-flowering and flowering stems 12–40 cm. Stems and leaves with very short hairs and setae up to 3 mm. Leaves ovate or elliptical, slightly cordate or rounded at base, the lower long-, the upper short-petiolate. Calyx 4–7 mm, lobed to $\frac{2}{3}-\frac{5}{8}$. Corolla (14–)15–18(–19) mm, pale yellow. Scales lingulate, rounded or sometimes emarginate at apex, the marginal papillae about as long as wide, the lower scattered, the upper dense. Stamens with filament $\frac{2}{3}$ as long as the anther or more. *Naturalized in hedges and grassland in S. & E. England.* [Br.] (*Caucasian region.*)

9. S. davisii Wickens, *Notes Roy. Bot. Gard. Edinb.* **29**: 168 (1969). Stock fusiform. Plant covered with rather soft hairs up to 4 mm and short setiform straight and hooked hairs, often more or less sericeous-villous. Stem 10–30 cm, unwinged. Lower leaves rather long-petiolate, the upper sessile, not decurrent. Calyx 9–13 mm, lobed to $\frac{2}{3}-\frac{3}{8}$. Corolla (17–)18–22 mm, white or white tinged with pink. Scales 5–6(–6·5) mm, lingulate, the marginal papillae not more than $1\frac{1}{2}$ times as long as wide, the lower scattered. Stamens 5–6(–6·5) mm; anthers (3–)3·5–4·2 mm, filament $\frac{1}{2}$ as long as or equalling anther. *Shady rocks.* ● *S. Aegean region (Amorgos).* Gr.

10. S. cycladense Pawł., *Fragm. Fl. Geobot.* **17**: 25 (1971). Plant with numerous basal leaves in a rosette, densely and rather softly villous-pubescent, with setae and hooked hairs. Stem 5–20 cm, simple or scarcely branched. Lower leaves ovate to ovate-oblong, shortly petiolate, greyish beneath. Calyx 6–7(–10) mm, lobed to $\frac{2}{3}-\frac{5}{8}$. Corolla 13–16·5 mm, ?white. Scales 1·6–4 mm, lingulate, marginal papillae not more than $1\frac{1}{2}$ times as long as wide, the lower more or less scattered. Stamens 2·6–4·3 mm; anthers 2–3 mm, 4–6 times as long as wide; filament $\frac{1}{3}-\frac{1}{2}$ as long as anther. $2n = 30$. ● *S. Aegean region (Sikinos).* Gr.

11. S. tauricum Willd., *Ges. Naturf. Freunde Berlin Neue Schr.* **2**: 120 (1799). Stock fusiform. Stem 20–60 cm, rather stout, much-branched, densely hairy with very short hooked hairs and setae up to 2(–3) mm. All cauline leaves except the uppermost petiolate, slightly cordate to truncate or rounded at the base, not decurrent, rather densely hairy on both surfaces, asperous. Calyx 4–7 mm, lobed to $\frac{2}{3}-\frac{5}{8}$. Corolla (8–)9–12 (–15) mm, pale yellow; lobes not recurved. Scales lingulate, rounded or sometimes emarginate at the apex, the marginal papillae up to twice as long as wide, scattered towards the base. Stamens with filament mostly shorter than anther; anthers 2–3 mm, $2\frac{1}{2}$–4(–5) times as long as wide. $2n = 18$. *Woods.* S.E. Europe. Bu Rm Rs (C, W, K, E).

12. S. orientale L., *Sp. Pl.* 136 (1753) (*S. tauricum* sensu Coste, non Willd.). Stock fusiform. Stem up to 70 cm or more, much-branched. Leaves ovate, subcordate or rounded or truncate at base, densely hairy on both surfaces, rather softly hispid, often subtomentose beneath; the lower long-, the upper short-petiolate, the uppermost sessile, not decurrent. Calyx 6–9

(–12) mm, lobed to $\frac{1}{4}-\frac{2}{5}$. Corolla (13–)14–18(–19) mm, white; lobes not recurved. Scales lingulate, the marginal papillae up to $2\frac{1}{2}$ times as long as wide. Stamens with filament c. $\frac{2}{3}$ as long as anther; anthers 2·5–3·5(–4) mm, $(2\frac{1}{2}-)3$–5(–5$\frac{1}{2}$) times as long as wide. *Damp, shady places. Around Istanbul; S.W. Ukraine; locally naturalized elsewhere.* Rs (W) Tu [Br Ga ?He It]. (*N.W. Anatolia.*)

13. S. bulbosum C. Schimper, *Flora (Regensb.)* **8**: 17 (1825) (incl. *S. zeyheri* C. Schimper). Rhizome slender, creeping, producing subglobose tubers. Stems 15–50 cm, simple or little-branched; stem and leaves with dense, very small hooked hairs and scattered setae up to 1·5(–2) mm. Lower leaves ovate to elliptic-lanceolate, gradually attenuate or abruptly contracted into a long petiole, the uppermost sessile, slightly decurrent. Calyx lobed to $\frac{1}{3}-\frac{6}{7}$. Corolla (excluding scales) (7–)8–11(–12) mm, pale yellow, with erect lobes which are $\frac{1}{4}-\frac{1}{3}$ as long as the tube. Scales (5–)5·5–9(–10) mm, exserted for 1–4(–5) mm, acute, lanceolate-subulate, rarely triangular-lanceolate; marginal papillae dense, about as long as wide. Stamens with filament $\frac{1}{5}-\frac{1}{2}$ as long as anther; anthers 2·5–4 mm, minutely apiculate. *S. Europe, eastwards from Corse.* Al Bu Co Ga Gr He It Ju Sa Si [Ge].

14. S. ottomanum Friv., *Flora (Regensb.)* **19**: 439 (1936). Stock fusiform. Stem 30–80 cm, branched. Leaves ovate or ovate-lanceolate, more or less rounded-cuneate at base, the lower long-petiolate, the upper sessile and slightly decurrent. Flowers small. Calyx 3–5 mm, lobed to $\frac{3}{5}-\frac{4}{5}$. Corolla (excluding scales) 5–7 mm, pale yellow; lobes c. $\frac{1}{3}$ as long as tube, more or less erect. Scales 5–9·5 mm, exserted for 2–5·5 mm, linear-lanceolate, more or less acute; marginal papillae dense, about as long as wide. Stamens with filament $\frac{1}{3}-\frac{3}{8}$ as long as anther; anthers 2–2·5 mm, minutely apiculate. $2n = 20$. *Woods.* ● *Balkan peninsula, Romania.* Al Bu Gr Ju Rm Tu.

19. Procopiania Guşuleac[1]

Like *Symphytum* but corolla-tube constricted at base and apex; lobes longer than the rest of the corolla; stamens long-exserted; filaments more than $1\frac{1}{2}$ times as long as anthers.

Literature: M. Guşuleac, *Bul. Fac. Şti. Cernăuţi* **2**: 434–438 (1928). H. Runemark, *Bot. Not.* **120**: 84–94 (1967) (sub *Symphytum*). B. Pawłowski, *Fragm. Fl. Geobot.* **17**: 39–58 (1971).

1 Corolla-tube 4·5–6 mm; lobes spirally contorted above
        3. circinalis
1 Corolla-tube 2–4·5 mm; lobes not contorted
2 Marginal papillae of corolla-scales not more than $1\frac{1}{2}$ times as long as wide   **1. cretica**
2 Lower marginal papillae of corolla-scales slender, at least 3 times as long as wide   **2. insularis**

1. P. cretica (Willd.) Guşuleac, *Bul. Fac. Şti Cernăuţi* **2**: 435 (1928) (*Trachystemon creticus* (Willd.) G. Don fil.). Perennial, more or less branched herb (5–)10–50 cm, hispid, with short and long, hooked and straight hairs. Leaves ovate, often subcordate, petiolate; the upper sessile, decurrent. Cymes many-flowered. Calyx 4·5–7(–8) mm, lobed to $\frac{1}{2}-\frac{4}{5}$; lobes acute. Corolla blue-violet, rarely white; tube (2–)2·5–4(–4·5) mm; lobes $2\frac{1}{3}-4\frac{1}{2}$ times as long as tube, patent, not contorted, only slightly recurved in the uppermost part. Scales narrowly lanceolate-subulate; marginal papillae not more than $1\frac{1}{2}$ times as long as wide. Stamens 9–11(–14) mm, glabrous or with minute lateral hairs at the base, without basal scale or rarely with a minute, unilateral, very shortly ciliate scale; anthers 1·5–3(–3·5) mm. ● *S. Greece and S. Aegean region.* Cr Gr.

¹ By B. Pawłowski.

2. P. insularis Pawł., *Fragm. Fl. Geobot.* **17**: 45 (1971). Like **1** but calyx-lobes usually obtuse; corolla-lobes 2–3 times as long as tube; corolla-scales in their lower part with long, slender papillae, at least 3 times as long as wide; base of filaments surrounded by a cup-shaped, lobed, rather long-ciliate scale. 2*n*= 28. ● *Islands of Aegean region.* Cr Gr.

3. P. circinalis (Runemark) Pawł., *loc. cit.* 48 (1971) (*Symphytum circinale* Runemark). Like **1** but corolla white; tube 4·5–6 mm; lobes 1½–2 times as long as tube, erecto-patent below, deflexed and spirally contorted above; lower marginal papillae of the corolla-scales 2–4 times as long as wide, the upper short and wide, deflexed. Stamens (9·5–)10–12(–13) mm, with 1 unilateral, lobed, rather long-ciliate scale or 2 opposite, laterally connected scales at the base; anthers 2–3·5 mm. 2*n*= 28. *Greece* (*Evvoia; not recently collected*). Gr. (*Islands of E. Aegean.*)

20. Brunnera Steven[1]

Perennial herbs. Flowers in ebracteate cymes in terminal panicles. Calyx lobed almost to the base, accrescent. Corolla blue or purple, very small, rotate, with campanulate tube, ovate-orbicular lobes and 5 ciliate scales wider than long closing the throat. Stamens included, inserted at the middle of the tube. Style included; stigma capitate. Nutlets oblong-obovoid, erect, rugose or ribbed, with a thickened collar-like ring at the base.

Literature: M. Guşuleac, *Bul. Fac. Şti. Cernăuţi* **2**: 412–418 (1928); *Feddes Repert.* **29**: 44–46 (1931).

1. B. macrophylla (Adams) I. M. Johnston, *Contr. Gray Herb.* nov. ser., **73**: 54 (1924). Appressed-pubescent throughout, and sparsely setose-hispid on stems and leaves; stems 20–50 cm, erect, branched only in inflorescence. Basal leaves 5–20 cm, cordate, long-petiolate; cauline leaves ovate, sessile. Pedicels 2–5 mm, up to 8 mm in fruit. Calyx *c.* 1 mm at anthesis, *c.* 2 mm in fruit; teeth linear-lanceolate, subacute. Corolla 3–4 mm; tube *c.* 1 mm; limb 3–4 mm in diameter, blue. Nutlets 2·5–4 mm, oblong-obovoid, slightly asymmetrical, rugose and somewhat longitudinally ribbed; collar-like basal ring ribbed. *Cultivated for ornament and perhaps locally naturalized.* [?Br ?Cz.] (*Caucasus.*)

21. Anchusa L.[1]

(incl. *Lycopsis* L.)

Annual, biennial or perennial herbs, rarely woody at the base. Flowers in axillary and terminal cymes, bracteate usually throughout. Calyx lobed from ⅓ to almost to the base. Corolla purple, blue, yellow or white, with cylindrical, straight or curved tube and rotate to campanulate limb divided into equal or unequal lobes, with 5 ovate or oblong, papillose or hairy scales in throat. Stamens included or slightly exserted, variously inserted. Style included; stigma capitate. Nutlets ovoid to reniform or hemispherical, erect or oblique, reticulate or rugose, usually more or less tuberculate, with a thickened collar-like ring at the base.

Literature: M. Guşuleac, *Bul. Fac. Şti. Cernăuţi* **1**: 72–123 (1927); *Feddes Repert.* **26**: 286–322 (1929). W. Greuter, *Candollea* **20**: 192–210 (1965).

1 Corolla yellow
  2 Bracts exceeding flowers; nutlets 4–5×2–3 mm, erect
      21. aegyptiaca
  2 Bracts shorter than flowers; nutlets *c.* 2×4 mm, oblique
    3 Leaves mostly more than 10 mm wide   **10. officinalis**
    3 Leaves mostly less than 10 mm wide   **3. ochroleuca**

1 Corolla blue, violet, purplish or whitish
  4 Corolla zygomorphic, the tube curved, the limb oblique, with 5 unequal lobes
    5 Filaments inserted at or below middle of corolla-tube; cymes bracteate throughout   **22. arvensis**
    5 Filaments inserted above middle of corolla-tube; cymes bracteate only at base
      6 Nutlets densely rugulose between the primary reticulations; lobes of style obliquely truncate   **23. cretica**
      6 Nutlets smooth between the primary reticulations; lobes of style ± acute   **24. variegata**
  4 Corolla actinomorphic
    7 Corolla-tube much shorter than limb
      8 Annual; inflorescence simple or once branched **16. macedonica**
      8 Perennial; inflorescence paniculate
        9 Caespitose; leaves up to 7 mm wide   **20. serpentinicola**
        9 Not caespitose; most leaves at least 10 mm wide **19. barrelieri**
    7 Corolla-tube equalling or longer than limb
      10 Calyx divided more or less to the base
        11 Nutlets longer than wide
          12 Nutlets *c.* 4 mm; corolla-limb 5–7 mm in diameter   **13. thessala**
          12 Nutlets at least 6 mm; corolla-limb (8–)10–15 mm in diameter   **11. azurea**
        11 Nutlets wider than long, or hemispherical
          13 Caespitose perennial with stems not more than 10 cm   **4. cespitosa**
          13 Annual, biennial or perennial, not caespitose, with stems more than 10 cm
            14 Style 2–3 times as long as calyx; corolla-scales with subglobose papillae   **17. stylosa**
            14 Style less than twice as long as calyx; corolla-scales hairy or with cylindrical papillae
              15 Corolla-tube 12–15 mm, about twice as long as calyx   **15. macrosyrinx**
              15 Corolla-tube less than 10 mm, not more than 1½ times as long as calyx
                16 Pedicels equalling or exceeding calyx at anthesis   **18. spruneri**
              16 Pedicels shorter than calyx at anthesis
                17 Cymes capitate, with lower bracts much exceeding calyx; nutlets ± hemispherical   **12. aggregata**
                17 Cymes ± elongate, with bracts not exceeding calyx; nutlets obliquely ovoid
                  18 Annual; cymes with *c.* 5 flowers   **14. pusilla**
                  18 Usually perennial; cymes with more than 5 flowers   **10. officinalis**
      10 Calyx lobed to not more than ⅔
        19 Calyx-teeth acute
          20 Most hairs on leaves tubercle-based   **7. granatensis**
          20 Most hairs on leaves not tubercle-based   **10. officinalis**
        19 Calyx-teeth obtuse
          21 Bracts equalling or exceeding calyx   **9. crispa**
          21 Bracts shorter than calyx
            22 Stamens inserted at top of corolla-tube and overlapping scales
              23 Plant with strongly tubercle-based setae   **8. calcarea**
              23 Plant pubescent and with setae which are mostly not tubercle-based   **5. undulata**
            22 Stamens inserted below top of tube and not or only slightly overlapping scales
              24 Plant subglabrous or with setae less than 1 mm; nutlets subreniform   **2. gmelinii**
              24 Plant with hairs or setae more than 1 mm; nutlets obliquely ovoid
                25 Plant with dense, stout, patent, white, mostly tubercle-based setae   **6. sartorii**
                25 Plant pubescent and with setae which are mostly not conspicuously white or tubercle-based
                  26 Stamens inserted at middle of tube and not nearly reaching scales   **5. undulata**
                  26 Stamens inserted above middle of tube and reaching base of scales   **1. leptophylla**

[1] By A. O. Chater.

1. A. leptophylla Roemer & Schultes, *Syst. Veg.* **4**: 90 (1819). Perennial, rarely biennial, more or less greyish with dense, short, soft, appressed, not tubercle-based hairs; stems 40–70 cm, erect. Leaves 50–100(–200) × 6–13 mm. Cymes short, usually several, crowded; pedicels up to 2 mm in fruit; bracts shorter than calyx. Calyx 5–7 mm, divided to $\frac{1}{3}$–$\frac{1}{2}$ into oblong, obtuse lobes. Corolla purple, becoming blue; tube 6–8 mm, $1\frac{1}{4}$–$1\frac{1}{2}$ times as long as calyx; limb 6–8 mm in diameter, with oblong lobes; scales oblong, obtuse, densely and shortly hairy. Stamens inserted in upper half of tube and not or scarcely overlapping scales. Nutlets *c.* 2·5 × 3·5 mm, obliquely ovoid. *S.E. Europe, from E. Bulgaria to Krym.* Bu Rm Rs (W, K).

2. A. gmelinii Ledeb., *Beitr. Naturk.* (*Dorpat*) **1**: 62 (1820). Perennial, rarely biennial, subglabrous or scabrid with sparse, stout, short, tubercle-based setae; stems 30–80 cm, erect. Leaves 50–150 × 3–8 mm. Cymes short, few or solitary; pedicels up to 3 mm in fruit; bracts shorter than calyx. Calyx 4–5 mm, up to 11 mm in fruit, divided to $\frac{1}{3}$–$\frac{1}{2}$ into oblong-lanceolate, obtuse lobes. Corolla violet or pale blue; tube 6–8 mm, $1\frac{1}{4}$–2 times as long as calyx; limb 7–10 mm in diameter, with ovate lobes; scales oblong, obtuse, densely and shortly hairy; stamens inserted at about middle of tube and not or scarcely overlapping scales. Nutlets *c.* 2 × 4 mm, subreniform. *S. part of U.S.S.R., extending to E. Romania; Turkey-in-Europe.* Rm Rs (C, W, E) Tu.

3. A. ochroleuca Bieb., *Fl. Taur.-Cauc.* **1**: 125, 421 (1808) (incl. *A. pseudochroleuca* Schost.). Perennial, rarely biennial, with dense, short, soft, appressed, usually not tubercle-based hairs and sometimes with some longer, stout hairs; stems 30–80 cm, erect. Leaves 40–80(–200) × 3–10(–25) mm. Cymes short, several, crowded; pedicels very short; bracts mostly shorter than calyx. Calyx 4–6(–8·5) mm, up to 9(–15) mm in fruit, divided to $\frac{1}{3}$–$\frac{1}{2}$ into linear-oblong, obtuse to acute lobes. Corolla pale yellow; tube 5–10 mm, $1\frac{1}{4}$–$1\frac{1}{2}$ times as long as calyx; limb 7–10(–15) mm in diameter. Nutlets *c.* 2 × 3·5 mm, obliquely ovoid. *S.E. & E.C. Europe.* Bu Hu Rm Rs (W, ?K, E) [Ho It].

4. A. cespitosa Lam., *Encycl. Méth. Bot.* **1**: 504 (1785). Caespitose perennial, scabrid-hispid throughout, with many of the hairs minutely tubercle-based; stems 1–10 cm. Leaves 20–50(–75) × 2·5–4 mm, rosulate. Cymes often subsessile, dense, with 3–5 flowers; pedicels very short. Calyx 4–6 mm, divided almost to the base into linear, obtuse lobes. Corolla deep blue; tube 6–7 mm, *c.* $1\frac{1}{2}$ times as long as calyx; limb 10–12 mm in diameter; scales oblong, ciliate; stamens inserted in upper half of tube and slightly overlapping scales. Nutlets *c.* 2·5 × 3·5 mm, obliquely ovoid. ● *Mountains of Kriti.* Cr.

5. A. undulata L., *Sp. Pl.* 133 (1753). Biennial or perennial, pubescent and hispid with both short and long hairs or setae which are usually mostly not tubercle-based; stems 10–50 cm, erect or ascending. Leaves 50–150 × 3–10(–25) mm, often sinuate-dentate and undulate. Cymes several, dense; pedicels very short; bracts shorter than calyx. Calyx 5–10 mm, up to 15(–20) mm in fruit, divided to *c.* $\frac{1}{2}$ into oblong-lanceolate, obtuse lobes. Corolla blue, violet or purple; tube 7–13 mm, $1\frac{1}{2}$–2 times as long as calyx; limb 3–8 mm in diameter. Nutlets *c.* 2 × 3–4 mm, obliquely ovoid. *Mediterranean region, Portugal.* Al Bu Co Cr Ga Gr Hs It Lu Sa Si Tu.

(a) Subsp. **undulata**: Bracts lanceolate or ovate-lanceolate. Corolla-tube 8–13 mm; limb 6–8 mm in diameter; stamens inserted at top of tube and overlapping scales. *Iberian peninsula.*

(b) Subsp. **hybrida** (Ten.) Coutinho, *Fl. Port.* 495 (1913)

(*A. hybrida* Ten.): Bracts ovate-lanceolate, cordate at base. Corolla-tube 5–8 mm; limb 3–5 mm in diameter; stamens inserted about middle of tube and not reaching scales. *Throughout the range of the species except the Iberian peninsula.*

A. subglabra A. Caballero, *Anal. Inst. Bot. Cavanilles* **5**: 509 (1945), is like **5**(a) but is glabrous or subglabrous, has bracts equalling the calyx and corolla-tube equalling or shorter than the calyx; it was described from W.C. Spain (near Hervás) and requires further investigation.

6. A. sartorii Heldr. ex Guşuleac, *Bul. Fac. Şti. Cernăuţi* **1**: 96 (1927). Like **5**(b) but covered with dense, white, patent, mostly tubercle-based setae without shorter hairs; stems procumbent. *Maritime sands.* ● *S. Aegean region.* Gr.

7. A. granatensis Boiss., *Voy. Bot. Midi Esp.* **2**: 430 (1841). Like **5**(a) but with few or no short hairs and numerous weakly tubercle-based setae; calyx-lobes acute; corolla-tube *c.* 6 mm, equalling or only just exceeding calyx. ● *C. & S. Portugal; mountains of S. Spain.* Hs Lu.

Plants from Portugal approach **5** in their indumentum.

8. A. calcarea Boiss., *op. cit.* 431 (1841). Like **5**(a) but usually perennial, with very strongly white-tubercle-based setae throughout; calyx up to 8 mm in fruit, divided to only *c.* $\frac{1}{3}$; corolla-tube 6·5–8 mm and slightly exceeding calyx. *Maritime sands.* ● *S. & W. parts of Iberian peninsula.* Hs Lu.

9. A. crispa Viv., *Fl. Cors., App.* **1**: 1 (1825). Annual or biennial, pubescent and hispid with both short and long hairs or setae; stems 10–35 cm, procumbent or ascending. Leaves 50–170 × 5–10 mm. Cymes several, long, lax; pedicels very short, up to 2 mm in fruit; bracts equalling or exceeding calyx. Calyx *c.* 5 mm, up to 8 mm in fruit, divided to $\frac{1}{3}$–$\frac{1}{2}$ into lanceolate, obtuse lobes. Corolla blue; tube 4–5 mm, equalling or shorter than calyx; limb *c.* 5 mm in diameter; stamens inserted at top of tube and overlapping scales. Nutlets *c.* 2 × 3 mm, obliquely ovoid. $2n=16$. *Maritime sands.* ● *Corse, Sardegna.* Co Sa.

10. A. officinalis L., *Sp. Pl.* 133 (1753) (incl. *A. procera* Besser ex Link). Perennial, rarely biennial, hispid with uniform hairs or setae; stems 20–80(–170) cm, erect. Leaves 50–120 × 10–20 mm. Cymes several, dense; pedicels very short, up to 5 mm in fruit; bracts equalling or shorter than calyx. Calyx 5–7 mm, up to 10 mm in fruit, divided to $\frac{1}{2}$ way or almost to the base into lanceolate, acute lobes. Corolla violet or reddish, rarely white or yellow; tube 5–7 mm, equalling or up to $1\frac{1}{2}$ times as long as calyx; limb 7–15 mm in diameter; stamens inserted in upper half of tube and reaching or partly overlapping scales. Nutlets *c.* 2 × 4 mm, obliquely ovoid. $2n=16+0-2B$. *Much of Europe, but absent from the extreme north, much of the west, and parts of the Mediterranean region.* Al Au Bl Bu Co Cz Da Ga Ge Gr He Ho Hu It Ju *No Po Rm Rs (*N, B, C, W, E) Su Tu [*Be *Fe].

A. davidovii Stoj., *Bull. Inst. Roy. Hist. Nat.* (*Sofia*) **6**: 210 (1933), from S.W. Bulgaria (Rila Planina), is a dwarf variant of **10** with smaller, pale blue flowers and corolla-tube shorter than the calyx.

A. velenovskyi (Guşuleac) Stoj., *op. cit.* 211 (1933), from the coasts of Romania and Bulgaria, has the calyx 3–5 mm, with obtuse teeth and large pale blue or purplish flowers; it requires further investigation.

11. **A. azurea** Miller, *Gard. Dict.* ed. 8, no. 9 (1768) (*A. italica* Retz.). Perennial, hispid with dense, patent, rigid or soft, often tubercle-based hairs; stems 20–150 cm, erect. Leaves (50–)100–300 × (10–)15–50 mm. Cymes many; pedicels 1–3 mm, up to 10(–15) mm in fruit; bracts shorter than calyx. Calyx 6–8 (–10) mm, up to 18 mm in fruit, divided almost to the base into linear, acute lobes. Corolla violet or deep blue; tube 6–10 mm, slightly exceeding or shorter than calyx; limb (8–)10–15 mm in diameter; stamens inserted at top of tube, overlapping scales. Nutlets (6–)7–10 × 2–3 mm, oblong or oblong-obovoid, erect. *S. & S.C. Europe; naturalized further north.* Al Bl Bu Co Cr Ga Gr Hs Hu It Ju Lu Rm Rs (W, K, E) Sa Si Tu [Au Cz Ge He].

12. **A. aggregata** Lehm., *Pl. Asperif.* 395 (1818) (*Hormuzakia aggregata* (Lehm.) Guşuleac). Annual, densely hispid with long, white, mostly tubercle-based setae; stems (5–)10–50 cm, ascending, branched. Leaves (50–)70–150 × (4–)7–15 mm, linear-lanceolate, entire. Cymes more or less capitate, with subsessile flowers, oblong and dense in fruit; at least the lower bracts greatly exceeding calyx. Calyx *c.* 5 mm in flower, *c.* 8 mm in fruit, divided almost to the base into linear, obtuse lobes. Corolla blue; tube 6–8 mm; limb *c.* 5 mm in diameter; stamens inserted at top of tube, overlapping scales. Nutlets *c.* 3 × 4 mm, hemispherical. *Maritime sands. S. Sicilia.* †Gr Si. (*N. Africa, Asia.*)

13. **A. thessala** Boiss. & Spruner in Boiss., *Diagn. Pl. Or. Nov.* 2(11): 99 (1849). Annual, hispid with short, strongly tubercle-based hairs; stems 10–40 cm, erect. Leaves 20–50 × 5–10 mm, lanceolate to oblong, entire. Cymes paired, short; pedicels very short; bracts about equalling calyx. Calyx 5–6 mm, divided almost to the base into linear, acute lobes. Corolla pale violet to whitish; tube *c.* 6 mm, equalling or up to 1½ times as long as calyx; limb 5–7 mm in diameter; scales with acute papillae; stamens inserted near top of tube and overlapping scales; style *c.* 1½ times as long as calyx. Nutlets *c.* 4 × 1–1·5 mm, oblong-obovoid, erect. 2*n* = 24. *S.E. Europe, from E.C. Greece to Krym.* Bu Gr Rm Rs (K) Tu.

14. **A. pusilla** Guşuleac, *Bul. Fac. Şti. Cernăuţi* 1: 109 (1927). Like 13 but leaves spathulate-oblanceolate, weakly dentate; calyx *c.* 7 mm; scales with long hairs; nutlets *c.* 3 × 4 mm, obliquely ovoid. *E. Krym.* Rs (K). (*Caucasus.*)

15. **A. macrosyrinx** Rech. fil., *Österr. Bot. Zeitschr.* **107**: 471 (1960). Like 13 but leaves linear-lanceolate, weakly repand-dentate; calyx 6–7 mm; corolla blue; tube 12–15 mm, twice as long as calyx; nutlets *c.* 3 × 4 mm, obliquely ovoid. ● *S. Greece (Attiki).* Gr.

16. **A. macedonica** Degen & Dörfler, *Denkschr. Akad. Wiss. Math.-Nat. Kl.* (*Wien*) **64**: 730 (1897). Annual or biennial, hirsute with tubercle-based hairs; stems up to 20 cm, ascending. Leaves 50–120 × 7–12 mm, linear-oblong or -lanceolate. Cymes solitary, short; pedicels 5–10 mm; bracts exceeding calyx. Calyx *c.* 5 mm, divided almost to the base into linear, subacute lobes. Corolla blue; tube 1–2 mm, much shorter than limb; limb 5–7 mm in diameter. Nutlets *c.* 2 × 4 mm, obliquely ovoid. ● *Macedonia.* Gr Ju.

17. **A. stylosa** Bieb., *Fl. Taur.-Cauc.* 1: 123 (1808). Annual, with sparse, stoutish, short, tubercle-based hairs; stems 10–40 cm, erect. Leaves 20–50(–100) × 4–15(–20) mm, oblong or oblong-lanceolate, often denticulate. Cymes paired, short; pedicels very short, up to 5(–10) mm in fruit; bracts shorter than calyx. Calyx *c.* 5 mm, up to 9 mm in fruit, divided almost to the base into linear-lanceolate, subobtuse lobes. Corolla violet-

blue; tube 8–10 mm, 1½–3 times as long as calyx; limb 7–10 mm in diameter; scales with subglobose papillae; style 2–3 times as long as calyx after anthesis. Nutlets *c.* 3 × 2 mm, obliquely ovoid. *S.E. Europe, from N. Greece and Albania to Krym.* Al Bu Gr Rm Rs (W, K) Tu.

18. **A. spruneri** Boiss., *Diagn. Pl. Or. Nov.* 2(11): 98 (1849). Like 17 but pedicels *c.* 5 mm, equalling or exceeding calyx at anthesis; corolla-tube shorter or up to 1¼ as long as calyx; scales with elongate, cylindrical papillae; style not more than 1½ times as long as calyx. ● *N.W. part of Aegean region.* Gr.

19. **A. barrelieri** (All.) Vitman, *Summa Pl.* 1: 388 (1789). Perennial, hispid with stout or slender, patent or appressed, often slightly tubercle-based hairs; stems (20–)50–80 cm, erect. Leaves 30–70(–120) × (5–)10–15(–30) mm, oblong-spathulate to oblong-lanceolate or almost linear, subentire. Cymes many, in a paniculate inflorescence; pedicels 2–5(–7) mm; bracts mostly shorter than calyx. Calyx 2–3 mm, up to 6 mm in fruit, divided to ½ or almost to the base into lingulate, very obtuse or subspathulate lobes. Corolla blue or bluish-violet; tube 1–1·5 mm, much shorter than limb; limb 7–10 mm in diameter; style very short; scales with short, cylindrical papillae. Nutlets 2–4 mm, ovoid, erect. *From Italy eastwards to C. Ukraine.* Al Bu Hu It Ju Rm Rs (W, ?K) [?Cz].

20. **A. serpentinicola** Rech. fil., *Österr. Bot. Zeitschr.* **107**: 472 (1960). Like 19 but caespitose; stems up to 25 cm; leaves up to 7 mm wide; corolla-limb *c.* 5 mm in diameter. *On basic rocks.* ● *Macedonia.* Gr Ju.

21. **A. aegyptiaca** (L.) DC., *Prodr.* 10: 48 (1846). Annual, hispid with stout, tubercle-based hairs; stems 5–30 cm, procumbent or ascending. Leaves 25–40 × 10–15 mm, oblong-ovate to -lanceolate, denticulate to sinuate-dentate. Cymes very lax, leafy; pedicels 2–3 mm, elongating and recurved in fruit; bracts leaf-like, exceeding calyx. Calyx *c.* 5 mm, divided almost to the base into linear-lanceolate, obtuse lobes. Corolla pale yellow; tube *c.* 4 mm, straight, slightly shorter than calyx; limb 3–5 mm in diameter, with 5 slightly unequal lobes; stamens inserted at about the middle of the tube, 2 higher than the other 3. Nutlets 4–5 × 2–3 mm, ovoid, erect. *S. part of Aegean region.* Cr Gr ?Si.

22. **A. arvensis** (L.) Bieb., *Fl. Taur.-Cauc.* 1: 123 (1808) (*Lycopsis arvensis* L.). Annual, hispid with mostly tubercle-based hairs; stems 10–60 cm, ascending. Leaves (15–)30–100(–150) × (3–)5–20(–25) mm, linear-lanceolate to broadly lanceolate, undulate, more or less dentate. Cymes several, short, bracteate throughout; pedicels very short, up to 10 mm in fruit; bracts longer or shorter than calyx. Calyx 4–5 mm, up to 10(–15) mm in fruit, divided almost to the base into linear-lanceolate, acute lobes. Corolla blue, rarely whitish; tube 4–7 mm, about equalling calyx, curved; limb 4–6 mm in diameter, with 5 unequal lobes; stamens inserted at or below middle of tube. Nutlets 1·5–2 × 3–4 mm, obliquely ovoid. 2*n* = 48. *Most of Europe.* Al Au Be Bl Br Bu Co Cz Da Fe Ga Ge Gr *Hb He Ho Hs Hu It Ju Lu No Po Rm Rs (N, B, C, W, K, E) *Sa Si Su [Fa].

(a) Subsp. **arvensis** (*Lycopsis arvensis* subsp. *occidentalis* Kusn.): Leaves linear-lanceolate or lanceolate, strongly undulate, dentate. Cymes remaining dense. Corolla-tube 5–7 mm, curved at about the middle. Nutlets sparsely tuberculate. 2*n* = 48. *Throughout the range of the species except for most of S.E. Europe.*

(b) Subsp. **orientalis** (L.) Nordh., *Norsk Fl.* 526 (1940) (*A. orientalis* (L.) Reichenb. fil., non L., *Lycopsis orientalis* L.): Leaves broadly lanceolate, scarcely undulate, subentire. Cymes becoming more or less lax at start of anthesis. Corolla-tube 4–5 mm, curved below the middle. Nutlets densely and strongly tuberculate. *S.E. Europe; naturalized in Spain and Portugal.*

23. A. cretica Miller, *Gard. Dict.* ed. 8, no. 7 (1768) (*Lycopsis variegata* auct., non L.). Annual, hispid with slightly tubercle-based hairs; stems 10–50 cm, ascending or erect. Leaves 30–100 × 5–20 mm, linear-lanceolate, subentire or repand-dentate. Cymes solitary, rarely paired, bracteate only at base; pedicels very short, up to 7 mm in fruit; bracts equalling or shorter than calyx. Calyx 4–5 mm, up to 10 mm in fruit, divided almost to the base into linear-lanceolate, acute lobes. Corolla purplish, becoming blue with white lines, rarely entirely white; tube 4–7 mm, about equalling calyx, curved; limb 6–9 mm in diameter, with 5 unequal lobes; stamens inserted above middle of tube; style with obliquely truncate lobes. Nutlets 1·5–2 × 3–4 mm, obliquely ovoid, ovate in dorsal view, densely rugulose between the primary reticulations. $2n=16$. ● *Italy, Sicilia; S. & W. parts of Balkan peninsula.* Al Gr It Ju Si.

24. A. variegata (L.) Lehm., *Pl. Asperif.* 223 (1818) (*Lycopsis variegata* L.). Like **23** but stems procumbent; cymes remaining more dense in fruit; corolla white or pale purplish, becoming pale blue with reddish markings; style with more or less acute lobes; nutlets suborbicular in dorsal view, smooth between the primary reticulations. $2n=16$. *S. Aegean region.* Cr Gr.

22. Pentaglottis Tausch[1]

(*Caryolopha* Fischer & Trautv.)

Perennial, hispid herbs. Flowers in axillary and terminal, bracteate cymes. Calyx lobed almost to the base, accrescent. Corolla blue, with short, infundibuliform or cylindrical tube and rotate limb, with 5 broadly ovate, hairy scales closing mouth of tube. Stamens included, inserted in upper part of tube. Style included; stigma capitate. Nutlets ovoid, erect, concave and with a slightly thickened collar-like ring and an excentric stalked attachment at the base.

1. P. sempervirens (L.) Tausch ex L. H. Bailey, *Man. Cult. Pl.* ed. 2, 837 (1949) (*Anchusa sempervirens* L., *Caryolopha sempervirens* (L.) Fischer & Trautv.). Hispid throughout; stems 30–100 cm, ascending or erect, branched. Basal leaves 10–40 cm, ovate-oblong or ovate, acute, narrowed into a long petiole; cauline sessile, acuminate. Cymes dense, with 5–15 flowers; each branch of the inflorescence subtended by a large bract. Calyx 2·5–5 mm at anthesis, c. 8 mm in fruit; lobes linear-lanceolate. Corolla-tube 4–6 mm; limb 8–10 mm in diameter, bright blue. Nutlets 1·5–2 mm, asymmetrically ovoid, rugose-reticulate, blackish. *Damp or shady places. S.W. Europe, from C. Portugal to S.W. France; widely cultivated, and naturalized in N.W. Europe and N. Italy.* Ga Hs Lu [Be Br Hb It].

23. Borago L.[1]

Annual or perennial herbs. Flowers in branched, lax, usually bracteate cymes. Calyx lobed almost to the base, accrescent. Corolla blue, pink or white, rotate to campanulate, the tube short or absent, with short, glabrous, emarginate, exserted scales. Stamens exserted, inserted near base of corolla; anthers connivent, mucronate; filaments with a long, narrow appendage at apex. Style included; stigma capitate. Nutlets obovoid, erect, rugose, concave and with a thickened collar-like ring at the base.

Erect, robust annual; calyx more than ½ as long as corolla
1. officinalis
Decumbent, slender perennial; calyx less than ½ as long as corolla
2. pygmaea

1. B. officinalis L., *Sp. Pl.* 137 (1753). Hispid annual; stems 15–70 cm, erect, robust, often branched. Basal leaves 5–20 cm, ovate to lanceolate, petiolate; upper cauline leaves sessile, amplexicaul. Pedicels 5–30 mm, stout, patent or deflexed after anthesis. Calyx 8–15 mm at anthesis, up to 20 mm in fruit; lobes linear-lanceolate, acute, connivent in fruit. Corolla rotate, bright blue, rarely white; tube very short or almost absent; lobes 8–15 mm, lanceolate, acute. Nutlets 7–10 mm, oblong-obovoid. $2n=16$. *Dry, often waste places. S. Europe; widely cultivated for ornament and as a pot-herb and for bees, naturalized in the warmer parts of C., E. & W. Europe and an occasional casual or short-lived escape further north.* *Az Bl Co Cr Ga Gr Hs It Ju Lu Sa Si Tu [Au Br Cz Ge He Ho Hu Po Rm Rs (C, W, K, E)].

2. B. pygmaea (DC.) Chater & W. Greuter, *Bot. Jour. Linn. Soc.* **65**: 261 (1972) (*Campanula pygmaea* DC., *Borago laxiflora* Poiret). Hispid perennial; stems 15–60 cm, decumbent, slender, branched. Lower leaves 5–20 cm, oblong to obovate, petiolate; upper leaves sessile, amplexicaul. Pedicels 10–40 mm, filiform, deflexed after anthesis. Calyx 4–6 mm at anthesis, up to 8 mm in fruit; lobes lanceolate, acute, not connivent in fruit. Corolla campanulate, clear blue; tube short; lobes 5–8 mm, ovate, acute. Nutlets 3–4 mm, obovoid. $2n=32$. *Damp places.* ● *Corse, Sardegna, Capraia.* Co It Sa.

24. Trachystemon D. Don[1]

Perennial herbs. Flowers in dense, bracteate cymes in a lax panicle. Calyx lobed to c. ½, accrescent. Corolla bluish-violet, with infundibuliform tube exceeding the calyx and limb divided to the base into revolute lobes; scales very short, villous, in 2 series of 5, the lower at about the middle, the upper at the apex of the tube. Stamens long-exserted, inserted between the 2 series of scales. Style included; stigma entire. Nutlets obliquely ovoid, carinate, rugose-reticulate, more or less flat and with a thickened collar-like ring at the base.

Literature: M. Guşuleac, *Bul. Fac. Şti. Cernăuţi* **2**: 431–434 (1928); *Feddes Repert.* **29**: 117–118 (1931).

1. T. orientalis (L.) G. Don fil., *Gen. Syst.* **4**: 309 (1837). Rhizomatous, pubescent and usually also sparsely hispid; stems 20–60 cm, erect, branched. Basal leaves 15–50 cm, ovate, sub-acute to acuminate, cordate at base, long-petiolate; cauline leaves ovate to lanceolate, sessile. Cymes with 5–15 flowers; pedicels long. Calyx 3–6 mm at anthesis, 6–9 mm in fruit; lobes ovate, obtuse. Corolla-tube 5–7 mm; limb 9–12 mm, bluish-violet. Nutlets c. 2 × 4 mm, rugose, reticulate; collar-like basal ring slightly ribbed. $2n=56$. *Shady places. E. Bulgaria, Turkey-in-Europe.* Bu Tu [Br]. (*N. Anatolia, W. Caucasus.*)

25. Mertensia Roth[2]

Perennial, usually glabrous herbs. Flowers in terminal cymes. Calyx lobed to the base, not accrescent. Corolla blue, pink or white, cylindrical or with campanulate limb, without scales but often with 5 invaginations in the throat. Stamens included or slightly exserted, inserted in upper part of tube. Style included or exserted; stigma very small, entire. Nutlets flattened-trigonous, fleshy, strongly angled.

1. M. maritima (L.) S. F. Gray, *Nat. Arr. Brit. Pl.* **2**: 354 (1821). Glabrous, glaucous; stems up to 60 cm, leafy, purple, pruinose, erect or decumbent, sometimes cushion-forming. Leaves 0·5–6 cm, spathulate, obovate or lanceolate, obtuse or apiculate, the lower petiolate, the upper sessile; upper surface papillose. Racemes often branched, with leaf-like bracts. Pedicels 2–10 mm, often recurved in fruit. Calyx-lobes *c.* 6 mm, ovate. Corolla *c.* 6 mm in diameter, pink, becoming blue and pink, with invaginations in the throat; limb campanulate, divided to *c.* ⅓ into ovate-triangular lobes. Nutlets *c.* 6 mm, flattened; outer coat fleshy, becoming papery. $2n=24$. *Maritime sands and shingle. Coasts of N. Europe, southwards to 54° (formerly 52° 30′ N.) in Britain and Ireland, but not in the Baltic region.* Br Da Fa Hb Is No Rs (N) Sb Su.

26. Amsinckia Lehm.[1]

Annuals. Flowers in usually ebracteate, terminal cymes. Calyx lobed almost to the base, accrescent, sometimes with some of the lobes connate. Corolla yellow or orange, with cylindrical tube and infundibuliform or campanulate limb, without scales. Stamens included, inserted in upper part of tube. Style included; stigma capitate. Nutlets ovoid-trigonous, granulate, rugose or smooth, attached to receptacle for only part of their length.

Native of America; the identity of the plants found in Europe is in some cases uncertain.

Literature: A. Brand in Engler, *Pflanzenreich* 97 (**IV. 252**): 204–217 (1931). P. M. Ray & H. F. Chisaki, *Amer. Jour. Bot.* **44**: 529–554 (1957).

Corolla-tube hairy inside at throat	**1. lycopsoides**
Corolla-tube glabrous inside at throat	**2. calycina**

1. A. lycopsoides (Lehm.) Lehm., *Del. Sem. Horto Hamburg.* **1831**: 3 (1831). Stem 20–50 cm, erect or ascending, often branched, with patent, stout hairs. Leaves 30–80 × 5–15 mm, linear to oblanceolate, with erecto-patent hairs on both surfaces. Inflorescence ebracteate or with one bract at base; flowers subsessile. Calyx 3–5 mm in flower, 6–11 mm in fruit. Corolla 5–8 mm, deep yellow, hairy inside at the throat; limb *c.* 2 mm. Nutlets 2–3 mm, slightly flattened, strongly muricate, not rugose or only weakly so near apex. *Waste places. Naturalized in N.E. England (Farne Islands); a frequent casual in France and Fennoscandia.* [Br.] (*E. North America.*)

2. A. calycina (Moris) Chater, *Bot. Jour. Linn. Soc.* **64**: 380 (1971) (*Lithospermum calycinum* Moris, *A. hispida* I. M. Johnston, nom. illegit., *A. angustifolia* Lehm.). Stem 15–50 cm, erect, often branched, with long, patent and short, crispate hairs. Leaves 50–150 × 4–10 mm, linear-lanceolate, with erecto-patent hairs on both surfaces. Inflorescence bracteate at base; pedicels up to 3 mm. Calyx 3–4 mm in flower, 4–5 mm in fruit. Corolla 4–8 mm, yellow, glabrous inside; limb 1–2 mm. Nutlets *c.* 2·5 mm, strongly muricate and weakly transversely rugose. *Waste places. A frequent casual in N. & W. Europe, and probably locally naturalized in France.* [Ga.] (*South America and S. North America.*)

The following species are frequent casuals in W. & N.W. Europe, and are perhaps becoming naturalized; all have the corolla-tube glabrous inside and can be distinguished from each other and from **2** thus:

1	Corolla 10–15 mm	**A. douglasiana**
1	Corolla less than 10 mm	
2	Nutlets muricate and strongly transversely rugose	**A. intermedia**
2	Nutlets muricate but not or weakly rugose	
3	Nutlets setulose; inflorescence ebracteate	**A. menziesii**
3	Nutlets glabrous; inflorescence bracteate at base	**A. calycina**

A. douglasiana A. DC. in DC., *Prodr.* **10**: 118 (1846) has leaves 15–40 × 2–5 mm, ebracteate racemes, sessile, orange flowers and weakly muricate nutlets which are not rugose.

A. intermedia Fischer & C. A. Meyer, *Ind. Sem. Hort. Petrop.* **2**: 26 (1836) has leaves 10–40 × 5–15 mm, usually ebracteate racemes and sessile, orange-yellow flowers.

A. menziesii (Lehm.) A. Nelson & Macbride, *Bot. Gaz.* **61**: 36 (1916) has leaves 30–80 × 5–15 mm, usually ebracteate racemes and sessile, pale yellow flowers.

27. Asperugo L.[2]

Annuals. Flowers solitary and axillary, or in axillary pairs. Calyx lobed almost to the base, the lobes leaf-like, dentate, accrescent and deltate in fruit to form a 2-lipped covering round the nutlets. Corolla purple or violet, infundibuliform, with 5 short scales in the throat. Stamens included, inserted in middle of tube. Style included; stigma capitate. Nutlets ovate, laterally compressed, attached to receptacle for only a part of their length at the margin.

1. A. procumbens L., *Sp. Pl.* 138 (1753). Hispid; stems up to 70 cm, procumbent or climbing, branched, with stiff, deflexed hairs. Leaves 20–75 mm, lanceolate, subacute to obtuse, entire or slightly dentate, the lower subopposite or verticillate, petiolate, the upper sessile. Corolla 2–3 mm, purplish, becoming dark violet; scales white. Nutlets *c.* 3 mm, densely and finely tuberculate. $2n=48$. *Cultivated fields, waste places and farmyards; nitrophile. N., E. & E.C. Europe; introduced or casual elsewhere.* Al Au Bu Cz Da Fe *Ge Gr *He Hu It Ju No Po Rm Rs (N, B, C, W, K, E) Su Tu [*Ga Ho Hs Sa].

28. Rochelia Reichenb.[3]

(*Cervia* Rodr. ex Lag.)

Annual or perennial herbs. Flowers in bracteate cymes forming lax, terminal panicles. Calyx lobed to the base, accrescent. Corolla usually blue, with cylindrical tube and infundibuliform limb, usually with 5 scales or invaginations in the throat. Stamens included, inserted at about middle of tube. Style included; stigma capitate. Ovary 2-locular. Nutlets 2, subpyriform, usually glochidiate, attached to receptacle for most of their length.

1. R. disperma (L. fil.) C. Koch, *Linnaea* **22**: 649 (1849) (*Cervia disperma* (L. fil.) Hayek). Annual; stem 5–20(–35) cm. Leaves 2–5 × 10–30 mm, elliptical to oblong, strigose. Bracts like the leaves but smaller; pedicels deflexed in fruit. Calyx 2–4 mm, with hooked or curved hairs; lobes linear, becoming patent then incurved in fruit. Corolla 2–4 mm, blue, with or without saccate invaginations in the throat; lobes ovate. Nutlets *c.* 3 mm, glochidiate. *Dry, stony ground. E. Romania, Greece and S. part of U.S.S.R.; C. & S. Spain.* Gr Hs Rm Rs (C, W, K, E).

(a) Subsp. **disperma**: Calyx with hooked hairs; corolla without saccate invaginations in the throat. *Romania, Greece.*

(b) Subsp. **retorta** (Pallas) E. Kotejowa in Pawł., *Fl. Polska* **10**: 217 (1963) (*R. retorta* (Pallas) Lipsky): Calyx with curved hairs; corolla with distinct saccate invaginations in the throat. *Spain; S. part of U.S.S.R.*

Intermediates occur in S.W. & S.C. Asia.

---

[1] By A. O. Chater.   [2] By A. Hansen.   [3] By L. F. Ferguson.

29. Myosotis L.[1]

Annual or perennial herbs. Flowers in usually paired cymes which are usually ebracteate or sometimes bracteate below. Calyx more or less accrescent in fruit, regularly 5-lobed to half-way to base or more. Corolla rotate; tube usually short; limb regularly 5-lobed, flat or slightly concave, usually blue (sometimes white, yellow or yellow and blue); scales 5, usually included, papillose, white or yellow. Stamens usually included, with a terminal lingulate appendage; filaments inserted about the middle of the tube. Style included; stigma capitate. Nutlets 4, ovoid, erect, more or less compressed, smooth and shiny, brown to black, often with a distinct rim; area of attachment usually small, sometimes with a spongy or lingulate appendage.

Literature: J. Grau, *Österr. Bot. Zeitschr.* **111**: 561–617 (1964); *Mitt. Bot. Staatssamm.* (*München*) **5**: 675–688 (1965); **6**: 517–530 (1967); **7**: 17–100 (1968). R. Schuster, *Feddes Repert.* **74**: 39–98 (1967). T. Vestergren, *Svensk Bot. Tidskr.* **24**: 449–467 (1930); *Ark. Bot.* **29** A (8): 1–39 (1938).

1 Calyx with equal and uniform, straight, appressed setae, pointing towards the apex
  2 Upper cauline leaves with erecto-patent hairs on upper surface; attachment of the nutlets not spongy; annuals of dry habitats
    3 Corolla *c.* 5 mm in diameter; stem up to the highest leaf with patent hairs   **2. cadmea**
    3 Corolla not more than 3 mm in diameter; lower part of stem with patent hairs
      4 Pedicels not thickened in fruit, usually patent; stems erect   **3. ucrainica**
      4 Pedicels thickened towards the apex in fruit; stems erect, ascending or procumbent
        5 Inflorescence bracteate   **4. pusilla**
        5 Inflorescence ebracteate, or only the lowermost flowers in the axils of bracts
          6 Calyces up to 3 mm in fruit, crowded and appressed to the stem   **5. litoralis**
          6 Calyces in fruit rarely appressed, less than 3 mm and not crowded   **1. incrassata**
  2 Upper cauline leaves with short appressed hairs on upper surface; attachment of the nutlets spongy; annuals or perennials of wet habitats
    7 Calyx persistent in fruit, divided at anthesis to less than ½ way; teeth broadly triangular
      8 Annual or biennial, without stolons; limb of corolla saucer-shaped; nutlets up to 1 mm, ellipsoid, brown   **33. sicula**
      8 Biennial or perennial, often with subterranean stolons; limb of corolla flat; nutlets more than 1·5 mm, ovoid, usually black   **(36–39). scorpioides group**
    7 Calyx often caducous at maturity, divided at anthesis to at least ½ way; teeth narrowly triangular
      9 Stems with patent hairs at least at the base
        10 Stolons absent   **31. welwitschii**
        10 Stolons present   **30. secunda**
      9 Stems with appressed hairs or glabrous at the base
        11 Stolons present   **32. stolonifera**
        11 Stolons absent
          12 Pedicels (except in the single central flower) not longer than calyx in fruit   **33. sicula**
          12 Pedicels (especially those of the lower flowers) longer than calyx in fruit
            13 Nutlets up to 1·2 mm, brown; main veins of calyx branched at about the middle of the calyx-tube; calyx-tube up to 3 mm at maturity   **34. debilis**
            13 Nutlets up to 2 mm, dark brown to black; main veins of calyx branched at base; calyx-tube up to 6 mm at maturity   **35. laxa**

1 Calyx with hooked hairs or with hairs of 2 kinds; some setiform, usually hooked and ±patent; some shorter, slender, straight or curved
  14 Perennial or biennial
    15 Calyx-teeth lanceolate, obtuse; hairs on calyx straight, those of the teeth pointing forwards, those of the tube backwards   **40. azorica**
    15 Calyx-teeth linear to broadly triangular; hairs on calyx all pointing forwards, or with hooked hairs at the base
      16 Nutlets obtuse; attachment-area large, elliptical to orbicular, often slightly but regularly reniform, with lateral folds; calyx narrowed at the base in fruit, never deciduous   **(21–29). alpestris group**
      16 Nutlets ±acute; attachment-area small, irregularly reniform, without lateral folds; calyx rounded at the base in fruit, usually deciduous
        17 Calyx closed in fruit; limb of corolla saucer-shaped   **6. arvensis**
        17 Calyx open in fruit; limb of corolla flat
          18 Nutlets 2–3 mm; calyx shorter than corolla-tube; hooked hairs of the calyx 0·4 mm, stiff
            19 Limb of corolla not more than 4 mm in diameter, yellowish-white to pale blue   **18. soleirolii**
            19 Limb of corolla up to 8 mm in diameter, usually bright blue
              20 Nutlets *c.* 2 mm; stem not woody at base   **19. decumbens**
              20 Nutlets *c.* 3 mm; stem woody at base   **20. latifolia**
          18 Nutlets not more than 1·8 mm; calyx equalling or longer than corolla-tube; hooked hairs of the calyx 0·2 mm, soft
            21 Pedicels not more than 2·5 mm in fruit; flowers yellowish-white to pale blue   **18. soleirolii**
            21 Pedicels at least 5 mm in fruit; flowers bright to deep blue   **17. sylvatica**
  14 Annual
    22 Lower surface of leaves, especially on the veins, and base of stem with hooked hairs; inflorescence-axis with some upwardly projecting or patent hairs
      23 Calyx divided to ⅓ in fruit, almost all deflexed towards the base of the stem; nutlets narrowly ellipsoid   **16. refracta**
      23 Calyx divided to ½ in fruit, all but the oldest patent or pointing towards the apex of the stem; nutlets ovoid
        24 Hairs on stem pointing upwards, many appressed; calyx not deciduous in fruit; nutlets without a longitudinal furrow   **13. stricta**
        24 Hairs on stem all patent; calyx deciduous in fruit; nutlets with a ±well-marked longitudinal furrow
          25 Leaves broadly ovate, the lower petiolate; calyx with hooked hairs only   **15. speluncicola**
          25 Leaves lanceolate, sessile; calyx with both hooked hairs and soft, straight, deflexed hairs   **14. minutiflora**
    22 Lower surface of leaves and base of stem without hooked hairs; hairs of the inflorescence always appressed
      26 Flowers yellow, at least in bud; corolla-tube often lengthening after anthesis; nutlets always with a rim
        27 Flowers changing from yellow or cream to blue   **9. discolor**
        27 Flowers remaining yellow
          28 Stems rarely more than 10 cm; calyx with many stiff hooked hairs; corolla *c.* 4 mm   **12. persoonii**
          28 Stems up to 20 cm; calyx with few, short hooked hairs; corolla *c.* 2 mm   **11. balbisiana**
      26 Flowers blue, rarely white; corolla-tube rarely lengthening after anthesis
        29 Pedicels of the lowest flowers much elongated and deflexed in fruit; nutlets with a distinct basal appendage   **41. sparsiflora**
        29 Pedicels rarely more than twice as long as calyx and not deflexed in fruit; nutlets without a distinct appendage
          30 Calyx closed in fruit; the lowermost pedicels longer than the calyx, usually directed upwards; nutlets black   **6. arvensis**
          30 Calyx open or half-open in fruit; the lowermost pedicels rarely longer than the calyx, usually patent; nutlets brown

---

[1] By J. Grau and H. Merxmüller.

31 Nutlets with a rim; corolla-tube lengthening after
anthesis **10. congesta**
31 Nutlets without a rim; corolla-tube not lengthening
after anthesis
32 Flowers irregularly spaced, often close together but
sometimes wide apart; stem often sharply bent
several times **8. ruscinonensis**
32 Stem straight, with regularly spaced flowers
7. ramosissima

1. M. incrassata Guss., *Fl. Sic. Syn.* **1**: 214 (1843) (*M. cretica* Boiss. & Heldr., *M. idaea* Boiss. & Heldr.). Annual, covered with straight, setiform hairs, patent at the base of the stem, pointing forwards on the leaves, otherwise appressed. Stem 5–20 cm, ascending to erect, often much-branched at the base. Basal leaves 4 × 1 cm, the cauline smaller, ovate-lanceolate. Inflorescence with numerous flowers; pedicels clavate in fruit, thickened towards apex. Calyces in fruit in 2 regular rows, usually pointing upwards, rarely appressed, often deciduous. Limb of corolla up to 3 mm in diameter, saucer-shaped, blue, rarely white; tube shorter than calyx. Nutlets up to 1·2 mm, brown, without rim; attachment-area with 2 marginal grooves. $2n = 24$. *Dry, open habitats. S. Europe, from Sicilia eastwards.* Al Bu Cr Gr It Ju Rs (K) Si Tu.

2. M. cadmea Boiss., *Diagn. Pl. Or. Nov.* **2**(11): 122 (1849). Like **1** but up to 25 cm, erect; hairs on stems patent: leaves up to 7 × 1·5 cm, oblanceolate; calyx patent in fruit, the pedicels not thickened; limb of corolla up to 6 mm in diameter, flat. $2n = 24$. *E. & C. parts of Balkan peninsula.* Gr Ju Tu.

3. M. ucrainica Czern., *Bull. Soc. Nat. Moscou* **18**(2): 133 (1845). Like **1** but up to 30 cm, erect, not branched at the base; pedicels in fruit up to 10 mm, patent, not thickened; limb of corolla *c.* 1·5 mm in diameter. *Forest-margins and river-banks. Ukraine and S.C. Russia.* Rs (C, W).

4. M. pusilla Loisel. in Desv., *Jour. Bot. Rédigé* **2**: 260 (1809). Annual. Stem 2–7 cm, procumbent, often much-branched and forming a cushion. Leaves narrowly lingulate, dark grey-green, thickly covered with straight hairs. Inflorescence usually bracteate throughout; pedicels 2·5 mm in fruit, uniformly thickened. Calyx *c.* 3·5 mm in fruit, patent or appressed, longer than the pedicels. Limb of corolla up to 2 mm in diameter, saucer-shaped, pale blue to white. Nutlets as in **1**. $2n = 24$. *Sandy places near the sea. S. France, Corse, Sardegna.* Co Ga Sa.

5. M. litoralis Steven ex Bieb., *Fl. Taur.-Cauc.* **3**: 118 (1819). Like **4** but ebracteate; pedicels somewhat clavate at maturity. Calyces very crowded in fruit, appressed to stem and pointing upwards. *Sandy places or rocks near the sea. Aegean region; Turkey-in-Europe; Krym.* Gr Rs (K) Tu.

6. M. arvensis (L.) Hill, *Veg. Syst.* **7**: 55 (1764) (*M. intermedia* Link). Biennial. Stem up to 60 cm, robust, often much-branched at the base. Basal leaves up to 8 × 1·5 cm, oblanceolate, not distinctly petiolate. Leafy part of stem with patent hairs, the leafless part with appressed, straight hairs. Inflorescence ebracteate; pedicels directed upwards in fruit, the lowest up to 1 cm, becoming shorter above. Calyx up to 7 mm in fruit, closed, with many patent, hooked hairs especially at its base, deciduous. Limb of corolla *c.* 3 mm in diameter, saucer-shaped, bright blue. Nutlets up to 2·5 × 1·2 mm, greenish-black to black, pointed, with a rim; attachment-area small. *Dry places, and as a ruderal. Throughout Europe.* All except Az Cr Sb.

Very variable, the habit depending greatly on the environment.

(a) Subsp. **arvensis**: Calyx not more than 5 mm in fruit; hooked hairs not more than 0·4 mm. Nutlets not more than 2 mm. $2n = 52$. *Throughout the range of the species.*

(b) Subsp. **umbrata** (Rouy) O. Schwarz, *Mitt. Thür. Bot. Ges.* **1**(1): 112 (1949): Calyx up to 7 mm in fruit; hooked hairs up to 0·6 mm. Nutlets up to 2·5 mm. $2n = 66$. ● *W. Europe.*

7. M. ramosissima Rochel in Schultes, *Österreichs Fl.* ed. 2, **1**: 366 (1814) (*M. collina* auct. plur., non Hoffm., *M. gracillima* Loscos & Pardo, *M. hispida* Schlecht.). Annual. Stem up to 40 cm. Rosette-leaves up to 4 × 1 cm, lanceolate, obtuse; leaves with soft, patent straight hairs. Stem soft, at base with patent, above with appressed, straight hairs. Inflorescence lax, ebracteate; pedicels in fruit scarcely longer than calyx, patent, more or less straight. Calyx up to 4 mm in fruit, divided to $\frac{1}{2}$, half-open, with many, often deflexed hooked hairs at the base, deciduous. Limb of corolla up to 3 mm in diameter, saucer-shaped, bright blue. Nutlets 1·2 × 0·7 mm, brown, without rim; attachment-area usually filled with spongy tissue. *Dry places. Most of Europe except the extreme north.* All except ?Az Fa ?Is Rs (N) Sb.

(a) Subsp. **ramosissima**: Flowers only on the upper part of the stem. Calyx up to 4 mm in fruit, elongated, segments narrowly triangular. Corolla scarcely exceeding calyx. Nutlets smooth. $2n = 48$. *Throughout the range of the species.*

(b) Subsp. **globularis** (Samp.) Grau, *Mitt. Bot. Staatssamm.* (*München*) **7**: 58 (1968) (*M. globularis* Samp.): Flowers extending almost to the base of the stem. Calyx scarcely more than 2 mm in fruit, almost spherical, segments broadly triangular. Corolla distinctly exceeding calyx. Nutlets with an indistinct rim at the apex. *Sandy places, especially near the sea.* ● *W. Europe, from N. Portugal to S. England.*

Intermediates between (a) and (b) have been described as subsp. *lebelii* (Godron) Blaise.

8. M. ruscinonensis Rouy, *Bull. Soc. Bot. Fr.* **38**: 377 (1891). Like **7** but low-growing; cymes patent; flowers usually irregularly arranged on the axis, often fused together; axis of inflorescence often sharply bent; pedicels up to 3 times as long as the calyces in fruit. Flowers white to bright blue, often bracteate. $2n = ?36, 48$. *Sandy places by the sea.* ● *S. France.* Ga.

9. M. discolor Pers., *Syst. Veg.* ed. 15, 190 (1797) (*M. collina* Hoffm., *M. versicolor* Sm.). Annual; stem up to 30 cm, slender, often branched at the base. Basal leaves up to 4 cm, lanceolate, obtuse; cauline leaves ovate-lanceolate, the upper acute. Hairs on the stem patent below, appressed above; those on the pedicels often patent. Inflorescence not leafy at the base, many-flowered, lax below, dense above, pedicels shorter than calyx in fruit. Calyx up to 4·5 mm in fruit, rarely deciduous, with soft deflexed, hooked hairs. Corolla up to 4 mm; limb *c.* 1·3 mm in diameter, saucer-shaped, pale yellow or cream at first, becoming pink, violet or even blue; tube pale blue to dark violet. Nutlets *c.* 1·2 × 0·8 mm, dark brown, with a wide rim. *Dry places; sometimes in wet meadows. Europe, eastwards to Latvia, N. Ukraine & C. Jugoslavia, and northwards to Iceland and S. Fennoscandia.* ?Al Au Az Be Br Co Cz Da Fa Ga Ge Hb He Ho Hu Is It Ju Lu No Po Rm Rs (B, C, W) Sa Su.

(a) Subsp. **discolor**: Flowers cream or yellow at first; corolla up to 4 mm; with at least one pair of opposite cauline leaves. $2n = 72, ?64$. *Throughout the range of the species.*

(b) Subsp. **dubia** (Arrondeau) Blaise, *Bot. Jour. Linn. Soc.* **65**: 261 (1972) (*M. dubia* Arrondeau): Flowers cream at first; corolla not more than 2 mm; with no opposite leaves. $2n = 24$. ● *W. Europe.*

Subsp. *canariensis* (Pitard) Grau, from Madeira and Islas Canarias, has been recorded from N.W. Europe but probably in error. It has shorter flowers and more condensed inflorescences than 9(b).

10. **M. congesta** R. J. Shuttlew. ex Albert & Reyn., *Coup d'Oeil Fl. Toulon Hyères* 16 (1891). Like **9** but not more than 25 cm, usually smaller and more slender; flowers always blue; calyx 3 mm in fruit, deciduous; corolla not more than 2 mm, limb 1 mm in diameter; nutlets 1 × 0·5 mm, dark brown. $2n=24$, ?32, 48. *Mediterranean region; very local.* Co Cr Ga Gr.

Often confused with 7(a); further investigation is needed.

11. **M. balbisiana** Jordan, *Pug. Pl. Nov.* 128 (1852). Like **9** but smaller (up to 20 cm) and more delicate; calyx not more than 3 mm in fruit, the hooked hairs few, short; corolla *c.* 2 mm; limb scarcely more than 1 mm in diameter, pale yellow. ● *S. France, N. Spain.* Ga Hs.

12. **M. persoonii** Rouy, *Fl. Fr.* 6: 327 (1900) (*M. lutea* (Cav.) Pers., non Lam.). Like **9** but low-growing, compact, scarcely more than 10 cm; inflorescence dense, the lowermost flowers with bracts; calyx 3 mm in fruit, enlarging only slightly, the hooked hairs numerous, stiff; corolla *c.* 4 mm, limb 2 mm in diameter when flat, bright yellow. $2n=48$. ● *N. & C. Spain, Portugal.* Hs Lu.

13. **M. stricta** Link ex Roemer & Schultes, *Syst. Veg.* 4: 104 (1819) (*M. micrantha* auct., non Pallas ex Lehm., *M. vestita* Velen.). Annual; stems up to 30 cm but often less, erect, usually much-branched at the base, leafy up to the inflorescence, with hooked hairs at the base. Basal leaves up to 2·5 × 0·5 cm, lanceolate; hairs on lower surface of leaves, especially the midrib, hooked, the others straight. Axis of inflorescence with patent and upwardly pointing to subappressed hairs; flowers crowded towards the apex; pedicels up to 1·5 mm, but usually shorter. Calyx with deflexed hooked hairs and appressed, straight hairs at the base, *c.* 4 mm in fruit, divided to ½, closed, not deciduous. Corolla scarcely exceeding the calyx; limb scarcely more than 1 mm in diameter, saucer-shaped, bright to pale blue. Nutlets 1·5 × 1 mm, brown, with a distinct rim, keeled at the apex. $2n=36$, 48. *Dry, sandy places. Most of Europe.* All except ?Az Bl Br Co Fa Hb Sa Sb Si.

14. **M. minutiflora** Boiss. & Reuter, *Pugillus* 80 (1852) (*M. rhodopea* Velen.). Annual; stem not more than 15 cm, delicate, branched at the base, with widely spaced leaves up to the lowest flowers, with a dense, patent indumentum above but also with hooked hairs below. Basal leaves up to 2 × 0·4 cm, oblanceolate, sessile, with hooked hairs beneath. Flowers very crowded, especially towards the apex of the inflorescence. Pedicels in lower part of inflorescence up to 3 mm in fruit, often bent and deflexed, in upper part 1 mm, patent or pointing upwards. Calyx up to 3·5 mm in fruit, divided to more than ½, open, deciduous, with deflexed, hooked hairs and appressed, straight hairs at the base. Corolla scarcely exceeding the calyx; limb scarcely 1 mm in diameter, saucer-shaped, pale blue. Nutlets 1·5 × 1 mm, brown, broadly ovate, with distinct rim and obliquely triangular attachment-area on the inner side of the keel which runs from apex to base. $2n=48$. *Mountains of S. Spain, N.E. Greece and S. Bulgaria.* Bu Gr Hs. (*S.W. Asia.*)

15. **M. speluncicola** (Boiss.) Rouy, *Naturaliste* (*Paris*) 1: 501 (1881). Like **14** but with broadly lanceolate to ovate leaves; basal leaves distinctly petiolate, often smaller than cauline; flowers less crowded, often bracteate; calyces in fruit without straight hairs at base, hooked hairs mainly restricted to the veins; limb of corolla 2 mm in diameter, shallowly saucer-shaped to flat, white. $2n=24$. *Mountain caves. S.E. France; C. Appennini, S. Jugoslavia.* Ga It Ju. (*S. Anatolia.*)

16. **M. refracta** Boiss., *Voy. Bot. Midi Esp.* 2: 433 (1841). Annual; stem up to 25 cm, often shorter, usually rigidly erect, leafy up to the lowest flowers, with patent, lanate indumentum above and hooked hairs below. Leaves 4 × 1 cm, narrowly to broadly lanceolate, rarely ovate, the lower surface with hooked hairs. Flowers lax below, crowded above. Pedicels scarcely more than 1 mm in fruit, almost always deflexed at maturity. Calyx up to 4·5 mm in fruit, divided to *c.* ⅓, open, deciduous, with strongly deflexed, hooked hairs and short, delicate straight hairs at the base. Limb of corolla up to 1·5 mm in diameter, flat, or rather smaller and saucer-shaped, pale to bright blue. Nutlets 2 × 1 mm, ellipsoid or narrowly obovate, brown, with a small, lateral attachment-area, distinct rim and, on the inner side, a furrow running from apex to base. *Mountains of S. Spain, E. Mediterranean region and Krym.* Cr Gr Hs Ju Rs (K).

1 Only the lowermost pedicels deflexed in fruit; leaves broadly lanceolate to ovate **(c) subsp. aegagrophila**
1 All pedicels deflexed in fruit; leaves narrowly to rather broadly lanceolate
2 Calyx-tube ± evenly covered with hooked hairs; calyces not appressed in fruit **(a) subsp. refracta**
2 Calyx-tube without hooked hairs in lower ⅓; calyces closely appressed in fruit **(b) subsp. paucipilosa**

(a) Subsp. **refracta**: Rigidly erect. Leaves narrowly to rather broadly lanceolate. Calyx-tube more or less evenly covered with deflexed hooked hairs; calyx usually not appressed when deflexed in fruit. Nutlets broadest at or below the middle, with furrow in the middle. $2n=44$. *Throughout the range of the species.*

(b) Subsp. **paucipilosa** Grau, *Mitt. Bot. Staatssamm.* (*München*) 7: 90 (1968): Rigidly erect. Leaves narrowly to rather broadly lanceolate. Calyx-tube without hooked hairs in lower ⅓; calyces closely appressed when deflexed in fruit. Nutlets widest above the middle, furrow lateral and coinciding with rim. $2n=20$. *S. Greece, Kriti.*

(c) Subsp. **aegagrophila** W. Greuter & Grau, *Candollea* 25: 8 (1970): Delicate, flexuous. Leaves broadly lanceolate to ovate. Calyx-tube more or less evenly covered with hooked hairs; only the lowermost calyces deflexed in fruit. Nutlets as in subsp. (a). ● *Kriti.*

17. **M. sylvatica** Hoffm., *Deutschl. Fl.* 61 (1791). Biennial to perennial; stem up to 50 cm, often much-branched, very leafy, with appressed hairs in inflorescence, otherwise with setiform hairs. Leaves *c.* 8 × 3 cm, broadly to narrowly ovate, or elliptical, the basal usually not distinctly petiolate. Pedicels up to 7(–15) mm, directed upwards. Calyx up to 5 mm in fruit, with short hooked hairs up to 0·2 mm, with linear to narrowly triangular teeth and rounded base, open, deciduous. Limb of corolla up to 8 mm in diameter but often smaller, flat, bright blue. Nutlets 1·7 × 1·2 mm, ovoid, acute, with rim; attachment-area very small. *Much of Europe, but absent from the south-west and much of the north.* Al Au Be Br Bu Cz Da Ga Ge Gr He Ho Hu It Ju *No Po Rm Rs (B, C, W) Su Tu [Fe].

1 Calyx with sparse hooked hairs almost disappearing at maturity; nutlets ovoid **(a) subsp. sylvatica**
1 Calyx with numerous hooked hairs; nutlets narrowly ovoid
2 Corolla-limb not more than 4 mm in diameter **(b) subsp. subarvensis**
2 Corolla-limb more than 4 mm in diameter
3 Pedicels not more than 7 mm **(c) subsp. cyanea**
3 Lowest pedicels at least 10 mm **(d) subsp. elongata**

(a) Subsp. **sylvatica**: Calyx green, with sparse hairs and few hooked bristles. Corolla-tube as long as the calyx. Nutlets ovoid. $2n=18(20)$. *Throughout the range of the species except the south.*

(b) Subsp. **subarvensis** Grau, *Österr. Bot. Zeitschr.* **111**: 571 (1964): Plant much-branched. Flowers small. Calyx with numerous hooked hairs. Corolla-tube equalling calyx. Nutlets narrowly ovoid. $2n=18$. ● *Italy, Jugoslavia.*

(c) Subsp. **cyanea** (Boiss. & Heldr.) Vestergren, *Ark. Bot.* **29A**(8): 10 (1938): Pedicels not more than 7 mm. Calyx silvery, with numerous hooked hairs. Corolla-tube often shorter than calyx. Nutlets narrowly ovoid. $2n=18$, 20. *S. Italy, Balkan peninsula.*

Plants transitional to subsp. (d) occur in Italy.

(d) Subsp. **elongata** (Strobl) Grau, *Österr. Bot. Zeitschr.* **111**: 574 (1964): Like (c) but stems often rather tall; pedicels up to 15 mm; calyx not so readily deciduous in fruit. $2n=20$, 22. ● *S. Italy, Sicilia.*

18. **M. soleirolii** Gren. & Godron, *Fl. Fr.* **2**: 534 (1853). Like **17** but pedicels up to 2·5 mm; calyx with dense hooked hairs, often shorter than the corolla-tube; corolla white to pale blue, limb not more than 4 mm in diameter. $2n=18$. ● *Corse.* Co.

19. **M. decumbens** Host, *Fl. Austr.* **1**: 228 (1827). Like **17** but always perennial, often with creeping rhizome; calyx-teeth narrowly to broadly triangular, divided to half-way at anthesis, with numerous hooked hairs up to 0·4 mm; limb of corolla up to 8 mm in diameter; corolla-tube usually longer than calyx, anthers sometimes exserted; nutlets *c.* $2\times1\cdot5$ mm; attachment-area broadly reniform. *N., C. & S.W. Europe, mainly in the mountains.* Au Br Cz Fe Ga Ge He Hs It Ju No Po Rm ?Rs (N) Su.

1 Corolla-tube twice as long as calyx; style longer than corolla-tube
  2 Anthers exserted   **(c) subsp. variabilis**
  2 Anthers not exserted   **(d) subsp. kerneri**
1 Corolla-tube only slightly longer than calyx; style not longer than corolla-tube
  3 Nutlets $2\times1\cdot1$ mm, narrowly ovoid   **(b) subsp. teresiana**
  3 Nutlets $2\times1\cdot5$ mm, broadly ovoid
    4 Calyx often more than 7 mm in fruit; calyx-teeth narrowly triangular   **(e) subsp. florentina**
    4 Calyx rarely more than 6 mm in fruit; calyx-teeth narrowly to broadly triangular   **(a) subsp. decumbens**

(a) Subsp. **decumbens**: Corolla-tube twice as long as calyx; anthers included; style not exserted. Calyx up to 6 mm in fruit. Nutlets broadly ovoid. $2n=32$. ● *N. Europe; C. & S. Alps.*

(b) Subsp. **teresiana** (Sennen) Grau, *Österr. Bot. Zeitschr.* **111**: 578 (1964): Corolla-tube less than twice as long as calyx; anthers included; style exserted. Calyx up to 5 mm in fruit. Nutlets narrowly ovoid. $2n=32$, 34. ● *Mountains of S.W. Europe.*

(c) Subsp. **variabilis** (M. Angelis) Grau, *op. cit.* 580 (1964) (*M. sylvatica* subsp. *variabilis* (M. Angelis) Nyman): Corolla-tube about twice as long as calyx; anthers and style exserted. Calyx-teeth and nutlets as in subsp. (b). $2n=32$. ● *S.E. Austria, Romania, Poland.*

Plants from Romania like subsp. (c) but without hooked hairs or bristles have been described as **M. transsylvanica** Porc., *Magyar Növ. Lapok* **9**: 130 (1885).

(d) Subsp. **kerneri** (Dalla Torre & Sarnth.) Grau, *Österr. Bot. Zeitschr.* **111**: 580 (1964): Like subsp. (c) but anthers not exserted. $2n=32$. ● *E.C. Europe.*

(e) Subsp. **florentina** Grau, *Mitt. Bot. Staatssamm. (München)*

8: 130 (1970): Like (a) but fruiting calyx often much larger and more than 7 mm, with more numerous hooked hairs; basal leaves often much larger. $2n=28$. ● *Appennini.*

20. **M. latifolia** Poiret in Lam., *Encycl. Méth. Bot.* **12**: 45 (1816). Like **19** but stems up to 70 cm, woody at base, often with large basal leaves; calyx more than 7 mm in fruit, with numerous hooked hairs up to 0·6 mm; corolla-tube twice as long as calyx, limb more than 10 mm in diameter, anthers included; style not exserted; nutlets up to 3×2 mm. *Açores.* Az. (*N.W. Africa, Islas Canarias.*)

(21–29). **M. alpestris group.** Perennial with short rhizome and numerous, often long and somewhat fleshy roots. Stems usually roughly hairy below, rarely glabrous; inflorescence with appressed hairs. Calyx densely hairy, more or less acute at the base in fruit, closed or slightly open, never deciduous. Corolla-tube shorter than calyx; limb flat. Nutlets often more than 2 mm, ovoid to ellipsoid, obtuse, black, shining; attachment-area wide, elliptical to reniform, usually extended laterally into two grooves.

A group of closely related species, many of which occur in the montane and alpine zones of the European mountains.

1 Calyx with few or no hooked bristles
  2 Basal leaves narrowly ovate, acute, with slender petioles; cauline leaves lanceolate to linear; hairs on calyx appressed
    3 Basal leaves glabrescent only beneath or not at all   **24. stenophylla**
    3 Basal leaves glabrescent on both surfaces   **25. asiatica**
  2 Basal leaves elliptical to oblong-lanceolate, sessile or stalked; cauline leaves ovate to lanceolate; hairs on calyx patent
    4 Calyx with dense, straight, white hairs, more or less swollen at base in fruit   **23. ambigens**
    4 Calyx rarely with dense, white hairs, not swollen at base in fruit   **21. alpestris**
1 Calyx with ± numerous hooked bristles
  5 Hooked bristles on pedicels as well as calyx; nutlets narrowly ovoid; basal leaves glabrescent beneath
    6 Basal leaves petiolate, rounded; plant up to 25 cm   **27. corsicana**
    6 Basal leaves not petiolate, lingulate; plant up to 12 cm, forming a mat   **26. alpina**
  5 Hooked bristles not present on pedicels; basal leaves hairy beneath, rarely glabrescent
    7 Cauline leaves uniform, narrowly linear, obtuse; nutlets narrowly ovoid   **29. lithospermifolia**
    7 Cauline leaves variable, broadly ovate to linear; nutlets ovoid to ellipsoid
      8 Nutlets up to 1·8 mm; rim much elongated at upper end   **22. gallica**
      8 Nutlets usually more than 1·8 mm; rim not much elongated
        9 Nutlets with complete rim; calyx with numerous hooked hairs; flowers deep blue   **28. suaveolens**
        9 Only upper part of nutlet with rim; calyx with fewer hooked hairs; flowers bright blue   **21. alpestris**

21. **M. alpestris** F. W. Schmidt, *Fl. Boëm.* **3**: 26 (1794) (*M. sylvatica* subsp. *alpestris* (F. W. Schmidt) Gams). Stems 5–35 cm. Basal leaves petiolate or sessile; cauline very variable, ovate, elliptical, spathulate or linear. Pedicels directed upwards in fruit, not longer than calyx. Calyx up to 7 mm and often much enlarged in fruit, densely hairy, with or without hooked bristles. Corolla-limb up to 9 mm in diameter, bright or deep blue. Nutlets up to 2·5 mm, ovoid to ellipsoid, blunt, with rim only in the distal half; attachment-area wide, elliptical, elongated laterally into 2 grooves, often extending towards the base. $2n=24$, 48, 72 (70). *Mountains of Europe, from Scotland and the Carpathians to N. Spain and Bulgaria.* Au Br Bu Cz Ga Ge He Hs It Ju Po Rm Rs (W).

Very variable, with many local forms. Small mat-forming plants with straight, appressed calyx-hairs from S. Jugoslavia have been described as **M. mrkvickana** Velen., *Rel. Mrkvick.* 22 (1922).

22. **M. gallica** Vestergren, *Ark. Bot.* 29A(8): 27 (1938). Like 21 but up to 20 cm; flowers bright blue; calyx always with hooked bristles at the base; nutlets up to 1·8 mm with distinct rim, much elongated at upper end. 2*n*=24. ● *S.E. France.* Ga.

23. **M. ambigens** (Béguinot) Grau, *Mitt. Bot. Staatssamm.* (*München*) 8: 127 (1970). Like 21 but not more than 12 cm; calyx with straight, very dense hairs, especially on the margins of the segments, and very few hooked bristles, swollen at the base in fruit; flowers whitish to bright blue. 2*n*=24. ● *C. Appennini.* It.

24. **M. stenophylla** Knaf in Berchtold & Opiz, *Ökon.-Techn. Fl. Böhm.* 2: 126 (1839). Stems 5–35 cm, branched at base. Basal leaves 6×0·5 cm, narrowly ovate, acute, with a distinct, slender petiole; cauline numerous, lanceolate, acute. Pedicels directed upwards in fruit, not longer than calyx. Calyx up to 7 mm and often much enlarged in fruit, with appressed or bent hairs and no hooked hairs. Corolla-limb up to 8 mm in diameter, bright or deep blue. Nutlets up to 2·5 mm, ovoid to ellipsoid, obtuse, without rim; attachment-area wide, elliptical, elongated laterally into 2 grooves often extending towards the base. 2*n*=48. *E. & E.C. Europe.* Au Cz Hu Po Rm Rs (C, W, K, E).

25. **M. asiatica** (Vestergren) Schischkin & Serg. in Krylov, *Fl. Zap. Sibir.* 9: 2272 (1937). Like 24 but low-growing, not more than 10 cm; leaves and often also the stem almost glabrous. *Arctic and subarctic Russia.* Rs (N).

26. **M. alpina** Lapeyr., *Hist. Abr. Pyr.* 85 (1813) (*M. pyrenaica* Pourret). Stems not more than 12 cm, forming a dense mat. Basal leaves almost always sessile, glabrous beneath; cauline variable. Pedicels up to 2 mm, directed upwards in fruit. Calyx up to 7 mm, with numerous hooked bristles at the base which extend to the very short pedicels, crowded and often very accrescent in fruit. Corolla-limb up to 8 mm in diameter, bright or deep blue. Nutlets 2×1·2 mm, narrowly ovoid, with rim only at apex; attachment-area wide, elliptical, elongated laterally into 2 grooves, often extending towards the base. 2*n*=24. *Rocks, screes and mountain pastures.* ● *Pyrenees.* Ga Hs.

27. **M. corsicana** (Fiori) Grau, *Mitt. Bot. Staatssamm.* (*München*) 8: 128 (1970). Stems up to 25 cm, almost glabrous. Basal leaves 3×1·5 cm, elliptical, often crenate towards apex, glabrous beneath, with a few bristles above; petiole up to 9 cm, distinct; cauline leaves ovate to lanceolate, glabrescent beneath. Pedicels up to 4 mm in fruit. Calyx, especially towards base, with dense, deflexed hooked bristles, which extend to the upper part of the pedicels, up to 5 mm in fruit. Limb of corolla up to 6 mm in diameter. Nutlets 2×1 mm, narrowly ovoid, with elliptical attachment-area and almost complete rim. 2*n*=24. *Alpine habitats.* ● *Corse.* Co.

28. **M. suaveolens** Waldst. & Kit. ex Willd., *Enum. Pl. Hort. Berol.* 176 (1809). Stems up to 40 cm, often bushy and much-branched. Basal leaves petiolate; cauline ovate to lanceolate. Calyx densely hairy, with hooked bristles, much enlarged in fruit up to 7 mm, and especially the lower on long pedicels. Corolla usually deep blue. Nutlets *c.* 2 mm, usually ellipsoid, with complete rim and wide, elliptical attachment-area which often extends towards the base. 2*n*=24, 48. *Alpine habitats.* ● *Balkan peninsula.* Al Bu Gr Ju.

29. **M. lithospermifolia** (Willd.) Hornem., *Hort. Hafn.* 1: 173 (1813). Stems up to 40 cm, often much-branched at the base, densely hairy, grey-green. Basal leaves up to 5×1 cm, obovate to narrowly elliptical, obtuse, gradually narrowed into a long petiole; cauline up to 4×0·5 cm, numerous, linear, obtuse. Inflorescence with numerous flowers; pedicels up to 5 mm, directed upwards. Calyx, especially at the base, with numerous patent or deflexed hooked bristles. Nutlets up to 2·2×1·2 mm, narrowly ovoid. *Mountain meadows.* Krym. Rs (K). (*Caucasus, N. Anatolia.*)

30. **M. secunda** A. Murray, *North. Fl.* 1: 115 (1836). Annual to biennial; stems 20–40 cm, usually slender, with rooting stolons emerging from the axils of the lowermost leaves, and with patent to upwardly pointing hairs in the leafy parts. Leaves 4×1·5 cm, elliptical. Inflorescence often much-branched, lax, with appressed or upwardly pointing hairs. Calyx divided to rather more than ½, open, saucer-shaped and up to 5 mm in diameter in fruit, easily deciduous, deflexed. Limb of corolla up to 8 mm in diameter, bright blue. Nutlets 1·8×1·2 mm, black, ovoid, acute, with rim, rhombic attachment-area and spongy appendage. 2*n*=24, 48. *Wet places; in mountains in the southern part of its range.* W. Europe. Az Br Fa Ga Hb Hs Lu.

31. **M. welwitschii** Boiss. & Reuter in Boiss., *Diagn. Pl. Or. Nov.* 3(3): 138 (1856). Like 30 but stems up to 60 cm, robust, never with rooting stolons but producing lateral flowering branches at the base; hairs at base of stem somewhat deflexed; leaves 7×1·5 cm; calyx up to 6 mm in fruit, campanulate; limb of corolla up to 10 mm in diameter. 2*n*=24. *Wet places, usually near the coast. Portugal.* Lu. (*Morocco.*)

Plants slightly hairy or almost glabrous at the base of the stem are known from two localities in Portugal and have been described as **M. lusitanica** Schuster, *Feddes Repert.* 74: 85 (1967). They also resemble 34, and further investigation is needed.

32. **M. stolonifera** (DC.) Gay ex Leresche & Levier, *Deux Excurs. Bot.* 83 (1880). Like 30 but smaller; hairs at base of stem appressed; stolons numerous in lower part of plant, often much-branched; leaves up to 2×1 cm; calyx up to 3 mm in fruit; limb of corolla 4–5 mm in diameter; nutlets 1·2×0·7 mm, ovoid, the rim only weakly developed. 2*n*=24. *Wet places; in mountains in the southern part of its range.* ● *W. Europe.* Br Hs Lu.

33. **M. sicula** Guss., *Fl. Sic. Syn.* 1: 214 (1843). Annual or biennial, stems either short and little-branched, or up to 50 cm and much-branched, erect to ascending; hairs on stems and leaves straight, appressed and pointing upwards. Basal leaves 6×0·8 cm, broadly linear, usually withering at maturity; cauline ovate-oblong. Inflorescence with flowers in two distinct rows; only the lowermost pedicels longer than the calyx at maturity. Calyx up to 6 mm in fruit; tube twice as long as the scarcely divergent teeth. Limb of corolla up to 3 mm in diameter, saucer-shaped, bright blue. Nutlets 1 mm, ovoid or ellipsoid, dark to medium brown, with appendage on the attachment-area. 2*n*=46, 46+2. *Wet places. S. & W. Europe, northwards to N.W. France.* Bu Co Ga Gr Hs It Ju Sa Si Tu.

34. **M. debilis** Pomel, *Nouv. Mat. Fl. Atl.* 298 (1875). Like 33 but basal leaves narrowly ovate; pedicels in lower and middle part of inflorescence longer than calyx at maturity; calyx divided to *c.* ½ in fruit, campanulate, with patent teeth, branches of the 5·

main veins arising at about the middle of the tube (not at the base); limb of corolla 4 mm in diameter; nutlets ovoid. 2n=48. *Wet places. Iberian peninsula.* Hs Lu.

35. **M. laxa** Lehm., *Pl. Asperif.* 83 (1818) (*M. lingulata* Lehm.). Annual or biennial; stems 20–50 cm, branched from the base, with straight, forwardly-directed, appressed hairs. Leaves up to 8 × 1·5 cm, lanceolate, extending up to the first flowers. Pedicels at the base of the inflorescence often elongated in fruit, up to 2·5 cm, and deflexed. Calyx divided to ½ or less, up to 8 mm in fruit, often deciduous; teeth narrowly triangular; branches of the main veins arising at the base of the tube. Limb of corolla up to 5 mm in diameter, bright blue, more or less flat. Nutlets up to 2 × 1·4 mm, dark brown, ovoid, obtuse, usually truncate at the base, with a spongy appendage on the attachment-area. *Wet places. Most of Europe, but rarer in the south.* All except Az Bl Cr Is Sa Sb Si.

(a) Subsp. **caespitosa** (C. F. Schultz) Hyl. ex Nordh., *Norsk Fl.* 529 (1940) (*M. caespitosa* C. F. Schultz, *M. scorpioides* subsp. *caespitosa* (C. F. Schultz) F. Hermann): Erect. Pedicels not more than 1 cm. Calyx not more than 5 mm. Nutlets not more than 1·5 × 1 mm. *Throughout the range of the species.*
(b) Subsp. **baltica** (Sam.) Hyl. ex Nordh., *loc. cit.* (1940) (*M. baltica* Sam., *M. scorpioides* subsp. *laxa* sensu Hegi): Ascending. Pedicels in lower part of inflorescence up to 2·5 cm. Calyx up to 8 mm. Nutlets up to 2 × 1·4 mm. 2n=?84, 88. ● *Fennoscandia, Estonia, Latvia.*

Subsp. *laxa* occurs in North America and may possibly occur in Europe; further investigation is needed.

(36–39). **M. scorpioides** group. Biennial or perennial, often with rhizome and stolons. Stems often with decurrent leaf-bases and angled; indumentum in lower part of stem variable, inflorescence-axis with upwardly-pointing appressed hairs. Calyx with short, straight, appressed bristles, divided to ½ or less, open in fruit, not deciduous; teeth broadly triangular. Limb of corolla flat; tube usually as long as the calyx. Nutlets ovoid-ellipsoid; rim absent or indistinct; attachment-area roundish with a spongy appendage.

Very variable, especially in C. Europe.

1 Stems not more than 10 cm in fruit, mat-forming   **37. rehsteineri**
1 Stems more than 10 cm in fruit, not mat-forming
 2 All or some of the hairs at the base of the stem deflexed
  3 Stem with deflexed hairs only at the base; usually without subterranean stolons   **38. nemorosa**
  3 Stem with deflexed, appressed hairs as far up as the upper leaves, rough; subterranean stolons present   **39. lamottiana**
 2 Stems glabrous at the base or with patent or upwardly-pointing hairs
  4 Limb of corolla not more than *c.* 5 mm in diameter; biennial, usually without stolons   **38. nemorosa**
  4 Limb of corolla more than 5 mm in diameter; perennial, with subterranean stolons
   5 Lower leaves hairy beneath, the hairs long, somewhat patent and pointing towards the base of the leaf   **38. nemorosa**
   5 Lower leaves glabrous beneath, or the hairs short, setiform and usually pointing towards the apex of the leaf   **36. scorpioides**

36. **M. scorpioides** L., *Sp. Pl.* 131 (1753) (*M. palustris* (L.) Hill, *M. scorpioides* subsp. *palustris* (L.) F. Hermann). Perennial; with creeping rhizome and runners; stems up to 100 cm, ascending to erect; base of stem glabrous or with patent hairs. Leaves up to 10 × 2 cm, oblong to oblong-lanceolate, usually with appressed forwardly-pointing hairs, rarely the lowermost with

deflexed bristles beneath, often glabrescent. Pedicels up to 10 mm in fruit. Calyx up to 6 mm in fruit, with a wide base. Limb of corolla up to 8 mm in diameter; style often protruding. Nutlets up to 1·8 × 1·2 mm, ovoid, with a faint rim above. 2n=66, ?64. *Wet places; often in water. C. & N. Europe.* Au ?Be Br Cz Da Fa Fe Ga Ge Hb He Ho Hu Is It Ju No Po Rm Rs (N, B, C, W, K, E) Sa Su.

37. **M. rehsteineri** Wartm., *Ber. Tät. St. Gall. Naturw. Ges.* **1884**: 276 (1884) (*M. scorpioides* subsp. *caespititia* (DC.) E. Baumann). Like **36** but smaller, mat-forming; stems not more than 10 cm, with appressed, upwardly-directed hairs; leaves smaller, not more than 2·5 × 1 cm; calyx not more than 5 mm in fruit, crowded; pedicels not more than 7 mm in fruit; corolla-limb not more than 10 mm, pink to bright blue. 2n=22. *Lake-margins.* ● *Alps.* Au Ge He It.

38. **M. nemorosa** Besser, *Enum. Pl. Volhyn.* 52 (1822). Like **36** but biennial (rarely perennial), usually without stolons; stems up to 50 cm, shining, the base glabrous or with deflexed, patent or upwardly pointing, hairs; lowermost leaves usually with deflexed, not appressed hairs beneath; calyx not more than 5 mm in fruit; corolla-limb not more than 6 mm in diameter; nutlets not more than 1·8 × 0·8 mm, narrowly ovoid. 2n=22, 44. *Wet meadows and woods. C. & E. Europe, rare in S. Europe.* All except Az Bl Cr Fa Fe Hs Is Lu Sb Tu.

More hairy, often more robust plants with long, subterranean stolons from S. & S.E. Europe have been distinguished as subsp. **orbelica** (Velen.) Schuster, *Feddes Repert.* **74**: 74 (1967). They have 2n=22, 44 and are perhaps related to **39**.

39. **M. lamottiana** (Br.-Bl. ex Chassagne) Grau, *Mitt. Bot. Staatssamm.* (*München*) **8**: 133 (1970). Perennial with subterranean stolons; stems up to 70 cm, erect, roughly hairy with appressed, deflexed hairs up to the uppermost leaves, shining, little-branched. Cauline leaves up to 11 × 2 cm, elliptical, acute, with appressed forwardly-pointing hairs above and patent, more or less backwardly-pointing hairs beneath. Pedicels up to 7 mm in fruit, slender, patent. Calyx up to 5 mm in fruit. Limb of corolla up to 6 mm in diameter. Nutlets 2 × 1 mm, narrowly ovoid. 2n=44. *Wet mountain meadows.* ● *Mountains of S.W. Europe, from S.C. France to N.W. Spain.* Ga Hs.

40. **M. azorica** H. C. Watson, *Bot. Mag.* **70**: t. 4122 (1844). Perennial; stems up to 60 cm, much-branched, the leafy parts densely hairy with deflexed hairs. Leaves up to 10 × 2 cm, narrowly obovate, with patent, backwardly-pointing hairs beneath and almost appressed, forwardly-pointing bristles above. Inflorescence dense, with upwardly-pointing, somewhat appressed hairs; pedicels not longer than calyx at maturity. Calyx up to 5 mm in fruit; teeth narrowly linear, twice as long as tube; hairs straight, deflexed at the base, apically directed above. Corolla-limb up to 6 mm in diameter, blue. Nutlets 1·5 × 1 mm, ovoid, obtuse, blackish-brown, with rim; attachment-area ovate, with apex directed towards the apex of the nutlet. *Wet rocks.* ● *Açores.* Az.

M. maritima Hochst. in Seub., *Fl. Azor.* 37 (1844), which occurs in the Açores, is said to differ in having larger flowers; further investigation is needed.

41. **M. sparsiflora** Mikan ex Pohl, *Bot. Taschenb.* **1807**: 74 (1807). Annual; stems up to 40 cm, slender, ascending to erect, often much-branched; leafy part of stems with deflexed bristles. Leaves up to 8 × 2·5 cm, ovate to elliptical, with scattered somewhat deflexed hairs beneath and appressed, forwardly-pointing

hairs above. Inflorescence very lax, especially at the base, with numerous leaves; pedicels, especially the lowermost, much elongated and up to 2 cm in fruit. Limb of corolla up to 2 mm in diameter, saucer-shaped, bright blue. Calyx strongly accrescent, up to 7 mm in fruit, divided to $\frac{1}{2}$, with narrowly triangular teeth; tube with patent, hooked hairs. Nutlets 2×1.5 mm, brown, with an indistinct rim; attachment-area broadly reniform, with a lingulate, white appendage. $2n = 18$. *Moist, shady places. E., S.E. & E.C. Europe.* Al Au Bu Cz Ge Gr Hu Ju Po Rm Rs (N, B, C, W, K, E) [Fe No Su].

30. Eritrichium Schrader ex Gaudin[1]

Caespitose perennials, sometimes woody at the base. Flowers in terminal, usually mostly ebracteate cymes. Calyx lobed almost to the base, accrescent or not. Corolla violet, blue or whitish, rotate with short tube, with 5 scales in the throat. Stamens included, inserted at about middle of tube, without appendages. Style included; stigma capitate. Nutlets ovoid-trigonous or turbinate, smooth or tuberculate, usually with a dentate-pectinate wing, attached to receptacle by only part of their length.

Literature: A. Brand in Engler, *Pflanzenreich* 97 (IV. 252): 187–201 (1931). L. Lechner-Pock, *Phyton (Austria)* 6: 98–206 (1956).

1 Pedicels 10–20 mm in fruit   **1. pectinatum**
1 Pedicels not more than 5 mm in fruit
  2 Sides of nutlets making an obtuse angle at point of attachment; corolla 5–7 mm   **2. nanum**
  2 Sides of nutlets making an acute angle at point of attachment; corolla 1·5–5·5 mm
    3 Leaves 4–8 mm wide; nutlets hairy above   **3. villosum**
    3 Leaves 1–3 mm wide; nutlets glabrous   **4. aretioides**

1. E. pectinatum (Pallas) DC., *Prodr.* 10: 127 (1846). Laxly caespitose, grey with dense, stout, appressed hairs. Stems 15–30 cm, erect, simple, from a branched, woody stock. Leaves 10–20 × 3–4 mm, linear-lanceolate to -oblong. Inflorescence usually branched, with 5–25 flowers; pedicels 2–5 mm in flower, 10–20 mm in fruit. Calyx *c.* 2 mm, scarcely accrescent. Corolla *c.* 5 mm; limb 6–8 mm in diameter, bright blue. Nutlets *c.* 1 × 1·5–2 mm, turbinate; upper surface smooth; margin with a dentate-pectinate wing; sides making an acute angle at the point of attachment of the nutlet. *Mountain rocks. On the border of Europe in C. Ural.* Rs (C). (*Siberia.*)

2. E. nanum (L.) Schrader ex Gaudin, *Fl. Helv.* 2: 57 (1828). Densely caespitose, villous. Stems 1–7·5 cm, simple, from a branched, woody stock. Leaves 4–10 × 1·5–4·5 mm, linear-oblong to -spathulate. Inflorescence simple or branched, with 3–7(–10) flowers; pedicels 1–3 mm. Calyx 2–3 mm, scarcely accrescent. Corolla 5–7 mm; limb 7–9 mm in diameter, pinkish-purple at first, becoming bright blue or rarely whitish. Nutlets 1–2 × *c.* 2·5 mm, ovoid-trigonous; upper surface smooth, glabrous; margin with a narrow, dentate wing; sides making an obtuse angle at the point of attachment of the nutlet. *Rocks and open, stony ground, usually above 2500 m. C. & S. Alps; E. & S. Carpathians.* Au Ga He It Ju Rm ?Rs (W).

(a) Subsp. **nanum**: Villous; stems 1–5 cm; leaves 1·5–3 mm wide. $2n = 46$. *Throughout the range of the species.*

(b) Subsp. **jankae** (Simonkai) Jáv. in Jáv. & Csapody, *Icon. Fl. Hung.* 409 (1932): More densely white-villous, with longer hairs than in subsp. (a); stems (2·5–)5–7·5 cm; leaves 3–4·5 mm wide. *E. & S. Carpathians.*

---

[1] By A. O. Chater.

3. E. villosum (Ledeb.) Bunge, *Mém. Sav. Étr. Pétersb.* 2: 531 (1835). Densely caespitose, grey with appressed, often sparse hairs. Stems 1–15 cm, simple, from a branched, woody stock. Leaves 10–20 × 4–8 mm, oblong to oblanceolate. Inflorescence usually simple, with 3–6 flowers; pedicels 1–3 mm. Calyx 2–3 mm, scarcely accrescent. Corolla 1·5–5 mm; limb (3–)5–8 mm in diameter, violet, becoming bright blue. Nutlets 1–1·3 × *c.* 2 mm, obovoid-turbinate; upper surface smooth or minutely tuberculate, shortly hairy; margin with a narrow, denticulate wing; sides making an acute angle at the point of attachment of the nutlet. *Gravelly places, rocks and sandy seashores. Arctic Russia.* Rs (N). (*N. Asia.*)

4. E. aretioides (Cham.) DC., *Prodr.* 10: 125 (1846). Forming large cushions or mats up to 25 cm in diameter, densely grey- or whitish-villous. Stems 1–8(–10) cm, erect, simple. Leaves 3–12 × 1–3 mm, elliptical to oblanceolate. Inflorescence simple, with 1–6 flowers; pedicels 1–3 mm. Calyx *c.* 2 mm in flower, *c.* 3 mm in fruit. Corolla 4–5·5 mm; limb 4–7·5 mm in diameter, violet, becoming bright blue. Nutlets *c.* 1·5 × 2 mm, obovoid-turbinate; upper surface smooth, glabrous; margin with a narrow, denticulate wing; sides making an acute angle at point of attachment of the nutlet. *Gravelly places and sandy seashores. Arctic Russia* (*Vajgač, N. Ural*). Rs (N). (*Arctic Siberia, Alaska.*)

31. Lappula Gilib.[1]

(*Echinospermum* Swartz ex Lehm.; incl. *Hackelia* Opiz, *Heterocaryum* A. DC.)

Usually annual or biennial herbs. Flowers in bracteate, terminal, often branched cymes. Calyx lobed almost to the base, accrescent. Corolla blue or white, with short tube and infundibuliform to almost rotate limb, with 5 short scales in the throat. Stamens included, variously inserted. Style included; stigma capitate. Nutlets ovoid-trigonous to oblong and flattened, with 1–3 rows of cylindrical, conical or flattened spines or glochidia on the sides, sometimes also on the back, attached to receptacle for usually only part of their length on the margin.

Literature: A. Brand in Engler, *Pflanzenreich* 97(IV. 252): 95–155 (1931).

1 Pedicels deflexed in fruit   **1. deflexa**
1 Pedicels erect or erecto-patent in fruit
  2 Nutlets convex on upper surface, with unbarbed spines   **2. spinocarpos**
  2 Nutlets flat on upper surface, with glochidia
    3 Nutlets oblong, attached to receptacle by their whole length   **3. echinophora**
    3 Nutlets ovoid-trigonous, attached to receptacle by only part of their length
      4 Nutlets with 1 row of glochidia   **4. marginata**
      4 Nutlets with 2–3 rows of glochidia for at least part of their length
        5 Plant usually green, with sparse, erecto-patent or patent as well as appressed hairs; limb of corolla 2–4 mm in diameter   **5. squarrosa**
        5 Plant grey, with dense, appressed hairs; limb of corolla (3–)5–8 mm in diameter   **6. barbata**

1. L. deflexa (Wahlenb.) Garcke, *Fl. Nord-Mittel-Deutschl.* ed. 6, 275 (1863) (*Echinospermum deflexum* (Wahlenb.) Lehm., *Hackelia deflexa* (Wahlenb.) Opiz). Annual or biennial, with sparse to dense, soft, erecto-patent hairs. Stem 20–90 cm, ascending or erect, branched. Leaves 20–80 × 3–15 mm, linear- to oblong-lanceolate, the lower petiolate, the upper sessile. Pedicels 2–5 mm, patent in flower, strongly deflexed in fruit. Calyx *c.* 1·5 mm in flower, 2–3 mm in fruit. Corolla *c.* 3 mm; limb

3–6 mm in diameter, campanulate, pale blue or white. Nutlets 3–5 mm, flattened-trigonous, smooth or minutely granulate and often hairy on upper surface; margin with 1 row of glochidia which are dilated and flattened towards the base. $2n=24$. *U.S.S.R. and Fennoscandia; mountains of C. Europe; E. Pyrenees.* Au Cz Fe Ga Ge He Hs It No Rm Rs (N, B, C, W, K, E) Su.

2. **L. spinocarpos** (Forskål) Ascherson ex O. Kuntze, *Acta Horti Petrop.* 10: 215 (1887) (*Sclerocaryopsis spinocarpos* (Forskål) Brand). Annual, densely appressed-grey-hairy. Stem 5–25 cm, decumbent or erect, branched. Leaves 20–40 × 1–3 mm, linear. Pedicels *c.* 1 mm, erect. Calyx 3–4 mm in flower, 6–8 mm in fruit. Corolla 3–4 mm; limb infundibuliform, blue. Nutlets 2·5–3 mm, ovoid-trigonous, convex on the upper surface; margin and upper surface with large, irregular, conical spines. *Dry clay soils. S.E. Russia (Volga delta).* Rs (E).

3. **L. echinophora** (Pallas) O. Kuntze, *op. cit.* 214 (1887) (*Heterocaryum echinophorum* (Pallas) Brand). Annual, hispid with erecto-patent hairs often arising from small tubercles. Stem 5–50 cm, erect, often branched. Leaves 15–50 × 1–10 mm, linear to linear-lanceolate, the lower petiolate, the upper sessile. Pedicels up to 5 mm, erect. Calyx *c.* 2 mm in flower, *c.* 5 mm in fruit. Corolla 2–2·5 mm; limb infundibuliform, pale blue or whitish. Nutlets unequal in size, the larger 5–7 mm, the smaller 4–5 mm, oblong, attached to the receptacle by their whole length; upper surface flat, sometimes with a row of glochidia down the centre; margin raised, with 1 row of glochidia. *Stony places and steppes.* ● *S.E. Russia, W. Kazakhstan.* Rs (E).

4. **L. marginata** (Bieb.) Gürke in Engler & Prantl, *Natürl. Pflanzenfam.* 4(3a): 107 (1895) (incl. *L. patula* (Lehm.) Menyh., *L. semiglabra* (Ledeb.) Gürke, *L. stricta* (Ledeb.) Gürke, *L. tenuis* (Ledeb.) Gürke, *Echinospermum patulum* Lehm.). Annual, grey with dense, erecto-patent hairs. Stem 5–70 cm, erect, simple or branched at base. Leaves 20–100 × 2–8 mm, oblong to linear, the lower petiolate, the upper sessile. Pedicels up to 1·5 mm, erect. Calyx 2–3 mm in flower, 3–4 mm in fruit. Corolla 3–4 mm; limb infundibuliform, blue, rarely white. Nutlets 2–4 mm, flattened-trigonous; upper surface smooth or tuberculate; margin with 1 row of glochidia which are dilated, flattened and united at the base. *Dry places. From Hungary and N. Greece to W. Kazakhstan; S.E. Spain; an occasional casual elsewhere.* Bu Gr Hs Hu Rm Rs (W, K, E) *Si.

5. **L. squarrosa** (Retz.) Dumort., *Fl. Belg.* 40 (1827). Annual or biennial, with sparse to dense, patent or erecto-patent as well as appressed hairs. Stem 10–70 cm, erect, often branched. Leaves 20–70 × 3–10 mm, oblong to linear-lanceolate, sessile or the lower shortly petiolate. Pedicels up to 4 mm, erect. Calyx 1·5–3 mm in flower, 4–5 mm in fruit. Corolla *c.* 4 mm; limb 2–4 mm in diameter, almost rotate, pale blue. Nutlets 2·5–4 mm, flattened-ovoid-trigonous; upper surface minutely tuberculate; margin with 2–3 rows of glochidia which are usually dilated and flattened at the base. $2n=48$. *Most of Europe except the extreme north and west, but probably only naturalized in the north.* Al Au ?Bl Bu Cz Ga Gr He Hs Hu It Ju Rm Rs (C, W, K, E) Su Tu [Da Fe *Ge Ho No Po Rs (N, B)].

(a) Subsp. **squarrosa** (*L. echinata* Fritsch, *L. myosotis* Moench, *Echinospermum lappula* (L.) Lehm.; incl. *Lappula consanguinea* (Fischer & C. A. Meyer) Gürke): Glochidia on nutlets 1–1·5 mm, all of subequal length, those of the inner row not confluent at base. *Cultivated fields and other dry disturbed habitats. Throughout the range of the species.*

(b) Subsp. **heteracantha** (Ledeb.) Chater, *Bot. Jour. Linn. Soc.* 64: 380 (1971) (*Echinospermum heteracanthum* Ledeb., *Lappula heteracantha* (Ledeb.) Gürke; incl. *L. semicincta* (Steven) M. Popov): Glochidia of the outer row very short, those of the inner row 2–3 mm, confluent at the base. *Scrub, grassland and dry, rocky places. S.C. Europe and S. part of U.S.S.R.*

6. **L. barbata** (Bieb.) Gürke in Engler & Prantl, *Natürl. Pflanzenfam.* 4(3a): 107 (1895) (*Echinophorum barbatum* (Bieb.) Lehm.). Annual or biennial, grey with dense, appressed hairs. Stem 10–70 cm, erect, often branched. Leaves 30–70 × 5–8 mm, linear to lanceolate, the lower petiolate, the upper sessile. Pedicels up to 5 mm, erect. Calyx 2–3 mm in flower, *c.* 4 mm in fruit. Corolla 3–5 mm; limb (3–)5–8 mm in diameter, almost rotate, white or blue. Nutlets 3–4 mm, flattened-trigonous; upper surface minutely tuberculate; margin with 2–3 rows of glochidia which are usually neither dilated nor flattened at the base. *Dry places. S.E. Europe, from S. Albania to E. Ukraine; E. Spain.* Al Bu Hs Rm Rs (W, K, E).

32. Trigonotis Steven[1]

Annual or perennial herbs. Flowers in terminal, ebracteate cymes. Calyx lobed to below the middle, accrescent. Corolla blue, infundibuliform to campanulate or rotate with short tube and with papillose or hairy scales in the throat. Stamens included, inserted about middle of tube. Style included; stigma capitate. Nutlets trigonous, erect, smooth, with a small attachment area, often stipitate.

1. **T. peduncularis** (Trev.) Bentham ex Baker & S. Moore, *Jour. Linn. Soc. London (Bot.)* 17: 384 (1879). Annual; stems 5–20 cm, erect or ascending, branched, scabrid-pubescent. Leaves 7–20 × 5–8 cm, spathulate to oblong or lanceolate, greyish with appressed hairs. Cymes secund, very lax in fruit; pedicels 3–6 mm in fruit and patent and thickened at the apex. Calyx *c.* 1 mm in flower, 2–2·5 mm in fruit, greyish with appressed hairs; lobes ovate-oblong, acute. Corolla *c.* 1·5 mm; limb 2–3 mm in diameter, bright blue, lobed to *c.* ¾; lobes ovate, obtuse. Nutlets *c.* 1 mm, minutely hairy on upper surface or glabrous. *S.E. Russia (near Astrakhan').* Rs(E). (*C. & E. Asia.*)

33. Omphalodes Miller[2]

Annual, biennial or perennial herbs. Flowers solitary and axillary or in mostly terminal, bracteate or ebracteate cymes. Calyx lobed to the base, accrescent. Corolla blue or white, subrotate or almost campanulate, with short tube and 5 prominent, saccate invaginations in the throat. Stamens included, inserted at about middle of tube. Style included; stigma capitate. Nutlets depressed-globose, smooth, sometimes hairy, winged, with the wing erect or incurved to form an umbilicus, attached to the receptacle for almost their whole length.

Literature: A. Brand in Engler, *Pflanzenreich* 78(IV. 252): 96–112 (1921).

1 Flowers solitary, axillary; corolla 3–4 mm **1. scorpioides**
1 Flowers in mostly terminal cymes; corolla 5–10 mm
  2 Perennial; flowers blue
    3 Basal leaves lanceolate **2. nitida**
    3 Basal leaves ovate
      4 Leaves acute; calyx with appressed hairs **3. verna**
      4 Leaves obtuse; calyx glabrous **4. luciliae**
  2 Annual; flowers white, rarely blue
    5 Wing of nutlets strongly incurved, crenate or dentate
      6 Inflorescence ebracteate **5. linifolia**
      6 Inflorescence bracteate **6. kuzinskyanae**

---

[1] By A. O. Chater.   [2] By L. F. Ferguson.

5 Wing of nutlets erect, entire
  7 Nutlets hirsute   **7. littoralis**
  7 Nutlets glabrous   **8. brassicifolia**

1. O. scorpioides (Haenke) Schrank, *Denkschr. Akad. Wiss. München* **3**: 222 (1812). Biennial; stems 12–40 cm, decumbent to ascending, branched, glabrous to sparsely hirsute. Leaves (10–)15–40 × 5–15 mm, lanceolate to spathulate, setose, the lower opposite, attenuate, petiolate, the upper alternate, sessile. Flowers axillary. Pedicels 7–13 mm, elongating in fruit. Calyx 2–3 mm, up to 5 mm in fruit. Corolla 3–4 mm, blue. Nutlets 2–3 mm, hirsute; wing entire, forming an umbilicus. 2n=24. *Damp, shady places. C. & E. Europe, from c. 46° to 57° N.* Au Cz Ge Hu Po Rm Rs (C, W, E).

2. O. nitida Hoffmanns. & Link, *Fl. Port.* **1**: 194 (1810) (*O. lusitanica* auct.). Perennial; stems 20–65 cm, erect, branched, sparsely setose. Basal leaves (7–)10–20(–30) × 1–3·5 cm, lanceolate, acute to acuminate, glabrous above, sparsely strigose beneath, ciliolate, long-petiolate; the cauline smaller, shortly petiolate to sessile. Inflorescence terminal, lax, bracteate at base. Pedicels 5–15(–55) mm. Calyx-lobes 2–4 mm, up to 7 mm in fruit, oblong to lanceolate, becoming ovate. Corolla 5–9 mm, blue, folds yellow. Nutlets 2·3 mm; wing long-dentate, forming an umbilicus. 2n=24. *Damp, shady places.* ● *N. Portugal, N.W. Spain.* Hs Lu.

3. O. verna Moench, *Meth.* 420 (1794). Perennial, stoloniferous; stems 5–20(–40) cm. Basal leaves 5–20(–35) × 2–6 cm, ovate to cordate, mucronate to acuminate, acute, sparsely hirsute, long-petiolate; the cauline smaller, ovate to elliptical, shortly petiolate to sessile. Inflorescence terminal, lax, bracteate at base. Pedicels 8–12 mm, up to 30 mm in fruit. Calyx 4 mm; lobes elliptical, hairy. Corolla 8–10 mm; lobes blue; folds yellow. Nutlets c. 2 mm, hirsute; wing ciliate, forming an umbilicus. 2n=48. *Damp, mountain woods.* ● *From the S.E. Alps southwards to the N. Appennini and C. Romania; cultivated for ornament and widely naturalized.* Au *Gr It Ju Rm [Be Br Cz Ga Ge He Ho Hu Po Rs (W)].

4. O. luciliae Boiss., *Diagn. Pl. Or. Nov.* **1**(4): 41 (1844). Like **3** but caespitose; leaves not distinctly mucronate or acuminate, usually obtuse; inflorescence more bracteate; calyx glabrous; corolla 7 mm; wing of nutlets narrow, not ciliate. *Rock-crevices. Mountains of S. & C. Greece.* Gr. (*Anatolia.*)

5. O. linifolia (L.) Moench, *Meth.* 419 (1794). Glaucous annual; stems 5–40 cm, erect, simple or branched from the base. Basal leaves 1–5(–10) × 0·1–1·5(–2) cm, spathulate to cuneiform, petiolate, sparsely strigose-ciliate; the cauline smaller, lanceolate to linear, sessile. Inflorescence terminal, 5- to 15-flowered, lax, ebracteate. Pedicels 1–5 mm, up to 20 mm in fruit. Calyx-lobes 1·5–3 mm, up to 7 mm in fruit, lanceolate, becoming ovate. Corolla 5–10 mm, white or bluish. Nutlets 4 mm, glabrous or hirsute; wing lobed, dentate or crenate, forming an umbilicus. 2n=28. *Dry, open habitats; somewhat calcicole. S.W. Europe.* Ga Hs Lu [?Rm Rs (W, K)].

6. O. kuzinskyanae Willk., *Österr. Bot. Zeitschr.* **39**: 318 (1889). Like **5** but stems 5–15 cm; cauline leaves elliptical to ovate; inflorescence bracteate, at least in part; corolla blue, rarely white; nutlets with rigid, hooked hairs. 2n=28. *Maritime rocks and sands.* ● *N.W. Spain, C. Portugal.* Hs Lu.

---

¹ By M. Kovanda.

7. O. littoralis Lehm., *Ges. Naturf. Freunde Berlin Mag.* **8**: 98 (1818). Annual; stems 4–15 cm. Basal leaves 1–2(–3) × 0·3–1·1 cm, spathulate, long-petiolate, more or less persistent; cauline smaller, lanceolate, sessile. Inflorescence lax, few-flowered, bracteate. Pedicels 1–4 mm, up to 10 mm in fruit. Calyx-lobes 3–4 mm, up to 6 mm in fruit, elliptical, becoming ovate. Corolla 6 mm, white. Nutlets hirsute; wing entire, ciliate, erect and narrow. *Maritime sands.* ● *W. coast of France.* Ga.

8. O. brassicifolia (Lag.) Sweet, *Hort. Brit.* 293 (1827) (*O. amplexicaulis* Lehm.). Like **7** but stems 10–15 cm; leaves up to 1–2(–3·5) cm, amplexicaul; inflorescence ebracteate; corolla 8 mm; nutlets glabrous, with very narrow wing. *Mountains of S. Spain.* Hs. (*N. Africa.*)

The record of *O. pavoniana* Boiss. from Spain is probably erroneous.

34. Cynoglossum L.¹

Biennial, perennial or rarely annual herbs. Flowers in usually ebracteate cymes. Calyx 5-lobed almost to the base, accrescent. Corolla with short, cylindrical to infundibuliform tube and rotate limb, with 5 scales closing the throat. Stamens included, inserted at or above the middle of the tube. Style included; stigma small, subcapitate. Nutlets ovoid to subglobose, glochidiate, the external surface convex, flat or slightly concave, sometimes with a distinct border and disc.

Literature: A. Brand in Engler, *Pflanzenreich* **78** (IV. 252): 114–153 (1921). H. Riedl, *Österr. Bot. Zeitschr.* **109**: 385–394 (1962).

C. glochidiatum Wall. ex Bentham in Royle, *Ill. Bot. Himal. Mount.* 306 (1836), a native of C. Asia and India, is cultivated for ornament and occurs occasionally as a garden escape in N. Europe.

1 Corolla-lobes villous   **11. clandestinum**
1 Corolla-lobes glabrous
  2 Mature nutlets flat or slightly concave outside, with a thickened border
    3 Cymes bracteate
      4 Biennial; calyx-lobes ovate; nutlets 5–8 mm   **9. cheirifolium**
      4 Perennial; calyx-lobes linear-lanceolate; nutlets 7–9 mm   **10. magellense**
    3 Cymes ebracteate (rarely bracteate at base)
      5 Annual; nutlets 7–10 mm in diameter   **3. columnae**
      5 Biennial; nutlets 5–8 mm in diameter
        6 Cauline leaves oblong to lanceolate; corolla dull purple; stamens inserted in upper part of corolla-tube   **1. officinale**
        6 Cauline leaves linear-lanceolate; corolla deep blue; stamens inserted in middle of corolla-tube   **2. dioscoridis**
  2 Mature nutlets convex outside, without a thickened border
    7 Leaves ± glabrous above   **4. germanicum**
    7 Leaves densely hairy to tomentose on both surfaces
      8 Corolla 7–9 mm, with distinct reticulate venation   **5. creticum**
      8 Corolla 4–6 mm, without reticulate venation
        9 Perennial; corolla-scales crescent-shaped   **6. sphacioticum**
        9 Biennial; corolla-scales ± square to trapeziform
          10 Scales reaching beyond the base of the corolla-lobes; pedicels in fruit shorter than calyx   **8. nebrodense**
          10 Scales about reaching the base of the corolla-lobes; pedicels in fruit longer than calyx   **7. hungaricum**

1. C. officinale L., *Sp. Pl.* 134 (1753). Biennial. Stems (20–)30–60(–90) cm, hirsute. Cauline leaves oblong to lanceolate, the lower shortly petiolate, the middle (10–)16–25(–30) mm wide, sessile to amplexicaul, shortly and softly hairy on both surfaces. Cymes ebracteate (rarely bracteate at base). Calyx-lobes up to

4 mm, ovate, hirsute. Corolla 5–6 mm, dull purple; tube cylindrical; limb shorter than tube. Stamens inserted in upper part of tube. Nutlets 5–8 mm in diameter, ovoid, with a distinct border; external face densely glochidiate. $2n=24, 48$. *Dry open habitats. Most of Europe except the extreme north and the extreme south.* Al Au Be Br Bu Co Cz Da Fe Ga Ge Gr Hb He Ho Hs Hu It Ju No Po Rm Rs (B, C, W, K, E) Sa Su.

Plants with bright reddish corolla 12–15 mm in diameter and with very short tube, occurring in W. Bulgaria, have been called **C. rotatum** Velen., *Sitz.-Ber. Böhm. Ges. Wiss. (Math.-Nat. Kl.)* 1893(37): 48 (1894); their taxonomic status is uncertain.

2. **C. dioscoridis** Vill., *Prosp. Pl. Dauph.* 21 (1779) (*C. loreyi* Jordan ex Lange). Biennial. Stems (15–)20–40(–50) cm, hirsute. Cauline leaves linear-lanceolate, the middle (5–)7–9(–12) mm wide, more or less amplexicaul, hairy on both surfaces. Cymes ebracteate. Calyx-lobes *c.* 3 mm, elliptical to oblong, hirsute. Corolla *c.* 5 mm, deep blue, campanulate; limb about as long as tube. Stamens inserted in middle of tube. Nutlets 5–6 mm in diameter, with a distinct border; external face densely glochidiate. *Wood-margins and rocky hillsides; calcicole.* ● *S. & E. France, N.E. Spain.* Ga Hs ?It.

3. **C. columnae** Ten., *Fl. Nap.* 1, *Prodr.*: 14 (1811). Annual. Stems (15–)25–45(–60) cm, hirsute. Cauline leaves oblong to lanceolate, sessile to amplexicaul, more or less tomentose on both surfaces. Cymes ebracteate. Calyx-lobes *c.* 5 mm, ovate, hirsute. Corolla 5–6 mm, deep blue; tube broadly cylindrical. Stamens inserted in upper part of tube. Nutlets 7–10 mm in diameter, with a distinct border; glochidia on disc shorter than those on margin. *Dry open habitats.* ● *C. & E. Mediterranean region.* Al Cr Gr It Ju Si.

4. **C. germanicum** Jacq., *Obs. Bot.* 2: 31 (1767) (*C. montanum* auct., non L.). Biennial. Stems (20–)30–50(–60) cm, hirsute. Cauline leaves lanceolate, subsessile to amplexicaul, hairy beneath, more or less glabrous and shining above. Cymes ebracteate. Calyx-lobes 4–5 mm, narrowly ovate, sparsely hirsute. Corolla 5–6 mm, reddish-violet; tube broadly cylindrical to infundibuliform; limb shorter than tube, with glabrous lobes. Stamens inserted in middle of tube. Nutlets 6–8 mm in diameter, ovoid, without a distinct border; external face convex, densely glochidiate. $2n=24$. *Woods; somewhat calcicole. W. & C. Europe, northwards to C. England and extending locally to S. Italy, Bulgaria and Ukraine.* Al Au Be Br Bu Cz Ga Ge ?Hs It Ju Po Rm Rs (C, W, K) Sa.

5. **C. creticum** Miller, *Gard. Dict.* ed. 8, no. 3 (1768) (*C. pictum* Aiton). Biennial. Stems (20–)30–60 cm, hirsute to tomentose. Cauline leaves oblong to lanceolate, shortly petiolate to amplexicaul, densely hairy on both surfaces. Cymes ebracteate. Calyx-lobes 6–8 mm, oblong, hirsute. Corolla 7–9 mm, deep blue, with distinct reticulate venation; tube broadly infundibuliform; limb about as long as tube, divided to *c.* ½, the lobes glabrous. Stamens inserted in lower part of tube. Nutlets 5–7 mm in diameter, ovoid, without a distinct border; external face convex, densely and unequally glochidiate. $2n=24$. *Open, usually dry habitats. S. Europe, extending northwards to N.C. France.* Al Az Bl Bu Co Cr Ga Gr Hs It Ju Lu ?Rm Rs (K) Sa Si Tu.

6. **C. sphacioticum** Boiss. & Heldr. in Boiss., *Diagn. Pl. Or. Nov.* 2(11): 125 (1849). Perennial. Stems 10–18 cm, hirsute. Cauline leaves linear-lanceolate, sessile, hairy on both surfaces. Cymes ebracteate. Calyx-lobes *c.* 3 mm, oblong, hirsute. Corolla

c. 4 mm, deep blue to violet; tube broadly infundibuliform; limb as long as tube or somewhat longer, divided to ⅔, the lobes glabrous; scales crescent-shaped. Stamens inserted in middle of tube. Nutlets 5–6 mm in diameter, ovoid, without a distinct border; external face convex, densely glochidiate. *Mountain screes.* ● *Kriti.* Cr.

7. **C. hungaricum** Simonkai, *Term. Füz.* 2: 151 (1878). Biennial. Stems (15–)20–40(–50) cm, hirsute. Cauline leaves lanceolate, sessile to semi-amplexicaul, hairy on both surfaces. Cymes ebracteate; pedicels in fruit longer than calyx. Calyx-lobes 4–5 mm, lanceolate, hirsute. Corolla 4–6 mm, dull red; tube broadly cylindrical; limb longer than tube, divided to ½, the lobes glabrous; scales more or less square to trapeziform. Stamens inserted in middle of tube. Nutlets 4–6 mm in diameter, without a distinct border; external face convex, densely glochidiate. $2n=24$. *Dry open habitats. S.E. & E.C. Europe.* Al Au Bu Cz Gr Hu Ju Rm Rs (W, K).

Extremely variable, particularly in indumentum and floral and fruit characters, and local variants have been given specific, subspecific or varietal rank. Variants transitional to **8** appear to be frequent in the Balkan peninsula and further work is required to understand the structure of this complex. The name *C. hungaricum* Simonkai was originally applied to plants from the northern edge of the range, where there is a tendency for the indumentum to be rather sparse.

8. **C. nebrodense** Guss., *Fl. Sic. Prodr.* 1: 216 (1827). Biennial. Stems (15–)20–40 cm, hirsute to tomentose. Cauline leaves lanceolate to narrowly lanceolate, shortly petiolate to semi-amplexicaul, more or less hairy or scabrid on both surfaces. Cymes ebracteate; pedicels in fruit shorter than calyx. Calyx-lobes *c.* 4 mm, lanceolate, hirsute. Corolla 4–6 mm, reddish-violet; tube broadly cylindrical; limb about as long as tube, divided to ½, the lobes glabrous; scales more or less square to trapeziform. Stamens inserted in middle of tube. Nutlets 4–5 mm in diameter, ovoid, without a distinct border; external face convex, densely glochidiate. *Mountain woods.* ● *Mediterranean region.* Gr Hs It ?Ju Si.

9. **C. cheirifolium** L., *Sp. Pl.* 134 (1753) (incl. *C. arundanum* Cosson, *C. heterocarpum* (G. Kunze) Willk.). Biennial. Stems (15–)25–40 cm, hirsute to tomentose. Cauline leaves lanceolate, sessile, whitish-tomentose on both surfaces. Cymes bracteate. Calyx-lobes 5–7 mm, ovate, hirsute. Corolla *c.* 8 mm, at first pale purplish, becoming purple, violet or deep blue; limb shorter than tube, the lobes glabrous. Stamens inserted at middle of tube. Nutlets 5–8 mm in diameter, with a distinct border; external face flat or concave, densely glochidiate to nearly smooth. *Dry open habitats. W. Mediterranean region, Portugal.* Bl Ga Hs It Lu Sa Si.

10. **C. magellense** Ten., *Fl. Nap.* 1, *Prodr.*: 66 (1811). Perennial. Stems (15–)20–35 cm, hirsute. Cauline leaves linear-lanceolate, sessile, densely hairy on both surfaces. Cymes bracteate, capitate at anthesis. Calyx-lobes *c.* 7 mm, linear-lanceolate, hirsute. Corolla *c.* 8 mm, reddish; tube broadly cylindrical; limb shorter than tube, the lobes glabrous. Stamens inserted in upper part of tube. Nutlets 7–9 mm in diameter, with a distinct border; external face flat or concave, sparsely glochidiate to smooth. *Mountain pastures.* ● *C. & S. Appennini.* It.

11. **C. clandestinum** Desf., *Fl. Atl.* 1: 159 (1798). Biennial. Stems (20–)30–50(–60) cm, hirsute. Cauline leaves lanceolate

to linear-lanceolate, shortly petiolate to amplexicaul, hairy on both surfaces. Cymes usually bracteate. Calyx-lobes 6–7 mm, ovate, hirsute. Corolla *c.* 6 mm, violet; tube broadly cylindrical; limb shorter than tube, the lobes villous. Stamens inserted in upper part of tube. Nutlets 5–7 mm in diameter, without a distinct border; external face more or less flat or concave, densely glochidiate. *Dry open habitats. S. part of Iberian peninsula, Sardegna, Sicilia.* Hs Lu Sa Si.

35. Solenanthus Ledeb.[1]

Biennial or short-lived perennial herbs. Flowers in mostly ebracteate cymes, forming a panicle. Calyx lobed to the base, lobes usually accrescent. Corolla purple or reddish-purple, cylindrical to infundibuliform, with 5 saccate invaginations in the throat. Stamens exserted or equalling corolla, inserted at about middle of corolla. Style exserted; stigma small. Nutlets depressed-ovoid to globose, glochidiate, with margins sometimes winged, attached to receptacle for most of their length.

Literature: A. Brand in Engler, *Pflanzenreich* 78(IV. 252): 153–160 (1921). W. B. Turrill, *Hook. Ic.* 33: sub t. 3278 (1935).

1 Basal and cauline leaves linear to linear-lanceolate   **1. reverchonii**
1 Basal and cauline leaves not linear or linear-lanceolate
   2 Upper cauline leaves ovate-rhombic   **6. albanicus**
   2 Upper cauline leaves not ovate-rhombic
     3 Stamens equalling or scarcely exceeding corolla   **3. apenninus**
     3 Stamens exceeding corolla by at least 2 mm
       4 Pedicels 0–4 mm in flower, scarcely elongating in fruit; corolla-lobes acute   **5. stamineus**
       4 Pedicels 4 mm or more, elongating in fruit; corolla-lobes obtuse
         5 Plant 30–60 cm; corolla (3–)5–6 mm   **2. biebersteinii**
         5 Plant 50–100 cm; corolla 7 mm   **4. scardicus**

1. S. reverchonii Degen, *Magyar Bot. Lapok* 2: 311 (1903). Biennial or perennial 30–70 cm. Stems simple. Basal leaves 20–30 × 1–2 cm, linear to linear-lanceolate, with 3(–5) longitudinal veins; cauline leaves smaller. Pedicels 0–2 mm, elongating in fruit. Calyx 4–6 mm; lobes accrescent, lanceolate to linear. Corolla 6–8·5 mm, with invaginations above the middle; lobes very small, lanceolate. Stamens exserted; filaments 7·5–9·5 mm. Nutlets 7–9 mm; external face sparsely glochidiate; margins with several rows of flattened glochidia. ● *S. Spain.* Hs.

2. S. biebersteinii DC., *Prodr.* 10: 165 (1846). Perennial 30–60 cm. Basal leaves 30–45 × 12–20 cm, elliptical to ovate, with 6–11 lateral veins; cauline leaves smaller; the lower broadly elliptical, the upper narrowly lanceolate. Pedicels *c.* 4 mm, elongating in fruit. Calyx 3–4 mm; lobes accrescent, linear-oblong. Corolla (3–)5–6 mm, with invaginations at or above the middle; lobes orbicular to obovate, obtuse. Stamens exserted; filaments 9–10 mm. Nutlets 5–6 mm; margins slightly winged; external face with few glochidia, other faces densely glochidiate. *Mountain woods. Krym.* Rs (K). (*Caucasus.*)

3. S. apenninus (L.) Fischer & C. A. Meyer, *Bull. Soc. Nat. Moscou* 11: 306 (1838). Biennial 60–120 cm. Basal leaves 30–50 × 5–9(–12) cm, elliptical to broadly lanceolate, with 3–6 lateral veins; cauline leaves smaller, the lower ovate-elliptical to elliptical, the upper narrowly obovate to attenuate, cuneate to amplexicaul-auriculate at base. Pedicels 8 mm, elongating in fruit. Calyx 8–10 mm; lobes accrescent, lanceolate. Corolla 7–9 mm, with invaginations at or below the middle; lobes obovate, obtuse. Stamens equalling or scarcely exceeding

corolla; filaments 8–10 mm. Nutlets 6–8 mm, sparsely glochidiate. *Mountain woods and pastures.* ● *C. & S. Italy, Sicilia.* ?Al ?Gr It Si.

4. S. scardicus Bornm., *Feddes Repert.* 17: 276 (1921). Perennial or biennial 50–100 cm. Stems branched. Basal leaves 25–35 × 7·5–11 cm, ovate to broadly lanceolate, with 3–5 lateral veins; cauline leaves smaller, narrowly elliptical to lanceolate, the upper sessile, sometimes amplexicaul. Pedicels 5–8 mm, elongating in fruit. Calyx 5–9 mm; lobes narrowly lanceolate to oblong-lanceolate, accrescent and becoming elliptic-lanceolate. Corolla 7 mm, with invaginations at or above the middle; lobes broadly elliptical to triangular, obtuse. Stamens exserted; filaments 7–9 mm. Nutlets 9 mm, densely glochidiate; margin winged. *Fagus-woods.* ● *E. Albania and W. Makedonija.* Al Ju.

5. S. stamineus (Desf.) Wettst., *Denkschr. Akad. Wiss. Math.-Nat. Kl. (Wien)* 50(2): 88 (1885). Perennial (10–)20–60(–100) cm. Stems simple. Basal leaves 15–30(–50) × 2–4(–7) cm, lanceolate, with (1–)3–5 longitudinal veins; cauline leaves smaller. Pedicels 0–4 mm. Calyx 3–5 mm; lobes oblong-lanceolate. Corolla 6–7 mm, with invaginations below the middle; lobes ovate, acute. Stamens exserted; filaments 12–14 mm. Nutlets 6–8 mm; margins with rows of flattened glochidia; external face sparsely, the other faces densely and shortly glochidiate. *Mountain rocks. S. Greece (Aroania Oros).* Gr. (*S.W. Asia.*)

6. S. albanicus (Degen & Bald.) Degen & Bald., *Magyar Bot. Lapok* 2: 315 (1903). Biennial 40–90 cm. Cauline leaves (3–)6–12 × (1·5–)3–3·5 cm, ovate-lanceolate to ovate-rhombic, sessile. Pedicels 0–4 mm. Calyx 4–6 mm; lobes lanceolate. Corolla (6–)7(–8) mm, with invaginations at or below the middle; lobes narrowly ovate, acute. Stamens exserted; filaments 10–11 mm. Nutlets 7 mm; glochidia large, interspersed with smaller ones. *Mountain rocks.* ● *N.W. Greece, S. Albania.* Al Gr.

36. Mattiastrum (Boiss.) Brand[1]

Perennial herbs. Flowers in terminal or axillary, ebracteate cymes. Calyx lobed almost to the base. Corolla blue, violet or purple, campanulate or infundibuliform; tube usually long, with 5 scales in the throat. Stamens included, inserted at top of tube. Style included or exserted; stigma small, capitate. Nutlets suborbicular to ovate, sometimes glochidiate, with a broad, flat wing on the margin, attached to the receptacle for their whole length.

Literature: A. Brand in Engler, *Pflanzenreich* 78(IV. 252): 54–66 (1921).

1. M. lithospermifolium (Lam.) Brand, *Feddes Repert.* 14: 155 (1915). Caespitose perennial; stems (2–)7–35(–50) cm, hispid to villous-pubescent. Basal leaves 3–5 × 0·5–1 cm, lanceolate to spathulate, petiolate; the cauline smaller, sessile. Inflorescence 1·5–3 cm, terminal, dense, elongating up to 6–10(–30) cm in fruit; pedicels 0–4 mm, elongating in fruit. Calyx 2–4 mm; lobes linear-oblong. Corolla 3–5 mm, campanulate, blue-violet. Style included. Nutlets 7–8 mm, suborbicular, sparsely glochidiate; wing spinulose-dentate. *Kriti.* Cr. (*W. Asia.*)

37. Rindera Pallas[1]

Perennial herbs. Flowers in terminal ebracteate cymes, forming panicles or umbels. Calyx lobed to the base, accrescent or not.

Corolla usually purple, cylindrical to infundibuliform, sometimes with 5 saccate invaginations in the throat. Stamens included or equalling corolla, inserted at or above middle of corolla. Style usually exserted; stigma small, capitate. Nutlets ovate to suborbicular, flattened, smooth, with a broad wing, attached to the receptacle for their whole length and adnate to style.

Literature: A. Brand in Engler, *Pflanzenreich* 78(IV. 252): 66–76 (1921).

1 Inflorescence paniculate; corolla lobed to *c.* ½; invaginations in corolla-tube indistinct **1. tetraspis**
1 Inflorescence umbellate; corolla lobed to *c.* ⅓; invaginations in corolla-tube distinct
2 Leaves green; stem 25–70 cm, leafy above **2. umbellata**
2 Leaves grey-lanate; stem 6–25 cm, not leafy above **3. graeca**

1. R. tetraspis Pallas, *Reise* 1: 486 (1771). Stems 18–50 cm, glabrous. Basal leaves (7–)12–18(–22) × (1–)2–3·5 cm, lanceolate to narrowly ovate, glabrous; cauline leaves smaller, the upper cuneate to cordate, sessile, with sub-lanate apices. Inflorescence paniculate. Pedicels 5–10 mm, elongating in fruit. Calyx 5–7 mm, villous. Corolla 12–14 mm, subconical, lobed to ½; lobes linear-oblong, purple; tube yellowish; invaginations small, at middle of corolla. Stamens included; anthers ½ as long as corolla-lobes. Nutlets 16 mm. *Stony slopes. S. part of U.S.S.R.* Rs (C, W, K, E).

2. R. umbellata (Waldst. & Kit.) Bunge, *Beitr. Kenntn. Fl. Russl.* 239 (1852). Stems 25–70 cm, hirsute. Basal leaves (7–)12–25 × 0·7–2 cm, linear-lanceolate to oblong-elliptical, sparsely sericeous; cauline leaves smaller, lanceolate, the upper narrowly ovate to oblanceolate. Inflorescence subumbellate. Pedicels 3–5 mm, elongating in fruit. Calyx 11–15 mm, sericeous. Corolla 11–15 mm, subconical, lobed to ⅓; lobes broadly oblong, yellow; invaginations above middle of corolla. Stamens equalling corolla. Nutlets 14 mm. *Dry, open habitats.* ● *From E. Jugoslavia to S.W. Ukraine.* Bu Ju Rm Rs (W).

3. R. graeca (A. DC.) Boiss. & Heldr. in Boiss., *Diagn. Pl. Or. Nov.* 1(7): 30 (1846). Caespitose, with appressed, sericeous hairs. Stems 6–25 cm, floccose. Basal leaves 3–12 × 0·2–0·9 cm, linear to oblanceolate, grey-lanate; cauline leaves smaller, narrowly linear, the upper rudimentary. Inflorescence subumbellate. Pedicels 0–4 mm, elongating up to 2–5 cm in fruit. Calyx 4–8 mm, sericeous. Corolla 10–12 mm, subcylindrical, lobed to ⅓; lobes oblong; invaginations above middle of corolla. Stamens equalling corolla. Nutlets 11–14 mm. *Mountain rocks.* ● *Greece.* Gr.

CXLIX. VERBENACEAE[1]

Herbs or shrubs with opposite or verticillate leaves. Flowers 4- to 5-merous, weakly zygomorphic. Calyx usually small. Corolla rotate. Ovary superior, initially 1-locular but becoming 2- to 4-locular by the development of septa; style terminal; ovules usually 1 in each loculus. Fruit usually a drupe, rarely a capsule or dividing into 2 or 4 1-seeded nutlets.

1 Herbs
2 Calyx with 5 nearly equal teeth; fruit of 4 nutlets **3. Verbena**
2 Calyx 2-lipped; fruit of 2 nutlets **4. Lippia**
1 Shrubs
3 Leaves digitate **1. Vitex**
3 Leaves simple
4 Leaves opposite; flowers yellow to red **2. Lantana**
4 Most leaves in whorls of 3; flowers lilac or white **4. Lippia**

1. Vitex L.[2]

Shrubs with digitate leaves. Flowers in cymes forming a terminal, paniculate or spike-like inflorescence. Corolla 2-lipped, the upper lip with 2, the lower with 3 lobes. Stamens exserted. Fruit a small drupe.

1. V. agnus-castus L., *Sp. Pl.* 638 (1753). 1–6 m, fragrant. Young twigs with 4 obtuse angles, grey-puberulent. Leaves petiolate, with 5–7 stipitate leaflets; leaflets 1·5–10 cm, linear-lanceolate, acuminate, entire, whitish-tomentose beneath, glabrous above. Calyx and outside of corolla tomentose. Corolla 8–10 mm, blue or pink. Drupe globose, reddish-black, slightly longer than calyx. *S. Europe.* Al Bl Bu Co Cr Ga Gr Hs It Ju Rs (K) Sa Si Tu.

2. Lantana L.[2]

Shrubs with simple leaves. Flowers in corymbs. Corolla weakly 2-lipped. Stamens included. Fruit a small drupe.

1. L. camara L., *Sp. Pl.* 627 (1753). Up to 1·5 m, with a strong, unpleasant smell. Young twigs 4-angled, sometimes prickly, pubescent. Leaves 5–8 cm, ovate, serrate, rugose, more or less hairy, petiolate. Inflorescences axillary; peduncles about equalling the subtending leaves. Corolla yellow or orange, usually changing to red; tube 7–8 mm; limb 4–5 mm in diameter. Drupe globose, black, distinctly longer than calyx. *Naturalized in Açores, and perhaps in parts of the Mediterranean region, where it is frequently cultivated for ornament.* [Az ?Hs Si]. (*Tropical and warm-temperate America; widely naturalized in the tropics and subtropics.*)

3. Verbena L.[3]

Annual or perennial herbs or small shrubs. Leaves opposite or rarely in whorls of 3. Flowers in bracteate spikes, usually arranged in panicles or corymbs. Calyx tubular, 5-ribbed and unequally 5-dentate. Corolla hypocrateriform, weakly 2-lipped, with obtuse or emarginate lobes. Stamens inserted at about the middle of the corolla-tube, included. Stigma unequally 2-lobed. Fruit separating at maturity into 4 nutlets.

Several American species, as well as their hybrids, are grown in European gardens and sometimes escape. Among them, **V. canadensis** (L.) Britton, *Mem. Torrey Bot. Club* 5: 276 (1894) (*V. aubletia* Jacq.), from E. North America, and **V. peruviana** (L.) Druce, *Rep. Bot. Exch. Club Brit. Is.* 3: 425 (1914) (*V. chamaedryfolia* Juss.), from South America, are locally naturalized in Germany and the latter also in E. France (near Strasbourg). *V. canadensis* has pinnatipartite to 3-fid leaves and shortly pedunculate spikes elongating in fruit, with crowded, large and showy flowers in shades of blue, purple or white. *V. peruviana* has crenate to incise-dentate leaves and long-pedunculate spikes with crowded, large and showy, bright scarlet to pink flowers; both are perennial herbs.

---

[1] Edit. T. G. Tutin. [2] By T. G. Tutin. [3] By J. do Amaral Franco.

1 Leaves dentate to incise-serrate, sessile, amplexicaul; spikes very dense
  2 Bracts distinctly longer than calyx; spikes pedunculate, not crowded **1. rigida**
  2 Bracts equalling or shorter than calyx; spikes mostly sessile and crowded **2. bonariensis**
1 At least the lower leaves deeply incised to 1- to 2-pinnatisect, petiolate; spikes slender, becoming lax
  3 Stem scabrid on the angles, usually erect; fruiting spikes 10–25 cm **3. officinalis**
  3 Stem strigulose, usually procumbent; fruiting spikes up to 8 cm **4. supina**

1. V. rigida Sprengel, *Syst. Veg.* **4**(2): 230 (1827) (*V. venosa* Gillies & Hooker). Scabrid-pubescent perennial with stiff, erect or ascending, little-branched stems up to 60 cm. Leaves 4–8 × 1–2 cm, oblong, acute, irregularly dentate, rigid, sessile, amplexicaul. Inflorescence a lax terminal corymb, with long branches each ending in a dense spike 2–5 cm in fruit; bracts longer than calyx, lanceolate-subulate, scabrid and ciliate; corolla reddish-purple, 2–3 times as long as calyx. *Roadsides and waste places. Naturalized in Açores.* [Az ?Ga.] (*Argentina and Brazil.*)

2. V. bonariensis L., *Sp. Pl.* 20 (1753). Perennial with stiff, erect, little-branched stems up to 200 cm, scabrid on the angles. Leaves 7–13 × 1–2·5 cm, oblong-lanceolate, regularly incise-serrate, rugose, scabrid and more or less villous, tomentose beneath, sessile, amplexicaul. Panicle terminal, composed of several long-pedunculate cymes, each cyme of dense, sub-cylindrical, sessile spikes up to 2 cm in fruit; bracts equalling or shorter than calyx, concave, lanceolate-acuminate, usually dark blue; corolla blue, twice as long as calyx. *Roadsides and gravelly waste places. Naturalized in W. Europe.* [Az ?Be ?Ga Lu.] (*Argentina and S. Brazil.*)

3. V. officinalis L., *Sp. Pl.* 20 (1753). Perennial; stems 30–60(100) cm, erect, quadrangular, longitudinally ribbed, scabrid on the angles and diffusely branched. Leaves more or less rhombic, strigulose, the lower 4–6 × 2–4 cm, petiolate, deeply incised, lyrate to 1- to 2-pinnatifid, the upper smaller, sessile and subentire or entire. Spikes 10–25 cm in fruit, terminal, long-pedunculate, solitary or in a very lax panicle; bracts ovate-acuminate, ciliate, up to half as long as calyx; corolla pale pink, twice as long as the calyx. Nutlets 1·5–2 mm, reddish-brown, with 4 to 5 longitudinal ribs on the back. $2n = 14$. *Europe, northwards to c. 54° N.; naturalized or casual further north.* All except Fa Fe Is No Rs (N, B) Sb Su; doubtfully native in Da.

4. V. supina L., *Sp. Pl.* 21 (1753). Annual; stems 10–40 cm, usually procumbent, much-branched from the base, strigulose,

quadrangular with rounded angles separated by narrow grooves. Leaves 2–4 × 1·5–2·5 cm, triangular, petiolate, 1- to 2-pinnatipartite or -pinnatisect, densely strigulose. Spikes up to 8 cm in fruit, terminal, shortly pedunculate, solitary, or sometimes paniculate, not very lax; bracts lanceolate, strigulose, usually half as long as the calyx; corolla lilac, about as long as calyx. Nutlets 2–2·5 mm, light brown, with one obtuse longitudinal rib on the back. *Waste places and damp or sandy ground. S. Europe, extending northwards to S. Hungary and S.C. Russia.* Bu Cr Gr Hs Hu It Ju Lu Rm Rs (W, K, E) Sa Si Tu.

4. Lippia L.[1]

Shrubs or herbs with simple leaves. Flowers in spikes. Corolla weakly 2-lipped. Stamens included. Fruit of 2 1-seeded nutlets.

1 Shrub 3–6 m; inflorescence a panicle of slender spikes **3. triphylla**
1 Herbs 15–30 cm, sometimes woody at base; inflorescence a short, stout spike
  2 Flowers white; calyx lobed almost to base **1. nodiflora**
  2 Flowers lilac; calyx lobed to not more than halfway **2. canescens**

1. L. nodiflora (L.) Michx, *Fl. Bor. Amer.* **2**: 15 (1803). Perennial herb with procumbent, non-flowering stems rooting at the nodes, and ascending flowering stems 15–30 cm. Leaves 1–2·5 cm, obovate to oblanceolate, with a long-cuneate base, remotely serrate in the distal half. Flowers in short, stout axillary spikes 5–7 mm in diameter; peduncles much exceeding the subtending leaves. Calyx lobed almost to the base. Corolla 2 mm wide, sparsely pubescent without, white; lobes subequal. *Wet, grassy places, usually near the sea. Mediterranean region.* Al Bl Co Cr Hs It ?Sa Si Tu.

2. L. canescens Kunth in Humb., Bonpl. & Kunth, *Nov. Gen. Sp.* **2**: 263 (1818). Like **1** but stems somewhat woody at base; spikes 9–12 mm in diameter; calyx lobed to not more than halfway; corolla 3 mm wide, densely pubescent without, lilac; lobes very unequal. *Locally naturalized in S.W. Europe; limits uncertain owing to confusion with* **1**. [Bl Co Ga Hs It Lu Sa.] (*South America.*)

3. L. triphylla (L'Hér.) O. Kuntze, *Rev. Gen. Pl.* **3**(2): 253 (1898). Deciduous shrub 3–6 m. Leaves 7–10 cm, lanceolate, acuminate, cuneate at base, usually in whorls of 3, lemon-scented when crushed. Inflorescence terminal, paniculate. Flowers in long, slender spikes. Corolla lilac. *Commonly cultivated for ornament and locally naturalized in the Mediterranean region.* [Al Ga Hs ?Ju.] (*Chile.*)

CL. CALLITRICHACEAE[2]

Monoecious herbs with opposite leaves. Flowers axillary. Perianth absent. Stamen 1. Ovary 4-locular, with 1 anatropous, pendent ovule in each loculus. Styles 2, filiform. Fruit separating into (2–)4 mericarps. Seeds with fleshy endosperm.

1. Callitriche L.[3]

Submerged, amphibious or terrestrial herbs, with axillary glandular scales, and sometimes with peltate cauline hairs, otherwise glabrous. Leaves simple. Flowers solitary, or 1 male and 1

female flower in the same leaf-axil. Bracteoles 0 or 2 (in European species), membranous. Anthers reniform. Mericarps with rounded, keeled or winged margin.

5 well-defined groups of species can be recognized in Europe, and hybridization between species of different groups is unknown. The amphibious species are very variable, and identification is made difficult because of the vegetative similarity of different species when growing in similar habitats.

Ripe fruits are essential for identification; the anatomical details of the mericarps can easily be observed (in European species) after stripping off the outer layer of cells. In **5–13**

---

[1] By T. G. Tutin.   [2] Edit. S. M. Walters.   [3] By H. D. Schotsman.

peltate hairs are present on the stems; their shape and the number of cells in the disc are of diagnostic importance. All species except **1–3** can occur in the terrestrial form on mud; the habitat given is for the aquatic form. The plants are usually annual, but the life-cycle depends on climatic factors, and most of the species can be perennial in some part of their range.

Literature: H. D. Schotsman in P. Jovet, *Flore de France* **1**. Paris. 1967.

1 Terrestrial
  2 Fruit *c.* 0·6 mm, wider than long
    3 Mericarps thicker at base than apex   **12. peploides**
    3 Mericarps of equal thickness throughout   **13. terrestris**
  2 Fruit 0·8 mm or more, suborbicular or longer than wide
    4 Mericarps not winged (sometimes weakly keeled)
      5 Fruit 1·5 mm, elliptical; mericarps with rounded, scarcely distinct margin; pollen-grains oblong-ellipsoid or slightly reniform   **6. obtusangula**
      5 Fruit 0·8–1·2 mm, suborbicular or slightly oblong; mericarps with distinct but obtuse margin; pollen-grains shortly ellipsoid or subglobose   **7. cophocarpa**
    4 Mericarps winged
      6 Mericarps winged only at apex; fruit obovate (rarely elliptical)   **9. palustris**
      6 Mericarps winged from base to apex; fruit suborbicular or slightly longer than wide
        7 Persistent styles deflexed and appressed to sides of fruit
          8 Fruit subsessile or very rarely shortly stalked   **10. hamulata**
          8 Fruit with long stalk up to 13 mm   **11. brutia**
        7 Persistent styles erect or recurved, not appressed to sides of fruit
          9 Styles recurved; leaves pale green, broadly elliptical or suborbicular; fruit pale brownish, with broadly winged mericarps   **5. stagnalis**
          9 Styles erect or patent; leaves dark green, elliptical; fruit brown, usually with narrowly winged mericarps   **8. platycarpa**
1 Aquatic
  10 All leaves submerged
    11 Leaves not transparent; wing of mericarp with 1 row of cells; persistent styles appressed to sides of fruit
      12 Leaves expanded at apex, with deep, crescentic emargination   **10. hamulata**
      12 Leaves not widened at apex, often with asymmetrical emargination   **11. brutia**
    11 Leaves transparent; wing of mericarp, if present, with at least 2 rows of cells; styles deciduous
      13 Fruit suborbicular   **1. hermaphroditica**
      13 Fruit wider than long
        14 Wing of mericarp narrow or absent; fruit *c.* 1·5 mm wide   **2. truncata**
        14 Wing of mericarp wide; fruit 1·8–2·2 mm wide
          15 Mericarps broadly reniform, with large seed; stellate thickenings present in cells of wall enclosing seed   **3. pulchra**
          15 Mericarps almost semicircular, with small seed; stellate thickenings not present in cells of wall enclosing seed   **4. lusitanica**
  10 Upper leaves spathulate, forming a floating or aerial rosette
    16 Rosette aerial, raised above surface of water; submerged linear leaves and basal part of rosette-leaves transparent; fruit wider than long   **4. lusitanica**
    16 Rosette floating; submerged leaves and basal part of rosette-leaves not transparent; fruit suborbicular or longer than wide
      17 Flowers submerged; pollen-grains colourless; persistent styles appressed to sides of fruit
        18 Usually robust; fruit *c.* 1·4 mm wide, suborbicular; mericarps narrowly winged   **10. hamulata**
        18 Slender; fruit 1–1·2 mm wide, usually longer than wide; mericarps often broadly winged   **11. brutia**
      17 Flowers aerial; pollen-grains yellow; persistent styles not appressed to side of fruit

    19 Mericarps unwinged, sometimes weakly keeled
      20 Fruit 1·5 mm, elliptical; mericarps with rounded, scarcely distinct margin; pollen-grains oblong-ellipsoid or slightly reniform   **6. obtusangula**
      20 Fruit 0·8–1·2 mm, suborbicular or slightly oblong; mericarps with distinct but obtuse margin; pollen-grains shortly ellipsoid or subglobose   **7. cophocarpa**
    19 Mericarps winged
      21 Mericarps winged only at apex; fruit 1(–1·5) mm, obovate (rarely elliptical), blackish   **9. palustris**
      21 Mericarps winged from base to apex; fruit 1·4–1·75 mm, suborbicular, brownish
        22 Leaves pale green, narrowly elliptical when submerged; mericarps broadly winged   **5. stagnalis**
        22 Leaves usually dark green, linear when submerged; mericarps narrowly winged   **8. platycarpa**

(*a*) Always submerged. Leaves more or less linear, transparent. Cauline hairs absent. Flowers solitary. Bracteoles absent. Stamen erect before and after dehiscence, scarcely elongating; anther-wall with thin-walled cells; pollen-grains colourless, without exine. Wing of mericarp, if present, composed of one or more rows of polygonal cells. Basic chromosome number 3, 4.

1. C. hermaphroditica L., *Cent. Pl.* **1**: 31 (1755) (*C. autumnalis* L.). Leaves widest at base, tapering towards emarginate apex. Styles patent or deflexed, deciduous. Fruit 1–3·3 mm, suborbicular, sessile or subsessile; mericarps usually broadly winged at least towards apex. $2n = 6$. *Lakes, canals and rivers. N. & E. Europe, southwards to c. 48° N. in U.S.S.R.* Br Da Fa Fe Ge Hb Ho Is No Po Rm Rs (N, B, C, W, E) Su.

Variants can be recognized, differing mainly in fruit size.

2. C. truncata Guss., *Pl. Rar.* 4 (1826). Leaves widest at base and tapering towards apex, or more or less elliptical, shallowly emarginate or truncate. Styles patent or deflexed, deciduous. Fruit 1–1·2 × 1·3–1·6 mm, wider than long, subsessile or stalked; mericarps not or only narrowly winged. ● *S. & W. Europe, northwards to England; S.E. Russia.* Be Br ?Co Ga Gr Hs It Ju Lu Rs (E) Sa Si.

1 Mericarps semicircular, unwinged; fruit subsessile; leaves often slightly elliptical   (c) subsp. **occidentalis**
1 Mericarps reniform, narrowly winged; fruit stalked; leaves tapering from base to apex
  2 Wing of mericarp composed of 1–3 rows of polygonal cells   (a) subsp. **truncata**
  2 Wing of mericarp consisting of a whitish, ciliate border, formed by the radial walls of the cells   (b) subsp. **fimbriata**

(a) Subsp. **truncata**: Fruit 1–1·2 × 1·3–1·4 mm, stalked. *Shallow pools. C. & E. Mediterranean region.*

(b) Subsp. **fimbriata** Schotsman in Jovet, *Fl. Fr.* **1**: 39 (1967): Fruit 1·2 × 1·5 mm, shortly stalked. *S.E. Russia.*

(c) Subsp. **occidentalis** (Rouy) Schotsman, *op. cit.* 36 (1967) (*C. truncata* race *occidentalis* Rouy): Fruit 1–1·2 × 1·4–1·6 mm, subsessile; sometimes 1–2 mericarps abortive. $2n = 6$. *Fresh or slightly brackish water. W. Europe, northwards to England.*

3. C. pulchra Schotsman, *op. cit.* 40 (1967). Leaves almost linear, slightly wider at base, emarginate. Styles with basal part recurved, upper part erect. Fruit 1·5–1·8 × 2–2·2 mm, stalked, often with only 2 mericarps; mericarps broadly reniform, with large seed and wide wing, the cells of the wall enclosing the seed with stellate thickenings. $2n = 8$. *Pools on calcareous soils. Gavdhos. Cr. (Libya.)*

(*b*) Two fruiting forms: completely submerged, with linear, transparent leaves; or upper part of stem aerial, forming a rosette with spathulate leaves raised above the surface of the water, the basal part of the spathulate leaves transparent. Cauline hairs absent. Flowers solitary, submerged or aerial. Bracteoles absent. Stamens of submerged flowers (in the European species) erect before and after dehiscence, those of aerial flowers recurved and elongating after dehiscence; anther-wall with thickened cells; pollen-grains pale yellowish, with thin exine. Wing of mericarp (in the European species) composed of several rows of large, transparent polygonal cells. Basic chromosome number 4.

4. C. lusitanica Schotsman, *Bol. Soc. Brot.* ser 2, **35**: 112 (1961). Submerged leaves tapering slightly towards emarginate apex. Styles patent. Fruit 1·2–1·4 × 1·8–2 mm, subsessile; mericarps almost semicircular, with small seed and wide wing, the cells of the wall enclosing the seed without stellate thickenings. 2*n*=8. *River-beds and small pools.* ● *E. Portugal, W. & S. Spain.* Hs Lu.

(*c*) Three forms: submerged and sterile; aquatic with floating rosette; and terrestrial. Submerged leaves not transparent. Disc of cauline hairs of 8–15 cells. Flowers solitary, or 1 male and 1 female flower in the same leaf-axil. Bracteoles 2. Stamen recurved and elongating after dehiscence; anther-wall with thickened cells; pollen-grains yellow, with exine. Wing of mericarp, if present, composed of a single row of radially elongated cells. Basic chromosome number 5.

5. C. stagnalis Scop., *Fl. Carn.* ed. 2, **2**: 251 (1772). Submerged leaves narrowly elliptical; floating rosettes with *c.* 6 broadly elliptical or suborbicular, pale green leaves. In terrestrial form, leaves small, broadly elliptical or suborbicular, pale green. Disc of cauline hairs orbicular, of 8–10(–12) cells. Flowers solitary, or 1 male and 1 female flower in the same leaf-axil. Pollen-grains subglobose. Styles in aquatic form erect or patent, in terrestrial form arcuate-recurved. Fruit 1·75 × 1·6–1·8 mm, suborbicular, pale brownish, deeply grooved between the divergent mericarps; mericarps broadly winged, radial thickening of cells of wing often weakly developed. 2*n*=10. *Springs and still or slow-moving water. Most of Europe westwards from Estonia, W. Ukraine and Greece.* All except Cr Fe Rm Rs (N, K, E) Tu.

6. C. obtusangula Le Gall, *Fl. Morbihan* 202 (1852). Floating rosettes with 12–20 often fleshy, rhombic (rarely obovate) leaves; in terrestrial form, leaves narrowly rhombic or elliptical, fleshy, often yellowish-green. Disc of cauline hairs orbicular or elliptical, of 8–10 cells. Flowers solitary. Pollen-grains oblong-ellipsoid or slightly reniform. Styles erect or patent. Fruit 1·5 × 1·2 mm, elliptical, brown, shallowly grooved between the parallel mericarps; mericarps with rounded, scarcely distinct margin. 2*n*=10. *Fresh or brackish, usually slow-moving water. S., W. & W.C. Europe, northwards to c. 54° N.* Au Be Br Co Ga Ge Gr Hb He Ho Hs It Ju Lu Sa Si.

7. C. cophocarpa Sendtner, *Veg. Südbayerns* 773 (1854) (*C. polymorpha* Lönnr.). Floating rosettes with 10–18 slightly rhombic or elliptical leaves; in terrestrial form, leaves elliptical. Disc of cauline hairs orbicular, of 8–12(–15) cells. Flowers solitary. Pollen-grains shortly ellipsoid or subglobose. Styles erect, often persistent. Fruit 0·8–1·2 × 0·9–1·1 mm, suborbicular or slightly oblong, brown, moderately deeply grooved between the parallel mericarps; mericarps unwinged, sometimes weakly keeled, with distinct but obtuse margin. 2*n*=10. *Slow-moving*

water. *Most of Europe, except the south-west and most of the islands.* Al Au Be Br Bu Cz Da Fa Fe Ga Ge He Ho Hu It Ju No Po Rm Rs (N, B, C, W, K, E) Su.

8. C. platycarpa Kütz. in Reichenb., *Pl. Crit.* **9**: 38 (1831). Submerged leaves linear; floating rosettes with elliptical, often dark green leaves; in terrestrial form, leaves elliptical, dark green. Disc of cauline hairs often asymmetrical, of 8–10 cells. Flowers solitary. Pollen-grains variously shaped (globose, ovoid, triangular etc.). Styles erect or patent. Fruit *c.* 1·5 mm, suborbicular, brown, sometimes with 1 or 2 mericarps abortive; mericarps narrowly winged, the cells of the wing thickened on radial walls and tapered at both ends. 2*n*=20. *Fresh (rarely brackish) flowing or still, often base-rich water.* ?● *N.W. & C. Europe; one locality in N. Spain.* Au Be Br Cz Da Fa Ga Ge Hb He Ho Hs Hu No Po ?Rm Su.

The distribution in S. & E. Europe is uncertain.

9. C. palustris L., *Sp. Pl.* 969 (1753) (*C. verna* L., *C. vernalis* Koch). Submerged leaves very narrowly linear; floating rosettes with elliptical or suborbicular leaves; in terrestrial form, leaves small, elliptical. Disc of cauline hairs orbicular, of 12–15 cells. Usually 1 male and 1 female flower in the same leaf-axil. Pollen-grains subglobose. Styles erect, caducous. Fruit 1(–1·5) mm, obovate, rarely elliptical, blackish; mericarps winged only at apex, the cells of the wing thickened on radial walls, not tapering. 2*n*=20. *Shallow, still water. Europe, southwards to the Pyrenees, Corse and the lower Volga, mainly in the mountains except in Fennoscandia & Russia.* Au †Be Br Bu Co Cz Da Fe Ga Ge He Ho ?Hs Hu Is It No Po Rm Rs (N, B, C, W, ?K, E) Su.

In the terrestrial form, the male flowers and styles are reduced or absent.

Variants with almost elliptical fruits and scarcely winged mericarps occur in the Pyrenees, with fruits up to 1·5 mm in E. Switzerland, and with broadly winged mericarps in E. Europe. Their status is uncertain.

(*d*) Three fruiting forms: submerged; aquatic with floating rosettes; and terrestrial. Submerged leaves not transparent. Disc of cauline hairs of (8–)12–20 cells. Flowers solitary. Bracteoles 2, deciduous, or absent. Stamen recurved towards one of the deflexed styles; anther-wall with thin-walled cells; pollen-grains colourless, without exine. Wing of mericarp composed of a single row of short cells with thickening on the radial and proximal walls. Basic chromosome number ?7.

10. C. hamulata Kütz. ex Koch, *Syn. Fl. Germ.* 246 (1835). Usually robust. Submerged leaves linear, with expanded, deeply emarginate apex; floating rosettes with elliptical or slightly obovate leaves; in terrestrial form, leaves elliptical, dark green. Disc of cauline hairs orbicular, of 10–15 cells. Pollen-grains globose. Styles deflexed, appressed to sides of fruit. Fruit 1·2–1·5 × 1·4 mm, suborbicular, subsessile or sometimes (in terrestrial form) very shortly stalked; mericarps narrowly winged. 2*n*=38, ?40. *Base-poor, cool, flowing water and lakes.* ?● *N.W. & C. Europe, extending eastwards to Finland and Latvia and southwards to Italy.* Au Be Br Bu Cz Da Fa Fe Ga Ge Hb He Ho Is It No Po Rm Rs (B) Su.

11. C. brutia Petagna, *Inst. Bot.* **2**: 10 (1787) (*C. pedunculata* DC.). Usually slender. Submerged leaves linear, not widened at apex, often irregularly emarginate; floating rosettes with elliptical or slightly obovate leaves; in terrestrial form, leaves elliptical, dark green. Disc of cauline hairs elliptical, often somewhat irregular in outline, of 8–16 cells. Pollen-grains subglobose. Styles deflexed, appressed to sides of fruit. Fruit 1–1·4 × 1–1·2 mm, suborbicular or slightly longer than wide, subsessile in aquatic form, with stalk up to 13 mm in terrestrial form; mericarps mostly broadly winged. $2n = 28$. *Still, often shallow water.* ● *W. & S. Europe, eastwards to Italy; one station in S. Sweden.* Be Bl Br Da Fa Ga Hb Hs Is It Lu No Sa Si Su.

Small aquatic forms of **10** and **11** are difficult to separate. **C. naftolskyi** Warburg & Eig, *Feddes Repert.* **26**: 84 (1929), from Israel, is doubtfully recorded from Sardegna; it has fruits with very long stalks and ripening in the soil, broadly winged mericarps with a complex wing-structure, and the disc of the cauline hairs composed of 12–19 cells.

(*e*) Always terrestrial. Leaves elliptical or spathulate. Disc of cauline hairs of 4–8 cells. 1 male and 1 female flower in the same leaf-axil. Bracteoles absent. Wing of mericarp composed of a single row of short cells with thickening on the radial and proximal walls. Basic chromosome number probably 5.

12. C. peploides Nutt., *Trans. Amer. Philos. Soc.* nov. ser., **5**: 141 (1835). Leaves elliptical or spathulate. Fruit 0·6 × 0·8 mm, subsessile; mericarps thicker and less rounded at base than at apex, very narrowly winged. *Naturalized in N. France (Seine-et-Oise).* [Ga.] (*North America.*)

13. C. terrestris Rafin., *Med. Reposit.* (*New York*) **5**: 358 (1808). Leaves elliptical. Fruit 0·6 × 0·8–0·9 mm; mericarps of equal thickness throughout and equally rounded at base and apex, narrowly winged. *Naturalized in S.W. France (Hautes-Pyrénées).* [Ga.] (*North America.*)

CLI. LABIATAE[1]

Herbs or shrubs, often glandular and aromatic. Leaves usually simple, exstipulate, opposite. Flowers zygomorphic, usually in contracted and modified cymes in the axils of opposite bracts or floral leaves, forming pseudowhorls called *verticillasters*, which in turn are arranged in simple or compound spike-like, cymose, corymbose, paniculate or capitate inflorescences; rarely in true cymes. Bracts leaf-like, or much reduced or modified (usually called *floral leaves* when conspicuous). Bracteoles usually small, sometimes absent. Calyx usually 4- or 5-lobed, often 2-lipped with the upper lip 3-toothed and the lower 2-toothed. Corolla sympetalous; limb usually 5-lobed, often 2-lipped with the upper lip 2-lobed and the lower lip 3-lobed, rarely all 5 lobes forming the lower lip. Stamens usually 4, didynamous, rarely 2. Ovary superior, 2-carpellate but appearing equally 4-lobed when mature due to further partition; style single, usually branched above and gynobasic. Fruit of four 1-seeded nutlets.

Sexual dimorphism occurs in several genera, with female flowers, which are normally smaller, occurring in addition to the usual hermaphrodite flowers on the same or on different plants.

The calyx may be entire or shallowly to deeply lobed or toothed, and in the key, the calyx-tube is regarded as extending from the base of the calyx up to the lowermost sinus. The number of veins refers to the lower part of the tube.

1 Corolla ± 1-lipped, the upper lip much reduced or absent
  2 Upper lip of corolla absent; corolla-tube glabrous inside
           2. Teucrium
  2 Upper lip of corolla present, entire or consisting of 2 small teeth; corolla-tube with a ring of hairs inside   **1. Ajuga**
1 Corolla ± actinomorphic, or 2-lipped with a distinct upper lip
  3 Fertile stamens 2
    4 Upper lip of corolla distinctly hooded (concave); stamens with 1 fertile cell   **40. Salvia**
    4 Upper lip of corolla flat or convex; stamens with 2 fertile cells
      5 Lower leaves pinnatifid   **34. Lycopus**
      5 Lower leaves entire
        6 Annual; upper lip of corolla entire or emarginate   **24. Ziziphora**
        6 Shrub; upper lip of corolla 2-fid   **37. Rosmarinus**
  3 Fertile stamens 4
    7 Upper lip of corolla distinctly hooded (concave), ± equalling to much exceeding the lower lip; calyx usually actinomorphic

    8 Calyx with an erect, dorsal projection   **3. Scutellaria**
    8 Calyx without a dorsal projection
      9 Calyx-tube ± hypocrateriform
        10 Calyx zygomorphic, curved   **14. Moluccella**
        10 Calyx ± actinomorphic, straight
          11 Corolla 10–18 mm; bracteoles present   **15. Ballota**
          11 Corolla 32–40 mm; bracteoles absent   **8. Eremostachys**
      9 Calyx-tube subglobose, campanulate, infundibuliform or tubular
        12 Style-branches distinctly unequal
          13 Calyx-teeth equal; tube with 5–10 veins   **9. Phlomis**
          13 Calyx 2-lipped, the lower teeth longer than the upper; tube with 13 veins
            14 Verticillasters of 2–6(8) equally pedicellate flowers; calyx gibbous at base, the tube constricted   **26. Acinos**
            14 Verticillasters many-flowered, or of pedunculate cymes; calyx not gibbous or constricted
              15 Calyx-tube straight; inflorescence of pedunculate cymes   **27. Calamintha**
              15 Calyx-tube curved; inflorescence of many-flowered verticillasters   **28. Clinopodium**
        12 Style-branches equal or subequal
          16 Calyx with 11–15 veins
            17 Calyx with 11–13 veins
              18 Stamens with anthers ± connate under upper lip of corolla; calyx with 13 veins   **29. Micromeria**
              18 Stamens divergent; calyx with 11–13 veins   **25. Satureja**
            17 Calyx with 15 veins
              19 Upper tooth of calyx wider than other 4 teeth   **20. Dracocephalum**
              19 Upper tooth of calyx not wider than other 4 teeth   **29. Micromeria**
         16 Calyx with 5–10 veins
           20 Lateral lobes of lower lip of corolla absent, or inconspicuous, sometimes reduced to small, acute teeth   **11. Lamium**
           20 Lateral lobes of lower lip of corolla distinct, ± obtuse
            21 Bracteoles pungent
              22 Annual; lower lip of corolla with 2 conical projections at base of lateral lobes   **10. Galeopsis**
              22 Shrub; lower lip of corolla without conical projections   **15. Ballota**
            21 Bracteoles not pungent, sometimes absent
              23 Calyx distinctly 2-lipped
                24 Calyx with throat closed in fruit when dry; filaments with a subulate appendage below the apex   **21. Prunella**
                24 Calyx with throat always open; filaments without an appendage

---

[1] Edit. V. H. Heywood and I. B. K. Richardson.

25 Corolla 17–20 mm; nutlets drupaceous **4. Prasium**
25 Corolla *c.* 8 mm; nutlets dry **25. Satureja**
23 Calyx not 2-lipped, the teeth equal or subequal
26 Nutlets hairy at apex **13. Leonurus**
26 Nutlets not hairy at apex
27 Verticillasters in secund spikes; corolla 3–4 mm **41. Elsholtzia**
27 Verticillasters not in secund spikes; corolla (4–)6–25 mm
28 Corolla bright yellow and brown; upper lip (8–)10–12 mm **12. Lamiastrum**
28 Corolla pale yellow, white, pink or purple; upper lip not more than 8 mm
29 Mostly dwarf shrubs with ascending stems; corolla usually not more than 11 mm; calyx usually less than 7 mm; stamens divergent **25. Satureja**
29 Mostly herbs with erect stems; corolla usually more than 11 mm; calyx usually more than 7 mm; stamens parallel
30 Calyx-tube expanded above middle; corolla-tube with a ring of hairs inside **15. Ballota**
30 Calyx-tube not or scarcely expanded above middle; corolla with or without a ring of hairs **16. Stachys**
7 Upper lip of corolla ± flat or convex, usually shorter than the lower lip, or corolla actinomorphic; calyx often 2-lipped
31 Upper tooth of calyx with a cordate to obovate apical appendage **38. Lavandula**
31 Upper tooth of calyx without appendage
32 Verticillasters aggregated into spicules **32. Origanum**
32 Verticillasters not aggregated into spicules
33 Corolla nearly actinomorphic, with 4 or 5 subequal lobes
34 Calyx zygomorphic
35 Calyx strongly 2-lipped, the throat glabrous inside **33. Thymus**
35 Calyx weakly 2-lipped, the throat hairy inside **35. Mentha**
34 Calyx ± actinomorphic
36 Flowers in pairs in a secund raceme; annual **36. Perilla**
36 Flowers in verticillasters; usually perennial
37 Inflorescence plumose **33. Thymus**
37 Inflorescence not plumose **35. Mentha**
33 Corolla 2-lipped
38 Flowers with corolla 25–40 mm and calyx 2-lipped, with the upper teeth much exceeding the lower
39 Calyx not more than 11 mm; bracteoles present **19. Glechoma**
39 Calyx more than 11 mm; bracteoles absent **7. Melittis**
38 Flowers either with corolla less than 25 mm, or with calyx not 2-lipped with upper teeth much exceeding the lower
40 Bracts pectinate-pinnatifid; style 4-fid **22. Cleonia**
40 Bracts not pectinate-pinnatifid; style 2-fid
41 Calyx with 5–10 veins
42 Corolla more than 11 mm
43 Calyx ± 2-lipped, the teeth unequal **33. Thymus**
43 Calyx not 2-lipped, the teeth ± equal
44 Flowers violet-blue; style not gynobasic **1. Ajuga**
44 Flowers rarely violet-blue; style gynobasic
45 Stamens included in corolla-tube; corolla-tube included in calyx; upper lip of corolla often oblong to linear and 2-fid
46 Bracteoles usually present; calyx 5- to 10-toothed; nutlets rounded at apex **5. Marrubium**
46 Bracteoles absent; calyx 10-toothed; nutlets truncate at apex **6. Sideritis**
45 Stamens usually not included in corolla-tube; corolla-tube usually exserted from calyx; upper lip of corolla usually ± ovate, entire or emarginate, rarely 2-fid
47 Calyx-tube hypocrateriform **15. Ballota**
47 Calyx-tube not hypocrateriform

48 Corolla 12(–14) mm; often dwarf shrubs with ascending stems; stamens divergent **25. Satureja**
48 Corolla 12–25 mm; usually herbs with erect stems; stamens parallel **16. Stachys**
42 Corolla not more than 11 mm
49 Calyx 2-lipped
50 Annual; stamens not exserted from corolla-tube, parallel **6. Sideritis**
50 Perennial; stamens exserted from corolla-tube, divergent
51 Fruiting calyx gibbous at base (S.C. Bulgaria) **25. Satureja**
51 Fruiting calyx not gibbous at base **33. Thymus**
49 Calyx not 2-lipped
52 Annual, without non-flowering shoots at flowering time
53 Corolla 3–4 mm; verticillasters 2-flowered, in crowded, secund spikes **41. Elsholtzia**
53 Corolla (4–)5–20 mm; verticillasters not in crowded, secund spikes
54 Calyx 3–4 mm; stamens divergent **25. Satureja**
54 Calyx 5–8 mm; stamens parallel
55 Corolla deep yellow or yellow and black **6. Sideritis**
55 Corolla pale yellow, white, pink, reddish or purple **16. Stachys**
52 Perennial, woody at base, with non-flowering shoots at flowering time
56 Bracteoles present
57 Calyx 7–12 mm; stamens parallel **16. Stachys**
57 Calyx 2·5–7(–9) mm; stamens divergent **25. Satureja**
56 Bracteoles absent
58 Stamens included in corolla-tube, parallel **6. Sideritis**
58 Stamens ± exserted from corolla-tube; divergent **25. Satureja**
41 Calyx with 11–15(–22) veins
59 Calyx-tube dorsally flattened, with 13 or 20–22 veins
60 Calyx with 20–22 veins **33. Thymus**
60 Calyx with 13 veins **30. Thymbra**
59 Calyx-tube not dorsally flattened, with 11–15 veins
61 Leaves 3(4)-foliolate **17. Cedronella**
61 Leaves not 3(4)-foliolate
62 Anther-cells at right angles to each other, those of the paired stamens arranged in the form of a cross; filaments parallel
63 Plant stoloniferous; anther-cells each opening by a separate slit **19. Glechoma**
63 Plant not stoloniferous; anther-cells opening by a common slit **18. Nepeta**
62 Anther-cells rarely at right angles to each other; filaments usually divergent
64 Style not gynobasic; corolla violet-blue **1. Ajuga**
64 Style gynobasic; corolla various
65 Calyx with 15 veins
66 Leaves mostly 2-pinnatisect **38. Lavandula**
66 Leaves not 2-pinnatisect
67 Upper tooth of calyx wider than other 4 teeth **20. Dracocephalum**
67 Upper tooth of calyx not wider than other 4 teeth
68 Middle lobe of lower lip of corolla usually concave, crenate; lateral lobes indistinct; stamens parallel **18. Nepeta**
68 Middle lobe of lower lip of corolla ± plane, usually entire or emarginate; lateral lobes distinct; stamens divergent or connivent
69 Verticillasters in secund spikes **31. Hyssopus**
69 Verticillasters not in secund spikes **29. Micromeria**

65 Calyx with (11–)13 veins
  70 Style-branches distinctly unequal
    71 Verticillasters of 2–6(8) equally pedicellate flowers; calyx usually gibbous at base, the tube constricted   **26. Acinos**
    71 Verticillasters many-flowered or of pedunculate cymes; calyx not gibbous or constricted
      72 Calyx-tube curved; inflorescence of many-flowered verticillasters   **28. Clinopodium**
      72 Calyx-tube straight; inflorescence of pedunculate cymes   **27. Calamintha**
  70 Style-branches equal or subequal
    73 Upper teeth of calyx similar to lower
      74 Inflorescence plumose   **33. Thymus**
      74 Inflorescence not plumose
        75 Stamens divergent   **25. Satureja**
        75 Stamens with anthers ±connate under upper lip of corolla   **29. Micromeria**
    73 Upper teeth of calyx distinct from lower
      76 Corolla-tube straight; leaves usually less than 2 cm
        77 Stamens connivent, curved; calyx-tube usually curved   **29. Micromeria**
        77 Stamens divergent, straight; calyx-tube straight   **33. Thymus**
      76 Corolla-tube distinctly curved; leaves at least 2 cm
        78 Corolla 8–15 mm, pale yellow, becoming white or pinkish   **23. Melissa**
        78 Corolla 17–21 mm, dark bluish-violet   **39. Horminum**

1. Ajuga L.[1]

Annual or perennial herbs. Calyx more or less actinomorphic, with 10 or more veins. Upper lip of the corolla usually very short; lower lip 3-lobed; tube with a ring of hairs inside. Stamens didynamous, all usually exserted. Style not gynobasic. Nutlets reticulate-veined or transversely rugose.

2, 3 and **4** are interfertile, and hybrids have been recorded from most areas where any two of these species grow together.

1 Leaves 3-partite with linear, sometimes 3-fid segments   **10. chamaepitys**
1 Leaves entire, toothed or shallowly lobed
2 Leaves 3–6(–8) mm wide, linear to linear-oblong   **9. iva**
2 Leaves 8–40(–50) mm wide, oblong to obovate-orbicular
  3 Stamens included in the corolla-tube; upper lip of corolla with 2 conspicuous lobes   **1. orientalis**
  3 Stamens ±exserted from the corolla-tube; upper lip of corolla entire or with 2 short teeth or lobes
    4 Acaulescent, or with stems not more than 5 cm   **5. tenorii**
    4 Caulescent, with stems more than 5 cm
      5 Flowers solitary in the axil of each bract
        6 Cauline leaves sessile, semiamplexicaul; corolla-tube shorter than the calyx   **7. laxmannii**
        6 Cauline leaves very shortly petiolate, not semiamplexicaul; corolla-tube exceeding the calyx
          7 Calyx 6–10(–11) mm; lower leaves 10–22 mm wide, entire or dentate   **6. salicifolia**
          7 Calyx 11–15 mm; lower leaves 25–45 mm wide, coarsely and remotely serrate-dentate   **8. piskoi**
      5 Flowers 4 or more in a whorl at each node
        8 Middle and upper part of stem pubescent on opposite faces, alternating at each node   **4. reptans**
        8 Middle and upper part of stem ± evenly hairy on all faces
          9 Stamens conspicuously exserted from the corolla-tube, the filaments hairy; upper bracts often shorter than the flowers   **2. genevensis**

          9 Stamens only slightly exserted from the corolla-tube, the filaments glabrous; upper bracts exceeding the flowers   **3. pyramidalis**

1. A. orientalis L., *Sp. Pl.* 561 (1753). Rhizomatous perennial; stems 10–60 cm, sparsely to densely lanate-villous. Lower leaves 30–90(–120) × 15–40(–50) mm, ovate to oblong, sinuate-crenate or crenate-dentate, sometimes shallowly lobed. Bracts ovate, usually shallowly lobed, tinged with blue, the upper shorter than the flowers. Verticillasters distant, usually 4- to 6-flowered. Calyx 5–8 mm, the teeth about as long as the tube. Corolla 12–16(–18) mm, violet-blue; tube exceeding the calyx; upper lip conspicuously 2-lobed. Stamens included in the corolla-tube. $2n=32$. *S. Europe, from Sicilia to Krym.* Al Gr It Rs (W, K) Si.

2. A. genevensis L., *Sp. Pl.* 561 (1753). Rhizomatous perennial; stems 10–40 cm, subglabrous to densely lanate-villous. Lower leaves 30–120 × 8–50 mm, obovate to oblong-obovate, crenate-dentate or shallowly lobed, the basal usually dead at anthesis. Bracts obovate, usually lobed and tinged with blue or violet, the upper often shorter than the flowers. Verticillasters distant or somewhat crowded, 6- to many-flowered. Calyx 4–6(–7) mm, the teeth about as long as the tube. Corolla 12–20 mm, bright blue, rarely pink or white; tube exceeding the calyx; upper lip with 2 teeth. Stamens exserted; filaments hairy. $2n=32$. *Europe except the south-west, the islands and most of the north; sometimes casual elsewhere.* Al Au Be Bu Cz Ga Ge Gr He Ho Hu It Ju Po Rm Rs (B, C, W, K, E) †Su ?Tu [Fe].

3. A. pyramidalis L., *Sp. Pl.* 561 (1753). Rhizomatous perennial without stolons; stems 5–30 cm, subglabrous or lanate-villous. Lower leaves 40–110 × 15–45 mm, obovate, entire or crenate-dentate, the basal usually persistent. Bracts ovate or obovate-orbicular, sometimes lobed, usually tinged with blue or violet, all exceeding the flowers. Verticillasters usually crowded, 4- to 8-flowered. Calyx 5–8 mm, the teeth as long as the tube. Corolla 10–18 mm, pale violet-blue, rarely pink or white; tube exceeding calyx; upper lip entire. Stamens slightly exserted; filaments glabrous. $2n=32$. *Europe, southwards to N. Portugal, N. Italy and Bulgaria, but absent from most of the U.S.S.R.* Al Au Be Br Bu ?Co Cz Da Fe Ga Ge Hb He Hs Is It Ju Lu No Po Rm Rs (B, C) Su.

Plants from Portugal have sometimes been treated as subsp. **meonantha** (Hoffmanns. & Link) R. Fernandes, *Bol. Soc. Brot.* ser. 2, **34**: 131 (1960), or as a distinct species, **A. occidentalis** Br.-Bl., *Agron. Lusit.* **18**: 89 (1956). They are densely villous with thick leaves, and the calyx has a narrow tube and narrow, acute teeth. These differences do not, however, seem to merit recognition at more than varietal rank.

A. rotundifolia Willk. & Cutanda, *Linnaea* **30**: 120 (1859), from mountain slopes of C. Spain, has stolons, spathulate lower leaves and lanceolate, acuminate calyx-lobes. It has sometimes been treated as a subspecies of **3** but its status is not clear.

4. A. reptans L., *Sp. Pl.* 561 (1753). Rhizomatous perennial with long stolons; stems 10–40 cm, pubescent on opposite faces, alternating at each node, sometimes glabrous at the base. Lower leaves 25–90 × 10–40 mm, ovate, entire or crenate. Bracts ovate, often tinged with blue, the upper shorter than the flowers. Verticillasters crowded, usually 6-flowered. Calyx 4–6 mm, the teeth about as long as the tube. Corolla 14–17 mm, blue, rarely pink or white; tube exceeding calyx; upper lip entire. Stamens exserted; filaments hairy. $2n=32$. *Most of Europe northwards to c. 61° N.* Al Au Be Br Bu Cz Da Ga Ge Gr Hb He Ho Hs Hu It Ju Lu No Po Rm Rs (N, B, C, W, K, E) Si Su Tu [Fe].

[1] By P. W. Ball.

5. A. tenorii C. Presl in J. & C. Presl, *Del. Prag.* 79 (1822). Rhizomatous perennial, acaulescent or with stem up to 5 cm; glabrous or sparsely hairy. Leaves 18–50 × 10–20 mm, oblong-obovate or spathulate, crenate-dentate. Bracts lanceolate, at least the upper shorter than the flowers. Calyx 5–6 mm, the teeth slightly longer than the tube. Corolla 16–25 mm, bright blue; tube exceeding calyx; upper lip entire or with 2 short teeth. Stamens exserted; filaments usually hairy. ● *Mountains of C. & S. Italy and Sicilia.* It Si.

6. A. salicifolia (L.) Schreber, *Pl. Vert. Unilab.* 26 (1773). Perennial; stem 20–30 cm, woody at base, lanate-villous. Lower leaves 25–60 × 10–22 mm, oblong-elliptical or lanceolate, entire or dentate; cauline leaves cuneate at base. Bracts similar to the leaves, exceeding the flowers. Flowers 2 at each node. Calyx 6–10(–11) mm, the teeth shorter than the tube. Corolla 18–25(–28) mm, yellow; tube slightly exceeding calyx; upper lip with 2 short lobes. Stamens exserted; filaments hairy. *S.E. Europe, from Turkey-in-Europe to Krym.* Bu Rm Rs (K) Tu.

(a) Subsp. **salicifolia**: Cauline hairs 0·2–1 mm, appressed to semipatent. Calyx (7–)8–11 mm. *Krym.*

(b) Subsp. **bassarabica** (Săvul. & Zahar.) P. W. Ball, *Bot. Jour. Linn. Soc.* 65: 258 (1972) (*A. oblongata* Bieb. subsp. *bassarabica* Săvul. & Zahar.): Cauline hairs dimorphic, some *c.* 0·2 mm, crispate, the rest 1–1·5 mm, straight, patent. Calyx 6–10 mm. *From E. Romania to Turkey-in-Europe.*

7. A. laxmannii (L.) Bentham, *Lab. Gen. Sp.* 697 (1835). Perennial; stem 20–50 cm, with long, more or less patent hairs. Middle cauline leaves 35–100 × 10–35 mm, ovate- to oblong-elliptical, entire to coarsely crenate-serrate; cauline leaves rounded at base, semiamplexicaul. Bracts similar to the leaves, exceeding the flowers. Flowers 2 at each node. Calyx (7–)9–15 mm, the teeth slightly shorter than the tube. Corolla (20–)25–35 mm, yellow or cream with purplish-brown veins; tube shorter than calyx; upper lip entire. Stamens exserted; filaments hairy. 2*n* = 62. *E. & E.C. Europe, northwards to N.W. Hungary and 51° N. in S.C. Russia.* Bu Gr Hu Ju Rm Rs (C, W, K, E) Tu.

8. A. piskoi Degen & Bald., *Egy Új Ajuga Fajról* 1 (1896). Perennial; stem 30–50 cm, hispid, the hairs mostly on the angles. Lower leaves 60–85 × 25–45 mm, ovate, coarsely and remotely serrate-dentate; cauline leaves cuneate at base. Bracts similar to the leaves, exceeding the flowers. Flowers 2 at each node. Calyx 11–15 mm, the teeth almost twice as long as the tube. Corolla 25–35 mm, pink with purple veins; tube slightly exceeding calyx; upper lip with 2 short teeth. Stamens exserted; filaments hairy. ● *S. Albania* (Nëmerçkë, S.W. of Leskovik). Al.

9. A. iva (L.) Schreber, *Pl. Vert. Unilab.* 25 (1773). Caespitose perennial; stem 5–20 cm, woody at base, usually much-branched, villous or lanate-villous. Leaves 14–35 × 3–6(–8) mm, linear or linear-oblong, entire or with 2–6 short lobes. Bracts similar to the leaves, exceeding the flowers. Flowers 2–4 at each node. Calyx 3·5–4·5 mm, the teeth as long as or shorter than the tube. Corolla 12–20 mm, purple, pink or yellow; tube exceeding calyx; upper lip entire. Stamens exserted; filaments hairy. 2*n* = *c.* 86. *S. Europe.* Bl Co Cr Ga Gr Hs It Ju Lu Sa Si.

10. A. chamaepitys (L.) Schreber, *op. cit.* 24 (1773). Annual or short-lived perennial; stem 5–30 cm, usually much branched, glabrous to densely lanate-villous. Leaves 3-partite with linear segments 0·5–3(–4) mm wide, the segments sometimes 3-fid.

---

¹ By T. G. Tutin and D. Wood.

Bracts similar to the leaves. Flowers 2–4 at each node. Calyx 4–6 mm, the teeth as long as or shorter than the tube. Corolla yellow with red or purple markings, rarely entirely purple; tube about equalling calyx; upper lip entire. Stamens exserted; filaments hairy. *Most of Europe except the north.* Al Au Be Bl Br Bu Cr Cz Ga Ge Gr He Ho Hs Hu It Ju Lu Po Rm Rs (C, W, K, E) Sa Si Tu.

(a) Subsp. **chamaepitys**: Annual. Leaf-segments 0·5–2 mm wide. Corolla (5–)7–15 mm. Nutlets 2·5–3 mm, reticulately veined. 2*n* = 28. *W. & C. Europe, Italy and Sicilia.*

(b) Subsp. **chia** (Schreber) Arcangeli, *Comp. Fl. Ital.* 560 (1882) (*A. chia* Schreber): Short-lived perennial. Leaf-segments 1·5–3(–4) mm wide. Corolla 18–25 mm. Nutlets 3–4 mm, transversely rugose. 2*n* = 30. *S.E. & E.C. Europe, extending northwards to c. 53° N. in E.C. Russia.*

These two subspecies are often treated as distinct species, but a wide range of intermediates can be found in much of C. & E. Europe, Italy and S. Spain. **A. suffrutescens** Lange, *Kong. Danske Vid. Selsk. Forhand.* **1893**: 196 (1893), from S. Spain, has the leaf-segments, corolla-size, and nutlet-size of subsp. *chamaepitys*, but the duration and nutlet-pattern of subsp. *chia*. **A. chamaepitys** var. *grandiflora* Vis. and **A. pseudochia** Schost., *Not. Syst.* (Leningrad) **8**: 147 (1940), are names given to plants like subsp. *chamaepitys* but with the corolla 12–20 mm and the nutlets 2·5–3·5 mm, sometimes more or less transversely rugose. These occur throughout much of E. & E.C. Europe, the Balkan peninsula, S. Italy and Sicilia. Plants showing various other combinations of characters are common in the Balkan peninsula.

2. Teucrium L.¹

Herbs or shrubs. Calyx tubular or campanulate, 2-lipped or actinomorphic, 5-toothed, the teeth equal or the upper the largest. Corolla with one 5-lobed lip; tube without a ring of hairs inside, often included in the calyx. Nutlets smooth or reticulate.

1 All leaves 1- to 2-pinnatifid or pinnatisect, the linear or oblong lobes extending almost or quite to the midrib
  2 Annual; calyx gibbous at base, the teeth shorter than the tube
        18. botrys
  2 Perennial; calyx not gibbous at base, the teeth as long as or longer than the tube
    3 Corolla *c.* 5 mm, not much longer than calyx
        5. campanulatum
    3 Corolla 10–15 mm, about twice as long as calyx
        4. pseudochamaepitys
1 Leaves entire, crenate or serrate, rarely some deeply lobed
  4 Branches ending in spines which are leafless or with very small leaves
    5 Annual herb; stems glandular and pubescent or villous
        16. spinosum
    5 Small shrub; stems grey- or white-tomentose, eglandular
        29. subspinosum
  4 Not spinose
    6 Calyx 2-lipped
      7 Verticillasters crowded in a dense, not or scarcely secund, spike-like inflorescence
        8 Corolla whitish
          9 Leaves cuneate to rounded at base; corolla 1½ times as long as calyx; stamens exserted   **6. arduini**
          9 Leaves cordate to truncate at base; corolla scarcely longer than calyx; stamens included   **7. lamifolium**
        8 Corolla pink, violet or purple
          10 Calyx (6–)7–8 mm; middle tooth of upper lip nearly twice as long as the lateral tooth   **9. francisci-werneri**
          10 Calyx not more than 5·5 mm; middle tooth of upper lip not or scarcely longer than the lateral teeth

11 Calyx 3·5–4·5 mm; teeth all nearly equal in length
10. heliotropifolium
11 Calyx (4–)5–5·5 mm; teeth of lower lip much longer than those of upper lip **8. halacsyanum**
7 Verticillasters forming a lax, secund inflorescence
12 Leaves gradually narrowed at base **14. asiaticum**
12 Leaves rounded to cordate or truncate at base
13 Corolla usually greenish-yellow **11. scorodonia**
13 Corolla pink or purplish
14 Fruiting calyx obscurely reticulate-veined **12. massiliense**
14 Fruiting calyx strongly reticulate-veined **13. salviastrum**
6 Calyx not 2-lipped
15 Inflorescence not capitate, longer than wide
16 Leaves entire
17 Calyx gibbous at base
18 Leaves linear-lanceolate to rhombic, ±acute; inflorescence dense **28. marum**
18 Leaves suborbicular, rounded at apex; inflorescence lax **10. heliotropifolium**
17 Calyx not gibbous at base
19 Leaves not densely tomentose above, often glabrescent **1. fruticans**
19 Leaves persistently grey-tomentose above
20 Twigs green or brown; leaves oblong to linear **2. brevifolium**
20 Twigs white-tomentose; leaves obovate to ovate-oblong **3. aroanium**
16 Leaves crenate, toothed or lobed
21 Herb, not woody at base
22 Perennial; corolla purplish **15. scordium**
22 Annual; corolla yellowish **17. resupinatum**
21 Dwarf shrub
23 Leaves at least 3 times as long as wide **20. webbianum**
23 Leaves at most twice as long as wide
24 Stems glabrous **22. lucidum**
24 Stems more or less pubescent or villous
25 Stems and leaves densely villous
26 Leaves truncate at base **32. rotundifolium**
26 Leaves cuneate at base
27 Inflorescence dense, spike-like; bracts (except the lowest) entire **30. compactum**
27 Inflorescence lax; bracts (except sometimes the uppermost) leaf-like, crenate or serrate
28 Verticillasters usually 4-flowered **26. fragile**
28 Verticillasters usually 2-flowered **25. intricatum**
25 Stems and leaves variously hairy to subglabrous, but not densely villous
29 Rhizomatous; stems woody at base only
30 Leaves subglabrous to moderately hairy with patent or irregularly arranged hairs **19. chamaedrys**
30 Leaves velutinous beneath **21. krymense**
29 Not rhizomatous; stems woody, except for ultimate branches
31 Leaves *c.* 5 mm **27. microphyllum**
31 Most leaves 10 mm or more
32 Corolla yellow; petiole as long as the width of the lamina **24. flavum**
32 Corolla pink or purple; petiole shorter than the width of the lamina **23. divaricatum**
15 Inflorescence capitate, the heads usually wider than long
33 Leaves entire or with fewer than 3 crenations on each side
34 Leaves in whorls of 4 **40. libanitis**
34 Leaves opposite
35 Leaves less than 10 mm
36 Leaves hairy on both surfaces **38. thymifolium**
36 Leaves glabrous above
37 Flowers in compound heads **44. turredanum**
37 Flowers in simple heads **43. pumilum**
35 Leaves more than 12 mm
38 Calyx with patent hairs at the base **42. aragonense**
38 Calyx glabrous, or uniformly hairy with appressed hairs

39 Flowers in compound heads
40 Leaves densely hairy above **39. cossonii**
40 Leaves glabrous or subglabrous above **44. turredanum**
39 Flowers in simple heads
41 Calyx 7–10 mm **37. montanum**
41 Calyx 4–5 mm **43. pumilum**
33 Leaves with at least 3 crenations on each side
42 Leaves ovate, obovate or orbicular
43 Stems with patent hairs
44 Leaves up to 25 mm; calyx 10–12 mm **31. pyrenaicum**
44 Leaves up to 15 mm; calyx less than 10 mm
45 Principal veins of calyx conspicuous **32. rotundifolium**
45 Principal veins of calyx obscured by indumentum **36. alpestre**
43 Stems with appressed hairs
46 Leaves at least 8 mm wide; indumentum very dense, greenish **35. cuneifolium**
46 Leaves less than 8 mm wide; indumentum grey
47 Leaves more densely hairy beneath than above **33. buxifolium**
47 Leaves equally hairy on both surfaces **34. freynii**
42 Leaves narrowly ovate to narrowly oblong or spathulate
48 Leaves in whorls of 3 or 4
49 Stems sparsely hairy with patent hairs **48. haenseleri**
49 Stems densely hairy with appressed hairs **45. polium**
48 Leaves opposite
50 Calyx 6–8 mm **41. carthaginense**
50 Calyx 2·5–5 mm
51 Hairs at leaf-apex predominantly glandular **49. charidemi**
51 Hairs at leaf-apex predominantly eglandular
52 Corolla creamy-yellow, the proximal lobes spathulate, widening slightly towards apex **36. alpestre**
52 Corolla white or red, the proximal lobes rounded or triangular, narrowing towards apex
53 Terminal flower-heads with more than 20 flowers
54 Leaves shortly petiolate; inflorescence narrowly oblong to subglobose **45. polium**
54 Leaves sessile; inflorescence subglobose **47. eriocephalum**
53 Terminal flower-heads with 3–20 flowers
55 Calyx evenly hairy; leaves with 2–5(–9) crenations **45. polium**
55 Calyx with hairs denser towards base; leaves with 5–9 crenations **46. gnaphalodes**

Sect. TEUCRIUM (Sect. *Teucris* Bentham). Shrubs or perennial herbs. Inflorescence not or rarely weakly secund. Calyx campanulate; tube straight, not gibbous at base; teeth equal or nearly so.

1. T. fruticans L., *Sp. Pl.* 563 (1753). Evergreen shrub up to 250 cm. Twigs 4-angled, white-tomentose. Leaves lanceolate to ovate, entire, flat, shortly petiolate, white- or reddish-tomentose beneath, thinly tomentose, often becoming glabrous and shining above. Verticillasters 2-flowered; bracts leaf-like. Calyx shortly campanulate, white-tomentose outside, glabrous within. Corolla 15–25 mm, blue or lilac; stamens long-exserted. *Dry, sunny places. W. Mediterranean region, extending to Portugal and the Adriatic islands.* Co Ga Hs It Ju Lu Sa Si.

T. creticum L., *Sp. Pl.* 563 (1753) (*T. rosmarinifolium* Lam.), which is like **1** but has linear leaves, has been recorded from Lampedusa; its occurrence there as a native is doubtful and it has not been seen for many years. It is a native of the E. Mediterranean region, but not its European part.

2. T. brevifolium Schreber, *Pl. Vert. Unilab.* 27 (1773). Like **1** but up to 60 cm; twigs puberulent, green or brown; leaves oblong or linear, revolute, grey-tomentose on both surfaces; corolla *c.* 10 mm. $2n = 30$. *Dry, rocky places near the sea. S. Aegean region.* Cr Gr.

3. T. aroanium Orph. ex Boiss., *Diagn. Pl. Or. Nov.* 3(4): 55 (1859) (*T. arvanicum* auct.). Like **1** but stems procumbent, rooting, much-branched; leaves obovate to ovate-oblong, sparsely tomentose above, densely tomentose beneath; calyx densely glandular; corolla 15–20 mm. *Mountain rocks.* ● *S. Greece (Khelmo, Akhaia).* Gr.

4. T. pseudochamaepitys L., *Sp. Pl.* 562 (1753). Villous, rarely glabrous perennial up to 40 cm, somewhat woody at the base. Lower leaves 1- to 2-pinnatisect, the upper usually 3-fid; all with linear, entire lobes. Inflorescence unbranched; verticillasters 2-flowered; bracts equalling or exceeding the pedicels. Calyx campanulate, glandular-pubescent; teeth longer than tube, triangular, acuminate. Corolla 10–15 mm, whitish or reddish; stamens long-exserted. *Dry places. S.W. Europe.* Ga Hs Lu.

5. T. campanulatum L., *Sp. Pl.* 562 (1753). Subglabrous perennial herb. Stems 10–30 cm, decumbent and rooting at the base, somewhat hairy above. Leaves 3-partite, the segments pinnatifid; lobes linear or linear-spathulate, sometimes dentate. Verticillasters 2- to 4-flowered; bracts leaf-like, longer than the flowers. Calyx campanulate, almost glabrous; teeth about as long to twice as long as the tube, acute or acuminate with a spinose point. Corolla *c.* 5 mm, whitish, only the lower lip exceeding the calyx; stamens shortly exserted. *Open habitats. Mediterranean region, from S.W. Spain to S.E. Italy.* Bl Hs It Si.

T. aristatum Pérez Lara, *Anal. Soc. Esp. Hist. Nat.* **18**: 90 (1889), described from S.W. Spain, is a variant of **5** with spinose calyx-teeth twice as long as tube. Plants intermediate between this and the typical ones occur in Italy and Sicilia.

Sect. STACHYOBOTRYS Bentham. Dwarf shrubs. Inflorescence not or weakly secund. Calyx campanulate; tube curved, gibbous at base; the upper tooth usually wider than the others.

6. T. arduini L., *Mantissa* 81 (1767). Stems *c.* 30 cm, erect or ascending, shortly villous. Leaves ovate, crenate-serrate, rounded or cuneate at base, puberulent and glandular beneath, petiolate. Inflorescence up to 16 cm in fruit, very dense, simple; bracts linear-lanceolate, equalling or longer than the calyx. Calyx villous and with short glandular hairs; teeth *c.* ⅓ as long as tube. Corolla 8–9 mm, glandular and villous, whitish; stamens long-exserted. *Rocky places.* ● *W. Jugoslavia, N. Albania.* Al Ju.

7. T. lamifolium D'Urv., *Mém. Soc. Linn. Paris* **1**: 320 (1822) (incl. *T. cordifolium* Čelak.). Like **6** but leaves coarsely crenate, cordate or truncate at base, shortly villous beneath; bracts subulate; corolla 6–7 mm; stamens included. *Scrub. Turkey-in-Europe, just extending into S.E. Bulgaria.* Bu Tu.

Sect. ISOTRIODON Boiss. Perennial herbs or dwarf shrubs. Inflorescence secund. Calyx tubular, curved, gibbous at base.

8. T. halacsyanum Heldr., *Österr. Bot. Zeitschr.* **29**: 241 (1879). Dwarf shrub up to 15(–50) cm. Stems lanate-villous, much-branched. Leaves broadly ovate to suborbicular, crenate, lanate-villous, rounded to cuneate at base, petiolate. Inflorescence dense. Bracts linear to lanceolate. Calyx (4–)5–5·5 mm, distinctly 2-lipped; upper lip with wide, acute, shallow teeth, the middle slightly but distinctly longer than the lateral; lower lip deeply bidentate, with ovate teeth. Corolla *c.* 7 mm, pubescent, violet; stamens long-exserted. 2*n* = 32. *Rocky places.* ● *W. Greece.* Gr.

9. T. francisci-werneri Rech. fil., *Phyton (Austria)* **1**: 207 (1949). Like **8** but calyx (6–)7–8 mm; middle tooth of upper lip nearly twice as long as the lateral, acuminate; lower lip deeply bidentate, with narrowly lanceolate teeth. ● *S. Greece.* Gr.

10. T. heliotropifolium W. Barbey, *Bull. Soc. Vaud. Sci. Nat.* ser. 3, 21: 223 (1886). Like **8** but leaves sometimes entire; inflorescence usually lax; calyx 3·5–4·5 mm; teeth all nearly equal in length; teeth of upper lip ovate, the middle widest, rounded or shortly cuspidate; lower lip with triangular or ovate, acute teeth. *Rocky places.* ● *Karpathos.* Cr.

Perhaps conspecific with **T. montbretii** Bentham, *Ann. Sci. Nat.* Ser. 2, 6: 56 (1836), from S. Anatolia and Syria.

Sect. SCORODONIA (Hill) Schreber. Perennial herbs or dwarf shrubs. Inflorescence secund, raceme-like. Calyx campanulate, 2-lipped; tube curved, gibbous at base; the upper tooth usually wider than the others.

11. T. scorodonia L., *Sp. Pl.* 564 (1753). Pubescent rhizomatous dwarf shrub. Stems 15–50(–100) cm, erect, branched. Leaves triangular-ovate, crenate, rugose, cordate at base, petiolate. Inflorescence up to 15 cm, simple or branched, rather lax; bracts ovate to lanceolate, much shorter than the calyx. Calyx villous and more or less glandular, strongly veined in fruit. Corolla 9 mm, villous, pale greenish-yellow, rarely white or reddish; stamens long-exserted. ● *S., W. & C. Europe, northwards to S. Norway and eastwards to W. Poland and N.W. Jugoslavia.* Au Be Br Co Cz Ga Ge Hb He Ho Hs Hu It Ju Lu No Po Sa Si [Da Su].

1 Corolla-tube scarcely exceeding calyx   **(c) subsp. baeticum**
1 Corolla-tube about twice as long as calyx
  2 Inflorescence-axis and calyx with short hairs and few or no glands; calyx 4·5–5·5 mm   **(a) subsp. scorodonia**
  2 Inflorescence-axis and calyx with long hairs, usually rather densely glandular; calyx 7–8 mm   **(b) subsp. euganeum**

(a) Subsp. **scorodonia**: Stems, petioles and leaves more or less appressed-pubescent. Leaves acute. Inflorescence-axis and calyx with short hairs and few or no glands. Calyx 4·5–5·5 mm. Corolla-tube about twice as long as calyx. 2*n* = 32, ?34. *Almost throughout the range of the species.*

(b) Subsp. **euganeum** (Vis.) Arcangeli, *Comp. Fl. Ital.* 558 (1882): Like subsp. (a) but inflorescence-axis and calyx with long hairs, usually rather densely glandular; calyx 7–8 mm. *Italy and Sicilia.*

(c) Subsp. **baeticum** (Boiss. & Reuter) Tutin, *Bot. Jour. Linn. Soc.* 65: 262 (1972) (*T. baeticum* Boiss. & Reuter): Stems, petioles and leaves with patent and somewhat crispate hairs. Leaves rounded at apex. Inflorescence-axis and calyx with abundant short, glandular hairs. Calyx 6–7 mm. Corolla-tube scarcely exceeding calyx. *S.W. Spain.*

12. T. massiliense L., *Sp. Pl.* ed. 2, 789 (1763). Like **11** but plant greyish-tomentose; leaves truncate or rounded at base; calyx glandular-pubescent, not strongly veined; corolla pink; tube included in the calyx. 2*n* = 32. ● *W. Mediterranean region; Kriti.* Co Cr Ga Hs Sa.

13. T. salviastrum Schreber, *Pl. Vert. Unilab.* 38 (1773). Dwarf shrub. Stems up to 30 cm. Leaves up to 15 mm, ovate-oblong, obtuse, rounded at base, crenate, puberulent above, grey-tomentose beneath, strongly rugose, petiolate. Inflorescence lax; bracts exceeding the pedicels. Calyx glandular and villous, strongly veined; teeth about half as long as tube, often with 1–2 lateral teeth, triangular with spinescent apex. Corolla *c.* 10 mm, purplish. ● *Mountains of C. Portugal.* Lu.

14. T. asiaticum L., *Mantissa* 80 (1767). Dwarf shrub with annual flowering stems *c.* 40 cm. Leaves up to 30 mm, narrowly oblong, dentate, gradually attenuate at base, acute, glandular and sparsely puberulent, shortly petiolate. Inflorescence very lax; lower bracts leaf-like and much longer than the flowers; upper bracts short. Calyx glandular but almost glabrous; teeth *c.* ½ as long as tube, broadly ovate, aristate. Corolla 10–13 mm, pink to purple. *Shady rocks.* ● *Islas Baleares.* Bl.

Sect. SCORDIUM (Miller) Bentham. Usually herbs. Inflorescence not secund. Calyx tubular, curved, gibbous at base; teeth subequal or the upper wider.

15. T. scordium L., *Sp. Pl.* 565 (1753). Softly hairy to subglabrous, stoloniferous perennial, smelling of garlic when crushed. Stems 10–60 cm, often freely branched. Leaves ovate to oblong, coarsely dentate, or crenate-dentate, sessile or subsessile. Flowers in the axils of leaf-like bracts which are usually longer than the flowers. Calyx villous, obscurely veined; teeth *c.* ⅓ as long as tube, triangular, acuminate. Corolla 7–10 mm, villous, purplish; tube slightly exceeding calyx. *Wet places. Europe, northwards to Ireland, Sweden and Estonia.* All except Az Fa Fe Is No Rs (N) Sb.

(a) Subsp. **scordium** (*T. scordium* subsp. *palustre* P. Fourn.): Stolons usually with normal foliage-leaves. Leaves of the main stem narrowed to the base; leaves of lateral branches cuneate at base and toothed in the upper part. *From N. Spain, N. Italy and Bulgaria northwards.*

(b) Subsp. **scordioides** (Schreber) Maire & Petitmengin, *Bull. Soc. Sci. Nancy* ser. 3, **9**: 411 (1908) (*T. scordioides* Schreber, *T. petkovii* Urum.): Stolons usually densely clothed with scale-leaves. Leaves of the main stem cordate, semiamplexicaul; leaves of lateral branches rounded at base, toothed all round. 2*n*=32. *S. Europe.*

16. T. spinosum L., *Sp. Pl.* 566 (1753). Pubescent or villous, glandular annual. Stems 30–50 cm, much branched, the branches spinose and more or less leafless at flowering time. Lower leaves oblong, narrowed at base, incise-serrate above; upper leaves small, entire. Flowers solitary or 4–6 together in the axils of the distant upper leaves. Calyx 6–7 mm, weakly veined; upper tooth broadly ovate, the others 2–2·5 mm, narrowly triangular to subulate, all spinescent. Corolla 6–8 mm, white, resupinate; tube included in the calyx. *Cultivated fields and sandy or gravelly ground. W. Mediterranean region, S. Portugal.* Hs It Lu Sa Si.

17. T. resupinatum Desf., *Fl. Atl.* **2**: 4 (1798). Like **16** but not spinose; all leaves, except those at the top of the inflorescence, crenate-serrate; upper calyx-tooth scarcely wider than the others, all 2·5–3 mm, broadly triangular or ovate-triangular, long-acuminate; corolla yellowish, with purple lateral lobes. *Cultivated, sandy ground. S.W. Spain.* Hs. (*N.W. Africa.*)

18. T. botrys L., *Sp. Pl.* 562 (1753). Villous annual up to 30 cm. Stems usually branched. Leaves petiolate, 1- to 2-pinnatisect; lobes oblong, usually entire. Inflorescence lax, usually occupying more than half the stem. Calyx villous and glandular, strongly gibbous, reticulate-veined; teeth *c.* ½ as long as tube, triangular, acuminate. Corolla 15–20 mm, purplish-pink, villous; tube included in the calyx. 2*n*=32. *Dry, stony places; somewhat calcicole. S., W. & C. Europe, northwards to S. England and S. Poland and eastwards to Romania.* Au Be Bl Br ?Bu Cz Ga Ge He Ho Hs Hu It Ju Po Rm ?Rs (W).

Sect. CHAMAEDRYS (Miller) Schreber. Dwarf shrubs. Inflorescence more or less secund, usually rather lax. Calyx tubular-campanulate, curved, oblique at base; teeth usually subequal.

19. T. chamaedrys L., *Sp. Pl.* 565 (1753) (incl. *T. pulchrius* Juz.). Rhizomatous dwarf shrub with annual flowering stems 5–50 cm, hairy. Leaves up to 2 × 1 cm, not more than twice as long as wide, oblong or oblong-obovate, entire to incise-serrate with obtuse teeth, or crenate-dentate, usually pubescent beneath. Inflorescence lax to dense, but always longer than wide. Calyx 5–8 mm, villous and glandular to glabrous; teeth *c.* ⅔ as long as tube, subequal, usually somewhat longer than wide, triangular, acuminate. Corolla 9–16 mm, pale to deep purple, rarely white, hairy. 2*n*=60. *Europe, northwards to the Netherlands, S. Poland and S.C. Russia.* Al Au Be Bl Bu Co Cz Ga Ge Gr He Ho Hs Hu It Ju Lu Po Rm Rs (C, W, K, E) Sa Si Tu [Br].

Very variable. Numerous subspecies have been recognized. For a detailed treatment of these see K. H. Rechinger, *Bot. Arch.* (*Berlin*) **42**: 344–389 (1941).

20. T. webbianum Boiss., *Elenchus* 78 (1838). Like **19** but always thinly grey-tomentose; leaves 4–6 times as long as wide, linear to oblong; calyx-teeth somewhat unequal, scarcely longer than wide. ● *Mountains of S.E. Spain.* Hs.

21. T. krymense Juz., *Not. Syst.* (*Leningrad*) **14**: 19 (1951) (incl. *T. fischeri* Juz.). Like **19** but leaves velutinous beneath, finely and deeply crenate-serrate; calyx-teeth 1½–2 times as long as wide. *Grassy mountain slopes.* ● *Krym.* Rs (K).

22. T. lucidum L., *Syst. Nat.* ed. 10, **2**: 1095 (1759). Like **19** but usually larger and almost glabrous; stems up to 60 cm, glabrous; leaves up to 4 × 3 cm, ovate-oblong, deeply toothed or shallowly lobed, teeth or lobes obtuse; inflorescence very lax; calyx-teeth *c.* ⅓ as long as tube, ciliolate on the margins; corolla 14–17 mm. *Dry, rocky hillsides.* ● *S.W. Alps.* Ga It.

23. T. divaricatum Sieber ex Boiss., *Fl. Or.* **4**: 816 (1879). Small, stout shrub 10–30(–50) cm. Leaves (6–)10–25 × (4–)8–15 mm, coriaceous, ovate, rather shallowly crenate; petiole shorter than the width of the lamina. Inflorescence usually lax. Calyx *c.* 8 mm, ciliolate. Corolla pink or purple. 2*n*=62, 64. *Aegean region.* Cr Gr.

1 Leaves nearly or quite glabrous; stems subglabrous to shortly crispate-hairy
2 Stems 30–50 cm; calyx with short, eglandular hairs only
        (d) subsp. athoum
2 Stems 15–30 cm; calyx with some long, glandular hairs
        (c) subsp. graecum
1 Whole plant ± villous
3 Stems with short, crispate, usually deflexed hairs
        (a) subsp. divaricatum
3 Stems with long, patent hairs
        (b) subsp. villosum

(a) Subsp. **divaricatum**: *E. Greece, Kriti.*
(b) Subsp. **villosum** (Čelak.) Rech. fil., *Bot. Arch.* (*Berlin*) **42**: 391 (1941): *Islands of S. Aegean region.*
(c) Subsp. **graecum** (Čelak.) Bornm., *Mitt. Thür. Bot. Ver.* nov. ser., **38**: 57 (1929): *E.C. Greece.*
(d) Subsp. **athoum** (Hausskn.) Bornm., *loc. cit.* (1929): *N.E. Greece* (*Athos*).

24. T. flavum L., *Sp. Pl.* 565 (1753). Like **23** but stems up to 50 cm, usually velutinous; leaves 10–40 × 10–25 mm; petiole as long as the width of the lamina; calyx 7–10 mm; corolla yellow. *Rocks and stony slopes. Mediterranean region.* Al Bl Co Cr Ga Gr ?Hs It Ju Sa Si Tu.

1 Calyx glabrous   **(d)** subsp. **gymnocalyx**
1 Calyx ± hairy
  2 Calyx *c.* 7 mm; teeth nearly as long as tube **(c)** subsp. **hellenicum**
  2 Calyx (8–)9–10 mm; teeth *c.* ½ as long as tube
    3 Leaves *c.* 20 mm, velutinous beneath   **(a)** subsp. **flavum**
    3 Leaves *c.* 10 mm, glaucous beneath   **(b)** subsp. **glaucum**

(a) Subsp. flavum: Plant usually velutinous; larger leaves usually 20–40 cm; calyx 8–10 mm; teeth *c.* ½ as long as tube. *From S.E. France to the W. part of the Balkan peninsula.*

(b) Subsp. glaucum (Jordan & Fourr.) Ronniger, *Verh. Zool.-Bot. Ges. Wien* **68**: (234) (1918): Stems with short, deflexed hairs. Leaves *c.* 10 mm, nearly glabrous, glaucous beneath. Calyx 9–10 mm; teeth *c.* ½ as long as tube. *From Islas Baleares to Greece.*

(c) Subsp. hellenicum Rech. fil., *Bot. Arch.* (*Berlin*) **42**: 397 (1941): Stems slender, velutinous. Leaves 10–15 mm, velutinous beneath, glabrescent. Calyx *c.* 7 mm; teeth nearly as long as tube. *Greece.*

(d) Subsp. gymnocalyx Rech. fil., *loc. cit.* (1941): Stems slender, often glabrescent. Leaves 10–20 mm, glabrous, glaucous beneath. Calyx *c.* 7 mm; teeth nearly as long as tube. ● *S.E. Greece.*

25. T. intricatum Lange, *Vid. Meddel. Dansk Naturh. Foren. Kjøbenhavn* **1863**: 21 (1863). Much-branched, spreading, greyish-pubescent dwarf shrub *c.* 20 cm. Leaves up to 8 mm, oblong or ovate, crenate to almost pinnatifid, obtuse, rather thick; margin recurved. Inflorescence lax; verticillasters usually 4-flowered. Bracts leaf-like. Calyx *c.* 7 mm, villous, gibbous at base; teeth ½ as long as tube, triangular. Corolla 15–18 mm, pubescent, pink. *Rocky places.* ● *S.E. Spain* (*Sierra de Gádor*). Hs.

26. T. fragile Boiss., *Elenchus* 77 (1838). Caespitose, softly and densely villous dwarf shrub. Stems up to 20 cm, but usually much less, slender. Leaves up to 6 mm, oblong to flabelliform, deeply serrate to pinnatifid, thin. Inflorescence rather lax; verticillasters usually 2-flowered. Bracts leaf-like. Calyx *c.* 7 mm, villous, not gibbous at base; teeth ½–⅓ as long as tube, triangular. Corolla 10–12 mm, pubescent, purple. *Mountain rocks and screes; calcicole.* ● *S. Spain.* Hs.

27. T. microphyllum Desf., *Ann. Mus. Hist. Nat.* (*Paris*) **10**: 300 (1807). Dwarf shrub 5–40 cm. Stems slender, appressed grey- or white-tomentose when young. Leaves *c.* 5 mm, oblanceolate to ovate, crenate-serrate, acute, white-tomentose beneath. Inflorescence rather lax; verticillasters 2-flowered. Lower bracts leaf-like, the upper small and entire. Calyx 6–7 mm, pubescent and glandular; teeth ½–¾ as long as tube, triangular, acute. Corolla 10–12 mm, villous, pink. *Rocky places. S. Aegean region.* Cr Gr.

28. T. marum L., *Sp. Pl.* 564 (1753). Small shrub up to 50 cm. Stems slender, white-tomentose. Leaves up to 10 mm, linear-lanceolate to rhombic, entire, rarely dentate, grey-tomentose beneath; margins often recurved. Inflorescence cylindrical, rather dense. Flowers 1–2 in the axils of leaf-like bracts. Calyx 6–7 mm, villous; teeth ⅓ as long as tube, triangular. Corolla 10–12 mm, villous, purplish. ● *Islands of W. Mediterranean region; also on one island* (*Murter*) *off N.W. Jugoslavia.* Bl Co Ga It Ju Sa.

Records from the mainland of Spain and Italy appear to be errors or to refer to naturalized plants.

29. T. subspinosum Pourret ex Willd., *Enum. Pl. Hort. Berol.* 596 (1809). Like **28** but main branches stout; lateral branches numerous, forming slender leafless spines; leaves 1–6 mm, very narrow, almost triangular; inflorescence lax, few-flowered; calyx 2–3 mm, puberulent; teeth *c.* ½ as long as tube, broadly triangular; corolla *c.* 8 mm, puberulent. ● *Mallorca.* Bl ?Sa.

30. T. compactum Clemente ex Lag., *Gen. Sp. Nov.* 17 (1816). Procumbent or ascending, villous dwarf shrub 10–30 cm. Leaves ovate, crenate, crenate-serrate or shallowly lobed, except for the cuneate base, sessile or subsessile. Inflorescence dense, spike-like. Flowers solitary or in pairs, sessile. Bracts lanceolate or linear-lanceolate, entire, about as long as the calyx. Calyx inflated, villous and glandular, strongly veined; teeth *c.* ⅓ as long as tube, triangular-acuminate. Corolla *c.* 12 mm, yellowish-white; tube included in the calyx. *Rocky places on mountains. S. Spain* (*Sierra Nevada*). Hs. (*N.W. Africa.*)

Sect. POLIUM (Miller) Schreber. Dwarf shrubs or herbs, often with branched hairs. Inflorescence simple or compoundly capitate. Calyx campanulate or tubular, actinomorphic.

31. T. pyrenaicum L., *Sp. Pl.* 566 (1753). Procumbent to ascending, villous herb 10–30 cm. Stem with patent hairs. Leaves up to 25 mm, suborbicular, crenate, with a cuneate base, shortly petiolate. Flowers in a terminal head. Bracts slightly shorter than the calyx; lower bracts leaf-like, upper bracts narrowly spathulate, entire. Calyx 10–12 mm, campanulate, sparsely hairy; teeth triangular, acuminate, ciliate; veins conspicuous. Corolla white, or white with purple, spathulate, proximal lobes. $2n = 26$. *Rocks, screes and mountain pastures; somewhat calcicole.* ● *N. Spain, S.W. France.* Ga Hs.

32. T. rotundifolium Schreber, *Pl. Vert. Unilab.* 42 (1773) (incl. *T. cinereum* Boiss., *T. granatense* (Boiss.) Boiss. & Reuter). Like **31** but more extensively branched, smaller in all its parts; leaves up to 15 mm, truncate at base; petiole almost equalling to longer than the lamina; calyx less than 10 mm; corolla purple or white. $2n = 26$. *Rock-crevices and screes; calcicole. Mountains of Spain.* Hs.

33. T. buxifolium Schreber, *loc. cit.* (1773). Much-branched dwarf shrub. Stems 7–25 cm, with dense, appressed white hairs. Leaves 5–12 mm, narrowly ovate to ovate, with 3–4 crenations, more densely hairy beneath than above; margins revolute. Flowers in a terminal head. Calyx 6–8 mm, campanulate, more or less hairy, often purple; veins distinct. Corolla pink; proximal lobes spathulate. *Sunny calcareous rocks.* ● *S.E. Spain.* Hs.

T. amplexicaule Bentham, *Lab. Gen. Sp.* 687 (1835) is a variant of **33** with glandular hairs on the calyx.

34. T. freynii Reverchon ex Willk., *Suppl. Prodr. Fl. Hisp.* 159 (1893). Dwarf shrub with persistent wiry branches. Stems up to 20 cm, with appressed white hairs. Leaves 2–4 mm wide, obovate, with 4–5 crenations, cuneate at the base, shortly petiolate, equally hairy on both surfaces. Flowers in a terminal head. Lower bracts leaf-like, upper spathulate. Calyx tubular, with a uniform indumentum of short crispate hairs; veins obscure. Corolla pink to brownish-cream; proximal lobes spathulate. ● *S.E. Spain* (*prov. Almería*). Hs.

35. T. cuneifolium Sibth. & Sm., *Fl. Graec. Prodr.* **1**: 395 (1809). Dwarf shrub. Stems 10–25 cm, densely green- to white-tomentose. Leaves 9–13 mm, at least 8 mm wide, ovate to orbicular, with 3–4 crenations, cuneate at the base, shortly petiolate, densely green- to white-tomentose. Flowers in terminal or axillary heads. Lower bracts leaf-like, upper entire, spathulate. Calyx 5–7 mm, hairy; veins obscure. Corolla cream; proximal lobes spathulate. *Calcareous rocks.* ● *Kriti.* Cr.

36. T. alpestre Sibth. & Sm., *loc. cit.* (1809). Dwarf, densely branched, hairy shrub up to 20 cm. Stem with patent hairs. Leaves 4–10 mm, obovate to linear, with 3–5 crenations, flat or with revolute margins, cuneate at the base, subsessile. Flowers in a terminal head, with occasional solitary, axillary flowers. Calyx campanulate, evenly hairy with dense short hairs; veins obscure. Corolla creamy yellow; proximal lobes spathulate. ● *Kriti and Karpathos.* Cr.

(a) Subsp. **alpestre**: Young stems and leaves with patent hairs; leaves flat or slightly revolute. $2n=78$. *Mountain rocks. Kriti.*
(b) Subsp. **gracile** (W. Barbey & Major) D. Wood, *Bot. Jour. Linn. Soc.* **65**: 261 (1972) (*T. gracile* W. Barbey & Major): Stems and leaves with appressed hairs; leaves strongly revolute. *Calcareous cliffs. Karpathos.*

37. T. montanum L., *Sp. Pl.* 565 (1753) (incl. *T. helianthemoides* Adamović, *T. jailae* Juz., *T. praemontum* Klokov, *T. pannonicum* A. Kerner). Dwarf shrub. Stems 10–25 cm, decumbent, with white, appressed, simple hairs. Leaves 13–30 mm, narrowly elliptical, entire, sessile, often glabrous above, hairy to densely hairy with prominent midrib beneath. Flowers in a terminal head. Bracts leaf-like. Calyx up to 10 mm, hairy or glabrous, often with setaceous teeth. Corolla cream; proximal lobes spathulate, constricted at the base. $2n=16, 26, 30, 60$. *Dry, rocky or stony ground; calcicole. S. & C. Europe, extending to the Netherlands and W. Ukraine.* Al Au Be Bu Cz Ga Ge Gr He Ho Hs Hu It Ju Po Rm Rs (W, K) Sa Si.

38. T. thymifolium Schreber, *Pl. Vert. Unilab.* 50 (1773). Dwarf shrub with appressed white hairs. Stems 5–10 cm, often decumbent. Leaves 5–9 mm, narrowly ovate to linear, entire, sessile, revolute, hairy on both surfaces; margins revolute. Flowers in a terminal head. Bracts leaf-like, equalling or shorter than the calyx. Calyx 7–9 mm, green or purple, evenly crispate-hairy; teeth acute; veins conspicuous. Corolla cream or red; proximal lobes spathulate. ● *S.E. Spain.* Hs.

39. T. cossonii D. Wood, *Bot. Jour. Linn. Soc.* **65**: 261 (1972) (*T. pulverulentum* (Barc.) Rouy, non (Jordan & Fourr.) F. W. Schultz & F. Winter). Dwarf shrub up to 30 cm. Stock very woody. Stem and leaves evenly covered with branched hairs. Leaves 15–30 mm, narrowly oblanceolate, slightly revolute, entire or shallowly crenate, subsessile. Flowers in terminal and lateral heads. Bracts leaf-like, the lower longer than, the upper equalling the calyx. Calyx 5–6 mm, evenly hairy. Corolla red; proximal lobes short, triangular. *Calcareous rocks.* ● *Mallorca.* Bl.

40. T. libanitis Schreber, *Pl. Vert. Unilab.* 48 (1773) (*T. verticillatum* Cav.). Dwarf shrub. Stems 15–25 cm, densely hairy. Leaves 10–15 mm, linear, entire, very densely hairy beneath, sessile, in whorls of 4; margins revolute. Flowers in a terminal subconical head. Bracts entire, spathulate. Calyx 4–5 mm; teeth very short, obtuse. Corolla yellowish; proximal lateral lobes triangular. *Dry hillsides; calcicole.* ● *S.E. Spain.* Hs.

41. T. carthaginense Lange, *Vid. Meddel. Dansk Naturh. Foren. Kjøbenhavn* **1881**: 97 (1882). Dwarf shrub. Stems 15–25 cm, hairy. Leaves 10–13 mm, spathulate, with 3–5 crenations, in small fascicles; margins revolute. Flowers in a terminal subcylindrical to globose head. Bracts entire, linear-lanceolate. Calyx 6–8 mm, strongly ribbed. Corolla white; proximal lobes narrowly triangular. *Rocky hillsides; calcicole.* ● *S.E. Spain.* Hs.

42. T. aragonense Loscos & Pardo, *Ser. Pl. Arag.* 85 (1863). Dwarf shrub. Stems 10–40 cm, hairy; flowering stems slender, straight. Leaves 15–20 mm, linear to spathulate, entire, green above, white-hairy beneath, sessile. Flowers in a simple or compound terminal head. Bracts leaf-like. Calyx 4–5 mm, subglabrous, with sericeous teeth. Corolla white or purple; proximal lobes triangular. ● *E. Spain.* Hs.

43. T. pumilum L., *Cent. Pl.* **1**: 15 (1755). Dwarf shrub. Stems 15–40 cm, densely hairy. Leaves 6–20 mm, linear, entire, glabrous or subglabrous above, sessile, ascending, rarely in fascicles; margins revolute. Flowers in a terminal, conical, simple head. Bracts longer than calyx, leaf-like. Calyx 4–5 mm, glabrous or hairy only at the base, glandular, with short obtuse or setaceous teeth. Corolla cream or purple; proximal lobes shortly spathulate. *Calcareous or gypsaceous soils.* ● *C. & S.E. Spain.* Hs.

(a) Subsp. **pumilum**: Stems 15–30 cm; bracts shorter than the flowers; calyx slightly hairy; teeth setaceous. *Throughout the range of the species.*
(b) Subsp. **carolipaui** (C. Vicioso ex Pau) D. Wood, *Bot. Jour. Linn. Soc.* **65**: 261 (1972) (*T. carolipaui* C. Vicioso ex Pau): Stems 25–40 cm; bracts much longer than the flowers; calyx glabrous; teeth obtuse. *S.E. Spain.*

44. T. turredanum Losa & Rivas Goday, *Anal. Inst. Bot. Cavanilles* **25**: 204 (1969). Like **43** but flowers always in compound heads; calyx densely hairy. *Calcicole.* ● *S.E. Spain (N.E. of Almería).* Hs.

45. T. polium L., *Sp. Pl.* 566 (1753). Dwarf shrub. Stems 6–45 cm, covered with white, greenish or golden branched hairs, often densely so. Leaves 7–27 mm, narrowly oblong to narrowly obovate, flat or with revolute margins, with 2–5(–9) crenations, often in fascicles. Flowers in a simple or compound head. Bracts leaf-like to entire. Calyx (2·5–)3–5 mm, densely and evenly hairy. Corolla white or red; proximal lobes rounded or triangular. $2n=26, 52, 78$. *Dry places. S. Europe, extending northwards to c. 51° N. in S.C. Russia.* Al Bl Bu Co Cr Ga Gr Hs It Ju Lu Rm Rs (C, W, K, E) Sa Si Tu.

1 Flowers in a simple head; lateral corolla-lobes always hairy
  2 Stem with golden hairs   **(b)** subsp. **aureum**
  2 Stem with white or grey hairs   **(a)** subsp. **polium**
1 Flowers in a compound head; lateral corolla-lobes glabrous or hairy
  3 Calyx less than 3 mm; lateral corolla-lobes always glabrous
                           (d) subsp. **pii-fontii**
  3 Calyx 3·5–7 mm; lateral corolla-lobes glabrous or hairy
    4 Flowering stems 10–25 cm; leaves less than 15 mm
                         (c) subsp. **capitatum**
    4 Flowering stems 26–45 cm; leaves more than 16 mm
                      (e) subsp. **vincentinum**

(a) Subsp. **polium**: Stems 6–17(–20) cm; hairs grey; margins of leaves more or less revolute; flower-heads simple; proximal corolla-lobes ciliate. ● *S., C. & E. Spain, S. France.*
(b) Subsp. **aureum** (Schreber) Arcangeli, *Comp. Fl. Ital.* 559 (1882) (*T. aureum* Schreber): Stems 15–30 cm; hairs golden; leaves flat or with revolute margins; flower-heads simple; proximal corolla-lobes ciliate. ● *Mountains of W. Mediterranean region.*
(c) Subsp. **capitatum** (L.) Arcangeli, *loc. cit.* (1882) (*T. capitatum* L.): Stems 10–25 cm, grey-hairy; leaves fasciculate, the margins more or less revolute; flower-heads compound; proximal corolla-lobes glabrous or hairy. *Almost throughout the range of the species.*

(d) Subsp. **pii-fontii** Palau, *Anal. Inst. Bot. Cavanilles* **11**(2): 487 (1953): Stems 15–35 cm; flowering stems straight, slender; leaves less than 10 mm, with axillary fascicles; flower-heads compound. ● *Spain, Islas Baleares.*

(e) Subsp. **vincentinum** (Rouy) D. Wood, *Bot. Jour. Linn. Soc.* **65**: 261 (1972) (*T. vincentinum* Rouy): Stems 25–45 cm, stout, with dense golden or grey hairs; leaves up to 28 mm; flower-heads compound; proximal corolla-lobes glabrous. *Maritime rocks and sands. S.W. Europe.*

46. T. gnaphalodes L'Hér., *Stirp. Nov.* 84 (1788). Dwarf shrub. Stems 15–30 cm, decumbent, with a dense woolly indumentum of branched hairs. Leaves 4–13 mm, oblong, with 5–9 crenations, sessile, in fascicles, often with golden hairs; margins strongly revolute. Flowers in terminal and axillary, 3- to 12-flowered, elongate heads. Calyx 4–5 mm, very hairy with hairs denser towards the base. Corolla white or red; proximal corolla-lobes triangular. *Dry places. C., S. & E. Spain.* Hs.

47. T. eriocephalum Willk., *Linnaea* **25**: 58 (1852). Dwarf shrub. Stems 13–30 cm, with a sparse to dense indumentum of branched hairs; flowering stems slender, straight, usually red. Leaves 7–11 mm, in fascicles, linear to oblong, with 4–6 crenations; margins revolute. Flowers in subglobose, compact, terminal and lateral, sessile or pedunculate heads. Bracts shorter than the calyx. Calyx densely or sparsely hairy; teeth acute. Corolla white or red; proximal lobes triangular. *Dry stony slopes near the coast.* ● *S. Spain.* Hs.

48. T. haenseleri Boiss., *Elenchus* 79 (1838) (incl. *T. reverchonii* Willk.). Dwarf shrub. Stems 10–35 cm, red or green, sparsely hirsute. Leaves 20–30 mm, narrowly oblong, with 3–12 crenations, often deflexed, in whorls of 3 or 4 and in axillary fascicles; margins revolute. Flowers in 3–10 subcylindrical to ovoid heads on each stem. Corolla whitish; proximal lobes triangular. ● *S. Spain, S. Portugal.* Hs Lu.

49. T. charidemi Sandwith, *Cavanillesia* **3**: 38 (1930). Dwarf shrub. Stems 8–30 cm, contorted, stems and leaves densely white-lanate. Leaves 5–19 mm, narrowly triangular, with 6–8 crenations, sessile; margins revolute. Flowers in terminal and lateral globose heads. Calyx 4 mm, evenly hairy. Corolla white; proximal lobes triangular. ● *S.E. Spain (Cabo de Gata).* Hs.

3. Scutellaria L.[1]

Rhizomatous perennials. Flowers in pairs, remote or in a dense oblong raceme. Calyx 2-lipped, the tube campanulate, with an erect dorsal scale; lips entire, closed in fruit. Corolla 2-lipped; tube long and more or less curved upwards from the base, glabrous inside. Stamens parallel.

Literature: R. von Bothmer, *Bot. Not.* **122**: 38–56 (1969). K. H. Rechinger fil., *Bot. Arch.* (*Berlin*) **43**: 1–70 (1941).

Measurements relating to inflorescences, bracts and flowers should be taken on parts that are at anthesis.

1 Bracts mostly leaf-like, though smaller towards the apex
  2 Petioles mostly 1–2 times as long as the lamina   **10. balearica**
  2 Petioles not more than ½ as long as the lamina
    3 Corolla 6–10 mm, almost straight, patent, pink   **13. minor**
    3 Corolla 10–23 mm, curved 45° or more, violet-blue, rarely pink
      4 Calyx glandular-pubescent; leaves usually ±hastate   **12. hastifolia**
      4 Calyx glabrous or eglandular-pubescent; leaves not hastate   **11. galericulata**

---

1 Bracts mostly distinct from the leaves
  5 Upper lip of corolla yellow or pink
    6 Leaves grey-tomentose beneath, deeply crenate to pinnatifid   **1. orientalis**
    6 Leaves green, glabrous to pubescent, serrate-crenate   **2. alpina**
  5 Upper lip of corolla purple, blue, reddish or white
    7 Corolla 18 mm or more
      8 Inflorescence ±4-angled; leaves not more than 3·5 cm, usually obtuse   **2. alpina**
      8 Inflorescence not 4-angled; leaves up to 8 cm, acute   **3. columnae**
    7 Corolla less than 18 mm
      9 Internodes of inflorescence *c.* 5 mm
        10 Leaves 2–4(–5·5) cm; bracts exceeding calyx   **8. sieberi**
        10 Leaves 1–2 cm; bracts not exceeding calyx   **9. hirta**
      9 Internodes of inflorescence (8–)9–12 mm
        11 Bracts 15–20 mm   **7. velenovskyi**
        11 Bracts 3–12 mm
          12 Flowers usually white; bracts usually more than 10 mm   **6. albida**
          12 Flowers usually bluish; bracts usually less than 10 mm
            13 Larger leaves at least 5 cm; calyx without or with only a few long, white, eglandular hairs   **4. altissima**
            13 Leaves not more than 4 cm; calyx usually with numerous long, white, eglandular hairs   **5. rubicunda**

1. S. orientalis L., *Sp. Pl.* 598 (1753). Stems up to 30 cm, decumbent, woody, sparsely tomentose. Leaves (0·5–)1–1·5(–3) × 0·5–2 cm, ovate-oblong to broadly ovate, deeply crenate-dentate to pinnatisect with linear-lanceolate or linear lobes, truncate at base, dark green and pubescent above, grey-tomentose beneath, petiolate; margins usually recurved. Inflorescence dense, oblong, usually 4-angled. Bracts 5–12 mm, usually exceeding the calyx, imbricate, ovate or rarely more or less lanceolate, entire, acute, somewhat scarious, purplish or yellow-green, glabrous to pubescent. Corolla 15–30(–35) mm, yellow, rarely pink; lower lip often reddish. *Dry, usually calcareous, rocky places. S.E. Europe, from Albania to Krym; mountains of S.E. Spain.* Al Bu Gr Hs Ju Rm Rs (K) Tu.

A variable species, especially in degree of leaf-dissection and in size and shape of the bracts. Many subordinate taxa have been described but they are not well enough defined to merit subspecific rank.

2. S. alpina L., *Sp. Pl.* 599 (1753). Stems (5–)25–35(–50) cm, ascending, simple or branched, more or less pubescent. Leaves 1·5–3 cm, ovate, rarely lanceolate, crenate-serrate, rarely subentire, usually obtuse, glabrous to pubescent; lower leaves petiolate, the upper more or less sessile; margins not recurved. Inflorescence dense, oblong, more or less 4-angled. Bracts 8–15 mm, exceeding the calyx, imbricate, sessile, ovate to ovate-lanceolate, entire, acute, somewhat scarious, purplish, rarely green. Corolla 20–25(–30) mm, the tube more or less pubescent outside. *Mountains of S. & S.C. Europe, and lowlands of Ukraine and S.C. Russia.* Al Bu Ga Ge Gr He Hs It Ju Rm Rs (C, W, E).

(a) Subsp. **alpina**: Leaves and bracts more or less pubescent. Corolla purplish; lower lip often whitish. 2*n*=22. *Calcareous rocks, screes and grassy slopes. From Spain to Greece and Romania.*

Dwarf variants occur relatively frequently in E. Spain, and have been called **S. javalambrensis** Pau, *Not. Bot. Fl. Esp.* **2**: 35 (1888).

(b) Subsp. **supina** (L.) I. B. K. Richardson, *Bot. Jour. Linn. Soc.* **65**: 262 (1972) (*S. supina* L.): Leaves and bracts more or less glabrous. Corolla yellow. *Meadows and steppes. U.S.S.R.*

[1] By I. B. K. Richardson.

3. **S. columnae** All., *Fl. Pedem.* **1**: 40 (1785). Stems 30–100 cm, erect, simple or branched. Leaves 4–8 × 2–5 cm, ovate, crenate-serrate, cordate at base, acute, subglabrous to densely pubescent. Flowers secund. Bracts 5–16 mm, shorter than the flowers, ovate to ovate-lanceolate, entire, acute, green. Corolla (18–)20–22(–28) mm, strongly curved, purplish; lower lip whitish; tube pubescent outside. 2*n* = 34. *Woods. Balkan peninsula, extending northwards to Hungary; Italy and Sicilia.* Al Bu Co Gr Hu It Ju Rm Si [Ga Ho].

(a) Subsp. **columnae**: Leaves subglabrous to sparsely pubescent. Bracts 5–8 mm. *Throughout the range of the species.*
(b) Subsp. **gussonii** (Ten.) Rech. fil., *Bot. Arch. (Berlin)* **43**: 56 (1941): Leaves densely pubescent. Bracts 12–16 mm. *S. Italy, Sicilia.*

4. **S. altissima** L., *Sp. Pl.* 600 (1753). Stems up to 100 cm, erect, simple or branched. Leaves 5–15 × 2–5 cm, ovate, serrate, glabrous or hairy beneath on the veins. Internodes of inflorescence 10–12 mm. Bracts 6–10 mm, shorter than the flowers, ovate to ovate-lanceolate, entire, acute, green. Calyx without or with only a few long, white, eglandular hairs. Corolla 12–16(–18) mm, bluish; lower lip whitish; tube subglabrous to pubescent outside. *Woods. E.C. & S.E. Europe; C. & S. Italy.* Al Bu Cz Gr Hu It Ju Rm Rs (C, W, K) [Au Be Br Ga Ge].

5. **S. rubicunda** Hornem., *Hort. Hafn.* **2**: 968 (1815) (*S. peregrina* L., nom. ambig.). Stems 5–40(–60) cm, procumbent to erect, usually branched. Leaves 1–3 cm, ovate-triangular to deltate, serrate to crenate, subcordate at base, acute to obtuse, glabrous to velutinous. Internodes of inflorescence (8–)9–11 mm. Bracts (3–)5–9(–10) mm, shorter than the flowers, ovate to ovate-lanceolate, entire. Calyx usually with numerous long, white, eglandular hairs. Corolla (9–)12–15 mm, usually purplish to bluish, the lower lip often whitish, rarely all white. *S. part of Balkan peninsula; C. & S. Italy and Sicilia.* Al Gr It Si.

1 Leaves ± glabrous, acute, ± serrate   **(a) subsp. linnaeana**
1 Leaves pubescent to velutinous, ± obtuse, crenate
  2 Upper lip of corolla white   **(d) subsp. geraniana**
  2 Upper lip of corolla bluish
    3 Flowering stems 15–30(–40) cm, ± erect; leaves 2–3 cm
         (b) subsp. rubicunda
    3 Flowering stems 5–10(–15) cm, ± procumbent; leaves 1–1·5(–2) cm   **(c) subsp. rupestris**

(a) Subsp. **linnaeana** (Caruel) Rech. fil., *Bot. Arch. (Berlin)* **43**: 22 (1941): Stems 25–40(–60) cm. Leaves 1·5–2(–3) cm, ovate-triangular, serrate to crenate-serrate, acute, more or less glabrous. Bracts (3–)5–8(–10) mm. *Rocky places.* ● *S. Italy, Sicilia.*
Large shade-forms resemble variants of **4**, but can be distinguished by the relatively dense pubescence of the corolla-tube. Plants intermediate between **5**(a) and **5**(b) occur in Greece.
(b) Subsp. **rubicunda**: Flowering stems 15–30(–40) cm, more or less erect. Leaves 2–3 cm, triangular to deltate, crenate, more or less obtuse, pubescent to velutinous. Bracts (5–)7–9(–10) mm. 2*n* = 34. *Rocky places. Albania, Greece.*
(c) Subsp. **rupestris** (Boiss. & Heldr.) I. B. K. Richardson, *Bot. Jour. Linn. Soc.* **65**: 262 (1972) (*S. rupestris* Boiss. & Heldr.): Flowering stems 5–10(–15) cm, more or less procumbent. Leaves 1–1·5(–2) cm, deltate, crenate, obtuse, pubescent. Bracts 6–7 mm. ● *Mountains of S. Greece.*
(d) Subsp. **geraniana** (Tuntas ex Halácsy) I. B. K. Richardson, *loc. cit.* (1972) (*S. sibthorpii* var. *geraniana* Tuntas ex Halácsy, *S. peregrina* var. *geraniana* (Tuntas ex Halácsy) Hayek): Stems 15–30 cm. Leaves 2–3 cm, triangular to deltate, subobtuse, crenate,

velutinous. Bracts *c.* 6 mm. Corolla white. *Mountains.* ● *S.E. Greece and Aegean Islands.*

This taxon and the following two species are difficult to separate, and many subordinate taxa, mostly of restricted distribution and based mainly on stem indumentum and bract characters, may be variously assigned to them.

6. **S. albida** L., *Mantissa Alt.* 248 (1771) (incl. *S. pallida* Bieb., *S. woronowii* Juz.). Stems 15–30(–40) cm, more or less erect, usually branched, with short appressed eglandular hairs. Leaves 2–3 cm, triangular to deltate, crenate, more or less obtuse, pubescent to velutinous. Internodes of inflorescence *c.* 10 mm. Bracts 10–12 mm, shorter than flowers, ovate to ovate-lanceolate, entire, green. Corolla (10–)12–16 mm, white. *Rocky places. S.E. Europe.* Bu Gr It Ju Rm Rs (K) Tu [Ga].

S. vacillans Rech. fil., *Bot. Arch. (Berlin)* **43**: 12 (1941), from Greece and the Aegean region, is like **6** but has the corolla often violet or bluish, and may be related to **5**. Further investigation is required.

7. **S. velenovskyi** Rech. fil., *Bot. Arch. (Berlin)* **43**: 9 (1941) (*S. pichleri* Velen.). Like **6** but stems 20–40(–60) cm, with both short and long patent eglandular hairs; leaves 4–6(–8) cm, ovate-triangular, pubescent; bracts 15–20 mm, equalling or exceeding the flowers, subobtuse. *Mountain pastures. C. part of Balkan peninsula, S.W. Romania.* Bu Ju Rm. (*S.W. Asia.*)

Subsp. **goulimyi** (Rech. fil.) Rech. fil., *Bot. Jahrb.* **80**: 388 (1961), from S.E. Greece (Evvoia), has the indumentum of **7** but is otherwise perhaps nearer to **6**.

S. naxensis Bothmer, *Bot. Not.* **122**: 46 (1969), described from the S. Aegean region (Naxos), has stems with long and short eglandular hairs, and large bracts. It is perhaps referable to **7**, and has 2*n* = 34.

8. **S. sieberi** Bentham in DC., *Prodr.* **12**: 420 (1848). Velutinous-tomentose. Stems 30–60 cm, more or less erect, branched, woody at base. Leaves 2–4(–5·5) cm, ovate-triangular, petiolate, crenate or crenate-serrate, truncate to subcordate at base, subacute. Internodes of inflorescence *c.* 5 mm. Bracts 7–9 mm, exceeding the calyx, ovate, entire, green. Corolla 10–14 mm, cream; upper lip reddish. *Rocky places.* ● *Kriti.* Cr.

9. **S. hirta** Sibth. & Sm., *Fl. Graec. Prodr.* **1**: 425 (1809). Densely pubescent. Stems 3–10(–20) cm, ascending, branched from a woody base. Leaves 1–2 cm, ovate to deltate, petiolate, crenate, more or less truncate at base, obtuse. Inflorescence few-flowered; internodes *c.* 5 mm. Bracts *c.* 5 mm, shorter than or equalling the calyx, entire, green. Corolla 9–10(–12) mm, cream; upper lip reddish. *Mountain screes.* ● *Kriti.* Cr.

10. **S. balearica** Barc., *Anal. Soc. Esp. Hist. Nat.* **6**: 399 (1877). Stems up to 8 cm, slender, decumbent, sometimes branched at base. Leaves 1(–1·5) cm, ovate to deltate, crenate, truncate to cordate at base, pubescent, often purplish beneath. Petiole mostly 1–2 times as long as lamina. Flowers few, in remote axillary pairs. Bracts like the leaves but smaller. Corolla *c.* 6 mm, purple. *Shady rocks.* ● *Islas Baleares (Mallorca).* Bl.

11. **S. galericulata** L., *Sp. Pl.* 599 (1753). Stems 7–70 cm, more or less erect, simple or branched. Leaves 1–5(–7) × 0·3–2 cm, ovate-elliptical to oblong-lanceolate, weakly crenate-serrate, more or less cordate at base, acute, glabrous to pubescent. Petiole up to ¼ as long as lamina. Flowers in remote

axillary pairs. Bracts like the leaves, usually exceeding the flowers, though smaller above. Calyx glabrous or eglandular-pubescent. Corolla 10–18 mm, violet-blue, rarely pink, whitish towards base of tube; tube curved through *c*. 45°. $2n = 31, 32$. *Damp places. Almost throughout Europe, except for some islands.* All except Az Bl Cr Fa Is Rs (K) Sb Si.

12. S. hastifolia L., *Sp. Pl.* 599 (1753) (incl. *S. dubia* Taliev & Širj.). Stems (5–)15–40(–50) cm, more or less erect, simple or branched, subglabrous to sparsely pubescent. Leaves (8–)15–25(–40) × (4–)5–12(–20) mm, lanceolate to ovate, entire at least in the distal half, sometimes with 1–3 small crenations proximally, more or less hastate, rarely subcordate, obtuse, more or less glabrous. Petioles up to ¼ as long as lamina. Flowers in remote or close axillary pairs. Bracts like the leaves but subsessile, shorter than the flowers, the upper truncate at base. Calyx 3–4 mm, glandular-pubescent. Corolla (10–)15–20 (–23) mm, violet-blue; tube curved through *c*. 70°. *Damp grassland. Much of Europe, but absent from the south-west, the extreme north and the islands.* Au Bu Cz Da Fe Ga Ge ?Gr Hu It Ju Po Rm Rs (B, C, W, E) Su Tu [?Be Br].

13. S. minor Hudson, *Fl. Angl.* 232 (1762). Like **12** but less robust, with stems 10–20(–50) cm; leaves 8–20(–40) × 4–10 (–20) mm, entire or with 2, rarely more, small crenations proximally, usually rounded to cordate at base; calyx 2–3 mm, eglandular-pubescent; corolla 6–10 mm, pink; tube almost straight, patent. *Damp places.* ● *W. Europe, extending eastwards to E. Germany and W. Italy; one station in S.W. Sweden.* Az Be Br Ga Ge Hb Ho Hs It Lu Su [Au].

S. × *hybrida* Strail (**11** × **13**) occurs with the parents.

4. Prasium L.[1]

Shrubs. Verticillasters reduced to a single flower (rarely 2), forming terminal racemes. Calyx campanulate, 2-lipped, 10-veined; upper lip 3-lobed; lower lip deeply 2-fid; all the lobes leaf-like. Corolla 2-lipped; tube with a ring of scale-like hairs inside; upper lip entire, hooded; lower lip 3-lobed. Stamens didynamous, parallel; anther-cells divergent. Style-branches subequal, subulate. Nutlets drupe-like.

1. P. majus L., *Sp. Pl.* 601 (1753). Stems up to 1 m, erect, divaricately branched, glabrous or glabrescent, rarely hirsute. Leaves 2–5 × 0·8–2 cm, ovate to ovate-lanceolate, acute, crenate or crenate-serrate, the lower cordate and the upper truncate at the base, usually glabrous; petiole 10–18 mm. Floral leaves like the cauline but smaller, or the upper entire. Calyx accrescent, up to 25 mm in fruit, glandular-puberulent or glabrous; lobes ovate to ovate-lanceolate, aristate. Corolla 17–20 mm, white or lilac; upper lip oblong, obtuse; lower lip with the middle lobe largest. Nutlets 3–4 mm, ovoid-trigonous, black. $2n = 34$. *Dry places, mainly near the sea. Mediterranean region, C. & S. Portugal.* Al Bl Co Cr Gr Hs It Ju Lu Sa Si.

5. Marrubium L.[2]

Perennial herbs. Bracteoles usually present, ascending from a deflexed base. Calyx narrowly obconical, usually 10-veined, with 5–10 teeth; tube densely hairy inside at the mouth. Corolla 2-lipped; upper lip straight, 2-fid; lower lip 3-lobed; tube included in calyx, with an uneven ring of hairs or glabrous inside. Stamens parallel, the outer pair the longer, all included in the corolla-tube; anther-cells diverging. Nutlets truncate at apex.

1 Calyx with 8–10 equal or unequal teeth
  2 Calyx-teeth straight; leaves oblong-ovate   **11. pestalozzae**
  2 Calyx-teeth hamate; leaves broadly ovate to orbicular
    3 Calyx-teeth 10, equal; petioles of lower leaves usually shorter than lamina   **10. vulgare**
    3 Calyx-teeth 10, 5 longer alternating with 5 shorter; petioles of lower leaves usually longer than lamina   **12. alternidens**
1 Calyx with 5 ± equal teeth
  4 Calyx exceeding corolla, with teeth usually at least as long as tube
    5 Calyx-teeth 1–1·5 mm wide at base, lanceolate, rigid in fruit; verticillasters with not more than 12 flowers; corolla purple   **8. alysson**
    5 Calyx-teeth less than 1 mm wide at base, subulate, not rigid in fruit; verticillasters many-flowered; corolla yellow
      6 Calyx-tube 4–5·5(–6) mm; leaves orbicular, regularly and lightly crenate, densely white-lanate beneath   **5. velutinum**
      6 Calyx-tube 6–8 mm; leaves oblong-ovate to elliptical, irregularly and rather deeply crenate, densely greyish hairy beneath   **6. friwaldskyanum**
  4 Calyx not exceeding corolla, with teeth shorter than tube
    7 Lower leaves reniform to reniform-orbicular, cordate at base, the petiole longer than lamina; corolla pink to lilac
      8 Leaves densely white-lanate on both surfaces; bracteoles conspicuous, plumose   **4. supinum**
      8 Leaves appressed-hairy above, more densely so beneath; bracteoles inconspicuous, not plumose   **7. leonuroides**
    7 Leaves oblong to ovate to suborbicular, cuneate at base, the petiole shorter than lamina; corolla white to yellow
      9 Stems with most branches patent; verticillasters with not more than 10 flowers   **9. peregrinum**
      9 Stems unbranched or the few branches erect; verticillasters with more than 10 flowers
        10 Corolla-tube narrowing gradually to base; lateral lobes of lower lip almost equalling the middle lobe   **3. incanum**
        10 Corolla-tube abruptly contracted below middle; lateral lobes of lower lip much smaller than the middle lobe
          11 Cauline leaves broadly ovate to suborbicular; corolla yellowish   **2. cylleneum**
          11 Cauline leaves oblong to ovate-oblong; corolla white   **1. thessalum**

1. M. thessalum Boiss. & Heldr. in Boiss., *Diagn. Pl. Or. Nov.* 3(4): 51 (1859). Stems up to 50 cm, lanate-tomentose, simple or with few, short, erect non-flowering branches. Leaves oblong to ovate-oblong or obovate, cuneate at base, crenate, tomentose above, white lanate-tomentose beneath; petiole shorter than lamina. Verticillasters many-flowered, globose. Bracteoles subulate, shorter than the calyx. Calyx-tube 4·5–5·5 mm, 10-ribbed, stellate-tomentose, especially on the ribs; teeth 5, 2·5–3 mm, shorter than corolla, equal, straight, erect, subulate, villous. Corolla up to 12 mm, white, pubescent outside; tube abruptly contracted below the middle; upper lip oblong, divided to ⅓–½ into 2 spathulate lobes; lower lip with lateral lobes much smaller than the middle. *Mountain pastures.* ● *E. Greece (Thessalia).* Gr.

2. M. cylleneum Boiss. & Heldr., *loc. cit.* (1859). Like **1** but leaves broadly obovate to suborbicular; indumentum yellowish; calyx-tube 5–6·5 mm, the teeth less densely villous; corolla yellowish. *Mountain rocks.* ● *S. Albania; S. Greece.* Al Gr.

Some specimens from Albania have 10 calyx-teeth; they may be hybrids between **2** and **10**, but their nature and status are obscure.

3. M. incanum Desr. in Lam., *Encycl. Méth. Bot.* 3: 716 (1792) (*M. candidissimum* auct., non L.). Stems up to 50 cm, densely white-tomentose, usually with many short, erect, vegetative branches. Leaves of the main stems oblong-ovate, cuneate

---

[1] By R. Fernandes.   [2] By J. Cullen.

at base, crenate-dentate, densely tomentose, grey-green above, whitish beneath; petiole shorter than lamina. Verticillasters many-flowered, globose. Bracteoles subulate, the longest almost equalling the calyx. Calyx-tube 6·5–7 mm, 10-ribbed, stellate-tomentose; teeth 5, 3–4 mm, shorter than corolla, equal, becoming patent, subulate, sparsely stellate-pubescent. Corolla white, pubescent outside; tube narrowing gradually to base; upper lip ovate, divided to ⅓ into 2 lanceolate or narrowly triangular lobes; lower lip with lateral lobes almost equalling the middle. *Rocky places. Italy and Sicilia; W. part of Balkan peninsula.* Al ?Hs It Ju Si.

4. **M. supinum** L., *Sp. Pl.* 583 (1753). Stems up to 45 cm, densely white-lanate below, lanate-tomentose above, usually simple. Leaves reniform or reniform-orbicular, cordate at base, deeply crenate, densely white-lanate on both surfaces; petiole longer than lamina. Verticillasters many-flowered, globose. Bracteoles conspicuous, subulate, villous-plumose. Calyx-tube 5–7 mm, weakly 10-ribbed, villous; teeth 5, shorter than corolla and calyx-tube, straight, erect or becoming patent, villous-plumose. Corolla pink to lilac, pubescent outside; upper lip divided to nearly ½ into 2 spathulate lobes; lower lip with lateral lobes much smaller than the middle. *Mountains of C. & S. Spain.* Hs.

5. **M. velutinum** Sibth. & Sm., *Fl. Graec. Prodr.* 1: 412 (1809). Stems up to 40 cm, yellowish-tomentose, simple or with short non-flowering branches. Leaves broadly ovate to orbicular, regularly and shallowly crenate-serrate, yellowish-appressed-tomentose above, white-tomentose beneath, shortly petiolate. Verticillasters many-flowered, globose. Bracteoles about as long as calyx-tube, subulate, more or less villous. Calyx-tube 4–5·5(–6) mm, 10-ribbed, yellowish-pubescent to -tomentose; teeth 5, 5–5·5 mm, exceeding the corolla, less than 1 mm wide at the base, subulate, not becoming rigid in fruit. Corolla yellow, pubescent outside. *Mountain rocks. N. & C. Greece.* Gr.

6. **M. friwaldskyanum** Boiss., *Diagn. Pl. Or. Nov.* 2(12): 74 (1853) (*M. velutinum* var. *friwaldskyanum* (Boiss.) Halácsy). Like 5 but stems usually longer; leaves oblong-ovate to -elliptical, irregularly and deeply crenate, densely hairy and greyish beneath, appressed-hairy and dark green above; calyx-tube 6–8 mm. *Stony ground. Bulgaria.* Bu.

7. **M. leonuroides** Desr. in Lam., *Encycl. Méth. Bot.* 3: 715 (1792). Stems up to 45 cm, green, tomentulose. Leaves reniform, cordate at base, deeply and obtusely incise-dentate, appressed-hairy above, more densely hairy beneath; petiole longer than lamina. Verticillasters many-flowered. Bracteoles inconspicuous, subulate, villous, more or less glabrous at apex. Calyx-tube white-lanate; teeth 5, somewhat shorter than corolla and calyx-tube, less than 1 mm wide at the base, subulate, not becoming rigid in fruit. Corolla pink to lilac. *Dry, stony ground. E. Krym.* Rs (K). (*Caucasus.*)

8. **M. alysson** L., *Sp. Pl.* 582 (1753). Stems up to 40 cm, densely white-lanate-tomentose, simple or with ascending to erect flowering branches. Leaves flabellate, long-cuneate, sub-sessile, deeply crenate distally, densely white-lanate, often glabrescent above. Verticillasters up to 12-flowered. Bracteoles absent or very small. Calyx-tube 4·5–5·5 mm, obscurely ribbed, densely lanate-tomentose; teeth 5, c. 3 mm, exceeding the corolla, 1–1·5 mm wide at the base, lanceolate, becoming rigid in fruit. Corolla purple, pubescent outside. *Dry, stony ground. Spain, Sardegna, S.E. Italy.* Hs It Sa.

---

¹ By V. H. Heywood, based mainly on information supplied by P. W. Ball.

9. **M. peregrinum** L., *Sp. Pl.* 582 (1753). Stems up to 60 cm, shortly whitish or yellowish appressed-tomentose, with patent branches from most axils. Leaves oblong or obovate, cuneate at base, crenate-serrate, whitish-tomentose above, more densely so beneath; petiole shorter than lamina. Verticillasters up to 10-flowered, distant, or the uppermost on the lateral branches rather crowded. Bracteoles short, subulate, or minute. Calyx-tube 3·5–5 mm, 10-ribbed, appressed-tomentose; teeth 5, 1·5–3 mm, lanceolate-subulate, erect or becoming patent. Corolla exceeding the calyx-teeth, white. 2n = 34. *Dry, open habitats. S.E. & E.C. Europe, northwards to c. 48° 30′ N.; naturalized or casual elsewhere.* Al Au Bu Cz Gr Hu Ju Rm Rs (W, K, E) Tu [Ga Ge It Po].

M. × paniculatum Desr. in Lam., *Encycl. Méth. Bot.* 3: 716 (1792) (*M. remotum* Kit.) is **9 × 10**. It is like 9 but has less regular branching, wider elliptical leaves, (5–)8–10 unequal, often slightly hamate calyx-teeth and, frequently, more flowers in the verticillasters. It has been recorded from stony and waste places in C. Europe and the Balkan peninsula.

10. **M. vulgare** L., *Sp. Pl.* 583 (1753). Stems up to 45 cm, white-lanate at least below, with many short non-flowering branches. Leaves orbicular to broadly ovate, subcordate or rounded at base, deeply and irregularly crenate, sparsely tomentose to subglabrous above, more densely tomentose beneath; petioles of lower leaves shorter than lamina. Verticillasters globose, many-flowered, distant. Bracteoles subulate, villous-plumose. Calyx-tube 3–4 mm, obscurely 10-striate, villous-pubescent; teeth 10, shorter than corolla, equal, patent, hamate, villous beneath, glabrous above. Corolla exceeding calyx-teeth, white. 2n = 34. *Waste places. Europe, from England, S. Sweden and C. Russia southwards.* All except Fa Fe Is No Rs (N) Sb; only as an alien in Hb.

In Spain intermediates (possibly hybrids) between 10 and 4 are found; these resemble 4, but have 10 calyx-teeth, 5 longer alternating with 5 shorter. They lack the plumose bracteoles of 4, and the calyx-teeth are sometimes slightly hamate. They have been called **M. × willkommii** Magnus ex Pau, *Bol. Soc. Ibér. Ci. Nat.* 25: 76 (1926).

11. **M. pestalozzae** Boiss., *Diagn. Pl. Or. Nov.* 3(4): 53 (1859) (*M. praecox* Janka). Like 10 but leaves oblong-ovate; calyx stellate-tomentose; teeth 8–10, straight, somewhat unequal. *Dry waste places. S.E. Europe, extending northwards to c. 53° N. in S.C. Russia.* Bu Gr Rm Rs (C, W, K, E).

Sometimes confused with *M. × paniculatum.*

12. **M. alternidens** Rech. fil., *Österr. Bot. Zeitschr.* 99: 37 (1952). Like 10 but less densely hairy; lower leaves broadly cuneate, with petiole as long as or longer than lamina; calyx densely lanate-tomentose; teeth unequal, 5 longer alternating with 5 shorter, all hamate. Corolla pinkish or white. *C. Albania (near Kruj); N.E. Greece (Thraki).* Al Gr. (*S.W. Asia.*)

Few collections have been made and further study is required.

6. Sideritis L.¹

Annual or perennial herbs, or small shrubs. Verticillasters 2- to many-flowered; bracteoles absent. Calyx campanulate, 10-veined, 5-toothed; teeth equal, or the upper larger than the 4 lower. Corolla usually yellow; tube not exceeding calyx; upper lip patent, more or less flat, entire to 2-fid; lower lip 3-lobed. Stamens included in corolla-tube. Nutlets rounded at apex.

A taxonomically difficult genus, requiring an extensive experimental investigation. Specific limits in Sect. *Sideritis*, and to a lesser extent in Sect. *Empedoclia*, are often obscure. Interspecific hybrids within Sect. *Sideritis* have been recorded from many localities in Spain and it seems probable that much of the taxonomic difficulty is due to the occurrence of these hybrids and possibly hybrid swarms.

Literature: P. Font Quer, *Trab. Mus. Ci. Nat. Barcelona* 5(4) (1924).

1 Calyx ±2-lipped, the upper tooth longer and wider than the 4 lower teeth; annual
  2 Stem with short glandular hairs; upper calyx-tooth narrowly lanceolate; corolla yellow with black lips   **26. lanata**
  2 Stem without glandular hairs; upper calyx-tooth ovate; corolla yellow, purple or white
    3 Calyx with prominent veins, not saccate at base; teeth patent in fruit   **27. romana**
    3 Calyx without prominent veins, saccate at base; teeth curved in fruit   **28. curvidens**
1 Calyx ±actinomorphic, the teeth all equal
  4 Annual
    5 Leaves 2–8 mm wide, oblong-lanceolate   **25. montana**
    5 Leaves 8–15 mm wide, ovate   **26. lanata**
  4 Shrub, or perennial herb woody at base
    6 Bracts entire
      7 Middle and lower leaves rounded or cordate at base, semi-amplexicaul and often perfoliate   **24. perfoliata**
      7 Middle and lower leaves attenuate at base, not semi-amplexicaul and never perfoliate
        8 Bracts narrowly ovate to triangular-cordate, not conspicuously acuminate or cuspidate; calyx 4–8 mm; corolla 8–10 mm
          9 Verticillasters distant; calyx without a ring of hairs inside   **19. incana**
          9 Verticillasters crowded; calyx with a ring of hairs inside
            10 Leaves narrowly lanceolate, glabrescent to villous; corolla yellow, sometimes tinged with purple   **8. hyssopifolia**
            10 Leaves oblong- to obovate-spathulate, densely white tomentose-lanate or sericeous-lanate; corolla purple   **18. stachydioides**
        8 Most bracts broadly ovate to suborbicular with a distinct acumen or cusp; calyx 7–12 mm; corolla (9–)10–15 mm
          11 Middle bracts usually equalling or shorter than flowers and often shorter than calyx; verticillasters usually distant   **21. syriaca**
          11 Middle bracts exceeding flowers; verticillasters usually crowded to form a dense spike
            12 Middle bracts with acumen 4–10 mm   **22. clandestina**
            12 Middle bracts with acumen 2–4 mm   **23. scardica**
    6 Bracts coarsely toothed and often spiny
      13 Most leaves spinose-dentate with pungent lateral teeth
        14 Plant with dense, short glandular papillae, at least on the upper part of the stem; verticillasters distant **16. ilicifolia**
        14 Plant glabrescent to villous-tomentose, sometimes also glandular, hairy; verticillasters crowded into dense spikes
          15 Middle cauline leaves with 2–6, usually very coarse, divaricate lateral spines   **14. spinulosa**
          15 Middle cauline leaves with at least 20 appressed lateral spines   **15. serrata**
      13 Leaves entire or toothed but the lateral teeth not pungent
        16 Calyx without a ring of long hairs inside
          17 All leaves entire, or rarely with few small teeth; bracts with up to 11 teeth   **19. incana**
          17 Lower leaves crenate-dentate; bracts incise-dentate, usually with more than 11 teeth   **20. lacaitae**
        16 Calyx with a ring of long hairs inside
          18 Corolla white, sometimes with reddish markings on the lips

          19 Erect shrub up to 100 cm; most verticillasters crowded into a dense spike; corolla *c.* 6 mm, not or scarcely exceeding calyx   **2. foetens**
          19 Herb or dwarf shrub up to 30 cm; most verticillasters distant; corolla 7–10 mm, distinctly exceeding calyx   **17. leucantha**
            20 Stem with at least some long, straight, patent hairs; lower bracts 8–20 mm wide   **4. hirsuta**
            20 Stem with short, crispate hairs; lower bracts 6–8 mm wide   **17. leucantha**
        18 Corolla yellow, pink or purple, sometimes with a white upper lip
          21 Calyx 12–17 mm; corolla 14–20 mm
            22 Leaves 45–90 mm; corolla pale yellow inside   **1. grandiflora**
            22 Leaves not more than 35 mm; corolla tinged with dark brown or purple inside   **5. endressii**
          21 Calyx usually not more than 12 mm; corolla usually not more than 13 mm
            23 Bracts shorter than calyx
              24 Leaves coarsely incise-dentate or -crenate
                25 Stem tomentose with crispate and appressed hairs   **6. scordioides**
                25 Stem hirsute with numerous ±straight hairs   **4. hirsuta**
              24 Leaves entire or with few small teeth
                26 Stem not more than 15 cm; verticillasters 1–3, usually distant   **7. glacialis**
                26 Stem up to 30 cm; verticillasters 5–15, mostly crowded   **8. hyssopifolia**
            23 Bracts equalling or exceeding calyx
              27 Verticillasters crowded into a dense spike (sometimes the lowest 1–2 distant)
                28 Stem and leaves densely white tomentose-lanate or sericeous-lanate; corolla purple   **18. stachydioides**
                28 Stem and leaves green or grey-green, glabrous to tomentose or villous; corolla yellow, sometimes tinged with purple
                  29 Lower leaves cordate or rounded at base, crenate-serrate; petiole as long as or almost as long as lamina   **9. ovata**
                  29 Lower leaves attenuate at base, entire or with few remote teeth or coarsely incise-dentate or -crenate; petiole absent or short
                    30 Shrub up to 2 m   **3. arborescens**
                    30 Perennial herb with woody base, not more than 80 cm
                      31 Leaves 1–4 mm wide
                        32 Verticillasters (2–)3–5; corolla dark brown inside   **12. javalambrensis**
                        32 Verticillasters 4–15; corolla yellow or purplish inside
                          33 Bracts with few aristate teeth   **8. hyssopifolia**
                          33 Bracts coarsely spinose-dentate with numerous teeth   **10. linearifolia**
                      31 Leaves more than 4 mm wide
                        34 Calyx 6–8 mm; leaves glabrescent to villous   **8. hyssopifolia**
                        34 Calyx 9–10 mm; leaves densely villous-tomentose   **15. serrata**
              27 Verticillasters ±distant, all except the uppermost with stem visible between each verticillaster
              35 Leaves linear or linear-lanceolate, mostly at least 5 times as long as wide
                36 Stem glabrous or sparsely hairy; calyx 7–9 mm   **3. arborescens**
                36 Stem tomentose; calyx 5–7 mm   **11. angustifolia**
                35 Leaves less than 5 times as long as wide
                37 Shrub up to 200 cm   **3. arborescens**
                37 Perennial herb with woody base, not more than 60 cm
                  38 Stem tomentose with crispate and appressed hairs
                  39 Verticillasters very numerous, many-flowered; leaves oblong-lanceolate; bracts broadly reniform   **12. reverchonii**

39 Verticillasters 3–12(–14), 2- to 6-flowered; leaves linear-oblong to obovate; bracts lanceolate- or ovate-cordate to suborbicular

40 Lower bracts 6–15 × 10–20 mm, similar to the upper; leaves incise-dentate or crenate **6. scordioides**

40 Lower bracts 4–10 × 6–8 mm, distinct from the upper; leaves entire to coarsely toothed or shallowly lobed **17. leucantha**

38 Stem glabrescent or with at least some ± straight long hairs

41 Calyx 9–13 mm; corolla tinged with dark brown or purple inside **5. endressii**

41 Calyx 5–8(–9) mm; corolla not dark-tinged inside **4. hirsuta**

Sect. SIDERITIS. Shrubs or perennial herbs with woody base. Bracts usually coarsely toothed and somewhat spinulose or spinose, not leaf-like.

1. S. grandiflora Salzm. ex Bentham, *Lab. Gen. Sp.* 577 (1834) (*S. baetica* Lange). Shrub with herbaceous, villous-tomentose flowering stems up to 50 cm. Leaves 45–90 × 12–25 mm, oblong-elliptical or lanceolate, crenate-serrate. Verticillasters 3–5, distant, many-flowered. Lower bracts 25–45 × 25–35 mm, broadly ovate-cordate, acuminate, coarsely spinose-serrate at base, remotely serrate or crenate-serrate at apex. Calyx 12–15 mm. Corolla 14–17 mm, pale yellow. *Dry grassland. S.W. Spain (near Cádiz).* Hs.

2. S. foetens Clemente ex Lag., *Gen. Sp. Nov.* 18 (1816) (*S. lasiantha* auct., vix Pers.). Fetid shrub up to 100 cm with erect, herbaceous, glabrescent flowering stems up to 100 cm. Leaves up to 20 × 1·5 mm, few, linear, entire. Verticillasters 10–25, c. 6-flowered, mostly crowded into long, dense spikes. Lower bracts 6–8 × 6–10 mm, about equalling calyx, broadly ovate-cordate, incise-dentate. Calyx c. 6 mm, with a ring of hairs inside. Corolla c. 6 mm, white with red lines on the lips. *Dry, saline soils.* ● *S.E. Spain (near Almería).* Hs.

3. S. arborescens Salzm. ex Bentham, *Lab. Gen. Sp.* 579 (1834). Shrub up to 100(–200) cm, with erect, herbaceous flowering stems up to 50 cm. Leaves 15–35 × 1–10 mm, oblong-lanceolate to elliptical, rarely linear, often remotely dentate. Verticillasters 4–20, usually distant, 6- to 10-flowered. Lower bracts 8–15 × 8–20 mm, about equalling calyx, cordate-ovate, incise-dentate, glandular. Calyx 7–9 mm, glandular, with a ring of hairs inside. Corolla 8–13 mm, yellow or pale pink. *Rocky and bushy places. S. Spain.* Hs.

A variable species which cannot always be clearly separated from **11**, **13** and large-flowered variants of **4**.

1 Leaves 1–2 mm wide, linear, entire or with few obscure teeth; corolla pale pink **(b) subsp. luteola**
1 Leaves 2–10 mm wide, linear-lanceolate to oblong-lanceolate or elliptical, entire to serrulate or with few, short lobes; corolla usually yellow
  2 Stem glabrous or sparsely hairy; leaves entire, crenate-dentate or serrulate; calyx 7–8 mm **(a) subsp. arborescens**
  2 Stem hairy, usually glandular; leaves with few coarse teeth or shallow lobes; calyx c. 9 mm **(c) subsp. paulii**

(a) Subsp. **arborescens**: *S.W. Spain (provs. Cádiz and Málaga).*
(b) Subsp. **luteola** (Font Quer) P. W. Ball ex Heywood, *Bot. Jour. Linn. Soc.* 65: 355 (1972) (*S. luteola* Font Quer): ● *S. Spain (prov. Granada).*
(c) Subsp. **paulii** (Pau) P. W. Ball ex Heywood, *op. cit.* (1972) (*S. paulii* Pau). ● *S.W. Spain (prov. Córdoba and Sevilla).*

Subsp. **(c)** is also recorded from a number of other areas in S. & S.C. Spain but the identity of these plants needs confirmation.

4. S. hirsuta L., *Sp. Pl.* 575 (1753) (incl. *S. hirtula* Brot.). Moderately to densely hairy perennial up to 50 cm, with at least some long, more or less straight, patent hairs on the stem. Leaves 10–20 × 4–10 mm, mostly obovate-oblong, incise-dentate or -crenate, green or grey-green. Verticillasters 4–15, usually c. 6-flowered, usually distant. Lower bracts 6–15 × 8–20 mm, about equalling or shorter than calyx, ovate-cordate, incise-dentate. Calyx 5–8(–9) mm, usually eglandular, with a ring of hairs inside. Corolla 8–10 mm, with white or pale yellow upper lip and pale or bright yellow lower lip, rarely entirely white. *Dry, open habitats.* ● *S.W. Europe.* Ga Hs It Lu.

Very variable in S. Spain, where many populations occur whose taxonomic status and position are uncertain.

S. bubanii Font Quer, *Butll. Inst. Catalana Hist. Nat.* 20: 141 (1920), from N.E. Spain (prov. Lérida), is like some variants of **4** but is tomentose, with many-flowered verticillasters, bracts equalling or longer than calyx, with subspinescent teeth, and corolla with the upper lip shorter than the lower (not longer as in **4**). It also shows affinities with **6** and **8** and its status is not clear.

5. S. endressii Willk., *Bot. Zeit.* 17: 276 (1859) (incl. *S. ruscinonensis* Timb.-Lagr., *S. angustinii* Sennen, *S. catalaunica* Sennen). Glabrescent to villous, green perennial up to 30 cm. Leaves 10–35 × 5–20 mm, elliptical to obovate or oblanceolate, sub-entire to crenate or crenate-dentate, sessile or shortly petiolate. Verticillasters 5–10(–15), 6-flowered, laxly crowded or more or less distant. Lower bracts 8–9 mm, ovate-cordate or -elliptical, prominently toothed, sometimes with only a few teeth. Calyx 9–17 mm, with a ring of hairs inside. Corolla 9–20 mm, yellow, usually dark brown or purplish-tinged inside. ● *S. France, E. Spain.* Ga Hs.

(a) Subsp. **endressii**: Leaves 10–25 × 5–15 mm. Bracts 8–9 mm. Calyx 9–12 mm. Corolla 9–13 mm. *Throughout the range of the species except S.E. Spain.*
(b) Subsp. **laxespicata** (Degen & Debeaux) Heywood, *Bot. Jour. Linn. Soc.* 65: 355 (1972) (*S. endressii* f. *laxespicata* Degen & Debeaux): Leaves up to 35 × 20 mm. Bracts up to 20 × 35 mm. Calyx 13–17 mm. Corolla 18–20 mm. *S.E. Spain (provs. Jaén and Albacete).*

Other variants from N.E. & S. Spain have usually distant verticillasters, larger and wider bracts, and variable indumentum (often densely hirsute or greenish and sparsely hairy); some of them are low-growing and coastal and may represent a further subspecies.

6. S. scordioides L., *Syst. Nat.* ed. 10, 2: 1098 (1759). Tomentose perennial up to 30 cm. Leaves 7–30 × 3–10 mm, linear-oblong to obovate, incise-dentate or -crenate. Verticillasters 3–10, c. 6-flowered, usually distant. Lower bracts 6–15 × 10–20 mm, usually about equalling calyx, ovate-cordate, incise-dentate. Calyx 6–9 mm, with a ring of hairs inside. Corolla 8–10 mm, yellow, sometimes with purplish markings. *Dry calcareous slopes.* ● *S. France, C. & E. Spain.* Ga Hs.

(a) Subsp. **scordioides**: Leaves usually sparsely tomentose, shallowly incise-dentate. *S. France, ?N.E. Spain.*
(b) Subsp. **cavanillesii** (Lag.) P. W. Ball ex Heywood, *Bot. Jour. Linn. Soc.* 65: 355 (1972) (*S. cavanillesii* Lag.; incl. *S. hirsuta* var. *chamaedryfolia* (Cav.) Willk.): Leaves densely grey-tomentose, deeply incise-dentate. *C. & E. Spain.*

7. S. glacialis Boiss., *Biblioth. Univ. Genève* ser. 2, 13: 410 (1838). Like **6** but not more than 15 cm, sparsely to densely tomentose; leaves 5–10 × c. 2 mm, lanceolate to obovate-spathulate, entire or with few rounded teeth at apex; verticillasters 1–3, more or less crowded; bracts 5–6 × 5–6 mm, shorter than calyx; corolla yellow with dark brown or purplish markings inside. $2n = 34$. *Rocks and screes. Mountains of S. Spain.* Hs.

Considered by some authors to be only a dwarf variant of **6**.

8. S. hyssopifolia L., *Sp. Pl.* 575 (1753). Glabrescent to villous perennial up to 40(–80) cm. Leaves $5–35 \times 2–10$ mm, linear to ovate or obovate, oblanceolate to obspathulate, entire or shallowly toothed or crenate, sessile or shortly petiolate. Verticillasters 5–15, *c.* 6-flowered, mostly crowded into a dense spike. Calyx 6–8 mm, with a ring of hairs inside. Corolla *c.* 10 mm, yellow sometimes tinged with purple. $2n = 30, 32 + 6B, 34$. *Rocky places, woods and pastures.* ● *S.W. Europe, mainly in the mountains, extending north-eastwards to the Swiss Jura.* Ga He Hs It Lu [Ge].

Extremely variable and often divided into a number of subspecies or varieties, of which the following are the most distinct, although in the Pyrenees intermediates between them occur with narrow leaves, shorter than the internodes but with the bracts of subsp. (a) and often with 2–3 verticillasters only.

(a) Subsp. **hyssopifolia**: Leaves mostly 4–10 mm wide, oblong-lanceolate or elliptical; lower bracts $9–12 \times 9–16$ mm, equalling or exceeding calyx, ovate to ovate-cordate, incise-dentate. *Throughout the range of the species except Portugal.*

(b) Subsp. **guillonii** (Timb.-Lagr.) Rouy, *Fl. Fr.* **11**: 261 (1909) (*S. guillonii* Timb.-Lagr.): Middle and upper leaves 2–4 mm wide, narrowly lanceolate; lower bracts $5–8 \times 3–4(–6)$ mm, usually shorter than calyx, linear to ovate, entire or with 1–3 teeth. *S.W. France, Pyrenees, C. Portugal.*

9. S. ovata Cav., *Icon. Descr.* **1**: 36 (1791). Stoloniferous, glabrous or sparsely hairy perennial 10–30 cm. Lower leaves $15–30 \times 12–18$ mm, oblong- or ovate-elliptical, crenate-serrate, cordate or rounded at base, long-petiolate. Verticillasters 4–8, *c.* 6-flowered, mostly crowded into a short dense spike. Lower bracts $8–15 \times 6–10$ mm, triangular-ovate to ovate-cordate, incise-dentate. Calyx 6–9 mm, with a ring of hairs inside. Corolla 8–10 mm, pale yellow. ● *N. Spain (Cordillera Cantábrica).* Hs.

S. lurida Gay in Durieu, *Pl. Astur. Exsicc.* no. 248 (1836), from the mountains of N. & W.C. Spain, is stoloniferous, with leaves $20–25 \times 3–4$ mm, subentire to shallowly dentate or serrate, bracts *c.* 7 mm, broadly ovate, dentate and verticillasters densely crowded; it has been variously placed but is probably nearest to **9** (which is also stoloniferous) and may be specifically separable from it.

10. S. linearifolia Lam., *Encycl. Méth. Bot.* **2**: 168 (1786) (*S. pungens* Bentham). Glabrescent perennial up to 60 cm. Stem green, 4-angled, glabrescent, the hairs short, often restricted to 2 opposite sides, curved forwards. Leaves $15–40 \times 1·5–3·5$ mm, linear or lanceolate, entire, spiny-tipped. Verticillasters 4–12, *c.* 6-flowered, usually crowded. Lower bracts $8–14 \times 9–16$ mm, cordate-ovate, equalling or exceeding calyx, very coarsely spinose-dentate with numerous teeth and a distinct, elongate, more or less entire terminal tooth. Calyx 5–7 mm, with a ring of hairs inside. Corolla 7–10 mm, yellow, often tinged with purple. *Scrub, cultivated fields and waste places. N., C. & E. Spain, S. Portugal.* Hs Lu.

Extremely variable in N. Spain. Dwarf, broader leaved variants are sometimes recognized as a separate variety, subspecies or species and some of them resemble **8(b)**. Plants from S. Portugal sometimes referred to this species, or to **12** or **3**, have greenish stems with forwardly curved hairs mainly on 2 opposite sides, leaves 20–30 mm, verticillasters crowded, and lowermost bracts with a distinct long terminal tooth.

S. giennensis [Pau ex] Font Quer, *Cavanillesia* **1**: 40 (1928) from S.E. Spain (Sierras de Cazorla, Cabrilla and del Cuarto) is somewhat intermediate between **8–10** and **5**. It has leaves $15–25 \times 2·5$ mm, linear-lanceolate or -spathulate; verticillasters 2–6, usually crowded; bracts with a large, distinct, entire central tooth; calyx *c.* 8 mm, about equalling the bracts; corolla 9–10 mm, yellow, darker inside. It may represent a distinct species but further material needs to be studied.

11. S. angustifolia Lag., *Gen. Sp. Nov.* 18 (1816) (*S. lagascana* Willk.). Like **10** but stem tomentose; leaves 10–18(–30) mm; verticillasters distant; lower, bracts $5–10 \times 4–14$ mm; spinose-dentate, without a distinct, elongate, terminal tooth. *E. & S. Spain.* Hs.

Plants from E. Spain often have more or less crowded verticillasters, longer leaves, smaller bracts (often with a distinct terminal tooth) and sometimes smaller flowers. They are somewhat intermediate between **10** and **11** and may correspond to **S. tragoriganum** Lag., *Gen. Sp. Nov.* 18 (1816) and **S. funkiana** Willk., *Bot. Zeit.* **17**: 282 (1859). Their status is not clear.

12. S. reverchonii Willk., *Suppl. Prodr. Fl. Hisp.* 156 (1893). Like **10** but stem tomentose; leaves $15–30 \times 5–7$ mm, oblong-lanceolate; verticillasters very numerous, usually many-flowered; lower bracts $6–12 \times 10–15$ mm, reniform, acuminate. *Calcareous sands.* ● *S.W. Spain (near Ronda).* Hs.

13. S. javalambrensis Pau, *Not. Bot. Fl. Esp.* **1**: 26 (1887). Stem up to 20 cm, greenish, more or less densely hairy with forwardly curved and semi-patent hairs. Leaves $(8–)10–14(–20) \times 1–2$ mm, linear to linear-lanceolate or -spathulate, entire or remotely toothed, more or less densely hairy with appressed forwardly directed hairs, often densely glandular-punctate. Verticillasters (2–)3–5, densely crowded. Lowermost bracts $(6–)10–11 \times 8–10(–12)$ mm, more or less equalling the calyx, usually with a distinct terminal tooth. Corolla yellow, dark brown inside. ● *E. Spain (Sierra de Javalambre).* Hs.

14. S. spinulosa Barnades ex Asso, *Enum. Stirp. Arag.* no. 113 (1784) (*S. spinosa* Lam.). Glabrescent to villous-tomentose perennial 10–30 cm. Leaves $10–20 \times 7–8$ mm, linear or lanceolate with 2–6 divaricate lateral spines. Verticillasters 5–12, *c.* 6-flowered, mostly crowded. Lower bracts $7–15 \times 12–20$ mm, exceeding the calyx, broadly ovate or cordate-ovate, very coarsely incise-dentate and spiny. Calyx 8–9 mm, with a ring of hairs inside. Corolla 8–9 mm, pale yellow, not exceeding calyx. $2n = 22, 28, 30$. ● *E. & C. Spain.* Hs.

15. S. serrata Cav. ex Lag., *Gen. Sp. Nov.* 18 (1816). Like **14** but more robust with stout woody stems; stem and leaves densely villous-tomentose and glandular; leaves $25–40 \times 5–8$ mm, entire or spinose-serrate with at least 20 lateral spines, the teeth at *c.* 45° to the margin; calyx 9–10 mm; corolla *c.* 10 mm. ● *S.E. Spain (Tobarra, S. of Albacete).* Hs.

16. S. ilicifolia Willd., *Enum. Pl. Hort. Berol.* 606 (1809). Glabrous or very sparsely hairy perennial 20–40 cm, with dense short glandular papillae at least on upper part of stem. Leaves $15–45 \times c.$ 8 mm, linear or lanceolate, with 6–10 divaricate lateral spines. Verticillasters 3–6, many-flowered, more or less distant. Lower bracts $10–15 \times 12–16$ mm, about equalling calyx, ovate to cordate-ovate, coarsely incise-dentate and spiny. Calyx 5–8 mm, with a ring of hairs inside. Corolla 8–9 mm, pale yellow, exceeding calyx. *Dry, rocky or gravelly soils.* ● *E. Spain.* Hs.

17. S. leucantha Cav., *Icon. Descr.* **4**: 2 (1797). Grey- or white-tomentose perennial up to 30 cm. Leaves $7–15 \times 2–6$ mm, linear to obovate, entire to coarsely toothed or shallowly lobed. Verti-cillasters 3–12(–14), 2- to 6-flowered, distant. Lower bracts 4–10 × 6–8 mm, about equalling the calyx, cordate-lanceolate to suborbicular, incise-dentate. Calyx 4–9 mm, with a ring of hairs inside; teeth green, pubescent to villous. Corolla 7–10 mm, white, very rarely pale or deep yellow, sometimes with dark brown or purplish markings inside. *Dry places; calcicole.* ● *S.E. Spain, extending locally westwards to Málaga.* Hs.

A variable species which has been divided into a series of variants from S. Spain (mainly in the coastal provinces) and from which **S. pusilla** (Lange) Pau, *Bull. Acad. Int. Géogr. Bot. (Le Mans)* **16** (*Mém.*): 77 (1906), from S. Spain (provs. Almería, Granada, Málaga) has been segregated. The latter differs mainly in having purplish calyx-teeth which are completely covered by villous hairs. Much further study of this complex is required.

18. S. stachydioides Willk., *Bot. Zeit.* **8**: 78 (1850). White tomentose-lanate or sericeous-lanate perennial up to 20 cm. Lower leaves 10–15 × 4–8 mm, oblong- to obovate-spathulate, entire. Verticillasters 5–10, *c.* 6-flowered, mostly crowded into a dense spike. Lower bracts 8–15 × 4–5 mm, exceeding calyx, ovate-oblong to cordate-triangular, subentire or with 4–6 teeth near the base. Calyx 7–8 mm, with a ring of hairs inside. Corolla 8–10 mm, purple. *Crevices of calcareous rocks.* ● *S.E. Spain (near Vélez Rubio).* Hs.

19. S. incana L., *Sp. Pl.* ed. 2, 802 (1763). Glabrous to white-tomentose perennial up to 60 cm. Leaves up to 15–40 × 1·5–4(–7·5) mm, linear to ovate-lanceolate or spathulate, entire or with few small teeth. Verticillasters (1–)2–10, *c.* 6-flowered, dis-tant. Lower bracts 3–10 × 2–8 mm, shorter than calyx, ovate to cordate-ovate, entire or with up to 11 teeth. Calyx 4–8 mm, without a ring of hairs inside. Corolla 8–10 mm, yellow or pink. *Dry places.* C., E. & S. Spain. Hs.

1 Plant glabrous or sparsely tomentose
2 Corolla yellow; lower bracts 4–10 mm, toothed (a) subsp. **incana**
2 Corolla pink; lower bracts 3–4 mm, entire or feebly toothed
                                          (d) subsp. **glauca**
1 Plant densely white-tomentose
3 Corolla yellow (b) subsp. **virgata**
3 Corolla pink (c) subsp. **sericea**

(a) Subsp. **incana**: Plant densely tomentose when young, becoming sparsely tomentose. Verticillasters 2–5. Lower bracts 4–10 × 6–8 mm, toothed. Calyx 5–6 mm. Corolla yellow. ● *C. Spain.*

(b) Subsp. **virgata** (Desf.) Malagarriga, *Collect. Bot. (Barcelona)* **7**: 681 (1968) (*S. virgata* Desf.): Like subsp. (a) but plant persistently densely white-tomentose; verticillasters up to 10; calyx 5–8 mm. *S. & S.E. Spain.*

(c) Subsp. **sericea** (Pers.) P. W. Ball ex Heywood, *Bot. Jour. Linn. Soc.* **65**: 355 (1972) (*S. sericea* Pers.): Like subsp. (a) but plant persistently densely white-tomentose; corolla pink. ● *S.E. Spain (Quesa, S.W. of Valencia).*

(d) Subsp. **glauca** (Cav.) Malagarriga, *Collect. Bot. (Barcelona)* **7**: 681 (1968) (*S. glauca* Cav.): Glabrous. Verticillasters 2–10. Lower bracts 3–4 × 2–2·5 mm, entire or feebly toothed. Calyx 4–5 mm. Corolla pink. ● *S.E. Spain (near Orihuela, N.E. of Murcia).*

Var. *edetana* Pau, from S.E. Spain (Sierra de Chiva, W. of Valencia), is intermediate between subspp. (a) and (d). It is glabrescent, with toothed bracts and a pink corolla.

20. S. lacaitae Font Quer, *Bol. Soc. Esp. Hist. Nat.* **24**: 208 (1924). More or less tomentose perennial 30–70 cm. Lower leaves 20–50 × 2–5 mm, narrowly oblanceolate, more or less crenate-dentate; upper leaves narrower, entire. Verticillasters 4–14, *c.* 6-flowered, usually distant. Lower bracts 6–7 × 8–10 mm, shorter than or about equalling calyx, suborbicular-cordate, incise-dentate, usually with more than 11 teeth. Calyx 6–7 mm, without a ring of hairs inside. Corolla *c.* 8 mm, pale yellow. ● *S.C. Spain (E. part of Sierra Morena).* Hs.

Similar plants from S.E. Spain (La Sagra), with stems 20–30 cm, leaves 10–20 × 1·5–2 mm, verticillasters 2–4, sometimes crowded, and calyx sometimes with a few hairs inside, may be referable to 7.

Sect. EMPEDOCLIA (Rafin.) Bentham. Perennial herbs with a woody base. Bracts entire, usually not leaf-like.

21. S. syriaca L., *Sp. Pl.* 574 (1753) (incl. *S. cretica* Boiss., non L., *S. raeseri* Boiss. & Heldr., *S. sicula* Ucria, *S. taurica* Stephan ex Willd.). Grey- or white-lanate perennial 10–50 cm. Lower leaves 10–60 × (5–)6–20 mm, oblong to narrowly obovate, entire, crenulate or denticulate; middle and upper leaves up to 80 × 18 mm, linear-lanceolate or oblong, entire. Verticillasters 5–20, 6- to 10-flowered, mostly distant, rarely all crowded. Middle bracts 6–12 mm (including acumen), usually shorter than or equalling flowers, suborbicular; acumen 2–3 mm. Calyx 7–12 mm; teeth 2·5–5 mm, half as long to almost as long as tube. Corolla 9–15 mm, yellow. 2*n* = 24. *Mountain rocks.* S. Europe, from Sicilia to Krym. Al Bu Cr Gr It Ju Rs (K) Si.

A variable species, usually divided into several species and subordinate taxa. Within a restricted part of its range it is often possible to distinguish many of the local populations from each other, but when the whole range of variation is taken into con-sideration no satisfactory subdivision seems possible.

22. S. clandestina (Bory & Chaub.) Hayek, *Prodr. Fl. Penins. Balcan.* **2**: 257 (1929). Yellowish- or grey-lanate perennial 15–40 cm. Lower leaves 25–50 × 8–20 mm, oblong-spathulate to obovate, entire or crenulate; middle and upper leaves 30–70 × 6–12 mm, linear to oblong-elliptical, entire. Verticillasters 4–20, many-flowered, crowded or the lower 1–3 distant. Middle bracts 15–25 mm (including acumen), exceeding the flowers, broadly ovate to suborbicular; acumen 4–10 mm, sparsely or densely lanate. Calyx 9–11 mm; teeth 3·5–4·5 mm, slightly shorter than tube. Corolla 10–15 mm, yellow. *Mountain rocks.* ● *S. Greece (Peloponnisos).* Gr.

Plants from Taïyetos (var. *clandestina*) have a grey indumen-tum, linear to linear-oblong middle and upper leaves, 10–20 verticillasters with the uppermost crowded, and densely lanate bracts with acumen 6–10 mm. Those from Killini (var. *cyllenea* (Heldr. ex Boiss.) Hayek) have a yellowish indumentum, oblong-elliptical middle and upper leaves, 4–10 distant verticillasters and sparsely lanate bracts with acumen 4–6 mm. The latter are more or less intermediate between var. *clandestina* and 23.

23. S. scardica Griseb., *Spicil. Fl. Rumel.* **2**: 144 (1844). Like **22** but usually densely white-lanate; lower leaves 40–80 × 6–20 mm, oblong-lanceolate; verticillasters crowded into a dense spike; middle bracts 12–20 mm, suborbicular-cordate, sparsely lanate, abruptly acuminate with acumen 2–4 mm; calyx 9–12 mm; calyx-teeth 3–4(–6) mm, usually about half as long as tube. 2*n* = 32. *Mountain rocks.* ● *C. part of Balkan peninsula.* Al Bu Gr Ju.

24. S. perfoliata L., *Sp. Pl.* 575 (1753). Sparsely hirsute or lanate and densely puberulent perennial 30–45 cm. Leaves 35–70 × 20–30 mm, lanceolate, entire or with few teeth, rounded or cordate at base, semiamplexicaul and often perfoliate. Verticillasters 10 or more, 6- to 15-flowered, distant. Middle bracts *c.* 25×15 mm, exceeding flowers, ovate, acuminate. Calyx 10–11 mm; teeth *c.* 5 mm, almost as long as tube. Corolla *c.* 14 mm, yellow. *Mountain rocks. N. Greece.* Gr. (*Anatolia.*)

Sect. HESIODIA Bentham. Annuals. Bracts more or less entire, leaf-like.

25. S. montana L., *Sp. Pl.* 575 (1753). Sparsely to densely villous-lanate annual up to 35 cm. Leaves 5–30 × 2–8 mm, oblong-lanceolate, dentate. Verticillasters *c.* 6-flowered, distant or more or less crowded. Calyx 6–8 mm; teeth 3–4 mm, ovate, aristate, the arista 1–2 mm. Corolla shorter than to about equalling calyx, yellow, or black with yellow lower lip. *S. & S.C. Europe; S. part of U.S.S.R.* Al Au Bu Cz Ga Gr Hs Hu It Ju Rm Rs (C, W, K, E) Tu [Ge].

1 Middle lobe of lower lip of corolla less than 1 mm
                       (c) subsp. **ebracteata**
1 Middle lobe of lower lip of corolla 1–2 mm
2 Plant sparsely to moderately villous-lanate; corolla black or brownish-black, usually with yellow lower lip
                       (a) subsp. **montana**
2 Plant densely villous-lanate; corolla yellow   (b) subsp. **remota**

(a) Subsp. **montana** (incl. *S. comosa* (Rochel) Stankov): Sparsely to moderately densely villous-lanate. Corolla black or brownish-black, usually with a yellow lower lip; middle lobe of lower lip 1–2 mm; tube pubescent to villous. 2*n*=16. *Almost throughout the range of the species except Spain.*

(b) Subsp. **remota** (D'Urv.) Heywood, *Bot. Jour. Linn. Soc.* 65: 355 (1972) (*S. remota* D'Urv.): Like subsp. (a) but densely villous-lanate; corolla yellow. *Aegean region.*

(c) Subsp. **ebracteata** (Asso) Murb., *Lunds Univ. Årsskr.* 34(7): 35 (1898): Sparsely villous-lanate. Corolla yellow; middle lobe of lower lip *c.* 0·8 mm; tube puberulent. *Spain.*

26. S. lanata L., *Fl. Palaest.* 22 (1756). Villous or hirsute annual 10–30 cm. Stem with short glandular hairs. Leaves 15–35 × 8–15 mm, ovate, crenate or crenate-dentate. Verticillasters *c.* 6-flowered, distant. Calyx 5 7 mm, more or less 2-lipped; upper tooth 2·5–4 mm, lower 1·5–3 mm, usually shorter than upper, rarely subequal, all lanceolate, mucronate. Corolla about as long as calyx, yellow with black lips. *Cultivated fields and waste places. Aegean region and S. part of Balkan peninsula.* Bu Gr Ju.

27. S. romana L., *Sp. Pl.* 575 (1753). Villous-lanate annual up to 30 cm. Stem eglandular. Leaves 10–25 × 5–12 mm, oblong-ovate, dentate or crenate-dentate. Verticillasters *c.* 6-flowered, distant. Calyx 6–10 mm, 2-lipped; upper tooth broadly ovate, lower lanceolate, all with a straight, pungent apex. Corolla 7–10 mm, yellow, white or purple, equalling or slightly exceeding calyx. *Dry places. W. & C. Mediterranean region, Portugal.* Al Bl Co Ga Gr Hs It Ju Lu Sa Si.

(a) Subsp. **romana**: Sparsely villous-lanate. Corolla yellow or white; upper lip 2–3 × 0·8–1 mm. 2*n*=28. *From N.W. Jugoslavia and Sicilia westwards.*

(b) Subsp. **purpurea** (Talbot ex Bentham) Heywood, *Bot. Jour. Linn. Soc.* 65: 355 (1972) (*S. purpurea* Talbot ex Bentham): Densely villous-lanate. Corolla purple or white; upper lip 4–5 × 1–2 mm. ● *W. part of Balkan peninsula.*

Plants intermediate between subspp. (a) and (b) occur in W. Jugoslavia.

28. S. curvidens Stapf, *Denkschr. Akad. Wiss. Math.-Nat. Kl.* (*Wien*) 50(2): 100 (1885) (*S. romana* subsp. *curvidens* (Stapf) Holmboe). Like **27**, but calyx strongly saccate at base, veins weak; teeth curved, acute, not pungent, usually ending in a recurved awn. 2*n*=28. *S. part of Balkan peninsula and Aegean region.* Cr Gr Tu.

7. Melittis L.[1]

Perennial herbs. Verticillasters 2- to 6-flowered, distant. Calyx 2-lipped; upper lip irregularly toothed or entire; lower lip usually 2-lobed, shorter than the upper. Corolla 2-lipped; upper lip weakly hooded; lower lip 3-lobed.

Literature: M. V. Klokov, *Not. Syst.* (*Leningrad*) 18: 183–217 (1957).

1. M. melissophyllum L., *Sp. Pl.* 597 (1753). Stems 20–70 cm, erect. Leaves 2–15 × 1–8 cm, oblong to ovate, cordate to truncate at base, coarsely crenate or dentate. Pedicels 4–10 mm. Calyx 12–25 mm. Corolla 25–40 mm, white, pink or purple or sometimes variegated; tube much exceeding the calyx. 2*n*=30. *Shady places.* ● *W., C. & S. Europe, extending eastwards to Lithuania and N.C. Ukraine.* Al Au Be Br Bu Co Cz Ga Ge Gr He Hs Hu It Ju Lu Po Rm Rs (B, C, W) Si.

A variable species. The following subspecific treatment is provisional.

1 Stem densely covered with small, stipitate glands (c) subsp. **albida**
1 Stem eglandular or sparsely glandular
2 Largest leaves 5–7(–9) cm, with not more than 20(–22) large teeth on each side   (a) subsp. **melissophyllum**
2 Largest leaves (6–)7–15 cm, with 20–32 large teeth on each side   (b) subsp. **carpatica**

(a) Subsp. **melissophyllum**: Stem eglandular or sparsely glandular. Largest leaves 5–7(–9) cm, with 10–20(–22) large teeth on each side. Corolla white, pink, purple or variegated. *W. & C. Europe.*

Plants with a purple corolla are most frequent in W. Europe, whilst those with a white corolla are most frequent in C. Europe.

(b) Subsp. **carpatica** (Klokov) P. W. Ball, *Bot. Jour. Linn. Soc.* 64: 71 (1971) (*M. carpatica* Klokov): Like subsp. (a) but largest leaves (6–)7–15 cm, with 20–32 large teeth on each side; corolla usually white, sometimes with pink or purple markings on the lower lip. *E.C. Europe, extending eastwards to W. Ukraine and S.W. White Russia.*

(c) Subsp. **albida** (Guss.) P. W. Ball, *loc. cit.* (1971) (*M. albida* Guss.): Stem densely covered with small, stipitate glands. Largest leaves 6–15 cm, with 20–30 large teeth on each side. Corolla white, often with pink or purple markings on the lower lip, rarely purple. *S. Italy, Sicilia, Balkan peninsula.*

8. Eremostachys Bunge[2]

Perennial herbs. Verticillasters (2)4- to 6-flowered. Calyx 5- to 10-veined; tube infundibuliform; limb subentire, with 5 rigid teeth. Corolla 2-lipped; upper lip hooded; lower lip 3-lobed. Stamens didynamous, parallel; filaments of upper stamens with an appendage at the base; anther-cells divergent. Style-branches unequal. Nutlets subobtuse, hairy at apex.

1. E. tuberosa (Pallas) Bunge in Ledeb., *Fl. Altaica* 2: 416 (1830). Perennial with a tuberous root. Stems 15–40 cm, branched. Leaves 6–14 × 6–9 cm, orbicular to ovate, serrate, hirsute. Bracteoles absent. Corolla 32–40 mm, whitish-pink; upper lip densely hairy inside; lower lip deflexed. *Dry steppes. S.E. Russia, W. Kazakhstan.* Rs (E). (*W.C. Asia.*)

---

[1] By P. W. Ball.   [2] By R. A. DeFilipps and I. B. K. Richardson.

9. Phlomis L.[1]

Herbs or shrubs. Verticillasters few- to many-flowered, crowded or distant. Calyx tubular, 5- to 10-veined, 5-toothed. Corolla 2-lipped; upper lip hooded, emarginate; lower lip patent, 3-lobed. Stamens included or exserted; anther cells divergent. Style-branches unequal. Nutlets trigonous, glabrous or pubescent.

Most species occur on dry, rocky ground.

Literature: F. Vierhapper, *Österr. Bot. Zeitschr.* **65**: 205–236, 252–257 (1915).

1 Corolla purple or pink, rarely white
2 Roots bearing tubers; upper lip of corolla straight, the inner margin ciliate with hairs 1 mm; nutlets pubescent at apex
1. tuberosa
2 Roots without tubers; upper lip of corolla curved, the margin not ciliate or with hairs not more than 0·5 mm; nutlets glabrous at apex
3 Stem and bracteoles with glandular hairs; calyx 18–25 mm
2. samia
3 Stem and bracteoles eglandular; calyx 8–20 mm
4 Herb not more than 70 cm; bracteoles subulate; leaves stellate-tomentose or sparsely stellate-pubescent beneath
3. herba-venti
4 Shrub up to 200 cm; bracteoles lanceolate, elliptical or oblong; leaves stellate-lanate beneath
5 Corolla 23–26 mm, stellate-tomentose; calyx-teeth 3–5 mm, subulate; bracteoles 2–5 mm wide **4. purpurea**
5 Corolla *c.* 20 mm, stellate-lanate; calyx-teeth 1–2 mm, broadly triangular; bracteoles not more than 2 mm wide
5. italica
1 Corolla yellow or brownish-yellow
6 Bracteoles and calyx-teeth uncinate **6. floccosa**
6 Bracteoles and calyx-teeth straight at apex
7 Bracteoles elliptic-lanceolate, oblanceolate, obovate or ovate
8 Shrub up to 55 cm; lower leaves 1·5–2·8 cm, broadly elliptical, oblong, obovate or suborbicular; calyx-teeth 0·5–1 mm **11. lanata**
8 Shrub up to 130 cm; lower leaves 3–9 cm, lanceolate or lanceolate-ovate; calyx-teeth 1–4 mm **12. fruticosa**
7 Bracteoles subulate, linear or narrowly lanceolate
9 Stem with glandular or clavate hairs **7. cretica**
9 Stem eglandular
10 Calyx stellate-puberulent or stellate-lanate
11 Calyx stellate-puberulent; dwarf shrub up to 45 cm
8. ferruginea
11 Calyx stellate-lanate; shrub up to 130 cm **12. fruticosa**
10 Calyx stellate-villous
12 Leaves linear, narrowly elliptical or spathulate, gradually tapering into an indistinct petiole **9. lychnitis**
12 Leaves ovate or lanceolate, abruptly contracted into a distinct petiole **10. crinita**

1. P. tuberosa L., *Sp. Pl.* 586 (1753). Herb up to 150 cm. Roots bearing small tubers. Stem eglandular or glandular. Basal leaves 5–25 cm, lanceolate-ovate to deltate, cordate or sagittate at base, crenate or dentate, membranous, sparsely to densely pubescent on both surfaces, the hairs simple above, stellate beneath, eglandular or glandular; petiole up to 30 cm. Floral leaves sessile or subsessile, lanceolate-ovate to deltate, obtuse to acuminate. Verticillasters 14- to 40-flowered. Bracteoles 8–13 mm, subulate, glabrous to densely hairy, ciliate. Calyx 8–13 mm, glabrous to hispid; calyx-teeth 2·5–3·5 mm, setaceous.

Corolla 15–20 mm, purple or pink; upper lip porrect, ciliate with hairs 1 mm. Nutlets pubescent. $2n=22$. *C. & S.E. Europe, westwards to E. Germany, and extending northwards to c. 55° N. in C. Russia.* Au Bu Cz Ge Gr Hu Ju Rm Rs (C, W, K, E).

A highly variable species, particularly in Russia, whence numerous taxa have been described based on minor differences in leaf-shape and -size, indumentum of leaves, bracts and calyx, and corolla-length. **P. glandulifera** Klokov in Kotov, *Fl. RSS Ucr.* **9**: 643 (1960), from S. Ukraine, appears to differ mainly in having glandular hairs on stems, leaves and floral leaves.

2. P. samia L., *Sp. Pl.* 585 (1753). Herb up to 100 cm. Roots without tubers. Stem glandular. Basal leaves 8–18 cm, lanceolate-ovate, cordate or sagittate at base, crenate or serrate, subcoriaceous, eglandular-stellate-tomentose above, whitish-stellate-tomentose with glandular hairs beneath; petiole up to 18 cm. Floral leaves shortly petiolate, ovate or lanceolate, acuminate. Verticillasters (6–)12- to 20-flowered. Bracteoles 20–26 mm, subulate, stellate-glandular-tomentose. Calyx 18–25 mm, stellate-glandular-tomentose; teeth 6–12 mm, subulate. Corolla (26–)30–35 mm, purple; upper lip galeate, not ciliate. Nutlets glabrous. *Greece, just extending to S. Jugoslavia.* Gr Ju.

3. P. herba-venti L., *Sp. Pl.* 586 (1753). Herb up to 70 cm. Roots without tubers. Stems greenish, stellate-hirsute with hairs 2–4 mm or whitish, stellate-tomentose with hairs up to 0·1 mm, sometimes with a secondary indumentum of hairs up to 1 mm, eglandular, rarely glabrous. Cauline leaves 7–17·5 cm, lanceolate or ovate, truncate, cuneate, rounded or cordate at base, entire, remotely denticulate, crenate or serrate, membranous, glabrous to hirsute or stellate-puberulent above, sparsely stellate-pubescent or whitish-stellate-tomentose beneath; petiole up to 8 cm. Floral leaves sessile to shortly petiolate, lanceolate or oblong, acute or acuminate. Verticillasters 2- to 14-flowered. Bracteoles 11–15 mm, subulate, appressed stellate-tomentose and with some longer hairs. Calyx 8–15 mm, appressed stellate-tomentose or stellate-hirsute; teeth 2–7 mm, subulate. Corolla 15–20(–25) mm, purple or pink; upper lip galeate, not ciliate. Nutlets glabrous. *S. & E. Europe.* Bu Ga Gr Hs It Ju Lu Rm Rs (C, W, K, E) Si Tu.

(a) Subsp. **herba-venti**: Stem greenish, stellate-hirsute with hairs 2–4 mm. Cauline leaves 9–17·5 cm, ovate or lanceolate, truncate, rounded or cordate at base, crenate or serrate, glabrous, subglabrous or shortly hirsute above. Verticillasters (6–)10- to 14-flowered. ● *Mediterranean region, Portugal.*

(b) Subsp. **pungens** (Willd.) Maire ex DeFilipps, *Bot. Jour. Linn. Soc.* **64**: 233 (1971) (*P. pungens* Willd.): Stem whitish, stellate-tomentose, the hairs up to 0·1 mm, sometimes with a secondary indumentum of hairs up to 1 mm, rarely glabrous. Cauline leaves 7–13 cm, lanceolate, cuneate at base, entire, remotely denticulate or serrate, stellate-puberulent or stellate-hispidulous above. Verticillasters 2- to 6–(10-)flowered. *S.E. Europe, extending locally westwards to Spain and northwards to c. 53° N. in C. Russia.*

Plants with a mixture of stellate-hirsute and whitish-stellate-tomentose indumentum on the stem, cauline leaves truncate to subcordate at the base, and corolla 21–25 mm, occur frequently in Krym, and have been described as **P. taurica** Hartwiss ex Bunge, *Mém. Acad. Sci. Pétersb.* ser. 7, **21(1)**: 77 (1873). They seem to represent variants of **3(b)**.

[1] By R. A. DeFilipps.

4. **P. purpurea** L., *Sp. Pl.* 585 (1753). Shrub up to 200 cm. Roots without tubers. Stem eglandular. Lower leaves 3·5–9 × 1·7–4 cm, lanceolate or oblong-lanceolate, cordate or truncate at base, crenate or crenulate, coriaceous, stellate-puberulent above, white-stellate-lanate beneath; petiole up to 2 cm. Floral leaves petiolate, lanceolate, acute. Verticillasters 10- to 12-flowered. Bracteoles 12–20 × 2–4(–5) mm, lanceolate, elliptical or oblong, acuminate, stellate-lanate. Calyx 15–20 mm, stellate-tomentose; teeth 3–5 mm, subulate. Corolla 23–26 mm, purple or pink, rarely white, stellate-tomentose, upper lip galeate, not ciliate. Nutlets glabrous. *S., C. & E. Spain, S. Portugal.* Hs Lu.

Plants from S.E. Spain (prov. Almería), with the wide bracts of 4, narrow leaves (7–19 mm wide) and calyx-teeth 1–3 mm, are intermediate between 4 and 5 and referable to **P. purpurea** var. **almeriensis** Pau, *Mem. Mus. Ci. Nat. Barcelona (Bot.)* 1(3): 29 (1925).

5. **P. italica** L., *Syst. Nat.* ed. 10, 2: 1102 (1759). Like 4 but leaves stellate-lanate above; bracteoles ½ as long as to equalling calyx, up to 2 mm wide; calyx-teeth 1–2 mm, broadly triangular; corolla *c.* 20 mm, stellate-lanate. ● *Islas Baleares.* Bl.

6. **P. floccosa** D. Don, *Bot. Reg.* 15: t. 1300 (1830). Dwarf shrub up to 35 cm. Stem eglandular. Lower leaves 3–5 cm, oblong-ovate or ovate-lanceolate, cordate or subcordate at base, crenate, coriaceous, stellate-lanate on both surfaces; petiole up to 3 cm. Floral leaves shortly petiolate, lanceolate, acute or acuminate. Verticillasters 4- to 8-flowered. Bracteoles 15–18 mm, linear, uncinate, stellate-lanate, ciliate with hairs 2–3 mm. Calyx 15–19 mm, stellate-lanate, ciliate; teeth 1–5 mm, subulate, uncinate. Corolla 25–32 mm, yellow. Nutlets glabrous. *Karpathos.* Cr. (*N. Africa.*)

7. **P. cretica** C. Presl in J. & C. Presl, *Del. Prag.* 84 (1822). Dwarf shrub up to 45 cm. Stem stellate-lanate with glandular or clavate hairs. Lower leaves 3–8 cm, lanceolate, oblong to oblong-lanceolate, cuneate, truncate or rounded at base, crenulate, coriaceous, stellate-tomentose above, white-stellate-lanate beneath; petiole up to 4 cm. Floral leaves petiolate, lanceolate, acute or acuminate. Verticillasters 14- to 30-flowered. Bracteoles 12–19 mm, linear to narrowly lanceolate, straight at apex, stellate-tomentose and ciliate. Calyx 13–19 mm, stellate-lanate with glandular hairs; teeth 1–5 mm, subulate, straight at apex. Corolla 25–27 mm, yellow. Nutlets glabrous. *Greece, Kriti.* Cr Gr.

P. viscosa Poiret in Lam., *Encycl. Méth. Bot.* 5: 271 (1804), from S.W. Asia, which differs from 7 in its membranous leaves, stellate-puberulent beneath, is widely cultivated and is perhaps naturalized in a few localities in S. France.

8. **P. ferruginea** Ten., *Fl. Nap.* 1, *Prodr.*: 35 (1811). Like 7 but hairs of stem and calyx not glandular or clavate; calyx stellate-puberulent. ● *S. Italy.* It.

9. **P. lychnitis** L., *Sp. Pl.* 585 (1753). Shrub up to 65 cm. Stem eglandular. Lower leaves 5–11 cm, linear, narrowly elliptical or spathulate, entire or with a few crenulations at apex, coriaceous, gradually tapering into an indistinct petiole, stellate-puberulent above, stellate-lanate beneath. Floral leaves sessile, broadly ovate or rhombic, cuspidate. Verticillasters 4- to 10-flowered. Bracteoles 12–20 mm, linear, straight at apex, stellate-villous. Calyx 15–20 mm, stellate-villous; teeth 2–4 mm, linear-lanceolate, straight at apex. Corolla 20–30 mm, yellow. Nutlets glabrous. $2n=20$. ● *S.W. Europe.* Ga Hs Lu.

In Spain, hybrid swarms with leaf-shapes transitional between 9 and 10 frequently occur; they have been discussed by C. Pau, *Mem. Mus. Ci. Nat. Barcelona (Bot.)* 1(1): 64, pl. 7, 8 (1922).

10. **P. crinita** Cav., *Icon. Descr.* 3: 25 (1795). Shrub up to 70 cm. Stem eglandular. Lower leaves 4–11·5 cm, ovate to lanceolate, cordate or rounded at base, entire, coriaceous, stellate-lanate on both surfaces, abruptly contracted into a distinct petiole up to 7 cm. Floral leaves sessile, ovate, gradually tapering into an acute or acuminate apex. Verticillasters 6- to 10-flowered. Bracteoles 13–15 mm, linear, straight at apex, stellate-villous. Calyx 13–17 mm, stellate-villous; teeth 2–5 mm, linear-lanceolate, straight at apex. Corolla 20–25 mm, brownish-yellow or yellow. Nutlets glabrous. *S. & E. Spain.* Hs.

11. **P. lanata** Willd., *Enum. Pl. Hort. Berol., Suppl.* 41 (1814). Shrub up to 55 cm. Stem eglandular. Lower leaves 1·5–2·8 cm, broadly elliptical, oblong, obovate or suborbicular, cuneate to rounded at base, crenulate, coriaceous, stellate-lanate on both surfaces; petiole up to 1 cm. Floral leaves subsessile, suborbicular, obtuse. Verticillasters 2- to 10-flowered. Bracteoles 6–10 × 3–5·5 mm, broadly elliptical, oblanceolate or obovate, mucronate, straight at apex, stellate-lanate. Calyx 10–12 mm, stellate-tomentose; teeth 0·5–1 mm, subulate. Corolla 20–23 mm, yellow. Nutlets pubescent. $2n=20$. ● *Kriti.* Cr.

12. **P. fruticosa** L., *Sp. Pl.* 584 (1753). Shrub up to 130 cm. Stem eglandular. Lower leaves 3–9 cm, elliptical, lanceolate or lanceolate-ovate, truncate or cuneate at base, entire or crenulate, coriaceous, shortly stellate-tomentose above, white-stellate-tomentose beneath; petiole up to 4 cm. Floral leaves sessile or petiolate, mostly lanceolate, obtuse. Verticillasters (6–)14- to 36-flowered. Bracteoles 10–20 × (2–)3–7 mm, obovate, ovate-lanceolate, oblanceolate or elliptical, acuminate, straight at apex, stellate-tomentose, ciliate or not, with hairs 2–3 mm. Calyx 10–19 mm, stellate-lanate, not ciliate; teeth 1–3·5(–4) mm, subulate. Corolla 23–35 mm, yellow. Nutlets glabrous or pubescent. $2n=20$. *Mediterranean region, westwards to Sardegna.* Al Cr Gr It Ju Sa Si [Az Br Ga Rs (K)].

Cultivated specimens often produce bracteoles which are ovate and wider than those of natural populations. Variants in which the bracteoles are not ciliate occur on Karpathos and Kasos, and have been distinguished as **P. pichleri** Vierh., *Österr. Bot. Zeitschr.* 65: 232 (1915).

10. Galeopsis L.[1]

Annuals. Verticillasters dense, the uppermost crowded, the lower distant; bracteoles subulate, pungent. Calyx tubular-campanulate, 5-toothed, the teeth spinose. Corolla distinctly 2-lipped; upper lip hooded; lower lip 3-lobed, with two bluntly conical protuberances at the base; tube about equalling to much exceeding the calyx, with a ring of hairs inside. Anther-cells fimbriate. Nutlets bluntly trigonous, rounded at apex.

A. Müntzing has studied the genetics of the genus intensively; natural hybrids have been reported between spp. 6–9, and artificial hybrids have been produced between most species in the genus.

All species grow in open habitats such as cultivated fields, roadsides, seashores, river-gravels, clearings in woodland, etc.

[1] By C. C. Townsend.

CLI *LABIATAE*

Literature: J. Briquet, *Mém. Cour. Acad. Roy. Sci. Belg.* **52**(9): 1–323 (1893). O. M. Gritsenko et al., *Ukr. Bot. Žur.* **27**: 241–248 (1970). J. T. Henrard, *Nederl. Kruidk. Arch.* **1918**: 158–188 (1919). A. Müntzing, *Hereditas* **10**: 241–260 (1928); **13**: 185–341 (1930); **16**: 73–154 (1932). C. C. Townsend, *Watsonia* **5**: 143–149 (1962). Z. Slavíková, *Act. Univ. Carol.* (*Biol.*) **1963**: 255–263 (1963); *Nov. Bot. Horti Bot. Reg. Univ. Carol. Prag.* **1963**: 39–43 (1963).

1 Plant with rigid, often yellowish, hairs; stem swollen at the nodes when fresh
2 Stem with ±dense appressed hairs on all 4 sides; calyx with setae confined to the teeth and sinuses   **7. pubescens**
2 Stem without appressed hairs or with hairs only on opposite sides; calyx-tube with setae at least on the ribs
3 Corolla (22–)27–34 mm, pale yellow, usually with large purple blotch on the lower lip   **6. speciosa**
3 Corolla 15–20(–28) mm, pink (rarely white or pale yellow) with darker markings
4 Corolla with the middle lobe of the lower lip broad and flat, entire   **8. tetrahit**
4 Corolla with the middle lobe of the lower lip narrower, convex with the margins ±deflexed, distinctly emarginate   **9. bifida**
1 Plant lacking rigid hairs; stem not swollen at the nodes
5 Leaves and calyx densely and softly sericeous; bracteoles of the lower verticillasters usually shorter than the calyx-tube
6 Corolla usually pale yellow, rarely pink with purple blotches; leaves lanceolate to lanceolate-ovate, attenuate into the petiole, subacute to acute   **1. segetum**
6 Corolla purple; leaves broadly ovate, truncate to subcordate at the base, obtuse   **2. pyrenaica**
5 Leaves and calyx pubescent to subglabrous, never sericeous; outer bracteoles of the lower verticillasters usually exceeding the calyx-tube
7 Stem subterete, pruinose, especially near the nodes, usually glabrous below the middle   **5. reuteri**
7 Stem quadrangular, not pruinose, hairy almost to the base
8 Calyx green, with patent, translucent, smooth or finely punctate hairs; leaves broadly ovate to ovate-lanceolate   **3. ladanum**
8 Calyx whitish, with closely appressed, opaque and ±densely papillose hairs and sometimes also with patent hairs; leaves linear to lanceolate (rarely ovate)   **4. angustifolia**

Subgen. **Ladanum** Reichenb. Vegetative parts, bracteoles and calyx with soft, crispate or patent hairs, and frequently also with glandular hairs, but without rigid yellowish hairs; stem 4-angled or terete, not swollen at the nodes.

1. G. segetum Necker, *Hist. Comment. Acad. Elect. Theod.-Palat.* **2**: 474 (1770) (*G. dubia* Leers). Stem up to 50 cm, 4-angled, with crispate hairs and glandular hairs. Leaves lanceolate to ovate, dentate, cuneate at the base, subacute. Calyx 7–10 mm, green, densely and softly patent-hairy with smooth hairs, glandular. Corolla (20–)25–30(–35) mm, pale yellow, rarely lilac, with purple blotches. $2n=16$. *Calcifuge.* ● *From Wales and N.E. Spain eastwards to Denmark and N.E. Italy.* Be Br Da Ga Ge He Ho Hs It [Au Cz Hu Ju Rm].

2. G. pyrenaica Bartl., *Ind. Sem. Horti Goetting.* **1848**: 4 (1848). Stems up to 40 cm, sometimes very dwarf, 4-angled, with soft crispate hairs and glandular hairs. Leaves ovate to deltate-ovate, dentate, truncate or subcordate at base; apex very obtuse. Calyx 8–10 mm, green, with dense, smooth, soft, patent hairs. Corolla 17–25 mm, pinkish-purple, usually with darker blotches. $2n=16$. *Sandy or gravelly ground by mountain streams; calcifuge.* ● *E. Pyrenees.* Ga Hs.

3. G. ladanum L., *Sp. Pl.* 579 (1753) (*G. intermedia* Vill.). Stem up to 40 cm, with crispate hairs and patent, glandular

hairs. Leaves ovate-lanceolate to broadly ovate, dentate, cuneate to subtruncate (rarely attenuate) at the base. Calyx 8–12 mm, green, with smooth or finely punctate, patent, translucent hairs. Corolla 15–28 mm, deep pink with yellow blotches. $2n=16$. *Somewhat calcifuge. Most of Europe except the islands and the extreme south.* Al Au Be Bu Co Cz Da Fe Ga Ge He Ho Hs Hu It Ju No Po Rm Rs (N, B, C, W, K, E) Su.

4. G. angustifolia Ehrh. ex Hoffm., *Deutschl. Fl.* ed. 2, **2**: 8 (1804) (*G. ladanum* subsp. *angustifolia* Gaudin, *G. ladanum* sensu Coste, non L.). Stem up to 40 cm, with crispate hairs, with or without glandular hairs. Leaves narrowly linear to lanceolate-ovate or rarely ovate, entire or dentate, rarely glabrous, cuneate at base. Calyx 8–13 mm; tube canescent to villous with strongly papillose, opaque hairs. Corolla 14–24 mm, deep reddish-pink with yellow blotches (rarely white). $2n=16$. *W., C. & S. Europe, eastwards to Poland and Bulgaria.* Au Be Br Bu Cz *Da Ga Ge He Ho Hs Hu It Ju Po Rm [Hb Su].

5. G. reuteri Reichenb. fil., *Icon. Fl. Germ.* **18**: 17 (1856). Stem up to c. 80 cm, wiry, glabrous or with very short, appressed hairs, especially above. Leaves lanceolate to almost linear, dentate, subacute, attenuate at base. Calyx 9–15 mm, appressed-hairy and sometimes shortly glandular. Corolla 16–20 mm, deep pink (rarely white), with yellow blotches. *Dry screes and rocks.* ● *Maritime Alps.* Ga It.

Perhaps only a subspecies of **4**.

Subgen. **Galeopsis**. Vegetative parts and usually also bracteoles and calyx with rigid, yellowish hairs and glandular hairs, sometimes with appressed hairs; stem 4-angled, more or less swollen at the nodes.

6. G. speciosa Miller, *Gard. Dict.* ed. 8, no. 3 (1768). Stem up to 100 cm, the angles setose; opposite sides, at least above, with appressed hairs. Leaves dentate, ovate to ovate-lanceolate, acuminate. Bracteoles setose, without soft hairs. Calyx 12–17 mm, with teeth, sinuses and ribs setose, otherwise subglabrous or with scattered smaller setae. Corolla (22–)27–34 mm, yellow, usually with a large purple blotch on lower lip. $2n=16$. *Europe, except most of the islands and much of the south.* Al Au Be Br Bu Cz Da Fe Ga Ge He Ho Hu It Ju No Po Rm Rs (N, B, C, W, E) Su [Hb].

7. G. pubescens Besser, *Prim. Fl. Galic.* **2**: 27 (1809). Stem up to 50 cm, with dense, appressed, whitish hairs on all 4 sides, the angles usually sparsely setose. Leaves ovate, acuminate, dentate. Bracteoles softly hairy, with weak setae at the margins. Calyx 7–11 mm. Corolla 20–25 mm, bright pinkish-red, usually with yellow blotches; central lobe of lower lip broad, rounded. $2n=16$. ● *C. Europe, extending to S.C. France, C. Italy, Albania and C. Ukraine.* Al Au ?Bu Cz Ga Ge He Hu It Ju Po Rm Rs (C, W) [Be Ho].

8. G. tetrahit L., *Sp. Pl.* 579 (1753). Stems up to 50 cm, the angles setose; opposite sides with appressed hairs. Leaves lanceolate to broadly ovate, acuminate, dentate. Bracteoles setose, with few or no soft hairs. Calyx 12–14 mm. Corolla 15–20(–28) mm, pink (rarely white or pale yellow), with darker markings; middle lobe of lower lip broad, entire, flat. $2n=32$. *Most of Europe, but rare in the south-east.* All except Az Bl Cr Gr Rs (K, E) Sa Sb Si Tu.

Plants from the French Alps with terete, glaucous stems have been called subsp. **glaucocerata** P. Fourn., *Quatre Fl. Fr.* 825 (1937); further information about them is required.

9. G. bifida Boenn., *Prodr. Fl. Monast.* 178 (1824). Like **8** but corolla rarely more than 15 mm; middle lobe of lower lip narrower, distinctly emarginate, the sides more or less deflexed so that the surface appears convex. $2n=32$. *Most of Europe, but absent from many of the islands and the south-west.* Au Be Br Bu Cz Da Fe Ga Ge Hb He Ho Hu It Ju No Po Rm Rs (N, B, C, W, K, E) Su.

11. Lamium L.[1]

Annual or perennial herbs. Verticillasters crowded. Calyx tubular or campanulate, 5-veined, with 5 equal or subequal teeth. Corolla white, pink or purple, 2-lipped; upper lip hooded; lower lip obcordate or broadly obovate with or without small lateral lobes. Anther-cells divaricate. Nutlets trigonous, truncate at apex.

Literature: P. Bernström, *Hereditas* **41**: 1–122 (1955) (Species 9–13 only).

Wiedemannia orientalis Fischer & Meyer, *Ind. Sem. Horti Petrop.* **4**: 51 (1838), native to S.W. Asia, has been recorded once from near Istanbul, but probably only as a casual. It is a purple-flowered annual, differing from species of *Lamium* mainly in the distinctly 2-lipped calyx.

1 Upper lip of corolla 2-fid
  2 Annual; corolla 12–25 mm; leaves mostly irregularly incised
          9. bifidum
  2 Perennial; corolla (22–)25–40 mm
    3 Lower leaves crenate or variously toothed **2. garganicum**
    3 Lower leaves deeply divided into (3–)5 segments which are themselves lobed **4. glaberrimum**
1 Upper lip of corolla entire, minutely toothed or emarginate
  4 Anthers glabrous
    5 Calyx 12–20 mm; corolla-tube straight, much longer than calyx **1. orvala**
    5 Calyx 7–12 mm; corolla-tube curved at base, shorter than or about as long as calyx **5. flexuosum**
  4 Anthers hairy
    6 Corolla-tube curved at base
      7 Annual; corolla with conspicuous, triangular lateral lobes **8. moschatum**
      7 Perennial; corolla with minute, linear or narrowly subulate lateral lobes
        8 Lateral lobes of the corolla composed of 1 tooth; corolla pink, purple or brownish-purple, rarely white **6. maculatum**
        8 Lateral lobes of the corolla composed of 2–3 teeth; corolla white **7. album**
    6 Corolla-tube straight
      9 Corolla 25–40 mm; perennial **2. garganicum**
      9 Corolla not more than 20(–25) mm
        10 Perennial; calyx-teeth c. ½ as long as tube **3. corsicum**
        10 Annual; calyx-teeth about as long as or longer than tube
          11 Calyx 8–12 mm, the teeth longer than the tube; lower lip of corolla c. 4 mm **12. moluccellifolium**
          11 Calyx 5–7 mm, the teeth not longer than the tube; lower lip of corolla 1·5–2·5 mm
            12 Bracts ±amplexicaul, usually wider than long **13. amplexicaule**
            12 Bracts not amplexicaul, longer than wide
              13 Leaves and bracts crenate or crenate-serrate, not decurrent along petiole **10. purpureum**
              13 Leaves and bracts incise-dentate, the upper ±decurrent along petiole **11. hybridum**

1. L. orvala L., *Syst. Nat.* ed. 10, **2**: 1099 (1759). Glabrous or sparsely hairy perennial 40–100 cm. Leaves 40–150 × 30–90(–120) mm, triangular-ovate, coarsely and irregularly toothed. Calyx 12–20 mm, the teeth shorter than to longer than tube.

---

[1] By P. W. Ball.

Corolla 25–45 mm, pink to dark purple, rarely white; tube 15–20 mm, straight, longer than calyx; upper lip 15–20 mm, irregularly toothed, the lateral lobes short, triangular; lower lip 15–20 mm, irregularly toothed. Anthers glabrous. ● *From N. Italy and W. Austria to W. Jugoslavia and S. Hungary.* Au Hu It Ju ?Rs (W).

2. L. garganicum L., *Sp. Pl.* ed. 2, 808 (1763). Perennial up to 50 cm. Leaves up to 70 × 40 mm, cordate-ovate to reniform, crenate to crenate-serrate. Calyx 7·5–18 mm, the teeth shorter than or rarely about as long as tube. Corolla 25–40 mm, pink to purple, rarely white; tube 15–25 mm, straight, much longer than calyx; upper lip 10–15 mm, 2-fid or irregularly toothed; lateral lobes short, triangular; lower lip 10–15 mm, obcordate. Anthers hairy. $2n=18$. *Mountain rocks. S. Europe, from Corse and S. France to the Aegean region.* Al Bu Co Cr Ga Gr It Ju Rm Tu.

1 Stem not more than 10 cm; leaves reniform or cordate-orbicular **(c) subsp. pictum**
1 Stem usually more than 10 cm; leaves cordate-ovate
  2 Stem, leaves and calyx densely hairy **(a) subsp. garganicum**
  2 Stem, leaves and calyx sparsely hairy to subglabrous **(b) subsp. laevigatum**

(a) Subsp. **garganicum** (incl. *L. garganicum* subsp. *striatum* (Sibth. & Sm.) Hayek): Stem up to 50 cm, densely hairy; leaves 15–70 × 10–40 mm, cordate-ovate, densely hairy; upper lip of corolla often 2-fid. *S. Italy; C. & S. parts of Balkan peninsula; Aegean region.*

(b) Subsp. **laevigatum** Arcangeli, *Comp. Fl. Ital.* 555 (1882) (incl. *L. garganicum* subsp. *glabratum* (Griseb.) Briq., *L. longiflorum* Ten.): Like subsp. (a) but sparsely hairy to subglabrous; upper lip of the corolla usually shallowly toothed, not 2-fid. *N. part of the range of the species.*

(c) Subsp. **pictum** (Boiss. & Heldr.) P. W. Ball, *Bot. Jour. Linn. Soc.* **65**: 350 (1972) (*L. pictum* Boiss. & Heldr.): Stem not more than 10 cm, usually glabrous; leaves 7–15 × 8–20 mm, reniform or cordate-orbicular, glabrous or sparsely hairy; upper lip of corolla 2-fid. *Mountains of C. & S. Greece.*

3. L. corsicum Gren. & Godron, *Fl. Fr.* **2**: 679 (1853). Like 2(a) but stem glabrescent; leaves not more than 40 × 30 mm, triangular-ovate, crenate; calyx c. 10 mm, the teeth about ½ as long as tube; corolla 15–20 mm. $2n=18$. *Mountain rocks and screes.* ● *Corse, Sardegna.* Co Sa.

This species may be better considered as another subspecies of 2.

4. L. glaberrimum Taliev, *Bull. Jard. Bot. Pétersb.* **2**: 136 (1902). Glabrous perennial 20–30 cm. Leaves 5–10(–16) × 10–20 mm, ovate to reniform, deeply divided into (3–)5 segments which are themselves lobed. Calyx 11–15 mm, the teeth shorter than tube. Corolla (22–)30–35 mm, pink; tube 17–25 mm, straight, much longer than calyx; upper lip c. 10 mm, 2-fid; lateral lobes subulate; lower lip obcordate. Anthers hairy. *Rocky hillsides and forest-margins.* ● *Krym.* Rs (K).

5. L. flexuosum Ten., *Fl. Nap.* 1, *Prodr.*: 34 (1811). Sparsely to densely hairy perennial 15–60 cm. Leaves 30–60 × 20–40 mm, cordate-ovate to lanceolate, crenate to serrate. Calyx 7–12 mm, the teeth about as long as tube. Corolla 15–20(–25) mm, white or pink; tube c. 10 mm, curved, equalling or shorter than calyx; upper lip 8–10 mm, entire; lateral lobes linear; lower lip c. 5 mm, obcordate. Anthers glabrous. *W. Mediterranean region.* Ga Hs It Si.

6. L. maculatum L., *Sp. Pl.* ed. 2, 809 (1763). Sparsely to densely hairy perennial 15–80 cm. Leaves 10–80 × 10–70 mm, triangular-ovate, coarsely and irregularly crenate-serrate. Calyx 8–15 mm, the teeth shorter to longer than tube. Corolla 20–35 mm, pinkish-purple, rarely white or brownish-purple; tube 10–18 mm, curved, equalling or longer than calyx; upper lip 7–14 mm, entire; lateral lobes subulate; lower lip 4–6 mm, obcordate. Anthers hairy. 2n=18. *Europe northwards to c. 54° N. in Germany and c. 59° N. in N.C. Russia, but absent from most of the islands.* Al Au Be Bu Cz Ga Ge Gr He Ho Hs Hu It Ju Lu Po Rm Rs (N, B, C, W, K, E) Sa Tu [Br Su].

A very variable species, especially in the shape and toothing of the leaves, length of the calyx-teeth and corolla-colour. It does not at the moment seem possible to recognize subspecies, as although there seems to be a good correlation of characters in some areas this is not maintained elsewhere.

7. L. album L., *Sp. Pl.* 579 (1753) (incl. *L. dumeticola* Klokov). Sparsely to densely hairy perennial 20–80 cm. Leaves 25–120 × 10–50 mm, ovate or ovate-oblong, coarsely serrate or crenate-serrate. Calyx 9–13(–15) mm, the teeth about as long as tube. Corolla (18–)20–25 mm, white; tube 9–14 mm, about as long as or slightly longer than calyx; upper lip 7–12 mm, entire or crenulate; lateral lobes with 2–3 small teeth; lower lip c. 5 mm, obcordate. Anthers hairy. 2n=18. *Most of Europe, but rare in the south and absent from many islands.* Au Be Br Bu Cz Da Fe Ga Ge Gr He Ho Hs Hu It Ju No Po Rm Rs (N, B, C, W, K, E) Su [Hb Is].

8. L. moschatum Miller, *Gard. Dict.* ed. 8, no. 4 (1768). Sparsely hairy annual 10–30 cm. Leaves 15–70 × 12–50 mm, ovate-orbicular to triangular-ovate, crenate to coarsely crenate-serrate. Calyx 7–14 mm, the teeth slightly longer than tube. Corolla (12–)16–25 mm, white; tube 6–12 mm, shorter than calyx; upper lip 10–12 mm, crenulate; lateral lobes triangular; lower lip 6–8 mm, obovate-orbicular. Anthers hairy. 2n=18. *E. part of Balkan peninsula, Aegean region.* ?Bu Cr Gr Tu.

9. L. bifidum Cyr., *Pl. Rar. Neap.* 1: 22 (1788). Glabrous or pubescent annual 10–40 cm. Leaves 10–40 × 8–25 mm, triangular-ovate, crenate or crenate-serrate, mostly irregularly incised. Calyx 8–10 mm, the teeth shorter than tube. Corolla 12–25 mm; tube 12–14 mm, longer than calyx; upper lip 6–8 mm, 2-fid; lateral lobes linear; lower lip c. 5 mm, obovate. Anthers hairy. *Open habitats. S. Europe, from Bulgaria westwards.* Bu Co Cr Gr It Ju Lu Rm Sa Si.

1 Corolla white   **(a) subsp. bifidum**
1 Corolla pink or purple
2   Corolla 20–25 mm, purple; lower leaves rounded at base
                      (b) subsp. balcanicum
2   Corolla 12–15 mm, pink; lower leaves cordate at base
                      (c) subsp. albimontanum

(a) Subsp. **bifidum**: Lower leaves incised, cordate to almost truncate at base. Calyx-teeth distinctly shorter than tube. Corolla 20–25 mm, white. *Mediterranean region from Corse and Sardegna eastwards; C. Portugal.*

(b) Subsp. **balcanicum** Velen., *Sitz.-Ber. Böhm. Ges. Wiss. (Math.-Nat. Kl.)* **1893**(37): 54 (1893): Lower leaves incised, rounded at base. Calyx-teeth slightly shorter than tube. Corolla 20–25 mm, purple. ● *Bulgaria, E. Jugoslavia, S. Romania.*

(c) Subsp. **albimontanum** Rech. fil., *Denkschr. Akad. Wiss. Math.-Nat. Kl. (Wien)* **105**(2, 1): 120 (1943): Lower leaves not incised, cordate at base. Calyx-teeth slightly shorter than tube. Corolla 12–15 mm, pink. ● *Kriti.*

---
¹ By P. W. Ball.

10. L. purpureum L., *Sp. Pl.* 579 (1753). Pubescent annual up to 40(–70) cm. Leaves 10–50 × 10–30 mm, ovate or ovate-orbicular, crenate or crenate-serrate. Lower bracts longer than wide, petiolate. Calyx 5–7 mm, the teeth about as long as tube. Corolla 10–18(–23) mm, pinkish-purple; tube 7–12 mm, straight, exceeding the calyx; upper lip 4–6 mm, entire; lateral lobes minute, linear; lower lip c. 2 mm, obcordate. Anthers hairy 2n–18. *Cultivated ground. Most of Europe.* All except Bl Cr Sb Si, but only naturalized in Is.

11. L. hybridum Vill., *Hist. Pl. Dauph.* 1: 251 (1786) (*L. hybridum* subsp. *dissectum* (With.) Gams). Like **10** but leaves and bracts irregularly incised, the upper leaves often somewhat decurrent along the petiole; corolla 10–15 mm; upper lip 3–4 mm. 2n=36. *Cultivated ground. Most of Europe except the southeast.* Au Be Br Co Da Fa Fe Ga Ge Hb He Ho Hs It Ju Lu No Po Rm Rs (N, B, C, W) Sa Su.

This appears to be an allopolyploid derived from **10** and probably **8** (Bernström, 1955).

12. L. moluccellifolium Fries, *Nov. Fl. Suec.* 72 (1819) (*L. hybridum* subsp. *intermedium* Gams). Pubescent annual up to 40 cm. Leaves 10–50 × 10–60 mm, ovate-orbicular, crenate or crenate-dentate. Lower bracts about as long as wide, distinctly petiolate. Calyx 8–12 mm, the teeth slightly longer than tube. Corolla 14–20 mm, pinkish-purple; tube 10–15 mm, straight, slightly exceeding the calyx; upper lip 4–5·5 mm, entire; lateral lobes minute, linear; lower lip c. 4 mm, obcordate. Anthers hairy. 2n=36. *Cultivated ground. N. Europe to c. 67° N. in Fennoscandia, and extending southwards to c. 52° 30′ in Germany.* Br Da Fa Fe Ge Hb No Rs (N, B) Su [Is Lu].

This appears to be an allopolyploid derived from **10** and **13** (Bernström, 1955).

13. L. amplexicaule L., *Sp. Pl.* 579 (1753). Sparsely to densely pubescent annual up to 30(–40) cm. Leaves 7–25 × 7–25 mm, orbicular or ovate-orbicular, crenate or lobed with obtuse lobes. Bracts up to 30 × 40 mm, usually wider than long, shallowly lobed, sessile or subsessile, cordate and somewhat amplexicaul. Calyx 5–7 mm, the teeth shorter or longer than tube. Corolla 14–20(–25) mm, pinkish-purple; tube 10–14 mm, straight, distinctly exceeding calyx; upper lip 3–5 mm, entire; lateral lobes minute or absent; lower lip 1·5–2·5 mm, obcordate. Anthers hairy. *Cultivated ground. Almost throughout Europe.* All except Fa Sb, but only naturalized in Is.

(a) Subsp. **amplexicaule** (incl. *L. rumelicum* Velen.): Sparsely to densely appressed-pubescent; leaves and bracts not divided or divided to less than ½-way. 2n=18. *Throughout the range of the species.*

(b) Subsp. **orientale** Pacz., *Florogr. Fitogeogr. Issled. Kalm. Step.* 103 (1892) (*L. paczoskianum* Vorosch.): Densely patent-pubescent; leaves and bracts divided to more than ½-way. *E. Ukraine, S.E. Russia.*

12. Lamiastrum Heister ex Fabr.¹

(*Galeobdolon* Adanson)

Like *Lamium* but corolla yellow; lower lip of the corolla with 3 more or less equal lobes, the middle lobe more or less triangular, acute.

Literature: S. Wegmüller, *Watsonia* 8: 277–288 (1971).

1. L. galeobdolon (L.) Ehrend. & Polatschek, *Österr. Bot. Zeitschr.* 113: 108 (1966) (*Galeobdolon luteum* Hudson, *Lamium galeobdolon* (L.) L.). Sparsely to densely hairy, erect perennial

15–60 cm. Leaves 3–8 × 2–6 cm, ovate to ovate-orbicular, coarsely toothed, rarely crenate, truncate to slightly cordate at base. Bracts similar to the leaves, the uppermost usually narrower and with fewer teeth. Calyx 7–10 mm, the teeth *c.* $\frac{1}{4}$ as long as the tube. Corolla 14–25 mm, bright yellow with brownish markings; tube straight, with a ring of hairs inside. *Woods and shady places. Most of Europe, but rare in the Mediterranean region and the north.* Al Au Be Br Bu Cz Da Ga Ge Gr Hb He Ho Hs Hu It Ju Po Rm Rs (N, B, C, W, E) Su Tu [Fe No].

1 Corolla 14–17 mm; lowest bracts usually more than twice as long as internode; not stoloniferous   **(c) subsp. flavidum**
1 Corolla 17–25 mm; lowest bracts usually less than twice as long as internode; usually stoloniferous
2 Verticillasters usually not more than 8-flowered; uppermost bracts 1–2 times as long as wide, crenate or with more or less obtuse teeth   **(a) subsp. galeobdolon**
2 Verticillasters 9- to 15-flowered; uppermost bracts $1\frac{1}{2}$–$3\frac{1}{2}$ times as long as wide, with acute or subobtuse teeth   **(b) subsp. montanum**

(a) Subsp. **galeobdolon** (*Lamium galeobdolon* subsp. *vulgare* (Pers.) Hayek): Usually stoloniferous. Lowest bracts shorter than or as long as internode; uppermost bracts 1–2 times as long as wide, crenate or with more or less obtuse teeth. Verticillasters up to 8(–10)-flowered. Corolla 17–21 mm. $2n = 18$. *Almost throughout the range of the species, but less frequent in the south.*

(b) Subsp. **montanum** (Pers.) Ehrend. & Polatschek, *op. cit.* 109 (1966) (*Lamium galeobdolon* subsp. *montanum* (Pers.) Hayek): Like subsp. (a) but lowest bracts usually as long as or up to twice as long as internode; uppermost bracts $1\frac{1}{2}$–$3\frac{1}{2}$ times as long as wide, with acute or subobtuse teeth; verticillasters (8–)10- to 15-flowered; corolla 18–25 mm. $2n = 36$. *Almost throughout the range of the species, but less frequent in the north.*

(c) Subsp. **flavidum** (F. Hermann) Ehrend. & Polatschek, *op. cit.* 110 (1966): Not stoloniferous. Lowest bracts at least twice as long as internode; uppermost bracts $1\frac{1}{2}$–4 times as long as wide, with acute patent teeth. Verticillasters 10- to 14-flowered. Corolla 14–17 mm. $2n = 18$. ● *Alps, Appennini, N.W. Jugoslavia.*

13. Leonurus L.[1]

Biennial or perennial herbs. Verticillasters many-flowered, crowded or more or less distant. Calyx campanulate, 5- or 10-veined, with 5 subequal spinose teeth. Corolla 2-lipped; tube shorter than calyx, not dilated at throat; upper lip hooded; lower lip 3-lobed. Nutlets trigonous, truncate and hairy at apex.

Literature: J. Holub, *Nov. Bot. Horti Bot. Univ. Carol. Prag.* **2**: 24–25 (1961). E. Ţopa, *Studii Cerc. Şti. Cluj* **3–4**: 1–10 (1951).

Leaves digitately 3- to 7-lobed; corolla 8–12 mm, distinctly exceeding calyx   **1. cardiaca**
Leaves entire, coarsely and irregularly crenate or toothed; corolla 5–7 mm, not or scarcely exceeding calyx   **2. marrubiastrum**

1. **L. cardiaca** L., *Sp. Pl.* 584 (1753) (incl. *L. glaucescens* Ledeb., *L. quinquelobatus* Usteri, *L. tataricus* L., *L. villosus* Desf. ex Sprengel). Glabrous to villous perennial 30–200 cm. Leaves 3–12 × 2–4 cm, the lower digitately 5- to 7-lobed, the lobes toothed or shallowly lobed, the upper shallowly to deeply 3-fid. Calyx 3·5–8 mm, the tube very prominently 5-veined; teeth almost as long as the tube, the lower deflexed. Corolla 8–12 mm, distinctly exceeding calyx, white or pale pink, sometimes with purple spots; upper lip densely villous on back; tube with a ring of hairs inside. Nutlets hispid at apex. $2n = 18$.

Most of Europe, but absent from the extreme north, the islands, and much of the Mediterranean region. Al Au Be Bu Cz Da Fe Ga Ge Gr He Ho Hs Hu It Ju No Po Rm Rs (N, B, C, W, K, E) Su Tu [Br].

Plants in the eastern part of the range usually have densely hairy stems with patent hairs 1–2 mm, leaves and bracts deeply divided and calyx 4·5–5(–8) mm, and are often regarded as a separate species or subspecies (subsp. **villosus** (Desf. ex Sprengel) Hyl., *Uppsala Univ. Årsskr.* **1945**(7): 273 (1945)); those in the central and western part usually have glabrous or sparsely hairy stems with recurved hairs 0·5 mm, leaves and bracts shallowly lobed and calyx 3·5–4(–5) mm. Sparsely hairy variants, however, are sometimes found in E. Europe and occasionally plants with a mixture of patent and recurved hairs occur.

2. **L. marrubiastrum** L., *Sp. Pl.* 584 (1753) (*Chaiturus leonuroides* Willd., *C. marrubiastrum* (L.) Ehrh. ex Spenner). Grey-pubescent biennial up to 125 cm. Leaves 2–6 × 1–3 cm, the lower ovate to suborbicular, coarsely and irregularly toothed or crenate, the middle and upper lanceolate to ovate, coarsely toothed. Calyx 5–7 mm, the tube with 10 indistinct veins; teeth shorter than tube, more or less patent. Corolla 5–7 mm, not or scarcely exceeding calyx, pale pink; upper lip pubescent on back; tube without a ring of hairs inside. Nutlets puberulent at apex. $2n = 24$. *C. & E. Europe, northwards to Czechoslovakia and C. Russia, and extending westwards to N.W. Italy; casual further north.* Au Bu Cz Ge Gr Hu It Ju Po Rm Rs (C, W, K, E) [Ga He].

14. Moluccella L.[1]

Annual or possibly short-lived perennial herbs. Verticillasters 6- to 10-flowered, distant; bracteoles subulate, spinose, deflexed. Calyx infundibuliform, more or less 2-lipped with greatly enlarged, reticulate-veined, membranous lips. Corolla 2-lipped; upper lip hooded; lower lip 3-lobed. Nutlets trigonous, truncate at apex.

1. **M. spinosa** L., *Sp. Pl.* 587 (1753). Glabrous annual or possibly short-lived perennial 30–100 cm. Leaves 2–6 × 2–6 cm, ovate to orbicular-ovate, coarsely serrate or incised. Calyx-tube *c.* 10 mm, with papilliform glands; lips 15–20 mm, the upper triangular, erect, with a single stout spine, the lower with 7–10 spines 5–10 mm long. Corolla 25–40 mm, white or pale pink, villous. *Mediterranean region; rare and apparently decreasing.* Gr Hs It Si [Bl].

15. Ballota L.[2]

Perennial herbs or small shrubs. Verticillasters few- to many-flowered. Bracteoles present. Calyx 10-veined; limb undulate, or with 5–16 crenations or teeth, the lobes more or less mucronate or gradually narrowed into an awn, rarely entire. Corolla-tube shorter than or equalling calyx, with a ring of hairs inside. Stamens parallel, the outer pair the longer; anther-cells diverging. Style-branches subequal. Nutlets oblong, rounded at apex.

Measurements relating to the calyx do not include awns or mucros.

Literature: A. Patzak, *Ann. Naturh. Mus.* (*Wien*) **62**: 57–86 (1958); **63**: 33–81 (1959); **64**: 42–56 (1961).

1 Small shrub; bracteoles pungent   **1. frutescens**
1 Herb; bracteoles not pungent
2 Verticillasters 6- to 12-flowered; calyx-limb undulate or irregularly crenate, smooth or mucronulate
3 Calyx-limb (12–)15–20 mm in diameter   **2. acetabulosa**
3 Calyx-limb 7–8 mm in diameter   **3. pseudodictamnus**

---

[1] By P. W. Ball.   [2] By A. Patzak.

2 Verticillasters usually more than 12-flowered; calyx-limb dentate
    4 Calyx-lobes 5, regular   **7. nigra**
    4 Calyx-lobes 6–many, ± irregular
      5 Calyx-limb 8–10 mm in diameter; lobes mucronulate
                                  4. hirsuta
      5 Calyx-limb 4–6 mm in diameter; lobes awned
        6 Calyx 8–10 mm; lobes triangular-lanceolate   **5. rupestris**
        6 Calyx 7–8 mm; lobes broadly triangular   **6. macedonica**

1. B. frutescens (L.) J. Woods, *Tourist's Fl.* 295 (1850). Shrub up to 60 cm. Stems numerous, divaricate, crispate-puberulent. Lower leaves 1·5–2·5 × 1–2 cm, ovate to ovate-oblong, obtuse, entire or crenate-denticulate; petiole 5–6 mm. Bracteoles 7–10 mm, 4 at each node, patent, pungent, glabrous. Verticillasters (2–)5- to 6-flowered. Calyx 12–15 mm, campanulate; limb 10–12 mm in diameter, irregularly dentate; lobes 5–10, 1·5–3 mm, broadly triangular, awned; awn *c.* 2·5 mm. Corolla 12–15 mm, white or lilac. *Rocky places.* ● *S.W. Alps.* Ga It.

2. B. acetabulosa (L.) Bentham, *Lab. Gen. Sp.* 595 (1834). Perennial up to 60(–80) cm, woody at base. Stems greyish-tomentose, with simple and stellate hairs, papillae and glands. Middle and upper cauline leaves 3–5 × 3–4 cm, broadly cordate at base, suborbicular, crenate-dentate; petiole 5–15 mm. Bracteoles 4–8 mm, linear to spathulate, membranous. Verticillasters 6- to 12-flowered. Calyx 12–15 mm, hypocrateriform; limb (12–)15–20 mm in diameter, the margin undulate, smooth or mucronulate. Corolla 15–18 mm, purple and white. 2*n* = 28. *S. & E. Greece and Aegean region.* Cr Gr ?Tu.

3. B. pseudodictamnus (L.) Bentham, *op. cit.* 594 (1834). Like **2** but plant 30–50 cm; stems yellowish-tomentose; middle cauline leaves 2–3 × 1·5–2 cm, subcordate at base, broadly ovate to suborbicular; petiole 2–4 mm; calyx 8–10 mm, infundibuliform; the limb 7–8 mm in diameter, irregularly crenate; lobes 5–10, the larger up to 2 mm, broadly triangular-acuminate, more or less mucronulate; corolla 14–15 mm. *S. Aegean region.* Cr Gr [It Si].

4. B. hirsuta Bentham, *op. cit.* 595 (1834) (*B. mollissima* Bentham, *B. hispanica* auct., non (L.) Bentham). Perennial 60–80 cm, woody at base. Stems hirsute, with glandular and eglandular simple and stellate hairs, papillae and glands. Lower and middle cauline leaves 3–6 × 3–5 cm, cordate or truncate at base, ovate or suborbicular, crenate; petiole of lower leaves 5–40 mm. Bracteoles 3–8 mm, linear-subulate (the outer oblanceolate), membranous. Verticillasters many-flowered. Calyx 10–12 mm, campanulate; limb 8–10 mm in diameter, irregularly dentate; lobes 10 or more, up to 2 mm, triangular-acuminate, sometimes dentate, mucronulate. Corolla 14–16 mm, purple or white. *Roadsides, walls and waste places.* C. & S. Portugal, C., S. & E. Spain, Islas Baleares. Bl Hs Lu.

5. B. rupestris (Biv.) Vis., *Fl. Dalm.* **2**: 216 (1847) (*B. acuta* Briq.). Perennial up to 70 cm, woody at base. Stems pubescent with glandular and eglandular simple and medifixed hairs. Lower and middle cauline leaves 4–6 × 4–5 cm, cordate at base, ovate or ovate-lanceolate, obtuse or subacute, and coarsely and irregularly crenate-dentate; petioles 10–25 mm. Bracteoles 5–8 mm, linear-subulate, membranous. Verticillasters many-flowered. Calyx 8–10 mm, subcylindrical; limb *c.* 6 mm in diameter, irregularly dentate; lobes 6–10, *c.* 2 mm, patent, triangular-lanceolate, awned; awn *c.* 1 mm. Corolla 12–14 mm, purple and white. *Rocky places.* C. & S. Italy, Sicilia; W. Jugoslavia, N. Albania. Al It Ju Si.

6. B. macedonica Vandas, *Magyar Bot. Lapok* **4**: 112 (1905) (*B. acuta* subsp. *macedonica* (Vandas) Hayek). Like **5** but middle cauline leaves 3–5·5 × 2–5 cm, cordate or truncate at base, ovate or triangular-ovate, acute; petiole 10(–15) mm; calyx 7–8 mm, the limb 4–5 mm in diameter; lobes 0·5–1 mm, broadly triangular; corolla 11–15 mm. *Rocky slopes, among scrub.* ● *W. part of Balkan peninsula from 39° 30′ to 42° N.* Al Gr Ju.

7. B. nigra L., *Sp. Pl.* 582 (1753). Perennial up to 130 cm. Stems pubescent to subglabrous. Lower cauline leaves 3–8 × 2–6 cm, cordate to cuneate at base, ovate or ovate-oblong; petiole usually short, rarely up to 80 mm. Bracteoles 3–9 mm, subulate or filiform, membranous. Verticillasters many-flowered. Calyx 7–13 mm; limb regularly dentate; lobes 5, awned or mucronate. Corolla 9–15 mm, pink, lilac or white. 2*n* = 22. *Europe, from England, S. Sweden and C. Russia southwards as a native or naturalized alien; casual or imperfectly naturalized further north.* All except Fa Fe Is No Rs (N, B) Sb.

Doubtfully native in much of N.W. & N.C. Europe; the limits of native distribution cannot be precisely ascertained.

A polymorphic species with 6 subspecies in Europe.

1 Calyx-lobes (3–)4–6·5 mm, triangular- to subulate-lanceolate, awned; awn 1·5–3 mm   **(a) subsp. nigra**
1 Calyx-lobes 1–3 mm, ovate to triangular-ovate, acuminate, mucronate; mucro 0·2–0·8(–1·7) mm
  2 Calyx-lobes erect or erecto-patent
    3 Calyx scarcely widened above; plant sericeous-tomentose, with shining, sessile glands   **(b) subsp. sericea**
    3 Calyx infundibuliform above; plant subglabrous or pubescent, with papillae and non-shining, sessile glands
                                   (c) subsp. foetida
  2 Calyx-lobes curved and patent
    4 Plant greyish-tomentulose-velutinous, with numerous sessile glands; stems not more than 30 cm, little-branched
                                   (d) subsp. velutina
    4 Plant pubescent to subglabrous, rarely sericeous-tomentulose, without sessile glands; stems up to 80 cm, much-branched
      5 Calyx 9–11 mm, the lobes with mucro 0·2–0·8 mm; corolla lilac   **(e) subsp. uncinata**
      5 Calyx 7–8 mm, the lobes with mucro 1–1·5 mm; corolla pink   **(f) subsp. anatolica**

(a) Subsp. **nigra** (*B. nigra* subsp. *ruderalis* (Swartz) Briq.): Pubescent to subglabrous, with papillae and non-shining glands. Stems up to 120 cm; lower internodes up to 10 cm, the upper shorter. Petiole up to 50 mm. Calyx 9–13 mm; lobes (3–)4–6·5 mm, erect, triangular- to subulate-lanceolate, awned; awn 1·5–3 mm. Corolla 12–14 mm, lilac. *Europe from Sweden, Germany and Albania eastwards; naturalized in N.W. Europe.*

(b) Subsp. **sericea** (Vandas) Patzak, *Ann. Naturh. Mus.* (*Wien*) **62**: 70 (1958): Sericeous, with numerous sessile, shining glands. Stems up to 70 cm, little-branched; lowermost internodes 3–4 cm, the upper shorter. Petiole very short. Calyx 8(–10) mm, scarcely widened above; lobes 1–1·5(–2) mm, broadly triangular, mucronate, patent; mucro 0·4–0·8 mm. Corolla 11–12 mm, lilac. *Albania, Macedonia.*

(c) Subsp. **foetida** Hayek, *Prodr. Fl. Penins. Balcan.* **2**: 278 (1929) (*B. borealis* Schweigger): Pubescent to subglabrous, with papillae and non-shining glands. Stems up to 100 cm; lower internodes up to 10 cm, the upper shorter. Petiole up to 40 mm. Calyx 7–12 mm, infundibuliform above; lobes (1·5–)2·5(–3) mm, erecto-patent, triangular-ovate, mucronate; mucro 0·2–0·5(–1·7) mm. Corolla 12–15 mm, lilac, rarely white. *Europe from Germany, Italy, and Albania westwards; naturalized in S. Scandinavia, Latvia and W. Estonia.*

(d) Subsp. **velutina** (Pospichal) Patzak, *Ann. Naturh. Mus.* (*Wien*) **62**: 67 (1958) (*B. velutina* Pospichal): Greyish-tomentulose-velutinous, with numerous sessile, shining glands. Stems up to 30 cm, little-branched. Petioles 30–50(–80) mm. Calyx 7–8(–9) mm; lobes (1·5–)2–2·5(–2·8) mm, recurved, ovate-lanceolate, mucronate; mucro 0·5–0·8(–1) mm. Corolla *c.* 14 mm, lilac. *W. Jugoslavia.*

(e) Subsp. **uncinata** (Fiori & Béguinot) Patzak, *op. cit.* 64 (1958): Pubescent to subglabrous, rarely sericeous-tomentulose, without sessile glands. Stems up to 80 cm, much-branched. Petiole short. Calyx 9–11 mm; lobes 1–2 mm, recurved, ovate or triangular-ovate, mucronate; mucro 0·2–0·8 mm. Corolla 12–14 mm, lilac. *S. Europe.*

(f) Subsp. **anatolica** P. H. Davis, *Notes Roy. Bot. Gard. Edinb.* **21**: 61 (1952): Hirsute, with a few sessile glands. Stems up to 70 cm, shortly branched above. Petiole up to 40(–50) mm. Calyx 7–8 mm; lobes 2–2·5 mm, recurved, triangular-lanceolate, mucronate; mucro 1–1·5 mm. Corolla 9–12 mm, pink. *Turkey-in-Europe.*

16. Stachys L.[1]

(incl. *Betonica* L.)

Annual or perennial herbs or small shrubs. Verticillasters 2- to many-flowered (rarely reduced to a solitary flower), in dense or lax spike-like, rarely capituliform, inflorescences. Bracteoles present or absent. Calyx tubular or campanulate, rarely weakly 2-lipped, with 5 equal teeth, 5- to 10-veined. Corolla 2-lipped; upper lip flat or hooded, entire to 2-fid; lower lip 3-lobed. Stamens didynamous, included or slightly exserted; anther-cells parallel or divergent. Style-branches subequal. Nutlets obovoid or oblong, rounded at apex.

1 Bracteoles as long as or longer than calyx-tube
  2 Annual; verticillasters in a dense capitulum, sometimes with 1 verticillaster distant   **58. serbica**
  2 Perennial; verticillasters in an elongate spike or all distant
    3 Upper part of stem and lower surface of leaves stellate-pubescent   **5. scardica**
    3 Plant without stellate hairs
      4 Stock with persistent rosettes of leaves; flowering branches axillary, subscapose with 1–4(–5) pairs of leaves
        5 Upper lip of corolla 2-fid or emarginate; corolla pale yellow   **1. alopecuros**
        5 Upper lip of corolla entire or crenulate; corolla white, pink or purple
          6 Calyx distinctly reticulate-veined; corolla 20–24 mm   **2. monieri**
          6 Calyx not reticulate-veined; corolla 12–18 mm
            7 Corolla purple, rarely pink or white; calyx-teeth ¼–¾ as long as tube   **3. officinalis**
            7 Corolla white; calyx-teeth about as long as tube   **4. balcanica**
      4 Stock without persistent rosettes of leaves; flowering stems terminal, usually not subscapose
        8 Verticillasters 2- to 4-flowered
          9 Lower branches not becoming spinose; leaves crenate, lanate   **35. mucronata**
          9 Lower branches persistent, becoming spinose; leaves entire, glabrescent but with sessile glands   **41. glutinosa**
        8 Verticillasters 6- to many-flowered
          10 Upper lip of corolla 2-fid   **17. acutifolia**
          10 Upper lip of corolla entire or emarginate
            11 Corolla pale yellow; verticillasters 2- to 12-flowered
              12 Corolla *c.* 15 mm; leaves oblong-lanceolate   **18. obliqua**
              12 Corolla *c.* 20 mm; leaves cordate-ovate   **19. decumbens**
            11 Corolla white, pink or purple
              13 Verticillasters *c.* 6-flowered   **39. euboica**
              13 Verticillasters 10- to many-flowered

[1] By P. W. Ball.

              14 Pedicels 1–2 mm in flower   **6. balansae**
              14 Flowers sessile or subsessile   **(7–16). germanica** group
1 Bracteoles absent, or not more than ½ as long as calyx-tube
  15 Annual
    16 Calyx-teeth ⅓–½ as long as tube
      17 Calyx with dense eglandular hairs and sessile glands   **53. annua**
      17 Calyx with glandular hairs but without sessile glands
        18 Stem ± scabrid on the angles; lower leaves cordate at base   **54. spinulosa**
        18 Stem smooth; lower leaves rounded or attenuate at base   **55. milanii**
    16 Calyx-teeth as long as or slightly shorter than tube
      19 Corolla 6–12 mm, scarcely exceeding calyx
        20 Calyx 5–7(–8) mm in flower; upper lip of corolla entire   **56. arvensis**
        20 Calyx 7–10 mm in flower; upper lip of corolla emarginate   **57. brachyclada**
      19 Corolla 12–20 mm, distinctly exceeding calyx
        21 Upper lip of corolla 2-fid; corolla yellow or white with a yellow lower lip   **51. ocymastrum**
        21 Upper lip of corolla entire; corolla pink   **52. marrubiifolia**
  15 Perennial herb or small shrub
    22 Most verticillasters 2-flowered, or flowers solitary
      23 Lower and middle leaves pinnatipartite with narrowly linear lobes; pedicels 2–5 mm   **32. angustifolia**
      23 Lower and middle leaves entire, linear to suborbicular; pedicels 0–2 mm
        24 Leaves suborbicular to ovate-orbicular, not more than 1½ times as long as wide   **50. corsica**
        24 Leaves linear to ovate, at least twice as long as wide
          25 Leaves ovate, cordate at base   **24. sylvatica**
          25 Leaves linear to lanceolate or oblong-spathulate, cuneate or rounded at base
            26 Calyx with sessile or shortly stipitate glands
              27 Plant with persistent, softly spinose branches; leaves entire, obtuse   **41. glutinosa**
              27 Plant without persistent, softly spinose branches; leaves crenate or crenate-dentate, acute   **53. annua**
            26 Calyx eglandular
              28 Calyx densely sericeous; stem appressed-hairy   **31. virgata**
              28 Calyx sparsely hairy; stem glabrous or with sparse patent hairs   **33. tetragona**
    22 Most verticillasters 3- to 16-flowered
      29 Small shrub with spinose branches   **42. spinosa**
      29 Herb or unarmed small shrub
        30 Corolla yellow, sometimes with white or purple markings
          31 Calyx eglandular
            32 Stem glabrous or with straight hairs
              33 Calyx-teeth less than 1 mm wide, linear-subulate, about as long as tube   **29. atherocalyx**
              33 Calyx-teeth more than 1 mm wide at base, triangular or triangular-lanceolate, shorter than tube
                34 Stems herbaceous at base; verticillasters 6- to 16-flowered, at least the upper crowded   **26. recta**
                34 Stems woody and much-branched at base; verticillasters 2- to 6-flowered, distant   **33. tetragona**
            32 Stem ± tomentose or lanate
              35 Stem and leaves green to grey-green, tomentose-hirsute; corolla 12–14 mm   **47. maritima**
              35 Stem and leaves white, densely tomentose or lanate-tomentose
                36 Corolla 15–20 mm; leaves linear-lanceolate to obovate-elliptical   **43. iva**
                36 Corolla 12–14 mm; leaves suborbicular to ovate-orbicular   **44. chrysantha**
          31 Calyx glandular
           37 Stock with persistent rosettes of leaves   **48. pubescens**
          37 Stock without persistent rosettes of leaves
            38 Stem and leaves with crispate hairs
              39 Lower leaves cuneate at base   **23. plumosa**
              39 Lower leaves cordate or truncate at base

40 Calyx 7–8 mm   **27. beckeana**
40 Calyx *c.* 10 mm   **21. anisochila**
38 Stem and leaves with ± straight hairs
  41 Calyx-teeth about as long as tube, all ± equal
    42 Calyx 10–15 mm; stem and leaves glandular
           23. plumosa
    42 Calyx 9–10 mm; stem and leaves eglandular
           28. parolinii
  41 At least the upper calyx-teeth shorter than tube
    43 Middle and lower leaves cordate at base **20. canescens**
    43 Middle and lower leaves cuneate to truncate at
      base
      44 Calyx 10–12 mm
        45 Stem and leaves densely glandular-puberulent;
           leaves 18–40 mm wide   **22. menthifolia**
        45 Stem and leaves eglandular or sparsely glandular;
           leaves not more than 20 mm wide   **26. recta**
      44 Calyx 5–10 mm
        46 Corolla 15–20 mm; verticillasters 6- to 16-
           flowered   **26. recta**
        46 Corolla 10–16 mm; verticillasters 2- to 6-flowered
                53. annua
30 Corolla white, pink or purple
  47 Leaves orbicular to ovate-orbicular, not more than 1½
    times as long as wide
    48 Stem and leaves densely hirsute, green or grey-green;
      leaves cordate at base   **40. circinata**
    48 Stem and leaves densely tomentose to sericeous-lanate,
      white; leaves rounded at base
      49 Calyx-teeth (2–)4 times as long as wide   **45. candida**
      49 Calyx-teeth 1½(–2) times as long as wide
              46. spreitzenhoferi
  47 Leaves linear-oblong to cordate-ovate, more than 1½
    times as long as wide
    50 Calyx with eglandular hairs
      51 Leaves cordate at base   **25. palustris**
      51 Leaves cuneate at base, rarely truncate
        52 Upper lip of corolla 2-fid   **49. arenaria**
        52 Upper lip of corolla entire or crenulate
          53 Middle and upper leaves linear-lanceolate, acute,
            serrate   **36. spruneri**
          53 Middle and upper leaves elliptical or ovate, sub-
            obtuse or obtuse, denticulate   **38. ionica**
    50 Calyx with glandular hairs
      54 Leaves cordate at base
        55 Leaves cordate-ovate, 2–3 times as long as wide,
          all petiolate   **24. sylvatica**
        55 Leaves oblong to oblong-lanceolate, usually at least
          4 times as long as wide, the lower shortly petiolate,
          the upper sessile   **25. palustris**
      54 Leaves cuneate, rarely truncate at base
      56 Leaves obtuse
        57 Stem and leaves ± lanate; pedicels 3–5 mm in flower;
          corolla scarcely exceeding calyx **35. mucronata**
        57 Stem and leaves sericeous-pubescent or hirsute;
          flowers sessile or subsessile; corolla 1½–2 times as
          long as calyx
          58 Upper leaves and bracts sessile, usually con-
            spicuously net-veined   **37. swainsonii**
          58 Upper leaves and bracts shortly petiolate, not
            conspicuously net-veined   **38. ionica**
      56 Leaves acute
        59 Calyx-teeth *c.* ½ as long as tube   **53. annua**
        59 Calyx-teeth about as long as tube
          60 Most leaves more than 10 mm wide, oblong-
            lanceolate to ovate, less than 4 times as long as
            wide   **23. plumosa**
          60 Middle and upper leaves usually not more than 10
            mm wide, linear-lanceolate to lanceolate, at least
            4 times as long as wide
            61 Stem with numerous sessile glands; leaves promi-
              nently veined beneath   **36. spruneri**
            61 Stem eglandular; leaves not prominently veined

    62 Calyx with moderately dense, long, appressed
      hairs; corolla purple, the lower lip 5–6 mm
               30. iberica
    62 Calyx glabrous or sparsely hirsute; corolla white
      or pale pink, the lower lip 8–9 mm
               34. leucoglossa

Sect. BETONICA (L.) Bentham. Perennial stock with persistent terminal rosettes of leaves; flowering stems lateral, usually subscapose. Verticillasters many-flowered; bracteoles as long as or longer than calyx-tube.

1. S. alopecuros (L.) Bentham, *Lab. Gen. Sp.* 531 (1834) (*S. jacquinii* Godron) Fritsch, *Betonica alopecuros* L., *B. jacquinii* Godron). Erect, hirsute perennial 20–60 cm. Leaves 30–100 × 30–70 mm, triangular- or ovate-cordate, coarsely crenate or crenate-dentate. Verticillasters in a dense spike, the lower sometimes distant. Calyx 7–8 mm, reticulate-veined, the teeth ⅓–½ as long as tube. Corolla 15–20 mm, pale yellow; tube not exceeding calyx; upper lip bifid. *Meadows, scrub and screes; calcicole. Mountains of C. & S. Europe, eastwards to E. Austria, C. Jugoslavia and S.E. Greece.* Al Au Ga Ge Gr He Hs It Ju.

2. S. monieri (Gouan) P. W. Ball, *Bot. Jour. Linn. Soc.* 65: 356 (1972) (*Betonica monieri* Gouan, *B. hirsuta* L., *S. densiflora* Bentham). Erect, hirsute perennial 10–40 cm. Leaves 20–80 × 10–30 mm, oblong, cordate at base, crenate. Verticillasters in a dense spike. Calyx 10–15 mm, reticulate-veined; teeth ⅓–½ as long as tube. Corolla 20–24 mm, pink; tube exceeding calyx; upper lip entire. *Dry pastures.* ● *Alps and Pyrenees.* Au Ga He Hs It Ju ?Rm.

3. S. officinalis (L.) Trevisan, *Prosp. Fl. Euganea* 26 (1842) (*S. betonica* Bentham, *Betonica officinalis* L.; incl. *S. bulgarica* (Degen & Nejc.) Hayek). Erect, subglabrous to densely hirsute perennial 15–100 cm. Leaves 30–120 × 15–50 mm, oblong to ovate-oblong, cordate at base, coarsely crenate or crenate-dentate. Verticillasters in a dense spike, sometimes interrupted below. Calyx 5–9(–12) mm; teeth ¼–¾ as long as tube. Corolla 12–18 mm, bright reddish-purple, rarely pink or white; tube exceeding calyx; upper lip entire. $2n = 16$. *Most of Europe northwards to C. Scotland, S. Sweden and N.W. Russia.* Al Au Be Br Bu Cz Da Ga Ge Gr Hb He Ho Hs Hu It Ju Lu Po Rm Rs (N, B, C, W, K, E) Su Tu [Fe No].

4. S. balcanica P. W. Ball, *Bot. Jour. Linn. Soc.* 65: 262 (1972) (*S. haussknechtii* (Uechtr. ex Hausskn.) Hayek, non Vatke, *Betonica haussknechtii* Uechtr. ex Hausskn.). Like **3** but corolla white; calyx-teeth about as long as tube. *Woods.* ● *C. Greece; S.E. Bulgaria.* Bu Gr.

5. S. scardica (Griseb.) Hayek, *Prodr. Fl. Penins. Balcan.* 2: 280 (1929). Erect, hirsute perennial 30–60 cm. Leaves 40–90 × 20–40 mm, the lower ovate-lanceolate, the middle and upper oblong-lanceolate, coarsely crenate-serrate or -dentate, slightly cordate at base, densely grey-stellate-tomentose beneath. Verticillasters in a lax spike. Calyx 9–15 mm; teeth *c.* ½ as long as tube. Corolla 15–18 mm, white, sometimes tinged with pink; tube about as long as calyx; upper lip entire. *Subalpine meadows and scrub.* ● *Balkan peninsula.* Al Bu Gr Ju.

Sect. ERIOSTOMUM (Hoffmanns. & Link) Dumort. Perennial stock without persistent terminal rosettes of leaves; flowering stems terminal. Verticillasters usually many-flowered; bracteoles as long as or longer than calyx-tube.

6. S. balansae Boiss. & Kotschy in Boiss., *Fl. Or.* 4: 722 (1879) (incl. *S. heterodonta* Zefirov). Erect, grey, patent- or crispate-hairy perennial 60–100 cm. Leaves 30–100 × 15–50 mm, oblong-ovate, truncate or more or less cordate at base, crenate-dentate, sparsely hairy above, tomentose beneath. Verticillasters many-flowered, mostly distant; pedicels 1–2 mm in flower. Calyx 8–9 mm; teeth unequal, the uppermost *c.* ½ as long as tube. Corolla *c.* 13 mm, pink, densely villous; upper lip entire. *Meadows and wood-margins. Krym.* Rs (K). (*Caucasian region.*)

(7–16). **S. germanica group.** Erect, grey- or white-hairy perennials (10–)20–100(–120) cm. Leaves usually crenate or crenate-serrate. Verticillasters many-flowered, usually distant; flowers sessile or subsessile. Calyx 6–12 mm; teeth equal or unequal. Corolla 15–25 mm, usually pink or purple, densely hairy; upper lip entire or emarginate.

A taxonomically difficult group in need of revision. Species **8–15** are often difficult to distinguish from each other and some authors have considered them to be variants of a single species.

1 Stem with glandular hairs   **7. alpina**
1 Stem without glandular hairs
  2 Most leaves ±cuneate at base
    3 Calyx-teeth recurved after flowering, usually glandular
                            12. thirkei
    3 Calyx-teeth not recurved, eglandular
      4 Upper surface of leaves moderately densely hairy, the surface more or less visible beneath the indumentum
                            14. cretica
      4 Upper surface of leaves densely sericeous-lanate, the surface completely obscured by the indumentum   **15. byzantina**
  2 Most leaves truncate or cordate at base
    5 Calyx-teeth 5–7 mm, with an arista 1·5–3 mm   **13. cassia**
    5 Calyx-teeth 3–5 mm, with an arista less than 1·5(–2) mm
      6 Leaves green beneath   **16. heraclea**
      6 Leaves grey-green to white beneath
        7 Whole plant very densely white-tomentose or lanate-tomentose; calyx-teeth all equal
          8 Stem usually more than 30 cm; upper bracts shorter than flowers; bracteoles lanceolate   **10. tournefortii**
          8 Stem 15–30 cm; upper bracts exceeding flowers; bracteoles subulate   **11. sericophylla**
        7 At least the upper surface of the leaves grey-green and only moderately densely tomentose; upper 2 calyx-teeth usually longer than the lower 3
          9 Calyx-teeth eglandular, not recurved in fruit   **8. germanica**
          9 Calyx-teeth glandular, usually ±recurved in fruit
                            9. tymphaea

7. S. alpina L., *Sp. Pl.* 581 (1753). Stems 30–100 cm, hirsute or hirsute-tomentose and with glandular hairs at least above. Leaves 50–180 × 30–90 mm, oblong-ovate or ovate, cordate at base, grey or grey-green. Calyx 6–12 mm, glandular; teeth unequal, the upper 2 *c.* ½ as long as tube. Corolla 15–22 mm, dull purple, rarely tinged with yellow. *Shady places, mainly in the mountains. W., C. & S. Europe, northwards to Wales.* Al Au Be Br Bu Cz Ga Ge Hs Hu It Ju Po Rm Rs (W).

8. S. germanica L., *Sp. Pl.* 581 (1753). Stems 30–100(–120) cm, tomentose or lanate-tomentose, eglandular. Leaves 30–120 × 10–50 mm, oblong to oblong-ovate, cordate, grey-tomentose beneath, green above. Calyx 6–12 mm, eglandular; teeth unequal, the upper 2 ⅓–½ as long as tube. Corolla 15–20 mm. $2n = 30$. *W., C. & S. Europe northwards to S. England, and extending eastwards to c. 38° E. in S.C. Russia.* Al Au Be Bl Br Bu Co Cz Ga Ge Gr He Hs Hu It Ju Lu Po Rm Rs (C, W, K) Sa Si Tu.

1 Lowest bracts triangular-ovate, cordate, widest at the base
                            (c) subsp. lusitanica
1 Lowest bracts cuneate at base, not cordate, widest above the base
  2 Calyx-teeth *c.* ½ as long as tube   **(a) subsp. germanica**
  2 Calyx-teeth *c.* ⅓ as long as tube   **(b) subsp. heldreichii**

(a) Subsp. **germanica**: Lowest bracts cuneate at base, widest above the base. Calyx-teeth *c.* ½ as long as tube. *Almost throughout the range of the species.*

(b) Subsp. **heldreichii** (Boiss.) Hayek, *Prodr. Fl. Penins. Balcan.* 2: 285 (1929): Like subsp. (a) but calyx-teeth *c.* ⅓ as long as tube. *S. part of Balkan peninsula.*

(c) Subsp. **lusitanica** (Hoffmanns. & Link) Coutinho, *Fl. Port.* 520 (1913): Lowest bracts triangular-ovate, cordate, widest at the base. *S. part of Iberian peninsula.*

9. S. tymphaea Hausskn., *Mitt. Thür. Bot. Ver.* Ser. 1, **5**: 70 (1886) (*S. reinertii* Heldr. ex Halácsy). Like **8** but stems 10–50 cm; leaves 30–60 × 10–35 mm, oblong-ovate or ovate; calyx-teeth glandular, usually recurved in fruit. ● *C. & S. Italy; W. part of Balkan peninsula.* Al Gr It Ju.

10. S. tournefortii Poiret in Lam., *Encycl. Méth. Bot.* 13: 227 (1817). Stems 30–100 cm, very densely white-lanate-tomentose, eglandular. Leaves 50–75 × 30–50 mm, ovate or ovate-oblong, cordate, densely white-tomentose. Upper bracts shorter than flowers; bracteoles lanceolate. Calyx 6–12 mm, eglandular; teeth equal, *c.* ⅓ as long as tube. Corolla 15–20 mm. *Dry, rocky places. Kriti.* Cr. (*Libya.*)

11. S. sericophylla Halácsy, *Consp. Fl. Graec.* **2**: 519 (1902). Like **10** but stems not more than 30 cm; all bracts exceeding the flowers; bracteoles subulate. *Mountain rocks.* ● *S. Albania.* Al.

12. S. thirkei C. Koch, *Linnaea* 21: 685 (1849). Stems 30–100 cm, densely white-tomentose, eglandular. Leaves 30–60 × 10–20 mm, oblong-lanceolate, cuneate at base, densely white-tomentose or sometimes grey-green and only moderately densely tomentose above. Verticillasters sometimes crowded into a short, dense spike. Calyx *c.* 10 mm; teeth slightly unequal, glandular, recurved after flowering, the uppermost *c.* ½ as long as tube. Corolla *c.* 15 mm. *Dry, rocky places. Balkan peninsula; S. Italy.* Bu Gr It Ju Tu.

13. S. cassia (Boiss.) Boiss., *Diagn. Pl. Or. Nov.* 2(13): 74 (1853). Stems 30–100 cm, tomentose or lanate-tomentose, eglandular. Leaves 30–120 × 15–60 mm, oblong-ovate to ovate, more or less cordate, green and sparsely pubescent or tomentose above, grey-green and moderately densely tomentose beneath. Calyx 10–12 mm; teeth eglandular, slightly unequal, almost as long as tube, with an arista 1·5–3 mm. Corolla 15–20 mm. *Dry, rocky places. C. part of Balkan peninsula.* Bu Gr Ju.

14. S. cretica L., *Sp. Pl.* 581 (1753) (*S. italica* Miller). Stems 20–80 cm, tomentose to lanate. Leaves 30–100 × 10–30 mm, oblong-ovate to ovate, cuneate at base, rarely the basal truncate or subcordate, tomentose and grey-green above, densely tomentose or lanate, grey or white beneath. Calyx 6–12 mm; teeth usually eglandular, slightly unequal, *c.* ½ as long as tube. Corolla 15–20 mm. *S. Europe from S. France eastwards.* Al Bu Cr Ga Gr It Ju Rs (K) Si Tu.

1 Calyx-teeth with arista 1–2 mm   **(c) subsp. bulgarica**
1 Calyx-teeth with arista not more than 1 mm
  2 Lowest leaves at least 3 times as long as wide, cuneate at base   **(a) subsp. cretica**
  2 Lowest leaves usually less than 3 times as long as wide, often rounded or subcordate at base   **(b) subsp. salviifolia**

(a) Subsp. **cretica**: Stems usually densely lanate or lanate-tomentose. Leaves at least 3 times as long as wide, cuneate at base. Verticillasters distant. Calyx-teeth with arista not more than 1 mm. *C. & S. Greece and Aegean region; Krym.*

(b) Subsp. **salviifolia** (Ten.) Rech. fil., *Ann. Naturh. Mus. (Wien)* 48: 170 (1937) (*S. germanica* subsp. *salviifolia* (Ten.) Gams): Stems densely tomentose or lanate-tomentose. Leaves usually less than 2–3 times as long as wide, the lowest often rounded or subcordate at base. Verticillasters crowded, or the lower distant. Calyx-teeth with arista not more than 1 mm. *C. Mediterranean region.*

(c) Subsp. **bulgarica** Rech. fil., *op. cit.* 171 (1937) (*?S. thracica* Davidov): Stems shortly tomentose. Leaves 3(–4) times as long as wide, attenuate at base, rarely the lowest subcordate. Verticillasters usually crowded. Calyx-teeth with arista 1–2 mm. *E. part of Balkan peninsula.*

15. **S. byzantina** C. Koch, *Linnaea* 21: 686 (1849) (*S. lanata* Jacq., non Crantz, *S. olympica* auct., vix Poiret). Stems 15–80 cm, densely white-lanate-tomentose. Leaves (6–)30–100 × (3–)15–40 mm, the lower oblong-spathulate, the upper elliptical, all attenuate at base and very densely white-sericeous-lanate. Calyx 8–12 mm; teeth *c.* ⅓ as long as tube. Corolla 15–25 mm. *Turkey-in-Europe (N. of Istanbul); widely cultivated for ornament and locally naturalized.* Tu [Ga]. (*S.W. Asia.*)

16. **S. heraclea** All., *Fl. Pedem.* 1: 31 (1785). Stems 20–60 cm, rather sparsely hirsute or lanate-hirsute, eglandular. Leaves 30–90 × 10–30 mm, oblong or lanceolate, subcordate or truncate, green and sparsely villous or lanate-villous above, more densely lanate-villous beneath. Calyx 9–14 mm; teeth unequal, *c.* ½ as long as tube, glandular. Corolla 15–20 mm. *Dry hillsides.* ● *W. Mediterranean region.* Co Ga Hs It Si.

17. **S. acutifolia** Bory & Chaub. in Bory, *Expéd. Sci. Morée* 3(2): 168 (1832). Erect, lanate-hirsute perennial 15–70 cm. Leaves 20–60 × 10–25 mm, ovate-oblong, rounded or slightly cordate at base, crenate-dentate, usually green above and grey-green beneath. Verticillasters many-flowered, distant. Calyx 9–12 mm; teeth slightly unequal, *c.* ½ as long as tube, sometimes glandular. Corolla 18–25 mm, pink, densely white-hirsute; upper lip 2-fid; lower lip white with pink markings. *C. & S. Greece.* Gr ?It.

18. **S. obliqua** Waldst. & Kit., *Pl. Rar. Hung.* 2: 142 (1803–1804). Erect, lanate-hirsute perennial 20–50 cm. Leaves 40–60 × 15–25 mm, oblong-lanceolate, rounded or slightly cordate at base, crenate, green above, green or grey-green beneath. Verticillasters 6- to 10-flowered, crowded or the lower distant. Calyx *c.* 10 mm; teeth *c.* ⅔ as long as tube. Corolla *c.* 15 mm, pale yellow, densely hairy; upper lip *c.* 6 mm; lower lip *c.* 8 mm. *Balkan peninsula.* Al Bu Gr Ju Tu.

Sect. STACHYS. Perennial herbs or small shrubs; stock without persistent terminal rosettes of leaves; flowering stems terminal. Verticillasters usually not more than 12-flowered; bracteoles usually absent or minute.

19. **S. decumbens** Pers., *Syn. Pl.* 2: 123 (1806). Decumbent perennial up to 40 cm, branched and woody at base; flowering stems lanate-hirsute, glandular. Leaves 25–35 × 20–25 mm, cordate-ovate, crenate-serrate, grey. Verticillasters 2- to 12-flowered, crowded into a dense, ovoid spike; bracteoles about as long as calyx-tube. Calyx 11–15 mm; teeth *c.* ½ as long as tube. Corolla *c.* 20 mm, pale yellow, densely hairy; upper lip *c.* 4 mm; lower lip *c.* 10 mm. *Cliffs.* ● *N.W. Greece, S. Albania.* Al Gr.

20. **S. canescens** Bory & Chaub. in Bory, *Expéd. Sci. Morée* 3(2): 167 (1832) (*S. messeniaca* Boiss.). More or less erect, hirsute or lanate-hirsute perennial 10–30 cm, woody at base.

Leaves 15–50 × 8–30 mm, elliptical to orbicular-ovate, cordate at base, crenate, hirsute. Verticillasters 4- to 6-flowered, in a dense spike. Calyx 7–12 mm, glandular-hairy; teeth *c.* ½ as long as tube, slightly unequal. Corolla 15–20 mm, pale yellow, pubescent; upper lip 5–6 mm; lower lip 8–10 mm. *Cliffs.* ● *S. Greece.* Gr.

21. **S. anisochila** Vis. & Pančić, *Mem. Ist. Veneto* 15: 13 (1870). Erect, densely deflexed-hirsute perennial *c.* 50 cm. Leaves 20–40 × 10–20 mm, oblong-ovate or ovate, cordate or truncate at base, crenate-serrate, pubescent. Verticillasters 4- to 10-flowered. Calyx *c.* 10 mm, with sessile glands; lower 2 teeth *c.* 5 mm, about as long as tube; upper 3 teeth 2–3 mm and connate at base. Corolla 15–20 mm, pale yellow, pubescent; upper lip 4–5 mm; lower lip 7–10 mm. *Rocky places.* ● *S. & E. Bulgaria; W. Jugoslavia and Albania.* Al Bu Ju.

22. **S. menthifolia** Vis., *Flora (Regensb.)* 12 (Erganz. 1): 14 (1829) (*S. grandiflora* Host). Erect, sparsely hairy and densely glandular-puberulent perennial 30–50 cm. Leaves 40–80 × 18–40 mm, lanceolate or ovate, rounded or cuneate at base, crenate-serrate, glandular-puberulent and sparsely pubescent. Verticillasters 6- to 10-flowered, crowded. Calyx 10–12 mm, glandular-puberulent and sparsely pubescent; lower 2 teeth 4–5 mm, *c.* ⅔ as long as tube; upper 3 teeth 2–3 mm, connate at base. Corolla 18–20 mm, pale yellow, pubescent; upper lip 5–6 mm; lower lip *c.* 10 mm. *Rocky places.* ● *W. part of Balkan peninsula.* Al Gr Ju.

23. **S. plumosa** Griseb., *Spicil. Fl. Rumel.* 2: 139 (1844) (incl. *S. freynii* Hausskn., *S. viridis* Boiss. & Heldr.). Erect, glandular-pubescent and sparsely to densely hirsute perennial 30–80 cm. Leaves 25–100 × 8–50 mm, ovate or oblong-ovate, acute, cuneate or rounded at base, crenate to serrate, glandular-puberulent and sparsely hirsute. Verticillasters 6- to 16-flowered, the upper crowded, the lower distant. Calyx 9–15 mm, sparsely to densely hairy and glandular-puberulent; teeth equal, about as long as tube. Corolla 14–20 mm, white or pale pink or reddish-brown with pink or purple markings, possibly also pale yellow with purple markings, pubescent; upper lip 4–5 mm; lower lip 7–10(–12) mm. *Dry pastures and mountain rocks.* ● *S. Jugoslavia, W. Bulgaria, N. & C. Greece.* Bu Gr Ju.

24. **S. sylvatica** L., *Sp. Pl.* 580 (1753). Erect, hirsute and glandular-pubescent perennial 30–120 cm. Leaves 40–140 × 20–80 mm, cordate-ovate, acute, crenate-serrate, sparsely hirsute, all petiolate. Verticillasters (2–)6(–8)-flowered, more or less crowded. Calyx 6–8 mm, glandular-pubescent; teeth equal, as long as or slightly shorter than tube. Corolla 13–18 mm, dull reddish-purple with white markings, rarely white or pale pink, puberulent; upper lip 4–5 mm; lower lip 6–7 mm. $2n = 66$. *Shady places. Most of Europe, but rare in the Mediterranean region.* All except Az Bl Cr Fa Is Sa Sb.

25. **S. palustris** L., *Sp. Pl.* 580 (1753) (incl. *S. maeotica* Postr., *S. wolgensis* Wilensky). Erect, sparsely to densely hairy, usually eglandular perennial 30–120 cm. Leaves 30–120 × 7–35 mm, oblong or oblong-lanceolate, acute, cordate at base, crenate, appressed-pubescent, the lower shortly petiolate, the upper sessile. Verticillasters usually 4- to 10-flowered, the upper crowded, the lower distant. Calyx 6–8 mm, usually eglandular; teeth equal, as long as or slightly shorter than tube. Corolla (11–)12–15 mm, purple, puberulent; upper lip 3–4 mm; lower lip 5–7 mm. $2n = 64, 102$. *Damp places, and as a weed in cultivated fields. Most of Europe, but rare in the Mediterranean region.* All except Az Bl Cr Fa Is Sa Sb Si.

Hybrids between **24** and **25** and hybrid swarms derived from these have been recorded from many parts of Europe.

26. S. recta L., *Mantissa* 82 (1767) (*S. czernjajevii* Schost.). Erect or ascending, subglabrous to sparsely hirsute, usually eglandular perennial 15–100 cm. Leaves 10–80 × 1–20 mm, the lower oblong to ovate, rounded or cuneate at base, the upper linear to ovate-oblong, glabrous to hirsute, entire to crenate-serrate. Verticillasters 6- to 16-flowered, crowded or the lower distant. Calyx 5–9(–11) mm; teeth shorter than tube. Corolla 15–20 mm, pale yellow, pubescent; upper lip 4–7 mm; lower lip 5–12 mm. *Dry places. Europe northwards to Belgium and C. Russia, but absent from the islands.* Al Au Be Bu Cz Ga Ge Gr He Hs Hu It Ju Po Rm Rs (C, W, K, E) Tu.

A very variable species containing a large number of subordinate taxa. The following subspecies appear to be the most widespread and distinct of these, but many local populations showing different combinations of characters also occur.

1 Calyx 5–7 mm; lower lip of corolla 5–7 mm   **(a) subsp. recta**
1 Calyx 7–9(–11) mm; lower lip of corolla 7–12 mm
2 Middle and upper leaves 7–10 mm wide, oblong to obovate, crenate or crenate-serrate   **(b) subsp. labiosa**
2 Middle and upper leaves 1–6 mm wide, linear or linear-lanceolate, entire or weakly crenate   **(c) subsp. subcrenata**

(a) Subsp. recta: Middle and upper leaves 20–80 × 5–20 mm, oblong to elliptical or oblanceolate, crenate or crenate-serrate. Calyx 5–7 mm, eglandular; teeth more or less equal. Lower lip of corolla 5–7 mm. 2*n* = 32, 34. *Throughout the range of the species.*

(b) Subsp. labiosa (Bertol.) Briq., *Lab. Alp. Marit.* 259 (1893): Middle and upper leaves 10–30 × 6–10 mm, oblong or elliptical to obovate, crenate or crenate-serrate. Calyx 7–10 mm, often glandular; teeth unequal. Lower lip of corolla 8–12 mm. 2*n* = 34. *Alps, Appennini and W. part of Balkan peninsula.*

(c) Subsp. subcrenata (Vis.) Briq., *op. cit.* 257 (1893): Middle and upper leaves 10–60 × 1–6 mm, linear or linear-lanceolate, entire or weakly crenate. Calyx 7–9(–11) mm, often glandular; teeth unequal. Lower lip of corolla 7–12 mm. *S.E. Europe.*

S. albanica Markgraf, *Ber. Deutsch. Bot. Ges.* **44**: 428 (1926), from C. Albania, may be a variant of **26**, **28** or **29**. It has the calyx *c*. 10 mm, with equal teeth about as long as tube, and corolla *c*. 20 mm, white, the upper lip *c*. 7 mm, the lower lip *c*. 10 mm.

27. S. beckeana Dörfler & Hayck, *Denkschr. Akad. Wiss. Math.-Nat. Kl.* (*Wien*) **94**: 186 (1918). Like **26** but stem and leaves sparsely lanate; leaves 15–25 × 6–12 mm, ovate, rounded or subcordate at base; calyx 7–8 mm, densely villous and with sessile glands; lower lip of corolla 8–10 mm. *Cliffs and screes.* ● *S.W. Jugoslavia, N. Albania.* Al Ju.

Possibly not specifically distinct from **26**.

28. S. parolinii Vis., *Mem. Ist. Veneto* **1**: 46 (1843). Like **26** but the stem with strongly deflexed hairs; leaves 20–40 × 7–20 mm, lanceolate or elliptical; verticillasters 4- to 6-flowered; calyx 9–10 mm, hirsute and with short, glandular hairs; teeth slightly unequal, upper 2 about as long as tube; upper lip of corolla *c*. 4 mm, the lower lip 9–10 mm. *Rocky places.* ● *W. Greece.* Gr.

29. S. atherocalyx C. Koch, *Linnaea* **21**: 691 (1849). Like **26** but leaves 50–60 × 7–10 mm, linear or lanceolate, crenate-serrate; calyx 10–13 mm, the teeth not more than 1 mm wide, linear-subulate, about as long as tube. *Dry places. Balkan peninsula, S. Romania.* Al Bu Gr Ju Rm.

30. S. iberica Bieb., *Fl. Taur.-Cauc.* **2**: 51 (1808). Erect, glabrous to sparsely appressed-pubescent perennial 25–60 cm. Leaves 25–30 × 6–10 mm, oblong to lanceolate, acute, cuneate at base, crenate, sparsely hairy. Verticillasters 6- to 10-flowered, mostly distant. Calyx 9–10 mm, densely appressed-hairy and with glandular hairs; teeth slightly shorter to slightly longer than tube. Corolla *c*. 15 mm, purple, hirsute; upper lip 4–5 mm; lower lip 5–6 mm. *Dry hillsides and mountain scrub. Krym.* ?Gr Rs (K). (*Caucasian region.*)

31. S. virgata Bory & Chaub. in Bory, *Expéd. Sci. Morée* **3**(2): 166 (1832). Erect, appressed-hairy perennial 60–90 cm. Leaves *c*. 25 × *c*. 5 mm, linear, the lower obtuse, cuneate at base, the upper acute, entire, appressed sericeous-lanate. Verticillasters 2-flowered, distant. Calyx *c*. 10 mm, densely sericeous; teeth *c*. ¼ as long as tube. Corolla *c*. 10 mm, white, sericeous. *Dry, rocky places.* ● *S. Greece.* Gr.

A little-known species requiring further investigation.

32. S. angustifolia Bieb., *Fl. Taur.-Cauc.* **2**: 52 (1808). Erect, glabrous perennial 30–60 cm, woody at base. Lower and middle leaves pinnatipartite with linear or filiform lobes; upper leaves simple, narrowly linear. Verticillasters 2-flowered, distant; pedicels 2–5 mm. Calyx 6–8 mm, glabrous or sparsely hairy; teeth *c*. ½ as long as tube. Corolla 12–16 mm, yellow tinged with pink, pubescent; upper lip 3–4 mm; lower lip 5–8 mm. *Dry places.* ● *S.E. Europe.* Bu Gr Rm Rs (?C, W, K) Tu.

33. S. tetragona Boiss. & Heldr. in Boiss., *Fl. Or.* **4**: 736 (1879). Erect, glabrous or sparsely hairy perennial up to 35 cm, much branched and woody at base. Lower leaves 20–40 × 8–10 mm, oblong-spathulate, obtuse, crenate-dentate; upper leaves 15–25 × 2–3 mm, linear; all appressed-hairy. Verticillasters 2- to 6-flowered, distant. Calyx 6–9 mm, sparsely hairy; teeth *c*. ⅓ as long as tube. Corolla *c*. 15 mm, pale yellow with purple markings; upper lip *c*. 6 mm; lower lip *c*. 8 mm. *Cliffs.* ● *E.C. Greece.* Gr.

34. S. leucoglossa Griseb., *Spicil. Fl. Rumel.* **2**: 140 (1844) (incl. *S. patula* var. *samothracica* Degen). Erect, pubescent perennial (10–)30–50 cm. Lower leaves 15–30 × 5–12 mm, broadly elliptical; middle and upper leaves 15–50 × 2·5–6 mm, lanceolate, acute, cuneate at base; all pubescent, crenate-serrate to subentire. Verticillasters usually 6-flowered, in a lax spike or the lower distant. Calyx 6–9 mm, glandular, glabrous or sparsely pubescent; teeth about as long as tube. Corolla 16–18 mm, white or pale pink with purple markings, pubescent; upper lip *c*. 5 mm; lower lip 8–9 mm. ● *C. part of Balkan peninsula.* Bu Gr Ju.

35. S. mucronata Sieber ex Sprengel, *Syst. Veg.* **2**: 733 (1825). Erect, sparsely lanate perennial 10–50 cm, branched and woody at base. Leaves 20–60 × 5–20 mm, oblong, obtuse, rounded at base, crenate, lanate. Verticillasters 2- to 4-flowered, distant or somewhat crowded; bracteoles about as long as calyx-tube; pedicels 3–8 mm. Calyx 9–12 mm, hairy and with sessile glands; teeth about as long as tube. Corolla 10–12 mm, pale pink, densely hairy; both lips *c*. 5 mm. 2*n* = 30. *Cliffs and other dry, open habitats.* ● *Kriti and Karpathos.* Cr.

36. S. spruneri Boiss. in DC., *Prodr.* **12**: 488 (1848). Erect, sparsely hairy and densely glandular perennial 15–30 cm, woody at base. Lower leaves *c*. 30 × *c*. 10 mm, oblong-elliptical; middle and upper leaves 15–50 × 3–8 mm, linear-lanceolate, acute, cuneate at base, serrate, sparsely hispid. Verticillasters 4- to

6-flowered, crowded into a dense, ovoid spike. Calyx 9–12 mm, hirsute, glandular; teeth about as long as tube. Corolla 16–18 mm, white; upper lip *c.* 5 mm; lower lip *c.* 10 mm. *Cliffs.* ● *S.E. Greece.* Gr.

37. **S. swainsonii** Bentham, *Lab. Gen. Sp.* 535 (1834). Procumbent to erect, sericeous-pubescent to lanate-hirsute perennial 10–20 cm, branched and woody at base. Leaves 10–30 × 5–18 mm, oblong or elliptical to obovate, obtuse, cuneate at base, usually conspicuously net-veined, crenate or denticulate; upper leaves and bracts sessile. Verticillasters 2- to 10-flowered, crowded. Calyx 6–10 mm, hirsute or lanate-hirsute and glandular-hairy; teeth as long as or slightly shorter than tube. Corolla 14–18 mm, white or pale pink with pink markings; upper lip 2·5–5 mm; lower lip 5–8 mm. *Rock-crevices.* ● *S. & S.C. Greece.* Gr.

1 Lanate or lanate-hirsute; lower leaves not more than 15 mm
                        (c) subsp. **scyronica**
1 Sericeous-pubescent or hirsute; lower leaves up to 30 mm
  2 Verticillasters 2- to 6-flowered; hairs on stem and leaves
    ± appressed   (a) subsp. **swainsonii**
  2 Verticillasters 4- to 10-flowered; hairs on stem and leaves
    ± patent   (b) subsp. **argolica**

(a) Subsp. **swainsonii**: Plant sericeous-pubescent or hirsute, the hairs more or less appressed. Lower leaves up to 30 mm. Verticillasters 2- to 6-flowered. $2n = 34$. *S.C. Greece (Parnassos).*
(b) Subsp. **argolica** (Boiss.) Phitos & Damboldt, *Ber. Deutsch. Bot. Ges.* **82**: 600 (1969) (*S. argolica* Boiss.): Like subsp. (a) but the hairs more or less patent. Verticillasters 4- to 10-flowered. $2n = 34$. *S.E. Greece (Argolis).*
(c) Subsp. **scyronica** (Boiss.) Phitos & Damboldt, *loc. cit.* (1969): Plant lanate-hirsute. Lower leaves not more than 15 mm. *S.E. Greece (near Megara).*

38. **S. ionica** Halácsy, *Consp. Fl. Graec., Suppl.* 85 (1908). Like 37 but lanate or lanate-hirsute; leaves ovate or ovate-elliptical, truncate or rounded at base, not conspicuously net-veined; upper leaves and bracts petiolate; calyx with mostly sessile glands. *Cliffs, rocks.* ● *W. Greece (Ionioi Nisoi).* Gr.

39. **S. euboica** Rech. fil., *Bot. Jahrb.* **80**: 391 (1961). Ascending, hirsute and densely glandular perennial 15–20 cm, branched and woody at base. Lower leaves up to 30 × 20 mm, obovate; middle and upper leaves 20–50 × 10–25 mm, oblong or elliptic-lanceolate; all obtuse or subacute, cuneate at base, crenate to serrate. Verticillasters *c.* 6-flowered, crowded; bracteoles about as long as calyx-tube. Calyx 11–14 mm; teeth as long as or slightly longer than tube. Corolla white. $2n = 34$. *Limestone cliffs.* ● *S.E. Greece (S. Evvoia).* Gr.

40. **S. circinata** L'Hér., *Stirp. Nov.* 51 (1786). Erect, densely hirsute perennial 20–45 cm. Leaves 20–50 × 15–50 mm, ovate-orbicular, obtuse, deeply cordate at base, crenate, densely hirsute. Verticillasters *c.* 6-flowered, more or less distant. Calyx 8–10 mm, villous and sparsely glandular; teeth *c.* ⅔ as long as tube. Corolla 15–20 mm, white to bright pinkish-purple, pubescent; upper lip 5–8 mm; lower lip 10–12 mm. *Cliffs and screes.* S. Spain. Hs.

41. **S. glutinosa** L., *Sp. Pl.* 581 (1753). Small shrub 10–60 cm, the lower branches persistent and becoming softly spinose; stems with sparse, patent hairs and sessile glands. Leaves 15–35 × 3–5 mm, oblong-spathulate to linear-lanceolate, obtuse, cuneate at base, entire, glabrescent but with dense, sessile glands. Verticillasters 2-flowered, or flowers solitary, remote. Calyx 7–11 mm, glabrescent but glandular; teeth *c.* ¾ as long as tube. Corolla 10–13 mm, white or violet, pubescent; upper lip

2·5–4 mm; lower lip 4–6 mm. $2n = 32$. *Dry places.* ● *Corse, Sardegna, Capraia.* Co It Sa.

42. **S. spinosa** L., *Sp. Pl.* 581 (1753). Tussock-forming perennial up to 30 cm; stems spinose, appressed-sericeous-lanate and often glandular. Leaves 10–35 × 3–8 mm, caducous, linear-oblong, entire or crenate, densely sericeous-lanate. Verticillasters 4- to 6-flowered, usually 2–3 crowded together. Calyx 6–9 mm, hirsute and with sessile glands; teeth *c.* ¾ as long as tube. Corolla *c.* 15 mm, pale pink, sericeous; upper lip *c.* 5 mm; lower lip 6–8 mm. *Dry, stony places and cliffs.* ● *S. Aegean region.* Cr Gr.

43. **S. iva** Griseb., *Spicil. Fl. Rumel.* **2**: 143 (1844). Ascending, white-lanate perennial 15–40 cm, woody at base. Leaves 15–60 × 4–15 mm, linear-lanceolate to obovate-elliptical, obtuse, cuneate at base, crenate, densely white-lanate. Verticillasters 4- to 6-flowered, crowded or the lower distant. Calyx 8–12 mm, white-tomentose; teeth slightly shorter to as long as tube. Corolla 15–20 mm, yellow, tomentose; upper lip 4–6 mm; lower lip 7–10 mm. *Stony slopes, rocks and cliffs; calcicole.* ● *S. Jugoslavia, N. Greece.* Gr Ju.

44. **S. chrysantha** Boiss. & Heldr. in Boiss., *Diagn. Pl. Or. Nov.* **1**(7): 56 (1846). Procumbent or ascending, white-tomentose perennial 10–20 cm, woody and branched at base. Leaves 10–25 × 8–15 mm, suborbicular or ovate-orbicular, white-lanate-tomentose, slightly crenate. Verticillasters 4- to 6-flowered, more or less crowded. Calyx 6–9 mm, white-tomentose, eglandular; teeth ½ as long to as long as tube, *c.* 3 times as long as wide. Corolla 12–14 mm, yellow, tomentose; upper lip 4–6 mm; lower lip 6–8 mm. *Cliffs and screes.* ● *S. Greece (Lakonia).* Gr.

45. **S. candida** Bory & Chaub. in Bory, *Expéd. Sci. Morée* **3**(2): 167 (1832). Like 44 but calyx-teeth (2–)4 times as long as wide; corolla 15–18 mm, white with purple spots. ● *S. Greece (Taïyetos).* Gr.

46. **S. spreitzenhoferi** Heldr., *Österr. Bot. Zeitschr.* **30**: 344 (1880). Like 44 but calyx sometimes glandular-hairy, the teeth 1½(–2) times as long as wide; corolla white with pink spots. *Cliffs.* ● *S. Greece.* Gr.

47. **S. maritima** Gouan, *Fl. Monsp.* 91 (1764). Erect or ascending, tomentose-hirsute perennial 10–30 cm, with persistent rosettes of leaves. Leaves 10–30 × 6–20 mm, oblong or elliptic-oblong to broadly ovate, obtuse, rounded or cuneate at base, crenate, villous-tomentose. Verticillasters 4- to 6-flowered, crowded. Calyx 5–10 mm, villous-tomentose; teeth *c.* ½ as long as tube. Corolla 12–14 mm, pale yellow, tomentose; upper lip 4–5 mm; lower lip 5–6 mm. *Maritime sands. Coasts of the Mediterranean and Black Sea.* Al Bu Co Ga Hs It Ju Tu.

48. **S. pubescens** Ten., *Fl. Nap.* 1, *Prodr.*: 34 (1811). Erect or ascending, appressed-pubescent perennial 20–40 cm, with persistent rosettes of leaves. Leaves *c.* 25 × 4–5 mm, oblong to oblong-ovate, obtuse, rounded or cuneate at base, entire, pubescent. Verticillasters 4- to 6-flowered, more or less distant. Calyx 7–10 mm, villous and with sessile glands; teeth ½–¾ as long as tube. Corolla *c.* 12 mm, pale yellow, pubescent; upper lip 2–3 mm; lower lip *c.* 4 mm. *Maritime sands. S. & E. Italy; Jugoslavia; Krym.* It Ju Rs (K).

49. **S. arenaria** Vahl, *Symb. Bot.* **2**: 64 (1791). Procumbent, sparsely hispid perennial 20–80 cm. Leaves 25–60 × 6–12 mm,

linear-oblong to oblanceolate, obtuse, cuneate at base, crenate-dentate, sparsely villous-lanate. Verticillasters 6- to 10-flowered, distant. Calyx 8–12 mm, densely villous and with sessile glands; teeth as long as or slightly shorter than tube. Corolla 15–18 mm, reddish-purple, pubescent; upper lip 6–8 mm, 2-fid; lower lip 7–8 mm. *Sandy ground near the sea. Italy and Sicilia.* ?Hs It Si. (*N. Africa.*)

Sect. OLISIA Dumort. Annuals, rarely short-lived perennials. Verticillasters not more than 6-flowered; bracteoles usually absent or minute.

50. S. corsica Pers., *Syn. Pl.* **2**: 124 (1806). Procumbent, sparsely hairy short-lived perennial, with slender creeping stems often rooting at nodes. Leaves 5–10(–17) mm, suborbicular to orbicular-ovate, with 3–7(–10) broad crenations. Verticillasters 2-flowered or flowers solitary, remote. Calyx 5–6 mm, densely hairy; teeth about as long as tube, obtuse, aristate or mucronate. Corolla 12–18 mm, white to purple; upper lip 4–6 mm, emarginate or 2-fid; lower lip 6–10 mm. $2n=16$. *Damp places.* ● *Mountains of Corse and Sardegna.* Co Sa.

51. S. ocymastrum (L.) Briq., *Lab. Alp. Marit.* 252 (1893) (*S. hirta* L.). Erect, hirsute annual 20–50 cm. Leaves 10–50 × 7–40 mm, oblong-ovate to broadly ovate, obtuse, slightly cordate at base, crenate-serrate, hirsute. Verticillasters 4- to 6-flowered, remote or the upper crowded. Calyx 7–10 mm, hirsute; teeth equal, as long as or longer than tube. Corolla 12–15 mm, yellow or white with yellow lower lip; tube shorter than calyx; upper lip *c.* 6 mm, 2-fid; lower lip *c.* 7 mm. $2n=18$. *S.W. Europe, Italy; once recorded from Kriti.* Bl Co Cr Ga Hs It Lu Sa Si.

52. S. marrubiifolia Viv., *Fl. Cors., App.* **1**: 2 (1825). Erect, hairy annual 10–50 cm. Leaves 20–60 × 15–45 mm, ovate-orbicular to suborbicular, obtuse, cordate at base, crenate or crenate-dentate. Verticillasters 4- to 6-flowered, the upper crowded, the lower distant. Calyx 8–10 mm, densely hairy; 2 upper teeth 4–4·5 mm, about as long as tube; 3 lower teeth 3–3·5 mm. Corolla *c.* 15 mm, pink; tube shorter than calyx; upper lip *c.* 5 mm, entire; lower lip *c.* 7 mm. $2n=16$. *Maritime sands. W. Italy, Corse.* Co It. (*N.W. Africa.*)

53. S. annua (L.) L., *Sp. Pl.* ed. 2, 813 (1763). Erect, pubescent and sometimes glandular annual, rarely short-lived perennial 10–40 cm. Leaves 10–60 × 5–20(–30) mm, lanceolate, acute, rounded or cuneate at base, crenate or crenate-dentate, glabrous or pubescent. Verticillasters 2- to 6-flowered, the upper crowded, the lower distant. Calyx 5–8 mm, hirsute and with sessile glands; teeth equal, *c.* ½ as long as tube. Corolla 10–16 mm, white or pale yellow, sometimes with red spots; tube exceeding calyx; upper lip 3–7 mm, entire; lower lip 4–7 mm. *Cultivated fields and other open habitats; somewhat calcicole. Most of Europe except the north, but doubtfully native in much of the centre and west.* Al Au Be Bu Cz Ga Ge Gr He Ho Hs Hu It Ju Po Rm Rs (B, C, W, K, E) Tu [Da No Su].

54. S. spinulosa Sibth. & Sm., *Fl. Graec. Prodr.* **1**: 410 (1809). Erect, hispid and glandular-pubescent annual 10–60 cm; stem scabrid on the angles. Leaves 20–60 × 10–40 mm, ovate, the lower cordate, the upper rounded at base, crenate, sparsely hairy. Verticillasters 4- to 6-flowered, the upper crowded, the lower distant. Calyx 9–11 mm, glandular-pubescent; teeth *c.* ½ as long as tube. Corolla 20–22 mm, white or pale yellow with purple spots, glandular-pubescent; tube exceeding calyx; upper

lip 5–8 mm, entire; lower lip 8–11 mm. *Balkan peninsula.* Al Cr Gr Ju.

55. S. milanii Petrović, *Scrin. Fl. Select.* (*Magnier*) **6**: 117 (1887). Erect, glandular-pubescent annual 15–50 cm. Leaves 20–45 × 10–25 mm, oblong to ovate, obtuse, rounded or cuneate at base, crenate-serrate, pubescent. Verticillasters 2- to 6-flowered, remote. Calyx 7–9 mm, densely glandular-pubescent; teeth *c.* ⅔ as long as tube. Corolla *c.* 20 mm, pale yellow; tube about as long as calyx; upper lip 5–6 mm, entire; lower lip *c.* 12 mm. *Grassland.* ● *E. Jugoslavia, S. & E. Bulgaria.* Bu Ju.

56. S. arvensis (L.) L., *Sp. Pl.* ed. 2, 814 (1763). Erect, hirsute annual 10–40 cm. Leaves 10–40 × 8–30 mm, cordate-ovate, obtuse, crenate, hirsute. Verticillasters 4- to 6-flowered, usually 6–12, the upper crowded, the lower distant. Calyx 5–7(–8) mm, hirsute; teeth about as long as tube. Corolla 6–8 mm, white, pale pink or purple, scarcely exceeding calyx; upper lip *c.* 2 mm, entire; lower lip 2–3 mm. $2n=10$. *Cultivated fields and sandy ground; usually calcifuge. S., W. & C. Europe eastwards to W. Poland, W. Jugoslavia and Kriti and northwards to S. Sweden; an occasional casual further north and east.* Al Au Az Be Bl Br Co Cr Da Ga Ge Gr Hb He Ho Hs It Ju Lu Po Sa Si Su.

A common weed almost throughout its range.

57. S. brachyclada De Noë ex Cosson, *Ann. Sci. Nat.* ser. 4, **1**: 226 (1854). Ascending, densely hirsute annual 10–40 cm. Leaves 15–30 × 12–20 mm, ovate-orbicular, cordate at base, crenate, hirsute. Verticillasters 2- to 6-flowered, usually more than 12, the upper crowded, the lower distant. Calyx 7–10 mm, densely hairy; teeth slightly unequal, the upper 2 about as long as tube. Corolla 7–12 mm, white or pale pink with purple markings, scarcely exceeding calyx; upper lip *c.* 2 mm, emarginate; lower lip *c.* 2 mm. ● *Coasts of W. Mediterranean region* (*very local*). Bl Ga Hs. (*N.W. Africa.*)

58. S. serbica Pančić, *Fl. Princ. Serb.* 564 (1874). Erect, hirsute annual 10–25 cm. Leaves 10–40 × 8–25 mm, oblong-ovate to broadly triangular-ovate, obtuse, cordate or rounded at base, crenate to crenate-serrate. Verticillasters 4- to 6-flowered, crowded into a capitulum, rarely with 1 verticillaster distant; bracteoles about as long as calyx. Calyx 7–14 mm, densely hirsute; teeth about as long as tube, recurved in fruit. Corolla 10–20 mm, reddish-purple; tube about as long as calyx; upper lip 4–5 mm, entire; lower lip 3·5–4 mm. *Grassland.* ● *Balkan peninsula.* Al Bu Gr Ju.

17. Cedronella Moench[1]

Perennial herbs, woody at the base. Leaves 3(4)-foliolate. Inflorescence dense, terminal, racemiform or capituliform, the lowest verticillaster often remote. Bracts simple. Calyx tubular-campanulate, 13- to 15-veined, with 5 equal teeth. Corolla infundibuliform, 2-lipped; upper lip 2-lobed; lower lip 3-lobed, the middle lobe the longest. Stamens equalling or slightly exceeding the corolla, curved under the upper lip. Style-branches subequal.

1. C. canariensis (L.) Webb & Berth., *Phyt. Canar.* **3**: 87 (1845). Aromatic. Stems up to 150 cm, glabrous. Leaves 3(4)-foliolate; leaflets 6–13 × 1·5–4·5 cm, lanceolate, acuminate, often asymmetrical at the base, petiolate, glabrous above, pubescent beneath, serrate, the central the longest. Bracts 5–6 mm, linear, ciliate. Calyx-tube 8–12 mm, puberulent; teeth 3·5–4 mm, acuminate. Corolla 16–20 mm, pinkish to lilac, rarely white, puberulent. Nutlets 1·5–2 mm, broadly ellipsoid, rounded at apex. *Açores.* [*Az.] (*Madeira, Canarias.*)

---

[1] By D. Bramwell.

18. Nepeta L.[1]

Perennial, rarely annual herbs. Flowers hermaphrodite or unisexual; verticillasters in spike-like inflorescences or in lax or dense, sometimes pedunculate cymes. Calyx cylindrical to ovoid, straight or curved, 15-veined, accrescent; teeth 5, subequal, the upper sometimes exceeding the lower. Corolla cylindrical-campanulate or infundibuliform, 2-lipped; tube slender, long, glabrous inside; upper lip patent, flat, 2-fid; lower lip 3-lobed. Stamens didynamous, parallel; anther-cells divergent, opening by a common slit. Nutlets smooth, tuberculate or rugose.

Measurements and shapes of leaves refer to cauline leaves.

Several species are cultivated in gardens for ornament and are occasionally naturalized. These include **N. mussinii** Sprengel ex Henckel, *Adumbr. Pl. Horti Hal.* 15 (1806), with decumbent stems, small, ovate-cordate, grey-green leaves and lax terminal racemes of blue flowers, native to the Caucasus and Caspian region; and more commonly the sterile hybrid of horticultural origin, **N. × faassenii** Bergmans ex Stearn, *Jour. Roy. Hort. Soc.* **75**: 403 (1950) (supposedly *N. mussinii × nepetella*) with narrower, lanceolate to oblong-ovate, truncate leaves.

All species usually grow in dry habitats, on rocky, hilly or disturbed ground.

Literature. J. Briquet, *Les Labiées des Alpes Maritimes* 359–373. Genève & Bale. 1893. G. P. De Wolf, *Baileya* **3**: 99–107 (1955).

1 Flowers unisexual, in diffusely branched, patent inflorescences
  2 Stems subglabrous or glabrescent
    3 Flowers in lax cymes; calyx 7–9 mm   **21. ucranica**
    3 Flowers in dense cymes; calyx 4–6 mm   **24. beltranii**
  2 Stems pubescent to lanate
    4 Leaves serrate, the lowest cordate at the base; corolla blue, shorter than calyx   **22. parviflora**
    4 Leaves crenate-serrate, the lowest rounded or rarely subcordate at base; corolla pink or white, exceeding calyx   **23. hispanica**
1 Flowers hermaphrodite, in spike-like or branched, erect inflorescences
  5 Outermost bracteoles equalling or exceeding the calyx
    6 Bracts at least 5 mm wide, broadly ovate
      7 Bracts membranous   **2. tuberosa**
      7 Bracts herbaceous   **4. scordotis**
    6 Bracts less than 5 mm wide, ovate-lanceolate to linear-lanceolate, rarely ovate
      8 Leaves 0·5–3(–4) cm, all petiolate
        9 Calyx-tube straight, the upper teeth not exceeding the lower   **(6–11). sibthorpii** group
        9 Calyx-tube slightly curved, the upper teeth slightly exceeding the lower   **12. italica**
      8 Leaves 3–8(–10) cm, the upper sessile
        10 Calyx-teeth at least as long as the tube; inner bracts 1–4 mm wide   **3. apuleii**
        10 Calyx-teeth shorter than the tube
          11 Corolla 10–12 mm; bracteoles 0·5–1·5 mm wide, linear   **1. multibracteata**
          11 Corolla 16–17 mm; bracteoles 1·5–3 mm wide, linear-lanceolate   **5. granatensis**
  5 Bracteoles usually shorter than the calyx
    12 Calyx-tube straight, the upper teeth not exceeding the lower
      13 Cauline leaves 1–3 cm, mostly petiolate   **(6–11). sibthorpii** group
      13 Cauline leaves 2–7(–8) cm, mostly sessile or subsessile
        14 Stem, densely pubescent; calyx 7–9 mm; corolla 8–11 mm   **19. latifolia**

        14 Stems glabrous or puberulent; calyx 4–6 mm; corolla 6–8(–10) mm   **20. nuda**
    12 Calyx-tube usually ± curved, the upper teeth usually exceeding the lower
      15 Inflorescence leafy throughout   **16. foliosa**
      15 Inflorescence not leafy, at least in upper part
        16 Calyx 8–11 mm; corolla 12–17 mm
          17 Stems glabrous or puberulent; leaves crenulate   **17. grandiflora**
          17 Stems pubescent to villous; leaves coarsely crenate   **18. melissifolia**
        16 Calyx 5–8 mm; corolla 7–12 mm
          18 Leaves ovate-cordate, with petioles up to 40 mm; corolla 7–10 mm, scarcely exserted from calyx   **13. cataria**
          18 Leaves lanceolate to oblong-lanceolate, rarely ovate, with petioles not more than 15 mm; corolla 10–12 mm, distinctly exserted from calyx
            19 Leaves pubescent to velutinous or lanate; corolla cylindrical-campanulate   **14. nepetella**
            19 Leaves glabrous or subglabrous; corolla infundibuliform   **15. agrestis**

Sect. PYCNONEPETA Bentham. At least the lowest leaves petiolate. Inflorescence spike-like, rarely branched; verticillasters many-flowered. Outer bracts usually equalling or exceeding the calyx, often rigid or scarious. Flowers hermaphrodite. Calyx tubular, straight or curved; upper teeth exceeding the lower. Nutlets tuberculate or rugose.

1. N. multibracteata Desf., *Fl. Atl.* **2**: 11 (1798). Stems 60–120 cm, erect, simple, pubescent. Leaves 3–6 cm, ovate to oblong-ovate, cordate at base, the upper sessile. Inflorescence spike-like, simple or branched; verticillasters somewhat interrupted below. Bracts 9–11 × 2–3(–5) mm, linear; bracteoles 5–10 × 0·5–1·5 mm. Calyx 7–10 mm; teeth narrowly lanceolate, c. ½ as long as the tube. Corolla 10–12 mm, bluish-violet. *S. Portugal.* Lu. (*N. Africa.*)

2. N. tuberosa L., *Sp. Pl.* 571 (1753). Rhizome tuberous. Stems 25–80 cm, simple, pubescent to lanate, sometimes viscid. Leaves 3–8 cm, ovate-lanceolate to oblong, cordate at base, subglabrous to villous, often sublanate beneath, the lowest petiolate. Inflorescence spike-like, simple. Bracts and bracteoles 8–16 × 3–8 mm, ovate to lanceolate, membranous, greenish-white, pink or reddish-purple. Calyx 8–11 mm, slightly curved, glandular; teeth 2·5–3 mm, the upper slightly exceeding the lower. Corolla 9–12 mm, purple or violet. *Spain and Portugal; Sicilia.* Hs Lu Si.

1 Bracts lanceolate, suberect, overlapping; calyx-teeth 0·5 mm wide at base, sublinear   **(a) subsp. tuberosa**
1 Bracts ovate-lanceolate to ovate, patent, not overlapping; calyx-teeth 1 mm wide at base, lanceolate
  2 Cauline leaves much exceeding the internodes; bracts entire   **(b) subsp. reticulata**
  2 Cauline leaves shorter or scarcely longer than the internodes; most bracts somewhat toothed   **(c) subsp. gienensis**

(a) Subsp. **tuberosa**: Cauline leaves usually exceeding the internodes. Inflorescence with most verticillasters densely crowded. Bracts 10–12 mm, lanceolate, entire, reddish-purple, suberect. Calyx-teeth 2·5–3 × 0·5 mm, sublinear. Corolla 12–13 mm, violet-blue. *Throughout the range of the species.*

(b) Subsp. **reticulata** (Desf.) Maire in Jahandiez & Maire, *Cat. Pl. Maroc* **3**: 632 (1934): Cauline leaves much exceeding the internodes, often acuminate. Inflorescence with verticillasters less crowded than in (a). Bracts 10–13 mm, ovate-lanceolate, usually entire, greenish-white, often pink-tinged at the margins. Calyx-teeth 2·5–3 × 1 mm, lanceolate. Corolla 11–12 mm, purplish. *S. Spain.* (*N. Africa.*)

---

[1] By C. Turner.

(c) Subsp. **gienensis** (Degen & Hervier) Heywood, *Bot. Jour. Linn. Soc.* 65: 262 (1972) (*N. gienensis* Degen & Hervier): Cauline leaves usually shorter than the internodes, not acuminate. Inflorescence with verticillasters less crowded than in (a). Bracts ovate, cuspidate and slightly toothed, patent, greenish-white, often pink-tinged. Calyx-teeth 2·5–3 × 1 mm, lanceolate. Corolla 9–11 mm, pale blue. ● *S. Spain.*

3. **N. apuleii** Ucria, *Nuovo Racc. Opusc. Aut. Sic.* 6: 252 (1793). Stems 30–80 cm, simple, puberulent. Leaves 3–6 cm, ovate to ovate-lanceolate, cordate at base, sparsely pubescent, the upper sessile. Inflorescence spike-like, sometimes branched, many-flowered. Bracts 9–12 × 3–5 mm, ovate to ovate-lanceolate, acute, rigid, reddish-purple; bracteoles 7–12 × 1–4 mm. Calyx 8–10 mm, deeply toothed; teeth lanceolate, as long as or longer than the tube. Corolla 12–14 mm, pink, exserted. *S. Spain; Sicilia.* Hs Si. (*N. Africa.*)

4. **N. scordotis** L., *Cent. Pl.* 2: 20 (1756). Stems 25–60 cm, ascending, simple, villous or lanate. Leaves 3–6 cm, ovate-cordate, rugose, villous-lanate, mostly petiolate. Inflorescence spike-like, verticillasters interrupted below. Bracts 10–13 × 7–13 mm, broadly ovate, acute, villous; bracteoles smaller, ovate-lanceolate. Calyx 8–9·5 mm, straight; teeth 3–3·5 mm, lanceolate-subulate, the upper not exceeding the lower. Corolla 13–16 mm, blue, distinctly exserted. ● *S. Aegean region.* Cr Gr.

5. **N. granatensis** Boiss., *Elenchus* 76 (1838). Stems 80–160 cm, erect, viscid-villous. Leaves 4–8(–10) cm, ovate-cordate, acute, coarsely serrate or crenate, viscid-pubescent, the upper sessile. Inflorescence spike-like, rarely branched; verticillasters mostly distant. Bracts 12–18 × 3–4 mm, reddish-purple at the apex, lanceolate, acuminate; bracteoles 9–12 × 1·5–3 mm, linear-lanceolate. Calyx 10–13 mm, slightly curved; teeth 4–5 mm, the upper not exceeding the lower. Corolla 16–17 mm, white, distinctly exserted. *S. Spain.* Hs.

(6–11). **N. sibthorpii** group. Stems 30–70 cm, erect or decumbent and ascending, branched. Leaves 0·5–3(–4) cm, ovate to oblong, cordate at base, rugose, grey-green, all shortly petiolate. Bracts and bracteoles rarely exceeding the calyx, green, with more or less scarious margins. Calyx 5–8 mm, tubular-campanulate, straight; upper teeth not exceeding the lower. Corolla 9–13 mm, white, slightly exserted.

A group of closely related species, taxonomically difficult and much in need of revision, occurring in the southern part of the Balkan peninsula and in Anatolia.

1 Outermost bracteoles at least 1 mm wide, with distinctly
    scarious margins
2 Indumentum of the stem appressed, at least above **6. sibthorpii**
2 Indumentum of the stem patent **7. parnassica**
1 Outermost bracteoles less than 1 mm wide, with very narrow
    scarious margins
3 All verticillasters distant
4 Indumentum of the stem and leaves patent, viscid
        10. camphorata
4 Indumentum of the stem and leaves appressed, not viscid
        11. heldreichii
3 At least the upper verticillasters crowded
5 Indumentum of the stem tomentose to lanate, patent; calyx-
    teeth c. ½ as long as tube **9. dirphya**
5 Indumentum of the stem velutinous, appressed; calyx-teeth
    about as long as tube **8. spruneri**

6. **N. sibthorpii** Bentham, *Lab. Gen. Sp.* 474 (1834). Stems finely grey-tomentose, the hairs appressed at least above. In-florescence spike-like, the lower verticillasters distant. Bracteoles 6–8 × 1–1·5 mm, oblong-lanceolate, with distinct scarious margins. Calyx-teeth as long as the tube. ● *S. Greece.* Gr.

7. **N. parnassica** Heldr. & Sart. in Boiss., *Diagn. Pl. Or. Nov.* 3(4): 22 (1859) (incl. *N. orphanidea* Boiss.). Stems tomentose to lanate, the hairs patent, often viscid. Inflorescence spike-like, the lower verticillasters distant. Bracteoles 5–10 × 0·5–1·5 mm, lanceolate, with distinct scarious margins. Calyx-teeth about as long as the tube. ● *S. Albania, Greece.* Al Gr.

N. sphaciotica P. H. Davis, *Notes Roy. Bot. Gard. Edinb.* 21: 136 (1953) is like 7 but differs in its dwarf habit (possibly environmentally determined) with stems 12–18 cm, very short spike, and broader, greener leaves with fewer, coarser crenations. It occurs in a single locality in W. Kriti (Levka Ori) and may be a sub-species of 7.

8. **N. spruneri** Boiss., *Diagn. Pl. Or. Nov.* 3(4): 23 (1859). Stems grey-tomentose, the hairs appressed. Leaves grey-green, somewhat tomentose. Inflorescence spike-like, only the lower verticillasters distant. Bracteoles 3·5–6 × 0·25–0·5 mm, linear-lanceolate, with very narrow scarious margins. Calyx-teeth about as long as the tube. ● *Albania, N.W. & C. Greece.* Al Gr.

9. **N. dirphya** (Boiss.) Heldr. ex Halácsy, *Consp. Fl. Graec.* 2: 538 (1902). Like 8 but stems tomentose to lanate, the hairs patent; leaves grey; calyx-teeth c. ½ as long as the tube; corolla white with purple spots. ● *E. Greece (Evvoia).* Gr.

10. **N. camphorata** Boiss. & Heldr. in Boiss., *Diagn. Pl. Or. Nov.* 1(7): 49 (1846). Stems crispate-villous, the hairs patent, viscid. Leaves lanate-villous, the hairs patent, viscid. Inflorescence with distant verticillasters. Bracteoles 2·5–4 × 0·25–0·5 mm, linear, with very narrow scarious margins, ½ as long as calyx. Calyx grey; teeth much shorter than the tube. Corolla white with purple spots. ● *S. Greece (Taïyetos).* Gr.

11. **N. heldreichii** Halácsy, *Consp. Fl. Graec.* 2: 539 (1902). Like 10 but indumentum of stems and leaves appressed, not viscid; stems white-tomentose; leaves white-tomentose; bracts ⅓–½ as long as calyx. ● *S. Greece (Taïyetos).* Gr.

12. **N. italica** L., *Sp. Pl.* 571 (1753). Stems grey-pubescent to villous. Leaves 0·5–3(–4) cm, ovate to oblong, cordate at base, shortly petiolate. Inflorescence spike-like, the lower verticillasters distant. Bracts and bracteoles 8–10(–12) × 0·5–1·5 mm, linear to linear-lanceolate, with narrow scarious margins, rigid, equalling or exceeding the calyx. Calyx 7–8 mm, slightly curved; teeth 3–3·5 mm, the upper slightly exceeding the lower. Corolla 11–13 mm, creamy white, exserted. *E.C. Italy; Turkey-in-Europe.* It Tu. (*S.W. Asia.*)

Sect. NEPETA. Most leaves petiolate. Inflorescence spike-like or branched, sometimes leafy below; verticillasters many-flowered. Bracteoles subulate, not rigid, usually much shorter than the calyx. Flowers hermaphrodite. Calyx usually cylindrical or ovoid, often curved; upper teeth usually exceeding the lower. Corolla-tube curved, dilated at the throat. Nutlets smooth or tuberculate.

13. **N. cataria** L., *Sp. Pl.* 570 (1753). Stems 40–100 cm, erect, branched, grey-pubescent to tomentose. Leaves 2–8 cm, ovate, acute, cordate at base, crenate or serrate, grey-tomentose beneath; petiole 0·5–4 cm. Inflorescence spike-like, the lower verticillasters distant. Bracts 1·5–3 mm, linear-subulate. Calyx 5–6·5 mm, ovoid; teeth 1·5–2·5 mm, linear-lanceolate, patent.

Corolla 7–10 mm, shortly exserted from calyx, white with small purple spots. 2n=36. *S., E., E.C. and perhaps W. Europe; formerly cultivated as a medicinal herb and widely naturalized in N. & W.C. Europe.* Al *Be Bl *Br Bu Co Ga Gr *Ho Hs Hu It Ju Rm Rs (C, W, K, E) Si [Au Cz Da Fe Ge Hb He Lu No Po Rs (B) Su].

14. N. nepetella L., *Syst. Nat.* ed. 10, 2: 1096 (1759) (incl. *N. amethystina* Poiret, *N. aragonensis* Lam., *N. boissieri* Willk., *N. murcica* Guirão ex Willk.). Stems 30–80 cm, branched, pubescent to velutinous. Leaves 1–4 × 0·5–2 cm, lanceolate to oblong-lanceolate, truncate, rarely ovate and cordate at base, crenate to deeply dentate, grey-green to green, pubescent to velutinous or lanate, usually glandular; petiole up to 15 mm. Inflorescence usually branched. Bracts leaf-like below; bracteoles 2–4 mm, linear-lanceolate. Calyx 5–8 mm, pubescent to villous; teeth triangular, often tinged with pink or blue. Corolla 10–12 mm, cylindrical-campanulate, distinctly exserted from calyx, curved, white, pink or bluish-violet. 2n=34. *S.W. Europe, extending eastwards to S. Italy.* Bl Ga Hs It.

This species is exceedingly polymorphic, particularly in Spain, from where numerous variants have been accorded specific status in the past. Its variation requires further critical study but appears in part to be linked with habitat and altitudinal factors. The most extreme variants have very small, deeply dentate leaves and slender, few-flowered inflorescences (**N. boissieri** Willk., *Bot. Zeit.* 15: 219 (1857)) or are completely covered with a lanate indumentum (**N. mallophora** Webb & Heldr. ex Nyman, *Consp.* 585 (1881)). Nevertheless these are linked by continuous variation with more normal plants. Plants with a bluish-violet corolla are restricted to the Iberian peninsula and North Africa and have been distinguished as **N. nepetella** subsp. **amethystina** (Poiret) Briq., *Lab. Alp. Marit.* 368 (1893). Plants with pink or white corolla are found throughout the range of the species in Europe.

15. N. agrestis Loisel., *Mém. Soc. Linn. Paris* 6: 417 (1827). Like **14** but leaves green, glabrous or subglabrous; corolla infundibuliform. 2n=18. ● *Corse.* Co.

16. N. foliosa Moris, *Stirp. Sard.* 3: 10 (1829). Stems 30–80 cm, viscid-villous. Leaves 1·5–3 cm, oblong-lanceolate, cordate or cuneate at base, coarsely serrate-dentate, densely pubescent. Inflorescence spike-like, leafy. Bracteoles 3–4 mm, lanceolate to linear-lanceolate. Calyx 8–9 mm; teeth 2–3 mm, triangular-lanceolate. Corolla 10–13 mm, pale bluish-violet. ● *Sardegna (Monte Oliena).* Sa.

17. N. grandiflora Bieb., *Fl. Taur.-Cauc.* 2: 42 (1808). Stems 40–80 cm, erect, branched, glabrous or puberulent. Leaves 2–6(–10) cm, ovate, cordate at base, crenulate, glabrous. Inflorescence spike-like. Bracteoles 2–3 mm, linear-subulate. Calyx 9·5–11 mm, often blue; teeth 1·5–2 mm, lanceolate, acute. Corolla 14–17 mm, blue, distinctly exserted. *A frequent casual in E. & E.C. Europe and locally naturalized.* [?Ge Rs (B, C, W, K).] (*Caucasus.*)

18. N. melissifolia Lam., *Encycl. Méth. Bot.* 1: 711 (1785). Stems 20–40 cm, ascending, little-branched, pubescent to villous. Leaves 1·5–3·5 cm, ovate, cordate at base, coarsely crenate, pubescent. Bracts 5–7 mm, lanceolate; bracteoles shorter, linear. Calyx 8–10 mm, curved; teeth 2–3·5 mm. Corolla 12–15 mm, blue with small red spots. 2n=18. ● *S. Aegean region.* Cr Gr.

Sect. ORTHONEPETA Bentham. Most leaves sessile or subsessile. Inflorescence spike-like or branched; verticillasters many-flowered. Flowers hermaphrodite. Bracts small, inconspicuous, much shorter than the calyx. Calyx straight, the upper teeth not exceeding the lower. Nutlets tuberculate.

19. N. latifolia DC. in Lam. & DC., *Fl. Fr.* ed. 3, 3: 528 (1805). Stems 50–120 cm, erect, simple or sparsely branched, densely pubescent, often bluish above. Leaves 3–7 cm, ovate to ovate-oblong, more or less cordate at base, crenate or serrate, puberulent, sessile or subsessile. Inflorescence spike-like. Bracteoles 3·5–5 mm, lanceolate. Calyx 7–9 mm; teeth 2–3 mm, lanceolate, often bluish. Corolla 8–11 mm, blue. ● *Pyrenees and Iberian peninsula.* Ga Hs Lu.

20. N. nuda L., *Sp. Pl.* 570 (1753). Stems 50–120 cm, glabrous or puberulent. Leaves 2–5(–8) cm, ovate to ovate-oblong, more or less cordate at base, crenate, the lower sessile or shortly petiolate, the upper sessile. Inflorescence branched, rarely spike-like. Bracteoles 2–3 mm, linear to linear-lanceolate. Calyx 4–6 mm; teeth 1–2 mm. Corolla 6–8(–10) mm, pale violet or white. *S., E. & E.C. Europe, northwards to C. Russia.* Al Au Bu Cz Ga Gr He Hs Hu It Ju Po Rm Rs (C, W, K, E) [Ge].

(a) Subsp. **nuda** (incl. *N. nuda* subsp. *pannonica* (L.) Gams, *N. pannonica* L.): Lower leaves distinctly petiolate. Inflorescence lax. Calyx-teeth linear-lanceolate, pale green or blue-tinged. Corolla pale violet. *Throughout the range of the species.*

(b) Subsp. **albiflora** Gams in Hegi, *Ill. Fl. Mitteleur.* 5(4): 2372 (1927): Lower leaves sessile or subsessile. Inflorescence compact. Calyx-teeth lanceolate, white-tinged. Corolla white. *S. Jugoslavia, N. Greece.* Gr Ju. (*S.W. Asia.*)

Sect. OXYNEPETA Bentham. Leaves petiolate. Cymes few-flowered, in diffusely branched, patent inflorescences. Bracts leaf-like; bracteoles shorter than the calyx. Calyx-tube cylindrical, straight, sparsely hairy within; upper teeth longer than the lower. Flowers unisexual; central flowers of each cyme female, the stamens represented by staminodes; outer flowers male, with a rudimentary pistil. Nutlets tuberculate.

21. N. ucranica L., *Sp. Pl.* 570 (1753). Stems 15–50 cm, erect, much-branched, subglabrous or puberulent, glabrescent. Leaves (1–)2–4 cm, oblong-lanceolate, crenate-serrate, glabrous, the lower subcordate at base; petiole up to 15 mm. Cymes 3- to 5-flowered, lax, numerous. Bracteoles 6–7 mm, linear, acute. Calyx 7–9 mm, often bluish; teeth linear or linear-lanceolate, longer than the tube. Corolla 7–9 mm, bluish-violet, slightly exserted. *Bulgaria and Romania; E.C. & S.E. Russia.* Bu Rm Rs (C, E).

22. N. parviflora Bieb., *Fl. Taur.-Cauc.* 2: 41 (1808) (*N. euxina* Velen.). Like **21** but stems tomentose to lanate, at least below; leaves serrate with larger teeth, the lowest cordate at base, distinctly petiolate; cymes dense; calyx 6–7 mm; corolla 5–6 mm, blue, shorter than the calyx. *S.E. Europe.* Bu Rm Rs (C, W, K, E).

23. N. hispanica Boiss. & Reuter in Boiss., *Diagn. Pl. Or. Nov.* 3(4): 26 (1859). Like **21** but stems pubescent to tomentose; leaves lanceolate, puberulent, the lowest rounded or rarely subcordate at base; cymes dense; calyx 6–7 mm; corolla 5–7 mm, pink or white, equalling or exceeding the calyx. *C. & S.E. Spain.* Hs.

24. N. beltranii Pau, *Bol. Soc. Aragon. Ci. Nat.* 11: 40 (1912). Like **21** but leaves lanceolate, subglabrous to puberulent, the lowest rounded or rarely subcordate at base; cymes dense; calyx 4–6 mm, the teeth shorter than or equalling the tube; corolla pink or white. ● *C. Spain (Vaciamadrid, near Madrid).* Hs.

19. Glechoma L.[1]

Perennial herbs. Verticillasters 2- to 5-flowered, secund. Calyx tubular-campanulate, 2-lipped; upper lip with 3, the lower lip with 2 teeth. Corolla 2-lipped; tube straight, widening towards the apex; upper lip flat; lower lip 3-lobed. Anther-cells at right angles to each other, each opening by a separate slit. Nutlets smooth.

Calyx 5–6·5 mm; teeth of upper lip of calyx $\frac{1}{5}$–$\frac{1}{3}$(–$\frac{1}{2}$) as long as tube, triangular-acuminate   **1. hederacea**
Calyx 7–11 mm; teeth of upper lip of calyx $\frac{1}{2}$ as long to about as long as tube, linear-lanceolate, acuminate   **2. hirsuta**

1. G. hederacea L., *Sp. Pl.* 578 (1753). Subglabrous or pubescent; flowering stems up to 50(–60) cm, ascending or erect; non-flowering stems creeping and rooting. Leaves 4–35(–80) × 6–40(–80) mm, reniform to suborbicular-cordate, obtuse or subacute, coarsely crenate. Bracteoles 1–1·5 mm, setaceous. Calyx 5–6·5 mm, shortly hispid or pubescent and sometimes with sessile, yellow glands; teeth of upper lip $\frac{1}{5}$–$\frac{1}{3}$(–$\frac{1}{2}$) as long as tube, triangular-acuminate. Corolla (6–)15–22(–25) mm, pale violet with purple spots on lower lip, rarely white or pink. Nutlets *c.* 2 mm. $2n = 18, 36$. *Woods, grassland and waste places, usually in damp soil. Almost throughout Europe.* All except Bl Cr Fa Is Sb Tu.

G. serbica Halácsy & Wettst., *Verh. Zool.-Bot. Ges. Wien* 38: 71 (1888) (*G. hederacea* subsp. *serbica* (Halácsy & Wettst.) Soó), from Jugoslavia (Srbija), is, according to V. Blečić (*Bull. Inst. Jard. Bot. Univ. Beograd* nov. ser., 3: 221–225 (1968)) probably an ecological modification of **1**. It is subglabrous, with calyx 4–6 mm, hirsute, the teeth almost as long as the tube and the corolla 20–27 mm, violet.

2. G. hirsuta Waldst. & Kit., *Pl. Rar. Hung.* 2: 124 (1802–1803). Like **1** but usually with denser and longer hairs; leaves (15–)25–50 × (15–)25–60 mm, the upper broadly ovate-triangular, subacute, crenate-serrate; bracteoles 2–4 mm; calyx 7–11 mm, the teeth of upper lip $\frac{1}{2}$ as long to nearly as long as tube, linear-lanceolate, acuminate; corolla 20–30 mm, pale blue with white spots on lower lip; nutlets 3–4 mm. $2n = 36$. *Woods. E.C. & S.E. Europe, extending westwards to Sardegna and northwards to c. 56° N. in E.C. Russia.* Al Au Bu Cz Gr Hu It Ju Po Rm Rs (C, W) Sa Si.

Intermediates, presumably hybrids, between **1** and **2** sometimes occur.

20. Dracocephalum L.[2]

Annual or perennial herbs or dwarf shrubs, rarely rhizomatous. Verticillasters in spikes. Calyx tubular or tubular-campanulate, 15-veined, more or less 2-lipped; upper lip 3-toothed, the middle tooth larger than the others. Corolla 2-lipped; upper lip 2-lobed, emarginate; lower lip 3-lobed with the middle lobe larger than the others, emarginate; tube narrow at the base, widening at the throat. Stamens equalling or rarely exceeding the corolla and curved under the upper lip. Style-branches equal.

All European species grow mainly in dry, more or less open habitats, and tend to become naturalized as weeds or ruderals outside their native territory. Many of them are cultivated for ornament, as honey-plants or as medicinal plants.

Lallemantia iberica (Bieb.) Fischer & C. A. Meyer, *Ind. Sem. Horti. Petrop.* 6: 53 (1840), from the Caucasian region, is culti-vated for ornament and may be locally naturalized in E. & E.C. Europe. It is characterized by 2 longitudinal folds within the upper lip of the corolla, in which the stamens are concealed.

1 Corolla not more than 15 mm, only slightly exserted from the calyx-tube   **2. thymiflorum**
1 Corolla at least 17 mm, distinctly exserted from the calyx-tube
  2 Leaves pinnatipartite; corolla 35–50 mm   **4. austriacum**
  2 Leaves simple, entire or crenate; corolla 17–28 mm
    3 Leaves linear-lanceolate, entire; anthers hairy   **3. ruyschiana**
    3 Leaves ovate to oblong-lanceolate, crenate or serrate-crenate; anthers glabrous
      4 Rhizomatous perennial; basal and lower cauline leaves ovate, cordate at base   **1. nutans**
      4 Annual; basal and lower cauline leaves oblong to lanceolate, cuneate at base   **5. moldavica**

1. D. nutans L., *Sp. Pl.* 596 (1753). Rhizomatous perennial; stems up to 70 cm, shortly hairy. Basal and lower cauline leaves 15–40 × 10–30 mm, ovate, cordate at base, coarsely crenate, glabrous, the petiole longer than the lamina; upper cauline leaves up to 70 × 40 mm, oblong-ovate, cuneate, the petiole shorter than or equalling the lamina; uppermost cauline leaves subsessile, shortly pubescent. Verticillasters many-flowered, forming an elongate inflorescence. Bracts elliptical, acute, entire. Corolla 17–22 mm, lilac-blue, rarely white. Anthers glabrous. *Naturalized in N. & E. Russia.* [Rs (N, C, E).] (*N. & C. Asia.*)

2. D. thymiflorum L., *Sp. Pl.* 596 (1753). Annual; stems up to 60 cm, erect, sparsely pubescent. Basal and lower cauline leaves 10–35 × 7–20 mm, ovate to ovate-lanceolate, cordate at base, serrate-dentate or serrate, glabrous, the petiole longer than the lamina; upper cauline leaves ovate, cuneate at base, the petiole shorter than the lamina. Verticillasters 6- to 12-flowered, forming an interrupted spike. Bracts elliptical, acute, entire. Corolla 7–9 mm, lilac-blue. $2n = c.\ 14$. *S., W. & C. parts of U.S.S.R., extending to N.E. Bulgaria; naturalized or casual in parts of N. & E.C. Europe.* *Bu Rm Rs (B, C, W, E) [Da Fe Po Rs (N) Su].

3. D. ruyschiana L., *Sp. Pl.* 595 (1753). Perennial; stems up to 60 cm, erect or ascending, glabrous or shortly hairy. Leaves 20–70 × 2–9 mm, linear-lanceolate, entire, obtuse, glabrous, the margins revolute; lower leaves shortly petiolate, the upper sessile. Verticillasters 2- to 6-flowered, forming a dense, terminal spike. Bracts ovate-lanceolate, entire. Corolla 20–28 mm, blue to violet, rarely pink or white. Anthers hairy. $2n = 14$. *U.S.S.R. and C. Europe, extending locally westwards to Norway and the Pyrenees.* Au Ga Ge He Hu It No Po Rm Rs (N, B, C, W, E) Su.

4. D. austriacum L., *Sp. Pl.* 595 (1753). Perennial herb or dwarf shrub; stems up to 60 cm, erect or ascending, densely leafy, velutinous. Cauline leaves 3- to 5(–7)-pinnatipartite, the segments 20–30 × 1–2·5 mm, linear to linear-lanceolate, entire, more or less velutinous, the margins revolute. Verticillasters 2- to 4(–6)-flowered, forming a more or less dense, ovoid to oblong spike. Bracts 3-fid, aristate. Corolla 35–50 mm, blue-violet. $2n = 14$. *S.C. Europe, extending westwards to S.E. France and eastwards to W.C. Ukraine.* Au Cz Ga He Hu It Rm Rs (C, W).

5. D. moldavica L., *Sp. Pl.* 595 (1753). Annual; stems up to 60 cm, erect, glabrous or sparsely hairy. Leaves 15–70 × 7–20 mm, oblong to lanceolate, serrate-crenate, obtuse, glabrous or subglabrous; lower leaves shortly petiolate. Verticillasters 6(–10)-flowered, in the axils of upper leaves, lax below, dense above. Bracts leaf-like. Corolla 20–25 mm, white or violet. Anthers glabrous. *Naturalized as a weed and ruderal in E. & E.C. Europe; an occasional casual elsewhere.* [Po Rm Rs (B, C, W, E).] (*Siberia and C. Asia.*)

---

[1] By R. Fernandes.   [2] By V. H. Heywood.

21. Prunella L.[1]

Perennial herbs. Verticillasters mostly 6-flowered, in dense, terminal, cylindrical spikes; bracts distinctly different from the leaves; bracteoles small or absent. Calyx tubular-campanulate, 2-lipped, closed in fruit; upper lip more or less 3-toothed; lower lip with 2 larger teeth. Corolla-tube exceeding calyx, straight, obconical, with a ring of hairs inside; upper lip distinctly hooded; lower lip denticulate. Filaments with a subulate appendage below the apex; anther-cells divergent. Nutlets oblong.

All four species are interfertile, and consequently hybrids are common where two or more of them grow together. There is a complete range of intermediates between the extremes of the different species in respect of nearly every character, and all of this variation may be due to hybridization.

1 Corolla usually more than 18 mm; inflorescence not subtended by leaves **2. grandiflora**
1 Corolla not more than 17 mm; inflorescence usually subtended by leaves
  2 Corolla yellowish-white, rarely pink or purplish; at least the upper leaves pinnatifid or lobed; teeth of lower lip of calyx linear-lanceolate **1. laciniata**
  2 Corolla violet, rarely white; leaves entire or crenulate; teeth of lower lip of calyx lanceolate
    3 Leaves ovate to rhombic-ovate, entire or crenulate, petiolate; middle tooth of upper lip of calyx wider than lateral teeth **3. vulgaris**
    3 Leaves linear-lanceolate to elliptic-lanceolate, entire, sessile; teeth of upper lip of calyx subequal **4. hyssopifolia**

1. P. laciniata (L.) L., *Sp. Pl.* ed. 2, 837 (1763) (*P. alba* Pallas ex Bieb.). Densely pubescent; stems up to 30 cm. Leaves 4–7 × 2–3 cm; at least the upper deeply pinnatifid or lobed; petiole 0–4 cm. Inflorescence subtended by leaves. Bracts 10 × 15 mm. Calyx 10 mm; upper lip more or less truncate, the teeth almost obsolete; teeth of lower lip 2–2·5 mm, linear-lanceolate, ciliate. Corolla 15–17 mm, yellowish-white, rarely rose-pink or purplish. $2n=28, 32$. *S., W. & C. Europe, northwards to Belgium and C. Poland; locally naturalized further north.* Al Au Be Bl Bu Co Cr Cz Ga Ge Gr He Hs Hu It Ju Lu Po Rm Rs (W, K) Sa Si Tu [Br].

2. P. grandiflora (L.) Scholler, *Fl. Barb.* 140 (1775). Sparsely pubescent; stems up to 60 cm. Leaves 5–9 × 3·5–4 cm, ovate to ovate-lanceolate, entire or crenulate; petiole (0–)4(–9) cm. Inflorescence not subtended by leaves. Bracts 15–20 × 15–20 mm. Calyx 15 mm, the teeth awned; teeth of upper lip subequal; teeth of lower lip 3–4 mm, broadly lanceolate, shortly ciliate. Corolla (18–)25–30 mm, the lips deep violet, the tube whitish. $2n=28$. *Europe except for most of the north and the islands.* Al Au Be Bu Cz Da Ga Ge Gr He Hs Hu It Ju Lu Po Rm Rs (B, C, W, K, E) Su Tu.

(a) Subsp. **grandiflora**: Leaves cuneate at base, not lobed; inflorescence not more than 5 cm. *Calcicole. Throughout the range of the species, except S. Portugal and S.W. Spain.*
(b) Subsp. **pyrenaica** (Gren. & Godron) A. & O. Bolós in A. Bolós, *Veg. Com. Barcelon.* 472 (1950) (*P. hastifolia* Brot.): Leaves hastate; inflorescence up to 8 cm. *Calcifuge. S.W. Europe.*

3. P. vulgaris L., *Sp. Pl.* 600 (1753). More or less pubescent; stems up to 50 cm. Leaves 4–5(–9) × 2(–4) cm, ovate to rhombic-ovate, entire or crenulate, cuneate; petiole 0·5–4 cm. Inflorescence usually subtended by leaves. Bracts 5–15 × 7–13 mm.

Calyx 8–9 mm, the teeth mucronate; middle tooth of upper lip wider than lateral teeth; teeth of lower lip 1·5–2 mm, lanceolate, shortly ciliate. Corolla (10–)13–15 mm, deep violet-blue, rarely white. $2n=28$. *Almost throughout Europe.* All except Bl Sb.

4. P. hyssopifolia L., *Sp. Pl.* 600 (1753). Glabrous or sparsely pubescent; stems up to 40 cm. Leaves (1–)3–6(–8) × (0·3–)0·5–1·3(–1·8) cm, linear-lanceolate to elliptic-lanceolate, entire, sessile, rarely the lowest pair shortly petiolate. Inflorescence subtended by leaves. Bracts 10–12 × 10–13 mm. Calyx 8–9 mm, the teeth mucronate or awned; teeth of upper lip subequal; teeth of lower lip 3–3·5 mm, lanceolate, ciliate. Corolla 15–17 mm, violet, rarely whitish. ● *S.W. Europe.* Co Ga Hs It.

22. Cleonia L.[2]

Annual. Verticillasters 6-flowered, in dense, subcapitate to oblong spikes; bracts pectinate-pinnatifid. Calyx 2-lipped, 10-veined; upper lip flat, subentire to 3-toothed; lower lip with 2 long, subulate, subspinescent teeth. Corolla-tube exceeding calyx, the limb dilated; upper lip erect, entire; lower lip 3-lobed. Stamens didynamous; filaments with a subulate appendage beneath the apex; anther-cells divergent. Style 4-fid. Nutlets ovoid-trigonous.

1. C. lusitanica (L.) L., *Sp. Pl.* ed. 2, 837 (1763). Hispid or appressed-pubescent; stems up to 40 cm, simple or branched from the base. Leaves 3–9 × 0·4–1·8 cm, sessile, oblong to linear, obtuse, usually coarsely crenate to deeply dentate or weakly pinnatifid, scabrid or shortly hispid. Bracts sessile, imbricate, pectinate-pinnatifid, white-ciliate, acuminate; segments long-aristate. Calyx 5–7 mm, glandular-hispid, the throat somewhat closed in fruit; teeth of the upper lip obsolete or more or less distinct, with an arista 0·5–3 mm. Corolla 2–3 cm, purplish-violet (rarely white), puberulent. Nutlets c. 2·5 mm, light brown. *Dry, open habitats. C. & S. Spain, C. & S. Portugal.* Hs Lu.

23. Melissa L.[2]

Perennial herbs. Verticillasters few- to many-flowered. Calyx campanulate, 2-lipped, 13-veined; upper lip flattened, 3-toothed; lower lip 2-toothed. Corolla 2-lipped, the tube curved and dilated above the middle, without a ring of hairs inside; upper lip erect or deflexed, sometimes slightly hooded, emarginate; lower lip 3-lobed. Stamens didynamous, included, convergent; anther-cells divergent. Style-branches subequal.

1. M. officinalis L., *Sp. Pl.* 592 (1753). Stems 20–150 cm, erect, branched, shortly glandular-puberulent, and with sparse or dense, long, patent, eglandular hairs, or glabrescent. Leaves 2–9 × 1·5–7 cm, broadly ovate to rhombic or oblong, obtuse or acute, more or less deeply crenate except at the base. Floral leaves crenate-serrate. Verticillasters 4- to 12-flowered. Bracteoles 2–5 mm, ovate to linear, entire. Calyx 7–9 mm, with long, patent, eglandular and short glandular hairs; teeth of lower lip lanceolate-triangular. Corolla 8–15 mm, pale yellow, becoming white or pinkish. Nutlets 1·5–2 mm. *Scrub and other shady places. S. Europe; widely cultivated for its aromatic foliage and naturalized northwards to England, Sweden and C. Russia.* ?Al Bl Bu Co Cr Ga Gr Hs It Ju Rm Sa Si [Au Az Be Br Cz Da Ge Hb He Lu Po Rs (W, K, E) Su ?Tu].

(a) Subsp. **officinalis**: Lemon-scented, up to 90 cm. Leaves glabrescent, or sparsely hairy above, glandular-puberulent and more or less sparsely hairy beneath; at least the upper cauline and floral leaves more or less cuneate at the base. Middle tooth of the

---

    [1] By A. R. Smith.   [2] By R. Fernandes.

upper lip of fruiting calyx conspicuous, broadly triangular. *Widely cultivated, and naturalized throughout Europe except parts of the north.*

Cultivated specimens have $2n=32$. The limits of the native range of this subspecies are uncertain, as the plants have been widely introduced.

(b) Subsp. **altissima** (Sibth. & Sm.) Arcangeli, *Comp. Fl. Ital.* ed. 2, 427 (1894): Fetid, up to 150 cm; indumentum dense throughout. Leaves greyish- or whitish-tomentose beneath, at least the upper cauline and lower floral leaves truncate or subcordate at the base. Middle tooth of the upper lip of fruiting calyx inconspicuous, truncate or emarginate. $2n=64$. *S. Europe.*

M. bicornis Klokov in Kotov, *Fl. RSS Ucr.* **9**: 659 (1960), described from Krym, has stems up to 150 cm; lower cauline leaves cordate-triangular, hirsute on both surfaces; upper lip of calyx abruptly deflexed, the teeth inconspicuous. It may not be distinct from 1(b).

24. Ziziphora L.[1]

Annuals. Verticillasters few- to many-flowered, distant or crowded, in heads or spike-like inflorescences. Calyx long-tubular, 13-veined, 2-lipped, the upper lip 3-toothed, the lower 2-toothed; tube hairy in the throat. Corolla weakly 2-lipped; upper lip porrect, entire or emarginate; lower patent, 3-lobed; tube scarcely exserted from the calyx. Lower 2 stamens fertile, ascending under the upper lip of corolla or shortly exserted; upper 2 stamens rudimentary or absent. Style-branches unequal, the lower subulate, the upper much reduced.

All species grow in dry, usually open habitats.

Z. rigida (Boiss.) H. Braun, *Verh. Zool.-Bot. Ges. Wien* **39**: 222 (1889), from S.W. Asia, has once been recorded from Bulgaria, probably erroneously. It is perennial, with stems glabrous, leaves lanceolate to ovate-lanceolate, glabrous, and verticillasters crowded into a terminal head.

```
1 Middle and upper leaves up to 10 mm, ovate to ovate-
      lanceolate, or obovate
  2 Stems up to 10 cm; leaves ovate to ovate-lanceolate, entire
                                                2. hispanica
  2 Stems up to 25 cm; leaves obovate, denticulate towards the
      apex                                      3. acinoides
1 Middle and upper leaves up to 40 mm, linear to lanceolate
  3 Verticillasters in globose, bracteate heads    1. capitata
  3 Verticillasters in ovoid to oblong, spike-like inflorescences
    4 Corolla 12–15 mm, twice as long as calyx      5. taurica
    4 Corolla 8–10 mm, slightly longer than calyx
      5 All leaves linear to linear-lanceolate; calyx 5–7·5 mm
                                                   4. tenuior
      5 Lowermost leaves ovate, the remainder lanceolate; calyx
          7–10 mm                                  6. persica
```

1. **Z. capitata** L., *Sp. Pl.* 21 (1753). Annual; stems up to 20 cm, simple or branched, scabrid. Leaves $3-35 \times 1\cdot5-7$ mm, linear-lanceolate to lanceolate, acute, petiolate, entire, appressed-hairy above, especially near the margins and on veins. Verticillasters few-flowered, in a terminal globose, bracteate head. Bracts broadly ovate to suborbicular, acuminate, ciliate. Calyx 6–11 mm, hispid. Corolla 10–15 mm, pink. Anthers not appendiculate. *S.E. Europe.* Al Bu Gr ?Hs *It Ju Rm Rs(K, E) Tu.

2. **Z. hispanica** L., *Cent. Pl.* **1**: 3 (1755). Annual; stems up to 10 cm, sometimes branched, puberulent, with recurved hairs.

Leaves $5-10 \times 2-5$ mm, ovate to ovate-lanceolate, obtuse, entire, petiolate, glabrous or scabrid-puberulent above, scabrid-puberulent beneath, ciliate. Verticillasters crowded in a dense spike-like inflorescence. Bracts broadly obovate-orbicular, mucronate, ciliate. Flowers subsessile. Calyx 6–7 mm, hispid. Corolla 9–10 mm, pink. Anthers not appendiculate. *C. & S.E. Spain.* Hs.

3. **Z. acinoides** L., *Sp. Pl.* 22 (1753). Like **2** but stems up to 25 cm; leaves obovate, denticulate towards the apex; verticillasters distant; flowers pedicellate. ● *E.C. Spain.* Hs.

4. **Z. tenuior** L., *Sp. Pl.* 21 (1753). Annual; stems 5–20 (–25) cm, simple or branched, with recurved, short hairs. Leaves on non-flowering shoots $7-25 \times 1-6$ mm, sparse, linear-lanceolate to lanceolate, acute or acuminate, shortly petiolate, glabrous, often ciliate. Verticillasters many-flowered, distant or crowded, in an oblong, spike-like inflorescence. Bracts like the cauline leaves, ciliate. Calyx 5–7·5 mm, narrowly cylindrical, patent-hairy; teeth ovate-triangular, obtuse. Corolla 8–10 mm, the limb usually with a small clavate appendage near the base. Anthers appendiculate. *S.E. Russia, E. Ukraine.* Rs (W, ?K, E). (*S.W. & C. Asia.*)

5. **Z. taurica** Bieb., *Fl. Taur.-Cauc.* **1**: 414 (1808). Annual; stems up to 30 cm, simple or branched, with recurved hairs. Leaves linear-lanceolate, subacute, petiolate, subglabrous or shortly and sparsely hairy, sometimes ciliate. Verticillasters few-flowered, in a dense or lax, spike-like inflorescence. Bracts like the cauline leaves. Calyx 5–7 mm, narrowly cylindrical, patent-hairy; teeth ovate. Corolla 12–15 mm, the limb with a corniculate appendage near the base. Anthers appendiculate. *Krym.* Rs (K).

6. **Z. persica** Bunge, *Mém. Acad. Sci. Pétersb.* ser. 7, **21**(1): 39 (1873). Annual; stems up to 30 cm, simple or branched, with deflexed hairs. Leaves $7-40 \times 1-10$ mm, sparse, the lowermost ovate, petiolate, the others lanceolate, subsessile, all acute or acuminate, glabrous, the uppermost sometimes ciliate. Verticillasters many-flowered, crowded, in a dense, ovoid to oblong, spike-like inflorescence. Bracts like the cauline leaves but widened at the base, the upper with a few distant teeth. Calyx 7–10 mm, narrowly cylindrical, patent-hairy; teeth narrowly triangular or lanceolate. Corolla 9 mm, the limb usually without appendage. Anthers appendiculate. *Krym* (*near Simferopol'*). Rs (K). (*S.W. Asia.*)

25. Satureja L.[2]

Annual or perennial herbs or dwarf shrubs. Flowers in verticillasters or in lax cymes. Calyx tubular or campanulate, 10(–13)-veined, with 5 more or less equal (rarely distinctly unequal) teeth; tube straight, rarely gibbous, more or less hairy in the mouth. Corolla 2-lipped, with straight tube. Stamens shorter than corolla, curved. Style-branches subequal, subulate.

All species occur in dry, sunny habitats, particularly on cliffs.

```
1 Calyx-teeth very unequal, the upper 3 c. ½ as long as the lower 2
                                                   4. rumelica
1 Calyx-teeth equal or slightly unequal
  2 Annual; at least the lower calyx-teeth much longer than tube
                                                  12. hortensis
  2 Dwarf shrub, or perennial herb woody at base; calyx-teeth
      shorter than to slightly longer than tube
    3 Calyx-teeth obtuse, though sometimes apiculate   2. salzmannii
    3 Calyx-teeth acute or acuminate
```

[1] By V. H. Heywood.   [2] By P. W. Ball and F. M. Getliffe.

4 Bracteoles numerous, oblong or lanceolate, about as long as
calyx **1. thymbra**
4 Bracteoles absent or short, rarely a few almost as long as
calyx
5 Dwarf shrub, with ±spiny branches **3. spinosa**
5 Perennial herb woody at base, without spiny branches
5–11. montana group

1. S. thymbra L., *Sp. Pl.* 567 (1753) (incl. *S. biroi* Jáv.).
Much-branched, usually grey-puberulent dwarf shrub 20–35 cm.
Leaves (5–)7–20 × (1–)2–9 mm, oblong to obovate, acute. Verticillasters usually many-flowered, globose, dense, distant; bracts
equalling the verticillasters; bracteoles numerous, oblong or
lanceolate, about equalling calyx. Calyx 4–7 mm, with long,
patent, white hairs, the teeth slightly shorter than tube, acuminate. Corolla 8–12 mm, bright pink or reddish-purple. *S. Aegean
region; S. coast of Sardegna.* Cr Gr Sa.

2. S. salzmannii P. W. Ball, *Bot. Jour. Linn. Soc.* **65**: 356 (1972)
(*S. inodora* Salzm. ex Bentham, non Host). Much-branched,
puberulent dwarf shrub up to 30 cm. Leaves 5–8 × *c.* 1 mm,
linear-spathulate, obtuse, glabrous except for ciliate margin.
Verticillasters usually 2-flowered, few, crowded; bracts about
equalling the verticillasters. Calyx 4–5 mm, glabrous, the teeth
slightly shorter than tube, oblong, obtuse, sometimes apiculate.
Corolla 6–8 mm, white. *S.W. Spain.* Hs.

3. S. spinosa L., *Cent. Pl.* **2**: 19 (1756). Much-branched,
pulvinate, puberulent dwarf shrub up to 20 cm, the branches
becoming spiny at apex. Leaves up to 10 × 3 mm, narrowly
obovate, acute, scabrid-puberulent. Verticillasters 2-flowered,
few; bracts shorter than or about as long as the verticillasters.
Calyx 2·5–4 mm, puberulent, the teeth slightly shorter than tube,
acute. Corolla 5–8 mm, white or pale lilac. ● *Kriti.* Cr.

4. S. rumelica Velen., *Fl. Bulg.* 466 (1891). Puberulent perennial with stout woody stock and flowering stems 10–20 cm.
Leaves *c.* 15 × 5 mm, spathulate, obtuse, sometimes mucronate,
scabrid-puberulent. Verticillasters 6- to 10-flowered, crowded
into a short raceme; bracts equalling or exceeding the verticillasters. Calyx 5–6 mm, puberulent, the lower 2 teeth setaceous,
about as long as tube, the upper 3 teeth *c.* ½ as long as the lower;
fruiting calyx gibbous at base. Corolla *c.* 8 mm. ● *S.C. Bulgaria.* Bu.

(5–11). S. montana group. Perennials with stout woody
stock. Leaves linear to obovate, usually acute. Verticillasters up
to 14-flowered. Calyx 3–7(–9) mm, the teeth shorter than to
longer than tube, acute. Corolla white to pale pink or bright
purple.

A variable group, in which specific limits are not very clear.
It is possible that **6–10** should be treated as subspecies of **5**.

1 Stem and leaves with numerous hairs at least 0·2 mm; flowering
stems usually less than 10 cm
2 Calyx-teeth ½ as long as tube; leaves sparsely glandular **8. athoa**
2 Calyx-teeth at least ¾ as long as tube; leaves densely glandular
3 Lower leaves obtuse; corolla 6–7 mm **6. parnassica**
3 Lower leaves acute or acuminate; corolla 8–12 mm **7. pilosa**
1 Stem and leaves with no or very few hairs exceeding 0·2 mm;
flowering stems usually more than 10 cm
4 Lower bracts 10–20 mm, exceeding the verticillasters;
verticillasters usually crowded **5. montana**
4 Lower bracts 3–10 mm, shorter than or about equalling the
verticillasters; verticillasters usually distant
5 Calyx-teeth glabrous; bracteoles ½ as long as to as long as
calyx **11. coerulea**
5 Calyx-teeth sparsely to densely puberulent; bracteoles not
more than *c.* ½ as long as calyx

6 Calyx distinctly punctate-glandular; leaves spathulate to
obovate, usually obtuse **10. obovata**
6 Calyx with a few minute glands; leaves narrowly elliptical
to spathulate, acute **9. cuneifolia**

5. S. montana L., *Sp. Pl.* 568 (1753). Flowering stems 10–
40(–70) cm, glabrous or puberulent. Leaves 5–30 × 1–5(–7) mm,
linear to oblanceolate, glabrous except for ciliate margins.
Verticillasters usually crowded; lower bracts 10–20 mm, exceeding the verticillasters. Corolla 6–12(–14) mm. *S. Europe.* Al
Bu Ga ?Gr Hs It Ju Rm Rs (K) Tu.

1 Stem puberulent all round
2 Calyx 4–6 mm, tubular; verticillasters dense, with the
peduncles and pedicels less than 5 mm **(a) subsp. montana**
2 Calyx 2·5–4(–4·5) mm, tubular-campanulate; verticillasters
usually lax, with at least some peduncles and pedicels more
than 5 mm **(b) subsp. variegata**
1 Stem glabrous, or puberulent on 2 sides only
3 Leaves mostly 4–7 mm wide **(d) subsp. kitaibelii**
3 Leaves 2–4 mm wide
4 Calyx 3·5–4 mm; corolla *c.* 8 mm **(e) subsp. taurica**
4 Calyx 5–9 mm; corolla 10–14 mm **(c) subsp. illyrica**

(a) Subsp. **montana**: Stem puberulent all round. Leaves 5–30 ×
1–5 mm. Verticillasters dense, with the peduncles and pedicels
less than 5 mm. Calyx 4–6 mm, tubular. Corolla 6–12 mm.
From Spain to S. Albania and N.W. Jugoslavia.

(b) Subsp. **variegata** (Host) P. W. Ball, *Bot. Jour. Linn. Soc.*
65: 352 (1972) (*S. variegata* Host): Like subsp. (a) but verticillasters lax, with at least some peduncles and pedicels more than
5 mm; calyx 2·5–4(–4·5) mm, tubular-campanulate; corolla
6–10 mm. ● *N.E. Italy, W. Jugoslavia.*

(c) Subsp. **illyrica** Nyman, *Consp.* 591 (1881) (*S. subspicata*
Bartl. ex Vis.): Stem glabrous, or puberulent on only 2 sides.
Leaves 10–30 × 2–3 mm. Verticillasters dense. Calyx 5–9 mm,
tubular. Corolla 10–14 mm. ● *W. Jugoslavia, Albania.*

(d) Subsp. **kitaibelii** (Wierzb.) P. W. Ball, *Bot. Jour. Linn. Soc.*
65: 352 (1972) (*S. kitaibelii* Wierzb.): Flowering stem 30–70 cm,
glabrous, or puberulent on 2 sides only. Leaves 15–30 × 4–7 mm.
Verticillasters usually lax and distant. Calyx 4–6 mm, tubular-
campanulate. Corolla 10–12 mm. 2n = 30. ● *N. part of Balkan
peninsula, S.W. Romania.*

(e) Subsp. **taurica** (Velen.) P. W. Ball, *op. cit.* (1972) (*S. taurica*
Velen.): Stem glabrous, or puberulent on 2 sides only. Leaves
15–25 × 2·5–4 mm. Verticillasters more or less distant. Calyx
3·5–4 mm, tubular-campanulate. Corolla *c.* 8 mm. ● *Krym.*

S. macedonica Form., *Verh. Naturf. Ver. Brünn* **37**: 186
(1899) (*S. montana* var. *pisidica* auct., non (Wettst.) Halácsy),
was described to cover plants from N. Greece which had previously been confused with *S. pisidica* Wettst., a species that does
not occur in Europe. It has stems glabrous or hairy on 2 sides
only, leaves 10–15 × 2–5 mm, verticillasters crowded, calyx
4–6 mm, the lips equalling or longer than the tube and corolla
8–10 mm. It may represent another subspecies of **5**.

6. S. parnassica Heldr. & Sart. ex Boiss., *Fl. Or.* **4**: 563
(1879). Stock usually stout and woody; flowering stems up to
10 cm, more or less hispid with numerous hairs at least 0·2 mm.
Leaves 5–10 × 2·5–4 mm, the lower obovate, obtuse, the upper
spathulate, acute, hispid, densely glandular. Verticillasters
crowded; bracts equalling or exceeding the verticillaster. Calyx
3·5–4 mm, hispid; teeth as long as or slightly shorter than tube.
Corolla 6–7 mm. ● *S. & S.C. Greece.* Gr.

7. S. pilosa Velen., *Sitz.-Ber. Böhm. Ges. Wiss.* (*Math.-Nat.
Kl.*) **1899**(40): 6 (1899). Like **6** but flowering stems up to

12(–20) cm; leaves 10–20 × 3–8 mm, oblong-oblanceolate to obovate, all acute or acuminate, usually sparsely hispid; calyx 4–5·5 mm, shortly hairy; teeth slightly shorter than the tube; corolla 8–12 mm. ● *Bulgaria.* Bu ?Ju.

8. S. athoa K. Malý, *Glasn. Muz. Bosni Herceg.* **22**: 690 (1910). Like **6** but leaves sparsely glandular; calyx-teeth ½ as long as tube. ● *N.E. Greece (Athos).* Gr.

9. S. cuneifolia Ten., *Fl. Nap.* **1**, *Prodr.*: 33 (1811). Puberulent to almost hispid; flowering stems 10–50 cm. Leaves 5–16 × 1·5–4 mm, the lower obovate, the upper linear-spathulate and often with inrolled margins, acute or subobtuse. Verticillasters distant; bracts 3–10 mm, usually shorter than the verticillaster; bracteoles not more than ½ as long as calyx. Calyx 3–5(–6) mm, sparsely to densely puberulent, with few, minute glands; teeth ¼ as long to as long as the tube. Corolla 5–10 mm. *Mediterranean region.* Al Hs It Ju.

Records for Bulgaria are referable to **7**; records for Greece are referable to **5**.

10. S. obovata Lag., *Gen. Sp. Nov.* 18 (1816). Like **9** but leaves up to 10 mm, spathulate to obovate, usually obtuse; calyx distinctly glandular-punctate; corolla 6–7 mm. ● *S. & E. Spain.* Hs.

Usually separable from **9**, although plants apparently referable to **10** have been found in Asiatic Turkey within the distributional range of **9** and somewhat intermediate specimens have been seen from Spain, Italy and the Balkan peninsula.

S. intricata Lange, *Vid. Meddel. Dansk Naturh. Foren. Kjøbenhavn* **1881**: 96 (1882), from the mountains of S.E. Spain, is similar to **10** but has ciliate leaf-margins like **5**; the plants are caespitose with flowering stems up to 10 cm and usually solitary flowers. They may be a high-altitude ecotype of **10**.

11. S. coerulea Janka in Velen., *Fl. Bulg.* 465 (1891). Puberulent; flowering stems 15–25 cm. Leaves 8–20 × 1·5–2(–3) mm, linear or linear-oblong, acute (the lower sometimes subobtuse), glabrous except for ciliate margin. Verticillasters 2- to 4-flowered, more or less distant; bracts up to 10 mm, shorter than or equalling verticillasters; bracteoles ½ as long to as long as calyx. Calyx 4–5 mm; tube hairy; teeth as long as tube, glabrous. Corolla 6–10 mm. ● *Bulgaria, E. Romania.* Bu Rm.

12. S. hortensis L., *Sp. Pl.* 568 (1753) (incl. *S. laxiflora* C. Koch, *S. pachyphylla* C. Koch). Puberulent annual 10–25 (–35) cm. Leaves 10–30(–40) × 1–4(–5) mm, linear or linear-lanceolate, obtuse. Verticillasters 2- to 5-flowered, crowded or lax, distant; bracts exceeding the verticillasters. Calyx 3–4 mm, puberulent; teeth slightly unequal, the upper about as long as tube, the lower distinctly longer than tube. Corolla 4–7 mm, white, pink or lilac. *Mediterranean region.* Al Ga Gr Hs It Ju [Bu].

Widely cultivated as a pot-herb.

26. Acinos Miller[1]

Like *Satureja* but flowers always in verticillasters; calyx tubular, 13-veined, the tube usually curved, constricted near the middle and gibbous at the base, the limb 2-lipped with the lower teeth longer than the upper; style-branches unequal, the upper subulate, the lower longer and wider.

---

[1] By P. W. Ball and F. M. Getliffe.

All species occur in dry, sunny habitats.

Intermediates between **2**, **3**, **4** and **5**, presumably of hybrid origin, can be found in those areas where two or more of these species occur.

1 Leaves with obscure lateral veins and a translucent or whitish margin **1. corsicus**
1 Leaves with conspicuous lateral veins and without a translucent or whitish margin
  2 Corolla (10–)12–20 mm; perennial
    3 Leaf-lamina mostly 1–2 times as long as wide, rounded or broadly cuneate at base, obtuse or acute **2. alpinus**
    3 Leaf-lamina mostly 2–3 times as long as wide, narrowly cuneate at base, acuminate or narrowly acute **3. suaveolens**
  2 Corolla 7–12 mm
    4 Leaf-lamina not more than 1½ times as long as wide, very prominently veined beneath, abruptly acuminate or mucronate **5. rotundifolius**
    4 Leaf-lamina usually at least 1½ times as long as wide, not prominently veined, obtuse or acute
      5 Perennial with branched woody stock; corolla 10–12 mm, usually distinctly exceeding the subtending bract **2. alpinus**
      5 Annual or short-lived perennial; corolla 7–10(–12) mm, not exceeding the subtending bract **4. arvensis**

1. A. corsicus (Pers.) Getliffe, *Bot. Jour. Linn. Soc.* **65**: 263 (1972) (*Thymus corsicus* Pers., *Calamintha corsica* (Pers.) Bentham). Procumbent, sparsely to moderately densely pubescent perennial up to 10 cm. Leaves 4–10 × 3–5 mm, obovate-spathulate to suborbicular, obtuse, entire with a translucent or whitish margin, the lateral veins very obscure. Verticillasters 2- to 4-flowered. Calyx 6–7 mm, the tube only slightly contracted in the middle. Corolla 12–16 mm, violet. *Mountain rocks and screes.* ● *Corse.* Co.

2. A. alpinus (L.) Moench, *Meth.* 407 (1794) (*Calamintha alpina* (L.) Lam., *Satureja alpina* (L.) Scheele). Sparsely to densely pubescent perennial up to 45 cm. Leaves 6–20 × 4–16 mm, elliptical to suborbicular, acute or obtuse, rounded or broadly cuneate at base, entire or denticulate towards the apex, the lateral veins conspicuous. Verticillasters 3- to 8-flowered, the flowers exceeding the bracts. Calyx 5–8 mm; lower lip 1·5–3 mm. Corolla 10–20 mm, violet with white markings on the lower lip. *C. & S. Europe, mainly in the mountains.* Al Au Bu Cr Cz Ga Ge Gr He Hs It Ju Lu Po Rm Rs (W) Sa Si ?Tu.

1 Upper lip of the calyx with teeth 0·5–1 mm; hairs on the calyx crispate **(c) subsp. meridionalis**
1 Upper lip of the calyx with teeth 1–2 mm
  2 Leaves without prominent veins beneath; hairs on the calyx usually straight **(a) subsp. alpinus**
  2 Leaves with prominent veins beneath, particularly near the margin; hairs on the calyx usually crispate **(b) subsp. majoranifolius**

(a) Subsp. **alpinus**: Stem usually not more than 30 cm. Leaves up to 15 × 11 mm, without prominent veins beneath. Calyx 6–8 mm, with usually straight hairs; upper lip with teeth 1–2 mm. Corolla 12–20 mm. $2n = 18$. *C. Europe and parts of S. Europe.*

(b) Subsp. **majoranifolius** (Miller) P. W. Ball, *Bot. Jour. Linn. Soc.* **65**: 344 (1972) (*Melissa majoranifolia* Miller, *Calamintha alpina* subsp. *majoranifolia* (Miller) Hayek and incl. subsp. *hungarica* (Simonkai) Hayek): Like subsp. (a) but stem up to 45 cm; leaves up to 20 × 16 mm, with prominent veins beneath, particularly near the margin; calyx usually with crispate hairs. *S.E. Europe, usually at lower altitudes than subsp.* (a).

Plants intermediate between subspp. (a) and (b) have been given subspecific status. Examples are **Calamintha alpina** subsp.

nomismophylla Rech. fil., *Denkschr. Akad. Wiss. Math.-Nat. Kl.* (*Wien*) 105(1): 529 (1943), and subsp. **elatior** (Griseb.) Rech. fil., *loc. cit.* (1943), both from N. Greece (Khalkidhiki). The latter is like subsp. (a) but has short hairs on the calyx. They may merit recognition but further study is required throughout the range of the species.

(c) Subsp. **meridionalis** (Nyman) P. W. Ball, *Bot. Jour. Linn. Soc.* 65: 344 (1972) (*Calamintha alpina* subsp. *meridionalis* Nyman, *C. granatensis* Boiss. & Reuter): Like subsp. (a) but calyx 5–7 mm, with short crispate hairs; upper lip with teeth 0·5–1 mm; corolla 10–14 mm. $2n=18$. *S. Europe.*

3. **A. suaveolens** (Sibth. & Sm.) G. Don fil. in Loudon, *Hort. Brit.* 239 (1830) (*Calamintha suaveolens* (Sibth. & Sm.) Boiss.). Like 2(a) but leaves 10–20 × 3·5–8 mm, lanceolate or elliptic-lanceolate, acuminate or narrowly acute, narrowly cuneate at base, remotely serrulate; lower lip of calyx 2·5–3·5 mm. *Balkan peninsula, S. & W. Romania, C. & S. Italy.* ?Al Bu Gr It Ju Rm ?Tu.

4. **A. arvensis** (Lam.) Dandy, *Jour. Ecol.* 33: 326 (1946) (*A. thymoides* Moench, *Calamintha acinos* (L.) Clairv., *Satureja acinos* (L.) Scheele). Pubescent annual, and perhaps rarely a short-lived perennial, up to 40 cm. Leaves 8–15 × 4–7 mm, lanceolate to ovate, obtuse or acute, entire or denticulate, conspicuously veined beneath, the veins not prominent. Verticillasters 3- to 8-flowered, the flowers usually not exceeding the bracts. Calyx 5–7 mm; lower lip 1·5–2·5 mm; upper lip with teeth 0·7–1·5 mm. Corolla 7–10(–12) mm, usually violet with white marks on the lower lip. $2n=18$. *Most of Europe except the extreme north and parts of the south.* All except Az Cr Fa Is Lu Sb ?Si.

5. **A. rotundifolius** Pers., *Syn. Pl.* 2: 131 (1806) (*A. graveolens* (Bieb.) Link, *A. fominii* Schost., *Calamintha rotundifolia* (Pers.) Bentham, *C. exigua* (Sibth. & Sm.) Hayek, *C. graveolens* (Bieb.) Bentham, ?*C. maritima* Bentham). Like 4 but leaves 7–10 mm wide, obovate-orbicular to orbicular, acuminate or obtuse and mucronate at apex, prominently veined beneath; calyx 5–8 mm; lower lip 1·5–3 mm. *S.E. Europe; C. Italy and Sicilia; C. & S. Spain.* Bu Cr Gr Hs It Ju Rm Rs (W, K, E) Si Tu.

27. Calamintha Miller[1]

Perennial herbs, sometimes woody at the base. Flowers in opposite, axillary, usually pedunculate cymes. Calyx tubular, 13-veined, more or less 2-lipped; upper lip 3-toothed; lower lip 2-toothed and longer than the upper; tube straight, not gibbous, hairy in the mouth. Corolla 2-lipped; upper lip entire or emarginate; lower lip 3-lobed, the middle lobe the largest; tube straight. Stamens included, curved, convergent; anthercells divergent. Style-branches unequal, the upper subulate, the lower longer and wider.

Measurements of the calyx refer to its length at anthesis; it is often considerably shorter in fruit.

1 Calyx (10–)12–16 mm; corolla 25–40 mm   **1. grandiflora**
1 Calyx 3–10 mm; corolla not more than 22 mm

[1] By P. W. Ball and F. Getliffe.

2 Calyx with dense, long, patent hairs; leaves densely grey-tomentose   **4. cretica**
2 Calyx with short, appressed or crispate hairs; or with sparse long, patent hairs
3 Leaves densely grey-tomentose on both surfaces; upper calyx-teeth less than 0·5 mm   **5. incana**
3 Leaves green and relatively sparsely hairy, at least on the upper surface; upper calyx-teeth 0·5–2 mm
4 Lower calyx-teeth 2–4 mm, usually densely long-ciliate; hairs in mouth of calyx ± included   **2. sylvatica**
4 Lower calyx-teeth 1–2 mm, without or with very few long cilia; hairs in mouth of calyx somewhat exserted   **3. nepeta**

1. **C. grandiflora** (L.) Moench, *Meth.* 408 (1794) (*Satureja grandiflora* (L.) Scheele). Sparsely hairy perennial 20–60 cm. Leaves 30–80 × 20–50 mm, ovate or ovate-oblong, acute, coarsely dentate or serrate, with 6 or more teeth on each side. Cymes 1- to 5-flowered. Calyx (10–)12–16 mm, subglabrous; upper teeth 2–3 mm; lower teeth 3·5–5 mm, long-ciliate. Corolla 25–40 mm, pink. $2n=22$. *S. & S.C. Europe, from N.E. Spain eastwards.* Al Au Bu Co Ga Gr He Hs It Ju Rm Rs (K) Si Tu.

2. **C. sylvatica** Bromf., *Phytologist* (*Newman*) 2: 49 (1845). Pubescent, stoloniferous perennial 30–80 cm. Leaves (15–)20–70 × 10–45 mm, ovate or orbicular-ovate, subacute to obtuse, subentire, coarsely dentate or crenate-serrate, with 5–10 teeth on each side. Cymes 3- to 9-flowered; peduncle up to 15 mm; secondary branches 0–5 mm. Calyx 6–10 mm, subglabrous or pubescent, the hairs in the mouth included; upper teeth 1·5–2 mm; lower teeth 2–4 mm, long-ciliate. Corolla 10–22 mm, pink or lilac with white spots on the lower lip. *W., S. & S.C. Europe.* Al Au Az *Be Bl Br Bu Co Cz Ga Ge Hb He Ho Hs Hu It Ju Lu Po Rm Rs (W) Si ?Tu.

(a) Subsp. **sylvatica** (*Satureja calamintha* subsp. *officinalis* sensu Gams). Plant 30–80 cm. Leaves 25–70 × 18–45 mm, the middle and upper ovate, coarsely dentate or crenate-serrate, with 6–10 teeth on each side, the lower suborbicular. Peduncle up to 15 mm. Lower calyx-teeth 3–4 mm. Corolla 15–22 mm. $2n=24$. *From S. England and France eastwards to S.W. Ukraine.*
(b) Subsp. **ascendens** (Jordan) P. W. Ball, *Bot. Jour. Linn. Soc.* 65: 346 (1972) (*C. ascendens* Jordan, *C. hirta* (Briq.) Hayek, *C. menthifolia* auct., non Host, *C. officinalis* auct. excl. Gams, non Moench, *Satureja calamintha* subsp. *ascendens* (Jordan) Briq. & subsp. *menthifolia* sensu Gams). Plant 30–60 cm. Leaves (15–)20–40(–50) × 10–35 mm, ovate or orbicular-ovate, subentire or shallowly crenate-serrate, with 5–8 teeth on each side. Peduncle 0·5(–10) mm. Lower calyx-teeth 2–3·5 mm. Corolla 10–16 mm. $2n=48$. *Almost throughout the range of the species.*

3. **C. nepeta** (L.) Savi, *Fl. Pis.* 2: 63 (1798). Sparsely to densely pubescent perennial 30–80 cm. Leaves 10–35(–45) × 8–25(–30) mm, broadly ovate, obtuse, subentire or shallowly to deeply crenate-serrate with up to 9 teeth on each side. Cymes 5- to 20-flowered; peduncle absent or up to 22 mm; secondary branches absent or up to 10 mm. Calyx 3–7 × 1–1·5(–2) mm, sparsely to densely puberulent or pubescent, the hairs in the mouth exserted; upper teeth 0·5–1·5 mm, narrowly or broadly triangular; lower teeth 1–2 mm, puberulent, rarely with a few long cilia. Corolla (9–)10–15 mm, white or lilac. *S., W. & S.C. Europe, northwards to E. England.* Al Au Bl Br Bu Co Cr Ga Gr He Hs Hu It Ju Rs (W, K) Sa Si Tu [Ge].

Cymes (5–)10- to 20-flowered, the peduncle 8–22 mm, the secondary branches 5–10 mm; leaves 20–35(–45) mm, with 5–9 teeth on each side   **(a) subsp. nepeta**

Cymes 5- to 11(–15)-flowered, the peduncle 0–5(–10) mm, the secondary branches 0–5 mm; leaves 10–20(–25) mm, subentire or with up to 5 teeth on each side   **(b) subsp. glandulosa**

(a) Subsp. **nepeta** (*C. nepetoides* Jordan, *C. thessala* Hausskn., *Satureja calamintha* subsp. *nepetoides* (Jordan) Br.-Bl.): Leaves 20–35(–45) × 12–25(–30) mm, crenate-serrate with 5–9 teeth on each side. Cymes (5–)10- to 20-flowered; peduncle 8–22 mm; secondary branches 5–10 mm. Upper calyx-teeth 0·7–1·5 mm, narrowly triangular. $2n = 24$. *Mountains of S. & S.C. Europe.*

(b) Subsp. **glandulosa** (Req.) P. W. Ball, *Bot. Jour. Linn. Soc.* **65**: 347 (1972) (*Thymus glandulosus* Req., *C. glandulosa* (Req.) Bentham, *C. officinalis* Moench, *Satureja calamintha* subsp. *glandulosa* (Req.) Gams, subsp. *nepeta* sensu Briq. & subsp. *subnuda* (Waldst. & Kit.) Gams): Leaves 10–20(–25) × 8–12 mm, subentire or shallowly crenate-serrate with up to 5 teeth on each side. Cymes 5- to 11(–15)-flowered; peduncle 0–5(–10) mm; secondary branches 0–5 mm. Upper calyx-teeth 0·5–1 mm, broadly triangular. $2n = 20, 24$. *S. & W. Europe.*

4. C. cretica (L.) Lam., *Fl. Fr.* **2**: 395 (1778). Densely grey-pubescent perennial 10–30 cm, woody at base. Leaves 6–15 × 6–10 mm, broadly ovate, obtuse, tomentose or villous-tomentose, subentire or shallowly crenate-serrate with up to 5 teeth on each side. Cymes 1- to 3(–6)-flowered; peduncle 0–4 mm, secondary branches usually absent. Calyx 4–5 mm, with dense, long, patent hairs, the hairs in the mouth exserted; upper teeth 0·7–1 mm; lower teeth *c.* 2 mm, densely patent-pubescent. Corolla *c.* 10 mm, white. *Stony places.* ● *Kriti.* Cr.

5. C. incana (Sibth. & Sm.) Boiss., *Pl. Atticae Heldr.* **1844** (1845). Densely grey-tomentose perennial 10–40 cm, woody at base. Leaves 5–12 × 5–12 mm, suborbicular, subentire. Cymes 2- to 6-flowered; peduncle 0–5 mm, secondary branches absent. Calyx 3·5–5 mm, with dense, short, appressed or crispate hairs, the hairs in the mouth exserted; upper teeth 0·2–0·4 mm; lower teeth 1–1·5 mm, puberulent, or with a few long cilia. Corolla 8–11 mm, pale pink. *S. Greece and Aegean region.* ?Cr Gr.

28. Clinopodium L.[1]

Like *Calamintha* but verticillasters dense, many-flowered; calyx-tube curved.

Literature: R. von Bothmer, *Bot. Not.* **120**: 202–208 (1967).

1. C. vulgare L., *Sp. Pl.* 587 (1753) (*Calamintha clinopodium* Bentham, *C. vulgaris* (L.) Halácsy, non Clairv., *Satureja vulgaris* (L.) Fritsch). Pubescent to densely villous perennial herb 30–80 cm. Leaves 20–65 × 10–30 mm, ovate-lanceolate or ovate, subobtuse, rounded or cuneate at base, shallowly and remotely crenate-serrate, petiolate. Verticillasters distant or sometimes the upper crowded; bracts linear, almost as long as the calyx, plumose. Corolla 12–22 mm, pinkish-purple. *Scrub and open woodland. Most of Europe, northwards to c. 66° N. in Norway.* All except Bl Fa Is Sb.

(a) Subsp. **vulgare**: Calyx 7–9·5 mm; lower teeth 2·5–4 mm; upper teeth up to 2·5 mm. $2n = 20$. *Almost throughout the range of the species, but absent from parts of S. Europe.*

(b) Subsp. **arundanum** (Boiss.) Nyman, *Consp.* 587 (1881): Calyx 9·5–12 mm; lower teeth 4–5·5 mm; upper teeth 2·5–4 mm. $2n = 20$. *S. Europe.*

---

[1] By P. W. Ball and F. Getliffe.
[2] By A. O. Chater and E. Guinea.

29. Micromeria Bentham[2]

Perennial herbs or dwarf shrubs. Calyx tubular, 13(15)-veined, sometimes actinomorphic and straight, sometimes zygomorphic and somewhat curved, scarcely gibbous and somewhat 2-lipped with rather unequal teeth, hairy or glabrous in the mouth. Corolla 2-lipped, with straight tube. Stamens shorter than corolla, curved, convergent. Style-branches subequal, subulate.

All species, unless another habitat is specified, grow on rocks, or occasionally on walls and in other dry, open habitats.

1 Calyx actinomorphic, straight, with ± equal teeth; leaves often crenate or serrate
  2 Calyx-teeth triangular, about as long as wide, ± obtuse
    3 Calyx villous in throat   **1. fruticosa**
    3 Calyx glabrous in throat
      4 Stems 20–50 cm, glabrous or sparsely puberulent; leaves glabrous   **2. thymifolia**
      4 Stems 4–10 cm, velutinous; leaves whitish-velutinous   **3. taygetea**
  2 Calyx-teeth lanceolate to subulate, much longer than wide, acute
    5 Middle and upper leaves linear, not more than 12 mm
      6 Flowers sessile; verticillasters often subsessile
        7 Calyx glabrous in throat, the teeth *c.* ½ as long as tube; nutlets acute   **14. juliana**
        7 Calyx villous in throat, the teeth *c.* ¼ as long as tube; nutlets obtuse   **15. myrtifolia**
      6 Flowers pedicellate; verticillasters pedunculate
        8 Leaves mostly overlapping   **16. cristata**
        8 Leaves mostly not overlapping   **17. cremnophila**
    5 Middle and upper leaves ovate to ovate-lanceolate, the larger more than 12 mm
      9 Calyx 2·5–3 mm; teeth *c.* ⅔ as long as tube   **4. dalmatica**
      9 Calyx 3·5–4 mm; teeth ⅓–½ as long as tube
        10 Stems erect, straight; corolla 7–9 mm   **5. pulegium**
        10 Stems ascending; corolla 5–6 mm   **6. frivaldszkyana**
1 Calyx zygomorphic, often curved, with unequal teeth; leaves entire
  11 Ericoid dwarf shrub with acicular leaves 2–3 × *c.* 0·5 mm   **21. inodora**
  11 Not an ericoid dwarf shrub; leaves, at least the floral, more than 3 mm
    12 The upper, or at least the floral, leaves linear-lanceolate or linear
      13 Flowers sessile; verticillasters often subsessile   **14. juliana**
      13 Flowers usually distinctly pedicellate; verticillasters pedunculate
        14 Nutlets acuminate
          15 Leaves mostly overlapping   **16. cristata**
          15 Leaves mostly not overlapping   **17. cremnophila**
        14 Nutlets obtuse (rarely obtuse and apiculate)
          16 Calyx with long, dense, patent hairs, the rest of the plant sparsely or more shortly and always much less conspicuously hairy   **13. nervosa**
          16 Whole plant more or less uniformly hairy
            17 Corolla 3–4 mm   **18. kerneri**
            17 Corolla more than 4 mm
              18 Peduncles ± patent, the verticillasters with 2–6 flowers and greatly exceeding the subtending leaves   **19. parviflora**
              18 Peduncles erect or erecto-patent, the verticillasters with (2–)6–18 flowers and usually not or only slightly exceeding the subtending leaves   **20. graeca**
    12 Upper leaves, including the floral, ovate or triangular (rarely oblong-lanceolate)
      19 Leaves 2–6 mm
        20 Lower leaves cordate at base   **7. filiformis**
        20 Lower leaves rounded or cuneate at base

21 Leaves oblong-lanceolate, obtuse   **9. acropolitana**
21 Leaves triangular-ovate to elliptical, acute
  22 Calyx 4–5 mm; plant densely patent-pubescent **10. hispida**
  22 Calyx 2·5–3·5 mm; plant subglabrous to ±sparsely
      pubescent   **8. microphylla**
19 At least some leaves more than 6 mm
  23 Calyx villous, with patent hairs more than 0·5 mm
      13. nervosa
  23 Calyx glabrous or pubescent, with hairs less than 0·5 mm
    24 Calyx c. 3 mm; corolla 5–9 mm   **2. thymifolia**
    24 Calyx 5–8 mm; corolla 10–16 mm
      25 Densely patent-pubescent; calyx with 15 veins
        12. croatica
      25 Puberulent or sparsely pubescent; calyx with 13 veins
        11. marginata

1. M. fruticosa (L.) Druce, *Rep. Bot. Exch. Club Brit. Is.* **3**: 421 (1914) (*M. marifolia* (Cav.) Bentham). Dwarf shrub 30–60 cm, minutely whitish-velutinous; branches divaricate. Leaves 5–20 × 2–12 mm, lanceolate to ovate-orbicular, entire to weakly sinuate or dentate, flat. Verticillasters with 10–30(–50) flowers, lax, greatly exceeding subtending leaves. Calyx (1·5–)2·5–3·5 mm, obovoid-tubular to obconic, densely villous in throat; teeth c. ¼ as long as tube, triangular, equal. Corolla 5–10 mm, whitish. Nutlets obtuse. *Usually calcicole. E. Spain; C. Italy; S.W. Jugoslavia; Krym.* ?Gr Hs It Ju Rs (K).

(a) Subsp. **fruticosa**: Leaves strongly punctate beneath. Calyx and pedicels velutinous; calyx obovoid-tubular. ● *E. Spain; C. Italy.*
(b) Subsp. **serpyllifolia** (Bieb.) P. H. Davis, *Kew Bull.* [6]: 77 (1951) (*M. serpyllifolia* (Bieb.) Boiss.): Leaves not punctate beneath. Calyx and pedicels puberulent; calyx obconic. *Krym; S.W. Jugoslavia.* (*S.W. Asia.*)

2. M. thymifolia (Scop.) Fritsch in A. Kerner, *Sched. Fl. Exsicc. Austro-Hung.* **8**: 119 (1899) (*Satureja thymifolia* Scop.). Stems 20–50 cm, erect, mostly branched, glabrous or minutely deflexed-puberulent. Leaves 5–20 × 3–12 mm, elliptical to ovate, obtuse, remotely crenate-dentate or subentire, glabrous, punctate beneath. Verticillasters with (2–)10–30(–40) flowers, dense, shortly pedunculate, the lower shorter than subtending leaves. Calyx c. 3 mm, glabrous; teeth ⅙–¼ as long as tube, triangular, equal or unequal. Corolla 5–9 mm, white and violet. Nutlets obtuse or subacute. ● *W. part of Balkan peninsula, extending to Hungary and N. Italy.* Al Hu It Ju.

3. M. taygetea P. H. Davis, *Kew Bull.* [4]: 110 (1949). Stems 4–10 cm, erect, simple or branched, whitish-velutinous. Leaves up to 10 × 6 mm, broadly ovate, obtuse, entire, whitish-velutinous, somewhat punctate. Verticillasters with 2–10 flowers, dense, shortly pedunculate, the lower shorter than the subtending leaves. Calyx 2–3 mm, puberulent-velutinous, glabrous in throat; teeth c. ⅓ as long as tube, triangular, equal. Corolla c. 6 mm, purplish-violet. Nutlets obtuse. *Calcicole.* ● *S. Greece (Taïyetos).* Gr.

4. M. dalmatica Bentham in DC., *Prodr.* **12**: 225 (1848). Stems 20–50 cm, ascending or erect, simple or branched, deflexed-puberulent or pubescent. Leaves 10–20 × 3–15 mm, ovate, crenate-dentate, puberulent, densely punctate beneath. Verticillasters with (2–)10–60 flowers, dense, shortly pedunculate, longer than the subtending leaves or the lowest shorter. Calyx 2·5–3 mm, puberulent, glabrous or minutely hairy in throat; teeth c. ⅔ as long as tube, triangular-acuminate to lanceolate-subulate, equal. Corolla 5–6 mm, whitish or pale lilac. Nutlets obtuse or acuminate. ● *Balkan peninsula, from C. Jugoslavia to N.E. Greece.* Bu Gr Ju.

(a) Subsp. **dalmatica**: Stems paniculately branched, puberulent. Leaves puberulent chiefly on the veins beneath. *C. Jugoslavia.*
(b) Subsp. **bulgarica** (Velen.) Guinea, *Bot. Jour. Linn. Soc.* **64**: 381 (1971) (*M. origanifolia* subsp. *bulgarica* Velen., *M. bulgarica* (Velen.) Hayek): Stems simple, often pubescent. Leaves densely puberulent or pubescent. *S. Bulgaria and N.E. Greece.*

5. M. pulegium (Rochel) Bentham, *Lab. Gen. Sp.* 382 (1834). Stems 20–50(–90) cm, erect, straight, usually simple, deflexed-puberulent or -pubescent. Leaves 10–30 × 5–20 mm, ovate, obtuse to acute, more or less crenate-dentate, shortly appressed-pubescent, densely punctate beneath. Verticillasters with 10–70 flowers, lax, the lowest shorter and the upper longer than the subtending leaves. Calyx 3·5–4 mm, shortly pubescent, sparsely hairy in throat; teeth ⅓–½ as long as tube, linear-lanceolate or -setaceous, equal. Corolla 7–9 mm, white or lilac. Nutlets obtuse. ● *S.W. Romania.* Rm.

6. M. frivaldszkyana (Degen) Velen., *Österr. Bot. Zeitschr.* **49**: 291 (1899). Like **5** but stems not more than 30 cm, ascending, flexuous; leaves subglabrous; verticillasters with 6–14 flowers; calyx subglabrous, sometimes glabrous in throat; corolla 5–6 mm. ● *Mountains of C. & S. Bulgaria.* Bu.

7. M. filiformis (Aiton) Bentham, *Lab. Gen. Sp.* 378 (1834). Subglabrous or sparsely puberulent (rarely pubescent) dwarf shrub 5–20 cm with filiform, ascending, often branched stems. Leaves 2–4·5 × 1·5–4 mm, ovate to triangular, the lower cordate and the upper rounded at the base, subobtuse, entire, flat, subsessile. Verticillasters with 2–4 patent flowers; pedicels or peduncles ascending, usually equalling or exceeding subtending leaves. Calyx 3–4 mm, somewhat villous in throat; teeth c. ⅔ as long as tube, subulate, unequal. Corolla 4·5–5·5 mm, white. 2n = 30. ● *Islas Baleares, Corse, Sardegna.* Bl Co Sa.

8. M. microphylla (D'Urv.) Bentham, *op. cit.* 377 (1834). Subglabrous to somewhat pubescent dwarf shrub 10–30 cm with more or less filiform, procumbent or ascending, often branched stems. Leaves 3–6 × 2–4 mm, triangular-ovate to elliptical, the upper sometimes narrowly elliptical, rounded or cuneate at base, acute, entire, flat, subsessile. Verticillasters with 1–6 usually erecto-patent flowers; peduncles or pedicels c. ½ as long as subtending leaves. Calyx 2·5–3·5 mm, patent-pubescent, villous in throat; teeth c. ½ as long as tube, lanceolate-acuminate to -subulate, unequal. Corolla 5–8 mm, purple. ● *S. Italy and Sicilia; Kriti and Karpathos.* ?Bl Cr It Si.

The plants from Kriti and Karpathos have been described as **M. sphaciotica** Boiss. & Heldr. ex Bentham in DC., *Prodr.* **12**: 220 (1848) (*M. carpatha* Rech. fil.); they have a distinctive habit, longer leaves (the upper often narrower) and calyx, fewer flowers, and calyx-teeth subulate from a lanceolate base, but there is some doubt about their taxonomic status.

9. M. acropolitana Halácsy, *Consp. Fl. Graec., Suppl.* 87 (1908). Like **8** but stems erect; leaves c. 5 × 1–2 mm, oblong-lanceolate, obtuse; verticillasters with up to 10 flowers; peduncles or pedicels as long as subtending leaves. ● *C. Greece.* Gr.

10. M. hispida Boiss. & Heldr. ex Bentham in DC., *Prodr.* **12**: 215 (1848). Like **8** but densely patent-pubescent throughout; calyx 4–5 mm; teeth c. ⅓ as long as tube. ● *Kriti.* Cr.

11. M. marginata (Sm.) Chater, *Bot. Jour. Linn. Soc.* **64**: 381 (1971) (*M. piperella* (Bertol.) Bentham, *Thymus marginatus*

Sm.). Dwarf shrub 10–20 cm, with ascending, mostly simple, puberulent stems. Leaves (3–)6–12 × 3–5 mm, broadly ovate, entire, obtuse, rounded or subcordate at base, glabrous or sparsely hairy beneath. Verticillasters with 2–12 flowers, lax, pedunculate, exceeding the subtending leaves. Calyx 5–8 mm, pubescent, 13-veined, villous in throat; teeth *c.* ½ as long as tube, subulate, unequal. Corolla 12–16 mm, purplish or violet. ● *Maritime Alps.* Ga It.

12. M. croatica (Pers.) Schott, *Österr. Bot. Zeitschr.* **7**: 93 (1857). Like **11** but most of plant densely patent-pubescent; leaves up to 8 mm wide, somewhat repand, subacute; calyx 15-veined; teeth *c.* ⅔ as long as tube; corolla 10–15 mm. ● *Jugoslavia.* Ju.

13. M. nervosa (Desf.) Bentham, *Lab. Gen. Sp.* 376 (1834). Dwarf shrub 10–40(–50) cm; stems many, erect or ascending, more or less simple, deflexed-puberulent. Leaves 7–10 × 4–5 mm, ovate to ovate-lanceolate, acute, very shortly petiolate. Verticillasters with 4–20 flowers, fairly dense, shortly pedunculate, about equalling subtending leaves; pedicels less than ½ as long as calyx. Calyx 3–4 mm, with long, dense patent hairs, villous in throat; teeth *c.* ⅔ as long as tube, lanceolate-subulate, unequal. Corolla 4–6 mm, purplish. 2*n* = 30. *Mediterranean region.* Bl Cr Gr ?Hs It Si.

M. tapeinantha Rech. fil., *Denkschr. Akad. Wiss. Math.-Nat. Kl.* (*Wien*) **105**(2, 1): 123 (1943), described from Kriti, requires further investigation; it sometimes has the densely hairy calyx of **13**, but in other respects is close to **20** and may perhaps be of hybrid origin.

14. M. juliana (L.) Bentham ex Reichenb., *Fl. Germ. Excurs.* 311 (1831). Dwarf shrub 10–40 cm, puberulent to pubescent throughout; stems many, erect, mostly simple. Leaves 3–8 × 1–2·5 mm, the lowest ovate, the upper linear-lanceolate or -oblong, obtuse, entire, with revolute margins. Verticillasters sessile or shortly pedunculate, with 4–20 strict, sessile flowers; bracts about equalling calyx. Calyx 2·5–3·5 mm, straight, glabrous in throat; teeth ½ as long as tube, subulate, rigid, slightly unequal. Corolla *c.* 5 mm, purplish. Nutlets acute. 2*n* = 30. *Mediterranean region, westwards to S.E. France; C. Portugal.* Al Bu ?Co Cr Ga Gr It Ju Lu Si Tu.

15. M. myrtifolia Boiss. & Hohen. in Boiss., *Diagn. Pl. Or. Nov.* **1**(5): 19 (1844). Like **14** but calyx villous in throat, the teeth *c.* ¼ as long as tube; nutlets obtuse. 2*n* = 30. *Greece and Aegean region.* Cr Gr Tu.

16. M. cristata (Hampe) Griseb., *Spicil. Fl. Rumel.* **2**: 122 (1844). Stems 5–20 cm, many, erect, slender, simple, shortly and densely patent-pubescent. Leaves 5–7(–12) × 0·5–1·5 (–3) mm, the lower oblong or elliptical, the rest linear, puberulent or appressed-pubescent, overlapping throughout most of the stem, with revolute margins. Verticillasters with (2–)6(–10) flowers, shortly pedunculate and pedicellate, slightly shorter than subtending leaves. Calyx 3–4 mm, pubescent, villous in throat; teeth ¼–½ as long as tube, lanceolate to linear-lanceolate, slightly unequal. Corolla 4–6 mm, purplish. Nutlets acuminate. *N. & C. parts of Balkan peninsula.* Al Bu Gr Ju.

17. M. cremnophila Boiss. & Heldr. in Boiss., *Fl. Or.* **4**: 570 (1879). Like **16** but stems puberulent; leaves mostly not overlapping; verticillasters mostly longer than subtending leaves; calyx 2·5–3·5 mm; corolla 3·5 mm. *C. & N.W. Greece, Albania.* Al Gr.

18. M. kerneri Murb., *Lunds Univ. Årsskr.* **27**(5): 53 (1892). Stems 20–30 cm, many, erect, slender, patent-pubescent. Leaves 3–9 × 2–5 mm, shortly pubescent at least beneath, the lowest triangular-cordate, the middle ovate or ovate-lanceolate, obtuse, flat, the upper lanceolate or linear with revolute margin. Verticillasters with 6–10 flowers, shortly pedunculate, equalling or longer than subtending leaves; bracts *c.* ½ as long as calyx. Calyx *c.* 3 mm, sparsely hairy on veins and in throat; teeth *c.* ½ as long as tube, lanceolate-subulate, unequal. Corolla 3–4 mm, purplish. Nutlets obtuse. ● *W. Jugoslavia.* Ju.

19. M. parviflora (Vis.) Reichenb., *Fl. Germ. Excurs.* 859 (1832). Stems 5–25 cm, many, ascending, slender, often branched, deflexed-puberulent. Leaves 4–10 × 1–5 mm, glabrous or sparsely puberulent, the lowest cordate-orbicular to ovate, the upper oblong to linear with somewhat revolute margin. Verticillasters long-pedunculate, with 2–6 usually patent, long-pedicellate flowers, much exceeding the subtending leaves; bracts less than ½ as long as calyx. Calyx 3·5–4 mm, glabrous or sparsely puberulent, shortly villous in throat; teeth ½–⅔ as long as tube, lanceolate, unequal. Corolla 5–7 mm, purplish. Nutlets obtuse. ● *S. Jugoslavia, Albania.* Al Ju.

20. M. graeca (L.) Bentham ex Reichenb., *op. cit.* 311 (1831) (*Satureja graeca* L.). Sparsely pubescent to hirsute or hispid dwarf shrub 10–50 cm with many erect or ascending, usually simple stems. Leaves 5–12 × 2–7 mm, the lower ovate, acute, the upper lanceolate or linear-lanceolate with revolute margins. Verticillasters shortly pedunculate with (2–)6–18 usually erecto-patent, shortly pedicellate flowers, usually about equalling the subtending leaves; bracts mostly not more than ½ as long as calyx. Calyx (2–)3–5 mm, villous in throat; teeth ⅓–¾ as long as tube, lanceolate-subulate, unequal. Corolla 6–8(–13) mm, purplish. Nutlets obtuse or shortly apiculate. *Mediterranean region, C. & S. Portugal.* Bl Co Cr Ga Gr Hs It Ju Lu Sa Si Tu.

1 Axillary fascicles of leaves present on stems; corolla-tube 4–6 mm longer than calyx-tube **(g) subsp. fruticulosa**
1 Axillary fascicles of leaves absent
  2 Flowering in autumn; lower leaves ovate, fugacious, the rest linear; corolla-tube 5–6 mm longer than calyx-tube; minutely scabrid-puberulent **(f) subsp. consentina**
  2 Flowering usually in spring; lower leaves ovate-lanceolate, the rest linear or linear-lanceolate, or all except the floral leaves ovate
    3 Verticillasters rather strict, mostly with 2–4 subsessile flowers; nutlets shortly apiculate **(e) subsp. tenuifolia**
    3 Verticillasters ± divaricate, mostly with 4–18 shortly pedicellate flowers; nutlets obtuse, not apiculate
     4 Pubescent; corolla-tube 2–4 mm longer than calyx-tube
      5 Middle and upper leaves ovate to linear-lanceolate; stems stout **(a) subsp. graeca**
      5 Middle and upper leaves linear; stems slender, filiform **(b) subsp. imperica**
     4 Hispid or tomentose; corolla-tube 4–6 mm longer than calyx-tube
      6 Tomentose; upper leaves linear; verticillasters lax, the peduncles distinctly divergent **(c) subsp. garganica**
      6 Hispid; upper leaves linear-lanceolate; verticillasters more dense, the peduncles suberect or erecto-patent **(d) subsp. longiflora**

(a) Subsp. graeca: Plant pubescent; stems stout, without axillary fascicles of leaves. Lower leaves ovate-lanceolate, the rest ovate to linear-lanceolate. Verticillasters with 4–18 shortly pedicellate flowers; peduncles more or less divaricate. Corolla-tube 2–4 mm longer than calyx-tube. Flowering in spring. Nutlets obtuse. 2*n* = 20. *Throughout the range of the species.*

(b) Subsp. imperica Chater, *Bot. Jour. Linn. Soc.* **64**: 381

(1971) (*M. thymoides* De Not., non (Solander ex Lowe) Webb & Berth.): Like subsp. (**a**) but stems slender, filiform; middle and upper leaves linear. ● *N.W. Italy* (*prov. Imperia*).

(**c**) Subsp. **garganica** (Briq.) Guinea, *loc. cit.* (1971) (*Satureja graeca* subsp. *garganica* Briq.): Plant tomentose; stems without axillary fascicles of leaves. Lower leaves ovate-lanceolate, the upper linear. Verticillasters lax, with more than 4 flowers; peduncles distinctly divaricate. Corolla-tube 4–6 mm longer than calyx-tube. Flowering in spring. Nutlets obtuse. ● *S.E. Italy.*

(**d**) Subsp. **longiflora** (C. Presl) Nyman, *Consp.* 590 (1881): Like subsp. (**c**) but hispid; upper leaves linear-lanceolate; verticillasters more dense, the peduncles suberect or erecto-patent. ● *C. & S. Italy, Sicilia.*

(**e**) Subsp. **tenuifolia** (Ten.) Nyman, *Consp.* 590 (1881): Plant shortly hirsute; stems without axillary fascicles of leaves. Lower leaves ovate-lanceolate, the rest linear or linear-lanceolate. Verticillasters mostly with 2–4 subsessile flowers; peduncles rather strict. Corolla-tube 2–3 mm longer than calyx-tube. Flowering in spring. Nutlets shortly apiculate. ● *Italy, Sardegna, Sicilia.*

(**f**) Subsp. **consentina** (Ten.) Guinea, *Bot. Jour. Linn. Soc.* **64**: 381 (1971) (*Satureja consentina* Ten.): Plant minutely scabrid-puberulent; stems without axillary fascicles of leaves. Lower leaves ovate, fugacious, the rest linear. Verticillasters lax, with 2–6 flowers; peduncles erecto-patent. Corolla-tube 5–6 mm longer than calyx-tube. Flowering in autumn. Nutlets obtuse. *Usually calcifuge. S. Italy, Sicilia.*

(**g**) Subsp. **fruticulosa** (Bertol.) Guinea, *loc. cit.* (1971) (*Thymus fruticulosus* Bertol.): Plant pubescent to hirsute; stems with conspicuous axillary fascicles of leaves. Lower leaves ovate, the upper linear. Verticillasters with 1–4 flowers; peduncles divaricate. Corolla-tube 4–6 mm longer than calyx-tube. Flowering in spring. Nutlets obtuse. *Calcicole. S. Italy, Sicilia.*

21. M. inodora (Desf.) Bentham, *Lab. Gen. Sp.* 375 (1834). Ericoid, much-branched, puberulent dwarf shrub 10–30 cm with crowded, erect branches. Leaves 2–3 × *c.* 0·5 mm, acicular, rigid, evergreen, very dense and fasciculate, with revolute margin. Verticillasters mostly with 2 flowers, crowded, the pedicellate flowers exceeding the subtending leaves. Calyx *c.* 4 mm, villous in throat; teeth *c.* ½ as long as tube, lanceolate-subulate, unequal. Corolla 8–10 mm, purplish. Flowering in winter. *Dry hillsides and waste ground. Islas Baleares, S. Spain.* Bl Hs. (*N.W. Africa.*)

30. Thymbra L.[1]

Small shrubs. Floral leaves imbricate; inflorescence dense, narrow, spike-like. Bracteoles linear to lanceolate. Calyx dorsally flattened, 2-lipped, 13-veined; upper lip 3-toothed; lower lip 2-toothed. Corolla 2-lipped, the tube straight; upper lip porrect, emarginate; lower lip 3-lobed. Stamens included; anther-cells parallel.

Literature: K. H. Rechinger fil., *Kulturpfl., Beih.* **3**: 64–69 (1962).

Leaves glabrous or ciliate; calyx 5–8·5 mm, the throat hirsute
  inside; corolla 12–15 mm   **1. spicata**
Leaves tomentose; calyx *c.* 3 mm, the throat glabrous inside;
  corolla 5·5–6 mm   **2. calostachya**

1. T. spicata L., *Sp. Pl.* 569 (1753). Stems up to 55 cm; branches erect, rigid, with hairs in two decussate lines, glabrescent. Leaves 15–23 × 2–3 mm, linear to linear-lanceolate, entire, glabrous or ciliate, sessile. Bracteoles 10 mm, lanceolate, purplish, white-ciliate. Calyx 5–8·5 mm, glandular-punctate, the teeth ciliate; throat hirsute inside. Corolla 12–15 mm, pink; upper lip much longer than the lower; tube twice as long as the calyx. *Dry, sunny hillsides. Turkey-in-Europe; a few isolated stations in Greece.* Gr Tu. (*Anatolia to Israel.*)

2. T. calostachya (Rech. fil.) Rech. fil., *Kulturpfl., Beih.* **3**: 64 (1962). Stems up to 40 cm; branches erect or ascending, densely puberulent, glabrescent. Leaves 5–12 × 1·5–3 mm, oblong-lanceolate, entire, densely and shortly grey-tomentose, sessile. Bracteoles 2 mm, linear, greyish, velutinous-tomentose. Calyx *c.* 3 mm, glandular-punctate, shortly tomentose; throat glabrous inside. Corolla 5·5–6 mm, white; upper lip shorter than the lower; tube only slightly longer than the calyx. *Calcareous rocks.* ● *E. Kriti (near Sitia).* Cr.

31. Hyssopus L.[2]

Aromatic, perennial herbs or dwarf shrubs. Verticillasters 4- to 16-flowered, in terminal, secund, spike-like inflorescences. Calyx tubular, 15-veined, with 5 equal teeth. Corolla infundibuliform, 2-lipped; upper lip erect, emarginate; lower lip patent, 3-lobed, the middle lobe the largest, emarginate. Stamens exserted, spreading; anther-cells divergent. Style-branches equal.

Literature: J. Briquet, *Les Labiées des Alpes Maritimes* 380–388. Genève & Bale. 1893.

1. H. officinalis L., *Sp. Pl.* 569 (1753). Stems 20–60 cm, numerous, erect, rarely decumbent. Leaves 10–50 × 1–10 mm, linear, lanceolate or oblong, obtuse to acuminate, entire, glabrous to villous, sessile or subsessile. Bracts linear, acuminate, not aristate or with an arista 1–3 mm. Calyx glabrous or puberulent, the tube 3–5 mm, the teeth 1–3 mm, acuminate, aristate or not. Corolla 7–12 mm, blue or violet, rarely white. Nutlets *c.* 2 mm. *Dry hills and rocky ground. S., S.C. & E. Europe; locally naturalized from gardens elsewhere.* Al Au Bu Cz Ga He Hs Hu It Ju Rs (B, C, W, K, E) [Be Ge Ho Po].

A polymorphic species, in which variants have been named at the specific, subspecific and varietal level, mainly on the basis of differences in indumentum, presence or absence of aristae on bracts and calyx-teeth, and length of calyx-teeth. Although their geographical range is sometimes discontinuous, the most distinctive variants are treated here as subspecies. Further research is needed to ascertain their proper status.

1 Plants greyish- or whitish-villous   (**a**) subsp. **canescens**
1 Plants glabrous or glabrescent
 2 Calyx-teeth 1 mm, triangular, not aristate; corolla exceeding
   calyx-teeth by 3 mm   (**b**) subsp. **montanus**
 2 Calyx-teeth 2–3 mm, lanceolate or ovate-lanceolate, with an
   arista 1–3 mm; corolla exceeding calyx-teeth by 4 mm
   3 Bracts aristate   (**c**) subsp. **aristatus**
   3 Bracts not aristate   (**d**) subsp. **officinalis**

(**a**) Subsp. **canescens** (DC.) Briq., *Lab. Alp. Marit.* 387 (1893) (*H. officinalis* var. *canescens* DC., *H. canescens* (DC.) Nyman, *H. cinerascens* Jordan & Fourr., *H. cinereus* Pau): *France and Spain.*

(**b**) Subsp. **montanus** (Jordan & Fourr.) Briq., *op. cit.* 386 (1893) (*H. montanus* Jordan & Fourr., *H. officinalis* subsp. *montanus* var. *wolgensis* Briq.; incl. *H. cretaceus* Dubjansky): *France; S. part of U.S.S.R.; probably elsewhere.*

(**c**) Subsp. **aristatus** (Godron) Briq., *op. cit.* 383 (1893) (*H. aristatus* Godron, *H. officinalis* var. *pilifer* Griseb. ex Pant.,

---

[1] By V. H. Heywood and R. A. DeFilipps.   [2] By R. A. DeFilipps.

subsp. *pilifer* (Griseb. ex Pant.) Murb.): *France and Spain; Balkan peninsula.*

(d) Subsp. **officinalis** (*H. angustifolius* Bieb., *H. decumbens* Jordan & Fourr., *H. officinalis* subsp. *angustifolius* (Bieb.) Arcangeli, var. *decumbens* (Jordan & Fourr.) Briq.): *Throughout the range of the species.*

32. Origanum L.[1]

Dwarf shrubs or annual, biennial or perennial herbs. Verticillasters few- to many-flowered, aggregated into short, terminal or lateral spicules; spicules arranged in paniculate, cymose or corymbiform inflorescences. Bracts distinct from the leaves, imbricate, often conspicuous, coloured. Calyx campanulate or turbinate, 2-lipped and either actinomorphic with 5 equal teeth, or entire, obliquely truncate at apex, or 1-lipped and deeply slit on one side. Corolla 2-lipped, the upper lip entire or emarginate; lower lip 3-lobed. Stamens didynamous, exserted or included; anther-cells divergent. Style-branches equal.

All species grow mainly in dry places, often on rocky slopes or in dwarf scrub.

1 Calyx 2-lipped
  2 Upper lip of calyx entire or subentire
    3 Stems densely leafy above; leaves lanate   **9. dictamnus**
    3 Stems ± naked above; leaves sparsely pubescent **10. tournefortii**
  2 Upper lip of calyx 3-toothed
    4 Spicules erect, grouped in panicles at the ends of branches
                                        8. majoricum
    4 Spicules nutant, grouped in cymes
      5 Caespitose dwarf shrub; leaves less than 5 mm, densely hairy   **13. vetteri**
      5 Rhizomatous perennial; leaves more than 11 mm, glabrous at least above
        6 Leaves cordate at base, glabrous or very sparsely scabrid beneath; bracts 7–8 mm wide, glabrous   **11. scabrum**
        6 Leaves rounded at base, sparsely pubescent beneath; bracts 3 mm wide, hairy   **12. lirium**
1 Calyx either with 5 equal teeth, or truncate and entire at apex, or 1-lipped and with a deep slit on one side
  7 Calyx entire, truncate   **7. microphyllum**
  7 Calyx with 5 equal teeth or 1-lipped and deeply slit
    8 Calyx 1-lipped, with a deep slit on one side
      9 Stems papillose and hirsute; spicules in dense terminal corymbs   **6. onites**
      9 Stems not papillose, glabrous to tomentose; spicules in panicles   **5. majorana**
    8 Calyx with 5 equal teeth
      10 Stems usually unbranched; calyx densely covered with purple glands; corolla 8–10 mm   **1. compactum**
      10 Stems branched at least above; calyx sparsely to densely covered with yellowish glands; corolla 4–7 mm
        11 Bracts twice as long as calyx, obovate to orbicular, membranous   **4. virens**
        11 Bracts as long as to nearly twice as long as calyx, ovate, herbaceous
          12 Bracts 2–3 mm, usually green, densely glandular on outer surface   **2. heracleoticum**
          12 Bracts 4–5 mm, usually purple, eglandular or sparsely glandular on outer surface   **3. vulgare**

Sect. ORIGANUM. Calyx with 5 more or less equal teeth. Bracts 2–5 mm, green or purplish.

1. O. compactum Bentham, *Lab. Gen. Sp.* 334 (1834). Dwarf shrub; stems up to 35 cm, usually unbranched, hirsute. Leaves 18–20 × 10–15 mm, ovate, entire, densely glandular-punctate on both surfaces, hairy on veins and at margins, petiolate.

Spicules 10–23 mm, densely fasciculate, forming a long, interrupted panicle. Bracts 4–5 mm, twice as long as calyx, ovate-oblong, acute, purplish. Calyx purple-glandular-punctate. Corolla 8–10 mm, 4 times as long as calyx-tube, pink or white. *S.W. Spain.* Hs. (*N. Africa.*)

2. O. heracleoticum L., *Sp. Pl.* 589 (1753) (*O. hirtum* Link). Woody, rhizomatous perennial; stems up to 60 cm, branched above, hirsute. Leaves 15–22 × 6–15 mm, ovate to oblong, entire or remotely serrate, sparsely hairy, glandular-punctate, petiolate. Spicules 5–20 mm, ovoid to oblong, crowded or lax and interrupted, forming a panicle. Bracts 2–3 mm, as long as to nearly twice as long as calyx, ovate, densely glandular, green or rarely purplish, glabrous or hairy. Calyx glabrous or hairy, yellow-glandular-punctate. Corolla 4–5 mm, white, rarely pink. $2n=30$. *S.E. Europe, from Sardegna to the Aegean region.* ?Al Bu Cr Gr It Ju Sa Tu.

Variable in colour and indumentum of bracts and calyx.

3. O. vulgare L., *Sp. Pl.* 590 (1753) (incl. *O. dilatatum* Klokov, *O. vulgare* subsp. *viride* (Boiss.) Hayek). Woody, rhizomatous perennial; stems up to 90 cm or more, usually branched above, pubescent, hirsute or velutinous, rarely glabrous. Leaves 10–40(–50) × 4–25 mm, ovate, entire or shallowly crenate-serrate, glabrous or hairy, glandular-punctate, petiolate. Spicules 5–30 mm, ovoid, oblong, or prismatic, forming a corymb or panicle. Bracts 4–5 mm, as long as to nearly twice as long as calyx, ovate, not apiculate, hairy or glabrous, eglandular or sparsely glandular-punctate, herbaceous, violet-purple or greenish. Calyx yellow-glandular-punctate, hairy or glabrous. Corolla 4–7 mm, white or purplish-red. $2n=30$, ?32. *Most of Europe.* All except Az Bl Cr Fa Is.

Extremely variable in colour and indumentum of bracts and calyx, shape and length of spicules, and colour of corolla. Many variants have been described as separate species or subspecies (cf. R. Soó & A. Borhidi, *Ann. Univ. Sci. Budapest. Rolando Eötvös (Biol.)* **9–10**: 361–364 (1968)), but they do not form distinctly separable populations. Some variants with prismatic spicules 12–20 mm are frequently cultivated as a pot-herb.

4. O. virens Hoffmanns. & Link, *Fl. Port.* **1**: 119 (1809). Like 3 but bracts twice as long as calyx, obovate to orbicular, apiculate, glandular-punctate, glabrous, membranous, pale green; calyx glabrous; corolla white. $2n=30$. *S.W. Europe.* Az Bl Hs Lu.

Sect. MAJORANA (Miller) T. Vogel. Calyx 1-lipped and deeply slit on one side, 2-lipped, or entire and truncate at apex. Bracts 3–4 mm, green.

5. O. majorana L., *Sp. Pl.* 590 (1753) (*Majorana hortensis* Moench). Annual, biennial or perennial herb; stems 20–60 cm, erect or ascending, glabrous to tomentose, not papillose. Leaves 5–20(–35) × 5–10(–15) mm, ovate to elliptic-spathulate, obtuse or acute, rounded or attenuate at base, entire, grey- or white-tomentose to sparsely pilose and green; veins somewhat raised; lower leaves with petiole up to 15 mm, the others subsessile. Spicules 5–12 mm, globose, ovoid or oblong, densely grouped at the end of short branches, forming a narrow, terminal panicle. Bracts *c.* 3–4 mm, orbicular, tomentose, glandular-punctate. Calyx *c.* 2·5 mm, 1-lipped, with a deep slit on one side, ciliate. Corolla *c.* 4 mm, white, pale lilac or pink. Nutlets 0·75–1 mm, light brown. *Cultivated as a pot-herb and locally naturalized in S. Europe.* [Co He Hs It Ju.] (*N. Africa, S.W. Asia.*)

---

[1] By R. Fernandes and V. H. Heywood.

O. dubium Boiss., *Fl. Or.* **4**: 553 (1879), from the E. Mediterranean region, differs from **5** in its always perennial, suffruticose habit, white-tomentose leaves (at least when young) and corolla 6–7 mm, but is doubtfully a separate species. Its only European record from the Kikladhes (Naxos) has not been verified since Boissier's original citation.

6. **O. onites** L., *Sp. Pl.* 590 (1753). Dwarf shrub; stems up to 60 cm, erect, densely papillose and hirsute. Leaves 5–22 × 4–12(–17) mm, ovate to orbicular-ovate, rounded or cordate at base, acute or shortly acuminate, entire or remotely serrate, papillose and villous, glandular-punctate, with the veins somewhat raised on lower surface; lower leaves petiolate, the others sessile. Spicules 4–10 mm, ovoid, oblong or prismatic, very compact, forming a terminal, dense corymb 2–8(–10) cm in diameter. Bracts 3–3·5 mm, slightly exceeding the calyx, the lower ovate, acute, the upper orbicular-ovate, obtuse, puberulent. Calyx 2·75–3 mm, 1-lipped, with a deep slit on one side, ovate-spathulate, entire or minutely 3-dentate, subglabrous, glandular-punctate, ciliate. Corolla 4–5·5 mm, white. Nutlets *c.* 1 mm, oblong-ellipsoid. 2*n*=30. *Mediterranean region.* Cr Gr ?Hs Ju Si.

7. **O. microphyllum** (Bentham) Boiss., *Fl. Or.* **4**: 552 (1879). Dwarf shrub; stems 20–50 cm, numerous, usually branched; branches slender, distinctly 4-angled, purplish, retrorsely hispidulous, glabrescent. Leaves on main branches 4–8 × 3–6 mm, ovate to oblong, obtuse, rounded or subcordate at base, puberulent to whitish-tomentose, with indistinct veins; petiole 1–1·5 mm. Spicules 4–11 mm, subglobose or obconical to oblong, white-lanate, 1–3 at the end of lateral branches, forming lax, terminal panicles. Bracts 3–4 mm, broadly spathulate, rounded. Calyx 2–2·25 mm, obliquely truncate, entire at apex, glabrous, densely ciliate at the mouth. Corolla *c.* 5 mm, purple. ● *Kriti.* Cr.

O. × minoanum P. H. Davis, *Notes Roy. Bot. Gard. Edinb.* **21**: 137 (1953) (*Majorana leptoclados* Rech. fil., non *Origanum leptocladum* Boiss.) (*O. heracleoticum × microphyllum*) occurs locally with the parents in Kriti.

8. **O. majoricum** Camb., *Mém. Mus. Hist. Nat.* (*Paris*) **14**: 296 (1827). Dwarf shrub or perennial herb; stems up to 60 cm or more, pubescent. Leaves up to 25 × 18 mm, ovate-elliptical to ovate-lanceolate, obtuse or acute, shortly villous; lower leaves with petiole up to 10 mm, the upper sessile. Spicules up to 3·5 cm, ovoid to oblong-prismatic, grouped in panicles at the ends of branches. Bracts 3–4 mm, ovate or ovate-spathulate, acute, villous, ciliate, somewhat exceeding the calyx. Calyx 2·5–3·5 mm, 2-lipped; tube campanulate, subglabrous; upper lip deeply 3-toothed, the lower 2-partite, all teeth ciliate. Corolla pink, slightly exceeding to twice as long as the calyx-tube. Sterile? ● *S.W. Europe; very local.* Bl Hs *Lu.

Possibly a hybrid between **4** and **5**.

Sect. AMARACUS Bentham. Calyx 2-lipped. Bracts up to 11 mm, purplish.

9. **O. dictamnus** L., *Sp. Pl.* 589 (1753) (*Amaracus dictamnus* (L.) Bentham). White-lanate dwarf shrub; stems up to 20 cm. Leaves 13–25 × 12–25 mm, broadly ovate to orbicular, entire, lanate, the veins raised, conspicuous; lower leaves shortly petiolate. Spicules in groups of 3–10, dense, ovoid or oblong, arranged in opposite pedunculate pairs, in lax panicles. Bracts 7–10 mm,

conspicuous, purple, longer than calyx. Upper lip of calyx subentire, the lower shallowly toothed. Corolla pink, the tube twice as long as the calyx-tube. ● *Kriti.* Cr.

10. **O. tournefortii** Aiton, *Hort. Kew.* **2**: 311 (1789) (*Amaracus tournefortii* (Aiton) Bentham). Lanate or glabrous dwarf shrub; stems up to 25 cm. Leaves 15–30 × 12–20 mm, broadly ovate to orbicular, entire, subsessile, glaucous, sparsely pubescent, ciliate, the veins not raised. Spicules in groups of 2–6, oblong, arranged in opposite pedunculate groups of 1–3, in lax panicles. Bracts 8–11 mm, conspicuous, purple, longer than the calyx. Upper lip of calyx entire, the lower 2-toothed. Corolla pink, the tube 3 times as long as the calyx-tube. 2*n*=30. ● *Kikladhes, Kriti.* Cr Gr.

11. **O. scabrum** Boiss. & Heldr. in Boiss., *Diagn. Pl. Or. Nov.* **1**(7): 48 (1846) (*Amaracus scaber* (Boiss. & Heldr.) Briq.). Rhizomatous perennial; stems up to 45 cm, erect, glabrous, branched above. Leaves 11–30 × 11–20 mm, ovate to suborbicular, cordate at the base, glabrous, or very sparsely scabrid on the veins beneath, sessile. Spicules ovoid, nodding, in lax panicles. Bracts 8–10 × 7–8 mm, conspicuous, ovate to ovate-elliptical, purplish, glabrous. Upper lip of calyx 3-toothed. Corolla pink, twice as long as the calyx-tube. ● *Mountains of S. Greece.* Gr.

(a) Subsp. **scabrum**: Leaf-margins scabrid. Bracts ovate. Upper lip of calyx divided to ⅓, 3 times as long as the lower. *Taïyetos.*
(b) Subsp. **pulchrum** (Boiss. & Heldr.) P. H. Davis, *Kew Bull.* [4]: 405 (1949) (*Amaracus scaber* subsp. *pulcher* (Boiss. & Heldr.) Hayek): Leaf-margins smooth. Bracts ovate-elliptical. Upper lip of calyx divided to ½, scarcely longer than the lower. *Evvoia.*

12. **O. lirium** Heldr. ex Halácsy, *Verh. Zool.-Bot. Ges. Wien* **49**: 192 (1899) (*Amaracus lirius* (Heldr. ex Halácsy) Hayek). Rhizomatous perennial; stems up to 40 cm, erect, branched above. Leaves 12–24 × 10–15 mm, ovate, rounded at base, glabrous above, sparsely puberulent on the veins beneath, sessile. Spicules ovoid, nodding, in lax panicles. Bracts *c.* 5 × 3 mm, ovate, purplish, hairy. Upper lip of calyx shallowly toothed, scarcely longer than the lower. Corolla pink, twice as long as the calyx-tube. ● *Mountains of S. Greece* (*Evvoia*). Gr.

Probably originated from hybridization between **11(a)** and **2**.

13. **O. vetteri** Briq. & W. Barbey in Stefani, Major & W. Barbey, *Karpathos* 124 (1895) (*Amaracus vetteri* (Briq. & W. Barbey) Hayek). Caespitose dwarf shrub. Leaves less than 5 mm, broadly ovate-triangular, obtuse, cordate at base, densely hairy, shortly petiolate. Spicules few-flowered, nodding, in lax panicles. Bracts obovate, hairy. Upper lip of calyx 3-toothed, twice as long as the lower. Corolla pink, twice as long as the calyx-tube. ● *Karpathos.* Cr.

33. Thymus L.[1]

Small shrubs or perennial herbs, woody at least at base. Verticillasters 1- to many-flowered, often crowded into a terminal, capitate inflorescence. Calyx cylindrical to campanulate, usually 2-lipped with the 3 upper teeth different from the lower; tube straight, not gibbous, hairy in the throat. Corolla 2-lipped, with straight tube. Stamens usually exserted, straight, divergent.

Literature: V. Borbás, *Math. Term. Közl.* **24**: 37–116 (1890). E. Huguet del Villar, *Cavanillesia* **6**: 104–125 (1934). J. Jalas, *Veröff. Geobot. Inst. Rübel* (*Zürich*) **43**: 186–203 (1970). M.

---

[1] By J. Jalas.

Klokov, *Not. Syst. (Leningrad)* **16**: 293–318 (1954). M. Klokov & N. Schostenko, *Trav. Inst. Bot. (Charkov)* **14**(3): 107–157 (1939). C. C. Lacaita, *Cavanillesia* **3**: 20–47 (1930). B. Pawłowski, *Fragm. Fl. Geobot.* **12**: 387–412 (1966). C. D. Pigott, *New Phytol.* **53**: 470–495 (1954). K. Ronniger, *Feddes Repert.* **31**: 129–157 (1932); *Deutsche Heilpfl.* **10**(5): 1–24 (1944). P. Schmidt, *Hercynia* **5**: 385–419 (1969). J. Velenovský, *Beih. Bot. Centr.* **19**(2): 271–287 (1906).

The flowering stems may be hairy on all sides, on two sides, or on the angles only. Leaf-measurements refer, unless otherwise stated, to leaves from the middle part of flowering stems; they include the petiole. At least the distal pairs of lateral veins of the leaves, when visible, may anastomose to form a marginal vein either at the apex of the leaf or along the whole length of the leaf. Leaf-venation is best seen in dried specimens. Gynodioecism is usual in the genus, pistillate flowers often having smaller calyx and corolla; measurements of calyx and corolla refer to hermaphrodite flowers.

Hybrids are frequently formed, even between members of different sections as well as between species at different levels of polyploidy.

1 Calyx-tube dorsally flattened, with 20–22 veins; leaves sub-triquetrous **1. capitatus**
1 Calyx-tube dorsally convex, with 10–13 veins; leaves not subtriquetrous
2 Upper calyx-teeth like lower (Iberian peninsula)
  3 Bracts similar to the leaves; upper calyx-teeth 2–3 mm **2. mastichina**
  3 Bracts not similar to the leaves; upper calyx-teeth 1–1·5 mm **3. tomentosus**
2 Upper calyx-teeth conspicuously different from lower
  4 Lower calyx-teeth almost leaf-like, triangular **4. caespititius**
  4 Lower calyx-teeth lanceolate-subulate
   5 Leaf-margins revolute
    6 Bracts not similar to leaves, at least twice as wide as leaves
     7 Leaves linear to narrowly lanceolate
      8 Creeping shoots absent
       9 Corolla 6–10 mm; bracts often sparsely dentate to shallowly lobed **8. villosus**
       9 Corolla *c.* 15 mm; bracts entire
        10 Bracts membranous, whitish; corolla whitish **10. membranaceus**
        10 Bracts coriaceous, purplish; corolla purple
         11 Inflorescence up to 4 cm, oblong-conical; leaves tomentose and with a few longer hairs, ciliate at base **7. cephalotos**
         11 Inflorescence not more than 2·5 cm, subglobose; leaves tomentose but without longer hairs, not or sparsely ciliate at base **9. longiflorus**
      8 Creeping shoots present
       12 Leaves glabrous to subglabrous; bracts purplish **12. mastigophorus**
       12 Leaves ± velutinous; bracts greenish
        13 Leaves shortly petiolate **13. dolopicus**
        13 Leaves sessile **15. parnassicus**
     7 Leaves ovate-lanceolate
      14 Leaves glabrous on upper surface **18. carnosus**
      14 Leaves tomentose on upper surface
       15 Inflorescence 6–8 mm in diameter; calyx 3–4 mm, the upper teeth rarely ciliate **16. capitellatus**
       15 Inflorescence 10–18 mm in diameter; calyx 4–6 mm, the upper teeth usually ciliate **17. camphoratus**
    6 Bracts ± similar to leaves
     16 Leaves more than 3 mm wide, ovate to rhombic; corolla-tube ± infundibuliform (Balkan peninsula) **6. teucrioides**
     16 Leaves up to 2·5 mm wide, linear to elliptical; corolla-tube cylindrical
      17 Leaves not ciliate **19. vulgaris**
      17 Leaves ± ciliate, at least at base

      18 Long, arcuate, procumbent or creeping shoots absent
       19 Calyx 5–7 mm, the tube cylindrical; verticillasters 2-flowered **11. antoninae**
       19 Calyx 3–4 mm, the tube campanulate; verticillasters 6- or more-flowered, or inflorescence capituliform
        20 Calyx shortly tomentose, the upper teeth not ciliate; inflorescence elongate, interrupted **21. zygis**
        20 Calyx pubescent, the upper teeth ciliate; inflorescence usually capituliform
         21 Leaves 3–5 mm, petiolate; corolla pinkish-purple **20. hyemalis**
         21 Leaves (4–)5–8(–10) mm, sessile; corolla creamy white **22. baeticus**
      18 Long, arcuate, procumbent or creeping shoots present
       22 Calyx *c.* 3 mm **23. hirtus**
       22 Calyx 3·5–5·5 mm
        23 Upper calyx-teeth 0·5–0·7 mm **25. serpylloides**
        23 Upper calyx-teeth 1–2 mm
         24 Leaves subglabrous, sparsely ciliate at the base (Spain) **24. loscosii**
         24 Leaves puberulent to velutinous or glabrous, ciliate to above the middle (S.E. Europe)
          25 Leaves 1–1·5 mm wide, shortly petiolate **13. dolopicus**
          25 Leaves 0·3–1 mm wide, sessile **14. cherlerioides**
   5 Leaf-margins not or only slightly revolute
    26 Stems with hairs on the angles only
     27 Leaves with marginal vein **60. bihoriensis**
     27 Leaves without marginal vein
      28 Leaves truncate to subcordate at base; upper calyx-teeth $\frac{1}{10}$–$\frac{1}{7}$ as long as calyx **63. oehmianus**
      28 Leaves cuneate to subtruncate at base; upper calyx-teeth at least $\frac{1}{5}$ as long as calyx
       29 Creeping, non-flowering, terminal shoots present **62. alpestris**
       29 Creeping, non-flowering, terminal shoots absent **61. pulegioides**
    26 Stems hairy all round or on 2 opposite sides
     30 Leaves not ciliate
      31 Branches erect **5. piperella**
      31 Branches procumbent, ascending
       32 Leaves oblong-elliptical **47. nitens**
       32 Leaves ovate to ovate-elliptical
        33 Leaves 5–7 × 2–3 mm; flowers not in a distinct inflorescence **48. willkommii**
        33 Leaves 7–12 × 3–6 mm; flowers in capitate inflorescences **49. richardii**
     30 Leaves ciliate at least at base
      34 Bracts ovate, not similar to the leaves
       35 Upper calyx-teeth not more than 0·7 mm
        36 Leaves acute, often denticulate, velutinous **26. holosericeus**
        36 Leaves obtuse, entire, glabrous
         37 Leaves *c.* 1 mm wide, fleshy; veins indistinct **39. zygioides**
         37 Leaves 1·7–2 mm wide, not fleshy; veins distinct **37. aznavourii**
       35 Upper calyx-teeth 0·8 mm or more
        38 Leaves puberulent to velutinous
         39 Calyx 6–8 mm; leaves velutinous **27. laconicus**
         39 Calyx 3·5–5 mm; leaves puberulent **30. aranjuezii**
        38 Leaves glabrous, sometimes ciliate
         40 Largest bracts at least 5 mm wide
          41 Leaves 7–10 mm, coriaceous; corolla 8–12 mm **29. granatensis**
          41 Leaves 10–17 mm, herbaceous; corolla 6–8 mm **28. bracteosus**
         40 Largest bracts not more than 4 mm wide
          42 Leaves 1·5–2 mm wide, herbaceous; calyx-tube distinctly shorter than lips **31. bracteatus**
          42 Leaves not more than 1·2 mm wide, coriaceous or ± fleshy; calyx-tube about as long as lips

43 Leaves 5–9 mm, narrowly spathulate, ±fleshy; veins indistinct, diverging **30. aranjuezii**

43 Leaves 8–20 mm, linear, ±coriaceous; veins distinct, parallel

44 Calyx 5–7 mm; bracts straw-coloured **33. atticus**

44 Calyx 3–5 mm

45 Corolla pink, becoming yellow; bracts straw-coloured **34. plasonii**

45 Corolla whitish, pink or purple; bracts usually purplish **35. striatus**

34 At least the lowermost bracts ±similar to the leaves (although sometimes wider and sometimes coloured)

46 Marginal vein present along the whole length of the leaf (Carpathians)

47 Creeping, non-flowering branches present **58. pulcherrimus**

47 Creeping, non-flowering branches absent **59. comosus**

46 Marginal vein absent or present in upper part of leaf only

48 Flowering stems usually more than 10 cm; non-flowering branches usually absent

49 Stems with hairs on two sides **64. alternans**

49 Stems usually with hairs all round

50 Upper calyx-teeth ciliate

51 Leaves elliptic-spathulate, petiolate, with two pairs of distinct lateral veins forming a marginal vein at the apex **51. binervulatus**

51 Leaves linear to lanceolate-elliptical, with lateral veins indistinct or disappearing towards the margin

52 Calyx 3·5–5 mm; upper teeth up to 1·5 mm

53 Bracts with distinct veins; inflorescence interrupted **40. comptus**

53 Bracts with indistinct veins; inflorescence capitate **36. spinulosus**

52 Calyx 2·5–3·5(–4) mm; upper teeth *c.* 1 mm

54 Inflorescence often branched; leaves coriaceous, with distinct venation **41. sibthorpii**

54 Inflorescence seldom branched; leaves herbaceous, with weak venation **42. pannonicus**

50 Upper calyx-teeth not ciliate

55 Leaves linear to lanceolate, sessile

56 Inflorescence interrupted; upper calyx-teeth up to 1·5 mm **40. comptus**

56 Inflorescence capitate; upper calyx-teeth *c.* 1 mm **45. pallasianus**

55 Leaves elliptical to rhombic-ovate, petiolate

57 Leaves acute, with indistinct veins **46. herba-barona**

57 Leaves subobtuse, with distinct veins **50. guberlinensis**

48 Flowering stems seldom more than 10 cm; long, creeping non-flowering branches present

58 Upper calyx-teeth 0·3–0·7 mm, not ciliate

59 Leaves ±sessile; calyx 4–4·5 mm; both sparsely glandular-punctate **37. aznavourii**

59 Leaves shortly petiolate; calyx 3–4 mm; both densely glandular-punctate **38. kirgisorum**

58 Upper calyx-teeth usually more than 0·7 mm, usually ciliate

60 Leaves with indistinct lateral veins

61 Leaves ±spathulate

62 Upper calyx-teeth 1–1·5 mm, not ciliate **32. leptophyllus**

62 Upper calyx-teeth 0·3–0·7 mm, ciliate **39. cygioides**

61 Leaves linear to elliptical, rarely obovate

63 Leaves less than 1 mm wide, ciliate at least in the basal half **14. cherlerioides**

63 Leaves at least 1 mm wide, ciliate only at base

64 Flowering stems with a basal cluster of small leaves; calyx 2·5–4(–5) mm

65 Calyx-tube about as long as upper lip **54. stojanovii**

65 Calyx-tube shorter than upper lip **55. longicaulis**

64 Flowering stems without a basal leaf-cluster; calyx 4–5 mm

66 Creeping shoots usually with a terminal inflorescence **43. glabrescens**

66 Creeping shoots without a terminal inflorescence **24. loscosii**

60 Leaves with distinct lateral veins

67 Leaves with lateral veins curved along the margin and anastomosing at apex **56. praecox**

67 Leaves with lateral veins disappearing towards the margin

68 Calyx-tube clearly shorter than upper lip

69 Flowering stems without a basal cluster of small leaves; calyx 5–6 mm **44. longedentatus**

69 Flowering stems with a basal cluster of small leaves

70 Calyx 4–6 mm, purplish **53. thracicus**

70 Calyx 2·5–4 mm, greenish or purple

71 Leaves coriaceous **52. ocheus**

71 Leaves herbaceous **55. longicaulis**

68 Calyx-tube about as long as upper lip

72 Leaves up to 1·5 mm wide; flowering stems hairy on 2 opposite sides only **57. nervosus**

72 Leaves mostly more than 1·5 mm wide; flowering stems usually hairy all round

73 Leaves sessile or nearly so

74 Upper calyx-teeth about as long as wide **65. serpyllum**

74 Upper calyx-teeth distinctly longer than wide

75 Leaves coriaceous; creeping shoots without a terminal inflorescence **35. striatus**

75 Leaves herbaceous; creeping shoots ending in an inflorescence **43. glabrescens**

73 Leaves distinctly petiolate

76 Petiole often up to ½ of the total leaf-length **66. talijevii**

76 Petiole not more than ⅓ of the total leaf-length **65. serpyllum**

Subgen. **Coridothymus** (Reichenb. fil.) Borbás. Leaves subtriquetrous, the margins flat. Calyx-tube dorsally flattened, with 20–22 veins.

1. T. capitatus (L.) Hoffmanns. & Link, *Fl. Port.* 1: 123 (1809) (*Satureja capitata* L., *Coridothymus capitatus* (L.) Reichenb. fil.). Dwarf shrub 20–50(–150) cm, with ascending to erect woody branches bearing axillary leaf-clusters (often the only leaves during the dry season). Leaves of long shoots 6–10 × 1–1·2 mm, sessile, linear, acute, subglabrous, sparsely ciliate at base; lateral veins not visible. Inflorescence oblong-conical; bracts *c.* 6 × 2 mm, imbricate, ovate to lanceolate, greenish, ciliate; bracteoles *c.* 6 mm, similar to the leaves. Calyx *c.* 5 mm, the upper lip shorter than the lower, all the teeth ciliate. Corolla up to 10 mm, purplish-pink, the upper lip bifid. $2n = 30$. *Calcicole. Mediterranean region, Portugal.* Al Bl Co Cr Gr Hs It Ju Lu Sa Si Tu.

Subgen. **Thymus.** Leaves not subtriquetrous, the margins flat or revolute. Calyx-tube dorsally convex, with 10–13 veins.

Sect. MASTICHINA (Miller) Bentham. Erect dwarf shrubs. Leaves distinctly petiolate, flat. Calyx almost actinomorphic; tube short; all teeth similar, more or less subulate, yellowish, patent in fruit. Corolla scarcely longer than calyx.

2. T. mastichina L., *Sp. Pl.* ed. 2, 827 (1763). Stems 20–50 cm, erect, with axillary leaf-clusters. Leaves 8–10 × 2–3 mm, narrowly ovate to elliptic-lanceolate, often more or less crenulate, tomentose to subglabrous, not ciliate at base. Inflorescence 10–20 mm in diameter, usually subglobose. Bracts similar to the leaves, greenish. Calyx 4–6 mm; upper teeth 2–3 mm; all teeth long-ciliate, giving the inflorescence a somewhat plumose appearance (resembling *Trifolium arvense*). Corolla whitish; tube scarcely exceeding calyx. 2n = 56. ● *Spain and Portugal.* Hs Lu.

3. **T. tomentosus** Willd., *Enum. Pl. Hort. Berol.* 626 (1809). Like 2 but leaves *c.* 5 × 1·5 mm, entire, tomentose; inflorescence *c.* 7 mm in diameter, denser and more globose; bracts not similar to the leaves; calyx 3–4 mm, the upper teeth 1–2 mm. ● *S. Portugal, S.W. Spain.* Hs Lu.

Sect. MICANTES Velen. Leaves fleshy, with almost obsolete nerves. Verticillasters pedunculate, with 2–20 pedicellate flowers. Calyx campanulate, with almost obsolete veins; lower teeth almost leaf-like. Corolla distinctly longer than calyx.

4. **T. caespititius** Brot., *Fl. Lusit.* 1: 176 (1804) (*T. micans* Solander ex Lowe). Caespitose, with woody, creeping shoots bearing erect flowering stems 2–7 cm and axillary clusters of small leaves. Leaves *c.* 8 × 1·7 mm, narrowly spathulate, glabrous, ciliate at base. Inflorescence lax, with 2-flowered verticillasters. Bracts similar to the leaves. Calyx 3–4 mm; upper teeth almost obsolete, not ciliate, the lower teeth flat, as long as wide. Corolla 6–14 mm, purplish-pink or whitish. *Portugal, N.W. Spain, Açores.* Az Hs Lu.

Sect. PIPERELLA Willk. Small shrubs. Leaves coriaceous, not ciliate, distinctly petiolate; margins flat. Flowers not forming a distinct inflorescence; at least lower cymes often pedunculate. Bracts similar to the leaves. Calyx-tube cylindrical; upper teeth as long as wide, not ciliate. Corolla distinctly longer than calyx, with narrow tube.

5. **T. piperella** L., *Syst. Nat.* ed. 12, 2: 400 (1767). Branches erect, tomentose, with axillary leaf-clusters. Leaves 5–6 × 3–4 mm, ovate, obtuse, glabrous. Inflorescence lax; verticillasters usually with 6–10 flowers. Calyx 4–6 mm. Corolla 7–10 mm, purple. ● *E. Spain.* Hs.

Sect. TEUCRIOIDES Jalas. Dwarf shrubs. Leaves herbaceous, distinctly petiolate; margins revolute. Flowers not forming a distinct inflorescence. Bracts similar to the leaves. Calyx-tube campanulate; upper teeth as long as wide. Corolla distinctly longer than calyx, with almost infundibuliform tube.

6. **T. teucrioides** Boiss. & Spruner in Boiss., *Diagn. Pl. Or. Nov.* 1(5): 15 (1844). Much-branched; branches erect, tomentose, with axillary clusters of small leaves. Leaves *c.* 10 × 4 mm, ovate to rhombic; margin subserrate. Verticillasters with 2–6 flowers. Calyx 4–6 mm. Corolla 9–12 mm, purple. ● *Mountains of Albania and Greece.* Al Gr.

Sect. PSEUDOTHYMBRA Bentham. Dwarf shrubs. Leaves tomentose at least beneath; margins usually revolute. Bracts usually not similar to the leaves, usually coloured and conspicuous. Calyx usually more than 5 mm, subcylindrical; upper teeth narrowly lanceolate. Corolla-tube narrow, distinctly longer than calyx.

7. **T. cephalotos** L., *Sp. Pl.* 592 (1753). Plant 15–30 cm, with ascending to erect woody branches bearing axillary leaf-clusters. Cauline leaves 7–10 × 0·5–1 mm, linear to linear-oblanceolate,

tomentose, ciliate; margins revolute. Inflorescence up to 4 cm, oblong-conical. Bracts up to 20 × 10 mm, ovate, entire, coriaceous, purplish. Calyx 5–7 mm, the tube cylindrical; upper teeth lanceolate. Corolla *c.* 15 mm, purple. ● *S. Portugal.* Lu.

8. **T. villosus** L., *Sp. Pl.* 592 (1753). Like 7 but lower part of bracts often greenish, often sparsely dentate to shallowly lobed; calyx 4–6 mm; corolla 6–10 mm. ● *S.W. Spain, S. Portugal.* Hs Lu.

Shows considerable variation in the length of corolla-tube and in the shape of bracts. Specimens with short corollas and more or less entire bracts have been called subsp. *lusitanicus* (Boiss.) Coutinho, *Bol. Soc. Brot.* 23: 87 (1907) but there seems to be a continuous series of character-combinations connecting the extremes.

9. **T. longiflorus** Boiss., *Elenchus* 75 (1838) (incl. *T. funkii* Cosson, *T. murcicus* Porta, *T. moroderi* Martinez). 10–30 cm, much-branched, with ascending to erect stems. Leaves (6–)8–12(–15) × (0·5–)0·8–1 mm, linear, velutinous-tomentose, not or sparsely ciliate at base, sessile to shortly petiolate; margins revolute. Inflorescence up to 2·5 cm, subglobose. Bracts up to 13 × 8 mm, broadly ovate to elliptical, acuminate, entire, usually ciliate, coriaceous, purplish. Calyx 5–7 mm, the tube cylindrical; upper teeth narrowly lanceolate. Corolla *c.* 15 mm, purple. ● *S.E. Spain.* Hs.

Variable, especially in the size of leaves and inflorescence and shape of bracts.

10. **T. membranaceus** Boiss., *loc. cit.* (1838). Like 9 but bracts membranous, whitish; corolla whitish. ● *S.E. Spain.* Hs.

Perhaps not specifically distinct from 9. Further investigation is needed.

11. **T. antoninae** Rouy & Coincy, *Bull. Soc. Bot. Fr.* 37: 165 (1890) (*T. portae* Freyn). 10–15 cm, with ascending to erect woody branches. Leaves 5–8 × 0·5 mm, linear, tomentose, sparsely ciliate at base; margin revolute. Inflorescence not differentiated, the verticillasters with 2 flowers only. Bracts similar to the leaves. Calyx 5–7 mm, the tube cylindrical; upper teeth narrowly lanceolate. Corolla 8–12 mm, purple; tube cylindrical. ● *S.E. Spain (prov. Albacete).* Hs.

Sometimes considered to be a hybrid between 9 and 21 but the morphological evidence does not seem to support this view.

12. **T. mastigophorus** Lacaita, *Cavanillesia* 3: 40 (1930) (*T. hirtus* auct., non Willd., *T. hispanicus* auct., non Poiret). Plant with woody, creeping primary branches bearing erect flowering stems 3–7 cm and axillary leaf-clusters. Cauline leaves 5–7 × 0·5–0·7 mm, linear, subacicular, somewhat fleshy, glabrous to subglabrous; margins revolute, ciliate. Inflorescence capitate. Bracts up to 8 × 3 mm, lanceolate-ovate, ciliate, purplish. Calyx 4–6 mm, subcylindrical; upper teeth narrowly lanceolate, ciliate. Corolla 6–7 mm, pink-purple. ● *N.C. Spain.* Hs.

Closely related to, and perhaps only a subspecies of the polymorphic **T. munbyanus** Boiss. & Reuter, *Pugillus* 96 (1852) (*T. ciliatus* (Desf.) Bentham, non Lam.) from N. Africa.

Similar plants from areas south of Madrid, usually referred to **12**, belong to **30**.

13. T. dolopicus Form., *Deutsche Bot. Monatsschr.* **15**: 75 (1897). Plant with creeping woody primary branches bearing axillary leaf-clusters and erect to ascending flowering stems up to 10 cm. Leaves 6–10 × 1–1·5 mm, narrowly lanceolate-spathulate, shortly petiolate, velutinous, the basal half ciliate; margins subrevolute. Inflorescence capitate, ovoid to globose. Bracts 1·5–4 mm wide, similar to the leaves or oblong-ovate, greenish. Calyx 4·5–5·5 mm; upper teeth *c.* 1·5 mm, lanceolate, ciliate. Corolla pink-purple, the tube scarcely exceeding calyx. ● *Mountains of N. & C. Greece.* Gr.

T. leucotrichus Haláscy, *Consp. Fl. Graec.* **2**: 561 (1902), a seldom-collected plant of the high mountains of C. & S. Greece, is like **13** but with leaves linear-lanceolate, sessile, covered by long and short hairs, and purplish bracts. Its status is uncertain, as is also that of *T. leucotrichus* var. *creticus* (Bald.) Ronniger from Kriti. Similar, possibly conspecific, taxa occur in Anatolia.

14. T. cherlerioides Vis., *Ill. Pi. Grecia*: 8 (1842) (*T. boissieri* Halácsy, *T. hirsutus* auct., non Bieb.; incl. *T. pseudo-humillimus* Klokov & Schost., *T. tauricus* Klokov & Schost.). Caespitose, with long, creeping branches bearing axillary leaf-clusters and erect flowering stems 1–8 cm. Leaves 4–10(–15) × 0·3–1 mm, linear to linear-lanceolate, sessile, glabrous or puberulent to velutinous, usually herbaceous; veins indistinct; margins more or less revolute, ciliate at least in basal half. Bracts up to 2·5 mm wide, more or less similar to the leaves, sometimes purple. Calyx 3·5–5 mm, tubular-campanulate; upper teeth 1–2 mm, lanceolate, usually ciliate and purple. Corolla 5–7 mm, pink; tube cylindrical. 2n=28. *Mountains of C. part of Balkan peninsula; Krym.* Al Bu Gr Ju Rs (K).

Variable in leaf- and bract-shape, and indumentum. Some local variants may deserve subspecific rank but a more thorough knowledge is needed of the variation pattern of the species in S.W. Asia.

Most plants from Krym differ from those from the Balkan peninsula in having almost flat leaf-margins. They have been treated as two different species on the basis of indumentum characters and of having ciliate (*T. pseudohumillimus*) or non-ciliate (*T. tauricus*) upper calyx-teeth. The chromosome number cited refers to plants from Krym.

15. T. parnassicus Halácsy, *Denkschr. Akad. Wiss. Math.-Nat. Kl. (Wien)* **61**: 254 (1894). Like **14** but leaves 6–12 × 0·5–1·5 mm, velutinous; bracts 6–12 × 2–4 mm, greenish; calyx greenish; corolla *c.* 7 mm. *Mountains of S. part of Balkan peninsula, northwards to Makedonija.* ?Al Gr Ju.

Perhaps only a subspecies of **14**.

Sect. THYMUS (Sect. *Vulgares* Velen., Sect. *Zygis* Willk.). Small shrubs. Leaves with prominent mid-vein and usually revolute margins, usually tomentose at least beneath. Bracts similar to leaves or not, mostly greenish. Calyx more or less campanulate; upper teeth more or less as long as wide. Corolla scarcely longer to distinctly longer than calyx.

16. T. capitellatus Hoffmanns. & Link, *Fl. Port.* **1**: 125 (1809). 20–40 cm, with erect to patent woody branches bearing axillary leaf-clusters. Cauline leaves 3–5 × 1–1·5 mm, petiolate, narrowly ovate, tomentose; margins revolute, not ciliate. Inflorescence compound, of subglobose, often pedunculate cymes 6–8 mm in diameter. Bracts broadly ovate, greenish. Calyx 3–4 mm, campanulate; upper teeth about as long as wide, not ciliate. Corolla whitish. ● *S. Portugal.* Lu.

17. T. camphoratus Hoffmanns. & Link, *op. cit.* 131 (1809) (*T. algarbiensis* Lange). Like **16** but inflorescence 10–18 mm in diameter, terminal; bracts often purplish; calyx 4–6 mm, the upper teeth triangular-subulate, usually ciliate. ● *S. Portugal.* Lu.

18. T. carnosus Boiss., *Voy. Bot. Midi Esp.* **2**: 490 (1841). Pulvinate, 20–40 cm, with erect to ascending woody branches bearing axillary leaf-clusters. Leaves 5–7 × 1–2 mm, petiolate, ovate-lanceolate, fleshy, glabrous above, tomentose beneath; margins revolute, ciliate at base. Inflorescence 1–3 cm. Bracts ovate, greenish. Calyx 3–4 mm, the tube campanulate; upper teeth as long as wide, not ciliate. Corolla whitish. *Maritime sands.* ● *S. Portugal.* Lu.

19. T. vulgaris L., *Sp. Pl.* 591 (1753) (incl. *T. aestivus* Reuter ex Willk., *T. ilerdensis* F. González ex Costa, *T. valentinus* Rouy, *T. webbianus* Rouy). 10–30(–50) cm, with erect to semi-patent woody branches. Leaves 3–8 × 0·5–2·5 mm, linear to elliptical, scarcely exceeding the axillary leaf-clusters, petiolate, tomentose; margins revolute, not ciliate. Inflorescence capituli-form or interrupted, many-flowered. Bracts similar to the leaves but somewhat wider and sometimes with almost flat margins, greyish-green. Calyx 3–4 mm, shortly hairy, the tube campanulate; upper teeth as long as wide, not ciliate. Corolla whitish to pale purple. 2n=30. ● *W. Mediterranean region, extending to S.E. Italy.* ?Co Ga Hs It [Bl He].

Very variable in shape of leaves, bracts and inflorescence. Widely cultivated as a pot-herb.

20. T. hyemalis Lange, *Vid. Meddel. Dansk Naturh. Foren. Kjøbenhavn* **1863**: 173 (1864). Like **19** but leaves 3–5 × 0·5–1 mm, puberulent to glabrescent, ciliate at base; calyx coloured, with longer patent hairs; upper teeth ciliate; corolla up to 7 mm, pink-purple. *S.E. Spain.* Hs.

T. glandulosus Lag. ex H. del Villar, *Cavanillesia* **6**: 105 (1934), non Req., from 2 or 3 localities in S.E. Spain (prov. Almería), and also from N. Africa, seems not to be specifically different from **20**. It has longer internodes, leaves 8–10 mm, and white to purplish-violet corolla. It is said to flower in spring (not in winter as **20**).

21. T. zygis L., *Sp. Pl.* 591 (1753) (incl. *T. sabulicola* Cosson, *T. sylvestris* Hoffmanns. & Link). 10–30 cm, with ascending to erect, puberulent, woody stems. Leaves 6–10 × 1 mm, usually exceeding the axillary leaf-clusters, linear, subacute, sessile, tomentose; margins revolute, sparsely ciliate at base. Inflorescence up to 10 cm, usually interrupted. Bracts similar to the leaves, mostly exceeding the verticillasters. Calyx 3–4 mm, shortly tomentose, greyish-green, the tube campanulate; upper teeth as long as wide, usually not ciliate. Corolla whitish. 2n=60. *Spain and Portugal.* Hs Lu.

22. T. baeticus Boiss. ex Lacaita, *Cavanillesia* **3**: 44 (1930) (*T. hirtus* auct., non Willd.). Like **21** but stems arcuate to erect, with somewhat longer and coarser indumentum; leaves (4–)5–8(–10) × 0·5–1(–1·5) mm, linear-lanceolate, subobtuse, puberulent, ciliate at base; bracts up to 3 mm wide; inflorescence usually many-flowered, capituliform, sometimes branched; calyx puberulent to hirsute, the lips about as long as tube; upper teeth up to 1 mm, usually lanceolate; corolla up to 6 mm, distinctly longer than calyx, creamy white. 2n=60. *Mountains of S. & S.E. Spain.* Hs.

Variable in development of indumentum, and in shape of bracts and inflorescence.

23. T. hirtus Willd., *Enum. Pl. Hort. Berol.* 623 (1809) (*T. diffusus* Salzm. ex Bentham). Plant with long arcuate to procumbent woody branches and sometimes axillary leaf-clusters. Erect flowering stems up to 5 cm; axillary leaf-clusters absent or poorly developed. Leaves up to 6 × 2 mm, lanceolate to elliptic-rhombic, petiolate, puberulent, sparsely glandular-punctate; margins revolute, ciliate at base; lateral veins indistinct. Inflorescence capituliform. Bracts up to 5 × 3 mm, similar to the leaves, with revolute margins at least distally. Calyx *c.* 3 mm, hirsute, the lips as long as or longer than tube, upper teeth as wide as long, ciliate. Corolla pink. *Gibraltar.* Hs. (*N.W. Africa.*)

24. T. loscosii Willk. in Willk. & Lange, *Prodr. Fl. Hisp.* 2: 401 (1868). Like **23** but flowering stems with well-developed axillary leaf-clusters except at the base; leaves (5–)7–10 × 1–2 mm, linear to narrowly lanceolate, densely glandular-punctate, subglabrous; margins revolute to almost flat; calyx 4–5 mm, upper teeth 1–1·5 mm, narrowly lanceolate; corolla up to 8 mm, whitish. ● *N.E. Spain.* Hs.

(a) Subsp. **loscosii**: Flowering stems puberulent. Leaves and bracts 0·5–1 mm wide, with more or less revolute margins. Inflorescence poorly differentiated, usually with 2-flowered verticillasters and pedicels up to 1 mm. Calyx subglabrous; upper teeth not ciliate. *Almost throughout the range of the species.*
(b) Subsp. **fontqueri** Jalas, *Bot. Jour. Linn. Soc.* 64: 250 (1971) (*T. angustifolius* auct., non Pers.): More robust. Flowering stems hirsute. Leaves and bracts 1–2 mm wide, with weakly revolute to almost flat margins. Inflorescence capituliform, usually with 6-flowered verticillasters and pedicels *c.* 2 mm. Calyx hirsute; upper teeth ciliate. 2n = 56. *Pobla de Segur (prov. Lérida).*

25. T. serpylloides Bory, *Ann. Gén. Sci. Phys.* (*Bruxelles*) 3: 16 (1820). Plant with slender arcuate to procumbent woody stems bearing axillary clusters of small leaves. Flowering stems 1–8 cm, borne in rows, pubescent, with a basal cluster of small leaves; axillary clusters absent. Leaves 5–8 × 0·5–1 mm, linear-lanceolate, densely glandular-punctate, puberulent to glabrous; margins revolute to subrevolute, ciliate at base. Inflorescence subcapitate, lax, with 2- to 6-flowered verticillasters; pedicels up to 3 mm. Bracts 1·5–2 mm wide, green, similar to the leaves, almost flat. Calyx 3·5–4·5 mm, pubescent; upper teeth 0·5–0·7 mm, as long as wide. Corolla 5–6 mm. *Mountains of S. Spain, mostly above 2000 m.* Hs.

(a) Subsp. **serpylloides**: Leaves sparsely pubescent to glabrous, green, with purple glands. Calyx purple; upper teeth not ciliate. Corolla purplish. ● *Sierra Nevada.*
(b) Subsp. **gadorensis** (Pau) Jalas, *Bot. Jour. Linn. Soc.* 64: 251 (1971) (*T. zygis* var. *gadorensis* Pau, *T. sylvestris* auct., non Hoffmanns. & Link): Leaves puberulent, pale greyish-green with yellowish glands. Calyx greenish; upper teeth ciliate. Corolla pale pink. *Sierra de Gádor.* (*N.W. Africa.*)

On some inland mountains of S. Spain (Sierra del Pinar, Serranía de Magina, Sierra de Alcaraz) plants occur which evidently are close to or conspecific with **25**. They tend to be more robust than subspp. (a) and (b), with larger leaves and more elongate to interrupted inflorescences. Some may correspond to **T. orospedanus** H. del Villar, *Cavanillesia* 6: 118 (1934), while others may be hybrids of **25** with **20, 21** or **22**. Further studies are needed.

Sect. HYPHODROMI (A. Kerner) Halácsy. Small shrubs. Stems hairy all round or on 2 opposite sides. Leaves coriaceous,

often more or less fleshy with a prominent midrib and weak lateral veins disappearing towards the margin, rarely tomentose-velutinous; margins flat, ciliate at least at base. Bracts similar to leaves or not, rarely coloured. Corolla scarcely longer to distinctly longer than calyx.

26. T. holosericeus Čelak., *Flora* (*Regensb.*) 66: 167 (1883). Up to 10 cm, caespitose, woody at the base. Flowering stems 3–6 cm, with internodes distinctly shorter than leaves, velutinous. Leaves 10–15 × 1·5–2 mm, narrowly oblanceolate, acute, often denticulate, velutinous, the basal half ciliate. Inflorescence a dense, oblong capitulum. Bracts *c.* 9 × 4 mm, ovate, acute, bluish-grey-green. Calyx 4–5 mm, the tube more or less cylindrical, villous-velutinous; upper teeth *c.* 0·5 mm, narrowly lanceolate, ciliate. Corolla 7–9 mm, pinkish-purple. ● *W. Greece* (*Kefallinia*). Gr.

27. T. laconicus Jalas, *Bot. Jour. Linn. Soc.* 64: 252 (1971) (*T. pastoralis* Turrill, non Iljin). Like **26** but leaves up to 11 × 1·5 mm, narrowly spathulate, obtuse to subacute; bracts up to 12 × 8 mm, ovate-elliptical, subacute, somewhat purplish; calyx 6–8 mm, with lips longer than tube, and upper teeth up to 1·5 mm; corolla *c.* 11 mm. ● *S. Greece* (*Akr. Malea*). Gr.

28. T. bracteosus Vis. ex Bentham, *Lab. Gen. Sp.* 346 (1834). Plant with long, procumbent, woody stems. Flowering stems *c.* 10 cm, pubescent to villous. Leaves 10–17 × 2–3 mm, oblong-spathulate, obtuse, herbaceous, glabrous, the basal half ciliate. Inflorescence capituliform. Bracts up to 13 × 6 mm, ovate, long-ciliate, often partly purplish. Calyx (4–)5–6 mm, the tube subcylindrical, shorter than the lips, hirsute at least ventrally; upper teeth 1·5–2 mm, narrowly lanceolate, ciliate. Corolla 6–8 mm, purple. ● *W. Jugoslavia.* Ju.

29. T. granatensis Boiss., *Elenchus* 74 (1838). Like **28** but leaves 7–10 × 1·5–2·5 mm, coriaceous; bracts 8–11 mm, usually purple; calyx 5–7 mm, the upper teeth 1–2 mm; corolla 8–12 mm, pink. ● *Mountains of S. Spain.* Hs.

Variable, especially in the length of the corolla.

30. T. aranjuezii Jalas, *Bot. Jour. Linn. Soc.* 64: 252 (1971) (*T. granatensis* var. *micranthus* Willk., *T. hirtus* auct., non Willd., *T. hispanicus* auct., non Poiret, *T. mastigophorus* auct., non Lacaita). Plant with short, woody, arcuate to procumbent branches bearing axillary clusters of small leaves. Flowering stems 3–10 cm, borne in rows, usually lacking axillary leaf-rosettes, pubescent to white-villous. Leaves 5–9 × 0·5–1 mm, narrowly spathulate, more or less fleshy, puberulent to glabrous, ciliate at base. Inflorescence a dense capitulum. Bracts (4–)6–9 × (2–)3–4 mm, ovate, narrowed above middle, obtuse, coriaceous, ciliate, usually not coloured. Calyx 3·5–5 mm, the tube about as long as lips; upper teeth 0·8–1·3 mm, ciliate. Corolla scarcely exceeding calyx, pink. 2n = 28. ● *C. & E. Spain.* Hs.

31. T. bracteatus Lange ex Cutanda, *Fl. Comp. Madrid* 538 (1861). Plant with long slender arcuate to procumbent, non-flowering branches bearing axillary clusters of small leaves. Flowering stems up to 12 cm, erect to ascending, pubescent; axillary clusters absent. Leaves 8–12 × (1·2–)1·5–2 mm, herbaceous, pale green, glabrous, ciliate at base, mostly petiolate. Inflorescence capitate to somewhat elongated with the lowest verticillasters distant. Bracts 2·5–4 mm wide, similar to the leaves, elliptical to narrowly ovate, mostly exceeding the verticillasters, pale green, ciliate. Calyx 4·5–6 mm, with campanulate to subcylindrical, pubescent tube distinctly shorter than lips;

upper teeth 1·2–1·8 mm, narrowly lanceolate, ciliate. Corolla 7–8 mm, whitish. 2*n* = 56. ● *C. & E. Spain.* Hs.

32. T. leptophyllus Lange, *Overs. Kong. Danske Vid. Selsk. Forh.* **1893**: 198 (1893–1894) (*T. angustifolius* auct., non Pers.). Like **31** but smaller in all parts; flowering stems 1·5–4 cm; leaves 5–8 × 1–1·5 mm, linear-oblanceolate to narrowly spathulate; bracts up to 2 mm wide; calyx 3·5–5 mm, the tube about as long as the lips, subglabrous, the upper teeth 1–1·5 mm, not ciliate except for short one-celled hairs. ● *C. & E. Spain.* Hs.

Although **32** tends to occur in drier areas and mostly at somewhat lower altitudes than **31**, intermediates between the two seem to occur in areas where they are sympatric. A specimen from prov. Murcia differs from **32** in having elliptical, petiolate leaves 1·5–2 mm wide, in which it resembles **T. atlanticus** (Ball) Roussine, *Nat. Monsp.* (*Bot.*) **16**: 165 (1965), from N. Africa.

33. T. atticus Čelak., *Flora* (*Regensb.*) **65**: 564 (1882). Plant with woody, procumbent or arcuate shoots bearing axillary leaf-clusters. Flowering stems up to 15 cm, with axillary leaf-clusters at least at the base, pubescent all round. Leaves 12–20 × 1 mm, linear, subobtuse, subcoriaceous, glabrous, ciliate at the base with veins rather indistinct, parallel. Bracts broadly ovate, narrowed from middle, obtuse, the lowermost usually 9 × 3 mm, coriaceous, straw-coloured, with distinct veins and shortly appressed-pubescent beneath, ciliate. Inflorescence a dense head. Calyx 5–7 mm, puberulent, with prominent veins; upper teeth 1–1·5 mm, lanceolate. Corolla whitish or pink. *E. half of Balkan peninsula.* Bu Gr Tu.

34. T. plasonii Adamović, *Österr. Bot. Zeitschr.* **57**: 200 (1907). Like **33** but smaller in all parts; stems patent-hirsute; leaves 8–12 × 0·7–1 mm, coriaceous, with more distinct veins, ciliate to above the middle; largest bracts *c.* 5 × 2 mm, glabrous; calyx 3·5–4 mm; corolla pink, becoming yellow. ● *N.E. Greece (near Thessaloniki).* Gr.

35. T. striatus Vahl, *Symb. Bot.* **3**: 78 (1794) (*T. acicularis* Waldst. & Kit., *T. comptus* auct., non Friv.; incl. *T. pseudo-atticus* Ronniger). Like **33** but usually less robust; stems sometimes hairy only on 2 opposite sides; bracts usually narrower and purplish, sometimes similar to leaves, glabrous or pubescent to villous; calyx 3–5 mm, purplish, the upper teeth lanceolate; corolla whitish, pink or purple. 2*n* = 26, 54. *Balkan peninsula; C. & S. Italy, Sicilia.* Al Bu Gr It Ju Si ?Tu.

Variable in hairiness, leaf- and calyx-size, and development of bracts. Variants with leaves 1·5 mm or more wide tend to occupy a more southern part of the range than those with narrow (*c.* 1 mm wide) leaves. The two chromosome numbers refer to morphologically extreme variants and further information is needed about the pattern of cytological and morphological variation to decide if subspecies can be recognized.

36. T. spinulosus Ten., *Fl. Nap.* **1**, *Prodr.*: 35 (1811) (*T. conspersus* Čelak.). Stems ascending from woody base, hairy all round; creeping shoots absent or short and few. Leaves 9–12 × 1·5–2·5 mm, lanceolate, subacute, often sparsely hirsute above; midvein prominent, the lateral less distinct. Inflorescence capitate. Bracts with indistinct lateral veins, herbaceous, greenish; lowest similar to the leaves, the upper *c.* 7 × 3 mm, more or less ovate. Calyx 3·5–5 mm, greenish, densely reddish-glandular-punctate; upper teeth up to 1·2 mm, lanceolate, ciliate. Corolla whitish. ● *Mountains of S. Italy and Sicilia.* It Si.

The plants from Sicilia described as **T. paronychioides** Čelak., *Flora* (*Regensb.*) **65**: 564 (1882), do not appear to be specifically different, but further material is required.

37. T. aznavourii Velen., *Sitz.-Ber. Böhm. Ges. Wiss.* (*Math.-Nat. Kl.*) **1903**(28): 17 (1904) (*T. sintenisii* auct., non Čelak.). Flowering stems 3–5(–10) cm. Leaves 9–10 × 1·7–2 mm, coriaceous, more or less sessile, oblanceolate, obtuse, entire, glabrous, ciliate at base, sparsely glandular-punctate above at apex; midrib prominent, the lateral veins less distinct. Inner bracts *c.* 5–7 × 2–2·5 mm, ovate. Calyx 4–4·5 mm, sparsely yellowish-glandular-punctate; upper teeth not more than 0·5 mm, not ciliate. ● *Turkey-in-Europe (Safraköy, near Istanbul).* Tu.

38. T. kirgisorum Dubjanski in O. & B. Fedtsch., *Consp. Fl. Turkestan.* **5**: 128 (1913) (incl. *T. calcareus* Klokov & Schost., *T. cretaceus* Klokov & Schost., *T. graniticus* Klokov & Schost., *T. kaljmijussicus* Klokov & Schost.). Like **37** but leaves (5–)7–13 × 0·8–1·5(–2) mm, linear-lanceolate to narrowly elliptical, shortly petiolate, densely glandular-punctate; bracts more or less similar to the leaves; calyx 3–4 mm, densely glandular-punctate. *S. part of U.S.S.R.* Rs (C, W, E).

T. dubjanskii Klokov & Schost., *Bull. Jard. Bot. URSS* **30**: 545 (1932), from S.E. Russia (Chvalynsk, S.W. of Kujbyšev), is a small, compact, woody plant with leaves 8–11 × 2–3 mm, and calyx 4–4·5 mm. The small amount of material seen is evidently heterogeneous, and one specimen from the type-locality closely resembles and may be identical with **T. rariflorus** C. Koch, *Linnaea* **21**: 666 (1848), from the Caucasus.

39. T. zygioides Griseb., *Spicil. Fl. Rumel.* **2**: 118 (1844). Flowering stems 2–6(–8) cm, in rows on long, creeping, woody branches which bear axillary leaf-clusters distally. Leaves 5–10(–15) × *c.* 1 mm, spathulate, obtuse, entire, fleshy, ciliate to above middle, otherwise glabrous; veins indistinct. Bracts shorter and wider, ovate to elliptical, often pubescent beneath. Inflorescence usually densely capitate. Calyx 3·5–4·5 mm; upper teeth 0·3–0·7 mm, as long as or longer than wide, ciliate. Corolla pink. *E. part of Balkan peninsula, extending to S.E. Romania.* Bu Gr Rm Tu.

T. eupatoriensis Klokov & Schost., *Acta Inst. Bot. Acad. Sci. URSS* (*Ser.* 1) **2**: 287 (1936), from Krym, differs, according to the description, by its larger leaves, and more leaf-like bracts, but material seen does not confirm this and suggests that it is conspecific with **39**. Another related plant from Krym is **T. liaculatus** Klokov in Kotov, *Fl. RSS Ucr.* **9**: 668 (1960). **T. moldavicus** Klokov & Schost., *Bull. Jard. Bot. Kieff* **16**: 17 (1932) may also belong here.

Sect. SERPYLLUM (Miller) Bentham. Small shrubs or perennial herbs. Stems usually herbaceous or woody at the base, hairy all round or on 2 opposite sides, rarely on the angles only. Leaves not tomentose or velutinous; margins usually more or less flat. Bracts more or less similar to the leaves. Calyx more or less campanulate.

40. T. comptus Friv., *Flora* (*Regensb.*) **19**: 439 (1836) (*T. glaucus* Friv. ex Podp.). Flowering stems up to 15 cm, ascending, woody at the base; non-flowering branches short, procumbent. Stems hairy all round, densely and usually patent-villous. Leaves 9–15 × 0·5–2(–3) mm, linear to lanceolate, more or less acute, usually ciliate to above middle and sparsely hairy; veins distinct, midrib prominent, the lateral veins almost parallel.

Inflorescence up to 10 cm, with remote, many-flowered verticillasters. Bracts somewhat similar to the leaves but shorter and wider, with distinct lateral veins. Calyx (3–)4–5 mm, the lips of equal length; upper teeth up to 1·5 mm, aristate, ciliate or with one-celled scabridities only. *E. part of Balkan peninsula, mainly near the coast.* Bu Gr Tu.

Contains several local variants, some of which may merit subspecific recognition.

41. **T. sibthorpii** Bentham, *Lab. Gen. Sp.* 345 (1834) (incl. *T. tosevii* Velen., *T. korthiaticus* Adamović, *T. macedonicus* (Degen & Urum.) Ronniger). Flowering stems usually 10–20(–30) cm, erect to ascending, often branched, shortly pubescent all round (rarely on 2 sides only), woody at base. Creeping, non-flowering stems absent. Leaves usually 10–15 × 3–5 mm, shortly petiolate, elliptic-lanceolate, acute, coriaceous, densely glandular-punctate; lateral veins distinct. Inflorescence often branched, long. Bracts similar to the leaves. Calyx 2·5–3·5(–4) mm, campanulate, usually greenish to straw-coloured; upper lip longer than tube; upper teeth *c.* 1 mm, triangular to lanceolate, ciliate. Corolla pale pink or red. *C. part of Balkan peninsula.* ?Al Bu Gr Ju ?Rm Tu.

This species shows considerable variation, especially in robustness, leaf-size and indumentum.

T. degenii H. Braun, *Mitt. Naturw. Ver. Steierm.* 54: 262 (1918) (*T. tosevii* subsp. *degenii* (H. Braun) Ronniger), which appears to be common in the mountains of S. Bulgaria and N.E. Greece (especially Rodopi) between 1000 and 2000 m, is less robust and more procumbent, with flowering stems usually less than 10 cm, smaller and less coriaceous leaves, and the inflorescence often globose and tinged with purple.

Another procumbent variant from Bulgaria and N. Greece was described as *T. heterotrichus* subsp. *cinerascens* Velen., *Sitz.-Ber. Böhm. Ges. Wiss.* (*Math.-Nat. Kl.*) 1903(28): 15 (1904). It has minutely puberulent leaves with weak lateral veins and somewhat revolute margins.

T. grisebachii Ronniger, *Beih. Bot. Centr.* 54B: 663 (1936), described from the border of Greece and Jugoslavia (Kaimak-čalan), has stems hairy on 2 sides only.

T. substriatus Borbás, *Verh. Naturf. Ver. Brünn* 38: 210 (1900), from Macedonia and Bulgaria, is like 41 but has flowering stems which are patent-villous, and somewhat longer calyx.

T. heterotrichus Griseb., *Spicil. Fl. Rumel.* 2: 116 (1844), from N. Greece (Makedhonia), is like 41 but has leaves 12–17 × 2–3 mm, somewhat rounded at apex, subsessile and sparsely glandular-punctate, inflorescence globose, calyx 4–5 mm with upper teeth short, not ciliate. Its relationship with 41 is not clear.

42. **T. pannonicus** All., *Auct. Syn. Meth. Stirp. Hort. Taurin.* 6 (1773) (incl. *T. dzevanovskyi* Klokov & Schost., *T. latifolius* (Besser) Andrz., *T. marschallianus* Willd., *T. serpyllum* subsp. *auctus* Lyka, subsp. *brachyphyllus* Lyka & subsp. *marschallianus* (Willd.) Nyman, *T. stepposus* Klokov & Schost.). Like 41 but more herbaceous; leaves subsessile, herbaceous with indistinct veins and sparsely glandular-punctate; inflorescence rarely branched. 2n=28. *E.C. & E. Europe, northwards to c. 57° N. in C. Russia.* Au Bu Cz Hu Ju Po Rm Rs (C, W, K, E) [Ge Rs (N)].

T. dimorphus Klokov & Schost., *Jour. Agr. Bot.* (*Kharkov*) 1(3): 122 (1927) (incl. *T. amictus* Klokov, *T. litoralis* Klokov & Schost.), from S.E. Ukraine and S.E. Russia, is said to differ from 42 in its distinctly petiolate middle and lower cauline leaves, long patent hairs especially below the nearly globose inflorescence, and calyx 3·5–5 mm. It combines characters of 42 with those of 43 and is of uncertain relationship. **T. pseudogranaticus** Klokov & Schost., *op. cit.* 124 (1927) may be a hybrid between *T. dimorphus* and 38.

T. bulgaricus (Domin & Podp.) Ronniger in Hayek, *Prodr. Fl. Penins. Balcan.* 2: 352 (1930), from Bulgaria, is more robust than *T. dimorphus*, and has leaves with prominent midribs and usually shorter, recurved stem-pubescence. It may be more closely related to 40 or 41.

43. **T. glabrescens** Willd., *Berlin. Baumz.* ed. 2, 507 (1811). Creeping branches mostly ending in a terminal inflorescence. Flowering stems 5–15(–30) cm, more or less herbaceous, erect to ascending, usually hairy all round, usually without basal leaf-clusters. Leaves mostly more than 1·5 mm wide, elliptic-lanceolate to obovate, mostly rounded at apex, herbaceous, ciliate at the base, at least the lower petiolate. Inflorescence globose or interrupted, often the lowest verticillasters remote. Bracts similar to the leaves. Calyx (3·5–)4–5 mm; tube as long as upper lip; upper teeth usually longer than wide, ciliate. *S.E. & E.C. Europe, extending locally westwards to S.E. France.* ?Al Au Bu Cz Ga He Hu It Ju Po Rm Rs (C, W, K, ?E) Tu.

This is a rather heterogeneous assemblage of plants probably of different allopolyploid origins. They are often difficult to separate from 42 and 55. The following treatment covers the major part of the variation but is largely provisional.

1 Leaves on flowering stems larger above, usually more than 3 mm wide, and the veins weak **(a)** subsp. **glabrescens**
1 Leaves on flowering stems mostly about equal, usually less than 3 mm wide, and with a prominent midrib
2 Leaves 4–10(–15) × 1·5–3 mm, elliptical, shortly cuneate at base, obtuse at apex; upper calyx-teeth often ±pungent **(b)** subsp. **decipiens**
2 Leaves 7–13 × 1·2–2·5 mm, linear-lanceolate, long-cuneate at base, subacute at apex; upper calyx-teeth not pungent **(c)** subsp. **urumovii**

(a) Subsp. **glabrescens** (*T. serpyllum* subsp. *glabrescens* (Willd.) Lyka, *T. austriacus* Bernh. ex Reichenb., *T. loevyanus* Opiz, *T. tschernjajevii* Klokov & Schost.): 2n=28, 32, 56, 58. *Throughout the range of the species, except the S. Alps.*
(b) Subsp. **decipiens** (H. Braun) Domin, *Pl. Čechosl. Enum.* 197 (1935) (*T. oenipontanus* H. Braun, *T. serpyllum* subsp. *decipiens* (H. Braun) Lyka): 2n=52. ● *Dry valleys of S. & E. Alps, mainly between 900 and 1400 m.*
(c) Subsp. **urumovii** (Velen.) Jalas, *Bot. Jour. Linn. Soc.* 64: 261 (1971) (*T. callieri* subsp. *urumovii* Velen.; incl. *T. callieri* Borbás ex Velen., *T. hirsutus* Bieb., nom. ambig., *T. jailae* (Klokov & Schost.) Stankov, *T. zelenetzkyi* Klokov & Schost.): 2n=56. *S.E. Europe.*

44. **T. longedentatus** (Degen & Urum.) Ronniger, *Feddes Repert.* 20: 335 (1924). Primary stems woody, creeping and at least partly subterranean, bearing large compact leaf-rosettes and flowering stems up to 10 cm, without a basal cluster of small leaves, villous all round. Leaves 12–25 × 1·5–3 mm, linear-oblanceolate, petiolate, long-cuneate at base, subacute to sub-obtuse at apex; midrib rather prominent, the lateral veins subparallel. Inflorescence capitate, elongated, sometimes with the lowest verticillasters remote. Pedicels and calyces usually densely stipitate-glandular. Bracts similar to the leaves. Calyx 5–6 mm, the upper lip somewhat longer than the lower;

upper teeth 1·5–2 mm, aristate, ciliate. *E. part of Balkan peninsula.* Bu ?Gr Tu.

T. conspersus var. **lycaonicus** Čelak., *Flora (Regensb.)* **66**: 156 (1883) (*T. zygioides* var. *lycaonicus* (Čelak.) Ronniger), from the E. Aegean region, may represent a subspecies of **44**. It differs in having leaves usually 10–12 × 1–1·5 mm, sparsely hairy, and calyx 3·5–4·5 mm, with upper teeth up to 1·2 mm. It has been recorded from Turkey-in-Europe (near Gelibolu), and intermediates are found in Bulgaria.

45. T. pallasianus H. Braun, *Österr. Bot. Zeitschr.* **42**: 337 (1892). Stems erect to arcuate or procumbent, woody; creeping, non-flowering branches absent. Flowering stems up to 15 cm, with axillary leaf-clusters and shortly hairy all round. Leaves 9–20 mm, linear to lanceolate, sessile, glabrous, usually finely scabrid; midrib prominent. Inflorescence capitate. Bracts similar to the leaves, the inner sometimes ovate-lanceolate. Calyx 3·5–4·5 mm; tube campanulate to subcylindrical, hirsute; upper teeth *c.* 1 mm, usually not ciliate. Corolla pale pink. *Sandy or clayey steppes. S. part of U.S.S.R.* Rs (C, W, E).

(a) Subsp. **pallasianus**: Leaves *c.* 1 mm wide, linear to narrowly spathulate, ciliate in lower half. Calyx purplish; upper teeth often with hyaline margins, the median up to 1 mm and usually exceeding the others. *Throughout the range of the species.*

(b) Subsp. **brachyodon** (Borbás) Jalas, *Bot. Jour. Linn. Soc.* **64**: 262 (1971) (*T. brachyodon* Borbás, *T. eltonicus* Klokov & Schost.; incl. *T. lanulosus* Klokov & Schost.): Leaves 2–3(–6) mm wide, lanceolate to elliptic-lanceolate, ciliate at base. Calyx greenish; upper teeth up to 0·7 mm, equal, as long as wide, the margins not hyaline. *S.E. Russia and N.W. Kazakhstan.*

T. borysthenicus Klokov & Schost., *Jour. Agr. Bot. (Kharkov)* **1**(3): 134 (1927), from sandy ground by the lower Dnepr in S. Ukraine, with a straggling habit, leaves *c.* 9 × 0·5 mm sometimes with recurved margins, and upper calyx-teeth not hyaline, may deserve recognition as a subspecies of **45**.

T. ciliatissimus Klokov in Kotov, *Fl. RSS Ucr.* **9**: 669 (1960), from similar habitats in S.W. Ukraine, has linear leaves ciliate to near the apex, and ciliate upper calyx-teeth. Its status is uncertain.

46. T. herba-barona Loisel., *Fl. Gall.* 360 (1807). Branches procumbent to ascending, woody; long, creeping, non-flowering branches absent. Flowering stems 5–10 cm, hairy all round. Leaves usually 6–9 × 2–4 mm, elliptical to rhombic-ovate, petiolate, acute, herbaceous, with flat to subrevolute margins, glabrous or puberulent, ciliate at base; veins indistinct. Inflorescence subcapitate, lax, sometimes with the lower verticillasters remote. Bracts similar to the leaves. Calyx 3–5 mm; upper teeth up to 1·5 mm, somewhat longer than wide, scabrid, not ciliate. Corolla up to 9 mm, pale purple. 2*n* = 56. ● *Corse, Sardegna.* Co Sa.

Variable in dimensions of leaf and calyx, and in indumentum.

47. T. nitens Lamotte, *Prodr. Fl. Centr. Fr.* **2**: 595 (1881). Like **46** but leaves up to 12 mm, oblong-elliptical, not ciliate at base; inflorescence up to 10 cm, more lax. 2*n* = 28. ● *S. France.* Ga.

48. T. willkommii Ronniger, *Feddes Repert.* **28**: 147 (1930). Woody at base, with long, slender, woody, arcuate to procumbent or creeping branches. Flowering stems 2–5 cm, pubescent all round. Leaves 5–7 × 2–3 mm, ovate to ovate-elliptical, glabrous, not ciliate at base; lateral veins somewhat indistinct. Flowers not forming a distinct inflorescence. Verticillasters 2- to

6-flowered, the cymes sometimes pedunculate; pedicels 1–2 mm. Bracts similar to the leaves. Calyx 4–5 mm; tube subcylindrical, about as long as lips, sparsely hairy; lips of about equal length; upper teeth *c.* 1 mm, lanceolate, not ciliate. Corolla 6–7 mm. ● *N.E. Spain.* Hs.

49. T. richardii Pers., *Syn. Pl.* **2**: 130 (1806). Like **48** but more robust; flowering stems 4–12 cm; leaves 7–12 × 3–6 mm, the lateral veins more distinct; inflorescence capitate; cymes 3- to many-flowered, pedunculate; calyx 4·5–6 mm; tube cylindrical, often longer than lips, the upper lip longer than the lower; upper teeth up to 1·5 mm; corolla 7–9 mm, purple. ● *Islas Baleares, Sicilia, W. Jugoslavia.* Bl Ju Si.

1 Calyx sparsely pubescent to subglabrous; margins of upper teeth without multicellular hairs   **(a) subsp. richardii**
1 Calyx hirsute; margins of upper teeth with multicellular hairs
2 Leaves more than twice as long as wide; calyx densely stipitate-glandular   **(b) subsp. nitidus**
2 Leaves less than twice as long as wide; calyx not stipitate-glandular   **(c) subsp. ebusitanus**

(a) Subsp. **richardii** (*T. aureopunctatus* (G. Beck) K. Malý): 2*n* = 28. *Islas Baleares (Mallorca), W. Jugoslavia (near Konjic).*
(b) Subsp. **nitidus** (Guss.) Jalas, *Bot. Jour. Linn. Soc.* **64**: 264 (1971) (*T. nitidus* Guss.): *W. Sicilia (Isola Marettimo).*
(c) Subsp. **ebusitanus** (Font Quer) Jalas, *loc. cit.* (1971) (*T. richardii* var. *ebusitanus* Font Quer): *Islas Baleares (Ibiza).*

50. T. guberlinensis Iljin, *Not. Syst. (Leningrad)* **1**(5): 1 (1920) (*T. zheguliensis* Klokov & Schost.). Stems ascending to arcuate, woody; long, creeping, non-flowering branches absent. Flowering stems up to 15 cm, shortly hairy all round. Leaves mostly 8–12 × 2·5–3·5 mm, ovate-spathulate to elliptical, subobtuse, petiolate, sparsely glandular-punctate; veins distinct. Inflorescence densely capitate. Bracts similar to the leaves. Calyx 3·5–4·5 mm, the tube campanulate; upper teeth not ciliate. Corolla pale violet. *S.E. Russia.* Rs (?C, E).

T. mugodzharicus Klokov & Schost., *Bull. Jard. Bot. URSS* **30**: 537 (1932), from E. Russia and W. Kazakhstan, with flowering stems 5–10 cm, and leaves 6–10 × 1·5–3 mm, elliptical to elliptic-lanceolate, subpetiolate, often with a recurved margin, is probably not specifically distinct from **50**.

T. bashkiriensis Klokov & Schost., *op. cit.* 533 (1932), from E. Russia (Baškirskaja A.S.S.R.), a smaller plant with flowering stems 3–6 cm, leaves 6–8 × 1·5–1·75 mm, oblanceolate to linear-spathulate, ciliate to about middle, and calyx 3·5 mm, with upper teeth up to 1 mm and lanceolate, is of uncertain status.

The plant recorded from a single locality in Turkey-in-Europe (Büyükdere, N. of Istanbul) as *T. punctatus* var. *subisophyllus* (Borbás) Ronniger probably belongs to *T. sipyleus* Boiss., from Asia Minor.

51. T. binervulatus Klokov & Schost., *Žur. Inst. Bot. URSR* **9**: 195 (1936). Plant with woody, slender, creeping stems with terminal inflorescences; flowering stems up to 8 cm, hirsute all round. Leaves 8–9 × 3–4 mm, elliptic-spathulate, subcoriaceous, with a distinct, ciliate petiole and two pairs of distinct lateral veins, fusing in the upper part to form a marginal vein. Inflorescence capitate. Bracts similar to the leaves. Calyx 3·5–4 mm, greenish; upper teeth up to 1 mm, lanceolate, sparsely ciliate. Corolla pale lilac. ● *Foothills of S. Ural (E. of Ufa).* Rs (C).

52. T. ocheus Heldr. & Sart. ex Boiss., *Diagn. Pl. Or. Nov.* **2**(4): 6 (1859) (*T. chaubardii* (Boiss. & Heldr. ex Reichenb. fil.)

Čelak.). Plants usually with long, creeping and woody non-flowering branches bearing axillary leaf-clusters; flowering stems up to 10 cm, each with a basal cluster of small leaves, hairy all round or on 2 opposite sides. Leaves up to 13 × 3–4 mm, elliptical to lanceolate, coriaceous; lateral veins distinct; marginal veins absent. Inflorescence globose to somewhat cylindrical. Bracts similar to the leaves. Calyx 3–4 mm, usually greenish, the upper lip longer than the tube; upper teeth *c.* 1 mm, pungent, usually ciliate. Corolla *c.* 6 mm, pale pink. *Mountains of Greece and Macedonia.* ?Bu Gr Ju.

53. T. thracicus Velen., *Österr. Bot. Zeitschr.* **42**: 16 (1892) (incl. *T. alsarensis* Ronniger, *T. longidens* (Velen.) Podp., *?T. nikolovii* (Degen & Urum.) Stoj. & Stefanov). Like **52** but leaves mostly 8–12 × (2–)3–4 mm, elliptic-spathulate, subobtuse; calyx 4–6 mm, tinged purple; upper lip somewhat longer than lower; corolla up to 8 mm, purple. *C. part of Balkan peninsula and N. Aegean region.* Bu Gr Ju Tu.

54. T. stojanovii Degen, *Magyar Bot. Lapok* **23**: 72 (1925). Caespitose, with woody, creeping, non-flowering stems. Flowering stems 5(–7) cm, with a basal cluster of small leaves, usually hairy all round, and often with axillary leaf-clusters. Leaves 5–7 × 1–2 mm, lanceolate-elliptical, somewhat fleshy, subacute, mostly exceeding the internodes, often hairy, ciliate at the base; lateral veins indistinct. Inflorescence globose to somewhat cylindrical. Bracts similar to the leaves. Calyx 3–4(–5) mm, purplish; upper lip equalling or slightly longer than tube; upper teeth 0·5–1(–1·5) mm, ciliate, subspinescent. Corolla scarcely exceeding calyx, purple. ● *Mountains of C. part of Balkan peninsula.* ?Al Bu Gr Ju.

55. T. longicaulis C. Presl, *Fl. Sic.* 37 (1826) (incl. *T. illyricus* Ronniger, *T. kosaninii* Ronniger, *T. lykae* Degen & Jáv., *T. malyi* Ronniger, *T. moesiacus* Velen., *T. rohlenae* Velen., *T. serpyllum* subsp. *dalmaticus* (Reichenb.) Nyman). Plant with long, somewhat woody, creeping branches, non-flowering or with a terminal inflorescence. Flowering stems borne in rows, each with a basal cluster of small leaves, slender, usually not more than 10 cm, hairy all round or on 2 opposite sides. Leaves at least 1 mm wide, linear-lanceolate to elliptical, herbaceous (sometimes fleshy), ciliate at the base; lateral veins more or less indistinct; marginal vein absent. Bracts similar to the leaves. Calyx 2·5–3·5 mm, with tube campanulate, usually shorter than upper lip; upper teeth longer than wide. Corolla purple. $2n = 30$, more than 50, 58. *S. Europe, westwards to S.E. France.* Al ?Au Bu Ga Gr ?He ?Hu It Ju Rm Si Tu [Ge].

The wide variation found within this species includes two different levels of polyploidy, but it is not yet known whether these are correlated with any morphological or geographical features. There seem to be no grounds so far for recognizing any of the variants at subspecific level. The species is connected by numerous, probably hybridogeneous, intermediates with **35**, **41**, **43**, **56** and **61**.

T. dolomiticus Coste, *Bull. Soc. Bot. Fr.* **40**: 130 (1893), from S. France, should probably be included in **55** as a subspecies. It is characterized by a diploid chromosome number ($2n = 28$), flowering stems hairy all round, leaves 5–7 × 1–1·5 mm, hairy, and calyx 3–4 mm, with tube shorter than lips.
A tetraploid plant ($2n = 54$), which is more widespread in S. France, characterized by stems incompletely hairy on opposite sides, leaves *c.* 11 × 2·5 mm, glabrous, subcoriaceous, with slightly more prominent venation than in **55**, and calyx 3·5–4·5 mm with almost cylindrical tube as long or slightly longer than lips, does not seem to differ significantly from **T. embergeri**

Roussine, *Rec. Trav. Montpellier* (*Sér. Bot.*) **5**: 79 (1952), also from France, for which, however, the chromosome number $2n = 48$ has been reported.

T. sintenisii Čelak., *Flora* (*Regensb.*) **67**: 537 (1884), from Turkey-in-Europe (Eceabat), may belong here. It has stems hairy all round; leaves 9–11 × 1·2–1·5 mm, shortly petiolate, linear-lanceolate, obtuse, with a weakly prominent midrib and almost invisible lateral veins; calyx *c.* 5 mm, with upper lip longer than tube and upper teeth 0·8–1 mm. The bracts and corolla are unknown.

T. adamovicii Velen., *Beih. Bot. Centr.* **19**(2): 282 (1906), from C. Jugoslavia (near Gornji Milanovac), also with stems hairy all round, has elliptical leaves up to 9 × 2·5 mm, with subrevolute margins and more prominent midrib, densely covered by very short and stiff, one-celled hairs, and the calyx *c.* 3 mm, broadly campanulate, with subulate, nearly herbaceous lower teeth and upper teeth *c.* 0·5 mm, as long as wide. It may also belong here.

56. T. praecox Opiz, *Naturalientausch* **6**: 40 (1824). Plant with long, somewhat woody, creeping branches, non-flowering or with a terminal inflorescence. Flowering stems borne in rows, each with a basal cluster of small leaves, slender, usually not more than 10 cm, hairy all round or on 2 opposite sides. Leaves mostly obovate, broadly spathulate to suborbicular, subcoriaceous, ciliate at the base; lateral veins prominent, marginal veins present in the upper part. Bracts similar to the leaves. Calyx 3–5 mm, with tube campanulate, about equalling upper lip; upper teeth longer than wide. Corolla purple. *S., W. & C. Europe.* Al Au Be Br Bu Cz Fa Ga Ge Gr Hb He Ho Hs Hu Is It Ju No Po Rm ?Si ?Tu.

Very variable in leaf-shape and indumentum. The following tentative treatment aims to summarize the main features of the geographical variation, although the subspecific boundaries are rather obscure in several places.

1 Stems hairy all round
  2 Basal leaves on flowering stems petiolate; corolla up to 6 mm
                                               (a) subsp. **praecox**
  2 Basal leaves on flowering stems sessile; corolla up to 8 mm
                                                 (b) subsp. **skorpilii**
1 Stems with hairs mostly on 2 sides only
  3 Cauline leaves larger above   **(c)** subsp. **polytrichus**
  3 Cauline leaves about equal
    4 Leaves 5–8 mm, mostly obovate, densely glandular-punctate; calyx 3–4 mm   **(e)** subsp. **arcticus**
    4 Leaves 8–14 mm, linear-spathulate, very sparsely glandular-punctate; calyx 4–5 mm   **(d)** subsp. **zygiformis**

(a) Subsp. **praecox** (*T. humifusus* Bernh., *T. serpyllum* subsp. *clivorum* Lyka, subsp. *hesperites* Lyka & subsp. *praecox* (Opiz) Vollmann): $2n = c.$ 50, *c.* 54, 56, 58. ● *W. & C. Europe, mainly lowland.*

(b) Subsp. **skorpilii** (Velen.) Jalas, *Bot. Jour. Linn. Soc.* **64**: 266 (1971) (*T. skorpilii* Velen., *T. jankae* Čelak., *T. eximius* Ronniger): *N. & C. parts of Balkan peninsula.*

(c) Subsp. **polytrichus** (A. Kerner ex Borbás) Jalas, *Veröff. Geobot. Inst. Rübel* (*Zürich*) **43**: 189 (1970) (*T. alpigenus* (A. Kerner ex H. Braun) Ronniger, *T. balcanus* Borbás, *T. kerneri* Borbás, *T. polytrichus* A. Kerner ex Borbás, *T. serpyllum* subsp. *polytrichus* (A. Kerner ex Borbás) Briq. & subsp. *trachselianus* (Opiz) Lyka): $2n = 28$, *c.* 50, 54, 55, 56. ● *Mountains of S. & S.C. Europe.*

(d) Subsp. **zygiformis** (H. Braun) Jalas, *Bot. Jour. Linn. Soc.* **64**: 267 (1971) (*T. albanus* H. Braun, *T. zygiformis* H. Braun): ● *W. part of Balkan peninsula;* ?*C. Appennini.*

(e) Subsp. **arcticus** (E. Durand) Jalas, *Veröff. Geobot. Inst. Rübel* (*Zürich*) **43**: 190 (1970) (*T. drucei* Ronniger): $2n = c.$ 50, 51, 54. *W. Europe.*

T. widderi Ronniger ex Machule, *Mitt. Thür. Bot. Ges.* **1**(4): 89 (1957), from the E. Alps, differs from **56(c)** in having marginal veins in the leaf.

T. doerfleri Ronniger, *Denkschr. Akad. Wiss. Math.-Nat. Kl. (Wien)* **99**: 189 (1924) (*T. hirsutus* var. *doerfleri* (Ronniger) Ronniger), from N.E. Albania (Koritnik), resembles **56(d)** but has stems hairy all round and leaves densely covered by long and short hairs. Its status is uncertain.

57. T. nervosus Gay ex Willk., *Suppl. Prodr. Fl. Hisp.* 144 (1893). Like **56(d)** but leaves *c.* 6 × 1–1·5 mm, subcoriaceous, linear-spathulate, with distinct subparallel lateral veins and no marginal veins; calyx 3–4 mm, the tube about as long as upper lip. $2n=28$. ● *C. & E. Pyrenees; S.E. France (Mont Ventoux).* Ga Hs.

58. T. pulcherrimus Schur, *Verh. Mitt. Siebenb. Ver. Naturw.* **10**: 140 (1859) (incl. *T. carpathicus* Čelak., *T. circumcinctus* Klokov, *T. serpyllum* subsp. *sudeticus* Lyka, non *T. sudeticus* Opiz ex Borbás). Plant with long, creeping, somewhat woody non-flowering branches. Flowering stems up to 10 cm, with a basal cluster of small leaves. Leaves larger above on the flowering stems, ovate to orbicular, petiolate, truncate to broadly cuneate at base, shallowly crenate, with distinct marginal vein from apex to base. Inflorescence globose to somewhat cylindrical. Bracts similar to the leaves. Calyx 3·5–5 mm; tube subcylindrical, about equalling lips; upper teeth up to 1·3 mm, longer than wide. Corolla up to 8(–9) mm. $2n=56$ (60). ● *Carpathians, extending to N.C. Czechoslovakia (Jeseniky).* Cz Po Rm Rs (W).

Variable in habit, stem-indumentum and calyx-teeth.

59. T. comosus Heuffel ex Griseb., *Arch. Naturgesch. (Berlin)* **18**(1): 328 (1852). Branches suberect to procumbent, woody at base. Flowering stems 5–15 cm, sometimes branched, hairy all round. Leaves up to 17 × 9 mm, broadly ovate or rhombic-ovate to suborbicular, petiolate, with distinct marginal veins, glabrous to slightly hairy, ciliate at base. Inflorescence up to 7 cm, capituliform to cylindrical. Bracts similar to the leaves. Calyx 4–5·5 mm, subcylindrical; upper teeth up to 2 mm, narrowly lanceolate, ciliate. Corolla 8–9 mm, purple. $2n=28$. ● *Mountains of W. & C. Romania.* Rm.

60. T. bihoriensis Jalas, *Bot. Jour. Linn. Soc.* **64**: 268 (1971) (*T. marginatus* A. Kerner, non Sm.). Like **59** but flowering stems 4-angled, with hairs on the angles only. ● *Mountains of N. & C. Romania.* Rm.

61. T. pulegioides L., *Sp. Pl.* 592 (1753) (*T. alpestris* auct., non Tausch ex A. Kerner, *T. chamaedrys* Fries, *T. enervius* Klokov, *T. froelichianus* Opiz, *T. montanus* Waldst. & Kit., non Crantz, *T. serpyllum* subsp. *carniolicus* (Borbás) Lyka, subsp. *chamaedrys* (Fries) Vollmann, subsp. *effusus* (Host) Lyka, subsp. *montanus* Arcangeli & subsp. *parviflorus* (Opiz ex H. Braun) Lyka, *T. ucrainicus* Klokov & Schost.). Plant suberect to procumbent, somewhat woody at base; long, creeping, non-flowering branches absent. Flowering stems up to 25(–40) cm, sometimes branched, 4-angled, hairy on the angles only. Leaves up to 18 × 10 mm, more or less equal, ovate or oblong-lanceolate, obtuse, usually cuneate and usually ciliate at base, petiolate; marginal vein absent. Inflorescence usually interrupted at least below. Bracts similar to the leaves. Calyx 3–4 mm, campanulate; upper teeth usually longer than wide, at least ⅛ as long as calyx, usually ciliate. Corolla *c.* 6 mm, pink-purple. $2n=28$, 30. *Europe, except parts of the north and east and many of the islands.* Al Au

Be Br Bu Cz Da Ga Ge Gr ?Hb He Ho Hs Hu It Ju Lu No Po Rm Rs (B, C, W) Si Su ?Tu [?Fa Fe ?Is ?Rs (N)].

Very variable. Several of the character-combinations given the rank of species or subspecies possess a distinct geographical distribution. However the total variation has been found to be so continuous and reticulate that attempts to make subspecific divisions seem premature, if not impossible.

62. T. alpestris Tausch ex A. Kerner, *Sched. Fl. Exsicc. Austro-Hung.* **1**: 56 (1881) (*T. chamaedrys* auct., non Fries, *T. serpyllum* subsp. *alpestris* (Tausch ex A. Kerner) Lyka; incl. *T. subalpestris* Klokov). Like **61** but creeping non-flowering branches present; leaves of flowering stems larger above; inflorescence capitate; calyx subglabrous; upper teeth not or sparsely ciliate. $2n=28$. *Mostly above 1000 m.* ● *Mountains of E.C. Europe.* Au Cz ?Ga Po Rm Rs (W).

Intermediates between this and **61** are frequent in areas where they grow together. A slight decrease in pollen fertility has been demonstrated for these putative hybrids. Most records of **62** from the Alps and mountains of the Balkan peninsula seem to refer to variants of **61** or **56(c)**.

63. T. oehmianus Ronniger & Soška, *Feddes Repert. (Beih.)* **100**: 171 (1938). Like **61** but flowering stems glabrous at base; leaves 8–12 × 7–10 mm, broadly ovate, truncate to subcordate at base, not ciliate; calyx 4–4·5 mm, glabrous, the tube subcylindrical, the upper teeth 0·4–0·5 mm, triangular, not ciliate; corolla *c.* 8 mm, pink. ● *S. Jugoslavia (gorge of river Treska, near Skopje).* Ju.

64. T. alternans Klokov, *Not. Syst. (Leningrad)* **16**: 293 (1954). Like **61** but flowering stems hairy on 2 opposite sides; leaves mostly lanceolate-elliptical, with shorter petiole, ciliate at base. ● *E. Carpathians.* ?Cz Rs (W).

This plant has the appearance of a hybrid between **42** and **61** but its range is outside that of **42**. Its status needs further investigation.

65. T. serpyllum L., *Sp. Pl.* 590 (1753). Plant with long, slender, creeping, non-flowering branches, woody at base and rooting at nodes, sometimes with a terminal inflorescence. Flowering stems rarely more than 10 cm, hairy all round. Leaves linear to elliptical, subsessile, ciliate at base; lateral veins disappearing towards margin. Inflorescence usually capitate. Calyx 3–5 mm, campanulate; upper teeth about as long as wide, usually ciliate. Corolla purple. ● *Europe northwards from N.E. France, N. Austria, and N. Ukraine.* Au Be Br Cz Da Fe Ga Ge Ho Hu No Po ?Rm Rs (N, B, C, W) Su.

(a) Subsp. **serpyllum** (*T. serpyllum* subsp. *angustifolius* (Pers.) Arcangeli & subsp. *rigidus* (Wimmer & Grab.) Lyka): Leaves 5–10 × (1–)2–3(–4) mm, more or less equal. Calyx 3–4 mm, the upper teeth mostly 0·5–0·8 mm. Corolla 6–7 mm. $2n=24$. *Throughout the range of the species.*

(b) Subsp. **tanaensis** (Hyl.) Jalas, *Acta Bot. Fenn.* **39**: 20 (1947) (*T. subarcticus* Klokov & Schost.): Leaves usually 7–13 × 3–5 mm, the upper larger. Calyx 4–5 mm; upper teeth up to 1·2 mm. Corolla 7–8 mm. $2n=24$. *N.E. Fennoscandia, N. Russia.*

66. T. talijevii Klokov & Schost., *Jour. Inst. Bot. Acad. Sci. Ukr.* **9**: 195 (1936) (incl. *T. hirticaulis* Klokov, *T. paucifolius* Klokov). Like **65(b)** but leaves more distinctly petiolate and more clearly increasing in size up the flowering stems. *E. Russia.* Rs (N, C).

34. Lycopus L.[1]

Perennial, odourless herbs with creeping rhizome. Verticillasters many-flowered, dense, distant. Calyx campanulate, 13-veined, with 5 equal teeth. Corolla-tube shorter than calyx, with 4 sub-equal lobes, the uppermost usually wider than others. Stamens 2, exceeding corolla; staminodes 0 or 2.

Leaves pinnatifid or pinnatisect at base, toothed or shallowly lobed at apex; bracts 3–5 mm; calyx-teeth *c.* 2 mm, spiny; staminodes minute or absent **1. europaeus**
Leaves pinnatifid or pinnatisect to the apex; bracts 6–9 mm; calyx-teeth *c.* 1·5 mm, not spiny; staminodes conspicuous, capitate **2. exaltatus**

1. L. europaeus L., *Sp. Pl.* 21 (1753) (incl. *L. mollis* A. Kerner). Stems 20–120 cm, sparsely to densely hairy. Leaves 3–10(–15) × 1–5 cm, ovate-lanceolate or elliptical, pinnatifid or pinnatisect at base, shallowly lobed or coarsely toothed at apex; upper leaves and bracts sometimes undivided but coarsely toothed. Bracts 3–5 mm. Calyx 3·5–4·5 mm; teeth (1·6–)2·1(–2·5) mm, about twice as long as tube, spiny. Corolla *c.* 4 mm, white with a few small purple spots; staminodes minute or absent. $2n = 22$. *Wet places. Europe, northwards to 63° 30′ N. in Fennoscandia.* All except Bl Cr Fa Is Sb.

A variable species, in which subordinate taxa based on leaf shape and indumentum have been recognized. Plants with leaves usually lanceolate, lobed, usually 3½ times as long as wide, often glabrous or sparsely hairy (subsp. *europaeus*), occur throughout much of Europe. Plants from C. Europe often have leaves ovate, lobed, usually 2½ times as long as wide, grey-tomentose, and have been treated as subsp. **mollis** (A. Kerner) Rothm. ex Skalický, *Acta Mus. Nat. Pragae* **24B**: 203 (1968), while those from the E. Mediterranean region and Balkan peninsula, often with leaves broadly ovate, crenate or crenate-lobed, as long as to twice as long as wide, more or less tomentose, have been recognized as subsp. **menthifolius** (Mabille) Skalický, *op. cit.* 206 (1968).

2. L. exaltatus L. fil., *Suppl.* 87 (1781). Like **1** but more robust, with stems 90–160 cm; leaves pinnatifid or pinnatisect to the apex; bracts 6–9 mm; calyx-teeth (1·3–)1·5(–2·0) mm, not spiny; staminodes conspicuous, capitate. *Wet places. C. & E. Europe, Italy.* Al Au Bu Cz Ge Gr Hu It Ju Po Rm Rs (C, W, K, E).

35. Mentha L.[2]

Perennial (rarely annual) herbs with creeping rhizomes and scented foliage. Flowers hermaphrodite or female, on the same or different plants, usually in dense, many-flowered verticillasters, sometimes forming a long spike-like inflorescence or a terminal head. Calyx actinomorphic or weakly 2-lipped, tubular or campanulate, 10- to 13-veined, with 5(4) subequal or rarely unequal teeth. Corolla weakly 2-lipped, with 4 subequal lobes, the upper lobe wider and usually emarginate; tube shorter than the calyx. Stamens about equal, divergent or ascending under the upper lip of corolla, exserted (except in *M. pulegium*, some hybrids and female flowers). Style-branches subequal. Nutlets smooth, reticulate or tuberculate.

A taxonomically complex genus. In Sect. *Mentha* identification is frequently difficult, since, in addition to much phenotypic plasticity and genetic variability, most of the species are capable of hybridization with each other. Hybrids are frequent in nature but can usually be recognized by their intermediate appearance and sterility, although fertile hybrid swarms occasionally occur. Several nothomorphs (nm., hybrid forms) in Sect. *Mentha* are given taxonomic recognition in the account.

Species delimitation in the *M. spicata* group is confused due to hybridization and chromosome doubling, with introgression occurring between some species in certain areas. The text includes most of the common variants, but the hybrids are too variable to be fully covered in a key.

In the following account, measurements of leaves refer to those from the middle of the main stem, and of the calyx, to its size at anthesis.

All species except 3 occur in damp or wet habitats.

Literature: J. Briquet, *Les Labiées des Alpes Maritimes* 18–97. Genève & Bale. 1891. J. Fraser, *Rep. Bot. Exch. Club Brit. Is.* **8**: 213–247 (1927). F. Petrak in K. H. Rechinger fil., *Flora Aegaea*, in *Denkschr. Akad. Wiss. Math.-Nat. Kl. (Wien)* **105(1)**: 541–548 (1943). N. Hylander, *Bot. Not.* **118**: 225–242 (1965).

1 Calyx-teeth 4; bracteoles digitately lobed **4. cervina**
1 Calyx-teeth 5; bracteoles simple
  2 Calyx hairy in throat, with distinctly unequal teeth
    3 Stems filiform, procumbent and mat-forming; leaves 2–7 mm **1. requienii**
    3 Stems not mat-forming, though sometimes procumbent; leaves 8–30 mm
      4 Perennial; leaves hairy at least beneath; corolla 4–6 mm **2. pulegium**
      4 Annual; leaves glabrous; corolla not more than 3·5 mm **3. micrantha**
  2 Calyx glabrous in throat, with ± equal teeth
    5 Bracts like the leaves; inflorescence terminated by leaves, or by very small upper verticillasters
      6 Plant hairy, green, usually fertile; calyx 1·5–2·5 mm, broadly campanulate, the teeth deltate or broadly triangular **5. arvensis**
      6 Plant glabrous or hairy, often tinged with red, usually sterile; calyx 2–4 mm, narrowly campanulate or tubular, the teeth narrowly triangular to subulate
        7 Calyx 2–3·5 mm, campanulate, the teeth rarely more than 1 mm; plant usually glabrous **7. × gentilis**
        7 Calyx 3·5–4 mm, tubular, or, if shorter, the teeth usually 1–1·5 mm and plant distinctly hairy
          8 Plant subglabrous, with a sweet scent; upper bracts usually suborbicular, cuspidate **8. × smithiana**
          8 Plant distinctly hairy, with a sickly scent; upper bracts ovate to ovate-lanceolate, not cuspidate **6. × verticillata**
    5 Bracts mostly small and inconspicuous, unlike the leaves; flowers in terminal spikes or heads
      9 Leaves sessile (the lower rarely shortly petiolate); flowers in a spike 5–15 mm in diameter **(11–14). spicata** group
      9 Leaves distinctly petiolate; flowers in a head or oblong spike 12–20 mm in diameter
        10 Flowers in an oblong spike; leaves usually lanceolate; plant sterile **10. × piperita**
        10 Flowers in a head, sometimes with one to three verticillasters below; leaves usually ovate
          11 Leaves and calyx-tube hairy; plant fertile **9. aquatica**
          11 Leaves and calyx-tube glabrous or subglabrous; plant sterile **10. × piperita**

Sect. AUDIBERTIA (Bentham) Briq. Bracts like the leaves. Verticillasters 2- to 6-flowered. Calyx turbinate-campanulate, weakly 2-lipped; throat hairy within. Corolla-tube straight.

1. M. requienii Bentham, *Lab. Gen. Sp.* 182 (1833). Glabrous or sparsely hairy perennial 3–12 cm, with pungent scent; stems

---

[1] By P. W. Ball.   [2] By R. M. Harley.

filiform, diffuse, usually procumbent, rooting at the nodes, mat-forming. Leaves and bracts 2–7 mm, petiolate, ovate-orbicular, entire or sinuate. Calyx 1–1·5(–2·5) mm; teeth triangular, subulate. Corolla scarcely exserted, pale lilac. Nutlets smooth, pale brown. $2n=18$. ● *Corse, Sardegna, Montecristo; cultivated elsewhere for ornament and locally naturalized in W. Europe.* Co It Sa [Br Hb Lu].

Sect. PULEGIUM (Miller) DC. Bracts like the leaves. Verticillasters many-flowered. Calyx tubular, weakly 2-lipped; throat hairy within. Corolla-tube gibbous.

2. **M. pulegium** L., *Sp. Pl.* 577 (1753) (*Pulegium vulgare* Miller). Subglabrous to tomentose perennial 10–40 cm, with pungent scent; stems procumbent to ascending. Leaves 8–30 × 4–12 mm, narrowly elliptical, attenuate at base, rarely suborbicular, shortly petiolate, entire or with up to 6 teeth on each side, hairy at least beneath. Bracts like the leaves, but usually smaller. Calyx (2–)2·5–3 mm; teeth ciliate, the lower subulate, the upper shorter and wider. Corolla (4–)4·5–6 mm, lilac. Stamens exserted or included; fertile anthers 0·4 mm. Nutlets 0·75 mm, pale brown. $2n=20$. *S., W. & C. Europe, northwards to Ireland and C. Poland, and extending to W. & S. Ukraine.* All except Da Fa Fe Is No Rs (N, B, C, E) Sb Su.

Very variable in habit, leaf-shape and indumentum; several varieties have been described.

3. **M. micrantha** (Bentham) Schost. in Schischkin, *Fl. Yugo-Vostoka* 6: 181 (1936). Like **2** but subglabrous annual with erect stems; leaves glabrous; calyx up to 2(?–2·5) mm; corolla up to 3·5 mm; fertile anthers 0·25 mm; nutlets 0·5 mm. *Dry steppes. S.E. Russia, W. Kazakhstan.* Rs (E).

Sect. PRESLIA (Opiz) Harley. Bracts like the leaves; bracteoles digitately lobed. Verticillasters many-flowered. Calyx tubular, with 4 equal teeth; throat hairy within. Corolla-tube straight.

4. **M. cervina** L., *Sp. Pl.* 578 (1753) (*Preslia cervina* (L.) Fresen.). Subglabrous perennial 10–40 cm, with pungent scent as in **2**; stems procumbent and rooting below, erect above. Leaves 10–25 × 1–4 mm, glabrous, sessile, linear-oblanceolate, attenuate at base, entire or obscurely toothed. Bracts like the leaves but wider. Calyx-teeth triangular, with a whitish apical spine. Corolla lilac or white. *S.W. Europe.* Ga Hs Lu.

Sect. MENTHA. Bracts variable. Verticillasters usually many-flowered. Calyx tubular or campanulate, with 5 more or less equal teeth; throat glabrous within. Corolla-tube straight.

5. **M. arvensis** L., *Sp. Pl.* 577 (1753) (incl. *M. lapponica* Wahlenb.). Hairy perennial or rarely annual up to 60 cm, with sickly scent; flowering stems ascending or erect. Leaves (15–)20–50(–70) × 10–30(–40) mm, elliptic-lanceolate to broadly ovate, usually elliptical with the base narrowing to a petiole, shallowly toothed. Flowers in remote sessile verticillasters, the inflorescence leafy at the apex. Bracts like the leaves, smaller above. Calyx 1·5–2·5 mm, broadly campanulate, hairy; teeth deltate or broadly triangular. Pedicels glabrous or hairy. Corolla lilac, white or rarely pink. Nutlets pale brown. $2n=24, 72, 90$. *Most of Europe, but absent from many of the islands.* Au Be Br Bu Cz Da Fe Ga Ge Gr Hb He Ho Hs Hu It Ju Lu No Po Rm Rs (N, B, C, W, K, E) ?Si Su.

The European plants form part of a widespread polymorphic complex. Many infraspecific taxa have been recognized, chiefly

on the basis of vegetative characters, but none seems very satisfactory. Among these is subsp. **austriaca** (Jacq.) Briq., *Lab. Alpes Marit.* 88 (1891) (*M. austriaca* Jacq.; incl. var. *parietariifolia* J. Becker), with leaves elliptic-lanceolate, attenuate at the base. The very similar **M. lapponica** Wahlenb., *Fl. Lapp.* 161 (1812), from N. Europe, has been recognized by a few authors. It differs from subsp. *austriaca* only in being subglabrous, with few-flowered verticillasters.

M. × muellerana F. W. Schultz, *Flora (Regensb.)* 37: 543 (1854) (*M. wohlwerthiana* F. W. Schultz, *M. carinthiaca* auct., vix Host, *M. malinvaldii* Camus, *M. stachyoides* auct., ? an Host) (*M. arvensis × suaveolens*) occurs occasionally in the vicinity of its parents, and is probably often confused with **5**. It is sterile and has $2n=60$.

M. × dalmatica Tausch, *Syll. Pl. Nov. Ratisbon. (Königl. Baier. Bot. Ges.)* 2: 249 (1828) (*M. carinthiaca* sensu Petrak, ? an Host, *M. haynaldiana* Borbás, *M. iraziana* Borbás, *M. cinerascens* H. Braun) (*M. arvensis × longifolia*), is a rare, sterile hybrid occasionally found in the vicinity of its parents. It is sometimes confused with hairy forms of **7**, but can generally be distinguished by its more slender habit, narrow bracts, smaller calyx and different scent.

6. **M. × verticillata** L., *Syst. Nat.* ed. 10, **2**: 1099 (1759) (*M. sativa* L.) (*M. aquatica × arvensis*). Like **5** but usually more robust and often hairier; calyx 2·5–3·5(–4) mm, tubular, the teeth more or less triangular, acute. $2n=42, 78, 84, 90, 96, 120, 132$. *Most of Europe, but absent from many of the islands.* Au ?Az Be Br Bu Cz Da Fe Ga Ge ?Gr Hb He Ho Hs Hu It Ju No Po Rs (B, C, W, ?K, ?E) Su.

Usually sterile; the occasional highly fertile plants are probably the result of back-crossing with the parents. Often confused with **5** or hairy variants of **7**.

7. **M. × gentilis** L., *Sp. Pl.* 577 (1753) (*M. sativa* var. *gentilis* (L.) Reichenb.) (*M. arvensis × spicata*). Perennial 30–90 cm, usually glabrous, often red-tinged, with sweet scent like **14**. Leaves (15–)30–70(–90) × (8–)15–40(–45) mm, ovate-lanceolate, lanceolate or elliptic-oblong, acute, narrowing to a short petiole, in cultivars sparsely hairy but otherwise often more densely so. Flowers usually in remote, sessile verticillasters. Bracts like the leaves or decreasing distinctly in size upwards, the uppermost sometimes shorter than the flowers. Pedicels usually glabrous. Calyx 2–3·5 mm, campanulate, glabrous below; teeth 0·5–1 mm, more or less subulate or rarely triangular, often conspicuously ciliate. Stamens usually included. $2n=54, 60, 84, 96, 108, 120$. *Frequently cultivated and often escaping.* *Be *Br *Co *It *Ga [Au Cz Da Fe Ge Hb He Ho Hs Hu Ju Lu No Po Rm Rs (C, W) Su].

Usually sterile, but sometimes back-crossing with its parents.

8. **M. × smithiana** R. A. Graham, *Watsonia* 1: 89 (1949) (*M. rubra* Sm., non Miller) (*M. aquatica × arvensis × spicata*). Robust perennial 50–150 cm, subglabrous and conspicuously red-tinged, with a sweet scent like **14**. Leaves 30–70(–90) × 20–40 mm, ovate, rounded at base, petiolate, sparsely hairy to subglabrous. Flowers usually in remote verticillasters, sometimes crowded at apex of stem. Bracts decreasing upwards, usually longer than the flowers, the upper suborbicular, cuspidate at apex, usually serrate. Calyx 3·5–4 mm, tubular, glabrous or sparsely hairy above; teeth (0·75–)1–1·5 mm, weakly ciliate. Stamens usually exserted. $2n=120$. *Spreading vegetatively and perhaps also arising spontaneously; rarely cultivated.* [Au Be Br Cz Ga Ge Hb He Ho Hu It Ju Po Rm.]

Usually sterile, but occasionally producing a few viable seeds which give rise to segregating progeny. Frequently confused with **7**.

9. M. aquatica L., *Sp. Pl.* 576 (1753) (*M. hirsuta* Hudson). Subglabrous to tomentose, often purplish perennial (10–)20–90 cm, with strong scent. Leaves (15–)30–90 × (10–)15–40 mm, ovate to ovate-lanceolate, usually truncate at base, petiolate, serrate. Inflorescence of 2–3 congested verticillasters with inconspicuous bracts, forming a terminal head up to 2 cm in diameter, sometimes with 1–3 distant verticillasters below, in the axils of leaf-like bracts. Calyx (2·5–)3–4 mm, tubular, the veins distinct; teeth subulate or narrowly triangular. Pedicels hairy. Corolla lilac. Nutlets pale brown. 2n = 96. *Europe except the extreme north.* All except Rs (N) Sb but only as a naturalized alien in Is.

Very variable in habit, leaf-shape and indumentum. Many infraspecific taxa have been recognized, at least some of which retain their characters in cultivation.

10. M. × piperita L., *Sp. Pl.* 576 (1753) (*M. nigricans* Miller) (*M. aquatica × spicata*). Perennial 30–90 cm, with a pungent scent, usually subglabrous but occasionally hairy to grey-tomentose, often purple-tinged. Leaves 40–80(–90) × 15–40 mm, ovate-lanceolate or lanceolate, rarely ovate, cuneate to subcordate and long-petiolate, usually serrate. Inflorescence usually of numerous congested verticillasters with inconspicuous bracts forming a terminal, oblong spike (30–)50–80 × 12–18 mm, often interrupted below, rarely a narrow elongate spike or subglobose head. Calyx 3–4 mm, tubular, the tube usually glabrous; teeth ciliate. Pedicels usually glabrous. Corolla lilac-pink. Sterile. 2n = 66, 72. *Widespread as a garden escape and also spontaneous.* [Au Az Be Bl Br Bu Cz Da Ga Ge Gr Hb He Ho Hs Hu It Ju Lu Po Rm Rs (C, W, E) Si.]

The subglabrous plants were formerly widely cultivated and have escaped to become well naturalized in many areas. They are usually easy to recognize. Hairy plants may have arisen spontaneously where both parents occur. They are much more variable in shape of leaf and inflorescence, and in scent, and are readily confused with *M. × dumetorum* (see below). Without field information and further cytological study the limits of the two are uncertain.

M. × piperita nm. **citrata** (Ehrh.) Boivin, *Naturaliste Canad.* **93**: 1061 (1966) (*M. odorata* Sole, *M. adspersa* Moench), a sterile cultivar with 2n = 84, 120, is widely cultivated and occasionally naturalized. It has a strong scent resembling Eau de Cologne, glabrous or subglabrous ovate, subcordate leaves, and an inflorescence like that of **9**, but smaller.

M. × maximilianea F. W. Schultz, *Pollichia* **12**: 34 (1854) (*M. rodriguezii* Malinv., *M. suavis* Guss., *M. schultzii* Bout.) (*M. aquatica × suaveolens*), with 2n = 60, 72–78, 120, occurs occasionally in the vicinity of both parents, and is intermediate between them and usually sterile, although fertile back-crosses are reported. It has been confused with *M. × dumetorum* and **10**.

M. × tutinii P. Silva in Palhinha, *Cat. Pl. Vasc. Açores* 103 (1966), from the Açores, which reputedly is **10 × 11**, appears to be a narrow-leaved variant of **9 × 11**.

M. × dumetorum Schultes, *Obs. Bot.* 108 (1809) (*M. pubescens* auct., ? an Willd., *M. nepetoides* Lej., *M. ayassei* Malinv., *M. hirta* Willd.) (*M. aquatica × longifolia*), occurs in many places in the vicinity of its parents, but, owing to confusion with hairy variants of **10**, its distribution and frequency are uncertain.

M. × pyramidalis Ten., *Fl. Nap.* **1**, *Prodr.*: 34 (1811) (*M. reverchonii* Briq.) (*M. aquatica × microphylla*), with 2n = c. 72, occurs in S. Italy and Kriti.

(11–14). M. spicata group. Leaves sessile or the lower very shortly petiolate. Flowers in slender spikes 5–15 mm in diameter. Nutlets red-brown to black, reticulate (except in glabrous variants of **14**).

In the key no attempt has been made to include the vast array of cultivated and wild hybrid variants; those involving *M. spicata* are usually recognizable by their sterility. Among the most valuable characters are the indumentum on the lower surface of the leaves and the length of the fertile anthers and nutlets. Glabrous plants or those with the sweet scent of **14** can unhesitatingly be referred to this species or one of its hybrids, although the absence of these characters is no positive indication to the contrary.

1 Stems and leaves glabrous or sparsely hairy   **14. spicata**
1 Stem and leaves hairy
  2 Leaves not more than 45 mm, with numerous branched hairs beneath
    3 Leaves 20–40 mm wide, ovate-oblong to suborbicular, obtuse, sometimes minutely apiculate, rarely subacute; verticillasters crowded, except at base   **11. suaveolens**
    3 Leaves 5–20 mm wide, ovate to ovate-lanceolate, usually acute; verticillasters distant   **13. microphylla**
  2 Leaves (30–)50–90(–110) mm, branched hairs absent, or if present usually few beneath
    4 Hairs all simple; leaves widest near the middle; fertile anthers 0·28–0·38 mm   **12. longifolia**
    4 Some branched hairs usually present on the leaves beneath; leaves widest near the base; fertile anthers 0·38–0·52 mm   **14. spicata**

11. M. suaveolens Ehrh., *Beitr. Naturk.* **7**: 149 (1792) (*M. rotundifolia* auct., non (L.) Hudson, *M. macrostachya* Ten., *M. insularis* Req.). Perennial 40–100 cm, with sickly sweet scent. Stem sparsely hairy to densely white-tomentose. Leaves (15–)30–45 × (10–)20–40 mm, sessile or very shortly petiolate, strongly rugose, ovate-oblong to suborbicular, obtuse, cuspidate or rarely acute, widest near the base, serrate, with 10–20 teeth, often apparently crenate due to the teeth being bent down towards the underside of the leaf, hairy above, usually grey- or white-tomentose to -lanate beneath, the hairs on the lower surface branched, with the basal cell 43–57 μm in diameter. Verticillasters many, usually congested, forming a terminal spike 40–90 × 5–10 mm, often interrupted below and usually branched. Calyx 1–2 mm, campanulate, hairy; teeth subequal. Pedicels hairy. Corolla whitish or pink. Fertile anthers 0·28–0·38 mm. Nutlets 0·57–0·75 mm. 2n = 24. *S. & W. Europe, northwards to the Netherlands; cultivated as a pot-herb, and naturalized in N. & C. Europe.* *Al Az Be Bl Br Co Cr Ga Ge Gr He Ho Hs It Lu Sa Si *Tu [Au Cz Da Hb Ju Po Rm Su].

Plants from mountains on some islands of the western part of the Mediterranean region have been described as **M. insularis** Req., *Gior. Bot. Ital.* **2**: 111, 115 (1846). They are sparsely hairy, with ovate-oblong to ovate-lanceolate, shortly petiolate or sessile, often acute leaves with an undulate margin. These characters however occur sporadically in other populations, so that the plants seem insufficiently distinct to merit subspecific status. They have 2n = 24.

M. × rotundifolia (L.) Hudson, *Fl. Angl.* 221 (1762) (*M. niliaca* Juss. ex Jacq., *M. amaurophylla* Timb.-Lagr., *M. nouletiana* Timb.-Lagr., *M. timbalii* Rouy, *M. villosa* auct., non Hudson)

(*M. longifolia* × *suaveolens*), with $2n = 24$, is extremely variable, showing a wide range of characters intermediate between the parents. It is a frequent and highly fertile hybrid which can form introgressive swarms where its parents grow together. It is very rarely found in cultivation but often persists in the absence of the parents. It is frequently confused with *M.* × *villosa*.

M. × **villosa** Hudson, *Fl. Angl.* ed. 2, 250 (1778) (*M. cordifolia* auct., ? an Opiz, *M. gratissima* Weber, *M. lamyi* Malinv., *M. mosoniensis* H. Braun, *M. nemorosa* Willd., *M. niliaca* auct., non Juss. ex Jacq., *M. scotica* R. A. Graham) (*M. spicata* × *suaveolens*), a highly sterile hybrid with $2n = 36$, is widely naturalized and also probably often arises spontaneously. It is morphologically extremely diverse, both hairy and glabrous plants being frequent, these ranging from plants almost indistinguishable from **14** to others very similar to **11**. Many of these have been given taxonomic recognition and some have spread clonally to become locally common.

M. × **villosa** Hudson nm. **alopecuroides** (Hull) (*M. alopecuroides* Hull, *M. velutina* Lej.), with $2n = 36$, is often confused with **11**, but it is more robust, 60–140 cm, and has a sweet scent like **14**; middle cauline leaves 40–80 × 30–60 mm, broadly ovate or orbicular, softly hairy, the margin with patent teeth; verticillasters forming a robust spike 10–12 mm wide; calyx 2 mm; corolla pink. It was formerly much cultivated and is widely naturalized in W. Europe.

12. M. longifolia (L.) Hudson, *Fl. Angl.* 221 (1762) (*M. sylvestris* L., *M. incana* Willd.). Perennial 40–120 cm, with a musty scent. Stem white- or grey-villous, sometimes sparsely hairy. Leaves (40–)50–90(–110) × (10–)20–40 mm, sessile or very shortly petiolate, smooth or slightly rugose near the base, usually oblong-elliptical, acute, serrate with usually irregular, patent teeth, green to grey-tomentose above, grey- or white-villous beneath; hairs simple, with the basal cell 18–33 μm in diameter. Verticillasters many, usually congested, forming a terminal, usually branched spike 40–100 × (9–)10–15 mm. Calyx (1–)1·5–3 mm, narrowly campanulate, hairy; teeth subequal. Pedicels hairy. Corolla lilac or white. Fertile anthers 0·28–0·38 mm. Nutlets 0·54–0·79 mm. $2n = 24$. *Most of Europe, from S. Sweden and N.C. Russia southwards.* Al Au Be Bu Cr Cz Ga Ge Gr He Hs Hu It Ju Lu Po Rm Rs (B, C, W, K, E) Si Su Tu.

Extremely variable in height, leaf-size and -shape, indumentum and inflorescence, and complicated by the occurrence of hybrids. There is clinal variation from west to east with reduction in size of leaves and calyx. Plants from Austria and eastwards with narrow leaves, narrow interrupted spikes and smaller calyx have been described as **M. longifolia** subsp. **grisella** (Briq.) Briq. in Engler & Prantl, *Natürl. Pflanzenfam.* 4(3a): 322 (1897). These, however, grade imperceptibly into **13**, with which there appears to have been very extensive hybridization. As a result, many populations, especially from Hungary, and to a lesser extent from the Balkan peninsula and S. Italy, are extremely complex.

The distribution in W. Europe is partly obscured owing to confusion with hairy plants of **14**, to which all records of **12** from N.W. Europe appear to be referable.

13. M. microphylla C. Koch, *Linnaea* 21: 648 (1849) (*M. sieberi* auct., ? an C. Koch). Like **12** but with leaves 10–45 × 5–20(–25) mm, broadly ovate to lanceolate, widest below the

middle, strongly rugose, tomentose and grey or grey-green on both surfaces, serrate with usually few salient teeth, undulate, the hairs on the lower surface both simple and branched with the basal cell 44–63 μm in diameter; verticillasters usually remote, forming a little-branched spike 60–120 × 5–10 mm; calyx 1–2 mm; fertile anthers 0·30–0·44 mm; nutlets 0·6–0·83 mm. $2n = 48$. *S. Italy, Sicilia; Balkan peninsula and Aegean region.* Al Bu Cr Gr It Ju ?Sa Si Tu.

Apparently an allotetraploid originally derived from **11** and **12**.

14. M. spicata L., *Sp. Pl.* 576 (1753) (*M. viridis* (L.) L., *M. crispa* L., *M. crispata* Schrader, *M. longifolia* auct., non (L.) Hudson, *M. sylvestris* auct., non L., *M. niliaca* auct., non Juss. ex Jacq., *M. cordifolia* auct., non Opiz). Perennial 30–100 cm, with strong, sweet scent, less frequently with pungent scent (as in **10**) or musty. Leaves (30–)50–90 × (7–)15–30 mm, lanceolate or lanceolate-ovate, smooth or rugose (rarely 40–60 × 25–40 mm, ovate or ovate-oblong, strongly rugose), widest near the base, serrate with regular teeth (rarely the whole leaf strongly crispate), glabrous to densely hairy, the hairs on the lower surface both simple and branched, the basal cell 30–47 μm in diameter. Inflorescence variable. Calyx 1–3 mm, campanulate, glabrous or hairy; teeth subequal. Corolla lilac, pink or white. Fertile anthers 0·38–0·52 mm. Nutlets 0·74–0·94 mm, reticulate in hairy plants, smooth in glabrous plants. $2n = 48$. *Widely cultivated as a pot-herb and for its aromatic oils, and naturalized throughout a large part of Europe.* [Al Au Az Be Bl Br Bu Cr Cz Da Ga Ge Gr Hb He Ho Hs Hu It Ju Lu No Po Rm Rs (W, K) Su Tu.]

The origin of this species is unknown, but it probably arose in cultivation. It behaves as a segmental allopolyploid derived by hybridization and chromosome doubling from **11** and **12**. It is propagated vegetatively in cultivation but often tends to segregate parental characters when selfed. This makes separation from its hybrids with **11** and **12** very difficult except on fertility criteria. Plants with broad, rugose, glabrous leaves from S.W. England and probably elsewhere have been known as *M. cordifolia* auct., ? non Opiz (*M. suaveolens* × *spicata*) but are fertile tetraploids and should be referred to **14**.

Many records of **12**, especially from lowland areas of Europe, are referable to **14** (or to its hybrids with **11** or **12**). These hairy plants are best distinguished from **12** by anther, nutlet and hair characters. Variants with crispate leaves occur in cultivation and this character can also be found in many hybrids of which *M. spicata* is a parent. Glabrous plants from Kriti and Turkey-in-Europe, similar to **14** but with small anthers, may be diploids; their taxonomic status is not clear.

M. × **villosonervata** auct., ? an Opiz (*M. longifolia* × *spicata*) is probably widespread as an escape, but is much confused with *M.* × *villosa*. It differs mainly in its narrower, usually patently toothed leaves with branched hairs few or absent. It is sterile and has $2n = 38$.

36. Perilla L.[1]

Annual. Flowers in pairs in a secund raceme. Calyx campanulate, 10-veined, with 5 subequal teeth in flower, 2-lipped in fruit. Corolla nearly actinomorphic, with 5 small lobes; tube shorter than calyx. Stamens subequal; anther-cells parallel, becoming divergent.

1. P. frutescens (L.) Britton, *Mem. Torrey Bot. Club* 5: 277 (1894) (*P. ocymoides* L.). Up to 1 m, pubescent. Leaves 4·5–8(–12) × 3–6(–7) cm, broadly ovate, deeply serrate-crenate, acuminate, cuneate at base, petiolate. Inflorescence 3–10 cm; floral leaves somewhat smaller than the foliage leaves. Calyx

---

[1] By I. B. K. Richardson.

3–3·5 mm, gibbous at base, glabrous inside. Corolla 3·5–4 mm, white, with a ring of hairs inside. *Cultivated in S.E. Europe for ornament and for its aromatic oil, and locally naturalized in E. Ukraine.* [Rs (W, ?E).] (*Himalayan region.*)

37. Rosmarinus L.[1]

Evergreen shrubs. Verticillasters few-flowered, in short axillary racemes. Calyx campanulate, 2-lipped; upper lip entire; lower lip 2-lobed. Corolla 2-lipped, exserted; upper lip strongly concave and 2-fid; lower lip 3-fid with a cochleariform middle lobe. Stamens 2, distinctly exserted, parallel, the filaments with a small, lateral, recurved tooth near the base; anthers 1-locular. Style long, incurved, unbranched.

Leaves 15–40 mm; inflorescence with stellate hairs only; branches brown **1. officinalis**
Leaves 5–15 mm; inflorescence with both stellate and long, glandular hairs; branches grey **2. eriocalix**

1. R. officinalis L., *Sp. Pl.* 23 (1753). Up to 2 m, with erect, ascending or rarely procumbent, brown branches; aromatic. Leaves 15–40 × 1·2–3·5 mm, linear, coriaceous, with revolute margins, bright green and rugulose above, white-tomentose beneath, sessile. Peduncle and pedicels stellate-tomentose; calyx 3–4 mm, green or purplish and sparsely tomentose when young, later 5–7 mm, subglabrous and distinctly veined. Corolla 10–12 mm, pale blue (rarely pink or white). Nutlets brown. *Dry scrub. Mediterranean region, extending to Portugal and N.W. Spain; cultivated elsewhere for ornament or for its aromatic oil.* 2n=24. Bl Co *Cr Ga Gr Hs It Ju Lu Sa Si [Al Az Bu He Rs (K)].

2. R. eriocalix Jordan & Fourr., *Brev. Pl. Nov.* **1**: 44 (1866) (*R. tournefortii* De Noë ex Turrill; incl. *R. tomentosus* Huber-Morath & Maire). Like **1** but usually procumbent; branches grey; leaves 5–15 × 1–2 mm, glabrous and green or greyish-tomentose; indumentum on peduncle, pedicels and calyx with both stellate and long, simple, glandular hairs. *Calcareous rocks. S. Spain.* Hs. (*N. Africa.*)

Variable in leaf-indumentum. Coastal plants tend to be densely stellate-tomentose while mountain plants are glabrous.

38. Lavandula L.[2]

Small shrubs. Verticillasters in lax or crowded spikes; bracts differing distinctly from the leaves. Calyx (8–)13(–15)-veined, the teeth small, subequal; uppermost tooth usually with a cordate to obovate appendage at the apex. Corolla 2-lipped, usually pale purple to blue-violet; upper lip 2-lobed; lower lip 3-lobed, the lobes equal.

All species occur in dry, usually sunny habitats.

Literature: D. A. Chaytor, *Jour. Linn. Soc. London* (*Bot.*) **51**: 153–204 (1937). A. Rozeira, *Brotéria* (*Ser. Ci. Nat.*) **18**: 5–84 (1949).

1 Leaves toothed to 2-pinnatisect
2 Leaves crenate-dentate to pectinate-pinnatifid; flowers 6–10 in each verticillaster **3. dentata**
2 Leaves mostly 2-pinnatisect; flowers 2 in each verticillaster **7. multifida**

1 Leaves entire
3 Upper bracts oblong-obovate, much longer than the flowers and the other bracts, and without flowers in their axils
4 Upper bracts white or purple; leaves tomentose **1. stoechas**
4 Upper bracts green; leaves shortly hirsute **2. viridis**
3 All bracts similar, not longer than the flowers and all with flowers in their axils
5 Bracts ovate-rhombic, cuspidate or acuminate; bracteoles minute or absent **4. angustifolia**
5 Bracts linear or lanceolate; bracteoles 2–5 mm, linear or setaceous
6 Leaves very shortly and densely white-tomentose when young, grey-green and less densely tomentose when mature; calyx 13-veined **5. latifolia**
6 Leaves densely and persistently white-lanate-tomentose; calyx 8-veined **6. lanata**

1. L. stoechas L., *Sp. Pl.* 573 (1753). Shrub up to 100 cm, tomentose. Leaves 10–40 mm, linear to oblong-lanceolate, entire, usually grey-tomentose. Spike usually 2–3 cm; fertile bracts 4–8 mm, rhombic-cordate, tomentose; upper bracts 10–50 mm, oblong-obovate, usually purple, without flowers in their axils; verticillasters 6- to 10-flowered. Calyx 4–6 mm, 13-veined, the upper tooth with an obcordate appendage at the apex 1–1·5 mm wide. Corolla 6–8 mm, usually dark purple. 2n=30. *Mediterranean region, Portugal.* Bl Co Cr Ga Gr Hs It Lu Sa Si Tu.

1 Peduncle shorter than spike **(a) subsp. stoechas**
1 Peduncle longer than spike
2 Calyx-appendage lobed **(f) subsp. cariensis**
2 Calyx-appendage entire or nearly so
3 Lower bracts exceeding the calyx, contiguous **(b) subsp. pedunculata**
3 Lower bracts not exceeding the calyx, not contiguous
4 Spike not more than twice as long as wide **(c) subsp. lusitanica**
4 Spike more than twice as long as wide
5 Lower bracts shorter than calyx, ±acute **(d) subsp. luisieri**
5 Lower bracts ±equalling the calyx, acute or obtuse, linear-elliptical **(e) subsp. sampaiana**

(a) Subsp. **stoechas**: Peduncle shorter than spike; lower bracts acute. *From N.E. Spain eastwards to Greece.*

(b) Subsp. **pedunculata** (Miller) Samp. ex Rozeira, *Brotéria* (*Ser. Ci. Nat.*) **18**: 72 (1949): Peduncle longer than spike; lower bracts exceeding the calyx, contiguous; calyx-appendage entire or nearly so. ● *C. Spain; mountains of N.E. Portugal.*

(c) Subsp. **lusitanica** (Chaytor) Rozeira, *Agron. Lusit.* **24**: 173 (1964): Peduncle longer than spike; spike not more than twice as long as wide; lower bracts not exceeding calyx, not contiguous; calyx-appendage entire or nearly so. *Sandy soils.* ● *C. & S. Portugal.*

(d) Subsp. **luisieri** (Rozeira) Rozeira, *loc. cit.* 173 (1964): Peduncle longer than spike; spike more than twice as long as wide; lower bracts shorter than the calyx, more or less deltate, acute, not contiguous; calyx-appendage entire or nearly so. *On schistose, loamy soils.* ● *C. & S. Portugal.*

(e) Subsp. **sampaiana** Rozeira, *Brotéria* (*Ser. Ci. Nat.*) **18**: 70 (1949): Like subsp. (d) but lower bracts about equalling the calyx, linear-elliptical, sometimes obtuse. *Mainly on granite soils.* ● *W. Spain, N. & C. Portugal.*

(f) Subsp. **cariensis** (Boiss.) Rozeira, *loc. cit.* (1949) (*L. cariensis* Boiss.): Peduncle longer than spike; calyx-appendage lobed. *Turkey-in-Europe* (*N. of Istanbul*). (*W. Anatolia.*)

2. L. viridis L'Hér., *Sert. Angl.* 19 (1789). Like **1** but stem and leaves shortly hirsute; peduncle as long as or longer than spike; upper bracts 8–20 mm, light green; calyx-appendage

---

[1] By J. do Amaral Franco and M. L. da Rocha Afonso.
[2] By E. Guinea.

2·5–3·5 mm wide, entire; corolla white. *S.W. Spain, S. Portugal.* Hs Lu.

3. L. dentata L., *Sp. Pl.* 572 (1753). Shrub up to 100 cm, grey-tomentose. Leaves 15–35 mm, oblong-linear or lanceolate, crenate-dentate to pectinate-pinnatifid with obtuse lobes, grey-tomentose beneath, grey-green above. Spike 2·5–5 cm; bracts 5–8 mm, rhombic-ovate to obovate-suborbicular, acuminate, sparsely tomentose; upper bracts 8–15 mm, ovate, purple, without flowers in their axils; verticillasters 6- to 10-flowered. Calyx 5–6 mm, 13-veined, the upper tooth with an obcordate appendage at the apex. Corolla *c.* 8 mm, dark purple. 2n=44. *S. & E. Spain, Islas Baleares.* Bl Hs [It Lu Si].

4. L. angustifolia Miller, *Gard. Dict.* ed. 8, no. 2 (1768) (*L. spica* L., nom. ambig.). Shrub up to 100(–200) cm, tomentose. Leaves 20–40(–50) mm, lanceolate, oblong or linear, entire, white-tomentose when young, becoming green. Spike 2–8 cm; bracts (1–)3–8 mm, usually broadly rhombic-obovate, acuminate; bracteoles absent or minute; verticillasters 6- to 10-flowered. Calyx 4·5–7 mm, 13-veined, the upper tooth with an obcordate appendage at the apex. Corolla 10–12 mm, purple. *Mediterranean region; widely cultivated for ornament and for perfume.* Co Ga Gr Hs It Ju Sa Si [Bu Rs (K)].

(a) Subsp. **angustifolia** (*L. officinalis* Chaix, *L. vera* DC.): Bracts usually shorter than calyx; calyx 4·5–6 mm, the appendage on the upper tooth obscure. *Throughout the range of the species.*

(b) Subsp. **pyrenaica** (DC.) Guinea, *Bot. Jour. Linn. Soc.* 65: 263 (1972) (*L. pyrenaica* DC.): Bracts usually exceeding calyx; calyx 6–7 mm, the appendage on the upper tooth distinct. ● *E. Pyrenees and N.E. Spain.*

5. L. latifolia Medicus, *Bot. Beobacht.* **1783**: 135 (1784) (*L. spica* auct., non L.). Like 4(a) but leaves grey-green and more densely tomentose; bracts linear or linear-lanceolate; bracteoles 2–3 mm, linear-subulate; corolla 8–10 mm. *Mediterranean region, Portugal.* ?Bl Ga Hs It Ju Lu Si.

6. L. lanata Boiss., *Elenchus* 72 (1838). Like 4(a) but stem and leaves persistently white-lanate-tomentose; leaves 35–50 mm, oblong-lanceolate or linear-spathulate; spike 4–10 cm; bracts 4–8 mm, linear or lanceolate, usually equalling or exceeding calyx; bracteoles 2–5 mm, setaceous; calyx 8-veined; corolla 8–10 mm, lilac. 2n=50. *Dry, calcareous rocks and screes.* ● *Mountains of S. Spain.* Hs.

7. L. multifida L., *Sp. Pl.* 572 (1753). Shrub up to 100 cm, grey-tomentose and sometimes also with long straight hairs. Leaves mostly 2-pinnatisect, green, sparsely puberulent. Spike 2–7 cm; bracts *c.* 5 mm, cordate-ovate, acuminate; verticillasters 2-flowered. Calyx *c.* 5 mm, 15-veined; upper tooth without an appendage. Corolla *c.* 12 mm, blue-violet. *W. part of Mediterranean region, S. Portugal.* Hs It Lu Si.

39. Horminum L.[1]

Rhizomatous perennials. Verticillasters (2–)4- to 6-flowered, secund. Bracts distinct from the leaves. Calyx 2-lipped, 13-veined; tube campanulate, the throat glabrous; upper lip 3-toothed; lower lip 2-fid. Corolla 2-lipped; tube curved upwards; upper lip erect, emarginate; lower lip with 3 unequal lobes.

[1] By V. H. Heywood.
[2] By I. C. Hedge.

1. H. pyrenaicum L., *Sp. Pl.* 596 (1753). Rhizome usually horizontal, bearing erect, unbranched, pubescent flowering stems up to 45 cm. Leaves 3–7 × 2–5 cm, mostly basal, ovate to ovate-orbicular or elliptical, obtuse, crenate, ciliate, long-petiolate. Bracts 4–11 mm, ovate-acute or -acuminate, entire, membranous, ciliate. Corolla 17–21 mm, dark bluish-violet, twice as long as calyx. 2n=12. *Grassy places in mountains; calcicole. Pyrenees, Alps.* Au Ga Ge He Hs It Ju.

40. Salvia L.[2]

Herbs or shrubs. Flowers in axillary verticillasters (rarely cymes). Calyx 2-lipped, the teeth unequal; lower lip deeply 2-fid or -dentate; upper lip 3-dentate, rarely subentire. Corolla 2-lipped; upper lip straight or falcate; lower lip 3-lobed, the middle lobe the largest; tube straight or invaginated, with or without a ring of hairs inside. Stamens 2; connective usually articulating with the filament, one arm with a fertile cell, the other more or less sterile, often expanded and flattened distally (dolabriform).

Most species grow on stony slopes or roadsides, in dry grassland, cultivated ground or similar habitats, and if no ecological information is given in the text a habitat of this type is to be presumed.

Many non-European species are cultivated for ornament.

1 Staminal connective shorter than or equalling the filament; arms subequal
  2 Leaves lobed or pinnate
    3 Stems ±leafless above
      4 Cymes not or scarcely pedunculate   **9. ringens**
      4 Cymes distinctly pedunculate
        5 Leaves with 1–3 pairs of segments   **8. brachyodon**
        5 Leaves with 1 pair of segments   **6. candelabrum**
    3 Stems leafy above
      6 Leaves with many pairs of linear or oblong lateral segments   **11. scabiosifolia**
      6 Leaves with 1–4 pairs of ovate or irregular lateral segments
        7 Leaves with both glandular and eglandular hairs; calyx 10–15 mm, urceolate, truncate   **10. pinnata**
        7 Leaves with only eglandular hairs; calyx 5–8 mm, campanulate, toothed   **5. triloba**
  2 Leaves simple, not lobed or pinnate
    8 Calyx enlarged in fruit; bracts exceeding the calyx   **12. pomifera**
    8 Calyx not enlarged in fruit; bracts shorter than calyx
      9 Stems ±leafless above
        10 Stems glabrous; pedicels 0–2 mm   **7. blancoana**
        10 Stems hairy at least below; pedicels 5–10 mm
          11 Leaves oblong-elliptical; corolla 30–40 mm   **6. candelabrum**
          11 Leaves oblong to oblong-linear; corolla 20–25 mm   **2. lavandulifolia**
      9 Stem leafy above
        12 Calyx 5–8 mm; stems appressed-white-tomentose   **5. triloba**
        12 Calyx 10–15 mm; stems glabrous or with patent hairs
          13 Leaves ±cuneate at base   **1. officinalis**
          13 Leaves cordate or rounded at base
            14 Corolla not more than 35 mm; leaves not more than 6·5 cm wide   **3. grandiflora**
            14 Corolla 40–50 mm; leaves up to 12(–15) cm wide   **4. eichlerana**
1 Staminal connective longer than filament; arms unequal
  15 Short arm of staminal connective subulate
    16 Upper lip of corolla narrowed at base; verticillasters with (8–)15–30 flowers   **35. verticillata**
    16 Upper lip of corolla not narrowed at base; verticillasters with 6–12 flowers   **36. napifolia**
  15 Short arm of staminal connective ±dolabriform or with a ±sterile cell
    17 Upper lip of calyx flat, not concave, in fruit; calyx tubular or campanulate

18 Bracts exceeding corolla
  19 Corolla 14–18 mm, the upper lip ± straight   **34. viridis**
  19 Corolla 20–30 mm, the upper lip strongly falcate **14. sclarea**
18 Bracts shorter than corolla
  20 Leaves pubescent or strigulose
    21 Annual or biennial; pedicels compressed, deflexed in fruit   **34. viridis**
    21 Perennial; pedicels terete, erect in fruit
      22 Corolla yellow with reddish-brown markings; hood emarginate; calyx 12–17 mm **18. glutinosa**
      22 Corolla violet-blue with white or yellow markings; hood 2-fid; calyx 5–12 mm **19. forskaohlei**
  20 At least the basal leaves lanate or densely tomentose beneath
    23 Bracts equalling or exceeding calyx, greyish-white   **13. phlomoides**
    23 Bracts shorter than calyx, green or violet-tinged
      24 Corolla 10–15 mm, the upper lip weakly falcate; stems eglandular **16. aethiopis**
      24 Corolla 15–35 mm, the upper lip strongly falcate; stems glandular above
        25 Bracts less than ⅓ as long as calyx **17. candidissima**
        25 Bracts ⅓–⅔ as long as calyx **15. argentea**
17 Upper lip of calyx concave-bisulcate in fruit; calyx campanulate
  26 Stamens long-exserted; corolla yellow-white **31. austriaca**
  26 Stamens included or slightly exserted; corolla usually violet-blue, never yellow-white
    27 Inflorescence nodding at apex before anthesis; stems leafless above **30. nutans**
    27 Inflorescence erect before anthesis; stems ± leafy above
      28 Leaves pinnate, with 4–6 pairs of narrow linear segments **32. jurisicii**
      28 Leaves simple, or pinnatifid with wide lobes
        29 Pedicels 8–10 mm; lower lip of corolla white, the hood coloured **20. bicolor**
        29 Pedicels less than 5 mm; lower lip and hood of corolla coloured, or corolla all white
          30 Bracts usually violet or purplish, imbricate in bud; leaves oblong
            31 Stems glandular; bracts ½–¾ as long as calyx **27. valentina**
            31 Stems eglandular; bracts as long as or longer than calyx
              32 Stems densely villous; verticillasters with 6–8 flowers **26. amplexicaulis**
              32 Stems pubescent; verticillasters with 2–6 flowers **25. nemorosa**
          30 Bracts green, not imbricate in bud; leaves ovate to oblong
            33 Stems glandular below **28. sclareoides**
            33 Stems eglandular below
              34 Leaves ± pinnatifid; corolla 6–10(–15) mm **29. verbenaca**
              34 Leaves simple; corolla 10–30 mm
                35 Leaves densely white-appressed-hairy beneath **24. transsylvanica**
                35 Leaves patent-hairy or subglabrous beneath
                  36 Inflorescence dense; cauline leaves few; plant 20–40 cm **23. teddii**
                  36 Inflorescence usually ± lax; cauline leaves few or many; plant up to 100 cm
                    37 Inflorescence with elongated, erecto-patent branches, rarely simple; corolla 11–20 mm **33. virgata**
                    37 Inflorescence ± shortly branched; corolla (10–)20–30 mm
                      38 Leaves ovate to ovate-oblong; corolla of hermaphrodite flowers (15–)20–30 mm **21. pratensis**
                      38 Leaves narrowly oblong; corolla of hermaphrodite flowers 10–20 mm **22. dumetorum**

Sect. SALVIA (Sect. *Eusphace* Bentham). Shrubs or perennial herbs. Calyx not or scarcely accrescent. Upper lip of corolla more or less straight; tube with a ring of hairs inside. Staminal connective shorter than or equalling the filament; arms subequal, one with a more or less sterile cell.

1. S. officinalis L., *Sp. Pl.* 23 (1753). Shrub up to 60 cm. Stems erect, with numerous patent-tomentose branches. Leaves simple, petiolate, oblong, more or less narrowed at base, rugose, white-pubescent beneath, greenish above, densely pubescent when young. Verticillasters with 5–10 flowers. Calyx 10–14 mm, pubescent, glandular-punctate. Corolla up to 35 mm, violet-blue, pink or white. ● *N. & C. Spain, S. France; W. part of Balkan peninsula. Widely cultivated as a pot-herb, and naturalized in parts of S. & S.C. Europe.* Al Ga ?Gr Hs Ju [Bl Bu *Co Cz He It Lu Rm Sa Si].

2. S. lavandulifolia Vahl, *Enum. Pl.* **1**: 222 (1804). Small shrub or herb up to 50 cm. Stems erect or ascending, pubescent. Leaves up to 50 mm, simple, petiolate, narrowly oblong or oblong-linear, crenulate, tomentose. Verticillasters with 6–8 flowers. Pedicels *c.* 5 mm. Calyx 8–12 mm, often reddish-purple, pubescent, glandular-punctate. Corolla 20–25 mm, blue or violet blue. ● *C., S. & E. Spain, just extending into S. France.* Ga Hs.

3. S. grandiflora Etlinger, *De Salvia* 17 (1777) (incl. *S. brachystemon* Klokov). Small shrub or herb up to 100 cm. Stems erect or ascending, patent-hairy. Leaves up to 6·5 cm wide, simple, long-petiolate, ovate or oblong, rounded or cordate at base, rugose, more or less eglandular-pubescent. Verticillasters with 4–10 flowers. Pedicels *c.* 10 mm, rarely 2–4 mm. Calyx 10–15 mm, often reddish-purple, glandular-viscid. Corolla up to 35 mm, lilac, pink or violet-blue, rarely white. $2n = 16$. *C. & S. parts of Balkan peninsula; Krym.* Al Bu Gr Ju Rs (K) Tu.

4. S. eichlerana Heldr. ex Halácsy, *Verh. Zool.-Bot. Ges. Wien* **49**: 191 (1899). Like **3** but up to 180 cm; leaves up to 12(–15) cm wide; corolla 40–50 mm. ● *N. & C. Greece.* Gr.

5. S. triloba L. fil., *Suppl.* 88 (1781) (*S. lobryana* Aznav.). Shrub up to 120 cm. Stems appressed white-tomentose. Leaves simple, or pinnate with 1–2 pairs of ovate lateral segments and a large oblong-elliptical terminal segment, petiolate, rugose, eglandular, greyish-white beneath, greenish above. Verticillasters with 2–6 flowers. Calyx 5–8 mm, campanulate, toothed, often purple, glandular or eglandular-pubescent. Corolla 16–25 mm, lilac or pink, rarely white. *C. & E. Mediterranean region, from Sicilia to Kriti.* Al Cr Gr It Si [Lu].

Polymorphic, especially in density of indumentum, as well as in height, shape of leaf, and length of spike. Some variants have been recognized in Portugal as subspecies (see A. R. Pinto da Silva, *Agron. Lusit.* **20**: 237–238 (1959)), but similar plants occur throughout the range of the species.

6. S. candelabrum Boiss., *Elenchus* 72 (1838). Herb up to 80 cm, woody at base. Stems erect, with few branches, hairy below, glabrous and leafless above. Leaves up to 90 mm, simple or with one pair of small lateral segments at base, petiolate, oblong-elliptical, rugose. Inflorescence lax, with 3- to 5-flowered pedunculate cymes. Pedicels *c.* 10 mm. Calyx 10–13 mm, glandular-viscid. Corolla 30–40 mm; upper lip white with violet markings; lower lip violet-blue. ● *S. Spain.* Hs.

S. × hegelmaieri Porta & Rigo, *Atti Accad. Agiati* **9**: 56 (1891), from S.E. Spain, is a hybrid between **6** and **2**.

7. S. blancoana Webb & Heldr., *Cat. Pl. Hisp. App.* (1850). Herb up to 100 cm, woody at base. Stems erect, simple, glabrous, more or less leafless above. Leaves up to 90 mm, simple, petiolate, oblong-elliptical, rugulose, appressed-pubescent. Verticillasters with 2–6 flowers. Pedicels 0–2 mm. Calyx 10–14 mm, glandular-viscid. Corolla up to 40 mm, violet-blue. *S.E. Spain.* Hs. (*Morocco.*)

8. S. brachyodon Vandas, *Österr. Bot. Zeitschr.* **39**: 179 (1889). Herb up to 70 cm, woody at base. Stems erect, not or scarcely branched, hairy below, glabrous and leafless above. Leaves pinnatisect, with 1–3 pairs of small lateral segments and a large oblong-elliptical terminal segment, petiolate, white-lanate, becoming pubescent. Inflorescence with 2- to 5-flowered, distinctly pedunculate cymes. Bracts shorter than calyx. Calyx 8–10 mm, glandular-viscid. Corolla 35–40 mm, violet-blue. ● *S.W. Jugoslavia.* Ju.

9. S. ringens Sibth. & Sm., *Fl. Graec. Prodr.* **1**: 14 (1806). Herb up to 60 cm, woody at base. Stems erect, scarcely branched, more or less leafless above. Leaves pinnatisect or pinnate with 3–6 pairs of small lateral segments and an ovate or elliptical terminal segment, petiolate, rugose, appressed-hairy. Verticillasters with 2–4 flowers, the cymes not or scarcely pedunculate. Calyx 10–17 mm, densely glandular-pubescent. Corolla up to 40 mm, violet-blue or blue. *Dry places. S. & E. parts of Balkan peninsula, just extending to S.E. Romania.* Al Bu Gr Ju Rm.

10. S. pinnata L., *Sp. Pl.* 27 (1753). Herb 40–100 cm. Stems procumbent or ascending, glandular-viscid. Leaves pinnate or pinnatisect, petiolate, with 2–4 pairs of irregular lateral segments and an ovate terminal segment, with glandular and eglandular hairs. Verticillasters with 4–6 flowers. Pedicels up to 15 mm. Calyx 10–15 mm, urceolate, truncate, often purple, densely glandular, slightly accrescent. Corolla 25–35 mm, lilac to pink. *Turkey-in-Europe.* Tu. (*E. Mediterranean region.*)

11. S. scabiosifolia Lam., *Jour. Hist. Nat.* (*Paris*) **2**: 44 (1792) (incl. *S. adenostachya* Juz., *S. demetrii* Juz.). Herb 20–50 cm, woody at base. Stems much-branched, ascending, patent-villous, eglandular. Leaves pinnate or pinnatisect, petiolate, with many pairs of linear or oblong lateral segments and a larger terminal segment. Verticillasters with 6–10 flowers. Calyx 10–15 mm, accrescent. Corolla 25–30 mm, violet-blue or violet. ● *Krym; one station in N. Bulgaria.* Bu Rs (K).

S. bracteata Banks & Solander in A. Russell, *Nat. Hist. Aleppo* ed. 2, **2**: 242 (1794), which is widespread in Anatolia, has been recorded once in Turkey-in-Europe from a newly made road. It is probably only a casual introduction.

Sect. HYMENOSPHACE Bentham. Like Sect. *Salvia* but shrubs; calyx strongly accrescent.

12. S. pomifera L., *Sp. Pl.* 24 (1753) (*S. calycina* Sibth. & Sm.). Shrub up to 100 cm, much-branched, canescent. Leaves simple, petiolate, ovate, rounded or cordate at base, rugose, densely canescent when young. Verticillasters with 2–4 flowers. Bracts large, exceeding the calyx, caducous. Calyx 10–12 mm, often reddish-purple, with eglandular hairs and sessile glands. Corolla *c.* 35 mm, violet-blue, the lower lip paler. 2n=14. *S. Greece and S. Aegean region.* Cr Gr.

Sect. AETHIOPIS Bentham. Biennial or perennial herbs. Calyx tubular or campanulate; upper lip straight in fruit. Upper lip of corolla more or less falcate; tube without a ring of hairs. Sta-

minal connective longer than filament; arms unequal, the shorter more or less dolabriform.

13. S. phlomoides Asso, *Intr. Oryctogr. Arag.* 158 (1779). Perennial 15–50 cm. Stems erect, usually simple, arachnoid-tomentose below, glandular-hairy and more or less leafless above. Leaves simple or irregularly lobed, shortly petiolate, oblong or oblong linear, lanate beneath, appressed-arachnoid above. Verticillasters with 6–10 flowers. Bracts as long as or longer than calyx, white-canescent. Calyx 15–19 mm, glandular-villous. Corolla 25–32 mm, white or pinkish to purplish. *C., S. & E. Spain.* Hs.

14. S. sclarea L., *Sp. Pl.* 27 (1753). Biennial or perennial up to 100 cm. Stems erect, much branched, glandular above. Leaves simple, petiolate, broadly ovate, cordate, pubescent. Verticillasters with 4–6 flowers. Bracts exceeding corolla, lilac or white. Calyx *c.* 10 mm, with spinose teeth, pubescent and glandular-punctate. Corolla 20–30 mm, lilac or pale blue; upper lip strongly falcate. 2n=22. *S. Europe.* Al Bl Bu Co Ga Gr Hs It Ju Lu Rm Rs (W, K) Sa Tu [Au Cz He].

15. S. argentea L., *Sp. Pl.* ed. 2, 31 (1762) (*S. verbascifolia* sensu Hayek, non Bieb.). Perennial 30–100 cm. Stems erect, much branched, eglandular-villous below, glandular and more or less leafless above. Leaves simple, ovate or oblong, irregularly lobed or toothed, lanate when young. Verticillasters with 4–8 flowers. Bracts ⅓–⅔ as long as calyx, green. Calyx 8–10 mm, glandular-viscid, with somewhat spinose teeth. Corolla 15–35 mm, white, tinged with pink or yellow; upper lip strongly falcate. 2n=20. *S. Europe, eastwards to Bulgaria.* Al Bu Gr Hs It Ju Lu Si.

Plants from the Iberian peninsula have been treated as subsp. **patula** (Desf.) Maire in Jahandiez & Maire, *Cat. Pl. Maroc* **3**: 642 (1934). They have the leaves usually cordate at the base (rarely rounded), and the calyx more zygomorphic. Similar plants occur elsewhere in the range, and they are not worth subspecific rank.

16. S. aethiopis L., *Sp. Pl.* 27 (1753). Biennial or perennial 30–100 cm. Stems robust, erect, white-lanate, eglandular, more or less leafless above. Leaves simple, petiolate, more or less ovate, acuminate, deeply and irregularly serrate, lanate when young. Verticillasters with 6–10 flowers. Bracts slightly shorter than calyx, green or violet-tinged. Calyx 5–8 mm, lanate. Corolla 10–15 mm, white; upper lip weakly falcate. 2n=24. *S. & S.E. Europe, extending northwards to S. Czechoslovakia and to c. 51° N. in S.C. Russia.* Au Bu Cz Ga Gr Hs Hu It Ju Lu Rm Rs (W, K) Tu [Ge Po].

17. S. candidissima Vahl, *Enum. Pl.* **1**: 278 (1804). Perennial up to 60 cm. Stems simple or little-branched, glandular above. Leaves simple, petiolate, ovate, cordate at base, acute, pannose. Verticillasters with 2–6 flowers. Bracts less than ⅓ as long as calyx, green; pedicels *c.* 2 mm. Calyx *c.* 10 mm, densely glandular-pubescent. Corolla 20–30 mm, white with a few purple spots on lower palate; upper lip strongly falcate. *S. Albania, W. Greece.* Al Gr. (*S.W. Asia.*)

S. sonklarii Pant., *Magyar Növ. Lapok* **5**: 151 (1881), with 2-flowered verticillasters and spiny calyx-teeth, has been described from C. Jugoslavia. Its status is uncertain.

Sect. DRYMOSPHACE Bentham. Like Sect. *Aethiopis* but perennial herbs; upper lip of corolla strongly falcate; shorter arm of staminal connective with a sterile cell.

18. S. glutinosa L., *Sp. Pl.* 26 (1753). Stems 50–100 cm, erect, simple or branched, hairy below, glandular-viscid above. Leaves simple, petiolate, ovate, cordate or hastate at base, serrate or crenate, pubescent. Verticillasters with 2–6 flowers. Calyx 12–17 mm, tubular-campanulate. Corolla 30–40 mm, yellow with reddish-brown markings; lower lip straight; upper lip emarginate. 2n=16. *Woods, mainly in mountain districts. From C. France and C. Russia southwards to N. Spain, S. Italy and C. Greece.* Al Au Bu Co Cz Ga Ge Gr He Hs Hu It Ju Po Rm Rs (C, W, K, E) Tu.

19. S. forskaohlei L., *Mantissa* 26 (1767). Stems 25–100 cm, erect, simple or little-branched. Leaves simple or lyrate, ovate, cordate at base, serrate or crenate, glandular-strigulose. Verticillasters with 4–8 flowers. Calyx (5–)8–12 mm, with somewhat spinose teeth. Corolla 20–30 mm, violet-blue with white or yellow markings; lower lip deflexed; upper lip 2-fid. *S.E. part of Balkan peninsula.* Bu Gr Tu. (*N. Anatolia.*)

Sect. PLETHIOSPHACE Bentham. Perennial or rarely biennial herbs. Calyx campanulate; upper lip concave-bisulcate in fruit. Upper lip of corolla straight or falcate; tube without a ring of hairs. Staminal connective longer than filament; arms unequal, the shorter dolabriform.

Hybrids between species of this section are frequent. Gynodioecism and cleistogamy occur in several species.

20. S. bicolor Lam., *Tabl. Encycl. Méth. Bot.* **1**: 69 (1791) (incl. *S. inamoena* Vahl). Sometimes biennial. Stems up to 100 cm, thick, erect, branched, pubescent below, glandular-viscid above, leafy. Leaves simple, broadly ovate, serrate, or pinnatifid with broad lobes, petiolate, cordate at base. Inflorescence branched, lax. Verticillasters with 4–6 flowers. Pedicels 8–10 mm. Calyx 8–10 mm, campanulate or infundibuliform, more or less deflexed in fruit. Corolla 20–30 mm; upper lip violet-blue, prominently falcate; lower lip white. *S.W. Spain.* Hs. (*N.W. Africa.*)

21. S. pratensis L., *Sp. Pl.* 25 (1753) (*S. tenorii* Sprengel, *S. bertolonii* Vis.). Stems up to 100 cm, erect, branched, eglandular-pubescent below, glandular above. Basal leaves simple, long-petiolate, ovate or ovate-oblong, cordate at base, crenate or serrate, patent-hairy or subglabrous beneath; cauline leaves smaller, few, sessile. Inflorescence more or less shortly branched, somewhat lax. Verticillasters with 4–6 flowers. Bracts less than ½ as long as calyx, green. Pedicels *c.* 2 mm. Flowers hermaphrodite or female. Calyx 7–11 mm, glandular-pubescent. Corolla of hermaphrodite flowers (15–)20–30 mm, violet-blue, rarely pink or white. 2n=18. *Most of Europe northwards to England, N. Germany and N.C. Russia.* Al Au Be Br Bu Co Cz Ga Ge Gr He Ho Hs Hu It Ju Po Rm Rs (C, W, K, E) Tu [*Rs (B) Su].

Extremely variable, particularly in shape and size of the corolla.

22. S. dumetorum Andrz. ex Besser, *Cat. Pl. Jard. Krzemien.* ed. 2, 288 (1811) (*S. stepposa* Schost.). Like **21** but leaves narrower, oblong or ovate-oblong; corolla of hermaphrodite flowers 10–20 mm. *Romania, S. part of U.S.S.R.* ?Po Rm Rs (C, W, E).

The taxonomic status of this plant is uncertain. It may be better treated as a subspecies of **21**.

23. S. teddii Turrill, *Kew. Bull.* **1937**: 82 (1937). Stems 20–40 cm, erect, simple, usually eglandular-pubescent. Basal leaves simple, petiolate, ovate or oblong-ovate, serrate or crenate, bullate, patent-hairy beneath, glabrous above; cauline leaves small, few, sessile. Inflorescence dense; verticillasters with 4–6 flowers. Bracts less than ½ as long as calyx, green; pedicels 2–3 mm. Calyx 7·5–10 mm, glandular- or eglandular-pubescent. Corolla 12–15 mm, violet-blue. ● *N.E. Greece.* Gr.

24. S. transsylvanica (Schur ex Griseb.) Schur, *Verh. Siebenb. Ver. Naturw.* **4** App.: 57 (1853) (*S. baumgartenii* Heuffel). Stems 60–100 cm, usually simple, eglandular below, glandular-pubescent above. Leaves simple, long-petiolate, ovate or oblong-lanceolate, cordate at base, acute, more or less regularly crenate, densely white-appressed-hairy beneath. Inflorescence lax; verticillasters with 3–6 flowers. Bracts less than ½ as long as calyx, green; pedicels 2–3 mm. Calyx 8–9 mm. Corolla 16–21 mm, blue or violet-blue. ● *N. & C. Romania.* Rm.

25. S. nemorosa L., *Sp. Pl.* ed. 2, 35 (1762) (*S. sylvestris* auct., non L.). Stems 30–60 cm, many, more or less appressed- or patent-pubescent, eglandular. Basal leaves simple, petiolate, oblong, cordate or rounded at base, attenuate at apex, regularly crenate, more or less pubescent; cauline leaves more or less sessile. Inflorescence usually dense; verticillasters with 2–6 flowers. Bracts as long as or longer than calyx, ovate, rounded at base, violet, imbricate in bud; pedicels *c.* 2 mm. Calyx 6–7 mm. Corolla 8–12(–14) mm, violet-blue, rarely pink or white. 2n=12. *C., S.E. & E. Europe, northwards to C. Russia, and extending to C. Italy.* Al Au Bu Cz Ge He Hu It Ju Po Rm Rs (C, W, K, E) ?Tu [Br Ga No Su].

(a) Subsp. **nemorosa**: Calyx with a very fine, closely appressed indumentum; corolla 8–12 mm. *Throughout the range of the species except E. part of U.S.S.R.*

(b) Subsp. **tesquicola** (Klokov & Pobed.) Soó, *Acta Bot. Acad. Sci. Hung.* **11**: 249 (1965) (*S. tesquicola* Klokov & Pobed.): Calyx with long, patent indumentum; corolla up to 14 mm. *E. Europe, from Bulgaria to E.C. Russia.*

S. × sylvestris L., *Sp. Pl.* 24 (1753) is a hybrid between **25** and **21** and occurs frequently.

26. S. amplexicaulis Lam., *Tabl. Encycl. Méth. Bot.* **1**: 68 (1791). Like **25** but stems up to 80 cm, densely patent-villous; leaves shortly petiolate or sessile, densely eglandular-pubescent beneath, subglabrous above; verticillasters with 6–8 flowers; bracts almost as long as calyx, cordate at base; calyx 6–8·5 mm, patent-pubescent; corolla 8·5–12 mm, violet. *Balkan peninsula.* Al Bu Gr Ju Tu.

27. S. valentina Vahl, *Enum. Pl.* **1**: 268 (1804). Stems up to 30 cm, erect or ascending, branched, densely glandular-villous. Basal leaves simple, petiolate, oblong, cordate at base, acute to acuminate, serrate, gland-dotted; cauline leaves sessile. Inflorescence racemose; verticillasters with 4–6 flowers. Bracts ½–¾ as long as calyx, purplish, imbricate in bud; pedicels *c.* 3 mm. Calyx 7–9 mm, purple, with eglandular hairs and punctate glands. Corolla *c.* 8 mm, violet-blue. ● *E. Spain.* Hs.

28. S. sclareoides Brot., *Fl. Lusit.* **1**: 17 (1804) (*S. bullata* Vahl, non Ortega). Stems 15–40 cm, erect, usually simple, glandular. Basal leaves simple, oblong-ovate, cordate at base, crenate or serrate, bullate; cauline leaves small, few, sessile. Inflorescence racemose; verticillasters with 3–6 flowers. Bracts shorter than calyx, green; pedicels 2–3 mm. Calyx 10–11 mm, glandular, pubescent. Corolla 15–20 mm, purplish or violet, rarely white. ● *Portugal, S.W. Spain.* Hs Lu.

29. S. verbenaca L., *Sp. Pl.* 25 (1753) (*S. clandestina* L., *S. horminoides* Pourret, *S. controversa* sensu Willk., non Ten.). Stems 10–80 cm, erect, simple or branched, eglandular-pubescent below, and more or less glandular above. Leaves more or less pinnatifid with wide lobes, oblong to ovate. Inflorescence dense or lax; verticillasters with 6–10 flowers. Bracts shorter than calyx, green; pedicels 2–3 mm. Flowers hermaphrodite or female, sometimes cleistogamous. Calyx 6–8 mm, glandular or eglandular, enlarging slightly in fruit. Corolla 6–10(–15) mm, blue, lilac or violet. 2*n*=42, 59, 60, 64. *S. & W. Europe, northwards to Scotland.* Al Bl Br Bu Co Cr Ga Gr Hb Hs It Ju Lu Rs (K) Sa Si Tu.

A polymorphic species with numerous local variants.

30. S. nutans L., *Sp. Pl.* 27 (1753). Stems up to 170 cm, erect, simple, appressed-hairy, eglandular, leafless above. Leaves simple, long-petiolate, ovate to oblong, cordate at base, irregularly crenate or serrate. Inflorescence nodding at apex before anthesis; verticillasters with 4–6 flowers. Calyx 5–8 mm, shortly villous and punctate-glandular. Corolla 12–16 mm, violet; upper lip suberect. ● *From C. Hungary and S.W. Bulgaria eastwards to E.C. Russia.* Bu Hu ?Ju Rm Rs (C, W, K, E) Tu.

S. cremenescensis Besser, *Enum. Pl. Volhyn.* 40 (1822), described from W. Ukraine, and **S. cernua** Czern. ex Schost., *Not. Syst.* (*Leningrad*) 8: 154 (1940), widespread in E. Ukraine, are of hybrid origin, with **30** as one parent.

31. S. austriaca Jacq., *Fl. Austr.* 2: 8 (1774). Stems up to 100 cm, erect, usually simple, glandular-hairy. Basal leaves simple, petiolate, irregularly toothed, pubescent beneath, glabrous above; cauline leaves few, sessile, or absent. Verticillasters with 4–9 flowers. Calyx 8–10 mm, glandular-pubescent. Corolla 12–17 mm, yellowish-white. Stamens long-exserted. *From S. Czechoslovakia and N. Ukraine southwards to N.E. Bulgaria.* Au Bu Cz Hu Ju Rm Rs (W, K, E).

32. S. jurisicii Košanin, *Glas Srpske Kralj. Akad.* 119: 26 (1926). Stems up to 60 cm, diffusely branched, with dense, long, eglandular hairs, leafy. Leaves pinnate with 4–6 pairs of narrow linear segments. Inflorescence lax, much-branched; verticillasters with 4–6 flowers. Pedicels 2–5 mm. Calyx 3–5 mm, shortly villous and glandular-punctate. Corolla 9–12 mm, violet-blue, resupinate. Stamens included or slightly exserted. ● *S. Jugoslavia* (*Makedonija*). Ju.

33. S. virgata Jacq., *Hort. Vindob.* 1: 14 (1770) (*S. sibthorpii* Sibth. & Sm., *S. similata* Hausskn.). Stems up to 100 cm, simple or branched, pubescent to villous, eglandular. Basal leaves simple, ovate-oblong, cordate, serrate, subglabrous beneath; cauline leaves few or many. Inflorescence of elongated erecto-patent branches, rarely simple; verticillasters with 4–6 flowers. Bracts less than ½ as long as calyx, green; pedicels *c.* 2 mm. Flowers hermaphrodite or female. Calyx 7·5–10(–13) mm, villous. Corolla of hermaphrodite flowers 11–20 mm, violet-blue, rarely white. 2*n*=18. *Italy; S. & E. parts of Balkan peninsula; Krym.* Al Bu Gr It Ju Rs (K) Tu.

Sect. HORMINUM Dumort. Annuals. Calyx tubular, deflexed in fruit. Upper lip of corolla more or less straight; tube without a ring of hairs inside. Staminal connective longer than filament; arms unequal, the shorter more or less dolabriform.

34. S. viridis L., *Sp. Pl.* 24 (1753) (*S. horminum* L.). Stems up to 50 cm, erect, simple or branched, eglandular- or glandular-hairy. Leaves simple, petiolate, ovate or oblong, rounded or cordate at base, obtuse, regularly crenate, pubescent. Verticillasters with 4–8 flowers, with or without a coma of violet, green, pink or white sterile bracts. Pedicels *c.* 5 mm, compressed, deflexed in fruit. Calyx 7–10 mm, pubescent. Corolla 14–18 mm, pink or violet. 2*n*=16. *S. Europe; occasionally casual in C. Europe.* Al Bu Cr Gr Hs It Ju Lu Rs (K) Si Tu [?Au].

Sect. HEMISPHACE Bentham. Perennial herbs. Calyx tubular or campanulate, deflexed in fruit. Upper lip of corolla more or less straight; tube with a ring of hairs inside. Staminal connective longer than and not articulating with filament; arms unequal, the shorter subulate.

35. S. verticillata L., *Sp. Pl.* 26 (1753) (*S. peloponnesiaca* Boiss. & Heldr.). Stems up to 80 cm, erect, often simple, eglandular-hairy. Leaves simple, or lyrate with 1–2 pairs of small lateral segments, petiolate, ovate-triangular, cordate to truncate at base, acute. Verticillasters with (8–)15–30 flowers. Pedicels 4–6 mm. Calyx *c.* 6 mm, with eglandular hairs and sessile glands. Corolla 8–15 mm, lilac-blue; tube slightly exserted; upper lip narrowed at base. 2*n* = 16. *S., E. & E.C. Europe; naturalized further north.* Al Au Bu Cz Ga Gr Hs Hu It Ju Po Rm Rs (C, W, K, E) Si Tu [Be Br Da Ge *He Ho No Rs (B) Su].

36. S. napifolia Jacq., *Hort. Vindob.* 2: 71 (1772). Stems up to 100 cm, branched above, glandular-pubescent below, eglandular-villous above. Leaves simple, or lyrate with 1–2 pairs of small lateral segments, petiolate, ovate-triangular, glandular-viscid. Verticillasters with 6–12 flowers. Pedicels 3–5 mm. Calyx 6–8 mm, often violet, with eglandular hairs and sessile glands. Corolla 10–16 mm, violet-blue; tube exserted; upper lip not narrowed at base. *Turkey-in-Europe (Istanbul region).* Tu. (*S.W. Asia.*)

41. Elsholtzia Willd.[1]

Annual. Verticillasters 2-flowered, in crowded, secund spikes; bracts differing distinctly from the leaves. Calyx campanulate, 5-veined, with 5 subequal teeth, accrescent. Corolla 2-lipped; tube straight; upper lip hooded; lower lip 3-lobed. Stamens divergent, slightly exceeding the upper lip of corolla. Nutlets smooth or verrucose-rugose.

1. E. ciliata (Thunb.) Hyl., *Bot. Not.* **1941**: 129 (1941) (*E. cristata* Willd., *E. patrinii* (Lepechin) Garcke). Stems 30–70 cm, erect, subglabrous or puberulent. Leaves 2–10 × 1–4 cm, ovate-elliptical, acute, crenate-serrate. Bracts 4–5(–7) mm, obovate-orbicular, cuspidate, entire, slightly exceeding flowers. Calyx 1·5–2 mm, pubescent. Corolla 3–4 mm, lilac. *Cultivated for ornament and naturalized in N., C. & E. Europe; casual elsewhere.* [Cz Da Ge Ju Po Rm Rs (N, C, W) Su Tu.] (*C. & E. Asia.*)

---

[1] By P. W. Ball.

CLII. SOLANACEAE[1]

Herbs or shrubs. Leaves simple to pinnate, exstipulate. Flowers actinomorphic or zygomorphic, hermaphrodite. Calyx (3–)5(–6)-lobed or -dentate. Corolla rotate to campanulate, infundibuliform or tubular, 5(6- or 10)-lobed, rarely subentire; lobes valvate or plicate in bud. Stamens 5(–8), adnate to the corolla-tube and alternating with the lobes; anthers usually with introrse longitudinal dehiscence. Ovary superior, with 2 (rarely more) loculi; style simple; stigma entire to 2-lobed. Fruit a capsule or berry, usually 2-locular. Seeds usually numerous (rarely 2–11).

Many members of this largely extra-European family are grown for their agricultural or horticultural value. On the information available it is often very difficult to distinguish between casual occurrence and true naturalization. Some species, therefore, have been mentioned which are probably no more than casual but, in some cases, may be in process of naturalization.

Some species and hybrids of **Petunia** Juss., from temperate South America, are widely cultivated for ornament. They are densely glandular-hairy herbs up to *c*. 50 cm, with the flowers solitary in the upper leaf-axils, a campanulate calyx, a white to pink or violet, infundibuliform corolla 2·5–7 cm, 5 included stamens and an ovoid, septicidal capsule 6–15 mm. **P. integrifolia** (Hooker) Schinz & Thell., *Viert. Naturf. Ges. Zürich* **60**: 361 (1915) (*P. violacea* Lindley), has been reported as locally naturalized in S. Europe but most modern cultivars are derived from hybrids between it and **P. axillaris** (Lam.) Britton, E. E. Sterns & Poggenb., *Prelim. Cat.* 38 (1888).

Salpiglossis sinuata Ruiz & Pavón, *Syst. Veg. Fl. Peruv.* 163 (1798), with an infundibuliform corolla and 4 unequal, included stamens, and **Schizanthus pinnatus** Ruiz & Pavón, *Fl. Peruv.* **1**: 13 (1798), with pinnate leaves, zygomorphic corolla, 2 long-exserted stamens and 2 staminodes, both viscid-pubescent annuals from Chile, are widely cultivated for ornament and may occasionally escape.

1 Anthers connivent
  2 Corolla campanulate   **6. Withania**
  2 Corolla rotate
    3 Anthers dehiscing by terminal pores   **10. Solanum**
    3 Anthers dehiscing by longitudinal slits
      4 Leaves interruptedly imparipinnate; corolla yellow   **11. Lycopersicon**
      4 Leaves entire; corolla whitish, greenish or purple **9. Capsicum**
1 Anthers not connivent
  5 Flowers in panicles or racemes
    6 Shrub; leaves glabrous
      7 Calyx not more than 7 mm; fruit a berry with not more than 11 seeds   **15. Cestrum**
      7 Calyx at least 10 mm; fruit a capsule with at least 20 seeds   **16. Nicotiana**
    6 Herb; leaves pubescent, often viscid
      8 Inflorescence usually branched; corolla tubular or narrowly infundibuliform; capsule dehiscing by apical valves   **16. Nicotiana**
      8 Inflorescence simple; corolla usually broadly infundibuliform; capsule circumscissile   **5. Hyoscyamus**
  5 Flowers solitary, or in pairs or small clusters, sometimes grouped into spikes
    9 Shrub, usually spiny   **2. Lycium**
    9 Herb, often woody at base
     10 Fruit a capsule

     11 Annual; calyx circumscissile, the base persistent; capsule with longitudinal or irregular dehiscence, usually spiny   **14. Datura**
     11 Perennial; calyx entirely persistent; capsule with circumscissile dehiscence, not spiny
      12 Plant glabrous; pedicels at least 2 cm   **4. Scopolia**
      12 Plant pubescent to villous, often viscid; pedicels not more than 1 cm   **5. Hyoscyamus**
    10 Fruit a berry
     13 Flowers in axillary pairs; berry with 4–8 seeds **13. Triguera**
     13 Flowers solitary, axillary; berry with numerous seeds
      14 Calyx divided almost to base
       15 Glabrous annual; corolla purple to blue; berry brown   **1. Nicandra**
       15 Pubescent, scrambling perennial; corolla white; berry creamy-white   **8. Salpichroa**
      14 Calyx divided to not more than ½-way
       16 Plant with stout, fleshy tap-root; leaves in a dense basal rosette   **12. Mandragora**
       16 Stout tap-root absent; leaves not in a dense rosette
        17 Stamens inserted near apex of corolla-tube; calyx enclosing and usually much exceeding the berry   **7. Physalis**
        17 Stamens inserted near base of corolla-tube; calyx not enclosing berry   **3. Atropa**

1. Nicandra Adanson[2]

Glabrous annuals. Leaves alternate, simple. Flowers actinomorphic, solitary, axillary. Calyx divided almost to base, strongly accrescent. Corolla campanulate. Stamens 5, included, inserted at base of corolla. Stigma capitate. Fruit a 3- to 5-locular, thin-walled, dry berry.

1. **N. physalodes** (L.) Gaertner, *Fruct. Sem. Pl.* **2**: 237 (1791). Stems 10–80(–200) cm. Leaves 4–15(–37) × 2–10(–30) cm, elliptic-lanceolate to ovate, obtuse to acute, cuneate at base, coarsely and irregularly sinuate-serrate or -dentate, sometimes somewhat lobed near base. Flowers 2–4 cm; pedicels 1·5–3 cm. Calyx 10–20 mm, up to 35 mm in fruit; teeth ovate, cuspidate-acuminate, green or scarious. Corolla 2–4 cm in diameter, lilac to blue. Berry *c*. 15 mm, globose, brown. *Widely cultivated for ornament and locally naturalized in C. & S.E. Europe, casual elsewhere.* [Au ?Az Bu He Hu ?It Ju ?Lu Rm Rs (K).] (*Peru.*)

2. Lycium L.[3]

Usually spiny shrubs. Leaves alternate or in clusters, simple, entire, shortly petiolate. Flowers solitary or in small clusters, axillary, pedicellate. Calyx cupuliform, regularly 5-dentate, or 2-lipped with 2–3 teeth connate. Corolla infundibuliform or sub-cylindrical. Stamens inserted in throat of corolla-tube. Stigma shallowly 2-lobed. Fruit a berry.

Literature: N. Feinbrun & W. T. Stearn, *Israel Jour. Bot.* **12**: 114–123 (1964). N. Feinbrun, *Collect. Bot.* (*Barcelona*) **7**: 359–379 (1968).

1 Leaves usually widest at or below the middle, the larger at least 10 mm wide; corolla-lobes about as long as or slightly longer than the tube
  2 Corolla-tube narrowly cylindrical at base for *c*. 2·5–3 mm; leaves usually widest at the middle   **4. barbarum**
  2 Corolla-tube narrowly cylindrical at base for *c*. 1·5 mm; leaves usually widest below the middle   **5. chinense**

---

[1] Edit. D. M. Moore.   [2] By D. M. Moore.   [3] By W. T. Stearn.

1 Leaves 0·5–10 mm wide, usually widest above the middle;
  corolla-lobes not more than ⅓ as long as the tube
  3 Calyx 5–7 mm; corolla 20–22 mm **6. afrum**
  3 Calyx 1·5–4 mm; corolla 8–18 mm
    4 Stamens included
      5 Leaves 20–50 mm; calyx 2–3 mm; corolla 11–13 mm
        1. europaeum
      5 Leaves 3–15 mm; calyx 1·5–2 mm; corolla 13–18 mm
        2. intricatum
    4 Stamens exserted
      6 Leaves 0·5–1·5 mm wide; filaments puberulent at base;
        fruit black **3. ruthenicum**
      6 Leaves 3–10 mm wide; filaments glabrous; fruit reddish
        1. europaeum

1. L. europaeum L., *Sp. Pl.* 192 (1753). Plant 1–4 m; branches rigid, very spiny; spines stout. Leaves 20–50 × 3–10 mm, usually oblanceolate. Flowers solitary or in clusters of 2(–3). Calyx 2–3 mm, 5-dentate or 2-lipped. Corolla 11–13 mm, narrowly infundibuliform, pink or white; lobes 3–4 mm. Stamens usually exserted; filaments glabrous, somewhat unequal. Fruit reddish. *Mediterranean region and Portugal; doubtfully native over much of its range.* Al Bl ?Co ?Cr Ga Gr Hs It Ju Lu Sa Si Tu.

There is one doubtful record from W. Kazakhstan of **L. depressum** Stocks, *Jour. Bot. Kew Gard. Misc.* 4: 179 (1852) (*L. turcomanicum* Turcz. ex Miers), a native of W. & C. Asia, which differs from **1** in having the flowers in clusters of 2–4 and with the corolla 8–10 mm.

2. L. intricatum Boiss., *Elenchus* 66 (1838). Plant 0·3–2 m, much-branched, very spiny; spines stout, rigid. Leaves 3–15 × 1–6 mm, oblanceolate. Flowers solitary or in clusters of 2–3. Calyx 1·5–2 mm, shallowly 5-dentate. Corolla 13–18 mm, narrowly infundibuliform, blue-violet, purple, lilac, pink or white; lobes 2–3 mm. Stamens included; filaments glabrous. Fruit orange-red or black. $2n = 24$. *S. Europe, from Kriti to Portugal; very local.* Cr Gr Hs Lu Si. (*N. Africa, S.W. Asia.*)

The plants with orange-red fruits from Portugal and Spain (*L. intricatum* sensu stricto) are connected with those with black fruits from Sicilia and the Aegean region, which have been called **L. schweinfurthii** Dammer, *Bot. Jahrb.* **48**: 224 (1912), by a series of populations extending along the coast of N. Africa into W. Asia. The relationships between these populations have yet to be clarified, particularly since the eastern plants have been considered a variety of **L. persicum** Miers, *Ann. Nat. Hist.* ser. 2, **14**: 12 (1854).

3. L. ruthenicum Murray, *Comment. Gotting.* 2, *Cl. Phys.*: 9 (1780). Plant 0·5–2 m; spines slender. Leaves 7–30 × 0·5–1·5 mm, very narrowly oblanceolate, somewhat fleshy. Calyx 3–4 mm, 2-lipped. Corolla 8–10 mm, narrowly infundibuliform, purplish above, whitish below; lobes 2–3 mm. Stamens exserted; filaments puberulent at base. Fruit black. *W. Kazakhstan.* Rs (E). (*C. & S.W. Asia.*)

4. L. barbarum L., *Sp. Pl.* 192 (1753) (*L. halimifolium* Miller, *L. vulgare* Dunal). Plant up to 2·5 m; stems arcuate; spines slender, few. Leaves 20–100 × 6–30 mm, very narrowly elliptical to narrowly lanceolate, usually widest at the middle. Calyx c. 4 mm, 2-lipped. Corolla c. 9 mm, infundibuliform, the tube narrowly cylindrical at base for 2·5–3 mm, purple, becoming brownish; lobes c. 4 mm. Stamens long-exserted; filaments with dense tuft of hairs at base. Fruit red. $2n = 24$. *Cultivated for hedges and naturalized in a large part of Europe.* [Au Be Br Bu

Co Cr Cz Da Ga Ge Hb He Ho Hs Hu It Ju Lu No Po Rm Rs (C, W, K, E) Su Tu.] (*China.*)

5. L. chinense Miller, *Gard. Dict.* ed. 8, no. 5 (1768) (*L. rhombifolium* (Moench) Dippel). Like **4** but leaves 10–140 × 5–60 mm, the lower much larger than the upper, lanceolate to ovate, usually widest below the middle; calyx 3 mm; corolla 10–15 mm, broadly infundibuliform, the tube narrowly cylindrical at base for c. 1·5 mm, the lobes 5–8 mm. *Occasionally cultivated for hedges and locally naturalized in W., C. & S. Europe.* [Br Cz Ga Ge Hb He Ho Hu It ?Ju Lu Rm Su.] (*China.*)

6. L. afrum L., *Sp. Pl.* 191 (1753). Plant 1–2 m; branches rigid, very spiny; spines stout. Leaves 10–23 × 1–2 mm, very narrowly oblanceolate. Calyx 5–7 mm, deeply 5-dentate. Corolla 20–22 mm, subcylindrical, purplish-brown; lobes *c.* 2 mm. Stamens included; filaments with dense tuft of hairs at base. Fruit purplish. *Cultivated and locally naturalized in W. part of Mediterranean region.* [Ga Hs It.] (*South Africa.*)

3. Atropa L.¹

Erect perennial herbs. Leaves alternate or opposite, simple, entire. Flowers solitary, axillary. Calyx campanulate, 5-lobed, somewhat accrescent. Corolla tubular-campanulate; limb short, 5-lobed. Stamens subequal, inserted at base of corolla; filaments tomentose at base. Ovary with annular receptacular disc at base; stigma peltate. Fruit a berry.

Corolla campanulate; lobes not more than ⅓ as long as tube;
  stamens and style included or slightly exserted **1. bella-donna**
Corolla infundibuliform; lobes about as long as tube; stamens and
  style long-exserted **2. baetica**

1. A. bella-donna L., *Sp. Pl.* 181 (1753). Plant green, glabrous to glandular-pubescent; stems 50–150(–200) cm, much-branched. Leaves up to 20 cm, not crowded, ovate, acuminate, cuneate at base; petiole short. Pedicels nodding. Calyx campanulate, with acuminate lobes, becoming stellate. Corolla 2·5–3 cm, campanulate, not more than 2½ times as long as the calyx, brownish-violet or greenish; lobes up to ⅓ as long as tube. Stamens and style included or slightly exserted; anthers ellipsoid, whitish. Berry 15–20 mm in diameter, globose, shining, black, rarely yellowish-green, the flesh usually reddish-purple, very poisonous. *Damp or shady places, mainly in the mountains. S., W. & C. Europe, northwards to N. England, and extending to W. Ukraine; long cultivated for its medicinal properties and naturalized elsewhere.* Al Au Be Br Bu Co Cz Ga Ge Gr He Hs Hu It Ju Lu Po Rm Rs (W, K) Sa Si Tu [Da Hb Rs (B, C) Su].

2. A. baetica Willk., *Linnaea* 25: 50 (1852). Plant yellowish-green, glabrous; stems 30–45 cm, simple or branched above, fleshy. Leaves densely crowded, ovate, shortly acuminate, cuneate at base, subcoriaceous; petiole long. Pedicels erect. Calyx-lobes apiculate. Corolla infundibuliform, twice as long as the calyx, green to yellow; lobes about as long as tube. Stamens exserted; anthers ovoid, pale yellow. Style long-exserted. Berry c. 10 mm in diameter, globose. *Shady, calcareous rocks and screes. S. Spain.* Hs. (*Morocco.*)

4. Scopolia Jacq.¹

Perennial herbs with a fleshy, horizontal rhizome. Leaves alternate, simple. Flowers solitary in leaf-axils. Calyx campanulate, shallowly 5-lobed, accrescent. Corolla cylindrical to campanulate, shallowly 5-lobed. Stamens 5, included, inserted at base of corolla. Stigma capitate. Fruit a circumscissile capsule, surrounded by the calyx.

1. S. carniolica Jacq., *Obs. Bot.* **1**: 32 (1764). Glabrous. Stems 20–60 cm, simple or branched above. Lower leaves scale-like, oblong-spathulate; upper leaves up to 20 × 8 cm, elliptical to ovate or obovate, somewhat acuminate, cuneate at base, entire, petiolate. Pedicels 2–4 cm, filiform, nodding. Calyx *c.* 1 cm. Corolla 1·5–2·5 cm, dark brownish-violet outside, yellowish to brownish-green within. Capsule *c.* 1 cm in diameter, globose. 2*n*=48. *C. & S.E. Europe, extending to Italy, C. Ukraine and Lithuania.* Au Cz Hu It Ju Po Rm Rs (B, W) [Da Ge].

5. Hyoscyamus L.[1]

Annual, biennial or perennial herbs, often viscid. Leaves alternate, simple. Flowers axillary, in bracteate spikes or racemes. Calyx campanulate-tubular, 5-dentate, 10- or more veined, accrescent. Corolla usually broadly infundibuliform, with 5 obtuse lobes. Stamens usually slightly exserted, inserted at base of corolla. Stigma capitate. Fruit a capsule, included in the calyx, circumscissile; seeds reniform to orbicular.

1 Cauline leaves sessile
2 Cauline leaves amplexicaul; corolla pale yellow, usually with purple veins; fruiting calyx ventricose **1. niger**
2 Cauline leaves not amplexicaul; corolla purple with darker reticulate veins; fruiting calyx tubular **2. reticulatus**
1 Leaves all petiolate
3 Corolla 1–1·5(–2) cm, usually slightly exceeding calyx **5. pusillus**
3 Corolla at least 2·5 cm, much exceeding calyx
4 Flowers sessile, except the lowest, in a dense spike; corolla usually yellowish-white **3. albus**
4 Flowers all shortly pedicellate, in a lax, few-flowered raceme; corolla golden yellow **4. aureus**

1. H. niger L., *Sp. Pl.* 179 (1753) (incl. *H. bohemicus* F. W. Schmidt). Viscid, velutinous, fetid annual or biennial; stem (10–)30–80 cm, erect, simple or branched. Leaves (5–)15–20 cm, ovate to ovate-oblong, entire to angled, coarsely dentate or deeply incised, the basal forming a rosette, petiolate, the cauline amplexicaul, sessile. Flowers subsessile, in dense, unilateral spikes; bracts leaf-like. Calyx 1–1·5 cm; teeth triangular, acute; fruiting calyx ventricose towards base, with pungent teeth. Corolla 2–3 cm in diameter, slightly zygomorphic, pale yellow, usually with purple veins. 2*n*=34. *Almost throughout Europe, but only as a casual in the extreme north.* All except Az Bl Cr Fa Is Sb.

2. H. reticulatus L., *Sp. Pl.* ed. 2, 257 (1762). Fetid annual or biennial, with glandular, patent pubescence; stem 30–60 cm, stout, erect, branched. Leaves 12–15 cm, oblong to lanceolate, entire to sinuate, coarsely dentate or pinnatisect, sessile to subsessile. Flowers sessile, in unilateral spikes; bracts lanceolate, entire. Calyx 1·5 cm; fruiting calyx 2·5 × 1 cm, tubular, with recurved, lanceolate, cuspidate, pungent teeth about ⅓ as long as tube. Corolla 3–3·5 cm in diameter, almost actinomorphic, purple with darker reticulate veins. *Turkey-in-Europe.* Tu. (*S.W. Asia, Egypt.*)

3. H. albus L., *Sp. Pl.* 180 (1753). Viscid annual, biennial or perennial; stem 30–90 cm, woody below when perennial, erect, branched, densely glandular-villous, the hairs patent. Leaves 4–10 × 3–8 cm, orbicular-ovate, obtuse, cuneate to cordate at base, incise-dentate with wide rounded teeth, sparsely glandular-

villous; petiole 2·5–5(–12) cm. Flowers sessile, except the lowest, in dense unilateral spikes; bracts leaf-like. Calyx 1–1·5 cm, densely glandular-villous; teeth *c.* 3 mm, broadly triangular; fruiting calyx 2–2·5 cm, ventricose below, hypocrateriform above, with short, triangular, pungent teeth. Corolla 3 cm, tubular-campanulate, somewhat zygomorphic, glandular-villous outside, usually yellowish-white, the throat greenish or purplish. Stamens included to slightly exserted. 2*n* = 68. *S. Europe.* Al Az Bl Bu Co Cr Ga Gr Hs It Ju Lu Rm Rs (W, K) Sa Si Tu.

4. H. aureus L., *Sp. Pl.* 180 (1753). Viscid, glandular-villous biennial to perennial; stems decumbent or pendent, weak, brittle. Leaves up to 6 × 5 cm, ovate to suborbicular, broadly cuneate to cordate at base, irregularly lobed and dentate or doubly dentate, densely glandular-villous; petiole *c.* (0·5–)2–3 cm. Flowers in lax, few-flowered racemes; pedicels short; bracts leaf-like. Calyx 1·5–2 cm, densely villous below, less so above; teeth triangular, acute; fruiting calyx up to 3 cm, infundibuliform, with triangular, acuminate, pungent teeth and brownish indumentum. Corolla up to 4·5 cm, 2·5 cm in diameter, infundibuliform, strongly zygomorphic, golden yellow with purplish throat. Stamens and style long-exserted. *Kriti; probably naturalized.* [*Cr.] (*S.W. Asia.*)

5. H. pusillus L., *Sp. Pl.* 180 (1753). Annual, with crispate, papilliform indumentum; stem 5–35 cm, procumbent to erect, branched at the base. Leaves 3–7·5 cm, oblong to lanceolate, acute, cuneate at base, gradually narrowing into a short petiole, sparsely pubescent, entire to sinuate, pinnatifid or pinnatisect with acute lobes. Flowers subsessile, in lax spikes. Calyx *c.* 1 cm; fruiting calyx 1·5–2 cm, patent or deflexed, with triangular, cuspidate, spine-tipped teeth. Corolla 1–1·5(–2) cm, usually slightly exceeding calyx, slightly zygomorphic, yellow with purple throat. Stamens included. *S.E. Russia, W. Kazakhstan.* Rs (E). (*S.W. & C. Asia.*)

6. Withania Pauquy[1]

Shrubs. Leaves alternate or opposite, simple, entire. Flowers solitary or in clusters, axillary. Calyx campanulate, 5-dentate, accrescent. Corolla campanulate, 5-lobed. Stamens 5, equal, included, inserted near base of corolla; anthers connivent. Ovary with nectary at base; stigma capitate. Fruit a globose berry, surrounded by inflated calyx; seeds subreniform.

Leaves cuneate at base, stellate-tomentose beneath; corolla *c.* 5 mm; fruiting calyx urceolate **1. somnifera**
Leaves cordate at base, glabrous to very sparsely pubescent; corolla 8–15 mm; fruiting calyx campanulate **2. frutescens**

1. W. somnifera (L.) Dunal in DC., *Prodr.* **13**(1): 453 (1852). Stem 60–120 cm, erect, branched, woody below, greyish stellate-tomentose. Leaves 3–10 × 2–7 cm, ovate to obovate or oblong, acute to obtuse, cuneate at base, subglabrous above, stellate-tomentose beneath; petiole 1–2 cm. Flowers in clusters of 4–6; pedicels up to 5 mm, patent. Calyx *c.* 5 mm, densely stellate-tomentose, with triangular teeth; fruiting calyx 10–20 × 8–10 mm, urceolate, 10-ribbed, with short teeth. Corolla *c.* 5 mm, yellowish-green. Anthers *c.* 1 mm, ovate. Berry 5–8 mm in diameter, shining, red. *Roadsides, scrub and waste places. Mediterranean region.* Bl Cr Gr Hs ?Lu Sa Si.

2. W. frutescens (L.) Pauquy, *De la Belladone* 15 (1825). Stem 100–180 cm, erect, much-branched, woody, glabrous, with grey, corky bark; mature branches much twisted and entangled; younger branches slender, straight, whitish to pale brownish-yellow. Leaves 1–5 × 1–4·5 cm, broadly ovate to ovate-orbicular,

[1] By J. G. Hawkes.

obtuse, cordate at base, glabrous to very sparsely pubescent; petiole 0·5–2 cm. Flowers solitary, rarely in clusters of 2–3; pedicels 10–15 mm, nodding. Calyx *c.* 5 mm, sparsely pubescent, with triangular teeth; fruiting calyx 15–25 × 10–18 mm, campanulate, 5-ribbed, with long, recurved, acuminate-cuspidate teeth. Corolla 8–15 mm, deeply lobed, greenish-yellow. Anthers *c.* 3 mm, lanceolate. Berry 7–8 mm in diameter, green. *Roadsides and waste places. S. & E. Spain, Islas Baleares.* Bl Hs.

7. Physalis L.[1]

Annual or perennial herbs. Leaves alternate, opposite or in whorls of 3, simple. Flowers solitary, axillary. Calyx campanulate, 5-dentate, accrescent. Corolla 5-lobed, rotate or broadly campanulate. Stamens 5, exserted, inserted near top of the short corolla-tube. Stigma capitate. Fruit a globose berry, surrounded and usually much exceeded by the inflated calyx; seeds suborbicular to reniform.

Literature: R. B. Fernandes, *Bol. Soc. Brot.* **44**: 343–366 (1970). S. J. Van Oostroom & T. J. Reichgelt, *Gorteria* **1**: 65–71 (1962). U. T. Waterfall, *Rhodora* **60**: 107–114, 128–142, 152–173 (1958); **69**: 82–120, 203–239, 319–329 (1967).

1 Corolla dirty white, rotate, the limb distinctly 5-lobed; fruiting calyx orange or red **1. alkekengi**
1 Corolla yellow, usually with greyish, purple or brownish marks at throat, broadly campanulate, the limb subentire; fruiting calyx green or yellowish-green
  2 Plant subglabrous or with a few short hairs on younger parts only **4. philadelphica**
  2 Plant with dense indumentum
    3 Leaves acuminate; anthers 3·5–4 mm **2. peruviana**
    3 Leaves acute; anthers 1–2 mm **3. pubescens**

1. P. alkekengi L., *Sp. Pl.* 183 (1753). Perennial, with horizontal rhizome; stem 25–60(–100) cm, erect, simple or branched above. Leaves 4–15 × 2–8 cm, broadly ovate, slightly acuminate, truncate to broadly cuneate at base, entire to sinuate or with a few coarse teeth, sparsely pubescent; petiole 1·5–6 cm. Pedicels 5–15 mm. Calyx 5–18 mm, tomentose; teeth *c.* ⅛ total length, attenuate; fruiting calyx 25–50 mm, inflated, red to orange. Corolla 15–25 mm in diameter, rotate, distinctly 5-lobed, dirty white. Anthers 2 mm. Berry 12–17 mm, red to orange. 2*n*=24. *Europe, northwards to N. France and C. Russia, but doubtfully native in the northern and western parts of this range.* Al Au †Be Bu *Co Cz *Ga Ge Gr He *Hs Hu It Ju Po Rm Rs (C, W, K, E) Tu [Br Ho].

2. P. peruviana L., *Sp. Pl.* ed. 2, 1670 (1763). Perennial, with dense indumentum; stem 30–100 cm, simple or branched above. Leaves (5–)7–11(–15) × 4–7(–9) cm, broadly ovate, acuminate, cordate at base, entire or slightly dentate; petiole 1–4 cm. Pedicels 7 mm. Calyx 8–9 mm; teeth ½ total length, acuminate; fruiting calyx 30–50 mm, green, pubescent. Corolla 15–25 mm in diameter, subentire, yellow with dark purple markings. Anthers 3·5–4 mm, purple; filaments purple. Berry 12–20 mm, yellow. *Cultivated for its edible fruits and locally naturalized in C. & S. Europe.* [Au Az Cz Hs It.] (*South America.*)

3. P. pubescens L., *Sp. Pl.* 183 (1753). Annual, usually softly villous, sometimes glandular-viscid; stem 10–90 cm, erect, branched. Leaves (4–)7–11(–15) × (2–)4–7(–9) cm, broadly ovate, acute, truncate to cordate at base, usually dentate, the teeth acute, unequal; petiole 2–7 cm. Pedicels 3–5 mm. Calyx 4–10 mm; teeth *c.* ½ total length, narrowly lanceolate; fruiting

calyx 20–30 mm, 5-angled, green, softly pubescent. Corolla 10–15 mm in diameter, subentire, yellow with dark purple markings at throat. Anthers 1–2 mm, purple; filaments 3–5 mm. Berry 10–15(–20) mm, yellow or green, sweet-scented. *Cultivated for its edible fruits in Ukraine, and on a small scale elsewhere, and locally naturalized.* [?Al ?Hu It Rm Rs (C, W).] (*North and South America.*)

4. P. philadelphica Lam., *Encycl. Méth. Bot.* **2**: 101 (1786). Annual, subglabrous but with a few short hairs on young shoots, leaves and calyces; stem 45–60 cm, erect, branched. Leaves 2–10 × 1–4 cm, ovate to ovate-lanceolate, somewhat acuminate, cuneate at base, entire or sinuate to somewhat dentate towards the base; petiole (1–)2–5 cm. Pedicels 5–10 mm. Calyx 4–10 mm; teeth ⅛–½ total length, ovate; fruiting calyx 30–50 mm, green, often purple-veined. Corolla 5–30 mm in diameter, subentire, yellow, with brownish-purple markings at throat. Anthers *c.* 1·25–4 mm, purple, curved after dehiscence; filaments purple. Berry 13–40(–60) mm, green to purple, filling and sometimes splitting calyx. 2*n*=24. *Cultivated for its edible fruits in Ukraine, and locally naturalized there and elsewhere.* [?Cz Lu Rs (C, W).] (*North and South America.*)

There is still some doubt as to whether the cultivated European and American plants, formerly called **P. ixocarpa** Brot. ex Hornem., *Hort. Hafn.*, *Suppl.* 26 (1819), are identical. In addition, Fernandes (*loc. cit.*) has distinguished the Portuguese plants with small flowers and fruits as *P. ixocarpa* in contrast to the cultivated plants with large flowers and fruits.

P. angulata L., *Sp. Pl.* 183 (1753), from Tropical America, which can be distinguished from **4** by the shorter anthers (1·5–2 mm) which are not curved after dehiscence, and the smaller, yellowish-green berry 10–12 mm, is also cultivated locally for its edible fruits and is found as an occasional casual.

8. Salpichroa Miers[1]

Scrambling, much-branched perennial herbs, with leaf-opposed branches. Leaves alternate or opposite, simple. Flowers solitary, axillary. Calyx 5-lobed. Corolla urceolate, with 5 deflexed lobes. Stamens 5, equal, somewhat exserted, inserted in corolla-tube; anthers dorsifixed. Ovary with nectary at base; style hairy below; stigma capitate. Fruit a berry; seeds strongly compressed.

1. S. origanifolia (Lam.) Baillon, *Hist. Pl.* **9**: 288 (1888). Stems flexuous, angled, pubescent, woody below. Leaves 1·5–2·5(–5) × 1–2(–4) cm, ovate-rhombic to suborbicular, entire, sparsely pubescent, narrowed into petiole of equal length. Pedicels nodding. Calyx 3–4 mm, divided almost to base; lobes subulate. Corolla 6–10 mm, whitish, densely lanate below insertion of stamens. Berry 1–1·5 cm, ovoid, creamy-white, edible; seeds 1·5–2 mm, hairy. *Locally naturalized in S. & S.W. Europe.* [Az Br Co Ga Hs It Lu.] (*E. temperate South America.*)

9. Capsicum L.[1]

Annual or biennial herbs or shrubs. Leaves alternate, in pairs or groups of 3, simple. Flowers solitary, in pairs or in groups of 3, axillary, extra-axillary or leaf-opposed. Calyx campanulate or tubular, 5(–6)-dentate, fleshy, slightly accrescent. Corolla rotate, with stellate, 5(–6)-lobed limb. Stamens 5(–6), exserted, inserted at top of corolla-tube; anthers connivent. Ovary 2- to 3-locular; stigma capitate. Fruit a rather dry berry, inflated, incompletely 2- to 3-locular; seeds subreniform, compressed.

---

[1] By J. G. Hawkes.

Literature: A. Terpó, *Feddes Repert.* **72**: 155–191 (1966). C. B. Heiser & B. Pickersgill, *Taxon* **18**: 277–283 (1969). C. B. Heiser & P. G. Smith, *Econ. Bot.* **7**: 214–227 (1953).

1. C. annuum L., *Sp. Pl.* 188 (1753). Annual or biennial herb 20–50(–100) cm. Leaves ovate to lanceolate, entire, glabrous; petiole 4–16 cm. Flowers solitary, rarely in pairs. Calyx campanulate. Corolla white, rarely purple, with whitish, green or violet blotches. Anthers ovate-lanceolate. Berry 5–15 cm, very variable in shape, usually red, often orange, yellow, green, brownish-purple or black. *Widely cultivated as a vegetable and condiment and occasionally occurring as a casual, especially in S., E.C. & E. Europe. (Tropical America.)*

C. frutescens L., *Sp. Pl.* 189 (1753), from Tropical America, is occasionally cultivated in S.W. Europe. It is a shrub 35–115 cm, which also differs from **1** in having elliptical leaves, the flowers in pairs or groups of 3, a tubular calyx and greenish-white corolla.

C. baccatum L. var. **pendulum** (Willd.) Eshbaugh, *Taxon* **17**: 52 (1968) is cultivated in parts of Germany, Italy and the U.S.S.R. It differs from **1** by the yellow blotches at the base of the petals and the deeply bifurcate base of the anthers.

10. Solanum L.[1]

Herbs or shrubs, sometimes scrambling or climbing, usually hairy, often with prickles. Leaves alternate or in pairs, simple to pinnate. Inflorescence axillary, extra-axillary or leaf-opposed, of 1 or more helicoid cymes usually becoming scorpioid, sometimes of cymose umbels, rarely reduced to a single flower. Calyx campanulate, usually 5-fid. Corolla rotate; limb orbicular or pentagonal to stellate, often recurved. Stamens exserted, inserted in throat of corolla-tube; anthers usually connivent, in a cylindrical to ovoid or conical column, dehiscing by 2 terminal pores, later splitting introrsely. Stigma capitate. Fruit a succulent to dry berry, 2- to 4-locular; seeds small, ovoid, compressed.

1 Indumentum of stellate or branched hairs
  2 Stems and leaves with many conspicuous prickles
    3 Corolla yellow; lower stamen much longer than the others   **14. cornutum**
    3 Corolla purple, blue or white; stamens equal   **12. sodomeum**
  2 Stems and leaves with few, inconspicuous prickles, or prickles absent
    4 Petioles with prominent, auricular false stipules at base   **13. mauritianum**
    4 Petioles without false stipules
      5 Leaves linear- to oblong-lanceolate, densely stellate-hairy above   **15. elaeagnifolium**
      5 Leaves ovate to lanceolate, sparsely stellate-pubescent above
        6 Corolla white or pale blue; berry not more than 10 mm, yellow   **11. bonariense**
        6 Corolla purplish to violet; berry at least 50 mm, dark purple to blackish-violet (rarely whitish or yellow)   **10. melongena**
1 Indumentum of simple, sometimes glandular hairs, or absent
  7 Plant glabrous or subglabrous
    8 Leaves (5–)10–15(–20) cm, often laciniate; corolla orbicular, purple, appearing 10-lobed   **9. laciniatum**
    8 Leaves not more than 10 cm, entire; corolla stellate, white, 5-lobed   **6. pseudocapsicum**
  7 Plant with simple, sometimes glandular hairs
    9 Leaves interruptedly imparipinnate; underground stolons and tubers present   **8. tuberosum**
    9 Leaves entire to deeply pinnatisect or imparipinnate; stolons and tubers absent

[1] By J. G. Hawkes and J. M. Edmonds.

  10 Scrambling perennial, woody at base; cymes much-branched, usually with at least 10 flowers; corolla usually dark purple   **7. dulcamara**
  10 Decumbent to erect annual or perennial, sometimes slightly woody; cymes usually unbranched, 2- to 10-flowered; corolla usually white
    11 Leaves deeply pinnatisect, with ± linear lobes   **5. triflorum**
    11 Leaves entire or sinuate-dentate
      12 Calyx strongly accrescent, enclosing at least lower half of berry; berry with stone-cells   **4. sarrachoides**
      12 Calyx slightly accrescent, deflexed or adhering to base of berry; berry usually without stone-cells
        13 Cymes 3- to 5-flowered; peduncles (4–)7–13(–19) mm; berry usually longer than wide, red, orange or yellow   **3. luteum**
        13 Cymes (3–)5- to 10-flowered; peduncles (10–)14–30 mm; berry globose or wider than long, purple, black or green
          14 Peduncle strongly deflexed in fruit; berry purple   **2. sublobatum**
          14 Peduncle usually erecto-patent in fruit; berry black or green   **1. nigrum**

1. S. nigrum L., *Sp. Pl.* 186 (1753). Subglabrous to villous annual up to 70 cm, with simple hairs; stems decumbent to erect. Leaves 2·5–7 × 2–4·5(–6) cm, ovate-rhombic to lanceolate, entire to sinuate-dentate. Cymes (3–)5- to 10-flowered, lax, solitary; peduncles (10–)14–30 mm, usually erecto-patent in fruit, the pedicels usually much shorter, recurved in fruit. Calyx 1–2·5 mm, slightly accrescent, deflexed or adhering to base of berry; lobes usually ovate. Corolla (8–)10–14(–18) mm in diameter, usually 1½–3 times as long as calyx, white. Anthers 1·5–2·5 mm, yellow. Berry 6–10 mm wide, usually wider than long, dull black or green; seeds 1·7–2·4 mm. *Disturbed and cultivated ground. Most of Europe, but doubtfully native in much of the north.* All except Fa Is Sb.

(a) Subsp. **nigrum** (*S. dillenii* Schultes, *S. judaicum* Besser, *S. suffruticosum* Schousboe ex Willd.): Plant subglabrous to pubescent, the hairs usually appressed, eglandular. $2n = 72$. *Throughout the range of the species.*

(b) Subsp. **schultesii** (Opiz) Wessely, *Feddes Repert.* **63**: 311 (1960) (*S. decipiens* Opiz): Plant villous, the hairs usually patent, glandular. $2n = 72$. *Drier parts of C., S. & E. Europe.*

S. melanocerasum All., *Auct. Syn. Stirp. Horti Taur.* 12 (1773), distinguished from **1** by the prominently and dentately winged stems, brown anthers and berry 15–17 mm in diameter, occasionally escapes from cultivation and is a casual in Belgium, England, Germany and Sweden.

S. americanum Miller, *Gard. Dict.* ed. 8, no. 5 (1768) (*S. nodiflorum* Jacq.), distinguished from **1** by its more or less umbellate cymes, corolla 5–9 mm in diameter, erect fruiting pedicels, globose, shiny black berry and seeds 1–1·5 mm, and with $2n = 24$, is a rare casual in parts of W., C. & S. Europe.

2. S. sublobatum Willd. ex Roemer & Schultes, *Syst. Veg.* **4**: 664 (1819) (*S. gracile* Dunal, non Sendtner). Like **1** but often perennial; cymes umbellate; peduncle strongly deflexed in fruit; berry ovoid, dull purple. $2n = 24$. *Locally naturalized in S.W. Europe.* [Ga He Hs ?Lu.] *(S.E. South America.)*

3. S. luteum Miller, *Gard. Dict.* ed. 8, no. 3 (1768). Subglabrous to villous annual up to 50 cm, with simple hairs; stems decumbent to erect. Leaves 2–7 × 1·5–4 cm, rhombic to ovate-lanceolate, entire to sinuate-dentate. Cymes 3- to 5-flowered, lax, solitary; peduncles (4–)7–13(–19) mm, the pedicels often longer and deflexed in fruit. Calyx 1–2·5 mm, slightly accrescent, deflexed or adhering to base of berry; lobes triangular. Corolla 8–16 mm in diameter, 3–5 times as long as calyx, white. Anthers 1·5–2·5 mm, yellow. Berry 6–10 mm in diameter, usually longer than wide, red, orange or yellow; seeds 1·6–2 mm. *Waste places and disturbed ground. Europe northwards to N. France and C. Russia; introduced further north.* Al Au Az Bl Br Bu Co Cr Cz Ga Ge Gr He Hs Hu It Ju Lu Po Rm Rs (C, W, K) Sa Si Tu [Be Da Ho ?No ?Rs (B) Su].

(a) Subsp. **luteum** (*S. villosum* Miller): Plant villous, the hairs usually patent, glandular; stems rounded, with smooth ridges. $2n=48$. *Throughout the range of the species, but most frequent in warm, dry areas.*

(b) Subsp. **alatum** (Moench) Dostál, *Květena ČSR* 1270 (1949) (*S. miniatum* Bernh. ex Willd., *S. zelenetzkii* Pojark.): Plant subglabrous to pubescent, the hairs usually appressed, eglandular; stems angled, with dentate ridges. $2n=48$. *Throughout the range of the species.*

4. **S. sarrachoides** Sendtner in C.F.P. Mart., *Fl. Brasil.* 10: 18 (1846). Annual up to 40 cm; stems decumbent to ascending, with dense, patent, glandular hairs. Leaves 2–5 × 1·5–3·5 cm, rhombic to ovate-lanceolate, sinuate-dentate, rarely subentire, glandular-pubescent. Cymes 3- to 8(–10)-flowered, solitary; peduncles 6–15 mm. Calyx (2·5–)3–4·5 mm, strongly accrescent and enclosing at least lower half of berry. Corolla 9–14 mm in diameter, $1\frac{1}{2}$–3 times as long as calyx, white. Anthers 1–2 (–2·5) mm, yellow. Berry 6–10 mm in diameter, green or black, with stone cells; seeds 1·8–2·2 mm. $2n=24$. *Locally naturalized as a weed or ruderal.* [Br Ga Ge.] (*Brazil.*)

The degree of leaf-indentation and accrescence of the calyx appear to be correlated with differences in habitat.

5. **S. triflorum** Nutt., *Gen. N. Amer. Pl.* 1: 128 (1818). Fetid annual with simple hairs; stem 15–100 cm, procumbent, branched, subglabrous. Leaves 2–3 cm, ovate to elliptical, deeply pinnatisect, decurrent on the short petiole; lobes more or less linear, obtuse, entire to dentate or lobed, subglabrous. Cymes usually 2- to 3-flowered, contracted, subumbellate, solitary. Calyx 4–5 mm; teeth triangular-ovate. Corolla *c.* 10 mm in diameter, slightly exceeding calyx, stellate, white. Berry 10 mm in diameter, more or less globose, marbled whitish and green. *Locally naturalized in N.W. Europe.* [Be Br.] (*W. North America.*)

Var. *ponticum* (Prodan) Borza, with subentire leaves, is recorded from Britain, the Netherlands and Romania.

6. **S. pseudocapsicum** L., *Sp. Pl.* 184 (1753). Glabrous annual or perennial; stem 30–120 cm, erect, branched. Leaves 1·5–10 × 0·5–2 cm, oblong- to linear-lanceolate, acuminate, cuneate at base, entire to somewhat sinuate, alternate or in unequal pairs; petiole 0·2–2 cm. Cymes 1- to 3-flowered, solitary, extra-axillary to leaf-opposed, sessile. Calyx *c.* 5 mm, deeply incised; teeth linear-lanceolate. Corolla 10–15 mm in diameter, stellate, white. Anthers 3 mm, orange. Berry 10–15 mm in diameter, globose, red, shining, solitary. *Widely cultivated for ornament and locally naturalized in S.W. Europe.* [Az Ga Lu.] (*E. South America.*)

S. capsicastrum Link ex Schauer, *Allgem. Gartenz.* 1: 228 (1833), from temperate E. South America, is widely cultivated for ornament and occasionally found as a casual. It has frequently been confused with 6, from which it is distinguished by the branched hairs on stem and leaves, the shorter stem, narrower leaves and the berry *c.* 20 mm in diameter.

7. **S. dulcamara** L., *Sp. Pl.* 185 (1753) (incl. *S. littorale* Raab). Scrambling, glabrous to pubescent or densely villous-tomentose perennial, with simple hairs; stems 30–200 cm, woody below. Leaves (3–)5–9 × (1·5–)2·5–5(–9) cm, ovate, acute to acuminate, truncate, cordate or hastate at base, entire or with 1–4 lobes or small, stipitate pinnae at base; petiole up to 3 cm. Cymes usually with 10–25 or more flowers, lax, terminal, becoming lateral, several on extra-axillary peduncles usually much-branched 10–40 mm above base. Calyx *c.* 3 mm; teeth 0·5–1 mm, rounded. Corolla 10–15(–20) mm in diameter, stellate, dark purple (rarely white); lobes 7–10 mm, recurved, each with 2 greenish spots at base. Anthers 5–7 mm, pale yellow. Berry 10–15 × 7·5–10 mm, ovoid, red, shining. $2n=24$. *Damp woods, hedges, river-banks and sea-shores. Most of Europe except the extreme north.* All except Az Cr Fa Is Sb; only casual in Rs (N).

A procumbent variant with fleshy leaves (**S. marinum** (Bab.) Pojark. in Schischkin & Bobrov, *Fl. URSS* 22: 16 (1955)) occurs on maritime sand and shingle.

S. persicum Willd. in Roemer & Schultes, *Syst. Veg.* 4: 662 (1819), from S.E. Russia (and W. Asia), is probably no more than a minor variant of **7**, from which it is said to differ by its tomentose stem, ovate-lanceolate, grey-tomentose leaves and subglobose berries.

8. **S. tuberosum** L., *Sp. Pl.* 185 (1753). Sparsely pubescent annual with simple, appressed hairs, with underground stolons bearing terminal tubers; stems 30–80(–150) cm, erect, branched, succulent, winged. Leaves interruptedly imparipinnate, with 3–5(–7) pairs of ovate, acuminate leaflets between which are smaller, unequal, ovate to orbicular leaflets; petiole with auricular or semi-lunate false stipules at base. Cymes many-flowered, 2–4 on axillary to extra-axillary branched peduncles 50–100 mm; pedicels 20–35 mm, usually articulated in middle $\frac{1}{3}$. Calyx 5–10 mm; teeth lanceolate, acuminate. Corolla 25–35(–40) mm in diameter, orbicular to pentagonal, white, purple or blue to violet or pinkish. Anthers 6–7 mm, yellow to orange. Berry 20–40 mm in diameter, more or less globose, succulent, greenish to purplish. $2n=48$. *Cultivated for its edible tubers (potatoes) almost throughout Europe.* (*South America.*)

The European cultivars belong to subsp. *tuberosum*.

9. **S. laciniatum** Aiton, *Hort. Kew.* 1: 247 (1789). Glabrous or subglabrous shrub; stem 100–150 cm, branched, purplish. Leaves entire or laciniate often on same plant; entire leaves (5–)10–15(–20) × 1·5–4·5 cm, narrow- to linear-lanceolate, long-acuminate, narrowly cuneate at base, decurrent on the short petiole; laciniate leaves with 1–3 pairs of narrowly triangular to linear-lanceolate, long-acuminate lobes. Cymes 3- to 12-flowered, solitary or rarely in pairs, axillary to extra-axillary, sessile, the axis elongating in fruit to *c.* 20 cm. Calyx 5–7 mm; teeth short, mucronulate. Corolla (30–)40–50(–60) mm in diameter, orbicular, purple; lobes appearing 10 by extension of interpetalar membranes for 3–5 mm on each side of lobe-apex. Anthers 3–5 mm, yellow. Berry *c.* 20 × 10 mm, ovoid, yellow to orange-yellow. *Cultivated in E. Europe and locally in C. & W. Europe for the foliage which is a source of steroid precursors.* (*Australia, New Zealand.*)

European material has long been confused with *S. aviculare* G. Forster from Australia and New Zealand.

10. **S. melongena** L., *Sp. Pl.* 186 (1753). Perennial or annual; stems 30–70(–120) cm, often somewhat woody, erect, branched, stellate-tomentose, sometimes with scattered prickles. Leaves 7–15 × 3–10 cm, ovate, entire to shallowly sinuate, sparsely

stellate-pubescent above, more densely so beneath; petiole 3–7 mm, densely stellate-tomentose, sometimes with few fine prickles. Cymes 1(–3)-flowered, solitary, extra-axillary, rarely leaf-opposed, sessile; pedicels deflexed, strongly accrescent in hermaphrodite flowers. Calyx 15–20 mm, 5- to 9-lobed, accrescent; lobes up to 12 mm, sometimes unequal, lanceolate, long-acuminate. Corolla 25–30(–50) mm in diameter, orbicular-pentagonal, 5- to 7(–8)-lobed, purplish to violet. Stamens 5–7(–8); anthers 6–8 mm. Berry 50–200(–300) × 50–100(–150) mm, oblong to ovoid or long-pyriform, rarely globose, blackish-violet, dark purple, yellow or whitish. *Widely cultivated for its edible fruits (aubergines) in S., E.C. & E. Europe. (India.)*

11. S. bonariense L., *Sp. Pl.* 185 (1753). Shrub; stem up to 200 cm, erect, stout, subglabrous to sparsely stellate-tomentose, usually with a few prickles when young, unarmed later; branches flexuous. Leaves 6–15 × 3·5–13 cm, ovate-oblong to -lanceolate or lanceolate, acute to acuminate, cuneate to auriculate at base, shallowly sinuate to entire, rarely with 1–9 obtuse lobes 2–3 × 3–4 cm, stellate-hairy; petiole 1·5–4 cm. Cymes few-flowered, 2–4 on terminal to lateral, extra-axillary peduncles branched c. 30–50 mm above base. Calyx c. 4–7 mm; teeth ovate, abruptly acuminate. Corolla 25–35 mm in diameter, orbicular-pentagonal, white or pale blue. Anthers 6–7 mm. Berry 7–10 mm, globose, yellow. *Locally naturalized in S. Europe.* [Bl Hs ?It.] *(Temperate South America.)*

12. S. sodomeum L., *Sp. Pl.* 187 (1753). Shrubby perennial; stem 50–300 cm or more, stout, erect, much-branched, sparsely stellate-pubescent, with many straight, wide-based pale yellow prickles up to 1·5 cm. Leaves 5–13(–18) × 4–9(–15) cm, more or less ovate, pinnatisect almost to midrib, with many prickles, sparsely stellate-hairy above, more densely so beneath; lobes rounded, sinuate; petioles 1–3(–5) cm. Cymes few-flowered, solitary, extra-axillary, sessile; pedicels c. 10 mm, elongating and recurving in fruit; upper flowers smaller, male. Calyx 5–7 mm, accrescent and with dense prickles in hermaphrodite flowers; lobes lanceolate. Corolla 25–30 mm in diameter, orbicular-pentagonal, pale violet. Anthers 5–6 mm, equal. Berry 20–30 mm in diameter, globose, yellow to brown, shining. *Maritime sands, roadsides and waste places, especially near the sea. Naturalized in S. Europe; an occasional casual elsewhere.* [Al Az Bl Bu Co Gr Hs It Ju Lu Sa Si.] *(Africa.)*

S. sisymbrifolium Lam., *Tabl. Encycl. Méth. Bot.* **2**: 25 (1794), from temperate South America, distinguished from **12** by its corolla 30–35 mm in diameter and red berry partly enclosed by the accrescent calyx, is occasionally found as a casual in C. Europe.

13. S. mauritianum Scop., *Delic. Fl. Insubr.* **3**: 16 (1788). Fetid, densely stellate-tomentose shrub; stem up to 450 cm. Leaves 20–30(–40) × 6–11(–20) cm, ovate-elliptical, long-acuminate, cuneate at base; petiole (2–)4–5 cm, with prominent, auricular false stipules at base. Cymes many-flowered, numerous; peduncles extra-axillary, terminal, becoming lateral, much-branched, undivided up to 120 mm above base. Calyx up to 7 mm, deeply divided; lobes oblong-elliptical. Corolla c. 20 mm in diameter, stellate to pentagonal, violet; lobes c. 7 × 6 mm, ovate-triangular. Anthers up to 3 mm, pale yellow. Berry c. 15 mm in diameter, globose, dull yellow. *Naturalized in Açores.* [Az.] *(Central America.)*

14. S. cornutum Lam., *Tabl. Encycl. Méth. Bot.* **2**: 25 (1794). Annual with dense tomentum of stipitate stellate and simple hairs intermixed with patent to recurved yellow prickles up to 1 cm; stem 30–60 cm, much-branched, sometimes woody at base. Leaves 6–12 × 4–8 cm, deeply pinnately lobed; lobes obovate to orbicular, sinuate; petiole 1–5 cm. Cymes 3- to 10-flowered, solitary, extra-axillary; peduncles 30–60 mm. Calyx 10 mm; lobes c. 5 mm, unequal, lanceolate, accrescent; tube with dense prickles, accrescent to enclose berry. Corolla 20–40 mm in diameter, somewhat zygomorphic, appearing 10-lobed by expansion of interpetalar membranes, yellow. Upper 4 anthers c. 4–6 mm, equal, yellow, the other c. 8–12 mm, purplish distally. Berry c. 10 mm in diameter, globose, dry. *Cultivated ground and waste places. Locally naturalized, mainly in E.C. & E. Europe.* [Bu Ga Ge Gr Hu Rs (C, W, K, E).] *(Mexico, S.W. United States.)*

S. heterodoxum Dunal, *Hist. Solanum* 235 (1813), and **S. citrullifolium** A. Braun, *Ann. Sci. Nat.* ser. 3 (Bot.), **12**: 356 (1849), from Mexico, which can be distinguished from **14** by their purple corolla, are perhaps naturalized in parts of S. Europe.

15. S. elaeagnifolium Cav., *Icon. Descr.* **3**: 22 (1795). Perennial herb or dwarf shrub with dense, whitish, stellate tomentum, usually with scattered, inconspicuous, usually reddish prickles on stem, mature leaves and calyx; stem 30–50 cm or more, erect, sparingly branched. Leaves 4–10(–16) × 1–2·5(–4) cm, linear- to oblong-lanceolate, obtuse to acute, rounded or truncate at base, entire to shallowly sinuate; petiole 5–20 mm. Cyme 1- to 5-flowered, solitary, extra-axillary; peduncle 5–20 mm. Calyx 5–7 mm, slightly accrescent, patent in fruit; lobes 2–4 mm, linear. Corolla 25–35(–40) mm in diameter, orbicular, purplish. Anthers 7–9 mm, yellow. Berry 10–13 mm in diameter, globose, dry, yellow. *Locally naturalized in Greece.* [Gr.] *(Temperate South America.)*

11. Lycopersicon Miller[1]

Erect to decumbent, branched annuals with glandular, aromatic indumentum. Leaves alternate, interruptedly imparipinnate. Flowers in one or more extra-axillary, often leaf-opposed, monochasial cymes. Calyx 5-lobed almost to base, accrescent, deflexed in fruit. Corolla rotate; limb 5(–6)-lobed. Stamens inserted in throat of corolla; filaments very short; anthers connivent with connective prolonged into a prominent sterile beak. Stigma capitate. Fruit a berry, with 2–3 or more loculi; seeds elliptical, compressed, enclosed in mucilage.

1. L. esculentum Miller, *Gard. Dict.* ed. 8, no. 2 (1768). Plant villous. Leaves at least 20 cm; leaflets ovate to ovate-lanceolate, irregularly incised-serrate to pinnatisect, somewhat glaucous beneath. Cymes 3- to 20-flowered; peduncles simple or dichotomously branched; pedicels articulated near the middle, deflexed in fruit. Corolla up to 2·5 cm in diameter, yellow. Berry 2–10 cm in diameter, globose, ovoid or pyriform, often depressed or irregularly lobed and ridged, densely glandular-villous when young, glabrescent, red to pink or yellowish. 2n=24. *Cultivated for the edible fruits (tomatoes) on a field scale throughout S. and parts of C. & E. Europe; a frequent casual, but nowhere truly naturalized. (South and Central America; Mexico.)*

Variants with globose berries 1·5–2(–3) cm in diameter are sometimes cultivated, and have been called subsp. **galeni** (Miller) Luckwill, *Gen. Lycopers.* 23 (1943), or var. **cerasiforme** (Dunal) Alef.

12. Mandragora L.[1]

Perennial herbs with stout, erect, often bifid, occasionally anthropomorphic, fleshy tap-root; acaulescent or with very short stem. Leaves in a dense basal rosette, simple. Flowers

---

solitary, axillary. Calyx campanulate, 5-lobed, accrescent. Corolla campanulate, 5-lobed, plicate between the lobes, persistent. Stamens 5, subexserted, inserted in lower half of corolla-tube; filaments villous below; anthers dorsifixed. Ovary surrounded at base by glandular disc; stigma capitate. Fruit a berry, becoming unilocular by obliteration of septum.

Literature: J. G. Hawkes, *Bot. Jour. Linn. Soc.* **65**: 356 (1972).

Corolla not more than 2·5 cm, greenish-white, with narrowly triangular lobes; berry globose   **1. officinarum**
Corolla (2·5–)3–4 cm, violet, with wide triangular lobes; berry ellipsoid   **2. autumnalis**

1. M. officinarum L., *Sp. Pl.* 181 (1753) (*M. acaulis* Gaertner, *M. vernalis* Bertol.). Leaves petiolate, ovate to ovate-lanceolate, entire, undulate, sparsely villous on veins at least when young. Pedicels usually shorter than leaves. Calyx slightly accrescent, much shorter than berry. Corolla not more than 2·5 cm, greenish-white, with narrowly triangular lobes. Berry globose, yellow. ● *N. Italy and W. Jugoslavia.* It Ju.

2. M. autumnalis Bertol., *Elench. Pl. Hort. Bot. Bon.* 6 (1820) (*M. officinarum* L. pro parte). Like **1** but leaves subglabrous; calyx strongly accrescent, usually as long as or longer than berry; corolla (2·5–)3–4 cm, violet, with wide, triangular lobes; berry ellipsoid, yellow to orange. $2n = 84$. *Mediterranean region, C. & S. Portugal.* Cr Gr Hs It Lu Sa Si [?Bl].

13. Triguera Cav.[1]

Erect or ascending, annual or perennial herbs. Leaves alternate, simple. Flowers slightly zygomorphic, in extra-axillary pairs, the pairs distant or in a dense terminal spike. Calyx campanulate, deeply and unequally 5-fid, accrescent. Corolla infundibuliform; tube short, globose; lobes short, unequal. Stamens subequal, included, inserted at top of corolla-tube; anthers connivent; filaments connate at base, forming an urceolate nectary. Stigma capitate. Fruit a dry, indehiscent, globose berry, with evanescent septum; seeds 4–8, reniform, compressed.

Flower-pairs in a dense terminal spike; calyx and pedicels glabrous; corolla yellowish-green, each lobe with two purple lines   **2. osbeckii**
Flower-pairs distant; calyx and pedicels densely tomentose; corolla violet with blackish throat, the lobes without purple lines
                    1. ambrosiaca

1. T. ambrosiaca Cav., *Monad. Class. Diss. Dec. 2, App.*: 2 (1786). Annual, smelling of musk. Stems 15–45 cm, subsulcate, glabrous. Leaves ovate to obovate, entire to deeply sinuate-dentate, glabrous or sparsely villous, the lower petiolate, the upper sessile and slightly decurrent. Flower-pairs distant; pedicels 10–15 mm, nodding, arising from small cupuliform glands on a short common peduncle, tomentose. Calyx *c.* 1 cm, densely tomentose. Corolla 1·5–2·5 cm, violet with blackish throat; lobes rounded, mucronate. Ovary glabrous, half enclosed by nectary. Berry *c.* 1 cm in diameter, surrounded by the accrescent calyx; seeds 5 mm, brownish-black. *Roadsides and waste places.* S. Spain. Hs. (*N. Africa.*)

T. inodora Cav., *op. cit.* 3 (1786), from S. Spain (near Córdoba), appears to have been collected only once. It is like **1** but is scentless, has entire, lanceolate, glabrous, sessile leaves, glabrous pedicels and calyx, the corolla has a pale violet throat and muticous lobes with 5 yellowish-white lines, and the ovary is completely enclosed by the nectary.

2. T. osbeckii (L.) Willk. in Willk. & Lange, *Prodr. Fl. Hisp.* **2**: 524 (1870). Like **1** but perennial, scentless, with a stout, fleshy, branched rhizome; stems pubescent; leaves ovate-elliptical to oblong, pubescent, the lower bullate-undulate; flower-pairs in a dense, terminal spike; pedicels *c.* 5 mm, glabrous; calyx glabrous; corolla *c.* 3·5 cm, subrotate, greenish at base, yellowish at margin, the lobes with two purple lines; ovary villous. S. Spain (*near Cádiz*). Hs. (*N. Africa.*)

14. Datura L.[2]

Erect annuals. Leaves alternate, simple, shortly petiolate. Flowers actinomorphic, solitary, axillary. Calyx tubular, often 5-angled, 5-dentate, circumscissile after flowering, the lower part persistent. Corolla tubular or infundibuliform; limb 5(10)-lobed. Stamens 5, equal, included or exserted, inserted near base of corolla. Stigma 2-lobed. Ovary 2-locular, sometimes 4-locular at base. Fruit a capsule, dehiscing regularly by 4 valves or irregularly.

Literature: A. G. Avery, S. Satina & J. Rietsema, *Blakeslee: The Genus* Datura. New York. 1959. W. E. Safford, *Jour. Washington Acad. Sci.* **11**: 173–189 (1921).

1 Corolla at least 11 cm; capsule nodding, dehiscing irregularly
                    3. innoxia
1 Corolla not more than 10 cm; capsule erect, dehiscing regularly
  2 Calyx-teeth usually 5–10 mm, unequal; capsule with rather slender, ±equal spines not more than 15 mm, rarely smooth   **1. stramonium**
  2 Calyx-teeth 3–5 mm, subequal; capsule with stout, conical, unequal spines 10–30 mm, the upper longer than the lower
                    2. ferox

1. D. stramonium L., *Sp. Pl.* 179 (1753) (*D. tatula* L.). Plant 50–200 cm, glabrous to puberulent. Leaves 5–18(–21) × 4–15 cm, ovate to elliptical, acute, cuneate to subcordate at base, sinuate-dentate to -lobed. Calyx 30–50 mm, angled; teeth (3–)5–10 mm, unequal. Corolla 5–10 cm, infundibuliform, white or purple. Capsule (2·5–)3·5–7 × (2–)3–5 cm, ovoid, erect, dehiscing regularly, densely covered with more or less equal, rather slender spines up to 15 mm, rarely smooth. $2n = 24$. *Cultivated ground, waste places and other open habitats. Naturalized in most of Europe except the extreme north, but in some regions very irregular in its appearances.* [All except Bl Fa Fe Hb Is Rs (N) Sb.] (*America.*)

2. D. ferox L., *Demonstr. Pl.* 6 (1753). Plant 50–150 cm, usually glabrous. Leaves 5–14 × 4–13 cm, broadly ovate, coarsely sinuate-dentate. Calyx 25–40 mm, angled; teeth 3–5 mm, subequal. Corolla 4–6 cm, infundibuliform, white. Capsule 5–8 × 4–6 cm, ovoid, erect, dehiscing regularly, with stout, conical spines 10–30 mm, the upper longer than the lower. *Cultivated for ornament and locally naturalized in the Mediterranean region.* [Ga Hs It Si.] (*E. Asia.*)

3. D. innoxia Miller, *Gard. Dict.* ed. 8, no. 5 (1768). Plant 30–200 cm, pubescent. Leaves 5–16 × 3·5–11 cm, ovate, unequally truncate to cuneate at base, entire or sinuate. Calyx 60–95 mm; teeth 11–25 mm, unequal, linear-triangular, acute. Corolla 11–19 cm, tubular, white, sometimes with a violet tint. Capsule 5·5–6·5 cm in diameter, ovoid, nodding, dehiscing irregularly, with long, slender spines. *Locally naturalized in the Mediterranean region.* [Ga Hs It Lu Sa Si.] (*Central America.*)

D. metel L., *Sp. Pl.* 179 (1753), which is like **3** but has glabrous or subglabrous stem, leaves and calyx, a 10-lobed corolla and very short spines or tubercles on the capsule, is frequently culti-

---

[1] By J. G. Hawkes.   [2] By D. M. Moore.

vated for ornament, often as *flore pleno* with a white, yellow or purple corolla. It may be locally naturalized in the Mediterranean region but most reports of this species appear to refer to **3**.

15. Cestrum L.[1]

Fetid shrubs. Leaves alternate, simple. Flowers actinomorphic, in axillary or terminal leafy racemes. Calyx tubular, 5-dentate, persistent. Corolla tubular-infundibuliform. Stamens 5, included, inserted near base of corolla. Stigma 2-lobed. Fruit a 2-locular berry; seeds 2–11.

1. C. parqui L'Hér., *Stirp. Nov.* 73 (1788). Stems 100–300 cm, branched. Leaves 3–14 × 1–4 cm, linear- to elliptic-lanceolate, acute, cuneate at base, entire, glabrous, shortly petiolate. Inflorescence 6–13 cm; flowers sessile or very shortly pedicellate. Calyx 4–7 mm; teeth 1–1·5 mm, pubescent distally. Corolla 18–25 mm, greenish-yellow to yellow; lobes 3–6 mm, ovate to ovate-lanceolate, usually acute, pubescent towards margins. Berry 7–10 mm, ovoid, blackish; seeds reddish-brown. *Locally naturalized in the Mediterranean region.* [Gr Hs It Si.] (*Warm-temperate South America.*)

16. Nicotiana L.[1]

Annual to perennial herbs, rarely shrubs. Leaves alternate, simple. Flowers actinomorphic or zygomorphic, pedicellate, in a terminal panicle or false raceme. Calyx subglobose to tubular, sometimes campanulate or cupuliform, 5-dentate, persistent, often somewhat inflated in fruit. Corolla tubular or infundibuliform; limb 5-lobed (rarely subentire). Stamens 5, subequal or 1 shorter, included (rarely slightly exserted). Stigma capitate. Fruit a capsule, dehiscing by 2 or 4 distal valves.

Literature: T. H. Goodspeed, *The Genus* Nicotiana. Waltham, Mass. 1954.

1 Shrub; leaves glabrous, glaucous   **1. glauca**
1 Herb; leaves viscid-pubescent, not glaucous
   2 Petiole not winged   **2. rustica**
   2 Petiole winged or absent
     3 Inflorescence a panicle; stamens inserted in basal half of corolla-tube   **3. tabacum**
     3 Inflorescence a false raceme; stamens inserted in distal half of corolla-tube   **4. alata**

1. N. glauca R. C. Graham, *Edinb. New Philos. Jour.* **5**: 175 (1828). Glabrous shrub (1–)2–6(–10) m. Leaves 5–25 cm, elliptical to lanceolate or ovate, acute, glaucous; petiole not winged. Flowers numerous in a lax panicle. Calyx 10–15 mm, tubular; teeth triangular, acute, equal. Corolla (25–)30–40(–45) mm, tubular, yellow; limb 2–4 mm, with short lobes.

Stamens subequal, inserted in basal ¼ of corolla-tube. Capsule 7–10 mm, ellipsoid. *Cultivated for ornament and widely naturalized on rocks, walls and roadsides in the Mediterranean region and Portugal.* [?Bl Co Cr Ga Gr Hs It Lu Sa Si.] (*Argentina, Bolivia.*)

2. N. rustica L., *Sp. Pl.* 180 (1753). Annual; stem 0·5–1·5 m, viscid-pubescent. Leaves 10–15(–30) cm, ovate to elliptical or rarely suborbicular, acute, puberulent; petiole not winged. Flowers numerous in a compact to lax panicle. Calyx 8–15 mm, tubular to cupuliform; teeth triangular, acute, unequal. Corolla 12–17 mm, tubular-infundibuliform, greenish-yellow; limb 3–6 mm, with short, obtuse, apiculate lobes. Stamens unequal, inserted in basal ¼ of corolla-tube. Capsule 7–16 mm, ellipsoid-ovoid to subglobose. *Formerly widely cultivated for tobacco and locally naturalized; now largely replaced by* **3**. [Al Au Be ?Bl Bu Co ?Cr Cz ?Da Ga Ge Gr He Ho Hs Hu It Ju Rm Rs (C, W, K, E) ?Sa ?Si ?Tu.] (*North America.*)

3. N. tabacum L., *Sp. Pl.* 180 (1753) (*N. latissima* Miller). Viscid annual or short-lived perennial 1–3 m. Leaves up to 50 cm or more, ovate to elliptical or lanceolate, acuminate, decurrent, sessile or with a short, winged petiole. Flowers numerous in a much-branched panicle. Calyx 12–20(–25) mm, tubular to tubular-campanulate; teeth triangular, acuminate, unequal. Corolla (30–)35–55 mm, infundibuliform, pale greenish-cream, often pinkish distally; limb 10–15 mm, with acuminate lobes, sometimes subentire. Stamens unequal, 4 sometimes slightly exserted, inserted in basal ¼–⅓ of corolla-tube. Capsule 15–20 mm, ellipsoid to globose. *Cultivated for tobacco throughout most of Europe except the north, and often naturalized.* [Al Au Az Be ?Bl Bu Co Cr Cz Da Ga Ge Gr He Ho Hs Hu It Ju Lu Rm Rs (C, W, K, E) ?Sa ?Si Tu.] (*? N.E. Argentina, Bolivia.*)

A very polymorphic species containing a multitude of cultivated forms used in most modern tobaccos. No satisfactory formal infraspecific classification can be applied here.

4. N. alata Link & Otto, *Icon. Pl. Rar. Horti Bot. Berol.* 63 (1830). Viscid, short-lived perennial 0·6–1·5 m. Lower leaves 20–25 cm, in a sparse rosette, spathulate to elliptical, obtuse to subacute, somewhat amplexicaul, sessile or with short, winged petiole; cauline leaves decurrent, ovate to elliptic-ovate, sessile, auriculate. Flowers few in a lax false raceme. Calyx 15–25 mm, cupuliform to campanulate; teeth subulate-acicular, equal or unequal. Corolla 50–100 mm, tubular to somewhat infundibuliform, pale greenish; limb 15–25 mm, with 2 long and 3 shorter acute lobes. Stamens unequal, inserted in distal ¼–⅓ of corolla-tube. Capsule 12–17 mm, ovoid. *Cultivated for ornament and locally naturalized in C. Europe.* [Au Cz Rm.] (*N.E. Argentina to S.E. Brazil.*)

CLIII. BUDDLEJACEAE[2]

Trees, shrubs or rarely herbs with small, sessile or stipitate glands and usually stellate hairs. Leaves opposite, very rarely alternate, simple. Flowers actinomorphic or weakly zygomorphic, usually 4-merous. Disk small or absent. Stamens epipetalous, alternating with the corolla-lobes. Ovary superior or rarely semi-inferior, 2-locular; ovules numerous; placentation axile. Style 1. Fruit a capsule.

In addition to the species described below, **Buddleja alternifolia** Maxim., *Bull. Acad. Imp. Sci. Pétersb.* **26**: 494 (1880), and **B. globosa** J. Hope, *Verh. Holland. Maatsch. Wetensch. Haarlem* **20**(2): 417 (1782), are commonly cultivated and may become naturalized. The former, from China, has alternate leaves and lilac flowers in sessile axillary clusters borne on the previous year's wood. The latter, from Chile and Peru, has opposite leaves and orange flowers in long-pedunculate globose heads.

[1] By D. M. Moore.   [2] Edit. T. G. Tutin.

1. Buddleja L.[1]

Shrubs or small trees with interpetiolar stipules often reduced to a ridge. Hairs stellate. Inflorescence cymose, capitate or a long panicle. Calyx campanulate. Corolla with a nearly cylindrical tube and patent limb. Stamens included or very shortly exserted. Capsule septicidal.

1 Leaves densely tomentose beneath; corolla-tube straight
  2 Corolla-tube 4–5 times as long as calyx; stamens inserted
    about the middle of the tube   **1. davidii**
  2 Corolla-tube 2–3 times as long as calyx; stamens inserted near
    the top of the tube   **2. albiflora**
1 Leaves sparsely pubescent beneath; corolla-tube curved
  3 Racemes erect; calyx-teeth as wide as long   **4. lindleyana**
  3 Racemes pendent; calyx-teeth distinctly longer than wide
      3. japonica

1. B. davidii Franchet, *Nouv. Arch. Mus. Hist. Nat. Paris* ser. 2, **10**: 65 (1887). Deciduous shrub 1–5 m. Twigs bluntly angled or almost terete, tomentose. Leaves 10–25 cm, ovate-lanceolate to lanceolate, acuminate, serrate, thinly pubescent to almost glabrous above, tomentose beneath, shortly petiolate. Flowers in dense many-flowered cymes, forming a terminal panicle. Young twigs, peduncles, pedicels and calyx tomentose. Corolla pale lilac to deep violet with an orange ring at the mouth, stellate-pubescent and glandular outside; tube 4–5 times as long as calyx, straight; stamens inserted about the middle of the tube. *Cultivated for ornament and commonly naturalized in parts of W. & C. Europe.* [Au Be Br Ga Ge Hb He Ho Hs It.] (*China.*)

2. B. albiflora Hemsley, *Jour. Linn. Soc. London* (*Bot.*) **26**: 118 (1889). Like **1** but up to 10 m; twigs glabrescent; corolla pale lilac, the tube 2–3 times as long as calyx; stamens inserted near the top of the tube. *Sometimes cultivated for ornament and occasionally naturalized in France.* [Ga.] (*W. China.*)

3. B. japonica Hemsley, *op. cit.* 119 (1889). Deciduous shrub up to 1·5 m. Twigs sharply 4-angled, with winged angles, stellate-pubescent. Leaves 8–20 cm, ovate-lanceolate, acuminate, remotely toothed or entire, sparsely pubescent or almost glabrous beneath. Flowers in narrow, pendent panicles up to 25 cm. Calyx-teeth longer than wide. Corolla lilac, the tube curved and densely pubescent outside. *Sometimes cultivated for ornament and occasionally naturalized in France.* [Ga.] (*Japan.*)

4. B. lindleyana Fortune ex Lindley, *Bot. Reg.* 30 (*Misc.*): 25 (1844). Like **3** but twigs not or narrowly winged; flowers in erect panicles; calyx-teeth as wide as long; corolla purple, the tube floccose outside. *Sometimes cultivated for ornament and occasionally naturalized in France.* [Ga.] (*China.*)

CLIV. SCROPHULARIACEAE[2]

Herbs, rarely shrubs. Leaves simple (though sometimes deeply lobed), exstipulate, alternate or opposite (rarely all basal, or the uppermost whorled). Flowers zygomorphic (though sometimes only very slightly), usually in bracteate spikes or racemes, less often solitary in the leaf-axils or in cymes. Calyx usually 4- or 5-lobed, sometimes 2-lipped. Corolla (4)5(8)-lobed or 2-lipped. Stamens usually 2 or 4, rarely 3, 5 or 6–8; staminodes sometimes present. Ovary superior, 2-locular (sometimes 1-locular in upper part and rarely throughout); style single; stigma usually capitate. Fruit a capsule (rarely indehiscent); seeds usually numerous.

1 Fertile stamens 2 (staminodes sometimes present as well)
  2 Corolla not 2-lipped
    3 Herb, sometimes woody at the base; capsule loculicidal
      (sometimes also septicidal), ±compressed at right angles
      to the septum   **21. Veronica**
    3 Shrub; capsule septicidal only, compressed parallel to the
      septum   **22. Hebe**
  2 Corolla ± distinctly 2-lipped, with upper lip entire or 2-lobed
    and lower lip 3-lobed
    4 Flowers solitary in leaf-axils
      5 Bracteoles present; calyx-lobes unequal; capsule loculicidal
        and septicidal   **1. Gratiola**
      5 Bracteoles absent; calyx-lobes equal; capsule septicidal only
        2. Lindernia
    4 Flowers in terminal spikes or racemes
      6 Calyx membranous, obscurely toothed, usually split on
        lower side; capsule with 1–2 seeds   **19. Lagotis**
      6 Calyx herbaceous, distinctly 5-lobed, not split on lower side;
        capsule with numerous seeds
        7 Leaves alternate or basal   **20. Wulfenia**
        7 Leaves opposite   **23. Paederota**
1 Fertile stamens (3)4–5(–8)
  8 Corolla not 2-lipped
    9 Leaves opposite
      10 Stems procumbent; corolla lilac-pink   **3. Bacopa**
      10 Stems erect; corolla yellow
        11 Leaves lyrate-pinnatifid; filaments with violet hairs
          7. Verbascum
        11 Leaves biserrate; filaments glabrous or with colourless
          hairs   **8. Scrophularia**
    9 Leaves alternate or basal
      12 Leaves all basal   **4. Limosella**
      12 Cauline leaves present
        13 Flowers in axils of foliage-leaves   **24. Sibthorpia**
        13 Flowers in terminal spikes, racemes or panicles
          14 Corolla-tube very short; stamens exserted   **7. Verbascum**
          14 Corolla-tube at least half as long as diameter of limb;
            stamens included
            15 Flowers in elongate racemes; corolla 9–15 mm, with ±
              porrect lobes; anther-lobes distinct   **17. Digitalis**
            15 Flowers in subcorymbose racemes; corolla 6–9 mm, with
              patent lobes; anther-lobes confluent   **18. Erinus**
  8 Corolla ± distinctly 2-lipped, with upper lip entire or 2-lobed
    and lower lip 3-lobed
    16 Corolla-tube with a spur, pouch or gibbosity at the base on
      the abaxial side
      17 Capsule dehiscing by longitudinal slits
        18 Leaves palmately veined, reniform to suborbicular,
          petiolate   **15. Cymbalaria**
        18 Leaves not palmately veined, usually narrow, ± sessile
          14. Linaria
      17 Capsule dehiscing by 2–3 apical pores
        19 Mouth of corolla-tube not closed by a palate
          20 Basal leaves alternate, forming a rosette; anther-lobes
            confluent; seeds without longitudinal ribs **9. Anarrhinum**
          20 Lower leaves usually opposite, not forming a definite
            rosette; anther-lobes distinct; seeds with longitudinal
            ribs   **13. Chaenorhinum**
        19 Mouth of corolla-tube ± closed by a raised palate at base
          of lower lip
          21 Corolla-tube with a narrow spur at the base   **16. Kickxia**

---

[1] By T. G. Tutin.   [2] Edit. D. A. Webb.

21 Corolla-tube with a broad pouch at the base
22 Annual; calyx unequally lobed, longer than corolla-tube **12. Misopates**
22 Perennial; calyx ± equally lobed, shorter than corolla-tube
23 Leaves palmately veined; loculi of capsule equal **11. Asarina**
23 Leaves not palmately veined; loculi of capsule unequal **10. Antirrhinum**
16 Corolla-tube symmetrical at the base, without spur, pouch or gibbosity
24 Parasitic, without chlorophyll; leaves reduced to fleshy, whitish or brownish scales **39. Lathraea**
24 Holophytic or hemiparasitic; chlorophyll present (rarely masked by anthocyanin); leaves not markedly fleshy
25 Capsule septicidal
26 Plant eglandular; flowers solitary in leaf-axils **2. Lindernia**
26 Plant glandular, at least on pedicels or corolla; flowers usually in spikes, cymes or panicles
27 Corolla-tube cylindrical; flowers in spikes **25. Lafuentea**
27 Corolla-tube subglobose; flowers pedicellate, usually in cymes **8. Scrophularia**
25 Capsule loculicidal or fruit indehiscent
28 Pedicel bearing 2 bracteoles
29 Flowers in terminal racemes; calyx 2-lipped; corolla purple **37. Siphonostegia**
29 Flowers in axils of lower leaves; calyx equally 5-lobed; corolla pale yellow **38. Cymbaria**
28 Pedicel without bracteoles
30 Anthers with one lobe pendent, the other medifixed (N. & E. Russia) **26. Castilleja**
30 Anthers with lobes ± parallel
31 Calyx 5-toothed or 2-lipped or obscurely lobed, not regularly 4-lobed
32 Upper leaves alternate
33 Leaves entire or remotely dentate; upper lip of corolla flat **6. Dodartia**
33 Leaves pinnatifid or pinnatisect (rarely crenate-serrate); upper lip of corolla galeate, laterally compressed **34. Pedicularis**
32 Upper leaves opposite or whorled
34 Leaves pinnatisect **34. Pedicularis**
34 Leaves entire to incise-dentate
35 Fruit 1-seeded **28. Tozzia**
35 Fruit a capsule with numerous seeds
36 Calyx-tube keeled; upper lip of corolla flat **5. Mimulus**
36 Calyx-tube not keeled; upper lip of corolla galeate or cucullate
37 Plant glandular-pubescent; flowers sessile; upper lip of corolla not rostrate **33. Bellardia**
37 Plant subglabrous; flowers distinctly pedicellate; upper lip of corolla conspicuously rostrate **36. Rhynchocorys**
31 Calyx distinctly and ± regularly 4-lobed
38 Capsule with 1–4 seeds
39 Annual; upper lip of corolla cucullate; capsule 4-seeded, longer than wide, usually mucronate or acuminate **27. Melampyrum**
39 Rhizomatous perennial; upper lip of corolla flat; capsule 1-seeded, globose **28. Tozzia**
38 Capsule with more than 4 seeds
40 Calyx-tube subglobose, inflated, especially in fruit; anther-lobes not mucronate; seeds discoid, with marginal wing **35. Rhinanthus**
40 Calyx-tube usually cylindrical or campanulate, scarcely inflated; anther-lobes mucronate; seeds not discoid
41 Lobes of upper lip of corolla recurved; lobes of lower lip conspicuously emarginate; one anther-lobe with a longer mucro than the other **29. Euphrasia**
41 Upper lip of corolla entire, or with flat lobes; lobes of lower lip entire or very slightly emarginate; anther-lobes equally mucronate
42 Seeds c. 0·5 mm, smooth or finely reticulate, not winged **32. Parentucellia**
42 Seeds 1–2 mm, winged or longitudinally ridged or striate
43 Rhizomatous perennial herb; seeds with 1 or more membranous, longitudinal wings **31. Bartsia**
43 Annual herb or dwarf shrub; seeds with longitudinal striae or low ridges
44 Inflorescence secund; leaves usually entire or subentire **30. Odontites**
44 Inflorescence not secund; leaves coarsely serrate **33. Bellardia**

1. Gratiola L.[1]

Perennial herbs; leaves opposite. Flowers solitary in the leaf-axils; pedicel bearing 2 bracteoles just below the calyx. Calyx deeply and unequally 5-lobed. Corolla with wide tube and somewhat 2-lipped limb; upper lip flat, 2-lobed, the lower slightly larger, 3-lobed. Stamens included, the 2 upper lateral fertile, the other 3 reduced to staminodes. Anther-lobes transverse, surrounded by a membranous expansion of the connective. Stigma 2-lobed. Capsule 4-valved (septicidal and loculicidal); seeds numerous.

1 Plant glabrous; pedicels shorter than subtending leaves **1. officinalis**
1 Plant glandular-puberulent at least above; pedicels as long as subtending leaves
2 Leaves linear; corolla 12–18 mm; lower staminodes conspicuous **2. linifolia**
2 Leaves lanceolate; corolla 8–12 mm; lower staminodes minute **3. neglecta**

1. G. officinalis L., *Sp. Pl.* 17 (1753). Glabrous. Stems 10–50(–80) cm, hollow, 4-angled above, erect from a creeping and rooting base. Leaves 20–50 mm, linear to lanceolate, serrate to subentire, semiamplexicaul, glandular-punctate. Pedicels shorter than subtending leaves. Corolla 10–18 mm, white, usually veined and tinged with purplish-red. Upper staminode minute, the 2 lower long and slender. *Ditches, river-banks and wet meadows.* $2n=32$. *Europe, northwards to the Netherlands, Estonia and S. Ural.* Al Au Be Bu Cz Ga Ge Gr He Ho Hs Hu It Ju Lu Po Rm Rs (B, C, W, K, E) Sa Tu.

2. G. linifolia Vahl, *Enum. Pl.* 1: 89 (1804). Like 1 but glandular-puberulent at least above; stems terete; leaves linear, more or less entire; pedicels at least as long as subtending leaves. *Ditches, river-banks and marshes.* ● *Portugal, S.W. Spain.* Hs Lu.

Plants intermediate between **1** and **2** occur in Portugal and N. Spain. They have been named **G. officinalis** subsp. **broteri** Nyman, *Consp.* 536 (1881), but are probably better regarded as hybrids.

3. G. neglecta Torrey, *Cat. Pl. New York* 89 (1819). Like 1 but glandular-puberulent above; stems 5–30 cm, not creeping at base; leaves lanceolate; pedicels at least as long as subtending leaves; corolla 8–12 mm; all staminodes minute. *Naturalized in flooded gravel-pits in E. France (Haut-Rhin).* [Ga.] (*E. North America.*)

2. Lindernia All.[2]

Glabrous annuals; leaves opposite. Flowers solitary in the leaf-axils. Calyx equally and deeply 5-lobed. Corolla with narrowly campanulate tube and 2-lipped limb; upper lip small, flat,

[1] By D. A. Webb.   [2] By D. A. Webb and D. Philcox.

erect, 2-lobed, the lower larger, patent, 3-lobed. Stamens 4, or 2 fertile with 2 staminodes, included; anther-lobes divergent. Stigma 2-lobed. Capsule septicidal; seeds numerous.

Fertile stamens 4; pedicels usually exceeding subtending leaf; flowers mostly cleistogamous, with corolla not exceeding calyx
1. procumbens

Fertile stamens 2, staminodes 2; pedicels usually shorter than subtending leaf; flowers mostly chasmogamous, with corolla distinctly exceeding calyx
2. dubia

1. L. procumbens (Krocker) Philcox, *Taxon* **14**: 30 (1965) (*L. pyxidaria* L. pro parte, *L. gratioloides* sensu Hayek, non Lloyd). Stems 5–18 cm, procumbent to ascending. Leaves up to 20 × 10 mm, elliptical to oblong, obtuse, entire or obscurely crenate-serrate. Pedicels 8–20 mm, slender, usually exceeding subtending leaf. Flowers usually cleistogamous, with corolla 2·5–4 mm, closed at the mouth and persistent in fruit, rarely chasmogamous with corolla 5–6 mm, open at the mouth and deciduous. Calyx 3–4 mm; lobes linear-oblong. Corolla pale pink. Stamens 4, all fertile, included. Capsule 3–5 mm, ellipsoid. *Wet, muddy or sandy places. C. & S.E. Europe, extending locally to N.W. Portugal, N. Italy and S.C. Russia.* Au Bu Cz Ga Ge He Hs Hu It Ju Lu Po Rm Rs (C, W, E).

2. L. dubia (L.) Pennell, *Scroph. E. Temp. N. Amer.* 141 (1935) (*L. gratioloides* Lloyd). Like **1** but stems often more or less erect; leaves up to 30 mm, ovate-oblong, often more distinctly crenate-serrate; pedicels 6–15 mm, usually stouter, usually shorter than subtending leaf; flowers mostly chasmogamous (a few of the later sometimes cleistogamous); calyx 5 mm; corolla 7–8 mm, lilac; fertile stamens 2; staminodes 2, spurred at the base. *Muddy or sandy river-banks, and in rice-fields. Naturalized in S.W. Europe, extending to N.W. France and N. Italy.* [Ga Hs It Lu.] (*E. North America.*)

3. Bacopa Aublet[1]

Glabrous perennial herbs; leaves opposite. Flowers solitary in the leaf-axils. Calyx deeply 5-lobed; lobes unequal in width. Corolla campanulate, more or less equally 5-lobed. Stamens 4, equal; anther-lobes parallel, contiguous. Stigma subentire. Capsule 4-valved (loculicidal and septicidal); seeds numerous.

1. B. monnieri (L.) Pennell, *Proc. Acad. Nat. Sci. Philad.* **98**: 94 (1946). Stems up to 60 cm, procumbent and rooting at the nodes. Leaves 7–20 mm, rather succulent, sessile, spathulate to cuneate-obcordate, obtuse, entire or obscurely crenate. Pedicels up to 15 mm, with 2 linear-lanceolate bracteoles *c.* 2 mm long just below the flower. Calyx 5–6 mm. Corolla 8–10 mm, pale lilac-pink. Capsule 5–8 mm. *River-banks and other wet places near the sea. Locally naturalized in N. Portugal and N.W. Spain.* [Hs Lu.] (*Widespread in tropics and subtropics.*)

4. Limosella L.[2]

Small, glabrous, annual or perennial herbs, usually stoloniferous; leaves simple, all basal, subtending 1-flowered scapes. Calyx campanulate, shortly and equally (4)5-lobed. Corolla campanulate, (4)5-lobed. Stamens 4, equal. Ovary with septum incomplete at top; stigma entire. Capsule septicidal; seeds numerous.

1 Larger leaves with elliptic-spathulate lamina **1. aquatica**
1 Leaves all reduced to linear-subulate phyllodes
  2 Corolla-tube shorter than calyx; seeds narrowly ellipsoid, 2–2½ times as long as wide **2. tenella**
  2 Corolla-tube longer than calyx; seeds broadly ellipsoid, *c.* 1½ times as long as wide **3. australis**

1. L. aquatica L., *Sp. Pl.* 631 (1753). Annual, sometimes perennating by stolons. Summer leaves with linear-oblong to elliptic-spathulate lamina 8–20 × 2–6(–10) mm and petiole 20–120 mm, usually preceded by spring leaves with very narrow lamina or reduced to linear-subulate phyllodes. Pedicels 3–20(–30) mm, erect in flower, usually deflexed in fruit. Corolla 3 × 2 mm, white, the lobes variably tinged or striped with pinkish-purple; tube about equalling lobes and shorter than calyx. Seeds narrowly ellipsoid (2–2½ times as long as wide). 2*n*=40. *Ditches, lake-shores and places subject to periodical flooding. Most of Europe except the Mediterranean region, but rare and local in many districts.* Au Be Br Cz Da Fe Ga Ge Hb He Ho Hs Hu Is It Ju Lu No Po Rm Rs (N, B, C, W, E) Su.

2. L. tenella Quézel & Contandr., *Candollea* **20**: 73 (1965). Like **1** but all leaves reduced to linear-subulate phyllodes, the largest *c.* 20 × 1 mm; pedicels erect, nearly equalling the leaves; flowers usually 4-merous; corolla *c.* 1·5 mm. 2*n*=40. *Muddy lake-shores.* ● *N.W. Greece (Timfi Oros).* Gr.

3. L. australis R. Br., *Prodr. Fl. Nov. Holl.* 443 (1810). Like **1** but with all leaves reduced to linear-subulate phyllodes, the largest 30–40 × 1·5 mm; corolla 3·5–4 × 3 mm, with white lobes and orange tube exceeding the calyx; seeds broadly ellipsoid (*c.* 1½ times as long as wide). 2*n*=20. *Places subject to periodical flooding. Britain (Wales).* Br. (*America, Africa, Australia.*)

A hybrid between **1** and **3** has been recorded from the European habitat of **3**.

5. Mimulus L.[2]

Perennial herbs; leaves opposite. Flowers solitary in the leaf-axils or in lax, leafy, terminal racemes. Calyx campanulate, shortly 5-toothed; tube keeled and plicate. Corolla with cylindrical-campanulate tube and 2-lipped limb; upper lip flat, erect, 2-lobed; lower lip 3-lobed. Stamens 4, didynamous, included; anther-lobes transverse. Stigma shortly 2-lobed. Capsule loculicidal; seeds numerous.

Literature: A. L. Grant, *Ann. Missouri Bot. Gard.* **11**: 99–388 (1924).

1 Calyx-teeth ± equal; corolla *c.* 2 × 1 cm; seeds tuberculate **3. moschatus**
1 Calyx-teeth conspicuously unequal; corolla *c.* 4 × 3 cm; seeds finely striate
  2 Inflorescence glandular-pubescent; pedicels 12–25 mm **1. guttatus**
  2 Inflorescence glabrous; pedicels 25–70 mm **2. luteus**

1. M. guttatus DC., *Cat. Pl. Horti Monsp.* 127 (1813) (*M. luteus* auct., non L.). Glabrous except for glandular-pubescent inflorescence. Stems up to 50 cm, hollow, stout, ascending to erect. Leaves broadly ovate, acute to obtuse, irregularly dentate, the lowest petiolate, the others sessile. Racemes 3- to 7-flowered; pedicels 12–25 mm. Calyx 15–20 mm, the upper tooth longer and wider than the others. Corolla *c.* 4 × 3 cm, bright yellow, usually spotted with red near the throat; mouth of tube nearly closed by 2 hairy ridges on lower lip. Seeds finely striate. 2*n*=48. *Cultivated for ornament and naturalized, chiefly by streams, over a large part of Europe.* [Au Be Br Bu Cz Da Fe Ga Ge Hb He Ho Ju No Po Rm Rs (B, C) Su.] (*W. North America.*)

2. M. luteus L., *Sp. Pl.* ed. 2, 884 (1763) (*M. smithii* auct., vix Paxton). Like **1** but entirely glabrous; stems often decumbent; leaves narrower, more acuminate, more regularly dentate; pedicels 25–70 mm; corolla usually with large red patches on the limb and with the mouth of the tube open. *Naturalized locally*

---

¹ By D. Philcox.   ² By D. A. Webb.

by streams in Scotland, and perhaps elsewhere in N.W. Europe. [Br.] (*Chile.*)

Hybrids between **1** and **2**, and other more complex hybrids involving *M. cupreus* Regel, are cultivated and are locally naturalized in N.W. Europe. Many of the records for **2** are, in fact, referable to such hybrids; they may be distinguished from **2** by the very short but dense glandular-puberulence of the inflorescence.

3. M. moschatus Douglas ex Lindley, *Bot. Reg.* **13**: t. 1118 (1828). Whole plant viscid-pubescent. Stems up to 35 cm, decumbent to ascending. Leaves ovate, denticulate to subentire, shortly petiolate. Flowers in most of the leaf-axils; pedicels 10–15 mm. Calyx 8–10 mm, with subequal teeth. Corolla *c.* 2 × 1 cm, rather pale yellow, sometimes striped with red at the throat; mouth of tube open. Seeds tuberculate. *Cultivated for ornament, and formerly for its scent; locally naturalized in damp or shady places in N., C. & W. Europe.* [Az Be Br Cz Ga Ge Ho It Lu Rm Rs (B, C).] (*W. North America.*)

6. Dodartia L.[1]

Perennial herbs; leaves opposite below, alternate above. Flowers in terminal racemes. Calyx equally 5-lobed. Corolla with cylindrical tube and 2-lipped limb; upper lip erect, flat, emarginate; lower lip 3-lobed. Stamens 4, didynamous, slightly exserted. Stigma 2-lobed. Capsule loculicidal; seeds numerous.

1. D. orientalis L., *Sp. Pl.* 633 (1753). Glabrous. Stems 15–50 cm, freely branched from the base; branches usually suberect, terete, junciform. Leaves up to 4 cm but mostly much smaller, oblong-lanceolate to linear, acute, sessile, entire or remotely dentate, the lower ones caducous. Racemes lax, with 3–8 flowers; pedicels very short. Calyx 4–5 mm, campanulate; teeth deltate, acute. Corolla 15–20(–25) mm, dark violet-purple; lower lip much longer than upper. Capsule globose, about equalling calyx. Seeds smooth. *Dry steppes and saline soils; also as a ruderal. S.E. Russia, W. Kazakhstan.* Rs (E) [Rs (W)]. (*S.W. & C. Asia.*)

7. Verbascum L.[2]

(incl. *Celsia* L.)

Herbs, very rarely small shrubs. Leaves simple, alternate (very rarely opposite), the basal forming a rosette. Flowers in terminal racemes, spikes or panicles. Calyx equally 5-lobed. Corolla usually yellow, with a very short tube and a rotate, nearly equally 5-lobed limb. Stamens 4 or 5 (sometimes 4 fertile and 1 staminode); filaments usually villous. Capsule septicidal; seeds numerous.

Two types of anther are found in the genus. Those of the 2 or 3 posterior (upper) stamens are always reniform and transversely medifixed; those of the 2 anterior (lower) stamens may be similar, or may be elongate, longitudinally inserted, and decurrent on the filament. These two types are referred to below as *reniform* and *decurrent* respectively. In a few species an intermediate condition is found, in which the anther is obliquely inserted.

Hybrids are very often found where two or more species grow together. They are intermediate between their parents in most morphological features, and appear to be always sterile. The species to which hybrid offspring have been most frequently attributed are **5, 27, 34, 58, 72, 82** and **85**.

Most of the species grow in dry, sunny, semi-open habitats such as roadsides, stony slopes, dry grassland, forest-clearings, etc., and this type of habitat may be assumed unless another is stated.

Literature: S. Murbeck, *Lunds Univ. Årsskr.* nov. ser., **22(1)**: 1–239 (1925); **29(2)**: 1–630 (1933); **32(1)**: 1–46 (1936); **35(1)**: 1–71 (1939). A. Huber-Morath & K. H. Rechinger, *Mitt. Thür. Bot. Ges.* **2(1)**: 42–55 (1960).

```
1  Spiny shrub                                              1. spinosum
1  Unarmed herb
  2  Plant ±acaulescent (stems not more than 3 cm); pedicels
     25–90 mm                                               23. acaule
  2  Plant caulescent, with stems at least 10 cm; pedicels 2–25 mm
    3  Each bract with a single flower in its axil
      4  Flowers subtended by bracteoles as well as by a bract
        5  Anthers all reniform
          6  Basal leaves coarsely incise-dentate to pinnatisect;
             filament-hairs white                           3. purpureum
          6  Basal leaves crenate; filament-hairs violet
                                                            4. adrianopolitanum
        5  Anthers of lower stamens decurrent or obliquely inserted
          7  Plant glandular-pubescent above                7. virgatum
          7  Plant eglandular                               2. ovalifolium
      4  Bracteoles absent
        8  Anthers all reniform
          9  Pedicels shorter than the subtending bract
            10  Annual; leaves mostly pinnatisect          26. orientale
            10  Perennial; leaves crenate to somewhat pinnatifid
                                                           24. pyramidatum
          9  Pedicels longer than the subtending bract
            11  Pedicels 2–5 mm                            25. zuccarinii
            11  Pedicels at least 7 mm
              12  Stems not more than 25 cm; largest basal leaves not
                  more than 2·5 cm wide; stamens 4          22. cylleneum
              12  Stems 30–100 cm; largest basal leaves at least 2·5 cm
                  wide; stamens usually 5
                13  Basal leaves deeply crenate or weakly pinnatifid,
                    crispate-villous                        21. xanthophoeniceum
                13  Basal leaves entire or slightly sinuate or weakly
                    crenate, glabrous or sparsely pubescent
                                                            20. phoeniceum
        8  Anthers of lower stamens decurrent or obliquely inserted
          14  Pedicels 2–10 mm
            15  Calyx 8–15 mm, with serrate lobes; capsule 9–15 mm
                                                            14. creticum
            15  Calyx 3–9 mm, with entire lobes; capsule 5–9 mm
              16  Axis of inflorescence glabrous; bracts 2–3 mm
                                                            19. hervieri
              16  Axis of inflorescence pubescent; bracts usually more
                  than 3 mm
                17  Stamens 4
                  18  Basal leaves entire or crenate       17. bugulifolium
                  18  Basal leaves lyrate to pinnatisect
                    19  Calyx 6–9 mm; style 22–30 mm        13. laciniatum
                    19  Calyx 3–4 mm; style 5–10 mm
                                                    (8–11). daenzeri group
                17  Stamens 5
                  20  Basal leaves ovate to ovate-oblong; petiole at least
                      ⅓ as long as lamina                  18. spectabile
                  20  Basal leaves oblong to lanceolate; petiole short or
                      absent
                    21  Corolla 15–18 mm in diameter; bracts 3–5 mm
                                                            6. siculum
                    21  Corolla 20–40 mm in diameter; bracts 7–22 mm
                      22  Plant glabrous below, ±glandular-pubescent
                          above; pedicel usually longer than calyx
                                                            5. blattaria
                      22  Plant usually glandular-pubescent throughout;
                          pedicel shorter than calyx        7. virgatum
          14  Pedicels at least 12 mm
            23  Style more than 15 mm
```

---

[1] By D. A. Webb.　　　　[2] By I. K. Ferguson.

24 Calyx *c.* 4 mm; corolla 30–35 mm in diameter
                                  12. barnadesii
24 Calyx 6–9 mm; corolla 35–50 mm in diameter
                                  13. laciniatum
23 Style less than 15 mm
25 Leaves tomentose or densely villous, at least beneath
                                  15. arcturus
25 Leaves glabrous, pubescent or sparsely villous
  26 Calyx 4–8 mm; fertile stamens 5   **5. blattaria**
  26 Calyx 3–5 mm; fertile stamens 4
    27 Basal leaves 5–11 cm wide; cauline leaves opposite
                              16. levanticum
    27 Basal leaves 1·5–5 cm wide; cauline leaves alternate
                      (8–11). daenzeri group
3 At least the lower bracts each with a cluster of several flowers in its axil
28 Anthers of lower stamens decurrent or obliquely inserted
29 Axis of inflorescence conspicuously glandular
30 Hairs of the inflorescence and cauline leaves simple
31 Bracts less than 7 mm, equalling the longer pedicels
                                43. anisophyllum
31 Bracts more than 7 mm, exceeding the longer pedicels
                                7. virgatum
30 Hairs of the inflorescence and cauline leaves branched
32 Upper cauline leaves broadly ovate to orbicular, usually crowded; inflorescence ±floccose with coarse white hairs interspersed with sessile glands
33 Inflorescence densely white-floccose; lower filaments 1½–2 times as long as the upper **44. eriophorum**
33 Inflorescence densely glandular but rather sparsely floccose; lower filaments equal in length to the upper **45. baldaccii**
32 Upper cauline leaves triangular, lanceolate or ovate-lanceolate, few and small; inflorescence pubescent or tomentose with fine, short, often yellow hairs, later glabrescent
34 Upper cauline leaves deltate-cordate **48. pelium**
34 Upper cauline leaves lanceolate to ovate-oblong, not cordate **46. epixanthinum**
29 Axis of inflorescence ±eglandular
35 Upper cauline leaves distinctly decurrent
36 Stigma capitate
37 Upper filaments hairy **34. thapsus**
37 Filaments all glabrous **35. litigiosum**
36 Stigma spathulate, decurrent on style
38 Bracts glabrous and scarious in fruit **31. vandasii**
38 Bracts persistently tomentose, not scarious in fruit
39 Indumentum very dense, hard and rough **33. macrurum**
39 Indumentum ±sparse, often somewhat floccose, soft and smooth
40 Hairs on inflorescence yellowish, crispate
                                28. densiflorum
40 Hairs on inflorescence white, sericeous **32. niveum**
35 Upper cauline leaves not or scarcely decurrent
41 Lower filaments glabrous
42 Longer pedicels at least 10 mm
43 Bracts ovate-lanceolate **27. phlomoides**
43 Bracts linear to linear-lanceolate **36. longifolium**
42 Longer pedicels not more than 9 mm
44 Filament-hairs white or yellow **27. phlomoides**
44 Filament-hairs violet
45 Petiole of basal leaves 2–5 cm; longer pedicels 1–4 mm; capsule 5–6 mm **47. euboicum**
45 Petiole of basal leaves 5–12 cm; longer pedicels 4–9 mm; capsule 6–10 mm **39. nevadense**
41 Lower filaments villous, at least in part
46 Bracts suborbicular-deltate, glabrescent, membranous
                              49. dieckianum
46 Bracts linear to ovate, usually with persistent tomentum, herbaceous
47 Mature leaves whitish-tomentose above; indumentum of inflorescence distinctly floccose
48 Longer pedicels 1–5 mm

49 Filament-hairs violet **37. boerhavii**
49 Filament-hairs white **40. lagurus**
48 Longer pedicels 5–16 mm
50 Inflorescence simple or with a few short branches
                              36. longifolium
50 Inflorescence a pyramidal panicle with numerous branches **38. argenteum**
47 Mature leaves ±green above; indumentum of inflorescence ±persistent
51 Bracts linear-lanceolate
52 Leaves entire or obscurely crenate; longer pedicels 10–16 mm **36. longifolium**
52 Leaves distinctly crenate; longer pedicels 2–9 mm
53 Corolla with pellucid glands; anthers of lower stamens 1·5–2·5 mm, shortly decurrent or obliquely inserted **42. nicolai**
53 Corolla without pellucid glands; anthers of lower stamens 3·5–4 mm, decurrent **41. georgicum**
51 Bracts ovate to ovate-lanceolate
54 Inflorescence branched **29. samniticum**
54 Inflorescence simple
55 Indumentum of calyx and bracts persistent, yellowish-grey, harshly tomentose; anthers of lower stamens 2–2·3 mm, shortly decurrent
                              30. guicciardii
55 Indumentum of calyx and bracts ±floccose, white, softly tomentose; anthers of lower stamens 3·5–4 mm, long-decurrent **41. georgicum**
28 Anthers all reniform
56 Basal leaves distinctly lobed
57 Basal leaves lobed for *c.* 85% of distance to midrib
58 Inflorescence usually simple; lobes of basal leaves entire or dentate **59. halacsyanum**
58 Inflorescence branched; lobes of basal leaves pinnately lobed **60. pinnatifidum**
57 Basal leaves lobed for not more than 60% of distance to midrib
59 Bracts 1–2 mm; axis of inflorescence glabrous
60 Bracteoles present **66. nobile**
60 Bracteoles absent
61 Leaves 5–8 cm wide; capsule subglobose
                              63. pseudonobile
61 Leaves 1·5–4 cm wide; capsule cylindrical
                              62. cylindrocarpum
59 Bracts 3–15 mm; axis of inflorescence ±hairy
62 Calyx 6–12 mm **56. undulatum**
62 Calyx 2–6 mm
63 Indumentum usually hard, yellowish-grey; filament-hairs violet **58. sinuatum**
63 Leaves glabrescent to softly tomentose; filament-hairs white or yellow
64 Leaves more or less glabrescent above, ±persistently and softly yellowish-grey-tomentose beneath **81. banaticum**
64 Leaves softly whitish-grey-tomentose on both surfaces **61. leucophyllum**
56 Basal leaves not lobed
65 Inflorescence, bracts and calyx glandular-pubescent
66 Filament-hairs violet
67 Basal leaves 15–30 cm, oblong-lanceolate to ovate-elliptical, subentire **46. epixanthinum**
67 Basal leaves 4–15 cm, ovate, crenate **77. durmitoreum**
66 Filament-hairs white or yellow
68 Glands on inflorescence black; corolla without pellucid glands **75. davidoffii**
68 Glands on inflorescence pale; corolla with pellucid glands
69 Glandular hairs on inflorescence partly long, partly short; calyx-lobes oblong to oblong-lanceolate; capsule 4–6 mm **74. glandulosum**
69 Glandular hairs on inflorescence all very short; calyx-lobes linear to linear-oblong; capsule 5–8 mm **76. jankaeanum**

65 Inflorescence, bracts and calyx eglandular
70 Basal and lower cauline leaves cordate, truncate or
   very shortly cuneate at base
71 Bracts and calyx ±glabrous
72 Inflorescence branched   **87. glabratum**
72 Inflorescence simple   **86. lanatum**
71 Bracts and calyx ±densely hairy
73 Calyx 5–8 mm; capsule 7–10 mm   **78. rotundifolium**
73 Calyx 2·5–5 mm; capsule 2·5–6 mm
74 Filament-hairs white or yellow
75 Leaves ovate, crenate to subentire, rounded to
   cordate at base; stem ±white-tomentose
        80. delphicum
75 Leaves oblong-obovate, deeply and irregularly
   crenate or lobed, shortly cuneate at base; stem
   ±glabrescent   **81. banaticum**
74 Filament-hairs violet
76 Basal leaves truncate or very shortly cuneate at
   base; pedicels about as long as calyx   **83. chaixii**
76 Basal leaves cordate; pedicels 2–3 times as long as
   calyx   **85. nigrum**
70 Basal and lower cauline leaves tapered gradually to the
   petiole
77 Leaves densely and persistently whitish- or greyish-
   tomentose on both surfaces
78 Stem and inflorescence-axis densely and persistently
   tomentose
79 Pedicels longer than calyx
80 Bracts linear; inflorescence simple or sparingly
   branched   **36. longifolium**
80 Bracts ovate-lanceolate or lanceolate; inflorescence freely branched
81 Filament-hairs violet   **54. adeliae**
81 Filament-hairs white or yellow   **53. speciosum**
79 Pedicels not longer than calyx
82 Flowers sessile
83 Inflorescence dense, cylindrical; pedicels adnate
   to axis of inflorescence
84 Calyx 5–7 mm; lobes broadly ovate and shortly
   cuspidate   **51. botuliforme**
84 Calyx 9–11 mm; lobes linear-lanceolate
        50. macedonicum
83 Inflorescence lax, with dense but widely spaced
   clusters of flowers; pedicels absent or very
   short, not adnate to axis
85 Calyx 3–5 mm   **71. mucronatum**
85 Calyx 6–12 mm   **56. undulatum**
82 Pedicels 2–8 mm
86 Filament-hairs violet
87 Bracts 5–7 mm, ovate to deltate; capsule
   5–6 mm   **52. dentifolium**
87 Bracts 8–15 mm, lanceolate, acuminate; capsule
   7–10 mm   **78. rotundifolium**
86 Filament-hairs white or yellow
88 Bracts broadly ovate, ovate-deltate to ovate-
   cordate
89 Calyx-lobes triangular-ovate   **55. lasianthum**
89 Calyx-lobes oblong-lanceolate to lanceolate
90 Indumentum floccose; inflorescence ±dense
        49. dieckianum
90 Indumentum ±persistent; inflorescence lax
        56. undulatum
88 Bracts ovate-lanceolate, lanceolate to linear
91 Bracts linear   **72. pulverulentum**
91 Bracts ovate-lanceolate to lanceolate
92 Inflorescence lax, with flower-clusters widely
   spaced   **71. mucronatum**
92 Inflorescence dense, with flower-clusters contiguous or overlapping
93 Calyx-lobes linear; corolla 15–20 mm in
   diameter   **79. reiseri**
93 Calyx-lobes ovate-lanceolate; corolla 25–40
   mm in diameter   **30. guicciardii**

78 Stem and inflorescence-axis ±glabrous when
   mature (sometimes floccose when young)
94 Calyx 5–12 mm
95 Filament-hairs violet   **78. rotundifolium**
95 Filament-hairs white or yellow
96 Bracts ovate-lanceolate to linear; calyx-lobes
   linear   **79. reiseri**
96 Bracts suborbicular-deltate, cordate or ovate-
   deltate; calyx-lobes oblong-lanceolate to
   lanceolate
97 Plant floccose, later glabrescent; inflorescence
   ±dense   **49. dieckianum**
97 Plant ±persistently tomentose; inflorescence
   lax   **56. undulatum**
94 Calyx 2–5 mm
98 Bracts and calyx-lobes ovate to ovate-orbicular
        68. herzogii
98 Bracts and calyx-lobes ovate-lanceolate to linear-
   oblong
99 Leaves distinctly crenate, the cauline at least
   shortly decurrent
100 Basal leaves 25–65 × 7–20(–25) cm; connectives
   of all anthers papillose   **71. mucronatum**
100 Basal leaves 7–15 × 3–8 cm; connectives of two
   lower anthers smooth   **70. decorum**
99 Leaves entire or obscurely crenate, the cauline
   not decurrent
101 Filament-hairs violet   **69. mallophorum**
101 Filament-hairs white or yellow
102 Inflorescence lax; capsule cylindrical, twice as
   long as wide; indumentum of leaves
   ±persistent   **67. graecum**
102 Inflorescence dense; capsule ovoid-globose,
   only slightly longer than wide; indumentum of leaves floccose   **72. pulverulentum**
77 Mature leaves sparsely tomentose to ±glabrous, green
   at least above
103 Upper bracts conspicuous, suborbicular-deltate,
   ±glabrescent   **49. dieckianum**
103 Upper bracts inconspicuous, ovate-lanceolate to
   linear, usually with persistent tomentum
104 Calyx 9–12 mm   **57. pentelicum**
104 Calyx 2–8 mm
105 Basal leaves not more than 5 cm wide
106 Inflorescence simple   **42. nicolai**
106 Inflorescence branched
107 Leaves coarsely crenate-dentate; bracts 6–10
   mm   **64. humile**
107 Leaves entire or finely crenate; bracts 2–4 mm
        65. adenanthum
105 Basal leaves more than 5 cm wide
108 Filament-hairs violet
109 Basal leaves orbicular to ovate-elliptical,
   petiolate; capsule 7–10 mm   **78. rotundifolium**
109 Basal leaves obovate-oblong, sessile; capsule
   3–6 mm   **84. bithynicum**
108 Filament-hairs white
110 Inflorescence simple
111 Basal leaves 15–50 × 6–15 cm, elliptic- to
   obovate-oblong; bracts ovate to ovate-
   lanceolate, acuminate-cuspidate
        73. gnaphalodes
111 Basal leaves 5–25 × 2–7 cm, oblong-ovate to
   -lanceolate; bracts linear to linear-lanceolate   **42. nicolai**
110 Inflorescence branched
112 Capsule oblong-cylindrical   **67. graecum**
112 Capsule ovoid-globose or ovoid-ellipsoid
113 Indumentum floccose; pedicels 2–5(–7) mm
        72. pulverulentum
113 Indumentum persistent; pedicels 6–11 mm
        82. lychnitis

A. Flowers solitary in the axil of each bract (except sometimes in 7).

1. V. spinosum L., *Cent. Pl.* **2**: 10 (1756). Freely-branched shrub up to 50 cm; branches ending in a spine; young shoots tomentose. Leaves 1·5–5 × 0·3–1 cm, oblong-lanceolate, irregularly toothed or lobed, whitish-grey-tomentose. Bracts minute, ovate; pedicels 3–10 mm. Calyx 2 mm; lobes oblong to oblong-lanceolate. Corolla 10–18 mm in diameter. Stamens 5; anthers all reniform; filament hairs short, lilac. Capsule 3–4 mm, ovoid. *Mountains and stony hillsides.* ● *Kriti.* Cr.

2. V. ovalifolium Donn ex Sims, *Bot. Mag.* **26**: t. 1037 (1807). Biennial; stem up to 100 cm. Basal leaves 5–30 × 2–10 cm, ovate to oblong-lanceolate, crenate, shortly petiolate, glabrous or sparsely hairy above, densely tomentose beneath. Bracts 10–14 mm, ovate-lanceolate to suborbicular, acuminate to cuspidate; bracteoles present; flowers sessile. Calyx 7–12 mm; lobes lanceolate, acuminate. Corolla 20–40(–50) mm in diameter, with pellucid glands. Stamens 5, the lower with decurrent anthers; filament-hairs yellow. Capsule 5–7 mm, globose to ovoid. *S.E. Europe, from Turkey to S. Russia.* Bu Gr Rm Rs (W, K, E) Tu.

(a) Subsp. **ovalifolium** (*V. crenatifolium* Boiss., *V. pulchrum* Velen.): Indumentum yellowish. Cauline leaves broadly ovate-cordate, cuspidate. Inflorescence simple or slightly branched, usually dense; bracts suborbicular, cuspidate. Filaments of lower stamens more or less glabrous. *Throughout the range of the species.*

(b) Subsp. **thracicum** (Velen.) Murb., *Lunds Univ. Årsskr.* nov. ser., **29**(2): 496 (1933) (*V. thracicum* Velen.): Indumentum grey. Cauline leaves ovate to oblong-lanceolate, acuminate. Inflorescence branched, usually lax; bracts ovate to ovate-lanceolate, acuminate. Filaments of lower stamens usually villous near the middle. ● *Bulgaria.*

V. formosum Fischer ex Schrank, *Pl. Rar. Hort. Monac.* 22 (1819), from the Caucasus, has once been recorded from Krym. It has suborbicular basal leaves, a deep violet patch in the centre of the corolla, and violet filaments connate in their basal part to form a tube.

3. V. purpureum (Janka) Huber-Morath, *Neue Denkschr. Schweiz. Naturf. Ges.* **87**: 125 (1971) (*V. glanduligerum* Velen.). Biennial; stem up to 100 cm, appressed-pubescent. Basal leaves 10–25 × 2–7 cm, oblong-lanceolate, incise-dentate to pinnatisect with linear-oblong lobes, petiolate, glabrescent above, arachnoid-tomentose beneath. Inflorescence branched, lax. Bracts 10–15 mm, lanceolate; bracteoles present; pedicels 1–3 mm. Calyx 4–6 mm; lobes lanceolate, acute. Corolla 20–30 mm in diameter. Stamens 5; anthers all reniform; filament-hairs pale yellow. Capsule 4–5 mm, ovoid-ellipsoid. ● *E. Bulgaria, extending to S.E. Romania and Turkey-in-Europe.* Bu Rm Tu.

4. V. adrianopolitanum Podp., *Verh. Zool.-Bot. Ges. Wien* **52**: 665 (1902) (*V. adenotrichum* sensu Hayek pro parte, non Halácsy). Like **3** but stem glandular-hirsute; basal leaves 5–10 × 1·5–2 cm, oblong-obovate, lanate when young, crenate but not lobed; filament-hairs violet. ● *S.E. Bulgaria.* Bu ?Gr ?Tu.

5. V. blattaria L., *Sp. Pl.* 178 (1753) (incl. *V. rhinanthifolium* Davidov, *V. carduifolium* Murb. ex Hayek). Biennial (or sometimes annual), glabrous below, with stalked or sessile glands above; stem 30–120 cm. Basal leaves (4–)8–25 × (1–)2·5–4 cm, oblong to lanceolate, sessile or shortly petiolate, crenate-sinuate to pinnatifid, glabrous. Inflorescence usually simple, lax.

Bracts 7–8(–22) mm, ovate, acuminate; bracteoles absent; pedicels 5–25 mm. Calyx (4–)5–8 mm. Corolla 20–30 mm in diameter, yellow, rarely white. Stamens 5, the lower with decurrent anthers; filament-hairs of the 2 lower stamens purple, of the 3 upper white and purple. Style *c.* 10 mm. Capsule 5–8 mm, globose. *Usually in rather damp places. Europe, northwards to the Netherlands and C. Russia.* Al Au ?Az Be Bu Co Cz Ga Ge Gr He Ho Hs Hu It Ju Po Rm Rs (C, W, K, E) Sa Si Tu [Br].

6. V. siculum Tod. ex Lojac., *Fl. Sic.* 2(2): 116 (1907). Like **5** but basal leaves 12–35 × 3–10 cm, obovate-elliptical, densely pubescent beneath with tomentose midrib; inflorescence branched; pedicels, bracts and calyx 3–5 mm; corolla 15–18 mm in diameter. ● *N.W. Sicilia, S.W. Italy.* It Si.

7. V. virgatum Stokes in With., *Arr. Brit. Pl.* ed. 2, **1**: 227 (1787). Like **5** but usually glandular-pubescent throughout; basal leaves (10–)15–33 × 4·5–9 cm; flowers in lower part of inflorescence often in fascicles of 2–5; bracts 8–20 mm, serrate-dentate; bracteoles usually present; pedicels 2–5(–7) mm; corolla 30–40 mm in diameter; lower stamens with obliquely inserted or shortly decurrent anthers. ● *W. Europe, northwards to S.W. England.* Az Bl Br Co Ga Hs *It Lu.

(8–11). V. daenzeri group. Basal leaves 9–22 × 1·5–5 cm, oblong- to obovate-oblanceolate, lyrate to pinnatisect, with 4–10 pairs of coarsely and irregularly incise-dentate lobes. Inflorescence lax; pedicels 5–35 mm. Calyx 3–4 mm. Fertile stamens 4, with or without a staminode, the lower with shortly decurrent anthers; filament-hairs violet. Style 5–10 mm.

1 Bracts dentate; lamina of basal leaves ± hairy beneath
2 Leaf-lobes acute; pedicels 15–30 mm; corolla 25–35 mm in
 diameter **8. daenzeri**
2 Leaf-lobes obtuse; pedicels 5–10(–15) mm; corolla 15–20 mm
 in diameter **9. rupestre**
1 Bracts entire; lamina of basal leaves glabrous
3 Stem sparsely glandular-pubescent above, glabrous below;
 capsule ellipsoid-globose, obtuse **10. roripifolium**
3 Stem minutely puberulent throughout; capsule oblong-ovoid
 or ovoid, acute **11. boissieri**

8. V. daenzeri (Fauché & Chaub.) O. Kuntze, *Rev. Gen. Pl.* **2**: 469 (1891) (*Celsia daenzeri* Fauché & Chaub.). Biennial; stem 50–150 cm. Leaves somewhat hairy beneath, irregularly pinnatisect or lyrate; lobes oblong to lanceolate, acute. Inflorescence sparsely glandular-pubescent, usually simple. Bracts triangular-ovate, acuminate, dentate; pedicels 15–30 mm. Corolla 25–35 mm in diameter. Staminode present. Capsule subglobose, obtuse. ● *S. Greece.* Gr.

V. friedrichsthalianum O. Kuntze, *op. cit.* 468 (1891) (*Celsia speciosa* Fenzl), from S. Greece, with numerous stems up to 60 cm, smaller corolla, and calyx-lobes of lower flowers large and leaf-like, is probably a variant of **8**.

9. V. rupestre (Davidov) I. K. Ferguson, *Bot. Jour. Linn. Soc.* **64**: 230 (1971) (*Celsia rupestris* Davidov). Perennial; stems 20–60 cm. Leaves glandular-villous; lobes ovate, obtuse. Inflorescence densely glandular-pubescent, simple or branched. Bracts ovate, acute, the lower dentate; pedicels 5–10(–15) mm. Corolla 15–20 mm in diameter. Staminode absent. Capsule ovoid-globose, obtuse. ● *Mountains of S. Bulgaria and N.E. Greece.* Bu Gr.

10. V. roripifolium (Halácsy) I. K. Ferguson, *loc. cit.* (1971) (*Celsia roripifolia* Halácsy). Biennial; stems 50–150 cm. Leaves glabrous; lobes oblong to lanceolate or triangular, acute. Inflorescence sparsely glandular-pubescent, usually branched. Bracts lanceolate to linear-lanceolate, entire; pedicels 15–35 mm. Staminode usually absent. Capsule ellipsoid-globose, obtuse. ● *S. Bulgaria and N.E. Greece.* Bu Gr.

11. V. boissieri (Heldr. & Sart. ex Boiss.) O. Kuntze, *Rev. Gen. Pl.* **2**: 469 (1891) (*Celsia boissieri* Heldr. & Sart. ex Boiss.). Perennial; stems several, 40–60 cm. Leaves glabrous, except for pubescent petiole; lobes incise-dentate with triangular, acuminate teeth. Inflorescence minutely glandular-puberulent, usually simple. Bracts ovate-lanceolate, entire; pedicels 10–25 mm. Corolla 22–25 mm in diameter. Staminode present. Capsule oblong-ovoid or ovoid, acute. ● *E. Greece.* Gr.

12. V. barnadesii Vahl, *Symb. Bot.* **2**: 39 (1791) (*Celsia barnadesii* (Vahl) G. Don fil.). Biennial, glandular-puberulent throughout; stem 50–90 cm. Basal leaves 7–13 × 1·5–2·5(–4) cm, narrowly lanceolate, pinnatifid to pinnatisect with 7–18 pairs of triangular-lanceolate, dentate, acute segments, shortly petiolate. Inflorescence usually simple, lax. Bracts 3–7 mm, triangular, acute; pedicels 12–20 mm. Calyx *c.* 4 mm; lobes obovate, somewhat unequal. Corolla 30–35 mm in diameter. Stamens 4, without staminode, the lower with decurrent anthers 5–7 mm long; filaments of upper with yellow and violet hairs, of lower glabrous. Style 17–23 mm. Capsule 6–9 mm, subglobose. ● *C. & S.W. Spain, C. & S. Portugal.* Hs Lu.

13. V. laciniatum (Poiret) O. Kuntze, *Rev. Gen. Pl.* **2**: 469 (1891) (*Celsia barnadesii* var. *baetica* Willk.). Like **12** but stem 70–180 cm; basal leaves oblong-ovate, with fewer, wider lobes and longer petiole; pedicels 8–16 mm; calyx 6–9 mm; corolla 35–50 mm in diameter; anthers of lower stamens 7–9 mm; style 22–30 mm; capsule 8–11 mm. ● *S.W. Spain.* Hs.

Celsia commixta Murb., *Lunds Univ. Årsskr.* nov. ser., **22**: 222 (1925) (*C. betonicifolia* sensu Willk., non Desf.), a species from N. Africa, like **13** but with lax inflorescence, pedicels 12–25 mm, bracts *c.* 7 mm, calyx 5–7 mm, corolla 30–35 mm in diameter and lower stamens with anthers curved at apex, has been recorded from S.W. Spain (near Cádiz), but perhaps only as a casual.

14. V. creticum (L.) Cav., *Elench. Pl. Horti Matrit.* 39 (1803) (*Celsia cretica* L.; incl. *C. sinuata* Cav.). Biennial, glandular-pubescent and also with somewhat longer eglandular hairs; stems 40–150 cm. Basal leaves 15–25 × 4–7 cm, oblong, crenate-dentate to somewhat pinnatifid. Inflorescence simple. Bracts 12–25 mm, ovate, acuminate, serrate; pedicels 3–6 mm. Calyx 8–15 mm; lobes serrate. Corolla 40–50 mm in diameter. Stamens 4, without staminode, unequal, the lower with decurrent anthers 5–6 mm long. Style 20–25 mm. Capsule 9–15 mm, ellipsoid-globose. *W. Mediterranean region.* Bl ?Co ?Ga Hs It Sa Si.

15. V. arcturus L., *Sp. Pl.* 178 (1753) (*Celsia arcturus* (L.) Jacq.). Perennial with woody stock; stems 30–70 cm, numerous. Basal leaves 8–15 × 2–5 cm, lyrate, with an ovate-oblong or oblong terminal lobe and 2–4 much smaller lateral ones, whitish-lanate beneath, green and sparsely lanate above, crenate. Cauline leaves alternate, few, much smaller. Inflorescence densely glandular-pubescent. Bracts 5–8(–11) mm, the lower ones entire; pedicels 15–20(–30) mm. Calyx 3–5 mm. Corolla 25–30 mm in diameter. Fertile stamens 4, with or without a staminode, the lower with shortly decurrent anthers; filament-hairs violet. Style 9–14 mm. Capsule 5–7 mm, subglobose. *Crevices in calcareous rocks.* ● *Kriti.* Cr.

16. V. levanticum I. K. Ferguson, *Bot. Jour. Linn. Soc.* **64**: 230 (1971) (*Celsia horizontalis* auct., non Moench). Like **15** but usually without a woody stock and often biennial or annual; leaves 8–25 × 5–11 cm, lyrate, with a broadly ovate or ovate-oblong terminal lobe and 3–7 ovate lateral lobes, more or less pubescent and glandular, villous on the veins beneath, acutely and often doubly dentate; cauline leaves opposite; lower bracts dentate. *Naturalized in C. Portugal.* [Lu.] (*S.W. Asia.*)

17. V. bugulifolium Lam., *Encycl. Méth. Bot.* **4**: 226 (1797) (*Celsia bugulifolia* (Lam.) Jaub. & Spach). Usually perennial; stem 20–75 cm, glandular-pubescent. Basal leaves 4–8(–12) × 1·5–3(–6) cm, triangular-ovate, subentire to crenate, subglabrous, but usually puberulent on the veins beneath, long-petiolate. Inflorescence simple. Bracts 7–14 mm, oblong-lanceolate, entire; pedicels 2–3 mm. Calyx 6–8 mm. Corolla 25–35 mm in diameter, yellowish- to bluish-green, the upper lobes with purple lines. Stamens 4, without staminode, the lower with decurrent anthers *c.* 5 mm long; filament-hairs violet. Style 6–10 mm. Capsule 5–8 mm, broadly ellipsoid. *Turkey-in-Europe, S.E. Bulgaria.* Bu Tu. (*N.W. Anatolia.*)

An isolated record from E. Jugoslavia (Zaječar) requires confirmation.

18. V. spectabile Bieb., *Fl. Taur.-Cauc.* **3**: 158 (1819). Biennial; stem 50–120 cm, lanate below, glandular-hirsute above. Basal leaves 8–20 × 5–8 cm, oblong-ovate, crenate-dentate, hairy on the veins beneath, elsewhere glabrous to sparsely pubescent, long-petiolate. Inflorescence simple. Bracts 6–15 mm, lanceolate to linear, long-acuminate; pedicels 5–10 mm. Calyx 5–7 mm. Corolla 35–40 mm in diameter, yellow, with purplish spots at base of upper lobes, glandular-pubescent outside. Stamens 5, the lower with decurrent anthers; upper filaments with violet hairs, the lower glabrous. Capsule 7–9 mm, ellipsoid. *Fagus-woods. Mountains of Krym.* Rs (K). (*Caucasus, N. Anatolia.*)

19. V. hervieri Degen, *Magyar Bot. Lapok* **5**: 5 (1906). Biennial; stem 150–300 cm, floccose-tomentose below. Basal leaves 25–50 × 7–18(–25) cm, obovate-oblong, crenate-dentate, silvery-sericeous-tomentose, shortly petiolate. Inflorescence simple, lax, glabrous. Bracts 2–3 mm, acuminate, entire; pedicels 3–10 mm. Calyx 3–4 mm. Corolla 20–25 mm in diameter, glabrous. Stamens 5, the lower with decurrent anthers; filament-hairs white. Capsule 5–6 mm, globose. ● *S.E. Spain.* Hs.

20. V. phoeniceum L., *Sp. Pl.* 178 (1753) (incl. *Celsia rechingeri* Murb.). Usually perennial; stem 30–100 cm, crispate-villous below, glandular-pubescent above. Basal leaves 4–17 × 2·5–9 cm, ovate, entire, sinuate to weakly crenate, sparsely pubescent or glabrous, petiolate; cauline leaves few, small. Inflorescence lax, usually simple. Bracts lanceolate, usually shorter than pedicels; pedicels 10–25 mm. Calyx 4–8 mm; lobes lanceolate to obovate-elliptical. Corolla 20–30(–35) mm in diameter, violet (rarely yellow). Stamens 4 or 5; anthers all reniform; filament-hairs violet. Style 5–8 mm. Capsule 6–8 mm, ovoid. $2n = 32, 36$. *S.E. & E.C. Europe, extending northwards to C. Russia and westwards to C. Germany; cultivated for ornament, and an occasional casual elsewhere.* Al Au Bu Cz Ge Gr Hu It Ju Po Rm Rs (*B, C, W, K, E) Tu [Ho].

Yellow-flowered plants from Macedonia have been distinguished as subsp. **flavidum** (Boiss.) Bornm., *Bot. Jahrb.* **61**, Beibl. **4**: 48 (1928), but differ from **20** only in this character. They have been confused with **21**.

21. V. xanthophoeniceum Griseb., *Spicil. Fl. Rumel.* **2**: 42 (1844). Like **20** but with deeply crenate or weakly pinnatifid, crispate-villous basal leaves; cauline leaves very small or absent; corolla yellow. *S.E. part of Balkan peninsula.* Bu Gr Tu.

22. V. cylleneum (Boiss. & Heldr.) O. Kuntze, *Rev. Gen. Pl.* **2**: 469 (1891) (*Celsia cyllenea* Boiss. & Heldr.). Perennial; stems 10–25 cm, several, glandular-puberulent throughout. Basal leaves 2–5·5 × 0·5–2·5 cm, ovate-oblong, coarsely dentate, glandular-puberulent at least beneath, with petiole 2–7·5 cm; cauline leaves very few, small. Inflorescence lax, simple. Bracts 3–5 mm, ovate to lanceolate, acute; pedicels 7–15 mm. Calyx 4–6 mm. Corolla 16–22 mm in diameter. Stamens 4, without staminode, the lower 2 much longer than the upper; anthers all reniform; filament-hairs violet. Style 6–8 mm. Capsule 5–6 mm, ovoid. *Shady mountain rocks.* ● *S. Greece (Killini Oros).* Gr.

23. V. acaule (Bory & Chaub.) O. Kuntze, *loc. cit.* (1891) (*Celsia acaulis* Bory & Chaub.). Like **22** but acaulescent or with stems not more than 3 cm; inflorescences shorter and denser, corymbose; bracts c. 10 mm, oblong-linear, subacute; pedicels 25–90 mm; lower stamens only slightly longer than the upper. ● *Higher mountains of S. Greece (Peloponnisos).* Gr.

24. V. pyramidatum Bieb., *Fl. Taur.-Cauc.* **1**: 161 (1808). Perennial, appressed-pubescent throughout and glandular-pubescent on upper surface of leaves; stem 60–150 cm. Basal leaves 12–30(–40) × 5–10(–15) cm, obovate-oblong, crenate-dentate to weakly pinnatifid. Inflorescence rather lax, with numerous arcuate-ascending branches, forming a narrowly pyramidal panicle. Bracts 6–10 mm, lanceolate, acuminate; pedicels 3–5(–8) mm. Calyx 3–5 mm. Corolla 20–30 mm in diameter. Stamens usually 5, but sometimes 4 in some flowers; anthers all reniform; filament-hairs violet. Capsule 4–5 mm, elliptic-ovoid. *Extreme S. part of U.S.S.R.* Rs (K, E). (*Caucasian region.*)

25. V. zuccarinii (Boiss.) I. K. Ferguson, *Bot. Jour. Linn. Soc.* **64**: 230 (1971) (*V. graecum* var. *zuccarinii* Boiss., *Celsia tomentosa* Zucc.). Biennial or perennial, floccose or lanate with white hairs below, glabrous in inflorescence. Basal leaves 2–8 × 1–5 cm, densely lanate, obovate-spathulate, subentire, long-petiolate. Inflorescence lax, freely branched. Bracts c. 1·5 mm, ovate; pedicels 2–5 mm. Calyx 3–4 mm. Corolla c. 20 mm in diameter. Stamens 4, without staminode; anthers all reniform; filament-hairs violet. ● *E. Greece (Evvoia).* Gr.

26. V. orientale (L.) All., *Fl. Pedem.* **1**: 106 (1785) (*Celsia orientalis* L.). Annual, puberulent to glabrescent below, glandular-puberulent above; stem 15–70 cm. Basal leaves obovate-oblong, petiolate, coarsely crenate to pinnatifid, often withered at time of flowering; cauline leaves usually bipinnatisect with linear to oblong segments, sessile. Inflorescence lax, simple. Bracts 6–17 mm, the lowest pinnatisect, the upper linear, entire; pedicels 1·5–5 mm. Calyx 5–7 mm. Corolla 14–20 mm in diameter, yellow with a few brownish spots. Stamens 4, without staminode; anthers all reniform; filament-hairs yellow. Style c. 5 mm. Capsule 4–7 mm, ellipsoid. *Balkan peninsula; Krym.* Bu Gr Ju Rs (K) Tu.

B. Flowers in fascicles in the axils of most of the bracts.

27. V. phlomoides L., *Sp. Pl.* 1194 (1753) (incl. *V. belasitzae* Stoj. & Stefanov). Grey-, white- or yellow-tomentose biennial; stem 30–120 cm. Basal leaves 5–30(–45) × 3–12 cm, oblong-elliptical, entire or crenate; upper cauline leaves ovate to lanceolate, acuminate, denticulate or crenulate, not or only very shortly decurrent. Inflorescence usually simple. Bracts 9–15 mm; longer pedicels 3–15 mm. Calyx 5–12 mm. Corolla 20–55 mm in diameter, tomentose outside. Stamens 5, the lower with decurrent anthers; upper filaments with white or yellow hairs, the lower glabrous. Style 6–15 mm, with spathulate, decurrent stigma. Capsule 5–8 mm, elliptic-ovoid. *Most of Europe except the north.* Al Au Bu Co Cz Ga Ge Gr He Hs Hu It Ju *Lu Po Rm Rs (C, W, K, E) Sa Si Tu [?Be Br ?Ho].

28. V. densiflorum Bertol., *Rar. Lig. Pl.* **3**: 52 (1810) (*V. thapsiforme* Schrader; incl. *V. macrantherum* Halácsy, *V. velenovskyi* Horák). Like **27** but cauline leaves long-decurrent; bracts 15–40 mm, usually long-acuminate, decurrent. *Most of Europe northwards to the Netherlands, S.E. Sweden and C. Russia, but rather local in the west.* Al Au Be Bu Cz Da Ga Ge Gr He Ho Hs Hu It Ju Po Rm Rs (*B, C, W, K, E) Si Su Tu.

29. V. samniticum Ten., *Fl. Nap.* **3**: 219 (1824–1829) (*V. phlomoides* subsp. *sartorii* (Boiss. & Heldr.) Nyman). Biennial; stem 60–150 cm, sparsely tomentose, usually more or less glabrescent. Basal leaves 15–30 × 4–10 cm, obovate-oblong, greenish above, grey-tomentose beneath, rugose, crenate or slightly lobed, subsessile or shortly petiolate. Inflorescence lax, with a few suberect branches, forming a narrow panicle. Bracts 10–12 mm, ovate to ovate-lanceolate, persistently and harshly yellowish-grey-tomentose; longer pedicels 4–6(–8) mm. Calyx 6–9 mm, persistently tomentose. Corolla 25–40 mm in diameter. Stamens 5, the lower with decurrent anthers; filament-hairs white. Capsule 5–6 mm, ovoid-globose. ● *W. part of Balkan peninsula; C. Italy.* Al Gr It Ju.

30. V. guicciardii Heldr. ex Boiss., *Diagn. Pl. Or. Nov.* **3**(6): 127 (1859). Like **29** but basal leaves sometimes wider and shorter, remotely crenate or subentire, and always with petioles 4–7 cm; inflorescence usually simple; longer pedicels 2–4 mm; calyx 4–7 mm; lower stamens with obliquely inserted or shortly decurrent anthers. ● *Greece and Albania.* Al Gr.

31. V. vandasii (Rohlena) Rohlena, *Acta Bot. Bohem.* **3**: 44 (1924) (*V. pannosum* auct., non Vis.). Biennial, with dense, greyish-white tomentum, persistent below, more or less floccose above; stem 60–200 cm. Basal leaves 15–30 × 5–11 cm, ovate-oblong, obscurely crenate, rugose beneath; cauline leaves decurrent. Inflorescence simple or branched. Bracts 8–20 mm, floccose-tomentose when young, later glabrescent and scarious; pedicels partly adnate to stem, the free part 1–5 mm. Calyx 5–8 mm; indumentum like that of the bracts. Corolla 25–40 mm. Stamens 5, the lower with decurrent anthers; filament-hairs white. Capsule 5–9 mm, elliptic-ovoid. ● *C. & S. Jugoslavia, just extending to S.W. Romania and N.E. Albania.* Al Ju Rm.

32. V. niveum Ten., *Fl. Nap.* **1**, *Prodr.*: 16 (1811). Biennial; stem 30–150 cm, sparsely to densely white-tomentose to -sericeous. Basal leaves ovate to oblong-lanceolate, entire to dentate, petiolate. Upper cauline leaves decurrent. Bracts acuminate, persistently tomentose; pedicels almost completely adnate to stem. Corolla 25–50 mm in diameter. Stamens 5, the lower with decurrent anthers; filament-hairs white. Stigma decurrent. Capsule 4–8 mm, ovoid to subglobose. ● *W. Jugoslavia and Albania; C. & S. Italy.* Al It Ju.

1 Bracts not more than 12 mm, not exceeding the unopened flowers **(c) subsp. garganicum**
1 Bracts usually more than 12 mm, exceeding the unopened flowers
  2 Basal leaves entire or finely crenate-dentate; inflorescence usually branched **(b) subsp. visianianum**
  2 Basal leaves coarsely and acutely dentate; inflorescence simple **(a) subsp. niveum**

(a) Subsp. **niveum**: Stem 35–150 cm. Basal leaves 8–36 × 2–8 cm, coarsely and acutely dentate, sparsely grey-tomentose. Inflorescence simple. Bracts 14–22 mm, lanceolate. Calyx 8–12 mm. Corolla 30–50 mm in diameter. *S. Italy.*

(b) Subsp. **visianianum** (Reichenb.) Murb., *Lunds Univ. Årsskr.* nov ser., 29(2): 101 (1933) (*V. macrurum* sensu Hayek pro parte, non Ten.): Stem 50–150 cm. Basal leaves 10–20 × 6–10 cm, entire to finely crenate-dentate, rather densely grey-tomentose. Inflorescence usually branched. Bracts 12–18 mm. Calyx 8–10 mm. Corolla 30–40 mm in diameter. *W. Jugoslavia, Albania.*

(c) Subsp. **garganicum** (Ten.) Murb., *op. cit.* 99 (1933): Stem 30–50 cm. Basal leaves 6–20 × 3–7 cm, entire or finely dentate, rather densely grey-tomentose. Inflorescence usually simple. Bracts 7–10(–12) mm. Calyx 7–9 mm. Corolla 25–30 mm in diameter. *C. Italy.*

33. V. macrurum Ten., *Fl. Neap. Prodr. App. Quinta* 9 (1826). Like **32** but indumentum dense, harsh, grey or yellowish; basal leaves oblong-oblanceolate, sessile; all cauline leaves decurrent. Basal leaves 20–50 × 5–13 cm. Inflorescence usually simple. Bracts (7–)10–14 mm, ovate-lanceolate. Calyx 7–12 mm; lobes ovate, somewhat acuminate. ● *S. Italy, Sicilia; S. Greece, Kriti.* Cr Gr It Si ?Tu.

34. V. thapsus L., *Sp. Pl.* 177 (1753) (incl. *V. simplex* Hoffmanns. & Link). Biennial, more or less densely greyish- or whitish-tomentose; stem 30–200 cm. Basal leaves 8–50 × 2·5–14 cm, elliptic- to obovate-oblong, obtuse, entire or finely crenate; upper cauline leaves decurrent. Inflorescence usually simple. Bracts 12–18 mm, ovate to lanceolate, acuminate; pedicels partly adnate to stem. Calyx (5–)8–12 mm; lobes lanceolate. Corolla 12–35 mm in diameter. Stamens 5, the upper with villous filaments, the lower with glabrous to villous filaments and decurrent anthers; filament-hairs white. Stigma capitate. Capsule 7–10 mm, elliptic-ovoid. $2n = 32, 36$. *Most of Europe except the extreme north and much of the Balkan peninsula.* Au Az Be Bl Br Co Cz Da Fe Ga Ge Hb He Ho Hs Hu It Ju Lu No Po Rm Rs (N, B, C, W, K, E) Sa Si Su.

1 Filaments of lower stamens glabrous or sparsely hairy **(a) subsp. thapsus**
1 Filaments of lower stamens densely villous
  2 Plant variably tomentose, but not white; lower cauline leaves scarcely decurrent **(b) subsp. crassifolium**
  2 Plant densely white-tomentose; lower cauline leaves long-decurrent **(c) subsp. giganteum**

(a) Subsp. **thapsus**: Basal leaves shortly petiolate; lower cauline leaves long-decurrent. Corolla-lobes ovate. Filaments of lower stamens glabrous or sparsely hairy. *Throughout the range of the species.*

(b) Subsp. **crassifolium** (Lam.) Murb., *Lunds Univ. Årsskr.* nov. ser., 29(2): 126 (1933) (*V. crassifolium* Lam.): Basal leaves petiolate; lower cauline leaves scarcely decurrent. Corolla-lobes oblong. Filaments of lower stamens densely villous. ● *Mountains of S. & C. Europe, from Portugal to the E. Alps.*

(c) Subsp. **giganteum** (Willk.) Nyman, *Consp.* 527 (1881): Basal leaves subsessile; lower cauline leaves long-decurrent.

Corolla-lobes ovate. Filaments of lower stamens densely villous. ● *Mountains of S. & S.E. Spain.*

Plants from S.E. Spain (Sierra de Cazorla) differ from subsp. (c) in their denser, harsher indumentum and sometimes branched inflorescence. They may warrant recognition as a distinct taxon when further information is available.

35. V. litigiosum Samp., *Lista Esp. Herb. Port.* 108 (1913). Like **34** but stem 20–100 cm; bracts 8–10 mm; filaments of all the stamens glabrous. Upper and middle cauline leaves decurrent. *Sandy places near the sea.* ● *C. Portugal.* Lu.

36. V. longifolium Ten., *Fl. Nap. 1, Prodr.*: 16 (1811) (incl. *V. pachyurum* Bornm., *V. samaritanii* Heldr. ex Boiss., *V. epirotum* Halácsy). Biennial or perennial, more or less densely whitish- or yellowish-tomentose, floccose above; stem 50–150 cm. Basal leaves 20–50 × 5–15 cm, elliptic-oblong to ovate-lanceolate, acute to acuminate, entire or slightly crenate; petiole 2–7 cm. Inflorescence usually simple. Bracts 12–20(–40) mm, linear; longer pedicels 10–16 mm. Calyx 5–8 mm; lobes linear-lanceolate or lanceolate. Corolla 25–35 mm in diameter. Stamens 5, the upper with densely villous filaments, the lower with glabrous to villous filaments and obliquely inserted or shortly decurrent anthers; filament-hairs white, yellow or pale violet. Stigma clavate to subspathulate, shortly decurrent. Capsule 5–8 mm, ellipsoid. ● *Balkan peninsula, C. & S. Italy.* Al Bu Gr It Ju.

Plants from Italy usually have a relatively lax, slender inflorescence and at least some of the filament-hairs violet; most, but not all plants from the Balkan peninsula have a stouter and denser inflorescence and white or yellow filament-hairs. There are, however, too many exceptions to allow subspecies to be effectively delimited.

A single plant from S.E. Jugoslavia (near the Bulgarian border, S.E. of Niš), with leaves green, not tomentose, pedicels *c.* 5 mm, filament-hairs violet, anthers of lower stamens long-decurrent and stigma long-decurrent, has been named **V. viridissimum** Stoj. & Stefanov, *Österr. Bot. Zeitschr.* 73: 281 (1924). Its status is obscure, but it seems to be most closely related to **36**.

37. V. boerhavii L., *Mantissa* 45 (1767). Biennial, white-tomentose all over, floccose at least above; stem 30–120 cm. Basal leaves 10–30 × 4–12 cm, ovate-elliptical, obtuse or subacute, crenate-dentate, sometimes slightly lobed at base; petiole 4–8 cm. Inflorescence usually simple. Bracts 10–25 mm, oblong to linear, acute; longer pedicels 1–4 mm. Calyx 6–9 mm; lobes oblong to linear. Corolla 22–32 mm in diameter. Stamens 5, the lower with shortly decurrent anthers; all filaments with violet hairs. Capsule 6–10 mm, ovoid. *W. Mediterranean region.* Bl Co Ga Hs It.

38. V. argenteum Ten., *Fl. Nap.* 3: 220 (1824–1829). Like **37** but basal leaves *c.* 40 × 12 cm, oblong to ovate-lanceolate, more or less entire; inflorescence freely branched, forming an ovoid-pyramidal panicle; pedicels 5–10 mm; corolla 18–25 mm in diameter. ● *S. Appennini (Monti del Matese).* It.

39. V. nevadense Boiss., *Voy. Bot. Midi Esp.* 2: 443 (1841). Usually perennial, more or less densely greyish-floccose-tomentose; stems 70–100 cm. Basal leaves 12–25 × 4–12 cm, oblong-ovate to -lanceolate, acute to acuminate, entire; petiole 5–12 cm. Inflorescence lax, simple. Bracts 6–15(–25) mm, linear; pedicels 4–9 mm. Calyx 4–8(–10) mm; lobes linear- to oblong-lanceolate. Corolla 30–40 mm in diameter. Stamens 5, the upper with densely violet-villous filaments, the lower with decurrent anthers and glabrous filaments. Capsule 6–10 mm, pyramidal-ovoid. ● *Mountains of S. & S.E. Spain*. Hs.

V. charidemi Murb., *Lunds Univ. Årsskr.* nov. ser., **29**(2): 161 (1933), described from a single plant from S.E. Spain (Cabo de Gata), with persistent white indumentum, triangular-lanceolate bracts, and corolla 18–20 mm in diameter, is doubtfully distinct from **39**.

40. V. lagurus Fischer & C. A. Meyer, *Ind. Sem. Hort. Petrop.* **5**: 42 (1839) (incl. *V. ponticum* Stefanov). Greyish-floccose-tomentose biennial; stems 60–120 cm. Basal leaves 6–25 × 2·5–10 cm, oblong-elliptical, more or less entire, petiolate. Inflorescence usually simple. Bracts 5–8 mm, lanceolate, acuminate; pedicels 3–5 mm. Calyx 4–7 mm; lobes lanceolate. Corolla 20–30 mm in diameter. Stamens 5, the lower with shortly decurrent anthers; filament-hairs white. Capsule 5–6 mm, ovoid. *Turkey-in-Europe, just extending to S.E. Bulgaria.* Bu Tu. (*N.W. Anatolia.*)

41. V. georgicum Bentham in DC., *Prodr.* **10**: 228 (1846). Biennial; stems up to 150 cm. Basal leaves 10–35 × 3–12 cm, obovate-oblong to oblong-lanceolate, more or less crenulate, glabrous or sparsely hairy above, more or less floccose-tomentose beneath; petiole 2–4·5 cm. Inflorescence simple; pedicels 2–5 mm. Bracts 10–14 mm, the lower ovate, long-acuminate, the upper linear-lanceolate to linear, more or less floccose, softly white-tomentose. Corolla 20–25 mm, without pellucid glands. Stamens 5, the lower with decurrent anthers 3·5–4 mm; filament-hairs yellowish-white. Capsule 5–7 mm, globose. *Turkey-in-Europe (N.E. of Çatalca).* Tu. (*Anatolia, Caucasus.*)

42. V. nicolai Rohlena, *Feddes Repert.* **3**: 148 (1906). Whitish-tomentose biennial or perennial, more or less glabrescent above; stems 30–80 cm. Basal leaves 5–25 × 2–7 cm, oblong-ovate to -lanceolate, more or less crenate, greenish above when mature; petiole 1–5 cm. Inflorescence simple; pedicels 4–9 mm. Bracts 8–11 mm, linear to linear-lanceolate. Calyx 5–8 mm; lobes ovate or oblong-lanceolate. Corolla 20–35 mm, with pellucid glands. Stamens 5, the lower with shortly decurrent anthers 1·5–2·5 mm; filament-hairs white. Capsule 5–8 mm, broadly ellipsoid. ● *S. Jugoslavia, N. Albania.* Al Ju.

V. scardicola Bornm., *Feddes Repert.* **18**: 136 (1922), from S.W. Jugoslavia (Šar Planina), is very like **42** except that the tomentum is yellow and more persistent, but according to some authors all the anthers are reniform. This point requires further investigation.

43. V. anisophyllum Murb., *Lunds Univ. Årsskr.* nov. ser., **29**(2): 177 (1933) (*V. heterophyllum* Velen., non J. Miller). Biennial; stem 40–100 cm. Basal leaves 5–10 × 2·5–4 cm, oblong-lanceolate to ovate-elliptical, crenate, densely greyish-yellow-tomentose; petiole 2–5 cm. Inflorescence lax, branched, glandular-pubescent. Bracts 3–6 mm, ovate-lanceolate, acuminate; longer pedicels 3–6 mm. Calyx 5–7 mm; lobes oblong- to linear-lanceolate. Corolla c. 30 mm in diameter. Stamens 5, the lower with shortly decurrent anthers; filament-hairs violet. Capsule 5–8 mm, ovoid-ellipsoid. ● *W. Bulgaria (Konjavska Planina, S.W. of Sofija).* Bu.

44. V. eriophorum Godron, *Mém. Acad. Montp., Sect. Méd.* **1**: 440 (1853) (*V. heteropogon* Pančić, *V. malacotrichum* Boiss. & Heldr.). Biennial, white-floccose-tomentose, often glabrescent; stems 100–250 cm. Basal leaves 25–45 × 9–15 cm, ovate-oblong, entire or crenate; petiole short or absent. Inflorescence simple or branched, lax; axis densely floccose-tomentose with coarse white hairs, and also bearing sessile glands. Bracts 8–18 mm, linear-lanceolate; longer pedicels 3–5 mm. Calyx 7–9 mm; lobes oblong-lanceolate, acute. Corolla 30–40 mm in diameter. Stamens 5, the lower with long-decurrent anthers and with densely hairy filaments 1½–2 times as long as those of the upper; filament-hairs violet. Capsule 5–10 mm, ovoid. ● *C. part of Balkan peninsula.* Bu Gr Ju.

45. V. baldaccii Degen, *Österr. Bot. Zeitschr.* **46**: 416 (1896). Like **44** but axis of inflorescence densely glandular and only sparsely white-floccose; basal leaves smaller, crenate to coarsely dentate; filaments of lower stamens sparsely hairy and about equal in length to those of the upper. ● *S.W. Jugoslavia, Albania, N.W. Greece.* Al Gr Ju.

46. V. epixanthinum Boiss. & Heldr. in Boiss., *Diagn. Pl. Or. Nov.* **1**(7): 39 (1846) (incl. *V. tymphaeum* Freyn & Sint., *V. parnassicum* Halácsy, *V. adenotrichum* Halácsy, *V. agrimonioides* Degen & Borbás). Biennial or perennial, densely to sparsely whitish- or yellowish-tomentose; stems 25–150 cm. Basal leaves 15–30 × 4–8(–10) cm, oblong-lanceolate to ovate-elliptical, entire or weakly crenate, more or less acute, petiolate; upper cauline leaves ovate-oblong to lanceolate, sessile. Inflorescence usually simple; axis densely glandular-pubescent and sparsely tomentose-floccose. Bracts 8–18 mm, linear; longer pedicels 5–10 mm. Calyx 5–10 mm; lobes usually linear-lanceolate. Corolla 25–35 mm in diameter. Stamens 5, the lower with obliquely inserted or shortly decurrent anthers 2–3 mm; filament-hairs violet. Capsule 5–8 mm, ovoid to ellipsoid. ● *Mountains of Greece.* Gr.

Very variable, especially in the density, colour and persistence of the tomentum.

V. foetidum Boiss. & Heldr. in Boiss., *Diagn. Pl. Or. Nov.* **3**(3): 141 (1856), from C. Greece (Parnassos), is like **46** but more robust, with oblong- to ovate-elliptical basal leaves up to 40 × 15 cm and anthers of lower stamens 3·5–4 mm.

47. V. euboicum Murb. & Rech. fil., *Lunds Univ. Årsskr.* nov. ser., **32**(1): 15 (1936). Like **46** but covered with dense, white tomentum, floccose above; inflorescence without glandular hairs; pedicels 1–4 mm, lower stamens with glabrous or subglabrous filaments; anthers 3–4 mm. ● *E. Greece (Evvoia).* Gr.

48. V. pelium Halácsy, *Verh. Zool.-Bot. Ges. Wien* **48**: 131 (1898). Like **46** but basal leaves 8–25 × 5–15 cm, broadly elliptic-ovate, truncate or subcordate at base, glandular-pubescent above but green and not tomentose; upper cauline leaves deltate-cordate. ● *E. Greece (Pilion).* Gr.

49. V. dieckianum Borbás & Degen, *Magyar Bot. Lapok* **4**: 82 (1905) (*V. luteoviride* Turrill). Biennial, white-tomentose-floccose at first, more or less glabrescent; stem 50–200 cm. Basal leaves 10–35 × 3–8(–12) cm, oblong-elliptical, crenate, sessile or with a short, winged petiole; upper cauline leaves ovate-cordate. Inflorescence freely branched. Bracts 8–10 mm, suborbicular-deltate, cordate, shortly cuspidate, yellowish-green, subglabrous and membranous when mature; longer pedicels 2–3(–7) mm. Calyx 5–8(–12) mm; lobes oblong-lanceolate. Corolla 20–30 mm in diameter. Stamens 5, the lower with reniform or obliquely inserted anthers; filament-hairs white. Capsule c. 5 mm, ellipsoid to subglobose. ● *Macedonia, N. Bulgaria.* Bu Gr Ju.

50. V. macedonicum Košanin & Murb., *Bull. Inst. Jard. Bot. Univ. Beograd* **1**: 220 (1930). Densely and persistently white-tomentose biennial; stem 100–150 cm. Basal leaves 8–20 × 4–8 cm, elliptical, crenate; petiole 4–9 cm. Cauline leaves decurrent. Inflorescence usually simple. Bracts *c.* 8 mm, linear-lanceolate, persistently tomentose at least beneath; pedicels almost completely adnate to axis of inflorescence. Calyx 9–11 mm; lobes linear-lanceolate. Corolla *c.* 25 mm in diameter. Stamens 5; anthers all reniform; filament-hairs white. *Calcareous rocks in a river-gorge.* ● *Makedonija (near Pološki Monaskir).* Ju.

51. V. botuliforme Murb., *Lunds Univ. Årsskr.* nov. ser., **29**(2): 288 (1933). Like **50** but larger, with compact, less densely tomentose inflorescence; bracts broadly ovate, cuspidate; calyx 5–7 mm; lobes broadly ovate, shortly cuspidate. ● *N. Greece.* Gr.

52. V. dentifolium Delile, *Sem. Hort. Bot. Monsp.* 28 (1837) (*V. granatense* Boiss.). Persistently greyish-tomentose biennial; stem 100–160 cm. Basal leaves 22–45 × 6–12 cm, oblong-oblanceolate, entire, crenate or dentate, petiolate. Inflorescence freely branched. Bracts 5–7 mm, ovate to deltate, cuspidate; longer pedicels 3–5 mm. Calyx 4–6 mm; lobes ovate-lanceolate. Corolla 22–28 mm in diameter. Stamens 5; anthers all reniform; filament-hairs violet, at least in part. Capsule 5–6 mm, sub-globose. *S. Spain (Sierra Nevada and adjoining region).* Hs. (*N.W. Africa.*)

53. V. speciosum Schrader, *Hort. Gotting.* 22 (1809). Persistently tomentose biennial; stem 50–200 cm. Basal leaves 12–40 × 3–14 cm, oblong-oblanceolate to obovate, entire, shortly petiolate. Inflorescence freely branched. Bracts (5–)8–15 (–20) mm, the lower ovate-lanceolate, the upper lanceolate; longer pedicels (3–)5–12 mm. Calyx 3–6 mm; lobes narrowly lanceolate. Corolla 18–30 mm in diameter. Stamens 5; anthers all reniform; filament-hairs white. Capsule 3–7 mm, ovoid-oblong. *S.E. & E.C. Europe, northwards to S.C. Czechoslovakia and westwards to Slovenija.* Al Au Bu Cz Gr Hu Ju ?Po Rm Rs (W) Tu.

(a) Subsp. **speciosum**: Tomentum of leaves greyish, rather harsh. Capsule 3–6 mm. *Throughout the range of the species except C. & S. Greece.*

(b) Subsp. **megaphlomos** (Boiss. & Heldr.) Nyman, *Consp.* 529 (1881): Tomentum of leaves white or yellowish, soft, thick. Capsule 5–7 mm. ● *Mountains of C. & S. Greece.*

54. V. adeliae Heldr. ex Boiss., *Diagn. Pl. Or. Nov.* 3(3): 145 (1856). Like **53**(b) but basal leaves obovate-oblong to ovate-elliptical with petiole 2–8 cm; bracts broadly ovate to orbicular, apiculate; calyx 2·5–4 mm; filament-hairs violet; capsule 3–4·5 mm. ● *Kikladhes (Naxos, Amorgos).* Gr.

55. V. lasianthum Boiss. ex Bentham in DC., *Prodr.* **10**: 234 (1846) (*V. pycnostachyum* var. *samothracicum* Degen). Like **53**(a) but basal leaves sometimes crenate; bracts broadly ovate to cordate-deltate; pedicels not more than 3 mm; calyx-lobes triangular-ovate. *Islands of Aegean region.* Gr. (*Anatolia.*)

A single plant from S. Greece (Lakonia), apparently intermediate between **53**(b) and **55**, but with nearly simple inflorescence, has been described as **V. orphanideum** Murb., *Lunds Univ. Årsskr.* nov. ser., **29**(2): 312 (1933).

56. V. undulatum Lam., *Encycl. Méth. Bot.* **4**: 221 (1797). White-, grey- or yellow-tomentose biennial or perennial; stems 30–120 cm, usually several. Basal leaves 8–18 × 5–8 cm, oblong, more or less entire, sinuate to pinnately lobed, undulate; petiole 3–7 cm. Inflorescence simple or sparingly branched, very lax, eglandular. Bracts 6–12 mm, ovate-deltate; longer pedicels 1–3 mm. Calyx 6–12 mm; lobes lanceolate. Corolla 25–50 mm in diameter, without or with few pellucid glands. Stamens 5; anthers all reniform; filament-hairs white. Capsule 4–6 mm, broadly ellipsoid. ● *S. & W. parts of Balkan peninsula.* Al Gr Ju.

Variable in colour and density of tomentum and in extent of lobing of leaves.

57. V. pentelicum Murb., *Lunds Univ. Årsskr.* nov. ser., **29**(2): 216 (1933). Like **56** but basal leaves plane, finely crenate-dentate, greenish and scarcely tomentose above; bracts and calyx glandular-hairy; corolla with numerous pellucid glands. ● *S. Greece (Pendelikon Oros).* Gr.

58. V. sinuatum L., *Sp. Pl.* 178 (1753). Biennial, shortly but densely grey- to yellow-tomentose, sometimes floccose, glandular in inflorescence; stems 50–100 cm. Basal leaves 15–35 × 6–15 cm, oblong-spathulate, sinuate-pinnatifid, often somewhat undulate, sessile or very shortly petiolate. Inflorescence freely branched, lax. Bracts 3–8 mm, cordate-deltate, shortly cuspidate; longer pedicels 2–5 mm. Calyx 2–4 mm; lobes lanceolate. Corolla 15–30 mm in diameter. Stamens 5; anthers all reniform; filament-hairs violet. Capsule 2·5–4 mm, subglobose. $2n = 30.$ *S. Europe.* Al Bl Bu Co Cr Ga Gr Hs It Ju Lu Rs (W, K) Sa Si Tu.

59. V. halacsyanum Sint. & Bornm. ex Halácsy, *Österr. Bot. Zeitschr.* **42**: 374 (1892). Biennial; stems 60–120 cm. Basal leaves 8–20 × 2·5–4 cm, narrowly oblong-oblanceolate, pinnatifid to pinnatisect, greyish-white-tomentose; segments entire to crenate-dentate but not lobed. Inflorescence simple or with a few short branches; axis more or less glabrescent; cymes persistently tomentose. Bracts 7–12(–15) mm; flowers sessile or subsessile. Calyx 7–11 mm; lobes lanceolate. Corolla 30–45 mm in diameter. Stamens 5; anthers all reniform; filament-hairs yellow. ● *N.E. Greece (near Kavalla).* Gr.

60. V. pinnatifidum Vahl, *Symb. Bot.* **2**: 39 (1791). Biennial, tomentose when young but usually more or less glabrescent; stems 30–50 cm, several, branched from near the base into numerous slender, simple, lax racemes. Basal leaves 6–17 × 2·5–6 cm, oblong, incise-dentate to deeply pinnatisect, the lobes entire, dentate or pinnatifid. Bracts 9–15 mm, triangular-ovate; flowers sessile. Calyx 5–7 mm; lobes linear-lanceolate. Corolla 25–30 mm in diameter. Stamens 5; anthers all reniform; filament-hairs white or yellow. Capsule 4–5 mm, ovoid-ellipsoid. *Aegean region; Krym.* Gr Rs (K) Tu.

61. V. leucophyllum Griseb., *Spicil. Fl. Rumel.* **2**: 46 (1844). Biennial; stems 20–70 cm. Basal leaves 4–15 × 2–5 cm, oblong-lanceolate, sinuate to pinnatifid (sometimes pinnatisect towards the base), more or less undulate, silvery-tomentose; petiole 2–5 cm. Inflorescence freely branched, lax, sparsely tomentose, eglandular. Bracts 3–7 mm, lanceolate; longer pedicels 2–5 mm. Calyx 4–6 mm; lobes linear-oblong. Corolla 20–30 mm in diameter; outer surface tomentose. Stamens 5; anthers all reniform; filament-hairs white. Capsule 5–6 mm, ellipsoid. ● *S. & S.W. parts of Balkan peninsula.* Al Gr Ju.

62. V. cylindrocarpum Griseb., *loc. cit.* (1844). Biennial; stems 30–80 cm. Basal leaves 8–25 × 1·5–4 cm, oblong-lanceolate, sinuate-pinnatifid with more or less crenate lobes, more or less densely grey-tomentose; petiole 1–4 cm. Inflorescence branched, lax, glabrous. Bracts 1–2 mm, ovate-lanceolate; usually two bracteoles in each fascicle of flowers; pedicels 2–5 mm. Calyx 2–3 mm; lobes oblong to lanceolate. Corolla 12–15 mm in diameter, glabrous. Stamens 5; anthers all reniform; filament-hairs white. Style *c.* 5 mm. Capsule 4–5 mm, cylindrical. ● *N. Aegean region* (*Thasos*). Gr.

63. **V. pseudonobile** Stoj. & Stefanov, *Jahrb. 20 Univ. Sofia Agron. Fak.* 2: 69 (1924). Like **62** but basal leaves 5–8 cm wide, sometimes sparsely tomentose; style 6–7 mm; capsule 3·5–4 mm, subglobose. ● *S.W. Bulgaria, N.E. Greece.* Bu Gr.

64. **V. humile** Janka, *Österr. Bot. Zeitschr.* 23: 241 (1873). Biennial or perennial, arachnoid-tomentose on younger parts but more or less glabrescent; stems 20–80 cm. Basal leaves (3–)10–20(–30) × 1–5 cm, lanceolate to narrowly elliptical, crenate, sometimes slightly lobed towards the base; petiole 1–10 cm. Inflorescence branched, lax. Bracts 6–10 mm, linear; longer pedicels 3–5 mm. Calyx 2–5 mm; lobes linear. Corolla 15–25 mm in diameter; outer surface tomentose. Stamens 5; anthers all reniform; filament-hairs white or pale violet. Capsule 3–5 mm, ellipsoid. ● *S.E. part of Balkan peninsula.* Bu Gr Tu.

65. **V. adenanthum** Bornm., *Feddes Repert.* 18: 137 (1922). Grey-green perennial, rather sparsely tomentose and often glabrescent above, densely tomentose below; stem 60–80 cm. Basal leaves 4–9 × 1–2·5 cm, oblong to elliptical, crenate to subentire; petiole 1–4 cm. Inflorescence freely branched, lax. Bracts 2–4 mm, linear-lanceolate; longer pedicels 4–8 mm. Calyx 2–3·5 mm; lobes linear-lanceolate to -oblong. Corolla 20–30 mm in diameter; outer surface sparsely tomentose. Stamens 5; anthers all reniform; filament-hairs white. Capsule 5–8 mm, cylindrical-ellipsoid. ● *Macedonia.* Bu Gr Ju.

66. **V. nobile** Velen., *Fl. Bulg., Suppl.* 209 (1898) (incl. *V. nikolovii* Stoj.). Stems 60–140 cm. Basal leaves 5–27 × 1–7·5 cm, oblong-oblanceolate, sinuate to pinnatifid or pinnatisect, with obtuse or subacute, remotely dentate segments, grey- or yellowish-tomentose at least beneath, distinctly petiolate. Inflorescence branched, lax, glabrous; fascicles with 1–3 flowers. Bracts 1–2 mm, ovate-lanceolate; bracteoles absent; longer pedicels 3–5 mm. Calyx 2–3(–4) mm; lobes oblong-linear to oblong-lanceolate. Corolla 20–35 mm in diameter, glabrous. Stamens 5; anthers all reniform; filament-hairs yellowish. Capsule *c.* 3 mm, globose. ● *S. & E. Bulgaria, N.E. Greece.* Bu Gr.

Smaller plants from N. Greece, with leaves deeply pinnatisect with pinnatifid to incise-dentate segments and corolla 15–20 mm in diameter, have been described as **V. dingleri** Mattf. & Stefanov, *Bull. Soc. Bot. Bulg.* 1: 101 (1926) (*V. parviflorum* auct. balcan., non Lam.).

67. **V. graecum** Heldr. & Sart. ex Boiss., *Diagn. Pl. Or. Nov.* 3(3): 148 (1856). Biennial, densely white-tomentose below, more or less floccose and often glabrescent above; stem 40–150 cm. Basal leaves 5–30 × 3–10 cm, ovate to lanceolate, entire or crenulate; petiole 1·5–10 cm. Inflorescence branched, lax. Bracts 2–4(–10) mm, linear to linear-lanceolate; longer pedicels 2–5(–15) mm. Calyx 2–5 mm; lobes linear-oblong. Corolla 15–30 mm in diameter. Stamens 5; anthers all reniform; filament-hairs white. Capsule 4–7 mm, oblong-cylindrical. ● *S. part of Balkan peninsula.* ?Al Gr Ju Tu.

V. haussknechtii Heldr. ex Hausskn., *Mitt. Geogr. Ges. Thür. Jena* 5 (*Bot. Ver.*): 71 (1887) (*V. degenii* Halácsy), from E. & N.E. Greece, and **V. dimoniei** Velen., *Sitz.-Ber Böhm. Ges. Wiss.* (*Math.-Nat. Kl.*) 1910(8): 10 (1910), from N. Greece, closely resemble **67**; the former has more or less glabrescent leaves, the latter has densely glandular bracts and calyx-lobes. Both are best, perhaps, treated as varieties.

68. **V. herzogii** Bornm., *Feddes Repert.* 19: 97 (1923). Like **67** but inflorescence simple or sparingly branched; leaves distinctly crenate; bracts deltate-orbicular; calyx-lobes ovate. ● *N.W. Macedonia.* Ju.

69. **V. mallophorum** Boiss. & Heldr. in Boiss., *Diagn. Pl. Or. Nov.* 1(7): 39 (1846). Biennial, persistently white-tomentose below, floccose and more or less glabrescent above. Basal leaves 15–45 × 5–15 cm, obovate-oblong, entire; petiole short or absent. Cauline leaves not decurrent. Inflorescence freely branched, moderately dense. Bracts 4–7 mm, linear; longer pedicels 3–7 mm. Calyx 2·5–4 mm; lobes linear, acute. Corolla 15–30 mm in diameter. Stamens 5; anthers all reniform; filament-hairs violet. Capsule 4–6 mm, cylindrical. ● *C. & S. Italy; C. & S. Greece, S. Albania.* Al Gr It.

70. **V. decorum** Velen., *Sitz.-Ber. Böhm. Ges. Wiss.* (*Math.-Nat. Kl.*) 1890(1): 56 (1890). More or less densely white-floccose biennial; stems 50–120 cm. Basal leaves 7–15 × 3–8 cm, ovate to obovate-oblong, crenate, often slightly lobed at the cuneate base; petiole 2–6 cm, winged. Cauline leaves numerous, the upper shortly decurrent. Inflorescence sparingly branched, rather lax. Bracts (4–)8–11(–15) mm, lanceolate to linear-lanceolate; longer pedicels up to 5 mm, but flowers sometimes subsessile. Calyx 2–4 mm; lobes oblong to linear-lanceolate. Corolla 20–30 mm in diameter. Stamens 5; anthers all reniform; connectives of two lower stamens smooth; filament-hairs pale yellow. Capsule 5–8 mm, pyramidal-ovoid. *Crevices of limestone rocks.* ● *Mountains of S. Bulgaria* (*C. Rodopi*). Bu.

71. **V. mucronatum** Lam., *Encycl. Méth. Bot.* 4: 218 (1797). Densely white-tomentose; stems up to 200 cm. Basal leaves 25–65 × 7–20(–25) cm, ovate or obovate to oblong-lanceolate, coarsely crenate to subundulate; petiole 2–8 cm. Cauline leaves ovate to ovate-lanceolate, sessile, long-decurrent. Inflorescence freely branched, rather lax; pedicels 0–3 mm. Bracts 5–7 mm, lanceolate or ovate-lanceolate, acuminate. Calyx 3–5 mm; lobes lanceolate. Corolla 20–32 mm in diameter. Stamens 5; anthers all reniform; connectives all papillose; filament-hairs white or yellow. Capsule 3–4 mm, globose. *Turkey-in-Europe.* ?Cr Tu. (*Anatolia and E. Aegean region.*)

72. **V. pulverulentum** Vill., *Prosp. Pl. Dauph.* 22 (1779) (incl. *V. acutifolium* Halácsy, *V. floccosum* Waldst. & Kit.). Densely white-floccose biennial, the older parts more or less glabrescent; stems 50–120 cm. Basal leaves 12–40 × 6–15 cm, obovate-oblong to oblanceolate, crenate to subentire, subsessile or shortly petiolate; cauline leaves usually not decurrent. Inflorescence branched, fairly dense. Bracts 3–5 mm, linear; longer pedicels 2–5(–7) mm. Calyx 2–3·5 mm; lobes linear-lanceolate. Corolla 18–25 mm in diameter. Stamens 5; anthers all reniform; filament-hairs white. Capsule 3–5(–8) mm, ellipsoid-globose. *W., S. & S.C. Europe, northwards to England.* Al Be Br Bu Co Ga Ge Gr He Hs Hu It Ju Lu Rm Sa Si [Au].

73. **V. gnaphalodes** Bieb., *Fl. Taur.-Cauc.* 3: 152 (1819). Densely white-floccose biennial, the older parts more or less glabrescent; stems 100–150 cm. Basal leaves 15–50 × 6–15 cm,

elliptic- to obovate-oblong, obscurely crenate, greyish-green above, white-tomentose beneath; petiole 5–10 cm. Inflorescence simple, dense, stout. Bracts 6–10 mm, ovate to ovate-lanceolate, acuminate-cuspidate; longer pedicels 5–10(–12) mm. Calyx 4–7 mm; lobes narrowly oblong-lanceolate. Corolla 15–30 mm in diameter. Stamens 5; anthers all reniform; filament-hairs white. Capsule 5–6 mm, ovoid-ellipsoid. *Krym.* Rs(K). (*Caucasian region.*)

74. V. glandulosum Delile, *Ann. Sci. Nat.* ser. 3 (Bot.), **12**: 367 (1849) (*V. gloeotrichum* Hausskn. & Heldr., *V. meteoricum* Hausskn.). Biennial; stems c. 100 cm, densely glandular-hairy with long and short hairs. Basal leaves 15–22 × 5–11 cm, ovate- to oblanceolate-oblong, crenate, gradually tapered to the winged petiole, grey-tomentose on both surfaces and glandular-pubescent above, but partly glabrescent. Inflorescence freely branched, rather lax, glandular-hairy. Bracts 5–15 mm, ovate-lanceolate; longer pedicels 3–8 mm. Calyx 3–7 mm; lobes oblong-lanceolate. Corolla 20–35 mm in diameter, with pellucid glands. Stamens 5; anthers all reniform; filament-hairs white. Capsule 4–6 mm, ovoid-ellipsoid. ● *N. & E. Greece, S. Jugoslavia.* Gr Ju.

75. V. davidoffii Murb., *Lunds Univ. Årsskr.* nov. ser., **29**(2): 302 (1933). Like **74** but glandular hairs on stem and inflorescence all short and with black glands; basal leaves 5–15 × 3–7 cm, subentire, less gradually tapered to petiole; calyx-lobes linear; corolla without pellucid glands; capsule (5–)6–9 mm, subglobose. ● *S.W. Bulgaria* (*Pirin Planina*). Bu.

76. V. jankaeanum Pančić, *Nov. Elem. Fl. Bulg.* 32 (1886). Like **74** but glandular hairs all short and confined to inflorescence; basal leaves 30–40 × 9–11 cm, more completely glabrescent above, subentire; longer pedicels 6–13 mm; capsule 5–8 mm. ● *Mountains of W. Bulgaria.* Bu.

77. V. durmitoreum Rohlena, *Feddes Repert.* **3**: 149 (1906). Sparsely yellowish-tomentose and glandular-pubescent biennial or perennial; stems 30–50 cm. Basal leaves 4–15 × 3–8 cm, ovate, crenate, acute, rounded to subcordate at base; petiole 2–7 cm. Inflorescence usually simple. Bracts 7–10 mm, linear to linear-lanceolate; longer pedicels 5–10 mm. Calyx 5–9 mm; lobes linear-oblong. Corolla 25–30 mm in diameter, yellow. Stamens 5; anthers all reniform; filament-hairs violet. Capsule 6–7 mm, ovoid-ellipsoid. ● *S.W. Jugoslavia* (*mountains of Crna Gora and Hercegovina*). Ju.

78. V. rotundifolium Ten., *Fl. Nap.* **1**, *Prodr.*: 66 (1811). White-tomentose biennial, more or less floccose above; stem 50–150 cm. Basal leaves 15–25 × 6–17 cm, ovate-elliptical to suborbicular, somewhat crenate or entire; petiole 7–15 cm. Inflorescence simple or sometimes branched. Bracts 8–15 mm, lanceolate, acuminate; longer pedicels 3–6 mm. Calyx 5–8 mm; lobes linear-lanceolate. Corolla 15–40 mm in diameter. Stamens 5, the lower 2 with filaments much longer than the others; anthers all reniform, or the 2 lower sometimes somewhat oblique; filament-hairs violet. Capsule 7–11 mm, ovoid to conical. *W. Mediterranean region.* Co Hs It Sa Si.

1 Leaves with ±persistent, rough tomentum; capsule conical
　　　　　　　　　　　　　　　　(c) subsp. conocarpum
1 Leaves with ±floccose, soft tomentum; capsule ellipsoid to ovoid
2 Basal leaves ovate to orbicular, obtuse　**(a) subsp. rotundifolium**
2 Basal leaves ovate-elliptical, ±acute　　**(b) subsp. haenseleri**

(a) Subsp. **rotundifolium**: Basal leaves ovate to orbicular, obtuse; leaves ±floccose with soft tomentum. Capsule ellipsoid to ovoid. *S. Italy, Sicilia.*

(b) Subsp. **haenseleri** (Boiss.) Murb., *Lunds Univ. Årsskr.* nov. ser., **29**(2): 401 (1933) (*V. haenseleri* Boiss.): Basal leaves ovate-elliptical, subacute; leaves with floccose, soft, tomentum. Capsule ellipsoid to ovoid. *C., S. & S.E. Spain.*

(c) Subsp. **conocarpum** (Moris) I. K. Ferguson, *Bot. Jour. Linn. Soc.* **65**: 269 (1972) (*V. conocarpum* Moris, *V. boerhavii* subsp. *conocarpum* (Moris) Nyman): Basal leaves ovate to ovate-elliptical, subacute; indumentum more or less persistent, rough. Capsule conical. ● *Corse and Sardegna.*

A single plant from S. Jugoslavia, resembling **78** in vegetative characters but **77** in its inflorescence, has been named **V. chrysanthum** Murb., *Lunds Univ. Årsskr.* nov. ser., **29**(2): 384 (1933).

79. V. reiseri Halácsy, *Verh. Zool.-Bot. Ges. Wien* **48**: 132 (1898). Densely white-tomentose biennial, only slightly floccose; stems 20–25 cm. Basal leaves 4–9 × 2·5–5 cm, obovate- to oblong-elliptical, obscurely crenate; petiole 2–6 cm. Inflorescence usually somewhat branched, dense. Bracts c. 10 mm, the lower ovate-lanceolate, the upper linear; longer pedicels 3–8 mm. Calyx 5–8 mm; lobes linear. Corolla 15–20 mm in diameter. Stamens 5; anthers all reniform; filament-hairs yellow. Capsule 6–8 mm, ovoid. ● *S.C. Greece* (*Vardhousia Ori*). Gr.

80. V. delphicum Boiss. & Heldr. in Boiss., *Diagn. Pl. Or. Nov.* **3**(3): 146 (1856). Biennial; stem 50–150 cm, more or less white-tomentose. Basal leaves 6–30 × 4–15 cm, ovate, crenate to subentire, rounded to cordate at base, sparsely hairy or subglabrous above, densely whitish-tomentose beneath; petiole 5–15 cm. Inflorescence simple or branched, fairly dense. Bracts 4–10 mm, ovate-lanceolate to lanceolate; longer pedicels 4–7 mm. Calyx 3–5 mm; lobes linear-lanceolate. Corolla 15–20 mm in diameter. Stamens 5; anthers all reniform; filament-hairs white. Capsule 3–5 mm, ovoid-ellipsoid. ● *E. Greece.* Gr.

81. V. banaticum Schrader, *Monogr. Verbasci* **2**: 28 (1823). Biennial; stems 50–100 cm, sparsely tomentose, more or less glabrescent. Basal leaves 10–30 × 3–12 cm, oblong-obovate, deeply and irregularly crenate and usually pinnately lobed towards the shortly cuneate base, persistently tomentose beneath, more or less glabrescent above; petiole 4–10 cm. Inflorescence freely branched, lax; fascicles with 1–6 flowers. Bracts 3–10 mm, the lower ovate-cordate, the upper lanceolate; longer pedicels 3–10 mm. Calyx 2·5–4 mm; lobes linear-lanceolate. Corolla 15–22 mm in diameter. Stamens 5; anthers all reniform; filament-hairs white or yellow. Capsule 3–4 mm, ellipsoid-subglobose. ● *S.E. Europe, eastwards to S.W. Ukraine.* Al Bu Gr Ju Rm Rs (W) Tu.

82. V. lychnitis L., *Sp. Pl.* 177 (1753). Sparsely grey-tomentose biennial; stems 50–150 cm. Basal leaves 15–30 × 6–15 cm, ovate to oblanceolate-oblong, coarsely crenate to subentire, cuneate at base, green above, whitish-tomentose beneath; petiole 0·5–5 cm. Inflorescence usually freely branched. Bracts 8–15 mm, linear to lanceolate; longer pedicels 6–11 mm. Calyx (1·5–)2·5–4 mm; lobes lanceolate. Corolla 12–20 mm in diameter, yellow or white. Stamens 5; anthers all reniform; filament-hairs yellow or white. Capsule 4–5 mm, ovoid-ellipsoid. *Europe, southwards from England and from c. 56° N. in C. Russia, but rare in the Mediterranean region.* Al Au Be Br Bu Co Cz Da Ga Ge He Ho Hs Hu It Ju Po Rm Rs (C, W, K, E) [Fe ?No Rs (B) Su].

83. V. chaixii Vill., *Prosp. Pl. Dauph.* 22 (1779). Perennial, more or less greyish-tomentose to -pubescent; stems 50–100 cm. Basal leaves 10–30 × 4–12 cm, ovate-oblong, crenate and sometimes slightly lobed towards the truncate to cuneate base; petiole 5–25 cm. Inflorescence usually branched. Bracts 2–5 mm, linear to linear-lanceolate; longer pedicels 3–6 mm. Calyx 3–5 mm; lobes linear to lanceolate. Stamens 5; anthers all reniform; filament-hairs violet. Capsule 3–6 mm, ellipsoid. *S., C. & E. Europe, westwards to N.E. Spain and northwards to S. Poland and C. Russia.* Au Bu Cz Ga Gr He Hs Hu It Ju Po Rm Rs (C, W, K, E) Si.

1 Filaments of lower stamens hairy throughout; corolla 20–25
   (–30) mm in diameter   **(c) subsp. orientale**
1 Filaments of lower stamens glabrous for a short distance below
   the anther; corolla 15–22 mm in diameter
2 Basal leaves usually slightly lobed towards the base, greyish-
   tomentose beneath; upper cauline leaves remotely dentate
                           (a) subsp. chaixii
2 Basal leaves not lobed, green beneath; upper cauline leaves
   finely crenulate-serrate   **(b) subsp. austriacum**

(a) Subsp. **chaixii**: Basal leaves usually slightly lobed towards the base, greyish-tomentose beneath; upper cauline leaves remotely dentate. Filaments of lower stamens glabrous for a short distance below the anther. Corolla 15–22 mm in diameter. ● *S. Europe, from Spain to Jugoslavia.*

(b) Subsp. **austriacum** (Schott ex Roemer & Schultes) Hayek, *Prodr. Fl. Penins. Balcan.* 2: 127 (1929) (*V. austriacum* Schott ex Roemer & Schultes): Basal leaves not lobed, green beneath; upper cauline leaves finely crenulate-serrate. Filaments of lower stamens glabrous for a short distance below the anther. Corolla 15–22 mm in diameter. *E.C. Europe and N. part of Balkan peninsula.*

(c) Subsp. **orientale** Hayek, *loc. cit.* (1929) (*V. orientale* Bieb., non (L.) All.): Basal leaves crenate, densely tomentose beneath. Filaments of lower stamens hairy right up to anther. Corolla 20–25(–30) mm in diameter. *S. & C. parts of U.S.S.R.; S. Romania.*

84. V. bithynicum Boiss., *Diagn. Pl. Or. Nov.* 1(4): 63 (1844). Like **83(c)** but basal leaves up to 15 cm wide, obovate-oblong, cuneate at the base, sessile or subsessile; longer pedicels 5–10 mm; corolla 10–17 mm in diameter. *Turkey-in-Europe (near Istanbul).* Tu. (*N.W. Anatolia.*)

85. V. nigrum L., *Sp. Pl.* 178 (1753). Rather sparsely pubescent, rarely tomentose perennial; stems 50–100 cm. Basal leaves 15–40 × 5–25 cm, ovate to oblong, cordate, crenate. Bracts 4–7(–15) mm, linear; longer pedicels 5–12(–15) mm. Calyx 2·5–4·5 mm; lobes linear. Corolla 18–25 mm in diameter, yellow (rarely white). Stamens 5; anthers all reniform; filament-hairs violet. Capsule 4–5 mm, ovoid-ellipsoid. *From England, C. Fennoscandia and N. Russia southwards to N. Spain, N. Italy, N. Greece and S. Ukraine.* Al Au Be Bl Br Bu Cz Da Fe Ga Ge Gr He Ho Hs Hu It Ju No Po Rm Rs (N, B, C, W, E) Su.

(a) Subsp. **nigrum**: Basal leaves 15–30 × 5–15 cm, with petiole 5–15 cm. Inflorescence simple or rarely with a few erect branches. *Throughout the range of the species except S. part of Balkan peninsula.*

(b) Subsp. **abietinum** (Borbás) I. K. Ferguson, *Bot. Jour. Linn. Soc.* 64: 232 (1971) (*V. abietinum* Borbás, *V. bornmuelleri* Velen.): Basal leaves 20–40 × 10–25 cm, with petiole 10–30 cm. Inflorescence branched with suberect or ascending branches. ● *Balkan peninsula, W. & C. Romania.*

---

¹ By I. B. K. Richardson.

V. hypoleucum Boiss. & Heldr. in Boiss., *Diagn. Pl. Or. Nov.* 3(3): 147 (1856) (*V. nigrum* subsp. *hypoleucum* (Boiss. & Heldr.) Nyman) was described from 2 plants from S. Greece (Killini Oros). They are like **85(b)** but with smaller leaves (as in **85(a)**), which are sinuately lobed with 4–9 deltate, subentire lobes on each side and are somewhat floccose-tomentose. Until more material is forthcoming their taxonomic status must remain doubtful.

86. V. lanatum Schrader, *Mongr. Verbasci* 2: 28 (1823). Perennial; stems 50–120 cm. Basal leaves 8–22(–35) × 5–13(–22) cm, broadly ovate to oblong, cordate or rarely truncate at base, singly or doubly crenate to incise-dentate, sparsely hairy and green above, more or less grey-tomentose beneath; petiole 5–13(–25) cm. Inflorescence simple. Bracts 6–15 mm, linear; longer pedicels 5–12 mm. Calyx 3–5 mm; lobes linear to lanceolate, acute, glabrous. Corolla 16–28 mm in diameter. Stamens 5; anthers all reniform; filament-hairs violet. Capsule 4–5 mm, ellipsoid. *Chiefly in mountain woods.* ● *E.C. Europe, N. Italy, N. Bulgaria.* Au Bu It Ju Rm Rs (W).

87. V. glabratum Friv., *Flora (Regensb.)* 19: 440 (1836). Perennial; stems 40–120 cm. Basal leaves 17–25(–32) × 7–12(–18) cm, broadly ovate to oblong, sparsely hairy above, more or less densely white-tomentose beneath, at least on the veins; petiole 15–22 cm, often densely white-tomentose or lanate. Inflorescence branched. Bracts 6–15 mm, linear; longer pedicels 5–18 mm. Calyx 3·5–4 mm; lobes lanceolate, glabrous. Corolla 16–22 mm in diameter. Stamens 5; anthers all reniform; filament-hairs violet. Capsule 4–5 mm, oblong- or subglobose-ellipsoid. ● *Balkan peninsula, Romania, E. Carpathians.* Al Bu Gr Ju Rm Rs (W).

1 Basal leaves coarsely crenate or lobed   **(c) subsp. bosnense**
1 Basal leaves coarsely dentate or serrate
2 Fascicles 2- to 7-flowered; pedicels of primary flowers 6–7 mm
                           (a) subsp. glabratum
2 Fascicles (1–)2- to 3-flowered; pedicels of primary flowers
   (6–)8–18 mm   **(b) subsp. brandzae**

(a) Subsp. **glabratum**: Basal leaves broadly ovate-cordate, coarsely and doubly dentate or serrate, incise-dentate towards the base. Cauline leaves ovate-cordate. Fascicles 2- to 7-flowered. Pedicels of primary flowers 6–7 mm. *From C. Greece to the Ukrainian Carpathians and C. Jugoslavia.*

(b) Subsp. **brandzae** (Franchet ex Brandza) Murb., *Lunds Univ. Årsskr.* nov. ser., 29(2): 445 (1933) (*V. rohlenae* K. Malý): Basal leaves ovate-oblong, coarsely incise-dentate, truncate or shortly cuneate at base. Cauline leaves elliptical to triangular-ovate. Fascicles (1–)2- to 3-flowered. Pedicels of primary flowers (6–)8–18 mm. *From Albania to S.W. Romania.*

(c) Subsp. **bosnense** (K. Malý) Murb., *op. cit.* 447 (1933) (*V. bosnense* K. Malý). Basal leaves oblong-elliptical, coarsely crenate or lobed, especially towards the base. Cauline leaves oblong to oblong-lanceolate, crenate or lobed. Fascicles 2- to 7-flowered. Pedicels of primary flowers 7–15 mm. *From C. Jugoslavia to Albania and N. Greece.*

8. Scrophularia L.¹

Herbs or small shrubs; stems erect or ascending. Leaves usually opposite, entire to 2-pinnatisect. Flowers in cymes (rarely reduced to a single flower) in the axils of leaf-like or reduced bracts, forming more or less distinct terminal panicles or racemes. Calyx equally 5-lobed. Corolla with a short, more or less globose tube and usually 2-lipped limb. Fertile stamens 4; the fifth (adaxial) stamen usually represented by a scale-like staminode, rarely absent. Capsule septicidal. Seeds numerous, ovoid, rugose.

Literature: H. Stiefelhagen, *Bot. Jahrb.* **44**: 406–496 (1910). A. Vaarama & H. Hiirsalmi, *Hereditas* **58**: 333–358 (1967).

1 Most of the leaves pinnatisect, 3-sect or deeply pinnatifid
  2 Corolla 3–5(–8) mm; staminode linear to lanceolate, or absent   **(25–28). canina** group
  2 Corolla (4–)6–20 mm; staminode orbicular to reniform
    3 Most of the pedicels shorter than or equalling calyx
      4 Perennial, usually with several stems
        5 Corolla 12–20 mm   **(10–13). sambucifolia** group
        5 Corolla 4–9 mm   **23. heterophylla**
      4 Biennial, usually with a single stem
        6 Leaf-lobes ovate to oblong; mucro of capsule *c.* 1 mm **21. lucida**
        6 Leaf-lobes ±linear; mucro of capsule 2–3 mm **22. myriophylla**
    3 Most of the pedicels longer than calyx
      7 Leaves densely glandular-pubescent   **9. grandiflora**
      7 Leaves ±glabrous
        8 Corolla at least 10 mm   **(10–13). sambucifolia** group
        8 Corolla not more than 9 mm
          9 Stems winged on the angles; leaves usually with 1 pair of basal lobes   **19. auriculata**
          9 Stems not or scarcely winged; leaves usually with more than one pair of basal lobes
            10 Most of the pedicels less than twice as long as calyx; capsule 3–6 mm; leaves often glaucous **23. heterophylla**
            10 Most of the pedicels at least twice as long as calyx; capsule 5–8 mm; leaves not glaucous   **(10–13). sambucifolia** group
1 Most of the leaves entire to incise-dentate or shallowly pinnatifid
  11 Corolla not 2-lipped, with ±equal lobes   **1. vernalis**
  11 Corolla 2-lipped
    12 Stems winged on the angles
      13 Rhizome nodular; scarious margin of calyx-lobes less than 0·5 mm wide   **18. nodosa**
      13 Rhizome not nodular; scarious margin of calyx-lobes more than 0·5 mm wide
        14 Staminode suborbicular, ±entire; leaves often cordate at base   **19. auriculata**
        14 Staminode wider than long, ±2-lobed; leaves not cordate at base   **20. umbrosa**
    12 Stems not winged
      15 Leaves glabrous or subglabrous
        16 Annual
          17 Capsule subglobose; calyx-lobes acute, without scarious margin   **5. peregrina**
          17 Capsule conical; calyx-lobes obtuse, with a narrow scarious margin   **6. arguta**
        16 Usually perennial; leaves sometimes divided
          18 Most of the pedicels shorter than or equalling calyx
            19 Staminode orbicular to reniform   **23. heterophylla**
            19 Staminode linear to ovate, or absent **(25–28). canina** group
          18 Most of the pedicels longer than calyx
            20 Bracts mostly leaf-like   **(10–13). sambucifolia** group
            20 Bracts mostly not leaf-like
              21 Leaves pubescent, at least on the veins
                22 Leaves mostly less than 9 cm, doubly dentate; cymes with not more than 7 flowers **15. scopolii**
                22 Leaves up to 15 cm, simply dentate; cymes with up to 30 flowers   **16. alpestris**
              21 Leaves glabrous
                23 Rhizome nodular; scarious margin of calyx-lobes less than 0·5 mm wide   **18. nodosa**
                23 Rhizome not nodular; scarious margin of calyx-lobes more than 0·5 mm wide   **14. laevigata**
      15 Leaves ±densely (though sometimes very shortly) puberulent to pubescent or lanate
        24 Leaves not more than 1·5 cm wide, minutely but densely glandular-puberulent; corolla 3–6(–10) mm (S. part of U.S.S.R.)

          25 Leaves 0·2–0·7 cm wide, ±linear-oblong   **29. cretacea**
          25 Leaves 0·6–1(–1·5) cm wide, oblong-lanceolate to ovate-elliptical   **30. rupestris**
        24 Larger leaves more than 1·5 cm wide, puberulent to lanate; corolla 7–18 mm
          26 Bracts mostly not leaf-like
            27 Most of the pedicels longer than calyx
              28 Scarious margin of calyx-lobes more than 0·5 mm wide; capsule narrowed abruptly to a mucro or beak   **15. scopolii**
              28 Scarious margin of calyx-lobes less than 0·5 mm wide; capsule acuminate   **17. herminii**
            27 Most of the pedicels shorter than or equalling calyx
              29 Leaves minutely glandular-puberulent   **23. heterophylla**
              29 Leaves pubescent   **24. taygetea**
          26 Bracts mostly leaf-like
            30 Scarious margin of calyx-lobes more than 0·5 mm wide
              31 Stems lanate; leaves ±orbicular   **7. pyrenaica**
              31 Stems pubescent; leaves ovate to lanceolate
                32 Calyx-lobes *c.* 4 mm, with scarious margins less than 1 mm wide; capsule 5–8 mm   **8. scorodonia**
                32 Calyx-lobes 5–6 mm, with scarious margin more than 1 mm wide; capsule 7–10 mm   **9. grandiflora**
            30 Scarious margin of calyx-lobes less than 0·5 mm wide
              33 Calyx hirsute in fruit   **3. aestivalis**
              33 Calyx glabrous in fruit
                34 Calyx-lobes pubescent in flower; pedicels glandular-pubescent   **2. bosniaca**
                34 Calyx-lobes glabrous in flower; pedicels glandular-lanate   **4. divaricata**

1. S. vernalis L., *Sp. Pl.* 620 (1753). Biennial or perennial; stems up to 100 cm, glandular-pubescent. Leaves 3–7(–15) cm, broadly ovate, acute to subobtuse, cordate to subtruncate at base, deeply doubly serrate, more or less pubescent. Bracts mostly leaf-like. Calyx-lobes oblong-lanceolate, obtuse to subacute, glandular-pubescent; margin not scarious. Corolla 6–8 mm, yellow, not 2-lipped, and with subequal lobes. Staminode absent. Capsule 8–10 mm, ovoid-conical, acuminate. 2*n* = 40. *Mountain woods; also as an alien in damp waste places. From S.C. Germany and C. Russia southwards locally to the Pyrenees, Sicilia and C. Jugoslavia.* Au Cz Ga Ge ?Hs Hu It Ju ?Po Rm Rs (C, W, E) Si [Be Br Da ?He Ho Su].

2. S. bosniaca G. Beck, *Ann. Naturh. Mus. (Wien)* **2**: 135 (1887). Perennial; stems 50–80 cm, more or less glandular-pubescent. Leaves up to 10 × 6 cm, ovate, rounded to subcordate at base, doubly dentate, puberulent. Cymes usually 6- to 9-flowered; bracts mostly leaf-like; pedicels (1–)3–5 times as long as the calyx, glandular-puberulent. Calyx-lobes ovate, obtuse, more or less pubescent in flower, glabrous in fruit; margin not or very narrowly scarious. Corolla *c.* 9 mm, brownish-green. Staminode suborbicular. Capsule 8–10 mm, ovoid, acuminate. *Mountain woods.* ● *S.W. Jugoslavia, N. Albania.* Al Ju.

3. S. aestivalis Griseb., *Spicil. Fl. Rumel.* **2**: 36 (1844). Like **2** but stems glandular-villous; leaves irregularly dentate, lanate to villous; cymes mostly 3- to 6-flowered; pedicels 2–3(–4) times as long as the calyx; calyx-lobes oblong, obtuse to subacute, hirsute in fruit; corolla *c.* 8 mm, pinkish- or greenish-yellow; capsule ovoid to subglobose, acuminate. *Shady places on mountains.* 2*n* = 44. ● *C. part of Balkan peninsula.* Al Bu Gr Ju.

4. S. divaricata Ledeb., *Icon. Pl. Fl. Ross.* **2**: 10 (1830). Like **2** but sometimes biennial; leaves truncate at base, pubescent; cymes 5- to many-flowered; pedicels glandular-lanate; calyx-lobes ovate to lanceolate, subacute, glabrous; staminode reniform; capsule *c.* 6 mm, subglobose, acuminate. *Calcareous hillsides. S.C. Russia (near Bogučar).* Rs (E). (*Caucasus.*)

5. S. peregrina L., *Sp. Pl.* 621 (1753). Glabrous or subglabrous annual; stems 15–90 cm. Leaves up to 10 × 6 cm, ovate, irregularly serrate, acute to subobtuse, cordate, truncate, or rarely rounded at base. Bracts mostly leaf-like; pedicels 2–3 times as long as the calyx, glandular. Calyx-lobes triangular- to ovate-lanceolate, acute; margin not scarious. Corolla 6–9 mm, dark red to purplish-brown. Staminode obovate-orbicular, obtuse. Capsule *c.* 6 mm, subglobose, subacute. $2n = 36$. *Scrub, cultivated ground and waste places. Mediterranean region, Portugal; an occasional casual further north.* Al Bl Co Cr Ga Gr Hs It Ju Lu Sa Si.

6. S. arguta Aiton, *Hort. Kew.* **2**: 342 (1789). Like **5** but stems more or less pubescent; leaves not more than 4 × 3 cm, doubly dentate, acute to subobtuse; pedicels 1–2 times as long as the calyx; calyx-lobes obtuse, with narrow, scarious margin; capsule 6–8 mm, conical. *Rocky places. S.E. Spain (Sierra de Gádor).* Hs. (*N. Africa, Islas Canarias.*)

7. S. pyrenaica Bentham in DC., *Prodr.* **10**: 306 (1846). Perennial; stems 10–80 cm, glandular-lanate. Leaves up to 10 × 10 cm, more or less orbicular, cordate to truncate at base, doubly dentate, densely glandular-pubescent. Bracts mostly leaf-like; pedicels 2–4 times as long as the calyx, glandular-pubescent. Calyx-lobes orbicular to ovate, obtuse; margin broadly scarious. Corolla 8–11 mm, yellowish, with reddish-brown upper lip. Staminode more or less reniform. Capsule *c.* 4 mm, ovoid-globose, acuminate. $2n = 58$. *Mountain rocks.* ● *Pyrenees.* Ga Hs.

8. S. scorodonia L., *Sp. Pl.* 620 (1753). More or less pubescent perennial; stems 25–100(–150) cm. Leaves up to 10 × 5 cm, ovate to lanceolate, obtuse to acute, cordate at base, doubly crenate-serrate. Bracts mostly leaf-like; pedicels 3–4 times as long as the calyx, glandular-pubescent. Calyx-lobes *c.* 4 mm, orbicular, more or less puberulent; margin broadly scarious. Corolla 8–12 mm, dull purple. Staminode suborbicular. Capsule 5–8 mm, ovoid-globose, subobtuse. $2n = 58, 60$. *Hedges, riversides and meadows. W. Europe, northwards to S. England.* Az Br Ga Hs Lu.

9. S. grandiflora DC., *Cat. Pl. Horti Monsp.* 143 (1813). Glandular-pubescent perennial; stems 30–80(–150) cm. Leaves up to 25 × 20 cm, ovate, the lower sometimes lyrate, truncate to cordate at base, more or less obtuse, doubly dentate. Bracts mostly leaf-like; pedicels 1–2 times as long as the calyx, densely glandular-pubescent. Calyx-lobes 5–6 mm, ovate-orbicular, sparsely pubescent; margin sinuate, undulate, broadly scarious. Corolla yellowish-green, with orange to purplish upper lip. Staminode more or less orbicular. Capsule 7–10 mm, ovoid, acuminate. ● *W.C. Spain, N.C. Portugal.* Hs Lu.

(a) Subsp. **grandiflora**: Densely glandular-pubescent. Lower leaves lyrate. Corolla 12–18 mm. $2n = 60$. *Woods, walls, and roadsides. C. Portugal.*

(b) Subsp. **reuteri** (Daveau) I. B. K. Richardson, *Bot. Jour. Linn. Soc.* **65**: 266 (1972) (*S. reuteri* Daveau, *S. herminii* sensu Lange, non Hoffmanns. & Link): More or less glandular-pubescent. Lower leaves usually undivided. Corolla 9–12(–14) mm. $2n = 58$. *Damp, rocky places. Mountains of W.C. Spain.*

(**10–13**). **S. sambucifolia** group. Glabrous perennials. Leaves pinnatifid to 2-pinnatisect or 3-sect, the uppermost sometimes undivided. Pedicels 1–3 times as long as the calyx. Calyx-lobes orbicular to ovate, with usually wide scarious margin. Corolla 6–20 mm. Staminode obovate to reniform. Capsule 5–10 mm, globose to ovoid or conical.

1 Most of the bracts leaf-like; corolla 12–20 mm
  2 Pedicels glandular; staminode obovate; capsule shortly acuminate   **10. sambucifolia**
  2 Pedicels usually eglandular; staminode reniform; capsule long-acuminate   **11. trifoliata**
1 Most of the bracts small, not leaf-like; corolla 6–14(–20) mm
  3 Leaves bipinnatisect   **12. sciophila**
  3 Leaves pinnatifid to pinnatisect, usually lyrate
    4 Calyx-lobes undulate; corolla 12–20 mm   **10. sambucifolia**
    4 Calyx-lobes plane; corolla 6–10 mm   **13. sublyrata**

10. S. sambucifolia L., *Sp. Pl.* 620 (1753). Stems 50–80(–150) cm, robust. Leaves 6–15(–20) cm, pinnatisect (the uppermost rarely undivided); lobes ovate to lanceolate, singly or doubly crenate or serrate. Bracts very variable in size; pedicels glandular. Calyx-lobes undulate. Corolla 12–20 mm, eglandular outside, greenish with brownish-red upper lip. Staminode obovate, entire or emarginate. Capsule 10–12 mm, ovoid-globose, shortly acuminate. $2n = 58$. *Wet places. C. & S. Portugal, S. Spain.* Hs Lu.

11. S. trifoliata L., *Syst. Nat.* ed. 10, **2**: 1114 (1759). Like **10** but stems up to 200 cm; leaves mostly lyrate-pinnatisect or 3-sect, with usually doubly serrate lobes; most of the bracts leaf-like; pedicels usually eglandular; calyx-lobes plane; corolla often glandular outside, purplish; staminode reniform; capsule ovoid-conical, long-acuminate. $2n = c. 84$. *Streamsides and other damp, shady places.* ● *Islands of W.C. Mediterranean region.* Co It Sa.

12. S. sciophila Willk., *Bot. Zeit.* **8**: 77 (1850) (*S. grenieri* Reuter ex Lange). Stems 25–60 cm. Leaves up to 9 cm, bipinnatisect. Bracts not leaf-like; pedicels more or less glandular. Calyx-lobes plane. Corolla 8–14 mm, yellowish-green, with reddish-purple upper lip. Staminode ovate to orbicular. Capsule 5–7 mm, subglobose to ovoid-acuminate. $2n = 58$. *Damp, rocky places, mainly in the mountains.* ● *S.E. Spain.* Hs.

13. S. sublyrata Brot., *Phyt. Lusit.* ed. 3, **2**: 156 (1827) (*S. ebulifolia* Hoffmanns. & Link, non Bieb.). Stems up to 80 cm. Leaves 2·5–10 cm, lyrate-pinnatifid (rarely 2-pinnatifid), with elliptical, irregularly serrate lobes; uppermost leaves sometimes entire. Bracts usually not leaf-like; pedicels glandular. Calyx-lobes plane. Corolla 6–10 mm, greenish-yellow with a purplish-brown upper lip. Staminode reniform. Capsule 6–8 mm, conical, acute. *Sandy or rocky places.* ● *N. & N.C. Portugal, W.C. Spain.* Hs Lu.

Plants with entire upper leaves and leaf-like bracts, which are sometimes annual, have been called **S. schousboei** Lange in Willk. & Lange, *Prodr. Fl. Hisp.* **2**: 553 (1870). They have $2n = 60$. They may be variants of **13**, but are close to many plants from North Africa which are usually referred to **14**.

14. S. laevigata Vahl, *Symb. Bot.* **2**: 67 (1791) (*S. laxiflora* Lange; incl. *S. valentina* Rouy). Glabrous perennial; rhizome not nodular. Stems up to 70 cm, 4-angled. Leaves up to 5 × 2 cm, entire, ovate to ovate-lanceolate, subacute, doubly dentate. Cymes 2- to 5-flowered; bracts mostly not leaf-like; pedicels 2–3 times as long as calyx, glandular. Calyx-lobes ovate-orbicular, entire; margin broadly scarious. Corolla *c.* 8 mm, green, the upper lip purple. Staminode orbicular. Capsule 6–8 mm, globose. *S. & S.E. Spain.* Hs ?Lu.

15. S. scopolii Hoppe in Pers., *Syn. Pl.* **2**: 160 (1806). More or less pubescent perennial; stems up to 100 cm. Leaves up to 9 cm, ovate to oblong-lanceolate, doubly dentate, acute to obtuse, cordate to rounded at base. Bracts mostly not leaf-like; peduncles 10–20 mm; pedicels mostly 2–3 times as long as calyx, usually glandular-pubescent. Cymes 4- to 7-flowered. Calyx-lobes ovate-orbicular, glabrous; margin broadly scarious. Corolla 7–12 mm, greenish, with purplish-brown upper lip. Staminode more or less reniform. Capsule 5–7 mm, ovoid-globose, mucronate. 2*n*=26. *Woods and damp places, mainly in the mountains. S.E. & E.C. Europe, extending to Italy and Sicilia; one isolated station in E. Russia; occasionally casual elsewhere.* Al Au Bu Cz Gr Hu It Ju Po Rm Rs (C, W, K) Si Tu.

16. **S. alpestris** Gay ex Bentham in DC., *Prodr.* 10: 307 (1846). Like 15 but leaves often up to 15 cm, simply dentate; cymes 3- to 30-flowered; peduncles 10–50 mm; pedicels 2–5(–7) times as long as calyx. 2*n*=68, 72. *Damp woods and by streams.* ● *Mountains of N. Spain and S. France.* Ga Hs.

17. **S. herminii** Hoffmanns. & Link, *Fl. Port.* 1: 266 (1813–1820) (*S. bourgaeana* Lange). Glandular-pubescent perennial. Stems 30–90 cm, not winged. Leaves up to 15×10 cm, ovate to lanceolate, truncate to cordate at base, serrate, the lower more or less obtuse, the upper acute. Cymes 3- to 6-flowered; bracts mostly not leaf-like; pedicels 3–4 times as long as calyx, glandular-hirsute. Calyx-lobes ovate-orbicular, entire; margin narrowly scarious. Corolla 8–9 mm, yellowish-green, with purplish upper lip. Staminode orbicular to reniform. Capsule *c.* 6 mm, globose to conical, acuminate. *Mountain rocks.* 2*n*=*c.* 52. ● *N. & C. Portugal; W.C. Spain.* Hs Lu.

18. **S. nodosa** L., *Sp. Pl.* 619 (1753). Glabrous perennial with a nodular rhizome; stems 30–80(–150) cm, 4-angled, sometimes narrowly winged. Leaves up to 12×7 cm, ovate to ovate-lanceolate, doubly (rarely singly) serrate, more or less acute, cordate to truncate at base. Cymes usually 5- to 7-flowered; bracts usually not leaf-like; pedicels 2–3 times as long as calyx, glandular. Calyx-lobes broadly ovate, entire; margin narrowly scarious. Corolla 7–10 mm, green, with purplish-brown upper lip. Staminode obovate, truncate or emarginate. Capsule *c.* 5 mm, broadly ovoid, shortly acuminate. 2*n*=36. *Usually in damp or shady places. Most of Europe.* Al Au Be Br Bu Cz Da Fe Ga Ge Gr Hb He Ho Hs Hu It Ju No Po Rm Rs (N, B, C, W, K, E) Su ?Tu.

19. **S. auriculata** L., *Sp. Pl.* 620 (1753) (*S. aquatica* auct., non L., *S. cretica* Boiss. & Heldr.). Glabrous, rarely somewhat pubescent perennial; rhizome not nodular; stem 50–100 cm, 4-angled, the angles produced into wings *c.* 1 mm wide. Leaves 5–12(–25) cm, simple or with one (rarely more) pair of small lobes at the base, ovate to elliptical, crenate, the lower obtuse, the upper subacute, truncate to subcordate at base. Bracts mostly linear, not leaf-like; pedicels up to twice as long as calyx, glandular. Calyx-lobes ovate-orbicular, serrate, glabrous; margin broadly scarious. Corolla (5–)7–9 mm, greenish, with purplish-brown upper lip. Staminode more or less orbicular, scarcely emarginate. Capsule 4–6 mm, subglobose, obtuse, mucronate. 2*n*=78, 80, 84, 86. *Riversides and other damp places. W. Europe, northwards to the Netherlands; Italy and Sicilia; Kriti.* Az Be Bl Br Co Cr Ga Ge Hb He Ho Hs It Lu Sa Si.

S. lyrata Willd., *Hort. Berol.* t. 55 (1805), from Spain and Portugal, is like 19 but has leaves with 2–4 pairs of lobes at the base and ovate bracts. It has 2*n*=58. Similar plants from Sardegna, with short peduncles and pedicels, have been described as **S. subverticillata** Moris, *Stirp. Sard. Elench.* 2: 8 (1827). The status of these taxa is obscure.

20. **S. umbrosa** Dumort., *Fl. Belg.* 37 (1827) (*S. alata* auct., *S. aquatica* L., nom. ambig., *S. neesii* Wirtgen, *S. samaritanii* Boiss. & Heldr. ex Halácsy). Like 19 but stem often more broadly winged; leaves without basal lobes, serrate or crenate-serrate, acute to subobtuse, more or less cuneate at base; bracts sometimes leaf-like; calyx-margin more or less dentate; staminode much wider than long, more or less 2-lobed. 2*n*=26, 52. *Damp, shady places. Europe, from Scotland, Denmark and Latvia southwards, but absent from much of the west.* Al Au Be Br Bu ?Co Cz Da Ga Ge Gr Hb He Ho Hu It Ju Po Rm Rs (B, C, W, K, E) Sa Si Tu.

21. **S. lucida** L., *Syst. Nat.* ed. 10, 2: 1114 (1759). Glabrous biennial; stem 10–100 cm, usually solitary. Leaves up to 9×5 cm, pinnatisect or bipinnatisect, rarely undivided; lobes ovate to oblong, more or less serrate or crenate. Bracts not leaf-like; pedicels *c.* ½ as long as calyx, glandular. Calyx-lobes more or less orbicular, entire or serrate; margin broadly scarious. Corolla 4–9 mm, greenish-brown. Staminode broadly orbicular. Capsule *c.* 5 mm, globose, with mucro *c.* 1 mm. 2*n*=24, 26. *Shady rocks. Greece and Aegean region; S.E. Italy; S.E. France.* Cr Ga Gr It.

22. **S. myriophylla** Boiss. & Heldr. in Boiss., *Diagn. Pl. Or. Nov.* 2(12): 39 (1853) (*S. laxa* Boiss. & Heldr.). Like 21 but stems up to 60 cm, often numerous; leaves up to 4×2 cm, all bipinnatisect, with very small, more or less linear, acute lobes; margins of calyx more or less dentate; capsule with mucro 2–3 mm. *Mountain rocks. S. Greece.* Gr.

23. **S. heterophylla** Willd., *Sp. Pl.* 3: 274 (1800). Glabrous or glandular, rarely puberulent, often glaucous perennial. Stems 10–70 cm, usually several. Leaves up to 5×3 cm, incise-dentate to bipinnatisect. Bracts not leaf-like; pedicels up to 1(–3) times as long as calyx, glandular. Calyx-lobes ovate to orbicular, entire to serrate; margin scarious. Corolla (4–)6–9 mm, reddish-purple to greenish. Staminode reniform to orbicular. Capsule 3–6 mm, globose. *Rocky places. Aegean region and Balkan peninsula, extending northwards to Romania and Krym.* Al Bu Cr Gr Ju Rm Rs (K).

(a) Subsp. **heterophylla**: Leaves often more or less fleshy, simple and crenate, or more or less pinnatifid with obtuse lobes. 2*n*=24. *Mainly in S. Greece and Aegean region.*
(b) Subsp. **laciniata** (Waldst. & Kit.) Maire & Petitmengin, *Bull. Soc. Sci. Nancy* ser. 3, 8: 178 (1907) (*S. laciniata* Waldst. & Kit., *S. variegata* auct. pro parte, non Bieb.; incl. *S. exilis* Popl.): Leaves thin, simple and serrate, or pinnatisect with more or less acute lobes. *Throughout the range of the species except the extreme south.*

S. olympica Boiss., *Diagn. Pl. Or. Nov.* 1(4): 69 (1844), from Anatolia and the Caucasus, has once been recorded from Krym, but requires confirmation. It is closely related to 23 and perhaps conspecific.

24. **S. taygetea** Boiss., *Diagn. Pl. Or. Nov.* 1(4): 68 (1844). Slender glandular-pubescent perennial; stems up to 40 cm. Leaves up to 4×2 cm, ovate, cuneate to truncate at base, more or less obtuse, simply or doubly serrate, sometimes pinnatifid towards base. Bracts sometimes leaf-like; pedicels about equalling the calyx, glandular. Calyx-lobes orbicular, more or less entire; margin scarious. Corolla 6–9 mm, greenish-yellow. Staminode reniform. Capsule *c.* 5 mm, globose, mucronate. *Rock-crevices.* ● *S. Greece (Taïyetos).* Gr.

(25–28). **S. canina** group. Glabrous perennial herbs or small shrubs. Leaves 1–8 cm, entire to bipinnatisect. Cymes 1- to

11(–25)-flowered; pedicels usually shorter than or equalling calyx, usually with sessile or subsessile glands, rarely glandular-pubescent. Calyx glabrous; lobes with wide scarious margin. Corolla 3–5(–8) mm, dark purplish-red. Staminode linear to lanceolate, or absent. Capsule 3–6 mm, more or less globose, mucronate to cuspidate.

A difficult group of taxa of uncertain status, sharing a distinctive facies but covering between them a wide range of variation. It might be possible, after a fuller study of the pattern of variation, to reduce the group to a single species with 5 or 6 subspecies, but it has been thought best, in the present state of our knowledge, to recognize the more distinct variants as independent species.

```
1  Cymes 1- to 2–(3-)flowered; pedicels spinescent
  2  All leaves incise-dentate to pinnatifid, the lowest opposite
                                          26. ramosissima
  2  Lower leaves 3-sect or pinnatisect, alternate    27. spinulescens
1  Cymes 3- to 11(–25)-flowered; pedicels usually not spinescent
  3  Leaves entire to serrate; lower bracts often leaf-like  25. frutescens
  3  At least the lower leaves pinnatifid to 2-pinnatisect; bracts
     not leaf-like                         28. canina
```

25. S. frutescens L., *Sp. Pl.* 621 (1753). Stems up to 60 cm, woody, sparingly branched. Leaves 1·5–4 × 0·6–2·5 cm, oblong-lanceolate to orbicular, entire to serrate or crenate, mostly opposite, coriaceous. Cymes 5- to 13-flowered; pedicels with subsessile glands; lower bracts often leaf-like. Corolla 4 mm. Capsule 5–6 mm. $2n=26$. *Maritime sands. W. half of Iberian peninsula.* Hs Lu ?Si.

26. S. ramosissima Loisel., *Fl. Gall.* 381 (1807). Stems 25–50 cm, with numerous erecto-patent, rigid branches. Leaves 1–2 × 0·2–0·4 cm, linear-oblong to oblanceolate, incise-dentate to pinnatifid, caducous. Cymes 1- to 2–(3-)flowered, forming long, narrow terminal panicles; bracts minute, not leaf-like; pedicels sparsely glandular, persistent, woody, spinescent. Corolla *c.* 4 mm. Capsule *c.* 4 mm. *Maritime sands and other dry places. W. Mediterranean region.* ?Bl Co Ga Sa ?Si.

Plants from Islas Baleares referred to this species are in many features transitional to **28(a)** and require further study.

27. S. spinulescens Degen & Hausskn., *Mitt. Thür. Bot. Ver.* nov. ser., **10**: 60 (1897). Stems up to 40 cm, somewhat woody, with numerous short, divaricate, somewhat spiny branches. Leaves 0·6–2 × 0·1–0·8 cm, all alternate, caducous, the lower pinnatisect, the upper dentate. Cymes 1-flowered, forming a lax, narrow raceme-like inflorescence; pedicels sparsely glandular, more or less spinescent. Corolla 3–4 mm; lobes with whitish margin. Capsule 3–3·5 mm. *Rocky slopes, c. 1000 m.* ● *N. Aegean region (Samothraki).* Gr.

28. S. canina L., *Sp. Pl.* 621 (1753). Stems 20–60 cm, usually sparingly branched. Leaves 1–8 cm, the lower pinnatifid to 2-pinnatisect, with linear to oblong, usually dentate lobes, the upper sometimes elliptic-oblong, undivided; at least the lowest opposite. Cymes (3–)5- to 11(–25)-flowered, forming a cylindrical panicle; bracts small, not leaf-like; pedicels with sessile glands or glandular-pubescent, seldom spinescent. Corolla 4–5(–8) mm. Capsule 4–5 mm. *S. & S.C. Europe, extending to C. France.* Al Au Bl Bu Co ?Cz Ga Ge Gr He Hs It Ju Lu Rs (K) Sa Si Tu.

```
1  Upper lip of corolla more than ½ as long as tube; pedicels
      usually glandular-pubescent            (c) subsp. hoppii
1  Upper lip of corolla c. ⅓ as long as tube; pedicels with sessile
      or subsessile glands
  2  Corolla concolorous                   (a) subsp. canina
  2  Corolla-lobes with conspicuous whitish margin (b) subsp. bicolor
```

(a) Subsp. canina: Leaves pinnatifid to 2-pinnatisect. Pedicels with sessile or subsessile glands. Upper lip of corolla *c.* ⅓ as long as tube; lobes without pale margin. *Waste places and stony ground.* $2n=24, 26$. *Throughout the range of the species.*

S. crithmifolia Boiss., *Voy. Bot. Midi Esp.* **2**: 447 (1841), from the mountains of S. & E. Spain, is like **28(a)** but usually more robust, with pinnatisect leaves up to 10 cm, corolla 6–8 mm, and no staminode. It has $2n=24$, and is perhaps more closely related to **23**.

(b) Subsp. bicolor (Sibth. & Sm.) W. Greuter, *Boissiera* **13**: 109 (1967) (*S. bicolor* Sibth. & Sm.): Leaves pinnatisect, with incise-dentate lobes. Pedicels with a few subsessile glands. Upper lip of corolla *c.* ⅓ as long as tube; lobes with conspicuous whitish margin. *Stony ground and waste places. S.E. Europe.*

S. pindicola Hausskn., *Mitt. Thür. Bot. Ver.* nov. ser., **10**: 59 (1897), from mountain rocks in C. Greece, has a yellow margin to the corolla-lobes, numerous, intricate branches, and cymes mostly 3-flowered.

(c) Subsp. hoppii (Koch) P. Fourn., *Quatre Fl. Fr.* 770 (1937) (*S. hoppii* Koch, *S. juratensis* Schleicher): Leaves mostly 2-pinnatisect. Pedicels usually glandular-pubescent. Upper lip of corolla ½–⅔ as long as tube; lobes without pale margin. $2n=24$, 26. ● *Jura, S. Alps, Appennini.*

Plants showing some approach to subsp. **(c)** are recorded from the Pyrenees and the mountains of Jugoslavia, but are probably best assigned to subsp. **(a)**.

29. S. cretacea Fischer ex Sprengel, *Syst. Veg.* **2**: 783 (1825). Usually whitish-glaucous perennial, densely but minutely glandular-puberulent throughout. Stems up to 50 cm, numerous, simple. Leaves 1·5–3 × 0·2–0·7 cm, linear-oblong to elliptic-oblanceolate, serrate or dentate. Cymes mostly 3-flowered, forming a narrow panicle; bracts not leaf-like; pedicels longer or shorter than calyx. Calyx-lobes with narrow scarious margin. Corolla 4–5 mm, dark brownish-red. Staminode oblong. Capsule 3–5 mm, globose, mucronate. *Chalky hillsides.* ● *S.E. Russia, E. Ukraine.* Rs (E).

30. S. rupestris Bieb. ex Willd., *Sp. Pl.* **3**: 274 (1800) (incl. *S. goldeana* Juz.). Like **29** but less glaucous; leaves 0·6–1(–1·5) cm wide, oblong-lanceolate to ovate-elliptical; cymes often 5-flowered; pedicels mostly shorter than calyx; calyx-lobes with somewhat wider scarious margin; corolla 4·5–6(–10) mm, sometimes yellowish towards the base; staminode cordate to ovate-oblong. *Dry, rocky or stony places. S.E. Russia, Krym.* Rs (K, E).

S. donetzica Kotov, *Ukr. Bot. Žur.* **1**(2): 298 (1940) and **S. sareptana** Kleopow in Majevski, *Fl. Sred. Ross.* ed. 7, 642 (1940), both from S.E. Russia, resemble **30** but have somewhat larger, dentate to pinnatifid leaves, cymes mostly 2- to 3-flowered, and smaller corolla and capsule.

9. Anarrhinum Desf.[1]
(*Simbuleta* Forskål)

Erect, biennial or perennial herbs. Leaves alternate, the basal undivided, forming a rosette, the cauline usually palmatisect. Flowers small, zygomorphic, in terminal, bracteate racemes or panicles. Calyx deeply and more or less equally 5-lobed. Corolla glabrous; tube cylindrical, usually with a conical spur at the base, sharply reflexed along the lower side of the tube (rarely reduced to an inconspicuous gibbosity); limb 2-lipped, the upper lip

2-lobed, the lower 3-lobed, more or less flat and without definite palate, leaving the mouth of the tube open. Stamens 4, didynamous, included; anther-lobes confluent. Capsule subglobose, with equal loculi, each dehiscing by an apical pore. Seeds numerous, tuberculate or papillose.

Measurements of leaf-length include the petiole, as it is not clearly defined from the lamina.

1 Plant glandular-hairy; corolla pale yellow or cream-coloured
                                       4. duriminium
1 Plant glabrous, though sometimes papillose; corolla lilac or bluish-violet, rarely white
2 Pedicels 5–11 mm in flower, exceeding the bracts; plant conspicuously papillose, at least in upper part
                             3. longipedicellatum
2 Pedicels not more than 3 mm in flower, equalling or shorter than the bracts; plant not or sparsely papillose
3   Corolla-tube gibbous at the base, but not spurred; seeds with low, rounded tubercles; most of the cauline leaves entire
                             5. corsicum
3   Corolla-tube spurred at the base; seeds with acute, conical papillae; most of the cauline leaves divided into 3–7 linear lobes
4   Cauline leaves sparse; racemes lax; calyx-lobes oblong to elliptical, obtuse, apiculate or shortly acute   **2. laxiflorum**
4   Cauline leaves crowded; racemes dense; calyx-lobes narrowly triangular-lanceolate, tapered to a long, slender point
                           1. bellidifolium

1. A. bellidifolium (L.) Willd., *Sp. Pl.* 3: 260 (1800). Glabrous biennial or perennial, sometimes sparsely papillose above; stems up to 80 cm. Basal leaves up to 8 × 3 cm, obovate-spathulate to elliptic-oblanceolate, irregularly dentate, obtuse, tapered to a petiole. Cauline leaves numerous, crowded, palmatisect with 3–5 linear to lanceolate, acute segments, the central 1–5 mm wide. Inflorescence simple or branched, dense; pedicels *c.* 1 mm, shorter than the bracts. Calyx-lobes 1–2 mm, narrowly triangular-lanceolate, tapered to a slender point. Corolla 4–5 mm, pale lilac to blue; spur 1·5–1·75 mm. Seeds covered with acute, conical papillae. 2n=18. *Calcifuge.* ● *S.W. Europe, extending locally to N.C. France, S.C. Germany and N. Italy.* Ga Ge Hs It Lu [He].

2. A. laxiflorum Boiss., *Elenchus* 71 (1838). Like **1** but basal leaves narrower; cauline leaves distant, with central segment not more than 2 mm wide; inflorescence lax, simple or with a few short branches; calyx-lobes 1·25–1·5 mm, oblong to elliptical, obtuse, apiculate or shortly acute; corolla 6–7 mm, pale blue to white. *Mountains of S. Spain.* Hs.

Some plants from S. Spain combine some characters of **1** with others of **2**. They may be hybrids, but require further investigation.

3. A. longipedicellatum R. Fernandes, *Bol. Soc. Brot.* ser. 2, 33: 14 (1959). Biennial, tinged with violet, especially in upper part, glabrous but densely papillose; stems up to 90 cm. Basal leaves 2–11 × 0·6–2·2 cm, obovate to oblanceolate-spathulate, crenate to doubly serrate, petiolate. Cauline leaves numerous, crowded, palmatisect with 3–7 lanceolate, acute segments. Inflorescence simple or branched, dense; pedicels 5–11 mm, very slender, exceeding the bracts. Calyx-lobes 2–3 mm, linear-lanceolate, acute. Corolla 5·5–6 mm, deep violet-blue; spur 1·5 mm; lower lip with two swellings at base, forming a low palate. Seeds covered with acute, conical papillae. ● *C. Portugal.* Lu.

4. A. duriminium (Brot.) Pers., *Syn. Pl.* 2: 159 (1806) (*A. hirsutum* Hoffmanns. & Link). Glandular-pubescent to -hirsute biennial or perennial; stems 15–90 cm. Basal leaves up to

17 × 4·5 cm, obovate-spathulate, crenate to doubly serrate, obtuse. Cauline leaves mostly 3-fid, with the central segment up to 17 mm wide, lanceolate to suborbicular, mucronate, the 2 lateral much smaller and linear-lanceolate. Inflorescence simple or branched, dense; pedicels *c.* 1·5 mm, shorter than the bracts. Calyx-lobes 2·5 mm, linear-lanceolate with subulate apex. Corolla *c.* 6 mm, pale yellow or cream-coloured. Seeds covered with acute, conical papillae. ● *N.W. part of Iberian peninsula.* Hs Lu.

5. A. corsicum Jordan & Fourr., *Brev. Pl. Nov.* 1: 41 (1866). Glabrous perennial; stems 20–50 cm, slender. Basal leaves up to 5 × 3 cm, oblong-spathulate, entire or weakly and distantly dentate. Cauline leaves distant, mostly linear to oblanceolate, entire, a few sometimes toothed or lobed. Inflorescence simple or branched, lax; pedicels *c.* 2 mm in flower, up to 5 mm in fruit, shorter than the bracts. Calyx-lobes 1·5–2 mm, lanceolate, acute. Corolla *c.* 4 mm, dull blue or violet; tube gibbous at the base but not spurred. Seeds covered with low, rounded tubercles. 2n=18. *Dry places.* ● *Corse.* Co.

10. Antirrhinum L.¹

Dwarf shrubs, or perennial herbs somewhat woody at the base; leaves simple, entire, pinnately veined, usually opposite below and alternate above. Flowers zygomorphic, in terminal, bracteate racemes, or solitary in the leaf-axils. Calyx deeply and more or less equally 5-lobed, shorter than corolla-tube. Corolla glandular-pubescent outside; tube cylindrical, wide, produced abaxially at the base into a short pouch; limb 2-lipped, the upper lip 2-lobed, the lower 3-lobed, with at its base a prominent palate which closes the mouth of the tube. Stamens 4, didynamous, included. Stigma capitate. Capsule with 2 unequal loculi, the adaxial longer, narrower above, and opening by a single apical pore, the abaxial shorter, wider above, and with 2 apical pores. Seeds numerous, reticulate-rugose.

The taxonomic difficulties of this genus are due, at least in part, to the fact that hybrids are very easily formed. An original pattern of narrow endemism in the Iberian peninsula has thus become complicated by hybridization with and introgression by a few wider-ranging species, some of them dispersed outside their original range by cultivation for ornament and subsequent naturalization.

Measurements of leaves refer to those near the middle of the stem, and of flowers to the largest on the plant.

Literature: W. Rothmaler, *Feddes Repert.* (*Beih.*) 136: 1–124 (1956).

1 Stem (excluding inflorescence) ±densely covered with eglandular hairs (rarely with a few glandular hairs as well)
2 Ovary and capsule glabrous (except sometimes for a few hairs at apex); lips of corolla widely divergent   **1. valentinum**
2 Ovary and capsule hairy all over; lips of corolla parallel for most of their length, divergent only in apical part
3   Nearly all the leaves opposite   **2. sempervirens**
3   Most of the upper leaves alternate
4   Hairs on calyx and capsule mostly glandular
5   Calyx-lobes 4–5 mm; corolla 18–25 mm   **6. charidemi**
5   Calyx-lobes 5–8 mm; corolla 25–35 mm   **7. molle**
4   Hairs on calyx and capsule mostly eglandular
6   Pedicel usually exceeding subtending leaf; leaves not more than 10 mm; petiole 1–2 mm   **5. microphyllum**
6   Pedicel usually shorter than subtending leaf; most of the leaves more than 10 mm; petiole 2–6 mm
7   Leaves soft, rounded at apex; corolla 20–25 mm
                             3. pulverulentum
7   Leaves coriaceous, truncate or slightly emarginate at apex; corolla 17–20 mm   **4. pertegasii**

1 Stem (excluding inflorescence) glabrous or glandular-hairy
  8 Leaves glabrous (except sometimes for a few hairs on the
     petiole)
    9 Corolla 17–30 mm
      10 Corolla pink, except for yellow palate   **13. barellieri**
      10 Corolla yellow
        11 Leaves 8–30 mm wide, ovate to elliptical **10. meonanthum**
        11 Leaves not more than 6 mm wide, linear to linear-oblong
                                      12. siculum
    9 Corolla 30–48 mm
      12 Bracts linear-lanceolate, exceeding flower-buds; calyx-
        lobes acute   **11. braun-blanquetii**
      12 Bracts ovate to ovate-lanceolate, not exceeding flower-
        buds; calyx-lobes obtuse
        13 Leaves more than 3 times as long as wide **17. majus**
        13 Leaves 1½–3 times as long as wide
          14 Leaves obtuse, ±truncate at base; corolla pale yellow;
            capsule 13–17 mm   **16. latifolium**
          14 Leaves subacute, ±cuneate at base; corolla purplish-
            pink; capsule 10–14 mm   **17. majus**
  8 Leaves glandular-hairy (sometimes only sparsely)
    15 Corolla white, purple or pink, except for yellow palate
      16 Upper leaves often opposite or whorled; corolla 35–45 mm
                                      15. australe
      16 Upper leaves alternate; corolla 20–32 mm
        17 Stems ascending or decumbent; corolla 20–25 mm
                                9. hispanicum
        17 Stems usually ±erect; corolla 25–32 mm **14. graniticum**
    15 Corolla yellow
      18 Capsule 7–10 mm
        19 Stems decumbent; corolla 30–35 mm **8. grosii**
        19 Stems erect; corolla 18–25 mm **10. meonanthum**
      18 Capsule 13–17 mm
        20 Leaves linear-oblong to elliptical; calyx-lobes ovate-
          lanceolate, acute; bracts exceeding flower-buds
                            11. braun-blanquetii
        20 Leaves broadly ovate; calyx-lobes broadly ovate, obtuse;
          bracts not exceeding flower-buds **16. latifolium**

1. A. valentinum Font Quer, *Ill. Fl. Occ.* 5 (1926). Eglandular-puberulent dwarf shrub; stems 20–30 cm, decumbent, intricately branched. Leaves 5–15 × 4–12 mm, elliptical to suborbicular, obtuse to slightly emarginate, mostly opposite; petiole 1–5 mm. Bracts similar to foliage-leaves. Pedicel 7–20 mm. Calyx-lobes 4–5 mm, oblong-lanceolate, subacute. Corolla 12–15 mm, white, with yellow palate and upper lip veined with pink; lips diverging at right angles to tube almost from their base. Capsule 7 mm, globose, usually glabrous. *Shady calcareous rocks.* ● *S.E. Spain (hills S. of Valencia).* Hs.

2. A. sempervirens Lapeyr., *Fig. Fl. Pyr.* 1: 7 (1795). Eglandular-puberulent to -villous dwarf shrub (sometimes also with glandular hairs in inflorescence). Stems up to 25 cm, procumbent. Leaves 12–30 × 5–17 mm, oblong to elliptical, mostly opposite; petiole 3–7 mm. Bracts similar to foliage-leaves. Pedicel 5–10 mm. Calyx-lobes 5–6 mm, lanceolate, acuminate. Corolla 20–25 mm, white or cream, often veined with purple and with a violet patch on upper lip; palate yellow or white. Capsule *c.* 6 mm, subglobose, glandular-pubescent. *Mountain rocks.* ● *C. Pyrenees and E.C. Spain.* Ga Hs.

3. A. pulverulentum Láz.-Ibiza, *Anal. Soc. Esp. Hist. Nat.* **29**: 164 (1901) (*A. sempervirens* var. *densiflorum* Lange ex Willk.). Like **2** but hairs rather longer; upper leaves alternate; corolla pale yellow; capsule eglandular-pubescent. *Calcareous rocks,* 1000–2000 m. ● *E. Spain (prov. Jaén to prov. Teruel).* Hs.

4. A. pertegasii Rothm., *Feddes Repert. (Beih.)* 136: 61 (1956). Like **2** but stems ascending to erect; leaves obovate to oblanceolate, coriaceous, those subtending the flowers alternate; calyx-lobes 4–5 mm; corolla 17–20 mm. *Calcareous rocks, c.* 1000 m. ● *E. Spain (near Tortosa).* Hs.

5. A. microphyllum Rothm., *loc. cit.* (1956). Like **2** but leaves 3–10 × 2–5 mm, ovate to orbicular; petiole 1–2 mm; pedicel 10–20 mm; calyx-lobes 3–5 mm, subacute; capsule oblong. *Calcareous rocks.* ● *E.C. Spain (near Sacedón, prov. Guadalajara).* Hs.

6. A. charidemi Lange, *Vid. Meddel. Dansk Naturh. Foren. Kjøbenhavn* **1881**: 99 (1882). Dwarf shrub; stems up to 35 cm, procumbent to ascending, intricately branched. Stems, leaves and pedicels eglandular-puberulent; calyx, corolla and capsule glandular-pubescent. Leaves 8–28 × 4–9 mm, elliptical to oblong-lanceolate, obtuse, mostly alternate; petiole 1–3 mm. Bracts similar to smaller foliage-leaves. Pedicel 3–12 mm. Calyx-lobes 4–5 mm, oblong, subacute. Corolla 18–25 mm, pink with red veins, rarely white; palate yellow. Capsule 8–9 mm, subglobose. *Schistose rocks.* ● *S.E. Spain (Cabo de Gata).* Hs.

7. A. molle L., *Sp. Pl.* 1198 (1753). Dwarf shrub; stems up to 40 cm, procumbent to ascending. Stems, leaves and pedicels eglandular-pubescent to -lanate; calyx, corolla and capsule glandular-pubescent. Leaves 8–22 × 5–20 mm, broadly ovate to elliptical, opposite below, alternate above; petiole 2–6 mm. Bracts similar to foliage-leaves. Pedicels 3–20 mm. Calyx-lobes 5–8 mm, lanceolate, acute. Corolla 25–35 mm, white or pale pink; palate yellow. Capsule 7–8 mm, subglobose. *Calcareous rocks and walls,* 500–1800 m. ● *E. & C. Pyrenees and adjoining mountains of N.E. Spain; N.E. Portugal.* Hs Lu.

The plants from Portugal have been distinguished as **A. lopesianum** Rothm., *Feddes Repert. (Beih.)* 136: 65 (1956), mainly on account of the longer hairs on their stems and leaves.

8. A. grosii Font Quer, *Bol. Soc. Esp. Hist. Nat.* **25**: 268 (1925). Glandular-pubescent dwarf shrub; stems 12–40 cm, decumbent. Leaves 20–30 × 12–22 mm, ovate, obtuse, opposite below, alternate above; petiole 2–3 mm. Lower bracts similar to foliage-leaves, the upper smaller and narrower. Pedicels 3–6 mm, shorter than bracts. Calyx-segments 7 mm, ovate-lanceolate, acute. Corolla 30–35 mm, pale yellow. Capsule 9–10 mm, subglobose, glandular-pubescent. *Mountain rocks.* ● *W.C. Spain (Sierra de Gredos).* Hs.

9. A. hispanicum Chav., *Monogr. Antirrh.* 83 (1833). Glandular-pubescent to -villous dwarf shrub; stems 20–60 cm, procumbent to ascending, freely branched. Leaves 5–35 × 2–20 mm, lanceolate to orbicular, opposite below, alternate above, or nearly all alternate. Bracts similar to foliage-leaves or much smaller. Pedicels 2–20 mm. Calyx-lobes 6–8 mm, ovate-lanceolate, subacute to subobtuse. Corolla 20–25 mm, white or pink; palate sometimes yellow. Capsule 6–9 mm, oblong to subglobose, glandular-pubescent. *Rocks and walls. S.E. Spain (Granada and Almería prov.).* Hs.

(a) Subsp. **hispanicum** (*A. glutinosum* Boiss. & Reuter, non Brot.): Leaves lanceolate to broadly ovate. Hairs on stem and leaves usually not more than 0·5 mm. *Mainly in the W. half of the range of the species.*

Very variable, especially in leaf-shape. The plants with widest leaves, which tend also to have longer hairs, and are therefore in some degree transitional to subsp. (b), have been distinguished as **A. rupestre** Boiss. & Reuter, *Pugillus* 82 (1852), but intermediates are so common that a separation even at subspecific level does not appear to be practicable.

(b) Subsp. mollissimum (Rothm.) D. A. Webb, *Bot. Jour. Linn. Soc.* **64**: 273 (1971) (*A. mollissimum* Rothm., *A. molle* auct., non L.): Leaves broadly ovate to orbicular. Hairs on stem and leaves up to 2 mm. *E. half of the range of the species.*

10. A. meonanthum Hoffmanns. & Link, *Fl. Port.* **1**: 261 (1813–1820) (incl. *A. ambiguum* Lange). Perennial herb, glandular-hairy at least in inflorescence and sometimes throughout. Stems up to 120 cm, erect, sparingly branched. Leaves 25–70 × 8–30 mm, lanceolate to ovate, acute or mucronate, glabrous or glandular-pubescent, opposite below, alternate above. Racemes 20–35 cm, with 30–60(–100) flowers. Bracts mostly 5–10 mm, linear-lanceolate, but the lowest transitional to foliage-leaves. Pedicels 1–4 mm. Calyx-lobes 5–8 mm, ovate-lanceolate, acute. Corolla 18–25 mm, pale yellow. Capsule 7–10 mm, oblong, glandular-hairy. ● *C. & N.W. Spain, N. Portugal.* Hs Lu.

11. A. braun-blanquetii Rothm., *Feddes Repert.* **56**: 280 (1954) (*A. meonanthum* auct., non Hoffmanns. & Link). Perennial herb, usually glabrous except for glandular-pubescent inflorescence. Stems up to 120 cm, erect, branched above. Leaves 25–60 × 4–15 mm, linear-oblong to elliptical, acute, opposite below, alternate above. Racemes 5–20 cm, with 5–30 flowers. Lower bracts similar to foliage-leaves, the upper smaller (6–12 mm), but exceeding the buds. Pedicels 3–6(–12) mm. Calyx-lobes 7–10 mm, ovate-lanceolate, acute. Corolla 30–40 mm, yellow. Capsule *c.* 15 mm, oblong, glandular-hairy. *Rocks and walls; calcicole.* ● *N.W. Spain, N.E. Portugal.* Hs Lu.

Plants with the leaves and lower part of stem glandular-pubescent have been distinguished as var. *oreophilum* Rothm. They may be hybrids with **10** or **14**.

12. A. siculum Miller, *Gard. Dict.* ed. 8, no. 6 (1768). Perennial herb, glabrous except for glandular-pubescent inflorescence. Stems 20–60 cm, erect, freely branched, often with many of the branches represented by short, axillary tufts of leaves. Leaves 20–60 × 2–6 mm, linear to narrowly elliptical, opposite below, alternate above. Lower bracts transitional to foliage-leaves, the upper *c.* 5 mm, about equalling pedicels. Calyx-lobes 5 mm, ovate-lanceolate, subacute. Corolla 17–25 mm, pale yellow, rarely veined with red. Capsule 10–12 mm, ovoid, glandular-pubescent. *Rocks and walls.* ● *Sicilia, Malta, ?S.W. Italy; cultivated for ornament and naturalized elsewhere in Mediterranean region.* *It Si [?Cr Ga Hs].

13. A. barrelieri Boreau, *Graines Recolt. Jard. Bot. Angers* 1854 [2] (1855). Perennial herb, usually glabrous except for glandular-pubescent inflorescence, but lower part of stem sometimes eglandular-villous. Stems 50–120 cm, erect; branches slender, numerous, often twining. Leaves 7–60 × 1–3(–9) mm, linear to linear-lanceolate, opposite below, alternate above. Bracts 2–10 mm, linear to broadly lanceolate. Pedicel 1–4 mm. Calyx-lobes 3–6 mm, lanceolate to ovate, acute to subobtuse. Corolla 20–30 mm, pink with yellow palate. Capsule 10–12 mm, ovoid-oblong, glandular-pubescent. *Hedges and rocky places.* *S. & E. Spain, S. Portugal.* Hs Lu.

Some plants in the northern part of the range (N.E. Spain) show evidence of introgression by **17(a)**; they correspond to *A. majus* subsp. *litigiosum* (Pau) Rothm., *Feddes Repert.* (*Beih.*) **136**: 99 (1956).

14. A. graniticum Rothm., *Bol. Soc. Brot.* ser. 2, **13**: 279 (1939) (*A. hispanicum* auct., non Chav.). Perennial herb, usually densely glandular-pubescent throughout. Stems up to 120 cm, ascending to erect; branches sometimes twining. Leaves 15–50 × 3–12(–20) mm, ovate to oblong-lanceolate, opposite below, alternate above. Bracts 3–10 mm. Pedicels 3–15 mm, usually exceeding bracts. Calyx-lobes (5–)7–10 mm, ovate, subacute. Corolla (22–)25–32 mm, pink or whitish, with yellow palate. Capsule 8–10 mm, oblong, glandular-pubescent. *Rocks, walls and stony hillsides; calcifuge.* ● *C. & S. Spain, N. & C. Portugal.* Hs Lu.

In most characters intermediate between **9(a)** and **17**, and perhaps of hybrid origin; it is, however, only in the southernmost part of its range that it is sympatric with either of these taxa, and over a considerable area of C. Spain and E. Portugal it is the only native representative of the genus. In S. Spain it tends to have less erect stems, longer bracts and more acute calyx-lobes; such plants (which approach more closely to **9(a)**) have been named **A. boissieri** Rothm., *Feddes Repert.* (*Beih.*) **136**: 88 (1956), but the correlation and constancy of these characters is poor.

15. A. australe Rothm., *Feddes Repert.* (*Beih.*) **136**: 91 (1956). Like **14** but stems erect, stout, sparingly branched; leaves ovate, mostly opposite or in whorls of 3 (the uppermost sometimes alternate); flowers often opposite or in whorls of 3; pedicels 1–5 mm, usually shorter than bracts; calyx-lobes broadly ovate, obtuse; corolla (35–)40–45 mm, bright pinkish-purple; capsule 11–14 mm. *Calcareous rocks and walls.* ● *S. & S.E. Spain.* Hs.

16. A. latifolium Miller, *Gard. Dict.* ed. 8, no. 4 (1768). Perennial herb, usually glandular-pubescent throughout, but sometimes only in inflorescence. Stems 60–100 cm, erect, sparingly branched. Leaves 20–70 × 8–32 mm, 1½–2½ times as long as wide, ovate, obtuse, more or less truncate at base, opposite below, alternate above. Bracts 5–12 mm, ovate. Pedicels 3–8(–15) mm. Calyx-lobes 7–9 mm, broadly ovate, obtuse. Corolla 33–48 mm, pale yellow. Capsule 13–17 mm, ovoid-oblong, glandular-pubescent. $2n = 16$. *Rocks and stony slopes.* ● *C. Italy to N.E. Spain.* Ga Hs It.

17. A. majus L., *Sp. Pl.* 617 (1753). Perennial herb, glabrous below, glabrous or glandular-pubescent in inflorescence. Stems up to 150(–200) cm, erect or straggling. Leaves 10–70 × 1–25 mm, linear to ovate, (1½–)2–12 times as long as wide, distinctly cuneate at base. Bracts 2–10 mm, ovate. Pedicels 2–10(–15) mm. Calyx-lobes 6–8 mm, ovate-oblong to suborbicular, obtuse. Corolla (27–)30–45 mm, usually pink or purple. Capsule 10–14 mm, oblong, glandular-pubescent or glabrous. *S.W. Europe, extending eastwards to Sicilia; widely cultivated for ornament and naturalized elsewhere.* Bl Ga Hs Lu Si [Al Au Be Br Cr Cz Ge Gr Hb He Ho *It Ju Rm Sa Tu].

The native plants may be divided into 4 subspecies, although there is a considerable overlap in most characters. Plants escaped from cultivation are mostly referable to subsp. (d) in the Mediterranean region and to subsp. (a) elsewhere, but some of them cannot be referred to any subspecies, and are perhaps of hybrid origin.

1 Leaves (6–)9–12 times as long as wide, the upper ones often opposite; inflorescence glabrous   **(d) subsp. tortuosum**
1 Leaves 1½–9 times as long as wide, the upper ones alternate; inflorescence glandular-pubescent
2 Leaves 1½–3 times as long as wide; lower pedicels longer than calyx   **(b) subsp. linkianum**
2 Leaves 4–9 times as long as wide; lower pedicels shorter than or equalling calyx

3 Cauline leaves 30–70 × 5–20 mm; stems erect, sparingly
   branched   **(a) subsp. majus**
3 Cauline leaves 20–40 × 4–8 mm; stems diffuse, freely branched
                                              (c) subsp. cirrhigerum

(a) Subsp. **majus**: Stems up to 120 cm, erect, sparingly branched. Leaves 30–70 × 5–20 mm, elliptic-lanceolate to linear-oblong, widest near the middle, the upper ones alternate. Inflorescence glandular-pubescent. Corolla 33–45 mm, purplish-pink with yellow palate, rarely (var. *striatum* (DC.) Rothm.) pale yellow veined with red. ● *E. Pyrenees, and mountains of S.C. France, N.E. Spain and Mallorca; widely naturalized elsewhere.*

(b) Subsp. **linkianum** (Boiss. & Reuter) Rothm., *Feddes Repert.* **54**: 19 (1944): Stems up to 80 cm, erect or ascending, sparingly branched. Leaves 20–55 × 8–20 mm, ovate to ovate-lanceolate, widest near the base, the upper ones alternate. Inflorescence glandular-pubescent. Corolla (27–)30–40 mm, purplish-pink with yellow palate. 2n=16. *Rocks and walls.* ● *W.C. Portugal.*

(c) Subsp. **cirrhigerum** (Ficalho) Franco, *Bot. Jour. Linn. Soc.* **64**: 275 (1971) (*A. latifolium* var. *cirrhigerum* Ficalho): Stems up to 200 cm, diffuse, freely branched, with branches often twining. Leaves 20–40 × 4–8 mm, linear-lanceolate, widest near the middle. Corolla 30–40 mm, purplish-pink with yellow palate. *Sandy ground near the sea. C. & S. Portugal, S.W. Spain.*

(d) Subsp. **tortuosum** (Bosc) Rouy, *Fl. Fr.* **11**: 59 (1909) (*A. tortuosum* Bosc): Stems erect or straggling, the branches sometimes twining. Leaves 10–60 × 1–5(–10) mm, linear to linear-oblong, the uppermost often opposite. Inflorescence glabrous or glandular-pubescent. Corolla 30–37 mm, purplish-pink, with white or yellow palate. *S. Spain, Sicilia; widely naturalized elsewhere in Mediterranean region.*

11. Asarina Miller[1]

Like *Antirrhinum* but leaves crenate-dentate or lobed, palmately veined; corolla glabrous except for palate; capsule with 2 equal loculi, each dehiscing by a single pore.

1. A. procumbens Miller, *Gard. Dict.* ed. 8, no. 1 (1768) (*A. lobelii* Lange, *Antirrhinum asarina* L.). Glandular-pubescent perennial; stems procumbent, woody at the base. Leaves up to 5 × 6 cm, ovate-cordate to reniform, crenate-dentate, sometimes slightly palmatifid; petiole about as long as lamina. Flowers solitary in leaf-axils; pedicel 12–20 mm. Calyx 10–13 mm, divided almost to the base; lobes lanceolate, acute, slightly unequal. Corolla 30–35 mm; tube whitish, slightly veined with purple; lips pale yellow; palate deep yellow. Capsule subglobose, glabrous, shorter than calyx. *Shady rocks, mostly in the mountains; calcifuge.* ● *S. France, N.E. Spain.* Ga Hs.

12. Misopates Rafin.[1]

Like *Antirrhinum* but annual; calyx-lobes conspicuously unequal, all longer than the corolla-tube; seeds somewhat flattened, with one face smooth, keeled, and produced into a narrow wing, the other finely tuberculate and with a wide, raised, sinuate, papillose border.

Corolla 10–17 mm, not exceeding the calyx, usually pink **1. orontium**
Corolla 18–22 mm, exceeding the calyx, white **2. calycinum**

1. M. orontium (L.) Rafin., *Autikon Bot.* 158 (1840) (*Antirrhinum orontium* L.). Stem (5–)20–50 cm, erect, sparingly branched, glabrous to sparsely hirsute below, glandular-pubescent

above. Leaves 20–50 × 2–7 mm, linear to oblong-elliptical, subacute, shortly petiolate, opposite below, often alternate above. Flowers in a lax, terminal raceme; bracts similar to leaves but decreasing in size upwards; pedicels very short in flower, up to 4 mm in fruit. Calyx 10–17 mm; lobes linear, the longest about 1½ times as long as the shortest. Corolla 10–15 mm, equalling or shorter than the calyx, pink (rarely white). Capsule 8–10 mm, ovoid, gibbous, glandular-hairy. 2n=16. *Cultivated ground; occasionally in other open habitats; somewhat calcifuge. S., W. & C. Europe, but doubtfully native in much of the centre and northwest; locally naturalized as a weed further north and east.* Al *Au Az *Be Bl *Br Bu Co Cr Ga Gr *Hb He Ho Hs *Hu It Ju Lu Rm Rs (W, K) Sa Si Tu [Cz Da Ge Po Rs (C) Su].

2. M. calycinum Rothm., *Feddes Repert.* (*Beih.*) **136**: 112 (1956). Like **1** but stem up to 80 cm, usually glabrous; raceme dense in flower, elongating in fruit; calyx 15–20 mm; corolla 18–22(–27) mm, exceeding the calyx, white; capsule 6–8 mm, sometimes glabrous. *Cultivated fields and other open habitats; usually calcicole. W. Mediterranean region, extending to Portugal and S.E. Italy.* Ga Hs It Lu Sa Si.

13. Chaenorhinum (DC.) Reichenb.[2]

Annual or perennial herbs. Leaves entire, shortly petiolate and usually opposite below, sessile and opposite or alternate above, those subtending flowers always alternate. Flowers in terminal, bracteate racemes or solitary in the leaf-axils. Calyx deeply and somewhat unequally 5-lobed. Corolla with more or less cylindrical tube, produced at the base into a straight spur; limb 2-lipped, the upper lip 2-lobed, the lower 3-lobed and with a low palate which does not close the mouth of the tube. Stamens 4, didynamous, included. Capsule obliquely ovoid to subglobose, with unequal loculi opening by pores. Seeds numerous, oblong-ellipsoid to truncate-conical, with smooth or variously ornamented, longitudinal ribs.

Literature: T. M. Losa España, *Anal. Inst. Bot. Cavanilles* **21**: 545–566 (1963).

1 Perennial
2 Non-flowering stolons, bearing only scale-leaves, present at
   base of plant   **2. glareosum**
2 Stolons absent
3   Corolla 6–8 mm, white   **4. tenellum**
3   Corolla at least 9 mm, blue, lilac, pink or yellow
4   Capsule ovoid, with loculi very unequal in length; spur of
       corolla 5·5–8 mm   **5. macropodum**
4   Capsule subglobose, with loculi not very unequal in length;
       spur of corolla 2–5(–6) mm
5   Stems densely glandular-villous throughout, with trans-
         lucent, yellowish hairs; pedicels recurved in fruit
                                                3. villosum
5   Stems glabrous below (rarely with short, white, eglandular
         hairs); pedicels usually erecto-patent in fruit
                                                1. origanifolium
1 Annual
6   Basal leaves usually in a rosette; spur of corolla acute,
     evenly tapered from the base
7   Spur at least as long as corolla-tube; lips of corolla widely
       divergent   **8. grandiflorum**
7   Spur usually shorter than corolla-tube; lips of corolla not
       widely divergent   **9. rubrifolium**
6   Basal leaves usually not in a rosette; spur of corolla obtuse to
     subacute, ± cylindrical or inflated in the middle
8   Corolla not more than 9 mm; spur 1·5–2·5 mm, ± cylindrical
                                                10. minus
8   Corolla at least 9 mm; spur 3–5 mm, somewhat inflated in the
       middle

9 Calyx-lobes not more than 1 mm wide; corolla 9–17 mm,
   lilac with darker veins   **6. serpyllifolium**
9 Calyx-lobes 1–2 mm wide; corolla 13–19 mm, with blue lips
   and yellowish palate, spur and tube   **7. robustum**

1. C. origanifolium (L.) Fourr., *Ann. Soc. Linn. Lyon* nov. ser.,
17: 127 (1869) (*Linaria origanifolia* (L.) Cav.). Perennial; stems
up to 35(–50) cm, numerous, ascending to erect, simple or
branched. Leaves lanceolate to suborbicular, acute to obtuse.
Racemes lax, at least in fruit; bracts linear to lanceolate; pedicels
usually erecto-patent in fruit. Calyx-lobes linear-spathulate to
oblanceolate. Corolla 9–20 mm; spur 2–5(–6) mm. Capsule
subglobose; loculi only slightly unequal. $2n = 14$. *Calcareous
rocks and walls. S.W. Europe.* Bl Ga Hs It Lu.

Very variable; it seems probable that further subspecies, in
addition to those described below, should be recognized.

1 Stems, petioles and pedicels covered with short, white, patent
   or deflexed, eglandular hairs   **(d) subsp. segoviense**
1 Plant glabrous below, more or less glandular-pubescent above
2 Leaves up to 10(–13) mm wide, ovate to suborbicular (rarely
   lanceolate); calyx 5–8 mm; capsule usually more than 3 mm
                (a) subsp. origanifolium
2 Leaves not more than 5 mm wide, narrowly elliptical to
   lanceolate; calyx 3–6 mm; capsule not more than 3 mm
3 Leaves usually thin; stems slender, flexuous or pendent
                 (b) subsp. cadevallii
3 Leaves thick and rigid; stems rigid   **(c) subsp. crassifolium**

(a) Subsp. origanifolium: Glabrous at least at the base. Leaves
up to 22(–28) × 10(–13) mm, suborbicular to lanceolate. Lower
pedicels up to 18(–30) mm in fruit. Calyx 5–8 mm. Corolla
10–20 mm, bluish-lilac with violet lines and pale yellow palate;
spur 2–5(–6) mm, cylindrical or somewhat inflated. Capsule
(2·5–)3–5 mm. Seeds 0·5–0·9 mm; ribs low or prominent, smooth
or denticulate. *Throughout the range of the species except for
parts of C. & S. Spain.*

(b) Subsp. cadevallii (O. Bolós & Vigo) Laínz, *Aportac. Fl.
Gallega* **6**: 27 (1968): Glabrous except in inflorescence; stems
long, slender, flexuous or pendent. Leaves not more than
17 × 5 mm, elliptical, usually thin and rather sparse. Lower
pedicels 10–17 mm in fruit. Calyx 3–5 mm. Corolla 9–14 mm,
lilac; spur 2–3·5 mm, contracted at both ends. Capsule *c.* 2 mm.
Seeds 0·5–0·6 mm; ribs low, smooth. ● *N.E. Spain.*

(c) Subsp. crassifolium (Cav.) Rivas Goday & Borja, *Anal.
Inst. Bot. Cavanilles* **19**: 451 (1961): Stems slender, rigid, papillose
but otherwise glabrous, except in inflorescence. Leaves 4–17 ×
2–5 mm, oblanceolate to spathulate, rarely ovate, thick and rigid,
usually crowded; margin inflexed. Lower pedicels 5–10(–17) mm
in fruit. Calyx 3–6 mm. Corolla 10–15 mm, pale pink or lilac
tinged with yellow; spur 2–3(–4·5) mm, obtuse, usually somewhat
inflated. Capsule 2–3 mm. Seeds as in subsp. (b). ● *S. & E.
Spain, Islas Baleares.*

(d) Subsp. segoviense (Willk.) R. Fernandes, *Bot. Jour. Linn.
Soc.* **64**: 220 (1971) (*C. segoviense* Willk.): Densely pubescent
throughout with white, patent or deflexed, eglandular hairs, and
with glandular hairs also in inflorescence; stems long, slender,
decumbent. Leaves 5–17 × 2–5 mm, elliptical, thin, rather sparse.
Racemes very lax; pedicels slender, up to 15(–30) mm in flower.
Calyx *c.* 5 mm. Corolla 11–13 mm, pink with darker lines; spur
2·5–4 mm, slightly constricted at base, obtuse. Seeds as in subsp.
(b). *Walls.* ● *C. Spain (Segovia).*

2. C. glareosum (Boiss.) Willk., *Ill. Fl. Hisp.* **2**: 29 (1886).
Perennial; stems up to 30 cm, numerous, ascending, the non-
flowering ones forming slender stolons with minute scale-leaves.

Leaves up to 12 × 7 mm, ovate to ovate-orbicular, obtuse, light
green, glabrous. Racemes few-flowered, subcorymbose; pedicels
up to 15 mm in fruit. Calyx 6·5–10 mm; lobes linear-spathulate,
obtuse. Corolla 17–24 mm; lips violet to lilac; palate yellow; tube
violet tinged with pink or yellow; spur 4–6 mm, constricted at the
base, yellowish. Seeds *c.* 0·6 mm, with prominent, smooth ribs.
$2n = 14$. *Schistose rocks and screes at c. 3000 m.* ● *S. Spain
(Sierra Nevada).* Hs.

3. C. villosum (L.) Lange in Willk. & Lange, *Prodr. Fl. Hisp.* **2**:
580 (1870). Perennial; densely viscid-villous throughout with
yellowish, translucent hairs; stems up to 35 cm, diffuse, pendent
or ascending. Leaves up to 25 × 13 mm, suborbicular to obovate
or rhombic, usually obtuse, rather thick. Pedicels up to 22 mm,
recurved in fruit. Calyx 5–8(–9·5) mm; lobes 1–4 mm wide,
linear-spathulate to obovate. Corolla 10–18 mm; lips lilac or
pale yellow with violet lines; palate yellow; spur 2–4 mm, obtuse.
Capsule about half as long as calyx; loculi only slightly unequal.
Seeds *c.* 0·5 mm, with low, smooth ribs. *Walls and rocks. S.
Spain, S.W. France.* Ga Hs.

4. C. tenellum (Cav.) Lange, *op. cit.* 581 (1870). Perennial,
villous with long white hairs. Stems up to 50 cm, numerous,
slender, procumbent. Leaves broadly ovate to orbicular, truncate
or cordate at base, usually sparse; lamina 3–14 × 3–14 mm, thin;
petiole 1–7 mm, slender. Pedicels up to 25 mm in fruit, very
slender. Calyx 3–5 mm; lobes linear-lanceolate to oblanceolate,
nearly equal. Corolla 6–8 mm, white; spur 2–3 mm, obtuse.
Seeds *c.* 0·6 mm, with low, smooth ribs. *Shady mountain rocks.*
● *E. Spain (near Ayora).* Hs.

5. C. macropodum (Boiss. & Reuter) Lange, *op. cit.* 579 (1870).
Perennial; stems 10–35 cm, erect or ascending, robust, glandular-
villous. Leaves up to 30 × 10 mm, the lowest lanceolate to elliptical, obtuse, subglabrous, the upper smaller, narrower and
villous. Pedicels up to 25(–35) mm in flower, more or less erect.
Calyx 7–9 mm, densely glandular-villous. Corolla 17–26 mm,
lilac with violet stripes, sometimes with yellow or white palate;
spur 5·5–8 mm, subacute. Capsule 5–6·5 mm, subacute, with
loculi very unequal in length. Seeds with denticulate or spinulose
ribs. *Mountain rocks and stony slopes.* ● *S. Spain.* Hs.

(a) Subsp. macropodum: Racemes lax throughout; pedicels
relatively stout. Calyx-lobes 2–3 mm wide, narrowly elliptical.
Seeds *c.* 1 mm. *Provinces of Córdoba and Granada.*

(b) Subsp. degenii (Hervier) R. Fernandes, *Bot. Jour. Linn. Soc.*
64: 222 (1971) (*C. robustum* f. *degenii* Hervier): Racemes dense
above; pedicels slender. Calyx-lobes 1–1·5 mm wide, spathulate. Seeds 0·5–0·7 mm. *Provinces of Jaén, Granada and Málaga.*

6. C. serpyllifolium (Lange) Lange in Willk. & Lange, *Prodr. Fl.
Hisp.* **2**: 578 (1870). Annual, glandular-villous, especially above.
Leaves up to 17 × 8 mm, the lower obovate to oblanceolate,
obtuse to subacute, subglabrous, the floral oblong, pubescent or
villous. Pedicels up to 20(–25) mm in fruit, erect. Calyx 3·5–6
mm; lobes not more than 1 mm wide, linear-spathulate, glandular-
villous, and also with long, white, eglandular hairs. Corolla lilac
with darker lines; spur 3–5 mm, somewhat inflated. Capsule
about equalling the calyx. Seeds 0·5–0·6 mm, with low, wide,
smooth ribs. ● *C. Spain; S.W. Portugal.* Hs Lu.

(a) Subsp. serpyllifolium: Stems 5–25(–40) cm, branched from
the base; branches ascending to erect. Calyx lobes *c.* 0·5 mm
wide, linear-spathulate. Corolla 9–15 mm, with more or less
cylindrical tube. *Cultivated ground and waste places. C. Spain.*

(b) Subsp. lusitanicum R. Fernandes, *Bot. Jour. Linn. Soc.* **64**:
223 (1971): Stems not more than 5 cm, erect, simple or sparingly

branched. Calyx-lobes up to 1 mm wide, oblong-elliptical. Corolla 13–17 mm, with the tube somewhat expanded above. *Calcareous sandstone rocks by the sea. S.W. Portugal.*

7. C. robustum Loscos, *Trat. Pl. Arag.* **1**: 14 (1876). Like **6(a)** but usually more robust; calyx 5·5–7 mm, with lobes up to 2 mm wide; corolla 13–19 mm, the lips bright blue, the upper with violet lines, the palate, tube and spur yellowish. *Stony or sandy soils.* ● *E. Spain.* Hs.

8. C. grandiflorum (Cosson) Willk., *Suppl. Prodr. Fl. Hisp.* 178 (1893). Annual; stems 5–20(–30) cm, erect or ascending, puberulent to glandular-pubescent, rarely glabrous. Leaves up to 25 × 12 mm, broadly ovate to lanceolate, obtuse to subacute, the basal forming a rosette. Racemes lax. Calyx 3–6 mm; lobes oblong to linear-spathulate, ciliate. Corolla 12–17 mm, bluish-violet; lips strongly divergent; spur 5–7 mm, conical-subulate, at least as long as tube and making an obtuse angle with it. Capsule *c.* 2·5 mm. Seeds *c.* 0·4 mm, conical, with minutely denticulate ribs. ● *S.E. Spain.* Hs.

9. C. rubrifolium (Robill. & Cast. ex DC.) Fourr., *Ann. Soc. Linn. Lyon* nov. ser., **17**: 127 (1869) (*Linaria rubrifolia* Robill. & Cast. ex DC.). Annual; stems up to 20(–40) cm. Leaves up to 20(–40) × 9(–15) mm, ovate to obovate, the lowest red beneath and usually forming a basal rosette, the upper smaller and green on both surfaces. Pedicels up to 16(–25) mm in fruit, erecto-patent. Calyx 4–7 mm; lobes linear-spathulate, 1–1·5 mm wide. Lips of corolla more or less approximated; spur conical, acute, usually shorter than tube. Seeds 0·4–0·5 mm. *Dry places. Mediterranean region, mainly in the west.* Bl Ga Gr Hs It Sa Si.

1 Corolla (13–)15–20 mm, yellow, tinged with violet on the
   tube; capsule 4–6 mm   (c) subsp. **raveyi**
1 Corolla usually less than 15 mm, blue, violet or lilac, with
   yellow palate; capsule 2·5–4·5 mm
2 Corolla usually less than 13 mm; seeds with rather prominent
   ribs   (a) subsp. **rubrifolium**
2 Corolla at least 13 mm; seeds with low ribs (b) subsp. **formenterae**

(a) Subsp. **rubrifolium**: Corolla-tube cylindrical, nearly closed at the mouth; spur 1–3·5(–4·5) mm, forming an obtuse angle with the tube; seeds with rather prominent, usually more or less tuberculate or spinulose ribs. *Throughout the range of the species.*
(b) Subsp. **formenterae** (Gand.) R. Fernandes, *Bot. Jour. Linn. Soc.* **64**: 227 (1971) (*C. formenterae* Gand.): Corolla-tube somewhat expanded towards the open mouth; spur 3–5 mm, usually continuing the line of the corolla-tube. Seeds with smooth, low ribs. ● *Islas Baleares* (*Ibiza, Formentera*).
(c) Subsp. **raveyi** (Boiss.) R. Fernandes, *loc. cit.* (1971) (*Linaria raveyi* Boiss.): Corolla-tube expanded towards the mouth; spur 4–6·5 mm, continuing the line of the tube. Seeds with prominent, tuberculate-spinulose ribs. *Damp, sandy places. S. Spain* (*mountains between Granada and Málaga*). (*N. Africa.*)

10. C. minus (L.) Lange in Willk. & Lange, *Prodr. Fl. Hisp.* **2**: 577 (1870) (*Linaria minor* (L.) Desf.). Annual, usually glandular-pubescent; stems up to 40 cm. Leaves 5–35 × 1–5(–8) mm, linear to oblong-lanceolate, obtuse. Pedicels 3–20 mm in fruit, erect or ascending. Calyx 2–5 mm; lobes linear-spathulate, obtuse. Corolla 5–9 mm; spur 1·5–2·5 mm, more or less cylindrical, obtuse. Capsule 3–6 mm. Seeds 0·5–1 mm, with smooth or minutely denticulate ribs. *Most of Europe, but doubtfully native in much of the north.* All except Az Bl ?Co Fa Is Sa ?Si Sb.

1 Calyx 2–2·5 mm; corolla 5 mm, pale yellow   (c) subsp. **idaeum**
1 Calyx 3–5 mm; corolla 6–9 mm, lilac with yellow palate
2 Pedicels 3–9 mm in fruit; capsule subglobose, usually shorter
   than calyx; seeds *c.* 1 mm   (b) subsp. **litorale**
2 Pedicels (5–)8–20 mm in fruit; capsule oblong-ovoid, about
   equalling calyx; seed 0·5–0·8 mm   (a) subsp. **minus**

(a) Subsp. **minus** (*C. viscidum* (Moench) Simonkai; incl. *C. klokovii* Kotov): Pedicels (5–)8–20 mm in fruit. Calyx 3–5 mm; corolla 6–9 mm, lilac with yellow palate. Capsule oblong-ovoid, about equalling the calyx; seeds 0·5–0·8 mm. *Cultivated fields, waste places and railway-lines. Throughout the range of the species.*
(b) Subsp. **litorale** (Willd.) Hayek, *Prodr. Fl. Penins. Balcan.* **2**: 146 (1929): Like subsp. (a) but pedicels 3–9 mm in fruit; capsule subglobose, shorter than the calyx; seeds *c.* 1 mm. $2n = 14$. ● *Shores of the Adriatic sea.*
(c) Subsp. **idaeum** (Rech. fil.) R. Fernandes, *Bot. Jour. Linn. Soc.* **64**: 229 (1971) (*C. idaeum* Rech. fil.): Not more than 5 cm, with divaricate branches. Calyx 2–2·5 mm; corolla 5 mm, pale yellow. Capsule subglobose, shorter than the calyx. ● *Kriti.*

14. Linaria Miller[1]

Annual to perennial herbs; leaves simple, entire, sessile, usually narrow, usually verticillate below and alternate above. Flowers in terminal, bracteate racemes or spikes, rarely solitary in the axils of foliage-leaves. Calyx deeply, often unequally 5-lobed, the adaxial lobe usually the longest, very rarely the shortest. Corolla usually glabrous except for the palate; tube cylindrical, produced at the base into a conical or cylindrical spur; limb 2-lipped, the upper lip 2-lobed, the lower 3-lobed, with at its base a more or less prominent, usually pubescent palate which usually closes the mouth of the tube. Stamens 4, didynamous, included. Capsule more or less globose, with equal loculi, dehiscing by several meridional fissures in the apical half. Seeds numerous.

A difficult genus, still requiring study. The four closely related groups comprising species **47–59** present particular difficulties, with a strongly reticulate pattern in the variation of those characters most commonly relied on for taxonomic differentiation. The species as here defined cannot be keyed or diagnosed as effectively as elsewhere in the genus, but they represent recurrent and recognizable morphological units, to which at least the majority of specimens can be referred. The only alternative treatment on the information at present available – the reduction of the whole complex to a small number of exceedingly variable species with probably ill-defined boundaries – would seem to be less useful.

The sections into which the genus has been traditionally divided are unsatisfactory; we have chosen to make only one division, based on seed-character.

Most species bear at their base short non-flowering shoots; in the annual species these may be hairy when the rest of the plant is glabrous, and bear verticillate leaves which are often wider than those of the flowering stems; descriptions of leaves refer to the largest of those on the flowering stems, unless the contrary is stated. Measurements of the calyx are taken to the apex of the longest lobe; those of the corolla from the apex of the spur to the apex of the upper lip; those of the spur from its apex to the insertion of the corolla in the receptacle; those of the capsule are taken before dehiscence; and those of the pedicel refer to the fruiting condition unless the contrary is stated.

Literature: B. Valdés, *Revisión de las Especies europeas de Linaria con Semillas aladas* (*Publ. Univ. Sevilla, Ser. Ci.* No. 7). Sevilla. 1970.

---

[1] By A. O. Chater, B. Valdés and D. A. Webb.

1 Seeds not discoid, without wing, or with angles very narrowly winged
2 Racemes glandular-puberulent to villous
3 Corolla mainly violet, purple or white (though often with yellow palate)
4 Corolla 9–17 mm; spur not more than 7 mm
5 Corolla 9–10 mm   **11. bipunctata**
5 Corolla 11–17 mm
6 Pedicels at least 15 mm in fruit   **2. pedunculata**
6 Pedicels not more than 4 mm in fruit
7 Leaves *c.* 1 mm wide; corolla violet; stigma bifid   **24. clementei**
7 Leaves 4–7 mm wide; corolla white or pale lilac; stigma clavate, ±entire   **29. nivea**
4 Corolla 16–30 mm; spur at least 8 mm
8 Stigma clavate, ±entire
9 Stems at least 20 cm, erect; racemes usually with at least 10 flowers   **22. elegans**
9 Stems less than 20 cm, flexuous-ascending; racemes with 3–8 flowers   **23. nigricans**
8 Stigma deeply bifid
10 Racemes usually with more than 10 flowers; lips of corolla widely divergent   **19. incarnata**
10 Racemes usually with not more than 10 flowers; lips of corolla ±approximated
11 Stems procumbent to ascending   **21. algarviana**
11 Stems erect
12 Racemes lax, at least in fruit; pedicels up to 15 mm, erecto-patent   **17. spartea**
12 Racemes rather dense, even in fruit; pedicels not more than 8 mm, erect   **18. viscosa**
3 Corolla mainly yellow (though sometimes with violet spur)
13 Leaves at least 4 mm wide
14 Pedicels and calyx eglandular-villous; stems glabrous   **7. tonzigii**
14 Pedicels and calyx glandular-pubescent; stems glandular-pubescent, at least immediately below the racemes
15 Stems erect; leaves oblong-lanceolate, 3–4 times as long as wide, mostly alternate   **13. hirta**
15 Stems decumbent to ascending; leaves ovate, about twice as long as wide, many of them opposite or verticillate   **8. cavanillesii**
13 Leaves not more than 3 mm wide
16 Pedicels mostly more than 3 mm in fruit
17 Racemes rather dense, even in fruit; capsules mostly overlapping   **18. viscosa**
17 Racemes very lax in fruit; capsules not overlapping
18 Corolla 15–30 mm; spur 9–18 mm   **17. spartea**
18 Corolla 13–15 mm; spur 6–7 mm   **20. hellenica**
16 Pedicels mostly less than 3 mm in fruit
19 Perennial; stems 25–70 cm; leaves 25–60 mm   **16. peloponnesiaca**
19 Annual or biennial; stems 5–30 cm; leaves not more than 22 mm
20 Spur 8–10 mm   **12. ficalhoana**
20 Spur 4–7 mm
21 Stems 5–12 cm, decumbent to ascending; calyx *c.* 4·5 mm   **10. huteri**
21 Stems 15–30 cm, usually ±erect; calyx 2·5–3 mm   **11. bipunctata**
2 Racemes glabrous
22 Pedicels strongly recurved in fruit   **1. reflexa**
22 Pedicels not recurved in fruit
23 Corolla mainly violet, purple or white (though often with yellow palate)
24 Leaves 10–25 mm wide   **14. triphylla**
24 Leaves not more than 8 mm wide
25 At least the longest pedicels 10–15 mm in flower   **2. pedunculata**
25 Pedicels less than 10 mm in flower
26 Spur 7–18 mm
27 Perennial; leaves up to 60 mm, crowded   **28. capraria**
27 Annual; leaves not more than 35 mm, distant
28 Stigma deeply bifid   **17. spartea**
28 Stigma clavate, ±entire
29 Spur curved; seeds 1–1·2 mm; corolla white   **25. chalepensis**
29 Spur straight; seeds 0·4–0·9 mm; corolla violet or bluish
30 Spur 7–8 mm; seeds 0·6–0·9 mm, strongly ruminate; leaves up to 4 mm wide   **4. pseudolaxiflora**
30 Spur 9–12 mm; seeds *c.* 0·4 mm, minutely rugulose-tuberculate; leaves 1–2 mm wide   **23. nigricans**
26 Spur 2–6 mm
31 Seeds less than 1 mm
32 Annual; corolla 10–15 mm; pedicels 2–6 mm in fruit   **26. canadensis**
32 Perennial; corolla 7–9 mm; pedicels not more than 2 mm in fruit   **31. microsepala**
31 Seeds at least 1 mm
33 Corolla white with lilac veins; spur straight, about half as long as remainder of corolla   **30. repens**
33 Corolla purplish-violet; spur curved, about as long as remainder of corolla   **27. purpurea**
23 Corolla mainly yellow (though sometimes with violet spur)
34 Annual; seeds 0·4–0·9 mm
35 Pedicels mostly less than 3 mm
36 Corolla 18–25 mm; racemes lax   **9. oligantha**
36 Corolla 10–14 mm; racemes capitate   **3. flava**
35 Pedicels all at least 3 mm
37 Spur 6–7 mm   **2. pedunculata**
37 Spur 9–18 mm   **17. spartea**
34 Perennial; seeds 1–1·5 mm
38 Corolla pale yellow with violet spur   **14. triphylla**
38 Corolla bright yellow throughout
39 Stems ±leafless for some distance below raceme; leaves not more than 3 mm wide   **16. peloponnesiaca**
39 Stems leafy up to base of raceme; leaves often more than 3 mm wide
40 Stems erect; leaves linear to ovate-lanceolate   **15. genistifolia**
40 Stems procumbent to ascending; leaves ovate to elliptical
41 Racemes mostly not more than 4 cm; capsule 2·5–3 mm   **5. cretacea**
41 Racemes mostly more than 4 cm; capsule *c.* 4 mm   **6. sabulosa**
1 Seeds discoid, with a marginal wing
42 Corolla mainly violet, purple, reddish or white (though often with yellow palate)
43 Corolla 35–55 mm   **32. triornithophora**
43 Corolla less than 35 mm
44 Wing of seed lacerate or fimbriate
45 Corolla 15–20 mm; pedicels at least 3 mm in flower; wing of seed deeply fimbriate   **33. pelisseriana**
45 Corolla 9–12 mm; pedicels not more than 2·5 mm in flower; wing of seed lacerate   **64. ricardoi**
44 Wing of seed entire
46 Corolla not more than 7 mm; spur shorter than remainder of corolla   **(68–70). arvensis** group
46 Corolla at least 8 mm; spur at least as long as remainder of corolla
47 Wing of seed swollen   **65. amethystea**
47 Wing of seed not swollen
48 Lower part of plant glandular or hairy
49 Corolla 8–15 mm; annual or biennial   **(57–59). diffusa** group
49 Corolla 15–27 mm; perennial
50 Corolla greyish-lilac with bluish-violet veins   **44. lilacina**
50 Corolla violet, purplish-brown or whitish-cream   **(47–50). tristis** group
48 Lower part of plant eglandular and glabrous
51 Disc of seed with long papillae
52 Annual; leaves 5–15 mm; racemes with not more than 4 flowers   **60. depauperata**
52 Perennial; leaves 13–27 mm; racemes usually with more than 4 flowers   **(47–50). tristis** group

51 Disc of seed smooth or with low tubercles
 53 Racemes sessile, overlapped at base by cauline leaves; bracts more than 10 mm, at least as long as cauline leaves   **63. glacialis**
 53 Racemes pedunculate, not overlapped by cauline leaves; bracts usually less than 10 mm, smaller than cauline leaves
  54 Stems ± erect; leaves often more than 3·5 mm wide
   55 Corolla deep violet with yellow palate; racemes lax in flower   **62. faucicola**
   55 Corolla lilac-grey with bluish-violet palate, or dark purple ± tinged with yellow, but without yellow palate; racemes dense in flower
    56 Corolla lilac-grey with bluish-violet palate; leaves elliptical to linear, flat   **43. anticaria**
    56 Corolla dark purple, ± tinged with yellow; leaves linear   **(47–50). tristis** group
  54 Stems decumbent to ascending; leaves not more than 3·5 mm wide
   57 Seeds 0·9–1·7 mm, strongly concavo-convex; racemes lax   **(57–59). diffusa** group
   57 Seeds 1·5–3·3 mm, usually flat or weakly concavo-convex; racemes usually dense
    58 Racemes ± glandular-pubescent; corolla with dark bluish-violet or purple palate
     59 Corolla greyish-lilac with bluish-violet veins; racemes sparsely hairy   **43. anticaria**
     59 Corolla dark purple, ± tinged with yellow; racemes densely hairy   **(47–50). tristis** group
    58 Racemes usually glabrous; corolla with yellow or orange palate
     60 Most leaves alternate   **(51–53). supina** group
     60 Most leaves verticillate
      61 Spur 8–10 mm   **61. alpina**
      61 Spur 10–15 mm   **(51–53). supina** group
42 Corolla mainly yellow (though sometimes with violet spur or partly tinged or veined with violet, purple or brown)
 62 Corolla not more than 9 mm
  63 Plant densely glandular-pubescent throughout   **67. arenaria**
  63 Plant glabrous, or glandular-pubescent only in inflorescence
   64 Stems erect   **(68–70). arvensis** group
   64 Stems procumbent to ascending   **(54–56). glauca** group
 62 Corolla at least 10 mm
  65 Plant entirely glabrous
   66 Leaves less than 2½ times as long as wide
    67 Spur 16–17 mm; leaves mostly alternate   **(47–50). tristis** group
    67 Spur 7–12 mm; leaves mostly verticillate
     68 Calyx 4–5 mm; pedicels less than 2 mm   **46. thymifolia**
     68 Calyx 7–11 mm; pedicels at least 2 mm   **42. platycalyx**
   66 Leaves at least 3 times as long as wide
    69 Wing of seed thickened, papillose   **65. amethystea**
    69 Wing of seed not thickened, smooth
     70 Many of the leaves alternate (the lowest often verticillate)
      71 Stems ± erect; corolla without purple or brownish markings   **(34–40). vulgaris** group
      71 Stems procumbent to ascending; corolla often tinged with purple or brown
       72 Spur not more than 9 mm   **(54–56). glauca** group
       72 Spur 10–15 mm   **(51–53). supina** group
     70 Leaves mostly verticillate or opposite (the uppermost usually alternate)
      73 Corolla 30–35 mm   **41. latifolia**
      73 Corolla 7–26 mm
       74 Leaves oblong to elliptical   **43. anticaria**
       74 Leaves usually linear
        75 Spur not more than 9 mm   **(54–56). glauca** group
        75 Spur 10–15 mm
         76 Seeds 0·9–1·2 mm   **9. oligantha**
         76 Seeds 1·5–2·8 mm   **(51–53). supina** group
  65 Plant pubescent or puberulent, at least in part

77 Wing of seed swollen   **65. amethystea**
77 Wing of seed not swollen
 78 Corolla tinged with violet, purple or brown
  79 Racemes sessile, overlapped at base by cauline leaves; bracts more than 10 mm, at least as long as cauline leaves   **63. glacialis**
  79 Racemes pedunculate, not overlapped by cauline leaves; bracts usually less than 10 mm, smaller than cauline leaves
   80 Lower lip of corolla with long lobes   **(51–53). supina** group
   80 Lower lip of corolla with short lobes
    81 Leaves linear to oblong-oblanceolate, mostly alternate, sometimes with revolute margins   **(47–50). tristis** group
    81 Leaves oblong to elliptical, mostly verticillate, flat   **45. verticillata**
 78 Corolla entirely yellow or orange
  82 Seeds 0·7–1·7 mm, with narrow wing
   83 Glabrous annual; leaves linear, sparse   **(54–56). glauca** group
   83 Subglabrous to hairy, usually perennial; leaves linear to oblong-elliptical, crowded   **66. saxatilis**
  82 Seeds at least 1·5 mm, with broad wing
   84 Leaves mostly verticillate (the uppermost usually alternate)
    85 Pedicels up to 17 mm in fruit, exceeding bracts   **42. platycalyx**
    85 Pedicels not more than 6 mm in fruit, usually not exceeding bracts
     86 Plant glandular-pubescent below   **45. verticillata**
     86 Plant glabrous below
      87 Leaves usually linear; seeds usually concavo-convex   **(51–53). supina** group
      87 Leaves oblong to elliptical; seeds flat   **43. anticaria**
   84 Many of the leaves alternate (the lowest often verticillate)
    88 More or less erect perennial   **(34–40). vulgaris** group
    88 Decumbent or ascending perennial or annual
     89 Pedicels usually longer than bracts; corolla 9–14 mm; wing of seed brown or black   **(54–56). glauca** group
     89 Pedicels shorter than bracts; corolla 13–27 mm; wing of seed white
      90 Lower lip of corolla with short lobes; leaves with revolute margins   **(47–50). tristis** group
      90 Lower lip of corolla with long lobes; leaves flat   **(51–53). supina** group

A. Seeds ovoid, reniform, trigonous or tetrahedral, sometimes somewhat compressed but never discoid, usually not winged (rarely with very narrow wings on the angles).

1. L. reflexa (L.) Desf., *Fl. Atl.* **2**: 42 (1798). Glabrous, slightly glaucous annual; stems 5–45 cm, procumbent, simple or sparingly branched. Leaves 6–25 × 2–10 mm, elliptic-oblanceolate to broadly elliptical, more or less acute, mostly alternate, the lowest verticillate. Racemes very lax, leafy; pedicels (7–)12–20 mm, scarcely elongating but strongly recurved in fruit, exceeding bracts. Calyx (2–)3–5 mm at anthesis, up to 7 mm in fruit; lobes lanceolate, acuminate, equal. Corolla (15–)20–30 mm, pale yellow, lilac or white; spur (8–)12–16 mm. Capsule 3·5–5 mm. Seeds 1–1·6 mm, oblong-reniform, strongly ruminate, grey-brown. *Dry, open habitats. C. Mediterranean region, mainly in the islands.* ?Al *Co *Ga ?Gr It Sa Si.

2. L. pedunculata (L.) Chaz., *Dict. Jard., Suppl.* **2**: 41 (1790). More or less glaucous annual (perhaps sometimes perennial), glabrous except sometimes for calyx; stems 10–30 cm, decumbent to erect, stoutish, branched above. Leaves 5–12 × 2–5 mm, oblong-lanceolate to elliptic-ovate, subacute, fleshy, mostly verticillate below, alternate above. Racemes lax, leafy; pedicels 10–15 mm at anthesis, up to 25 mm in fruit, erecto-patent. Calyx *c.* 4 mm, sometimes slightly glandular-hairy; lobes oblong, obtuse, equal. Corolla 11–16 mm, cream, usually with violet veins; spur 6–7 mm, deep violet. Capsule 4–5 mm. Seeds 0·5–0·7 mm, ovoid to subreniform, slightly flattened, smooth, black. *Maritime sands. S. Portugal, S.W. Spain. Hs Lu.*

3. **L. flava** (Poiret) Desf., *Fl. Atl.* **2**: 42 (1798). Glabrous annual; stems 5–22 cm, ascending, simple. Leaves 3–15 × 1–7 mm, elliptical to narrowly oblanceolate, subacute, verticillate below, alternate above. Racemes capitate, with 1–5 flowers; pedicels *c.* 1 mm. Calyx 3–4·5 mm; lobes narrowly oblong, subacute, unequal. Corolla 10–14 mm, yellow; spur 5–7 mm, straight; lobes of upper lip parallel. Capsule 3·5–5 mm. Seeds 0·6–0·9 mm, ovoid to subreniform, slightly flattened, rugulose, black. *Maritime sands. Corse, Sardegna. Co Sa.*

4. **L. pseudolaxiflora** Lojac., *Fl. Sic.* **2**(2): 132 (1907). Like **3** but stems decumbent; leaves not more than 10 × 4 mm; pedicels *c.* 6 mm, much longer than bracts; corolla 15–17 mm, mainly blue; spur 7–8 mm; seeds strongly ruminate. *Gozo, Linosa, Lampedusa, Malta. Si.*

L. laxiflora Desf., *Fl. Atl.* **2**: 45 (1798), from N.W. Africa, differs in having the corolla 18–23 mm, usually deep violet, the spur 10–14 mm, the lobes of the upper lip somewhat divergent, and pale brown seeds; it has been doubtfully recorded from Sicilia.

5. **L. cretacea** Fischer ex Sprengel, *Syst. Veg.* **2**: 791 (1825) (incl. *L. creticola* Kuprian.). Glabrous, glaucous perennial; stems 10–35 cm, procumbent to ascending, freely branched with usually strongly patent branches, leafy up to inflorescence. Leaves 7–13 × 5–9 mm, ovate to elliptical, acute, some verticillate, some alternate. Racemes up to 4(–5) cm, rather lax; pedicels 1–4 mm, usually about equalling bracts. Calyx 1·7–3 mm; lobes oblong to lanceolate, subacute, subequal. Corolla 11–14 mm, yellow; spur 4–6 mm. Capsule 2·5–3 mm. Seeds 1·2–1·5 mm, oblong-reniform, rugulose, black, without distinct angles or flanges. *Open habitats; calcicole.* ● *S.C. & S.E. Russia, E. Ukraine. Rs (C, E).*

6. **L. sabulosa** Czern. ex Klokov, *Bot. Žur.* **34**: 69 (1949). Like **5** but leaves sometimes narrower, mostly alternate; racemes 4–8 cm; corolla 13–17 mm; capsule *c.* 4 mm. *Maritime sands.* ● *W. coast of Krym. Rs (K).*

7. **L. tonzigii** Lona, *Natura* (*Milano*) **40**: 66 (1949). Glaucous perennial, glabrous except in inflorescence; stems 5–12(–20) cm, decumbent to ascending, usually simple. Leaves 8–22 × 5–8 mm, oblong-elliptical, more or less obtuse, mostly verticillate, more rarely opposite or alternate. Racemes capitate, villous; pedicels up to 3 mm, usually very short and shorter than bracts. Calyx 6–7 mm; lobes oblong-spathulate, obtuse, more or less unequal. Corolla 20–27 mm, pale yellow; spur 8–12 mm. Capsule 6–7 mm. Seeds 1·4–2 mm, angular, compressed, deeply foveolate-reticulate, blackish. *Dolomite screes, 2000–2500 m.* ● *Alpi Bergamasche. It.*

8. **L. cavanillesii** Chav., *Monogr. Antirrh.* 117 (1833). Glandular-hairy perennial; stems 15–40 cm, decumbent or ascending, simple or branched above. Leaves 15–22 × 9–14 mm, ovate to elliptic-ovate, acute, mostly verticillate or opposite, sometimes alternate above. Racemes dense; pedicels 1–4 mm, shorter than bracts. Calyx 6–9 mm; lobes linear-spathulate, obtuse, very unequal. Corolla 23–30 mm, yellow; spur 9–13 mm. Capsule *c.* 5 mm.

Seeds *c.* 1·2 mm, angular, compressed, densely rugulose-tuberculate, brownish-black. *Shady rock-crevices.* ● *S. & E. Spain. Hs.*

9. **L. oligantha** Lange, *Vid. Meddel. Dansk Naturh. Foren. Kjøbenhavn* **1881**: 100 (1882). Glabrous annual; stems 5–25 cm, decumbent or ascending, sparingly branched. Leaves 8–15 × 0·5–1 mm, linear, obtuse, opposite below, verticillate above, rarely alternate. Racemes lax, with 2–7 flowers; pedicels 0·5–1·5(–3) mm, shorter than bracts. Calyx *c.* 4 mm; lobes oblong-lanceolate, acute, subequal. Corolla 18–25 mm, yellow; spur 11–14 mm. Stigma entire. Capsule 4–5 mm. Seeds 0·9–1·2 mm, angular, compressed, blackish-brown, tuberculate, with smooth, acute angles which are sometimes almost wing-like. ● *S.E. Spain. Hs.*

10. **L. huteri** Lange, *op. cit.* 99 (1882). Glandular-puberulent annual; stems 5–12 cm, decumbent or ascending, simple or somewhat branched. Leaves 10–15 × 1–2 mm, linear to linear-lanceolate, acute, verticillate below, alternate above. Racemes short, dense; pedicels 1–3 mm, shorter than bracts. Calyx *c.* 4·5 mm; lobes linear-oblong, acute, unequal. Corolla 13–20 mm, yellow; spur 6–7 mm, usually pinkish-purple. Capsule *c.* 4 mm. Seeds 0·6–0·7 mm, reniform, minutely and densely tuberculate, blackish. *Rocks and stony places; calcicole.* ● *S. Spain. Hs.*

11. **L. bipunctata** (L.) Dum.-Courset, *Bot. Cult.* **2**: 93 (1802) (*L. filifolia* Lag.). Annual, glandular-puberulent at least above; stems 15–30 cm, diffuse or ascending to erect, usually much-branched. Leaves 6–20 × 0·5–1 mm, linear, obtuse, verticillate below, alternate above. Racemes short, rather dense at anthesis, lax in fruit; pedicels 1–3 mm. Calyx 2·5–3 mm; lobes oblong, subobtuse, unequal. Corolla 11–14 mm, yellow; spur 4–6 mm, often reddish. Capsule *c.* 3·5 mm. Seeds *c.* 0·6 mm, flattened-trigonous to reniform, obscurely and sparsely rugose or minutely tuberculate, with rounded angles. *Cultivated ground, waste places and maritime sands.* ● *Portugal, W. & C. Spain. Hs Lu.*

Plants from S.C. Spain (Sierra Morena) with slightly wider leaves, corolla 9–10 mm, perhaps sometimes pale lilac with yellow palate, and reniform, tuberculate seeds very narrowly winged on the margins have been called **L. intricata** Coincy, *Jour. Bot.* (*Paris*) **14**: 109 (1900) (*L. amoris* Pau, *L. diffusa* sensu Lange, non Hoffmanns. & Link); they are probably not specifically distinct from **11**.

12. **L. ficalhoana** Rouy, *Naturaliste* (*Paris*) **5**: 285 (1883). Densely glandular-pubescent annual or biennial; stems 3–15 cm, decumbent to erect, simple or branched above. Leaves 7–18 × 1–2·5 mm, linear-oblong to oblanceolate, obtuse, crowded, verticillate below, alternate above. Racemes short; pedicels 0·5–1·5 mm, much shorter than bracts. Calyx 3–4 mm; lobes elliptical to oblong-lanceolate, subobtuse, slightly unequal. Corolla 14–18 mm, yellow; spur 8–10 mm, reddish. Stigma clavate, entire. Capsule *c.* 4 mm. Seeds *c.* 0·6 mm, tetrahedral, finely rugulose. *Maritime sands.* ● *S.W. Portugal. Lu.*

13. **L. hirta** (L.) Moench, *Meth., Suppl.* 170 (1802). Annual, densely glandular-pubescent at least above; stems 15–80 cm, erect, usually simple, rather stout. Leaves 25–50 × 4–15 mm, oblong-lanceolate, acute to subobtuse, semiamplexicaul, mostly alternate, the lowest opposite. Racemes rather dense, elongating in fruit; flowers subsessile. Calyx 7–8 mm; lobes oblong-elliptical, obtuse, very unequal. Corolla 18–30(–40) mm, pale yellow; spur 10–16 mm. Capsule 5–6 mm. Seeds *c.* 1·2 mm, tetrahedral, strongly ruminate-alveolate, greyish-brown, with

acute, slightly winged angles. *Cultivated ground.* ● *C., S. & E. Spain, Portugal.* Hs Lu.

14. L. triphylla (L.) Miller, *Gard. Dict.* ed. 8, no. 2 (1768). Glabrous, somewhat glaucous annual; stems 10–45(–65) cm, usually single, erect, simple or branched above, stout. Leaves 15–35 × 10–25 mm, elliptical or obovate, truncate or semi-amplexicaul at base, mostly verticillate but some opposite or alternate. Racemes rather dense, greatly elongating in fruit; flowers subsessile. Calyx 9–12 mm; lobes ovate to lanceolate, obtuse to acute, unequal. Corolla 20–30 mm, white (rarely pale yellow or violet) with orange palate and violet spur; spur 8–11 mm. Capsule *c.* 8 mm. Seeds *c.* 1·5 mm, tetrahedral, strongly ruminate-alveolate, greyish-brown, with acute angles. *Cultivated ground. Mediterranean region.* Bl Co Cr Ga Gr Hs It Ju Sa Si.

15. L. genistifolia (L.) Miller, *op. cit.* no. 14 (1768). Glabrous perennial; stems (20–)30–100 cm, erect, branched, especially above, leafy up to inflorescence. Leaves 20–60 × 2–40 mm, linear to ovate, acute, more or less amplexicaul, alternate. Racemes (3–)10–20 cm, lax or dense; pedicels 1–13 mm, usually more or less equalling bracts. Calyx 2–12 mm; lobes linear-lanceolate to triangular-ovate or oblong-lanceolate, acute or acuminate, usually subequal. Corolla 13–50 mm, yellow; spur 4–25 mm. Capsule 3–7 mm. Seeds *c.* 1·2 mm, compressed-tetrahedral, rugulose, black, with a narrow, often whitish flange on the angles. 2*n* = 12. *S.E. & E.C. Europe, extending westwards to Italy and northwards to 54° N. in E.C. Russia.* Al Au Bu Cz Gr Hu It Ju Po Rm Rs (C, W, K, E) Tu [?Ge].

1 Corolla 20–50 mm; leaves mostly less than 4 times as long as wide   **(c) subsp. dalmatica**
1 Corolla 13–22 mm; leaves mostly at least 4 times as long as wide
  2 Leaves suberect, rigid   **(a) subsp. genistifolia**
  2 Leaves patent, flaccid   **(b) subsp. sofiana**

(a) Subsp. **genistifolia** (incl. *L. euxina* Velen., *L. pontica* Kuprian., *L. syspirensis* C. Koch): Leaves 3–15 mm wide, ovate- to linear-lanceolate, (3·5–)6–12 times as long as wide, suberect, rigid. Corolla 13–22 mm. 2*n* = 12. *Throughout the range of the species.*

(b) Subsp. **sofiana** (Velen.) Chater & D. A. Webb, *Bot. Jour. Linn. Soc.* **65**: 264 (1972) (*L. sofiana* Velen., *L. concolor* auct., ?an Griseb.): Leaves 2–6 mm wide, linear to oblong-lanceolate or oblanceolate, 5–15 times as long as wide, patent, flaccid. Corolla 13–20 mm. *E. Jugoslavia, W. Bulgaria.*

The identity of **L. concolor** Griseb., *Spicil. Fl. Rumel.* **2**: 21 (1844), from N.E. Greece, is uncertain; it is said to have more or less simple stems, rather obtuse calyx-lobes and the spur shorter than the rest of the corolla.

(c) Subsp. **dalmatica** (L.) Maire & Petitmengin, *Bull. Soc. Sci. Nancy* ser. 3, **9**: 403 (1908) (*L. dalmatica* (L.) Miller, *Antirrhinum dalmaticum* L.): Leaves up to 40 mm wide, ovate to lanceolate, 2–4(–5) times as long as wide, suberect, rigid. Corolla 20–50 mm. *Balkan peninsula, Romania; S. Italy; locally naturalized in C. Europe.*

16. L. peloponnesiaca Boiss. & Heldr. in Boiss., *Diagn. Pl. Or. Nov.* **3**(3): 163 (1856) (*L. sibthorpiana* Boiss. & Heldr.). Perennial, glabrous except for usually glandular-pubescent inflorescence. Stems 25–70(–120) cm, erect, sparingly branched above, usually leafless for some distance below inflorescence. Leaves 25–60 × 1·5–3 mm, linear, acute, alternate. Racemes 3–10 cm, very dense; pedicels 0·5–1(–1·5) mm, shorter than bracts. Calyx 1·5–4 mm; lobes ovate to oblong, obtuse to subacute, subequal. Corolla 13–20 mm, pale yellow; spur 7–8 mm. Capsule *c.* 4 mm.

Seeds 1–1·2 mm, compressed-tetrahedral, rugose-papillose, black, with a narrow flange on the margins. ● *S. & W. parts of Balkan peninsula.* Al Gr Ju ?Si.

17. L. spartea (L.) Willd., *Enum. Pl. Hort. Berol.* 640 (1809). Annual, glabrous below except for stems of non-flowering shoots, glabrous or glandular-pubescent in inflorescence; stems 15–60 cm, erect. Leaves 7–40 × 0·5–1 mm, linear, obtuse, rather remote, mostly alternate but the lowest opposite or verticillate. Racemes rather short and lax in flower, very lax in fruit; pedicels 3–15 mm, erecto-patent, greatly exceeding bracts. Calyx *c.* 4 mm; lobes oblong-lanceolate, obtuse to subacute, subequal, with wide scarious margins. Corolla (15–)18–30 mm, bright yellow (very rarely violet); lips closely approximated; spur 9–18 mm. Stigma deeply bifid. Capsule *c.* 4 mm. Seeds 0·6–0·7 mm, tetrahedral, coarsely rugose, black. *Dry, open habitats, especially on sandy soil.* ● *S.W. Europe, northwards to c. 46° N. in W. France.* Ga Hs Lu [Ge].

18. L. viscosa (L.) Dum.-Courset, *Bot. Cult.* **2**: 94 (1802). Like **17** but stems sometimes hairy at base; inflorescence always rather densely glandular-pubescent; leaves more crowded, especially below; racemes rather dense, even in fruit; calyx up to 6 mm, often with acuminate segments; pedicels not more than 8 mm, erect. 2*n* = 12. *Waste places and sandy fields. S. Italy and Sicilia; S. Spain, S. Portugal.* Hs It Lu Si.

The plants from Italy and Sicilia have been distinguished as **L. heterophylla** Desf., *Fl. Atl.* **2**: 48 (1798) (*L. stricta* (Sibth. & Sm.) Guss., non Hornem.), and differ from plants from the Iberian peninsula in more robust habit, leaves of non-flowering shoots in whorls of 6 (not 4) and fruiting pedicels shorter than calyx; they are best included in **18**.

Plants from S. Spain with decumbent to ascending stems and usually violet corolla have been described as **L. salzmannii** Boiss., *Voy. Bot. Midi Esp.* **2**: 456 (1841); they are probably not specifically distinct from **18**.

19. L. incarnata (Vent.) Sprengel, *Syst. Veg.* **2**: 796 (1825) (*L. bipartita* auct. eur., non (Vent.) Willd.). Like **17** but leaves up to 4 mm wide, all alternate; racemes usually with more than 10 flowers; calyx-lobes acuminate; corolla up to 22 mm, violet to reddish-purple, with lips widely gaping and lobes of the upper lip divergent; spur up to 11 mm. *Cultivated fields and dry grassland. Portugal, W. Spain; a frequent casual in parts of C. & N.W. Europe and perhaps locally naturalized.* Hs Lu. (*N.W. Africa.*)

20. L. hellenica Turrill, *Kew Bull.* [10]: 356 (1955). Annual, glabrous below, glandular-pubescent in inflorescence; stems *c.* 50 cm, erect, branched. Leaves 15–35 × 1–2·5 mm, linear to linear-oblong, obtuse, mostly alternate. Racemes with 10–20 flowers, very lax in fruit; pedicels up to 15 mm, erect. Calyx 4·5–5 mm; lobes oblong, subequal. Corolla 13–15 mm, yellow; spur 6–7 mm. Capsule *c.* 5 mm. Seeds *c.* 0·7 mm, reniform, strongly rugose, black. ● *S. Greece (S.E. Lakonia).* Gr.

Only once collected; it seems to be most closely related to **18**, but further material is required.

21. L. algarviana Chav., *Monogr. Antirrh.* 142 (1833). Annual, glabrous below, glandular-pubescent above; stems 10–30 cm, procumbent to ascending, simple. Leaves 4–8 × 0·6–3 mm, few, linear-oblong, subacute, alternate. Racemes lax, with 1–8 flowers; pedicels 5–12 mm, much exceeding bracts. Calyx 3–4·5 mm; lobes oblong-lanceolate, acuminate, slightly unequal, with white or violet scarious margin. Corolla 20–25 mm, violet, spotted with white or yellow on palate; spur 11–12 mm. Stigma deeply bifid. Capsule 3·5–4 mm. Seeds 0·6–0·8 mm, irregularly tetrahedral, strongly rugose, black. *Dry, sandy places.* ● *S.W. Portugal (W. Algarve).* Lu.

22. L. elegans Cav., *Descr. Pl.* 338 (1802) (*L. delphinoides* Gay ex Knowles & Westcott). Annual, glabrous below, glandular-pubescent above; stems 20–70 cm, erect, simple or branched. Leaves 7–35 × 0·5–4 mm, filiform to linear-lanceolate, obtuse, alternate. Racemes long, lax; pedicels 4–8 mm, much exceeding bracts, more or less erect. Calyx 3–5 mm; lobes lanceolate, acuminate, equal, with white or violet scarious margin. Corolla 17–25 mm, lilac to violet; spur 10–14 mm, longer than remainder of corolla, strongly curved; palate whitish; mouth of tube more or less open. Stigma clavate, entire to emarginate. Capsule 3–4 mm. Seeds *c.* 0·6 mm, tetrahedral, finely tuberculate, dark grey. *Dry, open habitats.* ● *N. & C. Spain, N. Portugal.* Hs Lu.

23. L. nigricans Lange in Willk. & Lange, *Prodr. Fl. Hisp.* **2**: 565 (1870) (*L. fragrans* Porta & Rigo). Annual, glabrous, or minutely glandular-puberulent in inflorescence; stems up to 17 cm, flexuous-ascending, simple or freely branched. Leaves 5–12 × 1–2 mm, linear-oblong to broadly elliptical, obtuse, blackening on drying, mostly alternate, the lowest verticillate. Racemes lax, with 3–8 flowers; pedicels 4–10 mm, patent to ascending, exceeding bracts. Calyx 3–4 mm; lobes oblong, subacute, equal. Corolla 16–20 mm, violet or lilac; spur 9–12 mm, straight. Capsule 3·5–4 mm. Seeds *c.* 0·4 mm, crescentic-trigonous, minutely rugulose-tuberculate, black. *Sandy places.* ● *S.E. Spain (Prov. Almería).* Hs.

24. L. clementei Haenseler ex Boiss., *Elenchus* 69 (1838). Glaucous perennial, glabrous except for glandular-pubescent inflorescence; stems 80–150 cm, erect, simple or branched above. Leaves 12–20(–30) × 1 mm, linear, subterete, obtuse, very sparse above, verticillate below, alternate above. Racemes subcapitate, laxer in fruit; pedicels 2–4 mm, slightly exceeding bracts. Calyx 2·5–3 mm; lobes lanceolate, acute, subequal. Corolla 13–17 mm, violet; spur *c.* 3·5 mm, straight. Stigma bifid. Capsule *c.* 3 mm. Seeds *c.* 0·6 mm, trigonous-reniform, strongly rugose, black. *Dry, sandy or calcareous places.* ● *S. Spain (Prov. Málaga).* Hs.

L. reverchonii Wittrock, *Acta Horti Berg.* **1**(4): 11 (1891), from the Sierra de Mijas, differs in its stems 40–110 cm, brownish and white-margined palate, filaments hairy above (not below) and capsule opening by 4 short teeth (not by 6 teeth to *c.* ½ way); its status is uncertain.

25. L. chalepensis (L.) Miller, *Gard. Dict.* ed. 8, no. 12 (1768). Glabrous annual; stems (10–)20–40 cm, usually single, simple or sparingly branched. Leaves 18–35 × 1–2 mm, linear, subobtuse, alternate. Racemes very lax; pedicels 1–2 mm at anthesis, up to 4 mm in fruit, mostly shorter than bracts. Calyx 7–10 mm; lobes linear, acute, unequal (the adaxial the shortest). Corolla 12–16 mm, white; spur 8–11 mm, curved; mouth of tube open. Capsule 4–5 mm. Seeds 1–1·2 mm, tetrahedral, strongly ruminate. *Dry places.* *Mediterranean region, from Islas Baleares eastwards.* Al Bl Bu Co Cr Ga Gr It Ju Sa Si Tu.

26. L. canadensis (L.) Dum.-Courset, *Bot. Cult.* **2**: 96 (1802). Glabrous, somewhat glaucous annual or biennial; stems (10–)25–80 cm, erect, usually simple. Leaves 15–30 × 1–2·5 mm, linear to linear-oblanceolate, obtuse, verticillate below, alternate above. Racemes rather dense at anthesis, lax in fruit; pedicels 1·5–3 mm at anthesis, up to 6 mm in fruit, exceeding bracts.

Calyx *c.* 3 mm; lobes linear-lanceolate, acute, equal. Corolla 10–15 mm, lilac to whitish; spur 4–6 mm, very slender, curved; palate rudimentary; lower lip very large. Capsule *c.* 3 mm. Seeds *c.* 0·5 mm, irregularly tetrahedral, with smooth, concave faces. *Cultivated for ornament and naturalized near Moskva.* [Rs (C).] (*North and South America.*)

27. L. purpurea (L.) Miller, *Gard. Dict.* ed. 8, no. 5 (1768). Glabrous, glaucous perennial; stems 20–60(–90) cm, ascending to erect, often branched above. Leaves 20–60 × 1–4(–8) mm, linear, subacute, verticillate below, alternate above. Racemes long, slender, rather dense; pedicels 1·5–4 mm, about equalling bracts. Calyx *c.* 3 mm; lobes linear-lanceolate, acute, equal. Corolla 9–12 mm, purplish-violet; spur *c.* 5 mm, curved. Capsule *c.* 3 mm. Seeds 1–1·3 mm, trigonous, rugose-ruminate, blackish. $2n = 12$. ● *C. & S. Italy, Sicilia; frequently cultivated for ornament and locally naturalized elsewhere.* It Si [Br ?Cz Hb].

28. L. capraria Moris & De Not., *Fl. Caprariae* 98 (1839). Like **27** but stems decumbent; racemes shorter, with fewer flowers; calyx *c.* 4·5 mm; corolla 15–17 mm; spur *c.* 8 mm; capsule *c.* 5 mm. ● *C. Italy (Arcipelago Toscana).* It.

29. L. nivea Boiss. & Reuter, *Diagn. Pl. Nov. Hisp.* 22 (1842). Like **27** but leaves 4–7 mm wide, linear-oblong to lanceolate; stems freely branched above; inflorescence glandular-pubescent; pedicels usually shorter than bracts; corolla 12–14 mm, white or pale lilac; spur 4–5 mm; capsule 4–5 mm. ● *Mountains of C. Spain.* Hs.

30. L. repens (L.) Miller, *Gard. Dict.* ed. 8, no. 6 (1768) (*L. striata* DC.; incl. *L. blanca* Pau, *L. monspessulana* (L.) Miller). Glabrous perennial; stems 30–120 cm, erect from a creeping rhizome, usually branched above. Leaves 15–40 × 1–2·5 mm, linear to linear-oblanceolate, acute, verticillate, sometimes alternate above. Racemes usually long and dense at anthesis, lax in fruit; pedicels 2–3 mm at anthesis, up to 4·5 mm in fruit; usually equalling bracts. Calyx 2–3 mm; lobes narrowly lanceolate, acute, subequal. Corolla 8–15 mm, white to pale lilac, with violet veins; spur 3–5 mm, conical, straight. Stigma capitate. Capsule 3–4 mm. Seeds 1·2–1·7 mm, ovoid-trigonous, strongly rugose, dark grey, with acute, almost winged angles. ● *From N. Spain and N.W. Italy to N.W. Germany; widely naturalized in N.W. & C. Europe.* Be ?Bl Ga Ge Hs It [Au Br Cz Da Fe Hb He Ho No Po Rs (B) Su].

L. × sepium Allman, *Proc. Roy. Irish Acad.* **2**: 405 (1843), (**30 × 34**) is the commonest hybrid in the genus. It is intermediate between the parents in most characters but, being partly fertile, is rather variable.

31. L. microsepala A. Kerner, *Sched. Fl. Exsicc. Austro-Hung.* **1**: 50 (1881). Like **30** but racemes denser in fruit; pedicels 0·5–1 mm at anthesis, *c.* 1·5 mm in fruit; calyx 1·5–2 mm, with oblong, obtuse lobes; corolla 7–9 mm; spur 2–3 mm; seeds *c.* 0·8 mm, coarsely tuberculate, black, with rounded angles. *Sandy and rocky places.* ● *Dalmatia.* Ju.

B. Seeds discoid, orbicular to reniform, surrounded by a conspicuous (rarely very narrow) marginal wing.

32. L. triornithophora (L.) Willd., *Enum. Pl. Hort. Berol.* 639 (1809). Glabrous, somewhat glaucous perennial; stems 50–130 cm, erect or diffuse, simple or branched above. Leaves 25–75 × 5–30 mm, lanceolate to ovate-lanceolate, acute, in remote whorls of 3(–5). Inflorescence up to 10 cm, lax, with 3–15 flowers, mostly in whorls; pedicels 10–30 mm, erect. Calyx 6–9 mm; lobes ovate-lanceolate, long-acuminate, equal. Corolla 35–55 mm, purple with yellow palate; spur 16–25 mm. Capsule 3–6 mm. Seeds 2–2·3 mm, suborbicular, dark brown; disc smooth or finely granular; wing rather narrow. *Hedges and thickets.* ● *N.W. & W.C. Spain, N. & C. Portugal.* Hs Lu.

33. L. pelisseriana (L.) Miller, *Gard. Dict.* ed. 8, no. 11 (1768). Glaucous, glabrous annual; stems 15–50 cm, erect, few, usually simple. Leaves 12–40 × *c.* 1 mm, linear, more or less erect, alternate or the lowest verticillate. Racemes dense, up to 5 cm in fruit; pedicels 3–6·5 mm, about equalling bracts. Calyx 4–5 mm; lobes linear-lanceolate, subequal. Corolla 15–20 mm, purplish-violet with whitish palate; upper lip very long, with parallel lobes; throat more or less closed; spur 7–9 mm. Capsule 2·5–3 mm. Seeds 1–1·2 mm, suborbicular, flat; disc smooth, black; wing deeply fimbriate, pale. 2*n* = 24. *Cultivated ground and waste places. Mediterranean region and W. Europe, northwards to the Channel Islands.* Al Bl Bu Co Cr Ga Gr Hs It Ju Sa Si Tu.

(34–40). L. vulgaris group. Perennials; stems 15–90 cm, erect or ascending. Leaves linear-filiform to narrowly elliptical, acute to subacute, mostly alternate. Racemes dense in flower, usually lax in fruit; flowers pedicellate. Calyx-lobes equal, or slightly unequal with the adaxial the longest. Corolla yellow; tube cylindrical, with closed mouth; spur as long as or shorter than rest of corolla. Capsule glabrous. Seeds 1·5–4 mm, orbicular to subreniform, flat; disc dark brown to black.

The main area of diversity is in S. Russia; it is probable that here at least hybridization occurs between various species of this group.

Part of the variation of the group has been interpreted as a result of hybridization with **15**, but no clear demonstration of this has yet been given.

1 Stems much-branched, the main axis usually not distinct; leaves usually distant; disc of seeds smooth
  2 Seeds 2–3·5 mm; pedicels 1–5 mm in flower, not more than 6 mm in fruit   **39. odora**
  2 Seeds 3–4 mm; pedicels 5–7 mm in flower, up to 12 mm in fruit   **40. loeselii**
1 Stems simple or branched, but the main axis always distinct; leaves crowded
  3 Disc of seeds smooth
    4 Capsule 9–14 mm, ovoid to oblong; calyx 5–8 mm **37. macroura**
    4 Capsule 5–6 mm, globose; calyx 4–5 mm   **38. debilis**
  3 Disc of seeds tuberculate
    5 Spur at least as long as lower lip of corolla; corolla 13–20 mm   **35. angustissima**
    5 Spur shorter than lower lip of corolla; corolla 19–33 mm
      6 Corolla 20–25(–28) mm; stems often hairy in lower half   **36. biebersteinii**
      6 Corolla (19–)25–33 mm; stems glabrous in lower half   **34. vulgaris**

34. L. vulgaris Miller, *Gard. Dict.* ed. 8, no. 1 (1768) (incl. *L. acutiloba* Fischer ex Reichenb.). Stems erect, simple, or branched only at base or in inflorescence, glabrous, or glandular-hairy above. Leaves 20–60 × 1–5(–15) mm, linear to narrowly elliptical or linear-oblanceolate, 1(3)-veined, crowded. Racemes with 5–30 flowers; pedicels 2–8 mm, often glandular-puberulent; bracts equalling or exceeding pedicels. Calyx 3–6 mm, often glandular-puberulent; lobes ovate to oblanceolate, subacute. Corolla (19–)25–33 mm, pale or bright yellow; spur 10–13 mm, stout, wider and shorter than lower lip. Capsule 5–11 × 5–7 mm, ovoid- or oblong-globose. Seeds 2–3 mm; disc tuberculate. 2*n* = 12. *Most of Europe except the extreme north and much of the Mediterranean region.* All except Az Bl Cr Fa Is Lu Sa Sb Si.

35. L. angustissima (Loisel.) Borbás, *Balaton Növényföldr.* 376 (1900) (*L. italica* Trev.). Like **34** but always glabrous; leaves 1–2(–5) mm wide; calyx-lobes linear-lanceolate, acuminate; corolla (13–)15–20 mm, pale yellow; spur (5–)7–10 mm, narrower than and at least as long as lower lip; capsule 4–6(–7·5) mm, ovoid. *S. & E.C. Europe, but absent from much of the Mediterranean region.* Al Au Bu Cz Ga ?Gr He Hs Hu It Ju Rm ?Rs (W).

L. bessarabica Kotov, *Ukr. Bot. Žur.* 11(4): 67 (1954), from S. Ukraine, has been equated with **35** by some Russian authors, but it has linear-oblong, very obtuse calyx-lobes and corolla up to 23 mm; it has been seldom collected and is perhaps of hybrid origin.

36. L. biebersteinii Besser, *Enum. Pl. Volhyn.* 25 (1822). Stems erect, simple or branched above, sparsely hairy. Leaves 25–55 × 1·5–7 mm, linear-lanceolate, 1- to 3-veined, acute, crowded. Racemes with (2–)5–30 flowers, rather short and dense; pedicels 1·5–7 mm, usually glandular-puberulent or -pubescent; bracts equalling or exceeding pedicels. Calyx 2·5–5 mm, often pubescent; lobes lanceolate, acute. Corolla 20–25(–28) mm, bright yellow; spur 5–10(–12) mm, stout, wider and shorter than lower lip of corolla. Capsule 6–9 mm, ovoid-oblong or -globose. Seeds 1·7–2·5 mm; disc tuberculate. *Steppes.* ● *S. part of U.S.S.R.* Rs (C, W, K, E).

Plants similar to **36** but entirely glabrous and with leaves 3–10 mm wide occur throughout its range, and also in Romania and Hungary. They have been called **L. × kocianovicii** Ascherson, *Österr. Bot. Zeitschr.* 15: 367 (1865) (*L. angustissima* subsp. *kocianovicii* (Ascherson) Sóo), and interpreted as hybrids between **15** and **34**; but they have also been interpreted as variants of **34** or **35**, or as an independent species (*L. ruthenica* Błoński, *L. rudis* Janka).

37. L. macroura (Bieb.) Bieb., *Cent. Pl.* 1: t. 27 (1810). Like **36** but glabrous, or sparsely glandular-pubescent only in inflorescence; leaves not more than 2 mm wide; calyx 5–8 mm, with ovate-oblong, very obtuse lobes; corolla (22–)28–33 mm, often with bluish veins; spur (10–)12–16 mm; capsule 9–14 mm, ovoid to oblong; seeds with smooth disc. *Steppes.* ● *S. Ukraine.* Rs (W, K, E).

L. incompleta Kuprian., *Sovetsk. Bot.* 4: 114 (1936), from S.E. Russia, is intermediate between **37** and **34** in most respects, but has most of its flowers abortive; it is perhaps of hybrid origin.

38. L. debilis Kuprian., *op. cit.* 4: 115 (1936). Like **36** but glandular-pubescent only in inflorescence; leaves *c.* 1 mm wide; bracts shorter than pedicels; calyx 4–5 mm; lobes ovate-oblong, subacute; corolla 24–29 mm; spur 12–16 mm; capsule 5–6 mm, globose; seeds with smooth disc. *S. Ural.* Rs (C, E).

39. L. odora (Bieb.) Fischer, *Cat. Jard. Gorenki* ed. 2, 25 (1812) (incl. *L. dolichocarpa* Klokov). Glabrous (except sometimes for calyx); stems ascending, much-branched, without a distinct main axis. Leaves 15–40(–60) × 1–1·5(–2·5) mm, linear to filiform, arising 1–3 cm apart and not crowded. Racemes with 7–12 flowers, terminating most branches; pedicels 1–5 mm in flower, up to 6 mm in fruit; bracts 1–3(–5) mm. Calyx 1·5–5 mm; lobes lanceolate to oblong-obovate, acute to subobtuse. Corolla 12–21 mm, pale yellow; spur 5–10 mm, slender, straight. Capsule 3–8 × 3–5 mm, subglobose to oblong-ellipsoid. Seeds 2–3·5 mm; disc smooth. *Sandy and stony places. S. part of U.S.S.R.; Turkey-in-Europe.* Rs (C, W, ?K, E) Tu.

Plants from Ukraine and the Don valley have subglobose capsules 3–5 mm and have been called **L. dulcis** Klokov, *Bot. Žur.*

34: 71 (1949); plants from the lower Volga have calyx 3–4 mm and corolla 17–21 mm and have been called **L. dolichoceras** Kuprian., *Acta Inst. Bot. Acad. Sci. URSS* (*Ser.* 1) **1**(2): 298 (1936); plants from S. Ural with corolla 14–20 mm and globose capsules 4–5 mm have been called **L. uralensis** Kotov, *Ukr. Bot. Žur.* **3**(3–4): 26 (1946). These are probably all conspecific with **39**.

40. **L. loeselii** Schweigger, *Königsb. Arch. Naturw.* **1**: 228 (1812). Like **39** but leaves 30–60 × 2–4 mm; pedicels 5–7 mm in flower, up to 12 mm in fruit; corolla 11–16 mm; spur 4·5–6 mm, usually curved; seeds 3–4 mm. *Maritime sands.* ● *Baltic coasts of Germany, Poland and U.S.S.R.* Ge Po Rs (B).

41. **L. latifolia** Desf., *Fl. Atl.* **2**: 40 (1798). Glabrous annual; stems 35–50 cm, erect, simple, stout. Leaves up to 65 × 15 mm, oblong-lanceolate, acute, mostly in whorls of 3 but the uppermost alternate. Racemes 7–15 cm, dense; flowers subsessile; bracts 10–20 mm, linear-lanceolate, patent. Calyx 7–10 mm; segments linear-lanceolate, somewhat unequal. Corolla 30–35 mm, yellow with greenish veins; spur 12–15 mm. Capsule *c.* 5 mm. Seeds *c.* 2 mm, reniform; disc smooth or tuberculate, black; wing whitish. *Cultivated ground.* S.W. Spain. Hs. (*N.W. Africa.*)

42. **L. platycalyx** Boiss., *Voy. Bot. Midi Esp.* **2**: 459 (1841). Annual or perennial, glabrous except for minutely ciliate bracts and calyx, somewhat glaucous; stems 10–40 cm, procumbent to erect, simple or branched above, rather stout. Leaves 3–35 × 2–20 mm, broadly elliptical to ovate, usually mucronate, mostly in whorls of 3, remote on the flowering stems. Racemes with 1–3 terminal flowers and often up to 3 others some distance below. Pedicels 2–10 mm in flower, up to 17 mm in fruit, exceeding the bracts. Calyx 7–11 mm; segments ovate, somewhat unequal. Corolla 20–27 mm, yellow; spur 9–12 mm. Capsule 4–8 mm. Seeds 2·3–3 mm, reniform-orbicular, flat; disc smooth or finely tuberculate, dark grey; wing pale yellow. *Shady limestone rocks.* ● *S.W. Spain.* Hs.

43. **L. anticaria** Boiss. & Reuter, *Pugillus* 86 (1852). Perennial, glabrous except sometimes for glandular-pubescent inflorescence, somewhat glaucous; stems 7–45 cm, procumbent to ascending. Leaves 7–35 × 1·5–10 mm, oblong to elliptical, acute, mostly in whorls of 4–6(–8) but the uppermost sometimes alternate. Racemes dense in flower, rather lax in fruit; bracts 2·5–10 mm; pedicels 1–2 mm. Calyx 4–7 mm; segments oblong-obovate, unequal. Corolla 18–26 mm, greyish-lilac with bluish-violet veins (very rarely entirely yellow); palate deep bluish-violet, sometimes with a yellow spot; spur 5–12 mm, usually slightly longer than rest of corolla. Capsule 3·5–7 mm. Seeds 1·7–3 mm, suborbicular, flat; disc tuberculate or smooth, blackish-brown; wing broad, paler. 2*n* = 12. *Shady limestone rocks.* ● *S. Spain.* Hs.

L. rossmaessleri Willk., *Linnaea* **30**: 118 (1859), from the Sierra de Almijara, with narrower leaves, which are often mostly alternate, and yellowish corolla-tube, is somewhat intermediate between **43** and **48**, but is best regarded as a variety of the former.

44. **L. lilacina** Lange, *Ind. Sem. Horto Haun.* **1854**: 24 (1854). Like **43** but glandular-pubescent throughout; corolla 17–27 mm, with orange palate; spur 8–12 mm, shorter than remainder of corolla, yellowish; seeds 1·5–2·6 mm, with acute, pale tubercles on disc. *Shady limestone rocks.* ● *S.E. Spain* (*prov. Jaén*). Hs.

45. **L. verticillata** Boiss., *Voy. Bot. Midi Esp.* **2**: 462 (1841). Perennial, glandular-pubescent below and usually also above; stems 5–35 cm, ascending to erect, simple or sparingly branched

above. Leaves (4–)6–25 × 1·5–6(–8) mm, oblong to elliptical, acute, in whorls of 3–5 below, usually alternate above. Racemes few-flowered, dense; bracts 3–10 mm; pedicels 0–4 mm. Calyx 3–6 mm; segments oblong to oblanceolate, obtuse, unequal. Corolla 15–35 mm, yellow, sometimes with violet or greenish stripes; spur 9–17 mm. Capsule 3·5–4·5 mm. Seeds 1·6–3·2 mm, suborbicular, concavo-convex; disc blackish-brown, smooth or minutely tuberculate; wing paler brown. *Mountain rocks.* ● *S. Spain* (*Sierra Nevada and adjacent ranges*). Hs.

46. **L. thymifolia** (Vahl) DC. in Lam. & DC., *Fl. Fr.* ed. 3, **3**: 587 (1805). Glaucous, glabrous annual, biennial or perennial; stems 6–30 cm, procumbent to suberect, numerous, usually freely branched. Leaves 3–15 × 1·5–10 mm, suborbicular-obovate to oblong, obtuse, somewhat fleshy, mostly in whorls of 3. Racemes subcapitate, with few, subsessile flowers. Calyx 4–5 mm; lobes broadly oblanceolate to oblong, unequal. Corolla 17–24 mm, yellow; tube with closed mouth; spur 7–11 mm. Capsule 5–6 mm, globose. Seeds 2·4–3·3 mm, suborbicular, concavo-convex; disc black, smooth; wing broad, black. *Maritime sands.* ● *W. coast of France, northwards to 46°N.* Ga.

(47–50). **L. tristis** group. Glaucous perennials, glabrous or glandular-puberulent below, usually glandular-pubescent in inflorescence; stems decumbent to ascending. Calyx-lobes unequal. Corolla 15–30 mm, usually yellow, variously tinged with purple or brown; tube rather wide; lobes of lower lip very short. Seeds reniform-orbicular, more or less concavo-convex, with broad wing.

1	Disc of seeds densely covered with papillae		**50. amoi**
1	Disc of seeds smooth or with low tubercles		
2	Leaves with revolute margins		**49. aeruginea**
2	Leaves flat		
3	Spur 16–17 mm; calyx-lobes oblong-spathulate		**47. lamarckii**
3	Spur 11–13 mm; calyx-lobes linear-oblanceolate		**48. tristis**

47. **L. lamarckii** Rouy, *Naturaliste* (*Paris*) **5**: 351 (1883). Glabrous, or sometimes glandular-pubescent in inflorescence; stems 20–40 cm, usually simple, leafless for *c.* 2 cm below inflorescence. Leaves (5–)7–15 × 3–6 mm, oblong-spathulate to obovate, flat, obtuse, somewhat fleshy, mostly alternate, sometimes verticillate at base. Racemes with up to 12 flowers, oblong, dense; pedicels 1–2 mm in flower, usually *c.* 3 mm in fruit, shorter than bracts. Calyx 6–7 mm; lobes oblong-spathulate. Corolla 25–31 mm, deep yellow; spur 16–17 mm, striped with orange-brown. Capsule 7–8 mm, ovoid. Seeds *c.* 3 mm; disc smooth, blackish; wing greyish-white. *Maritime sands.* ● *S. Portugal.* Lu.

48. **L. tristis** (L.) Miller, *Gard. Dict.* ed. 8, no. 8 (1768). Glabrous or glandular-puberulent below, more or less densely glandular-pubescent in inflorescence; stems 10–90 cm, usually simple, leafless for 1–8 cm below inflorescence. Leaves 8–40 × 0·5–4·5 mm, linear to oblong-lanceolate, flat, obtuse to subacute, alternate, often irregularly verticillate at base. Racemes with 2–15 flowers, dense or lax; pedicels 0·5–5 mm, shorter than bracts. Calyx 7–9 mm; lobes linear-oblanceolate, subobtuse. Corolla (18–)21–28 mm, yellow, variably tinged with purplish-brown; spur 11–13 mm, stout. Capsule 3·5–8 mm, globose. Seeds 1·7–3·2 mm; disc smooth or tuberculate, blackish; wing grey. 2*n* = 12. *Usually on limestone rocks.* ● *S. Spain; S. Portugal.* Hs Lu.

49. **L. aeruginea** (Gouan) Cav., *Elench. Pl. Horti Matrit.* 21 (1803) (*L. melanantha* Boiss. & Reuter). Perennial (rarely annual), glabrous or glandular-puberulent below, glandular-pubescent in inflorescence; stem 3–40(–70) cm, decumbent or ascending,

usually simple. Leaves 4–18(–32) × 0·3–1·7 mm, linear, with revolute margins, verticillate below, alternate above. Calyx 3–6 mm; lobes linear-lanceolate, subacute. Corolla 15–27 mm, usually yellow, variably tinged with purplish-brown, sometimes almost completely purplish-brown, violet, yellowish or whitish-cream; spur 5–11 mm. Capsule 4–6 mm, globose. Seeds 1·5–2·5(–3·5) mm; disc brown or blackish; wing pale brown or grey. 2n = 12. ● *Portugal, S. & E. Spain, Islas Baleares.* Bl Hs Lu.

(a) Subsp. **aeruginea**: Racemes with 2–35 flowers, dense; pedicels 0·5–3 mm, shorter than bracts; seeds tuberculate or smooth. *Portugal, S. & E. Spain.*

Plants from the Sierra Nevada at high altitudes have been named L. **nevadensis** (Boiss.) Boiss. & Reuter, *Pugillus* 87 (1852) (*L. supina* subsp. *nevadensis* (Boiss.) Nyman); they are consistently smaller in all their parts and have always a yellow corolla, but they do not merit more than varietal status.

(b) Subsp. **pruinosa** (Sennen & Pau) Chater & Valdés, *Bot. Jour. Linn. Soc.* 65: 264 (1972) (*L. supina* var. *pruinosa* Sennen & Pau): Racemes with 1–6 flowers, lax; pedicels 2–10 mm, usually longer than bracts; seeds smooth. *Islas Baleares.*

50. L. **amoi** Campo ex Amo, *Revista Progr. Ci. Exact. Fis. Nat.* 5: 56 (1855). Perennial, glabrous below, glandular-pubescent in inflorescence. Stems up to 30 cm, decumbent. Leaves 13–27 × 0·8–2·9 mm, linear, with revolute margins, verticillate below, alternate above, crowded. Racemes with 2–20 flowers, dense; pedicels 1–3 mm, shorter than bracts; Calyx 5–9 mm; lobes linear-oblanceolate, subacute. Corolla 20–28 mm, reddish-purple; spur 10–14 mm. Capsule 5–9 mm, globose. Seeds 2·2–3·2 mm; disc yellowish-black, densely covered with very prominent tubercles; wing yellowish-brown. 2n = 12. *Rocks and sandy places on dolomite.* ● *S. Spain (Sierra Tejeda, Sierra de Almijara).* Hs.

(51–53). L. **supina** group. Glaucous annuals, biennials or perennials, glabrous below, usually glandular-pubescent in inflorescence; stems procumbent to erect. Calyx 3–7 mm; lobes oblanceolate to linear-lanceolate or -oblong, subacute to obtuse, usually unequal. Corolla yellow, sometimes tinged with violet or red; tube narrow or wide; lobes of lower lip long. Seeds 1·5–2·8 mm, suborbicular, with broad wing.

1 Corolla yellow with reddish-brown stripes; seeds metallic-shiny
    51. caesia
1 Corolla without reddish-brown stripes, but sometimes violet-tinged; seeds not metallic-shiny
  2 Seed with grey or blackish wing; spur 10–15 mm   **52. supina**
  2 Seed with whitish wing; spur (4–)7–11 mm   **53. oblongifolia**

51. L. **caesia** (Pers.) DC. ex Chav., *Monogr. Antirrh.* 174 (1833). Annual, biennial or perennial, glabrous or sparsely glandular-hairy in inflorescence; stems 10–40 cm, usually simple. Leaves 5–30 × 0·5–2(–5) mm, linear-subulate to linear-oblong, alternate or the lower verticillate. Racemes with up to 15 flowers, dense in flower, lax in fruit; pedicels 1–2 mm, shorter than bracts. Corolla 19–25 mm, yellow with reddish-brown stripes; spur 9–12 mm. Capsule *c.* 5 mm, globose. Seeds 2–2·5 mm, flat to slightly concavo-convex; disc grey or blackish, metallic-shiny, smooth; wing pale brown or grey. 2n = 12. ● *W. half of Iberian peninsula.* Hs Lu.

Plants from the coast of N.W. Spain and C. & S. Portugal (var. *decumbens* Lange) differ from those of C. Spain in having glabrous inflorescences, wider leaves and wider and more obtuse calyx-lobes.

52. L. **supina** (L.) Chaz., *Dict. Jard., Suppl.* 2: 39 (1790). Annual, biennial or perennial, usually glandular-pubescent in inflorescence; stems 5–30 cm, simple. Leaves 5–20(–30) × 0·5–1 (–3·5) mm, linear to linear-oblanceolate, verticillate at least below, usually alternate above. Racemes with 2–5(–20) flowers, lax or dense; pedicels 1–2 mm in flower, up to 6 mm in fruit, shorter than bracts. Corolla 13–20(–27) mm, pale yellow, sometimes tinged with violet; spur 10–15 mm. Capsule 3–7 mm, globose. Seeds 1·7–2·8 mm, suborbicular, flat or somewhat concavo-convex; disc black, smooth or tuberculate; wing broad, grey or blackish. 2n = 12. *S.W. Europe, extending eastwards to c.* 10° 30′ *E. in N. Italy.* Ga Hs It Lu [*Br *Su].

Small plants from S.E. France (Var) referred to this species may be almost indistinguishable from 53(b).

53. L. **oblongifolia** (Boiss.) Boiss. & Reuter, *Pugillus* 86 (1852). Annual, glabrous to pubescent, usually glandular in inflorescence; stems 5–20 cm, simple. Leaves 3–17 mm, usually verticillate below, verticillate or alternate above. Racemes with 1–10 flowers, dense at anthesis, lax in fruit; pedicels up to 6 mm, longer or shorter than bracts. Corolla (12–)15–22 mm, yellow; spur (4–)7–11 mm, sometimes tinged with violet. Capsule 3·5–5 mm, globose. Seeds 1·5–2·2 mm, strongly concavo-convex; disc black, smooth or tuberculate; wing whitish, rarely blackish-brown. 2n = 12. ● *Mountains of S. Spain and Portugal.* Hs Lu.

(a) Subsp. **oblongifolia**: Leaves (1–)2–5 mm wide, oblong-lanceolate to ovate-elliptical. Corolla 15–22 mm. Pedicels shorter than bracts. *Mountains of S. Spain.*
(b) Subsp. **haenseleri** (Boiss. & Reuter) Valdés, *Rev. Esp. Eur. Linaria* 127 (1970) (*L. haenseleri* Boiss. & Reuter): Leaves 0·5–2 mm wide, linear to linear-oblong. Corolla 12–17 mm. Pedicels usually longer than bracts. *Throughout the range of the species.*

(54–56). L. **glauca** group. Glaucous annuals or perennials, glabrous below, sometimes glandular-puberulent in inflorescence; stems procumbent to nearly erect. Calyx 2·5–7 mm; lobes linear to linear-oblong or -spathulate, subequal or unequal. Corolla 7–20 mm, yellow, sometimes with purplish or bluish veins; tube narrow; lobes of lower lip long; spur usually wide. Seeds 1·2–3 mm, suborbicular, more or less flat.

1 Pedicels longer than bracts
  2 Corolla yellow; leaves 0·3–1 mm wide   **56. glauca**
  2 Corolla yellow with bluish or purplish veins; leaves 1–3·5 mm wide   **55. propinqua**
1 Pedicels shorter than bracts
  3 Racemes ± lax, often glandular-puberulent; corolla 7–12 mm
    56. glauca
  3 Racemes dense, glabrous; corolla 10–20 mm   **54. badalii**

54. L. **badalii** Willk., *Ill. Fl. Hisp.* 2: 33 (1887). Glabrous annual; stems 10–20 cm, several. Leaves 5–19 × 0·5–1·5 mm, linear, often with revolute margins, verticillate below, alternate above. Racemes dense, glabrous; pedicels 1–3 mm, shorter than bracts. Corolla 10–20 mm, yellow with purplish or bluish veins; spur 5–7·5 mm, shorter than lower lip. Capsule 3–5 mm, globose. Seeds (1·5–)2–3 × 1·5–2·7 mm, black; disc minutely tuberculate. *Calcareous and gypsaceous soils.* ● *N. Spain, extending southwards to Teruel.* Hs.

55. L. **propinqua** Boiss. & Reuter, *Pugillus* 88 (1852). Like **54** but leaves 1–3·5 mm wide, sometimes oblanceolate; racemes lax; pedicels longer than bracts; spur 7–9 mm, longer than lower lip; disc of seeds smooth to tuberculate. *Rocky or stony places; calcicole.* ● *N. Spain.* Hs.

56. L. glauca (L.) Chaz., *Dict. Jard., Suppl.* **2**: 39 (1790). Annual or perennial, glabrous or glandular-puberulent in inflorescence; stems 5–25 cm, usually several, procumbent to ascending-erect. Leaves 5–15 × 0·3–1 mm, linear, verticillate below, alternate above. Racemes with 3–7 flowers, rather lax, glabrous or glandular-pubescent; pedicels 1–4 mm, shorter or longer than bracts. Corolla 7–14 mm, yellow, rarely with bluish veins; spur 3–8 mm, stout. Capsule 3·5–5 mm, globose. Seeds 1·2–2·5 mm; disc black. ● *C. & E. Spain.* Hs.

1 Seeds not more than 1·5 mm, with black wing 0·2–0·4 mm wide
2 Annual; seeds with wing *c.* 0·2 mm wide **(a) subsp. glauca**
2 Perennial; seeds with wing 0·3–0·4 mm wide **(b) subsp. olcadium**
1 Seeds 1·5–2·5 mm, with wing 0·4–0·6 mm wide
3 Seeds with black wing; pedicels shorter than bracts
(d) subsp. bubanii
3 Seeds with white wing; pedicels usually longer than bracts
(c) subsp. aragonensis

(a) Subsp. **glauca**: Glabrous annual. Pedicels shorter than bracts. Seeds 1·2–1·5 mm; disc smooth or very slightly tuberculate; wing *c.* 0·2 mm wide, black. *C. Spain.*

(b) Subsp. **olcadium** Valdés & D. A. Webb, *Bot. Jour. Linn. Soc.* **65**: 265 (1972): Glabrous perennial. Pedicels shorter than bracts. Seeds *c.* 1·5 mm; disc slightly tuberculate; wing 0·3–0·4 mm wide, black. *S.E. Spain (near Albacete).*

(c) Subsp. **aragonensis** (Lange) Valdés, *Rev. Esp. Eur. Linaria* 177 (1970) (*L. diffusa* var. *aragonensis* Lange, *L. aragonensis* (Lange) Loscos): Annual, usually glandular-puberulent in inflorescence. Pedicels usually longer than bracts. Seeds 1·5–2·2 mm; disc smooth or with white tubercles; wing 0·4–0·6 mm wide, white. *E. Spain.*

(d) Subsp. **bubanii** (Font Quer) Valdés, *op. cit.* 178 (1970) (*L. bubanii* Font Quer): Glabrous annual or perennial. Pedicels shorter than bracts. Corolla with bluish veins. Seeds 1·5–2·5 mm; disc with dark tubercles; wing 0·4–0·6 mm wide, black. *N.E. Spain.*

(57–59). **L. diffusa** group. More or less glaucous annuals or biennials; stems procumbent to ascending. Calyx 3–4 mm; lobes linear or linear-lanceolate, subobtuse, slightly unequal. Corolla 8–15 mm, lilac; tube narrow; lobes of lower lip fairly long. Seeds 0·6–2 mm, suborbicular.

1 Plant glandular-pubescent at least in inflorescence 59. **satureioides**
1 Plant glabrous, but with sparse sessile glands throughout
2 Spur 6–10 mm; seeds 1·3–2 mm 57. **diffusa**
2 Spur *c.* 5 mm; seeds 0·6–0·8 mm 58. **coutinhoi**

57. L. diffusa Hoffmanns. & Link, *Fl. Port.* **1**: 257 (1813). Sparsely glandular; stems 10–40 cm, branched. Leaves 5–12 × 0·5–2 mm, linear to narrowly elliptical, verticillate below, alternate above. Racemes with few flowers, dense in flower, lax in fruit; pedicels 1·5–7·5(–18) mm, usually longer than bracts. Corolla 9–15 mm, bluish-lilac; spur 6–10 mm. Capsule 3–4 mm, oblong. Seeds 1·3–2 mm, black, flat; disc tuberculate; wing broad. ● *N. & C. Portugal, W. Spain.* Hs Lu.

58. L. coutinhoi Valdés, *Rev. Esp. Eur. Linaria* 183 (1970) (*L. multicaulis* sensu Coutinho, non (L.) Miller). Like **57** but corolla 8–9 mm; spur *c.* 5 mm; seeds 0·6–0·8 mm, with very narrow wing. ● *N. Portugal.* Lu.

59. L. satureioides Boiss., *Voy. Bot. Midi Esp.* **2**: 463 (1841). Glandular-pubescent at least in inflorescence; stems 5–25 cm, branched. Leaves 5–15(–25) × 0·5–1·2 mm, linear, verticillate below, alternate above. Racemes with 4–10 flowers, lax; pedicels

2·5–7 mm, longer than bracts. Corolla 8–14 mm, lilac with violet veins; spur 4–7 mm. Capsule 2–4 mm, globose. Seeds 1–1·7 mm, strongly concavo-convex; disc blackish, smooth; wing broad or narrow, white. *Stony or sandy ground; calcicole.* ● *S. Spain, S. Portugal.* Hs Lu.

60. L. depauperata Leresche ex Lange in Willk. & Lange, *Prodr. Fl. Hisp.* **2**: 569 (1870). Annual, glabrous except for glandular-pubescent inflorescence; stems 5–15 cm, ascending, simple or sparingly branched. Leaves 5–15 × 0·7–1·3 mm, linear, subacute, mostly alternate, glaucous. Racemes subcapitate, with (1–)2–4 flowers; bracts 3–7 mm; pedicels 1–2 mm. Calyx 4·5–5·5 mm; lobes linear-oblong, obtuse, somewhat unequal. Corolla 16–27 mm, violet to orange-red, with pale yellow spur and yellow palate; spur 8–12 mm, rather stout. Capsule 4·5–7·5 mm. Seeds 2·2–2·7 mm, orbicular-reniform, flat; disc densely covered with long, white papillae; wing broad, whitish towards the edge, darker near the disc. *Rocky places; calcicole.* ● *S.E. Spain.* Hs.

61. L. alpina (L.) Miller, *Gard. Dict.* ed. 8, no. 4 (1768) (incl. *L. petraea* Jordan). Glaucous annual, biennial or perennial, glabrous, or rarely glandular-hairy in inflorescence; stems 5–25 cm, decumbent or ascending, sometimes branched. Leaves 5–15 × 0·7–1·5(–2·5) mm, linear- to oblong-lanceolate, mostly verticillate. Racemes usually dense, with 3–15 flowers; pedicels 2–5 mm in flower, up to 13 mm in fruit. Calyx 3–5 mm; lobes oblong- to linear-oblanceolate, subobtuse, unequal. Corolla 13–22 mm, violet, usually with a yellow palate (rarely entirely yellow, whitish or pink); spur 8–10 mm. Capsule 3–5 mm. Seeds 1·7–2·5 mm, suborbicular, flat, black; disc smooth or tuberculate; wing broad. 2*n* = 12. *Screes, rocky slopes and river-gravels.* ● *Mountains of C. & S. Europe from the Jura and Carpathians to C. Spain, C. Italy and C. Greece.* Al Au Cz Ga Ge Gr He Hs It Ju Rm.

Plants from N.W. Spain with lax, few-flowered inflorescences and pedicels short in fruit have been called **L. filicaulis** Boiss. ex Leresche & Levier, *Jour. Bot.* (*London*) **17**: 200 (1879); they represent an extreme variant of a very variable species and may merit subspecific status.

62. L. faucicola Leresche & Levier, *Jour. Bot.* (*London*) **17**: 200 (1879). Glabrous annual or short-lived perennial; stems 10–27 cm, ascending, branched, slender. Leaves 7–20 × 1·5–5 mm, linear-oblong to oblanceolate, mostly in whorls of 4. Racemes with 2–4(–9) flowers, rather lax; pedicels 1·5–5(–7) mm, usually shorter than bracts. Calyx 4·5–6 mm; segments oblong, the adaxial the longest. Corolla (16–)22–27 mm, violet; spur 10–13 mm. Capsule 5 mm. Seeds 2–2·5 mm, suborbicular, flat, black; disc smooth or finely tuberculate; wing rather narrow. *Limestone rocks and screes.* ● *N.W. Spain (Picos de Europa).* Hs.

63. L. glacialis Boiss., *Elenchus* 70 (1838). Annual to perennial, glaucous, glabrous except for glandular-pubescent inflorescence; stems 5–15 cm, ascending, usually simple, leafless below, densely leafy above up to inflorescence. Leaves 10–15 × 2–6 mm, elliptical to linear-oblanceolate, mostly in whorls of 4. Racemes capitate, with 3–8 flowers; bracts 12–20 mm; pedicels 3–4 mm. Calyx 8–17 mm; segments linear to linear-oblanceolate. Corolla 20–27 mm, dull violet suffused with yellow; spur 8–15 mm, yellow with violet veins. Capsule 7–10 mm. Seeds 2·5–3 mm, suborbicular, flat; disc tuberculate or smooth, greyish-brown; wing paler brown. 2*n* = 12. *Rocks and screes above 2700 m.* ● *S. Spain (Sierra Nevada).* Hs.

64. L. ricardoi Coutinho, *Bol. Soc. Brot.* **22**: 131 (1906). Glabrous annual; stems 15–30 cm, decumbent to erect, slender, usually simple. Leaves 10–30 × 0·5 mm, linear, obtuse, mostly in whorls of 4. Inflorescence with 5–17 flowers, lax; pedicels 0·5–2·5 mm. Calyx 3·5–4·5 mm; segments oblanceolate, acute, slightly unequal. Corolla 9–12 mm, violet with yellow palate; spur 4–5 mm. Capsule 3–4·5 mm. Seeds 1–1·5 mm, reniform; disc black, with low, white papillae; wing lacerate, whitish. *Cornfields.* ● *S. Portugal.* Lu.

65. L. amethystea (Lam.) Hoffmanns. & Link, *Fl. Port.* **1**: 253 (1813). Annual, glabrous to glandular-puberulent below, usually pubescent (often with violet hairs) in inflorescence; stems 5–35 cm, ascending, usually simple. Leaves 4–20 × 0·3–2·5 mm, linear to oblanceolate, verticillate below, alternate above. Racemes with 2–5(–10) flowers, usually lax, often interrupted; pedicels 1–3 mm, usually shorter than bracts. Calyx 3–4 mm; lobes linear-oblong, somewhat unequal. Corolla 10–27 mm, bluish-violet, bright yellow or cream, variably spotted with purple; spur 4–15 mm, slender. Capsule 3–4·5 mm. Seeds 0·8–1·8 mm, sub-orbicular, flat, dark brown; disc tuberculate; wing finely papillose, thickened. *Cultivated ground and other open habitats. Spain and Portugal.* Hs Lu.

(a) Subsp. **amethystea** (incl. *L. broussonetii* (Poiret) Chav.): Corolla 10–22 mm, bluish-violet, rarely white or yellow with violet spur; spur 4–11 mm; racemes lax in fruit. *Throughout the range of the species.*

(b) Subsp. **multipunctata** (Brot.) Chater & D. A. Webb, *Bot. Jour. Linn. Soc.* **65**: 264 (1972) (*L. multipunctata* (Brot.) Hoffmanns. & Link, *Antirrhinum multipunctatum* Brot.): Corolla 19–27 mm, yellow, often with violet spur; spur 10–15 mm; racemes dense in fruit. ● *C. Portugal, S.W. Spain.*

Plants from S.W. Spain and S. Portugal (and N.W. Africa), similar to subsp. (a) but with calyx 4–5 mm, yellow corolla 8–13 mm, spur 4–6 mm and seeds with the wing thin or only slightly thickened and the disc sometimes smooth, have been called **L. munbyana** Boiss. & Reuter, *Pugillus* 89 (1852) (*L. pygmaea* Samp.). They resemble **69** in many characters, though not in the short spur. Their status is uncertain.

66. L. saxatilis (L.) Chaz., *Dict. Jard., Suppl.* **2**: 39 (1790) (*L. tournefortii* (Poiret) Steudel). Subglabrous to densely glandular-pubescent perennial (rarely annual); stems 7–50 cm, ascending to erect, simple or branched above. Leaves 4–20 × 1–5(–7) mm, linear to oblong-elliptical, mostly alternate, the lowest verticillate. Racemes with up to 10(–25) flowers; pedicels 0·5–3 mm, shorter than bracts. Calyx 3–4·5 mm; lobes linear-oblong, slightly unequal. Corolla 9–17 mm, yellow; spur 5–8 mm, slender. Capsule 2·5–4·5 mm. Seeds 0·6–1·7 mm, suborbicular, concavo-convex or flat, dark brown; disc tuberculate; wing thin, narrow. 2*n* = 12. *Dry, sandy or rocky places.* ● *N. & C. Spain, N. & C. Portugal.* Hs Lu.

Very variable in habit, indumentum, width of leaf and width of wing of seed.

67. L. arenaria DC., *Icon. Pl. Gall. Rar.* 5 (1808). Bushy annual, sometimes perennating by root-buds, densely glandular-puberulent throughout; stems 5–15 cm, ascending to erect, branched from the base. Leaves 5–14 × 1·5–2·5 mm, oblanceolate-elliptical, verticillate below, alternate above. Racemes short; pedicels 1–1·5 mm, much shorter than bracts. Calyx 4–5 mm; lobes oblanceolate, subequal. Corolla 4–7 mm, yellowish; spur 2–3 mm, sometimes violet. Capsule 3–5 mm. Seeds 1–1·5 mm, reniform-orbicular, flat, black; disc smooth; wing rather narrow. *Maritime sands.* ● *W. & N.W. France, from 45° to c. 49° 30′ N.; N.W. Spain.* Ga Hs [Br].

(68–70). L. arvensis group. Glaucous annuals, glabrous except in inflorescence; stems 10–50 cm, erect. Leaves linear to oblong-lanceolate, verticillate below, alternate above. Racemes glandular-pubescent, dense in flower, variably elongating in fruit; bracts linear-oblong to ovate-elliptical; pedicels 1–2 mm. Calyx 3–4 mm; lobes linear-oblong, subequal. Corolla 2·5–9 mm; spur shorter than or equalling rest of corolla. Capsule 4–6 mm. Seeds suborbicular to reniform, dark brownish-grey; wing broad.

1 Spur strongly curved   **68. arvensis**
1 Spur straight or slightly curved
  2 Corolla 5–9 mm, yellow, often with violet veins; spur 2–3·5 mm   **69. simplex**
  2 Corolla 2·5–5 mm, lilac-blue, spur *c.* 1 mm   **70. micrantha**

68. L. arvensis (L.) Desf., *Fl. Atl.* **2**: 45 (1798). Leaves 7–20 (–30) × 0·5–2 mm, linear. Racemes lax in fruit. Corolla 4–7 mm, pale lilac-blue; spur 1·5–3 mm, strongly curved. Seeds 1–1·5 mm; disc smooth or minutely tuberculate. *Cultivated ground and other open habitats. S., W. & C. Europe, northwards to N. France and Poland.* Al Au †Be Bl Bu Co Cz Ga Ge Gr Hs Hu It Ju Po ?Rm Sa Si [He].

69. L. simplex (Willd.) DC. in Lam. & DC., *Fl. Fr.* ed. 3, **3**: 588 (1805) (*L. parviflora* (Jacq.) Halácsy, non Desf.). Leaves 12–30 × 1–2·5 mm, linear to linear-lanceolate. Racemes lax in fruit. Corolla 5–9 mm, pale yellow, often with violet veins; spur 2–3·5 mm, straight. Seeds 1·5–2·3 mm; disc smooth or tuberculate. *Dry, open habitats. S. Europe.* Al Bl Bu Co Cr Ga Gr Hs It Ju Lu Rs (K) Sa Si Tu [He].

70. L. micrantha (Cav.) Hoffmanns. & Link, *Fl. Port.* **1**: 258 (1813). Leaves 15–30 × 2–10 mm, linear to oblong-lanceolate. Racemes fairly dense in fruit. Corolla 2·5–5 mm, lilac-blue; spur not more than 1 mm, straight or slightly curved. Seeds 1·3–1·8 mm; disc tuberculate. *Cultivated ground and waste places. Mediterranean region, S. Portugal.* Co Cr Ga Gr Hs It Ju Lu Sa.

15. Cymbalaria Hill[1]

Herbs, usually short-lived perennials but sometimes behaving as annuals; stems procumbent to decumbent. Leaves reniform to suborbicular, petiolate, palmately veined, entire to palmately lobed. Flowers solitary in the leaf-axils. Calyx deeply 5-lobed; segments somewhat unequal. Corolla, stamens and capsule as in *Linaria.* Seeds usually fairly numerous, variously tuberculate, ridged or alveolate.

Predominantly chasmophytes, confined as natives to the Mediterranean region and the S. Alps.

Literature: G. Cufodontis, *Arch. Bot. (Forlì)* **12**: 54–81, 135–158, 233–254 (1936); *Pflanzenareale* **4**: 69–71 (1938); *Bot. Not.* **1947**: 135–156 (1947). A. Chevalier, *Bull. Soc. Bot. Fr.* **83**: 638–653 (1937).

1 Stem and leaves glabrous at maturity
  2 Corolla not more than 15 mm; middle and upper internodes usually about equal
    3 Spur 1·5–3 mm, about as long as calyx; capsule glabrous; seeds remaining separate   **1. muralis**
    3 Spur 4–5 mm, much longer than calyx; capsule glandular-puberulent; seeds of each loculus concrescent to a single mass   **2. longipes**

¹ By D. A. Webb.

2 Corolla 15–30 mm; middle internodes usually much longer than upper
  4 Spur 6–9 mm, about twice as long as calyx   **5. pallida**
  4 Spur 4–5 mm, about as long as calyx   **6. hepaticifolia**
1 Stem and leaves with persistent hairs
5 Corolla 15–30 mm; stem puberulent to pubescent   **5. pallida**
5 Corolla not more than 15 mm; stem ± villous
  6 Capsule hairy
    7 Most of the leaves 7- to 9-lobed; calyx ± densely villous   **7. pilosa**
    7 Most of the leaves 3- to 5-lobed; calyx glabrous to sparsely hairy   **8. microcalyx**
  6 Capsule glabrous
    8 Stem and leaves puberulent; seeds covered with large, hemispherical tubercles   **4. muelleri**
    8 Stem and leaves pubescent to villous; seeds rugose or alveolate
      9 Most of the leaves 5- to 9-lobed   **1. muralis**
      9 Most of the leaves entire or 3-lobed, the larger sometimes 5-lobed   **3. aequitriloba**

1. C. muralis P. Gaertner, B. Meyer & Scherb., *Fl. Wetter.* 2: 397 (1800) (*Linaria cymbalaria* (L.) Miller). Glabrous to villous. Stems up to 60 cm; middle and upper internodes about equal. Leaves up to 55 × 65 mm, but more usually c. 12 × 15 mm, alternate, reniform to semicircular, rarely suborbicular, with 5–9 rounded to deltate, often mucronate lobes. Calyx 2–2·5 mm. Corolla 9–15 mm, lilac to violet with yellow palate, rarely white; spur 1·5–3 mm, about as long as calyx. Capsule glabrous, usually exceeding calyx. Seeds c. 1 mm, broadly ellipsoid to globose, black, ornamented with high, acute ridges and usually a few tubercles. *Shady rocks and woods; somewhat calcicole.* ● *Native in S. Alps, W. Jugoslavia, C. & S. Italy and Sicilia; cultivated for ornament and naturalized on walls (more rarely on stony ground or shingle) throughout most of S., W. & C. Europe.* He It Ju Si [Al Au Az Be Bl Br Bu Co Cr Cz Da Ga Ge Gr Hb Ho Hs Hu Lu No Po Rm Rs (?B, W, ?K) Su].

1 Plant glabrous at maturity   **(a) subsp. muralis**
1 Plant persistently villous
  2 Pedicel at anthesis about equalling subtending petiole; capsule c. 4 mm   **(b) subsp. visianii**
  2 Pedicel at anthesis much longer than subtending petiole; capsule c. 3 mm   **(c) subsp. pubescens**

(a) Subsp. **muralis**: Glabrous except sometimes for a few hairs on young leaves and stems. Pedicel at anthesis about equalling subtending petiole. Corolla violet. Capsule c. 4 mm, considerably exceeding calyx. Seeds coarsely rugose. *Throughout the range of the species except parts of Sicilia.*

(b) Subsp. **visianii** D. A. Webb, *Bot Jour. Linn. Soc.* 65: 265 (1972) (*Linaria cymbalaria* var. *pilosa* Vis., *L. pilosa* auct., non (Jacq.) DC.): Like (a) but villous throughout; corolla lilac. *C. & S. Italy; W. Jugoslavia; rarely naturalized elsewhere.*

(c) Subsp. **pubescens** (C. Presl) D. A. Webb, *loc. cit.* (1972) (*Linaria pubescens* C. Presl): Villous throughout. Pedicel at anthesis much longer than subtending petiole. Corolla lilac. Capsule c. 3 mm, equalling or slightly exceeding calyx. Seeds finely rugose. *N. & E. Sicilia.*

2. C. longipes (Boiss. & Heldr.) A. Cheval., *Bull. Soc. Bot. Fr.* 83: 641 (1937) (*Linaria cymbalaria* subsp. *longipes* (Boiss. & Heldr.) Hayek). Like **1(a)** but leaves not more than 15 × 20 mm; spur of corolla 4–5 mm, much longer than calyx; capsule 3–4 mm, minutely glandular-puberulent; seeds of each loculus concrescent when ripe, forming a coarsely alveolate and irregularly winged mass c. 3 × 2 mm. *Rocky or stony places by the sea. S. Aegean region.* Cr Gr. (*E. Mediterranean region.*)

3. C. aequitriloba (Viv.) A. Cheval., *op. cit.* 646 (1937) (*Linaria aequitriloba* (Viv.) Sprengel). Pubescent to villous. Stems up to 40 cm; middle and upper internodes usually about equal. Leaves up to 28 × 35 mm, but often very much smaller, mostly alternate but sometimes a few opposite, orbicular to reniform, entire to 3(–5)-lobed; lobes usually low and rounded, the central not much larger than the adjacent ones. Pedicel at anthesis longer than the subtending petiole. Calyx c. 2·5 mm. Corolla 8–13 mm; spur 2–3 mm. Capsule glabrous, slightly exceeding calyx. Seeds globose, alveolate. *Damp or shady places.* ● *Islands of W. Mediterranean, eastwards to Giglio.* Bl Co It Sa.

(a) Subsp. **aequitriloba**: Stems slender, not fragile; leaves not fleshy; calyx hairy to subglabrous; corolla violet; seeds 0·9–1·2 mm, rather coarsely alveolate. 2n = 56. *Throughout the range of the species.*

(b) Subsp. **fragilis** (Rodr.) D. A. Webb, *Bot. Jour. Linn. Soc.* 65: 265 (1972) (*Linaria fragilis* Rodr.): Stems relatively stout, fragile; leaves rather fleshy; calyx hairy; corolla pale blue to whitish; seeds 0·8–1 mm, finely alveolate. *Menorca (Barrancos de Algedar, S. of Ferrerias).*

4. C. muelleri (Moris) A. Cheval., *Bull. Soc. Bot. Fr.* 83: 645 (1937). Stems and leaves densely puberulent; stems c. 5 cm, crowded, caespitose. Leaves up to 15 × 18 mm, orbicular to reniform, entire or obscurely 3-lobed with rounded, subequal lobes. Pedicel c. 8 mm at anthesis, up to 30 mm in fruit. Calyx c. 2·5 mm. Corolla c. 10 mm, dull violet with reddish palate; spur 2 mm. Capsule glabrous, considerably exceeding calyx. Seeds 1–1·3 mm, black, globose, covered with large, hemispherical tubercles. *Rocks and walls.* ● *C. Sardegna (near Laconi).* ?Co Sa.

Very distinct in its condensed, caespitose habit (which is retained in cultivation) and the ornamentation of its seeds. Plants from two localities in Corse have been referred to this species, but until they have been rediscovered in the field and the structure of their seeds ascertained it is probably best to regard them as abnormal variants of 3(a).

5. C. pallida (Ten.) Wettst. in Engler & Prantl, *Natürl. Pflanzenfam.* 4(3b): 58 (1891). Stems and leaves usually puberulent to pubescent; more rarely (var. *beguinotii* (Cuf.) Cuf.) glabrous. Stems up to 20 cm; middle internodes very long, the upper much shorter, so that the long-petiolate leaves overtop the stem-apex. Leaves up to 25 × 30 mm, mostly opposite, suborbicular to semicircular-deltate, entire to 5-lobed with low and rounded to subacute, deltate lobes, the central usually much the largest. Calyx 3–4 mm, densely puberulent. Corolla 15–25(–30) mm, pale lilac-blue; spur 6–9 mm. Capsule glabrous, slightly exceeding calyx. Seeds c. 1 mm, ovoid, with acute, longitudinal ridges. *Rocks and screes.* ● *Mountains of C. Italy.* It [Br Cz].

6. C. hepaticifolia (Poiret) Wettst., *loc. cit.* (1891) (*Linaria hepaticifolia* (Poiret) Steudel). Like **5** but glabrous (except sometimes for sparsely hairy calyx); some of the leaves usually alternate; calyx 4–5 mm; corolla 15–18 mm, with spur 4–5 mm; capsule slightly shorter than calyx. 2n = 56. *Shady places and on rocks by streams.* ● *Corse.* Co.

7. C. pilosa (Jacq.) L. H. Bailey, *Gentes Herb.* 1: 136 (1923). Stems and leaves villous. Stems up to 35 cm; middle and upper internodes about equal. Leaves usually c. 10 × 12 mm, alternate, reniform, with (5–)7(–11) rounded or mucronate, subequal lobes. Pedicel at anthesis about equalling subtending petiole, elongating considerably in fruit. Calyx c. 2 mm, villous. Corolla 10–12 mm, pale lilac to white; spur c. 1 mm. Capsule pubescent, exceeding calyx. Seeds 0·6–0·8 mm, brown, subglobose, rather finely rugose.

Shady rocks and walls; calcicole. ● *C. & S. Italy, Sardegna.*
It Sa.

Much confused with **1(b)** and **1(c)**, but distinct in its very short spur and pubescent capsule.

8. C. microcalyx (Boiss.) Wettst. in Engler & Prantl, *Natürl. Pflanzenfam.* **4**(3b): 58 (1891) (*Linaria microcalyx* Boiss.). Stems and leaves villous. Stems up to *c.* 25 cm; middle and upper internodes about equal. Leaves up to 10 × 15 mm, mostly alternate, reniform to semicircular, entire or 3(–5)-lobed; lobes usually low and rounded. Calyx 1–2 mm, sparsely hairy to subglabrous. Corolla 9–13 mm, violet with yellow palate. Capsule pubescent to lanate, considerably exceeding calyx. *S. & W. parts of Balkan peninsula, S. Aegean region.* Al Cr Gr Ju.

1 Seeds *c.* 1 mm, ±equal, subglobose, regularly alveolate
2 Spur 1·5–2 mm, scarcely longer than calyx; capsule pubescent
        (b) subsp. minor
2 Spur 2–3 mm, much longer than calyx; capsule lanate
        (c) subsp. dodekanesi
1 Seeds often unequal, at least some of them 2 mm or more, furnished with ridges, wings or crests
3 Spur 1–2 mm; seeds triangular-ovate, with longitudinal ridges
        (a) subsp. ebelii
3 Spur 2–3 mm; seeds irregularly shaped, with sinuous, anastomosing crests and wings   **(d) subsp. microcalyx**

(a) Subsp. **ebelii** (Cuf.) Cuf., *Bot. Not.* **1947**: 151 (1947): Spur 1–2 mm, usually longer than calyx; seeds 2–2·5 × 1 mm, triangular-ovate, with narrow, mostly longitudinal ridges. ● *S.W. Jugoslavia, N. Albania.*

(b) Subsp. **minor** (Cuf.) W. Greuter, *Boissiera* **13**: 107 (1967): Spur 1·5–2 mm, scarcely longer than calyx; capsule pubescent; seeds *c.* 1 mm, subglobose, regularly alveolate. 2*n* = 28. ● *W. Greece.*

(c) Subsp. **dodekanesi** W. Greuter, *op. cit.* 108 (1967): Like (b) but spur 2–3 mm, about twice as long as calyx; ripe capsule densely lanate. *S. Aegean area.*

(d) Subsp. **microcalyx**: Spur 2–3 mm; seeds 1·5–3 mm, unequal, irregularly shaped, with prominent, sinuous, anastomosing crests and wings. ● *S. Greece.*

16. Kickxia Dumort.[1]
(*Elatinoides* (Chav.) Wettst.)

Annual or perennial herbs; stems usually procumbent or climbing. Leaves alternate (the lowest sometimes opposite), ovate, hastate or sagittate. Flowers solitary in the leaf-axils (rarely aggregated into spike-like racemes on lateral branches). Calyx with 5 subequal lobes. Corolla and stamens as in *Linaria*. Capsule more or less globose, with 2 equal loculi, each opening by a single large pore with detachable lid. Seeds alveolate or tuberculate.

1 All leaves rounded or cordate at the base, without lobes or auricles
2 Calyx-lobes 4·5–8 × 2–4 mm in fruit, ±ovate-cordate; corolla 10–15 mm   **4. spuria**
2 Calyx-lobes *c.* 3 × 1 mm in fruit, linear-lanceolate; corolla 8–11 mm   **5. lanigera**
1 At least some leaves hastate or sagittate
3 Corolla 4–6 mm; capsule 1·5–2 mm   **1. cirrhosa**
3 Corolla 7–15 mm; capsule at least 2·5 mm
4 Seeds tuberculate; wall of capsule rather thick and rigid; spur strongly curved   **2. commutata**
4 Seeds alveolate; wall of capsule thin, not rigid; spur ±straight   **3. elatine**

1. K. cirrhosa (L.) Fritsch, *Excursionsfl. Österr.* 492 (1897) (*Linaria cirrhosa* (L.) Cav.). Annual; stems up to 90 cm, trailing or climbing, very slender, pubescent to villous. Leaves glabrous or subglabrous, the lowest up to 3(–4·5) × 0·7(–1·4) cm, entire or with 1–2 teeth at the base, the middle and upper smaller, lanceolate-hastate to narrowly lanceolate-sagittate, acute; petiole equalling the lamina or shorter, somewhat twisted. Pedicels up to 3·5 cm in flower, 2–5 times as long as the subtending leaf, filiform, glabrous, patent. Calyx-lobes linear-lanceolate, acute. Corolla 4–6 mm, whitish, tinged and striped with violet. Capsule 1·5–2 mm, globose, thin-walled. Seeds *c.* 0·75 mm, tuberculate. *Damp or shady places, especially on sandy soil, usually near the sea. W. Mediterranean region and S.W. Europe.* Az Bl Co Ga Hs It ?Ju Lu Sa Si.

2. K. commutata (Bernh. ex Reichenb.) Fritsch, *loc. cit.* (1897) (*Linaria commutata* Bernh. ex Reichenb.). Perennial; stems 20–70 cm, numerous, procumbent, rooting, glandular-villous. Leaves villous, the lower broadly ovate, the middle and upper ovate-hastate to lanceolate-sagittate; petiole not more than half as long as lamina. Pedicels up to 3 cm in flower, filiform, glabrous, patent. Calyx 4–5 mm; lobes linear-lanceolate, very acute, hispid. Corolla 11–15 mm, whitish, with bluish-violet upper lip, yellow lower lip and purple-spotted palate; spur strongly curved. Capsule 2·5–4 mm, globose; wall rather thick and rigid. Seeds 0·75–1 mm, tuberculate. *Mediterranean region, extending to S.E. Bulgaria and W. France.* Al Bl Bu Co Cr Ga Gr Hs It Ju Sa Si Tu.

(a) Subsp. **commutata**: Middle leaves broadly ovate-hastate, obtuse, the upper somewhat narrower. Pedicels rarely more than twice as long as subtending leaf. *Throughout most of the range of the species, but rarer in the south-east.*

(b) Subsp. **graeca** (Bory & Chaub.) R. Fernandes, *Bot. Jour. Linn. Soc.* **64**: 74 (1971) (*Antirrhinum graecum* Bory & Chaub.): Branches longer and slenderer than in subsp. (a). Leaves acute, rather narrow, the uppermost and those of the branches lanceolate-sagittate, very small. Pedicels 2–6 times as long as subtending leaf. *S. part of Balkan peninsula, Aegean region.*

3. K. elatine (L.) Dumort., *Fl. Belg.* 35 (1827) (*Linaria elatine* (L.) Miller). Glandular-pubescent to -hirsute annual; stems branched from the base. Leaves ovate, sagittate or hastate; petiole not more than half as long as lamina. Calyx-lobes lanceolate, acuminate. Corolla 7–15 mm, yellowish or bluish, with violet upper lip. Capsule 4–4·5 mm, subglobose, thin-walled. Seeds alveolate. *Cultivated fields and other open habitats. S., W. & C. Europe, northwards to England and extending eastwards to Moldavia and Krym; naturalized or casual in some other regions.* All except Fa Fe Is No Rs (N, B, C, E), but not native in Da Hb Su.

(a) Subsp. **elatine**: Main branches decumbent, weak, often rather sparsely hairy; secondary branches few or none. Leaves acute to mucronate, mostly hastate, but the uppermost sagittate. Pedicels glabrous except just below the flower, 3–6 times as long as the calyx in flower, up to 3 cm in fruit. Corolla 7–10 mm. 2*n* = 36. *Throughout the range of the species except Portugal and parts of the Mediterranean region.*

(b) Subsp. **crinita** (Mabille) W. Greuter, *Boissiera* **13**: 108 (1967) (*K. elatine* subsp. *sieberi* (Reichenb.) Hayek, *Linaria sieberi* Reichenb., *L. crinita* Mabille): Main branches ascending, relatively stout, densely hairy, usually with several short, patent, flowering secondary branches. Leaves more or less obtuse, the lower ovate to indistinctly hastate, the middle and upper usually hastate. Pedicels hairy throughout, 2½–3 times as long as the calyx in flower, not more than 2 cm in fruit. Corolla up to 15 mm. *S. Europe.*

Some plants from Spain and Portugal, in other characters referable to subsp. (b), have long, subglabrous pedicels, as in subsp. (a).

K. caucasica (Mussin ex Sprengel) Kuprian. in Schischkin & Bobrov, *Fl. URSS* 22: 178 (1955), which occurs in S. Ukraine, should probably be regarded as an extreme variant of (b) in which even the middle leaves are scarcely hastate.

4. **K. spuria** (L.) Dumort., *Fl. Belg.* 35 (1827) (*Linaria spuria* (L.) Miller). Glandular-pubescent to -villous annual; stems 20–50 cm, decumbent. Leaves up to 7·5 cm, entire, or the lowest remotely denticulate, the lower broadly ovate, truncate or rounded at base, obtuse, the middle and upper ovate-lanceolate to suborbicular-cordate, mucronate; petiole very short. Pedicels 1–2 cm in flower, villous. Calyx-lobes accrescent, acute, more or less cordate at base in fruit. Corolla 10–15 mm, yellow, with deep purple upper lip; spur curved. Capsule depressed-globose. Seeds oblong-ellipsoid, alveolate. *Cultivated fields and other open habitats. S., W. & C. Europe, northwards to England, the Netherlands and S.W. Poland; naturalized or casual further north.* Al Au Az Be Bl Br Bu Co Cr Cz Ga Ge Gr He Ho Hs Hu It Ju Lu Po Rm Rs (K) Sa Si Tu [Da].

(a) Subsp. **spuria**: Sparsely to moderately hairy. Stems usually simple. Calyx-lobes 4–5 × 2–3 mm in flower, 5–8 × 2–4 mm in fruit, ovate-oblong, slightly cordate. Capsule 5 mm wide. Seeds 1–1·25 mm. *Throughout most of the range of the species, but rare in the Mediterranean region.*

(b) Subsp. **integrifolia** (Brot.) R. Fernandes, *Bot. Jour. Linn. Soc.* 64: 74 (1971) (*Antirrhinum spurium* var. *integrifolium* Brot.): Densely hairy throughout. Stems with slender, flexuous, small-leaved, lateral flowering branches. Calyx-lobes 3–4 × 1·5–2·5 mm in flower, 4·5–6 × 2–3 mm in fruit, ovate, distinctly cordate. Capsule 3–4 mm wide. Seeds 0·75–1 mm. *S. Europe.*

5. **K. lanigera** (Desf.) Hand.-Mazz., *Ann. Naturh. Mus. (Wien)* 27: 403 (1913) (*Linaria lanigera* Desf., *L. racemigera* sensu Rouy, non *L. spuria* var. *racemigera* Lange). Procumbent, densely glandular-villous annual; stems up to 90 cm, whitish, rigid. Leaves broadly ovate, the middle and upper often cordate, shortly acuminate (the lowest sometimes obtuse), entire to dentate. Flowers axillary, distant, long-pedicellate, or subsessile on short, bracteate, lateral branches. Calyx 2·5–3·5 mm; lobes linear-lanceolate, acute, not accrescent. Corolla 8–11 mm, whitish, with violet upper lip and blue-spotted palate; spur curved. Capsule *c.* 2·5 mm, depressed-globose, slightly emarginate. Seeds *c.* 0·75 mm, ovoid, alveolate. *Cultivated fields and waste places. S.W. Europe; Turkey-in-Europe.* Bl ?Ga Hs Lu ?Si Tu.

The distribution is not as disjunct as might appear, as the species is widespread in N. Africa and S.W. Asia.

17. Digitalis L.[1]

Biennial or perennial herbs, rarely small shrubs. Leaves simple, alternate, sometimes mostly basal. Flowers in terminal, bracteate, often secund racemes. Calyx equally 5-lobed, shorter than the corolla-tube. Corolla with cylindrical to inflated-globose tube, often constricted at the base; limb more or less 2-lipped; lips erecto-patent, the upper shorter than the lower. Stamens 4. Capsule ovoid to conical, septicidal. Seeds numerous.

Literature: L. I. Ivanina, *Acta Inst. Bot. Acad. Sci. URSS* (Ser. 1) 11: 198–302 (1955). K. Werner, *Bot. Jahrb.* 79: 218–254 (1960); *Wiss. Zeitschr. Univ. Halle (Math.-Nat. Kl.)* 13: 453–486 (1964).

---

[1] By V. H. Heywood.

1 Middle lobe of lower lip of corolla projecting far beyond the others
  2 Shrub; corolla-tube campanulate   **1. obscura**
  2 Herb, sometimes woody at the base; corolla-tube inflated-globose
    3 Axis of inflorescence glandular-villous or -velutinous; corolla whitish with brown or violet veins
      4 Bracts lanceolate; corolla 20–30 mm   **11. lanata**
      4 Bracts linear; corolla 10–20 mm   **12. leucophaea**
    3 Axis of inflorescence glabrous and eglandular; corolla yellow to brown
      5 Calyx-lobes obtuse, with a wide scarious margin
        10. ferruginea
      5 Calyx-lobes acute or acuminate, with scarious margin very narrow or absent   **9. laevigata**
1 Middle lobe of lower lip of corolla only slightly exceeding the others
  6 Corolla-tube slender, cylindrical or cylindrical-urceolate
    7 Calyx-lobes ovate, obtuse   **8. parviflora**
    7 Calyx-lobes linear to lanceolate, acute
      8 Plant pubescent or puberulent; corolla dull greenish-yellow, conspicuously veined   **7. viridiflora**
      8 Plant ±glabrous (except in inflorescence); corolla pale yellow to whitish, not conspicuously veined   **6. lutea**
  6 Corolla-tube stout, campanulate or campanulate-ventricose
    9 Corolla yellow; leaves ±smooth, usually glabrous above
      5. grandiflora
    9 Corolla purple, pink or white; leaves ±rugose, usually hairy above
      10 Whole plant covered with a yellowish, glutinous indumentum of glandular hairs   **3. thapsi**
      10 Plant greenish to whitish, not or only slightly glutinous, with an indumentum of glandular and eglandular hairs (rarely with eglandular hairs only or glabrous)
        11 Corolla glabrous outside   **2. purpurea**
        11 Corolla hairy outside
          12 Lower leaves crenate to serrate, long-petiolate   **2. purpurea**
          12 Lower leaves entire or shallowly dentate, shortly petiolate   **4. dubia**

Sect. FRUTESCENTES Bentham. Shrubs. Corolla orange-yellow or brown, with campanulate tube.

1. **D. obscura** L., *Sp. Pl.* ed. 2, 867 (1763). Glabrous shrub, 30–120 cm; stems leafless below, densely leafy above, decumbent to erect. Leaves linear-oblong to lanceolate, entire or serrate, coriaceous, shiny. Racemes elongate, somewhat secund. Corolla 20–30 mm, orange-yellow to brown, reticulate or spotted. Capsule conical, acute, exceeding the calyx. $2n = 56$. *Bushy and rocky places. E., C. & S. Spain.* Hs.

(a) Subsp. **obscura**: Leaves usually entire; sepals ovate-lanceolate. *E., C. & S.E. Spain.*

(b) Subsp. **laciniata** (Lindley) Maire in Jahandiez & Maire, *Cat. Pl. Maroc* 3: 688 (1934) (*D. laciniata* Lindley): Leaves usually deeply serrate (rarely entire); sepals lanceolate. *S. Spain.*

Sect. DIGITALIS. Herbs. Leaves more or less rugose. Corolla purple or white, with campanulate tube.

2. **D. purpurea** L., *Sp. Pl.* 621 (1753). Biennial or perennial, sometimes flowering in the first year, (25–)60–180 cm, pubescent to lanate, rarely subglabrous. Basal leaves ovate to lanceolate, long-petiolate, with an indumentum of multicellular eglandular hairs mixed with shorter glandular hairs in varying proportions, rarely eglandular only. Raceme simple or slightly branched, usually many-flowered. Bracts variable, sometimes minute. Pedicels shorter than, equalling or longer than the bracts and calyx. Calyx-lobes ovate-lanceolate to elliptical. Corolla 40–55 mm, purple, pale pink or white, usually spotted inside,

glabrous to pubescent or villous outside, ciliate. Capsule ovoid, obtuse, equalling or exceeding the calyx. *Usually calcifuge. W., S.W. & W.C. Europe; cultivated for ornament and as a medicinal plant and widely naturalized or casual further east.* *Az Be Br Co Cz Ga Ge Hb Hs Lu No Sa Su [*Au Da Ho Hu Po].

A complex polytypic species centred in the W. Mediterranean region, divisible into at least three subspecies, with several local variants.

1 Leaves greenish, glabrous or pubescent above **(a)** subsp. **purpurea**
1 Leaves whitish, tomentose or lanate
  2 Corolla white (rarely pale pink or pale yellow); whole plant densely lanate **(c)** subsp. **heywoodii**
  2 Corolla purplish; leaves and sometimes the whole plant tomentose
    3 Lower leaves broadly ovate, ±abruptly contracted at base; pedicel longer than calyx **(b)** subsp. **mariana**
    3 Lower leaves lanceolate to ovate-lanceolate, gradually tapered to the petiole; pedicel as long as or shorter than calyx **(a)** subsp. **purpurea**

(a) Subsp. purpurea: Lower leaves lanceolate to ovate, gradually tapered to the petiole, glabrous to tomentose (var. *tomentosa* (Hoffmanns. & Link) Brot.), greenish to whitish. Pedicels usually equalling or shorter than calyx. Corolla purple (rarely pink or white), glabrous or pubescent outside. $2n=56$. *Almost throughout the range of the species.*

Extremely variable in height, indumentum, size and shape of leaves and bracts, pedicels, calyx, colour and spotting of the corolla. Some local populations show characteristic combinations of these variable characters, but the pattern of variation is too complex and fluctuating to enable them to be treated as subspecies. Dwarf alpine ecotypes also occur. Superimposed on this local variation there is a regional differentiation between the populations of the Iberian peninsula and of the rest of Europe in habit, details of indumentum, branching of inflorescence and seed-colour.

(b) Subsp. mariana (Boiss.) Rivas Goday, *Farmacognosia (Madrid)* **5**: 144 (1946) (*D. mariana* Boiss.): Lower leaves broadly ovate, abruptly contracted at base, white-tomentose. Pedicels longer than calyx. Corolla purple, glabrous outside. *Rocky slopes and rock-crevices.* ● *S.C. Spain; N.E. Portugal.*
(c) Subsp. heywoodii P. & M. Silva, *Agron. Lusit.* **20**: 239 (1959): While plant densely white-lanate. Lower leaves ovate-lanceolate to lanceolate, tapered into the petiole or contracted at base. Pedicels equalling or longer than calyx. Corolla white, sometimes partly pale pink or pale yellowish, sparsely hairy outside. $2n=56$. *Granite rocks.* ● *S. Portugal (around Reguengos de Monsaraz).*

3. **D. thapsi** L., *Sp. Pl.* ed. 2, 867 (1763). Like 2(a) but whole plant covered with a yellowish indumentum of glandular hairs; inflorescence lax, usually branched at the base; pedicels about twice as long as calyx. Calyx-lobes ovate-lanceolate, acute. Corolla 25–50 mm, pubescent outside. $2n=56$. *Rocky slopes and uncultivated fields; calcifuge.* ● *E. Portugal, C. & W. Spain.* Hs Lu.

4. **D. dubia** Rodr., *Anal. Soc. Esp. Hist. Nat.* **3**: 45 (1874). Like 2(a) but not more than 50 cm; indumentum of very long, multicellular, eglandular hairs (30 cells or more) mixed with a few short glandular hairs; lower leaves entire or shallowly dentate, shortly petiolate; capsule shorter than calyx. Perennial; leaves mostly basal. Corolla 35–40 mm, pubescent outside. *Rocky slopes, woods and scrub; calcicole.* ● *Islas Baleares.* Bl.

Sect. MACRANTHAE Heywood. Herbaceous. Leaves more or less smooth. Corolla yellow, with campanulate-ventricose tube.

5. **D. grandiflora** Miller, *Gard. Dict.* ed. 8, no. 4, Corr. (1768) (*D. ambigua* Murray). Biennial to perennial (30–)60–100 cm. Leaves 70–250 × 20–60 mm, ovate-lanceolate, finely serrate, usually glabrous and shining green above, sparsely pubescent beneath. Calyx-lobes lanceolate, acute. Corolla 40–50 mm, yellow. $2n=56$. *Woods. E. & C. Europe, northwards to Estonia and southwards to N. Greece, and extending westwards to Belgium and S.C. France.* Al Au Be Bu Cz Ga Ge Gr He Hu It Ju Po Rm Rs (B, C, W, E) Tu.

Records for the Pyrenees and N. Spain appear to be erroneous.

Sect. TUBIFLORAE Bentham. Herbs. Corolla-tube cylindrical or cylindrical-urceolate.

6. **D. lutea** L., *Sp. Pl.* 622 (1753). Glabrous or slightly pubescent perennial 60–100 cm. Leaves oblong-oblanceolate, serrate to subentire. Racemes more or less dense, many-flowered. Calyx-lobes linear-lanceolate, acute, glandular-ciliate. Corolla 9–25 mm, pale yellow to whitish; lateral lobes of lower lip ovate; middle lobe 5–6 mm, ovate or elliptical. *W. & W.C. Europe, extending southwards to S. Italy.* Au Be Co Ga Ge He ?Ho Hs It [Cz ?Da Po].

(a) Subsp. lutea: Corolla 15–22(–25) mm; lateral lobes recurved. Racemes secund. $2n = 56$. ● *Throughout the range of the species except S. Italy and Corse.*
(b) Subsp. australis (Ten.) Arcangeli, *Comp. Fl. Ital.* 512 (1882) (*D. micrantha* Roth): Corolla 9–13(–15) mm; lateral lobes not recurved. Racemes not secund. $2n = 56$. ● *C. & S. Italy, Corse.*

D. × purpurascens Roth (*D. lutea × purpurea*) occurs occasionally with the parents.

7. **D. viridiflora** Lindley, *Digitalium Monogr.* 21 (1821). Puberulent to pubescent perennial 50–80 cm. Leaves oblong-lanceolate to -elliptical, denticulate. Racemes dense, many-flowered. Calyx-lobes lanceolate, acute. Corolla 11–20 mm, dull greenish-yellow, conspicuously veined; lower lip scarcely exceeding the upper. $2n=56$. *Woods. Balkan peninsula, from 39° 30′ to 43° 30′ N.* Al Bu Gr Ju Tu.

8. **D. parviflora** Jacq., *Hort. Vindob.* 1: 6 (1770). Perennial 30–60 cm. Stems very leafy, sparsely pubescent to glabrous below, more or less tomentose above when young. Leaves oblong-oblanceolate or -lanceolate, acute or more or less obtuse, entire or slightly dentate, coriaceous. Racemes long, dense, many-flowered, white-tomentose. Calyx-lobes ovate, obtuse. Corolla c. 12 mm, urceolate-tubular; tube reddish-brown with violet veins; limb white-ciliate, with middle lobe of lower lip $\frac{1}{4}$–$\frac{1}{3}$ as long as the tube, purple-brown. $2n=56$. ● *Mountains of N. Spain.* Hs.

Sect. GLOBIFLORAE Bentham. Herbs, sometimes woody at the base. Corolla-tube inflated-globose.

9. **D. laevigata** Waldst. & Kit., *Pl. Rar. Hung.* **2**: 171 (1803–1804). Glabrous perennial 60–100 cm. Leaves oblong-lanceolate to lanceolate, entire or denticulate. Racemes long, rather lax. Calyx-lobes ovate, acute or acuminate, without, or with a very narrow scarious margin. Corolla 15–35 mm, yellowish with purple-brown veins or markings; lobes of lower lip obtuse, the

middle one 5–15 mm. 2n = 56. *Scrub and woods.* ● *W. & C. parts of Balkan peninsula, extending northwards to Slovenija.* Al Bu Gr Ju.

(a) Subsp. **laevigata**: Corolla 25–35 mm; middle lobe of lower lip 9–15 mm. *From c. 40° 30′ N. northwards.*

(b) Subsp. **graeca** (Ivanina) Werner, *Bot. Jahrb.* **79**: 236 (1960): Corolla 15–25 mm; middle lobe of lower lip 5–10 mm. *From c. 40° 30′ N. southwards.*

D. macedonica Heywood, *Kew Bull.* [6]: 149 (1951), described from N.W. Greece (Smolikas Oros), resembles **9(b)** in habit and flower-colour but differs in its lanceolate to oblong-lanceolate calyx-lobes and scarcely ventricose corolla with acute lobes; in these features it agrees with **6**, which it also resembles vegetatively. It is regarded by Werner as a hybrid between **7** and **9(b)** but its status is not clear.

10. D. ferruginea L., *Sp. Pl.* 622 (1753). Perennial or biennial 30–120 cm; stem glabrous. Leaves oblong-lanceolate to lanceolate, glabrous or slightly pubescent beneath. Raceme usually long, dense and many-flowered; axis glabrous. Calyx-lobes ovate- or oblong-elliptical, obtuse, glabrous or pubescent, with a wide scarious margin. Corolla 15–35 mm, yellowish- or reddish-brown, with a reticulum of darker veins; middle lobe of lower lip 2–4 times as long as the lateral lobes. 2n = 56. *Woods and scrub. S. Europe, from Italy eastwards, and extending northwards to c. 46° in Hungary and Romania.* Al Bu Gr Hu It Ju Rm Tu.

11. D. lanata Ehrh., *Beitr. Naturk.* **7**: 152 (1792) (*D. orientalis* auct. balcan., non Lam.). Perennial or biennial 30–100 cm; stem usually solitary, often reddish-purple, glabrous or subglabrous. Leaves oblong-lanceolate to lanceolate, glabrous or sometimes ciliate. Racemes usually very dense; axis glandular-villous or -velutinous. Bracts lanceolate. Calyx-lobes lanceolate, acute, glandular-villous, without scarious margins. Corolla 20–30 mm, white or yellowish-white with brown or violet veins; middle lobe of lower lip 8–13 mm, oblong or ovate, whitish. 2n = 56. *Woods and scrub.* ● *Balkan peninsula, S. Hungary, S. & W. Romania.* Al Bu Gr Hu Ju Rm Tu [Au].

12. D. leucophaea Sibth. & Sm., *Fl. Graec. Prodr.* **1**: 439 (1809). Like **11** but bracts linear; corolla 10–16(–20) mm, the middle lobe of lower lip 4–7 mm, suborbicular, whitish with purple veins. *Woods and dry grassland.* ● *N.E. Greece (Ayion Oros, Thasos).* Gr.

18. Erinus L.[1]

Perennial herbs; leaves alternate, undivided. Flowers in terminal, bracteate racemes. Calyx equally 5-lobed. Corolla with cylindrical tube and 5 subequal, patent lobes. Stamens 4, didynamous, included. Stigma capitate, but with 2 lateral lobes. Capsule 4-valved (septicidal and loculicidal); seeds numerous.

1. E. alpinus L., *Sp. Pl.* 630 (1753). Usually more or less villous and somewhat viscid. Stems 5–20(–30) cm, simple, ascending to erect, forming a lax cushion. Leaves 5–20 mm, oblanceolate-cuneate, usually crenate-serrate at least towards the apex, the lower ones petiolate. Racemes subcorymbose in flower, elongating in fruit. Calyx lobed almost to base; lobes linear-oblong. Corolla 6–9 mm in diameter, bright purple (rarely white); lobes emarginate to truncate, the 2 upper narrower than the 3 lower. Seeds smooth. 2n = 14. *Rocks, screes and stony grassland; calcicole.* ● *Mountains of S. & S.C. Europe, from N. & E. Spain to C. Italy and W. Austria.* Au Bl Ga He Hs It Sa [Br Hb].

19. Lagotis Gaertner[1]

Perennial herbs; leaves undivided, alternate or opposite. Flowers zygomorphic, in dense, terminal, bracteate spikes. Calyx membranous, obscurely toothed, usually split along the lower side. Corolla with cylindrical tube, abruptly curved near the middle, and 2-lipped limb; upper lip erect, entire to emarginate; lower lip patent, 2(3)-lobed. Stamens 2, more or less exserted. Stigma capitate. Capsule usually indehiscent; ovules 1 in each loculus; seeds 1–2.

Filaments 2–4 mm, about equalling upper lip of corolla; cauline leaves alternate **1. minor**
Filaments not more than 1 mm, much shorter than upper lip of corolla; cauline leaves opposite **2. uralensis**

1. L. minor (Willd.) Standley, *Publ. Field Mus. Bot.* (*Chicago*) **8**: 325 (1931). Stem 10–30 cm, simple, erect. Basal leaves 2–4, elliptic to lanceolate, acute, dentate to subentire, with petiole about equalling lamina; cauline leaves smaller, sessile, alternate, passing gradually into bracts. Spike 1–8 cm. Calyx 4–6 mm. Corolla 8–10 mm, dull blue or whitish; upper lip about as long as wide, entire or slightly emarginate. Filaments 2–4 mm, about equalling upper lip of corolla. Capsule 6 × 2·5 mm. *Tundra. Arctic Russia.* Rs (N). (*N. & N.E. Asia.*)

2. L. uralensis Schischkin, *Not. Syst.* (*Leningrad*) **17**: 381 (1955). Like **1** but up to 40 cm; basal leaves ovate, obtuse, usually crenate; cauline leaves opposite; upper lip of corolla longer than wide, distinctly emarginate; filaments 0·5–1 mm, much shorter than upper lip of corolla. *C. & S. Ural.* Rs (C).

20. Wulfenia Jacq.[1]

Perennial herbs; leaves alternate or basal, undivided. Flowers in terminal, bracteate, spike-like racemes. Calyx equally 5-lobed. Corolla with obconical tube and short, obscurely 2-lipped limb; upper lip entire or emarginate; lower lip shortly 3-lobed. Stamens 2, inserted at apex of corolla-tube; filaments very short. Stigma capitate, emarginate. Capsule 4-valved (septicidal and loculicidal); seeds numerous.

Plant glabrous except for petioles and midribs; stem 20–40(–50) cm **1. carinthiaca**
Leaves hairy; flowers glandular-pubescent; stem not more than 20 cm **2. baldaccii**

1. W. carinthiaca Jacq., *Misc. Austr. Bot.* **2**: 60 (1781). Glabrous except for petioles and midribs of leaves. Stem 20–40(–50) cm, single, erect, arising from a horizontal rhizome, scapose, but with a few very small, alternate, scale-like leaves. Basal leaves 8–17 cm, oblanceolate to obovate, crenate, tapered to a short petiole. Raceme 6–10 cm in flower, dense, secund, elongating in fruit; pedicels very short. Calyx c. 8 mm, divided almost to the base; lobes linear-oblong. Corolla 12–15 mm, dark violet-blue. Capsule slightly shorter than calyx; seeds finely reticulate. 2n = 18. *Damp grassland, 1500–2000 m; calcifuge.* ● *S.E. Alps; E. Crna Gora.* ?Al Au It Ju.

2. W. baldaccii Degen, *Österr. Bot. Zeitschr.* **47**: 408 (1897). Like **1** but not more than 20 cm; leaves hairy; racemes fairly lax; calyx, corolla and capsule glandular-pubescent; corolla bright blue. *Shady mountain rocks.* ● *N. Albania.* Al.

21. Veronica L.[1]

(incl. *Pseudolysimachium* (Koch) Opiz)

Herbs, sometimes with a procumbent, woody stock. Leaves opposite below, the floral usually alternate. Flowers in terminal or axillary racemes, or solitary in the leaf-axils. Calyx deeply divided into 4(5) often unequal segments. Corolla rotate to campanulate; limb longer than tube, with 4 somewhat unequal lobes. Stamens 2, exserted. Fruit a loculicidal and sometimes also septicidal capsule, more or less flattened at right angles to the septum. Seeds usually numerous.

Measurements of pedicels refer to the fruiting condition unless the contrary is stated.

V. crista-galli Steven, *Trans. Linn. Soc. London* 11: 408 (1815), an annual species from the Caucasian region, distinguished by its large calyx in which the segments are united in pairs almost to the apex, was naturalized for nearly 80 years in one station in S. England (Sussex), but is probably extinct there now.

Literature: W. Kulpa, *Roczn. Nauk Roln.* **126 D**: 1–108 (1968) (morphology of seeds). H. Römpp, *Feddes Repert.* (*Beih.*) **50**: 1–172 (1928). G. Stroh, *Beih. Bot. Centr.* **61B**: 384–451 (1942).

1 Flowers in axillary racemes (which, if single and near the stem-apex, may appear terminal)
  2 Capsule both septicidal and loculicidal, dehiscing into four valves; seeds scarcely compressed, with excentric chalaza; plants of wet places, usually glabrous (Sect. *Beccabunga*)
    3 Flowering stems procumbent or ascending; leaves all petiolate
      4 Leaves thick; capsule scarcely compressed   **36. beccabunga**
      4 Leaves thin; capsule compressed in upper half   **37. scardica**
    3 Flowering stems ± erect; at least the upper leaves usually sessile
      5 Capsule elliptical; annual   **38. anagalloides**
      5 Capsule ovate to orbicular; often perennial
        6 Upper leaves ovate-rhombic, subsessile to shortly petiolate   **37. scardica**
        6 Upper leaves ovate-lanceolate to linear, sessile
          7 Pedicels patent; capsule usually wider than long   **40. catenata**
          7 Pedicels ± erect, at least in distal half; capsule not wider than long   **39. anagallis-aquatica**
  2 Capsule loculicidal only; seeds compressed, with ± central chalaza; plants seldom of wet places, and usually hairy (Sect. *Veronica*)
    8 Leaves 1- to 2-pinnatisect, with linear segments   **(13–17). austriaca** group
    8 Leaves entire to pinnatifid
      9 Racemes not more than 2 cm in fruit, 2- to 6(8)-flowered, corymbose or capitate in flower
        10 Racemes capitate, with pedicels not more than 3 mm; stems woody below; capsule at least as wide as long
          11 Leaves and capsule pubescent   **23. thymifolia**
          11 Leaves and capsule glabrous   **24. thessalica**
        10 Racemes corymbose, with pedicels at least 5 mm in fruit; stems herbaceous; capsule longer than wide
          12 Leafy stems procumbent; leaves pubescent; capsule 7 × 5 mm   **25. aphylla**
          12 Leafy stems ascending; leaves glabrous; capsule 5 × 3 mm   **26. baumgartenii**
      9 Racemes more than 2 cm in fruit, usually with more than 6 flowers, scarcely capitate or corymbose
        13 Petioles 7–15 mm; capsule *c.* 8 mm wide   **34. montana**
        13 Petioles not more than 7 mm; capsule not more than 6 mm wide
          14 Racemes usually longer than leafy stems, usually with lower bracts lobed or with a few leaves below the flowers (Bulgaria and Turkey)   **20. pectinata**

14 Racemes usually shorter than leafy stems, with entire bracts and bearing no leaves below the flowers
  15 Calyx at least 4 mm
    16 Pedicels 8–15 mm, recurved in fruit (S. part of U.S.S.R.)   **33. peduncularis**
    16 Pedicels not more than 9 mm, erect or erecto-patent in fruit
      17 Capsule distinctly wider than long
        18 Leaves 10–40 mm; corolla *c.* 10 mm in diameter, bright blue   **30. chamaedrys**
        18 Leaves 30–60 mm; corolla 5–8 mm in diameter, pale blue or whitish   **31. melissifolia**
      17 Capsule longer than wide, or about as long as wide
        19 Corolla white with pink centre   **32. micrantha**
        19 Corolla blue
          20 Stems flexuous; calyx-segments elliptic-oblong, obtuse   **22. aragonensis**
          20 Stems not flexuous; calyx-segments lanceolate to linear-oblong, acute   **(13–17). austriaca** group
  15 Calyx less than 4 mm
    21 Pedicels at least 6 mm in fruit
      22 Leaves linear-oblong, ± entire; racemes alternate   **35. scutellata**
      22 Leaves triangular-ovate, incise-serrate; racemes opposite   **12. urticifolia**
    21 Pedicels less than 6 mm in fruit
      23 Corolla pink
        24 Leaves *c.* 30 × 15 mm, glabrous   **29. dabneyi**
        24 Leaves *c.* 10 × 3 mm, crispate-pubescent   **21. rosea**
      23 Corolla blue or lilac
        25 Leaves pubescent or puberulent
          26 Leaves distinctly petiolate; calyx-segments 4, subequal   **27. officinalis**
          26 Leaves sessile or subsessile; calyx-segments usually 5, conspicuously unequal   **(13–17). austriaca** group
        25 Leaves glabrous or sparsely ciliate
          27 Sepals ciliate; leaves 2–5 mm wide   **19. rhodopaea**
          27 Sepals glabrous; leaves 5–13 mm wide
            28 Racemes more than 4 cm; capsule glabrous   **18. turrilliana**
            28 Racemes less than 4 cm; capsule pubescent   **28. allionii**
1 Flowers in terminal racemes (sometimes also in axillary racemes below the terminal) or solitary in leaf-axils
  29 Annual
    30 Flowers solitary in leaf-axils; stems ± procumbent
      31 Leaves 3- to 9-lobed; seeds at least 2 mm
        32 Calyx-segments cordate-ovate, acute; capsule glabrous   **57. hederifolia**
        32 Calyx-segments obovate, obtuse; capsule ciliate   **58. cymbalaria**
      31 Leaves crenate or serrate; seeds less than 2 mm
        33 Capsule keeled, with divergent lobes; corolla 8–12 mm in diameter   **55. persica**
        33 Capsule not or only slightly keeled, with sub-parallel or slightly divergent lobes; corolla less than 8 mm in diameter
          34 Capsule with sparse, long glandular hairs only; corolla usually whitish   **52. agrestis**
          34 Capsule with long glandular and shorter eglandular hairs; corolla blue
            35 Calyx-lobes oblong-lanceolate to oblanceolate, not overlapping near base   **54. opaca**
            35 Calyx-lobes broadly ovate, often overlapping near base   **53. polita**
  30 Flowers in racemes; stems usually ± erect
    36 Seeds ± cup-shaped (with one convex and one deeply concave face)
      37 Fruiting pedicels at least 4 times as long as calyx   **44. glauca**
      37 Fruiting pedicels less than 4 times as long as calyx
        38 Lower bracts and upper cauline leaves ± digitately lobed   **43. triphyllos**
        38 Bracts and leaves deeply crenate-serrate, but not lobed   **42. praecox**

[1] By S. M. Walters and D. A. Webb; spp. **36–40** based on data provided by N. G. Marchant.

36 Seeds slightly convex on one face, and flat or slightly concave on the other
39 Bracts not exceeding pedicels
40 Leaves entire or obscurely crenate   **41. acinifolia**
40 Leaves pinnatifid or laciniate   **45. aznavourii**
39 Bracts exceeding pedicels
41 Glabrous   **49. peregrina**
41 Pubescent
42 Corolla white (sometimes with blue veins)
43 Bracts mostly entire, linear; corolla exceeding calyx   **51. grisebachii**
43 Bracts mostly laciniate or digitate; corolla not exceeding calyx   **50. chamaepithyoides**
42 Corolla blue
44 Upper cauline leaves not pinnatifid   **46. arvensis**
44 Upper cauline leaves pinnatifid
45 Style c. 0·5 mm   **47. verna**
45 Style c. 1·5 mm   **48. dillenii**
29 Perennial
46 Flowers solitary in axils of foliage-leaves   **56. filiformis**
46 Flowers in ± distinct racemes
47 Corolla-tube longer than wide; racemes long and dense (Sect. *Pseudolysimachium*)
48 Pedicels longer than subtending bracts   **60. paniculata**
48 Pedicels shorter than subtending bracts
49 Leaves crenate, crenate-serrate, or subentire   **62. spicata**
49 Leaves biserrate
50 Leaves linear-lanceolate to lanceolate, with cuneate to truncate base; calyx-segments triangular-ovate   **59. longifolia**
50 Leaves triangular-ovate, with ± cordate base; calyx-segments linear-lanceolate   **61. bachofenii**
47 Corolla-tube much wider than long; racemes usually short or relatively lax (Sect. *Veronicastrum*)
51 Basal leaves larger than cauline, forming a ± distinct rosette
52 Racemes 6–30 cm; pedicels 8–20 mm; corolla white or pale blue with dark blue veins   **1. gentianoides**
52 Racemes 1–4 cm; pedicels 2–6 mm; corolla violet-blue or lilac, without conspicuous veins   **2. bellidioides**
51 Basal leaves not larger than cauline and not forming a rosette
53 Stems procumbent or decumbent; leaves usually not more than 8 mm (rarely up to 15 mm)
54 Leaves conspicuously crenate-serrate (S. Greece)   **10. erinoides**
54 Leaves ± entire
55 Lower pedicels at least 4 mm in fruit; racemes corymbose or ± elongate in fruit
56 Corolla blue; capsule 4–5 mm   **3. serpyllifolia**
56 Corolla pink; capsule 2–2·5 mm   **4. repens**
55 Pedicels not more than 2 mm in fruit; racemes ± capitate
57 Bracts similar in shape to foliage-leaves; calyx-segments 4, ciliate   **8. nummularia**
57 Bracts much narrower than foliage-leaves; calyx-segments usually 5, villous   **9. saturejoides**
53 Stems ascending or erect; larger leaves more than 8 mm
58 Capsule wider than long, not exceeding calyx **3. serpyllifolia**
58 Capsule usually longer than wide, exceeding calyx
59 Leaves conspicuously serrate; racemes up to 15 cm
60 Calyx 3·5–4 mm; fruiting pedicels arcuate-ascending to suberect   **11. ponae**
60 Calyx 2 mm; fruiting pedicels patent, but turned upwards sharply just below capsule   **12. urticifolia**
59 Leaves ± entire; racemes not more than 6 cm
61 Stems herbaceous; pedicels usually less than 5 mm; style less than ⅓ as long as capsule   **5. alpina**
61 Stems woody at base; pedicels usually more than 5 mm; style at least ½ as long as capsule
62 Corolla deep blue with reddish centre; hairs on inflorescence all eglandular and appressed or crispate   **6. fruticans**

62 Corolla pink (rarely deep blue); some hairs on inflorescence glandular and patent   **7. fruticulosa**

Sect. VERONICASTRUM Koch. Perennial. Leaves entire to crenate-serrate, sessile or shortly petiolate. Racemes terminal (rarely with 1 or 2 axillary racemes as well); at least the upper bracts markedly different from the foliage-leaves. Calyx-segments 4 or 5. Corolla-tube wider than long. Capsule dehiscing loculicidally, usually only at apex. Seeds strongly compressed, smooth or slightly tuberculate or rugose, with central chalaza.

1. **V. gentianoides** Vahl, *Symb. Bot.* **1**: 1 (1790). Stems 25–80 cm, erect, pubescent above. Leaves very variable in size but usually 20–60 × 4–15 mm, entire or obscurely crenate, glabrous or puberulent, the basal linear-oblanceolate to broadly obovate, shortly petiolate, forming a rosette, the cauline ovate to oblong-lanceolate, sessile, alternate or opposite. Racemes 6–30 cm; lower bracts similar to cauline leaves; pedicels 8–20 mm, usually exceeding the bracts. Calyx 4–5 mm; segments 4, triangular, pubescent, somewhat unequal. Corolla 10 mm in diameter, white or pale blue with dark blue veins. Capsule 5–6 × 4–5 mm, obovate, emarginate, glandular-hairy; style c. 5 mm. *Mountain grassland. Krym.* Rs (K) [Rs (C)]. (*S.W. Asia.*)

2. **V. bellidioides** L., *Sp. Pl.* 11 (1753). Pubescent throughout; stems 5–20 cm, a short basal part procumbent and rooting, densely leafy, the remainder erect, with 1–3 pairs of opposite leaves. Leaves 12–40 × 7–15 mm, oblong-obovate, crenate-serrate or subentire, subsessile. Racemes capitate to corymbose in flower, usually elongating to 2–5 cm in fruit; pedicels 2–6 mm; bracts c. 4 mm. Calyx 4–5 mm; segments 4, oblanceolate. Corolla 9–10 mm in diameter, lilac to violet-blue. Capsule 6–8 × 4·5–6 mm, elliptical to ovate, pubescent; style c. 4 mm. *Dry alpine pastures; somewhat calcifuge.* ● *Pyrenees and mountains of C. Europe and Balkan peninsula, southwards to Bulgaria.* Au Bu Cz Ga Ge He Hs It Ju Po Rm Rs (W).

(a) Subsp. **bellidioides**: Leaves subentire to obscurely crenate. Corolla deep violet-blue. Capsule only slightly longer than wide. $2n = 36$. *Throughout the range of the species.*
(b) Subsp. **lilacina** (Townsend) Nyman, *Consp., Suppl.* **2**(2): 237 (1890) (*V. lilacina* Townsend): Leaves distinctly crenate-serrate in apical half. Corolla lilac with white centre. Capsule at least 1½ times as long as wide. $2n = 18$. *Pyrenees, S.W. Alps.*

3. **V. serpyllifolia** L., *Sp. Pl.* 12 (1753). Stems 5–20(–30) cm, ascending. Leaves up to 25 × 12 mm, ovate, shortly petiolate, entire or obscurely crenate, glabrous or puberulent. Racemes sparsely to densely pubescent; pedicels 3–8 mm; bracts usually exceeding pedicels, the lower similar to the foliage-leaves. Calyx 4–6 mm; segments 4, oblong, slightly unequal. Corolla 6–10 mm in diameter. Capsule 4–5 mm, slightly wider than long, emarginate, pubescent; style c. 2 mm. *Almost throughout Europe.* All except Bl Cr Sb.

(a) Subsp. **serpyllifolia**: Stems procumbent for only a short distance at base, the greater part erect. Racemes elongate, with 20–40 flowers, subglabrous to eglandular-pubescent; pedicels about as long as calyx. Corolla 6–8 mm, white or pale blue with dark blue veins. $2n = 14$. *Throughout the range of the species, but mainly in the lowlands.*
(b) Subsp. **humifusa** (Dickson) Syme in Sowerby, *Engl. Bot.* ed. 3, **6**: 158 (1866) (*V. humifusa* Dickson; incl. *V. apennina* Tausch, *V. balcanica* Velen.): Stems often procumbent for a large part of their length. Racemes usually short, with 8–15 flowers, glandular-pubescent; pedicels considerably longer than calyx. Corolla 7–10 mm, bright blue. $2n = 14$. *Mountains, throughout the greater part of Europe.*

4. V. repens Clarion ex DC. in Lam. & DC., *Fl. Fr.* ed. 3, 3: 727 (1805). Glabrous below, pubescent in inflorescence; stems 5–10 cm, procumbent, rooting, branched, usually forming a dense mat from which arise ascending flowering branches. Leaves 4–8 mm, ovate or elliptical to suborbicular, shortly petiolate, entire or slightly crenate-serrate. Racemes short, often corymbose, with 3–6 flowers; pedicels 4–7 mm, usually longer than bracts. Calyx 3 mm; segments 4, elliptical, subequal. Corolla 10 mm in diameter, pink. Capsule 2–2·5 mm, wider than long, scarcely emarginate, glabrous; style *c.* 4 mm. 2*n*=14. *Damp places.* ● *Mountains of Corse and of S. & S.E. Spain.* Co Hs.

Some plants from the Sierra Nevada agree well with plants from Corse, but the majority of the former which are referred to *V. repens* appear to have blue flowers and to be transitional in varying degree to 3(b). It seems probable that at least some of them are hybrids.

5. V. alpina L., *Sp. Pl.* 11 (1753). Stems 5–15(–25) cm, ascending, herbaceous, glabrous or pubescent. Leaves 10–25 × 4–12 mm, ovate to elliptical, entire or weakly crenate-dentate, glabrous or subglabrous, sessile. Racemes subcapitate, elongating to *c.* 2 cm in fruit; pedicels 2–4(–6) mm, shorter than bracts. Calyx 3–4 mm; segments 4, elliptic-oblong, hairy at least on margins, subequal. Corolla *c.* 7 mm in diameter, deep blue. Capsule 6–7 × 4–5 mm, elliptical, rounded at apex, glabrous; style 1·5–2 mm. 2*n*=18, 36. *Arctic Europe; mountains of Europe southwards to S. Spain, S. Italy and S. Bulgaria.* Al Au Br Bu Co Cz Fa Fe Ga Ge He Hs Is It Ju No Po Rm Rs (N, W) Sb Su.

6. V. fruticans Jacq., *Enum. Stirp. Vindob.* 2, 200 (1762) (*V. saxatilis* Scop.). Stems 5–15 cm, simple, woody at the base, ascending from a procumbent, branched, woody stock. Leaves 8–20 × 2–7 mm, obovate to narrowly oblong, entire or slightly crenate-serrate, subsessile, glabrous or puberulent. Racemes short, sometimes more or less capitate in flower, with 4–10 flowers, elongating to 3–6 cm in fruit; pedicels *c.* 5 mm in flower, up to 20 mm in fruit; lower bracts similar to cauline leaves. Pedicels, calyx and capsule densely crispate- to appressed-pubescent, without glandular hairs. Calyx 5 mm; segments 4, elliptic-oblong. Corolla 11–15 mm in diameter, deep blue with reddish centre. Capsule 6–8 × 4–5 mm, ovate, with a narrow, scarcely emarginate apex; style 4 mm. 2*n*=16. *Rocks and stony grassland. N.W. Europe, extending very locally eastwards to N.W. Russia; mountains of C. & S. Europe southwards to N. Spain, Corse and C. Jugoslavia.* Au Br Co Cz Fa Fe Ga Ge He Hs Is It Ju No Po Rm Rs (N, W) Su.

7. V. fruticulosa L., *Sp. Pl.* ed. 2, 15 (1762). Like **6** but some of the hairs on pedicels, calyx and capsule patent and glandular; corolla 9–12 mm in diameter, pink (rarely deep blue); capsule 5–6·5 mm, suborbicular, with broader and distinctly emarginate apex. *Mountain rocks; calcicole.* ● *Higher mountains of W. & S.C. Europe, from the Vosges and Sierra Nevada eastwards to N.W. Jugoslavia.* Au Ga Ge He Hs It Ju.

The record from the Ukrainian Carpathians is erroneous.

In the eastern part of their joint range **6** and **7** are quite distinct, but in the Pyrenees and N. Spain, and to a smaller extent in the S.W. Alps, plants are found in which the distinctive characters of the species are combined in various different ways. Until the whole complex has been surveyed throughout its range it is difficult to provide a taxonomic scheme for these variants. The most distinct appears to be **V. mampodrensis** Losa & P. Monts., *Anal. Inst. Bot. Cavanilles* 11(2): 442 (1953), from the Cordillera Cantábrica. This resembles **6** in having eglandular hairs in the inflorescence and a capsule about 8 mm long; but the hairs are patent, as in **7**, which it also resembles in the shape of its capsule (very broadly elliptical, with wide, emarginate apex). The corolla is intermediate in colour (violet to purplish-pink). It has 2*n*=16.

8. V. nummularia Gouan, *Obs. Bot.* 1 (1773) (*V. nummulariifolia* auct.). Stems 5–15 cm, procumbent, matted, woody and leafless below, with short, ascending, crispate-pubescent flowering branches. Leaves 4–5 × 3–4 mm, broadly elliptical to ovate-orbicular, entire, ciliate towards the base, shortly petiolate. Racemes capitate, with 5–10 flowers, elongating to 1·5 cm in fruit; pedicels *c.* 1 mm; bracts similar to foliage-leaves but smaller. Calyx 3·5 mm; segments 4, oblong, ciliate. Corolla 6 mm in diameter, blue or pink; lobes narrow, suberect. Capsule 3·5–4 mm, suborbicular, slightly emarginate, ciliate; style 3·5 mm. 2*n*=16. *Damp rocks and screes above 1800 m.* ● *Pyrenees, Cordillera Cantábrica.* Ga Hs.

9. V. saturejoides Vis., *Fl. Dalm.* 2: 168 (1847). Stems 10–30 cm, procumbent to decumbent, woody and leafless below, pubescent above. Leaves 6–9(–15) mm, suborbicular to oblanceolate, entire, rather fleshy, crowded, more or less hairy on the margins. Racemes capitate in flower, up to 3 cm in fruit, with 6–12 flowers, eglandular-villous; pedicels 1 mm; bracts 3 mm, oblanceolate. Calyx 2·5–3 mm; segments (4–)5, oblong, the fifth very small. Corolla *c.* 7 mm in diameter, bright blue. Stamens exserted. Capsule 3·5–4 mm, obcordate, hairy; style 5–6 mm. *Mountain rocks and screes.* ● *Balkan peninsula, from N. Albania to S.W. Bulgaria.* Al Bu Ju.

10. V. erinoides Boiss. & Spruner in Boiss., *Diagn. Pl. Or. Nov.* 1(4): 78 (1844) (*V. thessalica* auct., non Bentham). Stems 10–25 cm, decumbent, appressed-pubescent, woody below. Leaves 4–8 × 2–4 mm, lanceolate, crenate-serrate, glabrous above, pubescent beneath. Racemes capitate, with 3–5 flowers, glandular-pubescent; pedicels *c.* 1 mm; bracts *c.* 1·5 mm. Calyx 4 mm; segments (4–)5, linear-oblong, the fifth very small. Corolla 7–9 mm in diameter, lilac. Stamens included. Capsule *c.* 7·5 × 6 mm, obcordate, glandular-pubescent; style 2·5 mm. 2*n*=16. *Screes and stony slopes.* ● *Mountains of S. Greece.* Gr.

A plant once collected in S. Greece (Killini Oros), like **10** but with 5-merous flowers, larger corolla, capsule glabrous except for apical cilia and 2*n*=32, has been described as **V. contandriopouli** Quézel, *Taxon* 16: 240 (1967). Until further material is available and the constancy of the 5-merous flowers is established its taxonomic status must remain obscure.

11. V. ponae Gouan, *Obs. Bot.* 1 (1773). Sparsely pubescent throughout. Stems 10–50 cm, erect or ascending from a creeping stock. Leaves 15–60 × 6–30 mm, oblong-lanceolate to broadly ovate, subsessile, serrate. Racemes up to 15 cm, lax, usually solitary and terminal but rarely also with 1–2 axillary racemes; pedicels 8–20 mm, arcuate-ascending to suberect, exceeding the bracts. Calyx 3·5–4 mm; segments 4, somewhat unequal. Corolla *c.* 10 mm in diameter, bluish-lilac. Capsule 5–6 mm, suborbicular, slightly emarginate, pubescent; style 5–6 mm. *Damp or shady places in the mountains.* ● *Spain, Pyrenees.* Ga Hs.

Specimens with axillary racemes can be mistaken for **12**; this differs, however, in its shorter sepals, capsule and style, and in the sharply geniculate fruiting pedicels.

Sect. VERONICA (Sect. *Chamaedrys* Griseb.). Perennial. Leaves entire to 2-pinnatisect, sessile or petiolate. Racemes axillary (sometimes pseudoterminal), very rarely with a small terminal raceme present as well. Bracts usually very different from foliage-leaves. Calyx-segments 4 or 5. Corolla-tube wider than long. Capsule dehiscing loculicidally, usually only at apex. Seeds strongly compressed, smooth or rugose, with central chalaza.

12. **V. urticifolia** Jacq., *Fl. Austr.* **1** : 37 (1773) (*V. maxima* auct., non Miller, *V. latifolia* auct., non L.). Sparsely hairy; stem 25–70 cm, erect. Leaves 40–80 × 20–40 mm, triangular-ovate, incise-serrate, sessile. Racemes 5–10 cm, opposite, lax, in the axils of the terminal 1–5 pairs of leaves, a terminal raceme sometimes also present; pedicels 6–8 mm, exceeding the bracts, glandular-pubescent, patent in fruit but turned sharply upwards just below the capsule. Calyx 2 mm; segments 4, elliptic-oblong, subequal. Corolla *c.* 7 mm in diameter, lilac. Capsule 3·5–4 × 4–5 mm, suborbicular but slightly wider than long, emarginate, ciliate. $2n = 18$. *Shady places; somewhat calcicole.* ● *S. & C. Europe, from the Jura and Carpathians southwards to N.E. Spain, C. Italy and S. Greece, mainly in mountain districts but below 2000 m; also in C. & S. Ural.* Al Au Bu ?Cz Ga Ge Gr He Hs It Ju Po Rm Rs (C, W).

(13–17). **V. austriaca** group. Stems procumbent to erect, more or less hairy. Leaves entire to 2-pinnatisect, puberulent or pubescent. Racemes usually opposite. Calyx-segments (4)5, narrow, usually very unequal. Corolla 5–13(–17) mm in diameter, blue.

In this group have been included such species as have been confused with *V. austriaca* or have been included in it by some authors. Other species, such as **18**, **19** and **20**, which are not included in the group, may be as closely related, but are in practice more easily distinguished.

In most Floras the number of calyx-segments for plants of this group is given as 5, and they are usually keyed out on this character. The small fifth segment may, however, be absent not merely from individual flowers, but from whole plants or even whole populations or varieties.

V. kindlii Adamović, *Denkschr. Akad. Wiss. Math.-Nat. Kl.* (*Wien*) **74** : 138 (1904), from the borders of Greece and Jugoslavia, should probably be placed in this group, but further material and information are required before its relationships can be determined.

1 Leaves pinnatisect
2 Capsule wider than long   **14. multifida**
2 Capsule suborbicular or longer than wide
  3 Stems decumbent to ascending; leaves 6–15(–20) mm; internodes usually less than 15 mm; racemes 0·5–2 cm in flower, not more than 5(–7) cm in fruit   **16. tenuifolia**
  3 Stems ± erect; leaves (12–)15–30 mm; internodes usually more than 15 mm; racemes 1·5–5 cm in flower, 4–12 cm in fruit   **13. austriaca**
1 Leaves entire to pinnatifid
4 Pedicels 5–9 mm; capsule wider than long   **15. orientalis**
4 Pedicels 1–5 mm; capsule suborbicular or longer than wide
  5 Calyx and capsule hairy   **13. austriaca**
  5 Calyx and capsule glabrous
    6 Stems procumbent to ascending   **17. prostrata**
    6 Stems ± erect   **13. austriaca**

13. **V. austriaca** L., *Syst. Nat.* ed. 10, **2** : 849 (1759) (*V. latifolia* L., nom. ambig., *V. maxima* Miller). Stems up to 100 cm, usually erect, sometimes procumbent to ascending. Leaves very variable in shape, broadly ovate to narrowly oblong-lanceolate in outline, entire to deeply 2-pinnatisect. Racemes dense to moderately lax; pedicels 2–5 mm, usually equalling or shorter

than bracts. Calyx (3–)4–6(–8) mm, glabrous or hairy. Corolla (8–)10–13(–17) mm in diameter, bright blue. Capsule 4–6 × 3–5 mm, suborbicular-cordate to broadly elliptical, usually longer than wide, glabrous or pubescent. *Most of Europe except Fennoscandia, the islands and parts of the extreme south.* Al Au Be Bu Cz Ga Ge Gr He Ho Hs Hu It Ju Po Rm Rs (N, B, C, W, K, E) Tu.

1 At least the middle cauline leaves pinnatisect or deeply pinnatifid   **(d) subsp. austriaca**
1 Leaves all entire to incise-dentate, rarely somewhat pinnatifid
2 Stems not more than 30 cm, procumbent to ascending; leaves usually 15–30 mm   **(b) subsp. vahlii**
2 Stems up to 100 cm, usually erect; leaves often up to 50 mm
  3 Leaves ± amplexicaul, with truncate or subcordate base, all ± crenate or serrate   **(a) subsp. teucrium**
  3 Leaves not amplexicaul, with truncate to cuneate base, at least the uppermost ± entire   **(c) subsp. dentata**

(a) Subsp. **teucrium** (L.) D. A. Webb, *Bot. Jour. Linn. Soc.* **65** : 267 (1972) (*V. teucrium* L.; incl. subsp. *crinita* (Kit.) Velen.): Stems 30–100 cm, erect to ascending. Leaves 20–70 × 6–45 mm, ovate to oblong, sessile, truncate or subcordate at base, crenate to incise-serrate. Racemes long. Calyx and capsule hairy or glabrous. $2n = 64$. *Throughout the range of the species except parts of W. Europe.*

(b) Subsp. **vahlii** (Gaudin) D. A. Webb, *loc. cit.* (1972) (*V. teucrium* subsp. *vahlii* Gaudin; incl. subsp. *orsiniana* (Ten.) Watzl): Stems 15–30 cm, procumbent to ascending. Leaves (10–)15–30 × 5–12 mm, lanceolate to ovate-oblong, rounded or cuneate at the base, sometimes very shortly petiolate, crenate to incise-serrate. Racemes short. Calyx and capsule hairy. $2n = 16$. ● *S. & W. Europe, northwards to Belgium and eastwards to Jugoslavia.*

(c) Subsp. **dentata** (F. W. Schmidt) Watzl, *Abh. Zool.-Bot. Ges. Wien* **5**(5) : 53 (1910) (*V. dentata* F. W. Schmidt): Stems 20–50 cm, usually erect. Leaves 20–50(–75) × 3–15 mm, ovate-oblong to linear-lanceolate, truncate to cuneate at the base, sometimes very shortly petiolate, the lower usually crenate-serrate, the uppermost usually subentire. Racemes fairly long. Calyx and capsule hairy or glabrous. $2n = 48, 64$. *C. Europe, Balkan peninsula, Ukraine and S.C. Russia.*

(d) Subsp. **austriaca** (subsp. *jacquinii* (Baumg.) J. Maly; incl. subsp. *orbiculata* (A. Kerner) K. Malý): Stems 25–50 cm, erect or ascending. Leaves usually blackening on drying, linear-lanceolate to suborbicular or deltate in outline, usually 1- or 2-pinnatisect with linear segments, but the uppermost and lowermost sometimes dentate to pinnatifid. Racemes usually long. Calyx and capsule hairy or glabrous. $2n = 48$. *E., E.C. & S.E. Europe, extending westwards to N. Italy.*

14. **V. multifida** L., *Sp. Pl.* 13 (1753). Like **13(d)** but smaller and more delicate; stems numerous, decumbent to ascending, not more than 30 cm; leaves *c.* 10 mm, mostly 2-pinnatisect, not blackening on drying; racemes not more than 5 cm in fruit; capsule wider than long. *S. Ukraine, S.E. Russia.* Rs (K, E). (*S.W. & C. Asia.*)

Much confused with **13(d)**, but probably more closely related to **15**.

V. caucasica Bieb., *Fl. Taur.-Cauc.* **1** : 13 (1808), usually regarded as endemic to the Caucasus, which is similar to **14** but with white flowers and longer pedicels, has been reported from Jugoslavia, but probably in error.

15. **V. orientalis** Miller, *Gard. Dict.* ed. 8, no. 10 (1768) (incl. *V. taurica* Willd.). Stems 10–25 cm, ascending, woody at the

base, sparsely pubescent below, more densely above. Leaves 10–20 × 2–4 mm, linear-oblong to oblanceolate, entire to irregularly pinnatisect. Racemes 1–4, up to 8 cm; pedicels 5–9 mm, exceeding the bracts. Calyx 4–6 mm; segments usually 4, unequal. Corolla 10–12 mm in diameter, blue (rarely pink). Capsule 4 × 5 mm, broadly obcordate, glabrous or glandular-pubescent. *Rocky hillsides. Krym.* ?Rm Rs (K). (*S.W. Asia.*)

16. **V. tenuifolia** Asso, *Syn. Stirp. Arag.* 2 (1779) (*V. assoana* (Boiss.) Willk.; incl. *V. commutata* Willk., non Seidl). Crispate-pubescent throughout. Stems 10–15 cm, decumbent to ascending, slender, woody at the base. Leaves 6–15(–20) mm, crowded, 1- to 2-pinnatisect with linear-oblong segments. Racemes 1–4, 0·5–2 cm in flower, up to 5(–7) cm in fruit; pedicels 4–5 mm; bracts 3 mm. Calyx 3·5 mm; segments 4 or 5, unequal. Corolla *c.* 10 mm in diameter, bright blue. Capsule 5 × 4·5 mm, obcordate, puberulent. *Dry grassland.* ● *N.E. Spain.* Hs.

Very similar to some variants of 13(d), but usually distinct in the much shorter internodes, especially in the upper part of the stem.

17. **V. prostrata** L., *Sp. Pl.* ed. 2, 17 (1762). Flowering stems up to 25(–40) cm, ascending; non-flowering stems short, procumbent. Leaves 8–25 × 2–12 mm, linear-oblong to ovate, very shortly petiolate, cuneate to truncate at the base, crenate-serrate to subentire; margin often revolute. Racemes usually 3–6 cm in fruit; pedicels 1–3 mm. Calyx 2–4 mm, glabrous; segments (4)5, linear-oblong, unequal. Corolla 6–11(–14) mm in diameter. Capsule 3·5–5 mm, suborbicular to obovate, usually slightly longer than wide, glabrous. *From the Netherlands and N.C. Russia southwards to N. Spain, C. Italy and Macedonia.* Al Au Be Bu Cz Ga Ge Gr He Ho Hs Hu It Ju Po Rm Rs (N, B, C, W, K, E) Tu.

(a) Subsp. **prostrata**: Leaves rather densely pubescent, ovate to oblong-lanceolate. Corolla 6–8(–11) mm in diameter, pale blue. 2n = 16. *From C. Germany, S.E. France and Italy eastwards.*
(b) Subsp. **scheereri** J. P. Brandt, *Bull. Soc. Neuchâtel. Sci. Nat.* **84**: 80 (1961): Leaves sparsely puberulent, lanceolate to linear-oblong. Corolla 8–11(–14) mm in diameter, deep blue. 2n = 32. *W. & W.C. Europe.*

18. **V. turrilliana** Stoj. & Stefanov, *Jour. Bot.* (*London*) **61**: 219 (1923). Stems up to 40 cm, decumbent, woody at the base, pubescent, usually with the hairs in 2 opposite lines. Leaves 12–20 × 6–10 mm, elliptical, obtuse, very shortly petiolate, crenate, coriaceous, glabrous. Racemes 2–8, 4–8 cm, opposite, rather lax; pedicels 2–3 mm. Calyx 2–2·5 mm; segments (4)5, oblong, obtuse, glabrous, the fifth, if present, very small, the others subequal. Corolla 8–10 mm in diameter, deep blue. Capsule 4–5 mm, suborbicular, glabrous. *Calcareous rocks.* ● *E. part of Balkan peninsula* (*Istrancadağlari*). Bu Tu.

19. **V. rhodopaea** (Velen.) Degen ex Stoj. & Stefanov, *Fl. Bălg.* 1010 (1925). Stems *c.* 5 cm, procumbent, pubescent, with the hairs usually in 2 opposite lines. Leaves 8–13 × 2–5 mm, elliptic- to linear-oblong, sessile, entire or minutely crenate, glabrous or minutely ciliate; margin revolute. Racemes 1–4, dense and subcapitate in flower, 3–5 cm in fruit, alternate. Calyx 3–4 mm; segments 4, ciliate, acute, unequal. Corolla 9 mm in diameter, deep blue. Capsule subglabrous. ● *Mountains of S. Bulgaria.* Bu.

Plants from W. Macedonia (on the borders of Greece and Jugoslavia) and from E. Albania have also been referred to this species, but their relationships are not clear.

20. **V. pectinata** L., *Mantissa* 24 (1767). Puberulent throughout. Stems 20–50 cm, decumbent. Leaves 10–25 × 4–15 mm, broadly elliptical to oblong, shortly petiolate, incise-serrate. Racemes up to 20 cm, glandular, often longer than the leafy stems, rather dense even in fruit, often with a few small leaves below the lowest flowers; lower bracts often pinnatifid; pedicels 2–5 mm. Calyx 4–5 mm; segments 4, linear, villous, subequal. Corolla 10–12 mm in diameter, blue. Capsule 5–6 × 4·5–5 mm, obcordate, pubescent. *E. part of Balkan peninsula.* Bu Tu. (*Anatolia.*)

Not recently recorded from Europe.

21. **V. rosea** Desf., *Fl. Atl.* **1**: 13 (1798). Crispate-pubescent throughout, but more sparsely in inflorescence; stems 10–20 cm, ascending, stout, woody at the base. Leaves 8–12 mm, oblanceolate, dentate to deeply pinnatifid distally. Racemes usually 2, up to 6(–10) cm in fruit, opposite, rather dense; pedicels 3–4 mm, equalling or slightly exceeding the bracts. Calyx 3·5 mm; segments 4(5). Corolla 10–12 mm in diameter, pale pink. Capsule 4 × 4·5 mm, broadly obcordate. *Dry, rocky ground. S. Spain* (*near Antequera*). Hs. (*N.W. Africa.*)

22. **V. aragonensis** Stroh, *Beih. Bot. Centr.* **61 B**: 398 (1942). Stems 10–25 cm, ascending or decumbent, weak and flexuous, crispate-pubescent. Leaves 7–10 × 2–4 mm, oblong to elliptical, entire or crenate, glabrous or pubescent. Racemes 1–2, 2–6 cm, rather lax, sometimes pseudo-terminal; pedicels 2–4 mm; bracts 3–5 mm. Calyx 5–6 mm; segments 4, oblong-elliptical, obtuse, glandular-pubescent, somewhat unequal. Corolla 8–10 mm in diameter, pale blue. Capsule 5 mm, about as wide as long, pubescent, emarginate. *Calcareous rocks and screes.* ● *C. Pyrenees.* Hs.

A little-known species requiring further study. It has been treated by some authors as related to *V. serpyllifolia*, but the structure of the seed as well as the position of the raceme justify its assignment to Sect. *Veronica*.

23. **V. thymifolia** Sibth. & Sm., *Fl. Graec. Prodr.* **1**: 6 (1806). Densely crispate-pubescent throughout; stems decumbent, matted, woody at the base. Leaves 4–6 × 1–2·5 mm, narrowly oblong to oblanceolate-spathulate, entire, with revolute margins. Racemes 1–2, very short and few-flowered; pedicels *c.* 2 mm, shorter than the bracts. Calyx 3 mm; segments 4, oblong, obtuse. Corolla 7–8 mm in diameter, blue, lilac or pink. Capsule 4 × 4 mm, obcordate, pubescent. *Mountain rocks.* ● *S. Greece and Kriti.* Cr Gr.

24. **V. thessalica** Bentham in DC., *Prodr.* **10**: 480 (1846). Stems 10–20 cm, procumbent, woody, glabrous. Leaves 5–8 × 2·5–4 mm, elliptical, entire or serrulate, glabrous except for ciliate petiole, crowded at stem-apex. Racemes usually 2, opposite, capitate, arising near the stem-apex and often approximated, so as to be mistaken for a single terminal raceme; axis villous; pedicels 2–3 mm; bracts 3–4 mm, linear-oblong, long-ciliate. Calyx 3 mm; segments 4–5, ciliate. Corolla 5 mm long, usually not opening widely. Capsule 2·5 × 3 mm, obcordate, glabrous. 2n = 16. *Mountain rocks and screes.* ● *E. Greece; N. Albania.* Al Gr.

Until recently included in Sect. *Veronicastrum*; in well-grown plants, however, the axillary position of the racemes is clear.

25. **V. aphylla** L., *Sp. Pl.* 11 (1753). Pubescent throughout; stems 1–4 cm, procumbent. Leaves 10–15 × 5–10 mm, elliptic-oblong to broadly obovate-spathulate, obscurely crenate, tapered to a short petiole, crowded near the stem-apex into a lax rosette.

Racemes 1–2, short, corymbose, with 2–6 flowers, arising near the stem-apex and sometimes pseudoterminal; peduncle 3–6 cm, erect, leafless; pedicels up to 10 mm, usually exceeding the bracts. Calyx 3 mm; segments 4, oblong, obtuse, subequal. Corolla 6–8 mm in diameter, deep blue (rarely pink). Capsule 7 × 5 mm, obovate, usually emarginate, purplish, pubescent. $2n = 18$. *Mountain rocks and stony alpine pastures.* ● *C. & S. Europe, from the Jura and the Carpathians southwards to the Pyrenees, S. Italy and C. Greece.* Al Au Cz Ga Ge Gr He Hs It Ju Po Rm Rs (W).

26. V. baumgartenii Roemer & Schultes, *Syst. Veg.* **1**: 100 (1817). Stems 4–12 cm, ascending, pubescent above. Leaves 8–18 × 3–8 mm, serrate to subentire, glabrous, subsessile, the lower ovate, the upper elliptic-oblong. Racemes 1–4, opposite, short, corymbose; pedicels up to 15 mm, pubescent, greatly exceeding the bracts. Calyx 2·5–3 mm; segments 4, elliptic-oblong, subequal. Corolla *c.* 9 mm in diameter, blue. Capsule 5 × 3 mm, elliptical, deeply emarginate, glabrous. *Mountain rocks and alpine pastures.* ● *E. & S. Carpathians; mountains of E. Jugoslavia and N. Bulgaria.* Bu Ju Rm Rs (W).

27. V. officinalis L., *Sp. Pl.* 11 (1753). Stems 10–50 cm, procumbent to ascending, hirsute to villous. Leaves 15–50 × 8–30 mm, ovate to elliptical, serrate, softly hairy; petiole 2–6 mm. Racemes 3–6(–10) cm, erect, opposite, dense, glandular-hairy; pedicels 1–3 mm; bracts 3–5 mm. Calyx 2·5–3 mm; segments 4, oblong, subequal. Corolla *c.* 8 mm in diameter, dull lilac-blue. Capsule *c.* 4 × 4 mm, obdeltate to slightly obcordate, pubescent. $2n = 18, 36$. *Woods and heaths. Almost throughout Europe.* All except Bl Cr Sb.

Variants with small, glabrous leaves, villous peduncles, deeply obcordate capsules, and sometimes with deep blue corolla are recorded from a few localities in N.W. & N.C. Europe. They are in some features transitional between **27** and **28**, and their status is not clear.

28. V. allionii Vill., *Prosp. Pl. Dauph.* 20 (1779). Like **27** but more or less glabrous except for crispate-puberulent inflorescence; stems 5–15 cm, woody at the base; leaves 8–20 × 5–13 mm, entire or minutely crenate; racemes shorter, usually solitary; calyx 2 mm; corolla deep blue; capsule 3 mm, obovate. $2n = 18$. *Alpine pastures.* ● *S.W. Alps.* Ga It.

29. V. dabneyi Hochst. in Seub., *Fl. Azor.* 39 (1844). Like **27** but glabrous except for pubescent inflorescence; leaves oblong, dark green, coriaceous; racemes up to 12 cm; pedicels 4–5 mm; corolla 10–12 mm in diameter, pale pink with deeper veins; capsule glabrous. ● *Açores.* Az.

30. V. chamaedrys L., *Sp. Pl.* 13 (1753). Stems 7–25 cm, ascending, sparsely to densely hairy. Leaves 10–40 × 6–30 mm, oblong-lanceolate to ovate-deltate, crenate-serrate to deeply and irregularly pinnatifid with linear-oblong lobes, sparsely hairy to subglabrous, sessile or shortly petiolate. Racemes 4–12(–16) cm, usually opposite but the lowest often unpaired; pedicels (2–)4–8 mm; bracts 3–7 mm. Calyx 4–6 mm; segments 4, lanceolate to linear-oblong, pubescent, somewhat unequal. Corolla *c.* 10 mm in diameter. Capsule 3–4 × 4–5 mm, obdeltate to obcordate, ciliate, shorter than calyx. *Grassland, thickets and hedges. Throughout Europe except for some islands and much of the Arctic.* All except Az Bl Co Cr Fa Sa Sb Si; only naturalized in Is.

(a) Subsp. **chamaedrys**: Leaves dark green, crenate-serrate. Stem often more or less hairy all round, at least above. Calyx

sparsely pubescent. Corolla bright blue. $2n = 32$. *Throughout the range of the species.*

(b) Subsp. **vindobonensis** M. Fischer, *Österr. Bot. Zeitschr.* **118**: 207 (1970): Leaves yellowish-green, incise-crenate to pinnatifid. Stem with hairs confined to 2 opposite lines throughout. Calyx densely glandular-puberulent. Corolla pale blue, rarely pink or white. $2n = 16$. *Dry grassland. E.C. & S.E. Europe.*

31. V. melissifolia Poiret in Lam., *Encycl. Méth. Bot.* **8**: 526 (1808). Like **30** but stems up to 50 cm; leaves 30–60 × 20–35 mm, triangular-ovate; pedicels 2–3 mm; corolla 5–8 mm in diameter, pale blue or whitish; capsule 4·5 × 5 mm, obcordate. *On the borders of Europe in S.E. Russia (near Krasnodar).* Rs (E). (*Anatolia, Caucasus.*)

32. V. micrantha Hoffmanns. & Link, *Fl. Port.* **1**: 286 (1813–1820). Like **30** but stems uniformly hairy all round; pedicels 1–2 mm; corolla *c.* 7 mm in diameter, white with a pink centre; capsule longer than wide. *Damp or shady places.* ● *N. Portugal.* Lu.

33. V. peduncularis Bieb., *Beschr. Länd. Terek Casp.* 126 (1800) (incl. *V. umbrosa* Bieb.). Sparsely pubescent throughout; stems 10–30 cm, ascending, diffuse. Leaves 7–25 × 3–15 mm, lanceolate to broadly ovate, serrate to pinnatifid; petiole 2–7 mm. Racemes 1–4, 5–10 cm, usually opposite, lax; pedicels 8–15 mm, greatly exceeding the bracts, recurved in fruit. Calyx 4–7 mm; segments 4, lanceolate, subequal. Corolla 10–12 mm in diameter, blue, pink or whitish. Capsule 2·5–4·5 × 3·5–5 mm, broadly obcordate, shorter than or very slightly exceeding calyx, pubescent. *Woods, thickets and stony slopes. S. & S.E. Ukraine, S.E. Russia.* Rs (K, E).

A single record for Romania requires confirmation.

34. V. montana L., *Cent. Pl.* **1**: 3 (1755). Softly hairy; stems 12–50 cm, decumbent. Leaves 20–35 × 15–30 mm, broadly ovate, coarsely crenate-serrate; petiole 7–15 mm. Racemes up to 10 cm, usually alternate, lax, flexuous; pedicels *c.* 8 mm, greatly exceeding the bracts. Calyx 4–5 mm; segments 4, elliptical, obtuse, subequal. Corolla 8–10 mm in diameter, pale lilac-blue. Capsule 6 × 8 mm, transversely elliptical to reniform, longer and wider than calyx, glandular-ciliate. $2n = 18$. *Woods. W., C. & S. Europe, northwards to N. Denmark, and extending eastwards to Latvia and W. Ukraine.* Al Au Be¦ Br Bu Co Cz Da Ga Ge Gr Hb He Ho Hs Hu It Ju Lu Po Rm Rs (B, C, W) Si Su ?Tu.

A record from the borders of Europe in S. Ural is probably erroneous.

35. V. scutellata L., *Sp. Pl.* 12 (1753). Usually glabrous, rarely pubescent. Stems up to 60 cm, ascending, weak. Leaves 15–50 × 3–7 mm, linear-oblong, entire or obscurely serrate, sessile, often reddish-brown. Racemes up to 7 cm, alternate, lax; peduncles very slender; pedicels 7–10 mm, filiform, greatly exceeding the bracts. Calyx 2 mm; segments 4, elliptical, subequal. Corolla 5–6 mm in diameter, pale pink or lilac. Capsule 4 × 5 mm, reniform. $2n = 18$. *Marshes and other wet places. Most of Europe, but rare in the Mediterranean region.* All except Az Bl Cr Fa Sa Sb Si.

Sect. BECCABUNGA (Hill) Dumort. Annuals or perennials of wet habitats, usually fleshy and glabrous. Flowers in opposite, axillary racemes. Calyx-segments 4. Corolla-tube wider than long. Capsule septicidal and loculicidal, dehiscing into 4 valves. Seeds not more than 0·75 mm, relatively thick, smooth, with convex back, flat chalazal face and excentric chalaza.

36. V. beccabunga L., *Sp. Pl.* 12 (1753). Glabrous perennial. Stems decumbent, rooting at the nodes. Leaves 10–40 mm, thick, orbicular, ovate or oblong, obtuse, petiolate, subentire or shallowly crenate-serrate. Racemes up to 3 times as long as subtending leaves; bracts from half as long to twice as long as pedicels. Calyx-segments more or less deflexed in fruit; corolla 5–7 mm in diameter, pale to dark blue (rarely pinkish). Capsule 2–3 × 2–4 mm, subglobose, scarcely emarginate; style *c.* 2 mm. $2n = 18, 36$. *Ditches, streams and marshes. Europe, southwards from c. 65°N.* All except Az Bl Cr ?Is Sb.

37. V. scardica Griseb., *Spicil. Fl. Rumel.* **2**: 31 (1844). Annual or perennial. Stem up to 20(–40) cm, erect or ascending from a procumbent, rooting base, glabrous or sparsely glandular-pubescent. Leaves 15–30 mm, thin, with distinct, usually reddish veins, scrrate in distal half; the lower ovate, shortly petiolate; the upper ovate-rhombic, usually subsessile and semi-amplexicaul. Racemes glabrous, up to 6 times as long as subtending leaves; bracts 2–5 mm; pedicels 5–10 mm, slender, patent. Calyx-segments closely appressed to mature capsule; corolla *c.* 4 mm, pale blue, or pinkish with blue veins. Capsule 2–3 × 2–3 mm, suborbicular, elliptical or obovate, compressed in upper half; style 1–1·5 mm. $2n = 18$. *Wet places. E.C. & S.E. Europe.* Al Au Bu Cz Gr Hu Ju Rm Rs (?C, W, ?K).

38. V. anagalloides Guss., *Pl. Rar.* 5 (1826) (incl. *V. poljensis* Murb.). Annual. Stem (5–)20–30 cm, erect, simple or with long branches from the base, usually glabrous. Leaves 15–40 mm, linear-lanceolate or lanceolate, sessile, subentire or serrate, glabrous. Racemes glabrous or glandular-pubescent, up to 6 times as long as subtending leaves; pedicels 3–7 mm, erecto-patent. Calyx-segments erect in fruit; corolla 3–5 mm in diameter, lilac or whitish. Capsule 2·5–3·5 × 2–2·5 mm, elliptical, not emarginate; style *c.* 1·5 mm. $2n = 18 + 0-2$ B. *Ditches, streamsides and wet mud. S., C. & E. Europe, northwards to Czechoslovakia and S.C. Russia.* Al Au Bl Bu Co Cz Ga Ge Gr Hs Hu It Ju ?Lu Rm Rs (C, W, K, E) Sa Si.

39. V. anagallis-aquatica L., *Sp. Pl.* 12 (1753) (incl. *V. lysima-chioides* Boiss.). Perennial (rarely annual). Stem (5–)30–60(–100) cm, erect, simple or much-branched at the base, glabrous or glandular-pubescent. Leaves 20–100 mm, light green, glabrous, the lower ovate, often petiolate, subentire, the upper ovate-lanceolate, more or less amplexicaul, irregularly serrate near the apex. Racemes glabrous or glandular-pubescent, up to 3 times as long as subtending leaves; pedicels 4–7 mm, erect, erecto-patent, or arcuate-ascending, usually at least as long as the linear bracts. Calyx-segments erect in fruit, but not closely appressed to capsule; corolla 5–10 mm in diameter, blue with violet veins. Capsule 2·5–4 × 2·5–3·5 mm, orbicular, ovate or elliptical; style 1·5–2 mm. $2n = 36$. *Streams, river-banks, and other wet places. Europe except the extreme north.* All except Fa Sb.

Plants with lax, usually glandular-pubescent racemes and arcuate-ascending pedicels have been distinguished as subsp. **divaricata** Krösche, *Allgem. Bot. Zeitschr.* **18**: 83 (1912). Such variants occur throughout the range of the species, and do not therefore merit subspecific distinction.

40. V. catenata Pennell, *Rhodora* **23**: 37 (1921) (*V. aquatica* Bernh., non S. F. Gray, *V. anagalloides* auct., non Guss). Like **39** but leaves dark green, linear to linear-lanceolate, all sessile; pedicels straight, patent, usually shorter than lanceolate bracts; calyx-segments usually patent or deflexed in fruit; corolla 3–5 mm in diameter, usually pink with reddish veins; capsule 2–3 × 2·5–3·5 mm, wider than long. $2n = 36$. *In or beside still or slow-moving water. Most of Europe from c. 58°N. southwards.* Al Au Az Be Br Bu Cz Da Ga Ge Gr Hb He Ho Hs Hu It Ju Lu Po Rm Rs (B, C, W, ?K, ?E) Su.

The vigorous hybrid between **39** and **40**, with $2n = 36$ (35), occurs throughout much of Europe, sometimes forming large clones in streams. Such plants have long, many-flowered racemes, but form very few 1- to 5-seeded capsules, and are most easily recognized by their sterility.

Sect. POCILLA Dumort. (Sect. *Alsinebe* Griseb.). Annual (except *V. filiformis*). Flowers solitary in leaf-axils, or in terminal racemes. Calyx-segments 4. Corolla-tube wider than long. Capsule dehiscing loculicidally and to a variable degree also septicidally. Seeds more or less compressed; chalazal face flat or excavate.

41. V. acinifolia L., *Sp. Pl.* ed. 2, 19 (1762) (incl. *V. perpusilla* Boiss.). Stem 5–15 cm, erect, simple or branched from the base, more or less glandular-pubescent throughout. Leaves 4–10 mm, shortly petiolate, ovate, entire or obscurely crenate, subglabrous to sparsely pubescent. Flowers in racemes; bracts shorter than or equalling pedicels, the upper oblanceolate. Calyx-segments ovate; corolla 2–3 mm in diameter, blue. Capsule *c.* 2 × 3 mm, broadly reniform with deep sinus, flat, glandular-hairy; style *c.* 1·5 mm, not or scarcely exceeding the sinus. Seeds *c.* 0·75 × 0·5 mm, broadly elliptical, flat. $2n = 14, 16$. *Cultivated ground and damp grassland. S., S.C. & W. Europe, northwards to N.C. France.* Al †Be Bu Co Cr Cz Ga Ge Gr He Hs Hu It Ju Lu Rm Rs (K) Tu.

42. V. praecox All., *Auct. Fl. Pedem.* 5 (1789). Glandular-pubescent. Stem 5–20 cm, erect, simple or branched from the base. Leaves 4–12 mm, shortly petiolate, ovate, deeply crenate-serrate. Flowers in narrow racemes, usually occupying most of stem; bracts shorter or longer than the arcuate-ascending pedicels, the lower similar to the leaves, the upper oblong-lanceolate. Calyx-segments oblong-ovate; corolla *c.* 3 mm in diameter, blue. Capsule *c.* 4 × 3–4 mm, obovate, with very shallow sinus, flat, glandular-hairy; style *c.* 2 mm, much exceeding the sinus. Seeds *c.* 1 × 0·75 mm, broadly elliptical, cup-shaped. $2n = 18$. *Dry, sandy or stony ground. S., C. & W. Europe, extending northwards locally to S.E. Sweden.* Al Au Be ?Bl Bu ?Co Cr Cz Da Ga Ge Gr He Ho Hs Hu It Ju Po Rm Rs (W, K, E) Si Su [Br].

V. kavusica Rech. fil., *Denkschr. Akad. Wiss. Math.-Nat. Kl. (Wien)* **105**(2, 1): 112 (1943), described from Kriti, differs in its glabrous, truncate or scarcely emarginate capsule, and seeds 1·5–2 mm. Some plants from Kriti, however, with large seeds and a more or less truncate capsule, have glandular hairs on the capsule as in **42**, and the specific distinction from the latter is not clear.

43. V. triphyllos L., *Sp. Pl.* 14 (1753). Glandular-pubescent. Stem 5–20 cm, suberect, simple or with decumbent to ascending branches from the base. Leaves 4–12 mm, mostly digitately lobed, with 3–7 spathulate or oblong lobes, but the lowest usually ovate, crenate. Flowers in lax racemes; uppermost bracts usually oblanceolate and not exceeding pedicels, the others usually lobed and often much larger. Calyx-segments ovate; corolla 3–4 mm in diameter, deep blue. Capsule 4·5–6 × 5–7 mm, reniform, glandular-hairy; style *c.* 1·5 mm, exceeding the sinus. Seeds *c.* 1·5 × 1·25 mm, broadly elliptical, cup-shaped. $2n = 14$. *Dry grassland, cultivated ground and waste places. From England, S. Sweden and Latvia southwards, but rather rare in the Mediterranean region.* Al Au Be Br Bu ?Co Cz Da Ga Ge Gr He Ho Hs Hu It Ju Lu Po Rm Rs (B, C, W, K, E) ?Si Su.

44. V. glauca Sibth. & Sm., *Fl. Graec. Prodr.* **1**: 9 (1806) (incl. *V. chaubardii* Boiss. & Reuter, *V. peloponnesiaca* Boiss. & Orph.). Stem 3–20(–35) cm, erect, simple or with decumbent to ascending branches near the base, usually crispate-pubescent and sometimes also with long, straight hairs, especially in inflorescence. Leaves 3–10(–18) mm, shortly petiolate, triangular-ovate, shallowly lobed to crenate-serrate, more or less pubescent. Flowers in racemes; pedicels very slender, more or less patent (sometimes recurved), much longer than bracts and at least 4 times as long as calyx. Calyx-segments 3–6(–7) mm, lanceolate, often unequal; corolla (7–)10–15 mm in diameter, deep blue. Capsule 3–4 × 4–5 mm, obovoid to subglobose, more or less inflated, emarginate to shallowly 2-lobed, usually glabrous; style 3–4 mm. Seeds 1·5–2 × 1–1·5 mm, suborbicular or broadly elliptical, cup-shaped. *Cultivated ground and dry grassland.* ● *S. part of Balkan peninsula.* Al ?Cr Gr.

Variable in hairiness, capsule-shape and angle of pedicels. *V. chaubardii* was described as having the capsule 2-lobed almost to the middle, but in type-material it is only shallowly 2-lobed.

45. V. aznavourii Dörfler, *Herb. Norm.* **48**: 244 (1907) (*V. pontica* Velen., non (Rupr. ex Boiss.) Wettst., *V. pumila* E. D. Clarke, non All.). Glandular-pubescent. Stem 5–15 cm, with suberect branches from the base. Leaves 3–12 mm, petiolate, the middle cauline deeply pinnatifid, with cuneate, distally toothed segments, the upper cauline and lower bracts more or less pinnatifid. Flowers in racemes; pedicels 1½–3 times as long as bracts, ascending. Calyx-segments 3–5 mm, ovate-lanceolate. Corolla 7–10 mm in diameter, blue. Capsule *c.* 4 × 5 mm, obcordate-reniform, deeply 2-lobed, glandular-hairy; style 3–4 mm, exceeding the sinus. Seeds 1·2–1·5 × 1–1·2 mm, broadly ovate, very shallowly cup-shaped or almost flat. *Dry grassland.* ● *E. part of Balkan peninsula.* Bu Gr Tu.

46. V. arvensis L., *Sp. Pl.* 13 (1753). Stem (2–)5–40 cm, simple and erect to much-branched and procumbent, pubescent, and often slightly glandular above. Leaves 2–15 mm, triangular-ovate with truncate or weakly cordate base, crenate-serrate, pubescent or subglabrous, the lower shortly petiolate, the upper sessile. Flowers in racemes; bracts longer than pedicels, the upper lanceolate, the lower ovate. Calyx-segments lanceolate; corolla 2–3 mm in diameter, blue. Capsule *c.* 3 × 3 mm, obcordate, flat, usually subglabrous except for marginal cilia; style *c.* 1 mm, scarcely exceeding the sinus. Seeds *c.* 1 × 0·75 mm, elliptical, flat. 2*n* = ?14, 16. *Cultivated ground, walls and other dry, open habitats. Almost throughout Europe.* All except Fa Sb; introduced in Is.

V. sartoriana Boiss. & Heldr. in Boiss., *Diagn. Pl. Or. Nov.* **3**(3): 171 (1856), is a dwarf plant (*c.* 5 cm), densely glandular-hairy, with longer pedicels, capsule *c.* 4 mm, and seeds *c.* 1·2 × 1·0 mm, broadly elliptical and slightly concave on one face. Such plants occur on mountains in Greece and Kriti, but do not seem sufficiently distinct from the very variable *V. arvensis* to merit specific rank.

47. V. verna L., *Sp. Pl.* 14 (1753) (*V. brevistyla* Moris). Stem 5–15 cm, erect, simple or branched from the base, pubescent in lower half, glandular above. Leaves 4–12 mm, more or less glandular-pubescent, the lower shortly petiolate, ovate to lanceolate, coarsely crenate, the upper sessile, pinnatifid with narrow lobes. Flowers in racemes; bracts exceeding pedicels, the upper lanceolate, the lower pinnatisect. Calyx-segments lanceolate; corolla *c.* 3 mm in diameter, blue. Capsule *c.* 3 × 4 mm, reniform, flat, glandular-hairy; style *c.* 0·5 mm. Seeds *c.* 1 × 0·75 mm, broadly elliptical, flat. 2*n* = 16. *Cultivated fields and other dry places. Much of Europe, but absent from the extreme north and*

parts of the west and of the Mediterranean region. Au Be Br Bu Co Cz Da Fe Ga Ge Gr He Ho Hs Hu It Ju No Po Rm Rs (N, B, C, W, K, E) Sa Si Su.

48. V. dillenii Crantz, *Stirp. Austr.* ed. 2, **2**: 352 (1769). Like **47** but up to 20 cm and more robust; corolla 4–5 mm in diameter, deep blue; style *c.* 1·5 mm; seeds *c.* 1·25 × 1 mm. 2*n* = 16. *Dry places. E. & E.C. Europe, northwards to the Baltic region, and extending locally westwards to N.E. Spain.* Au Cz Ga Ge Gr He Hs Hu It Ju Po Rm Rs (?B, C, W, K, E).

The distribution of this species is uncertain, as many authors do not separate it from **47**.

49. V. peregrina L., *Sp. Pl.* 14 (1753). Glabrous. Stem 5–25 cm, erect, simple or branched from the base. Leaves 5–20 mm, oblong to oblanceolate or spathulate, entire or obscurely serrate, the lower narrowed at base into a short petiole, the upper sessile. Flowers in racemes; bracts lanceolate, exceeding the subsessile flowers and capsules. Calyx-segments lanceolate; corolla 2–3 mm in diameter, whitish. Capsule *c.* 3 × 4 mm, obovate, shallowly emarginate; style less than 0·5 mm. Seeds *c.* 0·75 × 0·4 mm, narrowly elliptical, flat. *Cultivated ground, damp waste places and streamsides. Widely naturalized, mainly in W. & C. Europe.* [Au Az Be Br Cz Ga Ge Hb He Ho Hs Hu It Lu No Po Rm Rs (B) Su.] (*North and South America.*)

Rather variable in its persistence, and perhaps only casual in some of the territories listed.

50. V. chamaepithyoides Lam., *Tabl. Encycl. Méth. Bot.* **1**: 47 (1791) (*V. digitata* Vahl, non Lam.). Stem 5–15 cm, erect or suberect, often with procumbent to ascending branches from the base, pubescent and somewhat glandular. Leaves 3–8 mm, sparsely pubescent, petiolate, the lowest ovate, shallowly lobed, the others more or less deeply 3- to 5-lobed. Flowers in dense, narrow racemes, occupying most of stem; lower bracts like upper leaves, middle and upper bracts more or less pinnatifid with linear-subulate terminal lobe, much exceeding flower and fruit. Flowers subsessile; calyx-segments unequal, linear or linear-subulate; corolla *c.* 2 mm in diameter, white. Capsule *c.* 4 × 3 mm, narrowly obdeltate, with prominent veins, scarcely emarginate, sparsely hairy; style less than 1 mm. Seeds *c.* 1 × 0·5 mm, narrowly elliptical, flat or slightly concave on one face. *Cultivated ground.* ● *C. Spain.* Hs.

51. V. grisebachii Walters, *Bot. Jour. Linn. Soc.* **65**: 268 (1972) (*V. chamaepitys* Griseb., non Pers.). Like **50** but middle and upper bracts linear-subulate, entire; corolla *c.* 10 mm in diameter, with blue veins; style 2–3 mm. *S.E. part of Balkan peninsula.* Bu ?Gr Tu. (*Anatolia.*)

52. V. agrestis L., *Sp. Pl.* 13 (1753). Pubescent annual. Stems 5–30 cm, procumbent or decumbent. Leaves 5–15 mm, longer than wide, shortly petiolate, ovate with more or less truncate base, crenate or crenate-serrate, mostly alternate but the lowest opposite. Flowers solitary in leaf-axils; pedicels 5–15 mm, usually recurved in fruit. Calyx-segments 3–6 mm, ovate to ovate-lanceolate, subobtuse, not overlapping at base; corolla 3–6 mm in diameter, usually whitish with blue or pink upper lobe. Capsule 3–4 × 4–6 mm, 2-lobed with parallel lobes and deep, narrow sinus, not or slightly keeled, sparsely glandular-hirsute; style *c.* 1 mm, not exceeding the sinus. Seeds *c.* 1·5 × 1·25 mm, broadly elliptical, concave on one face. 2*n* = 28. *Cultivated ground; somewhat calcifuge. Most of Europe except the extreme north and parts of the south-east.* All except Bl Fa Is Rs (N, K) Sb Tu.

53. V. polita Fries, *Nov. Fl. Suec.* 63 (1819) (*V. didyma* auct., vix Ten.). Like **52** but leaves more regularly crenate-serrate, the lower often wider than long; calyx-segments broadly ovate, often overlapping near base; corolla blue; capsule with long glandular hairs and short, crispate eglandular hairs; style *c.* 1·5 mm, distinctly exceeding the wide sinus. 2*n*=14, 18. *Cultivated ground. Most of Europe except the Arctic, but probably not native in the north-east.* All except Az Fa Is Sb.

54. V. opaca Fries, *op. cit.* 64 (1819). Like **52** but leaves very shallowly crenate, the lower often wider than long; calyx-segments oblong-lanceolate to oblanceolate; corolla deep blue; capsule with long glandular hairs and short, crispate eglandular hairs; style 1–1·5 mm, usually more or less equalling the wide sinus. *Cultivated ground and waste places; somewhat calcicole. C. Europe, extending to E. France, C. Fennoscandia, C. Russia and Bulgaria.* Au Be Bu Cz Da Fe Ga Ge He Ho Hu It Ju No Po Rm Rs (B, C, W) Su.

55. V. persica Poiret in Lam., *Encycl. Méth. Bot.* **8**: 542 (1808) (*V. tournefortii* C. C. Gmelin pro parte, non Vill.). Pubescent. Stems (5)10–60 cm, procumbent. Leaves 5–25 mm, shortly petiolate, broadly ovate with more or less truncate base, mostly alternate but the lowest opposite. Flowers solitary in leaf-axils; pedicels 5–30 mm, recurved in fruit and much exceeding the subtending leaf. Calyx-segments (4–)6–7 mm, ovate-lanceolate, not or only slightly overlapping at base; corolla 8–12 mm in diameter, blue. Capsule 4–5 × 7–10 mm, 2-lobed, with very divergent lobes and wide, shallow sinus, strongly keeled, with eglandular hairs on sides and long glandular hairs on keel; style 2·5–3 mm, much exceeding the sinus. Seeds *c.* 1·75 × 1·25 mm, broadly elliptical, concave on one face. 2*n*=28. *Cultivated ground. Naturalized almost throughout Europe.* [All except Fa Rs (N) Sb.] (*S.W. Asia.*)

First recorded in Europe *c.* 1800; now in many regions the most abundant annual weed in the genus.

56. V. filiformis Sm., *Trans. Linn. Soc. London* **1**: 195 (1791). Procumbent, pubescent perennial, with slender stems up to 50 cm, freely rooting at the nodes, often mat-forming. Leaves 5–10 mm, petiolate, reniform, crenate, mostly opposite on non-flowering stems but alternate on flowering stems. Flowers solitary in leaf-axils; pedicels up to 40 mm, often deflexed after flowering. Calyx-segments 3–5 mm, oblanceolate to elliptical, obtuse; corolla 10–15 mm in diameter, pale lilac-blue. Capsule (rarely produced) *c.* 3 × 4 mm, 2-lobed, with subparallel lobes and rather wide sinus, subglabrous except for long hairs on keel; style *c.* 3 mm, much exceeding the sinus. Seeds *c.* 1·5 × 1 mm, elliptical, shallowly concave on one face. 2*n*=14. *Damp grassland. Originally cultivated for ornament; now widespread as a weed in gardens and lawns, and naturalized among native vegetation in several regions, mainly in N.W. and C. Europe.* [Au Be Br Cz Da Ga Ge Hb He Ho Po Rs (K) Su.] (*Caucasus and N. Anatolia.*)

The frequent sterility of this species is apparently associated with strong self-incompatibility and vigorous growth of single clones.

57. V. hederifolia L., *Sp. Pl.* 13 (1753). Pubescent. Stems 10–60 cm, procumbent. Leaves 5–15 mm, petiolate, suborbicular in outline, 3- to 7-lobed with large terminal lobe, mostly alternate, but the lowest opposite. Flowers solitary in leaf-axils; pedicels not or scarcely exceeding leaves. Calyx-segments 4–5 mm, cordate-ovate, acute, strongly ciliate, erect or suberect in fruit. Corolla 4–9 mm in diameter, blue or lilac. Capsule *c.* 4 × 4 mm, subglobose, shallowly 4-lobed, glabrous; style 0·3–1·1 mm.

Seeds *c.* 2·5 × 2·25 mm, suborbicular, cup-shaped. *Most of Europe except the Arctic, but doubtfully native in much of the north and north-east.* All except Az ?Cr Fa Is Rs (N) Sb.

1 Pedicels not more than 2½ times as long as calyx in fruit
  2 Middle lobe of leaf wider than long; pedicels usually glabrous
    except for an adaxial row of hairs   **(a) subsp. triloba**
  2 Middle lobe of leaf longer than wide; pedicels equally hairy
    all round   **(b) subsp. sibthorpioides**
1 Pedicels (2–)3–7 times as long as calyx in fruit
  3 Middle lobe of leaf wider than long; style 0·7–1 mm
          (d) subsp. hederifolia
  3 Middle lobe of leaf longer than wide; style less than 0·5 mm
          (c) subsp. lucorum

(a) **Subsp. triloba** (Opiz) Čelak., *Prodr. Fl. Böhm.* 333 (1871): Leaves thick, deeply 3(5)-lobed; middle lobe much wider than long. Pedicels not more than 2½ times as long as calyx, usually glabrous except for an adaxial row of short hairs. Calyx pubescent, shortly ciliate. Corolla 4–6 mm in diameter, dark blue. Style *c.* 1 mm. 2*n*=18. *Cultivated fields and other dry, open habitats. S. & S.C. Europe.*

(b) **Subsp. sibthorpioides** (Debeaux, Degen & Hervier) Walters, *Bot. Jour. Linn. Soc.* **65**: 269 (1972) (*V. sibthorpioides* Debeaux, Degen & Hervier): Like subsp. (a) but middle lobe of leaves longer than wide; pedicels equally hairy on all sides; corolla 2·5–3·5 mm in diameter; style 0·3–0·6 mm. *Shady, calcareous mountain rocks. S.E. Spain.* (*N. Africa.*)

(c) **Subsp. lucorum** (Klett & Richter) Hartl in Hegi, *Ill. Fl. Mitteleur.* ed. 2, **6**(1): 203 (1968): Leaves thin, very shallowly 5- to 7-lobed; middle lobe longer than wide. Pedicels 3½–7 times as long as calyx, with an adaxial row of short hairs and usually some patent hairs, especially in distal half. Calyx shortly ciliate, otherwise glabrous or sparsely pubescent. Corolla 4–6 mm in diameter, pale lilac. Style 0·3–0·5 mm. 2*n*=36. *Woods and hedges; also as a weed in cultivated ground. N., N.W. & C. Europe, from Britain and S.W. Finland southwards, and extending to N. part of Balkan peninsula.*

(d) **Subsp. hederifolia**: Leaves rather thick, 3- to 5-lobed; middle lobe wider than long. Pedicels (2)3–4 times as long as calyx, usually glabrous except for an adaxial row of rather long hairs. Calyx long-ciliate, otherwise glabrous. Corolla 6–9 mm in diameter, pale blue with white centre. Style 0·7–1 mm. 2*n*=54. *Cultivated ground. Throughout the range of the species.*

Subsp. (d) seems to have arisen as an allopolyploid hybrid between subspp. (a) and (c), and has now achieved, as a weed of cultivation, the widest distribution of all the subspecies.

58. V. cymbalaria Bodard, *Mém. Vér. Cymb.* 3 (1798). Like **57** but leaves 5- to 9-lobed with subequal lobes; pedicels much exceeding leaves; calyx-segments 3–4 mm, obovate, obtuse, patent in fruit; corolla 6–10 mm, white; capsule shallowly 2-lobed, ciliate; style *c.* 1·5 mm; seeds *c.* 2 × 2 mm. 2*n*=18, 36, 54. *Cultivated ground and other dry, open habitats. S. Europe.* Al Bl Bu Co Cr Ga Gr Hs It Ju Lu Rs (K) Sa Si.

V. panormitana Tineo ex Guss., *Fl. Sic. Prodr., Suppl.* 4 (1832), described from Sicilia, differs in its smaller fruit and flower, its shorter style and its subglabrous capsule. It may represent a distinct subspecies, but plants with some of these characters occur elsewhere in the Mediterranean region.

Sect. PSEUDOLYSIMACHIUM Koch. Perennial. Flowers in dense, many-flowered terminal racemes, spikes or panicles. Corolla-tube longer than wide. Capsule ovoid to suborbicular, only slightly compressed, scarcely emarginate, dehiscing loculicidally only at apex. Seeds *c.* 1 mm, relatively thick, smooth, with convex back, flat chalazal face, and excentric chalaza.

Most of the possible hybrids in this section have been artificially produced, and those from parents with the same chromosome number are highly fertile.

59. V. longifolia L., *Sp. Pl.* 10 (1753) (*Pseudolysimachium longifolium* (L.) Opiz; incl. *Veronica maritima* L.). Stem 40–120 cm, robust, erect, usually simple below, glabrous or puberulent. Leaves 30–120 × 5–20 mm, lanceolate to linear-lanceolate, acuminate, truncate to cuneate at base, acutely biserrate, glabrous or sparsely hairy, opposite or in whorls of 3 or 4; petiole usually less than 10 mm. Inflorescence up to 25 cm, very dense, usually with one or more small lateral branches; pedicels 1–2 mm, shorter than the linear-filiform bracts. Calyx-segments 2–3 mm, triangular-ovate, acute, unequal; corolla 6–8 mm in diameter, lilac or pale blue. Capsule *c.* 3 × 3 mm, broadly ovoid to subglobose, emarginate, glabrous; style 4–10 mm. 2*n* = 34, 68. *Riverbanks, woods, and other damp places. N., E. & C. Europe, extending westwards to Belgium and southwards to Bulgaria.* Au Be Bu Cz Da Fe Ga Ge Ho Hu It Ju No Po Rm Rs (N, B, C, W, E) Su.

Very variable, especially in size and leaf-shape.

60. V. paniculata L., *Syst. Nat.* ed. 10, **2**: 849 (1759) (*V. spuria* auct., non L., *Pseudolysimachium paniculatum* (L.) Hartl). Like **59** but more freely branched; leaves cuneate at base; inflorescence less dense, with several subequal branches; pedicels (2–)3–5 mm, longer than bracts; calyx-segments lanceolate, subequal; corolla 7–14 mm in diameter, blue; capsule 3–4 × 2–3 mm, broadly ovoid. 2*n* = 34. *Grassy and rocky places. C., E. & S.E. Europe.* Au Bu Cz †Ge Hu ?It Ju Po Rm Rs (N, B, C, W, E).

61. V. bachofenii Heuffel, *Flora (Regensb.)* **18**: 253 (1835) (*V. grandis* auct., vix Fischer). Like **59** but leaves triangular-ovate, cordate or subcordate at base, all opposite; inflorescence less dense; calyx-segments linear-lanceolate, subequal; capsule broadly obcordate. *Rocky woods. E. Carpathians and mountains of N. part of Balkan peninsula.* Bu Ju Rm Rs (W).

62. V. spicata L., *Sp. Pl.* 10 (1753) (*Pseudolysimachium spicatum* (L.) Opiz). Stems 5–60 cm, erect or ascending from a more or less extensive rhizome, usually hairy and glandular. Leaves 20–80 × 3–30 mm, usually shortly petiolate, linear- to ovate-lanceolate, cuneate at base, subentire to crenate, more or less hairy, opposite. Inflorescence up to 30 cm; pedicels usually less than 1 mm, much shorter than bracts. Calyx-segments usually ovate-elliptical, obtuse; corolla 4–8 mm in diameter, blue. Capsule 2–4 × 2–4 mm, emarginate, glabrous; style 4–10 mm. *Dry grassland and rocky slopes. Much of Europe, but rare in the west and absent from most of the islands.* Al Au Br Bu Cz Da Fe Ga Ge He Hs Hu It Ju No Po Rm Rs (N, B, C, W, K, E) Su Tu [Ho].

1 Plant densely greyish-white-tomentose **(b) subsp. incana**
1 Plant green, sparsely or moderately hairy
  2 Lower lobes of corolla linear, twisted **(d) subsp. orchidea**
  2 Lower lobes of corolla ovate-lanceolate, flat
    3 Petioles of lower leaves 10–20 mm; leaves glabrous or sparsely ciliate near the base **(e) subsp. crassifolia**
    3 Petioles of lower leaves rarely more than 10 mm; leaves usually more or less hairy
      4 Calyx glabrous except for stiff marginal cilia **(c) subsp. barrelieri**
      4 Calyx more or less hairy, usually glandular **(a) subsp. spicata**

(a) Subsp. **spicata** (incl. *V. hybrida* L.): Stem more or less densely pubescent, usually glandular (rarely glabrous or subglabrous). Leaves linear-lanceolate to ovate, subsessile or the lower with petioles not more than 10 mm. Calyx more or less

hairy, usually glandular; corolla clear blue, with ovate-lanceolate, flat lower lobes. 2*n* = 34, 68. *Throughout the range of the species.*

All plants from N. & N.W. Europe hitherto investigated are tetraploid, but diploid plants are known from C. & S.E. Europe. **V. euxina** Turrill, *Jour. Bot. (London)* **63**: 161 (1925), described from Bulgaria, is one of these diploids; it is densely glandular-hairy, with ovate-elliptical, subsessile lower leaves. It is not yet possible to distinguish the diploid taxa satisfactorily from the widespread and variable tetraploids.

Natural hybrids between **62(a)** and **59** are found occasionally in districts where both parents grow; they are particularly well-known in the Baltic area.

(b) Subsp. **incana** (L.) Walters, *Bot. Jour. Linn. Soc.* **65**: 269 (1972) (*V. incana* L.): Like subsp. (a) but whole plant densely greyish-white tomentose and eglandular. 2*n* = 68. *E. & E.C. Europe.*

(c) Subsp. **barrelieri** (Schott ex Roemer & Schultes) Murb., *Lunds Univ. Årsskr.* **27**(5): 73 (1892) (*V. barrelieri* Schott ex Roemer & Schultes; incl. *V. hololeuca* Juz., *V. andrasovszkyi* Jáv.): Like subsp. (a) but stem and leaves with long, stiff, whitish, eglandular hairs only (rarely glabrous); petiole of lower leaves up to 10 mm; calyx ciliate but otherwise glabrous. 2*n* = 34. *S.E. Europe, extending westwards to N. Italy.*

(d) Subsp. **orchidea** (Crantz) Hayek in Hegi, *Ill. Fl. Mitteleur.* **6**(1): 46 (1913) (*V. orchidea* Crantz, *Pseudolysimachium spicatum* subsp. *orchideum* (Crantz) Hartl): Stem subglabrous below, more or less glandular-hirsute above. Leaves thick and rather shiny, subglabrous or sparsely hairy, the lower ovate to ovate-lanceolate, with petiole *c.* 1 mm. Calyx glandular-hairy; corolla pale blue, with lower lobes linear, twisted. 2*n* = 34. *S.E. Europe, extending westwards to N. Italy and C. Czechoslovakia.*

(e) Subsp. **crassifolia** (Nyman) Hayek, *Prodr. Fl. Penins. Balcan.* **2**: 157 (1929): Like subsp. (d) but stem often glabrous except in inflorescence; leaves glabrous, the lower with petiole up to 20 mm; calyx ciliate but otherwise glabrous; corolla bright blue with ovate-lanceolate, flat lobes. ● *S. Carpathians and mountains of N. part of Balkan peninsula.*

V. urumovii Velen., *Österr. Bot. Zeitschr.* **51**: 31 (1901), from Bulgaria, is eglandular, with ovate-elliptical, setulose leaves, and calyx ciliate but otherwise glabrous. It seems to be most closely related to subsp. (e) but requires further investigation.

22. Hebe Commerson[1]

Like *Veronica* but erect shrubs; leaves opposite, entire or subentire; flowers in opposite axillary racemes; capsule septicidal, compressed parallel to the septum. Terminal buds large, without scale-leaves.

A large genus, almost confined to New Zealand; several species are cultivated in the milder parts of Europe. Some species not represented in Europe differ from the above description in several characters, and the genus is kept distinct from *Veronica* mainly on account of its characteristic habit.

1 Leaves narrowly lanceolate, 5–8 times as long as wide **1. salicifolia**
1 Leaves elliptical to obovate-oblong, 2–4 times as long as wide
  2 Young shoots and mature leaves glabrous; racemes dense, many-flowered **2. speciosa**
  2 Young shoots pubescent; mature leaves ciliate; racemes lax, with *c.* 10 flowers **3. elliptica**

1. H. salicifolia (G. Forster) Pennell, *Rhodora* **23**: 39 (1921). Up to 250 cm, of rather straggling habit. Young twigs glabrous. Leaves 5–15 × 1–2·5 cm, narrowly lanceolate, herbaceous, puberulent on midrib and petiole. Racemes 10–20 cm, slender, rather

---

[1] By D. A. Webb.

dense. Calyx 2 mm. Corolla white, sometimes tinged with lilac; lobes 3–4 mm, ovate-lanceolate, acute, suberect. Capsule 3·5 × 2·5 mm. *Cultivated for ornament and locally naturalized on walls and sea-cliffs in W. Europe.* [Az Br Hb.] (*New Zealand, Chile.*)

2. H. speciosa (R. Cunn. ex A. Cunn.) Andersen, *Trans. Proc. N.Z. Inst.* **56**: 693 (1926). Up to 150 cm, of compact habit. Young twigs glabrous. Leaves 5–10 × 2·5–5 cm, obovate-oblong, coriaceous to fleshy, glabrous. Racemes 7–10 cm, stout, very dense. Calyx 2 mm. Corolla violet to reddish-purple; lobes *c.* 4 mm, broadly ovate, obtuse, erecto-patent. Capsule 5–6 × 3–4 mm. *Cultivated for ornament, and for hedges near the sea; locally naturalized on sea-cliffs in Ireland.* [?Br Hb.] (*New Zealand.*)

Numerous cultivars exist, some of them probably of hybrid origin.

3. H. elliptica (G. Forster) Pennell, *Rhodora* **23**: 39 (1921) (*Veronica elliptica* G. Forster). Up to 80 cm, of compact habit. Young twigs pubescent. Leaves 1–3 × 0·5–1·5 cm, elliptical, coriaceous to fleshy, ciliate, pale greyish-green. Racemes 3–5 cm, lax, with *c.* 10 flowers. Calyx 2·5–4 mm. Corolla white with purple veins; lobes *c.* 5 mm, elliptical, subacute, more or less patent. Capsule 6–8 × 4–5 mm. *Naturalized on rocks in N.W. France (islands off Finistère).* [Ga.] (*New Zealand to Falkland Islands.*)

H. lewisii (Armstrong) Cockayne & Allan, *Trans. Proc. N.Z. Inst.* **57**: 21 (1927), which is perhaps naturalized in England, is probably a hybrid between **1** and **3**.

23. Paederota L.[1]

Perennial herbs; leaves opposite, undivided. Flowers in terminal, bracteate, spike-like racemes. Calyx somewhat unequally 5-lobed. Corolla with cylindrical tube and 2-lipped limb, usually somewhat shorter than the tube; upper lip entire (rarely 2-lobed), erect; lower lip 3-lobed, more or less patent. Stamens 2, exserted from corolla-tube, but sometimes shorter than limb. Stigma capitate. Capsule loculicidal, not laterally compressed; seeds numerous.

Flowers blue; leaves with not more than 9 teeth on each side
               1. bonarota
Flowers yellow; larger leaves with at least 10 teeth on each side **2. lutea**

1. P. bonarota (L.) L., *Sp. Pl.* ed. 2, 20 (1762) (*Veronica bonarota* L.). Stems 7–20 cm, numerous, simple, crispate-pubescent. Leaves 15–30 × 7–20 mm, subsessile or shortly petiolate, hirsute to subglabrous, with 3–6(–9), usually subobtuse teeth on each side, the lower broadly ovate to rhombic or suborbicular, the upper ovate to oblong-lanceolate. Racemes 2–4 cm in flower, dense, lengthening somewhat in fruit and usually glabrescent. Corolla 10–13 mm, violet-blue (rarely pink); tube slightly exceeding limb. Stamens equalling or slightly exceeding corolla. 2*n* = 36. *Crevices of limestone rocks.* ● *E. Alps, westwards to* 10° 15′ *E.* Au It Ju.

2. P. lutea Scop., *Annus Hist.-Nat.* **2**: 41 (1769) (*Veronica lutea* (Scop.) Wettst.). Like **1** but usually less hairy; stems sometimes longer; leaves narrowly ovate to lanceolate, rarely broadly ovate; teeth at least 10 on each side on larger leaves, usually more acute; corolla yellow, with tube distinctly longer than limb; stamens usually shorter than corolla (but sometimes longer). 2*n* = 36, *c.* 54. *Crevices of limestone rocks.* ● *E. Alps, westwards to* 11° 30′ *E.; mountains of W. Jugoslavia.* Au It Ju.

24. Sibthorpia L.[1]

Perennial herbs; stems procumbent, rooting at the nodes. Leaves alternate, petiolate, more or less orbicular. Flowers axillary, pedicellate, 4- to 8-merous. Corolla rotate, with short tube and entire, subequal lobes. Stamens equal in number to, or 1 or 2 less than the corolla-lobes. Capsule loculicidal. Seeds oblong-ovoid, reticulate or smooth.

Literature: O. Hedberg, *Bot. Not.* **108**: 161–183 (1955).

1 Pedicel more than 35 mm; flowers usually several in each leaf-axil; corolla at least 9 mm in diameter   **3. peregrina**
1 Pedicel not more than 35 mm; flowers solitary in leaf-axils; corolla not more than 7 mm in diameter
 2 Pedicel shorter than subtending petiole; corolla less than 3 mm in diameter, white, cream or pinkish **1. europaea**
 2 Pedicel usually longer than subtending petiole; corolla more than 3 mm in diameter, yellow **2. africana**

1. S. europaea L., *Sp. Pl.* 631 (1753) (*S. africana* auct., non L.). Stems up to 40 cm, slender, usually rather sparingly branched and far-creeping. Leaves 8–25 mm wide, reniform to orbicular, crenate, with 7–13 usually truncate crenations, more or less pubescent with hairs less than 1 mm; petiole (3–)5–40(–120) mm. Flowers solitary in leaf-axils; pedicel 1–5(–12) mm, straight. Corolla and calyx (4–)5-lobed; corolla 1·5–2·5 mm in diameter, white or cream, more or less tinged with pink, especially on lower lobes; lobes slightly unequal. Stamens (3–)4–5, the number variable on a single plant. Seeds 0·5 mm. 2*n* = 18. *Damp, shady places. W. Europe, northwards to* 52° N. *in Britain; mountains of Greece and Kriti.* Az Br Cr Ga Gr Hb Hs Lu.

2. S. africana L., *loc. cit.* (1753). Like **1** but stems usually more freely branched, so that the plant often forms a dense mat; leaves not more than 15 mm wide, with obtuse to subacute crenations; leaves and calyx grey-hirsute with hairs more than 1 mm; petiole 5–25(–50) mm; pedicel 8–20(–35) mm, usually coiled in fruit; corolla 4–7 mm in diameter, yellow; seeds 0·8 mm. *Shady walls and damp rock-crevices.* ● *Islas Baleares.* Bl.

The epithet *africana* was based on false information; the species which grows on the high mountains of tropical Africa is *S. europaea.*

3. S. peregrina L., *loc. cit.* (1753). Stems up to 80 cm, far-creeping. Leaves 15–50 mm wide, reniform to orbicular, with 17–35 usually subacute crenations, shortly hirsute with hairs more than 1 mm; petiole 15–65 mm. Flowers 1–3(–6) in each axil; pedicel 35–70 mm, usually somewhat coiled or curved in fruit. Calyx and corolla 5- to 8-lobed; corolla 9–13 mm in diameter, yellow; lobes more or less equal. Stamens equal in number to corolla-lobes. Seeds *c.* 1·2 mm. *Naturalized in Portugal (Sintra).* [Lu.] (*Madeira.*)

25. Lafuentea Lag.[1]

Perennial herbs; stem woody at the base; leaves opposite, simple. Flowers zygomorphic, in long, terminal, bracteate spikes. Calyx 5-lobed; corolla with long, cylindrical tube and 5-lobed, somewhat 2-lipped limb; upper lip 2-lobed, the lower 3-lobed. Stamens 4, didynamous, exserted. Capsule septicidal; seeds numerous.

1. L. rotundifolia Lag., *Gen. Sp. Nov.* 19 (1816). Plant densely greyish-pubescent to -hirsute, glandular, aromatic. Flowering stems erect, from a procumbent, woody stock. Leaves up to 3 × 3 cm, reniform to orbicular-deltate, irregularly crenate-dentate. Spikes up to 18 cm, with very numerous flowers; bracts

---

[1] By D. A. Webb.

linear-oblong, shorter than flowers. Calyx 4–6 mm, divided almost to the base, with linear-lanceolate lobes. Corolla 7–8 mm, white, striped with purple. Capsule 5 mm, ellipsoid, pubescent above; style persistent, equalling capsule. *Rock-crevices and under overhanging rocks.* ● *Lowlands of S. Spain, from Málaga to Orihuela.* Hs.

26. Castilleja Mutis ex L. fil.[1]

Perennial herbs, probably hemiparasitic, with alternate, simple, sessile leaves. Flowers zygomorphic, in dense, terminal, bracteate, spike-like racemes; bracts large, coloured. Calyx tubular, laterally compressed, 2- to 4-lobed. Corolla with a long tube and 2-lipped limb; lips subparallel, the upper laterally compressed, entire, the lower shorter, 3-lobed. Stamens 4, didynamous, included under upper lip of corolla; anthers with one lobe medifixed, the other pendent. Stigma capitate. Capsule loculicidal; seeds numerous.

Literature: O. V. Rebristaja, *Nov. Syst. Pl. Vasc.* (*Leningrad*) **1964**: 283–311 (1964).

1　Corolla 20–22 mm, the lower lip nearly as long as the upper; capsule 8–10 mm **2. schrenkii**
1　Corolla 22–30 mm, the lower lip *c.* ⅔ as long as the upper; capsule 10–15 mm
2　Densely grey-pubescent; stems 30–50 cm; bracts ovate-lanceolate, the lower usually ±entire **1. pallida**
2　Sparsely hairy, not grey; stems 10–30 cm; bracts ovate, all ±pinnatifid **3. lapponica**

1. C. pallida (L.) Sprengel, *Syst. Veg.* **2**: 774 (1825). Densely greyish-pubescent. Stems 30–50 cm, simple, erect. Leaves 30–70 × 2–7 mm, linear to lanceolate, acute, entire. Racemes 4–6 cm in flower, elongating to 9–12 cm in fruit; bracts up to 25 × 10 mm, ovate-lanceolate, pale yellow, the lower entire, the upper pinnatifid. Calyx 14–18 mm. Corolla 23–30 mm, pale yellow; lower lip much shorter than the upper. Style scarcely exserted. Capsule 12–15 mm, oblong-ovoid, acute, glabrous. *Meadows and forest-margins.* S. Ural. Rs (C). (*Siberia, N.E. Asia.*)

2. C. schrenkii Rebr., *Nov. Syst. Pl. Vasc.* (*Leningrad*) **1964**: 293 (1964). Like **1** but rather sparsely pubescent and not greyish; stems 20–30 cm; racemes 2·5–3 cm in flower, *c.* 5 cm in fruit; corolla 20–22 mm, with lower lip only slightly shorter than the upper; capsule 8–10 mm. *Tundra.* ● *N.W. Russia* (*near Lovozero, Arkangel'sk. Obl.*). Rs (N).

Perhaps only subspecifically distinct from **1**.

3. C. lapponica (Gand.) Rebr., *op. cit.* 291 (1964) (*C. arctica* auct. eur., non Krylov & Serg.). Sparsely hairy. Stems 10–30 cm, simple, ascending or erect. Leaves 30–50 × 4–15 mm, linear-lanceolate to ovate, entire or pinnatifid. Racemes 4–8 cm in flower, elongating to *c.* 12 cm in fruit; bracts 20–30 × 8–20 mm, ovate, pale yellow, all more or less pinnatifid. Calyx 15–17 mm, hirsute. Corolla 22–25 mm, pale yellow. Style distinctly exserted. Capsule 10–15 mm. $2n = 48$. *Tundra, river-banks and stony slopes.* Arctic Russia. Rs (N).

27. Melampyrum L.[2]

Hemiparasitic annuals. Leaves opposite, undivided, usually entire. Flowers in terminal, bracteate spikes or racemes. Calyx tubular, not inflated, with 4 acute, entire teeth. Corolla 2-lipped; tube cylindrical to obconical; upper lip cucullate, usually with margins more or less recurved; lower lip with 3 short, entire lobes. Stamens 4, didynamous, included under upper lip of corolla; anther-lobes equally mucronate. Capsule loculicidal, compressed at right angles to the septum; seeds 1–4, smooth.

In this and related genera the leaves on the main stem which lie between the uppermost branches and the lowest flowers of the terminal inflorescence are known as *intercalary leaves*.

The species of this genus, as of some other genera of hemiparasitic Scrophulariaceae, exhibit a characteristic type of variation, which, since it is to some extent correlated with variation in the season of germination and flowering, was originally described as seasonal dimorphism; but as many species have more than two variants, and since habitat-factors unconnected with seasonal variation have been shown to be relevant, it is now sometimes referred to as 'pseudoseasonal polymorphism'. The variation appears to be mainly eoctypic in nature; its seasonal aspect arises largely from the selection of early-flowering varieties in meadows which are cut for hay in midsummer. These variants have been very differently treated by different authors, being regarded sometimes as species, sometimes as subspecies, varieties or forms; and in many cases the information available is insufficient to distinguish between true ecotypic variation and purely phenotypic modification. The non-committal term 'ecotypic variant' is therefore used. Since these variants cut across the normal pattern of geographical speciation and subspeciation it is not possible – at least until more experimental work has been done – to accommodate them satisfactorily in any taxonomic categories; on the other hand the constant nature of the intraspecific variation throughout the genus demands some recognition. The principal recurrent variants are, therefore, designated and described below for the genus as a whole, and under each species or subspecies are listed such of these variants as have been recognized within it, together with the basionym (at whatever rank this may happen to be) under which each was first validly described. Such citation does not imply approval of the appropriateness of the rank in question.

In *Melampyrum* the recurrent ecotypic variants are as follows:

(1) *Autumnal.* Branches numerous, patent or arcuate; internodes numerous, short; intercalary leaves numerous. Late-flowering. Mainly in woods and scrub.

(2) *Aestival.* Branches 0–2 pairs, short, non-flowering, erect; internodes few, long; intercalary leaves absent; cotyledons often present at anthesis; flowers often larger than in (1). Early-flowering. Mainly in meadows.

(3) *Montane.* Branches 1–2 pairs, suberect, flowering; internodes numerous, fairly long; intercalary leaves 0–2 pairs. Leaves often fleshy. Mid-season-flowering. Mainly in mountain pastures.

(4) *Segetal.* Like (3) but with shorter internodes and not more than 1 pair of intercalary leaves. Mid-season-flowering. In cornfields.

Species which do not exhibit ecotypic variation usually conform in their morphology to that of the autumnal variant.

Literature: G. Beauverd, *Mém. Soc. Phys. Hist. Nat. Genève* **38**: 291–657 (1916). C. E. Britton, *Trans. Proc. Bot. Soc. Edinb.* **33**: 357–379 (1943). J. Jalas, *Ann. Bot. Fenn.* **4**: 486–526 (1967). A. J. E. Smith, *Watsonia* **5**: 336–367 (1963). R. de Soó, *Feddes Repert.* **23**: 159–176, 383–397; **24**: 127–193 (1926–1927).

1　Inflorescence dense, not secund; bracts densely imbricate, concealing the axis (Sect. *Spicata*)
2　Bracts cordate, folded along the midrib and with apex deflexed; inflorescence distinctly 4-angled **1. cristatum**
2　Bracts not cordate, ±flat, with apex seldom deflexed; inflorescence cylindrical or obscurely 4-angled

---

[1] By A. O. Chater and D. A. Webb.
[2] By R. de Soó and D. A. Webb.

3 Lower lip of corolla with upturned margins; throat of corolla-
 tube closed
 4 Calyx-tube glabrous except for strigulose veins; calyx-teeth
  conspicuously ciliate **2. ciliatum**
 4 Calyx-tube uniformly strigulose; calyx-teeth not conspicuously ciliate **3. arvense**
3 Lower lip of corolla with deflexed margins; throat of corolla-
 tube open
 5 Calyx densely villous **4. barbatum**
 5 Calyx glabrous or puberulent
  6 Teeth of bracts bearing long, rigid hairs; calyx-teeth about
   equalling the tube; corolla with white tube and pink
   lips **5. variegatum**
  6 Teeth of bracts pubescent; calyx-teeth 1½–2 times as long
   as the tube; corolla pink to violet, with white or yellow
   lower lip **6. fimbriatum**
1 Inflorescence ± lax, secund; bracts scarcely imbricate, not
 concealing the axis (Sect. *Melampyrum*)
 7 Bracts pectinate, with teeth ending in a flexuous arista
  8–10 mm **20. heracleoticum**
 7 Bracts entire to fimbriate, without long-aristate teeth
  8 Calyx with some hairs more than 0·5 mm, at least on the
   veins
   9 Calyx-tube with some long hairs between the veins, as well
    as on them
    10 Hairs uniformly oriented on all parts of calyx **7. nemorosum**
    10 Hairs on distal part of calyx directed towards the
     apex, those on proximal part deflexed
      (9–12). velebiticum group
   9 Calyx-tube with long hairs confined to the veins; remainder
    of surface puberulent or glabrous
    11 Calyx-teeth not more than 1½ times as long as tube
       14. subalpinum
    11 Calyx-teeth 2–3 times as long as tube
     12 Corolla 10–12(–17) mm, with tube distinctly curved or
      inflected **18. scardicum**
     12 Corolla (12–)16–24 mm; tube ± straight
        (15–17). bihariense group
  8 Calyx glabrous or puberulent, with hairs not more than 0·25
   mm
   13 At least the upper bracts violet-blue
    14 Hairs on calyx patent **8. polonicum**
    14 Hairs on distal part of calyx directed towards the apex,
     those on proximal part deflexed **(9–12). velebiticum** group
   13 Bracts green (rarely whitish)
    15 Calyx-teeth 11–14 mm **19. trichocalycinum**
    15 Calyx-teeth not more than 5 mm
     16 Throat of corolla usually ± closed; capsule dehiscing
      along one margin only **24. pratense**
     16 Throat of corolla open; capsule dehiscing along both
      margins
      17 Lips of corolla shorter than tube, the lower lip porrect
        13. bohemicum
      17 Lips of corolla equalling tube, the lower lip deflexed
        (21–23). sylvaticum group

Sect. SPICATA Wettst. Flowers in dense spikes, which are not
secund; bracts densely imbricate, concealing the spike-axis.
Leaves sessile.

1. M. cristatum L., *Sp. Pl.* 605 (1753). Stem 15–50 cm. Leaves
3–15 mm wide, linear-lanceolate to lanceolate, entire or toothed.
Spike 4-angled; bracts folded along midrib, with ovate-cordate,
pectinate base and long, entire, deflexed apex, yellowish-green,
the basal part variably tinged with purplish-red. Calyx 5–8 mm;
tube glabrous except for 2 hirsute veins; teeth 1·5–4 mm, unequal.
Corolla 12–16 mm, pale yellow, tinged at least on lower lip with
purple; palate deeper yellow; throat closed. Capsule dehiscing
along one margin only. $2n=18$. *Much of Europe, but absent from
parts of the north and south.* Al Au Be Br Bu Co Cz Da Fe Ga
Ge He Hs Hu It Ju No Po Rm Rs (N, B, C, W, E) Su.

Ecotypic variants:
 Autumnal: type of the species.
 Aestival: **M. solstitialis** Ronniger in Dörfler, *Herb. Norm.* **48**: 247
(1907).
 Montane: **M. cristatum** var. **majus** Baumg., *Enum. Stirp. Transs.* **2**:
197 (1816).

2. M. ciliatum Boiss. & Heldr. in Boiss., *Diagn. Pl. Or. Nov.*
3(3): 176 (1856). Stem up to 40 cm, branched. Leaves 2–5 mm
wide, linear to linear-lanceolate, entire. Spike cylindrical; bracts
green, ovate-lanceolate, laciniate, with lanceolate, acuminate,
strongly ciliate segments. Calyx *c.* 8 mm; tube glabrous except
for strigulose veins; teeth 4–5 mm, setaceous, ciliate. Corolla
20–25 mm, yellow, sometimes tinged with purple, the lower lip
spotted with orange; lower lip with upturned margins; throat
closed. *Evergreen scrub.* ● *E.C. Greece (Olimbos).* Gr.

3. M. arvense L., *Sp. Pl.* 605 (1753) (incl. *M. argyrocomum*
Fischer ex Steudel). Stem 15–50 cm. Leaves 2–15 mm wide,
lanceolate, entire or toothed. Spike cylindrical; bracts green,
whitish or reddish-pink, ovate-lanceolate, erect, pinnatifid, not
folded. Calyx 15–18 mm; tube uniformly and densely strigulose;
teeth 9–12 mm, equal. Corolla 20–25 mm, purplish-pink, marked
with yellow at the throat and sometimes on the lower lip; lower
lip with upturned margins; throat closed. $2n=18$. *Europe, from
England and S.W. Finland southwards to N. Spain, C. Italy and
Turkey.* Al Au Be Br Bu Co Cz Da Fe Ga Ge He Ho Hs Hu It Ju
Po Rm Rs (B, C, W, E) Su Tu.

Ecotypic variants:
 Autumnal: **M. pseudobarbatum** Schur, *Verh. Mitt. Siebenb. Ver.
Naturw.* **4**: 56 (1853).
 Aestival: **M. semleri** Ronniger & Poeverlein, *Allgem. Bot. Zeitschr.*
13: 179 (1907).
 Montane: **M. arvense** subsp. **schinzii** Ronniger in Schinz & R. Keller,
Fl. Schweiz ed. 3, **1**: 580 (1909).
 Segetal: type of the species.

4. M. barbatum Waldst. & Kit. ex Willd., *Sp. Pl.* **3**: 198 (1800).
Stem 15–40 cm. Leaves 3–15 mm wide, lanceolate, entire or
toothed. Spike cylindrical; bracts green, whitish or reddish-
purple, ovate-lanceolate, with long, narrow teeth towards the
base and entire, acuminate apex. Calyx 8–12 mm; tube densely
villous; teeth aristate. Corolla 20–30 mm; lower lip with deflexed
margins; throat open. ● *E.C. Europe, N. part of Balkan
peninsula, Italy.* Al Au Cz ?Gr Hu It Ju Rm [Po].

(a) Subsp. **barbatum**: Calyx-teeth about equalling the tube, with
arista 2–3 mm. Bracts usually whitish. Corolla 20–25 mm, pale
yellow to white. *From E. Austria to E. Romania and N.E.
Jugoslavia.*

Ecotypic variants:
 Autumnal: subsp. **filarszkyanum** (Soó) Soó, *Feddes Repert.* **24**: 138
(1927).
 Aestival: subsp. **kitaibelii** (Soó) Soó, *op. cit.* 139 (1927).
 Segetal: type of the species.

(b) Subsp. **carstiense** Ronniger, *Mitt. Naturw. Ver. Steierm.* **54**:
292 (1918): Calyx-teeth twice as long as the tube, with arista
3–4 mm. Bracts green or purplish. Corolla 27–30 mm, pale
yellow or purplish. *N. part of Balkan peninsula, extending to S.
Austria; E. Italy.*

5. M. variegatum Huter, Porta & Rigo, *Österr. Bot. Zeitschr.*
57: 246 (1907). Stem 20–40 cm, with arcuate branches. Leaves
5–10 mm wide, lanceolate. Spike cylindrical; bracts with long,
setiform teeth. Calyx-tube glabrous, except for hairy veins;
teeth aristate, as long as the tube. Corolla 20–25 mm, with
whitish tube and pink lips; lower lip with deflexed margins;
throat open. ● *C. & S. Italy.* It.

6. M. fimbriatum Vandas, *Österr. Bot. Zeitschr.* **39**: 53 (1889). Stem 30–50 cm, with arcuate branches. Leaves 5–8 mm wide, narrowly lanceolate. Spike cylindrical; bracts fimbriate, green, with pubescent teeth. Calyx-tube puberulent; teeth $1\frac{1}{2}$–2 times as long as the tube. Corolla 20–30 mm; tube pink; upper lip often purplish; lower lip white or pale yellow, with deflexed margins; throat open. ● *W. Jugoslavia, Albania.* Al ?Gr ?It Ju.

Sect. MELAMPYRUM. Flowers in lax, more or less secund spikes or racemes; bracts not concealing the inflorescence-axis. Leaves often shortly petiolate.

7. M. nemorosum L., *Sp. Pl.* 605 (1753). Stem 15–50 cm. Leaves 8–40 mm wide, narrowly lanceolate to ovate-cordate, usually entire. Bracts usually violet-blue, at least towards apex, but sometimes green, purple or whitish, deeply toothed, hairy at the base. Calyx 8–10 mm; tube usually densely and uniformly villous; teeth 4–5 × 1 mm, patent, triangular-lanceolate, separated by acute sinuses. Corolla 15–20 mm, bright yellow; throat closed. *N. & C. Europe, westwards to Denmark and Switzerland, and extending south-eastwards to S. Russia.* Au Cz Da Fe Ge He Hu It Ju Po Rm Rs (N, B, C, W, ?E) Su.

(a) Subsp. **nemorosum**: Leaves usually at least 15 mm wide. Bracts ovate-cordate to ovate-lanceolate, usually violet-blue. $2n = 18$. *Throughout the range of the species.*

Ecotypic variants:
Autumnal: type of the species.
Aestival: **M. moravicum** H. Braun, *Österr. Bot. Zeitschr.* **34**: 422 (1884).
Montane: **M. nemorosum** subsp. **silesiacum** Ronniger, *Viert. Naturf. Ges. Zürich* **55**: 314 (1910).

(b) Subsp. **debreceniense** (Rapaics) Soó, *Feddes Repert.* **24**: 148 (1927): Leaves 8–12 mm wide. Bracts narrowly lanceolate, green. *Sandy soils.* ● *Hungary, W. Romania.*

8. M. polonicum (Beauverd) Soó, *Feddes Repert.* **24**: 156 (1927). Like **7** but calyx-tube puberulent or subglabrous, except sometimes for long hairs on veins and margin; calyx-teeth longer and narrower; bracts lanceolate, violet-blue. ● *Poland and W. part of U.S.S.R.* ?Ge Po Rs (B, C, W).

Ecotypic variants:
Autumnal: type of the species.
Aestival: subsp. **hayekii** Soó, *loc. cit.* (1927).

(9–12). M. velebiticum group. Leaves linear-lanceolate to lanceolate. Bracts violet-blue. Calyx puberulent to hirsute, with hairs between the veins as well as on them, those on the distal part of the calyx directed towards the apex, those on the proximal part deflexed (rarely subglabrous except for hairy veins). Corolla 14–22 mm, yellow.

```
1  Calyx-teeth 4–6 mm, patent; corolla 16–22 mm
2    Calyx puberulent; bracts mostly entire        9. velebiticum
2    Calyx hirsute; bracts toothed                10. vaudense
1  Calyx-teeth 6–9 mm, porrect; corolla 14–16 mm
3    Calyx hirsute                                11. catalaunicum
3    Calyx puberulent                             12. italicum
```

9. M. velebiticum Borbás, *Magyar Ind. Akad. Értesitő* **1882**: 9 (1882). Stem 25–50 cm, branched. Leaves 5–25 mm wide, lanceolate, entire. Bracts mostly entire, only the uppermost toothed. Calyx 7–8 mm, puberulent (rarely subglabrous except for veins); teeth 4–5 mm, triangular-lanceolate, patent. Corolla *c.* 20 mm; throat open. $2n = 18$. ● *S. & S.W. Alps; mountains of E. & S. France and of N.W. Jugoslavia.* Ga He It Ju.

Ecotypic variants:
Autumnal: var. **meridionale** J. Murr, *Deutsche Bot. Monatsschr.* **17**: 100 (1899).
Aestival: type of the species.

M. degenianum Soó in Jáv., *Magyar Fl.* 1008 (1925), from N.W. Jugoslavia, differs in having the lower bracts hastate-dentate and the upper entire.

10. M. vaudense (Ronniger) Soó, *Feddes Repert.* **24**: 153 (1927). Stem 20–50 cm, branched. Leaves 12–30 mm wide, lanceolate. Bracts deeply toothed. Calyx 8–10 mm, hirsute; teeth 4–6 mm, triangular-lanceolate, patent. Corolla 16–20 mm; throat closed. ● *S.W. Alps, Jura.* Ga He.

11. M. catalaunicum Freyn, *Flora (Regensb.)* **67**: 680 (1884). Stem 20–40 cm, branched. Leaves 6–10 mm wide, linear-lanceolate. Bracts ovate-lanceolate, toothed. Calyx 10–13 mm, hirsute; teeth 6–9 mm, linear-lanceolate, porrect. Corolla 14–16 mm; throat closed. ● *N.E. Spain.* Hs.

12. M. italicum Soó, *Feddes Repert.* **24**: 152 (1927). Stem 20–50 cm, simple or branched. Leaves 5–25 mm wide, lanceolate. Bracts ovate to lanceolate, hastate-dentate. Calyx 7–10 mm, puberulent; teeth 6–9 mm, linear, porrect. Corolla 14–16 mm; throat closed. ● *Mountains of Italy.* It.

Ecotypic variants:
Autumnal: type of the species.
Aestival: subsp. **markgrafianum** Soó, *op. cit.* 153 (1927).

13. M. bohemicum A. Kerner, *Sched. Fl. Exsicc. Austro-Hung.* **1**: 35 (1881). Stem 20–25 cm, freely branched. Leaves 2 mm wide, linear. Bracts 1·5–3 mm wide, green, the lower entire, the upper hastate-dentate. Calyx *c.* 8 mm, subglabrous; teeth 4 mm, linear, more or less patent. Corolla *c.* 15 mm, yellow; lips shorter than the tube, the lower lip porrect; throat open. ● *Czechoslovakia and Austria.* Au Cz.

14. M. subalpinum (Juratzka) A. Kerner, *op. cit.* **1**: 32 (1881) (incl. *M. angustissimum* G. Beck). Stem 20–40 cm, simple or branched. Leaves (2–)14–18 mm wide, linear to ovate-lanceolate. Bracts ovate-lanceolate, deeply toothed, violet-blue. Calyx 8–10 mm, glabrous or puberulent, but with longer hairs on veins and on margin; teeth 4–6 mm, linear, porrect. Corolla 18–20 mm, yellow; throat open. ● *E. Alps.* Au ?Cz ?Ge.

Ecotypic variants:
Aestival: **M. grandiflorum** A. Kerner, *op. cit.* **2**: 114 (1882).
Montane: type of the species.

(15–17). M. bihariense group. Leaves 6–30 mm wide, lanceolate to ovate-lanceolate. Bracts ovate-cordate, toothed, usually violet-blue, rarely green. Calyx 9–13 mm, glabrous or puberulent but with longer hairs on veins and on margin; teeth 6–10 mm, narrowly lanceolate, porrect. Corolla 12–24 mm, yellow; throat half-closed.

```
1  Calyx-teeth usually not more than 8 mm, separated by ± obtuse
     sinuses; calyx-tube without tuft of hairs; corolla 18–24 mm
                                                      15. bihariense
1  Calyx-teeth usually at least 8 mm, separated by ± acute sinuses;
     calyx-tube with tuft of hairs on upper side near the base
2    Corolla 16–22 mm, with a tuft of short hairs inside between the
       insertion of the two filaments on either side, but otherwise
       glabrous                                    16. hoermannianum
2    Corolla 12–18 mm, with tufts of short hairs as above, but also
       with a tuft of longer hairs on each side at the insertion of
       longer filaments                              17. doerfleri
```

15. M. bihariense A. Kerner, *Sched. Fl. Exsicc. Austro-Hung.* **1**: 35 (1881). Stem 5–50 cm. Calyx-teeth 6–8(–10) mm; sinuses more or less obtuse; calyx-tube without basal tuft of hairs. Corolla 18–24 mm, without hairs at insertion of filaments. ● *Romania, extending to N.E. Hungary, C. Jugoslavia and C. Bulgaria.* Bu Hu Ju Rm.

Ecotypic variants:
Autumnal: f. **roemeri** Ronniger, *Denkschr. Akad. Wiss. Math.-Nat. Kl.* (*Wien*) **94**: 181 (1918).
Aestival: subsp. **kuemmerlei** (Soó) Soó, *Feddes Repert.* **24**: 159 (1927).
Montane: type of the species.

M. ambiguum Soó, *Feddes Repert.* **24**: 150 (1927), from Romania, resembles **15** in the size and form of its calyx, but resembles **7** in its indumentum.

16. M. hoermannianum K. Malý, *Glasn. Muz. Bosni Herceg.* **11**: 141 (1899). Calyx-teeth (6–)8–10 mm; sinuses more or less acute; calyx-tube with tuft of hairs on upper side near the base. Corolla 16–22 mm, with tufts of short hairs inside between insertion of longer and shorter filaments. ● *N.W. part of Balkan peninsula.* Ju.

Ecotypic variants:
Autumnal: **M. bosniacum** Ronniger, *Mitt. Naturw. Ver. Steierm.* **54**: 294 (1918).
Aestival: type of the species.
Montane: **M. hoermannianum** subsp. **beckianum** K. Malý, *Glasn. Muz. Bosni Herceg.* **31**: 77 (1919).

17. M. doerfleri Ronniger, *Denkschr. Akad. Wiss. Math.-Nat. Kl.* (*Wien*) **94**: 180 (1918). Like **16** but corolla 12–18 mm, with, in addition to the tufts of hairs between insertions of filaments, two longer tufts at insertion of longer filaments. ● *Crna Gora and N. Albania.* Al Ju.

Ecotypic variants:
Autumnal: subsp. **montenegrinum** Janchen, *Österr. Bot. Zeitschr.* **68**: 271 (1919).
Montane: type of the species.

18. M. scardicum Wettst., *Biblioth. Bot.* (*Stuttgart*) **26**: 81 (1892). Stem 10–40 cm. Leaves 5–15 mm wide, ovate-lanceolate. Bracts lanceolate, acutely toothed, violet-blue. Calyx *c.* 8 mm, puberulent, with longer hairs on the veins; teeth about 3 times as long as tube, curved upwards. Corolla 10–12(–17) mm, yellow; tube curved downwards; throat closed. ● *C. & S. Jugoslavia, W. & S. Bulgaria.* ?Al Bu Ju.

Ecotypic variants:
Autumnal: subsp. **serbicum** Ronniger, *Viert. Naturf. Ges. Zürich* **55**: 315 (1910).
Aestival: subsp. **wettsteinii** Ronniger, *loc. cit.* (1910).
Montane: type of the species.

19. M. trichocalycinum Vandas, *Sitz.-Ber. Böhm. Ges. Wiss.* (*Math.-Nat. Kl.*) 1890(1): 275 (1890). Stem 20–50 cm, freely branched. Leaves 2–5 mm wide, linear to linear-lanceolate. Bracts linear, entire or with 1–2 teeth, green. Calyx 13–18 mm, glabrous except for hairy veins; teeth 11–14 mm, filiform. Corolla *c.* 16 mm, yellow; throat closed. ● *Bosna.* Ju.

20. M. heracleoticum Boiss. & Orph. in Boiss., *Fl. Or.* **4**: 482 (1879). Stem 25–50 cm, freely branched. Leaves 3–10 mm wide, linear to narrowly lanceolate. Bracts green, pectinate in proximal part, the teeth ending in a flexuous filiform arista 8–10 mm. Calyx 12–14(–18) mm, glabrous or puberulent, with long hairs on the veins; teeth 8–10 mm, filiform. Corolla 16–18 mm, yellow; throat closed. ● *Albania and S. Jugoslavia.* Al Ju.

(21–23). M. sylvaticum group. Leaves 2–12 mm wide, lanceolate to elliptic-lanceolate. Bracts green, entire or with 1 or 2 teeth at the base. Calyx 6–7 mm, puberulent, without longer hairs on the veins; teeth *c.* 2·5 mm, lanceolate. Corolla 6–14 mm, white to yellow, often variegated with purple; throat open.

1	Anthers 1·5–2·2 mm	**21. sylvaticum**
1	Anthers 2·5–4 mm	
2	Bracts not more than 25 × 4 mm; corolla mainly pale yellow	**22. herbichii**
2	Bracts up to 40 × 7 mm; corolla mainly white	**23. saxosum**

21. M. sylvaticum L., *Sp. Pl.* 605 (1753). Stem 10–40 cm. Bracts 20–60 × 3–7 mm. Corolla (6–)8–10(–11) mm, usually deep golden-yellow (rarely pale yellow or whitish), the lower lip often spotted with purple. Anthers 1·5–2·2 mm. $2n = 18$. ● *N. Europe, extending southwards in the mountains to the Pyrenees, C. Italy and S. Bulgaria.* Au Br Bu Cz Da Fe Ga Ge Hb He Hs Hu Is It Ju No Po Rm Rs (N, B, C, W) Su.

Ecotypic variants:
Autumnal: type of the species.
Aestival: **M. aestivale** Ronniger & Schinz in Schinz & R. Keller, *Fl. Schweiz* ed. 3, **1**: 489 (1909).
Montane: **M. intermedium** Ronniger & Schinz, *loc. cit.* (1909).

A number of variants from the higher mountains of C. Europe, representing more extreme forms of the montane ecotype, have also been described.

22. M. herbichii Wołoszczak, *Spraw. Kom. Fizyogr. Krakow.* **21**: 133 (1887). Stem 12–30 cm. Bracts 10–25 × 1·5–4 mm. Corolla (7–)9–11(–14) mm, clear, pale yellow, the lower lip often spotted with purple. Anthers 2·5–4 mm. ● *E. & S. Carpathians.* Po Rm Rs (W).

Ecotypic variants:
Autumnal: **M. sylvaticum** f. **csatoi** Soó, *Feddes Repert.* **24**: 174 (1927).
Aestival: **M. sylvaticum** subsp. **moeszianum** Soó, *op. cit.* 190 (1927).
Montane: type of the species.

23. M. saxosum Baumg., *Enum. Stirp. Transs.* **2**: 199 (1816). Stem 10–40 cm. Bracts 25–40 × 5–7 mm. Corolla 8–13 mm, white, the lower lip marked with purple and sometimes with orange. Anthers 2·5–4 mm. ● *E. Carpathians and Transylvania.* ?Cz Po Rm Rs (W).

Ecotypic variants:
Autumnal: type of the species.
Aestival: subsp. **javorkae** (Soó) Soó, *Feddes Repert.* **24**: 176 (1927).
Montane: subsp. **baumgartenii** (Soó) Soó, *loc. cit.* (1927).

24. M. pratense L., *Sp. Pl.* 605 (1753) (incl. *M. laciniatum* Koshewn. & V. Zinger). Stem 5–60 cm. Leaves 1–35 mm wide, linear to ovate. Bracts green, ovate- to linear-lanceolate, entire or hastate-dentate. Calyx 7–8 mm, glabrous except for strigulose veins; teeth 4–5 mm, linear. Corolla 10–18 mm, whitish to bright yellow, rarely with tube and upper lip tinged with purple; throat usually closed. Anthers 1·5–3·2 mm. $2n = 18$. *Most of Europe.* Au Be Br Bu Cz Da Fe Ga Ge Hb He Ho Hs Hu It Ju Lu No Po Rm Rs (N, B, C, W, K, E) Su.

An extremely variable species, which has been treated differently by every author who has studied it closely. Several variants have been recognized as subspecies on the strength of differences in the colour of the corolla and of the degree of closure of its throat (e.g. subsp. **purpureum** (Hartman) Soó, *Feddes Repert.* **24**: 183 (1927), and subsp. **hians** (Druce) Beauverd in Schinz & R. Keller, *Fl. Schweiz* ed. 3, **2**: 309 (1914)). It has been shown, however, that at least in some parts of Europe these distinctive characters are distributed sporadically through populations which are otherwise typical, and show no certain correlation with other characters. Other variants recognized as subspecies (e.g. subsp. **commutatum** (Tausch ex A. Kerner) C. E. Britton, *Trans. Proc. Bot. Soc. Edinb.* **33**: 371 (1943), and subsp. **alpestre** Ronniger, *Viert. Naturf. Ges. Zürich* **55**: 322 (1910)), although they may show a fairly well-defined geographical range, are distinguished mainly or entirely by characters which are used to define the ecotypic variants, so that a very confusing taxonomic situation results. It is also apparent that a large part of the infraspecific variation is clinal (usually from north to south), without any appreciable discontinuities. It seems best, therefore, in the present state of our knowledge not to give formal recognition to any infraspecific taxa, but to refer the reader to the literature cited.

28. Tozzia L.[1]

Perennial, hemiparasitic herbs with scaly rhizome; leaves opposite, undivided. Flowers in bracteate, axillary and terminal racemes. Calyx unequally 3- to 5-toothed. Corolla-tube obconical; limb 5-lobed, somewhat 2-lipped. Stamens 4, didynamous, shortly exserted. Stigma capitate, small. Capsule loculicidal or indehiscent; ovules 4; seed 1.

1. **T. alpina** L., *Sp. Pl.* 607 (1753). Rhizome covered by fleshy, whitish, ovate, imbricate scales. Stems up to 50 cm, erect, branched, 4-angled, usually hairy on the angles. Leaves 10–25 mm, ovate, sessile, rather fleshy, usually coarsely serrate or crenate near the base. Racemes numerous, short; pedicels short, filiform, elongating in fruit. Calyx 1·5–3 mm, campanulate; teeth short and broad, often obscure. Corolla 4–10 mm, yellow, spotted with purple inside; upper lip erect, lower patent. Anthers apiculate at base. *Damp places, among coarse herbs, mostly from 1000 to 2250 m; somewhat calcicole.* ● *Mountains of Europe, from the Jura and the Carpathians southwards to the Pyrenees, N. Appennini and Bulgaria.* Au Bu Cz Ga Ge He Hs It Ju Po Rm Rs (W).

(a) Subsp. **alpina**: Corolla 6–10 mm, bright golden-yellow; anthers 1–1·6 mm. 2n=20. *From the Pyrenees to Austria and W. Jugoslavia.*

(b) Subsp. **carpathica** (Wołoszczak) Dostál, *Květena ČSR* 1318 (1949) (*T. carpathica* Wołoszczak): Corolla 4–7 mm, pale yellow; anthers 0·6–1 mm. *Carpathians; E. part of Balkan peninsula.*

29. Euphrasia L.[2]

Annual (rarely perennial) hemiparasitic herbs. Leaves opposite, crenate to incise-dentate. Flowers zygomorphic, in bracteate, terminal, spike-like racemes. Bracts (*floral leaves*) large and leaf-like; bracteoles absent. Calyx tubular or campanulate, 4-lobed. Corolla white to purple, marked with deep violet lines, and with yellow spots on lower lip and throat; limb 2-lipped, the upper lip galeate, the lower longer, flat, patent, 3-lobed with emarginate lobes. Stamens 4, didynamous; loculi parallel, spurred, one loculus with a much longer spur than the other. Stigma capitate. Capsule loculicidal; seeds numerous.

Most of the species are very variable and weakly differentiated, and they hybridize readily; the taxonomy of the genus is, therefore, difficult. Populations frequently occur in which one character falls outside the normal range of variation and hybrids sometimes occur independently of the parents, as well as commonly with them. The most frequent hybrids are mentioned below.

Although with sufficient experience it is possible to assign the

great majority of populations to a species, it is extremely difficult to provide rigidly diagnostic descriptions and consequently to construct a key. The keys given here are designed to be used with population-samples, not with individual plants. It is therefore necessary to have a representative sample of a population taken from a small area, and to ignore extreme states of each character.

The keys make no allowance for extreme and relatively rare variants or for hybrids; otherwise it would have been necessary to prefix 'usually' to almost every phrase. Determinations should, therefore, be checked by reference to a good herbarium.

The most important taxonomic characters are length of internodes; node at which the lowest flower is situated; number of branches; shape, size and indumentum of leaves; size of corolla; size, shape and indumentum of capsule. In this account the presence or absence of eglandular or short glandular hairs is not regarded as very important; the presence or absence of long glandular hairs is, however, an important character.

The nodes are counted from the base, excluding the cotyledonary node; measurements of leaves and internodes refer to those on the main stem; 'lower floral leaves' indicates those of the lowest four flowering nodes. The length of the corolla is measured from the base of the tube to the apex of the upper lip in its normal condition *in vivo*, *i.e.* with the lobes erect or recurved. Measurement of dried specimens in which the lobes of the upper lip have become fully extended in pressing adds 0·5–1 mm to the length.

The name *E. officinalis* L. is variously used by modern authors as a 'collective species' to comprise many (or even nearly all) of the species described below. The type of *E. officinalis* is a mixture of **3** and **17**. P. D. Sell and P. F. Yeo, *Bot. Jour. Linn. Soc.* **63**: 201 (1970) have selected the specimen of **3** as lectotype, but reject the name *E. officinalis* as a *nomen ambiguum*.

Literature: R. von Wettstein, *Monographie der Gattung Euphrasia.* Leipzig. 1896. A. Chabert, *Bull. Herb. Boiss.* ser. 2, **2**: 121–152, 265–280, 497–520 (1902). E. Joergensen, *Bergens Mus. Aarb.* **1916–1917**, *Naturvid. Række* **5**: 1–337 (1919). H. W. Pugsley, *Jour. Linn. Soc. London (Bot.)* **48**: 467–544 (1930); *Jour. Bot. (London)* **74**: 273–288 (1936). M. Smejkal, *Biol. Práce Slovensk. Akad. Vied (Bratislava)* **9**(9): 1 83 (1963).

1 Perennial; leaves all ±similar, with up to 12 teeth on each side (Açores)
  2 Cauline leaves suborbicular; calyx not more than 7·5 mm; capsule obcordate   **1. grandiflora**
  2 Cauline leaves deltate; calyx c. 9·5 mm; capsule acuminate   **2. azorica**
1 Annual; lower leaves markedly different in shape from upper, none with more than 9 teeth on each side
  3 Middle and upper leaves bearing glandular hairs with multicellular stalk c. 10–12 times as long as the gland
    4 Capsule more than twice as long as wide   **11. arctica**
    4 Capsule not more than twice as long as wide   **(3–8). rostkoviana** group
  3 Middle and upper leaves without glandular hairs, or glandular hairs with stalk of 1–2 cells, not more than 6 times as long as the gland
    5 Anthers light yellowish-brown
      6 Corolla 7–11 mm   **22. christii**
      6 Corolla not more than 6 mm
        7 Corolla 3–3·5 mm; lower floral leaves with obtuse teeth   **38. bottnica**
        7 Corolla 4·5–6 mm; lower floral leaves with acute teeth   **40. taurica**
    5 Anthers dark brown or purplish
      8 Capsule glabrous, or with a few small cilia; at least some leaf-teeth distant

[1] By D. A. Webb.   [2] By P. F. Yeo.

9 Floral leaves usually entire except for a pair of teeth situated near the apex, not more than 1·5 mm long, and shorter than the width of the remainder of the leaf (S.E. Alps) **46. tricuspidata**
9 Some floral leaves with at least 2 teeth on each side, which are either more than 1·5 mm long or longer than the width of the remainder of the leaf
  10 Lower floral leaves with 1–2(–3) teeth on each side
    11 Corolla at least 8 mm **45. cuspidata**
    11 Corolla not more than 7·5 mm
      12 Lower floral leaves with linear terminal lobe and long, narrowly falcate, acute but not aristate teeth **44. dinarica**
      12 Lower floral leaves with lanceolate terminal lobe and long, narrow, aristate (rarely deltate, acute) teeth **41. salisburgensis**
  10 Lower floral leaves with (3–)4 or more teeth on each side
    13 Corolla 8–9 mm **42. portae**
    13 Corolla not more than 7·5 mm
      14 Teeth of lower floral leaves much longer than width of remainder of leaf; corolla purplish **43. illyrica**
      14 Teeth of lower floral leaves not much longer than width of remainder of leaf; corolla white **41. salisburgensis**
8 Capsule ciliate with long, fairly numerous hairs; all leaf-teeth usually contiguous
  15 Corolla more than 7·5 mm
    16 Lowest flower at node 1–3
      17 Cauline and lower floral leaves narrowly obovate to trullate or ovate, usually crenate-serrate **(17–20). stricta** group
      17 Cauline and lower floral leaves elliptical to suborbicular, usually crenate **9. picta**
    16 First flower at node 4 or higher
      18 Basal teeth of lower floral leaves apically directed
        19 Capsule not more than 2½ times as long as wide
          20 Corolla less than 9 mm **16. confusa**
          20 Corolla 9–11 mm
            21 Cauline leaves ovate to oblong-ovate; lower floral leaves usually with more than 3 pairs of teeth **9. picta**
            21 Cauline leaves linear-oblanceolate to narrowly trullate; lower floral leaves usually with 2–3 pairs of teeth **10. marchesettii**
        19 Capsule more than 2½ times as long as wide
          22 Stem and branches flexuous **16. confusa**
          22 Stem erect, with straight or regularly curved branches
            23 Corolla not more than 10 mm; branches ±erect **(17–20). stricta** group
            23 Corolla at least 10 mm; branches ±patent
              24 Lowest flower at node 9 or lower; stem with not more than 4 pairs of branches; capsule usually not more than 3 times as long as wide **21. alpina**
              24 Lowest flower at node 10 or higher; stem with up to 13 pairs of branches; capsule at least 3½ times as long as wide **23. cisalpina**
      18 Basal teeth of lower floral leaves patent
        25 Lowest flower at node 8 or lower; capsule usually elliptical to obovate
          26 Capsule at least twice as long as wide; leaves often with short glandular hairs **11. arctica**
          26 Capsule less than twice as long as wide; leaves without glandular hairs
            27 Capsule not more than 5·5 mm **9. picta**
            27 Capsule at least 6 mm **11. arctica**
        25 Lowest flower at node 9 or higher; capsule oblong to elliptic-oblong
          28 Stem and branches usually flexuous; leaves near base of branches usually very small
            29 Corolla less than 9 mm **16. confusa**
            29 Corolla at least 9 mm **9. picta**
          28 Stem and branches usually not flexuous; leaves near base of branches not much smaller than the others

30 Teeth of floral leaves acute to acuminate; capsule usually slightly shorter than calyx; stem and branches relatively stout **13. nemorosa**
30 Teeth of floral leaves mostly aristate; capsule much shorter than calyx; stem and branches relatively slender
  31 Capsule usually not more than twice as long as wide; corolla at least 9 mm **9. picta**
  31 Capsule usually about 2½ times as long as wide; corolla not more than 9 mm **15. pseudokerneri**
15 Corolla not more than 7·5 mm
  32 Calyx-tube whitish and membranous, with prominent green veins
    33 Capsule not more than twice as long as wide; leaves thin **35. calida**
    33 Capsule more than twice as long as wide; leaves thick **(28–32). ostenfeldii** group
  32 Calyx-tube green and herbaceous
    34 Lowest flower at node 6 or higher
      35 Cauline internodes mostly 2–6 times as long as the leaves
        36 Basal teeth of lower floral leaves apically directed
          37 Teeth of lower floral leaves obtuse to acute, fairly short; corolla not more than 6·5 mm
            38 Leaves strongly tinged with purple, not darker beneath than above; corolla usually lilac to purple; capsule shorter than calyx **33. micrantha**
            38 Leaves weakly or moderately tinged with purple, often darker beneath than above; corolla usually white; capsule at least as long as calyx **34. scottica**
          37 Teeth of lower floral leaves acute to aristate, long; corolla usually at least 7 mm
            39 Basal teeth of lower leaves strongly apically directed **(17–20). stricta** group
            39 Basal teeth of lower floral leaves only slightly apically directed
              40 Capsule not more than 3 times as long as wide **11. arctica**
              40 Capsule at least 3 times as long as wide **(17–20). stricta** group
        36 Basal teeth of lower floral leaves patent
          41 Corolla at least 6·5 mm
            42 Lowest flower at node 9 or higher; leaves usually without glandular hairs; lower floral leaves smaller than upper cauline **13. nemorosa**
            42 Lowest flower at node 8 or lower; leaves usually with glandular hairs; lower floral leaves larger than upper cauline **11. arctica**
          41 Corolla not more than 6·5 mm
            43 Leaves densely hairy
              44 Lowest flower at node 8 or lower; stem not more than 15 cm; lower floral leaves about as wide as long **(28–32). ostenfeldii** group
              44 Lowest flower at node 9 or higher; stem up to 40 cm; lower floral leaves often longer than wide **13. nemorosa**
            43 Leaves sparsely hairy to subglabrous
              45 Stem and branches very slender, blackish; leaves strongly tinged with purple, not darker beneath than above; corolla usually lilac to purple **33. micrantha**
              45 Stem and branches either stout or lightly pigmented; leaves weakly or moderately tinged with purple; corolla usually white
                46 Lowest flower at node 7 or lower; stem slender; leaves usually light green above and purplish beneath; capsule usually longer than calyx **34. scottica**
                46 Lowest flower at node 8 or higher; stem robust; leaves not darker beneath than above; capsule usually shorter than calyx **13. nemorosa**

35 Cauline internodes mostly not more than twice as long as leaves
  47 Basal teeth of lower floral leaves apically directed
    48 Teeth of lower floral leaves not much longer than wide   **(28–32). ostenfeldii** group
    48 Teeth of lower floral leaves much longer than wide
      49 Stem and branches flexuous   **16. confusa**
      49 Stem and branches ±straight and erect
        50 Floral leaves with not more than 3 teeth on each side   **40. taurica**
        50 Some floral leaves with more than 3 teeth on each side   **(17–20). stricta** group
  47 Basal teeth of lower floral leaves patent
    51 Lowest flower at node 10 or higher
      52 Stem erect, stout, with stout, ascending branches; lower floral leaves mostly opposite   **13. nemorosa**
      52 Stem and branches slender and flexuous; lower floral leaves mostly alternate   **16. confusa**
    51 Lowest flower at node 9 or lower
      53 Leaves with numerous eglandular hairs   **(28–32). ostenfeldii** group
      53 Leaves with few eglandular hairs
        54 Stem branched from near the base; branches ascending, curved at base, terminating in long, dense spikes similar to that on main stem   **12. tetraquetra**
        54 Stem not branched much below the middle; branches straight, erecto-patent, not terminating in long, dense spikes   **11. arctica**
34 Lowest flower at node 5(–6) or lower
  55 Cauline internodes mostly at least 2½ times as long as the leaves
    56 Capsule broadly elliptical to obovate-elliptical
      57 Cauline internodes up to 9 times as long as the leaves; lower floral leaves ovate to rhombic, with teeth all apically directed   **36. saamica**
      57 Cauline internodes rarely more than 4 times as long as the leaves; lower floral leaves ± broadly ovate to suborbicular, with basal pair of teeth ±patent
        58 Teeth of lower floral leaves mostly subacute and not longer than broad; corolla 4·5–7 mm; lowest flower at node 2–4(–5)   **25. frigida**
        58 Teeth of lower floral leaves usually acute or acuminate and longer than broad; corolla at least 6·5 mm; lowest flower usually at node 4 or higher   **11. arctica**
    56 Capsule oblong to narrowly elliptical
      59 Lower floral leaves deeply serrate to pectinate; basal teeth apically directed   **(17–20). stricta** group
      59 Lower floral leaves crenate or shallowly serrate; basal teeth usually patent
        60 Upper cauline leaves elliptic-ovate to narrowly obovate
          61 Leaves usually light green above and purplish beneath; branches long; capsule usually more than twice as long as wide   **34. scottica**
          61 Leaves ±concolorous; branches short; capsule not more than twice as long as wide **14. coerulea**
        60 Upper cauline leaves suborbicular to broadly ovate or broadly obovate
          62 Lowest flower at node 4 or lower; lower floral leaves often considerably larger than the upper cauline
            63 Stem straight, erect; cauline leaves distinctly petiolate, with distal teeth incurved; corolla rarely more than 5·5 mm, the lower lip usually yellow   **24. minima**
            63 Stem often flexuous; cauline leaves obscurely petiolate, with distal teeth not incurved; corolla up to 7(–8) mm, the lower lip white or purple   **25. frigida**
          62 Lowest flower at node 4 or higher; lower floral leaves scarcely larger than the upper cauline

            64 Leaves densely hairy, without distinct petiole   **(28–32). ostenfeldii** group
            64 Leaves usually sparsely hairy to subglabrous, with short but distinct petiole   **24. minima**
  55 Cauline internodes mostly less than 2½ times as long as the leaves
    65 Corolla at least 6 mm
      66 Teeth of lower floral leaves usually very acute, all apically directed
        67 Stem and branches slender, flexuous; lower floral leaves mostly alternate   **16. confusa**
        67 Stem and branches usually stout, rigid; lower floral leaves mostly opposite   **(17–20). stricta** group
      66 Teeth of lower floral leaves subacute to acute, the basal pair patent
        68 Capsule at least as long as the calyx, usually emarginate   **25. frigida**
        68 Capsule shorter than the calyx, truncate or slightly emarginate   **12. tetraquetra**
    65 Corolla not more than 6 mm
      69 Lower floral leaves ovate to rhombic, with acute to aristate teeth, the basal pair apically directed
        70 Leaves with very short glandular hairs; floral leaves with not more than 3 teeth on each side; capsule apiculate (Krym)   **40. taurica**
        70 Leaves usually eglandular; floral leaves often with more than 3 teeth on each side; capsule scarcely apiculate
          71 Corolla deep purple (Færöer)   **37. atropurpurea**
          71 Corolla white to lilac
            72 Stem and branches flexuous   **16. confusa**
            72 Stem and branches straight, erect
              73 Lowest flower at node 4 or lower **39. willkommii**
              73 Lowest flower at node 5 or higher   **(17–20). stricta** group
      69 Lower floral leaves broadly ovate or deltate to suborbicular, with obtuse to subacute teeth, the basal pair patent
        74 Leaves with numerous eglandular hairs and without glandular hairs
          75 Lower floral leaves scarcely larger than the upper cauline   **(28–32). ostenfeldii** group
          75 Lower floral leaves considerably larger than the upper cauline
            76 Stem straight, erect, usually with 1–3 pairs of branches; cauline leaves distinctly petiolate, with distal teeth incurved   **24. minima**
            76 Stem often flexuous, with not more than 2 pairs of branches and often simple; cauline leaves obscurely petiolate, with distal teeth not incurved   **25. frigida**
        74 Leaves with few eglandular hairs, sometimes with short glandular hairs
          77 Capsule elliptical to obovate, emarginate; corolla not more than 5·5 mm, with very small lower lip   **27. cambrica**
          77 Capsule oblong to elliptic-oblong, usually truncate; corolla up to 7 mm, with conspicuous lower lip
            78 Distal teeth of lower floral leaves usually incurved; capsule usually exceeding calyx   **26. foulaensis**
            78 Distal teeth of lower floral leaves not incurved; capsule usually shorter than calyx **12. tetraquetra**

Sect. ATLANTICAE Pugsley. Perennial. Leaves all more or less similar. Corolla 13–17 mm; tube infundibuliform.

1. E. grandiflora Hochst. in Seub., *Fl. Azor.* 39 (1844). Stem up to 40 cm, erect or ascending. Leaves usually blackening on drying, mostly suborbicular, with up to 12 pairs of usually obtuse to subacute teeth, usually glabrous above, scabrid or hairy on the veins beneath; lower leaves hairier than upper. Calyx 5–7 mm, glabrous, or villous in the sinuses. Corolla 13–16 mm; lilac. Capsule 6–7·5 mm, obcordate. *Shallow soils near lava-flows on mountains.* ● *Açores (central group).* Az.

2. **E. azorica** H. C. Watson, *London Jour. Bot. (Hooker)* 3: 598 (1844). Like **1** but stem usually shorter; leaves not blackening on drying, mostly deltate, with subacute to acute teeth, hispid beneath, strigulose-scabrid above; upper leaves usually hairier than lower; calyx 9·5 mm; corolla 15–17 mm, white; capsule 10 mm, cuspidate, twisted at maturity. *Mountains.* ● *Açores (western group).* Az.

Sect. EUPHRASIA. Annual. Leaves rapidly changing in shape from below upwards. Corolla not more than 13·5 mm; tube cylindrical.

Subsect. *Ciliatae* Joerg. Capsule densely ciliate with long, fine hairs.

(3–8). **E. rostkoviana** group. Leaves with multicellular glandular hairs having the stalk about 10 times as long as the gland. Teeth of middle leaves usually with both edges convex. Capsule not more than twice as long as wide, usually shorter than the calyx, truncate or more or less emarginate.

1 Corolla not more than 7 mm
  2 Lowest flower at node 5–8; branches usually 1–4 pairs **5. anglica**
  2 Lowest flower at node 2–6(–7); branches usually 0–2 pairs
    3 Corolla not more than 6 mm
      4 Cauline leaves usually 4–9 mm, broadly ovate to elliptical, with rounded to cordate (rarely cuneate) base; glandular hairs usually numerous **7. hirtella**
      4 Cauline leaves usually 2·5–8 mm, elliptical to elliptic-obovate, with cuneate base; glandular hairs often sparse **8. drosocalyx**
    3 Corolla more than 6 mm
      5 Plant not more than 10(–15) cm; lower floral leaves 3–6(–7) mm, shorter than the flowers **4. rivularis**
      5 Plant up to 20 cm; lower floral leaves 4–12 mm, usually at least as long as the flowers **7. hirtella**
1 Corolla more than 7 mm
  6 Lowest flower at node 2–5(–6)
    7 Corolla 9–12·5 mm; lower floral leaves 5–12(–20) mm **3. rostkoviana**
    7 Corolla not more than 9 mm; lower floral leaves not more than 6(–7) mm **4. rivularis**
  6 Lowest flower at node 5 or above
    8 Leaves dull greyish-green, ± strongly flushed and punctate with dull violet or black; corolla usually lilac to deep reddish-purple **6. vigursii**
    8 Leaves light or dark green, usually without much purple pigment; corolla usually with at least the lower lip white
      9 Stem usually erect, with erect or divergent branches; lower floral internodes mostly 1½–3 times as long as the leaves; corolla 8–12 mm **3. rostkoviana**
      9 Stem usually flexuous, with flexuous or arcuate branches; lower floral internodes mostly less than 1½ times as long as the leaves; corolla 8 mm or less **5. anglica**

3. **E. rostkoviana** Hayne, *Darst. Beschr. Arzn. Gewächse* 9: t. 7 (1825). Stem up to 35(–50) cm, erect, with up to 8(–12) pairs of erect or divergent branches. Cauline internodes mostly much longer than the leaves. Corolla 8–12·5 mm; lower lip white, the upper frequently lilac. *Throughout a large part of Europe, but absent from parts of the south and east and many of the islands.* Al Au Be Br Bu Cz Da Fe Ga Ge Hb He Ho ?Hs Hu It Ju Po Rm Rs (N, B, C, W) Su.

In C. Europe intergradation with **9** takes place, giving large populations of plants some with and some without long glandular hairs, but similar in other characters.

1 Lowest flower at node 2–6; cauline internodes 2–6(–10) times as long as the leaves **(a) subsp. montana**
1 Lowest flower usually at node 6 or higher; cauline internodes mostly not more than 3 times as long as the leaves
  2 Lowest flower usually at node 6–10; lower floral leaves up to 15 mm **(b) subsp. rostkoviana**
  2 Lowest flower usually at node 9–16; lower floral leaves usually not more than 9 mm **(c) subsp. campestris**

(a) Subsp. **montana** (Jordan) Wettst., *Denkschr. Akad. Wiss. Math.-Nat. Kl. (Wien)* 70: 319 (1901) (*E. montana* Jordan): Stem with 0–3(–4) pairs of branches. Cauline internodes 2–6(–10) times as long as the leaves. Lowest flower at node 2–6. Lower floral leaves up to 12(–20) mm, orbicular to broadly deltate. Corolla 9–12·5 mm. Capsule 4·5–5 mm, often longer than the calyx, sometimes emarginate. $2n = 22$. *Meadows and pastures, mainly in the mountains.* ● *Throughout most of the range of the species.*

(b) Subsp. **rostkoviana**: Stem with up to 5(–12) pairs of branches. Cauline internodes mostly not more than 3 times as long as the leaves. Lowest flower at node (3–)6–10(–14). Lower floral leaves up to 15 mm, deltate to oblong-ovate. Corolla (6·5–)8–12 mm. Capsule 4–5 mm, not longer than the calyx. $2n = 22$. *Meadows and pastures. Throughout the range of the species.*

Plants from Finland and N. Russia which have rather small, oblong, closely-toothed leaves and often fewer branches and smaller corolla (6·5–9 mm) than plants from C. Europe, have been distinguished as **E. fennica** Kihlman, *Meddel. Soc. Fauna Fl. Fenn.* 24: 92 (1901). Early-flowering variants from the same area, also with small corollas, have been further separated as **E. onegensis** Cajander, *Meddel. Soc. Fauna Fl. Fenn.* 27: 101 (1901). There is continuous gradation between these plants and more normal representatives of subsp. *rostkoviana*.

(c) Subsp. **campestris** (Jordan) P. Fourn., *Quatre Fl. Fr.* 785 (1937): Stem with 3–8 pairs of branches. Cauline internodes not more than 2½ times as long as the leaves. Lowest flower at node 9–16. Lower floral leaves usually not more than 9 mm, broadly ovate to trullate. Corolla 8–11 mm. Capsule 4–5 mm, not longer than the calyx. *Dry grassland.* ● *From Belgium to Italy.*

Probably of hybrid origin between **3(b)** and **17**.

4. **E. rivularis** Pugsley, *Jour. Bot. (London)* 67: 225 (1929). Stems up to 10(–15) cm, slender, with 0–2 pairs of short branches. Cauline internodes shorter than or up to *c.* 4 times as long as leaves. Lowest flower at node (2–)3–5(–6). Leaves 2–6(–7) mm, orbicular to elliptical, often with few glandular hairs, the cauline with 1–4 pairs of very obtuse teeth; lower floral leaves 3–5(–6) pairs of obtuse to acute teeth. Corolla 6·5–9 mm; upper lip lilac; lower lip white or pale lilac. Capsule 3·5–5 mm, usually much less than twice as long as wide. $2n = 22$. *Damp mountain pastures.* ● *Wales and N.W. England.* Br.

5. **E. anglica** Pugsley, *loc. cit.* (1929). Stem up to 20 cm, usually flexuous, with (0–)1–4(–6) pairs of arcuate or flexuous branches, the lowest less than 2(–3) cm above the cotyledons. Cauline internodes up to 2½ times, lower floral up to 1½ times as long as the leaves. Lowest flower at node 5–8. Lower floral leaves 5–12 mm, suborbicular or deltate to oblong-ovate, with 4–7 pairs of obtuse to acute teeth. Corolla 6·5–8 mm, white or pale lilac. Capsule 4–5·5 mm, up to twice as long as wide. $2n = 22$. *Pastures and heaths.* ● *Britain and Ireland; ?N.W. France.* Br ?Ga Hb.

6. **E. vigursii** Davey, *Jour. Bot. (London)* 45: 219 (1907). Like **5** but stem and branches erect, dark purplish; lowest branches

(2–)3–7 cm above the cotyledons; corolla 7–8·5 mm, usually lilac to deep reddish-purple. 2*n*=22. *Heaths.* ● *S.W. England. Br.*

7. E. hirtella Jordan ex Reuter, *Compt. Rend. Soc. Hallér.* **4**: 120 (1856). Stem up to *c.* 20 cm, erect, with 0–2 pairs of erect, often short branches. Cauline internodes (1–)2–5 times, the lower floral (1–)1½–3½ times as long as the leaves. Lowest flower at node 2–6(–7). Upper cauline and lower floral leaves up to 12(–13·5) mm, suborbicular to ovate, cordate, rounded or broadly cuneate at base, with 3–6(–7) pairs of obtuse to acute or cuspidate teeth about as long as wide. Corolla 5·5–7 mm, white, or sometimes with lilac upper lip. Capsule 4–5·5 mm, often as long as or longer than the calyx. 2*n*=22. *Grassland, mainly in the mountains; somewhat calcifuge. From N.W. France and Austria southwards to N. Portugal, S. Bulgaria and S. Russia.* Al Au Bu Ga Ge He Hs It Ju Lu Rm Rs (?N, C, E).

8. E. drosocalyx Freyn, *Soc. Helv. Éch. Pl.* **16**: 9 (1885). Like **7** but stem usually not more than 10 cm; lowest flower at node 2–4; leaves often with few glandular hairs; upper cauline leaves with not more than 4 pairs of obtuse, more or less incurved teeth; lower floral leaves 4–11 mm, oblong-ovate to broadly ovate or trullate, with 2–5 pairs of teeth; corolla 4·5–6 mm. *Mountain pastures.* ● *Alps; S.W. Bulgaria.* ?Al Au Bu Ga Ge He Ju.

9. E. picta Wimmer, *Fl. Schles.* ed. 3, 407 (1857) (incl. *E. versicolor* A. Kerner). Stem up to 35 cm, erect or flexuous. Cauline internodes shorter than or up to 5(–8) times as long as leaves. Cauline leaves up to 13 mm, with 1–5 pairs of teeth; lower floral leaves 3–10(–12) mm, suborbicular to deltate or oblong-ovate, cuneate to rounded. Capsule usually shorter than calyx, 1¼–2½(–3) times as long as wide, truncate or emarginate. ● *Mountain districts of C. & S.E. Europe, from Switzerland to the E. Carpathians and Bulgaria.* Au Bu Cz ?Ga Ge He Hu It Ju Po Rm Rs (W).

(a) Subsp. **picta**: Stem with 0–3(–5) pairs of erect branches. Lowest flower at node 2–6(–7). Lower leaves often shortly petiolate, shortly hirsute above, the upper glabrous to setulose; lower floral leaves usually wide, with 2–4(–5) pairs of teeth. Corolla (6·5–)7·5–10(–11) mm, white with lilac upper lip, or entirely lilac, strongly violet-veined. 2*n*=22. *Alpine meadows and pastures. Mountains of C. Europe from the C. Alps to the E. Carpathians.*

(b) Subsp. **kerneri** (Wettst.) Yeo, *Bot. Jour. Linn. Soc.* **64**: 354 (1971) (*E. kerneri* Wettst.): Stem with (1–)7–12 pairs of ascending branches, which are often again branched. Lowest flower at node (6–)8–12(–17). Leaves minutely scabrid, occasionally with short glandular hairs; lower floral leaves usually rather narrow, with 3–5(–9) pairs of teeth. Corolla 9–11 mm, white, sometimes with lilac upper lip. *Limestone grassland, usually below 1500 m. Mountain districts of C. & S.E. Europe.*

Though the two subspecies are usually distinct, intermediate plants occur in some places.

E. pulchella A. Kerner, *Sched. Fl. Exsicc. Austro-Hung.* **1**: 48 (1881), from the S. & E. Alps, believed to be of hybrid origin between **9** and **24**, differs in its corolla only 7–8·5 mm and capsule 4·5–6·5 mm, 2–3 times as long as wide.

E. exaristata Smejkal, *Biol. Práce Slovensk. Akad. Vied* (*Bratislava*) **9**(9): 47 (1963), from the W. Carpathians (High Tatra), is like variants of **9**(a) with long internodes, but the leaves are even more distinctly petiolate, the lowest flower is at node 3–8, the corolla only 6·5–8 mm, and the capsule narrowly emarginate.

E. simonkaiana Degen & Lengyel ex Jáv., *Magyar Fl.* 1014 (1925) from N. Jugoslavia, is apparently also related to **9**. It has the lowest flower at node 4–6, branches short, erect, up to 4 pairs, lowest floral leaves with acute teeth, and corolla up to about 7·5 mm.

10. E. marchesettii Wettst. ex Marchesetti, *Fl. Trieste* 411 (1897). Like **9**(b) but cauline leaves linear-oblanceolate, narrowly oblong or narrowly trullate, with 1–3 pairs of usually shallow, obtuse to acuminate teeth; lower floral leaves lanceolate, narrowly oblong, trullate or rhombic, cuneate; teeth 2–3(–4) pairs. ● *Coastal regions of N.E. Italy.* It ?Ju.

Evidently of hybrid origin, the parents being **9**(b) and a species of Subsect. *Angustifoliae*.

11. E. arctica Lange ex Rostrup, *Bot. Tidsskr.* **1**(4): 47 (1870–1871). Stem up to 25(–35) cm. Lowest flower at node 3–9(–10). Leaves usually light green, the cauline with 1–6 pairs of obtuse to acute (rarely aristate) teeth, the upper ones usually broadly ovate; lower floral leaves usually with basal teeth patent. Corolla 6–11(–13) mm, deep lilac to white. Capsule 4–8 mm, usually 2–3 times as long as wide, truncate to emarginate. *Meadows, pastures and roadsides.* ● *N. Europe, extending southwards to the E. Carpathians.* Br Cz Da Fa Fe ?Ga Hb Is No Po ?Rm Rs (N, B, C, W) Su.

1 Capsule 4–5·5 mm, oblong; cauline internodes not more than 2(–3½) times as long as the leaves
  2 Leaves without glandular hairs, the lower ones deflexed; branches short, arcuate-ascending   **(c) subsp. minor**
  2 Leaves usually with glandular hairs, not deflexed; branches long, ascending or erect   **(d) subsp. borealis**
1 Capsule usually 5·5 mm or more, elliptical or elliptic-oblong; cauline internodes usually up to 4(–9) times as long as the leaves
  3 Stem usually flexuous, with 0–2(–5) pairs of long, often flexuous branches; lower floral leaves suborbicular or broadly ovate, with the teeth not much longer than wide; leaves without glandular hairs; corolla 7–11(–13) mm   **(b) subsp. arctica**
  3 Stem usually straight, with 0–5(–6) pairs of often short, usually straight, branches; lower floral leaves deltate or broadly to narrowly ovate; teeth sometimes much longer than wide; leaves usually with glandular hairs; corolla not more than 10 mm
    4 Lowest flower at node 3–6(–9); corolla (6·5–)8–10 mm; cauline internodes mostly 2½–6 times as long as the leaves   **(a) subsp. tenuis**
    4 Lowest flower at node (3–)4–9(–10); corolla 6–9(–10) mm; cauline internodes mostly 2–4 times as long as the leaves
      5 Branches frequently more than 3 pairs; lower floral leaves usually truncate or subcordate; capsule (4–)4·5–6·5(–7) mm   **(d) subsp. borealis**
      5 Branches rarely more than 3 pairs; lower floral leaves usually broadly cuneate; capsule 4·5–7(–8) mm   **(e) subsp. slovaca**

(a) Subsp. **tenuis** (Brenner) Yeo, *Bot. Jour. Linn. Soc.* **64**: 357 (1971) (*E. officinalis* var. *tenuis* Brenner, *E. tenuis* (Brenner) Wettst. 1896, non Wettst. 1893): Stem with 0–3(–4) pairs of usually rather short branches. Cauline internodes (1½–)2½–6(–9) times as long as leaves. Lowest flower at node 3–6(–9). Leaves often with glandular hairs, the cauline 3–14(–16) mm, mostly broadly ovate, trullate or oblong; lower floral leaves 5–12(–16) mm, broadly ovate to oblong-ovate, rounded to subcordate at base, with 3–6 pairs of subacute to aristate teeth. Corolla (6·5–)8–10 mm. Capsule 5–8 mm, at least as long as calyx, at least twice as long as wide, elliptic-oblong to obovate. 2*n*=44. *From Iceland to C. Russia.*

Plants approaching **11(b)** occur in Norway but have glandular hairs. In the Baltic region plants intermediate between **11(a)** and **17** are found.

(b) Subsp. **arctica** (*E. borealis* auct.): Stem flexuous, with 0–2(–5) pairs of long, often flexuous branches. Cauline internodes not more than 4(–5) times as long as leaves. Lowest flower at node (3–)4–6(–7). Leaves eglandular, the cauline 4–11(–16) mm, orbicular to oblong, the lower floral 5–10(–14) mm, suborbicular to broadly ovate, with 3–5(–6) pairs of obtuse to acuminate teeth not much longer than wide. Corolla 7–11(–13) mm, usually white with lilac upper lip. Capsule 5·5–8 mm, about as long as the calyx, sometimes less than twice as long as wide, elliptical to obovate. *Færöer, Zetland, Orkney.*

(c) Subsp. **minor** Yeo, *loc. cit.* (1971) (*E. borealis* auct. dan., non (Townsend) Wettst.): Stem with (0–)1–3(–6) pairs of short, stout, arcuate-ascending branches. Cauline internodes not more than 2(–3½) times as long as leaves. Lowest flower at node 5–9. Leaves not more than 10 mm, eglandular, the cauline deflexed, caducous, narrowly ovate to elliptical, the lower floral ovate to deltate, with 3–4(–5) pairs of subacute to acute teeth not much longer than wide. Corolla 6·5–8·5 mm. Capsule 4–5·5 mm, not exceeding calyx, more than twice as long as wide, oblong. *Denmark (around Skagen, N. Jylland).*

(d) Subsp. **borealis** (Townsend) Yeo, *op. cit.* 358 (1971) (*E. rostkoviana* f. *borealis* Townsend, *E. brevipila* auct., non Burnat & Gremli): Stem with 0–5(–6) pairs of usually long, erect or ascending branches. Cauline internodes 2–4(–7) times as long as leaves. Lowest flower at node 4–8(–10). Leaves often with short, and occasionally with long glandular hairs, the cauline 3–14 mm, ovate to oblong, the lower floral (4–)6–12(–14) mm, ovate to deltate, rounded to subcordate at base, with (3–)4–6(–7) pairs of subacute to acuminate teeth usually much longer than wide. Corolla 6–9(–10) mm. Capsule 4–7 mm, usually not exceeding calyx, at least twice as long as wide, elliptical to oblong. 2*n*=44. *Britain and Ireland; W. Norway.*

Plants resembling **11(d)** found in N.E. Europe are probably hybrids between **11(a)** and **17**.

(e) Subsp. **slovaca** Yeo, *loc. cit.* (1971) (*E. brevipila* auct., non Burnat & Gremli, *E. vernalis* auct., ? an List): Stem with 0–3(–6) pairs of usually erect branches. Cauline internodes 1½–5 times as long as leaves. Lowest flower at node (3–)4–9. Leaves usually with glandular hairs, the cauline 2–12(–16) mm, elliptical to broadly ovate, with obtuse to acuminate teeth, the lower floral 5–11(–14) mm, usually broadly ovate to trullate, broadly cuneate at base, with 3–6 pairs of subacute to aristate teeth about as long as wide or longer. Corolla 6·5–8·5(–9·5) mm. Capsule 4·5–7(–8) mm, not exceeding calyx, 2–3 times as long as wide, elliptical to oblong. *Carpathians, south-eastwards to c. 48° N.*

12. E. tetraquetra (Bréb.) Arrondeau, *Bull. Soc. Polymath. Morbihan* **1862**: 96 (1863). Stem up to 15(–20) cm, erect, stout, with 0–5(–8) pairs of erect or ascending branches; all internodes often shorter than the leaves and very rarely more than 1½ times as long. Lowest flower at node (3–)5–7(–9). Leaves fleshy, glabrous to setose, frequently also with short glandular hairs, the cauline 2–9(–11) mm, obovate to suborbicular, with 1–3(–4) pairs of obtuse to subacute teeth; floral leaves usually forming a dense, sometimes 4-angled spike, the lower 5–10(–14) mm, usually ovate to deltate, truncate at base, with 3–5(–6) pairs of subacute to acuminate teeth at least as long as wide. Corolla (4–)5–7(–8) mm, white to lilac. Capsule 4·5–5·5(–6) mm, usually shorter than calyx, at least twice as long as wide, oblong to elliptic-oblong, truncate or slightly emarginate. 2*n*=44. *Maritime sands and short grassland on sea-cliffs; rarely inland on calcareous pastures.* ● *Britain, Ireland, N.W. France.* Br Ga Hb.

Hybrids with **16** are frequent in Britain.

13. E. nemorosa (Pers.) Wallr., *Annus Bot.* 82 (1815) (incl. *E. curta* (Fries) Wettst.). Stem erect, up to 35(–40) cm, with (0–)1–9 pairs of ascending branches, often bearing secondary branches. Cauline internodes up to 4(–7) times, the lower floral up to 2(–3) times as long as the leaves. Lowest flower at node (5–)10–14. Leaves usually with deeply impressed veins, subglabrous to densely hairy, rarely with short glandular hairs, the cauline 2–12 mm, elliptical or oblong to ovate or deltate, with 1–5(–7) pairs of obtuse to acute teeth; lower floral leaves 4–9(–11) mm, ovate to deltate, usually rounded or truncate at base, with (3–)4–6(–9) pairs of subacute to aristate teeth usually slightly longer than wide, the basal teeth patent. Corolla 5–7·5(–8·5) mm, white to lilac. Capsule 4–5·5(–6) mm, usually slightly shorter than calyx, more than twice as long as wide, oblong to elliptic-oblong, truncate or slightly emarginate. 2*n*=44. *Grassy places.* ● N. & C. Europe, *extending southwards to N.E. Spain.* Au Be Br Cz Da Fe Ga Ge Hb He Ho Hs No Po Rs (N, B, C) Su.

Strongly hairy variants (*E. curta* (Fries) Wettst.) are increasingly common towards the north-eastern part of the range.

Hybrids with **16** and **17** are frequent.

14. E. coerulea Hoppe & Fürnrohr, *Flora* (Regensb.) 17: 445 (1834) (*E. uechtritziana* Junger & Engler). Like **13** but stem not more than 15 cm, with 0–1(–3) pairs of short branches; lowest flower at node 3–6; leaves not more than 8 mm, the cauline with not more than 4 pairs of teeth; lower floral leaves with up to 5 pairs of teeth; corolla purple to white with lilac upper lip; capsule not more than twice as long as wide. *Mountain grassland.* ● *Mountains of E.C. Europe.* Cz ?Ge Po Rm Rs (W).

15. E. pseudokerneri Pugsley, *Jour. Bot.* (*London*) **67**: 224 (1929). Like **13** but stem not more than 20(–30) cm, with patent or ascending branches; cauline internodes not more than 2½(–3) times, the lower floral not more than 1½(–2) times as long as the leaves; leaves subglabrous, the lower cauline oblong-ovate, with obtuse to subacute teeth, the upper cauline ovate with acute or acuminate teeth; lower floral leaves rounded to broadly cuneate at base; corolla (6–)7–9(–11) mm; capsule 3·5–5(–6) mm, much shorter than calyx, about 2½ times as long as wide. 2*n*=44. *Calcareous grassland.* ● *England; W. Ireland.* Br Hb.

16. E. confusa Pugsley, *Jour. Bot.* (*London*) **57**: 172 (1919). Stem up to 20(–45) cm, flexuous, with (0–)2–8(–10) pairs of usually long, slender, flexuous branches, usually bearing numerous secondary branches. Cauline internodes up to 2½ times, the lower floral up to 2(–3) times as long as leaves. Lowest flower at node (2–)5–12(–14). Leaves up to 10 mm, subglabrous to sparsely setose, sometimes also with short glandular hairs, the cauline ovate to ovate-lanceolate, with 1–5 pairs of obtuse to acute teeth, the lower floral 3·5–9(–10) mm, often alternate and with flowers only in alternate axils, usually ovate to trullate, cuneate, rounded, or occasionally truncate at base, with 2–6 pairs of subacute to acute teeth at least as long as wide. Corolla 5–9 mm, white to purple. Capsule 3·5–5·5(–6·5) mm, usually about as long as calyx, 2–3 times as long as wide, oblong to elliptic-oblong, truncate or emarginate. 2*n*=44. *Heathy pastures and sand-dunes.* ● *Ireland, England, Færöer; probably introduced in W. Norway.* Br Fa Hb [*No].

The characteristic profuse branching of this species is reduced in mountain habitats or by early starvation, and is generally less in N. Scotland and the Færöer. In S. England there is fairly frequent intergradation with **13**. Hybrids with **11, 33** and **34** are also frequent.

(17–20). E. stricta group. Stem and branches usually rigid and erect. Leaves glabrous to densely hairy, frequently also with short glandular hairs, usually narrowly ovate or elliptical to oblong-lanceolate; floral leaves usually appressed to the flower, with long, very acute to aristate teeth, the basal teeth apically directed. Capsule usually much shorter than calyx, $(1\frac{1}{2}-)2\frac{1}{2}-5$ times as long as wide, oblong to elliptic-oblong.

1 Corolla 4·5–6·5 mm   **20. liburnica**
1 Corolla (6–)7–11 mm
  2 Lowest flower usually at node 3 or lower; capsule emarginate
                                        18. hyperborea
  2 Lowest flower usually at node 4 or higher; capsule truncate or slightly emarginate
    3 Capsule usually 4–5·5 mm, $2\frac{1}{2}-3\frac{1}{2}$ times as long as wide; branches usually 2–6 pairs or more; lowest flower at node (3–)7–14(–18); plant usually strongly tinged with purple
                                        17. stricta
    3 Capsule usually 5–7 mm, 3–5 times as long as wide; branches 0–3(–6) pairs; lowest flower at node (4–)5–8(–9); plant not or slightly tinged with purple   **19. pectinata**

17. E. stricta D. Wolff ex J. F. Lehm., *Prim. Fl. Herbip.* 43 (1809) (*E. brevipila* Burnat & Gremli, *E. condensata* Jordan, *E. tatarica* auct., non Fischer ex Sprengel). Plant usually strongly tinged with purple. Stem up to 35 cm, with (0–)2–6(–13) pairs of branches. Cauline internodes up to 4 times, the lower floral up to 2(–3) times as long as leaves. Lowest flower at node (3–)7–14 (–18). Leaves glossy, the cauline 2–12(–16) mm, narrowly ovate to ovate-lanceolate, with 1–6 pairs of teeth, the lower floral 5–14 mm, ovate to trullate, rounded to cuneate at base, serrate or pectinate with 4–6 pairs of subacute to aristate teeth. Corolla (6–)7·5–10(–11) mm, lilac or white. Capsule 4–5·5(–7) mm, $2\frac{1}{2}-3\frac{1}{2}$ times as long as wide, truncate, or rarely emarginate. $2n=44$. *Meadows, dry grassland and scrub.* ● *Most of Europe, but absent from the extreme north and much of the west and south.* Al Au Be Bu Cz Da Fe Ga Ge Gr He Ho Hs Hu It Ju No Po Rm Rs (N, B, C, W, E) Su [Br].

In Fennoscandia, the Baltic region and the mountains of S. Europe glandular variants are frequent; in these areas, also, variants occur which are very like **11(a)**, but differ in their acutely toothed cauline leaves and sometimes in their capsules. Strongly branched variants with all the internodes shorter than the leaves occur in Denmark and Sweden, and at high altitudes in the Pyrenees. Strongly hairy variants occur in S.W. Europe.

Hybridizes with **11, 13, 21, 33** and **41**.

18. E. hyperborea Joerg., *Bergens Mus. Aarb.* **1916–1917**, *Naturvid. Række* 5: 255 (1919). Like **17** but not more than 25 cm; stems often flexuous, with 0–2(–3) pairs of branches; lowest flower at node (1–)2–3(–4); leaves rather large, strigose above, setose beneath, mainly on the veins, the cauline with 1–4 pairs of short, obtuse teeth; lower floral leaves with subacute to acute teeth; capsule 4–7 mm, $1\frac{1}{2}$ times to more than twice as long as wide, emarginate. *Damp meadows and* Betula-*groves.* ● *Arctic Europe.* No Rs (N).

19. E. pectinata Ten., *Fl. Nap.* 1, *Prodr.*: 36 (1811) (*E. tatarica* Fischer ex Sprengel). Plant with little purple pigment. Stem up to 35 cm, with 0–3(–6) pairs of branches. Lowest flower at node (4–)5–8(–9), the dense upper parts of the spikes often long. Cauline leaves 3–13 mm, ovate, elliptical, trullate, oblong or oblong-lanceolate, with 1–4(–6) pairs of subacute to acute teeth; lower floral leaves 5–13 mm, ovate, trullate or oblong-ovate, sometimes very wide, cuneate at base with 2–6(–7) pairs of acute to aristate teeth, often pectinate towards apex. Corolla 6·5–9(–10)

mm, usually white, sometimes lilac. Capsule 5–7 mm, 3–5 times as long as wide, truncate or slightly emarginate. $2n=44$. *Grassland, scrub and open woodland. S., S.C. & S.E. Europe.* Al Au Bu Co Cz Ga Gr He Hs Hu It Ju ?Lu Rm Rs (C, W, K, E).

20. E. liburnica Wettst., *Österr. Bot. Zeitschr.* **44**: 172 (1894) (? *E. irenae* Juz.). Like **19** but stem not more than 15(–25) cm; leaves sometimes purplish, narrower, the lower floral with 2–5 pairs of deltate or incurved, rarely aristate teeth; corolla 4·5–6·5 mm, usually lilac; capsule 4·5–6·5 mm, $2\frac{1}{2}-3\frac{1}{2}$ times as long as wide, usually shorter than or about as long as calyx. *Grassy places and open woods.* ● *Balkan peninsula and E. Carpathians.* Al Bu Gr ?It Ju Rm ?Rs (K).

21. E. alpina Lam., *Encycl. Méth. Bot.* **2**: 400 (1786). Plant usually strongly tinged with purple. Stem up to 20 cm, erect, with 0–4 pairs of patent branches. Cauline internodes up to $2\frac{1}{2}$ times as long as the leaves. Lowest flower at node (3–)5–8(–10). Leaves hirsute, puberulent or glabrous, the cauline obovate to ovate-lanceolate, with obtuse to aristate teeth, the lower floral 5–13 mm, ovate, ovate-lanceolate, trullate or rhombic, cuneate at base, with 3–6 pairs of subacute to aristate teeth mostly much longer than wide, the upper floral with aristate teeth, of which the distal sometimes exceed the aristate terminal lobe. Corolla 8–13·5 mm, usually deep lilac. Capsule 4·5–7·5 mm, usually much shorter than calyx, $2\frac{1}{2}-3(-4)$ times as long as wide. $2n=22$. *Alpine meadows and pastures, c. 1500–2500 m.; calcifuge.* ● *Pyrenees, Alps, N. & C. Appennini.* Ga He Hs It.

22. E. christii Favrat in Gremli, *Neue Beitr. Fl. Schweiz* 4: 27 (1887). Like **21** but with less purple pigment; stem not more than 8(–14) cm, with 0–1(–4) pairs of branches; cauline leaves with obtuse to acute teeth, the floral with shortly aristate teeth; corolla (7–)9–11 mm, yellow; anthers pale brown. ● *S. Switzerland, N. Italy.* He It.

Probably originated by hybridization between **21** and **24**.

23. E. cisalpina Pugsley, *Jour. Bot.* (*London*) 70: 262 (1932). Like **21** but stem up to 40 cm, with up to 13 pairs of branches; cauline internodes not more than $1\frac{1}{2}$ times as long as the leaves; lowest flower at node (8–)11–20; lower cauline leaves oblanceolate, the floral all with long-aristate teeth; capsule 4–5·5 mm, $3\frac{1}{2}-4$ times as long as wide. *Margins of* Castanea-*woods.* ● *N.W. Italy, S. Switzerland.* He It.

24. E. minima Jacq. ex DC. in Lam. & DC., *Fl. Fr.* ed. 3, 3: 473 (1805). Stem erect, with 0–3(–5) pairs of long, erect branches. Cauline internodes 1–4(–6) times, the lower floral $1-3\frac{1}{2}(-4\frac{1}{2})$ times as long as leaves. Lowest flower at node (1–)2–6. Leaves subglabrous to hairy, with some of the hairs sometimes glandular, the cauline 2–11(–13) mm, suborbicular to broadly elliptical or rhombic, shortly petiolate, with 1–4(–6) pairs of obtuse to subacute teeth, the lower floral 3–12(–15) mm, suborbicular to ovate, with 2–5(–6) pairs of obtuse to acute teeth not longer than wide. Corolla 4–6 mm; lower lip yellow, white or lilac, the upper usually lilac. Capsule 3·5–7 mm, elliptic-oblong, truncate to narrowly emarginate. *Grassy and stony places in high mountains.* ● *S. & C. Europe, from S.C. France and the Carpathians southwards to the Pyrenees, Appennini and Bulgaria.* Al Au Bu ?Co Cz Ga Ge Gr He Hs It Ju Po Rm Rs (W).

(a) Subsp. **minima**: Not more than 12(–20) cm. Lowest flower usually at node 2–5. Cauline leaves with the distal teeth incurved, the lower floral rounded to cuneate at base. Capsule about twice as long as wide. $2n=44$. *Almost throughout the range of the species.*

(b) Subsp. tatrae (Wettst.) Hayek in Hegi, *Ill. Fl. Mitteleur.* **6(1)**: 91 (1913) (*E. tatrae* Wettst.): Up to 25(–30) cm. Lowest flower often at node 6. Cauline leaves with the distal teeth seldom incurved, the lower floral subcordate to rounded at base. Capsule up to 3 times as long as wide. *E.C. Europe (Carpathians, Sudeten Mts.).*

E. mendoncae Samp., *Anais Fac. Ci. Porto* **22**: 50 (1937), known from one locality in N.E. Portugal, may represent a third subspecies. It has stems with up to 4 pairs of branches and cauline internodes 4–8 times as long as the leaves, the lowest flower at node 5–8 and the corolla 5·5–6·5 mm, yellow.

25. E. frigida Pugsley, *Jour. Linn. Soc. London (Bot.)* **48**: 490 (1930) (*E. minima* auct. scand., non DC.). Stem up to 20(–30) cm, flexuous or erect, with 0–2(–3) pairs of erect branches. Cauline internodes 1–5(–10) times, the lower floral 1–4(–5) times as long as the leaves. Lowest flower at node 2–4(–5). Cauline leaves 3–11(–13) mm, more or less setose, sometimes also with short glandular hairs, oblong to suborbicular, broadly cuneate or shortly petiolate at base, with 1–4(–5) pairs of obtuse to subacute teeth; lower floral leaves (3–)5–12(–18) mm, elliptical to deltate or suborbicular, cuneate to truncate at base, with 2–5(–8) pairs of obtuse to subacute teeth usually not longer than wide. Corolla 4–7(–8) mm, white to lilac, rarely purple. Capsule (4–)5–7 mm, about as long as or longer than the calyx, about twice as long as wide, obovate to broadly oblong, emarginate, or rarely truncate. 2*n*=44. *Grassy places or rock-ledges, mostly on mountains except in the extreme north.* N. Europe. Br ?Cz Fa Fe Hb Is No Rs (N, C) Sb Su.

Plants from Iceland referred to this species have usually densely hairy leaves and very small corolla, and may be difficult to distinguish from **28** or **34**.

26. E. foulaensis Townsend ex Wettst., *Monogr. Gatt. Euphrasia* 139, 299 (1896). Like **25** but more compact, with up to 3(–4) pairs of branches and lowest flower at node (2–)4–6; leaves small, dark green, becoming black on drying, not densely hairy, the lower floral with obtuse to acute teeth; corolla 4–6 mm; capsule 4·5–5·5(–7) mm, usually exceeding calyx, oblong to elliptic-oblong. 2*n*=44. *Cliff-tops and edges of salt-marshes.* ● *N. Scotland, Færöer.* Br Fa.

Strongly hairy plants referred to this species are probably hybrids with **28**, **29** or **30**.

27. E. cambrica Pugsley, *Jour. Bot. (London)* **67**: 224 (1929). Stem up to 8 cm, flexuous, with 0–2 pairs of flexuous branches. Internodes shorter than the leaves or up to 1½(–2) times as long. Lowest flower at node 2–4. Lower floral leaves 3·5–9 mm, broadly ovate to suborbicular or deltate, with 2–4 pairs of obtuse to subacute teeth and obtuse terminal lobe. Corolla 4–5·5 mm; lower lip very small, white or yellowish-white; upper white or lilac. Capsule (4–)5–7 mm, longer than the calyx, not more than twice as long as wide, elliptical or obovate, emarginate. *Mountain grassland.* ● *N. Wales.* Br.

(28–32). E. ostenfeldii group. Stem not more than 12(–15) cm. Lowest flower at node 9 or below. Leaves up to 11(–14) mm, rather densely eglandular-hairy, the lower floral with the teeth about as long as wide or less. Corolla 4–6(–7) mm. Capsule elliptic-oblong or oblong.

1　Leaves hairy mainly towards the apex; the cauline obovate to
　　narrowly ovate or elliptical　　　　　**32. campbelliae**
1　Leaves more or less uniformly hairy, usually suborbicular,
　　ovate, or ovate-oblong

2　Leaves with glandular hairs　　　　　　**31. dunensis**
2　Leaves without glandular hairs
3　Teeth of lower floral leaves mostly wider than long; branches
　　not more than 3 pairs　　　　　　　**30. rotundifolia**
3　Teeth of lower floral leaves mostly as long as wide; branches
　　up to 5 pairs
4　Corolla 5·5–7 mm; capsule usually more than twice as long
　　as wide　　　　　　　　　　　　**29. marshallii**
4　Corolla 3·5–6 mm; capsule not more than twice as long as
　　wide　　　　　　　　　　　　　**28. ostenfeldii**

28. E. ostenfeldii (Pugsley) Yeo, *Bot. Jour. Linn. Soc.* **64**: 359 (1971) (*E. curta* var. *ostenfeldii* Pugsley, *E. curta* auct., non (Fries) Wettst.). Stem erect or flexuous, with 0–4(–6) pairs of branches. Cauline internodes up to 3(–5) times as long as the leaves. Lowest flower at node (2–)3–7. Cauline leaves suborbicular to oblong-obovate, with not more than 4 pairs of teeth, the lower floral little larger than the upper cauline, suborbicular to oblong-ovate, with not more than 5 pairs of symmetrical, subacute to acute teeth. Corolla (3·5–)4·5–6 mm. Capsule 4–6 mm, as long as or longer than the calyx, not more than twice as long as wide, emarginate. *Grassy, stony or sandy places, often near the sea. N.W. Europe, from N. Wales to Iceland.* Br Fa Is.

E. davidssonii Pugsley, *Jour. Bot. (London)* **71**: 308 (1933), described from Iceland, with hairy, deeply toothed leaves, is probably related to **28**, but is insufficiently known. **E. eurycarpa** Pugsley, *Naturalist (Leeds)* **1945**: 42 (1945), described from W. Scotland (island of Rum), is like **28** but has longer internodes and very wide, deeply emarginate capsule. It is perhaps of hybrid origin.

29. E. marshallii Pugsley, *Jour. Bot. (London)* **67**: 224 (1929). Stem erect, with 1–5 pairs of rather long, erect branches. Cauline internodes up to 2½(–3) times as long as the leaves; upper floral internodes very short. Lowest flower at node (5–)7–9. Cauline leaves ovate, oblong-ovate or elliptical, with up to 5 pairs of teeth, the lower floral ovate or rhombic, with up to 5 pairs of teeth, the basal teeth patent or apically directed. Corolla 5·5–7 mm. Capsule 4·5–5·5(–6·5) mm, about as long as or shorter than the calyx, usually more than twice as long as wide, truncate. 2*n*=44. *Maritime grassland.* ● *N. Scotland.* Br.

30. E. rotundifolia Pugsley, *loc. cit.* (1929). Like **29** but with 0–3 pairs of short branches; cauline internodes 1–2(–2½) times as long as the leaves; cauline leaves orbicular to oblong-ovate, with obtuse teeth; lower floral leaves suborbicular to broadly ovate, rounded at base, with obtuse or subacute teeth usually wider than long; capsule 5·5–6 mm, slightly exceeding calyx, not more than twice as long as wide. *Maritime grassland.* ● *N. Scotland.* Br.

31. E. dunensis Wiinst., *Bot. Tidsskr.* **43**: 233 (1935). Like **29** but with 0–3 pairs of short erect branches; cauline internodes up to 1½(–3) times as long as the leaves; leaves with scattered short or long glandular hairs among the eglandular, the cauline caducous, with not more than 4 pairs of teeth; lower floral leaves broadly ovate or oblong-ovate. *Calcareous maritime grassland.* ● *N.W. Jylland (Bulbjerg).* Da.

Records from elsewhere in Denmark are erroneous. Hybrids with **17** are known.

32. E. campbelliae Pugsley, *Jour. Bot. (London)* **78**: 91 (1940). Stem up to 10 cm, erect, with 0–2 pairs of short erect branches. Cauline internodes 1–2½(–4) times as long as the leaves. Lowest flower at node 2–7. Leaves not more than 8 mm, sometimes purplish beneath and green above, the hairs mostly on the distal half, sparse to rather dense, the cauline obovate to elliptical, with not more than 3 pairs of teeth, the lower floral ovate, rounded or cuneate at base, shortly petiolate, with 2–5(–6) pairs of obtuse to acute teeth. Corolla 5·5–7 mm. Capsule 4·5–7·5 mm, shorter than calyx, more than twice as long as wide, more or less emarginate. *Heathy grassland near the sea.* ● *N.W. Scotland (Isle of Lewis); ?Zetland.* Br.

Possibly of hybrid origin, between **29** and **33** or **34**.

33. E. micrantha Reichenb., *Fl. Germ. Excurs.* 358 (1831) (*E. gracilis* (Fries) Drejer). Plant usually strongly tinged with purple. Stem up to 25 cm, erect, slender, with (0–)2–7(–10) pairs of slender, erect branches. Cauline internodes usually 2–4 times, the lower floral 1½–2½(–3) times as long as the leaves. Lowest flower at node (4–)6–14(–16). Leaves glabrous or minutely scabrid, not darker beneath, the cauline 2–8(–11) mm, narrowly ovate to obovate, with 1–6 pairs of usually subacute to acute teeth, the lower floral 3·5–7(–8) mm, ovate to rhombic, truncate or cuneate at base, with 3–6 pairs of acute to acuminate teeth. Corolla 4·5–6·5 mm, usually lilac to purple. Capsule 3–5(–6) mm, usually shorter than the calyx, more than twice as long as wide, oblong to elliptic-oblong, rounded to slightly emarginate. 2*n*=44. *Heaths, usually associated with* Calluna vulgaris. ● *N. & C. Europe, extending southwards to N. Spain.* Au Be Br ?Co Cz Da Fa Fe Ga Ge Hb Ho Hs It No Po Rm Rs (B) Su.

Hybrids with **11(d)**, **12**, **16**, and **17** are common.

Plants derived from the hybrid **33 × 17** have been described as **E. gratiosa** Wiinst., *Bot. Tidsskr.* **48**: 103 (1946).

Hairy plants from W. Scotland (island of Rum), similar to **33** but with very small leaves and corolla and more or less emarginate capsule, have been described as **E. rhumica** Pugsley, *Naturalist (Leeds)* **1945**: 41 (1945); they are possibly of hybrid origin.

34. E. scottica Wettst., *Monogr. Gatt. Euphrasia* 170 (1896). Stem up to 25 cm, erect, with 0–4 pairs of long, arcuate-erect branches. Cauline internodes 2–5 times, the lower floral 1½–3(–5) times as long as the leaves. Lowest flower at node (2–)3–6(–8). Leaves light green, often purple beneath, glabrous to moderately hairy, rarely with some glandular hairs. Cauline leaves ovate-elliptical to obovate, with up to 5(–6) pairs of shallow, obtuse or subacute teeth; lower floral leaves 4–10(–17) mm, often alternate, ovate, oblong or rhombic, with 3–5(–6) pairs of subacute to acute teeth. Corolla (3·5–)4·5–6·5 mm; lower lip small, white; upper white or lilac. Capsule 4–5·5(–9) mm, as long as or longer than the calyx, more than twice as long as wide, oblong to elliptic-oblong, more or less emarginate. 2*n*=44. *Wet moorland, fens and flushes.* ● *N. Europe.* Br Fa Fe Hb Is No ?Rs (C) Su.

Populations probably referable to the hybrid **16 × 34** are rather frequent in Britain, sometimes occurring independently of the presumed parents; these plants have more flexuous stems, more acutely toothed leaves and sometimes slightly larger flowers than **34**.

Specimens from Iceland often approach **28** in their hairy, broader leaves.

E. heslop-harrisonii Pugsley, *Naturalist (Leeds)* **1945**: 43 (1945), from N. Scotland, is like **34** but has usually flexuous stems, cauline internodes not more than 2(–3) times as long as the leaves, leaves not purple beneath, the cauline with not more than 4 pairs of teeth, the lower floral broadly ovate to trullate, and the capsules (4·5–)5·5–7 mm.

35. E. calida Yeo, *Bot. Jour. Linn. Soc.* **64**: 359 (1971). Like **34** but with cauline internodes not more than 3 times and lower floral internodes not more than 2½ times as long as the leaves; leaves thin, not purple beneath, the lower floral broadly ovate or broadly rhombic; calyx whitish and membranous except for the teeth and prominent veins; capsule 3·5–5(–6) mm, usually much shorter than the calyx and not more than twice as long as wide, broadly obovate to oblong. *Apparently associated with warm springs.* ● *Iceland.* Is.

36. E. saamica Juz., *Not. Syst. (Leningrad)* **17**: 365 (1955). Like **34** but with 0–1 pairs of branches; cauline internodes 2½–9 times, the lower floral 1½–5 times as long as the leaves; lowest flower at node 2–4; purple colouring of leaves not stronger beneath than above; cauline leaves with not more than 3 pairs of teeth, the lower floral with subacute to aristate teeth; corolla 5·5–7·5(–8) mm; capsule 5–7 mm, elliptical to obovate. ● *Arctic Fennoscandia.* ?Fe No Rs (N) Su.

37. E. atropurpurea (Rostrup) Ostenf. in Warming et al., *Bot. Faeroes* **1**: 55 (1901). Stem up to 17 cm, flexuous, with 0–3 pairs of arcuate-erect branches. Cauline internodes (1–)1½–3 times as long as the leaves. Lowest flower at node 3–6. Leaves up to 11 mm, green, often tinged with purple, the lower floral obovate-lanceolate to rhombic or ovate, with 2–4(–6) pairs of subacute to acute teeth. Corolla 5–6·5 mm, deep purple, at least in part. Capsule 4·5–6·5 mm, 1½–2½ times as long as wide, oblong or elliptic-oblong, truncate or emarginate. ● *Færöer.* Fa.

38. E. bottnica Kihlman, *Acta Soc. Fauna Fl. Fenn.* **13**(5): 20 (1896). Stem up to 20 cm, erect, with 0–3(–5) pairs of erect branches. Cauline internodes 2–4 times as long as the leaves. Lowest flower at node 2–5(–7). Leaves up to 11(–14) mm, more or less setulose, the lower floral with obtuse to subacute teeth. Corolla 3–3·5 mm. Anthers pale brown. Capsule 3–5 mm, not more than twice as long as wide, emarginate. *Sea-cliffs and salt-marshes.* ● *Around the Gulf of Bothnia.* Fe Su.

39. E. willkommii Freyn, *Flora (Regensb.)* **67**: 681 (1884). Stem up to 8 cm, with 0–2 pairs of erect or appressed branches; cauline internodes up to 2½(–3) times as long as the leaves. Lowest flower at node 2–4. Leaves up to 9 mm, subglabrous or with short glandular hairs, the cauline narrowly oblong to cuneate-obovate, with 1–3 pairs of obtuse teeth, the lower floral with (1–)2–5 pairs of linear and obtuse to deltate and acute teeth, the upper usually with aristate teeth, often nearly as long as the terminal lobe. Corolla 4·5–5·5 mm, yellowish with purplish upper lip. Capsule 3·5–5·5 mm, usually shorter than the calyx, 2–3 times as long as wide, elliptic-oblong, truncate or emarginate. *Grassy places, 1900–3000 m.* ● *S. Spain (Sierra Nevada).* Hs.

Plants from the Pyrenees referred to this species appear to resemble **20** more closely than **39**; their status is obscure.

40. E. taurica Ganeschin in Popl., *Spisok Rast. Sobr. Krymsk. Gos. Zapov.* 87 (1931). Stem up to 8 cm, erect, with short glandular and eglandular hairs, with (0–)1–3 pairs of erect branches. Cauline internodes 1–2(–2½) times as long as the leaves, the lower floral usually shorter than the leaves. Lowest flower at node (3–)4–7(–8). Leaves up to 8 mm, sparsely setulose and more or less densely covered with very short glandular hairs, usually trullate or rhombic, cuneate at base, with 2–3 pairs of narrowly triangular, subacute to aristate teeth. Corolla 4·5–6 mm, white. Capsule (4·5–)5–6 mm, shorter than or as long as the calyx, 2½–3 times as long as wide, elliptic-oblong, rounded or subtruncate, apiculate to cuspidate. *Rocky and grassy places.* ● *Mountains of Krym.* Rs (K).

Subsect. *Angustifoliae* (Wettst.) Joerg. Capsule glabrous, or with a few small hairs on the margin.

41. E. salisburgensis Funck, *Bot. Taschenb.* **1794**: 190 (1794). Stem up to 20(–25) cm, with 0–7 pairs of slender, erect or patent branches. Cauline internodes up to 3(–4) times as long as the leaves. Lowest flower at node 2–13(–16). Leaves 3–13(–17) mm, usually strongly tinged with purple, glabrous or minutely scabrid, with the teeth distant from the leaf-base and at least the two distal pairs distant from each other; cauline leaves up to 5 mm wide (excluding teeth), cuneate-obovate, to narrowly oblong-ovate, with 1–4 pairs of subacute to aristate teeth; lower floral leaves up to 6·5 mm wide (excluding teeth), broadly obovate to narrowly ovate or trullate, with 1–5(–6) pairs of acute to aristate teeth. Corolla 5–7·5(–8·5) mm, white (rarely purple); lower lip usually rather small. Capsule 4–6·5 mm, 2–3½ times as long as wide, oblong to elliptical, truncate to emarginate. 2*n*=44. *Stony grassland, screes and scrub, usually on basic soils. Mountains of C. and S. Europe; Fennoscandia; W. Ireland.* Al Au ?Br Bu Co Cr Cz Fe Ga Ge Gr Hb He Hs It Ju No Po Rm Rs (W, K) Su.

Plants from high altitudes or high latitudes are sparsely branched, with a low node of flowering and large, emarginate capsules; in C. & S.E. Europe such plants often have purplish corollas. Plants which also have relatively wide leaves with shallow, obtuse teeth are usually referred in Fennoscandia to **E. lapponica** T. C. E. Fries, *Ark. Bot.* **17**(6): 12 (1922) and in C. & S. Europe to **E. nivalis** G. Beck, *Verh. Zool.-Bot. Ges. Wien* **33**: 225 (1883).

Plants from S.E. Sweden (Gotland) approach **43** in leaf-shape, but the central portion of the leaf is wider and the teeth shorter.

42. E. portae Wettst., *Österr. Bot. Zeitschr.* **43**: 196 (1893). Like **41** but with 0–3(–5) pairs of branches; cauline internodes 1–2 times as long as the leaves; leaves usually green, the lower floral not more than 4·5 mm wide (excluding teeth), trullate to linear-lanceolate, with 2–4 pairs of teeth; corolla 7·5–9 mm, lower lip relatively large, usually white, upper lilac; capsule 2¾–4 times as long as wide. *Pastures and screes, 600–2300 m.* ● *S. Alps (Alpi Tridentine, Alpi dell'Ortles).* It.

Probably of hybrid origin, between **21** and **41**.

Dwarf, sparingly branched plants with a low node of flowering, from the C. Appennini, have been referred to a distinct species, **E. italica** Wettst., *Monogr. Gatt. Euphrasia* 242 (1896). Their status requires clarification.

43. E. illyrica Wettst., *Österr. Bot. Zeitschr.* **43**: 131 (1893). Like **41** but with lowest flower at node 9–15; lower floral leaves not more than 2 mm wide (excluding teeth), with 2–4 pairs of narrowly falcate, aristate teeth, the distal pair often short and embraced by the middle pairs; calyx minutely scabrid; corolla 6·5–8·5 mm, lilac to purple; capsule 2½–4 times as long as wide, elliptic-obovate, rounded, cuspidate or truncate. *Rocky places and stony grassland.* ● *N.W. Jugoslavia, N.E. Italy.* It Ju.

Records from Bulgaria require confirmation.

44. E. dinarica (G. Beck) Murb., *Lunds Univ. Årsskr.* **27**(5): 72 (1892). Stem up to 14 cm, erect, with 1–5(–8) pairs of erect branches. Cauline internodes up to 1½ times as long as the leaves. Leaves 3–18 mm, up to 1·5 mm wide (excluding teeth), with up to 3 pairs of distant teeth, those of the cauline leaves deltate to linear, with subobtuse, cartilaginous apex, those of the lower floral

leaves falcate, acute, several times as long as wide, overlapping so as to produce a latticed effect, the distal pair short, greatly exceeded by the linear terminal lobe. Corolla 5·5–6·5 mm, lilac or purplish. Capsule 4·5–6(–7·5) mm, 3–5 times as long as wide, elliptic-obovate. *Dry, stony grassland in mountains.* ● *W. Jugoslavia; S. Italy.* It Ju.

45. E. cuspidata Host, *Fl. Austr.* **2**: 186 (1831) (incl. *E. stiriaca* Wettst.). Stem up to 20(–25) cm, with 3–9(–12) pairs of branches. Cauline internodes not more than twice as long as the leaves. Lowest flower at node 6–13(–20). Leaves green or brown-tinged, up to 2·5(–3) mm wide (excluding teeth), obovate-lanceolate to linear-lanceolate; teeth not more than 2(–3) pairs, distant from the leaf-base and from each other, those of the lower floral leaves deltate to linear or falcate and aristate. Corolla 8–9·5(–10·5) mm, white or with upper lip lilac. Capsule 4·5–6(–6·5) mm, usually much shorter than the calyx, 2½–4 times as long as wide, elliptic-obovate, often asymmetrical. *Stony and rocky places; calcicole.* ● *S. & E. Alps and adjacent foothills.* Au Ge It Ju.

46. E. tricuspidata L., *Sp. Pl.* 604 (1753). Like **45** but with the leaves linear, not more than 2(–3·5) mm wide (including teeth); teeth shorter than the width of the remainder of the leaf, those of the lower floral leaves not more than 1(–2) pairs; corolla 9–12 (–13) mm, sometimes lilac. *Stony places and wood-margins.* ● *S.E. Alps.* It.

30. Odontites Ludwig[1]

(incl. *Dispermotheca* Beauverd, *Macrosyringion* Rothm. and *Odontitella* Rothm.)

Annual, hemiparasitic herbs or dwarf shrubs. Leaves opposite, sessile, usually narrow, entire or obscurely (rarely conspicuously) toothed. Flowers zygomorphic, in terminal, bracteate, spike-like, secund racemes. Calyx tubular-campanulate, 4-toothed. Corolla with a cylindrical tube and 2-lipped limb; upper lip entire to shortly 2-lobed, the lobes not recurved; lower lip 3-lobed, with entire or slightly emarginate lobes. Stamens 4, didynamous; anthers glabrous or hairy, equally mucronate at base. Stigma capitate. Capsule loculicidal; seeds 1–2 mm, rather few, with longitudinal striae or low ridges.

1 Corolla pink, purple or white, without any yellow tint
  2 Corolla glabrous
    3 Stem and leaves glandular-puberulent to villous, with ±patent hairs   **5. granatensis**
    3 Stem and leaves appressed-eglandular-puberulent   **6. purpurea**
  2 Corolla hairy
    4 Stems procumbent   **14. corsica**
    4 Stems erect
      5 Leaves lanceolate, serrate to crenate   **15. verna**
      5 Leaves linear, entire
        6 Bracts 6–12 mm, the lower often exceeding the flowers; calyx 4–5 mm; corolla cream, tinged with pink   **10. jaubertiana**
        6 Bracts not more than 5 mm, not exceeding the flowers; calyx 3 mm; corolla purplish-pink throughout   **11. kaliformis**
1 Corolla yellow, sometimes tinged with red or pink
  7 Shrub, with non-flowering shoots at the base
    8 Inflorescences 3–8 cm, dense, aggregated into ±compact panicles; lower bracts greatly exceeding flowers; immature capsule hairy at apex   **12. bocconei**
    8 Inflorescences 2–4 cm, ±lax, not aggregated into panicles; bracts not or only slightly exceeding the flowers; capsule glabrous   **13. linkii**
  7 Annual, often with woody stem, but without non-flowering shoots at the base
    9 Stems procumbent   **14. corsica**
    9 Stems erect

---

¹ By D. A. Webb and J. M. Camarasa.

10 Leaves 4–10 mm wide, lanceolate, usually crenate-serrate
    9. lanceolata
10 Leaves 1–4 mm wide, linear to narrowly oblong, entire,
    or rarely with 1–2 pairs of teeth
  11 Corolla 9–25 mm
    12 Corolla-tube 5–7 mm, shorter than calyx; inflorescences
      very dense; capsule glabrous except for apical cilia
      1. tenuifolia
    12 Corolla-tube 9–21 mm, considerably exceeding calyx;
      inflorescences lax, at least in fruit; capsule hairy
      13 Corolla 17–25 mm, with tube at least twice as long as
        calyx; bracts linear   **2. longiflora**
      13 Corolla 12–15 mm, with tube not more than 1½ times as
        long as calyx; bracts lanceolate **3. glutinosa**
  11 Corolla 5–9 mm
    14 Hairs on stem patent and usually glandular at least in
      part   **4. viscosa**
    14 Hairs on stem crispate or appressed, eglandular
      15 Bracts *c.* 5 mm; corolla glabrous; capsule 5–7 mm,
        narrowly oblong, with not more than 4 seeds
        7. rigidifolia
      15 Bracts 6–12 mm; corolla usually pubescent; capsule
        3–5 mm, elliptical, with more than 4 seeds
        16 Anthers widely exserted   **8. lutea**
        16 Anthers ±included under upper lip of corolla
        10. jaubertiana

1. O. tenuifolia (Pers.) G. Don fil., *Gen. Syst.* **4**: 611 (1838). Densely appressed-pubescent annual. Stem 10–40 cm, erect; branches ascending. Leaves up to 10 × 1·5 mm, linear, entire, few. Inflorescences *c.* 2 cm, very dense; flowers subsessile; bracts linear, shorter than calyx. Calyx 6–10 mm; teeth lanceolate, acute. Corolla 9–12 mm, yellow, glabrous; tube 5–7 mm. Anthers hirsute, somewhat exserted. Capsule shorter than calyx, glabrous except for apical cilia. *Heaths, pinewoods and maritime sands.* ● *Spain and Portugal.* Hs Lu.

2. O. longiflora (Vahl) Webb, *Iter Hisp.* 24 (1838). Glandular-puberulent annual. Stem 12–30 cm, erect, usually with erecto-patent branches. Leaves linear, entire, few, the lower up to 20 mm, but the upper usually *c.* 5 mm. Inflorescences lax, few-flowered, often interrupted; bracts 7–10 mm, linear; flowers subsessile. Calyx 6–8 mm; teeth linear, obtuse. Corolla 17–25 mm, yellow, puberulent; tube 14–21 mm, very slender; lower lip deflexed. Anthers with apical tuft of hairs, included. Capsule oblong, equalling the calyx, hairy. *Dry places in the mountains. Spain.* Hs.

3. O. glutinosa (Bieb.) Bentham in DC., *Prodr.* **10**: 549 (1846). Glandular-puberulent annual. Stem 12–30 cm, erect, simple or with a few suberect branches. Leaves 10–20 mm, linear, entire. Inflorescences up to 8 cm in fruit, dense at first, lax later; bracts 5–9 mm, lanceolate; pedicels 1–2 mm. Calyx 6–8 mm; teeth linear-lanceolate. Corolla 12–15 mm, yellow, sometimes tinged with brownish-purple, puberulent; tube 9–12 mm. Anthers with apical tuft of hairs, included. Capsule slightly exceeding calyx, hairy. *Stony mountain slopes. Balkan peninsula; Krym.* Al Bu Gr Ju Rs (K) [Ga].

O. pilatiana Ronniger ex Rohlena, *Mitt. Thür. Bot. Ver.* nov. ser., **51**: 353 (1944), differs only in being glabrous except in the inflorescence. It was described from a single collection on the borders of Albania and Jugoslavia (Kovat), and is best regarded as a mutant of **3**.

4. O. viscosa (L.) Clairv., *Man. Herb. Suisse* 207 (1811). Annual; more or less glandular-puberulent to -pubescent; sometimes also eglandular-hirsute. Stem 20–60 cm, erect; branches numerous, patent to erecto-patent. Leaves 15–40 × 1–4 mm, linear to oblong, entire or with 1–2 pairs of teeth. Inflorescences up to 6 cm, lax to moderately dense; pedicels 1–2 mm. Calyx 3 mm; teeth ovate-lanceolate, acute to obtuse. Corolla 5–6 mm, yellow, glabrous to pubescent; tube *c.* 3 mm. Anthers with apical tuft of hairs, more or less included. Capsule obovate, equalling the calyx, glabrous except for apical cilia. *Woods and dry hillsides. S.W. Europe, eastwards to S.W. Switzerland and N.W. Italy.* Ga He Hs It Lu.

(a) Subsp. **viscosa**: Bracts 5–7 mm, linear, not or scarcely ciliate. Stem and leaves densely glandular-pubescent. *Throughout the range of the species.*

(b) Subsp. **hispanica** (Boiss. & Reuter) Rothm., *Mitt. Thür. Bot. Ver.* nov. ser., **50**: 279 (1943) (*O. hispanica* Boiss. & Reuter): Bracts 3–4 mm, lanceolate, conspicuously white-ciliate. Stem and leaves rather sparsely glandular-pubescent and usually eglandular-hirsute. ● *Spain.*

5. O. granatensis Boiss., *Elenchus* 71 (1838). Glandular-puberulent to -villous annual. Stem 8–20 cm, erect; branches flexuous, ascending. Leaves 10–25 mm, linear-lanceolate, entire. Inflorescences up to 3 cm, with 4–9 flowers, rather lax; bracts 4–8 mm, linear-lanceolate; pedicels 1 mm. Calyx 4–5 mm; teeth triangular-oblong, subacute. Corolla 6–7 mm, dull purple, glabrous; tube 4 mm. Anthers hairy at apex, slightly exserted. Capsule 4 mm, obovate, ciliate. *Mountain scrub.* ● *N. & S. Spain (Cordillera Cantábrica, Sierra Nevada).* Hs.

6. O. purpurea (Desf.) G. Don fil., *Gen. Syst.* **4**: 611 (1838). Like **5** but sparsely eglandular-appressed-puberulent; stem up to 40 cm, with few, short branches; calyx 3–4 mm; anthers glabrous, included. *S. Spain.* Hs.

Many records of this species in Spain are errors for **11**.

7. O. rigidifolia (Biv.) Bentham in DC., *Prodr.* **10**: 550 (1846). Sparsely appressed-puberulent annual. Stem 15–40 cm, erect; branches long, ascending, few. Leaves 4–8(–20) mm, suberect, linear-oblanceolate, entire. Inflorescences up to 6 cm in fruit, moderately dense; bracts 5 mm, linear, flowers sessile. Calyx 5–7 mm; teeth lanceolate, acute. Corolla 7–9 mm, yellow tinged with red, glabrous. Anthers glabrous, included. Capsule 5–7 mm, narrowly oblong, sparsely puberulent. *Dry, open habitats. Sicilia.* Si.

8. O. lutea (L.) Clairv., *Man. Herb. Suisse* 207 (1811) (*Orthantha lutea* (L.) A. Kerner ex Wettst.). Appressed- or crispate-pubescent annual. Stem 10–40 cm, erect; branches ascending, numerous. Leaves 5–25 × 1–2 mm, linear, obtuse, entire or remotely and shortly toothed. Inflorescences 3–8(–12) cm, rather lax; bracts 6–8 mm, linear-lanceolate; pedicels *c.* 1 mm. Calyx 3–4 mm; teeth deltate to triangular, subacute. Corolla 5–8 mm, bright yellow, more or less pubescent; lower lip deflexed. Anthers glabrous, conspicuously exserted. Capsule 3–4 mm, elliptical, pubescent. $2n = 20$. *Dry grassland and scrub. S. & C. Europe, extending to N.C. France and to S.C. Russia.* ?Al Au Bu Co Cz Ga Ge He Hs Hu It Ju Po Rm Rs (C, W, K, E) Sa Si.

9. O. lanceolata (Gaudin) Reichenb., *Fl. Germ. Excurs.* 862 (1832). Like **8** but leaves 15–45 × 4–10 mm, lanceolate, crenate-serrate (rarely subentire); bracts 10–30 mm, lanceolate, resembling the leaves and greatly exceeding the flowers; calyx-teeth narrowly triangular; corolla 8–9 mm. *Meadows.* ● *E. Pyrenees, S.W. & S.C. Alps.* Ga Hs It.

Plants from the C. Pyrenees, which appear to correspond to **O. pyrenaea** (Bubani) Rothm., *Cavanillesia* **7**: 120 (1935), are intermediate in most respects between **8** and **9**, but have some glandular hairs in the inflorescence. They require further investigation.

Plants from Sicilia generally similar to **9** but with linear bracts and yellowish-pink corolla *c.* 10 mm have been called O. **verna** subsp. **sicula** (Guss.) P. D. Sell, *Watsonia* **6**: 303 (1967). They appear to be intermediate between **9** and **15** and require further investigation.

10. **O. jaubertiana** (Boreau) D. Dietr. ex Walpers, *Repert. Bot. Syst.* **3**: 401 (1844). Appressed- or crispate-pubescent annual. Stem 20–50 cm, erect, rather sparingly branched. Leaves 5–18 × 1–3 mm, linear to linear-lanceolate. Inflorescences 2–6 cm, rather lax; bracts 6–12 mm, linear-lanceolate; flowers subsessile. Calyx 4–5 mm; teeth triangular to deltate, subacute. Corolla 7–9 mm, cream or yellow, sometimes tinged with pink, puberulent; lower lip more or less porrect. Anthers more or less included, glabrous or slightly hairy. Capsule 3·5–4 mm, elliptical, pubescent. *Calcicole.* ● *W., C. & S. France.* Ga.

1 Stem and leaves glandular-pubescent; anthers purplish
(c) subsp. **cebennensis**
1 Stem and leaves eglandular-puberulent; anthers yellow
2 Corolla cream or pale yellow, tinged with red; leaves and bracts entire (a) subsp. **jaubertiana**
2 Corolla bright yellow; leaves and lower bracts usually with 1–2 pairs of teeth (b) subsp. **chrysantha**

(a) Subsp. **jaubertiana**: Eglandular. Branches patent to ascending, flexuous. Leaves and bracts linear, entire. Calyx-teeth shorter than tube. Corolla cream or pale yellow, tinged with red. Anthers yellow. *Almost throughout the range of the species.*

(b) Subsp. **chrysantha** (Boreau) P. Fourn., *Quatre Fl. Fr.* 787 (1937) (*O. chrysantha* Boreau): Eglandular. Branches ascending to suberect, more or less rigid, usually short. Leaves and bracts linear-lanceolate, usually with 1–2 teeth on each side. Calyx-teeth equalling tube. Corolla bright yellow. Anthers yellow. *Pastures and woodland-margins. N.C. France.*

(c) Subsp. **cebennensis** (Coste & Soulié) P. Fourn., *loc. cit.* (1937): Like (b) but stem and leaves glandular-pubescent; anthers purplish. *Rocky slopes. S. France* (*Causse Noir, near Millau*).

Plants from N.E. Spain described as *O. cebennensis* var. *roseiflora* Sennen appear to be transitional between subspp. (a) and (c).

11. **O. kaliformis** (Pourret) Pau, *Bol. Soc. Aragon. Ci. Nat.* **6**: 28 (1907) (*O. purpurea* sensu Lange pro parte, non (Desf.) G. Don fil.). Densely eglandular-appressed-puberulent annual. Stem 20–40 cm, erect, with short, ascending to suberect branches. Leaves 8–20 mm, linear, obtuse. Inflorescences 1–5 cm, fairly dense; bracts 3–5 mm, linear-oblong to ovate-lanceolate; pedicels 0·5–1 mm. Calyx 3 mm; teeth ovate-lanceolate, obtuse, about equalling the tube. Corolla *c.* 6 mm, purplish-pink, puberulent. Anthers glabrous, exserted. Capsule 2·5–3 mm, puberulent. *Dry places; calcicole.* ● *S. & E. Spain.* Hs.

12. **O. bocconei** (Guss.) Walpers, *Repert. Bot. Syst.* **3**: 400 (1844). Glabrous dwarf shrub. Stems branched from the base; branches 30–90 cm, up to 6 mm in diameter, pendent to erect; short, non-flowering shoots present. Leaves up to 45 × 7 mm, linear-oblong, acute. Inflorescences 3–8 cm, dense, aggregated to form fairly compact panicles; bracts up to 22 mm, leaf-like, at least the lowest greatly exceeding the flowers, usually recurved at apex; pedicels *c.* 1 mm. Calyx 6 mm; teeth triangular-lanceolate, acute. Corolla 8–10 mm, bright yellow; tube 5–6 mm.

Anthers sparsely hairy at apex, more or less included. Capsule 7 mm, oblong-obovate, hairy at apex when young. *Calcareous rocks.* ● *Sicilia.* Si.

13. **O. linkii** Heldr. & Sart. ex Boiss., *Diagn. Pl. Or. Nov.* 3(3): 177 (1856) (incl. *O. frutescens* Halácsy). Like **12** but sometimes sparsely pubescent; stems not more than 50 cm; leaves not more than 4 mm wide, linear; inflorescences rather lax, not aggregated into panicles; bracts 4–7 mm, not recurved at apex and scarcely exceeding the flowers; corolla 7 mm with tube 4·5 mm; anthers sometimes glabrous; capsule 5–6 mm, glabrous. *Mountain rocks.* ● *C. & S. Greece, S. Aegean region.* Cr Gr.

14. **O. corsica** (Loisel.) G. Don fil., *Gen. Syst.* **4**: 611 (1838). Scabrid-strigulose annual. Stems 8–15(–25) cm, procumbent, branched from the base. Leaves 6–10 mm, linear-oblong, obtuse. Inflorescences 1–2 cm, fairly dense; bracts 4–7 mm, linear-lanceolate; pedicels 1–2 mm. Calyx 4–5 mm; teeth triangular, obtuse. Corolla 5–6 mm, yellow (rarely white), pubescent; lower lip more or less porrect. Anthers glabrous, included. Capsule 4–5 mm, obovate, pubescent. *Rocky mountain pastures.* ● *Corse, Sardegna.* Co Sa.

15. **O. verna** (Bellardi) Dumort., *Fl. Belg.* 32 (1827) (*O. rubra* Besser). Appressed-pubescent to strigulose annual. Stem 10–50 cm, erect or ascending, somewhat 4-angled, simple or branched. Leaves 12–40 × 3–8 mm, lanceolate, crenate to serrate, rarely subentire. Inflorescences 3–10 cm, rather lax; flowers subsessile. Calyx 5–8 mm; teeth triangular, obtuse. Corolla 8–10 mm, pubescent, reddish-pink (rarely white); tube 5–6 mm; lower lip somewhat deflexed. Anthers glabrous, more or less exserted. Capsule 6–8 mm, oblong, pubescent. *Meadows, pastures, roadsides, cultivated fields and other disturbed habitats. Most of Europe, but absent from some of the islands.* All except Az Bl Cr Fa Is Sa Sb.

Very variable; the variation is, however, difficult to express adequately in infraspecific categories. There are local populations of distinct facies, and also some fairly well-marked ecotypes; apart from this the species is often cited as an illustration of seasonal dimorphism. The three best-known and most widespread variants are here treated as subspecies, but intermediate plants are fairly common, especially in N.C. Europe. Studies of cultivated plants and chromosome counts from different regions are required before the complex can be properly understood.

1 Bracts not exceeding flowers; branches usually long and ± patent (c) subsp. **serotina**
1 Bracts exceeding flowers; branches (if present) usually short and suberect
2 Stem usually simple; lower internodes longer than leaves
(b) subsp. **litoralis**
2 Stem usually with several suberect branches; lower internodes shorter than or equalling leaves (a) subsp. **verna**

(a) Subsp. **verna**: Stem 10–30 cm; lower internodes shorter than or equalling leaves. Branches fairly short, straight, suberect, the uppermost not more than 2 nodes below the terminal inflorescence. Leaves crenate to serrate, widest at the base. Bracts 10–15(–20) mm. Early-flowering. $2n=40$. ● *N. & C. Europe, eastwards to N.W. Russia, and extending southwards in the mountains to N. Portugal, C. Italy and N. Greece.*

Small plants from coastal regions of N.W. Europe, similar to (a) but late-flowering and with very short lower internodes, and thus transitional to (c), have been distinguished as subsp. **pumila** (Nordstedt) A. Pedersen, *Bot. Tidsskr.* **58**: 291 (1963).

(b) Subsp. litoralis (Fries) Nyman, *Consp.* 551 (1881) (*O. litoralis* Fries): Stem 10–35 cm, usually simple, rarely with 1–2 pairs of short, suberect branches immediately below the inflorescence; lower internodes longer than the leaves. Leaves crenate to serrate, widest at or near the base. Bracts 8–20 mm. Early-flowering. $2n=20$. *Meadows by the sea.* ● *Baltic region, extending to S. Norway and the Netherlands.*

(c) Subsp. serotina (Dumort.) Corb., *Nouv. Fl. Normand.* 437 (1894) (*O. serotina* Dumort.; incl. *O. salina* (Kotov) Kotov, *O. virgata* Lange): Stem 20–50 cm; lower internodes usually shorter than leaves. Branches long, more or less patent, often ascending near the apex, the uppermost usually 2–4 nodes below the terminal inflorescence. Leaves rather obscurely crenate to subentire, widest some distance above the base. Bracts 7–10 mm. Late-flowering. $2n=20$. *Almost throughout the range of the species, but rare in parts of the north.*

31. Bartsia L.[1]

Perennial, hemiparasitic herbs with a short, rhizomatous, subterranean stock. Leaves opposite, sessile, toothed. Flowers in terminal, bracteate, spike-like racemes, which are not secund. Calyx tubular-campanulate, 4-toothed. Corolla with a cylindrical tube and 2-lipped limb; upper lip entire or emarginate, longer than the lower; lower lip 3-lobed. Stamens 4, didynamous; anthers hairy, equally mucronate at the base. Stigma capitate. Capsule loculicidal; seeds *c.* 2 mm, rather few, with 1 or more longitudinal, membranous wings; hilum lateral.

1 All bracts ovate, crenate-serrate, exceeding the calyx **1. alpina**
1 At least the middle and upper bracts linear-lanceolate, entire,
    not exceeding the calyx
2 Lowest bracts ovate, serrate **2. spicata**
2 Lowest bracts linear-lanceolate, entire **3. aspera**

1. B. alpina L., *Sp. Pl.* 602 (1753). Glandular-hairy; stems 8–30(–40) cm, erect or ascending, simple. Leaves 10–25 × 6–15 mm, ovate, obtuse to subacute, crenate-serrate. Bracts like the leaves but decreasing in size upwards, dull purple, exceeding the calyx. Calyx 6–8 mm; teeth as long as tube or shorter, narrowly triangular, obtuse. Corolla 15–20 mm, dull, dark purple. Capsule distinctly longer than calyx. $2n=12, 24, 36$. *Damp places; usually on base-rich soils.* N. Europe, and southwards in the mountains to the Pyrenees, S. Alps and S.W. Bulgaria. ?Al Au Br Bu Cz Fa Fe Ga Ge He Hs Is It Ju No Po Rm Rs (N, W) Su.

2. B. spicata Ramond, *Bull. Soc. Philom. Paris* 2: 141 (1800). Like **1** but stems somewhat branched; middle and upper bracts linear-lanceolate, entire, not exceeding the calyx; calyx-teeth acute; capsule not or scarcely longer than calyx. *Mountain rocks; calcicole.* ● *C. Pyrenees, Cordillera Cantábrica.* Ga Hs.

3. B. aspera (Brot.) Lange in Willk. & Lange, *Prodr. Fl. Hisp.* 2: 614 (1870). Like **1** but appressed-hispid and scabrid; stems branched; all bracts linear-lanceolate, entire, shorter than the calyx; calyx-teeth acute; corolla yellow, changing to reddish-brown; capsule shorter than calyx. *Dry scrub.* ● *C. & S. Portugal; one station in S.W. Spain.* Hs Lu.

32. Parentucellia Viv.[1]

Like *Bartsia* but annual; corolla with the lower lip longer than the upper; seeds *c.* 0·5 mm, numerous, smooth or finely reticulate; hilum basal.

Calyx-teeth nearly as long as tube; corolla yellow or white,
    caducous **1. viscosa**
Calyx-teeth about half as long as tube; corolla reddish-purple
    (rarely white), persistent **2. latifolia**

1. P. viscosa (L.) Caruel in Parl., *Fl. Ital.* 6: 482 (1885) (*Bartsia viscosa* L.). Glandular-hairy; stem 10–50(–70) cm, erect, usually simple. Leaves 10–45 × 3–15 mm, oblong to lanceolate, acute to subacute, coarsely serrate. Bracts like the leaves, but decreasing in size upwards. Calyx 10–16 mm; teeth linear-lanceolate, nearly as long as tube. Corolla 16–24 mm, yellow (less often white), caducous. Capsule 7–9 mm, oblong, pubescent. Seeds *c.* 0·5 mm. $2n=48$. *Damp, grassy or sandy places. S. & W. Europe, northwards to S.W. Scotland; naturalized or casual further north.* Al Az Bl Br Co Cr Ga Gr Hb Hs It Ju Lu Sa Si Tu [Be Da Ho].

2. P. latifolia (L.) Caruel in Parl., *op. cit.* 480 (1885) (*Bartsia latifolia* (L.) Sibth. & Sm.). Like **1** but stem 5–30 cm; leaves 4–12 × 3–5 mm, triangular-lanceolate, very deeply toothed; calyx 6–10 mm, with teeth about half as long as tube; corolla 8–10 mm, reddish-purple (rarely white), persistent; capsule glabrous. *Sandy and stony places. S. Europe, extending northwards to N.W. France.* Al Bl Bu Co Cr Ga Gr Hs It Ju Lu Sa Si Tu.

33. Bellardia All.[1]

Like *Bartsia* but annual; calyx ventricose-campanulate, rather deeply cleft above and below into 2 shortly bidentate segments; lower lip of corolla longer than the upper; stigma clavate; seeds *c.* 1 mm, numerous, with several longitudinal ridges and fine transverse striations, not winged; hilum basal.

1. B. trixago (L.) All., *Fl. Pedem.* 1: 61 (1785) (*Bartsia trixago* L.). Glandular-pubescent; stems 15–70 cm, usually simple, erect. Leaves 15–90 × 1–15 mm, linear to linear-lanceolate, obtuse to subacute, coarsely and obtusely serrate. Bracts like the leaves, but decreasing in size upwards, the upper ovate, cordate, entire. Calyx 8–10 mm; teeth triangular, less than ¼ as long as tube. Corolla 20–25 mm, purple and white or yellow, more rarely entirely white. Capsule subglobose; seeds *c.* 0·7 mm, oblong to reniform. $2n=24$. *Stony or grassy places. S. Europe.* Al Az Bl Bu Co Cr Ga Gr Hs It Ju Lu Sa Si Tu.

34. Pedicularis L.[2]

Perennial (rarely annual or biennial), hemiparasitic herbs. Leaves usually alternate (rarely opposite or whorled), usually pinnatisect, the lower usually long-petiolate, the upper often subsessile. Flowers in terminal, bracteate spikes or racemes. Calyx tubular to campanulate, 5-toothed, often more or less 2-lipped. Corolla with cylindrical to obconical tube; limb 2-lipped, the upper lip galeate, laterally compressed, the lower 3-lobed, usually patent. Stamens 4, didynamous; anthers glabrous, the lobes equally mucronate at the base. Stigma capitate. Capsule not or moderately compressed, loculicidal; seeds numerous.

The descriptions of the leaves apply to those of the basal rosette; in most species the cauline leaves are smaller, less deeply divided, and with petiole shorter or absent.

Within the genus all intermediate states exist between virtually sessile flowers and those in which the pedicel is longer than the calyx. In this account *spike* is used to denote an inflorescence in which most of the pedicels are less than 2 mm.

Literature: G. Bonati, *Le genre Pedicularis.* Nancy. 1918. I. Klášterský, *Bull. Int. Acad. Tchèque Sci.* **29**: 202–230 (1928).

---

[1] By T. G. Tutin.   [2] By E. Mayer.

W. Limpricht, *Feddes Repert.* **20**: 161–265 (1924). H. Steininger, *Bot. Centr.* **28** (1886); **29** (1887); **30** (1887).

1 Corolla with obconical-campanulate tube and connivent lips, the lower yellow, bordered with red; capsule globose **2. sceptrum-carolinum**

1 Corolla with ±cylindrical tube and patent, ±concolorous lower lip; capsule ±ovoid, somewhat compressed laterally

2 Upper lip of corolla truncate or rounded, neither beaked nor toothed (rarely with 2 very small teeth directed upwards)

3 Acaulescent; flowers arising directly from the leaf-rosette; pedicels 10–30 mm **1. acaulis**

3 Caulescent; flowers in spikes or racemes; pedicels less than 10 mm

4 Upper cauline leaves and bracts in whorls of 3–4

5 Calyx-teeth entire; calyx-tube glabrous except for long hairs on the veins **15. verticillata**

5 Calyx-teeth serrate

6 Calyx glabrous; lower lip of corolla longer than the upper **16. amoena**

6 Calyx villous; lower lip of corolla shorter than or equalling the upper **17. arguteserrata**

4 Cauline leaves and bracts alternate

7 Corolla entirely yellow

8 Calyx split to half-way on lower side; teeth very short

9 Calyx pubescent to villous; corolla-tube hairy inside; bracts hairy beneath **5. hacquetii**

9 Calyx glabrous, or with a few hairs on the veins; corolla-tube glabrous inside; bracts glabrous **6. exaltata**

8 Calyx not split on lower side; teeth conspicuous

10 Upper lip of corolla arachnoid-villous; capsule slightly exceeding the calyx **3. foliosa**

10 Upper lip of corolla glabrous or pubescent; capsule nearly twice as long as calyx

11 Bracts exceeding the flowers; corolla up to 25 mm **4. hoermanniana**

11 Bracts shorter than the flowers; corolla not more than 15 mm **7. recutita**

7 Corolla red or pink, at least in part

12 Corolla mainly yellow, but with upper lip tinged with red

13 Stem usually more than 20 cm; leaves 25–30 mm wide **7. recutita**

13 Stem usually less than 20 cm; leaves 8–12 mm wide

14 Corolla 12–20 mm; calyx hairy; capsule *c.* 1½ times as long as calyx **13. oederi**

14 Corolla 10–12 mm; calyx glabrous; capsule at least twice as long as calyx **14. flammea**

12 Corolla entirely red or pink

15 Calyx ±glabrous; pedicels thickened distally **8. limnogena**

15 Calyx hairy; pedicels not thickened distally

16 Leaves deeply pinnatisect, with narrow rhachis

17 Leaves pinnatifid, with acute segments; calyx-teeth entire **9. rosea**

17 Leaves 2-pinnatisect, with obtuse segments; calyx-teeth serrate **10. orthantha**

16 Leaves pinnatifid or pectinate, with wide rhachis

18 Corolla 17–20 mm, hairy; calyx-teeth entire **11. dasyantha**

18 Corolla 10–15 mm, glabrous; calyx-teeth serrate to incise-dentate **12. hirsuta**

2 Upper lip of corolla terminating in a beak, or in 2 teeth directed forwards or downwards

19 Beak of corolla very short, sometimes ±obsolete, bidentate at apex

20 Lateral flowering branches present; plant usually annual or biennial

21 Corolla yellow, the upper lip becoming tinged with violet after anthesis **19. labradorica**

21 Corolla pink or red

22 Calyx not 2-lipped; lateral margins of upper lip of corolla entire **20. sylvatica**

22 Calyx 2-lipped; upper lip of corolla with a small tooth on each side some distance below the apex **18. palustris**

20 Lateral flowering branches absent; plant perennial

23 Corolla pink, red or purple

24 Calyx-teeth wider than long, obtuse **26. asparagoides**

24 Calyx-teeth longer than wide, acute

25 Calyx-teeth toothed or crenate **21. sudetica**

25 Calyx-teeth entire

26 Stem 2–5(–8) cm; bracts glabrous **36. ferdinandi**

26 Stem 7–30 cm; bracts whitish-lanate

27 Leaf-segments obtuse; bracts entire; filaments glabrous **32. dasystachys**

27 Leaf-segments apiculate; bracts serrate to pinnatisect; filaments ±hairy **35. petiolaris**

23 Corolla yellowish

28 Calyx split on lower side, the two lateral teeth on each side ±united **29. heterodonta**

28 Calyx with (4–)5 distinct teeth, not split on lower side

29 Calyx-teeth wider than long, usually obtuse

30 Calyx-teeth crenate **27. schizocalyx**

30 Calyx-teeth entire

31 All bracts pinnatisect, similar to the leaves **23. kaufmannii**

31 Upper bracts 3-fid or dentate, not pinnatisect

32 Inflorescence ±glabrous **22. comosa**

32 Inflorescence densely hairy

33 Capsule 16–18 mm, ±obtuse; lower lip of corolla usually glabrous **24. sibthorpii**

33 Capsule 9–12 mm, apiculate; lower lip of corolla ciliate **25. uralensis**

29 Calyx-teeth at least as long as wide, acute

34 Calyx-teeth about as long as wide **28. brachyodonta**

34 Calyx-teeth 2–3 times as long as wide

35 Inflorescence sparsely hirsute; calyx-teeth linear-spathulate, toothed near the apex only **30. leucodon**

35 Inflorescence lanate or villous; calyx-teeth lanceolate, with margin uniformly serrate, crenate or entire

36 Calyx glandular-villous **33. graeca**

36 Calyx eglandular-hairy

37 Stem arachnoid-lanate; upper bracts 3-lobed **31. physocalyx**

37 Stem subglabrous below; upper bracts entire **34. friderici-augusti**

19 Beak of corolla conspicuous, not bidentate at apex

38 Corolla yellow or cream-coloured

39 Calyx-teeth foliose, toothed or lobed

40 Stem uniformly lanate; calyx-teeth glabrous on margin and inner surface **39. tuberosa**

40 Stem with 2 rows of hairs; calyx-teeth pubescent on margin and inner surface

41 Bracts ciliate but otherwise glabrous; calyx-tube glabrous **40. elongata**

41 Bracts and calyx-tube densely villous **41. julica**

39 Calyx-teeth entire or obscurely serrate, not foliose

42 Calyx deeply split on lower side; leaves pinnatifid **38. lapponica**

42 Calyx not split on lower side; leaves pinnatisect

43 Calyx-teeth wider than long, obtuse **37. compacta**

43 Calyx-teeth at least as long as wide, acute

44 Bracts ±glabrous, with entire segments; calyx-tube glabrous **42. ascendens**

44 Bracts hairy, with serrate segments; calyx-tube villous **43. baumgartenii**

38 Corolla red, pink or purple

45 Corolla-tube at least 1½ times as long as calyx

46 Leaves crenate-serrate, not lobed; calyx strongly 2-lipped **54. resupinata**

46 Leaves pinnatifid or pinnatisect; calyx ±regularly toothed, not 2-lipped

47 Flowers subsessile; calyx densely lanate **50. gyroflexa**

47 Flowers distinctly pedicellate; calyx glabrous or slightly pubescent

48 Stem decumbent; calyx-teeth nearly as long as tube; corolla not more than 21 mm **51. elegans**

48 Stem erect; calyx-teeth about half as long as tube; corolla up to 25(–30) mm **52. portenschlagii**
45 Corolla-tube equalling or slightly exceeding calyx
49 Flowers distinctly pedicellate
50 Lower lip of corolla ciliate, at least at anthesis **45. rostratocapitata**
50 Lower lip of corolla glabrous
51 Stems procumbent to ascending; calyx glabrous or pubescent **46. kerneri**
51 Stems erect; calyx villous-lanate with reddish hairs **53. asplenifolia**
49 Flowers subsessile
52 Calyx glabrous; inflorescence capitate **47. pyrenaica**
52 Calyx hairy; inflorescence ± elongate
53 Petioles of basal leaves villous; cauline leaves 0–2 **49. cenisia**
53 All petioles glabrous; cauline leaves usually 3 or more
54 Bracts 3-lobed; calyx-teeth not foliose; corolla not more than 13 mm **44. rostratospicata**
54 Bracts pinnatifid; some of the calyx-teeth foliose; corolla up to 18 mm **48. mixta**

Sect. ANODONTAE (Bunge) Maxim. Perennial. Upper lip of corolla rounded at apex, neither rostrate nor toothed.

1. P. acaulis Scop., *Fl. Carn.* ed. 2, **1**: 439 (1771). Acaulescent; leaves and flowers arising from a cylindrical stock. Leaves ovate, pinnatisect with pinnatifid segments; petiole and rhachis pubescent; segments glabrous. Pedicels 10–30 mm, pubescent. Calyx hairy, with 5 lanceolate, foliose, serrate teeth about equalling the tube. Corolla up to 35 mm, white tinged with red; tube about equalling calyx; upper lip falcate; both lips ciliate. Capsule compressed-globose, obliquely apiculate, somewhat exceeding calyx. *Damp or shady grassland.* ● *Foothills of S. Alps and of mountains of N.W. Jugoslavia.* It Ju.

2. P. sceptrum-carolinum L., *Sp. Pl.* 608 (1753). Stem 15–80(–100) cm, erect, glabrous, often reddish. Basal leaves 12–20(–30) cm, forming a rosette, lanceolate, pinnatifid, glabrous or somewhat pubescent beneath; segments ovate, obtuse, crenate. Cauline leaves few. Flowers alternate or in whorls of 3, in a long, lax spike; bracts ovate. Calyx glabrous; teeth irregularly serrate. Corolla up to 32 mm, suberect, pale yellow with red margin to lower lip; tube obconical-campanulate; lips connivent, the upper falcate, ciliate, the lower with glabrous margin, but with 2 hairy ridges. Capsule globose, shortly mucronate, somewhat exceeding calyx. $2n=32$. *Fens, wet woods and river-banks. Fennoscandia and N. & C. parts of U.S.S.R., extending very locally in C. Europe to S. Germany and C. Romania.* †Au Cz Da Fe Ge No Po Rm Rs (N, B, C, W) Su.

3. P. foliosa L., *Mantissa* 86 (1767). Stem 20–50 cm, erect, pubescent, leafy mainly in upper half. Leaves broadly lanceolate, 2- to 3-pinnatisect, dull green, glabrous above, pubescent beneath. Flowers in a dense spike; bracts exceeding the flowers, the lower leaf-like, the upper less divided. Calyx membranous, hairy on the veins and margin; teeth unequal, broadly triangular, entire. Corolla up to 25 mm, pale yellow; tube exceeding the calyx; upper lip villous, nearly straight, obtuse. Capsule ovoid, shortly apiculate, somewhat exceeding calyx. $2n=16$. *Meadows, streamsides and scrub.* ● *Mountains of S. & S.C. Europe from the Vosges to N. Spain and S. Italy.* Au Ga Ge He Hs It.

4. P. hoermanniana K. Malý, *Glasn. Muz. Bosni Herceg.* **11**: 145 (1899) (*P. foliosa* auct. balcan., non L.). Like **3** but stem 50–100 cm; calyx coriaceous; upper lip of corolla shortly pubescent; capsule obtuse, almost twice as long as calyx. *Meadows and open scrub.* ● *Balkan peninsula, southwards to S.W. Bulgaria, and extending northwards to Slovenija.* ?Al Bu Ju.

5. P. hacquetii Graf, *Flora (Regensb.)* **17**: 40 (1834). Stem 30–120 cm, erect, glabrous or pubescent, leafy mainly in upper half. Leaves lanceolate, 3-pinnatisect, glabrous above, somewhat hairy beneath. Flowers in a rather dense spike; bracts exceeding the flowers, the lower leaf-like, the upper smaller, incise-dentate. Calyx coriaceous, pubescent to villous, split to half-way on lower side; teeth short but distinct, deltate, obtuse. Corolla up to 25 mm, pale yellow; tube nearly twice as long as calyx, hairy inside; upper lip straight, glabrous or sparsely ciliate. Capsule obliquely ovoid, shortly apiculate, somewhat exceeding calyx. *Mountain meadows.* ● *S.E. Alps, Carpathians, C. Appennini.* Au Cz It Ju Po Rm Rs (W).

6. P. exaltata Besser, *Flora (Regensb.)* **15** (*Beibl.* 2): 19 (1832). Like **5** but stem 100–150(–200) cm, leafy throughout; spike lax; bracts glabrous; calyx subglabrous, with very short, sometimes obsolescent teeth; corolla-tube glabrous inside. *Open woodland and scrub.* ● *Carpathians and adjacent hill-country, extending to W. White Russia and W. Ukraine.* Cz Po Rm Rs (C, W).

7. P. recutita L., *Sp. Pl.* 608 (1753). Glabrous except for calyx-teeth. Stem 20–60 cm, erect. Leaves lanceolate, pinnatisect, with lanceolate, incise-dentate segments, the lower leaves long-petiolate. Flowers in a raceme, dense at first, elongating later; bracts shorter than flowers, the lower pinnatisect, the upper 3-lobed or undivided. Calyx-teeth almost as long as tube, unequal, lanceolate, entire, ciliate. Corolla up to 15 mm, glabrous, greenish-yellow, usually tinged with dull crimson; upper lip almost straight, much longer than the lower. Capsule ovoid, apiculate, almost twice as long as calyx. $2n=16$. *Mountain meadows and damp or shady places.* ● *Alps.* Au Ga Ge He It Ju.

8. P. limnogena A. Kerner, *Österr. Bot. Zeitschr.* **13**: 362 (1863). Stem 8–20 cm, erect or ascending, glabrous except for calyx-teeth. Basal leaves numerous, lanceolate, pinnatisect with ovate-oblong, obtuse, pinnatifid segments; cauline leaves 2–4. Flowers in an elongate raceme; bracts oblong, toothed, shorter than flowers; pedicels thickened distally. Calyx membranous; teeth shortly triangular, acute, entire, sparsely ciliate. Corolla up to 12(–15) mm, purplish-red; upper lip almost straight, longer than the lower. Capsule ovoid-conical, with curved, obtuse apex, twice as long as calyx. *Streamsides, springs and wet grassland.* ● *Mountains of Transylvania and Makedonija.* Ju Rm.

9. P. rosea Wulfen in Jacq., *Misc. Austr. Bot.* **2**: 57 (1781). Stem 2–15 cm, erect, glabrous below, hairy and purple-tinted above. Leaves 5–8(–12) cm, including petiole, oblong-lanceolate, pinnatifid with more or less dentate, acute segments, glabrous. Flowers in a more or less capitate spike; bracts villous, longer than the flowers. Calyx lanate-villous; teeth unequal, nearly as long as tube, lanceolate, acute, entire. Corolla 12–18 mm, pink to lilac, glabrous; tube exceeding the calyx; upper lip nearly straight. Stigma clearly exserted. Capsule obliquely ovoid, apiculate, slightly exceeding calyx. $2n=16$. *Screes and stony grassland; calcicole. S.W., S. & E. Alps; two stations in the Pyrenees.* Au Ga It Ju.

(a) Subsp. **rosea**: Leaf-segments triangular, closely incise-dentate. Lower bracts dentate to somewhat laciniate, the upper entire. Filaments densely pubescent. *S. & E. Alps.*

(b) Subsp. **allionii** (Reichenb. fil.) E. Mayer, *Österr. Bot. Zeitschr.* **119**: 324 (1971) (*P. allionii* Reichenb. fil.): Leaf-segments lanceolate, obscurely and remotely dentate. Lower bracts deeply laciniate, the upper mostly bifid. Filaments sparsely pubescent to subglabrous. *S.W. Alps; Pyrenees.*

10. P. orthantha Griseb., *Spicil. Fl. Rumel.* **2**: 15 (1844). Stems 8–12 cm, erect or ascending, glabrous or sparsely crispate-hairy, blackish, leafy mainly in upper half. Leaves about as long as stem, ovate-lanceolate, 2-pinnatisect, with lanceolate, incise-dentate, obtuse segments, glabrous except for some hairs on petiole. Flowers in a capitate raceme; bracts crispate-hairy, exceeding the calyx, at least the lower pinnatifid. Calyx white-lanate, membranous; teeth about half as long as tube, lanceolate, obtuse, serrate. Corolla 10–12(–15) mm, purplish-red, glabrous; tube about twice as long as calyx; upper lip straight below, falcate at apex. Stigma slightly exserted. Capsule oblong-ovoid, obliquely acuminate, 1½ times as long as calyx. *Stony alpine pastures.* ● *Mountains of Bulgaria and S. Jugoslavia.* Bu Ju.

11. P. dasyantha Hadač, *Studia Bot. Čechica* **5**: 4 (1942). Stem 2–7 cm, erect, lanate, leafy. Leaves about as long as stem, glabrous except for lanate petiole, lanceolate, pinnatisect, with linear, pinnatifid segments. Flowers in a spike at first capitate, later cylindrical; bracts as long as flowers or shorter, the lower like the cauline leaves, the upper less divided and subglabrous. Calyx densely lanate; teeth narrowly triangular, acute, entire, half as long as tube. Corolla 17–20 mm, purplish-pink, hairy; upper lip straight except for curved and obscurely bidentate apex; lower lip ciliate. Capsule obliquely ovoid, curved towards apex, 1½ times as long as calyx. *Stony tundra. Arctic Europe.* Rs (N) Sb.

12. P. hirsuta L., *Sp. Pl.* 609 (1753). Stem 2–10(–15) cm, erect, subglabrous below, lanate above, leafy throughout. Leaves shorter than stem, linear to lanceolate, pinnatifid, with serrate segments. Flowers in a subcapitate, lanate raceme; bracts similar to cauline leaves, exceeding the calyx. Calyx lanate; teeth subequal, serrate or incise-dentate. Corolla 10–13(–15) mm, bright pink, glabrous; upper lip nearly straight, very obtuse, not toothed; lower lip nearly as long as upper. Capsule obliquely ovoid, apiculate, 1½ times as long as calyx. $2n=16$. *Damp, stony tundra, sea-shores and river-banks; calcicole. Arctic Europe.* Fe No Rs (N) Sb Su.

13. P. oederi Vahl in Hornem., *Dansk Oekon. Plantel.* ed. 2, 580 (1806). Stem 4–15(–20) cm, erect, usually glabrous below, hairy above, with few or no cauline leaves. Leaves shorter than stem, lanceolate, pinnatisect, with ovate to oblong, deeply crenate-dentate segments, glabrous. Flowers in a spike-like raceme dense at first, lax later; bracts lanceolate, hairy, deeply crenate-dentate, shorter than the flowers. Calyx hirsute to lanate; teeth unequal, lanceolate, ciliate. Corolla 12–20 mm, yellow, with crimson apex to upper lip, glabrous; tube exceeding calyx; upper lip straight except for falcate, obtuse apex. Capsule narrowly ovoid, obliquely rostrate, about 1½ times as long as calyx. $2n=16$. *Damp grassland or tundra. Arctic Russia; mountains of Europe from C. Fennoscandia and N. Ural to S.W. Alps and S.W. Bulgaria.* Au Bu Cz Ga Ge He It Ju No Po Rm Rs (N, C, W) Su.

14. P. flammea L., *Sp. Pl.* 608 (1753). Stem 2–10 cm, erect, glabrous, leafy. Basal leaves lanceolate, pinnatisect, with broadly ovate, acuminately dentate segments, glabrous, shorter than stem. Flowers in a lax, glabrous or subglabrous spike. Bracts linear-lanceolate, more or less toothed, sparsely ciliate, exceeding the calyx. Calyx glabrous, often red-spotted; teeth lanceolate, acute, serrate, unequal. Corolla 10–12 mm, yellow but with most of the upper lip dark red, glabrous; upper lip straight except for curved apex; lobes of lower lip acute. Capsule narrowly ovoid, obliquely rostrate, at least twice as long as calyx. $2n=16$. *Damp places; usually calcicole. Iceland; mountains of N. Fennoscandia.* Is No Su. (*Arctic and subarctic America.*)

15. P. verticillata L., *Sp. Pl.* 608 (1753). Stems 5–20(–30) cm, erect, glabrous to crispate-hairy. Basal leaves lanceolate, pinnatifid, with ovate, crenate segments; cauline leaves in whorls of (2–)3–4, similar to the basal but with shorter petioles. Flowers in whorls in a dense spike. Bracts lanceolate, often purplish, hirsute, the lower pinnatifid, the upper crenate. Calyx often purplish, hairy only on the veins; teeth short, entire. Corolla 12–18(–20) mm, purplish-red (rarely pink or white), glabrous; tube twice as long as calyx; upper lip almost straight; lobes of lower lip rounded, not ciliate. Capsule narrowly ovoid, apiculate, almost twice as long as calyx. $2n=12$. *Damp mountain grassland or tundra. Arctic and subarctic Russia; mountains of C. & S. Europe.* Al Au Bu Cz Ga Ge ?Gr He Hs It Ju Po Rm Rs (N, C, W).

16. P. amoena Adams ex Steven, *Mém. Soc. Nat. Moscou* **6**: 25 (1823). Stems 5–10(–20) cm, erect, usually crispate-hairy. Basal leaves linear-lanceolate, pinnatisect with linear, acutely toothed segments, glabrous or crispate-hairy; cauline leaves in whorls of (2–)3–4. Flowers in whorls in a raceme usually interrupted below. Bracts lanceolate, 3-fid with lobes toothed at the apex, crispate-hairy. Calyx glabrous; teeth unequal, the two larger half as long as the tube, more or less serrate. Corolla 15–20 mm, purple (rarely violet or white), glabrous; tube sharply inflected at the base; upper lip straight except for falcate apex; lower lip longer than the upper, with very broad lobes. Capsule obliquely ovoid, shortly apiculate, somewhat exceeding calyx. *Stony lichen-tundra. Arctic Russia.* Rs (N).

17. P. arguteserrata Vved. in Schischkin & Bobrov, *Fl. URSS* **22**: 809 (1955). Like **16** but stems up to 30 cm; leaves pinnatifid, with segments toothed or entire; bracts villous; calyx villous, usually tinged with violet; corolla pinkish-lilac, with lower lip shorter than or equalling the upper; capsule broadly oblong. *Open woods. C. Ural.* Rs (C). (*Siberia, Mongolia.*)

Sect. PEDICULARIS. Annual to perennial. Upper lip of corolla falcate, shortly and sometimes inconspicuously rostrate, terminating in 2 linear to subulate teeth.

18. P. palustris L., *Sp. Pl.* 607 (1753). Biennial, sometimes annual. Stem 5–70 cm, usually branched, glabrous or sparsely hairy. Leaves subglabrous, triangular-lanceolate to linear, pinnatisect, with oblong, crenate to pinnatifid segments; cauline leaves usually alternate, rarely opposite or whorled. Flowers in lax racemes, often interrupted below; bracts mostly leaf-like, the uppermost 3-lobed. Calyx ovate, inflated in fruit, strongly 2-lipped, without distinct teeth; lobes crenate. Corolla 15–25 mm, reddish-pink (rarely pale yellow or white); tube nearly twice as long as calyx; margin of upper lip with a tooth on each side some distance below the apex; lower lip ciliate. Capsule ovoid, obliquely apiculate, somewhat exceeding calyx. $2n=16$. *Damp meadows, fens and marshes. Europe, southwards to the Pyrenees, N. Italy, S. Bulgaria and S. Ural.* Au Be Br Bu Cz Da Fa Fe Ga Ge Hb He Ho Hu It Ju No Po Rm Rs (N, B, C, W) Su.

1 Corolla 18–25 mm   **(a) subsp. palustris**
1 Corolla 14–18 mm
   2 Stem freely branched, the branches longer than the main axis;
      leaf-segments pinnatifid   **(b) subsp. opsiantha**
   2 Stem simple, or with a few short branches; leaf-segments
      crenate   **(c) subsp. borealis**

(a) Subsp. **palustris**: Stem freely branched, with lower branches considerably longer than upper, giving a pyramidal outline to plant. Leaf-segments pinnatifid; rhachis of leaf *c.* 2 mm wide, flat. Corolla 18–25 mm. *Throughout the range of the species except parts of the north.*

(b) Subsp. **opsiantha** (E. L. Ekman) Almquist, *Acta Phytogeogr. Suec.* 1: 592 (1929) (incl. *P. karoi* Freyn): Stem freely branched, with all branches subequal in length, giving a cylindrical outline to plant. Leaf-segments pinnatifid; rhachis of leaf less than 1 mm wide, canaliculate. Corolla *c.* 15 mm. *N. & E. Europe.*

(c) Subsp. **borealis** (J. W. Zett.) Hyl., *Uppsala Univ. Årsskr.* **1945**(7): 293 (1945): Stem simple or with a few short branches. Leaves narrower than in other subspecies, often more or less linear. Leaf-segments crenate, scarcely lobed. Corolla 14–18 mm. *N. Russia, N. & C. Fennoscandia, Færöer.*

19. P. labradorica Wirsing, *Eclog. Bot.* t. 10 (1778). Biennial. Stem 10–15(–30) cm, usually with suberect branches from the base, crispate-pubescent. Leaves alternate, the lower and middle linear-lanceolate, pubescent beneath, pinnatisect with acutely toothed segments, the upper linear, villous, more or less entire. Flowers in racemes lax below, subcapitate above; bracts small, linear. Calyx coriaceous, glabrous or pubescent, 2-lipped. Corolla 17–19 mm, yellow, with upper lip usually tinged with red or purple; lower lip ciliate, shorter than the upper. Capsule linear-oblong, acuminate, $1\frac{1}{2}$ times as long as calyx. *Tundra. E. arctic Russia.* Rs (N). (*Siberia; arctic and subarctic America.*)

20. P. sylvatica L., *Sp. Pl.* 607 (1753). Perennial or biennial. Stems 5–25 cm, the central one erect, simple, the others usually longer, procumbent to ascending, simple or branched. Leaves lanceolate, 2-pinnatisect, glabrous or somewhat hairy. Flowers in short, rather lax racemes; bracts similar to leaves but smaller. Calyx membranous, inflated in fruit, scarcely 2-lipped; teeth distinct, unequal, foliose, lobed. Corolla 15–25 mm, pink or red (rarely white); tube twice as long as calyx; lower lip glabrous, equalling the upper. Capsule ovoid, shortly acuminate, shorter than calyx. *Bogs, heaths, moors and woods on peaty soil. W. & C. Europe, extending northwards to C. Sweden and eastwards to Lithuania and W. Ukraine.* Au Be Br Cz Da Ga Ge Hb He Ho Hs It Lu No Po Rm Rs (C, W) Su.

1 Calyx glabrous   (a) subsp. **sylvatica**
1 Calyx hairy, at least on the angles
  2 Calyx hairy on the angles only; central stem up to 25 cm
                         (b) subsp. **lusitanica**
  2 Calyx hairy all over; central stem not more than 10 cm
                         (c) subsp. **hibernica**

(a) Subsp. **sylvatica**: Stems glabrous or with 2 lines of hairs, the central one up to 15 cm. Calyx and pedicels glabrous. $2n = 16$. ● *Throughout the range of the species except Portugal.*

(b) Subsp. **lusitanica** (Hoffmanns. & Link) Coutinho, *Fl. Port.* 565 (1913) (*P. lusitanica* Hoffmanns. & Link): Stems more or less pubescent, the central one up to 25 cm. Pedicels glabrous or subglabrous; calyx hairy on the angles. *Portugal and S.W. Spain.*

(c) Subsp. **hibernica** D. A. Webb, *Watsonia* 3: 239 (1956): Stems pubescent, the central one not more than 10 cm. Calyx and pedicels villous. *Moors and bogs.* ● *N.W. Europe, from Ireland to Norway.*

21. P. sudetica Willd., *Sp. Pl.* 3: 209 (1800). Perennial. Stems 10–25 cm, ascending to erect, glabrous below, pubescent above. Leaves 3–8 cm, mostly basal, ovate-lanceolate, pinnatisect with pinnatifid segments. Flowers in a dense raceme; lower bracts leaf-like, the upper entire or serrate. Calyx lanate; teeth lanceolate, toothed or crenate. Corolla up to 20 mm, pink to purplish-red; tube nearly twice as long as calyx; lower lip glabrous, with sinuate-dentate lobes. Capsule ovoid, long-acuminate, twice as long as calyx. $2n = 16$. *Bogs, flushes and mossy tundra. Arctic Russia, N. & C. Ural; borders of Poland and Czechoslovakia (Sudeten Mts.).* Cz Po Rs (N, C). (*Arctic and subarctic Asia and America.*)

22. P. comosa L., *Sp. Pl.* 609 (1753). Perennial. Stem (10–) 20–85 cm, erect. Leaves up to 15 cm, pubescent, lanceolate, 2-pinnatisect with linear-lanceolate, toothed segments; cauline leaves usually numerous. Flowers in a spike, dense in flower, elongating in fruit; lower bracts leaf-like, exceeding the flowers, the upper linear-lanceolate, or 3-lobed, about equalling the calyx. Calyx membranous; teeth wider than long, more or less obtuse, ciliate. Corolla up to 25 mm, pale yellow; tube twice as long as calyx. Capsule ovoid, acuminate, nearly twice as long as calyx. *Alpine meadows and stony hillsides.* ● *Mountains of S. & S.C. Europe, from S.C. France, S.C. Czechoslovakia and E. Carpathians southwards to S. Spain, S. Italy and S. Bulgaria.* Al Bu Cz Ga Hs It Ju Rm.

(a) Subsp. **comosa**: Stem not more than 50 cm, crispate-pubescent. Calyx pubescent on the angles. Lower lip of corolla ciliate. *Almost throughout the range of the species.*

(b) Subsp. **campestris** (Griseb.) Soó, *Szek. Fl. Előmunk.* 113 (1940) (*P. campestris* Griseb. & Schenk): Stem up to 85 cm, subglabrous. Calyx and lower lip of corolla glabrous. *Balkan peninsula and Romania.*

23. P. kaufmannii Pinzger, *Progr. Sald. Realsch. Brandenb.* **1868**: 17 (1868). Like **22** but stem and leaves lanate; leaf-segments ovate-oblong; all bracts leaf-like and exceeding the flowers; calyx usually hairy; corolla 25–30 mm, with glabrous lower lip; capsule slightly shorter. *Meadows, steppes and thickets. S., C. & W. parts of U.S.S.R., just extending to S.W. Poland.* Po Rs (B, C, W, E).

24. P. sibthorpii Boiss., *Diagn. Pl. Or. Nov.* 1(4): 83 (1844). Perennial. Stem 20–30 cm, erect, lanate. Leaves glabrous above, villous beneath, lanceolate, 2-pinnatisect with ovate-oblong, coarsely serrate segments. Flowers in a dense to lax spike; bracts lanate, the lower leaf-like, the upper usually 3-lobed, exceeding the calyx. Calyx coriaceous, villous; teeth wider than long, acute to acuminate, entire. Corolla 25–28 mm, yellow; lower lip large, glabrous or sparsely ciliate. Capsule 16–18 mm, ovoid-oblong, obtuse, *c.* $1\frac{1}{2}$ times as long as calyx. *Open woods and grassy hillsides. Krym.* Rs (K). (*Caucasus, N. Anatolia.*)

25. P. uralensis Vved. in Schischkin & Bobrov, *Flora URSS* 22: 816 (1955). Like **24** but stem 30–80 cm; lower lip of corolla distinctly ciliate; capsule 9–12 mm, apiculate, about as long as calyx. *Meadows and open woods. E. & E.C. Russia.* Rs (N, C).

26. P. asparagoides Lapeyr., *Hist. Abr. Pyr.* 349 (1813). Perennial. Stem 20–50 cm, erect, pubescent. Leaves pubescent to subglabrous, lanceolate, 2-pinnatisect with linear-lanceolate, toothed segments. Flowers in a lax raceme; lower bracts leaf-like, the upper linear-lanceolate, serrate, scarcely exceeding the calyx. Calyx membranous with subglabrous tube; teeth wider than long, obtuse, entire, ciliate. Corolla up to 25 mm, purplish-red; lower lip glabrous. Capsule ovoid, acuminate, somewhat exceeding calyx. *Meadows and rocky hillsides.* ● *E. Pyrenees and adjoining ranges.* Ga Hs.

27. P. schizocalyx (Lange) Steininger, *Bot. Centr.* 29: 249 (1887). Like **26** but flowers in a dense spike, elongating in fruit; calyx villous, with crenate teeth; corolla yellow, with ciliate lower lip. *Rocky hillsides.* ● *Mountains of W. Spain.* Hs.

28. P. brachyodonta Schlosser & Vuk., *Syll. Fl. Croat.* 89 (1857). Perennial. Stem erect or ascending. Leaves pubescent, lanceolate, 2-pinnatisect with linear-oblong, apiculate, serrate segments. Bracts more or less hairy. Calyx with more or less hairy tube; teeth deltate, entire, ciliate. Corolla 20–25 mm, pale yellow; lower lip glabrous, with serrulate lobes and 2 tubercles near the base. Capsule ovoid, apiculate, somewhat exceeding calyx. *Mountain pastures and screes.* ● *N. & C. parts of Balkan peninsula.* Al Bu Gr Ju.

1 Stem 40–60 cm
  2 Stem hairy; bracts densely villous, the lower 2-pinnatisect, the upper 3-fid or undivided   **(a) subsp. brachyodonta**
  2 Stem glabrescent; bracts pubescent, the lower oblong, leaf-like, the upper palmatifid   **(b) subsp. moesiaca**
1 Stem not more than 30 cm
  3 Calyx-tube hairy all over; calyx-teeth subequal; spike elongate; bracts pinnatifid to 3-lobed   **(c) subsp. malyi**
  3 Calyx-tube hairy only on the veins; calyx-teeth unequal; spike capitate; bracts 3-lobed or undivided   **(d) subsp. grisebachii**

(a) Subsp. **brachyodonta**: Stem 40–60 cm, hairy. Spikes dense, cylindrical. Bracts villous, the lower 2-pinnatisect, the upper 3-fid or undivided. *W. Jugoslavia and N. Albania.*

(b) Subsp. **moesiaca** (Stadlm.) Hayek, *Prodr. Fl. Penins. Balcan.* 2: 193 (1929): Like subsp. (a) but stem glabrescent; bracts pubescent, the lower oblong, the upper palmatifid. *Macedonia and Bulgaria.*

(c) Subsp. **montenegrina** (Janka ex Nyman) D. A. Webb, *Bot. Jour. Linn. Soc.* 65: 357 (1972) (*P. friderici-augusti* subsp. *montenegrina* Janka ex Nyman). Stem not more than 30 cm, hairy. Bracts pinnatisect to 3-lobed. Calyx hairy all over; teeth subequal. *W.C. Jugoslavia and N. Albania.*

(d) Subsp. **grisebachii** (Wettst.) Hayek, *Prodr. Fl. Penins. Balcan.* 2: 193 (1929): Like subsp. (c) but bracts undivided to 3-lobed; calyx with tube hairy only on the veins and teeth unequal. *Albania to Bulgaria.*

29. P. heterodonta Pančić, *Fl. Princ. Serb., Addit.* 196 (1884). Perennial. Stem 12–30 cm, erect or ascending, hairy. Leaves subglabrous, lanceolate, 2-pinnatisect with lanceolate, toothed segments. Flowers in a spike dense in flower, elongating in fruit; bracts sparsely pubescent, triangular, pinnatifid to 3-lobed, equalling the calyx. Calyx with glabrous, membranous tube, deeply split along the lower side; teeth deltate, entire, ciliate, the upper one distinct, the two lateral on each side more or less united. Corolla 15–19 mm, yellowish-white; lower lip sparsely ciliate. Capsule ovoid, acuminate, somewhat exceeding calyx. *Meadows and stony hillsides.* ● *Mountains of C. & E. Jugoslavia.* Ju.

30. P. leucodon Griseb., *Spicil. Fl. Rumel.* 2: 17 (1844). Perennial. Stem 12–20(–30) cm, erect. Leaves more or less glabrous above, hairy beneath, oblong-lanceolate, 2-pinnatisect with lanceolate, toothed segments. Flowers in a lax spike, often interrupted below; bracts hairy, equalling or exceeding the calyx. Calyx sparsely hairy; teeth about as long as tube, linear-spathulate, toothed at apex, ciliate. Corolla 18–25 mm, yellowish-white; lower lip glabrous. Capsule ovoid, apiculate, nearly twice as long as calyx. *Alpine grassland and screes.* ● *C. part of Balkan peninsula.* Al Bu Ju.

(a) Subsp. **leucodon**: Stem leafless below, leafy above. Lower bracts leaf-like, the upper linear, entire or toothed. Calyx equally 5-toothed. *E. Albania, S.E. Jugoslavia.*

(b) Subsp. **occulta** (Janka) E. Mayer, *Österr. Bot. Zeitschr.* 119: 324 (1971) (*P. occulta* Janka): Stem equally leafy throughout. Bracts deltate, palmatisect, exceeding the calyx. Calyx usually unequally 4-toothed. *Bulgaria.*

31. P. physocalyx Bunge, *Bull. Sci. Acad. Imp. Sci. Pétersb.* 8: 252 (1841). Perennial. Stem 10–20 cm, ascending, arachnoid-lanate. Leaves subglabrous except for the villous petiole, lanceolate, 2-pinnatisect, with oblong, acute, toothed segments.

Flowers in a raceme dense in flower, elongating in fruit; bracts lanate, the lower leaf-like, the upper 3-lobed. Calyx lanate; teeth half as long as tube, lanceolate, acuminate, serrulate to subentire. Corolla 26–35 mm, yellow; lower lip glabrous. Capsule ovoid, mucronate, shorter than calyx. *Steppes. E. Russia.* Rs (C, E). (*W.C. Asia.*)

32. P. dasystachys Schrenk, *Bull. Phys.-Math. Acad. Pétersb.* 2: 195 (1844). Perennial. Stems 10–30 cm, erect, subglabrous below, hairy above. Leaves glabrous, oblong-lanceolate, pinnatisect, with ovate-lanceolate, obtuse, pinnatifid segments. Flowers in a capitate, whitish-lanate spike; bracts linear, entire, exceeding the calyx. Calyx lanate; teeth unequal, lanceolate, entire, half as long as tube. Corolla 22–25 mm, bright pinkish-red; lower lip glabrous, with denticulate lobes. Capsule ovoid, mucronate, shorter than calyx. *Saline or seasonally-flooded meadows. S.E. Russia, E. Ukraine, W. Kazakhstan.* Rs (C, W, E).

33. P. graeca Bunge, *op. cit.* 1: 376 (1843). Perennial. Stem 10–20 cm, erect, arachnoid-lanate. Leaves glabrous except for villous petiole, ovate-lanceolate, 2-pinnatisect with lanceolate, acuminate, toothed segments. Flowers in a raceme, capitate in flower, elongating in fruit. Bracts hairy, exceeding the calyx, the lower leaf-like, the upper linear-lanceolate, crenate or serrate. Calyx glandular-villous; teeth unequal, ovate-lanceolate, subentire. Corolla up to 25(–30) mm, pale yellow; lower lip glabrous, with denticulate lobes. Capsule ovoid, acuminate, equalling or slightly exceeding calyx. *Pastures and stony slopes.* ● *Mountains of Albania and Greece.* Al Gr.

34. P. friderici-augusti Tommasini, *Linnaea* 13: 74 (1839). Like 33 but stem glabrous or subglabrous below; petioles sparsely hairy; raceme eglandular-villous; upper bracts linear, entire, shorter than calyx. *Mountain grassland.* ● *Jugoslavia, N.E. Italy, N. Albania.* Al It Ju.

35. P. petiolaris Ten., *Fl. Nap.* 5, *Syll. App.* 4: 11 (1835). Perennial. Stem 7–25 cm, erect, pubescent to villous. Leaves glabrous except for villous petiole, 2-pinnatisect with ovate to lanceolate, apiculate, serrate segments. Flowers in a whitish-lanate, capitate spike; bracts villous, exceeding the calyx, the lower leaf-like, the upper pinnatifid to serrate. Calyx lanate; teeth lanceolate, acute. Corolla up to 22 mm, deep pinkish-red; lower lip glabrous. Capsule narrowly ovoid, acuminate, somewhat exceeding calyx. *Mountain meadows and stony hillsides.* ● *C. part of Balkan peninsula; C. & S. Appennini.* Al Bu ?Gr It Ju.

36. P. ferdinandi Bornm., *Notizbl. Bot. Gart. Berlin* 8: 213 (1922). Like 35 but stem 2–5(–8) cm; petioles glabrous or subglabrous; raceme very short, with 4–6(–8) flowers; bracts glabrous; calyx sometimes glabrescent; corolla 22–25 mm, pale pink. *Screes and stony slopes.* ● *S. Jugoslavia (Jakupica, S. of Skopje).* Ju.

Sect. RHYNCHOLOPHAE Maxim. Perennial. Upper lip of corolla more or less falcate, terminating in a distinct, usually long beak, without teeth.

37. P. compacta Stephan ex Willd., *Sp. Pl.* 3: 219 (1800). Stems (12–)25–50 cm, erect or ascending, glabrous or somewhat pubescent above. Leaves glabrous, lanceolate, pinnatisect with lanceolate, pinnatifid, apiculate segments. Flowers in a dense spike; bracts oblong-lanceolate, leaf-like, glabrous or hairy, the uppermost entire and shorter than the calyx. Calyx membranous, glabrous or pubescent; teeth unequal, broad, obtuse, entire.

Corolla 17–20 mm, yellow; lower lip as long as the upper, glabrous, with broad, denticulate lobes. Capsule ovoid, equalling calyx. *Meadows and wet tundra. N.E. Russia.* Rs (N, C). (*N. Asia.*)

38. P. lapponica L., *Sp. Pl.* 609 (1753). Stems (5–)10–25 cm, glabrous or somewhat pubescent above. Leaves glabrous, linear-lanceolate, pinnatifid, with wide rhachis and oblong, incise-dentate segments. Flowers few, fragrant, in a capitate, spike-like raceme; bracts leaf-like, equalling or exceeding the calyx. Calyx membranous, glabrous, split nearly halfway to base on lower side; teeth deltate, entire. Corolla 14–16 mm, very pale yellow; beak rather short; lower lip somewhat shorter than the upper, glabrous or sparsely ciliate. Capsule ovoid, curved, acuminate, $1\frac{1}{2}$ times as long as calyx, dehiscing along upper side only. $2n=16$. *Heaths and dry tundra. Arctic Europe and mountains of Fennoscandia.* Fe No Rs (N) Su. (*Arctic and subarctic Asia and America.*)

39. P. tuberosa L., *Sp. Pl.* 610 (1753). Stock somewhat tuberous. Stem 10–25 cm, ascending to erect, hairy, especially below. Petiole and sometimes rhachis hairy; leaves otherwise glabrous, pinnatisect, with pinnatifid segments. Flowers in a spike capitate at first, elongating in fruit; bracts pinnatifid, pubescent, slightly exceeding the calyx. Calyx pubescent to villous; teeth foliose, toothed, glabrous on inner surface, nearly as long as tube. Corolla up to 20 mm, pale yellow; tube exceeding calyx; beak long; lower lip glabrous. Capsule ovoid, apiculate, somewhat exceeding calyx. $2n=16$. *Open woods, alpine pastures and screes; calcifuge.* ● *Alps, N. & C. Appennini, Alpi Apuane; Pyrenees.* Au Ga He Hs It.

40. P. elongata A. Kerner, *Nov. Pl. Sp.* **1**: 14 (1870). Stem 15–35 cm, ascending to erect, glabrous except for 2 lines of hairs. Petiole and sometimes rhachis pubescent; leaves otherwise glabrous, oblong-lanceolate, 2-pinnatisect with lanceolate, subentire segments. Flowers in a spike rather dense at first, very lax in fruit; bracts glabrous, the lower pinnatisect, the upper pinnatifid to dentate. Calyx-tube glabrous; teeth foliose, toothed, ciliate, pubescent on inner surface, nearly as long as the tube. Corolla up to 16 mm, pale yellow; tube exceeding calyx; beak long; lower lip glabrous. Capsule ovoid, apiculate, slightly exceeding calyx. $2n=16$. *Dry pastures and screes; calcicole.* ● *S.E. Alps.* Au It Ju.

41. P. julica E. Mayer, *Phyton (Austria)* **9**: 301 (1961). Like **40** but leaf-segments toothed; bracts densely villous, the upper ones 3-lobed; calyx villous; corolla up to 18 mm; capsule equalling calyx. *Dry pastures and screes; calcicole.* ● *S.E. Alps.* Au It Ju.

42. P. ascendens Schleicher ex Gaudin in Murith, *Guide Bot. Valais* 83 (1810) (*P. barrelieri* Reichenb.). Stem 15–30 cm, erect, glabrous or with 1–2 lines of hairs. Leaves glabrous or slightly pubescent on rhachis, oblong-lanceolate, pinnatisect, with pinnatifid, toothed segments. Flowers in a spike, dense at first, becoming lax in fruit; bracts ovate-lanceolate, the lower pinnatisect, the upper 3- to 5-fid, with entire segments. Calyx-tube glabrous; teeth lanceolate, acuminate, entire or weakly toothed, usually ciliate, glabrous on inner surface, as long as the tube. Corolla up to 16 mm, dull yellow; tube exceeding calyx; lower lip glabrous. Capsule ovoid, shortly apiculate, twice as long as calyx. $2n=16$. *Mountain grassland; calcicole.* ● *S.W. & W.C. Alps.* Ga He It.

43. P. baumgartenii Simonkai, *Term. Füz.* **10**: 182 (1886). Like **42** but segments of bracts toothed, hairy; calyx villous, with deltate teeth $\frac{1}{3}$ as long as tube; corolla up to 20 mm; capsule only slightly exceeding calyx. *Dry, stony alpine pastures.* ● *S. Carpathians.* Rm.

44. P. rostratospicata Crantz, *Stirp. Austr.* ed. 2, **2**: 317 (1769) (*P. incarnata* Jacq., non L.). Stem 15–45 cm, ascending to erect, glabrous below, densely pubescent above. Leaves glabrous, lanceolate, pinnatisect with pinnatifid segments. Flowers in a rather lax spike; bracts hairy, with 3 linear-lanceolate lobes. Calyx hairy; teeth lanceolate, acute, entire or toothed. Corolla up to 13 mm, pink to purplish-red; tube about as long as calyx; beak long; lower lip glabrous. Capsule ovoid, apiculate, exceeding calyx. $2n=16$. *Alpine pastures.* ● *Alps; one station in E. Pyrenees.* Au Ga Ge He It Ju.

(a) Subsp. **rostratospicata**: Plant slender. Bracts and calyx arachnoid-tomentose. Calyx-teeth all entire. *E. Alps.*

(b) Subsp. **helvetica** (Steininger) O. Schwarz, *Mitt. Thür. Bot. Ges.* **1**(1): 114 (1949): Plant robust. Bracts and calyx villous. Calyx-teeth of lower flowers distinctly toothed. *C. & W. Alps; Pyrenees.*

45. P. rostratocapitata Crantz, *Stirp. Austr.* ed. 2, **2**: 320 (1769) (*P. rostrata* L. pro parte). Stem 5–20 cm, ascending, glabrous except for 1 or 2 lines of hairs. Leaves glabrous, often tinged with purple, oblong-lanceolate, 2-pinnatisect with oblong, entire or serrulate segments. Flowers in a capitate raceme; bracts similar to cauline leaves. Calyx glabrous or slightly pubescent on the veins; teeth foliose, lanceolate, crenate, recurved at apex. Corolla up to 25 mm, pink to purplish-red; tube slightly exceeding calyx; lower lip ciliate, at least at anthesis. Filaments of longer stamens sparsely pubescent to subglabrous. Capsule ovoid-oblong, apiculate, exceeding calyx. $2n=16$. *Alpine pastures and screes; usually calcicole.* ● *E. Alps, mountains of N.W. Jugoslavia.* Au Ge He It Ju.

Records for Romania appear to be erroneous.

(a) Subsp. **rostratocapitata**: Lower lip of corolla densely ciliate, at least at anthesis. *Throughout the range of the species.*

(b) Subsp. **glabra** H. Kunz, *Phyton (Austria)* **8**: 289 (1959): Lower lip of corolla glabrous. *Alpi Bergamasche.*

46. P. kerneri Dalla Torre, *Anleit. Beob. Bestimm. Alpenpfl.* 176 (1882) (*P. rhaetica* A. Kerner, *P. rostrata* L. pro parte). Like **45** but stem decumbent to ascending, sometimes pubescent; calyx glabrous or uniformly pubescent; corolla not more than 20 mm, with glabrous lower lip; filaments of longer stamens densely pubescent. $2n=16$. *Alpine pastures and screes; usually calcifuge.* ● *Alps, Pyrenees.* Au Ga He Hs It.

47. P. pyrenaica Gay, *Ann. Sci. Nat.* **26**: 210 (1832). Stem 10–25 cm, ascending to erect, glabrous except for 2 lines of hairs; cauline leaves few. Leaves glabrous, except sometimes for hairy petiole, 2-pinnatisect with ovate-lanceolate, toothed segments. Flowers in a short, subcapitate spike; bracts similar to leaves, ciliate, exceeding the calyx. Calyx glabrous; teeth nearly as long as tube, some foliose and lobed, some membranous, lanceolate, entire. Corolla up to 20 mm, dull pinkish-red; tube slightly exceeding calyx; beak long; lower lip glabrous. Capsule ovoid, slightly exceeding calyx. $2n=16$. *Alpine pastures and screes.* ● *Pyrenees.* Ga Hs.

48. P. mixta Gren. in Gren. & Godron, *Fl. Fr.* **2**: 617 (1853). Like **47** but stems ascending from a procumbent base; cauline leaves more numerous; spike lax; bracts and calyx lanate; corolla not more than 18 mm, pink with deep crimson upper lip.

Mountain grassland and screes. ● *Pyrenees, Cordillera Cantábrica.* Ga Hs.

49. P. cenisia Gaudin, *Fl. Helv.* **4**: 132 (1829). Like **47** but stem, bracts and calyx whitish-lanate; calyx-teeth all foliose; corolla pink with deep crimson upper lip. *Alpine pastures and screes.* ● *S.W. Alps, N. Appennini, Alpi Apuane.* Ga It.

50. P. gyroflexa Vill., *Hist. Pl. Dauph.* **2**: 426 (1787). Stem 10–35 cm, ascending to erect, lanate; cauline leaves few. Leaves 2-pinnatisect with lanceolate, incise-dentate segments. Flowers in a spike, dense in flower, becoming lax in fruit; bracts lanate, 3-lobed, exceeding the calyx. Calyx densely lanate; teeth lanceolate, foliose, pinnatifid, as long as tube or longer. Corolla up to 25 mm, pinkish-red; tube twice as long as calyx; beak short; lower lip ciliate. Capsule ovoid, slightly exceeding calyx. *Alpine pastures and screes; calcicole.* ● *S.W. & S. Alps; C. & S. Appennini; one station in E. Pyrenees.* Ga He It.

(a) Subsp. **gyroflexa**: Leaves puberulent, with crowded, acuminate segments. $2n = 16$. *Alps, Pyrenees.*

(b) Subsp. **praetutiana** (Levier ex Steininger) H. Kunz & E. Mayer, *Österr. Bot. Zeitschr.* **119**: 324 (1971) (*P. gyroflexa* var. *praetutiana* Levier ex Steininger): Leaves glabrous (except sometimes for petiole), with distant, obtuse segments. *Appennini.*

51. P. elegans Ten., *Fl. Nap.* **5**, *Syll. App.* **4**: 11 (1835). Stems 1–12 cm, decumbent, usually hairy, with 0–2 cauline leaves. Leaves glabrous, except sometimes for slightly pubescent rhachis, lanceolate, pinnatisect with oblong, pinnatifid, toothed segments. Flowers few, in a capitate raceme; bracts 2-pinnatifid, hairy, about equalling the calyx. Calyx strongly accrescent, sparsely pubescent; teeth lanceolate, foliose, nearly as long as tube. Corolla up to 21 mm, pinkish-red; tube distinctly exceeding calyx; beak short, conical; lower lip glabrous. Capsule ovoid, shortly apiculate, nearly twice as long as calyx. *Alpine pastures and screes.* ● *C. & S. Appennini.* It.

52. P. portenschlagii Sauter ex Reichenb., *Pl. Crit.* **5**: 1 (1827). Stem 2–10 cm, erect, glabrous except for 1 or 2 lines of hairs. Leaves glabrous, oblong-lanceolate, pinnatifid with ovate-lanceolate, dentate segments. Flowers 1–3(–5) in a capitate raceme; bracts pinnatifid, equalling or shorter than the calyx. Calyx glabrous or sparsely pubescent; teeth lanceolate, foliose, pinnatifid, recurved at apex, about half as long as tube. Corolla up to 25(–30) mm, pinkish- or purplish-red; tube nearly twice as long as calyx; beak short, truncate; lower lip glabrous. Capsule narrowly oblong, apiculate, exceeding calyx. *Alpine pastures and screes.* ● *N.E. Alps.* Au.

53. P. asplenifolia Floerke ex Willd., *Sp. Pl.* **3**: 208 (1800). Stem 2–8 cm, ascending to erect, glabrous below, lanate above with reddish hairs. Leaves glabrous, narrowly lanceolate, pinnatisect with ovate-lanceolate, crenate to pinnatifid segments. Flowers 2–5(–8) in a dense, capitate raceme; bracts pinnatifid, reddish-lanate, about as long as calyx. Calyx villous-lanate with reddish hairs; teeth short, foliose, pinnatifid. Corolla up to 17 mm, pinkish-red; tube equalling calyx; lower lip glabrous. Capsule oblong-ovoid, rostrate, exceeding calyx. $2n = 16$. *Alpine pastures and screes; calcifuge.* ● *E. Alps.* Au He It.

54. P. resupinata L., *Sp. Pl.* 608 (1753). Stems 30–60 cm, erect, subglabrous; cauline leaves numerous. Leaves hairy beneath, oblong-lanceolate, subcordate at base, crenate-serrate, not lobed; petiole short. Flowers in a spike, lax and interrupted below, dense

above; bracts pubescent, ovate, acuminate, crenate. Calyx pubescent, strongly 2-lipped; lips acute, entire. Corolla 20–25 mm, purplish-red; tube twice as long as calyx; beak short and broad; lower lip ciliate. Capsule oblong-ovoid, apiculate, 1½ times as long as calyx. *Meadows and open woods.* S. Ural. Rs (C). (*Siberia, E. Asia.*)

35. Rhinanthus L.[1]

(*Alectorolophus* Zinn)

Annual, hemiparasitic herbs. Leaves opposite, sessile, entire to dentate. Flowers in terminal, bracteate, spike-like racemes. Calyx laterally compressed, ovate to suborbicular, shortly 4-toothed, accrescent. Corolla mainly yellow, with a long tube and 2-lipped limb; upper lip galeate, laterally compressed, with 2 teeth slightly below the apex; lower lip slightly shorter than upper, 3-lobed. Stamens 4, didynamous, included under upper lip of corolla; anther-lobes hairy, not mucronate. Capsule loculicidal; seeds few, more or less discoid, usually with a marginal wing.

In this and related genera the leaves on the main stem which lie between the uppermost branches and the lowest flowers of the terminal spike are known as *intercalary leaves*.

'Glabrous', when used of the calyx in this genus, refers to the main surfaces; the margin may be scabrid or slightly puberulent.

The species of *Rhinanthus* show a recurrent pattern of variation similar to the 'pseudoseasonal polymorphism' described for *Melampyrum* (p. 253). Since, however, the principal habitat of most species of *Rhinanthus* is meadows cut for hay the seasonal aspect is less marked, and for this and other reasons the ecotypic variants are less easily distinguished than in *Melampyrum*. Nevertheless, the general pattern of this variation is so conspicuous and so constant in its general outline that it has been thought best to list, as for *Melampyrum*, the recurrent 'ecotypic variants' for those species in which they have been recognized, each being listed under the name by which it was first validly described, at whatever rank. Such citation does not imply approval of the appropriateness of the rank in question, and is designed primarily to lead the reader to a more detailed description.

Apart from this supposedly ecotypic variation, the pattern of variation of other characters in the genus is so reticulate and complex that there is great lack of agreement on the limits of species. The critical tradition of Central Europe has led to the recognition of a larger number of narrowly-defined species; in Northern Europe most recent authors have defined the species much more widely. This account attempts a compromise between the two traditions.

In *Rhinanthus* the recurrent ecotypic variants are as follows:

(1) *Autumnal.* Branches numerous, patent or arcuate; internodes numerous, short; intercalary leaves several. Late-flowering.

(2) *Aestival.* Branches usually 2–4 pairs, usually flowering; internodes numerous, usually short; intercalary leaves 0–2 pairs. Mid-season-flowering.

(3) *Vernal.* Branches few or absent, short, non-flowering; internodes few, long; intercalary leaves absent. Early-flowering.

(4) *Montane.* Rather dwarf; branches few or absent; internodes short, fairly numerous; intercalary leaves variable in number. Mid-season-flowering.

(5) *Alpine.* Dwarf; internodes numerous, short; intercalary leaves several. Mid-season- to late-flowering.

[1] By R. de Soó and D. A. Webb.

Literature: A. Chabert, *Bull. Herb. Boiss.* **7**: 497–517 (1899). D. J. Hambler, *Watsonia* **4**: 101–116 (1958). R. von Soó, *Feddes Repert.* **26**: 179–219 (1929); *Acta Bot. Acad. Sci. Hung.* **16**: 193–206 (1970). J. von Sterneck, *Abh. Zool.-Bot. Ges. Wien* **1**(2): 1–150 (1901).

1 Teeth of upper lip of corolla less than 1 mm, rounded; corolla-tube straight (Sect. *Rhinanthus*)
  2 Leaves and bracts deeply patent-dentate   **1. groenlandicus**
  2 Most of the leaves and bracts crenate-serrate, with apically directed teeth   **2. minor**
1 Teeth of upper lip of corolla at least 1 mm; corolla-tube usually ±curved
  3 Teeth of upper lip of corolla *c.* 1 mm long and wide, ±square (Sect. *Brevirostres*)
    4 Bracts densely glandular-hairy
     5 Leaves remotely patent-dentate; bracts much exceeding calyx, their upper teeth patent   **4. wettsteinii**
     5 Leaves crenate-serrate; bracts scarcely exceeding calyx, their upper teeth appressed   **5. pubescens**
    4 Bracts glabrous or subglabrous
     6 Throat of corolla open   **3. asperulus**
     6 Throat of corolla ±closed
      7 Corolla *c.* 18 mm   **6. pindicus**
      7 Corolla *c.* 15 mm   **7. antiquus**
  3 Teeth of upper lip of corolla 1·5–2·5 × 0·5–1 mm
    8 Corolla with straight tube and convex margins to upper lip   **8. dinaricus**
    8 Corolla with usually curved tube and concave margins to upper lip
     9 Corolla with open throat and ±porrect teeth on upper lip
      10 Teeth of bracts aristate, abruptly diminishing in length towards apex
       11 Bracts and calyx glabrous   **9. aristatus**
       11 Calyx and margin of bracts glandular-pubescent   **10. pampaninii**
      10 Teeth of bracts not aristate, gradually diminishing in length towards apex
       12 Bracts and calyx glabrous   **11. alpinus**
       12 Bracts and calyx glandular-hairy   **12. carinthiacus**
     9 Corolla with closed throat and patent teeth on upper lip
      13 Calyx villous, though partly glabrescent in fruit.   **25. alectorolophus**
      13 Calyx glabrous, puberulent or pubescent
       14 Calyx distinctly and ±densely glandular-hairy, at least on the margin
        15 Teeth at base of bracts longer and more crowded than those near apex   **13. burnatii**
        15 Teeth of bracts subequal and equally spaced
         16 Stem and upper bracts glandular-pubescent; surface of calyx glandular-pubescent even in fruit   **14. rumelicus**
         16 Stem and upper bracts subglabrous; surface of calyx glabrous in fruit   **15. wagneri**
       14 Calyx without glandular hairs, or rarely with a few on the margin
        17 Bracts and calyx eglandular-puberulent
         18 Teeth of bracts abruptly diminishing in length towards apex   **17. melampyroides**
         18 Teeth of bracts subequal, or gradually diminishing in length towards apex
          19 Stem with black streaks; teeth of bracts gradually but conspicuously diminishing in length towards apex   **16. mediterraneus**
          19 Stem without black streaks; teeth of bracts subequal in length   **24. freynii**
        17 Bracts and calyx glabrous or scabrid (rarely sparsely glandular-puberulent on margin)
         20 Teeth of bracts abruptly diminishing in length towards apex, the basal *c.* 8 mm
          21 Bracts narrowly triangular, acuminate, much exceeding calyx   **18. songeonii**
          21 Bracts triangular-ovate, acute, scarcely exceeding calyx   **19. ovifugus**
         20 Teeth of bracts gradually diminishing in length towards apex, the basal *c.* 5 mm
          22 Leaves acutely serrate or dentate
           23 Bracts about equalling calyx   **20. subulatus**
           23 Bracts much exceeding calyx   **21. borbasii**
          22 Leaves crenate-serrate
           24 Corolla 20–24 mm, golden-yellow; tube 5–7 mm wide   **22. halophilus**
           24 Corolla 16–20 mm, pale yellow; tube 2–3 mm wide   **23. angustifolius**

Sect. RHINANTHUS. Teeth of upper lip of corolla short, rounded.

1. R. groenlandicus (Ostenf.) Chab., *Bull. Herb. Boiss.* **7**: 511 (1899). Stem up to 30 cm, simple or slightly branched, stout, hairy on 2 opposite sides. Leaves rather fleshy, coarsely dentate, with patent, acute teeth, bright yellowish-green. Bracts exceeding the calyx, triangular, with long, patent teeth, glabrous or puberulent. Calyx glabrous or puberulent. Corolla *c.* 15 mm; tube straight; throat open. 2*n* = 22. *N. Europe, southwards to c.* 60° *N. in the mountains of Norway.* Fa Fe Is No Rs (N) Su.

2. R. minor L., *Amoen. Acad.* **3**: 54 (1756) (*Alectorolophus minor* (L.) Wimmer & Grab.; incl. *R. borealis* (Sterneck) Druce, *R. rusticulus* (Chab.) Druce, *R. nigricans* Meinsh.). Stem 5–50 cm, sometimes with black streaks. Leaves (2–)5–15 mm wide, ovate-oblong to linear-lanceolate, crenate-serrate (the uppermost sometimes dentate), rather dark green. Bracts longer than to slightly shorter than calyx, triangular, glabrous or somewhat scabrid. Calyx glabrous or puberulent. Corolla 13–15 mm; tube straight; throat more or less open. 2*n* = 22. *Most of Europe, but rare in the Mediterranean region.* All except Az Bl Co Cr Sa Sb Si Tu.

Ecotypic variants:
  Autumnal: var. **stenophyllus** Schur, *Enum. Pl. Transs.* 511 (1866).
  Aestival: var. **elatior** Schur, *loc. cit.* (1866).
  Vernal: type of the species.
  Montane: **R. crista-galli** var. **drummond-hayi** Buchanan-White, *Scott. Naturalist* **8**: 324 (1886).

In addition to the usual ecotypic variation this species shows much variation in other characters. Plants with puberulent calyx are commonest in northern areas or at high altitudes, and are sometimes separated as **R. borealis** (Sterneck) Druce, *Ann. Scott. Nat. Hist.* **1901**: 178 (1901). This character, however, shows little correlation with others, and as mixed populations of plants, some with pubescent and some with glabrous calyx, are common (as are, in some regions, plants with calyx puberulent when young but later glabrescent), taxonomic separation does not seem to be possible.

R. personatus (Behrendsen) Béguinot, *Sched. Fl. Ital. Exsicc.* ser. 3, **12**: 67 (1914), appears to differ from **2** only in having the throat of the corolla more or less closed.

Sect. BREVIROSTRES (Sterneck) Soó. Teeth of upper lip of corolla *c.* 1 × 1 mm, more or less square.

3. R. asperulus (Murb.) Soó, *Feddes Repert.* **26**: 206 (1929) (*Alectorolophus asperulus* Murb.). Stem up to 20 cm, simple or branched, without black streaks. Leaves oblong-lanceolate, acute, remotely dentate. Bracts equalling calyx, triangular-ovate, somewhat scabrid, with teeth somewhat diminishing in length towards apex. Calyx glabrous or puberulent. Corolla *c.* 15 mm; tube slightly curved; throat open. *W. Jugoslavia.* ?Gr Ju.

The plants with glabrous calyx have been separated as **R. illyricus** (G. Beck & Sterneck) Soó, *Izv. Bot. Inst.* (*Sofia*) **6**: 367 (1958).

Ecotypic variants:
Aestival: **Alectorolophus asperulus** subsp. **rohlenae** Sterneck, *Glasn. Muz. Bosni Herceg.* **31**: 64 (1919).
Montane: type of the species.

4. **R. wettsteinii** (Sterneck) Soó, *Feddes Repert.* **26**: 182 (1929). Stem 10–20 cm, simple or branched, densely glandular-pubescent, without black streaks. Leaves lanceolate, with remote, acute teeth. Bracts exceeding calyx, triangular, densely glandular-pubescent; teeth patent, the proximal nearly twice as long as the distal. Calyx densely glandular-pubescent. Corolla *c.* 18 mm; throat more or less closed. ● *C. & S. Appennini.* It.

5. **R. pubescens** (Sterneck) Boiss. & Heldr. ex Soó, *op. cit.* 206 (1929) (*Alectorolophus pubescens* Sterneck). Like **4** but stem up to 40 cm; leaves oblong, crenate-serrate; bracts scarcely exceeding calyx with appressed teeth. ● *Mountains of Greece.* Gr.

6. **R. pindicus** (Sterneck) Soó, *Izv. Bot. Inst. (Sofia)* **6**: 367 (1958) (*Alectorolophus pindicus* Sterneck). Like **4** but bracts and calyx glabrous, except for scabrid margin and teeth. ● *N.C. Greece.* Gr.

R. sintenisii (Sterneck) Soó, *loc. cit.* (1958) (*Alectorolophus sintenisii* Sterneck) from N.C. Greece (near Malakasi) is also like **4** but with bracts and calyx subglabrous except for densely glandular margins and teeth; but it has conical teeth on the upper lip of the corolla *c.* 2 mm long and its systematic position must, therefore, be considered doubtful.

7. **R. antiquus** (Sterneck) Schinz & Thell. in Schinz & R. Keller, *Fl. Schweiz* ed. 3, **2**: 315 (1914) (*Alectorolophus antiquus* Sterneck). Stem 5–15 cm, usually simple, without black streaks. Leaves ovate to lanceolate, acutely serrate. Bracts glabrous, triangular-ovate, acuminate; teeth abruptly diminishing in length towards apex, the basal up to 6 mm, acute but not aristate. Calyx glabrous. Corolla *c.* 15 mm; tube somewhat curved; throat more or less closed. ● *S. Alps.* He It.

Sect. ANOMALI (Sterneck) Soó. Upper lip of corolla with convex margins and with teeth 1·5–2·5 mm, 2–3 times as long as wide; throat open.

8. **R. dinaricus** Murb., *Lunds Univ. Årsskr.* **27**(5): 69 (1892) (*Alectorolophus dinaricus* (Murb.) Sterneck; incl. *R. praesignis* (G. Beck & Sterneck) Soó). Stem up to 30 cm, branched, without black streaks. Leaves lanceolate, remotely serrate. Bracts slightly exceeding calyx, glabrous, triangular-ovate; teeth acute, not aristate, gradually diminishing in length towards apex. Calyx glabrous or puberulent. Corolla *c.* 20 mm; tube straight. ● *W.C. Jugoslavia (Bosna, Hercegovina).* Ju.

Sect. ANOECTOLEMI Chab. Upper lip of corolla with concave margins and with teeth 1·5–2·5 mm, 2–3 times as long as wide; throat open.

9. **R. aristatus** Čelak., *Österr. Bot. Zeitschr.* **20**: 132 (1870) (*R. angustifolius* auct., non C. C. Gmelin, *Alectorolophus angustifolius* auct., non (C. C. Gmelin) Hcynh.; incl. *R. lanceolatus* (Kováts ex Neilr.) Chab.). Stem 5–50 cm, simple or branched, glabrous or subglabrous, with black streaks. Leaves linear to lanceolate, crenate-serrate. Bracts slightly exceeding calyx, glabrous, narrowly triangular, the basal 3–4 teeth subulate, aristate, the others much shorter, shortly aristate. Calyx glabrous. Corolla 15–18 mm; tube abruptly curved. 2*n*=22. ● *S.C.*

Europe, extending to S.E. France and S.W. Jugoslavia. Au Cz Ga Ge He It Ju.

Ecotypic variants:
Autumnal: type of the species.
Aestival: **Alectorolophus lanceolatus** var. **subalpinus** Sterneck, *Österr. Bot. Zeitschr.* **45**: 274 (1895).
Vernal: **Alectorolophus simplex** Sterneck, *Abh. Zool.-Bot. Ges. Wien* **1**(2): 89 (1901).
Montane: **R. alpinus** var. **lanceolatus** Kováts ex Neilr., *Nachtr. Fl. Wien* 213 (1851).
Alpine: **R. lanceolatus** var. **gracilis** Chab., *Bull. Herb. Boiss.* **7**: 509 (1899).

R. javorkae Soó, *Feddes Repert.* **26**: 207 (1929), from S.W. Bulgaria (Pirin Pl.), is intermediate between **17** and the alpine variant of **9**. It differs from **17** in its corolla not more than 18 mm, with abruptly curved tube and open throat, and from **9** in its puberulent bracts and calyx.

10. **R. pampaninii** Chab., *Nuovo Gior. Bot. Ital.* nov. ser., **12**: 199 (1905). Like **9** but stem 20–80 cm, glandular-pubescent; calyx and margin of bracts glandular-pubescent. ● *S. Alps.* It.

11. **R. alpinus** Baumg., *Enum. Stirp. Transs.* **2**: 194 (1816) (*Alectorolophus alpinus* (Baumg.) Sterneck). Stem up to 50 cm, simple or branched, with black streaks. Leaves narrowly lanceolate to oblong-lanceolate, crenate-serrate. Bracts equalling the calyx, pale green, glabrous, triangular-acuminate; teeth narrow but not aristate, gradually diminishing in length towards apex. Calyx glabrous. Corolla *c.* 15 mm, pale yellow, often tinged with purple; tube strongly and abruptly curved; throat widely open. ● *Mountains of E.C. & S.E. Europe, from W. Czechoslovakia to S.W. Bulgaria.* Au Bu Cz Po Rm Rs (W).

Ecotypic variants:
Autumnal: **Alectorolophus alpinus** var. **erectus** Sterneck, *Österr. Bot. Zeitschr.* **45**: 229 (1895).
Aestival: **Alectorolophus pulcher** var. **elatus** Sterneck, *op. cit.* 226 (1895).
Vernal: **R. pulcher** Sprengel, *Syst. Veg.* **2**: 772 (1825).
Montane: type of the species.

R. gracilis Schur, *Verh. Mitt. Siebenb. Ver. Naturw.* **10**: 176 (1859), from Romania and Bulgaria, differs from **11** only in its corolla up to 18 mm, with less abruptly curved tube and less widely open throat. It may be considered as intermediate between **11** and **23**.

12. **R. carinthiacus** Widder, *Carinthia II* **67**(147): 105 (1957). Like **11** but with bracts, calyx and upper part of stem densely glandular-hairy. ● *S.E. Austria (Kärnten).* Au.

Sect. CLEISTOLEMUS Chab. Upper lip of corolla with concave margins and with teeth 1·5–2·5 mm, 2–3 times as long as wide; throat closed.

13. **R. burnatii** (Chab.) Soó, *Feddes Repert.* **26**: 206 (1929) (*Alectorolophus burnatii* (Chab.) Sterneck). Stem 20–40 cm, branched, without black streaks. Leaves oblong-lanceolate, crenate-serrate. Bracts exceeding calyx, triangular-ovate, puberulent, glandular-hairy on margin; teeth long, shortly aristate, gradually diminishing in length towards apex. Calyx puberulent, with long glandular hairs on margin. Corolla *c.* 20 mm. 2*n*=22. ● *S. Alps and mountains of N.W. Jugoslavia.* Ga It Ju.

14. **R. rumelicus** Velen., *Sitz.-Ber. Böhm. Ges. Wiss. (Math.-Nat. Kl.)* **1887**: 455 (1887) (*Alectorolophus rumelicus* (Velen.) Borbás). Stem up to 60 cm, simple or branched, often with

black streaks, densely to sparsely glandular-hairy above. Leaves lanceolate to ovate. Bracts triangular-ovate, glandular-hairy (rarely subglabrous), with teeth scarcely diminishing in length towards apex. Calyx densely to sparsely glandular-hairy, the hairs persisting in fruit. Corolla up to 20 mm; tube slightly curved. *E.C. Europe and Balkan peninsula.* Al Bu Cz Hu Ju Rm Rs (W).

Ecotypic variants:
Autumnal: **R. major** var. **abbreviatus** Murb., *Lunds Univ. Årsskr.* **27**(5): 72 (1892).
Aestival: **R. rumelicus** subsp. **simonkaianus** Soó, *Feddes Repert.* **26**: 197 (1929).
Vernal: type of the species.
Montane: **Alectorolophus sagorskii** Semler, *Allgem. Bot. Zeitschr.* **14**: 118 (1908).

R. aschersonianus (M. Schulze) Soó, *Feddes Repert.* **26**: 198 (1929) (*Alectorolophus rumelicus* subsp. *aschersonianus* (M. Schulze) Hegi), from C. Germany (near Jena), and **R. oesilensis** (Ronniger & Saarson) Vassilcz. in Schischkin & Bobrov, *Fl. URSS* **22**: 684 (1955) (*R. rumelicus* subsp. *oesilensis* Ronniger & Saarson), from Estonia (island of Sarema), are scarcely distinguishable from **14**, though the latter is said to have narrower leaves.

15. R. wagneri Degen, *Österr. Bot. Zeitschr.* **44**: 39 (1894) (*Alectorolophus wagneri* (Degen) Sterneck). Like **14** but with stem and bracts subglabrous; calyx subglabrous except for glandular-hairy margin, the surface glabrescent in fruit. ● *E.C. Europe and N.C. parts of Balkan peninsula.* Bu ?Cz Hu Ju Rm.

Ecotypic variants:
Autumnal: type of the species.
Aestival: **R. rumelicus** subsp. **caroli-henrici** Soó, *Feddes Repert.* **26**: 198 (1929).
Vernal: **Alectorolophus anceps** Behrendsen, *Verh. Bot. Ver. Brandenb.* **45**: 201 (1904).
Montane: **Alectorolophus herzegovinus** Sagorski, *Österr. Bot. Zeitschr.* **59**: 81 (1909).

16. R. mediterraneus (Sterneck) Adamović, *Rad Jugosl. Akad. Znan. Umj.* **1913**: 63 (1913) (*Alectorolophus mediterraneus* Sterneck). Stem 5–40 cm, simple or branched, with black streaks, hairy on 2 opposite sides. Leaves ovate-oblong to oblong-lanceolate, with acute, patent teeth. Bracts exceeding calyx, triangular-ovate, puberulent; teeth aristate, gradually diminishing in length towards apex, the basal up to 8 mm. Calyx puberulent. Corolla *c.* 20 mm; tube somewhat curved. 2*n*=22. ● *W. & C. Mediterranean region.* Al Ga Hs It Ju ?Rs (K).

The identity of the plants from Krym referred to this species is subject to some doubt.

Ecotypic variants:
Autumnal: **R. ramosus** var. **arvernensis** Chab., *Bull. Herb. Boiss.* **7**: 499 (1899).
Aestival: type of the species.
Montane: **Alectorolophus behrendsenii** Sterneck ex Behrendsen, *Verh. Bot. Ver. Brandenb.* **45**: 202 (1904).
Alpine: **Alectorolophus pumilus** Sterneck, *Österr. Bot. Zeitschr.* **45**: 49 (1895).

17. R. melampyroides (Borbás & Degen) Soó, *Izv. Inst. Bot. (Sofia)* **6**: 363 (1958) (*Alectorolophus melampyroides* Borbás & Degen). Like **16** but stem without black streaks; leaves incise-dentate; teeth of bracts abruptly diminishing in length towards apex. ● *N. Albania.* Al ?Ju.

Ecotypic variants:
Autumnal: **Alectorolophus hayekii** Degen, *Magyar Bot. Lapok* **21**: 64 (1923).
Montane: type of the species.

18. R. songeonii Chab., *Bull. Herb. Boiss.* **7**: 497 (1899) (*Alectorolophus songeonii* (Chab.) Sterneck). Stem up to 70 cm, simple or branched, with black streaks, subglabrous or hairy on 2 opposite sides. Leaves narrowly lanceolate, acutely dentate, with patent, almost subulate teeth. Bracts much exceeding calyx, narrowly triangular, glabrous; teeth gradually diminishing in length towards apex, the basal very long and narrow, long-aristate, the upper remote. Calyx glabrous. Corolla up to 20 mm; tube slightly curved. ● *S. Alps; Albania.* Al Ga It.

Ecotypic variants:
Autumnal: type of the species.
Aestival: **R. ovifugus** subsp. **albanicus** Soó, *Feddes Repert.* **26**: 206 (1929).
Vernal: **Alectorolophus chabertii** Behrendsen, *Verh. Bot. Ver. Brandenb.* **45**: 204 (1904).

19. R. ovifugus Chab., *Bull. Herb. Boiss.* **7**: 501 (1899) (*Alectorolophus ovifugus* (Chab.) Sterneck). Stem up to 50 cm, simple or branched, with black streaks, hairy on 2 opposite sides. Leaves linear-lanceolate to ovate, subobtuse, with acute but not subulate teeth. Bracts slightly exceeding calyx, glabrous; teeth abruptly diminishing in length towards apex, the basal up to 8 mm, aristate, the upper remote, patent. Calyx glabrous. Corolla *c.* 20 mm; tube slightly curved. ● *S. Alps, Appennini, mountains of W. Jugoslavia.* Ga He It Ju.

Ecotypic variants:
Autumnal: **Alectorolophus divaricatus** Sterneck, *Abh. Zool.-Bot. Ges. Wien* **1**(2): 64 (1901).
Aestival: type of the species.
Vernal: **Alectorolophus laricetorum** Wilczek & Sterneck, *Ann. Bot. (Roma)* **12**: 9 (1913).
Montane: **Alectorolophus beyeri** Behrendsen, *Verh. Bot. Ver. Brandenb.* **45**: 47 (1903).
Alpine: **R. pulcher** var. **apenninus** Chab., *Bull. Herb. Boiss.* **7**: 507 (1899).

20. R. subulatus (Chab.) Soó, *Feddes Repert.* **26**: 182 (1929) (incl. *R. pectinatus* (Behrendsen) Vassilcz.). Stem up to 50 cm, simple or branched, with or without black streaks, subglabrous. Leaves linear to linear-lanceolate, acuminate, densely serrate-dentate. Bracts equalling calyx, triangular-ovate, glabrous; teeth subulate, gradually diminishing in length towards apex, the upper appressed. Calyx glabrous. Corolla 18–20 mm; tube slightly curved. *Krym.* Rs (K). (*Caucasus.*)

Ecotypic variants:
Autumnal: **Alectorolophus pectinatus** Behrendsen, *Verh. Bot. Ver. Brandenb.* **45**: 51 (1903).
Aestival: type of the species.

It is not certain that the aestival variant occurs in Europe.

21. R. borbasii (Dörfler) Soó, *Bot. Közl.* **36**: 311 (1939) (*Alectorolophus borbasii* Dörfler). Stem up to 60 cm, simple or branched, glabrous to moderately hairy, often with black streaks. Leaves linear-lanceolate, with patent, acute teeth. Bracts much exceeding calyx, broadly triangular, acuminate, glabrous; teeth gradually diminishing in length towards apex, the basal long, subaristate, the upper remote, patent. Calyx glabrous. Corolla-tube slightly curved. *S.E. part of C. Europe; S.E. part of U.S.S.R.* Au Cz Hu Ju Rm Rs (W, ?K, E).

(a) Subsp. **borbasii**: Leaves more or less patent. Corolla *c.* 20 mm. ● *S.E. part of C. Europe, from E. Austria to C. Romania.*

Ecotypic variants:
Autumnal: **Fistularia goniotricha** var. **interfoliata** Borbás, *Balaton Növényföld.* 379 (1900).
Aestival: **Rhinanthus major** subsp. **rapaicsianus** Soó, *Feddes Repert.* **26**: 201 (1929).
Vernal: type of the subspecies.

(b) Subsp. songaricus (Sterneck) Soó, *Acta Bot. Acad. Sci. Hung.* **13**: 303 (1967) (*R. songaricus* (Sterneck) B. Fedtsch.): Leaves suberect. Corolla *c*. 17 mm. *S.E. Russia, E. Ukraine.*

22. R. halophilus U. Schneider, *Wiss. Zeitschr. Univ. Greifswald* **11**: 155 (1962). Stem up to 25 cm, simple or slightly branched, glabrous. Leaves ovate-oblong, obtuse, rather fleshy. Bracts exceeding calyx, broadly triangular, glabrous; teeth gradually diminishing in length towards apex, the basal *c*. 5 mm, shortly aristate, the upper short, triangular, appressed. Calyx glabrous. Corolla 20–24 mm; tube 5–7 mm wide. *Alkaline soils.* ● *N. Germany (Mecklenburg).* Ge.

23. R. angustifolius C. C. Gmelin, *Fl. Bad.* **2**: 669 (1806) (*R. major* auct., non L., *Alectorolophus major* Reichenb.). Stem up to 60 cm, simple or branched, glabrous to moderately hairy, often with black streaks. Leaves linear- to ovate-lanceolate, crenate-serrate. Bracts exceeding calyx, glabrous; teeth gradually diminishing in length towards apex, the basal *c*. 5 mm, very shortly aristate, the upper short, triangular, appressed. Calyx glabrous. Corolla 16–20 mm; tube 2–3 mm wide, slightly curved. $2n = 22$. *Most of Europe, but absent from the Mediterranean region, the south-west and most of the islands.* Au Be Br Bu Cz Da Fe Ga Ge He Ho Hu ?It Ju No Po Rm Rs (N, B, C, W, K, E) Su Tu.

Very variable; by many authors the standard pattern of ecotypic variation has been combined with variation in other characters in the creation of infraspecific categories, so that the synonymy is somewhat confused. It seems best to recognize 3 subspecies, and, within two of them, a number of ecotypic variants.

1 Leaves 8–15 mm wide   **(c) subsp. grandiflorus**
1 Leaves 2–5(–8) mm wide
  2 Bracts narrowly triangular; seeds *c*. 4 mm
                                     (a) subsp. angustifolius
  2 Bracts ovate; seeds 2–2·5 mm   **(b) subsp. cretaceus**

(a) Subsp. angustifolius (*R. montanus* Sauter, *Alectorolophus major* subsp. *montanus* (Sauter) Wettst.): Leaves 2–5(–8) mm wide, linear-lanceolate. Internodes and branches numerous. Bracts narrowly triangular. Capsule 6–7 mm. Seeds *c*. 4 mm. *Throughout much of the range of the species, but absent from parts of the north and the south-east.*

Ecotypic variants:
Autumnal: type of the subspecies.
Montane: **R. major** subsp. **lykae** Soó, *Feddes Repert.* **26**: 201 (1929).

(b) Subsp. cretaceus (Vassilcz.) Soó, *Acta Bot. Acad. Sci. Hung.* **14**: 151 (1968) (*R. cretaceus* Vassilcz.): Like subsp. (a) but bracts ovate; capsule 7–8 mm; seeds 2–2·5 mm. ● *E. Ukraine and S.E. Russia (Donets basin).*

(c) Subsp. grandiflorus (Wallr.) D. A. Webb, *Bot. Jour. Linn. Soc.* **65**: 269 (1972) (*Alectorolophus grandiflorus* Wallr.; incl. *R. aestivalis* (Zinger) Schischkin & Serg., *R. vernalis* (Zinger) Schischkin & Serg.): Leaves 8–15 mm wide, lanceolate to ovate-lanceolate. Internodes and branches numerous or few. Bracts ovate. Capsule 10–12 mm. Seeds 3–5(–6) mm. *Throughout the range of the species.*

Ecotypic variants:
Autumnal: **R. major** var. **polycladus** Chab., *Mém. Herb. Boiss.* **8**: 12 (1900).
Aestival: **Alectorolophus aestivalis** Zinger, *Trudy Tiflissk. Bot. Sada* **12**: 184 (1913).
Vernal: **Alectorolophus vernalis** Zinger, *loc. cit.* (1913).
Montane: **Alectorolophus bosniacus** Behrendsen, *Verh. Bot. Ver. Brandenb.* **45**: 210 (1904).

A variant with unwinged seeds, distinguished as *R. major* var. *apterus* Fries (*R. apterus* (Fries) Ostenf.), occurs sporadically through the northern part of the range of this subspecies.

R. bosnensis (Behrendsen & Sterneck) Soó, *Omagiu Săvulescu* 739 (1959), is like **23**(c) but the bracts have teeth subequal in length and the throat of the corolla is incompletely closed. It is recorded from the E. Carpathians and N. Jugoslavia.

24. R. freynii (A. Kerner ex Sterneck) Fiori, *Nuovo Gior. Bot. Ital.* nov. ser., **19**: 576 (1912) (*Alectorolophus freynii* A. Kerner ex Sterneck). Stem up to 60 cm, simple or branched, hairy, without black streaks. Leaves ovate to oblong-lanceolate, crenate-serrate. Bracts rhombic-triangular, puberulent; basal teeth scarcely longer than the others. Calyx densely puberulent-velutinous with eglandular, unicellular hairs. Corolla *c*. 20 mm; tube slightly curved. ● *N. Italy, N.W. Jugoslavia.* It Ju.

Ecotypic variants:
Autumnal: **Alectorolophus sterneckii** Wettst., *Österr. Bot. Zeitschr.* **47**: 357 (1897).
Aestival: **R. freynii** subsp. **croaticus** Soó, *Feddes Repert.* **26**: 196 (1929).
Vernal: type of the species.
Montane: **R. bellunensis** Chab., *Allgem. Bot. Zeitschr.* **14**: 37 (1908).

R. helenae Chab. ex Pamp., *Nuovo Gior. Bot. Ital.* nov. ser., **14**: 606 (1907), from the S. Alps (around Trento), is like **24**, but with strongly and abruptly curved corolla-tube and throat of corolla more or less open. Its systematic position is doubtful.

25. R. alectorolophus (Scop.) Pollich, *Hist. Pl. Palat.* **2**: 177 (1777) (*R. major* L., nomen ambiguum, *Alectorolophus hirsutus* (Lam.) All.; incl. *R. patulus* (Sterneck) Schinz & Thell.). Stem up to 80 cm, simple or branched, without black streaks, hairy. Leaves lanceolate to ovate, crenate-serrate. Bracts rhombic-triangular, puberulent; basal teeth scarcely longer than the others. Calyx villous with long, white hairs, but partly glabrescent in fruit. Corolla *c*. 20 mm; tube slightly curved. $2n = 22$. ● *From N. France, the Netherlands and W.C. Russia southwards to N. Italy and N. Jugoslavia; occasionally casual elsewhere.* Au Be Cz Ga He Ho It Ju Po Rs (?B, C, W).

Variants with unwinged seeds are found throughout the range of the species, and show the same range of ecotypic variation as those with winged seeds.

Ecotypic variants:
Autumnal: **Alectorolophus patulus** Sterneck, *Österr. Bot. Zeitschr.* **47**: 433 (1897).
Aestival: **R. buccalis** Wallr., *Flora (Regensb.)* **25**: 504 (1842).
Vernal: type of the species.
Montane: **R. alectorolophus** var. **modestus** Chab., *Bull. Herb. Boiss.* **7**: 504 (1899).

R. facchinii Chab., *op. cit.* 506 (1899), from the Alps, is like **25**, but with corolla-tube strongly and abruptly curved and throat of corolla open. It thus bears the same relation to **25** as does *R. helenae* to **24**.

36. Rhynchocorys Griseb.[1]

Perennial, hemiparasitic herbs with opposite leaves. Flowers in lax, bracteate racemes; bracts leaf-like; bracteoles absent. Calyx laterally compressed; upper lip slightly emarginate; lower lip 2-lobed, exceeding the upper. Corolla-tube short, campanulate; upper lip galeate, rostrate; lower lip patent, 3-lobed. Stamens 4, didynamous; anthers oblique or transverse, glabrous, not mucronate. Capsule loculicidal; seeds striate.

---

[1] By I. B. K. Richardson.

1. R. elephas (L.) Griseb., *Spicil. Fl. Rumel.* **2**: 12 (1844). Stems up to 50 cm, erect or ascending, more or less branched, subglabrous. Leaves 10–30 × 5–15 mm, obovate to elliptical, shallowly crenate-serrate, obtuse, rounded to subcordate at base, glabrous; petioles 4–8 mm. Lower bracts like the leaves, the upper smaller, ovate, sessile; pedicels 4–8 mm. Corolla 10–15 mm, yellow; beak 6–10 mm, straight. Capsule suborbicular. *Mountain grassland. S. Italy, Sicilia; Balkan peninsula from C. Greece to C. Bulgaria.* Bu Gr It Si.

37. Siphonostegia Bentham[1]

Perennial, hemiparasitic herbs. Leaves opposite below, alternate above, subsessile. Flowers in a terminal, bracteate raceme. Calyx tubular, 2-lipped; upper lip 3-lobed, the lower 2-lobed. Corolla with a long tube and 2-lipped limb; upper lip truncate or emarginate, the lower shortly 3-lobed. Stamens 4, subequal; anthers not mucronate. Stigma subentire. Capsule loculicidal; seeds very small, numerous, with a loose, reticulate-rugose testa.

1. S. syriaca (Boiss. & Reuter) Boiss., *Fl. Or.* **4**: 471 (1879). Glandular-pubescent throughout; stems *c.* 60 cm, erect, rigid, simple or sparingly branched, densely leafy. Lower leaves 20–50 × 4–12 mm, oblong-lanceolate, entire. Bracts narrowly elliptical, equalling or exceeding the calyx. Calyx 17–20 mm; lips divided to their base into oblong-linear, subequal teeth. Corolla *c.* 30 mm, purple. Capsule included in the calyx. *Scrub. Mountains of E. Greece (Thessalia).* Gr. (*S. Anatolia*).

38. Cymbaria L.[2]

(incl. *Cymbochasma* (Endl.) Klokov & Zoz)

Perennial herbs, somewhat woody at the base, with undivided, opposite leaves. Flowers solitary, axillary, bracteolate. Calyx equally 5-lobed. Corolla with obconical tube and 2-lipped limb; lips subparallel, the upper entire, cucullate, the lower 3-lobed. Stamens 4, didynamous, somewhat exserted from upper lip of corolla. Stigma capitate. Capsule loculicidal; placentae large, 2-lobed; ovules numerous; seeds few.

1. C. borysthenica Pallas ex Schlecht. in Nees, *Horae Phys. Berol.* 109 (1820) (*Cymbochasma borysthenicum* (Pallas ex Schlecht.) Klokov & Zoz). Whole plant silvery-sericeous to greyish-villous. Stems 3–15 cm, simple, erect from a branched, woody stock. Leaves 5–20 × 1–3 mm, linear-lanceolate, acute, entire. Flowers few, in axils of lower leaves; pedicel 1–2 mm; bracteoles overlapping the calyx-tube. Corolla 25–35 mm, pale yellow; tube longer than lips. Seeds elliptical, flat, winged. *Steppes and other dry places.* ● *S. Ukraine, just extending to S. Russia.* Rs (W, E).

39. Lathraea L.[2]

Perennial herbs, wholly parasitic and without chlorophyll; leaves fleshy, scale-like. Flowers zygomorphic, in terminal racemes or corymbs. Calyx campanulate, unequally or subequally 4-lobed. Corolla with cylindrical tube and 2-lipped limb; upper lip entire, convex; lower entire or shortly 3-lobed. Stamens 4, didynamous, included under upper lip of corolla or slightly exserted from it; anther-lobes hairy, mucronate. Ovary 1-locular, with 2 large,

2-lobed, parietal placentae. Capsule dehiscing along midribs of carpels.

Included by some authors in Orobanchaceae, but the resemblance seems to be superficial (arising from the completely parasitic habit, which is, however, also found in Scrophulariaceae), and in many morphological features *Lathraea* more closely resembles such members of the Scrophulariaceae as *Bartsia* or *Rhinanthus*.

In all European species cleistogamous flowers may occur at the base of the inflorescence or underground. They have a corolla considerably shorter than that of normal flowers.

Literature: G. Beck von Mannagetta in Engler, *Pflanzenreich* 96(IV. 261): 317–326 (1930). E. Heinricher, *Monographie der Gattung* Lathraea. Jena. 1931.

1 Flowers arising from a mainly subterranean rhizome; corolla
    40–50 mm **3. clandestina**
1 Flowers borne on an erect, aerial stem; corolla 10–20 mm
2 Raceme secund; calyx *c.* 10 mm; corolla 14–17(–20) mm
    1. squamaria
2 Raceme cylindrical, not secund; calyx 5–7 mm; corolla *c.* 10
    mm **2. rhodopea**

1. L. squamaria L., *Sp. Pl.* 606 (1753). Plant white or cream below; upper part more or less tinged with lilac-pink. Stems 15–30 cm, stout, erect, arising from a branched, subterranean rhizome. Scale-leaves alternate, suborbicular, cordate-amplexicaul, entire, crowded and more or less squarrose below, more distant and appressed above; bracts similar but thinner. Inflorescence a rather dense, secund, spike-like raceme; flowers patent or somewhat drooping; pedicels 3–6(–10) mm. Calyx *c.* 10 mm, glabrous to glandular-pubescent. Corolla 14–17(–20) mm; lips subequal, only slightly divergent; apex of upper lip not cucullate; lower lip flat. Capsule *c.* 10 mm, subglobose; seeds 1–2 mm, subglobose, reticulate, numerous. 2*n* = 36. *Parasitic on the roots of various trees and shrubs, most frequently on species of* Alnus, Corylus *and* Fagus. *Much of Europe, but absent from parts of the north and south.* Au Be Br Bu Cz Da Fe Ga Ge Hb He †Ho Hs Hu It Ju No Po Rm Rs (B, C, W, K) Si Su Tu.

2. L. rhodopea Dingler, *Bot. Zeit.* **35**: 74 (1877). Like **1** but up to 50 cm; whole plant yellowish-brown; raceme not secund, usually longer and denser than in **1**; pedicels 1–2 mm; calyx 5–7 mm; corolla *c.* 10 mm, with upper lip slightly cucullate; seeds few. *Parasitic on the roots of various trees.* ● *S. Bulgaria, N.E. Greece.* Bu Gr.

3. L. clandestina L., *Sp. Pl.* 605 (1753). Stems almost or completely subterranean, freely branched so as to form large tufts, yellowish. Scale-leaves reniform, amplexicaul, alternate or opposite; bracts similar. Inflorescence a short, corymbose raceme with 4–8 flowers, the axis often largely subterranean; flowers erect; pedicels up to 30 mm. Calyx 18–20 mm, glabrous, pale lilac. Corolla 40–50 mm, violet, the lower lip reddish-purple; apex of upper lip somewhat cucullate; lower lip much shorter than upper, with margins deflexed. Capsule ovoid; seeds up to 5 mm, grey, reticulate-rugose, few. 2*n* = 42. *Parasitic on the roots of species of* Salix, Populus *and* Alnus, *or rarely of other trees.* ◉ *W. Europe, from Belgium to N. Spain, and extending eastwards to C. & S. Italy.* Be Ga Hs It [Br].

---

[1] By T. G. Tutin.   [2] By D. A. Webb.

CLV. GLOBULARIACEAE[1]

Perennial herbs or small shrubs. Leaves simple, alternate or in rosettes. Flowers 5-merous, zygomorphic, in dense capitula surrounded by an involucre. Corolla usually blue. Stamens 4, rarely 2. Ovary 1-locular; ovule 1, pendent. Fruit dry, enclosed in the persistent calyx.

Literature: O. Schwarz, *Bot. Jahrb.* **69**: 318–373 (1938).

1. Globularia L.[2]

Usually evergreen. Flowers in capitula. Corolla 2-lipped; upper lip usually 2-lobed, often small; lower lip 3-lobed, longer than the upper. Stamens and style exserted; stigma 2-lobed.

All species grow in dry places such as mountain rocks, grassland or stony slopes.

1 Shrub, with erect or procumbent woody branches
  2 Erect shrub; twigs leafy up to the inflorescence, never rooting at the nodes   **8. alypum**
  2 Procumbent shrub; twigs often rooting at the nodes; inflorescences usually on leafless peduncles arising from a leaf-rosette
    3 Leaves obovate to suborbicular, emarginate to rounded, sometimes mucronate at apex
      4 Involucral bracts abruptly cuspidate   **12. neapolitana**
      4 Involucral bracts gradually acuminate
        5 Subterranean stolons absent; calyx 2-lipped   **9. cordifolia**
        5 Subterranean stolons present; calyx almost actinomorphic   **13. stygia**
    3 Leaves lanceolate to oblanceolate, acute, very rarely some slightly emarginate
      6 Leaves 10–20(–25) × *c.* 1 mm, folded   **11. repens**
      6 Leaves 20–90 × 2–5 mm, flat   **10. meridionalis**
1 Herbs, often with a woody stock
  7 Leaves deciduous, not forming a well-marked rosette; cauline leaves mostly petiolate   **1. incanescens**
  7 Leaves evergreen, forming a well-marked rosette; cauline leaves absent or all sessile
    8 Flowering stem without or rarely with up to 3 small bracts; leaves erect or erecto-patent
      9 Stolons absent; leaves 6–12 cm   **14. nudicaulis**
      9 Stolons 1–5 cm; leaves 4–7 cm   **15. gracilis**
    8 Flowering stem with numerous small leaves; basal leaves ±horizontal
      10 Rosette-leaves with lateral veins distinctly visible on upper surface   **3. punctata**
      10 Rosette-leaves with lateral veins not or scarcely visible on upper surface
        11 Rosette-leaves with 2 or more long, patent, spinescent teeth on each side   **7. spinosa**
        11 Rosette-leaves entire, 3-dentate at apex or, rarely, with small, somewhat spinescent, forward-pointing teeth on the margin
          12 Stolons present; calyx-teeth 3–4 times as long as the tube   **2. trichosantha**
          12 Stolons absent; calyx-teeth not or little longer than the tube   **(4–6). vulgaris group**

1. G. incanescens Viv., *Fl. Ital. Fragm.* 2 (1808). Stems 3–10 cm, with slender rhizomes. Leaves orbicular to lanceolate, obtuse or emarginate, often mucronulate, more or less erect, the lower long-petiolate, most of the cauline more or less shortly but distinctly petiolate, the uppermost subsessile, all grey-green owing to calcareous secretions, deciduous. Capitula *c.* 1·5 cm in diameter; involucral bracts usually 5, linear-lanceolate, ciliate. Calyx-teeth *c.* 3 times as long as tube, aristate. Upper lip of corolla usually unlobed. $2n=16$. ● *N. Appennini and Alpi Apuane.* It.

2. G. trichosantha Fischer & C. A. Meyer, *Ind. Sem. Hort. Petrop.* **5**: 36 (1839). Stolons up to 30 cm; stems *c.* 20 cm. Lower leaves horizontal, forming a rosette, obovate, obtuse to emarginate, sometimes mucronate, petiolate; lateral veins not or scarcely visible above; cauline leaves elliptical, sessile. Capitula up to 2·5 cm in diameter; involucral bracts numerous, oblong-lanceolate, aristate. Calyx-teeth 3–4 times as long as tube. *E. part of Balkan peninsula; Krym.* Bu Rs (K) Tu.

3. G. punctata Lapeyr., *Hist. Abr. Pyr.* 57 (1813) (*G. aphyllanthes* auct., non Crantz, *G. willkommii* Nyman). Stolons absent; stems up to 30 cm. Lower leaves forming a rosette, obovate to spathulate, rounded or weakly emarginate, petiolate; lateral veins clearly visible above; cauline leaves lanceolate to oblong, sessile. Capitula *c.* 1·5 cm in diameter; involucral bracts numerous, lanceolate, acuminate. Calyx-teeth about as long as tube, linear-lanceolate. $2n=16$. *From N. France and Czechoslovakia southwards to N. Spain, S. Italy and N.E. Greece; S.C. & E. Russia.* Au Be Bu Cz Ga Ge Gr He Hs Hu It Ju Rm Rs (C, E) Tu.

(4–6). G. vulgaris group. Stolons absent. Leaves acute or 3-dentate with the middle tooth at least as long as the lateral, rarely with several, forwardly directed, somewhat spinescent teeth on the sides; lateral veins not or scarcely visible above; cauline leaves lanceolate, sessile. Involucral bracts numerous. Calyx-teeth not or little longer than tube.

1 Leaves lanceolate to elliptical, 3-dentate   **4. vulgaris**
1 Leaves obovate, acute or 3-dentate, with the margin of at least some leaves toothed or undulate-crenulate
  2 Petiole usually longer than lamina; margin of leaf undulate-crenulate   **5. valentina**
  2 Petiole usually shorter than lamina; margin of leaf flat, usually toothed   **6. cambessedesii**

4. G. vulgaris L., *Sp. Pl.* 96 (1753). Leaves lanceolate to elliptical, 3-dentate; margin flat; petiole usually longer than lamina. Flowering stems up to 20 cm. Capitula *c.* 2·5 cm in diameter. Involucral bracts lanceolate, acuminate. Calyx-teeth about as long as tube. $2n=32$. ● *Mountains of N.C. & E. Spain and S. France; Sweden (Öland & Gotland).* Ga Hs ?Rs (B) Su.

5. G. valentina Willk., *Rech. Glob.* 21 (1850). Like **4** but leaves obovate, acute or shortly acuminate; margin undulate-crenulate and sometimes serrate; flowering stems usually 20–30 cm; involucral bracts ovate, long-cuspidate; calyx-teeth rather longer than tube. ● *N.E. Spain.* Hs.

6. G. cambessedesii Willk., *Suppl. Prodr. Fl. Hisp.* 140 (1893). Like **4** but leaves obovate, acute or shortly acuminate, usually toothed; petiole usually shorter than lamina; flowering stems usually 20–30 cm; capitula *c.* 3·5 cm in diameter; calyx-teeth rather longer than tube. $2n=64$. ● *Near the N. coast of Mallorca.* Bl.

7. G. spinosa L., *Sp. Pl.* 96 (1753). Often with a very robust, woody stock. Stems up to 20 cm. Leaves ovate to spathulate, spinescent at apex and with 2 or more long, patent, spinescent

---

[1] Edit. V. H. Heywood.   [2] By T. G. Tutin.

teeth on each side, with abundant calcareous secretions when mature, petiolate; lateral veins not or scarcely visible above; cauline leaves lanceolate, sessile, the upper ciliate. Capitula 2–2·5 cm in diameter; involucral bracts numerous, lanceolate, acuminate. Calyx-teeth about as long as tube, triangular, acuminate. ● *S.E. Spain*. Hs.

8. **G. alypum** L., *Sp. Pl.* 95 (1753). Shrub up to *c.* 100 cm. Young twigs and leaves with abundant calcareous secretions. Leaves oblanceolate to obovate, mucronate or 3-dentate, very coriaceous, shortly petiolate, uniformly distributed on the main stems, fasciculate on short, non-flowering lateral branches. Capitula 1–2·5 cm in diameter, terminal, sometimes with sessile axillary capitula below them; involucral bracts broadly ovate, obtuse, sometimes mucronate, ciliate, densely imbricate. Calyx-teeth about twice as long as tube, setaceous, long-ciliate. $2n=16$. *Mediterranean region*. Al Bl Co Cr Ga Gr Hs It Ju ?Lu Sa Si Tu.

9. **G. cordifolia** L., *Sp. Pl.* 96 (1753). Dwarf shrub with creeping woody stems rooting at the nodes. Flowering stems 1–10 cm, leafless or with 1–3 small leaves. Rosette-leaves up to *c.* 25 × 8 mm, spathulate, usually emarginate, sometimes mucronate or 3-dentate, with the lateral teeth longer than the central one, flat, petiolate. Capitula 1–2 cm in diameter; involucral bracts numerous, ovate-lanceolate to ovate, acuminate, gradually narrowing upwards. Calyx 2-lipped; teeth lanceolate, acuminate. $2n=32$. ● *Mountains of C. & S. Europe, from the Carpathians to N.E. Spain and S. Bulgaria.* Al Au Bu Cz Ga Ge He Hs Hu It Ju.

10. **G. meridionalis** (Podp.) O. Schwarz, *Bot. Jahrb.* 69: 345 (1938) (*G. cordifolia* subsp. *bellidifolia* (Ten.) Hayek & subsp. *meridionalis* Podp.). Like **9** but more robust; leaves 20–90 × 2–5 mm, lanceolate to oblanceolate, acute, very rarely slightly emarginate; involucral bracts ovate, almost cuspidate; calyx-teeth linear-lanceolate. $2n=16$. ● *S.E. Alps; mountains of Balkan peninsula; C. & S. Appennini.* Al Au Bu Gr It Ju.

11. **G. repens** Lam., *Fl. Fr.* 2: 325 (1778) (*G. nana* Lam.). Like **9** but smaller; leaves 10–20(–25) × *c.* 1 mm, acute, folded; capitula

usually subsessile; involucral bracts lanceolate, shortly acuminate; calyx-teeth lanceolate, acuminate. $2n=16$. ● *Mountains of S.W. Europe.* Ga Hs It.

12. **G. neapolitana** O. Schwarz, *Bot. Jahrb.* 69: 348 (1938). Like **9** but with partly subterranean stolons; leaves suborbicular, mucronate, very rarely slightly emarginate; capitula *c.* 1·5 cm; involucral bracts abruptly contracted into an awn. ● *S. Italy (around Napoli).* It.

13. **G. stygia** Orph. ex Boiss., *Diagn. Pl. Or. Nov.* 3(4): 60 (1859). Dwarf shrub with slender, creeping woody stems rooting at the nodes, and with subterranean stolons. Rosette-leaves suborbicular, obtuse, rarely weakly emarginate, petiolate. Capitula subsessile; involucral bracts numerous, oblong-lanceolate, acuminate. Calyx almost actinomorphic; teeth narrowly lanceolate, subulate at apex. ● *Mountains of N. Peloponnisos.* Gr.

14. **G. nudicaulis** L., *Sp. Pl.* 97 (1753). Caespitose herb, without stolons. Stems up to 30 cm. Leaves 6–12 cm, more or less erect, crowded at base of stem, oblanceolate to obovate, obtuse, gradually narrowed into the petiole. Scape without or rarely with up to 3 small bracts. Capitula 1·5–3 cm in diameter; involucral bracts numerous, lanceolate to ovate, acuminate. Calyx-teeth shorter than tube, acute. $2n=16, 24$. *Alps, Pyrenees, mountains of N. Spain.* Au Ga Ge He Hs It Ju.

G. × fuxeensis Giraud., *Bull. Soc. Étud. Sci. Angers* 17: 53 (1889) (*G. nudicaulis × repens*) occurs locally with the parents. It is intermediate between them and is reported to be fertile, with $2n=16$.

15. **G. gracilis** Rouy & J. Richter in Rouy, *Ill. Pl. Eur. Rar.* 10: 81 (1898). Like **14** but smaller and shortly stoloniferous; leaves 4–7 cm; capitula usually less than 1·5 cm in diameter; calyx-teeth about equalling tube, acuminate. ● *Pyrenees.* Ga Hs.

CLVI. ACANTHACEAE[1]

Herbs or shrubs. Leaves simple, opposite, exstipulate. Flowers zygomorphic, in spike-like inflorescences. Bracts often conspicuous, coloured. Calyx 4-lobed, 2-lipped, or campanulate, 5-fid. Corolla sympetalous, zygomorphic, with a short tube; limb 2-lipped with the upper lip subentire and the lower 3-fid, or 1-lipped with the upper lip absent. Stamens 2 or 4. Ovary superior, 2-locular, with axile placentae; style single, unequally 2-lobed. Fruit a loculicidal capsule with indurated funicles (jaculators) which eject the seeds.

Basal leaves pinnatifid to pinnatisect; corolla 1-lipped   **1. Acanthus**
Basal leaves entire; corolla 2-lipped   **2. Justicia**

1. Acanthus L.[2]

Robust perennial herbs or small shrubs. Stems simple, terete, erect. Leaves mostly basal, pinnatifid to pinnatisect. Flowers in dense, terminal, cylindrical spikes. Bracts large, spinose-dentate;

bracteoles entire, lanceolate to linear. Calyx 4-lobed; upper and lower lobes large, the lateral small. Corolla 1-lipped (upper lip absent), 3-lobed; tube short. Stamens 4, included; anthers 1-celled, connate in pairs. Fruit a capsule.

1 Basal leaves spinose-dentate   **3. spinosus**
1 Basal leaves not spinose-dentate
  2 Leaf-lobes not narrowed at base; lower lip of calyx glabrous at apex   **1. mollis**
  2 Leaf-lobes narrowed at base; lower lip of calyx pubescent at apex   **2. balcanicus**

1. **A. mollis** L., *Sp. Pl.* 639 (1753). Stems 25–100(–200) cm. Leaves glabrous to puberulent; basal leaves with the lamina 20–60 × 5–15 cm, ovate, pinnatifid, the lobes not narrowed at the base, incise-dentate, long-petiolate; upper cauline leaves 1–3 cm, more or less ovate, spinose-dentate, more or less sessile. Bracts *c.* 4 cm, ovate, usually glabrous. Calyx 4–5 cm, glabrous. Corolla 3·5–5 cm, whitish with purple veins. $2n=56$. *Shady places and roadsides. W. & C. Mediterranean region, Portugal.* Co Ga Hs It Ju Lu Sa Si [Az Bl].

---

[1] Edit. V. H. Heywood.
[2] By V. H. Heywood and I. B. K. Richardson.

Records from Macedonia and Thrace probably refer to **2**.

2. A. balcanicus Heywood & I. B. K. Richardson, *Bot. Jour. Linn. Soc.* **65**: 357 (1972) (*A. longifolius* Host, non Poiret). Like **1** but basal leaves usually pinnatisect; leaf-lobes narrowed at base; upper cauline leaves more or less petiolate; lower lip of calyx pubescent at apex. *Woodland, scrub and stony hillsides.* ● *Balkan peninsula, extending to S.W. Romania and N.W. Jugoslavia.* Al Bu Gr Ju Rm ?Tu.

3. A. spinosus L., *Sp. Pl.* 639 (1753) (incl. *A. spinosissimus* Pers., *A. caroli-alexandri* Hausskn.). Like **1** but whole plant sometimes pubescent; stems 20–80 cm; basal leaves oblong, pinnatifid or pinnatisect, spinose-dentate. *Woodland and meadows. S.E. Italy; Balkan peninsula and Aegean region.* Al Bu Cr Gr It Ju.

2. Justicia L.[1]

Shrubs with opposite, entire leaves. Inflorescences axillary, pedunculate, dense and spike-like. Calyx campanulate, 5-fid. Corolla with a short tube and large, 2-lipped limb; upper lip subentire, lower 3-fid. Stamens 2, appressed to the upper lip; filaments adnate to the corolla-tube; anthers with the loculi attached at different levels, mucronate at base.

1. J. adhatoda L., *Sp. Pl.* 15 (1753). Stems 1–3 m, puberulent when young, swollen at the nodes. Leaves elliptical, obtuse to acuminate, puberulent. Inflorescence 3–8 cm. Bracts 10–15 mm, ovate to obovate. Corolla *c.* 30 mm, white; lower lip with transverse, pink bands. Capsule *c.* 25 mm. *Naturalized in Sicilia (Messina).* [Si.] (*Tropical Asia.*)

CLVII. PEDALIACEAE[2]

Herbs with mucilage-glands. Flowers solitary or in small axillary cymes. Calyx 5-fid. Corolla 5-lobed, weakly 2-lipped. Stamens 4, the 5th represented by a staminode. Ovary superior, 4-locular, with axile placentae. Fruit a loculicidal, 2-valved capsule.

1. Sesamum L.[1]

Flowers solitary. Corolla campanulate, the lower lip longer than the upper. Stamens didynamous; anthers parallel, dorsifixed.

1. S. indicum L., *Sp. Pl.* 634 (1753). Erect, pubescent annual 30–60 cm, simple or with erecto-patent branches. Leaves *c.* 10 cm, petiolate, the lower usually lobed or 3-sect, opposite, the upper oblong to linear-lanceolate, entire, alternate. Flowers axillary; corolla *c.* 3 cm, whitish, often with purplish or yellow markings. Capsule *c.* 25 × 5 mm, oblong, erect, scabrid-velutinous. *Cultivated in S.E. Europe for the oil obtained from the seeds, and locally naturalized.* [Bu Gr Hs Ju Rs (W, K, E) ?Tu.] (*Probably native of S.E. Asia.*)

CLVIII. MARTYNIACEAE[2]

Herbs. Inflorescence a terminal raceme. Calyx 5-fid. Corolla 5-lobed, weakly 2-lipped. Stamens 4, the 5th represented by a staminode. Ovary superior, 1-locular, with parietal placentae. Fruit a 2-valved capsule, terminating in a long, curved, horn-like process, splitting longitudinally when mature.

1. Proboscidea Schmidel[1]

Calyx split to the base ventrally. Corolla-tube widening upwards; limb nearly rotate. Stamens didynamous; anther-loculi divaricate.

1. P. louisianica (Miller) Thell., *Mém. Soc. Nat. Sci. Cherbourg* **38**: 480 (1912). Decumbent, viscid-pubescent annual 50–80 cm. Leaves 10–30 cm, suborbicular, obliquely cordate, entire or sinuate; petiole up to 20 cm. Corolla *c.* 5 cm, white or yellowish, spotted with purple. Capsule 6–10 cm, with horns 8–20 cm. *Naturalized as a weed in Portugal and S.E. Russia.* [Lu Rs (E).] (*S. North America.*)

CLIX. GESNERIACEAE[3]

Perennial, acaulescent herbs, with a basal rosette of simple, exstipulate leaves. Flowers solitary or in small umbels on axillary scapes, hermaphrodite, zygomorphic to almost actinomorphic. Corolla rotate to tubular, sometimes somewhat 2-lipped. Stamens equal in number to the corolla-lobes, or one of them reduced to a staminode or absent. Ovary superior, 1-locular, with 2 parietal placentae. Fruit a septicidal capsule.

A large family, mainly tropical and subtropical. The European representatives are very isolated geographically, their nearest relatives being in S. & E. Asia; they are usually regarded as

Tertiary relicts. The description for the family given above applies only to the European genera; in extra-European genera (among other differences) the stem is often developed, the leaves may be opposite, the ovary more or less inferior, and the fruit a berry.

1 Anthers connate in pairs, much shorter than the filaments;
  corolla 2-lipped, with tube longer than lobes   **1. Haberlea**
1 Anthers free, at least as long as the filaments; corolla almost
  actinomorphic, with tube equalling or shorter than lobes
2 Corolla-tube much shorter than lobes   **2. Ramonda**
2 Corolla-tube almost as long as lobes   **3. Jankaea**

---

[1] By T. G. Tutin.   [2] Edit. T. G. Tutin.   [3] Edit. D. A. Webb.

1. Haberlea Friv.[1]

Scape with 2 linear-lanceolate bracts at base of pedicels. Calyx with 5 lobes about equalling the tube; corolla with more or less cylindrical tube and 5-lobed, unequally 2-lipped limb, the tube longer than at least the upper lip. Stamens 4, included; filaments curved; anthers much shorter than filaments, connate in pairs. Staminode 1, small. Stigma 2-lobed. Capsule about equalling calyx.

1. **H. rhodopensis** Friv., *Magyar Tudós Társaság Évkőnyvei* (*Budapest*) 2: 249 (1835). Leaves 3–8 × 2–4 cm, obovate- to ovate-oblong, obtuse, coarsely crenate, softly hirsute, tapered to a short, broad petiole. Scapes 6–10 cm, with 1–5 flowers. Corolla 15–25 mm, pale bluish-violet, hairy within. $2n = 38, 48$. *Rock-crevices.* ● *Mountains of C. & S. Bulgaria and N.E. Greece.* Bu Gr.

H. ferdinandi-coburgii Urum., *Period. Spis. Bălg. Kniž. Druž.* 63: 2 (1902), known only from one locality in C. Bulgaria, has leaves which are almost glabrous above; other supposed differences seem to be less constant. It should probably be regarded as a variety of **1**.

2. Ramonda L. C. M. Richard[1]

Leaves rugose, hirsute above with short, white hairs, villous beneath and on petiole with long, brown hairs. Scape without bracts. Flowers usually 5-merous, but with some inconstancy in the number of parts. Calyx-lobes equal, longer than the tube; corolla more or less rotate, nearly actinomorphic. Stamens as many as corolla-lobes, exserted, all fertile; anthers about as long as filaments, free. Stigma capitate, entire. Capsule much exceeding calyx.

1 Style 5–7 mm; anthers 3–4 mm, mucronate   **1. myconi**
1 Style *c.* 3 mm; anthers 2·5–3 mm, obtuse
   2 Leaves truncate at base; corolla rotate, with patent lobes; anthers yellow   **2. nathaliae**
   2 Leaves cuneate at base; corolla cup-shaped, with erecto-patent lobes; anthers dark violet-blue   **3. serbica**

1. **R. myconi** (L.) Reichenb., *Fl. Germ. Excurs.* 388 (1831) (*R. pyrenaica* Pers.). Leaves 2–6 × 1–5 cm, rhombic-orbicular to ovate, obtuse, crenate-dentate, tapered to a short petiole. Scapes 6–12 cm, glandular-pubescent, with 1–6 flowers. Flowers usually 5-merous; corolla 3–4 cm in diameter, rotate, deep violet with a yellow centre. Anthers 3–4 mm, shortly mucronate, yellow. Style 5–7 mm; capsule *c.* 15 mm. $2n = 48$. *Shady crevices in limestone rocks.* ● *C. & E. Pyrenees and adjoining mountains of N.E. Spain.* Ga Hs.

2. **R. nathaliae** Pančić & Petrović in Petrović, *Fl. Agri Nyss.* 574 (1882). Leaves 3–5 × 1–3 cm, ovate to suborbicular, more or less truncate at the base, entire or obscurely crenate; petiole 1–4 cm. Scapes 6–8 cm, glandular-pubescent, with 1–3 flowers. Calyx-lobes 4–6; corolla 3–3·5 cm in diameter, rotate, with 4(–5) patent lobes, lilac to violet, with an orange-yellow centre. Anthers 2·5–3 mm, obtuse, yellow (sometimes slightly tinged with blue). Style *c.* 3 mm; capsule *c.* 9 mm. $2n = 48$. *Shady rock-crevices.* ● *S. Jugoslavia and N. Greece.* Gr Ju.

3. **R. serbica** Pančić, *Fl. Princ. Serb.* 498 (1874). Like **2** but leaves narrowly obovate, cuneate at base, irregularly and sometimes deeply crenate or dentate; petiole usually shorter; flowers usually 5-merous; corolla 2·5–3 cm in diameter, cup-shaped, with erecto-patent lobes; anthers dark violet-blue. $2n = c.\ 96$. *Shady crevices in limestone rocks.* ● *Albania, S. Jugoslavia, N.W. Bulgaria, N.W. Greece.* Al Bu Gr Ju.

3. Jankaea Boiss.[1]

Like *Ramonda* but leaves densely white-villous above; corolla usually 4-lobed, broadly campanulate, with tube about as long as lobes. Stamens 4, included.

1. **J. heldreichii** (Boiss.) Boiss., *Diagn. Pl. Or. Nov.* 1: 5 (1875). Leaves 2–4 cm, obovate, obtuse, entire. Scapes *c.* 4 cm, with 1–2 flowers. Calyx 5-lobed; corolla 4(5)-lobed, pale lilac. Anthers yellow. Capsule *c.* 7 mm. $2n = 56$. *Shady crevices in limestone rocks*, 1000–2000 *m*. ● *C. Greece* (*Olimbos*). Gr.

CLX. OROBANCHACEAE[2]

Perennial (rarely annual) herbs without chlorophyll, parasitic on the roots of other phanerogamic plants (usually herbaceous dicotyledons). Stems erect, usually simple. Leaves alternate, scale-like, often succulent at first. Flowers in a terminal spike or raceme, rarely in a panicle or solitary. Calyx tubular, cup-shaped or 2-lipped. Corolla 5-lobed, 2-lipped or almost regular. Stamens 4, didynamous. Ovary superior, 1-locular, with 2–4 parietal, often deeply lobed placentae; style single; stigma more or less 2-lobed. Fruit a loculicidal capsule; seeds small, numerous.

It is impossible to delimit this family satisfactorily from the Scrophulariaceae, and it would seem that nothing but tradition maintains its separate status.

Literature: G. Beck von Mannagetta in Engler, *Pflanzenreich* 96(**IV. 261**): 1–348 (1930).

1 Flowers solitary   **4. Phelypaea**
1 Flowers in racemes, spikes or panicles
   2 Corolla ± regular, 5-lobed   **1. Cistanche**
   2 Corolla distinctly 2-lipped
     3 Upper lip of corolla much longer than the lower; stamens exserted   **3. Boschniakia**
     3 Upper lip of corolla equalling or shorter than the lower; stamens included   **2. Orobanche**

1. Cistanche Hoffmanns. & Link[1]

Stems stout, simple. Leaves numerous. Flowers numerous, in dense spikes. Bracteoles 2, adnate to calyx. Calyx campanulate, with 5 equal, obtuse lobes. Corolla scarcely 2-lipped and nearly regular, with campanulate or obconical tube and 5 subequal, patent lobes. Stamens included. Placentae 4, not deeply lobed.

Bracts and bracteoles glabrous; the whole corolla bright yellow
    1. phelypaea
Bracts and bracteoles with villous margins; corolla with pale yellow tube and usually violet limb   **2. salsa**

---

[1] By E. M. Rix and D. A. Webb.
[2] Edit. D. A. Webb.

1. C. phelypaea (L.) Coutinho, *Fl. Port.* 571 (1913) (*C. tinctoria* (Forskål) G. Beck, *Phelypaea lutea* Desf., *P. tinctoria* (Forskål) Brot.). Glabrous. Stems 20–100 cm. Leaves *c.* 2 cm, ovate-lanceolate, obtuse, brown, with scarious, more or less denticulate margin. Spike 10–20 cm. Bracts 15–20 mm, irregularly crenulate; bracteoles oblong-lanceolate, slightly shorter than calyx. Calyx 13–18 mm, with short, ovate-orbicular, crenulate lobes. Corolla 30–40(–60) × 20–30 mm, bright shining yellow; tube usually curved; lobes ovate-orbicular, erecto-patent. Anthers and filament-bases hairy. Capsule *c.* 12 mm, ovoid. $2n=40$. *On woody Chenopodiaceae. S. Spain, S. Portugal; Kriti; one record from N.W. Spain.* Cr Hs Lu. (*N. Africa, Arabia.*)

2. C. salsa (C. A. Meyer) G. Beck in Engler & Prantl, *Natürl. Pflanzenfam.* 4(3b): 129 (1893). Like **1** but stems 10–40 cm, usually tomentose; bracts, bracteoles and sometimes calyx with villous margins; corolla 25–35 mm, with pale yellow tube and usually violet limb. *On woody Chenopodiaceae* (Atriplex, Anabasis, Kalidium). *S.E. Russia, W. Kazakhstan.* Rs (E). (*C. & S.W. Asia.*)

2. Orobanche L.[1]

Perennial, biennial or annual. Stems stout or slender, simple or branched. Leaves numerous. Flowers in usually dense spikes or racemes. Bracteoles, if present, adnate to calyx. Calyx with cylindrical to campanulate tube and 4(–5) teeth, or divided, usually deeply, into 2 lateral segments, which may be entire or bifid. Corolla strongly 2-lipped; lower lip 3-lobed, at least as long as the upper. Stamens included. Placentae 4, variably lobed.

For most species there is no reliable information as to annual or perennial habit.

The intrinsic taxonomic difficulties of this genus are greatly increased by the fact that important differential characters can be observed only with difficulty, or not at all, on dried specimens. This applies especially to the colour of the corolla and, to a lesser extent, of other parts. All Floras lay great stress on the shape of the corolla, especially the nature of the curvature of its 'dorsal line'. The shape of the corolla, however, can vary widely between flowers on the same spike, and it is often distorted by pressing; we have not used it, therefore, except in a few species where it is constant and distinctive.

Many species are becoming rarer, especially in N. Europe. A few, however, especially those which parasitize crop-plants, are extending their range.

There is much unconfirmed and probably erroneous information in the literature on the range of hosts of many species; if the parasite is growing in closed vegetation it may require very careful dissection of root-systems for the identity of the host to be revealed. We have tried to mention only those host-plants for which there is a good weight of concordant evidence.

It has often been suggested that the variation in some species is dependent, at least in part, on the species of host, but there appears to be no reliable experimental evidence on this point.

To facilitate identification a table is appended of the species of *Orobanche* most frequently associated with certain hosts. The list must not, however, be regarded as exhaustive, as some species show a very wide range of hosts, and for others our information on this subject is incomplete.

---

[1] A. O. Chater and D. A. Webb.

Individual host genera:
Berberis **39**; *Cistus* **30, 42**; *Digitalis* **21(b)**; *Euphorbia* **23**; *Hedera* **29**; *Helleborus* **16**; *Rubus* **?39**; *Thalictrum* **34**

Host families:

Compositae:	
Achillea	**9**
Anthemis	**5**
Artemisia	**7, 8, 10, 11, 12, 18, 26**; see also **9**
Carduus	**15**
Centaurea	**34**
Cirsium	**15**
Petasites, Tussilago, Adenostyles	**38**
Picris	**26**
Other genera	**9, 11, 19, 20, 21, 22, 24, 26, 34**
Dipsacaceae	**15, 17**
Labiatae:	
Ballota	**21(a)**
Rosmarinus	**30**; see also **3**
Salvia	**40**
Teucrium	**32**
Thymus	**14**
Other genera	**14**
Leguminosae:	
Lotus	**45**
Medicago	**33**
Psoralea	**3**
Trifolium	**27, 33**
Other genera	**13, 20, 25, 41, 42, 43, 44**
Rubiaceae	**28, 31**
Umbelliferae:	
Daucus	**21(a), 26**
Eryngium	**21(a)**
Laserpitium	**37**
Opoponax	**36**
Orlaya	**26**
Other genera	**6, 19, 35**
Cultivated plants	**1, 2, 11, 13**

1 Each flower subtended by 2 bracteoles (± adnate to calyx), as well as by a bract
  2 Anthers glabrous, or sparsely hairy at the base
    3 Stem usually branched; corolla 10–20(–22) mm
      4 Lobes of lower lip of corolla acute to acuminate   **5. oxyloba**
      4 Lobes of lower lip of corolla rounded to subacute   **1. ramosa**
    3 Stem usually simple; corolla (18–)20–25(–30) mm
      5 Inflorescence arachnoid-villous   **7. caesia**
      5 Inflorescence minutely puberulent   **9. purpurea**
  2 Anthers ± densely hairy
    6 Corolla (20–)22–35 mm
      7 Stem usually branched; bracts shorter than calyx   **2. aegyptiaca**
      7 Stem usually simple; bracts equalling or exceeding calyx   **8. arenaria**
    6 Corolla 16–22 mm
      8 Calyx-teeth not more than 1¼ times as long as tube   **3. lavandulacea**
      8 Calyx-teeth 2–3 times as long as tube
        9 Leaves *c.* 20 mm; bracteoles filiform; lobes of lower lip of corolla rounded   **4. trichocalyx**
        9 Leaves 5–10 mm; bracteoles linear-lanceolate; lobes of lower lip of corolla acute to acuminate   **6. schultzii**
1 Bracteoles absent
  10 Stigma purple, orange or dark red at anthesis
    11 Lower lip of corolla glandular-ciliate
      12 Stamens inserted not more than 3 mm above base of corolla
        13 Many of the hairs on the corolla dark at least at base or apex; on Labiatae   **14. alba**
        13 Hairs on corolla colourless or pale yellow
          14 Corolla 20–32 mm, pink or yellow, tinged with purple; on Rubiaceae   **31. caryophyllacea**
          14 Corolla 15–22 mm, yellow tinged with red; on *Berberis*   **39. lucorum**
      12 Stamens inserted at least 3 mm above base of corolla

15 Upper lip of corolla deeply 2-lobed; bracts 20–30 mm; on *Helleborus*   **16. haenseleri**
15 Upper lip of corolla entire or emarginate
 16 Calyx 10–17 mm, the segments usually entire; bracts 17–25 mm; on Rubiaceae   **31. caryophyllacea**
 16 Calyx not more than 12 mm, the segments bifid; bracts 12–20 mm; on *Teucrium*   **32. teucrii**
11 Lower lip of corolla not ciliate
 17 Corolla shining dark red inside, mostly dark red or purple outside
  18 Filaments inserted 3–7 mm above base of corolla; corolla glandular-pubescent outside   **44. foetida**
  18 Filaments inserted 1·5–2 mm above base of corolla; corolla glabrous or subglabrous outside   **45. sanguinea**
 17 Corolla not shining dark red inside, usually white or pale yellow outside at least towards the base
  19 Calyx-segments connate for *c.* ½ their length, with broadly triangular teeth   **28. clausonis**
  19 Calyx-segments free or slightly connate at base, with linear-lanceolate to filiform teeth
   20 Corolla villous (S.E. Europe and Sicilia)   **19. pubescens**
   20 Corolla glabrous to pubescent
    21 Corolla 22–30 mm; on *Helleborus*   **16. haenseleri**
    21 Corolla 10–23 mm
     22 Corolla broadly cylindrical-campanulate; filaments glabrous or subglabrous at base, usually glandular-puberulent near apex   **15. reticulata**
     22 Corolla tubular or narrowly campanulate; filaments hairy below (sometimes sparsely), subglabrous above   **(21–27). minor** group
10 Stigma yellow or white (rarely pink) at anthesis
 23 Corolla shining dark red inside, usually dark purplish-red or bright yellow outside
  24 Lower lip of corolla with middle lobe twice as big as the others   **43. variegata**
  24 Lower lip of corolla with subequal lobes
   25 Filaments inserted not more than 2 mm above base of corolla; lower lip of corolla ciliate; flowers fragrant   **42. gracilis**
   25 Filaments inserted 3–7 mm above base of corolla; lower lip of corolla not ciliate; flowers fetid   **44. foetida**
 23 Corolla not shining dark red inside, usually pale yellow, white or bluish outside, at least towards the base
  26 Corolla-tube narrowed to the mouth; on *Hedera*   **29. hederae**
  26 Corolla-tube cylindrical or campanulate, or constricted some distance below the mouth
   27 Hairs on corolla dark purplish-brown
    28 Stems usually 35–50 cm; lower lip of corolla with middle lobe the largest; usually on Dipsacaceae   **17. pancicii**
    28 Stems not more than 15 cm; lower lip of corolla with equal lobes; on *Artemisia alba*   **18. serbica**
   27 Hairs on corolla colourless or pale yellow
    29 Corolla predominantly white, cream, bluish or violet
     30 Lower lip of corolla glandular-ciliate
      31 Filaments inserted 5–7 mm above base of corolla; corolla mainly brownish-violet; on *Laserpitium siler*   **37. laserpitii-sileris**
      31 Filaments inserted 3–4 mm above base of corolla; corolla cream or pale yellow, tinged with pink; on *Opoponax chironium*   **36. chironii**
     30 Lower lip of corolla not ciliate
      32 Bracts, calyx and corolla arachnoid-villous   **10. coerulescens**
      32 Bracts, calyx and corolla subglabrous to glandular-pubescent
       33 Corolla becoming conspicuously inflated, scarious and shining at base   **11. cernua**
       33 Corolla not conspicuously inflated, scarious and shining at base
        34 Corolla constricted below middle, the upper part infundibuliform; filaments inserted 6–18 mm above base of corolla   **12. amoena**

        34 Corolla tubular to campanulate, not constricted below middle; filaments inserted 2–6 mm above base of corolla
         35 Corolla 20–30 mm with large, strongly divergent lips; bracts villous   **13. crenata**
         35 Corolla 10–23 mm with small lips, the upper usually porrect; bracts glandular-pubescent   **(21–27). minor** group
    29 Corolla predominantly yellow or reddish
     36 Lower lip of corolla glandular-ciliate
      37 Stems not or scarcely swollen at base; on Compositae, *Berberis* or *Salvia*
       38 Upper lip of corolla porrect; stamens inserted 2–3 mm above base of corolla   **39. lucorum**
       38 Upper lip of corolla erect or recurved at maturity; stamens inserted 4–6 mm above base of corolla
        39 Upper lip of corolla 2-lobed; style glabrous, becoming exserted and convolute   **38. flava**
        39 Upper lip of corolla entire or emarginate; style densely glandular-pubescent, not exserted or convolute   **40. salviae**
      37 Stems distinctly swollen at base; on Leguminosae or Umbelliferae
       40 Filaments inserted not more than 2 mm above base of corolla
        41 Corolla tubular, scarcely campanulate; on Umbelliferae   **35. alsatica**
        41 Corolla campanulate; on Leguminosae   **41. rapum-genistae**
       40 Filaments inserted more than 2 mm above base of corolla
        42 Upper lip of corolla deeply 2-lobed; corolla broadly campanulate   **37. laserpitii-sileris**
        42 Upper lip of corolla emarginate or shallowly 2-lobed; corolla narrowly campanulate
         43 Corolla cream or pale yellow, tinged with pink   **36. chironii**
         43 Corolla bright yellow, tinged with purple or brown
          44 Calyx 10–17 mm; on Leguminosae   **33. lutea**
          44 Calyx 6–11 mm; on Umbelliferae   **35. alsatica**
     36 Lower lip of corolla not ciliate
      45 Stamens inserted 7–15 mm above base of corolla; anthers villous; bracts ± glabrous   **30. latisquama**
      45 Stamens inserted not more than 6 mm above base of corolla; anthers glabrous; bracts glandular-pubescent
       46 Calyx-segments connate for *c.* ½ their length, with broadly triangular teeth   **28. clausonis**
       46 Calyx-segments free or slightly connate at base, with linear-lanceolate to filiform teeth
        47 Corolla at least 20 mm
         48 Stamens inserted not more than 2 mm above base of corolla   **41. rapum-genistae**
         48 Stamens inserted at least 3 mm above base of corolla
          49 Leaves 20–30 mm; calyx 10–17 mm; on Leguminosae   **33. lutea**
          49 Leaves 10–20 mm; calyx 6–11 mm; on Compositae   **34. elatior**
        47 Corolla usually not more than 20 mm
         50 Filaments with wide, ± deltate base; spike not more than 3 times as long as wide   **20. densiflora**
         50 Filaments gradually and only slightly widened towards base; spike (2–)4–8 times as long as wide
          51 Calyx-segments connate at base   **34. elatior**
          51 Calyx-segments free
           52 Stems conspicuously whitish-villous   **(21–27). minor** group
           52 Stems ± glandular-pubescent
            53 Bracts 10–12 mm; on *Euphorbia esula*   **(21–27). minor** group
            53 Bracts 13–20 mm; on Compositae or *Salvia*

54 Style becoming exserted and strongly convolute; upper lip of corolla 2-lobed **38. flava**
54 Style neither exserted nor convolute; upper lip
of corolla entire or emarginate **40. salviae**

Sect. TRIONYCHON Wallr. Stems simple or branched. Flowers pedicellate or sessile, with 2 bracteoles adnate to calyx. Calyx with cylindrical to campanulate tube and 4 subequal teeth, rarely with a fifth, much smaller tooth. Corolla white, cream, blue or violet, constricted below the middle near the insertion of the filaments; upper lip 2-lobed.

1. O. ramosa L., *Sp. Pl.* 633 (1753) (*Phelypaea ramosa* (L.) C. A. Meyer). Stems 5–30(–40) × 0·15–0·4(–0·6) cm, swollen at the base, simple or branched, glandular-pubescent. Leaves 3–8(–10) mm, ovate to ovate-lanceolate, acute. Inflorescence 2–25 cm, lax or dense, glandular-pubescent; bracts 6–8(–10) mm, ovate-lanceolate; bracteoles linear-lanceolate, about equalling calyx; pedicels 0–5(–8) mm. Calyx 6–8 mm. Corolla 10–22 mm, glandular-pubescent, suberect and inflated at base, erecto-patent and tubular-infundibuliform distally, whitish at base, cream, blue or violet distally (rarely white throughout); lower lip somewhat deflexed with suborbicular to elliptical, entire or denticulate lobes. Filaments glabrous, or hairy at the base, inserted 3–6 mm above base of corolla; anthers glabrous, or sparsely hairy at the base. Stigma white, cream or pale blue. Capsule 6–7(–10) mm. $2n = 24$. *On a wide variety of hosts. S. & S.C. Europe; naturalized further north.* Al *Au Bl Bu Co Cr Ga Gr He Hs *Hu It Ju Lu Rm Rs (*C, W, K, E) Sa Si Tu [Be Br Cz Ge Ho Po].

Very variable; within the limits of the species as here defined more than 20 variants have been described under binomials. Analysis of the literature, however, shows that (apart from a few extreme variants only once collected) there is no discontinuity in variation of the characters relied on for the discrimination of species; nor are they always constant within a gathering. Although differentiation seems to have proceeded along ecological rather than geographical lines, it seems best to recognize 3 rather ill-defined subspecies.

1 Corolla (15–)18–22 mm   **(c) subsp. mutelii**
1 Corolla 10–15(–17) mm
2 Calyx-teeth acuminate, usually shorter than tube; lobes of
    lower lip of corolla obtuse   **(a) subsp. ramosa**
2 Calyx-teeth filiform-subulate at apex, usually as long as tube;
    lobes of lower lip of corolla acute   **(b) subsp. nana**

(a) Subsp. **ramosa**: Stem usually branched. Flowers numerous. Calyx-teeth acuminate, usually shorter than tube. Corolla 10–13(–17) mm, whitish to pale blue or lilac; lobes of lower lip obtuse. *Usually on cultivated plants, especially* Cannabis sativa, Nicotiana tabacum *and* Solanum *spp. S. & S.C. Europe; introduced further north.*

(b) Subsp. **nana** (Reuter) Coutinho, *Fl. Port.* 566 (1913) (*O. nana* (Reuter) Noë ex G. Beck, *Phelypaea nana* (Reuter) Reichenb. fil.): Stem usually simple. Flowers few. Calyx-teeth triangular, with filiform-subulate apex, about equalling the tube. Corolla 11–15(–17) mm, usually bright blue or violet; lobes of lower lip acute. *S. Europe.*

(c) Subsp. **mutelii** (F. W. Schultz) Coutinho, *loc. cit.* (1913) (*O. mutelii* F. W. Schultz, *Phelypaea mutelii* (F. W. Schultz) Reuter): Stem usually branched. Flowers numerous. Calyx-teeth acuminate, about equalling the tube. Corolla (15–)18–22 mm, pale to bright blue or violet; lobes of lower lip obtuse. *S. Europe.*

2. O. aegyptiaca Pers., *Syn. Pl.* **2**: 181 (1806) (*Phelypaea aegyptiaca* (Pers.) Walpers). Like **1** but stems 15–50 × 0·4–0·6

cm, usually branched; leaves 5–12 mm; lower pedicels usually 10–20 mm; corolla 20–35 mm, straight or slightly curved, conspicuously infundibuliform in distal part; filaments usually hairy at base and sometimes above; anthers villous. $2n = 24$. *On* Gossypium *spp.,* Solanum *spp. and other cultivated plants. S.E. Europe; once recorded from Sicily.* Bu Rs (K, E) ?Si Tu. (*C. & S.W. Asia, Egypt.*)

O. coelestis (Reuter) Boiss. & Reuter ex G. Beck, *Biblioth. Bot.* (*Stuttgart*) **19**: 114 (1890) is doubtfully distinct from **2**; it appears to differ only in its stout, usually simple stem, corolla not more than 26 mm, and anthers sometimes glabrous. It occurs in S.W. Asia and N. Africa, and has once been recorded from the S. Aegean region (Kasos, near Karpathos).

3. O. lavandulacea Reichenb., *Pl. Crit.* **7**: 48 (1829) (*Phelypaea lavandulacea* (Reichenb.) Reuter). Stems 15–60 × 0·4–0·7 cm, swollen at base, simple or branched, glandular-pubescent. Leaves 7–12 mm, ovate-lanceolate, acuminate. Inflorescence (6–)12–30 cm, lax or dense, glandular-hairy; bracts 7–10(–12) mm, ovate-lanceolate; bracteoles linear-lanceolate, slightly shorter than calyx. Calyx 6–8 mm, usually blue; teeth equalling or slightly longer than tube, triangular with subulate apex. Corolla 16–22 mm, glandular-hairy, suberect at base, patent and narrowly campanulate distally with strongly divergent lips, white at base, bright blue distally; lobes of lower lip suborbicular, denticulate. Filaments glabrous or sparsely hairy, inserted *c.* 6 mm above base of corolla; anthers hairy. Stigma white or yellowish. Capsule 6–7 mm. *On a variety of herbs, most commonly* Psoralea bituminosa. *Mediterranean region.* Bl Co Ga Gr Hs It Ju Sa Si.

O. rosmarina G. Beck, *Österr. Bot. Zeitschr.* **70**: 243 (1921), resembles **3** in its deep violet-blue inflorescence and its sharply inflected corolla, but in its subglabrous anthers and shorter stem it comes near to **1(c)**. It has been recorded as parasitic on *Rosmarinus* from the C. Mediterranean region and Portugal.

4. O. trichocalyx (Webb & Berth.) G. Beck, *Biblioth. Bot.* (*Stuttgart*) **19**: 107 (1890) (*Phelypaea trichocalyx* Webb & Berth.). Like **3** but stems not more than 40 × 0·5 cm; leaves *c.* 20 mm; bracteoles filiform; calyx-teeth linear-subulate, 2–3 times as long as tube; corolla lilac distally. ? *On* Pteridium aquilinum. ● *S. Portugal, S.W. Spain.* Hs Lu.

5. O. oxyloba (Reuter) G. Beck in L. Koch, *Entw.-Gesch. Orob.* 209 (1887) (incl. *O. dalmatica* (G. Beck) Tzvelev). Stems 12–30 × 0·2–0·5 cm, scarcely swollen at base, simple or branched, glandular-pubescent. Leaves 5–15 mm, ovate-lanceolate. Inflorescence 10–15 × *c.* 3 cm, lax or dense, glandular-hairy; bracts 6–9 mm, ovate-lanceolate; bracteoles linear-subulate, shorter than calyx; lower flowers shortly pedicellate. Calyx 7–9 mm; teeth triangular-subulate, about equalling tube. Corolla 15–20 mm, glandular-puberulent, suberect at base, erecto-patent to patent and narrowly campanulate above, white at base, lilac distally; lobes of lower lip ovate, acute, crenate to subentire. Filaments glabrous, or somewhat hairy above, inserted *c.* 6 mm above base of corolla; anthers glabrous, or hairy at the base. Stigma white or pale blue. Capsule *c.* 7 mm. *On* Anthemis chia *and other herbs. S.E. Europe.* Bu Cr Gr Ju Rs (K) Tu.

6. O. schultzii Mutel, *Fl. Fr.* **2**: 429 (1835) (*Phelypaea schultzii* (Mutel) Walpers). Stems 10–60 × 0·4–0·8 cm, swollen at base, simple or branched, glandular-pubescent. Leaves 5–10 mm, numerous, ovate to lanceolate. Inflorescence 7 × 3–4 cm, dense above, lax below, glandular-hairy; bracts 8–12 mm, lanceolate;

bracteoles linear-lanceolate, slightly shorter than calyx; most of the flowers shortly pedicellate. Calyx 8–12 mm; teeth 2–3 times as long as tube, linear-lanceolate with subulate apex. Corolla 16–21 mm, glandular-hairy, erecto-patent, tubular to narrowly campanulate, nearly straight, white at base, blue to violet distally; lobes of lower lip elliptical, acute to acuminate. Filaments glabrous or subglabrous, inserted 4–5 mm above base of corolla; anthers hairy. Stigma white. Capsule 7–8 mm. *On various herbs, most commonly Umbelliferae. Mediterranean region.* ?Cr Hs Sa Si.

O. **rechingeri** Gilli, *Osterr. Bot. Zeitschr.* **113**: 214 (1966), described from a single locality in C. Greece, differs in its stems not more than 10 cm, inflorescence whitish-villous as in **7**, calyx-teeth 3–6 times as long as tube, and corolla and stigma pale yellow. It requires further investigation.

7. **O. caesia** Reichenb., *Pl. Crit.* **7**: 48 (1829). Stems 10–35 × 0·3–0·5 cm, scarcely swollen at base, simple, glandular-pubescent. Leaves 8–17 mm, ovate to lanceolate, numerous. Inflorescence 5–20 × 2·5–4 cm, usually dense, arachnoid-villous with white hairs; bracts 8–14 mm, ovate-lanceolate; bracteoles linear-lanceolate, shorter than calyx; flowers sessile or subsessile. Calyx 9–14 mm; teeth lanceolate, almost equalling or longer than tube. Corolla 20–25 mm, glandular-pubescent, suberect at base, campanulate and more or less patent distally, lilac to bluish-violet; lobes of lower lip oblong to ovate, obtuse to acuminate. Filaments glabrous, or slightly hairy at base, inserted 6–8 mm above base of corolla; anthers glabrous. Stigma white. Capsule 7–9 mm. *On Artemisia spp. E.C. Europe, Romania, S. & E. parts of U.S.S.R.* Au Cz Hu Rm Rs (C, W, K, E).

O. **nowackiana** Markgraf, *Ber. Deutsch. Bot. Ges.* **44**: 429 (1926), recorded from two localities in C. Albania, has been included in **7**, which it resembles in its simple stem, arachnoid-villous inflorescence and subglabrous stamens. It appears, however, to differ from all other species of the section in its very dwarf stature, subequally 5-toothed calyx, predominantly yellow corolla and purple stigma. Its systematic position must remain doubtful until more material is available.

8. **O. arenaria** Borkh., *Neues Mag. Bot.* (*Roemer*) **1**: 6 (1794) (*Phelypaea arenaria* (Borkh.) Walpers). Stems 15–60 × 0·3–0·6 cm, slightly swollen at base, usually simple, glandular-hairy with long, appressed hairs. Leaves up to 20(–25) mm, lanceolate, numerous. Inflorescence 5–15(–25) cm, dense, glandular-pubescent; bracts 15–20 mm, lanceolate; bracteoles linear, somewhat shorter than calyx; lower flowers shortly pedicellate. Calyx 12–18 mm; teeth lanceolate-acuminate, equalling or slightly longer than tube. Corolla 25–35 mm, glandular-pubescent, erecto-patent, campanulate, more or less straight, bluish-violet; lobes of lower lip broadly ovate, obtuse. Filaments glabrous or subglabrous, inserted 6–7 mm above base of corolla; anthers hairy. Stigma white. Capsule 8–10 mm. *On Artemisia spp. and perhaps on other herbs. From N.E. France and C. Russia southwards to S. Spain, C. Italy and S. Bulgaria.* Au Bu Cz Ga Ge He Hs Hu It Ju Lu Po Rm Rs (C, W, K, E).

O. **bungeana** G. Beck, *Biblioth. Bot.* (*Stuttgart*) **19**: 119 (1890), from S.W. Asia, has once been recorded from C. Greece, but requires confirmation. It appears to differ from **8** only in its shorter bracts and usually subcapitate inflorescence.

9. **O. purpurea** Jacq., *Enum. Stirp. Vindob.* 108, 252 (1762) (*Phelypaea caerulea* (Vill.) C. A. Meyer). Stems 15–60 × 0·3–0·8 cm, slightly swollen at base, usually simple; whole plant minutely glandular-puberulent, often with a greyish tinge. Leaves 7–20

mm, narrowly lanceolate, usually sparse above. Inflorescence 5–20 cm, rather dense; bracts 8–15 mm, lanceolate; bracteoles linear, shorter than calyx; lower flowers usually subsessile, rarely with pedicels up to 15 mm. Calyx 8–15 mm; teeth lanceolate, about equalling the tube. Corolla 18–25(–30) mm, erecto-patent, or inflected and more or less patent distally, narrowly campanulate, white below, bluish-violet distally with deep violet veins; lobes of lower lip ovate-elliptical, rounded to apiculate. Filaments glabrous or subglabrous, inserted *c*. 7 mm above base of corolla; anthers glabrous or subglabrous. Stigma white or pale blue. Capsule 8–9 mm. $2n=24$. *On Achillea spp. and other Compositae. Europe, northwards to C. England, S.E. Sweden and S.C. Russia.* Au Be ?Bl Br Bu Co ?Cr Cz Da Ga Ge Gr He Ho Hs Hu It Ju ?Lu Po Rm Rs (C, W, K, E) Su Tu.

O. **uralensis** G. Beck, *Biblioth. Bot.* (*Stuttgart*) **19**: 132 (1890), from E. Russia, differs only in having slightly longer hairs throughout and the corolla without deep violet veins. It grows on *Artemisia* spp.

Sect. OROBANCHE (Sect. *Osproleon* Wallr.). Stems simple. Flowers sessile, without bracteoles. Calyx split above and below almost or completely to the base, and thus divided into 2 lateral segments, which may be entire or equally or unequally bifid. Corolla variously shaped, usually tinged with yellow, brown or red; upper lip 2-lobed to entire.

10. **O. coerulescens** Stephan in Willd., *Sp. Pl.* **3**: 349 (1800) (incl. *O. korshinskyi* Novopokr.). Stems 10–40 cm, somewhat swollen at the base, arachnoid-villous above, usually glabrescent below, yellowish. Leaves 10–17(–25) mm, lanceolate. Spike up to 10(–18) × 2·5–3·5 cm, dense, arachnoid-villous; bracts about equalling corolla, lanceolate-acuminate. Calyx 8–13 mm; segments free, more or less bidentate. Corolla (10–)14–23 mm, arachnoid-villous, pale violet-blue, suberect and somewhat inflated at base, constricted and inflected near middle, the distal part patent; lobes of lower lip subequal, not ciliate. Filaments almost glabrous, inserted 6–7 mm above base of corolla; anthers usually glabrous. Stigma pale yellow. *On Artemisia spp. E. & C. Europe, from C. Germany, N. Ukraine and S. Ural southwards to W. Kazakhstan.* Au ?Bu Cz Ge Hu ?It Po Rm Rs (C, W, K, E).

11. **O. cernua** Loefl., *Iter. Hisp.* 152 (1758) (incl. *O. cumana* Wallr.). Stems up to 40 cm, only slightly swollen at base, more or less glandular-pubescent, yellowish. Leaves 5–10 mm, ovate-lanceolate. Spike up to 25 × 2·5–4 cm, usually dense, glandular-puberulent, brown, variably tinged with blue; bracts 7–12 mm, ovate to lanceolate. Calyx 7–12 mm; segments free, more or less bidentate. Corolla 12–20 mm, sparsely glandular-puberulent, suberect, conspicuously inflated and whitish-scarious at base, constricted and inflected near middle, the distal part patent or somewhat deflexed, violet-blue, sparsely pubescent; lobes of lower lip subequal, not ciliate. Filaments glabrous or subglabrous, inserted 4–6 mm above base of corolla; anthers glabrous. Stigma whitish. Capsule 8–10 mm. $2n=24$. *On Artemisia spp. and Helianthus annuus, rarely on other herbs. S. Europe, extending northwards to S. Czechoslovakia and to c. 54° N. in E. Russia.* Br Co Cr Cz Ga Gr Hs Hu It Ju Rm Rs (C, W, K, E) Si Tu.

12. **O. amoena** C. A. Meyer in Ledeb., *Fl. Altaica* **2**: 457 (1830). Stems 15–40 mm, slightly swollen at base, sparsely glandular-pubescent, yellow or purplish. Leaves *c*. 10 mm, ovate-lanceolate. Spike 6–15 × *c*. 4 cm, sparsely glandular-pubescent; bracts 7–18 mm, oblong-lanceolate. Calyx 7–15 mm; segments free, more or less bidentate, pale yellow. Corolla 20–40

mm, subglabrous, erecto-patent, slightly curved, constricted below the middle, the upper part infundibuliform, violet-blue except for whitish base; lobes of lower lip equal, not ciliate. Filaments hairy at least at apex and base, inserted 6–18 mm above base of corolla; anthers puberulent. Stigma whitish. Capsule 10–11 mm. *On* Artemisia *spp. S.E. Russia, W. Kazakhstan.* Rs (E).

13. O. crenata Forskål, *Fl. Aegypt.* 113 (1775) (incl. *O. speciosa* DC.). Stems up to 50(–80) × 1·2 cm, slightly swollen at the base, sparsely villous, usually yellowish. Leaves 12–25 mm, linear-lanceolate, dense below, sparse above. Spike up to 20(–50) × 4 cm, dense above, often lax below, villous; bracts 15–25 mm, linear- to ovate-lanceolate, long-acuminate; flowers fragrant. Calyx 10–20 mm; segments free, more or less bidentate. Corolla 20–30 mm, subglabrous, white, the lips often with lilac veins, erecto-patent, straight or slightly curved, campanulate; lips divergent, the lower with large, suborbicular lobes, not ciliate. Filaments hairy, inserted 2–4 mm above base of corolla; anthers glabrous. Stigma white, yellow or pinkish. Capsule 10–12 mm. 2*n* = 38. *Usually on leguminous crop-plants, rarely on other herbs. S. Europe; a rare casual elsewhere.* ?Al Az Bl Bu Co Cr Ga Gr Hs It Ju Lu Rs (K) Sa Si Tu.

14. O. alba Stephan ex Willd., *Sp. Pl.* **3**: 350 (1800) (*O. epithymum* DC.). Stems up to 35(–70) × 1 cm, slightly swollen at the base, glandular-pubescent, usually purplish-red. Leaves 12–20 mm, lanceolate. Spike 5–10(–20) × 2–4 cm, lax to moderately dense, glandular-pubescent; bracts 12–25 mm, lanceolate, acuminate; flowers fragrant. Calyx 8–16 mm; segments free, usually entire, with 1–3 conspicuous veins. Corolla 15–25 mm, glandular-puberulent especially distally, purplish-red, yellow or whitish, erecto-patent, broadly cylindrical-campanulate, slightly curved; upper lip entire or 2-lobed; lower lip glandular-ciliate, the middle lobe usually much the largest. Filaments densely hairy at least at the base, inserted 1–3 mm above the base of corolla; anthers subglabrous. Stigma red or purple. Capsule 10–12 mm. *On* Thymus *and other Labiatae. Europe northwards to Scotland, Belgium and C. Russia; isolated stations in the Baltic islands.* Al Au Be Br Bu Cr Cz Ga Ge Gr Hb He Hs Hu It Ju Po Rm Rs (C, W, K, E) Si Su Tu.

15. O. reticulata Wallr., *Orob. Gen.* 42 (1825) (incl. *O. pallidiflora* Wimmer & Grab.). Like **14** but often taller and more robust; flowers scarcely fragrant; calyx 7–12 mm, with indistinct veins; lower lip of corolla with equal lobes, not ciliate; filaments glabrous or subglabrous at the base, inserted 2–4 mm above base of corolla. 2*n* = 38. *Usually on species of* Cirsium, Carduus, Knautia *or related genera. Europe northwards to England and Estonia, but absent from most of the islands.* Al Au Br Bu Co Cz Da Ga Ge Gr He Ho Hs Hu It Ju Po Rm Rs (B, C, W, E) Su.

16. O. haenseleri Reuter in DC., *Prodr.* **11**: 22 (1847). Like **14** but stems 30–55 cm; spikes 10–35 cm; bracts 20–30 mm, linear-lanceolate; calyx-segments usually deeply bifid; corolla 22–30 mm, dull orange-red, the upper lip deeply 2-lobed, lower lip usually not ciliate; filaments inserted (1–)3–6 mm above base of corolla. *On* Helleborus foetidus (*? and on* Ononis *spp.*). ● *S. Spain.* Hs.

17. O. pancicii G. Beck, *Ann. Naturh. Mus.* (*Wien*) **2**: 148 (1887). Stems (15–)35–50 × 0·5–0·7 cm, somewhat swollen at the base, glandular-pubescent, yellowish, more rarely purple. Leaves up to 30 mm, lanceolate, dense below, sparse above. Spike 6–15 × 3–4 cm, usually dense, densely glandular-pubescent; bracts about as long as corolla, lanceolate; flowers fragrant. Calyx

9–14 mm; segments free or slightly connate, usually bidentate. Corolla 15–28 mm, glandular-pubescent, pale yellow or whitish, tinged with lilac distally, erecto-patent, broadly cylindrical-campanulate, slightly curved; upper lip emarginate or shortly 2-lobed; lower lip more or less glandular-ciliate, the middle lobe the largest. Filaments usually hairy, inserted 2–3 mm above base of corolla; anthers subglabrous. Stigma yellow. Capsule 9–14 mm. *Usually on Dipsacaceae.* ● *Bulgaria, C. Jugoslavia, N. Albania.* Al Bu Ju.

18. O. serbica G. Beck & Petrović in Petrović, *Add. Fl. Agri Nyss.* 146 (1885). Like **17** but stems not more than 15 cm; calyx *c.* 8 mm; corolla not more than 20 mm, with upper lip subentire; lobes of lower lip equal, glandular-ciliate; filaments inserted 3–4 mm above base of corolla. *On* Artemisia alba. ● *W. Bulgaria, S. Jugoslavia.* Bu Ju.

19. O. pubescens D'Urv., *Enum.* 76 (1822) (*O. versicolor* F. W. Schultz). Stems 15–50 × 0·4–0·7 cm, variably swollen at base, glandular-pubescent, pale yellow, often tinged with pink. Leaves 10–25 mm, oblong to linear-lanceolate. Spike 7–22 × 2–4 cm, rather lax, glandular-pubescent to villous; bracts 10–20 mm, lanceolate, acuminate. Calyx 8–16 mm; segments free or almost free, usually equally bifid. Corolla 10–20 mm, villous, pale yellow, tinged with violet above, erecto-patent, tubular, more or less uniformly curved; upper lip more or less entire; lower lip not or sparsely ciliate, the middle lobe the largest. Filaments hairy below, usually glabrous above, inserted 3–4 mm above base of corolla; anthers glabrous. Stigma violet. Capsule 8–10 mm. *On Compositae and Umbelliferae, perhaps also on Leguminosae. S.E. Europe, extending locally westwards to S.E. France, but doubtfully native in the west.* Bu Cr *Ga Gr *It Ju Rs (K) Si Tu.

20. O. densiflora Salzm. ex Reuter in DC., *Prodr.* **11**: 19 (1847). Stems 15–40 × 0·6–0·8 cm, more or less strongly swollen at base, glandular-pubescent but more or less glabrescent below, reddish. Leaves 20–30 mm, oblong-lanceolate, acute. Spike 5–10 × 3–35 cm, dense, glandular-pubescent to more or less villous; bracts 15–20 mm, lanceolate, acuminate. Calyx 7–12 mm; segments free or shortly connate at base, unequally bifid. Corolla 10–18 mm, glandular-puberulent to subglabrous, yellowish, erecto-patent to patent, tubular, slightly curved; upper lip emarginate to 2-lobed; lobes of lower lip subequal, not ciliate. Filaments sparsely hairy below, more or less glabrous above, inserted 1·5–4 mm above base of corolla, with wide, more or less deltate base; anthers glabrous. Stigma yellow. Capsule 8–9 mm. *On Leguminosae; perhaps also on Compositae. W.C. Portugal, S.W. Spain.* Hs Lu ?Sa ?Si.

(21–27). O. minor group. Stems 10–70(–100) × 0·3–1 cm, variably swollen at base, glandular-pubescent to villous, yellowish, more or less tinged with purple, red or brown. Leaves 10–25 mm. Spike 3–30 × 2–4 cm, usually lax below, more or less dense above, glandular-pubescent; bracts 7–22 mm, lanceolate, acuminate. Calyx 7–16 mm; segments free. Corolla 10–23 mm, erecto-patent to more or less patent, tubular or narrowly campanulate, curved or nearly straight; lower lip not ciliate. Filaments hairy below (sometimes only sparsely), subglabrous above, inserted 1·5–6 mm above base of corolla; anthers glabrous. Capsule 8–10 mm.

1 Stigma yellow
2 Stamens inserted (4–)5–6 mm above base of corolla; on *Euphorbia esula*   **23. esulae**
2 Stamens inserted 2–4·5 mm above base of corolla
  3 Corolla pale yellow throughout   **27. minor**
  3 Corolla tinged with red, violet or brown

4 Leaves ovate; corolla yellow tinged with red   **22. canescens**

4 Leaves linear-lanceolate to -oblong; corolla white or cream, usually tinged with violet, pink or brown   **21. amethystea**

1 Stigma red or purple

  5 Corolla tinged with pink, red or brown

    6 Calyx-segments 3- to 4-veined, not filiform distally   **21. amethystea**

    6 Calyx-segments 1-veined, filiform distally   **24. calendulae**

  5 Corolla tinged with purple or violet

    7 Bracts 7–15 mm; stamens inserted 2–3 mm above base of corolla   **27. minor**

    7 Bracts 10–22 mm; stamens inserted 3–5 mm above base of corolla

      8 Corolla rather sharply inflected near the base, the upper lip deeply bifid   **21. amethystea**

      8 Corolla not inflected near the base, the upper lip emarginate or slightly bifid

        9 Corolla 12–15(–20) mm, glandular-pubescent on tube, glabrous distally   **25. grisebachii**

        9 Corolla 14–22 mm, subglabrous or glandular-puberulent (rarely glandular-pubescent) throughout   **26. loricata**

21. O. amethystea Thuill., *Fl. Paris* ed. 2, 317 (1800). Leaves linear-lanceolate to -oblong, acute. Bracts 12–22 mm. Calyx 9–15 mm; segments unequally bidentate or entire. Corolla 15–25 mm, sparsely glandular-pubescent, white or cream, usually tinged with violet, pink or brown distally; lobes of lower lip subequal. Filaments inserted 3–4 mm above base of corolla. Stigma purple or yellow. *On various dicotyledonous herbs. S., W. & W.C. Europe northwards to S. England.* Al Br Bu Co Ga Ge Gr He Hs It Ju Lu Sa Si.

(a) Subsp. **amethystea** (incl. *O. attica* Reuter): Calyx-segments 1-veined, filiform above. Corolla often inflected near base, more or less straight above, usually tinged with violet; upper lip deeply 2-lobed. Stigma purple or yellow. *Usually on species of Eryngium, Daucus, Ballota and Compositae. Throughout the range of the species.*

(b) Subsp. **castellana** (Reuter) Rouy, *Fl. Fr.* **11**: 185 (1909) (*O. castellana* Reuter): Calyx-segments 3- to 4-veined, not filiform. Corolla uniformly curved throughout, usually tinged with pink or brown; upper lip entire to 2-lobed. Stigma purple. *Usually on Digitalis spp. S.W. Europe.*

22. O. canescens C. Presl in J. & C. Presl, *Del. Prag.* 72 (1822). Leaves ovate. Bracts 10–20 mm. Calyx 8–16 mm; segments usually unequally bifid. Corolla 12–18 mm, glandular-pubescent to subglabrous, yellow tinged with red; upper lip emarginate or 2-lobed; lower lip with subequal lobes. Filaments inserted 3–4·5 mm above base of corolla. Stigma yellow. *On Compositae and perhaps other plants. C. & E. parts of Mediterranean region.* Co Cr Gr It Sa Si.

23. O. esulae Pančić, *Fl. Princ. Serb. Addit.* 194 (1884). Bracts 10–12 mm. Calyx 7–9 mm; segments entire or variably bifid. Corolla (12–)15–18 mm, glandular-pubescent, yellow, tinged with brown or purple distally; upper lip 2-lobed; lower lip with the middle lobe the largest. Filaments inserted (4–)5–6 mm above base of corolla. Stigma yellow. *On Euphorbia esula.* ● *E. Jugoslavia, C. & S. Bulgaria.* Bu Ju.

24. O. calendulae Pomel, *Nouv. Mat. Fl. Atl.* 110 (1874) (*O. mauretanica* G. Beck). Leaves oblong-lanceolate. Bracts 15–20 mm. Calyx 10–15 mm; segments equally bifid. Corolla 12–20 mm, glandular-pubescent, pale yellow, often veined and tinged with brownish-red; upper lip 2-lobed, rarely emarginate; lobes of the lower lip equal or the middle the largest. Filaments inserted 1·5–4·5 mm above base of corolla. Stigma purple or reddish. *On Compositae and various other plants. S. Portugal, S.W. Spain.* Hs Lu.

25. O. grisebachii Reuter in DC., *Prodr.* **11**: 28 (1847). Leaves ovate-lanceolate to lanceolate. Bracts 10–20 mm. Calyx 8–16 mm; segments entire or unequally bifid. Corolla 12–15(–20) mm, glandular-pubescent on tube, glabrous distally, whitish, tinged with violet above; upper lip entire or emarginate; lower lip with subequal lobes. Filaments inserted 3–4 mm above base of corolla. Stigma purple. *On Leguminosae and perhaps also Compositae. Greece and Aegean region.* Cr Gr.

26. O. loricata Reichenb., *Pl. Crit.* **7**: 41 (1829) (incl. *O. picridis* F. W. Schultz ex Koch). Leaves narrowly lanceolate. Bracts 12–20 mm. Calyx 10–15 mm; segments unequally bifid, rarely entire. Corolla 14–22 mm, glandular-pubescent or -puberulent or subglabrous, white or pale yellow, tinged and veined with violet; upper lip emarginate to bifid; lobes of the lower lip subequal or the middle the largest. Filaments inserted 3–5 mm above base of corolla. Stigma purple. $2n = c.$ 38. *On species of* Artemisia, Picris *and other Compositae; also on species of* Daucus *and* Orlaya. *S., W. & C. Europe, extending to Denmark.* Al Au Be Bl Br Bu Co Cr Cz Da Ga Ge Gr He Ho Hs Hu It Ju Lu Po Rm Rs (W) Sa Si.

O. fuliginosa Reuter ex Jordan, *Obs. Pl. Crit.* **3**: 225 (1846) (incl. *O. angelifixa* Péteaux & St-Lager), recorded from Greece and S.E. France, appears to differ from **26** mainly in its yellow corolla tinged with brownish-red; it has been subsequently recorded from other localities in the C. & E. parts of the Mediterranean region, but there is little agreement among authors as to its differential characters, and some of the plants so named may be more closely related to **19** or **27**.

27. O. minor Sm. in Sowerby, *Engl. Bot.* **6**: t. 422 (1797). Leaves ovate to linear-lanceolate. Bracts 7–15 mm. Calyx 7–12 mm; segments equally or unequally bidentate or entire. Corolla 10–18 mm, glandular-pubescent or subglabrous, pale yellow, usually tinged with dull violet distally; upper lip emarginate to 2-lobed; lower lip with subequal lobes or the middle the largest. Filaments inserted 2–3 mm above base of corolla. Stigma purple, rarely yellow. $2n = 38$. *Most frequently on* Trifolium *spp., but also on a wide variety of other plants. W., S. & S.C. Europe; introduced further north.* Al Au *Az Be *Br Bu Co Cz Ga Ge Gr He *Ho Hs Hu It Ju Lu Rm Rs (W, K) Sa Si Tu [Da Hb Po Su].

O. maritima Pugsley, *Jour. Bot.* (*London*) **78**: 110 (1940) (*O. amethystea* auct., non Thuill.), with $2n = 38$, recorded from S.W. England, W. France and S. Spain, is probably best regarded as a variety of **27**, though in some characters it resembles **26**. **O. ozanonis** F. W. Schultz ex G. Beck, *Biblioth. Bot.* (*Stuttgart*) **19**: 249 (1890) (*O. minor* subsp. *ozanonis* (F. W. Schultz ex G. Beck) Rouy), once collected in S.E. France (Hautes-Alpes), was distinguished by its shorter calyx, more densely hairy corolla and stamens, and less distinctly lobed upper lip of corolla; it also is best included in **27**.

28. O. clausonis Pomel, *Nouv. Mat. Fl. Atl.* 107 (1874). Stems 15–40 × 0·3–0·8 cm, slightly swollen at base, glandular-pubescent, yellowish. Leaves 5–15 mm, ovate to lanceolate, acute. Spike dense, becoming more or less lax, glandular-pubescent; bracts 8–15 mm, ovate to lanceolate, acuminate. Calyx 5–8 mm; segments connate for about half their length, equally bifid with broadly triangular, acuminate teeth. Corolla (10–)15–17 mm, glandular-pubescent, yellow, erecto-patent, somewhat curved; upper lip more or less entire; lobes of the lower lip subequal, not ciliate. Filaments sparsely hairy below, glabrous above, inserted (1–)3 mm above base of corolla; anthers glabrous. Stigma orange to purple. *On Rubiaceae. Spain and Portugal; Malta.* Hs Lu Si.

29. **O. hederae** Duby, *Bot. Gall.* 1: 350 (1828). Stems 15–60 × 0·3–0·8 cm, usually strongly swollen at base, glandular-pubescent, yellow to reddish-purple. Leaves 12–30 mm, oblong to lanceolate, acute. Spike 10–40 × 2·5–4 cm, lax, glandular-pubescent; bracts 12–22 mm, lanceolate, acuminate. Calyx 10–15 mm; segments free, entire or unequally bifid. Corolla 10–22 mm, subglabrous, rarely glandular-pubescent, dull cream tinged distally with reddish-purple, erecto-patent to more or less patent; tube somewhat inflated below, gradually narrowed to the mouth, nearly straight; lips patent, the upper entire to emarginate, the lower not ciliate, with middle lobe usually the largest. Filaments more or less glabrous, rarely somewhat hairy below, inserted 3–4 mm above base of corolla; anthers glabrous. Stigma yellow. Capsule 10–12 mm. 2*n* = 38. *On Hedera spp. W., S. & S.C. Europe, northwards to Ireland.* Au Be Bl Br Co Ga Ge Gr Hb He Ho Hs Hu It Ju Lu Rs (K) Sa Si Tu [Po].

30. **O. latisquama** (F. W. Schultz) Batt. in Batt. & Trabut, *Fl. Algér., Dicot.* 659 (1890) (*O. macrolepis* G. Beck, non Turcz., *Boulardia latisquama* F. W. Schultz, *Ceratocalyx macrolepis* Cosson; incl. *C. fimbriatus* Lange). Stems 20–40 × 0·4–0·9 cm, slightly swollen at base, glabrous, usually deep reddish-purple. Leaves 20–30 mm, dense, ovate-oblong, the lower obtuse, the upper acute. Spike 6–20 × 3–4 cm, usually dense, glabrous or with subsessile glands; bracts 17–25 mm, oblong-ovate, acuminate, purplish. Calyx 11–16 mm; segments connate in proximal quarter, entire or bifid. Corolla 22–35 mm, glabrous but with subsessile glands, pale yellowish, veined or tinged with purple towards the lips, sometimes entirely purple, suberect, somewhat inflated at base, constricted at or below middle, campanulate above; upper lip subentire; lower lip equally 3-lobed, not ciliate. Filaments hairy, inserted 7–10(–15) mm above base of corolla; anthers villous. Stigma yellow. Capsule 10–12 mm. *On Rosmarinus officinalis, Cistus spp. and other shrubs. E. & S.E. Spain, Islas Baleares.* Bl Hs.

31. **O. caryophyllacea** Sm., *Trans. Linn. Soc. London* 4: 169 (1798) (*O. vulgaris* Poiret). Stems 15–50 × 0·4–0·8 cm, slightly swollen at base, glandular-hairy, yellowish or purplish. Leaves 15–30 mm, narrowly triangular-lanceolate, acute. Spike 6–20 (–30) × 2·5–4·5 cm, usually lax, glandular-pubescent to villous; bracts 17–25 mm, lanceolate; flowers fragrant. Calyx 10–17 mm; segments free, or sometimes connate in lower half, entire or bifid. Corolla 20–32 mm, glandular-pubescent, pink or pale yellow, variously tinged with dull purple, erecto-patent, narrowly campanulate, uniformly curved; upper lip emarginate; lower lip with subequal, glandular-ciliate lobes. Filaments hairy, inserted 1–3(–5) mm above base of corolla; anthers glabrous. Stigma purple. Capsule 9–12 mm. 2*n* = 38. *On Rubiaceae. W., C. & S. Europe.* Al Au Be Br Bu Cz Ga Ge Gr He Ho Hs Hu It Ju No Po Rm Rs (C, W, K, E) Si.

32. **O. teucrii** Holandre, *Fl. Moselle* 322 (1829). Like **31** but leaves not more than 15 mm; spikes not more than 10 cm; bracts 12–20 mm; calyx not more than 12 mm, with segments bifid; corolla curved at base and in upper lip, but more or less straight in the middle; filaments inserted 3–5 mm above base of corolla. *On Teucrium spp. From Belgium and the Carpathians southwards to the Pyrenees, N. Italy and C. Jugoslavia.* Au Be Bu ?Cz Ga Ge He Hs Hu It Ju Rm Rs (W).

33. **O. lutea** Baumg., *Enum. Stirp. Transs.* 2: 215 (1816) (*O. rubens* Wallr.). Like **31** but stems usually more strongly swollen at base; spike sometimes dense; corolla usually yellowish- or reddish-brown, with lower lip not or only sparsely ciliate; filaments inserted 3–6 mm above base of corolla; stigma yellow or whitish; capsule 10–14 mm. 2*n* = 38. *On species of* Medicago *and* Trifolium *and other Leguminosae. Europe northwards to the Netherlands and N. Poland, but rare in the Mediterranean region.* Al Au ?Be Bu Cz Ga Ge Gr He Ho Hs Hu It Ju Po Rm Rs (W, K, E) Sa Si.

34. **O. elatior** Sutton, *Trans. Linn. Soc. London* 4: 178 (1798) (*O. major* L. pro parte; incl. *O. krylowii* G. Beck). Stems 20–70 × 0·3–0·9 cm, slightly swollen at base, usually glandular-pubescent. Leaves 10–20 mm, triangular-lanceolate. Spike 6–20(–30) × 2·5–4·5 cm, usually dense at least above, glandular-pubescent; bracts 15–25 mm, lanceolate. Calyx (6–)9–11 mm; segments connate at base, usually unequally bidentate. Corolla (12–)18–25 mm, glandular-pubescent, yellow, often tinged with pink, erecto-patent to patent, rarely suberect, narrowly campanulate, uniformly curved; upper lip entire to shallowly 2-lobed; lower lip with subequal lobes, not ciliate. Filaments hairy, inserted 3–6 mm above base of corolla; anthers glabrous. Stigma yellow. Capsule 8–10 mm. 2*n* = 38. *On Centaurea spp., other Compositae and* Thalictrum. *From England and Estonia southwards to E. Spain, N. Italy, Bulgaria and Krym.* Al Au Br Bu Cz Da Ga Ge Gr Hs Hu It Ju Po Rm Rs (B, C, W, K, E) Su.

O. borbasiana G. Beck, *Biblioth. Bot. (Stuttgart)* 19: 173 (1890), from N.W. Jugoslavia, is said to differ in its shorter stems, villous leaves and bracts and filaments glabrous above, but does not seem to be specifically distinct from **34**.

35. **O. alsatica** Kirschleger, *Prodr. Fl. Alsace* 109 (1836) (*O. cervariae* Kirschleger ex Suard). Like **34** but stems distinctly swollen at base; leaves up to 30 mm; bracts 10–20 mm; corolla often tinged with purple or brown, narrower and scarcely campanulate, with the lower lip ciliate and the middle lobe often the largest; filaments inserted 1–7 mm above base of corolla. *On Umbelliferae. C. Europe, extending to E. France and C. Jugoslavia; W., C. & S. parts of U.S.S.R.* Au ?Be Cz Ga Ge He Hu ?It Ju Po Rm Rs (B, C, W, K, E).

Plants from N.W. Italy (Liguria), described as **O. caudata** De Not., *Repert. Fl. Ligust.* 306 (1844), appear to resemble **35** but require further investigation.

36. **O. chironii** Lojac., *Contr. Fl. Sic.* 12 (1878). Like **34** but stems not more than 45 cm, distinctly swollen at base; spike not more than 15 cm; calyx up to 15 mm, the segments free; corolla cream or pale yellow, tinged with pink, the lower lip ciliate; filaments inserted 3–4 mm above base of corolla. *On Opoponax chironium.* ● *Sicilia.* Si.

37. **O. laserpitii-sileris** Reuter ex Jordan, *Obs. Pl. Crit.* 3: 223 (1846). Like **34** but more robust; stems distinctly swollen at base; leaves up to 30 mm; calyx 11–15 mm; corolla 22–30 mm, mainly brownish-violet, yellowish towards base and on lower lip, more broadly campanulate; upper lip deeply 2-lobed; lower lip ciliate, the middle lobe often the largest; filaments inserted 5–7 mm above base of corolla. *On Laserpitium siler.* ● *Alps, Balkan peninsula; very local.* Au ?Bu Ga He Ju.

38. **O. flava** C. F. P. Mart. ex F. W. Schultz, *Beitr. Kenntn. Deutsch. Orob.* 9 (1829). Stems 20–65 × 0·5–0·9 cm, scarcely swollen at base, glandular-pubescent, yellowish or brownish. Leaves 15–25 mm, lanceolate, dense below, sparser above.

Spike 8–20(–25) × _c._ 3 cm, usually rather lax, glandular-pubescent; bracts 13–20 mm, oblong to lanceolate, acute to acuminate. Calyx 8–15 mm; segments free, entire or unequally bidentate. Corolla 15–22 mm, glandular-pubescent, yellow, slightly tinged with red, erecto-patent to patent, tubular, slightly inflated in the middle, uniformly curved; upper lip erect or recurved at maturity, 2-lobed; lower lip usually with middle lobe the largest, glabrous or minutely ciliate. Filaments hairy, inserted 5–6 mm above base of corolla; anthers glabrous. Stigma pale yellow; style becoming exserted and strongly convolute. Capsule 10–11 mm. $2n=38$. _On Petasites, Tussilago and_ Adenostyles _spp. Mountains of C. Europe and of N. & C. Jugoslavia._ Au Cz Ga He Hu It Ju Po Rm Rs (W).

39. O. lucorum A. Braun in Röhling, _Deutschl. Fl._ ed. 3, **4**: 456 (1833). Like **38** but stems 15–50 cm; upper lip of corolla porrect, usually emarginate; lower lip with subequal lobes, conspicuously ciliate; filaments inserted 2–3 mm above base of corolla; stigma yellow at first, but reddish or purplish later; style neither exserted nor strongly convolute; capsule up to 15 mm. _On_ Berberis vulgaris. ● _E. Alps._ Au Ge He It ?Rm.

Plants from S.E. France (Var) (var. _rubi_ (Duby) G. Beck), parasitic on _Rubus_ spp., have been referred to this species. **O. denudata** Moris, _Stirp. Sard., App._ [1] (1828), from Sardegna, also growing on _Rubus_ spp., is very like **38** but has the lower lip of the corolla densely ciliate. Both require further investigation.

40. O. salviae F. W. Schultz ex Koch in Röhling, _Deutschl. Fl._ ed. 3, **4**: 458 (1833). Like **38** but stems 12–50 cm; leaves 10–20 mm; upper lip entire or emarginate; lower lip with subequal lobes, usually ciliate; filaments inserted 3–4(–5) mm above base of corolla; stigma yellow, but orange or brownish later; style neither exserted nor convolute. _On_ Salvia glutinosa _and_ ?S. pratensis. _Alps, mountains of N. & C. Jugoslavia and S.W. Romania._ Au Ga Ge He It Ju Rm.

41. O. rapum-genistae Thuill., _Fl. Paris_ ed. 2, 317 (1800) (_O. major_ L. pro parte, _O. rapum_ auct.). Stems 20–80 × 0·5–2 cm, strongly swollen at base, pale yellow, sometimes tinged with red. Leaves 15–25(–60) mm, ovate to linear-lanceolate, rather dense. Spike 15–35 × 3–4 cm, dense; bracts 15–50 mm, linear-lanceolate; flowers fetid. Calyx 8–15 mm; segments free or slightly connate at base, bidentate, usually entire. Corolla 20–25 mm, suberect to erecto-patent, campanulate, slightly curved; lower lip with middle lobe usually the larger, minutely to conspicuously ciliate. Filaments inserted not more than 2 mm above base of corolla; anthers glabrous. Stigma yellow. Capsule 10–11(–14) mm. $2n=38$. _On various leguminous shrubs. W. Europe, northwards to S. Scotland and extending eastwards to C. Germany and S. Italy._ Be Br Co Ga Ge Hb He Ho Hs It Lu Sa Si.

1 Bracts, corolla, style and filaments glabrous or subglabrous
                                       (c) subsp. rigens
1 Bracts, corolla, style and upper part of filaments glandular-pubescent
2 Bracts at least 1½ times as long as corolla; upper lip of corolla distinctly 2-lobed, more or less erect   **(b) subsp. benthamii**
2 Bracts shorter than to slightly exceeding corolla; upper lip of corolla subentire, porrect   **(a) subsp. rapum-genistae**

(a) Subsp. rapum-genistae: Bracts shorter than to slightly exceeding corolla; bracts, corolla, style and upper part of filaments glandular-pubescent; corolla pale yellow to reddish, the upper lip subentire, porrect. _Throughout the range of the species._

(b) Subsp. benthamii (Timb.-Lagr.) P. Fourn., _Quatre Fl. Fr._ 796 (1937): Like subsp. (a) but bracts at least 1½ times as long as corolla; upper lip of corolla distinctly 2-lobed, more or less erect. _S.W. Europe and N. Italy._

(c) Subsp. rigens (Loisel.) P. Fourn., _loc. cit._ (1937) (_O. rigens_ Loisel.): Not more than 40 cm. Bracts shorter than to slightly exceeding corolla; bracts, corolla, style and filaments glabrous or subglabrous; corolla dark red, the upper lip shortly 2-lobed, more or less erect. _Corse, Sardegna, ?Sicilia._

42. O. gracilis Sm., _Trans. Linn. Soc. London_ **4**: 172 (1798) (_O. cruenta_ Bertol.). Stems 15–60 × 0·2–0·7 cm, somewhat swollen at base, glandular-pubescent, yellow or reddish. Leaves 7–15(–20) mm, ovate-deltate to lanceolate. Spike 3–20 × 2·5–4 cm, glandular-pubescent, lax at least below; bracts 10–20 mm, triangular, acuminate; flowers fragrant. Calyx 8–15 mm; segments free or connate at base, usually unequally 2-lobed. Corolla 15–25 mm, glandular-pubescent, yellow outside, usually with red veins and reddish towards the lips, shining dark red inside, erecto-patent, broadly tubular, slightly constricted near the middle, slightly curved; upper lip emarginate; lobes of lower lip subequal, ciliate. Filaments hairy at least below, inserted not more than 2 mm above base of corolla; anthers glabrous. Stigma yellow. Capsule 12–15 mm. _On various Leguminosae, rarely on_ Cistus _spp. S., W. & C. Europe, northwards to c. 49° N. in N.W. France, C. Germany and E. Carpathians._ Al Au Bl Bu Co Cr Cz Ga Ge Gr He Hs Hu It Ju Lu Po Rm Rs (W) Sa Si.

43. O. variegata Wallr., _Orob. Gen._ 40 (1825). Like **42** but stems 0·4–1 cm wide; leaves 15–30 mm, oblong-ovate to linear-lanceolate; spike 5–30 cm, usually denser; flowers fetid; bracts 15–25 mm; calyx 10–20 mm; corolla sometimes brownish-red outside, the lower lip with the middle lobe deflexed and twice as big as the others; filaments often hairy only above, inserted 2–4 mm above base of corolla. _On leguminous shrubs. W.C. part of Mediterranean region._ Co Ga ?Hs It Sa Si.

A single record for S. Jugoslavia requires confirmation; it is probably an error for **42**.

44. O. foetida Poiret, _Voy. Barb._ **2**: 195 (1789). Like **42** but stems 0·4–1 cm wide; spike usually denser; flowers fetid; calyx-segments very long-acuminate; corolla 12–23 mm, dark purplish-red, narrower, nearly straight; upper lip 2-lobed with erect lobes; lower lip not ciliate; filaments inserted 3–7 mm above base of corolla; stigma sometimes purple. _On various Leguminosae, mostly herbaceous. Iberian peninsula, Islas Baleares._ Bl Hs Lu.

45. O. sanguinea C. Presl in J. & C. Presl, _Del. Prag._ 71 (1822) (incl. _O. crinita_ Viv.). Stems 10–40 × 0·3–0·7 cm, swollen at base, glandular-pubescent, yellow or purplish-red. Leaves 12–35 mm, narrowly lanceolate to linear-oblong, dense especially below. Spike 5–15 × _c._ 2·5 cm, glandular-pubescent, dense; bracts 8–15 mm, triangular-lanceolate, acuminate; flowers scentless. Calyx 7–10 mm; segments shortly connate at base, unequally 2-lobed. Corolla 10–15(–17) mm, glabrous or subglabrous, dark red or purple, yellow at base (rarely entirely yellow) outside, shining dark red inside, more or less patent, tubular, curved or nearly straight; upper lip 2-lobed; lower lip small, not ciliate, with subequal lobes. Filaments glabrous above and usually also below, inserted 1·5–2 mm above base of corolla; anthers glabrous. Stigma purple. _On_ Lotus _spp. Mediterranean region._ Bl Co Cr Gr Hs It Ju ?Lu Sa Si.

3. Boschniakia C. A. Meyer[1]

Stems stout, simple. Flowers in dense, spike-like racemes; bracteoles absent. Calyx short, cupuliform, obscurely 3- to 5-lobed. Corolla 2-lipped; upper lip much longer than the lower. Stamens exserted. Placentae 2, deeply lobed.

1. **B. rossica** (Cham. & Schlecht.) B. Fedtsch. in B. Fedtsch. & Flerow, *Fl. Evr. Ross.* 896 (1910). Glabrous, except for minutely ciliate bracts, calyx and corolla. Stems 15–40 cm, red or brown, usually in groups of 2–8. Leaves 5–10 mm, ovate to deltate. Raceme 15–20 cm, dense, narrow. Bracts *c.* 8 mm, ovate; pedicels very short in flower, elongating in fruit. Calyx 3–4 mm. Corolla 10–12 mm, red or brownish; tube subglobose; lower lip equally 3-lobed; upper lip subentire. Stamens inserted 2–3 mm above base of corolla-tube. Capsule 6–7 mm. *On species of* Alnus. *N. Ural (Ilyč basin). Rs (N). (N. Asia and coasts of Bering Sea.)*

4. Phelypaea L.[1]

Stems slender, simple. Leaves few, amplexicaul. Flowers solitary (very rarely 2), terminal; bracteoles absent. Calyx tubular, 5-lobed, more or less 2-lipped. Corolla more or less 2-lipped, with 5 large, patent lobes. Stamens included. Placentae 4, deeply lobed.

Literature: O. Stapf, *Kew Bull.* **1915**: 285–295 (1915).

Corolla-lobes obovate-orbicular, not overlapping; anthers glabrous   **1. coccinea**
Corolla-lobes orbicular, overlapping; anthers hairy   **2. boissieri**

1. **P. coccinea** Poiret in Lam., *Encycl. Méth. Bot.* **5**: 268 (1804) (incl. *P. helenae* Popl.). Stem 30–40 cm, reddish or orange, glandular-puberulent, at least above. Leaves *c.* 2 cm, ovate-oblong, obtuse, confined to lower half of stem. Calyx 15–30 mm, puberulent; lobes 5, longer than the tube. Corolla *c.* 30 × 40 mm, yellow outside, bright red or orange-yellow inside; lobes obovate-orbicular, not overlapping. Filaments and anthers glabrous. Stigma large, reniform. Capsule *c.* 15 mm, ovoid. *On* Psephellus *spp. Krym. Rs (K). (Caucasus.)*

2. **P. boissieri** (Reuter) Stapf, *Kew Bull.* **1915**: 291 (1915). Like **1** but stem shorter and stouter; corolla always red inside, with overlapping, orbicular lobes; anthers hairy. *On various species of* Centaurea. *S. Jugoslavia (C. Makedonija, around Gradsko). Ju. (S. & E. Anatolia.)*

CLXI. LENTIBULARIACEAE[2]

Small, carnivorous perennial herbs. Leaves alternate or in a basal rosette. Flowers solitary or in a short raceme. Corolla 2-lipped, spurred. Stamens 2, epipetalous. Ovary superior, 1-locular; placenta free-central. Fruit a capsule; seeds numerous.

Leaves entire, in a basal rosette   **1. Pinguicula**
Leaves divided into filiform lobes, alternate   **2. Utricularia**

1. Pinguicula L.[3]

Leaves in a basal rosette, entire, soft and fleshy, clothed with viscid glands above; margin usually more or less involute. Flowers solitary, on naked pedicels. Calyx persistent, 2-lipped; upper lip 3(–7)-, lower 2-lobed. Corolla 2-lipped, spurred, open at throat, with a hairy and often spotted palate; upper lip 2-, lower 3-lobed. Capsule opening by 2 valves. Seeds numerous, very small.

In the descriptions the terms *spring* and *summer leaves* indicate that the rosettes of some species undergo a regular change of form during their development; this is often not evident in dried specimens. The measurements of the corolla and leaf include the spur and the petiole respectively.

Literature: S. J. Casper, *Feddes Repert.* **66**: 1–148 (1962); *Biblioth. Bot. (Stuttgart)* **127/128**: 1–209 (1966). A. Ernst, *Bot. Jahrb.* **80**: 145–194 (1961). J. Schindler, *Österr. Bot. Zeitschr.* **57**: 409–421, 458–469 (1907); **58**: 13–18, 61–69 (1908).

1 Corolla white, with one or more yellow spots on the palate; roots stout   **3. alpina**
1 Corolla violet, lilac, blue or pinkish; roots slender
 2 Lobes of the lower lip of the corolla emarginate; plant overwintering as a rosette
  3 Corolla 7–11 mm; tube cylindrical; spur 2–4 mm   **1. lusitanica**
  3 Corolla 16–32 mm; tube infundibuliform; spur 6–13 mm   **2. hirtiflora**
 2 Lobes of the lower lip of the corolla entire; plant overwintering as a bud
  4 Pedicels with glandular hairs, the stalks at least 10 times as long as the gland; leaves elliptical to suborbicular, strongly involute; corolla 6–11 mm   **4. villosa**
  4 Pedicels glandular, the stalk not more than twice as long as the gland; leaves not or only slightly involute; corolla at least 12 mm
   5 Corolla 25–40 mm; spur 9–24 mm, more than half as long as the rest of the corolla
    6 Leaves oblong to obovate-oblong, horizontal; lobes of the lower lip of the corolla about as wide as long   **11. grandiflora**
    6 Leaves elliptic-ovate to ligulate or linear-lanceolate, suberect; lobes of lower lip of corolla much longer than wide
     7 Spring leaves sessile, elliptic-ovate; summer leaves 100–260 mm, ligulate, acute; lobes of upper lip of calyx ovate-oblong to ovate-suborbicular   **9. vallisneriifolia**
     7 Leaves 60–175 mm, the lowest elliptical, the others lanceolate, obtuse; lobes of upper lip of calyx elliptic-ligulate to linear   **10. longifolia**
   5 Corolla 12–30 mm; spur 4–10 mm, usually less than half as long as the rest of the corolla
    8 Leaves not much longer than wide
     9 Lobes of upper lip of calyx about as long as wide   **7. balcanica**
     9 Lobes of upper lip of calyx much longer than wide   **8. nevadensis**
    8 Leaves distinctly longer than wide
     10 Lobes of lower lip of corolla oblong, divergent, not overlapping or touching   **12. vulgaris**
     10 Lobes of lower lip of corolla suborbicular, overlapping or touching
      11 Lobes of upper lip of calyx about as long as wide   **7. balcanica**
      11 Lobes of upper lip of calyx distinctly longer than wide

---

[1] By E. M. Rix and D. A. Webb.
[2] Edit. T. G. Tutin.   [3] By S. J. Casper.

12 Lower lip of calyx divided nearly to base; lobes
   divergent   **6. leptoceras**
12 Lower lip of calyx divided halfway to base; lobes
   not divergent   **5. corsica**

1. P. lusitanica L., *Sp. Pl.* 17 (1753). Overwintering as a rosette. Leaves 5–12, 10–24(–29) × 3–8 mm, oblong-ovate, greyish; margin strongly involute. Pedicels 1–8, 60–150 (–250) mm, glandular-pubescent, very slender. Corolla 7–9(–11) mm. Calyx-lobes ovate. Corolla-lips subequal, pinkish to pale lilac, yellow at the throat; lobes suborbicular, emarginate. Tube cylindrical; palate entire. Spur 2–4 mm, subcylindrical, deflexed. Capsule subglobose. 2*n* = 12. *Bogs and wet heaths. W. Europe, eastwards to 5° E. in S. France.* Br Ga Hb Hs Lu.

2. P. hirtiflora Ten., *Fl. Nap.* **1**, *Prodr.*: 6 (1811) (incl. *P. louisii* Markgraf). Overwintering as a rosette. Leaves 6–9, 20–60(–80) × 10–25(–35) mm, elliptic-oblong to ovate-oblong, truncate or emarginate at apex; margin scarcely involute. Pedicels 1–3, 50–110(–140) mm, slender. Corolla 16–25(–32) mm. Upper lip of the calyx with spathulate lobes, those of the lower lip obovate, very short. Corolla-lips unequal, pink to pale blue, yellow at the throat; lobes of upper lip obovate, weakly emarginate; those of lower lip cuneate, emarginate. Tube infundibuliform; palate divided. Spur 6–10(–13) mm, subulate, straight or somewhat curved. Capsule subglobose. 2*n* = 16. *Wet rocks in the mountains. S. & W. parts of Balkan peninsula; C. & S. Italy.* Al Gr It Ju.

3. P. alpina L., *Sp. Pl.* 17 (1753). Overwintering as a bud. Roots long and relatively thick, brownish. Leaves 5–8, (15–)25–45(–60) × 8–14 mm, elliptic-oblong to lanceolate-oblong, pale yellowish-green; margin involute. Pedicels 1–8, 50–110 (–130) mm, sparsely glandular. Corolla 10–16(–21) mm. Upper lip of the calyx with triangular lobes; lower lip lobed for ⅓ its length, the lobes obovate. Corolla-lips unequal, white with one or more yellow spots on the palate. Tube short, thick. Spur 2–3(–5) mm, curved, yellowish. Capsule ovoid-oblong. 2*n* = 32. *Bogs and wet places. Arctic and subarctic Europe; mountains of Fennoscandia; islands of Baltic sea; mountains and uplands of C. Europe; Pyrenees.* Au †Br Cz Fe Ga Ge He Hs Hu It Ju No Po Rm Rs (N, B, ?C, W) Su.

4. P. villosa L., *Sp. Pl.* 17 (1753). Overwintering as a bud. Leaves 1–5, 8–10(–13) × 4–6(–7) mm, elliptical to suborbicular; margins strongly involute, often touching each other. Pedicels usually solitary, 30–60(–95) mm, glandular-hairy with the stalk at least 10 times as long as the gland. Corolla 6–9(–11) mm. Calyx-lobes lanceolate, purple. Corolla-lips unequal, pale violet, with yellow stripes in the throat; lobes broadly cuneate. Spur 2–3(–5) mm, straight, cylindric-conical. Capsule obovoid, much longer than the persistent calyx. 2*n* = 16. Sphagnum-*bogs. Fennoscandia, southwards to 61° N. in Sweden; N.W. Russia.* Fe No Rs (N) Su.

5. P. corsica Bernard & Gren. in Gren. & Godron, *Fl. Fr.* **2**: 443 (1853). Overwintering as a bud. Leaves 5–9, 25–35(–40) × 9–16 mm, ovate to obovate-oblong, petiolate. Pedicels 40–90(–150) mm, glandular. Corolla 16–25(–30) mm. Upper lip of the calyx with narrowly oblong-lanceolate lobes; lower lip lobed for at least ½ its length; lobes not divergent. Corolla-lips unequal, pale blue to pink; lobes obovate, rounded, overlapping. Tube infundibuliform, about as long as upper lip of the corolla. Spur 4–6(–9) mm, straight, cylindrical-subulate. Capsule subglobose. 2*n* = 16. *Wet places in the mountains.* ● *Corse.* Co.

6. P. leptoceras Reichenb., *Pl. Crit.* **1**: 69 (1823). Overwintering as a bud. Leaves 5–8, 25–40(–65) × 10–16(–22) mm, oblong to ovate-oblong; margin scarcely involute. Pedicels 1–6, 40–100(–130) mm, glandular. Corolla 16–25(–30) mm. Upper lip of the calyx with 3(–7) triangular-acuminate to oblong-obtuse lobes, the middle lobe often much larger than the lateral and irregularly dentate or truncate; lower lip divided nearly to the base, the lobes lanceolate and usually divergent. Corolla-lips unequal, blue; lobes of the upper lip obovate-obtuse, those of the lower lip broadly obovate to suborbicular, overlapping, white-spotted at the base. Tube infundibuliform, somewhat wider than long. Spur 4–6(–9) mm, straight, cylindrical. Capsule ovoid. 2*n* = 32. *Wet places in the mountains.* ● *Alps, Alpe Apuani, N. Appennini.* Au Ga He It.

7. P. balcanica Casper, *Feddes Repert.* **66**: 105 (1962) (*P. leptoceras* sensu Hayek, non Reichenb.). Overwintering as a bud. Spring leaves 20–30 × 15–20 mm, broadly ovate; summer leaves 20–50 × 10–20 mm, elliptic-oblong. Pedicels 1–5, 40–80 (–100) mm, glandular. Corolla 14–19(–23) mm. Calyx-lobes broadly ovate, about as long as wide, rarely longer than wide; lower lip divided for up to ⅓ its length; lobes not divergent. Corolla-lips unequal, blue-lilac, the lobes of the upper lip broadly ligulate, those of the lower lip obovate-oblong, contiguous or overlapping, the middle lobe often larger, white-spotted at its base. Spur 3–5(–7) mm, straight, cylindrical. Capsule ovoid. 2*n* = 32. *Wet places in the mountains.* ● *Balkan peninsula.* Al Bu Gr Ju.

8. P. nevadensis (Lindb.) Casper, *op. cit.* 112 (1962). Overwintering as a bud. Leaves 5–8, 15–20 × 6–12 mm, little longer than wide. Pedicels 1–3, 40–55(–90) mm, glandular. Corolla 12–16 mm. Calyx-lobes linear, obtuse; lower lip divided for up to ⅓ its length; lobes usually not divergent. Corolla-lips lilac with white lobes; those of the upper lip ovate, obtuse, those of the lower lip obovate-oblong, overlapping, the two lips subequal. Tube infundibuliform, about as long as the lips. Spur 3–4 (–5) mm, straight. *Bogs and wet places in the mountains.* ● *S. Spain (Sierra Nevada, Sierra de Alfacar).* Hs.

9. P. vallisneriifolia Webb, *Otia Hisp.* ed. 2, 48 (1853). Overwintering as a bud. Spring leaves 30–40(–55) × 15–20 (–25) mm, sessile, elliptic-ovate, the margin not involute; summer leaves 100–200(–260) × 8–20(–30) mm, suberect, ligulate, undulate, acute, petiolate. Pedicels 1–8, 100–150(–175) mm, glandular. Corolla 25–35(–40) mm. Calyx-lobes ovate-oblong to obovate; lower lip divided for up to ½ its length. Corolla-lips unequal, violet, white-spotted at the throat; lobes obovate, overlapping. Tube *c.* 5 mm, broadly infundibuliform. Spur 10–14(–20) mm, cylindric-subulate, straight or curved. 2*n* = 32. *Shady, wet rocks.* ● *Mountains of S.E. & E.C. Spain.* Hs.

10. P. longifolia Ramond ex DC. in Lam. & DC., *Fl. Fr.* ed. 3, **3**: 728 (1805). Overwintering as a bud. Leaves 5–11, 60–130(–175) × 10–20(–25) mm, the lowest elliptical, the others linear-lanceolate, suberect, slightly undulate, obtuse, petiolate. Pedicels 1–5, 70–130(–150) mm, glandular. Corolla 22–40 (–46) mm. Calyx-lobes conspicuous, elliptic-ligulate to linear, obtuse, the lower divided for up to ½ its length or sometimes nearly to the base. Corolla-lips unequal, lilac to pale blue; lobes of the upper lip ovate, obtuse, those of the lower lip obovate-cuneate, much longer than wide, overlapping, white-spotted at the base. Tube short, broadly infundibuliform. Spur 10–16(–24) mm, cylindric-subulate, straight or curved. Capsule ovoid-subglobose. *Wet rocks.* ● *Mountains of S. Europe, from S.C. France to the Pyrenees and C. Appennini.* Ga Hs It.

1 Lobes of upper lip of calyx linear, sometimes more than 3
                                         (c) subsp. **reichenbachiana**
1 Lobes of upper lip of calyx wider, never more than 3
2   Corolla 30–46 mm   **(a)** subsp. **longifolia**
2   Corolla 22–35 mm   **(b)** subsp. **caussensis**

(a) Subsp. **longifolia**: Lobes of the upper lip of the calyx elliptic-ligulate; lower lip divided for up to ½ its length. Corolla 30–46 mm. *C. Pyrenees.*

(b) Subsp. **caussensis** Casper, *Feddes Repert.* **66**: 70 (1962): Lobes of the upper lip of the calyx ovate-lanceolate to linear, obtuse. Corolla 22–35 mm. *Mountains of S.C. France.*

(c) Subsp. **reichenbachiana** (Schindler) Casper, *op. cit.* 71 (1962): Lobes of the upper lip of the calyx linear, acute or obtuse, often irregular, sometimes more than 3; lower lip divided nearly to the base. $2n = 32$. *Mountains of C. & N.W. Italy, just extending to S.E. France.*

11. P. grandiflora Lam., *Encycl. Méth. Bot.* **3**: 22 (1789). Overwintering as a bud. Leaves 5–8, 30–45(–65) mm, oblong to obovate-oblong. Pedicels 1–5, 60–150(–230) mm, glandular. Corolla 25–35(–40) mm. Lower lip of calyx divided for up to ⅓ its length; lobes obovate. Corolla-lips unequal, violet to pinkish or pale lilac, white at the throat; lobes of the upper lip suborbicular or obovate, those of the lower lip suborbicular, about as long as wide, somewhat undulate, overlapping. Spur 10–12(–14) mm, straight, sometimes slightly bifid. Capsule subglobose. $2n = 32$. *Bogs and wet rocks.* ● *S.W. Ireland; mountains of S.W. Europe, from the Cordillera Cantábrica to the Swiss Jura.* Ga Hb He Hs [Br].

(a) Subsp. **grandiflora**: Upper lip of calyx divided nearly to the base; lobes ligulate. Corolla violet to pinkish; spur rather stout. *Throughout the range of the species.*

(b) Subsp. **rosea** (Mutel) Casper, *Feddes Repert.* **66**: 85 (1962): Like (a) but smaller in all its parts; lobes of the upper lip of the calyx short, elliptical, about as long as wide; corolla pinkish to pale lilac; spur subulate. *S.E. France (Savoie).*

12. P. vulgaris L., *Sp. Pl.* 17 (1753) (incl. *P. norica* G. Beck). Overwintering as a bud. Leaves 5–11, 20–45(–90) × 14–20 (–26) mm, oblong to obovate-oblong; margin somewhat involute. Pedicels 1–6, 75–180(–270) mm, glandular. Corolla 15–22 (–30) mm. Calyx-lobes triangular to ovate or oblong, acuminate or obtuse, broadly based, the lower lip divided for up to ½ its length. Corolla-lips unequal, violet, usually white at the throat; lobes of the upper lip oblong, those of the lower lip oblong, much longer than wide, divergent, not overlapping. Spur 3–6(–10) mm, cylindric-subulate, straight. Capsule ovoid. $2n = 64$. *Bogs, wet heaths and wet rocks. N., W. & C. Europe, extending eastwards to W. Ukraine.* Au Be Br Cz Da Fa Fe Ga Ge Hb He Ho Hs Hu Is It Ju Lu No Po Rm Rs (N, B, C, W) Su.

Hybridization between **11** and **12** almost certainly occurs and a hybrid between **12** and **3** has been reported.

2. Utricularia L.[1]

Annual or perennial rootless herbs with horizontal stems and erect racemose inflorescences; terrestrial in damp places, with usually inconspicuous simple leaves, or aquatic with numerous conspicuous leaves which are divided into linear to filiform segments; leaves bearing small bladders which trap minute organisms. Calyx 2-lipped, divided almost to the base, usually accrescent; corolla 2-lipped, yellow; lower lip more or less spurred, with a gibbous palate; upper lip usually smaller. Capsule usually globose, circumscissile, indehiscent or dehiscing by pores or slits. Seeds usually many.

The genus is predominantly tropical and most of the species native in Europe (3–6) are atypical in the development of winter buds (turions). The vegetative parts are usually very variable and the identification of non-flowering material is often impossible; many of the published records are, therefore, open to doubt.

1 Terrestrial; lower lip of corolla 3-lobed; bracts peltate; leaves
     entire   **1. subulata**
1   ± Aquatic; lower lip of corolla entire; bracts not peltate; leaves
     divided into linear to capillary segments
2   Seeds with an irregular wing; upper lip of corolla 3–4 mm
     wide   **2. gibba**
2   Seeds not winged; upper lip of corolla at least 6 mm wide
3   Spur saccate or broadly conical, obtuse, not longer than
     wide; leaf-segments without bristles   **3. minor**
3   Spur with at least the apical part narrowly cylindrical or
     subulate, usually ± acute, longer than wide; leaf-segments
     with bristles
4   Stems dimorphic, with either green ± palmately divided
     leaves without bladders or much reduced colourless leaves
     with bladders   **4. intermedia**
4   Stems all with green, ± palmately divided leaves with
     bladders
5   Lower lip of corolla with deflexed margin; pedicels 2–3
     times as long as bract; glands inside the apex of the spur
     on abaxial surface only   **5. vulgaris**
5   Lower lip of corolla ± flat, with undulate margin; pedicels
     3–5 times as long as bract; glands inside the apex of the
     spur on both adaxial and abaxial surfaces   **6. australis**

1. U. subulata L., *Sp. Pl.* 18 (1753). Terrestrial; stems capillary, subterranean; leaves 10–20 × 0·5 mm, often decayed at anthesis; bladders 0·2–0·5 mm. Inflorescence 3–20 cm; flowers 2–8; scape capillary; bracts *c.* 1 mm, peltate; pedicels 3–7 mm. Corolla 8–9 mm; upper lip 4 mm, broadly ovate; lower lip longer, with prominent 2-gibbous palate and 3-lobed limb, about as long as the subulate spur. Capsule globose; seeds ovoid, striate, unwinged. *Damp sand. Portugal (Beira Litoral), perhaps native.* *Lu. (North and South America, tropical Africa and Asia.)*

2. U. gibba L., *Sp. Pl.* 18 (1753). Aquatic or subaquatic; stems radiating from the base of the scape; leaves unequally forked from the base, the segments capillary and again 1–3 times forked; bladders 1–1·5 mm. Inflorescence 2–8 cm; flowers 1–3(–4); scape filiform; bracts 1 mm, scarcely auriculate; pedicels 2–10 mm, erect in fruit. Corolla 5–7 mm; upper lip 3–4 mm wide, broadly ovate, obtuse, obscurely 3-crenate; lower lip with prominent 2-gibbous palate, entire, about as long as the narrowly conical spur. Capsule globose; seeds winged. *Shallow water or liquid mud. W. Portugal, S.W. Spain.* Hs Lu. *(Africa, S. Asia, Australia.)*

The above description refers to subsp. **exoleta** (R. Br.) P. Taylor, *Mitt. Bot. Staatssamm. (München)* **4**: 101 (1961) (*U. exoleta* R. Br.). Subsp. **gibba**, from America and tropical Africa, has been recorded from Hungary as a casual (as *U. biflora* Lam.).

3. U. minor L., *Sp. Pl.* 18 (1753). Aquatic, more or less anchored to substratum; stems up to 25 cm, slender, more or less dimorphic; bearing either green, orbicular, more or less palmately lobed leaves 3–10 mm long, with glabrous, entire,

---

[1] By P. Taylor.

narrowly linear-filiform segments with few or no bladders; or bearing colourless, much reduced leaves with bladders *c.* 2 mm. Turions 1·5–5 mm, globose, glabrous. Inflorescence 4–15 cm; flowers 2–6; bracts *c.* 1·5 mm, auriculate; pedicels more or less recurved in fruit. Corolla 6–8 mm, pale yellow; upper lip *c.* 6 mm wide; spur saccate or obtusely conical, not longer than wide. Capsule 1·5–2 mm. Seeds unwinged. $2n=c.$ 40. *Shallow water and bogs. Most of Europe, but rare in the Mediterranean region.* Al Au Be Br Bu Cz Da Fe Ga Ge Gr Hb He Ho Hs Hu Is It Ju No Po Rm Rs (N, B, C, W, E) Su.

Plants, mainly from C. Europe, have been distinguished as **U. bremii** Heer in Koelliker, *Verz. Phan. Gew. Zürich* 142 (1839). They differ from **3** in being larger, with longer leaves and stouter, taller inflorescences bearing more, slightly larger, brighter yellow flowers, with more acute, longer spurs. They are probably best regarded as a variety of **3** since intermediates occur.

4. **U. intermedia** Hayne in Schrader, *Jour. für die Bot.* **1800**(1): 18 (1800). Aquatic or subaquatic, more or less anchored to substratum; stems 10–25 cm, slender, strongly dimorphic; bearing either distichous green, orbicular, more or less palmately lobed, bladderless leaves 4–15 mm long, with linear denticulate segments bearing 1–2 bristles on each tooth; or bearing colourless, much reduced leaves with bladders 3–4 mm. Turions 5–15 mm, ovoid, densely bristly. Inflorescence 10–20 cm; flowers 2–4; bracts *c.* 2 mm, auriculate. Corolla 10–15 mm; spur longer than wide, subulate from a short conical base. Capsule rarely produced. *Shallow water and bogs. Europe southward to S.W. France, N. Italy and S.C. Russia.* Au Be Br Cz Da Fe Ga Ge Hb He Ho It Ju No Po Rm Rs (N, B, C, W, E) Su.

U. ochroleuca R. Hartman, *Bot. Not.* **1857**: 30 (1857), is like **4** but with narrower, more acute segments to the green leaves, which may also bear some bladders, and with smaller flowers with a shorter spur; it is perhaps a variant of **4** or possibly **3 × 4**. It has $2n=c.$ 40 and occurs in N. & C. Europe.

5. **U. vulgaris** L., *Sp. Pl.* 18 (1753). Aquatic, floating submerged; stems up to 100 cm, slender; internodes 3–10 mm; leaves up to 3 cm, 2-lobed from the base; lobes more or less broadly ovate in outline, pinnately divided; segments capillary, denticulate, bearing solitary or fasciculate bristles on each tooth; bladders up to 3 mm. Inflorescence 10–30 cm; scape erect in fruit; flowers 4–10; bracts 5 mm, more or less auriculate; pedicels 6–15 mm, 2–3 times as long as the bract, strongly recurved in fruit. Corolla 12–18 mm, deep yellow; upper lip about as long as the conspicuous gibbous palate; lower lip with deflexed margin; spur more or less broadly conical at base, rather abruptly narrowed to a short, narrowly cylindrical, more or less acute apex with internal glands on the abaxial surface only. Capsule 3–5 mm, globose, usually freely produced. $2n=c.$ 40. *Still water. Most of Europe, but rare in the south.* Al Au Be Br Bu ?Cr Cz Da Fa Fe Ga Ge Gr Hb He Ho Hs Hu It Ju No Po Rm Rs (N, B, C, W, K, E) Si Su.

6. **U. australis** R. Br., *Prodr. Fl. Nov. Holl.* 430 (1810) (*U. neglecta* Lehm., *U. major* auct., *U. jankae* Velen.). Like **5** but internodes 8–20 mm; scapes more slender, often longer, sinuous after anthesis; pedicels 10–25 mm, 3–5 times as long as the bract, erect or patent in fruit; corolla lemon-yellow; upper lip longer than palate; lower lip more or less flat, with undulate margin; spur narrowly conical, gradually tapering from base to an obtuse apex, with internal glands on both adaxial and abaxial surfaces; capsule apparently never produced. $2n=c.$ 40. *Still water. Most of Europe except the extreme north and most of the U.S.S.R.* Au Be Br Bu ?Cr Cz Da Fe Ga Ge Hb He Ho Hs Hu It Ju Lu No Po Rm Rs (C) Su Tu.

5 and **6** have often been confused and many records of both are therefore doubtful.

CLXII. MYOPORACEAE[1]

Shrubs or small trees. Leaves simple, exstipulate. Flowers 5-merous. Calyx deeply lobed. Corolla with a short tube and weakly zygomorphic limb. Stamens 4–5, inserted on the corolla-tube and alternating with the lobes. Ovary superior, 2- to 10-locular; style filiform; ovules usually 1 or 2 in each loculus. Fruit a drupe.

1. Myoporum Solander ex G. Forster[2]

Evergreen trees or shrubs. Leaves alternate, rarely opposite, entire or serrate, with numerous pellucid glands. Flowers small, axillary, solitary or clustered in cymes. Calyx 5-lobed. Corolla campanulate-hypocrateriform with a short tube and 5 patent lobes, often bearded. Stigma obtuse. Fruit a scarcely fleshy drupe.

Leaves lanceolate, usually entire; stamens 4   **1. tenuifolium**
Leaves obovate, serrate in distal half; stamens 5   **2. tetrandrum**

1. **M. tenuifolium** G. Forster, *Fl. Ins. Austral. Prodr.* 44 (1786) (incl. *M. acuminatum* R. Br.). Small round-topped tree up to 8 m; bark brownish-grey, minutely fissured. Leaves 4·5–10(–17) × 1·5–3(–5) cm, lanceolate, acuminate, cuneate at base, usually entire, bright glossy green above; petiole 5–10 mm. Flowers in dense (1–)5- to 9-flowered cymes; pedicels up to 12 mm, slender. Calyx 2–3 mm, divided to half-way into ovate-lanceolate, acute lobes. Corolla white, with small purple spots; tube 4–5 mm; limb 10–12 mm in diameter. Stamens 4. Fruit 7–9 mm, ovoid, blackish-purple when ripe. *Planted for shelter in S.W. Europe and locally naturalized.* [Bl Hs Lu.] (*E. Australia.*)

2. **M. tetrandrum** (Labill.) Domin, *Mém. Soc. Sci. Bohême* **1921–1922**(2): 109 (1923) (*M. serratum* R. Br.). Like **1** but usually a shrub; leaves 2·5–5 × 1–2·5 cm, obovate, dark green above, obtuse and mucronate at apex, serrate in distal half; calyx-lobes lanceolate; stamens 5; fruit 4–6 mm, globose. *Planted for shelter in Portugal, mainly near the coast, and locally naturalized.* [Lu.] (*Australia, Tasmania.*)

---

[1] Edit. T. G. Tutin.   [2] By J. do Amaral Franco.

APPENDICES

NOTE TO APPENDICES I–III

Considerable variation is found in the orthography of the names of many authors, especially of the earlier ones and of those whose names are transliterated from Cyrillic script. Variant spellings are given here only if they are likely to give rise to doubts about identity.

The initials used by some authors vary according to whether the vernacular or latinized form of a Christian name is used (e.g. *Karl* or *Carolus*); the form most frequently used by the author is adopted in these lists.

The dates given for books and periodicals indicate, as far as can be ascertained, the date of effective publication; where this differs from dates on the title-page or elsewhere in the work itself, there is usually a reference to explain the dates given.

Certain publications are of a character intermediate between books and periodicals (e.g. seed-lists, *schedae*). The assignment of these to Appendix II or Appendix III is inevitably somewhat arbitrary.

In Appendix III there is normally no attempt made to indicate whether one periodical is a continuation of another, unless there is some continuity between them in the numbering of the volumes or series.

APPENDIX I
KEY TO THE ABBREVIATIONS OF AUTHORS' NAMES

Abromeit   J. Abromeit (1857–1946)
Acht.   B. Achtarov (1885–1959)
Adamović   L. Adamović (1864–1935)
Adams   M. F. Adams (J. F. Adam) (1780–1838)
Adanson   M. Adanson (1727–1806)
Ade   A. Ade (1876–1968)
Aellen   P. Aellen (b. 1896)
Agardh   C. A. Agardh (1785–1859)
Agardh, J.   J. G. Agardh (1813–1901)
Ahlfvengren   F. E. Ahlfvengren (1862–1921)
Ahti   T. Ahti (b. 1933)
Aichele   D. Aichele (b. 1928)
Airy Shaw   H. K. Airy Shaw (b. 1902)
Aiton   W. Aiton (1731–1793)
Aiton fil.   W. T. Aiton (1766–1849)
Albert   A. Albert (1836–1909)
Albov   N. M. Albov (Alboff) (1866–1897)
Alechin   V. V. Alechin (1884–1946)
Alef.   F. G. C. Alefeld (1820–1872)
Alexeenko   M. I. Alexeenko (Alexejenko) (b. 1905)
All.   C. Allioni (1728–1804)
Allan   H. H. B. Allan (1882–1957)
Alleiz.   C. d'Alleizette (1884–1967)
Allman   G. J. Allman (1812–1898)
Almq.   S. O. I. Almquist (1844–1923)
Alpers   F. Alpers (1841–1912)
Alston   A. H. G. Alston (1902–1958)
Ambrosi   F. Ambrosi (1821–1897)
Amo   Mariano del Amo y Mora (1809–1894)
Andersen   J. C. Andersen (b. 1873)
Anderson, E.   E. S. Anderson (1897–1969)
Anderson, G.   G. Anderson (d. 1817)
Andersson, N. J.   N. J. Andersson (1821–1880)
Andrae   C. J. Andrae (1816–1885)
Andrasovszky   J. Andrasovszky (1889–1943)
Andreas   C. H. Andreas (b. 1898)
Andrews   H. C. Andrews (d. 1830)
Andrz.   A. L. Andrzejowski (1785–1868)
Angells, M.   M. von Angelis (1805–1894)
Ångström   J. Ångström (1813–1879)
Antoine   F. Antoine (1815–1886)
Arcangeli   G. Arcangeli (1840–1921)
Ard.   P. Arduino (1728–1805)
Ardoino   H. J. P. Ardoino (1819–1874)
Aresch., F.   F. W. C. Areschoug (1830–1908)
Armstrong   J. B. Armstrong (1850–1926)
Arnold   (possibly a pseudonym; fl. 1785)
Arnott   G. A. W. Arnott (1799–1868)
Arrh., A.   J. I. A. Arrhenius (1858–1950)
Arrondeau   E. T. Arrondeau (d. 1882)
Artemczuk   I. V. Artemczuk (b. 1898)
Arvat   A. Arvat (1890–1950)
Arvet-Touvet   J. M. C. Arvet-Touvet (1841–1913)
Ascherson   P. F. A. Ascherson (1834–1913)
Aspegren   G. C. Aspegren (1791–1828)
Asso   I. J. de Asso y del Río (1742–1814)
Aublet   J. B. C. F. Aublet (1720–1778)
Aucher   P. M. R. Aucher-Eloy (1792–1838)

Avé-Lall.   J. L. E. Avé-Lallemant (1803–1867)
Avr.   N. A. Avrorin (b. 1906)
Aznav.   G. V. Aznavour (1861–1920)
Bab.   C. C. Babington (1808–1895)
Badaro   G. B. Badaro (1793–1831)
Bagnall   J. E. Bagnall (1830–1918)
Bailey, L. H.   L. H. Bailey (1858–1954)
Baillon   H. E. Baillon (1827–1895)
Bailly   E. Bailly (1829–1894)
Baker   J. G. Baker (1834–1920)
Baker fil.   E. G. Baker (1864–1949)
Baksay   L. Baksay (b. 1915)
Balansa   B. Balansa (1825–1891)
Balbis   G. B. Balbis (1765–1831)
Bald.   A. Baldacci (1867–1950)
Balf.   J. H. Balfour (1808–1884)
Balk.   B. E. Balkovsky (b. 1899)
Ball   J. Ball (1818–1889)
Ball, P. W.   P. W. Ball (b. 1932)
Banks   J. Banks (1743–1820)
Barbarich   A. I. Barbarich (b. 1903)
Barbaz.   F. Barbazita (fl. 1826)
Barbey, W.   W. Barbey-Boissier (1842–1914)
Barc.   F. Barceló y Combis (1820–1889)
Barkley, F. A.   F. A. Barkley (b. 1908)
Barkoudah   Y. I. Barkoudah (b. 1933)
Barn.   F. M. Barnéoud (b. 1821)
Barnades   M. Barnades (d. 1771)
Barrandon   A. Barrandon (1814–1897)
Bartal.   B. Bartalini (1746–1822)
Bartl.   F. G. Bartling (1798–1875)
Bartlett   H. H. Bartlett (1886–1960)
Barton, W.   W. P. G. Barton (1786–1856)
Basil.   N. A. Basilevskaja (Bazilevskaja) (b. 1902)
Basiner   T. F. J. Basiner (1817–1862)
Bässler   M. Bässler (b. 1935)
Bast.   T. Bastard (1784–1846)
Batsch   A. J. G. C. Batsch (1761–1802)
Batt.   J. A. Battandier (1848–1922)
Baudo   F. Baudo (fl. 1843)
Baum   B. R. Baum (b. 1937)
Baumann, E.   E. Baumann (1868–1933)
Baumg.   J. C. G. Baumgarten (1765–1843)
Baxter   W. Baxter (1787–1871)
Bean   W. J. Bean (1863–1947)
Beauv.   A. M. F. J. Palisot de Beauvois (1752–1820)
Beauverd   G. Beauverd (1867–1942)
Becherer   A. Becherer (b. 1897)
Bechst.   J. M. Bechstein (1757–1822)
Beck, G.   G. Beck von Mannagetta (1856–1931)
Becker, A.   A. Becker (1818–1901)
Becker, W.   W. Becker (1874–1928)
Beger   H. K. E. Beger (b. 1889)
Béguinot   A. Béguinot (1875–1940)
Behrendsen   W. Behrendsen (d. 1923)
Bellardi   C. A. L. Bellardi (1741–1826)
Belli   S. C. Belli (1852–1919)
Bellot   F. Bellot Rodríguez (b. 1911)

Bell Salter   T. Bell Salter (1814–1858)
Beltrán   F. Beltrán Bigorra (1886–1962)
Benj.   L. Benjamin (b. 1825)
Benn., A. W.   A. W. Bennett (1833–1902)
Benn., Ar.   Arthur Bennett (1843–1929)
Benson, L.   L. D. Benson (b. 1909)
Bentham   G. Bentham (1800–1884)
Berchtold   F. von Berchtold (1781–1876)
Berger, A.   A. Berger (1871–1931)
Bergeret, J. P.   J. P. Bergeret (1751–1813)
Berggren, Jakob   Jakob Berggren (1790–1868)
Bergius   P. J. Bergius (1730–1790)
Bergmans   J. Bergmans (b. 1892)
Bernard   P. F. Bernard (1749–1825)
Bernh.   J. J. Bernhardi (1774–1850)
Bernis   F. Bernis (fl. 1955)
Berth.   S. Berthelot (1794–1880)
Bertol.   A. Bertoloni (1775–1869)
Bertram   F. W. W. Bertram (1835–1899)
Bertsch, F.   F. Bertsch (1910–1944)
Bertsch, K.   K. Bertsch (1878–1965)
Besser   W. S. J. G. von Besser (1784–1842)
Beyer   R. Beyer (1852–1932)
Beyrich   H. C. Beyrich (1796–1834)
Biasol.   B. Biasoletto (1793–1859)
Biatzovsky   J. Biatzovsky (c. 1802–1863)
Bicknell, E. P.   E. P. Bicknell (1859–1925)
Bieb.   F. A. Marschall von Bieberstein (1768–1826)
Bigelow   J. Bigelow (1787–1879)
Bihari   J. Bihari (b. 1889)
Billot   P. C. Billot (1796–1863)
Binz   A. Binz (1870–1963)
Biria   J. A. J. Biria (b. 1889)
Biroli   G. Biroli (1772–1825)
Bischoff   G. W. Bischoff (1797–1854)
Bitter   F. A. G. Bitter (1873–1927)
Biv.   A. de Bivona-Bernardi (1774–1837)
Blaise   S. Blaise (fl. 1970)
Blakelock   R. A. Blakelock (1915–1963)
Blakeslee   A. F. Blakeslee (1874–1954)
Blanc   — Blanc (fl. 1866)
Blanche   E. Blanche (1824–1908)
Blanco   F. M. Blanco (1778–1845)
Blečić   V. Blečić (b. 1911)
Błocki   B. Błocki (1857–1919)
Błoński   F. Błoński (1867–1910)
Bloxam   A. Bloxam (1801–1878)
Bluff   M. J. Bluff (1805–1837)
Blume   C. L. von Blume (1796–1862)
Blytt   M. N. Blytt (1789–1862)
Bobrov   E. G. Bobrov (b. 1902)
Böcher   T. W. Böcher (b. 1909)
Bodard   P. H. H. Bodard (fl. 1798–1810)
Boedijn   K. B. Boedijn (b. 1893)
Boehmer   G. R. Boehmer (1723–1803)
Boenn.   C. M. F. von Boenninghausen (1785–1864)
Boguslaw   I. A. Boguslaw (fl. 1846)
Boiss.   P. E. Boissier (1810–1885)
Boivin   J. R. B. Boivin (b. 1916)
Bolle, F.   F. Bolle (b. 1905)
Bolós, A.   A. de Bolós (b. 1889)
Bolós, O.   O. de Bolós (b. 1924)
Bolton   J. Bolton (c. 1758–1799)
Bolus, L.   L. H. M. Bolus (Mrs F. Bolus) (1877–1970)
Bong.   H. G. von Bongard (1786–1839)
Bonjean   J. L. Bonjean (1780–1846)

Bonnet   E. Bonnet (1848–1922)
Bonnier   G. E. M. Bonnier (1853–1922)
Bonpl.   A. J. A. Bonpland (1773–1858)
Boos, J.   J. Boos (1794–1879)
Borbás   V. von Borbás (1844–1905)
Bord.   H. Bordère (1825–1889)
Bordzil.   E. I. Bordzilowski (1875–1949)
Boreau   A. Boreau (1803–1875)
Borhidi   A. Borhidi (b. 1932)
Boriss.   A. G. Borissova-Bekrjaševa (1903–1970)
Borja   J. Borja Carbonell (b. 1903)
Borkh.   M. B. Borkhausen (Borckhausen) (1760–1806)
Börner   C. J. B. Börner (b. 1880)
Bornm.   J. F. N. Bornmüller (1862–1948)
Boros   Á. Boros (b. 1900)
Borrer   W. Borrer (1781–1862)
Bory   J. B. G. M. Bory de Saint-Vincent (1778–1846)
Borza   A. Borza (1887–1971)
Borzi   A. Borzi (1852–1911)
Bosc   L. A. G. Bosc (1759–1828)
Bosse   J. F. W. Bosse (1788–1864)
Bothmer   S. R. von Bothmer (b. 1943)
Botsch.   V. P. Botschantzev (b. 1910)
Bouché   C. D. Bouché (1809–1881)
Boulay   N. J. Boulay (1837–1905)
Bourgeau   E. Bourgeau (1813–1877)
Bout.   J. F. D. Boutigny (1820–1884)
Boutelou   E. Boutelou (1776–1813)
Bouvet   G. Bouvet (1874–1929)
Br., N. E.   N. E. Brown (1849–1934)
Br., R.   R. Brown (1773–1858)
Brackenr.   W. D. Brackenridge (1810–1893)
Bradshaw, M. E.   M. E. Bradshaw (b. 1926)
Brand   A. Brand (1863–1931)
Brandt, J. P.   J. P. Brandt (1921–1963)
Brandza   D. Brandza (1846–1895)
Braun, A.   A. C. H. Braun (1805–1877)
Braun, G.   G. Braun (1821–1882)
Braun, H.   H. Braun (1851–1920)
Braun, J.   J. Braun (later J. Braun-Blanquet) (b. 1884)
Br.-Bl.   J. Braun-Blanquet (b. 1884)
Bréb.   L. A. de Brébisson (1798–1872)
Breistr.   M. Breistroffer (b. 1910)
Brenan   J. P. M. Brenan (b. 1917)
Brenner   M. M. W. Brenner (1843–1930)
Brewer   W. H. Brewer (1828–1910)
Briganti   V. Briganti (1766 1836)
Brign.   G. de Brignoli di Brunnhoff (1774–1857)
Briot   P. L. Briot (1804–1888)
Briq.   J. I. Briquet (1870–1931)
Britten   J. Britten (1846–1924)
Brittinger   C. C. Brittinger (1795–1869)
Britton   N. L. Britton (1859–1934)
Britton, C. E.   C. E. Britton (1872–1944)
Brocchi   G. B. Brocchi (1772–1826)
Bromf.   W. A. Bromfield (1801–1851)
Brot.   F. Avellar Brotero (1744–1828)
Brouss.   P. M. A. Broussonet (1761–1807)
Browicz   K. Browicz (b. 1925)
Brügger   C. G. Brügger (1833–1899)
Brumh.   P. Brumhard (b. 1879)
Brummitt   R. K. Brummitt (b. 1937)
Bruno   — Bruno (fl. 1760)
Bruun   H. G. Bruun (b. 1897)
Bubani   P. Bubani (1806–1888)
Buchanan-White   F. Buchanan-White (1842–1894)

Buchegger   J. Buchegger (b. 1886)
Buchholz   J. T. Buchholz (1888–1951)
Buchinger   J. D. Buchinger (1803–1888)
Buckn.   C. Bucknall (1849–1921)
Buffon   G. L. L. de Buffon (1707–1788)
Buhse   F. A. Buhse (1821–1898)
Buia   A. Buia (1911–1964)
Bunge   A. A. von Bunge (1803–1890)
Burgsd.   F. A. L. von Burgsdorff (1747–1802)
Burm. fil.   N. L. Burman (N. L. Burmannus) (1734–1793)
Burnat   E. Burnat (1828–1920)
Burtt, B. L.   B. L. Burtt (b. 1913)
Busch, N.   N. A. Busch (1869–1941)
Buschm.   A. Buschmann (b. 1908)
Buser   R. Buser (1857–1931)
Bush   B. F. Bush (1858–1937)
Butcher   R. W. Butcher (1897–1971)
Butkov   A. Y. Butkov (b. 1911)
Caballero, A.   A. Caballero (1877–1949)
Cadevall   J. Cadevall i Diars (1846–1910)
Cajander   A. K. Cajander (1879–1943)
Caldesi   L. Caldesi (1821–1884)
Calestani   V. Calestani (b. 1882)
Camb.   J. Cambessedes (1799–1863)
Campd.   F. Campderá (1793–1862)
Campo   P. del Campo (fl. 1855)
Camus, A.   A. Camus (1879–1965)
Camus   E. G. Camus (1852–1915)
Cañigueral   J. Cañigueral Cid (b. 1912)
Cannon   J. F. M. Cannon (b. 1930)
Cariot   A. Cariot (1820–1883)
Carrière   E. A. Carrière (1818–1896)
Caruel   T. Caruel (1830–1898)
Casav.   J. Ruíz Casaviella (1835–1897)
Casper   S. J. Casper (b. 1929)
Cast.   J. L. M. Castagne (1785–1858)
Cav.   A. J. Cavanilles (1745–1804)
Cavara   F. Cavara (1857–1929)
Ceballos   L. Ceballos Fernández de Córdoba (1896–1967)
Čelak.   L. J. Čelakovsky (1834–1902)
Cesati   V. de Cesati (1807–1883)
Chab.   A. C. Chabert (1836–1916)
Chaix   D. Chaix (1730–1799)
Cham.   L. A. von Chamisso (L.C.A. Chamisseau de Boncourt) (1781–1838)
Charrel   L. Charrel ('Abd-ur-Raḣmān-Nadji) (fl. 1888)
Chassagne   M. Chassagne (fl. 1904–1960)
Chater   A. O. Chater (b. 1933)
Chaub.   L. A. Chaubard (1785–1854)
Chav.   E. L. Chavannes (1805–1861)
Chaytor   D. A. Chaytor (fl. 1937)
Chaz.   L. M. Chazelles de Prizy (fl. 1790)
Chenevard   P. Chenevard (1839–1919)
Cheval., A.   A. J. B. Chevalier (1873–1956)
Chevall.   F. F. Chevallier (1796–1840)
Chiaje   S. delle Chiaje (1794–1860)
Chiarugi   A. Chiarugi (1901–1960)
Ching, R.-C.   Ren-Chang Ching (Jên-ch'ang Ch'in) (b. 1899)
Chiov.   E. Chiovenda (1871–1940)
Chitrowo   V. N. Chitrowo (1879–1949)
Chodat   R. H. Chodat (1865–1934)
Choisy   J. D. Choisy (1799–1859)
Chopinet   R. Chopinet (b. 1914)
Chowdhuri   P. K. Chowdhuri (b. 1923)
Chr., C.   C. F. A. Christensen (1872–1942)
Christ   H. Christ (1833–1933)

Christm.   G. F. Christmann (b. 1752)
Chrshan.   V. G. Chrshanovski (b. 1912)
Chrtek   J. Chrtek (b. 1931)
Clairv.   J. P. de Clairville (1742–1830)
Clapham   A. R. Clapham (b. 1904)
Clarion   ?J. Clarion (1776–1844)
Clarke, E. D.   E. D. Clarke (1779–1822)
Claus   K. Claus (1796–1864)
Clavaud   A. Clavaud (1828–1890)
Cleland   R. E. Cleland (b. 1892)
Clemente   S. de Rojas Clemente y Rubio (1777–1827)
Clementi, G. C.   G. C. Clementi (1812–1873)
Clerc   O. E. Clerc (1845–1920)
Cockayne   L. Cockayne (1855–1934)
Coincy   A. de Coincy (1837–1903)
Coleman   W. H. Coleman (?1816–1863)
Coleman, J. R.   J. R. Coleman (b. 1934)
Colla   L. A. Colla (1766–1848)
Collett   H. Collett (1836–1901)
Colmeiro   M. Colmeiro y Penido (1816–1901)
Commerson   P. Commerson (1727–1773)
Comolli   G. Comolli (1780–1859)
Conr.   P. Conrath (b. 1892)
Constance   L. Constance (b. 1909)
Contandr.   J. Contandriopoulos (b. 1922)
Conti, P.   P. Conti (1874–1898)
Coombe, D. E.   D. E. Coombe (b. 1927)
Copel.   E. B. Copeland (1873–1964)
Corb.   L. Corbière (1850–1941)
Cornaz   E. Cornaz (1825–1911)
Corr.   C. F. J. E. Correns (1864–1933)
Cosent.   F. Cosentini (1769–1840)
Cosson   E. S. C. Cosson (1819–1889)
Costa   A. C. Costa y Cuxart (1817–1886)
Coste   H. J. Coste (1858–1924)
Coulter   J. M. Coulter (1851–1928)
Court.   R. J. Courtois (1806–1835)
Coust.   P. Cousturier (d. 1921)
Coutinho   A. X. Pereira Coutinho (1851–1939)
Covas   G. Covas (b. 1915)
Coville   F. V. Coville (1867–1937)
Craib   W. G. Craib (1882–1933)
Crantz   H. J. N. von Crantz (1722–1799)
Crépin   F. Crépin (1830–1903)
Cristofolini   G. Cristofolini (b. 1939)
Crome   G. E. W. Crome (1780–1813)
Csapody   V. Csapody (fl. 1930)
Cuatrec.   J. Cuatrecasas (b. 1903)
Cuf.   G. Cufodontis (b. 1896)
Cullen   J. Cullen (b. 1936)
Cunn., A.   A. Cunningham (1791–1839)
Cunn., R.   R. Cunningham (1793–1835)
Curtis   W. Curtis (1746–1799)
Cusson   P. Cusson (1727–1783)
Cutanda   V. Cutanda (1804–1865)
Cyr.   D. Cyrillo (1739–1799)
Czecz.   H. Czeczott (b. 1888)
Czefr.   Z. V. Czefranova (b. 1923)
Czern.   V. M. Czernajew (Czernjaew) (1796–1871)
Czernov   E. G. Czernov (b. 1908)
Czernova   N. M. Czernova (b. 1901)
Czetz   A. Czetz (1801–1865)
Dahl, O. C.   O. C. Dahl (1862–1940)
Dalby   D. H. Dalby (b. 1930)
Dalla Torre   K. W. von Dalla Torre (1850–1928)
Damanti   P. Damanti (b. 1858)

Damboldt   J. Damboldt (b. 1937)
Dammer   C. L. U. Dammer (1860–1920)
Dandy   J. E. Dandy (b. 1903)
Danilov   A. D. Danilov (b. 1903)
Danser   B. H. Danser (1891–1943)
Darlington, W.   W. Darlington (1782–1863)
Darracq   U. Darracq (d. 1872)
Daveau   J. A. Daveau (1852–1929)
Davey   F. H. Davey (1868–1915)
Davidov   B. Davidov (1870–1927)
Davies   H. Davies (1739–1821)
Davis, P. H.   P. H. Davis (b. 1918)
DC.   A. P. de Candolle (1778–1841)
DC., A.   A. L. P. P. de Candolle (1806–1893)
DC., C.   A. C. P. de Candolle (1836–1918)
De Bary   H. A. de Bary (1831–1888)
Debeaux   J. O. Debeaux (1826–1910)
Déchy   M. Déchy (b. 1851)
Decken   C. C. von der Decken (1833–1865)
Decker   P. Decker (b. 1867)
Decne   J. Decaisne (1807–1882)
DeFilipps   R. A. DeFilipps (b. 1939)
Degen   A. von Degen (1866–1934)
Dehnh.   F. Dehnhardt (1787–1870)
Delarbre   A. Delarbre (1724–1841)
De la Soie   G. A. de la Soie (1818–1877)
De Lens   — De Lens (fl. 1828)
Delile   A. R. Delile (1778–1850)
Delponte   G. B. Delponte (1812–1884)
Dematra   Dematra (1742–1824)
Dennst.   A. W. Dennstedt (1776–1826)
De Noé   F. de Noé (fl. 1855)
De Not.   G. de Notaris (1805–1877)
Déséglise   P. A. Déséglise (1823–1883)
Des Etangs   N. S. C. des Etangs (1801–1876)
Desf.   R. L. Desfontaines (c. 1751–1833)
Desmoulins   C. Desmoulins (1797–1875)
Desportes   N. H. F. Desportes (1776–1856)
Desr.   L. A. J. Desrousseaux (1753–1838)
Desv.   A. N. Desvaux (1784–1856)
Deville   L. Deville (fl. 1859)
De Wild.   É. de Wildeman (1866–1947)
Dickson   J. Dickson (1738–1822)
Diels   F. L. E. Diels (1874–1945)
Dietr., A.   A. Dietrich (1795–1856)
Dietr., D.   D. N. F. Dietrich (1800–1888)
Dietr., F. G.   F. G. Dietrich (1768–1850)
Dingler   H. Dingler (1846–1935)
Dippel   L. Dippel (1827–1914)
Dobrocz.   D. N. Dobroczaeva-Kovalczuk (b. 1916)
Dode   L. A. Dode (1875–1943)
Döll   J. C. Döll (1808–1885)
Dolliner   G. Dolliner (1794–1872)
Domac   R. Domac (b. 1918)
Domin   K. Domin (1882–1952)
Domokos   J. Domokos (b. 1904)
Don, D.   D. Don (1799–1841)
Don, G.   G. Don (1764–1814)
Don fil., G.   G. Don (1798–1856)
Donadille   P. Donadille (b. 1936)
Donn   J. Donn (1758–1813)
Dörfler   I. Dörfler (1866–1950)
Dorthes   J. A. Dorthes (1759–1794)
Dostál   J. Dostál (b. 1903)
Douglas   D. Douglas (1798–1834)
Downar   N. V. Downar (fl. 1855–1862)

Drejer   S. T. N. Drejer (1813–1842)
Drenowski   A. K. Drenowski (Drenovsky) (1879–1967)
Dreves   J. F. P. Dreves (1772–1816)
Druce   G. C. Druce (1850–1932)
Drude   C. G. O. Drude (1852–1933)
Düben   M. W. von Düben (1814–1845)
Dubjansky   V. A. Dubjansky (1877–1962)
Dubois, F.   F. N. A. Dubois (1752–1824)
Duby   J. E. Duby (1798–1885)
Duchartre   P. E. S. Duchartre (1811–1894)
Duchesne   A. N. Duchesne (1747–1827)
Dudley, T. R.   T. R. Dudley (b. 1936)
Dufour   J.-M. L. Dufour (1780–1865)
Duh.   H. L. Duhamel de Monceau (1700–1781)
Düll   R. Düll (b. 1932)
Dum.-Courset   G. L. M. Dumont de Courset (1746–1824)
Dumort.   B. C. J. Dumortier (1797–1878)
Dunal   M. F. Dunal (1789–1856)
Dupont   — Dupont (fl. 1825)
Durand, B.   B. M. Durand (b. 1928)
Durand, E.   E.-M. (later E.) Durand (1794–1873)
Durande   J. F. Durande (1732–1794)
Durieu   M. C. Durieu de Maisonneuve (1796–1878)
Duroi   J. P. Duroi (1741–1785)
D'Urv.   J. S. C. D. D'Urville (1790–1842)
Duthie   J. F. Duthie (1845–1922)
Du Tour   — Du Tour de Salvert (fl. 1803–1815)
Duval-Jouve   J. Duval-Jouve (1810–1883)
Dvořáková   M. Dvořáková (b. 1940)
Dyer   W. T. Thiselton-Dyer (1843–1928)
Ecklon   C. F. Ecklon (1795–1868)
Edgew.   M. P. Edgeworth (1812–1881)
Edmondston   T. Edmondston (1825–1846)
Ehrenb.   C. G. Ehrenberg (1795–1876)
Ehrend.   F. Ehrendorfer (b. 1927)
Ehrh.   J. F. Ehrhart (1742–1795)
Eichw.   K. E. von Eichwald (1794–1876)
Eig   A. Eig (1894–1938)
Ekman, E. L.   E. L. Ekman (1883–1931)
Ekman, Elis.   H. M. E. A. E. Ekman (1862–1936)
Elias   Frère H. Elias (fl. 1907–1944)
Elkan   L. Elkan (1815–1851)
Elliott   S. Elliott (1771–1830)
Emberger   L. Emberger (1897–1969)
Enander   S. J. Enander (1847–1928)
Endl.   S. L. Endlicher (1804–1849)
Engelm.   G. Engelmann (1809–1884)
Engler   H. G. A. Engler (1844–1930)
Engler, V.   V. Engler (1885–1917)
Ern   H. Ern (b. 1935)
Eshbaugh   W. H. Eshbaugh (b. 1936)
Eschsch.   J. F. G. von Eschscholz (1793–1831)
Esteve   F. Esteve Chueca (b. 1919)
Etlinger   A. E. Etlinger (fl. 1777)
Evers   G. Evers (?1837–?1916)
Exell   A. W. Exell (b. 1901)
Fabr.   P. C. Fabricius (1714–1774)
Facch.   F. Facchini (1788–1852)
Farwell   O. A. Farwell (1867–1944)
Fasano   A. Fasano (fl. 1787)
Fauché   M. Fauché (fl. 1832)
Fauconnet   C. I. Fauconnet (1811–1876)
Favarger   C. P. E. Favarger (b. 1913)
Favrat   L. Favrat (1827–1893)
Fedde   F. K. G. Fedde (1873–1942)
Fedorov   An. A. Fedorov (b. 1908)

Fedtsch., B.   B. A. Fedtschenko (1872–1947)
Fedtsch., O.   O. A. Fedtschenko (1845–1921)
Fée   A. L. A. Fée (1789–1874)
Feinbrun   N. Feinbrun (b. 1900)
Fenzl   E. Fenzl (1808–1879)
Ferguson, I. K.   I. K. Ferguson (b. 1938)
Fernald   M. L. Fernald (1873–1950)
Fernandes, A.   A. Fernandes (b. 1906)
Fernandes, R.   R. Fernandes (b. 1916)
Ferrarini   E. Ferrarini (b. 1919)
Fiala   F. Fiala (1861–1898)
Ficalho   F. M. de Mello Breyner de Ficalho (1837–1903)
Fieschi   V. Fieschi (b. c. 1910)
Fil.   N. Filarszky (1858–1941)
Fingerh.   K. A. Fingerhuth (1802–1876)
Fiori   A. Fiori (1865–1950)
Fischer   F. E. L. von Fischer (1782–1854)
Fischer, M.   M. Fischer (b. 1942)
Fischer von Wald.   A. A. Fischer von Waldheim (1803–1884)
Fisher   T. R. Fisher (b. 1921)
Fitschen   J. Fitschen (1869–1947)
Fleischm.   A. Fleischmann (1805–1867)
Flerow   A. F. Flerow (1872–1960)
Fletcher   H. R. Fletcher (b. 1907)
Flod., B.   B. G. O. Floderus (1867–1941)
Floerke   H.-G. Floerke (1764–1835)
Flügge   J. Flügge (1775–1816)
Focke   W. O. Focke (1834–1922)
Foggitt   W. Foggitt (1835–1917)
Fomin   A. V. Fomin (1869–1935)
Font Quer   P. Font Quer (1888–1964)
Form.   E. Formánek (1845–1900)
Forrest   G. Forrest (1873–1932)
Forskål   P. Forskål (1732–1763)
Forster, E.   E. Forster (1765–1849)
Forster, G.   J. G. A. Forster (1754–1794)
Forster, J. R.   J. R. Forster (1729–1798)
Forster, T. F.   T. F. Forster (1761–1825)
Fortune   R. Fortune (1813–1880)
Fouc.   J. Foucaud (1847–1904)
Foug.   A. D. Fougeroux de Bondaroy (1732–1789)
Fourn., E.   E. P. N. Fournier (1834–1884)
Fourn., P.   P.-V. Fournier (1877–1964)
Fourr.   J. P. Fourreau (1844 1871)
Franchet   A. R. Franchet (1834–1900)
Franco   J. do Amaral Franco (b. 1921)
Franklin   J. Franklin (1786–1847)
Fraser, Neill   P. Neill Fraser (1830–1905)
Freitag   H. Freitag (b. 1932)
Fresen.   J. B. G. W. Fresenius (1806–1866)
Freyc.   L. C. Desaulses de Freycinet (1779–1842)
Freyer   H. Freyer (1802–1866)
Freyn   J. F. Freyn (1845–1903)
Frid.   K. N. Friderichsen (1853–1932)
Friedrich   H. Friedrich (b. 1925)
Fries   E. M. Fries (1794–1878)
Fries, T. C. E.   T. C. E. Fries (1886–1930)
Fries, Th.   T. M. Fries (1832–1913)
Fritsch   K. Fritsch (1864–1934)
Fritze   R. Fritze (fl. 1870)
Friv.   E. Frivaldszky von Frivald (I. Frivaldszky) (1799–1870)
Frodin   D. G. Frodin (b. 1940)
Froelich   J. A. von Froelich (1766–1841)
Fröhlich, A.   A. Fröhlich (b. 1882)
Fröhner   S. E. Fröhner (b. 1941)
Fuchs, H. P.   H. P. Fuchs (b. 1928)

Funck   H. C. Funck (1771–1839)
Fürnrohr   A. E. Fürnrohr (1804–1861)
Fuss   M. Fuss (1814–1883)
Gaertner   J. Gaertner (1732–1791)
Gaertner fil.   C. F. von Gaertner (1772–1850)
Gaertner, P.   P. G. Gaertner (1754–1825)
Gagnebin   A. Gagnebin (1707–1800)
Gaill.   C. Gaillardot (1814–1883)
Galeotti   H. G. Galeotti (1814–1858)
Gamajun.   A. P. Gamajunova (b. 1904)
Gams   H. Gams (b. 1893)
Gand.   M. Gandoger (1850–1926)
Ganeschin   S. S. Ganeschin (1879–1930)
García   J. G. García (1904–1971)
Garcke   F. A. Garcke (1819–1904)
Gariod   C. H. Gariod (1836–1892)
Gars.   F. A. de Garsault (1691–1776)
Gartner, H.   H. Gartner (fl. 1939)
Gasparr.   G. Gasparrini (1804–1866)
Gauckler   K. Gauckler (b. 1898)
Gaud.-Beaup.   C. Gaudichaud-Beaupré (1789–1854)
Gaudin   J. F. A. T. G. P. Gaudin (1766–1833)
Gaussen   H. Gaussen (b. 1891)
Gaut.   G. Gautier (1841–1911)
Gavioli   O. Gavioli (1871–1944)
Gawłowska   M. J. Gawłowska (b. 1910)
Gay   J. E. Gay (1786–1864)
Gay, C.   C. Gay (1800–1873)
Gáyer   G. Gáyer (1883–1932)
Geiger   P. L. Geiger (1785–1836)
Gelert   O. C. L. Gelert (1862–1899)
Genev.   L. G. Genevier (1830–1880)
Genn.   P. Gennari (1820–1897)
Genty   P. A. Genty (1861–1955)
Georgescu   C. C. Georgescu (1898–1968)
Georgi   J. G. Georgi (1729–1802)
Georgiev   T. Georgiev (b. 1883)
Gérard   L. Gérard (1733–1819)
Germ.   J. N. E. Germain de Saint-Pierre (1815–1882)
Getliffe   F. M. Getliffe (b. 1941)
Gibbs, P.   P. E. Gibbs (b. 1938)
Gibelli   G. Gibelli (1831–1898)
Gilib.   J. E. Gilibert (1741–1814)
Gill   J. Gill (b. 1936)
Gillet   C. C. Gillet (1806–1896)
Gilli   A. Gilli (b. 1903)
Gillies   J. Gillies (1747–1836)
Gillot   F. X. Gillot (1842–1910)
Gilmour   J. S. L. Gilmour (b. 1906)
Ging.   F. C. J. Gingins de Lassaraz (1790–1863)
Ginzberger   A. Ginzberger (1873–1940)
Girard   F. de Girard (fl. 1844)
Giraud.   L. Giraudias (1848–1922)
Gled.   J. G. Gleditsch (1714–1786)
Glück   C. M. H. Glück (1868–1940)
Gmelin, C. C.   C. C. Gmelin (1762–1837)
Gmelin, J. F.   J. F. Gmelin (1748–1804)
Gmelin, J. G.   J. G. Gmelin (1709–1755)
Gmelin, S. G.   S. G. Gmelin (1744 or 1745–1774)
Godet   C. H. Godet (1797–1879)
Godman   F. Du Cane Godman (1834–1919)
Godron   D. A. Godron (1807–1880)
Goffart   J. Goffart (1864–1954)
Goiran   A. Goiran (1835–1909)
Goldie   J. Goldie (1793–1886)
Golitsin   S. V. Golitsin (1897–1968)

Gontsch.   N. F. Gontscharov (1900–1942)
González, F.   F. González (fl. 1877)
González-Albo   J. González-Albo (fl. 1935)
Goodding   L. N. Goodding (b. 1880)
Gordon   G. Gordon (1806–1879)
Gorodkov   B. N. Gorodkov (1890–1953)
Gorschk.   S. G. Gorschkova (1889–1972)
Görz, R.   R. Görz (1879–1935)
Gouan   A. Gouan (1733–1821)
Goulimy   C. N. Goulimy (Goulimis) (1886–1963)
Goupil   C. J. Goupil (1784–1858)
Govoruchin   V. S. Govoruchin (1903–1970)
Grab.   H. E. Grabowski (1792–1842)
Graebner   K. O. P. P. Graebner (1871–1933)
Graells   M. de la P. Graells (1809–1898)
Graf   ?S. Graf (1801–1838)
Graham, R. A.   R. A. Graham (1915–1958)
Graham, R. C.   R. C. Graham (1786–1845)
Gram, K.   K. J. A. Gram (1897–1961)
Grande   L. Grande (1878–1965)
Grau   H. R. J. Grau (b. 1937)
Grauer   S. Grauer (1758–1820)
Gray, A.   A. Gray (1810–1888)
Gray, S. F.   S. F. Gray (1766–1828)
Grec.   D. Grecescu (1841–1910)
Greene, E. L.   E. L. Greene (1843–1915)
Greenman   J. M. Greenman (1867–1951)
Gregory, E. S.   E. S. Gregory (1840–1932)
Gremli   A. Gremli (1833–1899)
Gren.   J. C. M. Grenier (1808–1875)
Greuter, W.   W. R. Greuter (b. 1938)
Grev.   R. K. Greville (1794–1866)
Grigoriev   J. S. Grigoriev (b. 1905)
Grimm   J. F. K. Grimm (1737–1821)
Grinţ., G.   G. P. Grinţescu (1870–1947)
Griseb.   A. H. R. Grisebach (1814–1879)
Gröntved   J. Gröntved (1882–1956)
Gross, H.   H. Gross (b. 1888)
Grosser   W. C. H. Grosser (b. 1869)
Grosset   H. E. Grosset (b. 1903)
Grossh.   A. A. Grossheim (1888–1948)
Groves   H. Groves (1835–1891)
Gruner   L. F. Gruner (b. 1838)
Grynj   F. A. Grynj (b. 1902)
Gueldenst.   J. A. von Gueldenstaedt (1745–1781)
Guépin   J. P. Guépin (1779–1858)
Guérin   J. X. B. Guérin (1775–1850)
Guersent   L. B. Guersent (1776–1848)
Guicc.   G. Guicciardi (fl. 1855)
Guimar.   J. de Ascensão Guimarães (1862–1922)
Guimpel   F. Guimpel (1774–1839)
Guinea   E. Guinea (b. 1907)
Guinier   P. Guinier (b. 1876)
Guirão   A. Guirão y Navarro (d. 1890)
Guittonneau   G. Guittonneau (b. 1934)
Gulia   G. Gulia (1835–1889)
Gunnarsson   J. G. Gunnarsson (1866–1944)
Gunnerus   J. E. Gunnerus (1718–1773)
Günther   C. C. Günther (1769–1833)
Gürke   A. R. L. M. Gürke (1854–1911)
Guss.   G. Gussone (1787–1866)
Guşuleac   M. Guşuleac (1887–1960)
Guterm.   W. Gutermann (b. 1935)
Guthnick   H. J. Guthnick (1800–1870)
Habl.   C. von Hablitz (1752–1821)
Hacq.   B. A. Hacquet (1739–1815)

Hadač   E. Hadač (b. 1914)
Haenke   T. Haenke (1761–1817)
Haenseler   F. Haenseler (1766–1841)
Hagerup   O. Hagerup (1889–1961)
Hahne   A. Hahne (1873–1942)
Halácsy   E. von Halácsy (1842–1913)
Hall, W.   W. Hall (1743–1800)
Haller   A. von Haller (1708–1777)
Haller fil.   A. von Haller (1758–1823)
Halliday   G. Halliday (b. 1933)
Hallier   E. Hallier (1831–1904)
Hamet   Raymond-Hamet (fl. 1906–1960)
Hampe   G. E. Hampe (1795–1880)
Hand.-Mazz.   H. von Handel-Mazzetti (1882–1940)
Hanry   H. Hanry (1807–1893)
Hara   H. Hara (b. 1911)
Harley   R. M. Harley (b. 1936)
Harrison, H.-   J. Heslop-Harrison (b. 1920)
Hartig   H. J. A. R. Hartig (1839–1901)
Hartinger   A. Hartinger (1806–1890)
Hartl   D. Hartl (fl. 1955)
Hartman   C. J. Hartman (1790–1849)
Hartman fil.   C. Hartman (1824–1884)
Hartman, R.   R. W. Hartman (1827–1891)
Hartmann, F. X.   F. X. von Hartmann (1737–1791)
Hartweg   K. T. Hartweg (1812–1871)
Hartwiss   N. von Hartwiss (1791–1860)
Harvey   W. H. Harvey (1811–1866)
Harz, C. O.   C. O. Harz (1842–1906)
Hasselq.   F. Hasselquist (1722–1752)
Hassk.   J. C. Hasskarl (1811–1894)
Hausskn.   H. K. Haussknecht (1838–1903)
Haw.   A. H. Haworth (1768–1833)
Hayek   A. von Hayek (1871–1928)
Haynald   S. F. L. Haynald (1816–1891)
Hayne   F. G. Hayne (1763–1832)
Häyrén   E. F. Häyrén (1878–1957)
Hazsl.   F. A. Hazslinszky von Hazslin (1818–1896)
Hedberg   K. O. Hedberg (b. 1923)
Hedl.   J. T. Hedlund (1861–1953)
Hedley   G. W. Hedley (1871–1941)
Heer   O. Heer (1809–1883)
Hegelm.   C. F. Hegelmaier (1834–1906)
Hegetschw.   J. J. Hegetschweiler (1789–1839)
Hegi   G. Hegi (1876–1932)
Heimans   J. Heimans (b. 1889)
Heimerl   A. Heimerl (1857–1942)
Heister   L. Heister (1683–1758)
Heldr.   T. von Heldreich (1822–1902)
Heller   F. X. Heller (1775–1840)
Helm   G. F. Helm (fl. 1809–1828)
Hemsley   W. B. Hemsley (1843–1924)
Henckel   L. V. F. Henckel von Donnersmarck (1785–1861)
Hendrych   R. Hendrych (b. 1926)
Henriq.   J. A. Henriques (1838–1928)
Henry, A.   A. Henry (1857–1930)
Henry, Louis   Louis Henry (1853–1913)
Hepper   F. N. Hepper (b. 1929)
Herbert   W. Herbert (1778–1847)
Herbich   F. Herbich (1791–1865)
Hermann, F.   F. Hermann (b. 1873)
Herrmann, J.   J. Herrmann (1738–1800)
Herter   W. G. Herter (1884–1958)
Hervier   J. Hervier-Basson (b. 1846)
Hess, H.   H. Hess (b. 1920)
Heuffel   J. Heuffel (1800–1857)

Heukels   H. Heukels (1854–1936)
Heynh.   G. Heynhold (fl. 1828–1850)
Heywood   V. H. Heywood (b. 1927)
Hicken   C. M. Hicken (1875–1933)
Hiern   W. P. Hiern (1839–1925)
Hieron.   G. H. E. W. Hieronymus (1846–1921)
Hiitonen   H. I. A. Hiitonen (b. 1898)
Hildebr.   F. H. G. Hildebrand (1835–1915)
Hill   J. Hill (1716–1775)
Hill, A. W.   A. W. Hill (1875–1941)
Hitchc., A. S.   A. S. Hitchcock (1865–1935)
Hitchc., E.   E. Hitchcock (1793–1864)
Hladnik   F. Hladnik (1773–1844)
Hochst.   C. F. Hochstetter (1787–1860)
Hoffm.   G. F. Hoffmann (1761–1826)
Hoffm., O.   O. Hoffmann (1853–1909)
Hoffmanns.   J. C. von Hoffmannsegg (1766–1849)
Hofmann, E.   E. Hofmann (fl. 1839–1856)
Hohen.   R. F. Hohenacker (1798–1874)
Holandre   J. J. J. Holandre (1773–1857)
Holl   F. Holl (fl. 1820–1842)
Holm   T. Holm (1880–1943)
Holmberg   O. R. Holmberg (1874–1930)
Holmboe   J. Holmboe (1880–1943)
Holub, J.   J. Holub (b. 1930)
Holuby   J. L. Holuby (1836–1923)
Holzm.   T. Holzmann (b. 1843)
Honckeny   G. A. Honckeny (1724–1805)
Hooker   W. J. Hooker (1785–1865)
Hooker fil.   J. D. Hooker (1817–1911)
Hope   J. Hope (1725–1786)
Hoppe   D. H. Hoppe (1760–1846)
Horák   B. Horák (fl. 1900)
Hormuzaki   K. Hormuzaki (1863–1937)
Hornem.   J. W. Hornemann (1770–1841)
Hornsch.   C. F. Hornschuch (1793–1850)
Hornung   E. G. Hornung (1795–1862)
Horvatić   S. Horvatić (b. 1899)
Horvátovszky   S. Horvátovszky (fl. 1770)
Hose, J. A. C.   J. A. C. Hose (d. 1800)
Hossain   M. Hossain (b. 1928)
Host   N. T. Host (1761–1834)
House   H. D. House (1878–1949)
Houtt.   M. Houttuyn (1720–1798)
Houtzagers   G. Houtzagers (1888–1957)
Howard   H. W. Howard (b. 1913)
Howell   T. J. Howell (1842–1912)
Hruby   J. Hruby (1882–1964)
Hubbard   F. T. Hubbard (1875–1962)
Huber, J. A.   J. A. Huber (1867–1914)
Huber-Morath   A. Huber-Morath (b. 1901)
Hudson   W. Hudson (1730–1793)
Hudziok   G. W. Hudziok (b. 1929)
Huet   A. Huet du Pavillon (1829–1907)
Hull   J. Hull (1761–1843)
Hülphers   K. A. Hülphers (1882–1948)
Hülsen   R. Hülsen (1837–1912)
Hultén   E. O. G. Hultén (b. 1894)
Humb.   F. H. A. von Humboldt (1769–1859)
Hussenot   L. C. S. L. Hussenot (1809–1845)
Huter   R. Huter (1834–1909)
Huth   E. Huth (1845–1897)
Hy   F. C. Hy (1853–1918)
Hyl.   N. Hylander (1904–1970)
Iljin   M. M. Iljin (Ilyin) (1889–1967)
Iljinsky, A.   A. P. Iljinsky (1885–1945)

Ingram   C. Ingram (b. 1880)
Insenga   G. Insenga (1815 or 1816–1887)
Ionescu   M. A. Ionescu (b. 1900)
Irmisch   J. F. T. Irmisch (1816–1879)
Irmscher   E. Irmscher (1887–1968)
Ivanina   L. I. Ivanina (b. 1917)
Ivaschin   D. S. Ivaschin (b. 1912)
Iversen   J. Iversen (1904–1971)
Ives   E. Ives (1779–1861)
Jackson, A. B.   A. B. Jackson (1876–1947)
Jackson, B. D.   B. D. Jackson (1846–1927)
Jacq.   N. J. von Jacquin (1727–1817)
Jacq. fil.   J. F. von Jacquin (1766–1839)
Jaeger   H. Jaeger (1815–1890)
Jäggi   J. Jäggi (1829–1894)
Jahandiez   E. Jahandiez (1876–1938)
Jakowatz   A. Jakowatz (b. 1872)
Jalas   J. Jalas (b. 1920)
Jameson   W. Jameson (1796–1873)
Jan   G. Jan (1791–1866)
Janchen   E. Janchen (1882–1970)
Jancz.   E. Janczewski von Glinka (1846–1918)
Janisch.   D. E. Janischewsky (1875–1944)
Janka   V. Janka von Bulcs (1837–1890)
Jaquet   F. Jaquet (1858–1933)
Jardine, N.   N. Jardine (b. 1943)
Jasiewicz   A. Jasiewicz (fl. 1970)
Jaub.   H. F. Jaubert (1798–1874)
Jáv.   S. Jávorka (1883–1961)
Jeanb.   E. M. J. Jeanbernat (1835–1888)
Jensen, G.   J. G. K. Jensen (1818–1886)
Jermy   A. C. Jermy (b. 1932)
Joerg.   E. H. Joergensen (1863–1938)
Joh., K.   K. Johansson (1856–1928)
Johnston, I. M.   I. M. Johnston (1898–1960)
Jones, B. M. G.   B. M. G. Jones (b. 1933)
Jordan   A. Jordan (1814–1897)
Jordanov   D. Jordanov (b. 1893)
Jovet   P. A. Jovet (b. 1896)
Junge   P. Junge (1881–1919)
Junger   E. Junger (fl. 1891)
Juratzka   J. Juratzka (1821–1878)
Jurišić   Z. J. Jurišić (1863–1921)
Juss.   A. L. de Jussieu (1748–1836)
Juss., A.   A. H. L. de Jussieu (1797–1853)
Juz.   S. V. Juzepczuk (1893–1959)
Kalela   A. Kalela (b. 1908)
Kalenicz.   J. O. Kaleniczenko (1805–1876)
Kalm   P. Kalm (1715–1779)
Kaltenb.   J. H. Kaltenbach (1807–1876)
Kanitz   Á. Kanitz (1843–1896)
Kar.   G. S. Karelin (1801–1872)
Kárpáti   Z. E. Kárpáti (1909–1972)
Karsten   G. K. W. H. Karsten (1817–1908)
Kaschm., B.   B. F. Kaschmensky (d. 1909)
Kauffm.   N. N. Kauffmann (Kaufman) (1834–1870)
Kaulfuss   G. F. Kaulfuss (1786–1830)
Kazim.   T. Kazimierski (b. 1924)
Keissler   K. von Keissler (b. 1872)
Keller, J. B.   J. B. von Keller (1841–1897)
Keller, R.   R. Keller (1854–1939)
Kellerer, J.   J. Kellerer (fl. 1905)
Kenyon   W. Kenyon (fl. 1847)
Ker-Gawler   J. B. Ker (J. Gawler) (1764–1842)
Kerner, A.   A. J. Kerner von Marilaun (1831–1898)
Kerner, J.   J. Kerner (1829–1906)

Kiffm. R. Kiffmann (fl. 1952)
Kihlman A. O. Kihlman (Kairamo) (1858–1938)
Kindb. N. C. Kindberg (1832–1910)
Kir. I. P. Kirilow (1821 or 1822–1842)
Kirby M. Kirby (1817–1893)
Kirchner G. Kirchner (1837–1885)
Kirschleger F. R. Kirschleger (1804–1869)
Kit. P. Kitaibel (1757–1817)
Kitagawa M. Kitagawa (b. 1909)
Kitanov B. Kitanov (b. 1912)
Kittel M. B. Kittel (1798–1885)
Klásková, A. A. Klásková (later A. Skalická) (b. 1932)
Klášt. I. Klášterský (b. 1901)
Klebahn H. Klebahn (1859–1942)
Kleopow J. D. Kleopow (1902–1942)
Klett G. T. Klett (d. 1827)
Klika J. Klika (1888–1957)
Klinggr. K. J. von Klinggraeff (1809–1879)
Klink. M. Klinkowski (b. 1904)
Klokov M. V. Klokov (b. 1896)
Klotzsch J. F. Klotzsch (1805–1860)
Kluk K. Kluk (1739–1796)
Knaben G. Knaben (b. 1911)
Knaf J. Knaf (1801–1865)
Knerr E. B. Knerr (1861–1942)
Knight J. Knight (1781–1855)
Knobl. E. F. Knoblauch (1864–1936)
Knoche H. Knoche (1870–1945)
Knowles G. B. Knowles (fl. 1829–1852)
Knuth, R. R. G. P. Knuth (1874–1957)
Koch W. D. J. Koch (G. D. I. Koch) (1771–1849)
Koch, C. C. H. E. Koch (1809–1879)
Koch, L. L. K. A. Koch (b. 1850)
Koch, Walo Walo Koch (1896–1956)
Koehler J. C. G. Koehler (1759–1833)
Koehne B. A. E. Koehne (1848–1918)
Koelle J. L. C. Koelle (1763–1797)
Koelliker R. A. von Koelliker (1817–1905)
Koerte F. Koerte (1782–1845)
Komarov V. L. Komarov (1869–1945)
Kondrat. E. N. Kondratjuk (b. 1914)
König, D. D. König (b. 1909)
Korsh. S. I. Korshinsky (1861–1900)
Košanin N. Košanin (1874–1934)
Koshewn. D. A. Koshewnikow (1858–1882)
Kos.-Pol. B. M. Koso-Poliansky (1890–1957)
Kossko I. N. Kossko (1924–1956)
Kossych V. M. Kossych (b. 1931)
Kostel. V. F. Kosteletzky (1801–1887)
Kotejowa, E. E. Kotejowa (fl. 1963)
Kotov M. I. Kotov (b. 1896)
Kotschy T. Kotschy (1813–1866)
Kotula, A. A. Kotula (1822–1891)
Kovanda M. Kovanda (b. 1936)
Kováts J. Kováts von Szentlelek (1815–1873)
Kožuharov S. I. Kožuharov (b. 1933)
Kralik J. L. Kralik (1813–1892)
Krašan F. Krašan (1840–1907)
Krasch. H. M. Krascheninnikov (1884–1947)
Krause, E. H. L. E. H. L. Krause (1859–1942)
Krause, K. K. Krause (fl. 1958)
Krecz., V. V. I. Kreczetowicz (1901–1942)
Krendl F. Krendl (b. 1926)
Kress A. A. H. L. Kress (b. 1932)
Kreyer G. K. Kreyer (1887–1942)
Křísa B. Křísa (b. 1936)

Krocker A. J. Krocker (1744–1823)
Krok T. O. B. N. Krok (1834–1921)
Krösche E. Krösche (fl. 1912)
Krylov P. N. Krylov (1850–1931)
Krysht. A. N. Kryshtofovicz (1885–1953)
Kühlew. P. E. Kühlewein (1798–1870)
Kuhn M. F. A. Kuhn (1842–1894)
Kulcz. S. Kulczyński (b. 1895)
Kümmerle J. B. Kümmerle (1876–1931)
Kunth C. S. Kunth (1788–1850)
Kuntze, O. K. E. O. Kuntze (1843–1907)
Kunz, H. H. Kunz (b. 1904)
Kunze, G. G. Kunze (1793–1851)
Kupcsok S. Kupcsok (1850–1914)
Kupffer K. R. Kupffer (1872–1935)
Kuprian. L. A. Kuprianova (b. 1914)
Kurtz, F. F. Kurtz (1854–1920)
Kusn. N. I. Kusnezow (Kuznetzov) (1864–1932)
Kütz. F. T. Kützing (1807–1893)
Kuzen. O. I. Kuzeneva (b. 1887)
Kuzinský P. A. von Kuzinský (fl. 1889)
L. C. von Linné (C. Linnaeus) (1707–1778)
L. fil. C. von Linné (1741–1783)
Labill. J. J. H. de Labillardière (1755–1834)
Lacaita C. C. Lacaita (1853–1933)
Laest. L. L. Laestadius (1800–1861)
Lag. M. Lagasca y Segura (1776–1839)
Lagerh. N. G. von Lagerheim (1860–1926)
Lagger F. Lagger (1799–1870)
Lagrèze-Fossat A. R. A. Lagrèze-Fossat (1818–1874)
Laicharding J. N. von Laicharding (1754–1797)
Laínz M. Laínz (b. 1923)
Laínz, J. M. J. M. Laínz (b. 1900)
Lam. J. B. A. P. Monnet de la Marck (1744–1829)
Lamb. A. B. Lambert (1761–1842)
Lamotte M. Lamotte (1820–1883)
Landolt E. Landolt (b. 1926)
Láng, A. F. A. F. Láng (1795–1863)
Lang, O. F. O. F. Lang (1817–1847)
Lange J. M. C. Lange (1818–1898)
Langsd. G. H. von Langsdorff (1774–1852)
Lapeyr. P. Picot de Lapeyrouse (1744–1818)
Lapierre J. M. Lapierre (1754–1834)
La Pylaie A. J. M. B. de la Pylaie (1786–1856)
Larsen, K. K. Larsen (b. 1926)
Lasch W. G. Lasch (1787–1863)
Lasebna A. M. Lasebna (b. 1922)
Latourr. M. A. L. Claret de Latourrette (1729–1793)
Lauche W. Lauche (1827–1882)
Lauth T. Lauth (1758–1826)
Lavrenko E. M. Lavrenko (b. 1900)
Lawalrée A. Lawalrée (b. 1921)
Lawrance M. Lawrance (fl. 1790–1831)
Lawrence G. H. M. Lawrence (b. 1910)
Lawson, C. C. Lawson (1794–1873)
Lawson, P. P. Lawson (d. 1820)
Laxm. E. Laxmann (1737–1796)
Layens G. de Layens (1834–1897)
Laza M. Laza Palacios (b. 1901)
Láz.-Ibiza Blas Lázaro Ibiza (1858–1921)
Lebel J. E. Lebel (1801–1878)
Lecoq H. Lecoq (1802–1871)
Lecoyer C.-J. Lecoyer (1835–1899)
Ledeb. C. F. von Ledebour (1785–1851)
Leers J. D. Leers (1727–1774)
Lees E. Lees (1800–1887)

Lefèvre   L. V. Lefèvre (b. 1810)
Le Gall   N. J. M. le Gall (1787–c. 1860)
Le Grand   A. le Grand (1839–1905)
Lehm.   J. G. C. Lehmann (1792–1860)
Lehm., C. B.   C. B. Lehmann (fl. 1860)
Lehm., J. F.   J. F. Lehmann (fl. 1809)
Leins   P. Leins (b. 1937)
Lej.   A. L. S. Lejeune (1779–1858)
Le Jolis   A. F. le Jolis (1823–1904)
Lemaire   C. A. Lemaire (1801–1871)
Léman   D. S. Léman (1781–1829)
Lemke   W. Lemke (b. 1893)
Lengyel   G. Lengyel (1884–1965)
Lepechin   I. I. Lepechin (1737 or 1740–1802)
Leresche   L. Leresche (1808–1885)
Lesp.   G. Lespinasse (1807–1876)
Less.   C. F. Lessing (1810–1862)
Lester-Garland   L. V. Lester-Garland (1860–1944)
Lestib.   T. G. Lestiboudois (1797–1876)
Letendre   J. B. P. Letendre (1828–1886)
Léveillé   A. A. H. Léveillé (1863–1918)
Levier   E. Levier (1838–1911)
Lewis, P.   P. Lewis (b. 1924)
Ley, A.   A. Ley (1842–1911)
Leybold   F. Leybold (1827–1879)
L'Hér.   C. L. L'Héritier de Brutelle (1746–1800)
Libert   M. A. Libert (1782–1865)
Lid   J. Lid (1886–1971)
Liebl.   F. K. Lieblein (1744–1810)
Liebm.   F. M. Liebmann (1813–1856)
Liljeblad   S. Liljeblad (1761–1815)
Liljefors   A. W. Liljefors (b. 1904)
Lincz.   I. A. Linczevsky (b. 1908)
Lindb., H.   H. Lindberg (1871–1963)
Lindblad   M. A. Lindblad (1821–1899)
Lindblom   A. E. Lindblom (1807–1853)
Lindeb.   C. J. Lindeberg (1815–1900)
Lindem.   E. von Lindemann (1825–1900)
Lindley   J. Lindley (1799–1865)
Lindman   C. A. M. Lindman (1856–1928)
Lindt.   V. H. Lindtner (1904–1965)
Lingelsh.   A. von Lingelsheim (1874–1937)
Link   J. H. F. Link (1767–1851)
Linton, E. F.   E. F. Linton (1848–1928)
Lipsky   V. I. Lipsky (1863–1937)
List   ?F. L. List (fl. 1828–1837)
Litard.   R. V. de Litardière (1888–1957)
Litv.   D. I. Litvinov (Litwinow) (1854–1929)
Lloyd   J. Lloyd (1810–1896)
Loddiges   G. Loddiges (1784–1846)
Loefl.   P. Loefling (1729–1756)
Loesener   L. E. T. Loesener (1865–1941)
Loisel.   J. L. A. Loiseleur-Deslongchamps (1774–1849)
Lojac.   M. Lojacono-Pojero (1853–1919)
Lona   F. Lona (fl. 1949)
Londes   F. W. Londes (1780–1807)
Lönnr.   K. J. Lönnroth (1826–1885)
Lonsing   A. Lonsing (fl. 1939)
Lorent   J. A. von Lorent (1812–1884)
Loret   H. Loret (1810–1888)
Losa   M. Losa España (1893–1966)
Loscos   F. Loscos y Bernál (1823–1886)
Losinsk.   A. S. Losina-Losinskaya (1903–1958)
Loudon   J. C. Loudon (1783–1843)
Lour.   J. de Loureiro (1717–1791)
Löve, Á.   Á. Löve (b. 1916)

Löve, D.   D. Löve (b. 1918)
Lowe   R. T. Lowe (1802–1874)
Lucand   J.-L. Lucand (1821–1896)
Lucé   J. W. L. von Lucé (fl. 1823)
Luckwill   L. C. Luckwill (b. 1914)
Lüdi   W. Lüdi (1888–1968)
Ludwig   C. G. Ludwig (1709–1773)
Luerssen   C. Luerssen (1843–1916)
Luizet   D. Luizet (1851–1930)
Lund, N.   N. Lund (1814–1847)
Lundström, E.   E. Lundström (b. 1882)
Lyka   K. Lyka (b. 1869)
Lynch   R. I. Lynch (1850–1924)
Lynge   B. A. Lynge (1884–1942)
Maack   R. Maack (1825–1886)
Mabille   P. Mabille (1835–1923)
Macbride   J. F. Macbride (b. 1892)
Macfadyen   J. Macfadyen (1798–1850)
Mach.-Laur.   B. Machatschki-Laurich (fl. 1926)
Machule   M. Machule (b. 1899)
Mackay   J. T. Mackay (1775–1862)
Mackenzie   K. K. Mackenzie (1877–1934)
Magne   J. H. Magne (1804–1885)
Magnier   C. Magnier (fl. 1883)
Magnus   P. W. Magnus (1844–1914)
Maillefer   A. Maillefer (b. 1880)
Maire   R. C. J. E. Maire (1878–1949)
Majevski   P. F. Majevski (1851–1892)
Major   C. J. F. Major (1843–1923)
Makino   T. Makino (1862–1957)
Malagarriga   Hermano Teodoro (Ramón de Peñafort Malagarriga) (b. 1904)
Malbr.   A. F. Malbranche (1818–1888)
Malinovski   E. Malinovski (b. 1885)
Malinv.   L. J. E. Malinvaud (1836–1913)
Malladra   A. Malladra (1865–1944)
Malte   M. O. Malte (1880–1933)
Maly, F.   F. de Paula Maly (1823–1891)
Maly, J.   Joseph Karl Maly (1797–1866)
Malý, K.   Karl Malý (1874–1951)
Manden.   I. P. Mandenova (b. 1907)
Mansfeld   R. Mansfeld (1901–1960)
Manton   I. Manton (b. 1904)
Marchesetti   C. de Marchesetti (1850–1926)
Marcos   A. Marcos Pascual (b. 1900)
Marès   P. Marès (1826–1900)
Margot   H. Margot (fl. 1838)
Mariz   J. de Mariz (1847–1916)
Markgraf   F. Markgraf (b. 1897)
Marsden-Jones   E. M. Marsden-Jones (1887–1960)
Marshall   H. Marshall (1722–1801)
Marshall, E. S.   E. S. Marshall (1858–1919)
Marsson   T. F. Marsson (1816–1892)
Mart., C. F. P.   C. F. P. von Martius (1794–1868)
Mart., H.   H. von Martius (1781–1831)
Martelli, U.   U. Martelli (1860–1934)
Martens, M.   M. Martens (1797–1863)
Martin   B. A. Martin (1813–1897)
Martínez   M. Martínez Martínez (1907–1936)
Martrin-Donos   J. V. de Martrin-Donos (1801–1870)
Martyn   T. Martyn (1736–1825)
Marzell   H. Marzell (1885–1970)
Massara   G. F. Massara (1792–1839)
Masters   M. T. Masters (1833–1907)
Máthé   I. Máthé (b. 1911)
Mattei   G. E. Mattei (1865–1943)

Mattf. J. Mattfeld (1895–1951)
Mattuschka H. G. von Mattuschka (1734–1779)
Maurer W. Maurer (b. 1926)
Mauri E. Mauri (1791–1836)
Maxim. K. J. Maximowicz (1827–1891)
Maxon W. R. Maxon (1877–1948)
Mayer, E. E. Mayer (b. 1920)
Mayer, J. J. C. A. Mayer (1747–1801)
McClell. J. McClelland (1805–1883)
McMillan C. McMillan (1867–1929)
McNeill J. McNeill (b. 1933)
Medicus F. C. Medicus (Medikus) (1736–1808)
Medv. J. S. Medvedev (1847–1923)
Meerb. N. Meerburgh (1734–1814)
Meikle R. D. Meikle (b. 1923)
Meinsh. K. K. Meinshausen (1819–1899)
Meissner C. F. Meissner (1800–1874)
Mela A. J. Mela (1846–1904)
Melderis A. Melderis (b. 1909)
Melville R. Melville (b. 1903)
Mendes E. J. S. M. Mendes (b. 1924)
Menéndez Amor J. Menéndez Amor (b. 1916)
Menyh. L. Menyhárth (1849–1897)
Mérat F. V. Mérat (1780–1851)
Merc. E. Mercier (1802–1863)
Merino P. B. Merino y Román (1845–1917)
Merr. E. D. Merrill (1875–1956)
Mert. F. K. Mertens (1764–1831)
Merxm. H. Merxmüller (b. 1920)
Metsch J. C. Metsch (1796–1856)
Mett. G. H. Mettenius (1823–1866)
Metzel. A. Metzelova-Kropáčova (b. 1922)
Metzger J. Metzger (1789–1852)
Meyen F. J. F. Meyen (1804–1840)
Meyer, B. B. Meyer (1767–1836)
Meyer, C. A. C. A. von Meyer (1795–1855)
Meyer, D. E. D. E. Meyer (b. 1926)
Meyer, E. H. F. E. H. F. Meyer (1791–1858)
Meyer, G. F. W. G. F. W. Meyer (1782–1856)
Michalet E. Michalet (1829–1862)
Micheletti L. Micheletti (1844–1912)
Michx A. Michaux (1746–1802)
Michx fil. F. A. Michaux (1770–1855)
Middendorff A. T. von Middendorff (1815–1894)
Miégeville Abbé Miégeville (1814–1901)
Miers J. Miers (1789–1879)
Mikan J. C. Mikan (1743–1814)
Mikan fil. J. C. Mikan (1769–1844)
Milde C. A. J. Milde (1824–1871)
Miller P. Miller (1691–1771)
Miller, J. J. M. Miller (d. 1796)
Millsp. C. F. Millspaugh (1854–1923)
Min. N. A. Miniaev (b. 1909)
Miq. F. A. W. Miquel (1811–1871)
Mirbel C. F. B. Mirbel (1776–1854)
Mitterp. L. Mitterpacher (1734–1818)
Moench C. Moench (1744–1805)
Moessler J. C. Moessler (fl. 1805–1815)
Moesz G. Moesz (1873–1946)
Mohr D. M. H. Mohr (1779–1808)
Moldenke H. N. Moldenke (b. 1909)
Molina J. I. Molina (1740–1829)
Molinier R. Molinier (b. 1899)
Monnard J. P. Monnard (b. 1791)
Monnier, P. P. C. J. Monnier (b. 1922)
Montandon P. J. Montandon (fl. 1856)

Montbret G. Coquebert de Montbret (1805–1837)
Montelucci G. Montelucci (b. 1899)
Monts., P. P. Montserrat Recoder (b. 1920)
Moore, S. S. Le Marchant Moore (1850–1931)
Moq. C. H. B. A. Moquin-Tandon (1804–1863)
Morariu I. Morariu (b. 1905)
Moretti G. Moretti (1782–1853)
Mori A. Mori (1847–1902)
Moric. M. E. Moricand (1779–1854)
Moris G. G. Moris (1796–1869)
Moritzi A. Moritzi (1806–1850)
Morot M. L. Morot (fl. 1885)
Morren C. J. E. Morren (1833–1886)
Morton, C. V. C. V. Morton (b. 1905)
Möschl W. Möschl (b. 1906)
Moss C. E. Moss (1872–1930)
Mössler J. C. Mössler (fl. 1814–1835)
Motelay L. Motelay (1831–1917)
Mouillefert P. Mouillefert (1845–1903)
Mueller, F. F. H. J. von Mueller (1825–1896)
Mueller, P. J. P. J. Mueller (1832–1889)
Muenchh. O. Muenchhausen (1716–1774)
Muhl. G. H. E. Muhlenberg (1753–1815)
Müller Arg. J. Müller of Aargau (Argoviensis) (1828–1896)
Munby G. Munby (1812–1876)
Münch E. Münch (1876–1946)
Munz P. A. Munz (b. 1892)
Murb. S. S. Murbeck (1859–1946)
Murith L. J. Murith (1742–1816)
Murr, J. J. Murr (1864–1932)
Murray J. A. Murray (1740–1791)
Murray, A. A. Murray (c. 1798–1838)
Murray, E. A. E. Murray (b. 1935)
Murray, R. P. R. P. Murray (1842–1908)
Muschler R. Muschler (b. 1883)
Mussin A. A. Mussin-Puschkin (1760–1805)
Mutel A. Mutel (1795–1847)
Mutis J. C. Mutis (1732–1808)
Mygind F. Mygind (1710–1789)
Nakai T. Nakai (1882–1952)
Nasarow M. I. Nasarow (1882–1942)
Nath. A. G. Nathorst (1850–1921)
Naudin C. V. Naudin (1815–1899)
Necker N. J. de Necker (1730–1793)
Nees C. G. D. Nees von Esenbeck (1776–1858)
Nees, T. T. F. L. Nees von Esenbeck (1787–1837)
Neilr. A. Neilreich (1803–1871)
Nejc. I. Nejceff (1870–1913)
Nelson, A. A. Nelson (1859–1952)
Nenukow S. S. Nenukow (1906–1942)
Nestler C. G. Nestler (1778–1832)
Neuman L. M. Neuman (1852–1922)
Neumann, A. A. Neumann (fl. 1960)
Neumayer, H. H. Neumayer (1887–1945)
Neves, J. J. de Barros Neves (b. 1914)
Nevski S. A. Nevski (1908–1938)
Newbould W. W. Newbould (1819–1886)
Newman E. Newman (1801–1876)
Neygenf. F. W. Neygenfind (fl. 1821)
Nicotra L. Nicotra (b. 1846)
Niedenzu F. J. Niedenzu (1857–1937)
Nikif. N. B. Nikiforova (b. 1912)
Nobre A. Nobre (b. 1865)
Noë W. Noë (d. 1858)
Nolte E. F. Nolte (1791–1875)
Nordborg G. Nordborg (b. 1931)

Nordh.   R. Nordhagen (b. 1894)
Nordm.   A. von Nordmann (1803–1866)
Nordstedt   C. F. O. Nordstedt (1838–1924)
Norrlin   J. P. Norrlin (1842–1917)
Norton   J. B. Norton (1877–1938)
Notø   A. Notø (1865–1948)
Noulet   J. B. Noulet (1802–1890)
Novák   F. A. Novák (1892–1964)
Novopokr.   I. V. Novopokrovsky (1880–1951)
Nowacki   E. K. Nowacki (b. 1930)
Nutt.   T. Nuttall (1786–1859)
Nyárády, A.   A. Nyárády (b. 1920)
Nyárády, E. I.   E. I. Nyárády (1881–1966)
Nyl., F.   F. Nylander (1820–1880)
Nyl., W.   W. Nylander (1822–1899)
Nyman   C. F. Nyman (1820–1893)
Oborny   A. Oborny (1840–1924)
Ockendon   D. J. Ockendon (b. 1940)
Oeder   G. C. Oeder (1728–1791)
Ohwi   J. Ohwi (b. 1905)
Oken   L. Oken (1779–1851)
Oliver   D. Oliver (1830–1916)
Olivier   G. A. Olivier (1756–1814)
Onno   M. Onno (b. 1903)
Opiz   P. M. Opiz (1787–1858)
Opperman   P. A. Opperman (d. 1942)
Orlova   N. I. Orlova (b. 1921)
Orph.   T. G. Orphanides (1817–1886)
Örsted   A. S. Örsted (1816–1872)
Ortega   C. Gómez Ortega (1740–1818)
Ortmann   J. Ortmann (b. 1814)
Osbeck   P. Osbeck (1723–1805)
Óskarsson   I. Óskarsson (b. 1892)
Ostenf.   C. E. H. Ostenfeld (1873–1931)
Otth   K. A. Otth (1803–1839)
Otto   C. F. Otto (1783–1856)
Ovcz.   P. N. Ovczinnikov (b. 1903)
Pacher   D. Pacher (1817–1902)
Pacz.   I. K. Paczoski (1864–1942)
Padmore   P. A. Padmore (b. 1929)
Paegle   B. Paegle (fl. 1927)
Palassou   P. B. Palassou (1745–1830)
Palau   P. Palau i Ferrer (1881–1956)
Palhinha   R. T. Palhinha (1871–1957)
Palitz   R. Palitz (fl. 1935)
Pallas   P. S. Pallas (1741–1811)
Pamp.   R. Pampanini (1875–1949)
Pančić   J. Pančić (1814–1888)
Pangalo   K. I. Pangalo (1883–1965)
Pant.   J. Pantocsek (1846–1916)
Panţu   Z. C. Panţu (1866–1934)
Paol.   G. Paoletti (1865–1941)
Pardo   J. Pardo y Sastrón (1822–1909)
Parl.   F. Parlatore (1816–1877)
Parodi   L. R. Parodi (1895–1966)
Parry   W. E. Parry (1790–1855)
Pasquale, C. A.   C. (G.) A. Pasquale (1820–1893)
Passer.   G. Passerini (1816–1893)
Patrin   E. L. M. Patrin (1742–1815)
Patzak   A. Patzak (b. 1930)
Patze   C. A. Patze (1808–1892)
Pau   C. Pau (1857–1937)
Paulin   A. Paulin (1853–1942)
Paulsen   O. V. Paulsen (1874–1947)
Pauquy   C. L. C. Pauquy (1800–1854)
Pavlov   N. V. Pavlov (1893–1971)

Pavón   J. Pavón (1750–1844)
Pawł.   B. Pawłowski (1898–1971)
Pawł., S.   S. Pawłowska (b. 1905)
Pax   F. A. Pax (1858–1942)
Paxton   J. Paxton (1803–1865)
Pedersen, A.   A. Pedersen (b. 1920)
Pennell   F. W. Pennell (1886–1952)
Pénzes   A. Pénzes (b. 1895)
Peola   P. Peola (b. 1869)
Pérard   M. Pérard (1835–1887)
Pérez Lara   J. M. Pérez Lara (1841–1918)
Perf.   I. A. Perfiljew (1882–1942)
Perr.   E. Perrier de la Bâthie (1825–1916)
Pers.   C. H. Persoon (c. 1762–1836)
Personnat   V. Personnat (fl. 1854–1870)
Petagna   V. Petagna (1734–1810)
Péteaux   J. C. J. Péteaux (1840–1896)
Péterfi   M. Péterfi (1875–1922)
Peterm.   W. L. Petermann (1806–1855)
Petitmengin   M. G. C. Petitmengin (1881–1908)
Petrak   F. Petrak (b. 1886)
Petri   F. Petri (1837–1896)
Petrov   V. A. Petrov (1896–1955)
Petrović   S. Petrović (1839–1889)
Petzold   C. E. A. Petzold (1815–1891)
Philcox   D. Philcox (b. 1926)
Philippe   X. Philippe (1802–1866)
Phillips, E. P.   E. P. Phillips (1884–1967)
Phipps, C. J.   C. J. Phipps (1744–1792)
Phitos   D. Phitos (b. 1930)
Pierrat   D. Pierrat (1835–1895)
Pignatti   S. Pignatti (b. 1930)
Pilger   R. K. F. Pilger (1876–1953)
Piller   M. Piller (1733–1788)
Pinzger   P. Pinzger (fl. 1868)
Pio   G. B. Pio (fl. 1813)
Piré   L. A. H. J. Piré (1827–1887)
Pires de Lima   A. Pires de Lima (b. 1886)
Pirona   G. A. Pirona (1822–1895)
Pissjauk.   V. V. Pissjaukowa (b. 1906)
Pitard   C. J. Pitard (1873–1927)
Planchon   J. E. Planchon (1823–1888)
Planellas   J. Planellas Giralt (1821–1888)
Pobed.   E. G. Pobedimova (b. 1898)
Podl.   D. Podlech (b. 1931)
Podp.   J. Podpěra (1878–1954)
Poech   J. Poech (1816–1846)
Poeverlein   H. Poeverlein (1874–1957)
Poggenb.   J. F. Poggenburg (1840–1893)
Pohl   J. B. E. Pohl (1782–1834)
Poiret   J. L. M. Poiret (1755–1834)
Poirion   L. P. Poirion (b. 1901)
Poiteau   P. A. Poiteau (1766–1854)
Pojark.   A. I. Pojarkova (b. 1897)
Polatschek   A. Polatschek (b. 1932)
Pollich   J. A. Pollich (1740–1780)
Pollini   C. Pollini (1782–1833)
Pomel   A. Pomel (1821–1898)
Popl.   G. I. Poplavskaja (Poplawska) (1885–1956)
Popov, M.   M. G. Popov (1893–1955)
Porc.   F. Porcius (1816–1907)
Porsch   O. Porsch (1875–1959)
Porsild, A. E.   A. E. Porsild (b. 1901)
Porta   P. Porta (1832–1923)
Portenschl.   F. E. von Portenschlag-Ledermayer (1772–1822)
Pospichal   E. Pospichal (1838–1905)

APPENDIX I

Post   G. E. Post (1838–1909)
Postr.   S. A. Postrigan (b. 1891)
Pourret   P. A. Pourret de Figeac (1754–1818)
Pozd.   N. G. Pozdeeva (b. 1913)
Praeger   R. L. Praeger (1865–1953)
Prantl   K. A. E. Prantl (1849–1893)
Presl, C.   C. (K.) B. Presl (1794–1852)
Presl, J.   J. S. Presl (1791–1849)
Price   W. R. Price (b. 1886)
Pritchard   N. M. Pritchard (b. 1933)
Pritzel, G. A.   G. A. Pritzel (1815–1874)
Privalova   L. A. Privalova (b. 1919)
Proctor, M. C. F.   M. C. F. Proctor (b. 1929)
Prodan   J. Prodan (1875–1959)
Progel   A. Progel (1829–1889)
Prokh.   J. I. Prokhanov (1902–1964)
Prolongo   P. Prolongo y García (1806–1885)
Puel   T. Puel (1812–1890)
Puget   F. Puget (1829–1880)
Pugsley   H. W. Pugsley (1868–1947)
Pulliat   V. Pulliat (1827–1866)
Purkyně   E. Purkyně (1831–1882)
Pursh   F. T. Pursh (1774–1820)
Putterlick   A. Putterlick (1810–1845)
Quézel   P. Quézel (b. 1926)
Raab   W. Raab (fl. 1819)
Rabenh.   G. L. Rabenhorst (1806–1881)
Racib.   M. Raciborski (1864–1917)
Raddi   G. Raddi (1770–1829)
Radius   J. W. M. Radius (1797–1884)
Rafin.   C. S. Rafinesque-Schmaltz (1783–1840)
Rafn   C. G. Rafn (1769–1808)
Ramond   L. F. E. Ramond de Carbonnières (1753–1827)
Rapaics   R. Rapaics (1885–1953)
Rapin   D. Rapin (1799–1882)
Rau   A. Rau (1784–1830)
Raulin   V. F. Raulin (1815 or 1819–1905)
Raunk.   C. Raunkiær (1860–1938)
Räuschel   E. A. Räuschel (fl. 1772–1797)
Răvărut   M. Răvărut (b. 1907)
Raven, P. H.   P. H. Raven (b. 1936)
Rayss   T. Rayss (1890–1965)
Re, G. F.   G. F. Re (1772–1833)
Rebr.   O. V. Rebristaya (b. 1930)
Rech.   K. Rechinger (1867–1952)
Rech. fil.   K. H. Rechinger (b. 1906)
Rees   A. Rees (1743–1825)
Regel   E. A. von Regel (1815–1892)
Rehder   A. Rehder (1863–1949)
Rehmann   A. Rehmann (1840–1917)
Reichard   J. J. Reichard (1743–1782)
Reichenb.   H. G. L. Reichenbach (1793–1879)
Reichenb. fil.   H. G. Reichenbach (1824–1889)
Rendle   A. B. Rendle (1865–1938)
Renner   O. Renner (1883–1960)
Req.   E. Requien (1788–1851)
Resvoll-Holmsen   H. Resvoll-Holmsen (1873–1943)
Retz.   A. J. Retzius (1742–1821)
Reuss, G.   G. Reuss (1818–1861)
Reuter   G. F. Reuter (1805–1872)
Revel   J. Revel (1811–1887)
Reverchon   E. Reverchon (1835–1914)
Reyn.   A. Reynier (1845–1932)
Ricci   A. M. Ricci (1777–1850)
Richard, A.   A. Richard (1794–1852)
Richard, L. C. M.   L. C. M. Richard (1754–1821)

Richardson, I. B. K.   I. B. K. Richardson (b. 1940)
Richter   H. E. F. Richter (1808–1876)
Richter, J.   J.-A. Richter (1821–1910)
Richter, K.   K. Richter (1855–1891)
Riddelsd.   H. J. Riddelsdell (1866–1941)
Riedl   H. Riedl (b. 1936)
Rigo   G. Rigo (1841–1922)
Rikli   M. A. Rikli (1868–1951)
Rink   H. J. Rink (1819–1893)
Ripart   J. B. M. J. S. E. Ripart (1814–1878)
Risso   J. A. Risso (1777–1845)
Rittener   T. Rittener (fl. 1887)
Rivas Goday   S. Rivas Goday (b. 1905)
Rivas Martínez   S. Rivas Martínez (b. 1935)
Rix   E. M. Rix (b. 1943)
Robert   — Robert (fl. 1838)
Roberts, J.   J. Roberts (1912–1960)
Robill.   L. M. A. Robillard d'Argentelle (d. 1828)
Robinson   B. L. Robinson (1864–1935)
Robson   E. Robson (1763–1813)
Robson, N. K. B.   N. K. B. Robson (b. 1928)
Robyns   W. Robyns (b. 1901)
Rocha Afonso   M. da Luz de Oliveira Tavares Monteiro da Rocha Afonso (b. 1925)
Rochel   A. Rochel (1770–1847)
Rodr.   J. D. Rodriguez (1780–1846)
Roemer   J. J. Roemer (1763–1819)
Roemer, M. J.   M. J. Roemer (fl. 1835–1846)
Roemer, R. de   R. de Roemer (fl. 1852)
Rogow.   A. S. Rogowicz (1812–1878)
Rohde   M. Rohde (1782–1812)
Rohlena   J. Rohlena (1874–1944)
Röhling   J. C. Röhling (1757–1813)
Rohrb.   P. Rohrbach (1847–1871)
Ronniger   K. Ronniger (1871–1954)
Rose   J. N. Rose (1862–1928)
Rosellini   ?F. Rosellini (1817–1873)
Ross, J.   J. Ross (1777–1856)
Ross, R.   R. Ross (b. 1912)
Rossi   M. L. Rossi (1850–1932)
Rössler   W. Rössler (b. 1909)
Rostański   K. Rostański (b. 1930)
Rostock   M. Rostock (fl. 1884)
Rostrup   F. G. E. Rostrup (1831–1907)
Roth   A. W. Roth (1757–1834)
Rothm.   W. Rothmaler (1908–1962)
Rottb.   C. F. Rottboel (Rottbøll) (1727–1797)
Rouleau   E. Rouleau (b. 1916)
Roussine   N. Roussine (formerly N. A. Schostenko) (1889–1968)
Rouy   G. C. C. Rouy (1851–1924)
Roxb.   W. Roxburgh (1751–1815)
Royle   J. F. Royle (1779–1858)
Rozan.   M. A. Rozanova (1885–1957)
Rozeira   A. D. F. Rozeira (b. 1912)
Rudolph, J. H.   J. H. Rudolph (1744–1809)
Rudolphi   K. A. Rudolphi (1771–1832)
Ruiz   H. Ruiz López (1754–1815)
Runemark   H. Runemark (b. 1927)
Rupr.   F. J. Ruprecht (1814–1870)
Russell, A.   A. Russell (?1715–1768)
Russell, P.   P. G. Russell (b. 1889)
Ruthe   J. F. Ruthe (1788–1859)
Rydb.   P. A. Rydberg (1860–1931)
Rylands   T. G. Rylands (1818–1900)
Sa'ad   F. Sa'ad (b. 1925)
Saarson   B. Saarson (later B. Saarsoo) (1899–1969)

Sabine   J. Sabine (1770–1837)
Sabr.   H. Sabransky (1864–1916)
Sadler   J. Sadler (1791–1849)
Sageret   A. Sageret (1763–1851)
Sagorski   E. Sagorski (1847–1929)
Salis   C. Ulysses von Salis-Marschlins (1760–? 1818)
Salisb.   R. A. Salisbury (1761–1829)
Salmon   C. E. Salmon (1872–1930)
Salzm.   P. Salzmann (1781–1851)
Sam.   G. Samuelsson (1885–1944)
Sambuk   F. V. Sambuk (1900–1942)
Samp.   G. A. da Silva Ferreira Sampaio (1865–1937)
Sanadze   K. S. Sanadze (fl. 1946)
Sándor   I. Sándor (b. 1853)
Sandwith   N. Y. Sandwith (1901–1965)
Sanguinetti   P. Sanguinetti (1802–1868)
Santi, G.   G. Santi (1746–1822)
Sapjegin   A. A. Sapjegin (1883–1946)
Sarato   C. Sarato (1830–1893)
Sarg.   C. S. Sargent (1841–1927)
Sarnth.   L. von Sarntheim (1861–1914)
Sart.   G. B. Sartorelli (1780–1853)
Sauer   F. W. H. Sauer (1803–1873)
Saunders   W. W. Saunders (1809–1879)
Sauter   A. E. Sauter (1800–1881)
Sauvage   C. P. F. Sauvage (b. 1909)
Sauzé   C. Sauzé (1815–1889)
Savi   G. Savi (1769–1844)
Savi fil.   P. Savi (1798–1871)
Savigny   M. J. C. Lelorgne de Savigny (1777–1851)
Săvul.   T. Săvulescu (1889–1963)
Scaling   W. Scaling (fl. 1863–1882)
Schaeffer   J. C. Schaeffer (1718–1790)
Schaeftlein   H. Schaeftlein (b. 1886)
Schaffner   W. Schaffner (d. 1882)
Schagerström   J. A. Schagerström (1818–1867)
Schauer   J. K. Schauer (1813–1848)
Scheele   G. H. A. Scheele (1808–1864)
Schellm.   C. Schellmann (fl. 1938)
Schenk   J. A. Schenk (1815–1891)
Schenk, E.   E. Schenk (b. 1880)
Scherb.   J. Scherbius (1769–1813)
Scheutz   N. J. W. Scheutz (1836–1889)
Schiffner   V. F. Schiffner (1862–1944)
Schimper, C.   C. F. Schimper (1803–1867)
Schindler   J. Schindler (b. 1881)
Schinz   H. Schinz (1858–1941)
Schipcz.   N. V. Schipczinski (1886–1955)
Schischkin   B. K. Schischkin (1886–1963)
Schkuhr   C. Schkuhr (1741–1811)
Schlecht.   D. F. L. von Schlechtendal (1794–1866)
Schleicher   J. C. Schleicher (1768–1834)
Schlosser   J. C. Schlosser (1808–1882)
Schmalh.   I. F. Schmalhausen (1849–1894)
Schmeil   O. Schmeil (1860–1943)
Schmid, E.   E. Schmid (b. 1891)
Schmidel   C. C. Schmidel (1718–1792)
Schmidely   A. I. S. Schmidely (1838–1918)
Schmidt, A.   A. Schmidt (b. 1932)
Schmidt, Franz   Franz Schmidt (1751–1834)
Schmidt, F. W.   Franz Willibald Schmidt (1764–1796)
Schmidt, W. L. E.   W. L. E. Schmidt (1804–1843)
Schmidt Petrop., Friedrich   Friedrich Schmidt of St Petersburg (1832–1908)
Schneider, C. K.   C. K. Schneider (1876–1951)
Schneider, U.   U. Schneider (fl. 1962)

Schnittspahn   G. F. Schnittspahn (1810–1865)
Schnizlein   A. C. F. H. C. Schnizlein (1814–1868)
Scholler   F. A. Scholler (1718–1785)
Scholz, H.   H. Scholz (b. 1928)
Scholz, J. B.   J. B. Scholz (fl. 1900)
Schönheit   F. C. H. Schönheit (1789–1870)
Schönl.   S. Schönland (1860–1940)
Schost.   N. A. Schostenko (Desjatova-Schostenko) (later N. Roussine) (1889–1968)
Schotsman   H. D. Schotsman (b. 1921)
Schott   H. W. Schott (1794–1865)
Schousboe   P. K. A. Schousboe (1766–1832)
Schouw   J. F. Schouw (1789–1852)
Schrader   H. A. Schrader (1767–1836)
Schrank   F. von Paula von Schrank (1747–1835)
Schreber   J. C. D. von Schreber (1739–1810)
Schrenk   A. G. von Schrenk (1816–1876)
Schrödinger   R. Schrödinger (1857–1919)
Schroeter   C. Schroeter (1855–1939)
Schultes   J. A. Schultes (1773–1831)
Schultes fil.   J. H. Schultes (1804–1840)
Schultz, C. F.   C. F. Schultz (1765–1837)
Schultz, F. W.   F. W. Schultz (1804–1876)
Schultze, W.   W. Schultze (fl. 1894)
Schulz, A.   A. A. H. Schulz (1862–1922)
Schulz, O. E.   O. E. Schulz (1874–1936)
Schulze, M.   C. T. M. Schulze (1841–1915)
Schum., K.   K. M. Schumann (1851–1904)
Schummel   T. E. Schummel (1785–1848)
Schur   P. J. F. Schur (1799–1878)
Schuster   R. Schuster (b. 1935)
Schwantes   G. Schwantes (1881–1960)
Schwarz, A.   A. Schwarz (1852–1915)
Schwarz, O.   O. Schwarz (b. 1900)
Schwegler   H. W. Schwegler (b. 1929)
Schweigger   A. F. Schweigger (1783–1821)
Schweinf.   G. A. Schweinfurth (1836–1925)
Schwertschl.   J. Schwertschleger (1853–1924)
Scop.   G. A. Scopoli (1723–1788)
Sebastiani   A. Sebastiani (1782–1821)
Sebeók   A. Sebeók de Szent-Miklós (fl. 1780)
Seem.   B. C. Seemann (1825–1871)
Seemen   K. O. von Seemen (1838–1910)
Séguier   J. F. Séguier (1703–1784)
Seidl   W. B. Seidl (1773–1842)
Selin   G. Selin (1813–1862)
Sell, P. D.   P. D. Sell (b. 1929)
Semen., N.   N. Z. Semenova-Tjan-Schanskaja (1906–1960)
Semler   C. Semler (1875–1955)
Sendtner   O. Sendtner (1813–1859)
Sennen   Frère Sennen (E. M. Grenier-Blanc) (1861–1937)
Ser.   N. C. Seringe (1776–1858)
Serg.   L. P. Sergievskaja (1897–1970)
Serg., E.   E. V. Sergievskaja (C. V. Sergievskaja) (b. 1926)
Serres   J. J. Serres (d. 1858)
Sesler   L. Sesler (d. 1785)
Seub.   M. A. Seubert (1818–1878)
Shivas   M. G. Shivas (b. 1926)
Shull   G. H. Shull (1874–1954)
Shuttlew., R. J.   R. J. Shuttleworth (1810–1874)
Sibth.   J. Sibthorp (1758–1796)
Sieber   F. W. Sieber (1789–1844)
Siebert   A. Siebert (1854–1923)
Siebold   P. F. von Siebold (1796–1866)
Siegfr.   H. Siegfried (1837–1903)
Sikura   J. J. Sikura (fl. 1960)

Silliman   B. Silliman (1779–1864)
Silva, M.   M. da Silva (b. 1916)
Silva, P.   A. R. Pinto da Silva (b. 1912)
Sim, R.   R. Sim (1791–1878)
Simkovics   L. Simkovics (later L. von Simonkai) (1851–1910)
Simmler   G. Simmler (b. 1884)
Simmons   H. G. Simmons (1866–1943)
Simon primus, E.   E. Simon (1848–1924)
Simon secundus, E.   E. Simon (1871–1967)
Simon, T.   T. Simon (b. 1926)
Simonkai   L. von Simonkai (1851–1910)
Sims   J. Sims (1749–1831)
Sint.   P. E. E. Sintenis (1847–1907)
Širj.   G. I. Širjaev (Schirjaev) (1882–1954)
Skalická   A. Skalická (b. 1932)
Skalický   V. Skalický (b. 1930)
Skeels   H. C. Skeels (1873–1934)
Skvortsov, A.   A. K. Skvortsov (b. 1920)
Slavíková   Z. Slavíková (b. 1935)
Slosson   M. Slosson (b. 1873)
Sm.   J. E. Smith (1759–1828)
Sm., A. R.   A. R. Smith (b. 1938)
Sm., G. E.   G. E. Smith (1805–1881)
Sm., H.   K. A. H. Smith (b. 1889)
Sm., W. W.   W. W. Smith (1875–1956)
Small   J. K. Small (1869–1938)
Smejkal   M. Smejkal (b. 1927)
Smirnov   P. A. Smirnov (b. 1896)
Snogerup   S. E. Snogerup (b. 1929)
Soczava   V. B. Soczava (b. 1905)
Soják   J. Soják (b. 1936)
Solac.   T. Solacolu (1876–1940)
Solander   D. C. Solander (1733–1782)
Sole   W. Sole (c. 1739–1802)
Solemacher   J. V. L. A. G. Solemacher-Antweiler (b. 1889)
Solms-Laub.   H. M. C. L. F. Solms-Laubach (1842–1915)
Soltok.   M. Soltoković (fl. 1901)
Sommer.   I. Sommerauer (d. 1854)
Sommerf.   S. C. Sommerfelt (1794–1838)
Sommier   C. P. S. Sommier (1848–1922)
Sonder   O. W. Sonder (1812–1881)
Song.   A. Songeon (1826–1905)
Soó   R. de Soó (b. 1903)
Soška   T. Soška (fl. 1938)
Sosn., D.   D. I. Sosnowsky (1885–1952)
Soulié   J. A. Soulié (1868–1930)
Sowerby   J. Sowerby (1757–1822)
Soyer-Willemet   H. F. Soyer-Willemet (1791–1867)
Spach   E. Spach (1801–1879)
Speg.   C. Spegazzini (1858–1926)
Spenner   F. K. L. Spenner (1798–1841)
Sprengel   K. P. J. Sprengel (1766–1833)
Spribille   F. J. Spribille (1841–1921)
Spring   F. A. Spring (1814–1872)
Spruner   W. von Spruner (1805–1874)
Sprygin   I. I. Sprygin (1873–1942)
Stace   C. A. Stace (b. 1938)
Stadlm.   J. Stadlmann (b. 1881)
Standley   P. C. Standley (1884–1963)
Stankov   S. S. Stankov (1892–1962)
Stapf   O. Stapf (1857–1933)
Stearn   W. T. Stearn (b. 1911)
Stebbins   G. L. Stebbins (b. 1906)
Steele   W. E. Steele (1816–1883)
Stefani   C. de Stefani (1851–1924)
Stefanov   B. Stefanov (b. 1894)

Stefánsson   S. Stefánsson (1863–1921)
Steinh.   A. Steinheil (1810–1839)
Steininger   H. Steininger (1856–1891)
Stephan   C. F. Stephan (1757–1814)
Stern, F. C.   F. C. Stern (1884–1967)
Sternb.   C. M. von Sternberg (1761–1838)
Sterneck   J. von Sterneck (1864–1941)
Sterns, E. E.   E. E. Sterns (1846–1926)
Steudel   E. G. von Steudel (1783–1856)
Steven   C. Steven (1781–1863)
St-Hil.   A. C. F. P. de Saint-Hilaire (1779–1853)
Stiefelhagen   H. Stiefelhagen (fl. 1910)
St-Lager   J. B. Saint-Lager (1825–1912)
Stocks   J. E. Stocks (1822–1854)
Stoj.   N. Stojanov (1883–1968)
Stokes   J. Stokes (1755–1831)
Störk   A. Störk (1741–1803)
Strail   C. A. Strail (1808–1893)
Strempel   J. K. F. Strempel (1800–1872)
Strobl   P. G. Strobl (1846–1910)
Stroh   G. Stroh (b. 1864)
Stur   D. Stur (1827–1893)
Sturm   J. Sturm (1771–1848)
Suard   V. Suard (fl. 1839)
Suckow, G.   G. A. Suckow (d. 1867)
Sudre   H. Sudre (1862–1918)
Sudworth   G. B. Sudworth (1864–1927)
Suess.   K. Suessenguth (1893–1955)
Suk.   V. N. Sukaczev (Sukatschew) (1880–1967)
Sumnev.   G. P. Sumnevicz (1909–1947)
Sünd.   F. Sündermann (1864–1946)
Suter   J. R. Suter (1766–1827)
Sutton   C. Sutton (1756–1846)
Sutulov   A. N. Sutulov (fl. 1914)
Svob.   B. Svoboda
Swartz   O. P. Swartz (1760–1818)
Sweet   R. Sweet (1783–1835)
Swingle   W. T. Swingle (1871–1952)
Syme   J. T. I. Boswell Syme (formerly Boswell) (1822–1888)
Symons   J. Symons (1778–1851)
Szabó   Z. Szabó (1882–1944)
Szafer   W. Szafer (1886–1970)
Szov.   A. J. Szovits (d. 1830)
Szysz.   I. Szyszylowicz (1857–1910)
Tacik, T.   T. Tacik (b. 1926)
Talbot   W. H. F. Talbot (1800–1877)
Taliev   V. I. Taliev (1872–1932)
Tanfani   E. Tanfani (1848–1892)
Tardieu-Blot   M. L. Tardieu-Blot (b. 1902)
Taubert   P. H. W. Taubert (1862–1897)
Tausch   I. F. Tausch (1793–1848)
Taylor, P.   P. G. Taylor (b. 1926)
Temesy   E. Temesy (fl. 1957)
Ten.   M. Tenore (1780–1861)
Tepl.   F. A. Teplouchow (1845–1905)
Terechov   A. F. Terechov (b. 1890)
Terpó   A. Terpó (b. 1925)
Terracc., N.   N. Terracciano (1837–1921)
Tesseron   Y.-A. Tesseron (1831–1925)
Texidor   J. Texidor y Cos (1836–1885)
Teyber   A. Teyber (1846–1913)
Thell.   A. Thellung (1881–1928)
Thév.   A. V. Théveneau (1815–1876)
Thib.   ?E. Thibaud (fl. 1785)
Thielens   A. Thielens (1833–1874)
Thomas   E. Thomas (1788–1859)

Thommen   E. Thommen (1880–1961)
Thomson   T. Thomson (1817–1878)
Thore   J. Thore (1762–1823)
Thouars   L. M. A. Aubert du Petit-Thouars (1758–1831)
Thouin   A. Thouin (1747–1824)
Thuill.   J. L. Thuillier (1757–1822)
Thunb.   C. P. Thunberg (1743–1828)
Timb.-Lagr.   P. M. E. Timbal-Lagrave (1819–1888)
Timm   J. C. Timm (1734–1805)
Tineo   V. Tineo (1791–1856)
Tiss.   P. G. Tissière (1828–1868)
Tod.   A. Todaro (1818–1892)
Tolm.   A. I. Tolmatchev (b. 1903)
Tommasini   M. G. S. de Tommasini (1794–1879)
Top.   S. Topali (fl. 1938)
Țopa   E. Țopa (b. 1900)
Topitz   A. Topitz (b. 1857)
Torrey   J. Torrey (1796–1873)
Tourlet   E.-H. Tourlet (1843–1907)
Townsend   F. Townsend (1822–1905)
Trabut   L. Trabut (1853–1929)
Tratt.   L. Trattinick (1764–1849)
Trautv.   E. R. von Trautvetter (1809–1889)
Travis   W. G. Travis (1877–1958)
Trelease   W. Trelease (1857–1945)
Trev.   L. C. Treviranus (1779–1864)
Trevisan   V. B. A. Trevisan de Saint-Léon (1817–1897)
Trew   C. J. Trew (1695–1769)
Tropea   C. Tropea (fl. 1910)
Trotzky   P. Kornuch-Trotzky (1803–1877)
Truchaleva   N. A. Truchaleva (b. 1927)
Tryon jun., R. M.   R. M. Tryon jun. (b. 1916)
Tubilla   T. Andrés y Tubilla (1859–1882)
Tuntas   B. Tuntas (b. 1871)
Turcz.   N. S. Turczaninow (1796–1864)
Turesson   G. W. Turesson (1892–1970)
Turner, D.   D. Turner (1775–1858)
Turpin   P. J. F. Turpin (1775–1840)
Turra   A. Turra (1730–1796)
Turrill   W. B. Turrill (1890–1961)
Tutin   T. G. Tutin (b. 1908)
Tuzson   J. Tuzson (1870–1941)
Tzvelev   N. N. Tzvelev (b. 1925)
Ucria   Bernadino da Ucria (Michelangelo Aurifici) (1739–1796)
Uechtr.   R. F. C. von Uechtritz (1838–1886)
Ugr.   K. A. Ugrinsky (fl. 1920)
Uhrová   A. Hrabětová-Uhrová (b. 1900)
Ujhelyi   J. Ujhelyi (b. 1910)
Ulbr.   E. Ulbrich (1879–1952)
Underw.   J. Underwood (d. 1834)
Unger   F. J. A. N. Unger (1800–1870)
Ung.-Sternb.   F. Ungern-Sternberg (d. 1885)
Urban   I. Urban (1848–1931)
Urum.   I. K. Urumoff (1856–1937)
Utinet   —Utinet (fl. 1839)
Vacc.   L. Vaccari (1873–1951)
Vahl   M. H. Vahl (1749–1804)
Vahl, J.   J. L. M. Vahl (1796–1854)
Valck.-Suringar   — Valckenier-Suringar (1865–1932)
Valdés   B. Valdés Castrillón (b. 1942)
Valentine   D. H. Valentine (b. 1912)
Vandas   K. Vandas (1861–1923)
Vandelli   D. Vandelli (1735–1816)
Van den Bosch   R. B. van den Bosch (1810–1862)
Van Hall   H. C. van Hall (1801–1874)
Van Houtte   L. B. van Houtte (1810–1876)

Van Ooststr.   S. J. van Ooststroom (b. 1906)
Vasc.   J. de Carvalho e Vasconcellos (1897–1972)
Vassil., V.   V. N. Vassiliev (b. 1890)
Vassilcz.   I. T. Vassilczenko (b. 1903)
Vatke   G. K. W. Vatke (1849–1889)
Vaucher   J. P. E. Vaucher (1763–1841)
Velen.   J. Velenovský (1858–1949)
Vendr.   X. Vendrely (fl. 1895)
Vent.   E. P. Ventenat (1757–1808)
Vent, W.   W. Vent (b. 1920)
Verdcourt   B. Verdcourt (b. 1925)
Verlot   J.-B. Verlot (1825–1891)
Verlot, B.   P. B. L. Verlot (1836–1897)
Vest   L. C. von Vest (1776–1840)
Vestergren   J. T. C. Vestergren (1875–1930)
Vicioso, C.   C. Vicioso Martínez (b. 1887)
Vidal   L. M. Vidal
Vierh.   F. Vierhapper (1876–1932)
Vig.   L. G. A. Viguier (1790–1867)
Vigineix   G. Vigineix (d. 1877)
Vigo   J. Vigo Bonada (b. 1937)
Vill.   D. Villars (Villar) (1745–1814)
Villar, H. del   E. Huguet del Villar (1871–1951)
Vilmorin   P. L. F. L. de Vilmorin (1816–1860)
Vilmorin, R. de   R.-P.-V. de Vilmorin (b. 1905)
Vindt   J. Vindt (b. 1915)
Vis.   R. de Visiani (1800–1878)
Vitman   F. Vitman (1728–1806)
Viv.   D. Viviani (1772–1840)
Vogel   B. C. Vogel (1745–1825)
Vogel, T.   J. R. T. Vogel (1812–1841)
Vogler   J. A. Vogler (1746–1816)
Voigt   J. O. Voigt (1798–1843)
Volk.   A. Volkart (1873–1951)
Vollmann   F. Vollmann (1858–1917)
Vorosch.   V. N. Voroschilov (b. 1908)
Voss   A. Voss (1857–1924)
Vuk.   L. F. Vukotinović (1813–1893)
Vved.   A. I. Vvedensky (b. 1898)
Wagner, H.   J. Wagner (H. Wagner) (1870–1955)
Wagner, R.   R. Wagner (fl. 1887)
Wahlberg   P. F. Wahlberg (1800–1877)
Wahlenb.   G. Wahlenberg (1780–1851)
Waisb.   A. Waisbecker (1835–1916)
Waldst.   F. A. von Waldstein-Wartemberg (1759–1823)
Wale   R. S. Wale (d. 1952)
Walker, S.   S. Walker (b. 1924)
Wall.   N. Wallich (1786–1854)
Wallr.   K. F. W. Wallroth (1792–1857)
Walpers   W. G. Walpers (1816–1853)
Walsh   R. Walsh (1772–1852)
Walter   T. Walter (1740–1789)
Walters   S. M. Walters (b. 1920)
Wangenh.   F. A. J. von Wangenheim (1747–1800)
Wangerin   W. L. Wangerin (1884–1938)
Warburg   O. Warburg (1859–1938)
Warburg, E. F.   E. F. Warburg (1908–1966)
Warming   J. E. B. Warming (1841–1924)
Wartm.   F. B. Wartmann (1830–1902)
Watson, H. C.   H. C. Watson (1804–1881)
Watson, S.   S. Watson (1826–1892)
Watson, W. C. R.   W. C. R. Watson (1885–1954)
Watt   D. A. P. Watt (1830–1917)
Watzl   B. Watzl (b. 1886)
Webb   P. B. Webb (1793–1854)
Webb, D. A.   D. A. Webb (b. 1912)

Weber   G. H. Weber (1752–1828)
Weber fil.   F. Weber (1781–1823)
Weddell   H. A. Weddell (1819–1877)
Weevers   T. Weevers (1875–1952)
Weigel   C. E. von Weigel (1748–1831)
Weihe   K. E. A. Weihe (1779–1834)
Weiller   M. Weiller (1880–1945)
Wein, K.   K. Wein (1883–1968)
Weinm.   J. A. Weinmann (1782–1858)
Weiss   E. Weiss (1837–1870)
Welden   F. L. von Welden (1782–1853)
Welw.   F. Welwitsch (1806–1872)
Wendelberger   G. Wendelberger (b. 1915)
Wendelbo   P. E. B. Wendelbo (b. 1927)
Wenderoth   G. W. F. Wenderoth (1774–1861)
Wendl.   J. C. Wendland (1755–1828)
Wendl. fil.   H. L. Wendland (1792–1869)
Wenzig   T. Wenzig (1824–1892)
Werner   K. Werner (b. 1928)
Wesmael, A.   A. Wesmael (1832–1905)
Wessely   I. Wessely (fl. 1960)
Westcott   F. Westcott (d. 1861)
Weston   R. Weston (1733–1806)
Wettst.   R. von Wettstein (1863–1931)
Wettst., F.   F. von Wettstein (1895–1945)
Wheldon   J. A. Wheldon (1862–1924)
White   J. White (c. 1750–1832)
Whitehead   F. H. Whitehead (b. 1913)
Wibel   A. W. E. C. Wibel (1775–1814)
Wibiral   E. Wibiral (1878–1950)
Wichura   M. E. Wichura (1817–1866)
Wickens   G. E. Wickens (b. 1927)
Widder   F. Widder (b. 1892)
Widmer   E. Widmer (1862–1952)
Wieg.   K. McK. Wiegand (1873–1942)
Wierzb.   P. Wierzbicki (1794–1847)
Wiesb.   J. Wiesbaur (1836–1906)
Wight   R. Wight (1796–1872)
Wiinst.   K. J. F. Wiinstedt (1878–1964)
Wikstr.   J. E. Wikström (1789–1856)
Wilce   J. H. Wilce (b. 1931)
Wilczek   E. Wilczek (1867–1948)
Wilensky   D. G. Wilensky (1892–1959)
Willd.   C. L. Willdenow (1765–1812)
Williams, F. N.   F. N. Williams (1862–1923)
Willk.   H. M. Willkomm (1821–1895)
Wilmott   A. J. Wilmott (1888–1950)
Wilson, E. H.   E. H. Wilson (1876–1930)
Wimmer   C. F. H. Wimmer (1803–1868)
Winge   Ö. Winge (1886–1964)
Winter, F.   F. B. Winter (1795–1869)
Winter, N.   N. A. Winter (1898–1934)
Winterl   J. J. Winterl (1739–1809)
Wirsing   A. L. Wirsing (1734–1797)

Wirtgen   P. W. Wirtgen (1806–1870)
Wissjul.   E. D. Wissjulina (b. 1902)
With.   W. Withering (1741–1799)
Wittm.   M. C. L. Wittmack (1839–1929)
Wittrock   V. B. Wittrock (1839–1914)
Wohlf.   R. Wohlfahrt (1830–1888)
Wolf, N. M.   N. M. von Wolf (1724–1784)
Wolf, T.   F. T. Wolf (1841–1921)
Wolff, D.   D. Wolff (fl. ?1809)
Wolff, H.   H. Wolff (1866–1929)
Wolfner   W. Wolfner (fl. 1858)
Wollaston   G. B. Wollaston (1814–1899)
Wolley-Dod   A. H. Wolley-Dod (1861–1948)
Wolny   A. R. Wolny (d. ? 1829)
Wołoszczak   E. Wołoszczak (1835–1918)
Wood, D.   D. Wood (b. 1939)
Wood, W.   W. Wood (1745–1808)
Woods, J.   J. Woods (1776–1864)
Woodson   R. E. Woodson (1904–1963)
Wormsk.   M. Wormskiold (1783–1845)
Woronow   J. N. Woronow (Voronov) (1874–1931)
Woynar   H. K. Woynar (1865–1917)
Wulf   E. V. Wulf (E. W. Wulff, E. V. Vul'f) (1885–1941)
Wulfen   F. X. von Wulfen (1728–1805)
Wünsche   J. G. Wünsche (fl. 1804)
Yeo   P. F. Yeo (b. 1929)
Yuncker   T. G. Yuncker (1891–1964)
Zabel   H. Zabel (1832–1912)
Zahar.   C. Zahariadi (b. 1901)
Zahlbr.   J. Zahlbruckner (1782–1850)
Zamels   A. Zamels (Zamelis) (1897–1943)
Zanted.   G. Zantedeschi (1773–1846)
Zapał.   H. Zapałowicz (1852–1917)
Zawadzki   A. Zawadzki (1798–1868)
Zefirov   B. M. Zefirov (1915–1957)
Zelen.   N. M. Zelenetzky (1859–1923)
Zenari   S. Zenari (b. 1896)
Zerov   D. K. Zerov (1895–1971)
Žertová   A. Chrtková-Žertová (b. 1930)
Zett., J. W.   J. W. Zetterstedt (1785–1874)
Zeyher   C. L. P. Zeyher (1799–1858)
Zimm., W.   W. Zimmermann (b. 1892)
Zimmeter   A. Zimmeter (1848–1897)
Zinger, N.   N. Zinger (1866–1923)
Zinger, V.   V. J. Zinger (1836–1907)
Zinn   J. G. Zinn (1727–1759)
Zinserl.   Y. D. Zinserling (1894–1938)
Ziz   J. B. Ziz (1779–1829)
Zodda   G. Zodda (1877–1968)
Zoega   J. Zoega (1742–1788)
Zoz   I. G. Zoz (b. 1903)
Zsák   Z. Zsák (b. 1880)
Zucc.   J. G. Zuccarini (1797–1848)
Zuccagni   A. Zuccagni (1754–1807)

APPENDIX II

KEY TO THE ABBREVIATIONS OF TITLES OF BOOKS CITED IN VOLUME 3

Aiton, *Hort. Kew.*
W. Aiton, *Hortus kewensis, or a Catalogue of the Plants culti-vated in the Royal Botanic Garden at Kew.* Ed. 1. London. 1789. (1–3 in 1789.) Ed. 2, by W. T. Aiton. London. 1810–1813. (1 in 1810; 2 & 3 in 1811; 4 in 1812; 5 in 1813. Cf. F. A. Stafleu, *Taxonomic Literature* 4 (1967).)

Albert & Reyn., *Coup d'Oeil Fl. Toulon Hyères*
A. Albert & A. Reynier, *Coup d'Oeil sur la Flore de Toulon et d'Hyères.* Draguignan. 1891.

All., *Auct. Fl. Pedem.*
C. Allioni, *Auctarium ad Floram pedemontanam cum Notis et Emendationibus.* Augustae Taurinorum. 1789.

All., *Auct. Syn. Stirp. Horti Taur.*
C. Allioni, *Auctarium ad Synopsim methodicam Stirpium Horti reg. taurinensis.* [Torino.] 1773. (Cf. J. E. Dandy, *Taxon* 19: 617–626 (1970) & F. A. Stafleu, *Taxonomic Literature* 6 (1967).)

All., *Fl. Pedem.*
C. Allioni, *Flora pedemontana, sive Enumeratio methodica Stirpium indigenarum Pedemontii.* Augustae Taurinorum. 1785. (1–3 in 1785.)

Andrz., *Enum. Pl. Podol.*
A. L. Andrzejowski, Исчисленіе Растеній Подольской Губерніи и смежныхъ съ нею Мѣстъ [*Isčislenie Rastenij Podol'skoj Gubernii i smežnikh s neju Mest*]./*Enumeratio Plantarum sponte in Gubernico Podolico et Locis adjacentibus crescentium.* Kiev. 1860–1862.

Arcangeli, *Comp. Fl. Ital.*
G. Arcangeli, *Compendio della Flora italiana, ossia Manuale per la Determinazione delle Piante che trovansi selvatiche od inselvatichite nell'Italia e nelle Isole adiacenti.* Ed. 1. Torino. 1882. Ed. 2. Torino & Roma. 1894.

Asso, *Enum. Stirp. Arag.*
I. J. de Asso y del Rio, *Enumeratio Stirpium in Aragonia noviter detectarum.* Aragoniae. 1784.

Asso, *Intr. Oryctogr. Arag.*
I. J. de Asso y del Rio, *Introductio in Oryctographiam, et Zoologiam Aragoniae.* [Madrid.] 1784.

Asso, *Syn. Stirp. Arag.*
I. J. de Asso y del Rio, *Synopsis Stirpium indigenarum Aragoniae.* Massiliae. 1779.

Bab., *Man. Brit. Bot.*
C. C. Babington, *Manual of British Botany.* Ed. 1. London. 1843. Ed. 2. London. 1847. Ed. 3. London. 1851. Ed. 4. London. 1856. Ed. 5. London. 1862. Ed. 6. London. 1867. Ed. 7. London. 1874. Ed. 8. London. 1881. Ed. 9. London. 1904. Ed. 10, edit. A. J. Wilmott. London. 1922.

Bailey, L. H., *Man. Cult. Pl.*
L. H. Bailey, *Manual of cultivated Plants.* Ed. 1. New York. 1924. Ed. 2. New York. 1949.

Baillon, *Hist. Pl.*
H. E. Baillon, *Histoire des Plantes.* Paris. 1866–1895. (1: pp. 1–88 in 1866; pp. 89–288 in 1868; pp. 289–488, i–xi in 1869; 2: pp. 1–71 in 1869; pp. 73–428 in 1870; pp. 429–512 in 1872; 3: pp. 1–292 in 1871; pp. 293–545 in 1872; 4: pp. 1–264 in 1872; pp. 265–520 in 1873; 5: pp. 1–256 in 1874; pp. 257–516 in 1874

or 1875; 6: pp. 1–92 in 1875; pp. 93–216 in 1875 or 1876; pp. 217–304 in 1876; pp. 305–523 in 1877; 7: pp. 1–256 in 1879; pp. 257–546 in 1880; 8: pp. 1–316 in 1882; pp. 317–515 in 1885; 9: pp. 1–80 in 1886; pp. 81–224 in 1887; pp. 225–491 in 1888; 10: pp. 1–112 in 1888; pp. 113–220 in 1889; pp. 221–402 in 1890; pp. 403–476 in 1891; 11: pp. 1–304 in 1891; pp. 305–494 in 1892; 12: pp. 1–134 in 1892; pp. 135–334 in 1893; pp. 335–611 in 1894; 13: pp. 1–244 in 1894; pp. 245–523 in 1895. Cf. F. A. Stafleu, *Taxonomic Literature* 12–13 (1967).)

Barbarich, *Vyzn. Rosl. Ukr.*
A. Barbarich, Визначник Рослин України [*Vyznačnyk Roslyn Ukrajiny*]. Ed. 1. Kyjiv. 1950. Ed. 2, by D. K. Zerov *et al.* (edit.). Kyjiv. 1965.

Bartl., *Ind. Sem. Horti Goetting.*
F. G. Bartling, *Index Seminum Horti goettingensis.* Göttingen. By F. G. Bartling in 1848.

Barton, W., *Veg. Mat. U.S.*
W. P. G. Barton, *Vegetable Materia medica of the United States.* Philadelphia. 1817–1819. (1(1 & 2): pp. 1–148, tt. 1–12 in 1817; 1(3 & 4): pp. 149–273, tt. 13–24 in 1818; 2(1): pp. 1–74, tt. 25–30 in 1818; 2(2): pp. 75–239, tt. 31–50 in 1819. Cf. F. A. Stafleu, *Taxonomic Literature* 15 (1967).)

Bast., *Essai Fl. Maine Loire*
T. Bastard, *Essai sur la Flore du Département de Maine et Loire.* Angers. 1809.

Batt. & Trabut, *Fl. Algér.*
J. A. Battandier & L. Trabut, *Flore de l'Algérie. Ancienne Flore d'Alger transformée, contenant la Description de toutes les Plantes signalées jusqu'à ce Jour comme spontanées en Algérie.* ***Dicot.****, Dicotylédones* by J. A. Battandier. Alger & Paris. 1888–1890. (Pp. 1–184 in 1888; pp. 185–576 in 1889; pp. 577–825 in 1890. Cf. W. T. Stearn, *Jour. Soc. Bibl. Nat. Hist.* 1: 145 (1938).) ***Monocot.****, Monocotylédones* by J. A. Battandier & L. Trabut. Alger & Paris. 1895.

Baumg., *Enum. Stirp. Transs.*
J. C. G. Baumgarten, *Enumeratio Stirpium magno Transsilvaniae Principatui...indigenarum.* Vindobonae. 1816. (1–3 in 1816.)

Beck, G., *Fl. Nieder-Österr.*
G. Beck von Mannagetta, *Flora von Nieder-Österreich.* Wien. 1890–1893. (1: pp. 1–430 in 1890; 2(1): pp. 431–[894] in 1892; 2(2): *Allgem. Th.* pp. 1–74 & 895–1396 in 1893.)

Bentham, *Lab. Gen. Sp.*
G. Bentham, *Labiatarum Genera et Species.* London. 1832–1836. (Pp. 1–60 in 1832, pp. 61–322 in 1833; pp. 323–644 in 1834; pp. 645–783, i–lxviii in 1835; Preface in 1836. Cf. F. A. Stafleu, *Taxonomic Literature* 26–27 (1967).)

Berchtold & Opiz, *Ökon.-Techn. Fl. Böhm.*
F. von Berchtold & P. M. Opiz, *Ökonomisch-technische Flora Böhmens.* Prag. 1836–1843. (1 in 1836; 2(1) in 1838; 2(2) in 1839; 3(1) in 1841; 3(2) in 1843.) 1 by F. von Berchtold & W. B. Seidl.

Berchtold & J. Presl, *Rostlinář*
F. von Berchtold & J. S. Presl, *O Přirozenosti Rostlin aneb Rostlinář.* W Praze. 1823–1835. (1 in 1823; 2 in 1825; 3 in 1830–1835.)

Bertol., *Amoen.*

A. Bertoloni, *Amoenitates italicae, sistentes Opuscula ad Rem herbariam et Zoologiam Italiae spectantia.* Bononiae. 1819.

Bertol., *Elench. Pl. Hort. Bot. Bon.*

A. Bertoloni, *Elenchus Plantarum vivarum quas cum aliis vivis Plantis commutandas exhibet Hortus botanicus Archigymnasii bononiensis Anno MDCCCXX.* Bononiae. 1820.

Bertol., *Fl. Ital.*

A. Bertoloni, *Flora italica, sistens Plantas in Italia et Insulis circumstantibus sponte nascentes.* Bononiae. 1833–1854. (1–10; for dates cf. F. A. Stafleu, *Taxonomic Literature* 32–33 (1967) & V. Giacomini, *Regn. Veg.* **71**: 85–98 (1970).)

Bertol., *Rar. Lig. Pl.*

A. Bertoloni, *Rariorum Liguriae Plantarum Decas* 1 [2, 3]. Genuae, Pisis. 1803–1810. (**1**, Genuae in 1803; **2**, Pisis in 1806; **3**, Pisis in 1810. **2** & **3** titled *Rariorum Italiae Plantarum…*)

Bertol., *Virid. Bon.*

A. Bertoloni, *Viridarii bononiensis Vegetabilia cum aliis Vegetabilibus commutanda ad Annum* 1824. Bononiae. 1824.

Besser, *Cat. Pl. Jard. Krzemien.*

W. S. J. G. von Besser, *Catalogue des Plantes du Jardin botanique de Krzemieniec en Volhynie.* Ed. 1. [Krzemieniec.] 1810. Ed. 2, titled *Catalogue des Plantes du Jardin botanique du Gymnase de Volhynie à Krzemieniec.* Krzemieniec. 1811.

Besser, *Enum. Pl. Volhyn.*

W. S. J. G. von Besser, *Enumeratio Plantarum hucusque in Volhynia, Podolia, Gub. kiioviensi, Bessarabia cis-tyraica et circa Odessam collectarum, simul cum Observationibus in Primitias Florae Galiciae austriacae.* Vilnae. 1822.

Besser, *Prim. Fl. Galic.*

W. S. J. G. von Besser, *Primitiae Florae Galiciae austriacae utriusque.* Viennae. 1809. (**1** & **2** in 1809.)

Bieb., *Beschr. Länd. Terek Casp.*

F. A. Marschall von Bieberstein, *Beschreibung der Länder zwischen den Flüssen Terek und Kur am caspischen Meere. Mit einem botanischen Anhang.* Frankfurt. 1800.

Bieb., *Cent. Pl.*

F. A. Marschall von Bieberstein, *Centuria Plantarum rariorum Rossiae meridionalis, praesertim Tauriae et Caucasi, Iconibus Descriptionibusque illustrata.* Charkoviae & Petropoli. 1810–1843. (**1**, Charkoviae in 1810; **2**, Petropoli in 1832–1843.)

Bieb., *Fl. Taur.-Cauc.*

F. A. Marschall von Bieberstein, *Flora taurico-caucasica, exhibens Stirpes phaenogamas in Chersoneso taurica et Regionibus caucasicis sponte crescentes.* Charkoviae. 1808–1819. (**1** & **2** in 1808; **3**, *Supplementum*, in 1819.)

Bischoff, *Beitr. Fl. Deutschl.*

G. W. Bischoff, *Beiträge zur Flora Deutschlands und der Schweiz.* Heidelberg. 1851.

Biv., *Stirp. Rar. Sic. Descr.*

A. de Bivona-Bernardi, *Stirpium rariorum minusque cognitarum in Sicilia sponte provenientium Descriptiones nonnullis Iconibus auctae.* Panormi. 1813–1816. (**1** in 1813; **2** in 1814; **3** in 1815; **4** in 1816.)

Bodard, *Mém. Vér. Cymb.*

P. H. H. Bodard, *Mémoire sur la Véronique cymbalaire.* Pise. 1798.

Boenn., *Prodr. Fl. Monast.*

C. M. F. von Boenninghausen, *Prodromus Florae monasteriensis Westphalorum.* Monasterii. 1824.

Boiss., *Diagn. Pl. Or. Nov.*

P. E. Boissier, *Diagnoses Plantarum orientalium novarum.* Lipsiae & Parisiis. 1843–1859. (Ser. 1, **1**(1–3) in 1843; **1**(4–5) in 1844; **1**(6–7) in 1846 or 1847; **2**(8–11) in 1849; **2**(12–13) in 1853; ser. 2, **3**(1) in 1853; **3**(2–3) in 1856; **3**(4) in 1859; **3**(5) in 1856; **3**(6) in 1859. Cf. F. A. Stafleu, *Taxonomic Literature* 42 (1967).)

Boiss., *Elenchus*

P. E. Boissier, *Elenchus Plantarum novarum minusque cognitarum, quas in Itinere hispanico legit…* Genevae. 1838.

Boiss., *Fl. Or.*

P. E. Boissier, *Flora orientalis.* Basileae, Genevae & Lugduni. 1867–1884. (**1** Basileae & Genevae; **2–5** Basileae, Genevae & Lugduni. **1** in 1867; **2** in 1872; **3** in 1875; **4**: pp. 1–280 in 1875; pp. 281–1276 in 1879; **5**: pp. 1–428 in 1882; pp. 429–868 in 1884.) *Suppl.*, Supplementum. Basileae, Genevae & Lugduni. 1888.

Boiss., *Pl. Atticae Heldr.*

P. E. Boissier, *Plantae atticae Heldreichianae.* [*Exsiccata.*] 1844–1846.

Boiss., *Pl. Or. Nov.*

P. E. Boissier, *Plantarum orientalium novarum Decas prima* [*secunda*]. Genevae. 1875. (**1** & **2** in 1875.)

Boiss., *Voy. Bot. Midi Esp.*

P. E. Boissier, *Voyage botanique dans le Midi de l'Espagne pendant l'Année 1837.* Paris. 1839–1845. (**1**: pp. 1–40 in 1839; pp. 41–96 in 1844; pp. 97–248, i–x in 1845; **2**: pp. 1–96 in 1839; pp. 97–352 in 1840; pp. 353–544 in 1841; pp. 545–640 in 1842; pp. 641–710 in 1844; pp. 711–757 in 1845. Cf. B. R. Baum, *Taxon* **17**: 720–724 (1968).)

Boiss. & Reuter, *Diagn. Pl. Nov. Hisp.*

P. E. Boissier & G. F. Reuter, *Diagnoses Plantarum novarum hispanicarum, praesertim in Castella nova lectarum.* Genevae. 1842.

Boiss. & Reuter, *Pugillus*

P. E. Boissier & G. F. Reuter, *Pugillus Plantarum novarum Africae borealis Hispaniaeque australis.* Genevae. 1852.

Bolós, A., *Veg. Com. Barcelon.*

A. de Bolós, *Vegetación de las Comarcas barcelonesas.* Barcelona. 1950.

Bonnier, *Fl. Compl. Fr.*

G. E. M. Bonnier, *Flore complète illustrée en Couleurs de France, Suisse et Belgique, comprenant la Plupart des Plantes d'Europe.* Paris, Neuchâtel & Bruxelles. 1911–1935. (**1–8** Paris, Neuchâtel & Bruxelles; **9** Paris & Bruxelles; **10–13** Paris. **1**: pp. 1–48, tt. 1–24 in 1911; pp. 49–120, tt. 25–60 in 1912; **2** in 1913; **3** in 1914; **4** in 1921; **5** in 1922; **6** in 1923; **7** in 1924; **8** in 1926; **9** in 1927; **10** in 1929; **11**: pp. 1–80, tt. 601–630 in 1930; pp. 81–161, tt. 631–660 in 1931; **12**: pp. 1–24, tt. 661–672 in 1932; pp. 25–80, tt. 673–696 in 1933; pp. 81–133, tt. 697–721 in 1934; **13** in 1935. **7–13** partly by R. Douin. Cf. F. A. Stafleu, *Taxonomic Literature* 44–45 (1967).)

Borbás, *Balaton Növényföldr.*

V. von Borbás, *A Balaton Tavának és Partmellékének Növényföldrajza és Edényes Növényzete.* Budapest. 1900. (Comprising *A Balaton tudományos Tanulmányozásának Eredményei* 2; *A Balaton Flórája* 2.)

Borcau, *Fl. Centre Fr.*

A. Boreau, *Flore du Centre de la France.* Ed. 1. Paris. 1840. (**1** & **2** in 1840.) Ed. 2. Paris. 1849. (**1** & **2** in 1849.) Ed. 3. Paris. 1857. (**1** & **2** in 1857.)

Boreau, *Graines Recolt. Jard. Bot. Angers*

A. Boreau, *Graines recoltées au Jardin botanique de la Ville d'Angers en 1854.* Angers. 1855.

Börner, *Fl. Deutsche Volk*

C. J. B. Börner, *Eine Flora für das deutsche Volk.* Leipzig. 1912.

Bory, *Expéd. Sci. Morée*

J. B. G. M. Bory de Saint-Vincent, *Expédition scientifique de Morée.* 3(2) *Botanique.* Paris. 1832–1833. (Pp. 1–336 in 1832; pp. 337–367 in 1833.)

Br., R., *Prodr. Fl. Nov. Holl.*

R. Brown, *Prodromus Florae Novae Hollandiae et Insulae Van*

Diemen. Ed. 1. Londini. 1810. Ed. 2. Londini. 1821. Ed. 3, titled *Editio secunda...* by C. G. D. Nees von Esenbeck. Norimbergae. 1827.

Brandza, *Prod. Fl. Române*
D. Brandza, *Prodromul Floreï române*. Bucuresci. 1879–1883. (Pp. vii–lxx, 1–128 in 1879 or 1880; pp. i–vi, lxxi–lxxxiv, 129–568 in 1883. Cf. W. T. Stearn, *Jour. Bot.* (*London*) 79: 191 (1941).)

Briq., *Lab. Alp. Marit.*
J. I. Briquet, *Les Labiées des Alpes maritimes*. Genève & Bâle. 1891–1895. (Pp. 1–184 in 1891; pp. [185]–408 in 1893; pp. [409]–587 in 1895.)

Briq., *Prodr. Fl. Corse*
J. I. Briquet, *Prodrome de la Flore corse*. Genève, Bâle, Lyon & Paris. 1910–1955. (1, Genève, Bâle & Lyon in 1910; 2(1), Genève, Bâle & Lyon in 1913; 2(2), Paris in 1936; 3(1), Paris in 1938; 3(2), Paris in 1955.) 2(2) & 3(1 & 2) by R. V. de Litardière.

Britton, E. E. Sterns & Poggenb., *Prelim. Cat.*
Preliminary Catalogue of Anthophyta and Pteridophyta reported as growing spontaneously within one hundred Miles of New York City. Compiled by the following Committee of the Torrey Botanical Club... The Nomenclature revised and corrected by N. L. Britton, E. E. Sterns and Justus F. Poggenburg. New York. 1888.

Brot., *Fl. Lusit.*
F. Avellar Brotero, *Flora lusitanica*. Olisipone. 1804. (1 & 2 in 1804.)

Brot., *Phyt. Lusit.*
F. Avellar Brotero, *Phytographia Lusitaniae selectior*. Ed. 1. Olisipone. 1800. Ed. 2. Olisipone. 1801. Ed. 3. Olisipone. 1816–1827. (1 in 1816; 2 in 1827.)

Bunge, *Beitr. Kenntn. Fl. Russl.*
A. A. von Bunge, *Beitrag zur Kenntniss der Flor Russlands und der Steppen Central-Asiens*. [St. Petersburg.] 1852. (For date cf. F. A. Stafleu, *Taxonomic Literature* 68 (1967).)

Bunge, *Enum. Pl. Chin. Bor.*
A. A. von Bunge, *Enumeratio Plantarum quas in China boreali collegit...* Petropoli. 1833. (For date cf. W. T. Stearn, *Jour. Bot.* (*London*) 79: 63–64 (1941).)

Cav., *Descr. Pl.*
A. J. Cavanilles, *Descripcion de las Plantas, que D. Antonio Josef Cavanilles demonstró en las Lecciones públicas del Año 1801, precedida de los Principios elementales de la Botánica*. Ed. 1. Madrid. 1802–1803. (Pp. i–cxxxvi in 1802; pp. 1–284, titled *Géneros y Especies de Plantas demonstradas en las Lecciones públicas del Año* 1801, in 1802; pp. 285–625, titled *... del Año* 1802, in 1802 or 1803.) Ed. 2. Madrid. 1827.

Cav., *Elench. Pl. Horti Matrit.*
A. J. Cavanilles, *Elenchus Plantarum Horti regii matritensis. Anno M. DCCC. III*. Matriti. 1803.

Cav., *Icon. Descr.*
A. J. Cavanilles, *Icones et Descriptiones Plantarum, quae aut sponte in Hispania crescunt, aut in Hortis hospitantur*. Matriti. 1791–1801. (1 in 1791; 2 in 1793; 3 in 1795; 4: pp. 1–36 in 1797; pp. 37–82 in 1798; 5 in 1799; 6: pp. 1–40 in 1800; pp. 41–97 in 1801. Cf. F. A. Stafleu, *Taxonomic Literature* 79–80 (1967).)

Cav., *Monad. Class. Diss. Dec.*
A. J. Cavanilles, *Monadelphiae Classis Dissertationes decem*. Matriti. 1785–1790. (Pp. 1–48 in 1785; pp. 49–106+4 in *app.* in 1786; pp. 107–266 in 1787; pp. 267–354 in 1788; pp. 355–414 in 1789; pp. 415–464 in 1790. Cf. F. A. Stafleu, *Taxonomic Literature* 79 (1967).)

Čelak., *Prodr. Fl. Böhm.*
L. J. Čelakovsky, *Prodromus der Flora von Böhmen*. (Published as *Arch. Naturw. Landes. Böhm.* 1, *Bot. Abth.*: 1–112 (1867); 2(2), *Bot. Abth.*: 113–388 (1871); 3, *Bot. Abth.*: 389–691

(1875); 4, *Bot. Abth.*: 693–955 (1881). Cf. J. Futák & K. Domin, *Bibliografia k Flóre ČSR* 138–139 (1960).)

Cesati, *Sagg. Geogr. Bot. Fl. Lomb.*
V. de Cesati, *Saggio su la Geografia botanica e su la Flora della Lombardia*. Milano. 1844.

Chav., *Monogr. Antirrh.*
E. L. Chavannes, *Monographie des Antirrhinées*. Paris & Lausanne. 1833.

Chaz., *Dict. Jard., Suppl.*
L. M. Chazelles de Prizy. *Dictionnaire des Jardiniers. Supplément*. Paris, Metz. 1790. (1, Paris; 2, Metz.)

Clairv., *Man. Herb. Suisse*
J. P. de Clairville, *Manuel d'Herborisation en Suisse et en Valais*. Winterthur. 1811.

Clapham, Tutin & E. F. Warburg, *Fl. Brit. Is.*
A. R. Clapham, T. G. Tutin & E. F. Warburg, *Flora of the British Isles*. Ed. 1. Cambridge. 1952. Reprinted 1957 & 1958 with some corrections. Ed. 2. Cambridge. 1962.

Corb., *Nouv. Fl. Normand.*
L. Corbière, *Nouvelle Flore de Normandie*. Caen. 1894.

Coutinho, *Fl. Port.*
A. X. Pereira Coutinho, *Flora de Portugal (Plantas vasculares)*. Ed. 1. Paris, Lisboa, Rio de Janeiro, S. Paulo & Bello Horizonte. 1913. Ed. 2, by R. T. Palhinha. Lisboa. 1939.

Crantz, *Stirp. Austr.*
H. J. N. von Crantz, *Stirpes austriacae*. Ed. 1. Viennae & Lipsiae. 1762–1767. (1 in 1762; 2 in 1763; 3 in 1767.) Ed. 2. Viennae. 1769. (1 & 2, with continuous pagination, in 1769.)

Curtis, *Fl. Lond.*
W. Curtis, *Flora londinensis; or, Plates and Descriptions of such Plants as grow wild in the Environs of London*. London. 1746–1799. For detailed account and dates of plates cf. A. Stevenson, *Catalogue of botanical Books in the Collection of Rachel McMasters Miller Hunt*. 2(2): 389–412 (1961) & F. A. Stafleu, *Taxonomic Literature* 90–92 (1967).

Cutanda, *Fl. Comp. Madrid*
V. Cutanda, *Flora compendiada de Madrid y su Provincia*. Madrid. 1861.

Cyr., *Pl. Rar. Neap.*
D. Cyrillo, *Plantarum rariorum Regni Neapolitani Fasciculus primus [secundus]*. Neapoli. 1788–1792. (1 in 1788; 2 in 1792.)

Dalla Torre, *Anleit. Beob. Bestimm. Alpenpfl.*
K. W. von Dalla Torre, *Anleitung zur Beobachtung und zum Bestimmen der Alpenpflanzen* (constitutes Vol. 2 of C. von Sonklar *et al.*, *Anleitung zu wissenschaftlichen Beobachtungen auf Alpenreisen*). Wien. 1882.

DC., *Cat. Pl. Horti Monsp.*
A. P. de Candolle, *Catalogus Plantarum Horti botanici monspeliensis*. Monspelii. 1813.

DC., *Icon. Pl. Gall. Rar.*
A. P. de Candolle, *Icones Plantarum Galliae rariorum*. Parisiis. 1808.

DC., *Prodr.*
A. P. de Candolle, *Prodromus Systematis naturalis Regni vegetabilis*. Parisiis. 1824–1874. (1 in 1824; 2 in 1825; 3 in 1828; 4 in 1830; 5 in 1836; 6 in 1838; 7(1) in 1838; 7(2) in 1839; 8 in 1844; 9 in 1845; 10 in 1846; 11 in 1847; 12 in 1848; 13(1) in 1852; 13(2) in 1849; 14: pp. 1–492 in 1856; pp. 493–706 in 1857; 15(1) in 1864; 15(2): pp. 1–188 in 1862; pp. 189–1286 in 1866; 16(1) in 1869; 16(2): pp. 1–160 in 1864; pp. 161–691 in 1868; 17 in 1873. Index 1–4 in 1843; 5–7(1) in 1840; 7(2)–13 in 1858–1859; 14–17 in 1874. Cf. F. A. Stafleu, *Taxonomic Literature* 74–75 (1967).)

Déchy, *Kaukasus*
M. Déchy, *Kaukasus. Reisen und Forschungen im kaukasischen Hochgebirge*. Berlin. 1905–1907. (1 in 1905; 2 in 1906; 3 in 1907.)

Degen, *Fl. Veleb.*
A. von Degen, *Flora velebitica*. Budapest. 1936–1938. (**1** in 1936; **2** in 1937; **3** & **4** in 1938.)

Degen, *Egy Új Ajuga Fajról*
A. von Degen, *Egy új* Ajuga *Fajról*. Budapest. 1896.

Delarbre, *Fl. Auvergne*
A. Delarbre, *Flore d'Auvergne, ou Recueil des Plantes de cette ci-devant Province*. Ed. 1. Clermont-Ferrand. 1795. Ed. 2. Riom & Clermont-Ferrand. 1800.

Delile, *Sem. Hort. Bot. Monspel.*
A. R. Delile, *Semina Anni* 1836 *quae Hortus botanicus regius monspeliensis pro mutua Commutatione offert*. Monspelii. 1837.

De Not., *Repert. Fl. Ligust.*
G. de Notaris, *Repertorium Florae ligusticae*. Taurini. 1844. (For date cf. F. A. Stafleu, *Taxonomic Literature* 339 (1967).)

Desf., *Fl. Atl.*
R. L. Desfontaines, *Flora atlantica, sive Historia Plantarum, quae in Atlante, Agro tunetano et algeriensi crescunt*. Parisiis. 1798–1799. (**1** in 1798; **2**: pp. 1–160 in 1798; pp. 161–458 in 1799. Cf. F. A. Stafleu, *Taxonomic Literature* 103 (1967).)

Desv., *Jour. Bot. Appl.*
A. N. Desvaux (edit.), *Journal de Botanique, appliquée à l'Agriculture à la Pharmacie, à la Médecine et aux Arts*. Paris. 1813–1816. (**1** & **2** in 1813; **3**(1 & 2) in 1814; **3**(3) in 1816; **3**(4) in 1814; **3**(5) in 1816; **4** in 1814.)

Desv., *Jour. Bot. Rédigé*
A. N. Desvaux *et al.* (edit.), *Journal de Botanique, rédigé par une Société de Botanistes*. Paris. 1808–1809. (**1**: pp. 1–192 in 1808; pp. 193–384 in 1809; **2** in 1809. Cf. F. A. Stafleu, *Taxonomic Literature* 104–105 (1967).)

Domin, *Pl. Čechosl. Enum.*
K. Domin, *Plantarum Čechoslovakiae Enumeratio Species vasculares indigenas et introductas exhibens*. Praha. 1935. Reprinted as *Preslia* **13–15** (1936).

Don fil., G., *Gen. Syst.*
G. Don, *A general System of Gardening and Botany*. Also titled *A general History of dichlamydeous Plants*. London. 1831–1838. (**1** in 1831; **2** in 1832; **3** in 1834; **4** in 1837–1838. Cf. F. A. Stafleu, *Taxonomic Literature* 111 (1967).)

Dörfler, *Herb. Norm.*
I. Dörfler, *Herbarium normale. Conditum a F. Schultz, dein continuatum a K. Keck, nunc editum per I. Dörfler*. Vindobonae. 1894–1915. (**31–56** in 1894–1915.)

Dostál, *Květena ČSR*
J. Dostál, *Květena ČSR*. Praha. 1948–1950. (Preface etc., pp. 1–64 in 1950; pp. 1–800 in 1948; pp. 801–1488 in 1949; pp. 1489–2269 in 1950.)

Druce, *Brit. Pl. List*
G. C. Druce, *List of British Plants*. Ed. 1. Oxford. 1908. Ed. 2, titled *British Plant List*. Arbroath. 1928.

Druce, *Fl. Berks.*
G. C. Druce, *The Flora of Berkshire*. Oxford. 1898.

Duby, *Bot. Gall.*
J. E. Duby, *Aug. Pyrami de Candolle Botanicon gallicum, seu Synopsis Plantarum in Flora gallica descriptarum. Editio secunda. Ex Herbariis et Schedis candollianis propriisque digestum a J. E. Duby*. Paris. 1828–1830. (**1** in 1828; **2** in 1830.) This is Ed. 2 of Lam. & DC., *Syn. Pl. Fl. Gall.*

Duchartre, *Rev. Bot.*
P. E. S. Duchartre (edit.), *Revue botanique. Recueil mensuel renfermant l'Analyse des Travaux publiés en France et à l'Étranger sur la Botanique et sur ses Applications à l'Horticulture, l'Agriculture, la Médecine etc*. Paris. 1845–1847.

Dum.-Courset, *Bot. Cult.*
G. L. M. Dumont de Courset, *Le Botaniste Cultivateur*. Ed. 1.

Paris. 1802–1805. (**1–4** in 1802; **5** in 1805.) Ed. 2. Paris. 1811–1814. (**1–6** in 1811; **7** in 1814.)

Dumort., *Fl. Belg.*
B. C. J. Dumortier, *Florula belgica, Operis majoris Prodromus. Staminacia*. Tornaci Nerviorum. 1827.

Dunal, *Hist. Solanum*
M. F. Dunal, *Histoire naturelle, médicale et économique des* Solanum. Paris, Strasbourg & Montpellier. 1813.

Durieu, *Pl. Astur. Exsicc.*
M. C. Durieu de Maisonneuve, *Plant. select. hispano lusit. Sect. Ia asturicae. Anno* 1835 *collecta*. (*Exsicc.*) 1836.

D'Urv., *Enum.*
J. S. C. D. D'Urville, *Enumeratio Plantarum, quas in Insulis Archipelagi aut Littoribus Ponti euxini Annis* 1819 *et* 1820 *collegit atque detexit....* Parisiis. 1822.

Ehrh., *Beitr. Naturk.*
J. F. Ehrhart, *Beiträge zur Naturkunde*. Hannover & Osnabrück. 1787–1792. (**1** in 1787; **2** & **3** in 1788; **4** in 1789; **5** in 1790; **6** in 1791; **7** in 1792. Cf. F. A. Stafleu, *Taxonomic Literature* 126–127 (1967).)

Eichw., *Pl. Nov. It. Casp.-Cauc.*
K. E. von Eichwald, *Plantarum novarum vel minus cognitarum quas in Itinere caspio-caucasico observavit....* Vilnae & Lipsiae. 1831–1833. (Pp. 1–18 in 1831; pp. 19–42 in 1833.)

Engler, *Pflanzenreich*
H. G. A. Engler (edit.), *Das Pflanzenreich. Regni vegetabilis Conspectus*. (For authors and dates of individual volumes cf. M. T. Davis, *Taxon* **6**: 161–182 (1957) & F. A. Stafleu, *Taxonomic Literature* 133–146 (1967).) In citation the number of the *Heft* is placed first, followed by the systematic numbers (both Roman and Arabic) in brackets.

Engler & Prantl, *Natürl. Pflanzenfam.*
H. G. A. Engler & K. A. E. Prantl, *Die natürlichen Pflanzenfamilien nebst ihren Gattungen und wichtigeren Arten, insbesondere den Nutzpflanzen*. Ed. 1. Leipzig. 1887–1915. (**2**(1): pp. 1–172 in 1887; pp. 173–262 in 1889; **2**(2): pp. 1–96 in 1887; pp. 96–130 in 1888; **2**(3): pp. 1–144 in 1887; pp. 145–168 in 1889; **2**(4): pp. 1–48 in 1887; pp. 49–78 in 1888; **2**(5): pp. 1–144 in 1887; pp. 145–162 in 1888; **2**(6): pp. 1–144 in 1888; pp. 145–244 in 1889; **3**(1): pp. 1–48 in 1887; pp. 49–144 in 1888; pp. 145–289 in 1889; **3**(1a): pp. 1–48 in 1892; pp. 49–130 in 1893; **3**(1b) in 1889; **3**(2): pp. 1–96 in 1888; pp. 97–144 in 1889; pp. 145–281 in 1891; **3**(2a): pp. 1–48 in 1890; pp. 49–142 in 1891; **3**(3): pp. 1–48 in 1888; pp. 49–112 in 1891; pp. 113–208 in 1892; pp. 209–256 in 1893; pp. 257–396 in 1894; **3**(4): pp. 1–94 in 1890; pp. 95–362 in 1896; **3**(5): pp. 1–96 in 1890; pp. 97–128 in 1891; pp. 129–224 in 1892; pp. 225–272 in 1893; pp. 273–416 in 1895; pp. 417–468 in 1896; **3**(6): pp. 1–96 in 1890; pp. 97–240 in 1893; pp. 241–340 in 1895; **3**(6a): pp. 1–96 in 1893; pp. 97–254 in 1894; **3**(7): pp. 1–48 in 1892; pp. 49–241 in 1893; **3**(8): pp. 1–48 in 1894; pp. 49–144 in 1897; pp. 145–274 in 1898; **4**(1): pp. 1–96 in 1889; pp. 97–144 in 1890; pp. 145–183 in 1891; **4**(2): pp. 1–48 in 1892; pp. 49–310 in 1895; **4**(3a): pp. 1–48 in 1891; pp. 49–96 in 1893; pp. 97–224 in 1895; pp. 225–320 in 1896; pp. 321–384 in 1897; **4**(3b): pp. 1–96 in 1891; pp. 97–144 in 1893; pp. 145–240 in 1894; pp. 241–378 in 1895; **4**(4) in 1891; **4**(5): pp. 1–80 in 1889; pp. 81–272 in 1890; pp. 273–304 in 1892; pp. 305–368 in 1893; pp. 369–402 in 1894; *Gesamtregister* pp. 1–320 in 1898; pp. 321–462 in 1899; *Nachtrag* 1 in 1897; *Nachtrag* 2 in 1900; *Nachtrag* 3: pp. 1–192 in 1906; pp. 193–288 in 1907; pp. 289–379 in 1908; *Nachtrag* 4: pp. 1–192 in 1914; pp. 193–381 in 1915. Cf. F. A. Stafleu, *Taxonomic Literature* 146–149 (1967).) Ed. 2. Leipzig and Berlin. 1925→. (**13, 14a, 14e, 15a, 16b, 16c, 17b, 18a, 19a, 19b I, 19c, 20b** and **21** at Leipzig; **14d, 17a II** and **20d** at Berlin. **13** in 1926; **14a** in 1926; **14d** in 1956; **14e** in 1940;

15a in 1930; **16b** in 1935; **16c** in 1934; **17a II** in 1959; **17b** in 1936; **18a** in 1930; **19a** in 1931; **19b I** in 1940; **19c** in 1931; **20b** in 1942; **20d** in 1953; **21** in 1925.) Photographic reprint. Berlin. 1960.

Etlinger, *De Salvia*
A. E. Etlinger, *De Salvia Dissertatio inauguralis.* Erlangae. 1777.

Fedtsch., O. & B., *Consp. Fl. Turkestan.*
O. A. & B. A. Fedtschenko, *Conspectus Florae turkestanicae./* Перечень Растеній, дико растущихъ въ Русскомъ Туркестанѣ [*Perečen' Rastenij, diko rastuščikh v Russkom Turkestaně*]. Sanktpeterburg & Jur'ev. 1906–1916. (**1,** Sanktpeterburg; **2–6,** Jur'ev. **1** in 1906; **2 & 3** in 1909; **4** in 1911; **5** in 1913; **6** in 1916.)

Fedtsch., B. & Flerow, *Fl. Evr. Ross.*
B. A. Fedtschenko & A. F. Flerow, Флора Европейской Россіи [*Flora Evropejskoj Rossii*]. S.-Peterburg. 1908–1910. (Pp. 1–286 in 1908; pp. 287–710 in 1909; pp. 711–1204 in 1910.)

Fiori & Béguinot, *Sched. Fl. Ital. Exsicc.*
A. Fiori & A. Béguinot, *Schedae ad Floram italicam exsiccatam.* Florentiae.

Fiori & Paol., *Fl. Anal. Ital.*
A. Fiori & G. Paoletti, *Flora analitica d'Italia.* Padova. 1896–1909. (1: pp. i–c in 1908; pp. 1–256 in 1896; pp. 257–610 in 1898; 2: pp. 1–224 in 1900; pp. 225–304 in 1901; pp. 305–493 in 1902; 3: pp. 1–272 in 1903; pp. 273–524 in 1904; 4: pp. 1–217, index pp. 1–16 in 1907; index pp. 17–330 in 1908; 5 in 1909.)

Fischer, *Cat. Jard. Gorenki*
F. E. L. von Fischer, *Catalogue du Jardin des Plantes du Comte Alexis de Razoumoffsky à Gorenki près de Moscou.* Ed. 1. Moscou. 1808. Ed. 2. Moscou. 1812.

Fischer & C. A. Meyer, or **Fischer, C. A. Meyer & Avé-Lall.,** *Ind. Sem. Hort. Petrop.*
F. E. L. von Fischer & C. A. von Meyer, or F. E. L. von Fischer, C. A. von Meyer & J. L. E. Avé-Lallemant, or J. L. E. Avé-Lallemant, *Index Seminum, quae Hortus botanicus imperialis petropolitanus pro mutua Commutatione offert.* Petropoli. 1835–1850. (For dates cf. F. A. Stafleu, *Taxonomic Literature* 153 (1967).)

Fomin, *Fl. RSS Ucr.*
A. V. Fomin, *Flora RSS Ucr./Flora Reipublicae Sovieticae Socialisticae Ucrainicae.* Ed. 1. Kioviae. 1936–1965. (**1,** edit. A. V. Fomin in 1936; **2,** edit. E. I. Bordzilowski & E. M. Lavrenko in 1940. Titled Флора УРСР [*Flora URSR*]: **3,** edit. M. I. Kotov & A. I. Barbarich in 1950; **4,** edit. M. I. Kotov in 1952; **5,** edit. M. V. Klokov & E. D. Wissjulina in 1953; **6,** edit. D. K. Zerov in 1954; **7,** edit. M. V. Klokov & E. D. Wissjulina in 1955; **8,** edit. M. I. Kotov & A. I. Barbarich in 1958; **9,** edit. M. I. Kotov in 1960; **10,** edit. M. I. Kotov in 1961; **11,** edit. E. D. Wissjulina in 1962; **12,** edit. E. D. Wissjulina in 1965.) Ed. 2. Kioviae. 1938. (**1,** edit. E. I. Bordzilowski in 1938.)

Font Quer, *Ill. Fl. Occ.*
P. Font Quer, *Illustrationes Florae occidentalis, quae ad Plantas Hispaniae, Lusitaniae et Mauritaniae novas vel imperfecte cognitas, spectant auctore....* Barcelona. 1926.

Font Quer, *Iter Maroc. (Sched.)*
P. Font Quer, *Iter maroccanum (Schedae).* 1928–1930.

Forskål, *Fl. Aegypt.*
P. Forskål, *Flora aegyptiaco-arabica.* Hauniae. 1775.

Forster, G., *Fl. Ins. Austral. Prodr.*
J. G. A. Forster, *Florulae Insularum australium Prodromus.* Gottingae. 1786.

Fourn., P., *Quatre Fl. Fr.*
P.-V. Fournier, *Les quatres Flores de la France, Corse comprise.* Poinson-les-Grancey. 1934–1940. (Pp. 1–64 in 1934; pp. 65–256 in 1935; pp. 257–576 in 1936; pp. 577–832 in 1937;

pp. 833–896 in 1938; pp. 897–992 in 1939; pp. 993–1092 in 1940.) Photographic reprint with additions and corrections at end. Paris. 1946. Reprint with more additions and corrections. Paris. 1961.

Fries, *Nov. Fl. Suec.*
E. M. Fries, *Novitiae Florae suecicae.* Ed. 1. Lundae. 1814–1823. (Pp. 1–40 in 1814; pp. 43–60 in 1817; pp. 63–80 in 1819; pp. 83–122 in 1823.) Ed. 2. Londini Gothorum. 1828. *Mantissa* 1. Lundae. 1832. *Mantissa* 2. Upsaliae. 1839. *Mantissa* 3. Upsaliae. 1842.

Fritsch, *Exkursionsfl. Österr.*
K. Fritsch, *Excursionsflora für Österreich (mit Ausschluss von Galizien, Bukowina und Dalmatien).* Ed. 1. Wien. 1897. Ed. 2. Wien. 1909. Ed. 3, titled *Exkursionsflora für Österreich und die ehemals österreichischen Nachbargebiete.* Wien & Leipzig. 1922.

Froelich, *Gent. Diss.*
J. A. von Froelich, *De Gentiana Libellus./De Gentiana Dissertatio.* Erlangae. 1796.

Fuss, *Mantissa*
M. Fuss, *Joh. Christ. Gottl. Baumgarten Enumerationis Stirpium Transilvaniae indigenarum Mantissa* I. Cibinii. 1846.

Gaertner, *Fruct. Sem. Pl.*
J. Gaertner, *De Fructibus et Seminibus Plantarum.* Stuttgardiae, Tuebingae & Lipsiae. 1788–1807. (**1,** Stuttgardiae; **2,** Tuebingae; **3,** Lipsiae. **1** in 1788; **2:** pp. 1–184 in 1790; pp. 185–504 in 1791; pp. 505–520 in 1792; **3:** pp. 1–128 in 1805; pp. 129–256 in 1807. Cf. F. A. Stafleu, *Taxonomic Literature* 162 (1967).) **3** by C. F. von Gaertner.

Gaertner, P., B. Meyer & Scherb., *Fl. Wetter.*
P. G. Gaertner, B. Meyer & J. Scherbius, *Oekonomisch-technische Flora der Wetterau.* Frankfurt. 1799–1803. (**1** in 1799; **2** in 1800; **3(1)** in 1801; **3(2)** in 1803.)

Garcke, *Fl. Nord- Mittel-Deutschl.*
F. A. Garcke, *Flora von Nord- und Mittel-Deutschland.* Ed. 1. Berlin. 1849. Ed. 2. Berlin. 1851. Ed. 3. Berlin. 1854. Ed. 4. Berlin. 1858. Ed. 5. Berlin. 1860. Ed. 6. Berlin. 1863. Ed. 7. Berlin. 1865. Ed. 8. Berlin. 1867. Ed. 9. Berlin. 1869. Ed. 10. Berlin. 1871. Ed. 11. Berlin. 1873. Ed. 12. Berlin. 1875. Later editions with different titles exist.

Gaudin, *Fl. Helv.*
J. F. A. T. G. P. Gaudin, *Flora helvetica.* Turici. 1828–1833. (**1–3** in 1828; **4 & 5** in 1829; **6** in 1830; **7** in 1833.)

Georgi, *Bemerk. Reise*
J. G. Georgi, *Bemerkungen einer Reise in russischen Reich in Jahre* 1772. St. Petersburg. 1775. (**1 & 2** in 1775, with continuous pagination.)

Gmelin, C. C., *Fl. Bad.*
C. C. Gmelin, *Flora badensis, alsatica et Confinium Regionum cis et transrhenana, Plantas a Lacu Bodamico usque ad Confluentem Mosellae et Rheni sponte nascentes exhibens.* Carlsruhae. 1805–1826. (**1** in 1805; **2** in 1806; **3** in 1808; **4** in 1826.)

Gmelin, J. F., *Onomat. Bot. Compl.*
J. F. Gmelin, *Onomatologia botanica completa, oder vollständiges botanisches Wörterbuch.* Frankfurt & Leipzig. 1772–1778. (**1 & 2** in 1772; **3–5** in 1773; **6** in 1774; **7** in 1775; **8** in 1776; **9** in 1777; **10** in 1778.)

Gmelin, J. F., *Syst. Nat.*
Cf. L., *Syst. Nat.*

Gouan, *Fl. Monsp.*
A. Gouan, *Flora monspeliaca.* Lugduni. 1764. (For date cf. F. A. Stafleu, *Taxonomic Literature* 176 (1967).)

Gouan, *Obs. Bot.*
A. Gouan, *Illustrationes et Observationes botanicae, ad Specierum Historiam facientes, seu rariorum Plantarum indigenarum, pyrenaicarum, exoticarum Adumbrationes.* Tiguri. 1773.

Grauer, *Pl. Min. Cogn. Dec.*
S. Grauer, *Plantarum minus cognitarum Decuria.* Kiloniae. 1784.

Gray, A., *Man. Bot.*
A. Gray, *A Manual of Botany of the Northern United States.* Ed. 1. Boston & Cambridge. 1848. Ed. 2. New York. 1856. Later editions and revisions exist.

Gray, S. F., *Nat. Arr. Brit. Pl.*
S. F. Gray, *A natural Arrangement of British Plants.* London. 1821. (**1** & **2** in 1821. Cf. F. A. Stafleu, *Taxonomic Literature* 181 (1967).)

Gremli, *Beitr. Fl. Schweiz*
A. Gremli, *Beiträge zur Flora der Schweiz.* Aarau. 1870.

Gremli, *Neue Beitr. Fl. Schweiz*
A. Gremli, *Neue Beiträge zur Flora der Schweiz.* Aarau. 1880–1890. (**1** in 1880; **2** in 1882; **3** in 1883; **4** in 1887; **5** in 1890.)

Gren. & Godron, *Fl. Fr.*
J. C. M. Grenier & D. A. Godron, *Flore de France, ou Description des Plantes qui croissent naturellement en France et en Corse.* Paris. 1847–1856. (**1**: pp. 1–330 in 1847; pp. 331–762 in 1849; **2**: pp. 1–392 in 1850; pp. 393–760 in 1853; **3**: pp. 1–384 in 1855; pp. 385–779 in 1856. Cf. F. A. Stafleu, *Taxonomic Literature* 181–182 (1967).)

Griseb., *Spicil. Fl. Rumel.*
A. H. R. Grisebach, *Spicilegium Florae rumelicae et bithynicae.* Brunsvigae. 1843–1845. (**1** in 1843; **2**: pp. 1–160 in 1844; pp. 161–548 in 1845 or 1846. Cf. F. A. Stafleu, *Taxonomic Literature* 184 (1967).)

Grossh., *Fl. Kavk.*
A. A. Grossheim, Флора Кавказа [*Flora Kavkaza*]. Ed. 1. Tiflis & Baku. 1928–1934. (**1** in 1928; **2** in 1930; **3** in 1932; **4** in 1934. **1** & **2** at Tiflis. **3** & **4** at Baku.) Ed. 2. Baku, Moskva & Leningrad. 1939 → . (**1** in 1939; **2** in 1940; **3** in 1945; **4** in 1950; **5** in 1952; **6** in 1962; **7** in 1967. **1–3** at Baku; **4–6** at Moskva & Leningrad; **7** at Leningrad.)

Grossh., *Opred. Rast. Kavk.*
A. A. Grossheim, Определитель Растений Кавказа [*Opredelitel' Rastenij Kavkaza*]. Moskva. 1949.

Guss., *Cat. Pl. Boccad.*
G. Gussone, *Catalogus Plantarum, quae asservantur in Regio Horto ser. Fr. Borbonii Principis Juventutis in Boccadifalco prope Panormum.* Neapoli. 1821.

Guss., *Fl. Sic. Prodr.*
G. Gussone, *Florae siculae Prodromus.* Neapoli. 1827–1828. (**1** in 1827; **2** in 1828.) *Suppl., Supplementum.* 1832–1834. (Pp. 1–166 in 1832; pp. 167–242 in 1834.)

Guss., *Fl. Sic. Syn.*
G. Gussone, *Florae siculae Synopsis.* Neapoli. 1843–1845. (**1** in 1843; **2**: pp. 1–668 in 1844; pp. 669–903 in 1845. Cf. F. A. Stafleu, *Taxonomic Literature* 188 (1967).)

Guss., *Ind. Sem. Horto Boccad.*
G. Gussone, *Index Seminum in Horto Boccadifalci.* By G. Gussone in 1825 & 1826.

Guss., *Pl. Rar.*
G. Gussone, *Plantae rariores quas in Itinere per Oras jonii ac adriatici Maris et per Regiones Samnii ac Aprutii collegit....* Neapoli. 1826.

Hacq., *Pl. Carn.*
B. A. Hacquet, *Plantae alpinae carniolicae.* Viennae. 1782.

Halácsy, *Consp. Fl. Graec.*
E. von Halácsy, *Conspectus Florae graecae.* Lipsiae. 1900–1904. (**1**: pp. 1–576 in 1900; pp. 577–825 in 1901; **2** in 1902; **3** in 1904. Cf. F. A. Stafleu, *Taxonomic Literature* 189 (1967).) *Suppl., Supplementum* 1. Lipsiae. 1908. (*Supplementum* 2 published as *Magyar Bot. Lapok* **11**: 114–202 (1912).)

Hassk., *Cat. Horto Bogor.*
J. C. Hasskarl, *Catalogus Plantarum in Horto botanico bogo-*
riensi cultarum alter./Tweede Catalogus der in's Lands Plantentuin te Buitenzorg gekweekte gewassen. Batavia. 1844.

Hayek, *Prodr. Fl. Penins. Balcan.*
A. von Hayek, *Prodromus Florae Peninsulae balcanicae.* (Published as *Feddes Repert.* (*Beih.*) 30.) (**1**: pp. 1–352 in 1924; pp. 353–672 in 1925; pp. 673–960 in 1926; pp. 961–1193 in 1927; **2**: pp. 1–96 in 1928; pp. 97–336 in 1929; pp. 337–576 in 1930; pp. 577–1152 in 1931; **3**: pp. 1–208 in 1932; pp. 209–472 in 1933. Cf. F. A. Stafleu, *Taxonomic Literature* 194–195 (1967).) **2** & **3** edit. F. Markgraf.

Hayne, *Darst. Beschr. Arzn. Gewächse*
F. G. Hayne, *Getreue Darstellung und Beschreibung der in der Arzneykunde gebräuchlichen Gewächse, wie auch solcher, welche mit ihnen verwechselt werden können.* Berlin. **1–14.** 1805–1843. (**1** in 1805; **2** in 1809; **3** in 1813; **4** in 1816; **5** in 1817; **6** in 1819; **7** in 1821; **8** in 1822; **9** in 1825; **10** in 1827; **11** in 1830; **12** in 1833; **13** in 1837; **14**: tt. 1–12 in 1843; tt. 13–24 in 1846.) **12** by F. G. Hayne, J. F. Brandt & J. T. C. Ratzeburg. **13** continued by J. F. Brandt & J. T. C. Ratzeburg.

Hegi, *Ill. Fl. Mitteleur.*
G. Hegi, *Illustrierte Flora von Mitteleuropa.* Ed. 1. München. 1906–1931. (**1**: pp. 1–72 in 1906; pp. 73–312 in 1907; pp. 313–412 in 1908; **2**: pp. 1–128 in 1908; pp. 129–408 in 1909; **3**: pp. 1–36 in 1909; pp. 37–328 in 1910; pp. 329–472 in 1911; pp. 473–608 in 1912; **4(1)**: pp. 1–96 in 1913; pp. 97–144 in 1914; pp. 145–192 in 1916; pp. 193–320 in 1918; pp. 321–491 in 1919; **4(2)**: pp. 497–540 in 1921; pp. 541–908 in 1922; pp. 909–1112a in 1923; **4(3)**: pp. 1113–1436 in 1923; pp. 1437–1748 in 1924; **5(1)**: pp. 1–316 in 1924; pp. 317–674 in 1925; **5(2)**: pp. 679–994 in 1925; pp. 995–1562 in 1926; **5(3)**: pp. 1567–1722 in 1926; pp. 1723–2250 in 1927; **5(4)**: pp. 2255–2632 in 1927; **6(1)**: pp. 1–112 in 1913; pp. 113–304 in 1914; pp. 305–352 in 1915; pp. 353–400 in 1916; pp. 401–496 in 1917; pp. 497–544 in 1918; **6(2)**: pp. 549–1152 in 1928; pp. 1153–1386 in 1929; **7** in 1931.) *Nachtr., Nachträge, Berichtigungen und Ergänzungen* (to **4(1)** and **5(1–4)**). München. 1968. Ed. 2. München. 1936 → . (**1** in 1936; **2** in 1939; **3(1)**: pp. 1–240 in 1957; pp. 241–452 in 1958; **3(2)**: pp. 453–532 in 1959; pp. 533–692 in 1960; pp. 693–772 in 1961; pp. 773–852 in 1962; pp. 853–932 in 1969; **3(3)**: pp. 1–80 in 1965; **4(1)**: pp. 1–80 in 1958; pp. 81–160 in 1959; pp. 161–320 in 1960; pp. 321–480 in 1962; pp. 481–548 in 1963; **4(2)**: pp. 1–80 in 1961; pp. 81–224 in 1963; pp. 225–304 in 1964; pp. 305–384 in 1965; pp. 385–448 in 1966; **6(1)**: pp. 1–80 in 1965; pp. 81–160 in 1966; pp. 161–240 in 1968; pp. 241–320 in 1969; **6(2)**: pp. 1–96 in 1966; pp. 97–176 in 1970; **6(3)**: pp. 1–80 in 1964; pp. 81–160 in 1965; pp. 161–240 in 1966; pp. 241–320 in 1968.) Ed. 3. München. 1966 → . (**2(1)**: pp. 1–80 in 1966; pp. 81–160 in 1968; pp. 161–240 in 1969.)

Heldr., *Sched. Herb. Graec. Norm.*
T. von Heldreich, *Schedae Plantarum ad Herbaria graeca normalia.*

Henckel, *Adumbr. Pl. Horti Hal.*
L. V. F. Henckel von Donnersmarck, *Adumbrationes Plantarum nonnullarum Horti halensis academici selectarum.* Halae. 1806.

Hill, *Veg. Syst.*
J. Hill, *The Vegetable System.* London. 1759–1775. (**1** in 1759, reissued in 1770; **2** in 1761, reissued in 1771; **3** in 1761; **4** in 1762; **5** in 1763; **6** in 1764; **7** in 1764; **8** in 1765; **9** in 1765; **10** in 1765; **11** in 1767; **12** in 1767; **13** in 1768; **14** in 1769; **15** in 1769; **16** in 1770; **17** in 1770; **18** in 1771; **19** in 1771; **20** in 1772; **21** in 1772; **22** in 1773; **23** in 1773; **24** in 1774; **25** in 1774; **26** in 1775.)

Hoffm., *Deutschl. Fl.*
G. F. Hoffmann, *Deutschlands Flora, oder botanisches Taschenbuch für das Jahr* 1791. Ed. 1. Erlangen. 1791. *Kryptogamie...*

für das Jahr 1795. Erlangen. 1795. Ed. 2. ...*für das Jahr* 1800. Erlangen. 1800. (For dates cf. F. A. Stafleu, *Taxonomic Literature* 206 (1967).)

Hoffmanns. & Link, *Fl. Port.*
J. C. von Hoffmannsegg & J. H. F. Link, *Flore portugaise.* Berlin. 1809–1840. (**1**: tt. 1–20 in 1809; tt. 21–25 in 1810; tt. 26–35 in 1811–1813; tt. 36–50 in 1813; tt. 51–78 in 1813–1820; **2**: tt. 79–102 in 1820–1828; tt. 103–106 in 1834; tt. 107–109 in 1840. Cf. F. A. Stafleu, *Taxonomic Literature* 206–207 (1967).)

Holandre, *Fl. Moselle*
J. J. J. Holandre, *Flore de la Moselle.* Ed. 1. Metz. 1829. *Supplément.* Metz. 1836. Ed. 2, titled *Nouvelle Flore de la Moselle.* Metz & Paris. 1842.

Honckeny, *Vollst. Syst. Verz.*
G. A. Honckeny, *Vollständiges systematisches Verzeichnis aller Gewächse Teutschlandes.* Leipzig. 1782.

Hooker & Arnott, *Bot. Beech. Voy.*
W. J. Hooker & G. A. W. Arnott, *The Botany of Captain Beechey's Voyage.* London. 1830–1841. (Pp. 1–48 in 1830; pp. 49–144 in 1832; pp. 145–192 in 1833; pp. 193–240 in 1837; pp. 241–336 in 1838; pp. 337–384 in 1839; pp. 385–432 in 1840; pp. 433–485 in 1841. Cf. F. A. Stafleu, *Taxonomic Literature* 217 (1967).)

Hornem., *Dansk Oekon. Plantel.*
J. W. Hornemann, *Forsøg til en dansk oeconomisk Plantelære.* Ed. 1. Kjøbenhavn. 1796. Ed. 2. Kjøbenhavn. 1806. Ed. 3. Kjøbenhavn. 1821–1837. (**1** in 1821; **2** in 1837.)

Hornem., *Fl. Dan.*
Cf. Oeder, *Fl. Dan.*

Hornem., *Hort. Hafn.*
J. W. Hornemann, *Hortus regius botanicus hafniensis, in Usum Tironum et Botanophilorum.* Hauniae. 1813–1815. (**1** in 1813; **2** in 1815.) *Suppl., Supplementum.* 1819.

Host, *Fl. Austr.*
N. T. Host, *Flora austriaca.* Viennae. 1827–1831. (**1** in 1827; **2** in 1831.)

Host, *Syn. Pl. Austr.*
N. T. Host, *Synopsis Plantarum in Austria Provinciisque adjacentibus sponte crescentium.* Vindobonae. 1797.

Hudson, *Fl. Angl.*
W. Hudson, *Flora anglica.* Ed. 1. London. 1762. Ed. 2. London. 1778. Ed. 3. London. 1798.

Hull, *Brit. Fl.*
J. Hull, *The British Flora.* Ed. 1. Manchester. 1799. Ed. 2. Manchester. 1808.

Hultén, *Fl. Aleut. Isl.*
E. O. G. Hultén, *Flora of the Aleutian Islands.* Ed. 1. Stockholm. 1937. Ed. 2. Weinheim, Codicote & New York. 1960.

Humb., Bonpl. & Kunth, *Nov. Gen. Sp.*
F. H. A. von Humboldt, A. J. A. Bonpland & C. S. Kunth, *Nova Genera et Species Plantarum quas in Peregrinatione Orbis novi collegerunt, descripserunt, partim adumbraverunt Amatus Bonpland et Alexander de Humboldt. Ex Schedis autographis Amati Bonpland in Ordinem digessit Carolus Siegesmund Kunth.* Lutetiae Parisiorum. 1816–1825. (**1–7**; for dates cf. F. A. Stafleu, *Taxonomic Literature* 225–226 (1967).)

Jacq., *Collect. Bot.*
N. J. von Jacquin, *Collectanea ad Botanicam, Chemiam et Historiam naturalem spectantia.* Vindobonae. 1787–1797. (**1** in 1787; **2** in 1789; **3** in 1790; **4** in 1791; **5** (*Supplementum*) in 1797. Cf. F. A. Stafleu, *Taxonomic Literature* 232 (1967) & *Taxon* **12**: 63–64 (1963).)

Jacq., *Enum. Stirp. Vindob.*
N. J. von Jacquin, *Enumeratio Stirpium plerarumque, quae sponte crescunt in Agro vindobonensi, Montibusque confinibus.* Vindobonae. 1762.

Jacq., *Fl. Austr.*
N. J. von Jacquin, *Florae austriacae, sive Plantarum selectarum in Austriae Archiducatu sponte crescentium Icones.* Viennae. 1773–1778. (**1** in 1773; **2** in 1774; **3** in 1775; **4** in 1776; **5** in 1778.)

Jacq., *Hort. Vindob.*
N. J. von Jacquin, *Hortus botanicus vindobonensis.* Vindobonae. 1770–1776. (**1** in 1770; **2** in 1772; **3** in 1776.)

Jacq., *Misc. Austr. Bot.*
N. J. von Jacquin, *Miscellanea austriaca ad Botanicam, Chemiam et Historiam naturalem spectantia.* Vindobonae. 1779–1781. (**1** in 1779; **2** in 1781.)

Jacq., *Obs. Bot.*
N. J. von Jacquin, *Observationum botanicarum Iconibus ab auctore delineatis illustratarum Pars* 1[–4]. Vindobonae. 1764–1771. (**1** in 1764; **2** in 1767; **3** in 1768; **4** in 1771.)

Jahandiez & Maire, *Cat. Pl. Maroc*
E. Jahandiez & R. C. J. E. Maire, *Catalogue des Plantes du Maroc* (*Spermatophytes et Ptéridophytes*). Alger. 1931–1941. (**1** in 1931; **2** in 1932; **3** in 1934; **4** (*Supplément*) by L. Emberger & R. C. J. E. Maire, in 1941.)

Jaub. & Spach, *Ill. Pl. Or.*
H. F. Jaubert & E. Spach, *Illustrationes Plantarum orientalium, ou Choix de Plantes nouvelles ou peu connues de l'Asie occidentale.* Paris. 1842–1857. (**1–5**; for dates cf. F. A. Stafleu, *Taxonomic Literature* 233–234 (1967).)

Jáv., *Magyar Fl.*
S. Jávorka, *Magyar Flóra (Flora hungarica). Magyaroszág Virágos és Edényes Virágtalan Növényeinek Meghatározó Kézikönyve.* Budapest. 1924–1925. (Pp. 1–800 in 1924; pp. 801–1307, i–cii in 1925.)

Jáv. & Csapody, *Icon. Fl. Hung.*
S. Jávorka & V. Csapody, *A Magyar Flóra Képekben. / Iconographia Florae hungaricae.* Budapest. 1929–1934. (Pp. 1–96, tt. 1–7 in 1929; pp. 97–224, tt. 8–16 in 1930; pp. 225–288, tt. 17–20 in 1931; pp. 289–448, tt. 21–25 in 1932; pp. 449–576, tt. 26–30 in 1933; tt. 31–40 in 1934.)

Jordan, *Obs. Pl. Crit.*
A. Jordan, *Observations sur plusieurs Plantes nouvelles, rares ou critiques de la France.* Paris & Leipzig. 1846–1849. (**1–4**, Paris & Leipzig in 1846; **5**, Paris in 1847; **6**, Leipzig in 1847; **7**, Paris in 1849.)

Jordan, *Pug. Pl. Nov.*
A. Jordan, *Pugillus Plantarum novarum praesertim gallicarum.* Paris. 1852.

Jordan & Fourr., *Brev. Pl. Nov.*
A. Jordan & J. P. Fourreau, *Breviarium Plantarum novarum, sive Specierum in Horto plerumque cultura recognitarum Descriptio contracta ulterius amplianda.* Parisiis. 1866–1868. (**1** in 1866; **2** in 1868.)

Jovet, *Fl. Fr.*
P. A. Jovet (edit.), *Flore de France.* Paris. 1967→. (**1** by H. D. Schotsman.)

Kerner, A., *Monogr. Pulm.*
A. J. Kerner von Marilaun, *Monographia Pulmonariarum.* Oeniponte. 1878.

Kerner, A., *Nov. Pl. Sp.*
A. J. Kerner von Marilaun, *Novae Plantarum Species.* Innsbruck. 1870. (**1 & 2** in 1870.)

Kerner, A., *Sched. Fl. Exsicc. Austro-Hung.*
A. J. Kerner von Marilaun, *Schedae ad Floram exsiccatam austro-hungaricam.* Vindobonae. 1881–1913. (**1** in 1881; **2** in 1882; **3** in 1884; **4** in 1886; **5** in 1888; **6** in 1893; **7** in 1896; **8** in 1899; **9** in 1902; **10** in 1913.) **8 & 9** by K. Fritsch; **10** by R. de Wettstein.

Kirschleger, *Prodr. Fl. Alsace*

F. R. Kirschleger, *Prodrome de la Flore d'Alsace.* Strasbourg, Colmar & Mulhausen. 1836. *Appendice.* Strasbourg. 1838.

Koch, *Syn. Deutsch. Fl.*

W. D. J. Koch, *Synopsis der deutschen und schweizer Flora.* Ed. 1. Frankfurt. 1838. Ed. 2. Leipzig. 1847. Ed. 3. Leipzig. 1890–1907. (**1**: pp. 1–320 in 1890; pp. 321–640 in 1891; pp. 641–997 in 1892; **2**: pp. 999–1110 in 1892; pp. 1111–1270 in 1893; pp. 1271–1430 in 1895; pp. 1431–1590 in 1897; pp. 1591–1750 in 1900; pp. 1751–1910 in 1901; pp. 1911–1990 in 1902; **3**: pp. 1991–2070 in 1902; pp. 2071–2390 in 1903; pp. 2391–2710 in 1905; pp. 2711–3094 in 1907.)

Koch, *Syn. Fl. Germ.*

W. D. J. Koch, *Synopsis Florae germanicae et helveticae.* Ed. 1. Francofurti a. M. 1835–1837. (Pp. 1–352 in 1835; pp. 353–844 in 1837.) Ed. 2. Lipsiae. 1843–1845. (Pp. 1–452 in 1843; pp. 451 *bis*–964 in 1844; pp. 965–1164 in 1845.) Ed. 3. Lipsiae. 1857.

Koch, C., *Dendrologie*

C. H. E. Koch, *Dendrologie. Bäume, Sträucher und Halbsträucher, welche in Mittel- und Nordeuropa im Freien kultivirt werden.* Erlangen. 1869–1873. (**1** in 1869; **2(1)** in 1872; **2(2)** in 1873. Cf. F. A. Stafleu, *Taxonomic Literature* 241 (1967).)

Koch, L., *Entw.-Gesch. Orob.*

L. K. A. Koch, *Die Entwicklungsgeschichte der Orobanchen.* Heidelberg. 1887.

Koelliker, *Verz. Phan. Gew. Zürich.*

R. A. von Koelliker, *Verzeichniss der phanerogamischen Gewächse des Cantons Zürich.* Zürich. 1839.

Komarov, *Fl. URSS*

V. L. Komarov *et al.* (edit.) Флора СССР [*Flora SSSR*]./*Flora URSS.* Leningrad & Mosqua. 1934–1964. (**1–3**, Leningrad; **4–30**, Mosqua & Leningrad. **1** in 1933; **2** in 1934; **3** & **4** in 1935; **5** & **6** in 1936; **7** in 1937; **8** & **9** in 1939; **10** & **11** in 1941; **12** in 1946; **13** in 1948; **14** & **15** in 1949; **16** in 1950; **17** in 1951; **18** in 1952; **19** in 1953; **20** & **21** in 1954; **22** in 1955; **23** in 1958; **24** in 1957; **25** in 1959; **26** in 1961; **27** in 1962; **28** in 1963; **29** in 1964; **30** in 1960. **11** reprinted in 1945.) **1–13** edit. V. L. Komarov; **14, 15, 18, 22, 24, 26, 27** & **30** edit. B. K. Schischkin & E. G. Bobrov; **16, 17, 19, 21, 23** & **25** edit. B. K. Schischkin; **20** edit. B. K. Schischkin & S. V. Juzepczuk; **28** edit. E. G. Bobrov & S. K. Tcherepanov; **29** edit. E. G. Bobrov & N. N. Tzvelev.

Kotov or **Kotov & Barbarich, *Fl. RSS Ucr.***

Cf. Fomin, *Fl. RSS Ucr.*

Krocker, *Fl. Siles.*

A. J. Krocker, *Flora silesiaca.* Vratislaviae. 1787–1823. (**1** in 1787; **2(1** & **2)** in 1790; **3** in 1814; **4** (*Supplementum*) in 1823. Cf. F. A. Stafleu, *Taxonomic Literature* 245–246 (1967).)

Krylov, *Fl. Zap. Sibir.*

P. N. Krylov, Флора западной Сибири [*Flora zapadnoj Sibiri*]. Tomsk. 1927–1949. (**1** in 1927; **2** in 1928; **3** in 1929; **4** in 1930; **5** & **6** in 1931; **7** in 1933; **8** in 1935; **9** in 1937; **10** in 1939; **11** in 1949; **12(1)** in 1961; **12(2)** in 1964.)

Kuntze, O., *Revis. Gen.*

K. E. O. Kuntze, *Revisio Generum Plantarum vascularium omnium.* Leipzig. 1891–1898. (**1** & **2** in 1891; **3**: pp. i–cccx in 1893; pp. 1–576 in 1898. Cf. F. A. Stafleu, *Taxonomic Literature* 250 (1967).)

Kusn., N. Busch & Fomin, *Fl. Cauc. Crit.*

N. I. Kusnezow, N. A. Busch & A. V. Fomin, *Flora caucasica critica.* Матеріалы для Флоры Кавказа [*Materialy dlja Flory Kavkaza*]. Jurjev. 1901–1916. (**1(1)**: pp. 1–96 in 1911; pp. 97–224 in 1912; pp. 225–248, i–xlvi in 1913; **2(1)**: pp. 1–43 in 1911; **2(4)**: pp. 1–64 in 1906; pp. 65–144 in 1912; pp. 145–176 in 1913; **2(5)**: pp. 1–32 in 1916; **3(3)**: pp. 1–32 in 1901; pp. 33–112 in 1902; pp. 113–256, i–xix in 1903; **3(4)**: pp. 1–16

in 1904; pp. 17–64 in 1905; pp. 65–144 in 1907; pp. 145–304 in 1908; pp. 305–544 in 1909; pp. 545–820, i–lxxiv in 1910; **3(5)**: pp. 1–32 in 1913; pp. 33–48 in 1915; **3(7)**: pp. 1–80 in 1908; pp. 81–96 in 1910; pp. 97–112 in 1912; **3(8)**: pp. 1–16 in 1911; pp. 17–48 in 1912; **3(9)**: pp. 1–64 in 1906; pp. 65–224 in 1909; pp. 225–288 in 1910; pp. 289–320 in 1912; pp. 321–352 in 1913; **4(1)**: pp. 1–128 in 1901; pp. 129–208 in 1902; pp. 209–352 in 1903; pp. 353–384 in 1904; pp. 385–464 in 1905; pp. 465–512 in 1906; pp. 513–560 in 1907; pp. 561–590, i–lxii in 1908; **4(2)**: pp. 1–32 in 1912; pp. 33–208 in 1913; pp. 209–256 in 1914; pp. 257–320 in 1915; pp. 321–400 in 1916; **4(3)**: pp. 1–48 in 1916; **4(6)**: pp. 1–16 in 1903; pp. 17–48 in 1904; pp. 49–80 in 1905; pp. 81–144 in 1906; pp. 145–157, i–xviii in 1907.)

L., *Amoen. Acad.*

C. von Linné, *Amoenitates academicae.* Holmiae. 1749–1768. (**3** in 1756; **4** in 1759; **5** in 1760; **6** in 1763; **7** in 1768. Cf. F. A. Stafleu, *Taxonomic Literature* 281 (1967).)

L., *Cent. Pl.*

C. von Linné, *Centuria Plantarum.* Upsaliae. 1755–1756. (**1** in 1755; **2** in 1756.)

L., *Demonstr. Pl.*

C. von Linné, *Demonstrationes Plantarum in Horto upsaliensi* 1753.... Upsaliae. 1753.

L., *Diss. Erica*

C. von Linné, *Dissertationem botanicam de Erica....* Upsaliae. 1770.

L., *Fl. Monsp.*

C. von Linné, *Flora monspeliensis.* Upsaliae. 1756.

L., *Fl. Palaest.*

C. von Linné, *Flora palaestina.* Upsaliae. 1756.

L., *Mantissa*

C. von Linné, *Mantissa Plantarum.* (Pp. 1–142.) Holmiae. 1767.

L., *Mantissa Alt.*

C. von Linné, *Mantissa Plantarum altera.* (Pp. 143–587.) Holmiae. 1771.

L., *Sp. Pl.*

C. von Linné, *Species Plantarum.* Ed. 1. Holmiae. 1753. Ed. 2. Holmiae. 1762–1763. (Pp. 1–784 in 1762; pp. 785–1684 in 1763.)

L., *Syst. Nat.*

C. von Linné, *Systema Natura.* Ed. 10. Botany in **2**. Holmiae. 1759. Ed. 11. Botany in **1** & **2**. Lipsiae. 1762. Ed. 12. Botany in **2** & **3**. Holmiae. 1767–1768. (**2** in 1767; **3** in 1768.) Ed. 13, by J. F. Gmelin, Botany in **2**. Lipsiae. 1791–1792. (Pp. 1–884 in 1791; pp. 885–1661 in 1792. Cf. F. A. Stafleu, *Taxonomic Literature* 172 (1967).) Titled *Syst. Veg., Systema Vegetabilium.* Ed. 13, edit. J. A. Murray. Gottingae & Gothae. 1774. Ed. 15, by C. H. Persoon. Gottingae. 1797. Ed. nov. (15), by J. J. Roemer & J. A. Schultes, Studtgardtiae. 1817–1830. (**1** & **2** in 1817; **3** in 1818; **4** in 1819; **5** in 1819 or 1820; **6** in 1820; **7** by J. A. & J. H. Schultes: pp. i–xliv, 1–754 in 1829; pp. xlv–cviii, 755–1816 in 1830. Cf. F. A. Stafleu, *Taxonomic Literature* 397–398 (1967).) Ed. 16, by C. P. J. Sprengel. Gottingae. 1824–1828. (**1** in 1824; **2** in 1825; **3** in 1826; **4(1** & **2)** in 1827; **5** & *Suppl.* in 1828.)

L., *Syst. Veg.*

Cf. L., *Syst. Nat.*

L. fil., *Suppl.*

C. von Linné fil., *Supplementum Plantarum Systematis Vegetabilium Editionis XIII, Generum Plantarum Editionis VI, et Specierum Plantarum Editionis II.* Brunsvigae. 1781.

Labill., *Icon. Pl. Syr.*

J. J. H. de Labillardière, *Icones Plantarum Syriae rariorum.* Parisiis. 1791–1812. (**1** & **2** in 1791; **3** in 1809; **4** & **5** in 1812. Cf. F. A. Stafleu, *Taxonomic Literature* 252 (1967).)

Lag., Gen. Sp. Nov.
M. Lagasca y Segura, *Genera et Species Plantarum quae aut novae sunt aut nondum recte cognoscuntur.* Matriti. 1816.

Laínz, Aportac. Fl. Gallega
M. Laínz, *Aportaciones al Conocimiento de la Flora gallega* 6. Madrid. 1968. (Earlier volumes published as follows: **1** as *Brotéria (Ser. Ci. Nat.)* **24**: 108–151 (1955); **2** as *Anal. Inst. Bot. Cavanilles* **14**: 529–554 (1956); **3** as *Brotéria (Ser. Ci. Nat.)* **26**: 90–97 (1957); **4** as *Anal. Inst. Forest. Invest. Exper.* **10**: 299–334 (1966); **5** as *op. cit.* **12**: 1–51 (1967).)

Lam., Encycl. Méth. Bot.
J. B. A. P. Monnet de la Marck, *Encyclopédie méthodique. Botanique.* Paris. 1783–1817. (**1**: pp. 1–344 in 1783; pp. 345–752 in 1785; **2**: pp. 1–400 in 1786; pp. 401–774 in 1788; **3**: pp. 1–360 in 1789; pp. 361–759 in 1792; **4**: pp. 1–400 in 1797; pp. 401–764 in 1798; **5** in 1804; **6**: pp. 1–384 in 1804; pp. 385–786 in 1805; **7** in 1806; **8** in 1808; *Suppl.* **1**(= **9**): pp. 1–400 in 1810; pp. 401–761 in 1811; *Suppl.* **2**(= **10**): pp. 1–384 in 1811; pp. 385–876 in 1812; *Suppl.* **3**(= **11**): pp. 1–368 in 1813; pp. 369–780 in 1814; *Suppl.* **4**(= **12**): pp. 1–368 in 1816; pp. 369–731 in 1817; *Suppl.* **5**(= **13**) in 1817. Cf. F. A. Stafleu, *Taxonomic Literature* 255–256 (1967).) **5–13** by J. L. M. Poiret.

Lam., Fl. Fr.
J. B. A. P. Monnet de la Marck, *Flore française.* Ed. 1. Paris. 1778. (**1–3** in 1778.) Ed. 2. Paris. 1795. (**1–3** in 1795.) Ed. 3 by J. B. A. P. Monnet de la Marck & A. P. de Candolle. Paris. 1805–1815. (**1–4** in 1805; **5** in 1815. **4** issued in two parts with continuous pagination, the second called '*Tome quatrième. Seconde partie*' and '*Vol. V*'; **5** called '*Tome cinquième, ou sixième volume*' and '*Vol. VI*'; **1–4** also reprinted in 1815.)

Lam., Tabl. Encycl. Méth. Bot.
J. B. A. P. Monnet de la Marck, *Tableau encyclopédique et méthodique des trois Règnes de la Nature. Botanique.* Paris. 1791–1823. (**1**: pp. 1–200 in 1791; pp. 201–440 in 1792; pp. 441–496 in 1793; **2**: pp. 1–48 in 1794; pp. 49–72 in 1796; pp. 73–136 in 1797; pp. 137–551 in 1819; **3**: pp. 1–728 in 1823. Cf. F. A. Stafleu, *Taxonomic Literature* 256 (1967).)

Lam. & DC., Fl. Fr.
Cf. Lam., *Fl. Fr.*

Lamotte, Prodr. Fl. Centr. Fr.
M. Lamotte, *Prodrome de la Flore du Plateau central de la France.* Paris. 1877–1881. (**1**: pp. 1–355 in 1877; **2**: pp. 351–628 in 1881.)

Lange, Fl. Dan.
Cf. Oeder, *Fl. Dan.*

Lange, Ind. Sem. Horto Haun.
J. M. C. Lange, *Index Seminum in Horto academico hauniensi collectorum.* Hauniae. 1854, 1855, 1857 & 1861.

Lapeyr., Fig. Fl. Pyr.
P. Picot de Lapeyrouse, *Figures de la Flore des Pyrénées, avec des Descriptions, des Notes critiques et des Observations.* Paris. 1795–1801. (Tt. 1–10 in 1795; tt. 11–43 in 1801.)

Lapeyr., Hist. Abr. Pyr.
P. Picot de Lapeyrouse, *Histoire abrégée des Plantes des Pyrénées et Itinéraire des Botanistes dans ces Montagnes.* Toulouse. 1813. *Supplément.* 1818.

Lavrenko & Soczava, Descr. Veg. URSS
E. M. Lavrenko & V. B. Soczava (edit.), *Descriptio Vegetationis URSS./*Растительный Покров СССР [*Rastitel'nyj Pokrov SSSR*]. Leningrad. 1956. (**1** & **2** in 1956.)

Ledeb., Fl. Altaica
C. F. von Ledebour, *Flora altaica.* Berolini. 1829–1834. (**1** in 1829; **2** in 1830; **3** in 1831; **4** in 1833; Index in 1833 or 1834.)

Ledeb., Icon. Pl. Fl. Ross.
C. F. von Ledebour, *Icones Plantarum novarum vel imperfecte cognitarum Floram rossicam, imprimis altaicam, illustrantes.* Rigae. 1829–1834. (**1** in 1829; **2** in 1830; **3**: tt. 201–250 in 1831; tt. 251–300 in 1832; **4** in 1833; **5** in 1834.)

Ledeb., Ind. Sem. Horti Dorpat.
C. F. von Ledebour, *Index Seminum Horti botanici dorpatensis.* Dorpat. 1818–1835.

Le Gall, Fl. Morbihan
N. J. M. le Gall, *Flore du Morbihan.* Vannes. 1852.

Lehm., Del. Sem. Horto Hamburg.
J. G. C. Lehmann, *Delectus Seminum quae in Horto Hamburgiensium botanico e Collectione Anni* 1831 *mutuae Commutationi offeruntur.* Hamburgi. 1831.

Lehm., Pl. Asperif.
J. G. C. Lehmann, *Plantae e Familiae Asperifoliarum nuciferae.* Berolini. 1818.

Lehm., J. F., Prim. Fl. Herbip.
J. F. Lehmann, *Primae Lineae Florae herbipolensis.* Herbipoli. 1809.

Lej., Fl. Spa
A. L. S. Lejeune, *Flore des Environs de Spa.* Liège. 1811–1813. (**1** in 1811; **2** in 1813.)

Leresche & Levier, Deux Excurs. Bot.
L. Leresche & E. Levier, *Deux Excursions botaniques dans le Nord de l'Espagne et le Portugal en* 1878 *et* 1879. Lausanne. 1880.

L'Hér., Sert. Angl.
C. L. L'Héritier de Brutelle, *Sertum anglicum.* Parisiis. 1789–1792. (Pp. 1–36, tt. 1–2 in 1789; tt. 3–12 in 1790; tt. 13–34 in 1792. Cf. F. A. Stafleu, *Taxonomic Literature* 267 (1967).)

L'Hér., Stirp. Nov.
C. L. L'Héritier de Brutelle, *Stirpes novae aut minus cognitae.* Parisiis. 1785–1805. (Pp. i–vi, 1–20 in 1785; pp. vii–viii, 21–40 in 1785 or 1786; pp. ix–x, 41–62 in 1786; pp. xi–xii, 63–102 in 1788; pp. xiii–xiv, 103–134 in 1789; pp. xv–xvi, 135–184 in 1791; tt. 85–124 in 1805. Cf. F. A. Stafleu, *Taxonomic Literature* 266 (1967).)

Lindley, Digitalium Monogr.
J. Lindley, *Digitalium Monographia.* Londini. 1821.

Link, Enum. Hort. Berol. Alt.
J. H. F. Link, *Enumeratio Plantarum Horti regii botanici berolinensis altera.* Berolini. 1821–1822. (**1** in 1821; **2** in 1822.)

Link & Otto, Icon. Pl. Rar. Horti Bot. Berol.
J. H. F. Link & C. F. Otto, *Icones Plantarum rariorum Horti regii botanici berolinensis./Abbildungen neuer und seltener Gewächse des königlichen botanischen Gartens zu Berlin.* Berlin. 1828–1831. (Pp. 1–36 in 1828; pp. 37–60 in 1829; pp. 61–72 in 1830; pp. 73–96 in 1831. Cf. F. A. Stafleu, *Taxonomic Literature* 274 (1967).)

Loefl., Iter Hisp.
P. Loefling, *Iter Hispanicum, eller Resa til Spanska Länderna.* Stockholm. 1758.

Loisel., Fl. Gall.
J. L. A. Loiseleur-Deslongchamps, *Flora gallica.* Ed. 1. Lutetiae. 1806–1807. (Pp. 1–336 in 1806; pp. 337–742 in 1807.) Ed. 2. Paris. 1828.

Loisel., Not. Pl. Fr.
J. L. A. Loiseleur-Deslongchamps, *Notice sur les Plantes à ajouter à la Flore de France* (Flora gallica) *avec quelques Corrections et Observations.* Paris. 1810.

Lojac., Contr. Fl. Sic.
M. Lojacono-Pojero, *Contributi alla Flora di Sicilia.* Palermo. 1878.

Lojac., Fl. Sic.
M. Lojacono-Pojero, *Flora sicula, o Descrizione delle Piante vascolari spontanee o indigenate in Sicilia.* Palermo. 1888–1909. (**1**(1) in 1888; **1**(2) in 1891; **2**(1) in 1902; **2**(2) in 1907; **3** in 1909.)

Loscos, *Trat. Pl. Arag.*
    F. Loscos y Bernál, *Tratado de Plantas de Aragón.* 1876–1886. (**1** in 1876–1877; **2** & *Suppl.* 1–4 in 1878; **3** & *Suppl.* 5–8 in 1883–1886. Cf. S. F. Blake, *Geographical Guide to Floras of the World* **2**: 487 (1961).)

Loscos & Pardo, *Ser. Pl. Arag.*
    F. Loscos y Bernál & J. Pardo y Sastrón, *Series inconfecta Plantarum indigenarum Aragoniae praecipue meridionalis, e Lingua castellana in latinam vertit, recensuit, emendavit, Observationibus suis auxit atque edendam curavit M. Willkomm.* Dresdae. 1863.

Loudon, *Hort. Brit.*
    J. C. Loudon, *Hortus britannicus; a Catalogue of all the Plants indigenous, cultivated in, or introduced into Britain.* Ed. 1. London. 1830–1832. (**1** & **2** in 1830; *Suppl.* in 1832.) Ed. 2 & *Suppl.* London. 1832. Ed. 3 & *Suppl.* London. 1839.

Luckwill, *Gen. Lycopers.*
    L. C. Luckwill, *The Genus* Lycopersicon. (Aberdeen University Studies No. 120.) Aberdeen. 1943.

Mackay, *Fl. Hibern.*
    J. T. Mackay, *Flora hibernica.* Dublin. 1836.

Majevski, *Fl. Sred. Ross.*
    P. F. Majevski, Флора средней Россіи [*Flora srednej Rossii*]. Ed. 1. 1892. Ed. 2 by S. I. Korshinsky. 1895. Ed. 3 by B. A. Fedtschenko. 1902. Ed. 4 by D. I. Litvinov. 1912. Ed. 5 by D. I. Litvinov. 1917. Ed. 6 by B. A. Fedtschenko. 1933. Ed. 7, titled Флора средней Полосы европейской Части СССР [*Flora srednej Polosy evropejskoj Časti SSSR*] by V. L. Komarov. 1940. Ed. 8 by B. K. Schischkin. Moskva & Leningrad. 1954. Ed. 9 by B. K. Schischkin. Leningrad. 1964.

Marchesetti, *Fl. Trieste*
    C. de Marchesetti, *Flora di Trieste e de' suoi Dintorni.* Trieste. 1896–1897.

Marshall, *Arbust. Amer.*
    H. Marshall, *Arbustum americanum, the American Grove, or an alphabetical Catalogue of Forest Trees and Shrubs, Natives of the American United States.* Philadelphia. 1785.

Mart., C. F. P., *Fl. Brasil.*
    C. F. P. von Martius, *Flora brasiliensis.* Monachii. 1840–1906. (For dates etc. cf. I. Urban in C. F. P. Mart., *Fl. Brasil.* **1**(1): 212–267 (1906).)

Meyer, C. A., *Verz. Pfl. Cauc.*
    C. A. von Meyer, *Verzeichniss der Pflanzen, welche während der, auf allerhöchsten Befehl, in den Jahren 1829 und 1830 unternommenen Reise im Caucasus und in den Provinzen am westlichen Ufer des caspischen Meeres gefunden und eingesammelt worden sind.* St Petersburg. 1831.

Michx, *Fl. Bor. Amer.*
    A. Michaux, *Flora boreali-americana.* Parisiis & Argentorati. 1803. (**1** & **2** in 1803.)

Miller, *Gard. Dict.*
    P. Miller, *The Gardeners Dictionary.* Ed. 8. London. 1768. Ed. 9, by T. Martyn. London. 1797–1804. (**1** in 1797; **2** in 1804.)

Moench, *Meth.*
    C. Moench, *Methodus Plantas Horti botanici et Agri marburgensis a Staminum Situ describendi.* Marburgi Cattorum. 1794. *Suppl., Supplementum.* Marburgi Cattorum. 1802.

Moessler, *Handb.*
    J. C. Moessler, *Gemeinnütziges Handbuch der Gewächskunde.* Ed. 1. Altona. 1815. Ed. 2, by H. G. L. Reichenbach. Altona. 1827–1829. (**1** in 1827; **2** in 1828; **3** in 1829.) Ed. 3, by H. G. L. Reichenbach. Altona. 1833–1834. (**1** in 1833; **2** in 1834.)

Moris, *Stirp. Sard.*
    G. G. Moris, *Stirpium sardoarum Elenchus.* Carali. 1827–1829. (**1** & **2** in 1827; **3** in 1829.) *App., Appendix.* Carali. 1828.

Moris & De Not., *Fl. Caprariae*
    G. G. Moris & G. de Notaris, *Florula Caprariae.* Torino. 1839.

Murith, *Guide Bot. Valais*
    L. J. Murith, *Le Guide du Botaniste qui voyage dans le Valais.* Lausanne. 1810.

Murray, A., *North. Fl.*
    A. Murray, *The Northern Flora; or, a Description of the wild Plants belonging to the North and East of Scotland.* Edinburgh. 1836.

Mutel, *Fl. Fr.*
    A. Mutel, *Flore française destinée aux Herborisations.* Paris. 1834–1838. (**1** in 1834; **2** in 1835; **3** in 1836; **4** in 1837; **5** in 1838.)

Nees, *Horae Phys. Berol.*
    C. G. D. Nees von Esenbeck (edit.), *Horae physicae berolinenses.* Bonnae. 1820.

Neilr., *Diagn. Ung. Slav. Gefässpfl.*
    A. Neilreich, *Diagnosen der in Ungarn und Slavonien bisher beobachteten Gefässpflanzen welche in Koch's* Synopsis *nicht enthalten sind.* Wien. 1867.

Neilr., *Nachtr. Fl. Wien.*
    A. Neilreich, *Nachträge zur Flora von Wien.* Wien. 1851.

Nordh., *Norsk Fl.*
    R. Nordhagen, *Norsk Flora.* Oslo. 1940.

Nutt., *Gen. N. Amer. Pl.*
    T. Nuttall, *The Genera of North-American Plants, and a Catalogue of the Species to the Year 1817.* Philadelphia. 1818. (**1** & **2** in 1818.)

Nyman, *Consp.*
    C. F. Nyman, *Conspectus Florae europaeae.* Örebro. 1878–1882. (Pp. 1–240 in 1878; pp. 241–493 (part) in 1879; pp. 493–677 in 1881; pp. 677–858 in 1882; p. 256 *errata* in 1881; p. 493 (part) and p. 677 published twice. Cf. W. T. Stearn, *Jour. Bot.* (*London*) **76**: 113 (1938).) *Suppl., Supplementum* **1** in 1883–1884; **2**(1): pp. 1–224 in 1889; **2**(2): pp. 225–404 in 1890. *Addit., Additamenta* by E. Roth, in 1886.

Oeder, *Fl. Dan.*
    G. C. Oeder, *Icones Plantarum sponte nascentium in Regnis Daniae et Norvegiae, in Ducatibus Slesvici et Holsatiae, et in Comitatibus Oldenburgi et Delmenhorstiae: ad illustrandum Opus de iisdem Plantis, regio Jussu exarandum, Florae danicae Nomine inscriptum.* Hauniae. 1762–1883. (**1**: tt. 1–60 in 1762; tt. 61–120 in 1763; tt. 121–180 in 1764; **2**: tt. 181–240 in 1765; tt. 241–300 in 1766; tt. 301–360 in 1767; **3**: tt. 361–420 in 1768; tt. 421–480 in 1769; tt. 481–540 in 1770; **4**: tt. 541–600 in 1771; tt. 601–660 in 1775; tt. 661–720 in 1777; **5**: tt. 721–780 in 1778; tt. 781–840 in 1780; tt. 841–900 in 1782; **6**: tt. 901–960 in 1787; tt. 961–1020 in 1790; tt. 1021–1080 in 1792; **7**: tt. 1081–1140 in 1794; tt. 1141–1200 in 1797; tt. 1201–1260 in 1799; **8**: tt. 1261–1320 in 1806; tt. 1321–1380 in 1808; tt. 1381–1440 in 1810; **9**: tt. 1441–1500 in 1813; tt. 1501–1560 in 1816; tt. 1561–1620 in 1818; **10**: tt. 1621–1680 in 1819; tt. 1681–1740 in 1821; tt. 1741–1800 in 1823; **11**: tt. 1801–1860 in 1825; tt. 1861–1920 in 1827; tt. 1921–1980 in 1829; **12**: tt. 1981–2040 in 1830; tt. 2041–2100 in 1832; tt. 2101–2160 in 1834; **13**: tt. 2161–2220 in 1836; tt. 2221–2280 in 1839; tt. 2281–2340 in 1840; **14**: tt. 2341–2400 in 1843; tt. 2401–2460 in 1845; tt. 2461–2520 in 1849; **15**: tt. 2521–2580 in 1849; tt. 2581–2640 in 1858; tt. 2641–2700 in 1861; **16**: tt. 2701–2760 in 1867; tt. 2761–2820 in 1869; tt. 2821–2880 in 1871; **17**: tt. 2881–2940 in 1877; tt. 2941–3000 in 1880; tt. 3001–3060 in 1883. **1**–**4**: t. 600 edit. G. C. Oeder; **4**: t. 601–5 edit. O. F. Mueller; **6** & **7** edit. M. H. Vahl; **8**–**13** edit. J. W. Hornemann; **14**: tt. 2341–2400 edit. J. F.

Schouw & J. L. M. Vahl; **14**: t. 2401–**15**: t. 2580 edit. F. M. Liebmann; **15**: tt. 2581–2640 edit. J. Steenstrup & J. M. C. Lange; **15**: t. 2641–**17** edit. J. M. C. Lange.)

Opiz, *Naturalientausch*
P. M. Opiz, *Naturalientausch*. Prag. 1823–1830. (**1–5** in 1823; **6–8** in 1824; **9–10** in 1825; **11** in 1826–1828; **12**, titled *Beiträge zur Naturgeschichte*, in 1828–1830. Cf. J. Futák & K. Domin, *Bibliografia k Flóre ČSR* 441 (1960).)

Opiz, *Seznam*
P. M. Opiz, *Seznam Rostlin Květeny České*. V Praže. 1852.

Palhinha, *Cat. Pl. Vasc. Açores*
R. T. Palhinha, *Catálogo das Plantas vasculares dos Açores*. Lisboa. 1966.

Pallas, *Fl. Ross.*
P. S. Pallas, *Flora rossica*. Petropoli. 1784–1831. (**1**(1) in 1784; **1**(2) in 1788 or 1789; **2**(1) in 1831. Cf. F. A. Stafleu, *Taxonomic Literature* 347 (1967).)

Pallas, *Reise*
P. S. Pallas, *Reise durch verschiedene Provinzen des russischen Reichs*. Ed. 1. St Petersburg. 1771–1776. (**1** in 1771; **2** in 1773; **3** in 1776.) Ed. 2. St Petersburg. 1801.

Pančić, *Fl. Princ. Serb.*
J. Pančić, Флора Кнежевине Србије [*Flora Kneževine Srbije*]./ *Flora Principatus Serbiae*. Beograd. 1874. *Addit.*, Додатак [*Dodatak*]./*Additamenta*. Beograd. 1884.

Pančić, *Nov. Elem. Fl. Bulg.*
J. Pančić, Нова Грађа за Флору Кнежевине Бугарске [*Nova Graća za Floru Kneževine Bugarske*]./*Nova Elementa ad Floram Principatus Bulgariae*. U Beogradu. 1886.

Parl., *Fl. Ital.*
F. Parlatore, *Flora italiana*. Firenze. 1848–1896. (**1**: pp. 1–96 in 1848; pp. 97–568 in 1850; **2**: pp. 1–220 in 1852; pp. 221–638 in 1857; **3**: pp. 1–160 in 1858; pp. 161–690 in 1860; **4**: pp. 1–288 in 1868; pp. 289–623 in 1869; **5**: pp. 1–320 in 1873; pp. 321–671 in 1875 or 1876; **6**: pp. 1–336 in 1884; pp. 337–656 in 1885; pp. 657–971 in 1886; **7** in 1887; **8**: pp. 1–176 in 1888; pp. 177–773 in 1889; **9**: pp. 1–232 in 1890; pp. 233–624 in 1892; pp. 625–1085 in 1893; **10** in 1894; **11** in 1896. **6–11** by T. Caruel. Cf. F. A. Stafleu, *Taxonomic Literature* 349–350 (1967).)

Pau, *Not. Bot. Fl. Esp.*
C. Pau, *Notas botánicas á la Flora española*. Madrid & Segorbe. 1887–1895. (**1, 2, 4** 6 Madrid; **3** Segorbe. **1** in 1887; **2** & **3** in 1889; **4** in 1891; **5** in 1893; **6** in 1895.)

Pauquy, *De la Belladone*
C. L. C. Pauquy, *De la Belladone*. Paris. 1825.

Pawł., *Fl. Polska*
Cf. Racib. & Szafer, *Fl. Polska*

Pennell, *Scroph. E. Temp. N. Amer.*
F. W. Pennell, *The Scrophulariaceae of eastern temperate North America*. (The Academy of Natural Sciences of Philadelphia. Monographs, 1.) Philadelphia. 1935.

Pers., *Syn. Pl.*
C. H. Persoon, *Synopsis Plantarum*. Ed. 1. Paris & Tuebingae. 1805–1807. (**1** in 1805; **2**: pp. 1–272 in 1806; pp. 273–657 in 1807.) Ed. 2, titled *Species Plantarum*. Petropoli. 1817–1822. (**1** in 1817; **2** & **3** in 1819; **4** & **5** in 1821; **6** in 1822.)

Pers., *Syst. Veg.*
Cf. L., *Syst. Nat.*

Petagna, *Inst. Bot.*
V. Petagna, …*Institutiones botanicae*. Neapoli. 1785–1787. (**1** in 1785; **2** in 1787.)

Petrović, *Add. Fl. Agri Nyss.*
S. Petrović, Додатак Флори околине Нишa [*Dodatak Flori Okoline Niša*]./*Additamenta ad Floram Agri nyssani*. U Beogradu. 1855.

Petrović, *Fl. Agri Nyss.*
S. Petrović, Флора Околине Ниша [*Flora Okoline Niša*]./*Flora Agri nyssani*. Beograd. 1882.

Poiret, *Voy. Barb.*
J. L. M. Poiret, *Voyage en Barbarie, ou Lettres écrites de l'ancienne Numidie, pendant les Années 1785 et 1786; avec un Essai sur l'Histoire naturelle de ce Pays*. Paris. 1789.

Pollich, *Hist. Pl. Palat.*
J. A. Pollich, *Historia Plantarum in Palatinatu electorali sponte nascentium incepta*. Mannhemii. 1776–1777. (**1** in 1776; **2** & **3** in 1777.)

Pomel, *Nouv. Mat. Fl. Atl.*
A. Pomel, *Nouveaux Matériaux pour la Flore Atlantique*. Paris. 1874–1875. (Pp. i–iii, 1–260 in 1874; pp. [257–]261–399 in 1875.)

Popl., *Spisok Rast. Sobr. Krymsk. Gos. Zapov.*
G. I. Poplavskaja, Список Растений, собранных в Крымском государственном Заповеднике [*Spisok Rastenij, sobrannykh v Krymskom gosudarstvennom Zapovednike*]. Moskva & Leningrad. 1931.

Presl, C., *Fl. Sic.*
C. B. Presl, *Flora sicula*. Pragae. 1826.

Presl, J. & C., *Del. Prag.*
J. S. & C. B. Presl, *Deliciae pragenses, Historiam naturalem spectantes*. Pragae. 1822.

Presl, J. & C., *Fl. Čechica*
J. S. & C. B. Presl, *Flora Čechica/Kwětena Česká*. Pragae. 1819.

Pursh, *Fl. Amer. Sept.*
F. T. Pursh, *Flora Americae septentrionalis*. Ed. 1. London. 1814. (**1** & **2**, with continuous pagination, in 1814.) Ed. 2. London. 1816. (**1** & **2** with, continuous pagination, in 1816.)

Racib. & Szafer, *Fl. Polska*
M. Raciborski & W. Szafer (edit.), *Flora Polska*. Kraków & Warszawa. 1919 →. (**1–6, 9** & **10** Kraków; **7** & **8** Warszawa. **1** in 1919; **2** in 1921; **3** in 1927; **4** in 1930; **5** in 1935; **6** in 1947; **7** in 1955; **8** in 1959; **9** in 1960; **10** in 1963; **11** in 1967.) **2–6** edit. W. Szafer; **7–9** edit. W. Szafer & B. Pawłowski; **10** & **11** edit. B. Pawłowski.

Radius, *Diss.*
J. W. M. Radius, *Dissertatio de Pyrola et Chimophila*. Lipsiae. 1821.

Rafin., *Autikon Bot.*
C. S. Rafinesque-Schmaltz, *Autikon botanikon, or botanical Illustrations of 2500 new, rare or beautiful Trees, Shrubs, Plants, Vines, Lilies, Grasses, Ferns etc., of all Regions, but chiefly North America, with Descriptions etc. and 2500 self Figures or Specimens*. Philadelphia. 1840. (Only one vol., describing 1500 species. No illustrations actually published.)

Rafn, *Danm. Holst. Fl.*
C. G. Rafn, *Danmarks och Holsteens Flora systematisk, physisk og oeconomisk bearbeydet*. Kiöbenhavn. 1796–1800. (**1** in 1796; **2** in 1800.)

Rech. fil., *Fl. Iran.*
K. H. Rechinger, *Flora iranica. Flora des Iranischen Hochlandes und der umrahmenden Gebirge*. Graz. 1963 → .

Rees, *Cyclop.*
A. Rees, *The Cyclopaedia; or, universal Dictionary of Arts, Sciences and Literature*. London. 1802–1820. (For dates of parts cf. F. A. Stafleu, *Taxonomic Literature* 380–382 (1967).)

Reichenb., *Fl. Germ. Excurs.*
H. G. L. Reichenbach, *Flora germanica excursoria*. Lipsiae. 1830–1832. (Pp. i–viii, 1–136 in 1830; pp. 137–438 in 1831; pp. ix–xlviii, 435–878 in 1832.)

Reichenb., *Icon. Fl. Germ.*
H. G. L. Reichenbach, *Icones Florae germanicae et helveticae.*

Lipsiae & Gerae. 1834–1914. (1–19(1), **20** & **21** Lipsiae; **19**(2), **22**, **24** & **25**(1 & 2) Lipsiae & Gerae; **23** Gerae. **1**, titled *Agrostographia germanica, sistens Icones Graminearum et Cyperoidearum quas in Flora germanica recensuit Auctor* (= *Pl. Crit.* **11**) in 1834; **2** in 1837–1838; **3** in 1838–1839; **4** in 1840; **5** in 1841; **6** in 1844; **7** in 1845; **8** in 1846; **9** in 1847; **10** in 1848; **11** in 1849; **12** in 1850. By H. G. Reichenbach, **13** & **14** in 1851; **15**: tt. 1–80 in 1852; tt. 81–160 in 1853; **16**: tt. 1–100 in 1853; tt. 101–150 in 1854; **17**: pp. 1–32 in 1854; pp. 33–113 in 1855; **18**: pp. 1–32 in 1856; pp. 33–72 in 1857; pp. 73–103 in 1858; **19**(1): tt. 1–60 in 1858; tt. 61–150 in 1859; tt. 151–260 in 1860. By G. Beck von Mannagetta, **19**(2): pp. 1–10 in 1904; pp. 11–48 in 1905; pp. 49–104 in 1906; pp. 105–152 in 1907; pp. 153–184 in 1908; pp. 185–240 in 1909; pp. 241–288 in 1910; pp. 289–324 in 1911. By H. G. Reichenbach, **20**: pp. 1–48 in 1861; pp. 49–125 in 1862; **21**: tt. 1–70 in 1863; tt. 71–110 in 1864; tt. 111–150 in 1865; tt. 151–190 in 1866; tt. 191–220 in 1867; **22**: pp. 1–96 in ?1869. By G. Beck von Mannagetta, **22**: pp. 97–230 in 1903. By F. G. Kohl, **23**: pp. 1–32 in 1896; pp. 33–44 in 1897; pp. 45–68 in 1898; pp. 69–84 in 1899. By G. Beck von Mannagetta, **24**: pp. 1–24 in 1903; pp. 25–48 in 1904; pp. 49–64 in 1905; pp. 65–88 in 1906; pp. 89–112 in 1907; pp. 113–152 in 1908; pp. 153–160 in 1909; pp. 161–216 in ?1909; **25**(1): pp. 1–12 in 1909; pp. 13–32 in 1910; pp. 33–48 in 1911; pp. 49–72 in 1912; **25**(2): pp. 1–24 in 1913; pp. 25–40 in 1914.)

Reichenb., *Pl. Crit.*
H. G. L. Reichenbach, *Iconographia botanica, seu Plantae criticae*. Lipsiae. 1823–1832. (**1** in 1823; **2** in 1824; **3** in 1825; **4** in 1826; **5** in 1827; **6** in 1828; **7** in 1829; **8** in 1830; **9** in 1831; **10** in 1832. For **11** cf. Reichenb., *Icon. Fl. Germ.*)

Reichenb. fil., *Icon. Fl. Germ.*
Cf. Reichenb., *Icon. Fl. Germ.*

Retz., *Obs. Bot.*
A. J. Retzius, *Observationes botanicae*. Lipsiae. 1779–1791. (**1** in 1779; **2** in 1781; **3** in 1783; **4** in 1786 or 1787; **5** in 1788; **6** in 1791. Cf. F. A. Stafleu, *Taxonomic Literature* 389 (1967).)

Richard, A., *Tent. Fl. Abyss.*
A. Richard, *Tentamen Florae abyssinicae*. Parisiis. 1847–1851. (**1** in 1847; **2** in 1851.)

Roemer & Schultes, *Syst. Veg.*
Cf. L., *Syst. Nat.*

Röhling, *Deutschl. Fl.*
J. C. Röhling, *Deutschlands Flora*. Ed. 1. Bremen. 1796. Ed. 2. Frankfurt a. M. 1812–1813. (**1** & **2** in 1812; **3** in 1813.) Ed. 3 by F. K. Mertens & W. D. J. Koch. Frankfurt a. M. 1823–1839. (**1** in 1823; **2** in 1826; **3** in 1831; **4** in 1833; **5** in 1839.)

Roth, *Bot. Abh.*
A. W. Roth, *Botanische Abhandlungen und Beobachtungen*. Nürnberg. 1787.

Roth, *Catalecta Bot.*
A. W. Roth, *Catalecta botanica, quibus Plantae novae et minus cognitae describuntur atque illustrantur*. Lipsiae. 1797–1806. (**1** in 1797; **2** in 1800; **3** in 1806.)

Rouy & Fouc., or **Rouy & Camus** or **Rouy, *Fl. Fr.***
G. C. C. Rouy, *Flore de France*. Asnières, Paris & Rochefort. 1893–1913. (**1** in 1893; **2** in 1895; **3** in 1896; **4** in 1897; **5** in 1899; **6** in 1900; **7** in 1901; **8** in 1903; **9** in 1905; **10** in 1908; **11** in 1909; **12** in 1910; **13** in 1912; **14** in 1913.) **1–3** in collaboration with J. Foucaud, **6** & **7** with E. G. Camus.

Rouy, *Ill. Pl. Eur. Rar.*
G. C. C. Rouy, *Illustrationes Plantarum Europae rariorum*. Paris. 1895–1905. (**1–4** in 1895; **5–8** in 1896; **9** & **10** in 1898; **11** & **12** in 1899; **13** & **14** in 1900; **15** & **16** in 1901; **17** & **18** in 1902; **19** in 1904; **20** in 1905.)

Royle, *Ill. Bot. Himal. Mount.*
J. F. Royle, *Illustrations of the Botany and other Branches of the natural History of the Himalayan Mountains and of the Flora of Cashmere*. London. 1833–1840. (For dates cf. F. A. Stafleu, *Taxonomic Literature* 403–404 (1967).)

Ruiz & Pavón, *Fl. Peruv.*
H. Ruiz López & J. Pavón, *Flora peruviana et chilensis*. Matriti. 1798–1802. (**1** in 1798; **2** in 1799; **3** in 1802; **4** in ?1802. Cf. F. A. Stafleu, *Taxonomic Literature* 406 (1967).)

Ruiz & Pavón, *Fl. Peruv. Chil. Prodr.*
H. Ruiz López & J. Pavón, *Florae peruvianae, et chilensis Prodromus*. Ed. 1. Madrid. 1794. Ed. 2. Romae. 1797.

Ruiz & Pavón, *Syst. Veg. Fl. Peruv.*
H. Ruiz López & J. Pavón, *Systema Vegetabilium Florae peruvianae et chilensis*. Madrid. 1798.

Russell, A., *Nat. Hist. Aleppo*
A. Russell, *The natural History of Aleppo*. Ed. 1. London. 1756. Ed. 2. London. 1794. (**1** & **2** in 1794.)

Salisb., *Prodr.*
R. A. Salisbury, *Prodromus Stirpium in Horto ad Chapel Allerton vigentium*. Londini. 1796.

Samp., *Lista Esp. Herb. Port.*
G. A. da Silva Ferreira Sampaio, *Lista das Espécies representadas no Herbário português*. Porto. 1913. *Ap.* **1**, *Apêndice à Lista…*. Porto. 1914. *Ap.* **2**, *Segundo Apêndice à Lista…* Porto. 1914. *Ap.* **3**, *Terceiro Apêndice à Lista…*. Porto. 1914.

Sauvage & Vindt, *Fl. Maroc*
C. P. F. Sauvage & J. Vindt, *Flore du Maroc*. Tanger. 1952–1954. (**1** in 1952; **2** in 1954.)

Savi, *Fl. Pis.*
G. Savi, *Flora pisana*. Pisa. 1798. (**1** & **2** in 1798.)

Săvul., *Fl. Rep. Pop. Române*
T. Săvulescu, *Flora Republicii Populare Române*./*Flora Reipublicae popularis romanicae*. Bucureşti. 1952 → . (**1** in 1952; **2** in 1953; **3** in 1955; **4** in 1956; **5** in 1957; **6** in 1958; **7** in 1960; **8** in 1961; **9** in 1964; **10** in 1965; **11** in 1966. **3–10** titled *Flora Republicii Populare Romîne*./*Flora Reipublicae popularis romanicae*. **11** titled *Flora Republicii Socialiste România*./*Flora Reipublicae socialisticae România*.)

Schinz & R. Keller, *Fl. Schweiz*
H. Schinz & R. Keller, *Flora der Schweiz*. Ed. 1. Zürich. 1900. Ed. 2. Zürich. 1905. (**1** *Exkursionsflora* in 1905; **2** *Kritische Flora* in 1905.) Ed. 3. Zürich. 1909–1914. (**1** *Exkursionsflora* in 1909; **2** *Kritische Flora*, edit. H. Schinz & A. Thellung, in 1914.) Ed. 4. Zürich. 1923. (**1** *Exkursionsflora*, edit. H. Schinz & A. Thellung, in 1923.)

Schischkin, *Fl. Jugo-Vostoka*
B. K. Schischkin (edit.), Флора Юго-востока Европейской Части СССР [*Flora Jugo-vostoka Evropejskoj Časti SSSR*]./ *Flora Rossiae austro-orientalis*. **6**. Moskva & Leningrad. 1936. (**1–5** published in *Acta Horti Petrop*. 1927–1931.)

Schischkin, or **Schischkin & Bobrov**, or **Schischkin & Juz., *Fl. URSS***
Cf. Komarov, *Fl. URSS*.

Schlosser & Vuk., *Syll. Fl. Croat.*
J. C. Schlosser & L. F. Vukotinović, *Syllabus Florae croaticae*. Zagrabiae. 1857.

Schmidt, F. W., *Fl. Boëm.*
F. W. Schmidt, *Flora boëmica inchoata*. Pragae. 1793–1794. (**1** in 1793; **2–4** in 1794.)

Scholler, *Fl. Barb.*
F. A. Scholler, *Flora barbiensis*. Lipsiae. 1775. *Supplementum* by J. J. Bossart. Barbii. 1787.

Schott, *Sippen Österr. Primeln*
H. W. Schott, *Die Sippen der österreichischen Primeln*. Wien. 1851.

Schott, *Wilde Blendl. Österr. Primeln*

H. W. Schott, *Wilde Blendlinge österreichischer Primeln.* Wien. 1852.

Schott, Nyman & Kotschy, *Analect. Bot.*

H. W. Schott, C. F. Nyman & T. Kotschy, *Analecta botanica.* Vindobonae. 1854.

Schrader, *Hort. Gotting.*

H. A. Schrader, *Hortus gottingensis, sive Plantae novae et rariores Horti regii botanici gottingensis descriptae et Iconibus illustratae.* Goettingae. 1809.

Schrader, *Jour. für die Bot.*

H. A. Schrader, *Journal für die Botanik.* Göttingen. 1799–1804. (**1799**(1) in 1799; **1799**(2): pp. 1–200 in 1799; pp. 201–500 in 1800; **1800**(1): pp. 1–220 in 1800; pp. 225–446 in 1801; **1800**(2) in 1802; **1801**: pp. 1–272 in 1803; pp. 273–504 in 1803–1804. Cf. F. A. Stafleu, *Taxonomic Literature* 431–432 (1967).)

Schrader, *Monogr. Verbasci*

H. A. Schrader, *Monographia Generis Verbasci.* Gottingae. 1813–1823. (**1** in 1813; **2** in 1823.)

Schrank, *Pl. Rar. Horti Monac.*

F. von Paula von Schrank, *Plantae rariores Horti academici monacensis.* Monachii. 1817–1822. (**1**: tt. 1–20 in 1817; tt. 21–50 in 1819; **2**: tt. 51–60 in 1820; tt. 61–80 in 1821; tt. 81–100 in 1822. Cf. F. A. Stafleu, *Taxonomic Literature* 433 (1967).)

Schreber, *Pl. Vert. Unilab.*

J. C. D. von Schreber, *Plantae verticillatae unilabiatae.* Erlangae. 1773. Reissue, titled *Plantarum verticillatarum unilabiatarum Genera et Species.* Lipsiae. 1774. (For date cf. F. A. Stafleu, *Taxonomic Literature* 434 (1967).)

Schultes, *Obs. Bot.*

J. A. Schultes, *Observationes botanicae in Linnei Species Plantarum ex Editione C. L. Willdenow.* Oeniponti. 1809.

Schultes, *Österreichs Fl.*

J. A. Schultes, *Östreichs Flora.* Ed. 1. Wien. 1794. Ed. 2, titled *Österreichs Flora.* Wien. 1814.

Schultz, F. W., *Arch. Fl. Eur.*

F. W. Schultz, *Archives de la Flore d'Europe.* Wissembourg. 1872–1874. (Pp. 1–18 in 1872; pp. 21–? in 1874.)

Schultz, F. W., *Arch. Fl. Fr. Allem.*

F. W. Schultz, *Archives de la Flore de France et d'Allemagne.* Bitche, Haguenau & Deux-Ponts. 1842–1855. (Pp. 1–48 in 1842; pp. 49–76 in 1844; pp. 77–98 in 1846; pp. 99–154 in 1848; pp. 155–166 in 1850; pp. 167–194 in 1851; pp. 195–258 in 1852; pp. 259–282 in 1853; pp. 283–326 in 1854; pp. 327–350 in 1855.)

Schultz, F. W., *Beitr. Kenntn. Deutsch. Orob.*

F. W. Schultz, *Beitrag zur Kenntniss der deutschen Orobanchen.* München. 1829.

Schur, *Enum. Pl. Transs.*

P. J. F. Schur, *Enumeratio Plantarum Transsilvaniae.* Wien. 1866.

Scop., *Annus Hist.-Nat.*

G. A. Scopoli, *Annus* I[–V] *Historico-naturalis.* Lipsiae. 1769–1772. (**1–3** in 1769; **4** in 1770; **5** in 1772. Cf. F. A. Stafleu, *Taxonomic Literature* 440 (1967).)

Scop., *Delic. Fl. Insubr.*

J. A. Scopoli, *Deliciae Florae et Faunae insubricae* Pavia. 1786–1788. (**1 & 2** in 1786; **3** in 1788.)

Scop., *Fl. Carn.*

J. A. Scopoli, *Flora carniolica.* Ed. 2. Viennae. 1771–1772. (**1** in 1771; **2** in 1772. Cf. F. A. Stafleu, *Taxonomic Literature* 440 (1967).)

Sendtner, *Veg. Südbayerns*

O. Sendtner, *Die Vegetations-Verhältnisse Südbayerns.* München. 1854.

Sennen, *Diagn. Pl. Esp. Maroc.*

Frère Sennen, *Diagnoses des Nouveautés parues dans les Exsiccata* Plantes d'Espagne et du Maroc *de 1928 à 1935.* 1936.

Seub., *Fl. Azor.*

M. A. Seubert, *Flora azorica, quam ex Collectionibus Schedisque Hochstetteri Patris et Filii elaboravit....* Bonnae. 1844.

Sibth. & Sm., *Fl. Graec. Prodr.*

J. Sibthorp & J. E. Smith, *Florae graecae Prodromus: sive Plantarum omnium Enumeratio, quas in Provinciis aut Insulis Graeciae invenit Johannes Sibthorp...Characteres et Synonyma omnium cum Annotationibus elaboravit Jacobus Edvardus Smith.* Londini. 1806–1816. (**1**: pp. 1–218 in 1806; pp. 219–442 in 1809; **2**: pp. 1–210 in 1813; pp. 211–422 in 1816. Cf. F. A. Stafleu, *Taxonomic Literature* 445 (1967).)

Simonkai, *Enum. Fl. Transs.*

L. von Simonkai, *Enumeratio Florae transsilvanicae vasculosae critica./Erdély edényes Flórájának helyesbitett Foglalata.* Budapest. 1887.

Soó, *Omagiu Săvulescu*

R. de Soó *et al., Omagiu lui T. Săvulescu.* 1959.

Soó, *Szék. Fl. Előmunk.*

R. de Soó, *A Székelyföld Flórájának Előmunkálatai./Prodromus Florae Terrae Siculorum.* Koloszvár. 1940.

Sowerby, *Engl. Bot.*

J. Sowerby, *English Botany.* Ed. 1. *Text* by J. E. Smith. London. 1790–1814. (**1–36** in 1790–1814.) *Supplement* by various authors. 1831–1863. (**1–5** in 1831–1863.) Ed. 2. London. 1831–1853. (**1–12** in 1831–1853.) Ed. 3, with new descriptions by J. T. I. Boswell Syme. London. 1863–1886. (For dates cf. G. Sayre, *Dates Publ. Musci* 77 (1959).)

Sprengel, *Ind. Sem. Horti Halensis*

Index Seminum in Horto botanico halensis collectorum. Halae. By K. P. J. Sprengel in 1809–1811, 1813 & 1814, 1818 & 1828.

Sprengel, *Syst. Veg.*

Cf. L., *Syst. Nat.*

Stefani, Major & W. Barbey, *Karpathos*

C. de Stefani, C. J. F. Major & W. Barbey-Boissier, *Karpathos. Étude geologique, paléontologique et botanique.* Lausanne. 1895.

Steudel, *Nomencl. Bot.*

E. G. von Steudel, *Nomenclator botanicus.* Ed. 1. Stuttgardtiae & Tubingae. 1821. (**1 & 2**, with continuous pagination, in 1821 & 1824.) Ed. 2. Stuttgartiae & Tubingae. 1840–1841. (**1** in 1840; **2**: pp. 1–48 in 1840; pp. 49–176 in 1840 or 1841; pp. 177–810 in 1841. Cf. F. A. Stafleu, *Taxonomic Literature* 458 (1967).)

Stoj. & Stefanov, *Fl. Bălg.*

N. Stojanov & B. Stefanov, Флора на България [*Flora na Bălgarija*]. Ed. 1. Sofija. 1924–1925. Ed. 2. Sofija. 1933. Ed. 3. Sofija. 1948. Ed. 4, by N. Stojanov, B. Stefanov & B. Kitanov. Sofija. 1966–1967. (**1** in 1966; **2** in 1967.)

Sturm, *Deutschl. Fl.*

J. Sturm, *Deutschlands Flora.* Ed. 1. Nürnberg. 1798–1855. Ed. 2. Stuttgart. 1900–1907. (For dates cf. F. A. Stafleu, *Taxonomic Literature* 459–460 (1967).)

Sweet, *Hort. Brit.*

R. Sweet, *Hortus britannicus.* Ed. 1. London. 1826–1827. (Pp. 1–240 in 1826; pp. 241–492 in 1827.) Ed. 2. London. 1830. Ed. 3, by G. Don. London. 1839.

Ten., *Fl. Nap.*

M. Tenore, *Flora napolitana.* Napoli. 1811–1838. (**1** in 1811–1815, including *Prodr., Prodromo della Flora napolitana* followed by *Supplimento primo* and *Supplimento secondo* with continuous pagination; **2** in 1820, including *Prodr. Suppl. 3, Prodromo della Flora napolitana, Supplimento terzo*; **3** in 1824–1829, including *Prodr. Suppl. 4, Prodromo della Flora napolitana, Supplimento quarto*; **4** in 1830, including *Syll., Florae* *neapolitanae Sylloge* followed by *Addenda et Emendanda* and *Addenda et Emendanda altera* with continuous pagination, and *Syll. App.* **3**, *Ad Florae neapolitanae plantarum vascularium Syllogem Appendix tertia*; **5** in 1835–1838, including *Syll. App.* **4**, *Ad Florae neapolitanae Syllogem Appendix quarta*.) The included works were in most cases reprinted separately later, with different pagination.

Ten., *Fl. Neap. Prodr. App. Quinta*
M. Tenore, *Ad Florae neapolitanae Prodromum Appendix quinta*. Neapoli. 1826. This work was never included in the volumes of *Flora napolitana*.

Thore, *Essai Chlor. Land.*
J. Thore, *Essai d'une Chloris du Département des Landes*. Dax. 1803.

Thuill., *Fl. Paris*
J. L. Thuillier, *La Flore des Environs de Paris*. Ed. 1. Paris. 1790. Ed. 2. Paris. 1800. (For date cf. F. A. Stafleu, *Taxonomic Literature* 467 (1967).)

Thunb., *Fl. Jap.*
C. P. Thunberg, *Flora japonica*. Lipsiae. 1784.

Torrey, *Cat. Pl. New York*
J. Torrey, *A Catalogue of Plants growing spontaneously within thirty Miles of the City of New-York*. Albany. 1819.

Torrey & A. Gray, *Fl. N. Amer.*
J. Torrey & A. Gray, *A Flora of North America*. New York. 1838–1843. (**1**: pp. 1–360 in 1838; pp. 361–712 in 1840; **2**: pp. 1–184 in 1841; pp. 185–392 in 1842; pp. 393–504 in 1843. Cf. F. A. Stafleu, *Taxonomic Literature* 471 (1967).)

Tratt., *Ausgem. Taf. Arch. Gewächsk.*
L. Trattinick, *Ausgemalte Tafeln aus dem Archiv der Gewächskunde*. Wien. 1812–1814. (**1** in 1812; **2** in 1813; **3** & **4** in 1814.)

Trevisan, *Prosp. Fl. Euganea*
V. B. A. Trevisan de Saint-Léon, *Prospetto della Flora Euganea*. Padova. 1842.

Urban, *Symb. Antill.*
I. Urban, *Symbolae antillanae, seu Fundamenta Florae Indiae occidentalis*. Berolini, Lipsiae, Parisiis & Londini. 1898–1928. (**1** at Berolini, Parisiis & Londini; **2–7** at Lipsiae, Parisiis & Londini; **8** & **9** at Lipsiae. **1**: pp. 1–192 in 1898; pp. 193–471 in 1899; pp. 472–536 in 1900; **2**: pp. 1–336 in 1900; pp. 337–508 in 1901; **3**: pp. 1–352 in 1902; pp. 353–546 in 1903; **4**: pp. 1–192 in 1903; pp. 193–352 in 1905; pp. 353–528 in 1910; pp. 529–771 in 1911; **5**: pp. 1–176 in 1904; pp. 177–352 in 1907; pp. 353–555 in 1908; **6**: pp. 1–432 in 1909; pp. 433–721 in 1910; **7**: pp. 1–160 in 1911; pp. 161–432 in 1912; pp. 433–580 in 1913; **8**: pp. 1–480 in 1920; pp. 481–860 in 1921; **9**: pp. 1–176 in 1923; pp. 177–272 in 1924; pp. 273–432 in 1925; pp. 433–568 in 1928.) *A cumulative Index*, by E. Carroll & S. Sutton. Jamaica Plain, Mass. 1965.

Urban & Graebner, *Festschr. Ascherson*
I. Urban & K. O. P. P. Graebner, *Festschrift zur Feier des siebzigsten Geburtstages des... Dr. P. Ascherson*. Leipzig. 1904.

Vahl, *Enum. Pl.*
M. H. Vahl, *Enumeratio Plantarum*. Hauniae. 1804–1805. (**1** in 1804; **2** in 1805. Cf. F. A. Stafleu, *Taxonomic Literature* 480 (1967).)

Vahl, *Symb. Bot.*
M. H. Vahl, *Symbolae botanicae*. Hauniae. 1790–1794. (**1** in 1790; **2** in 1791; **3** in 1794. Cf. F. A. Stafleu, *Taxonomic Literature* 480 (1967).)

Valdés, *Rev. Esp. Eur. Linaria*
B. Valdés Castrillón, *Revisión de las Especies Europeas de Linaria con Semillas aladas*. (*Publ. Univ. Sevilla, Ser. Ci*. No. 7.) Sevilla. 1970.

Velen., *Fl. Bulg.*
J. Velenovský, *Flora bulgarica*. Pragae. 1891. **Suppl.**, *Supplementum*. Pragae. 1898.

Velen., *Rel. Mrkvičk.*
J. Velenovský, *Reliquiae mrkvičkanae*. Pragae. 1922.

Verlot, *Cat. Pl. Dauph.*
J.-B. Verlot, *Catalogue raisonné des Plantes vasculaires du Dauphiné*. Grenoble. 1872.

Vill., *Hist. Pl. Dauph.*
D. Villars, *Histoire des Plantes de Dauphiné*. Grenoble. 1786–1789. (**1** in 1786; **2** in 1787; **3**(1): pp. 1–580 in 1788; **3**(2): pp. 581–1092 in 1789. Cf. F. A. Stafleu, *Taxonomic Literature* 484 (1967).)

Vill., *Prosp. Pl. Dauph.*
D. Villars, *Prospectus de l'Histoire des Plantes de Dauphiné*. Grenoble. 1779.

Vis., *Fl. Dalm.*
R. de Visiani, *Flora dalmatica*. Lipsiae. 1842–1852. (**1** in 1842; **2** in 1847; **3** in 1850–1852.) *Supplementum*. Lipsiae. 1872. *Supplementum alterum*. Lipsiae. 1877–1882. (**1** in 1877; **2** in 1882.)

Vis., *Ill. Pi. Grecia*
R. de Visiani, *Illustrazione di alcune Piante della Grecia e dell' Asia Minore*. Venezia. 1842.

Vitman, *Summa Pl.*
F. Vitman, *Summa Plantarum*. Mediolani. 1789–1792. (**1–3** in 1789 or 1790; **4** in 1790; **5** in 1791; **6** in 1792. Cf. F. A. Stafleu, *Taxonomic Literature* 485 (1967).)

Viv., *Fl. Cors.*
D. Viviani, *Florae corsicae Specierum novarum, vel minus cognitarum Diagnosis*. Genuae. 1824. *App.* **1**, *Appendix*. Genuae. 1825. *App.* **2**, *Appendix altera*. Genuae. 1830.

Viv., *Fl. Ital. Fragm.*
D. Viviani, *Florae italicae Fragmenta, seu Plantae rariores vel nondum cognitae in variis Italiae Regionibus detectae*. Genuae. 1808.

Viv., *Fl. Lib.*
D. Viviani, *Florae libycae Specimen, sive Plantarum Enumeratio Cyrenaicam, Pentapolim, Magnae Syrteos Desertum et Regionem tripolitanum incolentium*. Genuae. 1824.

Vollmann, *Fl. Bayern*
F. Vollmann, *Flora von Bayern*. Stuttgart. 1914.

Wahlenb., *Fl. Lapp.*
G. Wahlenberg, *Flora Lapponica*. Berolini. 1812.

Waldst. & Kit., *Pl. Rar. Hung.*
F. A. von Waldstein-Wartemberg & P. Kitaibel, *Descriptiones et Icones Plantarum rariorum Hungariae*. Viennae. 1799–1812. (**1**: tt. 1–10 in 1799; tt. 11–30 in 1800; tt. 31–50 in 1800–1801; tt. 51–70 in 1801; tt. 71–90 in 1801–1802; tt. 91–100 in 1802; **2**: tt. 101–130 in 1802 or 1803; tt. 131–170 in 1803 or 1804; tt. 171–190 in 1804; tt. 191–200 in 1805; **3**: tt. 201–220 in 1806 or 1807; tt. 221–240 in 1807; tt. 241–250 in 1808 or 1809; tt. 251–260 in 1809; tt. 261–270 in 1810 or 1811; tt. 271–280 in 1812. Cf. F. A. Stafleu, *Taxonomic Literature* 489 (1967).)

Wall., *Pl. Asiat. Rar.*
N. Wallich, *Plantae Asiaticae rariores, or Descriptions and Figures of a select Number of unpublished East Indian Plants*. London. 1829–1832. (**1**: pp. 1–22 [= part 1] in 1829; pp. 22–84 [= parts 2–4] in 1830; **2**: pp. 1–? [= part 5] in 1830; pp. ?–86 [= parts 6–8] in 1831; **3**: pp. 1–? [= part 9] in 1831; pp. ?–117 [= parts 10–12] in 1832. Cf. F. A. Stafleu, *Taxonomic Literature* 490–491 (1967).)

Wallr., *Beitr. Bot.*
K. F. W. Wallroth, *Beiträge zur Botanik*. Leipzig. 1842–1844. (**1**: pp. 1–123 in 1842; pp. 125–252 in 1844.)

Wallr., *Orob. Gen.*
K. F. W. Wallroth, *Orobanches Generis* ΔΙΑΣΚΕΥΗ. Francofurti ad Moenum. 1825.

Wallr., *Annus Bot.*
K. F. W. Wallroth, *Annus botanicus, sive Supplementum tertium ad Curtii Sprengelii* Floram Halensem. Halae. 1815.

Wallr., *Sched. Crit.*
K. F. W. Wallroth, *Schedulae criticae de Plantis Florae Halensis selectis.* Halae. 1822.

Walpers, *Repert. Bot. Syst.*
W. G. Walpers, *Repertorium Botanices systematicae.* Lipsiae. 1842–1848. (**1**: pp. 1–768 in 1842; pp. 769–947 in 1843; **2** in 1843; **3**: pp. 1–768 in 1844; pp. 769–1002 in 1845; **4**: pp. 1–192 in 1845; pp. 193–576 in 1847; pp. 577–821 in 1848; **5**: pp. 1–192 in 1845; pp. 193–982 in 1846; **6**: pp. 1–384 in 1846; pp. 385–834 in 1847. Cf. F. A. Stafleu, *Taxonomic Literature* 491–492 (1967).)

Warming *et al.,* *Bot. Færöes*
J. E. B. Warming *et al., Botany of the Færöes.* Copenhagen, Christiania & London. 1901–1908. (Pp. 1–338 in 1901; pp. 339–532 in 1902; pp. 533–681 in 1903; pp. 683–834 & Appendix in 1905; pp. 835–864 in 1907; pp. 867–1070 in 1908.)

Webb, *Iter Hisp.*
P. B. Webb, *Iter hispaniense.* Paris & London. 1838.

Webb, *Otia Hisp.*
P. B. Webb, *Otia hispanica.* Ed. 1. Paris & London. 1839. Ed. 2. Paris & London. 1853.

Webb & Berth., *Phyt. Canar.*
P. B. Webb & S. Berthelot, *Phytographia canariensis.* Vol. 3(2) of *Histoire naturelle des Îles Canaries.* Paris. 1836–1850. (For dates cf. F. A. Stafleu, *Taxonomic Literature* 495 (1967).)

Weber, *Pl. Min. Cogn. Dec.*
G. H. Weber, *Plantarum minus cognitarum Decuria.* Kiloniae. 1784.

Wettst., *Monogr. Gatt. Euphrasia*
R. von Wettstein, *Monographie der Gattung* Euphrasia. Leipzig. 1896.

Widmer, *Eur. Arten Primula*
E. Widmer, *Die europäischen Arten der Gattung* Primula. München. 1891.

Willd., *Berlin. Baumz.*
C. L. Willdenow, *Berlinische Baumzucht, oder Beschreibung der in den Gärten um Berlin im Freien ausdauernden Bäume und Sträucher, für Gartenliebhaber und Freunde der Botanik.* Berlin. 1796.

Willd., *Enum. Pl. Horti Berol.*
C. L. Willdenow, *Enumeratio Plantarum Horti regii botanici berolinensis.* Berolini. 1809. *Suppl., Supplementum* by D. F. von Schlechtendal. Berolini. 1814. (For dates, cf. W. T. Stearn, *Jour. Bot.* (*London*) **75**: 234 (1937).)

Willd., *Hort. Berol.*
C. L. Willdenow, *Hortus berolinensis, sive Icones et Descriptiones, Plantarum rariorum vel minus cognitarum, quae in Horto regio botanico berolinensi excoluntur.* Berolini. 1803–1816. (**1(1)**: tt. 1–12 in 1803; **1(2–3)**: tt. 13–36 in 1804; **1(4–5)**: tt. 37–60 in 1805; **1(6)** & **2(7)**: tt. 61–84 in 1806; **2(8)**: tt. 85–96 in 1809; **2(9)**: tt. 97–108 in 1812. Cf. F. A. Stafleu, *Taxonomic Literature* 504 (1967).)

Willd., *Sp. Pl.*
C. L. Willdenow, ed. 4 of C. von Linné, *Species Plantarum.* Berolini. 1797–1806. (**1(1)**: pp. 1–495 in 1797; **1(2)**: pp. 496–1968 in 1798; **2(1 & 2)** in 1799; **3(1)**: pp. 1–847 in 1800; **3(2)**: pp. 848–1474 in 1802; **3(3)**: pp. 1475–2409 in 1803; **4(1)**: pp. 1–629 in 1805; **4(2)**: pp. 630–1157 in 1806. Cf. F. A. Stafleu, *Taxonomic Literature* 503–504 (1967).)

Willk., *Ill. Fl. Hisp.*
H. M. Willkomm, *Illustrationes Florae Hispaniae Insularumque Balearium.* Stuttgart. 1881–1892. (**1(1)**: pp. 1–12, tt. 1–9 in 1881; **1(2)**: pp. 13–28, tt. 10–18 in 1881; **1(3)**: pp. 29–40, tt. 19–28 in 1881; **1(4–6)**: pp. 44–88, tt. 29–56 in 1882; **1(7 & 8)**: pp. 89–120, tt. 57–74 in 1883; **1(9)**: pp. 121–136, tt. 75–83 in 1884; **1(10)**: pp. i–vii, 137–157, tt. 84–92 in 1885; **2(11)**: pp. 1–16, tt. 98–101 in 1886; **2(12)**: pp. 17–32, tt. 102–110 in 1886; **2(13)**: pp. 33–48, tt. 111–119 in 1887; **2(14)**: pp. 49–64, tt. 120–127 in 1888; **2(15 & 16)**: pp. 65–98, tt. 128–146 in 1889; **2(17)**: pp. 99–112, tt. 147–155 in 1890; **2(18)**: pp. 113–126, tt. 156–164 in 1891; **2(19)**: pp. 127–140, tt. 165–173 in 1892; **2(20)**: pp. i–vii, 141–156, tt. 174–183 in 1892. Cf. F. A. Stafleu, *Taxonomic Literature* 506 (1967).)

Willk., *Rech. Glob.*
H. M. Willkomm, *Recherches sur l'Organographie et la Classification des Globulairiées.* Leipsick. 1850.

Willk., *Suppl. Prodr. Fl. Hisp.*
H. M. Willkomm, *Supplementum Prodromi Florae hispanicae.* Stuttgartiae. 1893.

Willk. & Lange, *Prodr. Fl. Hisp.*
H. M. Willkomm & J. M. C. Lange, *Prodromus Florae hispanicae.* Stuttgartiae. 1861–1880. (**1**: pp. 1–192 in 1861; pp. i–xxx, 193–316 in 1862; **2**: pp. 1–272 in 1865; pp. 273–480 in 1868; pp. 481–680 in 1870; **3**: pp. 1–240 in 1874; pp. 241–512 in 1877; pp. 513–736 in 1878; pp. 737–1144 in 1880. Cf. F. A. Stafleu, *Taxonomic Literature* 506–507 (1967).)

Wimmer, *Fl. Schles.*
C. F. H. Wimmer, *Flora von Schlesien.* Ed. 1. Berlin. 1832. Ed. 2. Breslau. 1841. *Ergänzungsband.* Breslau. 1845. Ed. 3. Breslau. 1857.

Wirsing, *Eclog. Bot.*
A. L. Wirsing, *Eclogae botanicae e Dictionario Regni vegetabilis Buc'hodziano collectae.* Norimbergae. 1778.

With., *Arr. Brit. Pl.*
W. Withering, *A botanical Arrangement of all the Vegetables naturally growing in Great Britain, with Descriptions of the Genera and Species.* Ed. 1. Birmingham & London. 1776. Ed. 2, titled *A botanical Arrangement of British Plants.* Birmingham & London. 1787–1792. (**1 & 2** in 1787; **3(1)** in 1789; **3(2)** in 1792.) Ed. 3, titled *An Arrangement of British Plants* by J. Stokes. London. 1796. Ed. 4, titled *A systematic Arrangement of British Plants* by W. Withering fil. London. 1801. Ed. 5. Birmingham. 1812. Ed. 6, titled *An Arrangement of British Plants.* London. 1818. Ed. 7. London. 1830.

Woods, J., *Tourist's Fl.*
J. Woods, *The Tourist's Flora: a descriptive Catalogue of the flowering Plants and Ferns of the British Islands, France, Germany, Switzerland, Italy, and the Italian Islands.* London. 1850.

APPENDIX III

KEY TO THE ABBREVIATIONS OF TITLES OF PERIODICALS AND ANONYMOUS WORKS CITED IN VOLUME 3

Abh. Naturw. Ver. Bremen
  Abhandlungen herausgegeben vom naturwissenschaftlichen Verein zu Bremen. Bremen. 1868→ .

Abh. Zool.-Bot. Ges. Wien
  Abhandlungen der kaiserlich-königliche zoologisch-botanischen Gesellschaft in Wien. Wien. 1901→ .

Acta Bot. Acad. Sci. Hung.
  Acta botanica Academiae Scientiarum hungaricae. Budapest. 1954→ .

Acta Bot. Bohem.
  Acta botanica bohemica. Praha. 1922–1947.

Acta Bot. Fenn.
  Acta botanica fennica. Helsingforsiae. 1925→ .

Acta Bot. Neerl.
  Acta botanica neerlandica. Amsterdam. 1952→ .

Acta Geobot. Barcinon.
  Acta geobotanica barcinonensia. Barcelona. 1964→ .

Acta Geobot. Hung.
  Acta geobotanica hungarica/Debreceni Tisza István tudományos Társaság III. Matematikai-természettudományi Osztályának Munkái. Debrecen. 1936–1949.

Acta Horti Berg.
  Acta Horti bergiani. / Meddelanden från Kongl. Svenska Vetenskaps-Akademiens Trädgård, Bergielund. Stockholm. 1890→ .

Acta Horti Bot. Univ. Latv.
  Acta Horti botanici Universitatis latviensis. Riga. **1–14**, 1926–1944.

Acta Horti Petrop.
  Acta Horti petropolitani. Peterburgi. **1–43**, 1871–1931. (With alternative title Труды Имп. С.-Петербургскаго ботаническаго Сада [*Trudy Imp. S.-Peterburgskago botaničeskago Sada*] 1871–1918, and Труды главнаго ботаническаго Сада [*Trudy glavnago botaničeskago Sada*] 1918–1931.)

Acta Inst. Bot. Acad. Sci. URSS (Ser. 1)
  Труды ботаническаго Института Академии Наук СССР. Серия 1. Флора и Систематика высших Растений. [*Trudy botaničeskago Instituta Akademii Nauk SSSR. Serija 1. Flora i Sistematika vysšikh Rastenij.*]./*Acta Instituti botanici Academiae Scientiarum URSS. Ser.* 1. Leningrad & Mosqua. 1933→ .

Acta Mus. Nat. Pragae
  Sborník národního Musea v Praze. / Acta Musei nationalis Pragae. Praha. 1938→ .

Acta Phytogeogr. Suec.
  Acta phytogeographica suecica. Uppsala. 1929→ .

Acta Phytotax. Barcinon.
  Acta phytotaxonomica barcinonensia. Barcelona. 1969→ .

Acta Soc. Bot. Polon.
  Acta Societatis Botanicorum Poloniae. / Organ Polskiego Towarzystwa botanicznego. / Publication de la Société botanique de Pologne. Warszawa. 1923→ .

Acta Soc. Fauna Fl. Fenn.
  Acta Societatis pro Fauna et Flora fennica. Helsingforsiae. 1875→ .

Acta Univ. Carol. (Biol.)
  Acta Universitatis carolinae. Biologica. Praha. 1954→ .

Actes Soc. Linn. Bordeaux
  Actes de la Société linnéenne de Bordeaux. Bordeaux. 1830→ .

Agron. Lusit.
  Agronomia lusitana. Sacavém. 1939→ . (**25**→ , 1966→ , at Oeiras.)

Allgem. Bot. Zeitschr.
  Allgemeine botanische Zeitschrift für Systematik, Floristik, Pflanzen-geographie. Karlsruhe. 1895–1927.

Allgem. Gartenz.
  Allgemeine Gartenzeitung. Berlin. 1833–1856.

Amer. Jour. Bot.
  American Journal of Botany. Lancaster, Pennsylvania. 1914→ . (**26–36**, 1939–1949, at Burlington, Vermont; **37**→ , 1950→ , at Baltimore, Maryland.)

Amer. Midl. Nat.
  American Midland Naturalist. Notre Dame, Indiana. 1909→ .

Anais Fac. Ci. Porto
  Cf. *Ann. Sci. Acad. Polyt. Porto*

Anal. Ci. Nat.
  Anales de Historia natural. Madrid. **1–2**, 1799–1800; titled *Anales de Ciencias naturales*, **3–7**, 1801–1804.

Anal. Inst. Bot. Cavanilles
  Anales del Jardín botánico de Madrid. Madrid. **1–9**, 1941–1950; titled *Anales del Instituto botánico A. J. Cavanilles*, **10**→ , 1950→ .

Anal. Soc. Esp. Hist. Nat.
  Anales de la Sociedad española de Historia natural. Madrid. **1–30**, 1872–1902 (**21–30** also called Ser. 2, **1–10**).

Ann. Bot. Fenn.
  Annales botanici fennici. Helsinki. 1964→ .

Ann. Bot. (Roma)
  Annali di Botanica. Roma. 1903→ .

Ann. Gén. Sci. Phys. (Bruxelles)
  Annales générales des Sciences physiques. Bruxelles. **1–8**, 1819–1821.

Ann. Hist.-Nat. Mus. Hung.
  Annales historico-naturales Musei nationalis hungarici. / A Magyar nemzeti Muzeum Természetrajzi Osztályainak Folyóirata. Budapest. Ser. 1, **1–41**, 1903–1948 (**33**→ , 1940→ , Hungarian title reads *Az Országos Magyar Természettudományi Muzeum Folyóirata.*) Nov. ser., **1–9**, 1951–1958. Vol. nos. then revert to ser. 1, **51**→ , 1959→ . (With several variations of Hungarian title.)

Ann. Missouri Bot. Gard.
  Annals of the Missouri Botanical Garden. St Louis, Missouri. 1914→ .

Ann. Mus. Hist. Nat. (Paris)
  Annales du Muséum d'Histoire naturelle. Paris. **1–20**, 1802–1813.

Ann. Nat. Hist.
  Annals of natural History; or, Magazine of Zoology, Botany, and Geology. London. Ser. 1, **1–5**, 1838–1840; titled *The Annals and Magazine of natural History, including Zoology, Botany and Geology*, **6–20**, 1841–1842. Ser. 2, **1–20**, 1843–1857. Ser. 3, **1–20**, 1858–1867. Many later series.

Ann. Naturh. Mus. (Wien)
Annalen des k.k. naturhistorischen Hofmuseums. Wien. **1–31**, 1886–1917; titled *Annalen des naturhistorischen Hofmuseums*, **32**, 1918; titled *Annalen des naturhistorischen Museums in Wien*, **33→** , 1919→ .

Ann. Rep. Missouri Bot. Gard.
Annual Report. Missouri Botanical Garden. St Louis, Missouri. **1–23**, 1890–1912.

Ann. Sci. Acad. Polyt. Porto
Annaes scientificos da Academia polytechnica do Porto. Coimbra. **1–14**, 1905–1922; titled **Anais Fac. Ci. Porto**, *Anais da Faculdade de Sciências do Porto*, **15–46**, 1927–1963; titled *Anais da Faculdade de Ciências. Universidade do Porto*, **47→** , 1964→ . (**15→** at Porto.)

Ann. Sci. Nat.
Annales des Sciences naturelles. Paris. Ser. 1, **1–30**, 1824–1833. Ser. 2, **1–20**, 1834–1843. Ser. 3, **1–20**, 1844–1853. Ser. 4, **1–20**, 1854–1863. Ser. 5, **1–20**, 1864–1874. Ser. 6, **1–20**, 1875–1885. Ser. 7, **1–20**, 1885–1895. Ser. 8, **1–20**, 1895–1904. Ser. 9, **1–20**, 1905–1917. Ser. 10, **1–20**, 1919–1938. Ser. 11, **1–20**, 1939–1959. Scr. 12, **1→** , 1960→ . (Ser. 2–10, *Botanique*. Ser. 11→ , *Botanique et Biologie végétale*.)

Ann. Sci. Phys. Nat. Agric. Industr.
Annales des Sciences physiques et naturelles, d'Agriculture et d'Industrie. Lyon, Paris & Londres. Ser. 1, **1–11**, 1838–1849. Ser. 2, **1–8**, 1849–1856. Ser. 3, **1–11**, 1857–1867. Later series with other titles.

Ann. Scott. Nat. Hist.
The Annals of Scottish Natural History. Edinburgh. 1892–1911.

Ann. Soc. Linn. Lyon
Annales de la Société linnéenne de Lyon. Lyon. Ser. 1, **1836–1850/52**, 1836–1852. Nov. ser., **1–80**, 1853–1937.

Ann. Univ. Sci. Budapest. Rolando Eötvös
Annales Universitatis Scientiarum budapestinensis de Rolando Eötvös nominatae. Sectio biologica. Budapest. **1→** , 1957→ .

Annu. Cons. Jard. Bot. Genève
Annuaire de Conservatoire et du Jardin botaniques de Genève. Genève. **1–21**, 1897–1922.

Anzeig. Akad. Wiss. (Wien)
Anzeiger der kaiserlichen Akademie der Wissenschaften. Mathematisch-naturwissenschaftliche Classe. Wien. **1–55**, 1864–1918; titled *Anzeiger. Akademie der Wissenschaften in Wien. Mathematisch-naturwissenschaftliche Klasse*, **56 83**, 1919–1946; titled *Anzeiger. Österreichische Akademie der Wissenschaften. Mathematisch-naturwissenschaftliche Klasse*, **84→** , 1947→ .

Årbok Univ. Bergen
Bergens Museums Aarsberetning. Bergen. 1884–1892; titled *Bergens Museums Aarbog/Aarbok/Årbok*, 1893–1948; titled *Universitetet i Bergen Årbok*, 1949–1959; titled *Årbok for Universitetet i Bergen*, 1960→ .

Arboretum Kórnickie
Arboretum Kórnickie. Poznań. 1955→ .

Arch. Apothekerver. Nördl. Teutschl.
Archiv des Apothekervereins im nördlichen Teutschland. Hannover. **1–39**, 1821–1831.

Arch. Bot. (Forlì)
Archivio botanico. Forlì. 1925→ .

Arch. Bot. (Roemer)
Archiv für die Botanik. (Herausgegeben von D. Johann Jacob Römer.) Leipzig. **1–3**, 1796–1805.

Arch. Naturgesch. (Berlin)
Archiv für Naturgeschichte. Berlin. 1835→ .

Ark. Bot.
Arkiv för Botanik. Uppsala. 1903→ .

Atti Accad. Agiati
Atti della i.r. Accademia roveretana di Scienze, Lettere ed Arti degli Agiati. Rovereto.

Atti Riun. Sci. Ital.
Atti della Riunione degli Scienziati Italiani. 1839–1875. Various places of publication.

Baileya
Baileya. A quarterly Journal of horticultural Taxonomy. Ithaca, New York. 1953→ .

Bauhinia
Bauhinia. Zeitschrift der basler botanischen Gesellschaft. Basel. 1955→.

Beih. Bot. Centr.
Beihefte zum botanischen Centralblatt. Cassel, Jena & Dresden. 1891–1943.

Beitr. Naturk. (Dorpat)
Beiträge zur Naturkunde aus den Ostseeprovinzen Russlands. Dorpat. **1**, 1820.

Ber. Bayer. Bot. Ges.
Berichte der bayerischen botanischen Gesellschaft zur Erforschung der heimischen Flora. München. 1891→ .

Ber. Deutsch. Bot. Ges.
Berichte der deutschen botanischen Gesellschaft. Berlin. 1883→ . (**59–61**, 1941–1944, at Jena; **62→** , 1949→ , at Stuttgart.)

Ber. Tät. St. Gall. Naturw. Ges.
Bericht über die Tätigkeit der St. Gallischen Naturwissenschaftlichen Gesellschaft. St. Gallen. 1858–1901 and 1939→ .

Bergens Mus. Aarb.
Bergens Museums Aarbog/Aarbok/Årbok. Bergen. 1893-1948.

Biblioth. Bot. (Stuttgart)
Bibliotheca botanica. Abhandlungen aus dem Gesammtgebiete der Botanik. Cassel. 1886→ . (**28→** , 1894→, at Stuttgart.)

Biblioth. Univ. Genève
Bibliothèque universelle des Sciences, Belles-lettres, et Arts, faisant suite à la Bibliothèque britannique. Partie des Sciences. Genève. Ser. 1, **1–60**, 1816–1835. Ser. 2, **1–60**, 1836–1845.

Biol. Práce Slovensk. Akad. Vied (Bratislava)
Sekcie Slovenskej Akadémie Vied. Séria biologická. Práce II. Bratislava. **1**, 1955; titled *Biologické Práce, Edicia Sekcie biologických a Lekárskych Vied. Slovenskej Akadémie Vied*, **2→** , 1956→ , with minor changes of title.

Boissiera
Boissiera. Genève. 1936→ . (Supplement of *Candollea*.)

Bol. Inst. Estud. Astur. (Supl. Ci.)
Boletin del Instituto de Estudios asturianos (Suplemento de Ciencias). Oviedo. 1960→ .

Bol. Soc. Aragon. Ci. Nat.
Boletin de la Sociedad aragonesa de Ciencias naturales. Zaragoza. **1–17**, 1902–1918; titled **Bol. Soc. Ibér. Ci. Nat.**, *Boletin de la Sociedad ibérica de Ciencias naturales*, **18–33**, 1919–1934.

Bol. Soc. Brot.
Boletim da Sociedade broteriana. Coimbra. Ser. 1, **1–28**, 1880–1920. Ser. 2, **1→** , 1922→ .

Bol. Soc. Esp. Hist. Nat.
Boletin de la real Sociedad española de Historia natural. Madrid. 1901→ .

Bol. Soc. Geogr. Lisboa
Boletim da Sociedade de Geographia de Lisboa. Lisboa. 1876→ .

Bol. Soc. Ibér. Ci. Nat.
Cf. *Bol. Soc. Aragon. Ci. Nat.*

Bot. Arch. (Berlin)
Botanisches Archiv. Berlin. 1922→ .

Bot. Beobacht.
Botanische Beobachtungen. Mannheim. **1782–1783**, 1783–1784.

Bot. Centr.
Botanisches Centralblatt. Jena & Dresden. 1880→ .

Bot. Gaz.
Botanical Bulletin. Hanover, Indiana, etc. **1**, 1875–1876; titled Botanical Gazette, **2**→ , 1876→ . (**22**→ , 1896→ , at Chicago, Illinois.)

Bot. Jahrb.
Botanische Jahrbücher für Systematik, Pflanzengeschichte und Pflanzengeographie. Leipzig. 1880→ . (**69**→ , 1938→ , at Stuttgart.)

Bot. Jour. Linn. Soc.
Cf. Jour. Linn. Soc. London (Bot.).

Bot. Közl.
Cf. Növ. Közl.

Bot. Mag.
The botanical Magazine or Curtis's botanical Magazine. London. 1793→ . (For publication dates of **1**–**6** cf. F. A. Stafleu, Taxonomic Literature 92–97 (1967).) Continuous volume numbers are used in citations and series are ignored.

Bot. Not.
Botaniska Notiser. Lund. 1839→ .

Bot. Reg.
The botanical Register. London. **1**–**14**, 1815–1829; titled Edwards's botanical Register, **15**–**33**, 1829–1847. (For dates of publication cf. F. A. Stafleu, Taxonomic Literature 124–125 (1967).)

Bot. Taschenb.
Botanisches Taschenbuch für die Anfänger dieser Wissenschaft und der Apothekerkunst. Regensburg. 1790–1811.

Bot. Tidsskr.
Botanisk Tidsskrift. Kjøbenhavn. 1866→ .

Bot. Zeit.
Botanische Zeitung. Berlin. **1**–**68**, 1843–1910. (**14**–**68**, 1856–1910, at Leipzig.)

Bot. Žur.
Журналъ Русскаго ботаническаго Общества [Žurnal Russkago botaničeskago Obščestva]./Journal de la Société botanique de Russie. Petrograd. **1**–**16**, 1916–1931; titled Ботаничиский Журнал СССР [Botaničeskij Žurnal]./Journal botanique de l'URSS, Leningrad, Moscou, **17**–**32**, 1932–1947; titled Ботанический Журнал Акад. Наук СССР [Botaničeskij Žurnal Akad. Nauk SSSR], Leningrad, **33**→ , 1948→ .

Brotéria (Bot.)
Brotéria. Lisboa. 1902→ . Série botânica, **6**–**25**, 1907–1931. Série Ciências naturais, **1**→ , 1932→ .

Bul. Fac. Şti. Cernăuţi
Buletinul Facultăţii de Ştiinţe din Cernăuţi. Cernăuţi. **1**–**8**, 1927–1937.

Bull. Acad. Imp. Sci. Pétersb.
Bulletin de l'Académie impériale des Sciences de St.-Pétersbourg. St.-Pétersbourg. Ser. 1, **1**–**32**, 1860–1888. Ser. 2, **1**–**2**, 1889–1892. Different series and titles later.

Bull. Acad. Int. Géogr. Bot. (Le Mans)
Cf. Monde Pl.

Bull. Acad. Sci. URSS
Bulletin de l'Académie des Sciences de l'Union des Républiques Soviétiques Socialistes./Известия Академии Наук Союза Советских Социалистических Республик [Izvestija Akademii Nauk Sojuza Sovetskikh Socialističeskikh Respublik]. Leningrad. Ser. 7, 1931–1936. Other series with other titles.

Bull. Alp. Gard. Soc.
Bulletin of the Alpine Garden Society (of Great Britain). Various places of publication. 1930→ .

Bull. Herb. Boiss.
Bulletin de l'Herbier Boissier. Genève & Bâle. Ser. 1, **1**–**7**, 1893–1899. Ser. 2, **1**–**8**, 1900–1909.

Bull. Inst. Jard. Bot. Univ. Beograd
Гласник ботаничког Завода и Баште Универзитета у Београду [Glasnik botaničkog Zavoda i Bašte Univerziteta u Beogradu]. / Bulletin de l'Institut et du Jardin botaniques de l'Université de Beograd. Beograd. **1**–**4**, 1928–1937.

Bull. Inst. Roy. Hist. Nat. (Sofia)
Bulletin des Institutions royales d'Histoire Naturelle. Sofia. 1928–1943.

Bull. Int. Acad. Tchèque Sci.
Bulletin international. Académie des Sciences de l'Empereur François Joseph. Classe des Sciences mathématiques et naturelles. Prague. **1**–**20**, 1895–1916; titled Bulletin international. Académie des Sciences, **21**–**22**, 1919–1920; titled Bulletin international. Académie tchèque des Sciences, **23**–**53**, 1923–1953.

Bull. Jard. Bot. Bruxelles
Bulletin du Jardin botanique de l'État à Bruxelles. Bruxelles. **1**–**36**, 1920–1966. (**15**–**36**, 1938–1966, with alternative title Bulletin van den [de] Rijksplantentuin Brussel); titled **Bull. Jard. Bot. Nat. Belg.**, Bulletin du Jardin botanique national de Belgique. Bulletin van de nationale Plantentuin van Belge, **37**→ , 1967→ .

Bull. Jard. Bot. Kieff
Bulletin du Jardin botanique de Kieff./Вісник Київського ботанічного Саду [Visnyk Kyjivs'kogo botaničnogo Sadu]. Kieff. 1924–1934.

Bull. Jard. Bot. Pétersb.
Bulletin du Jardin impérial botanique de St.-Pétersbourg. / Извѣстія императорскаго С.-Петербургскаго ботаническаго Сада [Izvěstija imperatorskago S.-Peterburgskago botaničeskago Sada]. S.-Peterburg. **1**–**12**, 1901–1912; titled Bulletin du Jardin impérial botanique de Pierre le Grand. / Извѣстія императорскаго ботаническаго Сада Петра Великаго [Izvěstija imperatorskago botaničeskago Sada Petra Velikago], **13**–**17**, 1912–1917; titled **Bull. Jard. Bot. URSS**, Bulletin du principal Jardin botanique de la République Russe. / Известия главного ботанического Сада Р.С.Ф.С.Р. [Izvestija glavnogo botaničeskogo Sada R.S.F.S.R.], **18**–**22**, 1918–1923; titled Bulletin du Jardin botanique principal de la République Russe./ [Russian title as before], **23**–**24**, 1924–1925; titled Bulletin du Jardin Botanique principal de l'U.R.S.S. / Известия главного ботанического Сада С.С.С.Р. [Izvestija glavnogo botaničeskogo Sada S.S.S.R.], **25**–**30**, 1926–1932.

Bull. Jard. Bot. URSS
Cf. Bull. Jard. Bot. Pétersb.

Bull. Mus. Nat. Hist. Nat. (Paris)
Bulletin du Muséum d'Histoire naturelle. Paris. **1**–**12**, 1895–1906; titled Bulletin du Muséum national d'Histoire naturelle, **13**→ , 1907→ .

Bull. Phys.-Math. Acad. Pétersb.
Bulletin de la Classe physico-mathématique de l'Académie impériale des Sciences de St. Pétersbourg. St. Pétersbourg & Leipzig. **1**–**17**, 1843–1859.

Bull. Res. Counc. Israel
Bulletin of the Research Council of Israel. Jerusalem. **1**–**11** (from **5** onwards Botany in Section D), 1951–1963; titled Israel Jour. Bot., Israel Journal of Botany, **12**→ , 1963→ .

Bull. Sci. Acad. Imp. Sci. Pétersb.
Bulletin scientifique publié par l'Académie impériale des Sciences de Saint-Pétersbourg. Saint-Pétersbourg & Leipzig. **1**–**10**, 1836–1842.

Bull. Soc. Bot. Belg.
Bulletin de la Société royale de Botanique de Belgique. Bruxelles. 1862→ .

Bull. Soc. Bot. Bulg.
Извѣстия на българското ботаническо Дружество [Izvěstija

na bălgarskoto botaničesko Družestvo]. | *Bulletin de la Société botanique de Bulgarie*. Sofija. **1–9**, 1926–1943.

Bull. Soc. Bot. Fr.
Bulletin de la Société botanique de France. Paris. 1854→ .

Bull. Soc. Bot. Genève
Bulletin des Travaux de la Société botanique de Genève. Genève. Ser. 1, **1–11**, 1879–1905. Ser. 2, titled *Bulletin de la Société botanique de Genève*, **1–43**, 1909–1952.

Bull. Soc. Étud. Sci. Angers
Bulletin de la Société d'Études scientifiques d'Angers. Angers. **1–89** (=nov. ser., **2**), 1871–1960; titled *Bulletin de la Société d'Études scientifiques d'Anjou*, **90** (=nov. ser., **3**)→, 1962→.

Bull. Soc. Hist. Nat. Afr. Nord
Bulletin de la Société d'Histoire naturelle de l'Afrique du Nord. Alger. 1909→ .

Bull. Soc. Hist. Nat. Savoie
Bulletin de la Société d'Histoire Naturelle de Savoie. Chambéry. 1850–1886. Ser. 1, **1–7**, 1887–1895. Ser. 2, **1–23**, 1895–1935.

Bull. Soc. Nat. Moscou
Bulletin de la Société impériale des Naturalistes de Moscou. Section biologique. Moscou. Ser. 1, **1–62**, 1829–1886. Nov. ser., **1**→ , 1887→ . (Nov. ser., **31**→ , 1922→ , with alternative title in Russian Бюллетень Московскаго Общества испытателей Природы [*Bjulleten' Moskovkago Obščestva ispytatelej Prirody*]. Nov. ser., **52(5)**→ , 1947→ , with Russian title only.)

Bull. Soc. Neuchâtel. Sci. Nat.
Cf. *Bull. Soc. Sci. Nat. Neuchâtel*.

Bull. Soc. Philom. Paris
Bulletin de la Société philomathique de Paris. Paris. 1791→ .

Bull. Soc. Polymath. Morbihan
Bulletin de la Société polymathique du Morbihan. Vannes. 1861–1894.

Bull. Soc. Sci. Nancy
Bulletin de la Société des Sciences de Nancy. Nancy. Ser. 2, **1–16**, 1874–1900. Ser. 3, **1–15**, 1900–1915. Ser. 4, **1–3**, 1920–1929. Ser. 5, **1–19**, 1936–1960.

Bull. Soc. Sci. Nat. Neuchâtel
Bulletin de la Société des Sciences naturelles de Neuchâtel. Neuchâtel. **1–25**, 1847–1897; titled **Bull. Soc. Neuchâtel. Sci. Nat.**, *Bulletin. Société Neuchâteloise des Sciences naturelles*, **26**→ , 1898→ .

Bull. Soc. Vaud. Sci. Nat.
Bulletin de la Société vaudoise des Sciences naturelles. Lausanne. 1842→ .

Bull. Torrey Bot. Club
Bulletin of the Torrey botanical Club. New York. 1870→. (**73–78**, 1946–1951, titled *Bulletin of the Torrey botanical Club and Torreya*.)

Butll. Inst. Catalana Hist. Nat.
Butlletí de la Institució catalana d'Història natural. Barcelona. 1901→ .

Canad. Jour. Bot.
Canadian Journal of Botany. Ottawa. **29**→ , 1951→ . (Formerly *Canadian Journal of Research, Sect. C, Botanical Sciences*.)

Candollea
Candollea. Organe du Conservatoire et du Jardin botaniques de la Ville de Genève. Genève. 1922→ .

Carinthia II
Carinthia II/Mitteilungen des Vereines naturkundlicher Landesmuseum für Kärnten/Mitteilungen des naturhistorichen Landesmuseums für Kärnten. Klagenfurt. **81**→ , 1891→ .

Cavanillesia
Cavanillesia. Rerum botanicarum Acta. Barcinone. **1–8**, 1928–1938.

Collect. Bot. (Barcelona)
Collectanea botanica a barcinonensi botanico Instituto edita. Barcinone. 1946→ .

Comment. Gotting.
Commentarii Societatis regiae Scientiarum gottingensis. **1–4**, 1751–1754. Continued as **Novi Comment. Gotting.**, *Novi Commentarii Societatis regiae Scientiarum gottingensis*. **1–8**, 1771–1778. Continued as **Comment. Gotting.**, Ser. 2. *Commentationes Societatis regiae Scientiarum gottingensis*. **1–16**, 1779–1801. Continued as **Comment. Gotting. Recent.**, *Commentationes Societatis regiae Scientiarum gottingensis Recentiores*. **1–8**, 1811–1841.

Compt. Rend. Soc. Hallér.
Compte-rendu des Travaux de la Société hallérienne. Genève. **1–4**, 1852–1856. (**1**: pp. 1–12 in 1852–1853; **2**: pp. 13–76 in 1853–1854; **3**: pp. 77–90 in 1854–1855; **4**: pp. 93–184 in 1854–1856.)

Contr. Gray Herb.
Contributions from the Gray Herbarium of Harvard University. Cambridge, Massachusetts. Nov. ser., **1**→, 1891→.

Denkschr. Akad. Wiss. Math.-Nat. Kl. (Wien)
Denkschriften der kaiserlichen Akademie der Wissenschaften. Mathematisch-naturwissenschaftlichen Classe. Wien. **1–95**, 1850–1918; titled *Denkschriften. Akademie der Wissenschaften in Wien. Mathematisch-naturwissenschaftlichen Klasse*, **96–107**, 1919–1943; titled *Denkschriften. Österreichische Akademie der Wissenschaften. Mathematisch-naturwissenschaftliche Klasse*, **108**→ , 1947→ .

Denkschr. Akad. Wiss. München
Denkschriften der Akademie der Wissenschaften zu München. München. **1–9**, 1809–1825.

Denkschr. Schweiz. Naturf. Ges.
Cf. *Neue Denkschr. Schweiz. Ges. Naturw.*

Deutsche Bot. Monatsschr.
Deutsche botanische Monatsschrift. Sondershausen. **1–22**, 1883–1912. (**6–22** at Arnstadt.)

Deutsche Heilpfl.
Deutsche Heilpflanze. Stollberg i. E. 1934→ .

Dict. Sci. Nat.
Dictionnaire des Sciences naturelles. Paris. **1–60**, 1804–1830.

Econ. Bot.
Economic Botany. New York. 1947→ .

Edinb. New Philos. Jour.
The Edinburgh new philosophical Journal, exhibiting a View of the progressive Improvements and Discoveries in the Sciences and the Arts. Edinburgh. Ser. 1, **1–57**, 1826–1854. Nov. ser. **1–19**, 1855–1864.

Evolution
International Journal of organic Evolution. Lancaster, Pennsylvania. 1947→ .

Farmacognosia (Madrid)
Farmacognosia. Anales del Instituto José Celestino Mutis. Madrid. 1942→ .

Feddes Repert.
Repertorium Specierum novarum Regni vegetabilis. Berlin. **1–51**, 1905–1942; titled *Feddes Repertorium Specierum novarum Regni vegetabilis*, **52**→ , 1943. (This work includes *Repertorium europaeum et mediterraneum*, indicated by dual pagination.) **Beih.**, *Beihefte*. Berlin. **1**→ , 1914→ .

Flora (Regensb.)
Flora, oder allgemeine botanische Zeitung. Regensburg. 1818→ . (**72–95** at Marburg; **96**→ at Jena.)

Folia Geobot. Phytotax. (Praha)
Folia geobotanica & phytotaxonomica bohemoslovaca. Praha. 1966→ .

Fragm. Fl. Geobot.
Fragmenta floristica et geobotanica. Kraków. 1945→ .

Gard. Chron.
Gardeners' Chronicle and agricultural Gazette. London. Ser. **1**, **1841–1873**, 1841–1873. Ser. 2, titled *The Gardeners' Chronicle. A weekly illustrated Journal of Horticulture and allied Subjects*, **1–26**, 1874–1886. Ser. 3, **1**→ , 1887. (**140–154**, 1956–1963, titled *Gardeners' Chronicle and Gardening illustrated*; **155**→ , 1964→ , titled *Gardeners' Chronicle, Gardening illustrated and the Greenhouse.*)

Gentes Herb.
Gentes Herbarum. Occasional Papers on the Kinds of Plants. Ithaca, N.Y. 1920→ .

Ges. Naturf. Freunde Berlin Neue Schr.
Der Gesellschaft naturforschender Freunde zu Berlin, neue Schriften. Berlin. 1795–1803.

Gior. Bot. Ital.
Giornale botanico italiano. Firenze. **1–2**, 1844–1852.

Gior. Fis. (Brugnat.)
Giornale di Fisica, Chimica e Storia naturale, ossia Raccolta di Memorie sulle Scienze, Arti e Manifatture ad esse relative di L. Brugnatelli. Pavia. Ser. 1, **1–5**, 1808–1812; titled *Giornale di Fisica, Chimica, Storia naturale, Medicina ed Arti del Regno italico*, **6–10**, 1813–1817. Ser. 2, titled *Giornale di Fisica, Chimica e Storia naturale, Medicina ed Arti*, **1–10**, 1818–1827.

Glas Srpske Kralj. Akad.
Глас Српске Краљевске Академије [*Glas Srpske Kraljevske Akademije*]. Beograd. 1888–1941.

Glasn. Muz. Bosni Herceg.
Гласник Земаљског Музеја у Босни и Хеьцеговини [*Glasnik Zemaljskog Muzeja u Bosni i Hercegovini*]. Sarajevo, **1–52**, 1889–1940.

God. Sof. Univ. (Agron. Fak.)
Годишник на Софийския Университет. Агрономически Факултет [*Godišnik na Sofijskija Universitet. Agronomičeski Fakultet*]./Annuaire de l'Université de Sofia. Faculté Agronomique. Sofija. 1923→ .

Gorteria
Gorteria. Leiden. 1961→ .

Hercynia
Hercynia. Abhandlungen der Botanischen Vereinigung Mitteldeutschlands. Halle. 1937→ .

Hereditas
Hereditas. Genetiskt Arkiv. Lund. 1920→ .

Hist. Comment. Acad. Elect. Theod.-Palat.
Historia et Commentationes. Academia electoralis Scientiarum et elegantiorum Literarum Theodoro-Palatina. Mannheim. 1766–1794.

Hook. Ic.
Hooker's Icones Plantarum. London. 1836→ .

Illinois Biol. Monogr.
Illinois biological Monographs. Urbana, Illinois. **1**→, 1914→.

Israel Jour. Bot.
Cf. *Bull. Res. Counc. Israel.*

Izv. Bot. Inst. (Sofia)
Известия на ботаническия Институт [*Izvestija na botaničeskija Institut*]./Bulletin de l'Institut botanique. Sofia. 1950→ . (From **8**, 1961, alternative title is *Mitteilungen des botanischen Instituts.*)

Izv. Kazakhsk. Fil. Akad. Nauk SSSR (Ser. Bot.)
Известия Казахского Филиала, Академия Наук СССР. Серия ботаническая [*Izvestija Kazakhskogo Filiala, Akademija Nauk SSSR. Serija botaničeskaja*]. Alma Ata. 1944→ .

Jour. Agr. Bot. (Kharkov)
Journal of agricultural Botany. / Труди сільско-господарської Ботаніки [*Trudy sil'sko-gospodarskoji Botaniki*]. Kharkiv. 1926–1929.

Jour. Arnold Arb.
Journal of the Arnold Arboretum. Cambridge, Massachusetts. 1919→ . (**2(3)–13**, 1921–1932, at Lancaster, Pennsylvania; **14–26(2–3)**, 1933–1955, at Jamaica Plain, Massachusetts.)

Jour. Bot. Kew Gard. Misc.
Hooker's Journal of Botany and Kew Garden Miscellany. London. **1–9**, 1849–1857.

Jour. Bot. (London)
The Journal of Botany, British and foreign. London. **1–80**, 1863–1942.

Jour. Bot. (Paris)
Journal de Botanique. Paris. Ser. 1, **1–20**, 1887–1906. Ser. 2, **1–3**, 1907–1925.

Jour. Ecol.
The Journal of Ecology. Cambridge. 1913→ . (**44**→ , 1956→ , at Oxford.)

Jour. Hist. Nat. (Paris)
Journal d'Histoire naturelle. Paris. **1–2**, 1792.

Jour. Hort. Soc.
The Journal of the Horticultural Society. London. 1846–?1864.

Jour. Inst. Bot. Acad. Sci. Ukr.
Journal de l'Institut botanique de l'Académie des Sciences de la R.S.S. d'Ukraine./Журнал Інституту ботаніки Акад. Наук УРСР. [*Žurnal Instytutu botaniky Akad. Nauk URSR*]. Kieff. 1933–1939.

Jour. Linn. Soc. London (Bot.)
The Journal of the Proceedings of the Linnean Society. Botany. London. **1–7**, 1856–1864; titled *The Journal of the Linnean Society. Botany*, **8–46**, 1865–1924; titled *The Journal of the Linnean Society of London. Botany*, **47–61**, 1925–1968; titled *Bot. Jour. Linn. Soc., Botanical Journal of the Linnean Society*, **62**→ , 1969→ .

Jour. Roy. Hort. Soc.
The Journal of the Royal Horticultural Society. London. 1865→ .

Jour. Washington Acad. Sci.
Journal of the Washington Academy of Sciences. Washington, D.C. 1911→.

Kew Bull.
Bulletin of miscellaneous Information. Royal Gardens, Kew. London. **1887–1941**, 1887–1942; titled *Kew Bulletin*, **1**→, 1946→.

Kong. Danske Vid. Selsk. Forhand.
Oversigt over det Kongelige danske videnskabernes Selskabs Forhandlinger. København. 1806–1931.

Königsb. Arch. Naturw.
Königsberger Archiv für Naturwissenschaft und Mathematik. Königsberg. 1812→ .

Kulturpfl.
Die Kulturpflanze. Berichte und Mitteilungen aus dem Institut für Kulturpflanzenforschung der deutschen Akademie der Wissenschaften zu Berlin. Berlin. 1953→ . **Beih.**, *Beihefte*. 1956→.

Kungl. Svenska Vet.-Akad. Handl.
Kongl. svenska Vetenskaps Academiens Handlingar. Stockholm. Ser. 1, **1–40**, 1739–1779 (*svenska* omitted after 1755). Nov. ser., titled *Kongl. Vetenskaps Academiens nya Handlingar*, **1–33**, 1780–1812; titled *Kongl. Vetenskaps Academiens Handlingar*, **1813–1846**, 1813–1846; titled *Kongl. Vetenskaps-Akademiens Handlingar*, **1846–1858**, 1846–1858. Nov. ser., titled *Kongliga svenska Vetenskaps-Akademiens Handlingar. Ny Földj*, **1–35**, 1855–1902; titled *Kungliga svenska Vetenskaps-Akademiens Handlingar*, **36–63**, 1902–1923. Ser. 3, **1–25**, 1924–1948. Ser. 4, **1**→ , 1951→ .

Linnaea
Linnaea. Ein Journal für die Botanik in ihrem ganzen Umfange. Berlin. **1–43**, 1826–1882. (**9–34**, 1834–1866, at Halle.) (**1** in

1826; **2** in 1827; **3** in 1828; **4** in 1829; **5** in 1830; **6** in 1831; **7** in 1832; **8** in 1833; **9**: pp. 1–402 in 1834; pp. 403–758 in 1835; **10**: pp. 1–368 in 1835; pp. 369–758 in 1836; **11** in 1837; **12** in 1838; **13** in 1839; **14**: pp. 1–528 in 1840; pp. 529–728 in 1841; **15** in 1841; **16** in 1842; **17**: pp. 1–640 in 1843; pp. 641–764 in 1844; **18**: pp. 1–512 in 1844; pp. 513–774 in 1845; **19**: pp. 1–512 in 1846; pp. 513–765 in 1847; **20** in 1847; **21** in 1848; **22** in 1849; **23** in 1850; **24**: pp. 1–640 in 1851; pp. 641–804 in 1852; **25**: pp. 1–256 in 1852; pp. 257–772 in 1853; **26**: pp. 1–384 in 1854; pp. 385–807 in 1855; **27**: pp. 1–128 in 1855; pp. 129–799 in 1856; **28**: pp. 1–256 in 1856; pp. 257–640 in 1857; pp. 641–767 in 1858; **29**: pp. 1–384 in 1858; pp. 385–764 in 1859; **30**: pp. 1–256 in 1859; pp. 257–640 in 1859–1860; pp. 641–779 in 1861; **31** in 1861–1862; **32** in 1863; **33** in 1864–1865; **34** in 1865–1866; **35**: pp. 1–512 in 1867–1868; pp. 513–637 in 1868; **36**: pp. 1–256 in 1869; pp. 257–790 in 1870; **37**: pp. 1–544 in 1872; pp. 545–663 in 1873; **38**: pp. 1–144 in 1873; pp. 145–753 in 1874; **39** in 1875; **40** in 1876; **41**: pp. 1–117 in 1876; pp. 118–576 in 1877; pp. 577–655 in 1878; **42**: pp. 1–192 in 1878; pp. 193–667 in 1879; **43**: pp. 1–66 in 1880; pp. 67–252 in 1881; pp. 253–554 in 1882. Cf. R. C. Foster, *Jour. Arnold Arb.* **43**: 400–409 (1962).)

London Jour. Bot. (Hooker)
The London Journal of Botany. London. **1–7**, 1842–1848. (**1** in 1842; **2** in 1843; **3** in 1844; **4** in 1845; **5** in 1846; **6** in 1847; **7** in 1848.) Edit. W. J. Hooker.

Lunds Univ. Årsskr.
Lunds Universitets Årsskrift. | *Acta Universitatis lundensis.* Lund. Ser. 1, **1–40**, 1864–1902. Nov. ser., **1–59**, 1905–1963. *Sectio 2, Medica, Mathematica, Scientiae Rerum naturalium*, **1→** , 1964→ .

Magyar Bot. Lapok
Magyar botanikai Lapok. | *Ungarische botanische Blätter.* Budapest. **1–33**, 1902–1934.

Magyar Növ. Lapok
Magyar növénytani Lapok. Kolozsvár. **1–15**, 1877–1892.

Magyar Tudós Társaság Évkönyvei (Budapest)
A 'Magyar tudós Társaság' Évkönyvei. Budapest. **1–7**, 1831–1846; titled *A Magyar Tudom. Akadémia Évkönyvei*, **8–17**, 1847–1889.

Malpighia
Malpighia. Rassegna mensuale di Botanica. Messina. **1–34**, 1886–1937. (**3–23**, 1889–1909, at Genova; **24–29**, 1911–1923, at Catania; **30–31**, 1927–1928, at Palermo; **32–34**, 1932–1937, at Bologna.)

Mat. Étude Fl. Géogr. Bot. Or.
Matériaux pour servir à l'Étude de la Flore et de la Géographie botanique de l'Orient (Missions du Ministère de l'Instruction publique en 1904 et en 1906). Nancy. **1–7**, 1906–1922.

Math. Term. Közl.
Mathematikai és természettudományi Közlemények, vonatkozólag a hazai Viszonyokra. Budapest. 1861–1944.

Med. Reposit. (New York)
The medical Repository. New York. Ser. 1, **1–15**, 1797–1812. Nov. ser., **1–15**, 1812–1826.

Meddel. Grønl.
Meddelelser om Grønland, af Kommissionen for Ledelsen af de geologiske og geographiske Undersøgelsen in Grønland. København. 1879→ .

Meddel. Soc. Fauna Fl. Fenn.
Meddelanden af Societas pro Fauna et Flora fennica. Helsingfors. **1–50**, 1876–1925.

Mélang. Philos. Math. Soc. Roy. Turin (Misc. Taur.)
Miscellanea philosophica-mathematica Societatis privatae taurinensis. Turin. **1**, 1759; titled *Mélanges de Philosophie et de Mathématique de la Société royale de Turin. Miscellanea taurinensia*, **2–5**, 1760–1774.

Mém. Acad. Montp. (Sect. Méd.)
Mémoires de l'Académie des Sciences et Lettres de Montpellier. Sect. Médecine. Montpellier. 1849–1910.

Mém. Acad. Sci. Pétersb.
Записки имп. Академіи Наукъ (по физико-математическому Отдѣленію) [*Zapiski imp. Akademii Nauk (po fiziko-matematičeskomu Otděleniju)*]. | *Mémoires de l'Académie impériale des Sciences de St. Pétersbourg (Classe des Sciences physiques et mathématiques).* St. Pétersbourg. Ser. 5, **1–11**, 1809–1830. Ser. 6, **1–10**, 1831–1859. Ser. 7, **1–42**, 1859–1897.

Mém. Acad. Sci. Toulouse
Histoire et Mémoires de l'Académie royale des Sciences, Inscriptions et Belles-lettres de Toulouse. Toulouse. Ser. 1, **1–4**, 1782–1792. Ser. 2, **1–6**, 1827–1843. Ser. 3, titled *Mémoires de l'Académie royale des Sciences, Inscriptions et Belles-lettres de Toulouse*, **1–6**, 1844–1850. A further 9 series, with numerous minor changes of title, follow.

Mém. Cour. Acad. Roy. Sci. Belg.
Mémoires sur les Questions proposées par l'Académie royale des Sciences et Belles-Lettres de Bruxelles. Bruxelles. **1–5**, 1818–1826; titled *Mémoires couronnés par l'Académie royale des Sciences et Belles-Lettres de Bruxelles*, **6–15**, 1827–1841; titled *Mémoires couronnés et Mémoires des Savants étrangers publiés par l'Académie royale des Sciences et Belles-Lettres de Bruxelles*, **16–18**, 1844–1845; titled *Mémoires couronnés et Mémoires des Savants étrangers publiés par l'Académie royale des Sciences, des Lettres et des Beaux-Arts de Belgique*, **19–62**, 1845–1904.

Mém. Herb. Boiss.
Mémoires de l'Herbier Boissier. Genève. 1900→ .

Mem. Ist. Veneto
Memorie dell' i. r. Istituto Veneto di Scienze, Lettere ed Arti. Venezia. **1–13**, 1843–1866; titled *Memorie del reale Istituto Veneto di Scienze, Lettere ed Arti*, **14→** , 1868→ .

Mem. Mus. Ci. Nat. Barcelona (Bot.)
Memòries del Museu de Ciències naturals de Barcelona. Sèrie botànica. | *Memorias del Museo de Ciencias naturales de Barcelona. Serie botanica.* Barcelona. **1**, 1922–1925. (**1**(1) in 1922; **1**(2) in 1924; **1**(3) in 1925.)

Mém. Mus. Hist. Nat. (Paris)
Mémoires du Muséum d'Histoire naturelle. Paris. **1–20**, 1815–1832.

Mém. Sav. Étr. Pétersb.
Mémoires des Savants étrangers. Mémoires présentés à l'Académie impériale des Sciences de St. Pétersbourg par divers Savants et lus dans ses Assemblées. St. Pétersbourg. **1–9**, 1831–1859.

Mém. Soc. Linn. Paris
Mémoires de la Société linnéenne de Paris. **1–6**, 1822–1828.

Mém. Soc. Nat. Moscou
Mémoires de la Société impériale des Naturalistes de l'Université de Moscou. Moscou. **1–6**, 1806–1823.

Mém. Soc. Nat. Sci. Cherbourg
Mémoires de la Société nationale des Sciences naturelles et mathématiques de Cherbourg. Cherbourg. 1852→ .

Mém. Soc. Phys. Hist. Nat. Genève
Mémoires de la Société de Physique et d'Histoire naturelle de Genève. Genève. 1821 → .

Mém. Soc. Sci. Bohême
Cf. *Sitz.-Ber. Böhm. Ges. Wiss.*

Mem. Torrey Bot. Club
Memoirs of the Torrey botanical Club. New York. 1889 → . (**18–19**, 1931–1941, at Menasha, Wisconsin; **20**, 1943–1954, at Lancaster, Pennsylvania; **21→** , 1958→ at Durham, North Carolina.)

Mitt. Bot. Staatssamm. (*München*)
Mitteilungen aus der botanischen Staatssammlung. München. 1950→ .

Mitt. Deutsch. Dendrol. Ges.
Mitteilungen der deutschen dendrologischen Gesellschaft. Berlin. 1893→ . (1894–1912 at Poppelsdorf-Bonn; 1913–1932 at Thyrow; 1933 at Berlin; 1934 at Berlin & Dortmund; 1935–1942 at Dortmund; 1950→ at Darmstadt.)

Mitt. Geogr. Ges. Thür. Jena
Mitteilungen der geographischen Gesellschaft (für Thüringen) zu Jena. Jena. 1882→ .

Mitt. Naturw. Ver. Steierm.
Mitteilungen des naturwissenschaftlichen Vereins für Steiermark. Graz. 1863→ .

Mitt. Naturw. Ver. Wien
Mitteilungen des naturwissenschaftlichen Vereins an der Universität Wien. Wien. **1–12**, 1894–1914.

Mitt. Thür. Bot. Ges.
Mitteilungen der thüringischen botanischen Gesellschaft. Weimar. **1–2**, 1949–1960.

Mitt. Thür. Bot. Ver.
Mitteilungen des thüringischen botanischen Vereins. Weimar. Ser. 1, **1–9**, 1882–1890. Nov. ser., **1–51**, 1891–1944.

Monde Pl.
Le Monde des Plantes. Revue mensuelle de Botanique. Organe de l'Académie internationale de Géographie botanique. Le Mans. **1–8**: p. 56, 1891–1898. Continued as *Le Monde des Plantes*, **1**→ , 1899→ (later published elsewhere). Also continued as **Bull. Acad. Int. Géogr. Bot.** (*Le Mans*), *Bulletin de l'Académie internationale de Géographie botanique*, **8**: p. 47 [57]–**19**, 1899–1910; titled *Bulletin de Géographie botanique. Organe mensuel de l'Académie internationale de Botanique*, **21–27**, 1911–1919. (**16–20** at Paris.)

Nat. Monsp. (*Bot.*)
Naturalia monspeliensia. Série botanique. Montpellier. 1955→ .

Natura (*Milano*)
Natura. Milano. **1**→ , 1909→ .

Naturalist (*Leeds*)
The Naturalist. London. 1864→ . (Later vols. published at Leeds or Huddersfield.)

Naturaliste Canad.
Le Naturaliste Canadien. Québec. **1**→ , 1869→ .

Naturaliste (*Paris*)
Le Naturaliste. Journal des Échanges et des Nouvelles. Paris. 1879–1910.

Nederl. Kruidk. Arch.
Nederlandsch kruidkundig Archief. Leiden. Ser. 1, **1–5**, 1846–1870. Ser. 2, **1–6**, 1871–1895. Ser. 3, **1–2**, 1896–1904. Years are then used as vol. nos., **1904–1932**, 1904–1932. Vol. nos. then revert to ser. 1, **43–57**, 1933–1951. (Ser. 2, ser. 3 & **1904–1913**(1) at Nijmegen; **1913**(2)–**1919** at Groningen; **1920–1921** at Utrecht; **1922**–57 at Amsterdam.)

Neue Denkschr. Schweiz. Ges. Naturw.
Neue Denkschriften der allgemeinen schweizerischen Gesellschaft für die gesammten Naturwissenschaften. / *Nouveaux Mémoires de la Société helvétique des Sciences naturelles.* Neuchâtel. **1–40**, 1837–1906; titled *Neue Denkschriften der schweizerischen naturforschenden Gesellschaft*, same French title, **41–54**, 1906–1918; titled **Denkschr. Schweiz. Naturf. Ges.**, *Denkschriften der schweizerischen naturforschenden Gesellschaft.* / *Mémoires de la Société helvétique des Sciences naturelles*, **55**→, 1920→. (**8–9**, 1847, at Neuenburg; **10**, 1849, at Neuchâtel; **11**→, 1850→, at Zürich.)

Neues Mag. Bot. (*Roemer*)
Neues Magazin für die Botanik in ihrem ganzen Umfange. Herausgegeben von J. J. Roemer. Zürich. **1**, 1794.

New Phytol.
The new Phytologist. London. 1902→ . (**29–54**, 1930–1955, at Cambridge; **55**→ , 1956→ , at Oxford.)

Not. Syst. Herb. Inst. Bot. Zool. Acad. Uzbek.
Notulae systematicae ex Herbario Instituti botanici et zoologici Academiae Scientiarum uzbekistanicae./Ботанические Материалы Гербария Института ботаники и зоологии Академии Наук УзССР [*Botaničeskie Materialy Gerbarija Instituta botaniki i zoologii Akademii Nauk UzSSR*]. Taschkent. With minor changes of title.

Not. Syst. (*Leningrad*)
Ботанические Материалы Гербария главного ботанического Сада Р.С.Ф.С.Р. [*Botaničeskie Materialy Gerbarija glavnogo botaničeskogo Sada R.S.F.S.R.*]. / *Notulae systematicae ex Herbario Horti botanici petropolitani.* Mosqua & Leningrad. **1–4**, 1919–1923; titled the same in Russian and *Notulae systematicae ex Herbario Horti botanici Reipublicae rossicae*, **5**, 1924; titled Ботанические Материалы Гербария главного ботанического Сада СССР [*Botaničeskie Materialy Gerbarija glavnogo botaničeskogo Sada SSSR*]. / *Notulae systematicae ex Herbario Horti botanici URSS*, **6**, 1926; titled Ботанические Материалы Гербария ботанического Института Академии Наук СССР [*Botaničeskie Materialy Gerbarija botaničeskogo Instituta Akademii Nauk SSSR*]. / *Notulae systematicae ex Herbario Instituti botanici Akademiae Scientiarum URSS*, **7–8**(3), 1937–1938; titled Ботанические Материалы Гербария Института имени В. Л. Комарова Академии Наук СССР [*Botaničeskie Materialy Gerbarija botaničeskogo Instituta imeni V. L. Komarova Akademii Nauk SSSR*]. / *Notulae systematicae ex Herbario Instituti botanici nomine V. L. Komarovii Academiae Scientiarum U.R.S.S.*, **8**(4)–**22**, 1940–1963.

Notes Roy. Bot. Gard. Edinb.
Notes from the Royal Botanic Garden, Edinburgh. Edinburgh. 1900→ .

Notizbl. Bot. Gart. Berlin
Notizblatt des königl. botanischen Gartens und Museums zu Berlin. Leipzig. **1–15**, 1895–1944; with minor changes of title, becoming *Notizblatt des botanischen Gartens und Museums zu Berlin-Dahlem* from 1920 to 1944. (**6–7**, 1913–1920, at Leipzig & Berlin; **8–15**, 1921–1944, at Berlin-Dahlem.)

Nouv. Arch. Mus. Hist. Nat. Paris
Nouvelles Archives du Muséum d'Histoire naturelle de Paris. Paris. Ser. 1, **1–10**, 1865–1874. Ser. 2, **1–10**, 1878–1888. Ser. 3, **1–10**, 1889–1898. Ser. 4, **1–10**, 1899–1908. Ser. 5, **1–6**, 1909–1914.

Növ. Közl.
Növénytani Közlemények. Budapest. **1–7**, 1902–1908; titled **Bot. Közl.**, *Botanikai Közlemények*, **8**→ , 1909 → .

Nov. Syst. Pl. Vasc. (*Leningrad*)
Новости систематики высших Растений [*Novosti sistematiki vysšikh Rastenij*]. / *Novitates systematicae Plantarum vascularium.* Moskva & Leningrad. 1964→ .

Nova Acta Acad. Leop.-Carol.
Nova Acta (Leopoldina) Academiae Caesareae Leopoldino-Carolinae germanicae Naturae curiosorum. Norimbergae; later volumes elsewhere. Ser. 1, 1757–1928. Nov. ser., 1934→ . Titled also in German (in some volumes only in German) *Verhandlungen der kaiserlichen Leopoldino-Carolinischen deutschen Akademie der Naturforscher.* The name of the Academy has varied greatly; any of the adjectives *Leopoldinisch-Carolinische*, *kaiserliche* and *deutsche* may or may not appear.

Nova Acta Acad. Sci. Petrop.
Nova Acta Academiae Scientiarum imperialis petropolitana. Petropoli. **1–15**, 1787–1806.

Novit. Bot. Horti Bot. Univ. Carol. Prag.
Novitates botanicae et Delectus Seminum, Fructuum, Sporar-

umque Anno...collectorum, quae Praefectus Horti botanici Universitatis carolinae pragensis libentissime pro mutua Commutatione offert. Praga. 1960→ . (With minor variations of title up to 1964, but thereafter titled *Novit. Bot. Inst. Bot. Univ. Carol. Prag., Novitates botanicae cum Delectu Seminum, Fructuum, Sporarum Plantarumque quas Institutum botanicum et Hortus botanicus Universitatis carolinae pragensis in Anno... libentissime pro mutua Commutatione offert.*)

Nuovo Gior. Bot. Ital.
Nuovo Giornale botanico italiano. Firenze. Ser. 1, **1–25**, 1869–1893. Nov. ser., **1–68**, 1894–1961; titled *Giornale botanico italiano*, **69**→ , 1962→ .

Nuovo Racc. Opusc. Aut. Sic.
Nuovo Raccolta di Opuscoli di Autori siciliani. Palermo. 1788–?.

Österr. Bot. Wochenbl.
Österreichisches botanisches Wochenblatt. Wien. **1–7**, 1851–1857; titled *Österr. Bot. Zeitschr., Österreichische botanische Zeitschrift*, **8**→ , 1858→ . (**92–93**, 1943–1944, titled *Wiener botanische Zeitschrift.*)

Österr. Bot. Zeitschr.
Cf. *Österr. Bot. Wochenbl.*

Overs. Kong. Danske Vid. Selsk. Forh.
Oversigt over det kongelige danske videnskabernes Selskabs Forhandlinger. Kiöbenhavn. 1806→ .

Period. Spis. Bǎlg. Kniž. Druž.
Периодическо Списание на българското книжовно Дружество [*Periodičesko Spisanie na bǎlgarskoto knižovno Družestvo*]. Sofija. 1870–1910.

Pflanzenareale
Die Pflanzenareale. Sammlung kartographischer Darstellungen von Verbreitungsbezirken der lebenden und fossilen Pflanzen-Familien, -Gattungen und -Arten. Jena. 1926→ .

Phyton (Austria)
Phyton. Annales Rei botanicae. Horn, Austria. 1948→ .

Pollichia
Jahresbericht der Pollichia, eines naturwissenschaftlichen Vereins der bayerischen Pfalz. Landau. **1–32**, 1843–1874. (**15–32**, 1857–1874, titled *...der Rheinpfalz.*)

Preslia
Preslia. Věstník české botanické Společnosti. Praha. **1**, 1914; titled *Preslia. Věstník československé botanické Společnosti. / Bulletin de la Société botanique tchécoslovaque à Prague. / Reports of the Czechoslovak botanical Society of Prague*, **2–15**, 1923–1936; titled *Preslia. Věstnik čs. botanické Společnosti. / Bulletin de la Société botanique tchéque à Prague. / Reports of the Czech botanical Society of Prague*, **16–17**, 1939; titled *Preslia. Věstník české botanické Společnosti*, **18–21**, 1940–1942; titled *Preslia. Věstník československé botanické Společnosti v Praze*, **22–23**, 1948; titled *Preslia. Časopis československé botanické Společnosti*, **24**→ , 1952→ .

Proc. Acad. Nat. Sci. Philad.
Proceedings of the Academy of natural Sciences of Philadelphia. Philadelphia, Pennsylvania. **1**→, 1841→.

Proc. Bot. Soc. Brit. Is.
Proceedings of the botanical Society of the British Isles. London. **1–7**, 1954–1969.

Proc. Linn. Soc. London
Proceedings of the Linnean Society of London. London. 1838→ .

Proc. Roy. Irish Acad.
Proceedings of the Royal Irish Academy. Dublin. 1930→ .

Progr. Sald. Realsch. Brandenb.
Programm der Saldern'schen Realschule zu Brandenburg a.d.H. Brandenburg.

Publ. Field Mus. Bot. (Chicago)
Publications of the Field Columbian Museum. Botanical Series.
Chicago. **1**, 1895–1902; titled *Publications of the Field Museum of natural History. Botanical Series.* **2–8**, 1903–1932; titled *Publications. Field Museum of natural History. Botanical Series*, **9–23**-1941–1947; titled *Fieldiana, Fieldiana: Botany*, **24**→ , 1958→ .

Rad Jugosl. Akad. Znan. Umj.
Rad jugoslavenske Akademije Znanosti i Umjetnosti. Zagreb. 1867→ . (**270–272**, 1941, titled *Rad hrvatske Akademije Znanosti i Umjetnosti.*)

Rapports Comm. Congr. Internat. Bot.
Rapports et Communications. Congrès International de Botanique.

Rec. Trav. Montpellier (Sér. Bot.)
Recueil des Travaux de l'Institut botanique. Montpellier. **1–4**, 1944–1949; titled *Recueil des Travaux des Laboratoires de Botaniques, Géologie et Zoologie de la Faculté des Sciences de Montpellier. Série botanique*, **5**→ , 1952→ .

Rendic. Accad. Sci. (Napoli)
Rendiconto delle Adunanze e de' Lavori dell'Accademia delle Scienze. Sezione della Società reale borbonica di Napoli. Napoli. Ser. 1, **1–9**, 1842–1850. Ser. 2, with minor changes of title, **1–5**, 1852–1856. Ser. 3, with minor changes of title, **1859** & **1(1861)**, 1860–1861.

Rep. Bot. Exch. Club Brit. Is.
Report of the botanical Society and Exchange Club of the British Isles. Manchester. **1–13**, 1867–1947. (**2(1903)–3(4)**, 1904–1913, at Oxford; **3(5)–13**, 1913–1947, at Arbroath.) With minor changes of title.

Rev. Roum. Biol.
Revue de Biologie. Bucarest. **1–8**, 1954–1963; titled *Revue Roumaine de Biologie*, **9**→ , 1964→ .

Revista Progr. Ci. Exact. Fis. Nat.
Revista de los Progresos de la Ciencias exactas, fisicas y naturales. Madrid. **1–12**, 1850–1862.

Rhodora
Rhodora. Journal of the New England botanical Club. Boston, Massachusetts. 1899→.

Roczn. Nauk Roln.
Roczniki Nauk Rolniczych. Kraków,Poznań&Warszawa.1913→ .

Sborn. Bǎlg. Akad. Nauk.
Сборникъ на Българскага Академия на Наукитѣ [*Sbornik na Bǎlgarskaga Akademija na Naukitě*]. Sofija. 1911→ .

Schr. Ges. Naturf. Freunde Berlin
Schriften der berlinischen Gesellschaft naturforschender Freunde. Berlin. **1–6**, 1781–1785; titled *Schriften der Gesellschaft naturforschender Freunde zu Berlin*, **7–11**, 1787–1794.

Scott. Naturalist
The Scottish Naturalist. Perth. **1–11**, 1871–1891. (**5 & 6** at Edinburgh.) 1912→ , at Edinburgh.

Scrin. Fl. Select. (Magnier)
Scrinia Florae selectae (Directeur: Charles Magnier). Saint-Quentin. **1–15**, 1882–1896. (**1**: pp. 1–48 in 1882; **2**: pp. 49–56 in 1883; **3**: pp. 57–72 in 1884; **4**: pp. 73–88 in 1885; **5**: pp. 89–104 in 1886; **6**: pp. 105–120 in 1887; **7**: pp. 121–136 in 1888; **8**: pp. 137–156 in 1889; **9**: pp. 157–176 in 1890; **10**: pp. 177–196 in 1891; pp. 197–228 in ?1891; **11**: pp. 229–262 in 1892; **12**: pp. 263–298 in 1893; **13**: pp. 299–336 in 1894; **14**: pp. 337–364 in 1895; **15**: pp. 365–383 in 1896.)

Sitz.-Ber. Akad. Wiss. Wien
Sitzungsberichte der kaiserlichen Akademie der Wissenschaften. Mathematisch-naturwissenschaftliche Classe. / Sitzungsberichte der mathematisch-naturwissenschaftlichen Classe der kaiserlichen Akademie der Wissenschaften. Wien. **1–123**, 1848–1914; titled *Sitzungsberichte. Kaiserliche Akademie der Wissenschaften in Wien. Mathematisch-naturwissenschaftliche Klasse*, **124–155**, 1915–1947 (*Kaiserliche* dropped from **127**, 1918, onwards); titled *Sitzungsberichte. Österreichische Akademie der Wissenschaften. Mathematisch-naturwissenschaftliche Klasse*, **156**→ , 1947→ .

Sitz.-Ber. Böhm. Ges. Wiss.
Sitzungsberichte der königl. böhmischen Gesellschaft der Wissenschaften in Prag. Prag. 1859–1873; titled Zprávy o Zasedání královské české Společnosti Nauk v Praze. | Sitzungsberichte der königl. böhmischen Gesellschaft der Wissenschaften in Prag, 1874–1885; titled Zprávy o Zasedání královské české Společnosti Nauk. Třída mathematicko-přírodovědecká. | Sitzungsberichte der königl. böhmischen Gesellschaft der Wissenschaften. Mathematisch-naturwissenschaftliche Classe, 1886; titled Věstník královské české Společnosti Nauk. Třída mathematicko-přírodovědecká. | Sitzungsberichte der königl. böhmischen Gesellschaft der Wissenschaften. Mathematisch-naturwissenschaftliche Classe, 1887–1915; titled **Mém. Soc. Sci. Bohême**, Věstník královské české Společnosti Nauk. Třída mathematicko-přírodovědecká. | Mémoires de la Société royale des Sciences de Bohême. Classe des Sciences, 1919–1935; titled Věstník královské české Společnosti Nauk. Třída mathematicko-přírodovědecká. | Mémoires de la Société royale des Lettres et des Sciences de Bohême. Classe des Sciences, 1936–1953. (1940–1944 lacks French title.)

Soc. Helv. Éch. Pl.
Société helvétique pour l'Échange des Plantes./Helvetischer Verein für den Austausch von Pflanzen. Neuchâtel. 1–?, 1871–?.

Sovetsk. Bot.
Советская Ботаника [Sovetskaja Botanika]. Moskva & Leningrad. 1933–1947.

Spraw. Kom. Fizyogr. Krakow.
Sprawozdania Komisyi fizyograficznej c. k. Towarzystwa naukowego krakowskiego. Kraków. 1–6, 1867–1872; titled Akademia Umiejetnósci w Krakowie. Sprawozdania Komisye fizyografiznej, 7–52, 1873–1918; titled Polska Akademia Umiejetnósci. Sprawozdania Komisji fizjograficznej, 53–73, 1920–1939.

Studia Bot. Čech.
Studia botanica čechoslovaca. Pragae. 1–12, 1938–1951. (2–6, 1939–1943 titled Studia botanica čechica.)

Studii Cerc. Şti. Cluj
Studii şi Cercetari ştiinţifice. Filiala Cluj, Academia RPR. Bucureşti. 1950–1954.

Svensk Bot. Tidskr.
Svensk botanisk Tidskrift. Stockholm. 1907→ . (16→ , 1922→ , at Uppsala.)

Syll. Pl. Nov. Ratisbon. (Königl. Baier. Bot. Ges.)
Sylloge Plantarum novarum itemque minus cognitarum a praestantissimis Botanicis adhuc viventibus collecta et a Societate regia botanica ratisbonensi edita. (Königlich-baierische botanische Gesellschaft in Regensburg.) Ratisbonae. 1–2, 1824–1828. (1 in 1824; 2 in 1828.)

Taxon
Taxon. Official News Bulletin of the International Association for Plant Taxonomy. Utrecht. 1–16, 1951–1967; titled Taxon. Journal of the International Association for Plant Taxonomy, 17→, 1968→.

Term. Füz.
Természetrajzi Füzetek. Budapest. 1877–1902.

Trab. Mus. Ci. Nat. Barcelona
Trabajos del Museo de Ciencias naturales de Barcelona/Treballs del Museu de Ciències naturals de Barcelona/Musei barcinonensis Scientiarum naturalium Opera. Barcelona. 1917→ .

Trans. Amer. Philos. Soc.
Transactions of the American Philosophical Society. Philadelphia, Pennsylvania. 1–6, 1771–1809. Nov. ser., 1→, 1818→.

Trans. Linn. Soc. London
Transactions of the Linnean Society. London. Ser. 1, 1–6, 1791–1802; titled Transactions of the Linnean Society of London, 7–30, 1804–1875. Ser. 2, Botany, 1→ , 1875→ . From 1939 Botany and Zoology are combined.

Trans. Proc. Bot. Soc. Edinb.
Transactions of the Botanical Society. Edinburgh. 1–11, 1844–1873; titled Transactions and Proceedings of the Botanical Society, 12–19, 1873–1893; titled Transactions and Proceedings of the Botanical Society of Edinburgh, 20→ , 1894→ .

Trans. Proc. N.Z. Inst.
Transactions and Proceedings of the New Zealand Institute. Wellington. 1–63, 1869–1934.

Trans. Roy. Soc. Edinb.
Transactions of the Royal Society of Edinburgh. Edinburgh. 1783→ .

Trav. Inst. Bot. (Charkov)
Travaux de l'Institut botanique. / Труди Институту ботаникі [Trudy Ynstytutu botanyki]. Charkov. 1936–1938.

Trav. Inst. Sci. Chérif.
Travaux de l'Institut scientifique chérifien. Série botanique. Tanger. 1952→ . (1956→ at Rabat.)

Trav. Mus. Bot. Acad. Pétersb.
Travaux du Musée botanique de l'Académie impériale des Sciences de St. Pétersbourg [de Russie]. St. Pétersbourg. 1902–1932.

Trudy Peterb. Obšč. Estestv.
Труды Санкт-Петербуьгскаго Общества Естествоиспытателей [Trudy Sankt-Peterburgskago Obščestva Estestvoispytatelej]. S.-Peterburg. 1–18, 1870–1887; with additional title Travaux de la Société des Naturalistes de St.-Pétersbourg, 19–25, 1888–1895; titled Труды императорскаго С.-Петербургскаго Общества Естествоиспытателей [Trudy imperatorskago S.-Peterburgskago Obščestva Estestvoispytatelej]. / Travaux de la Société impériale des Naturalistes de St.-Pétersbourg, 26–43(2), 1896–1912; titled Труды Петроградскаго Общества Естествоиспытателей [Trudy Petrogradskago Obščestva Estestvoispytatelej]. / Travaux de la Société des Naturalistes de Pétrograd, 47(7–8)–53, 1916–1922; titled Труды Ленинградского Общества Естествоиспытателей [Trudy Leningradskogo Obščestva Estestvoispytatelej]. / Travaux de la Société des Naturalistes de Leningrad, 54, 1925 (French title ceases at 70, 1959; from 71(3)→ , 1969→ , with alternative title Transactions of the Leningrad Society of Naturalists.)

Trudy Tiflissk. Bot. Sada
Труды Тифлисскаго ботаническаго Сада [Trudy Tiflisskago botaničeskago Sada]. Tbilissi.

Ukr. Bot. Žur.
Українськиї ботанічниї Журнал [Ukrajinskyji botaničnyji Žurnal]. / The Ukrainian botanical Review. Kiev. 1–5, 1922–1929. Continued as Žur. Inst. Bot. URSR, Журнал Інституту ботаніки АН УРСР [Žurnal instytutu botaniky AN URSR]. / Journal de l'Institut botanique de l'Académie des Sciences de la RSS d'Ukraine. 1–23, 1934–1940. Continued as Ukr. Bot. Žur., Ботанічний Журнал [Botaničnyj Žurnal]. / Journal botanique de l'Académie des Sciences de la RSS d'Ukraine. 1–12, 1940–1955. Titled Український ботанічний Журнал [Ukrajinskyji botaničnyji Žurnal], 13→ , 1956→ .

Univ. Calif. Publ. Bot.
University of California Publications in Botany. Berkeley, California. 1902→ .

Uppsala Univ. Årsskr.
Uppsala Universitets Årsskrift. Uppsala. 1861→ .

Verh. Bot. Ver. Brandenb.
Verhandlungen des botanischen Vereins der Provinz Brandenburg. Berlin. 1859→ .

Verh. Holland. Maatsch. Wetensch. Haarlem
Verhandelingen uitgegeeven door de Hollandse Maatschappye der Wetenschappen te Haarlem. Haarlem. 1–30, 1754–1793.

Verh. Mitt. Siebenb. Ver. Naturw.
Verhandlungen und Mitteilungen des siebenbürgischen Vereins für Naturwissenschaften in Hermannstadt. Hermannstadt. 1850–1938.

Verh. Naturf. Ges. Basel
Verhandlungen der naturforschenden Gesellschaft in Basel. Basel. 1854→ .

Verh. Naturf. Ver. Brünn
Verhandlungen des naturforschenden Vereines in Brünn. Brünn. **1–75**, 1862–1944.

Verh. Zool.-Bot. Ges. Wien
Verhandlungen der k.k. zoologisch-botanischen Verein in Wien. Wien. 1852→ . From 1858, *Verein* is replaced by *Gesellschaft.*

Veröff. Geobot. Inst. Rübel (Zürich)
Veröffentlichungen des geobotanischen Institutes Rübel. Zürich. 1924→ .

Vid. Meddel. Dansk Naturh. Foren. Kjøbenhavn
Videnskabelige Meddelelser fra den naturhistoriske Forening i Kjöbenhavn. Kjöbenhavn. **1–63**, 1849–1912; titled *Videnskabelige Meddelelser fra dansk naturhistorisk Forening i Kjøbenhavn,* **64**→ , 1913→ . Years used as vol. nos. until 1912.

Viert. Naturf. Ges. Zürich
Vierteljahrsschrift der naturforschenden Gesellschaft in Zürich. Zürich. 1856→ .

Watsonia
Watsonia. Journal of the botanical Society of the British Isles. Arbroath. 1949→ . (**3** → ,1953 → , at London.)

Webbia
Webbia. Raccolta di Scritti botanici. Firenze. 1905→ .

Wiss. Zeitschr. Univ. Greifswald
Wissenschaftliche Zeitschrift der Universität Greifswald. Mathematisch-naturwissenschaftliche Reihe. Greifswald. **1–3**, 1951–1954; titled *Wissenschaftliche Zeitschrift der Ernst Moritz Arndt-Universität Greifswald,* **4**→ , 1954→ .

Wiss. Zeitschr. Univ. Halle
Wissenschaftliche Zeitschrift der Martin-Luther-Universität Halle-Wittenberg. Mathematisch-naturwissenschaftliche Reihe. Halle. 1951 → .

Wrightia
Wrightia. A botanical Journal. Dallas, Texas. 1945→ . (**2**→ , 1959→ , at Renner, Texas.)

Zeitschr. Naturwiss. (Halle)
Zeitschrift für die gesammten Naturwissenschaften. Halle. **1–54**, 1853–1881; titled *Zeitschrift für Naturwissenschaften,* **55–95**, 1882–1941. (Some volumes published at Berlin, Leipzig or Stuttgart.)

Žur. Inst. Bot. URSR
Cf. *Ukr. Bot. Žur.*

APPENDIX IV

GLOSSARY OF TECHNICAL TERMS

The number of technical terms used in *Flora Europaea* has been kept as low as is consistent with a reasonable standard of accuracy and brevity. Most of them are used in well-established traditional senses, and their meanings may be ascertained by reference to glossaries such as H. I. Featherly, *Taxonomic Terminology of the Higher Plants* (Ames, Iowa, U.S.A., 1954). No term is used in a sense inconsistent with that given by Featherly.

Experience has shown, however, that some useful terms are liable to misinterpretation, and others, which can be used in a wider sense, are used in a restricted sense in *Flora Europaea*. This glossary is intended simply to indicate without ambiguity the sense in which these potentially ambiguous terms are employed.

Certain technical terms, which are restricted to descriptions in particular families or genera, are explained under the family or genus concerned.

ABOVE Used to indicate both the upper surface of a normally horizontal organ and the upper part of an organ or of the whole plant.

ACHENE A small, dry, 1-seeded, indehiscent fruit, whether derived from a superior or from an inferior ovary.

ALTERNATE Arising singly at a node; includes regularly spiral, as well as distichous arrangements.

ANNUAL Completing its life-cycle from seed to seed in less than 12 months; includes 'overwintering' annuals, which germinate in autumn and flower the following year.

BELOW Used to indicate the basal part of a plant, stem or inflorescence; cf. *beneath*.

BENEATH Used to indicate the lower surface of a normally horizontal organ; cf. *below*.

BIDENTATE With two teeth.

BISERRATE Serrate, with the teeth themselves serrate.

CADUCOUS Falling unusually early.

CILIATE With hairs on the margin.

DECIDUOUS Of leaves: falling in autumn; of other organs: falling before the majority of adjacent or associated organs.

ERECTO-PATENT Diverging at an angle of 15–45° from the axis on which the structure is borne.

FLOCCOSE Clothed with woolly hairs, which are disposed in tufts or tend to rub off and adhere in small masses.

GLABRESCENT Becoming glabrous with increasing age or maturity. For structures very slightly but persistently hairy the term *subglabrous* is used.

HIRSUTE Covered with long, moderately stiff and not interwoven hairs.

HISPID Covered with stiff hairs or bristles.

LANATE Covered with soft, flexuous, intertwined hairs.

PELTATE Denotes an organ of which the stalk is attached to a more or less flat surface, and not to the margin; the attachment is not, however, necessarily central.

PUBERULENT With very short hairs.

PUBESCENT With soft, short hairs.

PYRENE A small stone, consisting of one or few seeds with a hard covering, enclosed in fleshy tissue, e.g. *Arctostaphylos*, *Corema*.

SEMI-PATENT Between patent and appressed.

SERICEOUS With silky, appressed hairs.

SETOSE Covered with stout, rigid bristles.

SIMPLE HAIR Indicates an unbranched hair; it may or may not bear a gland.

STOCK The persistent, usually somewhat woody base of an otherwise herbaceous perennial.

STOLON A short-lived, horizontal stem, either above or below the surface of the ground, rooting at one or more nodes.

STRIGOSE With stiff, appressed, straight hairs.

TERETE More or less cylindrical, without grooves or ridges.

TOMENTOSE With hairs compacted into a felty mass.

TUBERCULATE Covered with smooth, knob-like elevations.

VELUTINOUS With a dense indumentum of fine, soft, straight hairs.

VERRUCOSE Covered with rough, wart-like elevations.

VILLOUS Covered with long, soft, straight hairs.

APPENDIX V

VOCABULARIUM ANGLO-LATINUM

IN USUM LECTORUM LINGUAE ANGLICAE MINUS PERITORUM CONFECTUM

N.B. Plurimi termini ad descriptionem botanicam in lingua anglica usurpati aequi-pollentibus latinis persimiles sunt, e.g. *ovate* (ovatus), *inflorescence* (inflorescentia). Talia verba omnia sunt omissa.

above insuper, supra, super
all omnes
almost fere, paene
always semper
arable fields arva
around circum
arranged dispositus
attached affixus
awn arista
back dorsum
backward(s) retro
bank ripa
barbed pilis hamatis obsitus
bare nudus
bark cortex
basin-shaped pelviformis
beak rostrum
bearded barbatus
become fieri
below infra, sub
beneath infra, subtus
bent inflexus
berry bacca
between inter
bind colligare, firmare
bitter amarus
black niger, ater
bloom pruina
blotch macula
blue caeruleus
boat navicula
border margo
borne prolatus
branch ramus
breadth latitudo
bright laete
bristle seta
broad latus
bronze aeneus
brown fuscus, brunneus
bud gemma
bundle fasiculus
bushy spisse et iteratim ramosus
casual fortuitus
catkin amentum
chaffy paleaceus

chamber loculus
chequered cancellatus
chestnut castaneus
chief principalis
claw unguis
cliff rupes
climbing scandens
close propinquus, affinis
closed clausus
clothed vestitus
cluster glomerulus
coarse crassus, grossus
coast litus, ora
coat tunica
common vulgaris
completely omnino, ex toto
compound compositus
cone strobilus
corner angulus
cornfield seges
covered obtectus
cream ochroleucus, albido-flavescens
crest crista
crevice fissura
crimson kermesinus, sanguineus; ut flos *Paeoniae officinalis* coloratus
crowded confertus
cultivated cultus, sativus
curled crispus
cushion pulvinus
damp humidus
dark obscure
dead emortuus
decay dissolutio
deep profundus; intense
developed evolutus
die mori
docks navalia
downwards deorsum
downy lanuginosus
dry siccus
dull opace; impolitus
dwarf nanus
early prius, mox, praecoce
eastern orientalis
eastwards orientem versus

edge margo
edible edulis
either...or aut...aut
end pars terminalis
enlarge crescere, augere
entire integer
entirely omnino
equal aequalis, aequans
escape evadere; planta ex horto elapsa
established subspontaneus
evening vesper
evergreen sempervirens
exceeding superans
face facies
fan-shaped flabellatus
female femineus, pistillatus
feebly debiliter, perleviter
few pauci
finely subtiliter
first primus
flap valva, ligula
flat planus
flattened compressus, applanatus
flax *Linum usitatissimum*
flesh-coloured carneus, pallide et opace roseus
fleshy carnosus
floating natans
flooded inundatus
flower flos
fodder bestiarum pabulum
fold plica
following sequens
food cibus
forest silva magna
forwards porro
free liber
fringe fimbriae
fruit fructus
furnished munitus
furrow sulcus
garden hortus
glossy nitidus
golden aureus
grassy graminosus
gravelly glareosus

graze pascere
green viridis
grey cinereus
grooved canaliculatus, sulcatus
ground solum
group grex
grow crescere, habitare
hair pilum
hairy pilis munitus
half dimidium
hard durus
head caput, capitulum
heath ericetum, callunetum
hedge saepes
helmet galea
hill collis
hoary incanus
hollow fistulosus, cavus; cavum, excavatio
hood cucullus
hooked uncinatus
inner interior, internus
inside intus, intra; pagina vel pars interior
introduced inquilinus, allatus
jagged argutus
jointed articulatus
juice succus
keel carina
key clavis
lake lacus
late sero
later postea
leaf folium
leafless foliis carens
leaflet foliolum
length longitudo
less minus
level altitudo, gradus
lid operculum
light clare
limestone calx
lip labium
locally hic inde
low humilis, pusillus
lower inferior
lowland campestris, planitiem incolens
main principalis
male masculus, stamineus
many multi
marbled marmoratus
marsh palus
mat stratum e ramulis procumbentibus intertextis compositum
mauve malvinus
meadow pratum
mealy farinosus
medicinal officinalis
middle pars centralis; medius
midrib costa, folii nervus principalis

milky lacteus
mistake error
more plus, magis
most plerique, pars major
mountain mons
mouth os
much multo, multum
naked nudus
narrow angustus
native indigenus
naturalized inquilinus
near prope
nearly paene, fere
neither...nor nec...nec
net reticulum
never numquam
nodding nutans, cernuus
none nulli
northern borealis
northwards septentrionem versus
notch incisio
nut nux
often saepe
oil oleum
old vetus, antiquus
open apertus
orange aurantiacus
ornament decus
other alius, alter
otherwise aliter
outer exterior, externus
outside extra; pagina vel pars exterior
overlapping imbricatus
pale pallidus
papery chartaceus
pasture pascuum
patch macula
peat-bog turbarium
pink roseus
pitted foveolatus
planted cultus
point acumen
pond stagnum
pool stagnum
poor egens
prickle aculeus
pricklet aculeolus
purple purpureus
quarter pars quarta
rank ordo
rarely raro
ray radius
red ruber
related affinis
remains reliquiae
rest ceteri
rib costa
rice-field oryzetum
rich abundans
ridge carina

rind fructus cortex
ring anulus
ripe maturus
river flumen
road via
rock saxum, rupes
root radix
rosette rosula
rough asper
rounded rotundatus
rust-coloured ferrugineus
salt-marsh palus salsa
sand arena
scale squama
scanty exiguus
scar cicatrix
scarcely vix
scarlet laete et clare ruber, paullulo aurantiaco affectus; ut flos *Salviae splendentis* coloratus
scattered sparsus
scented fragrans
scree clivus alpestris, saxis deorsum conjectis copertus
scrub dumetum, fruticetum
sea mare
seed semen
seldom raro
several nonnulli, complures
shady umbrosus
shallow haud profundus
shape forma
sharply acute
sheath vagina
shelter tegmen contra ventum
shingle glarea maritima vel fluviatilis
shiny nitidus
shoot caudiculus, surculus
shore litus, ora
short brevis
shoulder angulus obtusus
shrub frutex
side latus, pagina
silky sericeus
silvery argenteus
slender tenuis, gracilis
slightly leviter, paullo
slipper calceolus
slit rima, foramen longum sed angustum
slope clivus, declivitas
small parvus
smell odor
smooth laevis
snow-patch locus in montibus ubi nix sero perdurat
soft mollis
soil solum
sometimes interdum
southern australis
southwards meridiem versus

spikelet spicula
spot punctum, macula
spreading patens, divaricatus
spring ver
spur calcar
square quadratus
stalk stipes
standard vexillum
stem caulis
stiff rigidus
stock caudex
stony lapidosus
stout crassus, robustus
straight rectus
streak linea
stream rivulus
stripe vitta
strong robustus, validus
suddenly abrupte
summer aestas
sunk(en) immersus
surface superficies, pagina
sweet dulcis
swollen tumidus, inflatus
tall altus
taste sapor
tawny fulvus
teeth dentes
thick crassus, densus, spissus

thicket dumetum
thin tenuis
third pars tertia
timber materia; lignum ad usum hominum aptum
tinged suffusus
tip apex
tipped ad apicem munitus vel tinctus
tooth dens
top vertex
tough lentus
tree arbor
true verus
tufted in fasciculos dispositus, caespitosus
twice bis
twig ramulus, virga
twining volubilis
twisted contortus
unarmed inermis
uncertain incertus, dubius
undivided indivisus
unequal inaequalis
united conjunctus, connatus
upper superior
uppermost supremus
upwards sursum
usually plerumque
vegetable olus

veil velum
vein nervus
velvety velutinus
vessel vas
violet violaceus
wart verruca
waste incultus
weak debilis, flaccidus
well bene
western occidentalis
westwards occidentem versus
wet madidus
white albus, candidus
whorled verticillatus
wide latus
widespread late diffusus
width latitudo
wing ala
winter hiems
wiry filo ferreo similis
withered marcidus
without sine
wood silva; lignum
woody lignosus
woolly lanatus
wrinkled rugosus
yellow flavus, luteus
young juvenis

INDEX

This index is intended to serve two purposes: to enable the reader to find the page on which any plant is mentioned, and to cite and explain names relegated to synonymy which occur in 'Standard Floras', but are not in sufficiently wide currency to justify their citation in the text (see p. xix).

Generic names adopted in *Flora Europaea* are printed in **bold-faced** type; specific and subspecific epithets adopted are printed in ordinary type. (This applies not only to numbered species and genera, but also to those mentioned incidentally in observations, or in the introductory descriptions of families or genera.) All synonyms are printed in *italic* type, and are followed by a page-reference (also in *italics*); for those not cited in the text the page number is followed by a further number or numbers in parentheses to indicate the species (and, where necessary, subspecies, genus and family) on that page to which the synonym is referable. Among these numbers roman numerals denote the family, arabic numerals in ordinary type the genus, arabic numerals in **bold-faced** type the species, and a small letter (also in **bold-faced** type) following the species number the subspecies. Thus,

Erica

mediterranea L., *7* (**14**)

indicates that the name is regarded as a synonym (partial or complete) of the species on p. 7 which is numbered 14, namely **E. herbacea.** Similarly,

Nymphoides

orbiculata Druce, *68* (CXLI, 2, **1**)

indicates that this name is regarded as a synonym of species 1 (**peltata**) in genus 2 (**Nymphoides**) of family CXLI (**MENYANTHACEAE**) on p. 68; because more than one family and genus are treated on the page, citation of genus and family is necessary to avoid ambiguity.

Synonyms of taxa mentioned in notes following a numbered species are indexed as being synonyms of that species. In cases where this procedure would be ambiguous or misleading, the synonym in question has been inserted in the text.

Some names of hybrids are similarly indexed with page and number references to their parent species.

All infraspecific taxa are arranged alphabetically, regardless of rank, under the species with which they are combined.

MAP I

To illustrate the boundaries of Europe for the purposes of *Flora Europaea*, and its division into 'territories' which are indicated by two-letter abbreviations after the summary of geographical distribution for each species. These abbreviations are derived from the Latin name of the territory concerned.

Al   Albania
Au   Austria, with Liechtenstein
Az   Açores
Be   Belgium, with Luxembourg
Bl   Islas Baleares
Br   Britain, including Orkneys, Zetland and Isle of Man; excluding Channel Islands and Northern Ireland
Bu   Bulgaria
Co   Corse
Cr   Kriti (*Creta*), with Karpathos, Kasos and Gavdhos
Cz   Czechoslovakia
Da   Denmark (*Dania*), including Bornholm
Fa   Færöer
Fe   Finland (*Fennia*), including Ahvenanmaa (Aaland Islands)
Ga   France (*Gallia*), with the Channel Islands (Îles Normandes) and Monaco; excluding Corse
Ge   Germany (both eastern and western republics)
Gr   Greece, excluding those islands included under Kriti (*supra*) and those which are outside Europe as defined for *Flora Europaea*
Hb   Ireland (*Hibernia*); both the republic and Northern Ireland
He   Switzerland (*Helvetia*)
Ho   Netherlands (*Hollandia*)
Hs   Spain (*Hispania*), with Gibraltar and Andorra; excluding Islas Baleares
Hu   Hungary
Is   Iceland (*Islandia*)
It   Italy, including the Arcipelago Toscano; excluding Sardegna and Sicilia as defined *infra*
Ju   Jugoslavia
Lu   Portugal (*Lusitania*)
No   Norway
Po   Poland
Rm   Romania
Rs   U.S.S.R. (*Rossia*). This has been subdivided as follows, using the floristic divisions of Komarov's *Flora U.R.S.S.*; in a few places, however, our boundaries deviate slightly from those of Komarov.
        Rs (N) *Northern division:* Arctic Europe, Karelo-Lapland, Dvina-Pečora
        Rs (B) *Baltic division:* Estonia, Latvia, Lithuania, Kaliningradskaja Oblast'
        Rs (C) *Central division:* Ladoga-Ilmen, Upper Volga, Volga-Kama, Upper Dnepr, Volga-Don, Ural
        Rs (W) *South-western division:* Moldavia, Middle Dnepr, Black Sea, Upper Dnestr
        Rs (K) *Krym* (Crimea)
        Rs (E) *South-eastern division:* Lower Don, Lower Volga, Transvolga
        White Russia falls entirely within Rs (C). Ukraine is largely in Rs (W), but partly in Rs (K), Rs (C) and Rs (E). The European part of Kazakhstan is in Rs (E)
Sa   Sardegna
Sb   Svalbard, comprising Spitsbergen, Björnöya (Bear Island) and Jan Mayen
Si   Sicilia, with Pantelleria, Isole Pelagie, Isole Lipari and Ustica; also the Malta archipelago
Su   Sweden (*Suecia*), including Öland and Gotland
Tu   Turkey (European part), including Imroz

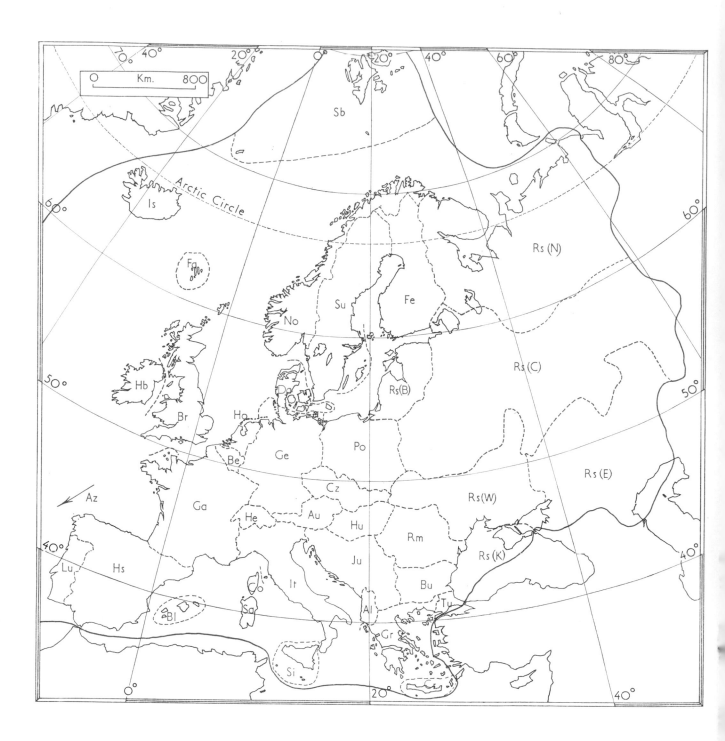

Km. 800

Arctic Circle

70° 40° 20° 20° 40° 60° 80°

60°

Is

Fₒ

Sb

Rs (N)

No

Su

Fe

Rs (C)

50°

Hb

Rs(B)

Br

Ho

Dₐ

Be

Ge

Po

Rs(E)

Az

Ga

He

Au

Cz

Rs(W)

Hu

Rs (K)

40°

Lu

Hs

Ju

Rm

Bu

It

Al

Tu

Bl

Sa

Gr

Si

50°

60°

50°

40°

40°

20°

MAP II

To illustrate the boundary between Europe and Asia in the Aegean region.

The boundary is based largely on the proposals of K. H. Rechinger, 'Grundzüge der Pflanzenverbreitung in der Aegäis', *Vegetatio* **2**: 55 (1949). His northern, western and Kikladhes divisions are regarded as entirely in Europe and his eastern division as entirely in Asia; it was, however, necessary to divide his southern and north-eastern divisions.

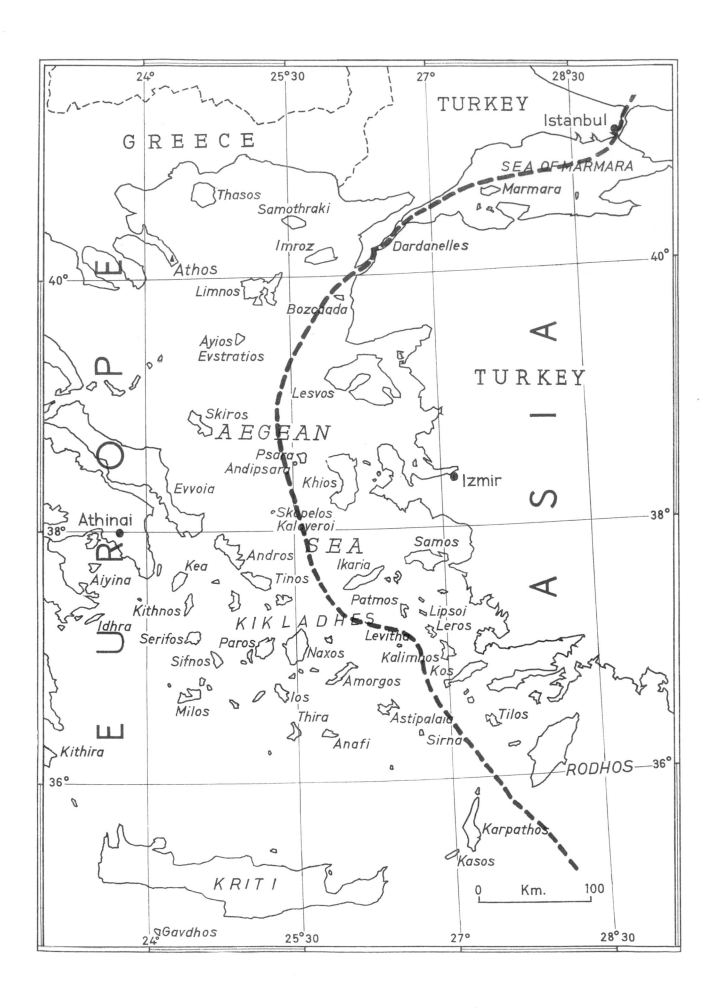

MAP III

To illustrate the boundary between Europe and Asia in the southern part of the U.S.S.R.

The southern boundary of Europe between the Caspian and Black Seas is defined for *Flora Europaea* as running up the Terek River westwards to 45° E.; thence along the eastern and northern boundaries of the Stavropol'skij Kraj (as marked in *The Times Atlas*) to meet the Kuban River a short distance east of Kropotkin; thence down the Kuban River to its more southerly mouth.

The eastern boundary of Europe is defined as running in the Arctic Ocean between Novaja Zemlja and Vajgač; up the Kara River to 68° N.; thence along the crest of the Ural Mountains (following the administrative boundaries) to 58° 30′ N.; thence by an arbitrary straight line to a point 50 km E. of Sverdlovsk, and by another arbitrary straight line to the head-waters of the Ural River (S. of Zlatoust); thence along the Ural River to the Caspian Sea.

The following administrative districts of the Russian S.F.S.R. near the eastern or southern boundary of Europe are regarded as entirely in Europe:

Arkhangel'skaja Obl.	Volgogradskaja Obl.
Komi A.S.S.R.	Astrakhanskaja Obl.
Permskaja Obl.	Kalmyckaja A.S.S.R.
Kujbyševskaja Obl.	Rostovskaja Obl.
Saratovskaja Obl.	

The following are regarded as partly in Europe, partly in Asia:

Russian S.F.S.R.
  Sverdlovskaja Obl.
  Čeljabinskaja Obl.
  Baškirskaja A.S.S.R. (only the extreme N.E. corner
    being in Asia)
  Orenburgskaja Obl.

Dagestanskaja A.S.S.R.
Čečeno-Inguškaja A.S.S.R.
Krasnodarskij Kraj
Kazakhstan
  Zapadno-Kazakhstanskaja Obl.
  Gur'jevskaja Obl.

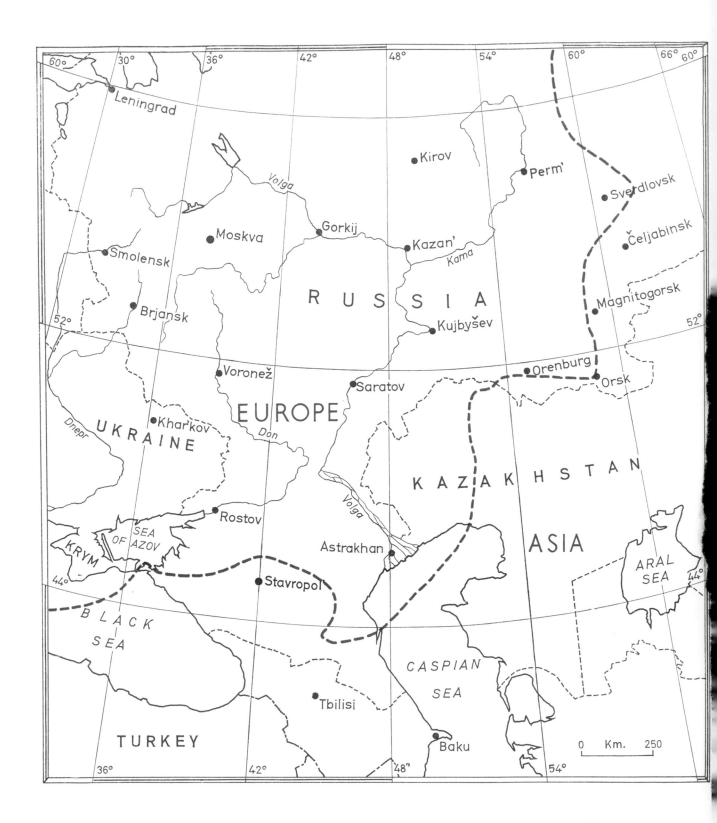

60°   30°   36°   42°   48°   54°   60°   66°   60°

● Leningrad

● Kirov

● Perm'

● Sverdlovsk

Volga

● Gorkij   ● Kazan'   *Kama*   ● Čeljabinsk

● Moskva

● Smolensk

R U S S I A

● Magnitogorsk

● Brjansk

52°   ● Kujbyšev   52°

● Voronež   ● Orenburg

● Saratov   ● Orsk

E U R O P E

Dnepr   ● Kharkov

U K R A I N E   *Don*

K A Z A K H S T A N

● Rostov

ASIA

KRYM   SEA OF AZOV   *Volga*

ARAL SEA

● Astrakhan

44°   44°

● Stavropol

B L A C K SEA

CASPIAN SEA

● Tbilisi

T U R K E Y

● Baku

0   Km.   250

36°   42°   48°   54°

MAP IV

To illustrate the meaning to be attached to certain phrases used in summaries of geographical distribution.

W. Europe: Açores, Portugal, Spain, Islas Baleares, France, Ireland, Britain, Færöer, Iceland, S.W. Norway, Netherlands, Belgium, N.W. Germany, W. Denmark (Jylland), Corse, Sardegna, and small parts of N.W. Italy and W. Switzerland

E. Europe: N.E. Greece and the Aegean islands, Bulgaria, S. & E. Romania, Finland, U.S.S.R.

N. Europe: Svalbard, Iceland, Færöer, Ireland, Britain (excluding S. England), Denmark, Fennoscandia, U.S.S.R. north of a line running through Minsk–Tula–Penza–Orsk

S. Europe: Europe south of a line running through Bordeaux–Chambéry–Aosta–Locarno–Riva–Udine–Zagreb–Beograd–Ploesti–Odessa–Rostov–Astrakhan'.

– – – – – – – eastern boundary of *W. Europe*   – · – · – · – southern boundary of *N. Europe*

○ ○ ○ ○ ○ ○ western boundary of *E. Europe*   × × × × × northern boundary of *S. Europe*

For the definition and illustration of the meaning of S.W., N.W., S.E., N.E. and C. Europe, and of certain other geographical phrases, see map v.

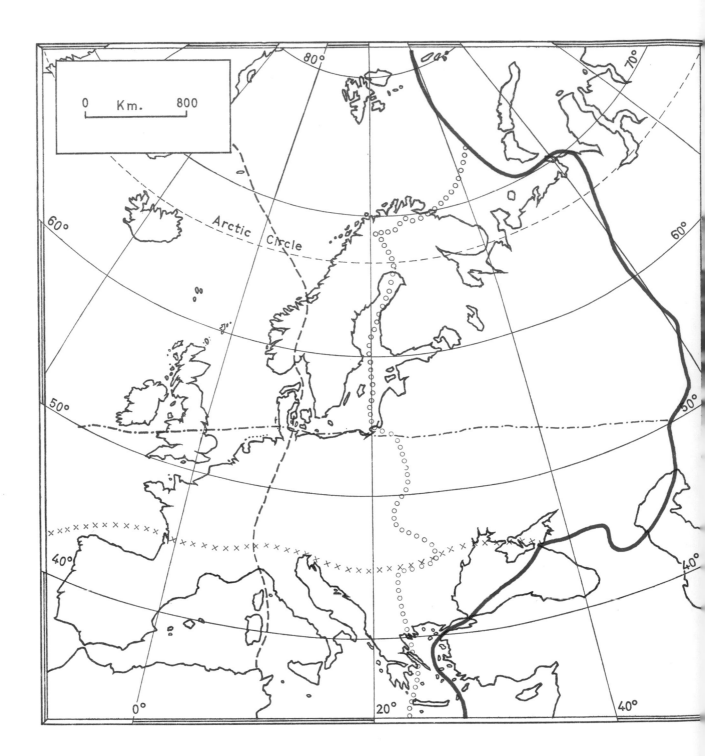

Arctic Circle

0 Km. 800

80°
70°
60°
60°
50°
50°
40°
40°
40°
0°
20°
40°

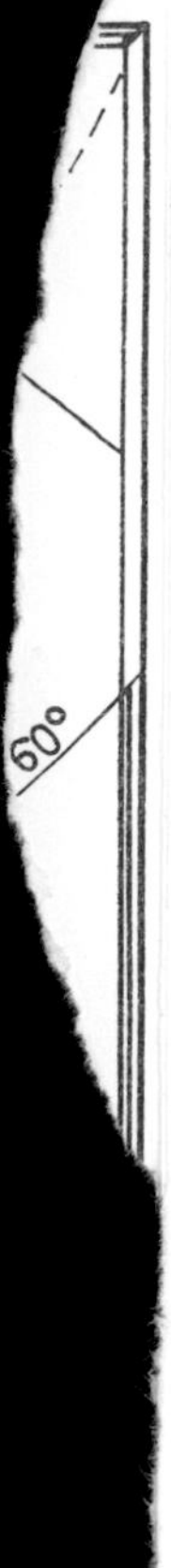

MAP V

To illustrate the meaning to be attached to certain phrases used in summaries of geographical distribution.

S.W. Europe: Açores, Portugal, Spain, Islas Baleares, Corse, Sardegna, S. France, N.W. Italy

N.W. Europe: Iceland, Færöer, Britain, N. France, Belgium, Netherlands, N.W. Germany, W. Denmark (Jylland), Norway

S.E. Europe: The Balkan peninsula, Aegean islands, S.E. Italy, S. & E. Romania, U.S.S.R. south of about 48° N.

N.E. Europe: U.S.S.R. north of a line from Vilnius to Sverdlovsk, Finland, E. Sweden, a small part of N.E. Norway.

C. Europe: Alsace and Lorraine, Germany, Switzerland, Austria, the Italian Alps from Monte Bianco eastwards, Hungary, Czechoslovakia, Poland, the Ukrainian Carpathians, N., W. & C. Romania, Jugoslavia north of the Danube–Sava–Kupa line.

Maps IV and V are intended merely to give precision to certain geographical phrases which are commonly used, but used in various senses in different parts of Europe. They do not purport to divide Europe into phytogeographical regions, as is apparent from the fact that along parts of their boundaries these regions overlap, and along other parts they are not contiguous.

Certain other phrases used in the summaries of geographical distribution, but not illustrated in the maps, may be briefly defined as follows:

Alps: Separated from the Appennini at 8° 15′ E. (above Savona); bounded on the east by the line Semmering–Graz–Maribor–Ljubljana–Trieste. Divided into three major divisions: *eastern*, *central*, and *south-western*, by the lines Arlberg–St Moritz–Chiavenna–Como and Genève–Chamonix–Aosta–Ivrea.

Arctic: This term is used to designate all territories north of the Arctic Circle, and is not restricted to those which have only 'arctic' vegetation.

Carpathians: Divided into *western*, *eastern* and *southern* divisions at the pass of Łupków (22° E.) and the Oituz Pass (46° 05′ N.). The western division is in Czechoslovakia and Poland, the southern entirely in Romania, the eastern extends from Czechoslovakia and Poland through Ukraine to Romania.

Pyrenees: Includes the subsidiary chains within 50 km of the main watershed, and extends westwards to Bilbao and Vitoria. Divided into *eastern*, *central* and *western* divisions at the Pont du Roi (0° 45′ E.) and the Col du Somport (0° 30′ W.).

Balkan peninsula: Jugoslavia south of the Danube–Sava–Kupa line, Bulgaria, Albania, Greece (including islands close to the mainland) and Turkey-in-Europe.

Fennoscandia: Norway, Sweden, Finland and part of N.W. Russia (Murmanskaja Oblast' and Karelskaja A.S.S.R.).

Mediterranean region: All European territories within 100 km of the Mediterranean Sea (including the Adriatic, but not the Black Sea), and including also all Italy except the Alpine region and all Spain except the west and north-west. It is divided into *eastern* and *western* divisions by a line following the main watershed of Italy and running east of Sicilia. *Central Mediterranean* indicates the region between 8° E. and 20° E.

Aegean region: All islands in the Aegean Sea which come within the scope of the *Flora*, and those parts of Greece and Turkey-in-Europe which drain into the Aegean Sea or the Dardanelles.

Macedonia: Comprises the Jugoslav republic of Makedonija, the Greek province of Makedhonia, and the Bulgarian province of Blagoevgrad.

0 Km. 800

80°

70°

Arctic Circle

60°

N.W.
EUROPE

N.E.
EUROPE

50°

50°

C.
EUROPE

40°

40°

S.W.
EUROPE

S.E.
EUROPE

0°

20°

40°